Cropping Systems and Agronomic Management Practices of Field Crops

Cropping Systems and Agronomic Management Practices of Field Crops

Editors

Umberto Anastasi
Aurelio Scavo

MDPI

Basel • Beijing • Wuhan • Barcelona • Belgrade • Novi Sad • Cluj • Manchester

Editors
Umberto Anastasi
University of Catania
Catania
Italy

Aurelio Scavo
University of Catania
Catania
Italy

Editorial Office
MDPI
St. Alban-Anlage 66
4052 Basel, Switzerland

This is a reprint of articles from the Special Issue published online in the open access journal *Agronomy* (ISSN 2073-4395) (available at: https://www.mdpi.com/journal/agronomy/special_issues/Cropping-Systems).

For citation purposes, cite each article independently as indicated on the article page online and as indicated below:

Lastname, A.A.; Lastname, B.B. Article Title. *Journal Name* **Year**, *Volume Number*, Page Range.

ISBN 978-3-7258-0063-6 (Hbk)
ISBN 978-3-7258-0064-3 (PDF)
doi.org/10.3390/books978-3-7258-0064-3

Contents

About the Editors

Umberto Anastasi

Umberto Anastasi, Ph.D., is an Associate Professor in Agronomy and Field Crops in the Department of Agriculture, Food and Environment (Di3A) of the University of Catania, Italy. At present, his research activities focus on the agronomic management models based on field crops adopted in the Mediterranean farming systems of southern Italy, mainly industrial crops and grain legumes. Moreover, his attention has been focused on the use of self-seeding leguminous species in farming systems based on field crops and orchards (ley-farming and cover-cropping). He also carries out research on cereal crops addressed especially to the varietal innovation and characterization of landraces germplasm, which is useful in organic and low-input farming systems. He is or has been involved in a number of national and international research projects, and has published more than 100 articles, including papers in peer-reviewed journals, reports in scientific and technical journals, and conference papers. He participates in various scientific conferences and acts as reviewer for different ISI/Scopus peer-reviewed journals. U. Anastasi teaches in various bachelor's and master's degree courses in agriculture, food and environmental science offered by the Di3A.

Aurelio Scavo

Aurelio Scavo is a Postdoctoral Fellowship Researcher in Agronomy and Field Crops in the Department of Veterinary Sciences at the University of Messina (UM), Italy. He has lectured for master's students and supervised B.Sc./M.Sc. theses. He is or has been involved in national and international research projects, and has received several scientific awards for his research activities. Currently, his main research lines are (1) the set-up of sustainable weed control practices; (2) the role of allelopathy in weed management, both in laboratory and field conditions; (3) the role of cover crops in agroecosystems; (4) the effects of mycorrhization on field crops; and (5) agronomical and technical practices to improve crop productivity in the Mediterranean area. He regularly participates in various national/international scientific conferences and acts as a reviewer for many ISI/Scopus peer-reviewed journals. He is the author/co-author of more than 37 papers in international scientific peer-reviewed journals (H-index 15 in the Scopus database), one book chapter, and more than 40 contributions to national or international congresses.

agronomy
MDPI

Editorial

Cropping Systems and Agronomic Management Practices of Field Crops

Umberto Anastasi and Aurelio Scavo *

Department of Agriculture, Food and Environment (Di3A), University of Catania, 95123 Catania, Italy; umberto.anastasi@unict.it
* Correspondence: aurelio.scavo@unict.it

Citation: Anastasi, U.; Scavo, A. Cropping Systems and Agronomic Management Practices of Field Crops. Agronomy 2023, 13, 2328. https://doi.org/10.3390/agronomy13092328

Received: 31 August 2023
Accepted: 4 September 2023
Published: 6 September 2023

1. Introduction

Agriculture is facing the challenge of a transition to sustainability to meet the growing demands for food, feed, and several other renewable nonfood raw materials under a changing climatic scenario. Research on innovative agronomic practices can help to guide this change, and can benefit the understanding of the complexity of agroecosystems. The optimization of the spatiotemporal combination of plants in farming systems (crop sequence, cover cropping, and intercropping), the reduction in the dependence on external energy input (soil tillage, agrochemicals, and mineral fertilizers), the set-up of innovative agronomic practices, and the increase in the use efficiency of native resources (radiation and rainfall, N_2, CO_2, H_2O, etc.) represent the driving forces behind this paradigmatic change. This approach will ensure the enhancement of the territorial vocation in productive and qualitative terms, also promoting several ecosystem services, from carbon sequestration to landscape ecology.

2. Overview of the Special Issue

In this Special Issue, we focus on the recent advancements in the wide scientific area of field crops in order to identify strategies and tactics calibrated site-by-site for eco-friendly and efficient agronomic management. It is a compilation of thirty-seven research articles and two reviews, where five are Editor's choice articles, and one is a feature paper. For simplicity, these original papers can be grouped into five groups:

1. Crop adaptation;
2. Weed management;
3. Fertilization;
4. Crop diversification;
5. Innovative cropping systems and agronomic practices.

2.1. Crop Adaptation

Climate change is nowadays affecting agricultural production in many areas worldwide. Consequently, it is of key importance to not only understand the impact of climate change on soil and atmosphere components of an agroecosystem, but also the study of the suitability of crops (i.e., plant species and cultivars) to changing climatic conditions and agronomic management. The studies of Mahmud et al. [1], Tuttolomondo et al. [2], and Ismael et al. [3] follow this direction, investigating the adaptability of orange fleshed sweet potatoes (*Ipomoea batatas* (L.) Lam.) to the riverbank inhabitants of the Gaibandha and Rangpur districts of Bangladesh, the productive and qualitative characteristics of three Sicilian *Salvia sclarea* L. populations, and the dynamics of rice (*Oryza sativa* L.) farming systems in Southern Mozambique to guide smallholder farmers. In another article, Zhou et al. [4] studied the effects of temperature and solar radiation on milling and the appearance quality of a number of rice varieties sowed at different times in the lower reaches of Huai River (Jiangsu, China). The authors found that temperature, compared to solar radiation, was the main environmental factor affecting the milling and appearance quality of rice in the studied area, and indicated the optimal thermal ranges and sowing dates to obtain a relatively high yield as well as good milling and appearance quality of rice in the lower reaches of Huai River. Similarly, Wang et al. [5] explored the characteristics of heat occurrence during maize flowering in the Huaibei Plain (Anhui, China) in orde to advise summer maize cropping strategies in the studied region and in other semiarid cropping systems.

2.2. Weed Management

The increasing intensification of weed control practices has posed serious environ mental issues, such as the leaching (and consequently water contamination) of glyphosate herbicide and its main degradation product aminomethylphosphonic acid (AMPA), as evaluated by Milan et al. [6]. The search for sustainable weed management techniques is o outstanding importance to reduce the negative effects associated with weed control and increase the resilience of cropping systems. This Special Issue involves two review articles that discuss two important and innovative aspects of weed management, i.e., the exploita tion of allelopathy [7] and the use of encapsulated herbicides in organic formulations [8 Of course, the choice of the weed management practice used should be associated with the context of the cropping system. For instance, Nazir et al. [9] found that the lowest rate o nutrients removed via weeds in rice across temperate climates depended on the combina tion between the establishment method (transplanting, direct seeding, or a system of rice intensification) and the adopted weed management practice. Furthermore, to contrast the increasing infestations of indigenous and exotic weeds in temperate regions, Iqbal et al. [10 studied the potential fit of forage cowpea (*Vigna unguiculata* (L.) Walp) in the temperate Himalayan region of Pakistan by dissecting the interactive effect of genetic potential and row configuration on weed density, growth attributes, biomass yield, and the nutritiona quality of the crop.

2.3. Fertilization

The Special Issue published five research articles on this topic. In the first article Singh et al. [11] applied the soil test crop response (STCR) in the fertilizer approach instead of the generally recommended dose (GRD) methodology to markedly enhance the pro ductivity, profitability, and nutrient use efficiency of rice. In the second article, 17-year-old integrated nutrient management under a maize–wheat cropping system was studied by Dhaliwal et al. [12] for the buildup of organic carbon, microbial communities, and soi nutrient status. Nevertheless, this Special Issue dealt with technical aspects of fertilization such as the subsurface application of mineral fertilizers to decrease the accumulation of nutrients in the top soil layers under no-tillage systems [13], the application of compos by microdosing to double the fertilized area and improve sorghum (*Sorghum bicolor* [L. Moench) productivity in Southern Mali [14], or the use of plasma-treated cattle slurry to produce nitrogen-enriched organic fertilizers [15].

In addition, in the present Special Issue, attention has also been paid to the combined effect of fertilization with weed management [16], irrigation [17,18], the crop establishment method [19], crop variety [20], sowing density [21,22], seed priming [23], and fungicide application [24]. For instance, with regard to the fertilization–irrigation combination Abdou et al. [17], in light of the water shortage caused by climate change, proposed a new agro-management practice (deficit irrigation and higher nitrogen fertilizer) for lowland rice in semiarid conditions as an alternative to the flooding system. In a similar experiment, Bhatt et al. [18] suggested the optimal potassium application rate for sugarcane (*Saccharum officinarum* L.) cropping systems under the potassium-deficient water-stressed conditions of Northern India.

2.4. Crop Diversification

The optimization of the spatiotemporal arrangement between crops through cover cropping, intercropping, or crop rotation is an ancient practice that has been gaining popularity in recent years by virtue of its numerous ecosystem services [25]. Five research articles have been published in this Special Issue on this topic. In the study of Johnson and co-authors [26], the interseeding of winter camelina (*Camelina sativa* (L.) Crantz) and winter rye (*Secale cereale* L.) into soybean (*Glycine max* (L.) Merr.) was investigated in the northern plains of the USA. Abbas et al. [27] demonstrated that maize–green gram intercropping is a sustainable agronomic practice to increase maize production and reduce weed infestations for smallholder farmers in semiarid environments. Kumar and co-workers [28], analyzing different crop diversification schemes under jute (*Corchorus olitorius* L.)-based cropping systems, suggested a jute–rice–baby corn scheme for system productivity and a jute–rice–pea scheme for system sustainability, with both productivity and sustainability being higher when the recommended doses of fertilizers were applied with crop residue incorporation. The management of cover crops shows outstanding importance in increasing their efficacy in agroecosystems [25]. The works of Salama and Abdel-Moneim [29] and Cottney et al. [30] deal with cover crop management. The former article evaluated the manipulation of sowing schedule and maize harvest regime in a soybean–fodder–maize intercropping system in Northern Egypt, whereas in the latter one, the choice of cover crop genotype and sowing date and their effects on the subsequent cash crop were studied in Northern Ireland.

2.5. Innovative Cropping Systems and Agronomic Practices

In order to contrast the harmful effects of climate change and to meet the needs of a growing global population, in this Special Issue, several innovative cropping systems and agronomic practices have been proposed. The aim of the study by Larkin et al. [31] was to examine four different potato cropping systems designed to address specific management goals (soil conservation, soil improvement, disease suppression, and a status quo standard rotation) for potato crop growth and yield characteristics under both irrigated and rainfed conditions in Maine (USA). Bunyangha et al. [32], comparing two paddy rice farming pathways (smallholder and large-scale commercial) and an adjacent natural wetland in the Mpologoma catchment (Uganda), highlighted that large-scale commercial paddies not only had higher richness and diversity than natural wetland and smallholder paddies, but also underscored the role of soil in influencing the macroinvertebrate community in rice paddies. In light of the production increase in quinoa worldwide (*Chenopodium quinoa* Willd.), Alvar-Beltrán et al. [33] quantified, for the first time, greenhouse gas emissions (CO_2, CH_4, and N_2O) and crop productivity (yields and biomass) under conventional (urea) and organic (digestate) fertilization. Analyzing agronomic management (tillage, weed control, growth regulation, rate of nitrogen and sulfur fertilizers) in the production of winter oilseed rape (*Brassica napus* L.), the most important oilseed crop in the temperate climates, Sokólski and co-authors [34] found that the chemical components of seeds were differently affected by tillage systems, that an increase in the N rate application enhanced the total protein content and decreased the crude fat content, and that sulfur fertilization increased glucosinolate concentrations. In another work, Akinseye et al. [35], adopting the Agricultural Production Systems sIMulator (APSIM) model in the three major agroecologies of North-Eastern Nigeria, identified the optimal sowing time and cultivar choice for sorghum productivity to overcome the low soil fertility and early terminal drought in the studied zone.

Regarding the innovative agronomic practices proposed here, Sun and co-workers [36] studied the border effects (in terms of dry matter, photosynthetic characteristics, and yield components) of winter wheat under hole sowing cultivation, a new wheat agricultural technology integrating rain, drought resistance, and the efficient utilization of light and heat resources. Madala et al. [37] evaluated, for the first time, the effects of planting pregerminated buds on stand establishment in sugarcane, while Li et al. [38] assessed the effect of mechanized transplanting properties on sweet potato growth and yield. Investigati the combination between soil management systems (conventional and no-tillage) a pre-harvest desiccation on the physiological quality of soybean seeds, Silva et al. [3 reported that the use of desiccant is dependent on the soil management system and th soybean seed longevity was higher in the no-tillage system, although desiccant applicati reduced it. Investigating the optimization of planting density in alpine mountain strawber cultivation (South Tyrol, Italy), Soppelsa et al. [40] indicated that a middle planting densi can be a fair compromise in terms of plant growth, yield, and farm profit.

3. Conclusions

In summary, the manuscripts collected in this Special Issue provided a relevant know edge contribution to the cropping systems and agronomic management practices of fie crops under a climate change scenario. We sincerely thank all of the contributing autho and reviewers, as well as the Academic Editors and the Managing Editor Amanda Li, f the time that they have dedicated to this successful Special Issue.

Author Contributions: U.A. and A.S. contributed equally to this article. All authors have read a agreed to the published version of the manuscript.

Conflicts of Interest: The authors declare no conflict of interest.

References

1. Mahmud, A.-A.; Alam, M.J.; Heck, S.; Grüneberg, W.J.; Chanda, D.; Rahaman, E.H.M.S.; Molla, M.S.H.; Anwar, M.M Talukder, M.A.-A.H.; Ali, M.A.; et al. Assessing the Productivity, Quality and Profitability of Orange Fleshed Sweet Po toes Grown in Riverbank of the Tista Floodplain Agro-Ecological Zone of Bangladesh. *Agronomy* **2021**, *11*, 2046. [CrossRef]
2. Tuttolomondo, T.; Virga, G.; Licata, M.; Iacuzzi, N.; Farruggia, D.; Bella, S.L. Assessment of Production and Qualitati Characteristics of Different Populations of *Salvia sclarea* L. Found in Sicily (Italy). *Agronomy* **2021**, *11*, 1508. [CrossRef]
3. Ismael, F.; Mbanze, A.A.; Ndayiragije, A.; Fangueiro, D. Understanding the Dynamic of Rice Farming Systems in Southe Mozambique to Improve Production and Benefits to Smallholders. *Agronomy* **2021**, *11*, 1018. [CrossRef]
4. Zhou, N.; Wei, H.; Zhang, H. Response of Milling and Appearance Quality of Rice with Good Eating Quality to Temperature a Solar Radiation in Lower Reaches of Huai River. *Agronomy* **2021**, *11*, 77. [CrossRef]
5. Wang, Z.; Sun, W.; Liu, X.; Li, Y.; Collins, B.; Ullah, N.; Song, Y. Analysis on Heat Characteristics for Summer Maize Cropping ir Semi-Arid Region. *Agronomy* **2022**, *12*, 1435. [CrossRef]
6. Milan, M.; Vidotto, F.; Fogliatto, S. Leaching of Glyphosate and AMPA from Field Lysimeters. *Agronomy* **2022**, *12*, 328. [CrossRe
7. Scavo, A.; Mauromicale, G. Crop Allelopathy for Sustainable Weed Management in Agroecosystems: Knowing the Present with View to the Future. *Agronomy* **2021**, *11*, 2104. [CrossRef]
8. Rodríguez-Mejías, F.J.; Scavo, A.; Chinchilla, N.; Molinillo, J.M.G.; Schwaiger, S.; Mauromicale, G.; Macías, F.A. Perspectiv and Advances in Organic Formulations for Agriculture: Encapsulation of Herbicides for Weed Control. *Agronomy* **2023**, *13*, 18 [CrossRef]
9. Nazir, A.; Bhat, M.A.; Bhat, T.A.; Fayaz, S.; Mir, M.S.; Basu, U.; Ahanger, S.A.; Altaf, S.; Jan, B.; Lone, B.A.; et al. Comparati Analysis of Rice and Weeds and Their Nutrient Partitioning under Various Establishment Methods and Weed Manageme Practices in Temperate Environment. *Agronomy* **2022**, *12*, 816. [CrossRef]
10. Iqbal, M.A.; Ahmed, A.; Imran, M.; Ahmed, H.E.; Hafez, R.M.; Hamad, A.A. Genetic Divergence and Spatial Configurati Influence the Weed Spectrum, Herbage Yield and Nutritive Quality of Temperate Cowpea. *Agronomy* **2022**, *12*, 1323. [CrossRe
11. Singh, V.K.; Gautam, P.; Nanda, G.; Dhaliwal, S.S.; Pramanick, B.; Meena, S.S.; Alsanie, W.F.; Gaber, A.; Sayed, S.; Hossain, A. S Test Based Fertilizer Application Improves Productivity, Profitability and Nutrient Use Efficiency of Rice (*Oryza sativa* L.) und Direct Seeded Condition. *Agronomy* **2021**, *11*, 1756. [CrossRef]
12. Dhaliwal, S.S.; Sharma, S.; Sharma, V.; Shukla, A.K.; Walia, S.S.; Alhomrani, M.; Gaber, A.; Toor, A.S.; Verma, Randhawa, M.K.; et al. Long-Term Integrated Nutrient Management in the Maize–Wheat Cropping System in Alluvi Soils of North-Western India: Influence on Soil Organic Carbon, Microbial Activity and Nutrient Status. *Agronomy* **2021**, *11*, 22 [CrossRef]
13. Kraska, P.; Andruszczak, S.; Gierasimiuk, P.; Rusecki, H. The Effect of Subsurface Placement of Mineral Fertilizer on Some S Properties under Reduced Tillage Soybean Cultivation. *Agronomy* **2021**, *11*, 859. [CrossRef]
14. Guindo, M.; Traoré, B.; Birhanu, B.Z.; Coulibaly, A.; Tabo, R. Microdosing of Compost for Sustainable Production of Improv Sorghum in Southern Mali. *Agronomy* **2022**, *12*, 1480. [CrossRef]

5. Cottis, T.; Mousavi, H.; Solberg, S.Ø. Plasma Treated Cattle Slurry Moderately Increases Cereal Yields. *Agronomy* **2023**, *13*, 1549. [CrossRef]
6. Abbas, R.N.; Iqbal, A.; Iqbal, M.A.; Ali, O.M.; Ahmed, R.; Ijaz, R.; Hadifa, A.; Bethune, B.J. Weed-Free Durations and Fertilization Regimes Boost Nutrient Uptake and Paddy Yield of Direct-Seeded Fine Rice (*Oryza sativa* L.). *Agronomy* **2021**, *11*, 2448. [CrossRef]
7. Abdou, N.M.; Abdel-Razek, M.A.; Abd El-Mageed, S.A.; Semida, W.M.; Leilah, A.A.A.; Abd El-Mageed, T.A.; Ali, E.F.; Majrashi, A.; Rady, M.O.A. High Nitrogen Fertilization Modulates Morpho-Physiological Responses, Yield, and Water Productivity of Lowland Rice under Deficit Irrigation. *Agronomy* **2021**, *11*, 1291. [CrossRef]
8. Bhatt, R.; Singh, J.; Laing, A.M.; Meena, R.S.; Alsanie, W.F.; Gaber, A.; Hossain, A. Potassium and Water-Deficient Conditions Influence the Growth, Yield and Quality of Ratoon Sugarcane (*Saccharum officinarum* L.) in a Semi-Arid Agroecosystem. *Agronomy* **2021**, *11*, 2257. [CrossRef]
9. Mohanta, S.; Banerjee, M.; Malik, G.C.; Shankar, T.; Maitra, S.; Ismail, I.A.; Dessoky, E.S.; Attia, A.O.; Hossain, A. Productivity and Profitability of Kharif Rice Are Influenced by Crop Establishment Methods and Nitrogen Management in the Lateritic Belt of the Subtropical Region. *Agronomy* **2021**, *11*, 1280. [CrossRef]
0. Liava, V.; Karkanis, A.; Danalatos, N.; Tsiropoulos, N. Effects of Two Varieties and Fertilization Regimes on Growth, Fruit, and Silymarin Yield of Milk Thistle Crop. *Agronomy* **2022**, *12*, 105. [CrossRef]
1. Sun, Y.; Yang, W.; Wu, Y.; Cui, Y.; Dong, Y.; Dong, Z.; Hai, J. The Effects of Different Sowing Density and Nitrogen Topdressing on Wheat Were Investigated under the Cultivation Mode of Hole Sowing. *Agronomy* **2023**, *13*, 1733. [CrossRef]
2. Dong, S.; Wang, G.; Li, X.; Kang, Y. A Trade-Off between the Growing Performance and Sowing Density of Sunflower (*Helianthus annuus* L.) under Fertigation in an Arid Saline Area. *Agronomy* **2023**, *13*, 179. [CrossRef]
3. Banerjee, P.; Venugopalan, V.K.; Nath, R.; Chakraborty, P.K.; Gaber, A.; Alsanie, W.F.; Raafat, B.M.; Hossain, A. Seed Priming and Foliar Application of Nutrients Influence the Productivity of Relay Grass Pea (*Lathyrus sativus* L.) through Accelerating the Photosynthetically Active Radiation (PAR) Use Efficiency. *Agronomy* **2022**, *12*, 1125. [CrossRef]
4. Gaile, Z.; Bankina, B.; Pluduma-Paunina, I.; Sterna, L.; Bimsteine, G.; Svarta, A.; Kaneps, J.; Arhipova, I.; Sutka, A. Performance of Winter Wheat (*Triticum aestivum*) Depending on Fungicide Application and Nitrogen Top-Dressing Rate. *Agronomy* **2023**, *13*, 318. [CrossRef]
5. Scavo, A.; Fontanazza, S.; Restuccia, A.; Pesce, G.R.; Abbate, C.; Mauromicale, G. The role of cover crops in improving soil fertility and plant nutritional status in temperate climates. A review. *Agron. Sustain. Dev.* **2022**, *42*, 93. [CrossRef]
6. Johnson, K.L.; Kandel, H.J.; Samarappuli, D.P.; Berti, M.T. Interseeding Camelina and Rye in Soybean with Varying Maturity Provides Soil Cover without Affecting Soybean Yield. *Agronomy* **2021**, *11*, 353. [CrossRef]
7. Abbas, R.N.; Arshad, M.A.; Iqbal, A.; Iqbal, M.A.; Imran, M.; Raza, A.; Chen, J.-T.; Alyemeni, M.N.; Hefft, D.I. Weeds Spectrum, Productivity and Land-Use Efficiency in Maize-Gram Intercropping Systems under Semi-Arid Environment. *Agronomy* **2021**, *11*, 1615. [CrossRef]
8. Kumar, M.; Mitra, S.; Mazumdar, S.P.; Majumdar, B.; Saha, A.R.; Singh, S.R.; Pramanick, B.; Gaber, A.; Alsanie, W.F.; Hossain, A. Improvement of Soil Health and System Productivity through Crop Diversification and Residue Incorporation under Jute-Based Different Cropping Systems. *Agronomy* **2021**, *11*, 1622. [CrossRef]
9. Salama, H.S.A.; Abdel-Moneim, M.H. Maximizing Land Use Efficiency and Productivity of Soybean and Fodder Maize Intercrops through Manipulating Sowing Schedule and Maize Harvest Regime. *Agronomy* **2021**, *11*, 863. [CrossRef]
0. Cottney, P.; Black, L.; Williams, P.; White, E. How Cover Crop Sowing Date Impacts upon Their Growth, Nutrient Assimilation and the Yield of the Subsequent Commercial Crop. *Agronomy* **2022**, *12*, 369. [CrossRef]
1. Larkin, R.P.; Honeycutt, C.W.; Griffin, T.S.; Olanya, O.M.; He, Z. Potato Growth and Yield Characteristics under Different Cropping System Management Strategies in Northeastern U.S. *Agronomy* **2021**, *11*, 165. [CrossRef]
2. Bunyangha, J.; Muthumbi, A.W.; Gichuki, N.N.; Majaliwa, M.J.G.; Egeru, A. Soil Macroinvertebrate Response to Paddy Rice Farming Pathways in Mpologoma Catchment, Uganda. *Agronomy* **2022**, *12*, 312. [CrossRef]
3. Alvar-Beltrán, J.; Dalla Marta, A.; Vivoli, R.; Verdi, L.; Orlandini, S. Greenhouse Gas Emissions and Yield Production from an Organic and Conventional Fertilization on Quinoa. *Agronomy* **2022**, *12*, 1015. [CrossRef]
4. Sokólski, M.; Załuski, D.; Szatkowski, A.; Jankowski, K.J. Winter Oilseed Rape: Agronomic Management in Different Tillage Systems and Seed Quality. *Agronomy* **2023**, *13*, 524. [CrossRef]
5. Akinseye, F.M.; Ajeigbe, H.A.; Kamara, A.Y.; Omotayo, A.O.; Tofa, A.I.; Whitbread, A.M. Establishing Optimal Planting Windows for Contrasting Sorghum Cultivars across Diverse Agro-Ecologies of North-Eastern Nigeria: A Modelling Approach. *Agronomy* **2023**, *13*, 727. [CrossRef]
6. Sun, Y.; Yang, C.; Liang, H.; Yang, Y.; Bu, K.; Dong, Y.; Hai, J. The Border Effects of Dry Matter, Photosynthetic Characteristics, and Yield Components of Wheat under Hole Sowing Condition. *Agronomy* **2023**, *13*, 766. [CrossRef]
7. Madala, H.V.; Lesmes-Vesga, R.A.; Odero, C.D.; Sharma, L.K.; Sandhu, H.S. Effects of Planting Pre-Germinated Buds on Stand Establishment in Sugarcane. *Agronomy* **2023**, *13*, 1001. [CrossRef]
8. Li, H.; Wang, B.; Shi, S.; Zhou, J.; Shi, Y.; Liu, X.; Liu, H.; He, T. Response of Crop Performance and Yield of Spring Sweet Potato (*Ipomoea batatas* [L.] Lam) as Affected by Mechanized Transplanting Properties. *Agronomy* **2023**, *13*, 1611. [CrossRef]

39. Silva, G.F.D.; Chamma, L.; Luperini, B.C.O.; Chaves, P.P.N.; Calonego, J.C.; Nakagawa, J.; Silva, E.A.A.D. Physiological Quality o Soybean Seeds as a Function of Soil Management Systems and Pre-Harvest Desiccation. *Agronomy* **2023**, *13*, 847. [CrossRef]
40. Soppelsa, S.; Gasser, M.; Zago, M. Optimizing Planting Density in Alpine Mountain Strawberry Cultivation in Martell Valley Italy. *Agronomy* **2023**, *13*, 1422. [CrossRef]

agronomy
MDPI

ticle

'he Effects of Different Sowing Density and Nitrogen :opdressing on Wheat Were Investigated under the :ultivation Mode of Hole Sowing

tao Sun [1,†], Wenlong Yang [1,†], Yilun Wu [1,†], Youhe Cui [1], Yongli Dong [1], Zhoujia Dong [2] and Jiangbo Hai [1,*]

1 College of Agronomy, Northwest A&F University, Yangling, Xianyang 712100, China; sunyitao@nwafu.edu.cn (Y.S.); yangwenlong2021@126.com (W.Y.); wuyilun21@163.com (Y.W.); cuiyouhe1999@163.com (Y.C.); 2008114433@nwafu.edu.cn (Y.D.)
2 Agricultural and Animal Husbandry Comprehensive Service Center, Tongren 811399, China; dzj678@163.com
* Correspondence: haijiangbo@126.com; Tel.: +86-133-8922-1092
† These authors contributed equally to this work.

Abstract: Hole sowing is a new and efficient cultivation method with few studies. This study investigated the effects of different sowing densities and nitrogen topdressing at the jointing stage on dry matter, quality, and yield under wheat hole sowing to provide a theoretical basis for integrating wheat fertilizer and density-supporting technology. In this study, a two-factor split-plot design was used. The sowing density was the main plot, and four levels were set: D1, D2, D3, and D4 (238, 327, 386, and 386 suitable seeds·m^{-2}). The four sowing levels were sown according to 8 grains/hole, 11 grains/hole, 13 grains/hole, and 16 grains/hole, respectively, with a row spacing of 25 cm and a hole spacing of 13.5 cm; the amount of nitrogen fertilizer applied at the jointing stage was the sub-area, with four levels: N1, N2, N3, and N4 (0, 60, 120, and 180 kg·ha^{-1}). After two years of experimental research, the following main conclusions are drawn: the use of high sowing density and nitrogen topdressing is helpful to improve the dry matter quality of wheat spikes at the maturing stage; the sowing density had significant or highly significant effects on protein content, starch content, and sedimentation value. The yield from 2018–2019 reached a maximum of 8448.67 kg·ha^{-1} under D4N4 treatment, and the yield from 2019–2020 reached a maximum of 10,136.40 kg·ha^{-1} under D4N3 treatment. Therefore, the combination of 225 kg·ha^{-1} sowing density and 120–180 kg·ha^{-1} nitrogen topdressing at the jointing stage can be used in field production, which can help improve wheat production potential. Similarly, understanding the interaction between wheat hole sowing and different sowing densities and nitrogen topdressing amounts provides a practical reference for high-yield wheat cultivation techniques.

Keywords: wheat; *Triticum aestivum* L.; hole sowing; cultivation techniques; yield

ation: Sun, Y.; Yang, W.; Wu, Y.; , Y.; Dong, Y.; Dong, Z.; Hai, J. The ects of Different Sowing Density l Nitrogen Topdressing on Wheat re Investigated under the ltivation Mode of Hole Sowing. onomy 2023, 13, 1733. https:// .org/10.3390/agronomy13071733

ademic Editors: Umberto Anastasi l Aurelio Scavo

ceived: 30 May 2023
vised: 25 June 2023
cepted: 26 June 2023
olished: 28 June 2023

1. Introduction

With the growth of the population, food security has become a severe problem for the world. In 2015, among the world's 7.3 billion people, an estimated 654 million people were malnourished [1–3]. By 2019, 864 million people were considered malnourished. In order to meet global food demand, food production needs to increase by 70~100% by 2050 [4–6]. Wheat is an important food source for humans worldwide, with 20% of the world's wheat consumption by 50% of the world's poorest people [7–10]. More than 50% of the world's wheat comes from developing countries, and more land is planted for wheat than for any other crop in the world [11,12].

In wheat cultivation, sowing density and nitrogen fertilizer are critical factors affecting wheat population structure and yield formation [13–16]. Suitable sowing density can make wheat make full use of water, nutrients, and light energy [17,18], alleviate the competition between populations and individuals, and help to construct a reasonable populatic structure [19,20]. Rational use of nitrogen fertilizer can promote the healthy growth wheat, improve grain quality, increase yield, and achieve sustainable development agriculture [15,21,22]. Many experts and scholars have carried out much research on tl level of nitrogen supply in crops. If the application of chemical fertilizer is stopped, will half the total global crop yield [23–25]. In addition, the unreasonable use of nitrog fertilizer will also lead to environmental problems such as groundwater pollution [2 greenhouse effect, soil acidification [27], and so on. Therefore, the rational use of nitrog fertilizer while achieving high yield and quality of wheat is significant for wheat productic

As a new cultivation technology, wheat hole sowing is an efficient agricultural tec nology integrating rain, drought resistance, and efficient utilization of light and he resources [28,29]. Due to the characteristics of wheat hole sowing cultivation, each ho has a noticeable border effect. The outer wheat of each hole has more solar energy, bett ventilation, and less nutrient competition than the inner wheat [30]. Therefore, in tl actual field production, the boundary advantage of hole sowing itself helps to impro productivity and bring more economic benefits and value to people.

In this study, from 2018 to 2020, through wheat cultivation with the hole sowi method, its border effect was measured. Different amounts of nitrogen fertilizer we applied according to different sowing densities and jointing stages to explore the effec of different sowing densities and nitrogen topdressing amounts and their interaction c the dry matter, quality, and yield of wheat. We assumed that different sowing densi nitrogen topdressing, and their interaction would have different effects on dry matter wheat spikes, grain quality, and yield. The objectives of this study were to: (1) explore tl effects of different sowing density and nitrogen topdressing on dry matter of wheat spik (2) evaluate the effects of different sowing density and nitrogen topdressing on gra quality; (3) evaluate the effects of different sowing density and nitrogen topdressing c yield and components. This study's results will help provide new ideas and references f future research on wheat hole sowing to help scholars quickly lock in relevant knowled and insights in the field.

2. Materials and Methods

2.1. Test Designs

This experiment was conducted at the Doukou Crop Experimental Demonstratic Station of Northwest A & F University from 2018 to 2020. The experimental demonstratic station is located in Xinglong Village, Yunyang Town, Jingyang County, Xianyang Ci Shaanxi Province, China, 108°52′ E, 34°37′ N. The precipitation during the two-year grow period of wheat was 84.57 mm and 122.68 mm, and the average temperature was 9.53 ° and 10.65 °C, respectively (Figures 1 and 2). The soil in the test field was loam. Befo sowing, 0–40 cm soil samples were randomly drilled at 5 points. After air drying, grindin and screening, the soil's basic nutrient content was determined: organic matter conte (potassium dichromate method) 18.02 g·kg^{-1}, total nitrogen content (inorganic and organ semi-micromethod of Kay's fixed nitrogen) 1.39 g·kg^{-1}, available nitrogen content (nitr acid powder test method) 86.8 mg·kg^{-1}, available phosphorus content (ultraviolet spe trophotometry colorimetry) 16.83 mg·kg^{-1}, available potassium content (flare photomete 232.07 mg·kg^{-1}, pH value 7.93, with medium fertility.

The 'XN805' wheat variety was selected as the experimental material. The variety is semi-winter mid-early-maturity variety, with semi-stowing seedlings, dark green leave medium tillering ability, high panicle rate, medium winter cold resistance, medium la spring cold resistance, and average plant height of 66.9 cm and of a compact plant type. Tl main area was sowing density, and four sowing density levels were set: D1 (238 suitab seeds·m^{-2}), D2 (327 suitable seeds·m^{-2}), D3 (386 suitable seeds·m^{-2}), D4 (475 suitab seeds·m^{-2}). The sub-area was the amount of nitrogen topdressing at the jointing sta (P, K fixed), and four nitrogen fertilizer application levels were set. The nitrogen fertiliz (nitrogen content 46.4%) and the base fertilizer were wheat special slow-release fertiliz

(N: P_2O_5: K_2O mass fraction 24: 15: 5) 750 kg·ha^{-1}, and the base fertilizer was applied once during rotary tillage. Nitrogen fertilizer without basal fertilizer was applied at the jointing stage: N1 (no urea), N2 (urea 60 kg·ha^{-1}), N3 (urea 120 kg·ha^{-1}), N4 (urea 180 kg·ha^{-1}). The sowing method was hole sowing. After calculating the four sowing density levels, the sowing was carried out according to 8 grains/hole, 11 grains/hole, 13 grains/hole, and 16 grains/hole, respectively. The row spacing was 25 cm and the hole spacing was 13.5 cm. Each plot was 3.5 m × 2 m = 7 m^2. Sowing was carried out manually on 5 October 2018 and 1 October 2019, weeding and pest control were carried out at different crop growth stages throughout the wheat growing season, and other management measures were taken to ensure consistency with local high-yielding farmland. During the experiment, the wheat was sown for 10 days, in mid-November, March, and May of the second year, and irrigated according to the actual situation in the field. The two-year processing was consistent and it was harvested on 4 June 2019 and 1 June 2020.

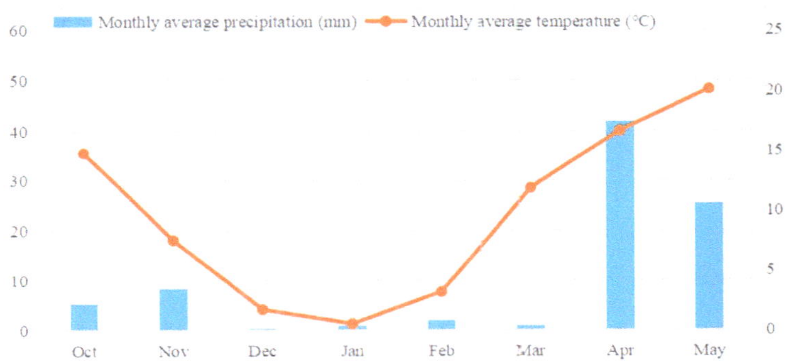

Figure 1. Total precipitation and monthly mean temperature during wheat growth stage from October 2018 to June 2019.

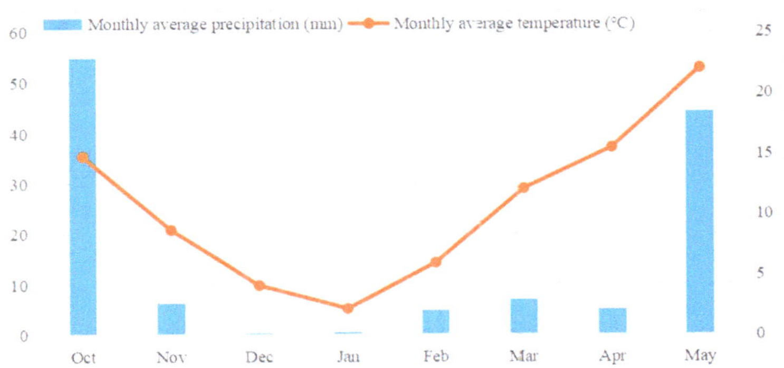

Figure 2. Total precipitation and monthly mean temperature during wheat growth stage from October 2019 to June 2020.

2.2. Determination Items and Methods

Dry matter of wheat spikes: 20 plants with uniform growth were randomly selected in each hole at the booting stage, heading stage, flowering stage, filling stage, and maturing stage. On the same day as harvesting, the wheat spikes were baked in an oven at 105 °C for 30 min, then the temperature was reduced to 60–80 °C, and the drying was continued for about 8 h so that it was quickly dried and then removed. Finally, the sample continued to dry for 4 h, with weighing again until the weight was constant, then the final weight was measured.

Wheat grain quality: After two months of physiological after-ripening, the protein content, stability time, starch content, and sedimentation value of wheat grain samples after harvest were measured by the Danish FOSS Infratec TM 1241 (Manufactured by FOSS China Co., Ltd., Beijing, China) near-infrared grain quality analyzer.

Yield composition statistics: After the wheat matured, the effective panicles in the 1 m double-row sample section of each plot were counted. After harvest, the samples were sun-dried to remove impurities and a few plates of 1000 grains were weighed, which was repeated 3 times and the average value was taken for the 1000-grain weight. In each plot 20 plants with uniform growth were randomly selected, and the grains per spike were counted to obtain the average value. Due to the small area of the plot, the hole sowing had an obvious border effect. In order to eliminate the influence of the border effect on the yield, 1 m^2 wheat was randomly taken from each plot in the middle and threshed with a thresher, dried in the sun, and weighed with an electronic balance to calculate the grain yield ($kg \cdot ha^{-1}$).

2.3. Statistical Analysis of Data

Microsoft Office Excel 2021 and SPSS 26.0 were used for statistical analysis. RStudio was used for linear regression analysis, correlation analysis, and figure drawing. The significance level ($p < 0.05$) was used to judge the average difference by the minimum significant difference test.

3. Results

3.1. Effects of Different Sowing Density and Nitrogen Topdressing on Dry Matter of Wheat Spikes

The dry matter of wheat spikes at different stages (booting stage, heading stage flowering stage, filling stage, and maturing stage) was measured and analyzed (Table 1). It can be seen from Table 1 that the effect of sowing density on the dry matter under different treatments designed in this experiment was very significant in the heading stage flowering stage, filling stage, and maturing stage from 2018–2019 and 2019–2020. The effect of nitrogen topdressing amount on the dry matter under different treatments was highly significant at the filling stage and maturing stage from 2018–2019 and 2019–2020, and there were also significant differences between the heading stage and the flowering stage from 2019–2020. There was no significant difference in the dry matter of wheat spikes in different years; there were significant differences in heading stage, flowering stage, filling stage, and maturing stage of wheat between different years and sowing densities.

Table 1. Effects of different sowing density and nitrogen topdressing on dry matter of wheat spikes in different stages.

Year	Sowing Density	Nitrogen Topdressing	Booting Stage	Heading Stage	Flowering Stage	Filling Stage	Maturing Stage
2018–2019	D1	N1	1.12 ab	1.35 cde	1.65 def	1.68 g	1.75 k
		N2	1.06 ab	1.42 abcde	1.78 bcdef	1.89 fg	2.08 jk
		N3	1.13 ab	1.56 abc	1.79 bcdef	2.21 bcd	2.35 hij
		N4	1.12 ab	1.39 bcde	1.74 cdef	2.18 bcd	2.42 ghij
	D2	N1	0.98 ab	1.23 de	1.62 ef	1.92 efg	2.24 ij
		N2	1.03 ab	1.27 de	1.58 f	1.92 efg	2.29 hij
		N3	1.07 ab	1.38 bcde	1.74 cdef	2.01 def	2.59 defgh
		N4	0.96 ab	1.19 e	1.93 abcd	2.17 bcde	2.53 efghi
	D3	N1	1.12 ab	1.48 abcd	1.92 abcd	2.07 cdef	2.44 fghi
		N2	0.94 b	1.26 de	1.99 abc	1.99 def	2.77 cdef

Table 1. *Cont.*

Year	Sowing Density	Nitrogen Topdressing	Booting Stage	Heading Stage	Flowering Stage	Filling Stage	Maturing Stage
		N3	0.88 b	1.42 abcde	1.88 abcde	2.13 cdef	2.81 cde
		N4	1.22 a	1.36 bcde	1.92 abcd	2.39 b	3.07 abc
	D4	N1	0.93 b	1.57 abc	1.91 abcde	2.19 bcd	2.75 cdefg
		N2	0.97 ab	1.57 abc	1.91 abcde	2.3 bc	2.92 bcd
		N3	1.06 ab	1.62 ab	2.04 ab	2.99 a	3.19 ab
		N4	1.22 a	1.66 a	2.13 a	3.09 a	3.39 a
	F value	FD	1.117	14.17 ***	10.608 ***	54.44 ***	214.04 ***
		FN	1.95	1.95	2.423	39.23 ***	79.96 ***
		FD × FN	2.15	0.893	1.209	4.109 *	0.047
2019–2020	D1	N1	0.98 cd	1.17 f	1.45 i	1.83 f	1.90 h
		N2	1.06 bcd	1.34 def	1.66 ghi	1.77 f	2.04 gh
		N3	1.13 bcd	1.44 cd	1.73 fghi	1.89 ef	2.14 gh
		N4	1.16 abc	1.42 cde	1.8 defgh	1.88 ef	2.16 g
	D2	N1	0.98 cd	1.41 cde	1.75 efghi	1.96 ef	2.07 gh
		N2	1.03 bcd	1.34 def	1.59 hi	1.83 f	2.08 gh
		N3	1.07 bcd	1.41 cde	1.62 hi	1.86 f	1.91 gh
		N4	0.99 cd	1.48 cd	1.73 fghi	2.09 de	2.45 f
	D3	N1	1.12 bcd	1.54 bc	1.96 cdefg	2.24 cd	2.52 ef
		N2	0.94 cd	1.24 ef	1.81 defgh	2.27 cd	2.71 cde
		N3	0.91 d	1.31 def	2.01 cdef	2.43 c	2.57 ef
		N4	1.25 ab	1.46 cd	2.06 bcde	2.41 c	2.88 bcd
	D4	N1	0.93 cd	1.37 cdef	2.13 bcd	2.34 c	2.64 def
		N2	1 cd	1.49 cd	2.35 ab	2.84 b	2.95 bc
		N3	1.09 bcd	1.73 ab	2.24 bc	2.96 ab	3.08 ab
		N4	1.39 a	1.81 a	2.67 a	3.08 a	3.3 a
	F value	FD	2.876 *	8.932 ***	45.67 ***	91.733 ***	105.9 ***
		FN	0.728	5.437 **	15.16 ***	6.903 ***	15.8 ***
		FD × FN	1.093	5.205 ***	1.206	6.291 ***	3.128 **
		FY	1.058	1.428	1.878	2.234	2.467
		FY × FD	ns	**	*	**	*
		FY × FN	ns	ns	ns	ns	ns
		FY × FD × FN	ns	ns	ns	ns	ns

Note: Y, D, and N represent different years, sowing density, and nitrogen topdressing, respectively. Different letters in the same column mean significant difference at 0.05. ns, not significant at 0.05 probability level; *, **, and *** refer to significant differences at 0.05, 0.01, and 0.001 level, same as Tables 2 and 3.

At the late filling stage, the assimilates of wheat plants are transported to the grains in large quantities, and their dry matter reaches the maximum at the maturity stage, eventually affecting their yield. Therefore, compared with other growth stages, the dry matter of the wheat maturity stage has a more significant impact. With the increase in sowing density, the overall performance of dry matter in the maturing stage was N4 > N3 > N2 > N1; with the increase in the amount of nitrogen, the overall performance of dry matter in the maturing stage was D4 > D3 > D2 > D1. As the dry matter of wheat spikes in the maturing stage was the most prominent, different sowing densities and nitrogen topdressing amounts were significantly different at this stage. In our study, we further explored the effects of sowing density (Figure 3) on the dry matter of wheat spikes at the maturing stage from 2018–2020 by linear regression analysis. Through the analysis of the two-year experiment, it can be found that the sowing density has a significant influence on the dry matter of wheat spikes, and it is significant.

The results showed that, at the maturity stage, the dry matter weight of wheat spikes treated with D4N4 was higher than that of other treatments, and the dry matter weight of wheat spikes treated with D1N1 was lower than that of other treatments.

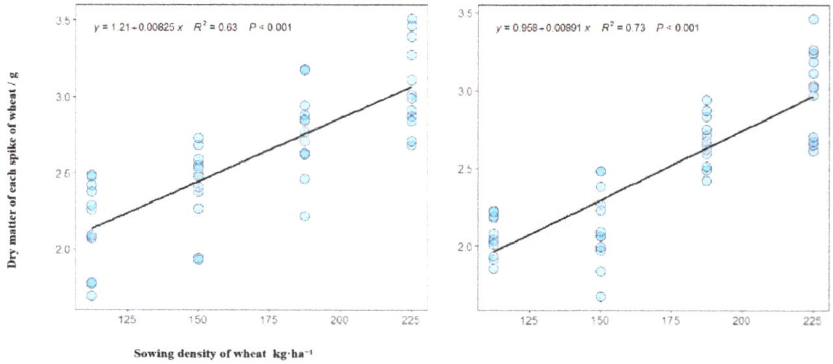

Figure 3. Linear regression analysis of different sowing density on dry matter of wheat spikes at t maturity stage from 2018–2020.

3.2. *Effects of Different Sowing Density and Nitrogen Topdressing on Grain Quality*

By analyzing the effects of different sowing densities and nitrogen topdressing c grain quality (Table 2), from 2018 to 2019, in terms of protein content, D2N4 treatme had the largest, 14.65%, and D4N3 treatment had the smallest, 13.46%; with the increa in sowing density, the protein content increased first and then decreased. The prote content was the highest at the D2 sowing density level, which was 14.42%, and the D2 lev was significantly higher than the D3 level. The starch content increased with increase sowing density and nitrogen topdressing amount. Under the conditions of different sowir densities, the settlement value of grains increased first and then decreased with the increa in sowing density. Compared with D1, D3, and D4, the value of D2 increased by 3.50 8.93%, and 13.22% respectively.

From 2019 to 2020, the protein content increased first and then decreased with t increase in sowing density, and the specific performance was D3 > D2 > D1 > D4. Compare with D2, D1, and D4, the value of D3 increased by 0.20%, 6.73%, and 5.00%, respective The level of D3 was significantly higher than that of D4. The effect of nitrogen topdressir on protein content was D4 > D2 > D3 > D1, but there was no significant difference amor different levels. The stabilization time increased first and then decreased with the increa in sowing density, and reached the maximum at the D2 level, and the stabilization tin decreased with the increase in nitrogen topdressing. The starch content increased with t increase in sowing density and nitrogen topdressing amount. The sedimentation value grains decreased with the increase in sowing density, and the value D1 was significant higher than those of the other three sowing density levels, reaching 67.22 mL.

Table 2. Effects of different sowing density and nitrogen topdressing on grain quality of wheat at t maturity stage.

Year	Sowing Density	Nitrogen Topdressing	Protein Content (%)	Stabilization Time (min)	Starch Content (%)	Settlement Value (mL)
2018–2019	D1	N1	14.22 abc	6.85 ab	67.5 d	50.77 abcd
		N2	14.37 abc	4.86 bcd	68.47 abcd	51.61 abc
		N3	14.49 ab	3.37 d	67.68 cd	52.06 abc
		N4	14.53 ab	2.97 d	67.67 cd	51.73 abc
	D2	N1	14.56 ab	8.10 a	67.63 cd	53.84 a
		N2	14.38 abc	4.31 bcd	68.11 bcd	50.66 abcd
		N3	14.08 abc	3.52 cd	67.83 cd	48.65 abcd
		N4	14.65 a	3.31 d	67.41 d	53.73 ab

Table 2. *Cont.*

Year	Sowing Density	Nitrogen Topdressing	Protein Content (%)	Stabilization Time (min)	Starch Content (%)	Settlement Value (mL)
	D3	N1	13.76 abc	8.26 a	68.39 abcd	47.25 abcd
		N2	13.49 c	3.17 d	68.46 abcd	43.60 d
		N3	14.04 abc	3.98 bcd	68.76 abc	49.51 abcd
		N4	14.09 abc	3.81 bcd	68.48 abcd	49.59 abcd
	D4	N1	13.57 c	7.25 abc	68.77 abc	46.25 bcd
		N2	13.65 bc	3.84 bcd	69.38 a	46.15 cd
		N3	13.46 c	2.25 d	68.79 abc	45.05 cd
		N4	13.57 c	2.89 d	69.17 ab	45.28 cd
	F value	FD	9.731 *	0.513	13.854 ***	7.398 *
		FN	0.596	17.575 **	2.012	0.663
		FD × FN	2.397 *	3.945 **	0.508	2.090 *
2019–2020	D1	N1	14.46 ab	5.80 ab	66.69 fg	64.71 cd
		N2	14.73 ab	3.21 de	67.38 defg	67.35 ab
		N3	15.41 ab	3.23 de	66.34 g	67.85 ab
		N4	15.58 a	2.47 de	67.57 def	68.97 a
	D2	N1	14.43 ab	7.90 a	66.97 efg	63.63 cd
		N2	15.56 ab	4.47 bcd	67.79 cde	62.94 d
		N3	15.41 ab	4.63 bcd	68.17 bcd	64.22 cd
		N4	15.47 ab	3.04 de	67.74 cdef	65.88 bc
	D3	N1	15.22 ab	8.03 a	68.72 abc	63.22 cd
		N2	15.49 ab	4.00 cde	67.79 cde	64.07 cd
		N3	15.25 ab	3.45 de	68.43 bcd	63.48 cd
		N4	15.02 ab	3.04 de	68.14 bcd	63.75 cd
	D4	N1	14.49 ab	6.59 abc	69.1 ab	62.61 d
		N2	14.67 ab	3.95 cde	69.05 ab	63.18 cd
		N3	14.28 b	1.52 e	69.13 ab	63.48 cd
		N4	14.66 ab	2.47 de	69.5 a	63.41 cd
	F value	FD	3.193 *	2.107	28.464 ***	19.318 **
		FN	1.68	23.404 **	0.761	3.835 *
		FD × FN	1.552	5.425 **	2.524 *	5.343 **
		FY	14.533 ***	4.392	68.157	56.828 ***
		FY × FD	ns	ns	ns	ns
		FY × FN	ns	***	ns	ns
		FY × FD × FN	ns	ns	ns	ns

The results showed that sowing density had significant effects on protein content, starch content, and settlement value but did not significantly affect stabilization time from 2018 to 2020. The amount of nitrogen topdressing had a significant effect on stabilization time. The interaction between sowing density and nitrogen topdressing significantly affected protein content, stabilization time, and settlement value from 2018–2019. From 2019–2020, it significantly impacted stabilization time, starch content, and settlement value. Different years had significant differences in protein content and settlement value, and the interaction between different years and nitrogen topdressing showed significant differences in stabilization time.

3.3. *Effects of Different Sowing Density and Nitrogen Topdressing on Yield and Components*

By analyzing the effects of different sowing densities and nitrogen topdressing rates on yield (Figures 4 and 5) and yield components (Table 3), it was observed that sowing density had a significant effect on grain per spike, effective spikes, and yield from 2018–2020. The amount of nitrogen topdressing only had a significant effect on grain per spike and yield from 2018–2019. The interaction between sowing density and nitrogen application rate had significant effects on grain per spike, effective spikes, and yield.

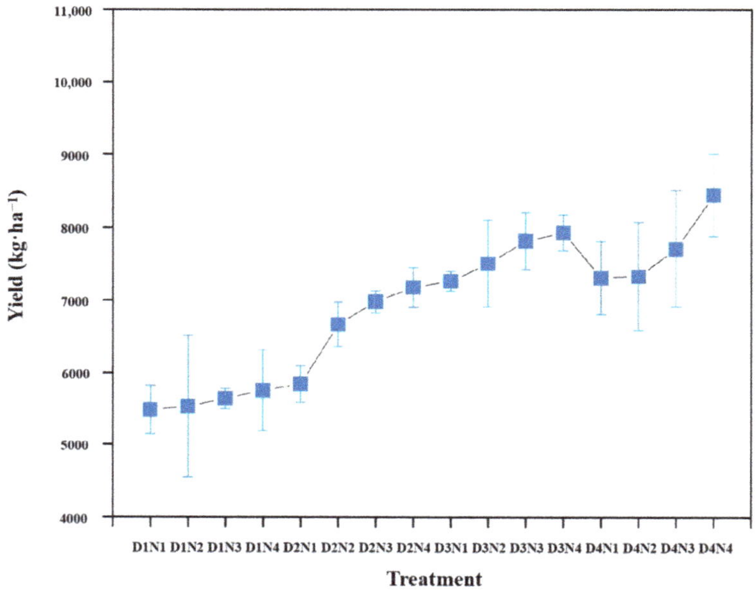

Figure 4. The dynamic changes in wheat yield under different sowing density and nitrogen topdressing from 2018–2019.

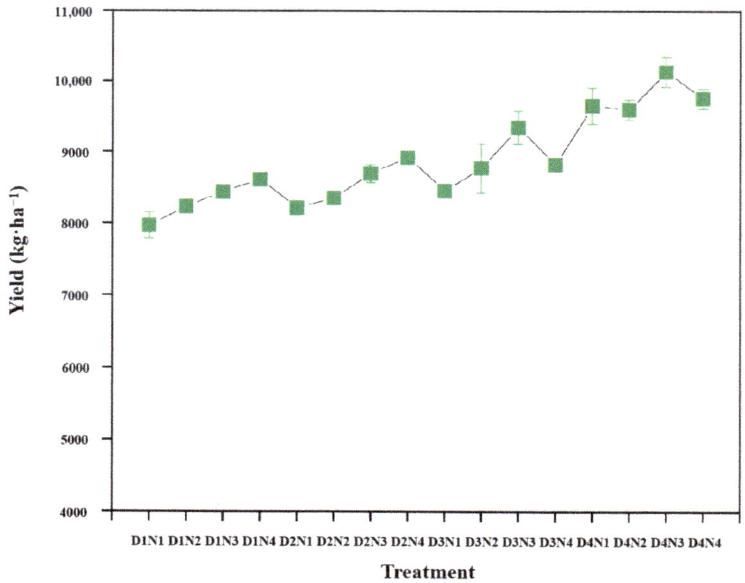

Figure 5. The dynamic changes in wheat yield under different sowing density and nitrogen topdressing from 2019–2020.

Table 3. Effects of different sowing density and nitrogen topdressing on yield and yield components.

Year	Sowing Density	Nitrogen Topdressing	Grain per Spike	Thousand-Grain Weight (g)	Effective Spikes ($\times 10^4 \cdot ha^{-1}$)	Yield ($kg \cdot ha^{-1}$)
2018–2019	D1	N1	37.30 bc	47.05	462.52 d	5491.63 e
		N2	38.46 ab	46.57	462.01 d	5536.10 de
		N3	38.78 a	46.21	471.09 d	5647.27 de
		N4	38.82 a	46.51	476.74 d	5758.43 de
	D2	N1	37.18 bc	46	516.04 cd	5847.37 de
		N2	37.37 bc	45.98	515.59 cd	6670.00 cd
		N3	37.66 abc	46.44	518.48 cd	6981.27 bc
		N4	37.84 abc	46.47	521.59 bc	7181.379 bc
	D3	N1	36.63 cd	47.74	578.96 bc	7270.30 abc
		N2	36.87 cd	45.99	580.29 bc	7514.872 abc
		N3	37.18 bc	45.62	594.01 ab	7826.17 abc
		N4	37.26 bc	46.27	604.87 ab	7937.40 ab
	D4	N1	35.81 d	46.4	624.76 ab	7314.60 abc
		N2	35.78 d	45.02	657.93 a	7336.83 abc
		N3	35.80 d	45.33	634.60 ab	7714.97 abc
		N4	36.85 cd	45.81	643.18 ab	8448.67 a
	F value	FD	22.766 **	1.887	51.404 **	30.285 **
		FN	4.104 *	1.357	1.298	4.165 *
		FD × FN	5.749 **	1.554	14.061 **	7.199 **
2019–2020	D1	N1	33.17 ab	46.64 ab	526.06 d	7972.82 f
		N2	33.76 a	46.13 ab	553.73 cde	8235.37 ef
		N3	33.93 a	48.13 a	536.27 cd	8441.39 cdef
		N4	33.56 a	47.51 ab	551.61 cd	8614.47 bcdef
	D2	N1	32.13 abc	47.82 ab	562.62 bcde	8220.78 ef
		N2	32.26 abc	47.12 ab	558.95 bcde	8356.01 def
		N3	31.57 abcd	46.92 ab	565.78 abcde	8695.35 bcdef
		N4	31.52 abcd	46.88 ab	574.29 abcde	8916.96 abcdef
	D3	N1	30.29 abcd	46.10 ab	578.07 abcde	8457.04 cdef
		N2	30.88 abcd	45.51 b	591.50 abcde	8772.18 bcdef
		N3	30.50 abcd	45.27 b	595.80 abcd	9346.59 abcde
		N4	29.43 bcd	45.50 b	603.10 abcd	8821.08 bcdef
	D4	N1	28.32 d	46.85 ab	631.22 a	9656.50 abc
		N2	29.20 cd	46.24 ab	621.45 abc	9599.63 abcd
		N3	32.64 abc	46.351 ab	626.31 ab	10136.40 a
		N4	31.95 abcd	46.69 ab	622.16 abc	9758.71 ab
	F value	FD	7.556 **	5.796 **	7.122 **	12.187 **
		FN	0.822	0.759	0.172	1.943
		FD FN	2.362 **	2.230 *	2.267 *	2.987 **
		FY	34.397 ***	46.409	567.549	7889.954 ***
		FY × FD	ns	ns	ns	ns
		FY × FN	ns	ns	ns	ns
		FY × FD × FN	ns	ns	ns	ns

From 2018 to 2019, with the increase in sowing density, the number of grains per spike and 1000-grain weight decreased gradually, and the number of effective spikes increased continuously. In terms of the number of grains per spike, the high sowing density (D4) significantly decreased it by 6.32% compared with the low sowing density (D1). There was no significant difference in 1000-grain weight among different levels. The number of effective spikes in the D4 treatment increased significantly by 6.65%, 23.17%, and 39.13%, respectively, compared with D3, D2, and D1. With the increase in nitrogen topdressing, the number of grains per spike increased gradually, and the specific performance of 1000-grain weight was N1 > N4 > N2 > N3. The number of effective spikes increased first and then decreased, and the number of effective spikes under N2 treatment was the highest, which was 525.86 $kg \cdot ha^{-1}$. From 2019 to 2020, with the increase in sowing density, the number of grains per spike decreased gradually, and the level of D1 was significantly higher than that of other levels. The 1000-grain weight performance was D2 > D1 > D4 > D3. The effecti spike number of the D4 treatment was significantly higher than that of D1 by 17.70%, a there was no significant difference among the other three sowing density levels.

From 2018 to 2019, the yield increased with the increase in sowing density. Compar with D3, D2, and D1, the value of D4 treatment increased by 66.59 kg·ha 1033.77 kg·ha^{-1}, and 2095.41 kg·ha^{-1}, respectively. The yield increased with the increase nitrogen application. Compared with N3, N2, and N1, the value of N4 treatment increas by 289.05 kg·ha^{-1}, 567.02 kg·ha^{-1}, and 850.49 kg·ha^{-1}, respectively. From 2019 to 20 there was a positive correlation between sowing density and yield. Compared with D1, D and D3, the value of D4 increased by 17.70%, 14.53%, and 10.51%, respectively. There w no significant difference among D1, D2, and D3 levels. The yield increased first and th decreased with the increase in nitrogen application. D4N3 treatment reached the maximu value of 10136.40 kg·ha^{-1}.

3.4. Correlation Analysis of Different Indexes of Wheat

Correlation analysis of different wheat indicators from 2018 to 2020 was carried o (Figures 6 and 7). It can be seen from Figure 6 that the sowing density was significant positively correlated with effective spikes, starch content, and yield from 2018 to 20 Sowing density was significantly negatively correlated with grain per spike, protein conte and settlement value. Nitrogen topdressing was significantly positively correlated wi grain dry matter. It was significantly negatively correlated with stabilization time. Fro 2019 to 2020, sowing density was significantly positively correlated with effective spik starch content, yield, and grain dry matter. The sowing density was significantly negative correlated with settlement value and grain per spike. Nitrogen topdressing was on significantly negatively correlated with stabilization time.

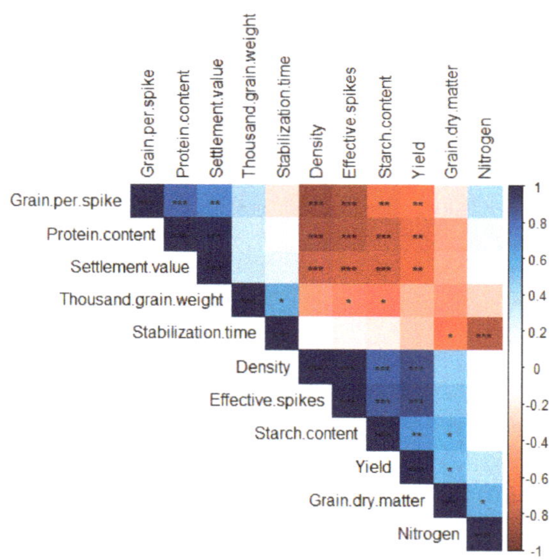

Figure 6. Correlation analysis of different wheat indexes from 2018–2019 (different colors in t figure represent positive and negative correlation, and color depth represents the correlation si The bluer the color, the greater the positive correlation coefficient; the redder the color, the greater t negative correlation coefficient. X axis and Y axis represent different indexes, r values in the figu are in different colors, *: $p \leq 0.05$, **: $p \leq 0.01$, ***: $p \leq 0.001$, the same as Figure 7).

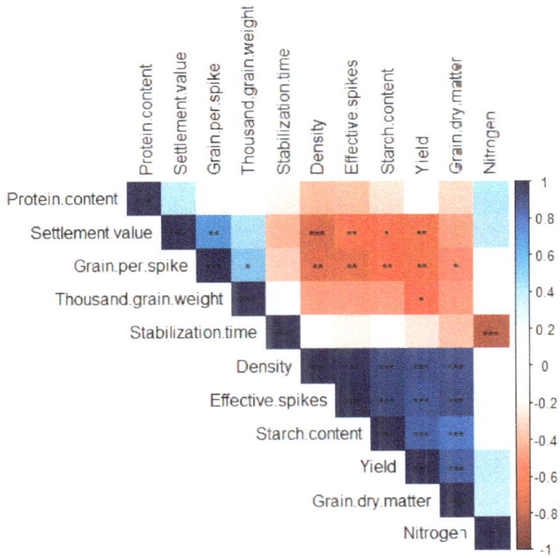

Figure 7. Correlation analysis of different wheat indexes from 2019–2020.

In summary, the increase in sowing density mainly promoted effective spikes, starch content, and yield and inhibited settlement value and grain per spike. The increase in nitrogen topdressing amount mainly inhibited the stabilization time.

4. Discussion

Sowing density is a limiting factor for plants to obtain environmental resources [31]. It is considered to be one of the most influential cultivation methods for grain yield and other agronomic traits. Changes in sowing density are particularly important in wheat crops and have a direct impact on grain yield and its components [14]. The dynamics of nitrogen and its loss trend create a challenging environment for the effective management of this nutrient in topdressing [32], which is mainly due to various reactions and instability in the soil. The low efficiency of nitrogen is attributed to the volatilization, leaching, and surface runoff of ammonia [20]. Some studies have found the effects of sowing density and nitrogen on crops. For example, Kanwal et al. [33], by evaluating the effects of different sowing densities and nitrogen doses on oat forage yield, found that the interaction of sowing density and nitrogen amount significantly changed the yield and quality attributes of oat green forage. The sowing rate of forage oat crops should be 90 kg·ha^{-1} and supplemented with 120 kg·ha^{-1} nitrogen, producing a higher yield, better quality, and better return.

Our research group has previously proved that hole sowing has an excellent effect on the growth characteristics of wheat by comparing the wheat hole sowing method with the traditional sowing method. Wu et al. [28] studied the effects of different sowing methods (drill sowing, wide sowing, and hole sowing) on the yield and quality of wheat. It was found that the hole sowing treatment increased the flag leaf area of wheat, the nitrogen application increased the dry matter quality of the above-ground part of the hole sowing treatment, and the actual yield of the hole sowing treatment was the highest. However, most of the field experiments on wheat sowing density and nitrogen topdressing in the past were carried out by drilling technology and the influence of the hole sowing cultivation method was not explored [34–37]. Under the conditions of this experiment, the density had a very significant effect on the number of effective spikes and yield. Increasing the sowing density would reduce the number of grains per spike and thousand grain weight, significantly increase the number of effective spikes per unit area, and expand the number of populations, which could compensate for the lack of individuals. Increasing the amoun of topdressing nitrogen had little effect on protein content and settlement value, which may be due to the high nutrient content in the soil before sowing in this experiment, so topdressing had little effect on the experiment. The results of our experiment were also slightly different between years. Different years had significant effects on dry matter of wheat spikes, protein content, settlement value, grain per spike, and yield. The dry matte of wheat spikes, the number of effective spikes per unit area, yield, protein content, and settlement value of each treatment from 2018–2019 were lower than those from 2019–2020 The main reason may be that, from 2019–2020, the precipitation and average temperature during the wheat growth period were higher than from 2018–2019, and abundant rainfall and suitable temperature were conducive to crop growth and development.

5. Conclusions

After two years of research on the use of different sowing densities and nitrogen top dressing amounts of wheat under hole sowing conditions, we found that field production can use a combination of a sowing density of 475 suitable seeds·m^{-2} and 120–180 kg·ha^{-1} of nitrogen topdressing at the jointing stage, which can fully tap the production potential o wheat. The experimental results fill the gap in wheat research on the cultivation method o hole sowing and provide valuable references and help for future researchers. In addition there are still some limitations and deficiencies in this experimental study. The experimen was only over a two-year research period, and due to the significant difference in climatic conditions between the two years, although the overall trend is consistent, the regularity and universality of individual index changes are not strong. It is necessary to further carry out long-term positioning experiments to more accurately grasp and lay a theoretical basis and technical support for fully tapping wheat's high-quality and high-yield potential under hole sowing conditions.

Author Contributions: Y.S.: Investigation, Methodology, Writing—original draft, W.Y.: Writing—original draft, Y.W.: Methodology, Y.C.: Writing—review and editing, Y.D.: Writing—review and editing, Z.D.: Methodology, J.H.: Project administration, Funding acquisition. All authors have read and agreed to the published version of the manuscript.

Funding: The research was financially supported by the Ecological Security and Bioremediation Mechanism of Saline-alkali Soil Improvement in the Middle Yellow River (No. DL2021172002L).

Institutional Review Board Statement: Not applicable.

Data Availability Statement: Not applicable.

Acknowledgments: The authors express their special gratitude to the funding source for the financia assistance and are also thankful to the anonymous reviewers for their constructive and valuable comments on earlier versions of this research article.

Conflicts of Interest: The authors declare no conflict of interest.

References

1. Clugston, G.A.; Smith, T.E. Global nutrition problems and novel foods. *Asia Pac. J. Clin. Nutr.* **2002**, *11*, S100–S111. [CrossRef]
2. Rahman, S.; Zheleva, B.; Cherian, K.M.; Christenson, J.T.; Doherty, K.E.; de Ferranti, D.; Gauvreau, K.; Hickey, P.A.; Kumar, R.K. Kupiec, J.K.; et al. Linking world bank development indicators and outcomes of congenital heart surgery in low-income and middle-income countries: Retrospective analysis of quality improvement data. *BMJ Open* **2019**, *9*, e028307. [CrossRef] [PubMed
3. Bongaarts, J. The State of Food Security and Nutrition in the World 2020. Transforming food systems for affordable healthy diets *Popul. Dev. Rev.* **2021**, *47*, 558. [CrossRef]
4. Shiferaw, B.; Smale, M.; Braun, H.-J.; Duveiller, E.; Reynolds, M.; Muricho, G. Crops that feed the world 10. Past successes and future challenges to the role played by wheat in global food security. *Food Secur.* **2013**, *5*, 291–317. [CrossRef]
5. Gong, F.; Wu, X.; Zhang, H.; Chen, Y.; Wang, W. Making better maize plants for sustainable grain production in a changing climate. *Front. Plant Sci.* **2015**, *6*, 835. [CrossRef]
6. Xu, H.; Twine, T.E.; Girvetz, E. Climate Change and Maize Yield in Iowa. *PLoS ONE* **2016**, *11*, e0156083. [CrossRef]
7. Li, Q.L. The produce and consumption of wheat and wheat flour in China. *Flour Milling* **2006**, *5*, 9–14.

El Serafy, S. System of Environmental-Economic Accounting 2012-Central Framework, United Nations, European Commission, Food and Agriculture Organisation of the United Nations, International Monetary Fund, Organisation for Economic Co-operation and Development and World Bank. *Ecol. Econ.* **2015**, *112*, 161–163. [CrossRef]
Eser, A.; Kassai, K.M.; Kato, H.; Kunos, V.; Tarnava, A.; Jolankai, M. Impact of nitrogen topdressing on the quality parameters of winter wheat (*Triticum aestivum* L.) Yield. *Acta Aliment.* **2020**, *49*, 244–253. [CrossRef]
Mohi-Ud-Din, M.; Rohman, M.M.; Alam, M.A.; Hasanuzzaman, M.; Islam, T. Wheat variety carrying 2N(v)S chromosomal segment provides yield advantage through lowering terminal heat-induced oxidative stress. *Protoplasma* **2023**, *260*, 53–76. [CrossRef]
Fischer, T.; Ammar, K.; Monasterio, I.O.; Monjardino, M.; Singh, R.; Verhulst, N. Sixty years of irrigated wheat yield increase in the Yaqui Valley of Mexico: Past drivers, prospects and sustainability. *Field Crop. Res.* **2022**, *283*, 108528. [CrossRef]
Mottaleb, K.A.; Kruseman, G.; Snapp, S. Potential impacts of Ukraine-Russia armed conflict on global wheat food security: A quantitative exploration. *Glob. Food Secur.-Agric. Policy Econ. Environ.* **2022**, *35*, 100659. [CrossRef]
Sawaguchi, A.; Sato, M. Cultivation by early winter seeding of spring wheat in central Hokkaido-Suitable seeding density. *Jpn. J. Crop Sci.* **2001**, *70*, 505–509. [CrossRef]
Valerio, I.P.; Felix de Carvalho, F.I.; de Oliveira, A.C.; Benin, G.; de Souza, V.Q.; Machado, A.d.A.; Bertan, I.; Busato, C.C.; da Silveira, G.; Rob Fonseca, D.A. Seeding density in wheat genotypes as a function of tillering potential. *Sci. Agric.* **2009**, *66*, 28–39. [CrossRef]
Shah, F.; Wu, W. Soil and Crop Management Strategies to Ensure Higher Crop Productivity within Sustainable Environments. *Sustainability* **2019**, *11*, 1485. [CrossRef]
Gao, Y.; Zhang, M.; Yao, C.; Liu, Y.; Wang, Z.; Zhang, Y. Increasing seeding density under limited irrigation improves crop yield and water productivity of winter wheat by constructing a reasonable population architecture. *Agric. Water Manag.* **2021**, *253*, 106951. [CrossRef]
Zhang, D.Y.; Sun, G.J.; Jiang, X.H. Donald s ideotype and growth redundancy: A game theoretical analysis. *Field Crop. Res* **1999**, *61*, 179–187. [CrossRef]
Ma, S.-C.; Wang, T.-C.; Guan, X.-K.; Zhang, X. Effect of sowing time and seeding rate on yield components and water use efficiency of winter wheat by regulating the growth redundancy and physiological traits of root and shoot. *Field Crop. Res* **2018**, *221*, 166–174. [CrossRef]
Hiltbrunner, J.; Streit, B.; Liedgens, M. Are seeding densities an opportunity to increase grain yield of winter wheat in a living mulch of white clover? *Field Crop. Res.* **2007**, *102*, 163–171. [CrossRef]
Soofizada, Q.; Pescatore, A.; Guerrini, L.; Fabbri, C.; Mancini, M.; Orlandini, S.; Napoli, M. Effects of Nitrogen plus Sulfur Fertilization and Seeding Density on Yield, Rheological Parameters, and Asparagine Content in Old Varieties of Common Wheat (*Triticum aestivum* L.). *Agronomy* **2022**, *12*, 351. [CrossRef]
Chung, R.S.; Chen, H.T.; Wu, P.C. Effect of split applications of nitrogen fertilizer on the growth and nitrogen composition of Chinese mustard. *Commun. Soil Sci. Plant Anal.* **1999**, *30*, 307–316. [CrossRef]
Liu, W.-Y.; Wu, C.-C.; Wang, S.-Y.S. Forest Management and Adaptation Strategies in Response to Climate Change by the Taiwanese Public. *Atmosphere* **2021**, *12*, 1056. [CrossRef]
Ju, X.T.; Kou, C.L.; Zhang, F.S.; Christie, P. Nitrogen balance and groundwater nitrate contamination: Comparison among three intensive cropping systems on the North China Plain. *Environ. Pollut.* **2006**, *143*, 117–125. [CrossRef]
Cordovil, C.M.d.s.; Cabral, F.; Coutinho, J. Potential mineralization of nitrogen from organic wastes to ryegrass and wheat crops. *Bioresour. Technol.* **2007**, *98*, 3265–3268. [CrossRef]
Fang, Q.; Ma, L.; Yu, Q.; Malone, R.W.; Saseendran, S.A.; Ahuja, L.R. Modeling Nitrogen and Water Management Effects in a Wheat-Maize Double-Cropping System. *J. Environ. Qual.* **2008**, *37*, 2232–2242. [CrossRef]
Jin, L.; Hu, K.; Deelstra, J.; Li, B.; Wei, D.; Wang, H. Evaluation of nitrogen fate, water and nitrogen use efficiencies of winter wheat in North China Plain based on model approach. *Acta Agric. Scand. Sect. B-Soil Plant Sci.* **2014**, *63*, 127–138. [CrossRef]
Guo, R.; Nendel, C.; Rahn, C.; Jiang, C.; Chen, Q. Tracking nitrogen losses in a greenhouse crop rotation experiment in North China using the EU-Rotate_N simulation model. *Environ. Pollut.* **2010**, *158*, 2218–2229. [CrossRef]
Wu, B.; Dong, Y.; Hai, J. Effect of Hole Seeding Way on Yield and Quality of Winter Wheat. *Acta Agric. Boreali-Occident. Sin* **2019**, *28*, 906–913. (In Chinese)
Qi, H.; Dong, Y.; Li, C.; Angelique Twizerimana Ren, H.; Hai, J. Effects of Sowing Methods and Seeding Rates on Yield and Quality of Winter Wheat Variety 'Xinong 20'. *Acta Agric. Boreali-Occident. Sin.* **2021**, *30*, 32–40. (In Chinese)
Sun, Y.T.; Yang, C.; Liang, H.J.; Yang, Y.Y.; Bu, K.M.; Dong, Y.L.; Hai, J.B. The Border Effects of Dry Matter, Photosynthetic Characteristics, and Yield Components of Wheat under Hole Sowing Condition. *Agronomy* **2023**, *13*, 766. [CrossRef]
Lloveras, J.; Manent, J.; Viudas, J.; Lopez, A.; Santiveri, P. Seeding rate influence on yield and yield components of irrigated winter wheat in a Mediterranean climate. *Agron. J.* **2004**, *96*, 1258–1265. [CrossRef]
Fageria, N.K.; Baligar, V.C. Enhancing nitrogen use efficiency in crop plants. *Adv. Agron.* **2005**, *88*, 97–185.
Kanwal, A.; Zubair, D.; Rehman, R.M.u.; Ibrahim, M.; Bashir, M.A.; Maqbool, M.M.; Imran, M.; Rehman, U.u.; Nasif, O.; Ansari, M.J. The Impact of Seeding Density and Nitrogen Rates on Forage Yield and Quality of *Avena sativa* L. *Biomed Res. Int.* **2022**, *2022*, 8238634. [CrossRef] [PubMed]

34. Banach, J.K.; Majewska, K.; Zuk-Golaszewska, K. Effect of cultivation system on quality changes in durum wheat grain and flo produced in North-Eastern Europe. *PLoS ONE* **2021**, *16*, e0236617. [CrossRef] [PubMed]
35. Boiek, K.S.; Zuk-Golaszewska, K.; Bojarczuk, J.; Golaszewski, J. The Effect of Different Nitrogen Fertilizer Rates, Sowing Dens and Plant Growth Regulator Application on the Quality and Milling Value of *Triticum durum* Desf. Grain. *Agronomy* **2022**, *12*, 16 [CrossRef]
36. Chen, M.; Zhu, Y.H.; Ren, M.J.; Jiang, L.; He, J.; Dong, R. Does Belt Uniform Sowing Improve Winter Wheat Yield under Hi Sowing Density? *Agronomy* **2022**, *12*, 2936. [CrossRef]
37. Marinho, J.D.; Silva, S.R.; Fonseca, I.C.D.; Zucareli, C. Nitrogen fertilization and sowing density on yield and physiologic quality of wheat seeds. *J. Seed Sci.* **2022**, *44*, e202244013. [CrossRef]

agronomy

MDPI

Article

Response of Crop Performance and Yield of Spring Sweet Potato (*Ipomoea batatas* [L.] Lam) as Affected by Mechanized Transplanting Properties

Hui Li [1], Baoqing Wang [2], Song Shi [1,2], Jilei Zhou [1,2], Yupeng Shi [1], Xuechuan Liu [1], Hu Liu [1,3,*] and Tengfei He [1,*]

1 Shandong Academy of Agricultural Machinery Sciences, Jinan 250100, China; lihuictrc@shandong.cn (H.L.); shisong@shandong.cn (S.S.); zhoujilei@shandong.cn (J.Z.); shiyupeng1997@163.com (Y.S.); xuechuan_13@163.com (X.L.)
2 Shandong Academy of Agricultural Sciences, Jinan 250100, China; xb970607@163.com
3 Huang Huai Hai Key Laboratory of Modern Agricultural Equipment, Ministry of Agriculture and Rural Affairs, Jinan 250100, China
* Correspondence: liuhu0725@163.com (H.L.); hetengfei.1@163.com (T.H.); Tel.: +86-531-88617528 (H.L. & T.H.)

Abstract: The sweet potato transplanters of diverse transplanting configurations have been shown to produce various planting properties in relation to different raised bed cropping systems, thus affecting crop growth and yield in sweet potato cultivation. In Shandong Province, a field experiment assessed the effects of three treatments (RB1, mulched raised beds with a finger-clip type transplanter; RB2, bare raised beds with a finger-clip type transplanter; and RB3, bare raised beds with a clamping-plate type transplanter) on soil temperature, plant growth, yield, and economic benefits. With the lowest coefficient variation of plant spacing and planting depth, the RB1 with the finger-clip type transplanter had 6.4% and 6.0% higher temperature at 5–10 cm soil layer by using the plastic-mulch for rapid early slips growth as compared with the RB2 and the RB3, respectively. Consequently, the leaf area index in the RB1 was increased by 5.6% and 6.4% as compared to the RB2 and the RB3, separately. This finally contributed to 57.5–70.8% greater fresh vines weight and 23.8–33.8% higher tubers yield in the RB1 compared with both the RB2 and the RB3 treatments, respectively. In general, in the mulched raised bed system of the Huang-Huai-Hai region of China, the finger-clip type transplanter could be a suitable option for the transplanting of sweet potato slips. In the bare raised bed system, meanwhile, the clamping-plate type transplanter has the potential to increase the production of sweet potatoes.

Keywords: crop performances; planting properties of sweet potato transplanter; planting system; yield

Citation: Li, H.; Wang, B.; Shi, S.; Zhou, J.; Shi, Y.; Liu, X.; Liu, H.; He, T. Response of Crop Performance and Yield of Spring Sweet Potato (*Ipomoea batatas* [L.] Lam) as Affected by Mechanized Transplanting Properties. *Agronomy* **2023**, *13*, 1611. https://doi.org/10.3390/agronomy13061611

Academic Editors: Umberto Anastasi and Aurelio Scavo

Received: 16 May 2023
Revised: 10 June 2023
Accepted: 12 June 2023
Published: 15 June 2023

1. Introduction

Food security is one of the greatest challenges facing humankind [1]. Agriculture is at the forefront of these challenges [2]. Sweet potato (*Ipomoea batatas* [L.] Lam) is one of the five most important crops in the world, rich in carbohydrates, and can serve as a source of protein, carotenoid, and essential vitamins for the survival needs of mankind [3,4]. This crop is widely cultivated from tropical to temperate regions, such as Asia, Africa, and Latin America [5,6]. There is an increasing need to produce more sweet potatoes on existing arable land given the challenges of both labor scarcity and population growth [7].

Sweet potato yields can vary significantly due to factors such as the soil, weather, crop variety, and cultivation management [8,9]. Under certain soil, weather, and sweet potato variety conditions, many efforts have been made to find cultivation modes that are more effective at enhancing productivity. Parwada et al. [10] established the proper ridging height and planting orientation in order to enhance constant reliable root yield and vine length among sweet potato producing farmers in Zimbabwe. Chagonda et al. [11] proposed that the horizontal vine orientation provided a significant storage root diameter, whil there was no significant difference between the ridge tillage and mound tillage system Abdallah et al. [12] evaluated the performance of sweet potato clones under differen watering strategies in the coastal lowlands of Kenya. Ribeiro et al. [13] conducted a study t evaluate the plant growth, yield, uptake, and removal of N by sweet potato plants fertilize with N and treated with paclobutrazol during two planting seasons. Pepó [14] showed tha a 0.75 m row spacing was more favourable than a 1.0 m one in Hungary.

China is the largest producer of sweet potatoes in the world [15]. Sweet potatoes ar widely cultivated in over half of the globe's poor counties due to their wide ecologica adaptation, strong tolerance to drought, and low requirement of soil fertilizer [16]. Th cultivation areas for sweet potatoes in China are generally divided into the northern Chin area, the Yangtze River area, the southern China area, etc., which are distinguished b climatic conditions, cultivation systems, and soil conditions [17]. As shown in Figure 1 the Huang-Huai-Hai region of China is one of the most important traditional sweet potat production regions in China, accounting for 30% of national sweet potato production [18,19 Many studies have shown that sweet potato cultivation on raised beds mulched with plasti film can be beneficial to sweet potato yield because it improves soil water moisture, soi bulk density, and soil porosity [20,21]. At present, farmers plant sweet potato on bare raise beds or raised beds mulched with plastic film in this area [17].

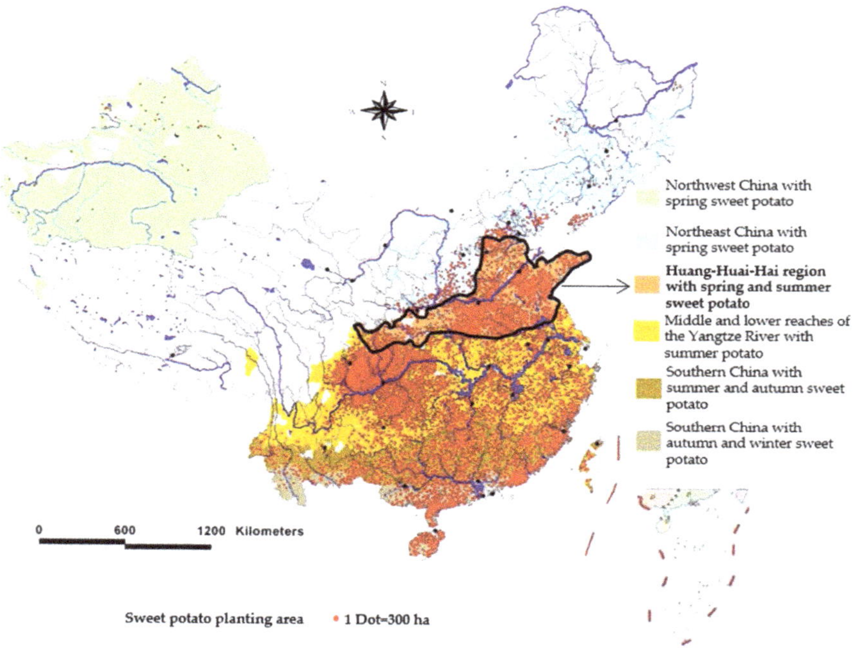

Figure 1. Traditional regional distribution and planting area of sweet potato cultivation in China.

However, most sweet potato production in the Huang-Huai-Hai region of China still occurs by the use of manual transplanting, which has caused this area to suffer from a labo shortage [22]. Sustainably producing the sweet potato crop in this region is thus a grea challenge. There were, indeed, not even any special transplanters for transplanting swee potato slips until Chen et al. [23] and Hu et al. [24] modified and improved the commercia clip-on-chain type transplanter for the horizontal transplanting of sweet potatoes in bare raised beds. These are mainly applicable for the bare raised bed cultivation system in rain-fed farming areas that have high soil moisture and that are rainy, so no additional watering is required. The two machines cannot be used to mechanically transplant the sweet potato slips in drought-affected areas, though, due to the lack of a timely watering function. Supplementary irrigation is thus required at the time of planting for proper sprouting and establishment, although the prolific root system of sweet potato does make it a drought-tolerant crop [25,26]. Since the Huang-Huai-Hai region of China has limited water resources, there is a need for sweet potato transplanters in this area to accomplish the planting operation for raised beds mulched with plastic film system and the bare raised bed system. After several years of development, sweet potato transplanters with a slip taking-planting mechanism have been developed [27,28], and some have now been manufactured commercially. These transplanters have encouraged the development and extension of sweet potato production in Huang-Huai-Hai region of China, but the literature contains little information about their impact on planting properties and crop performance [29,30]. This paper compares two of the most widely used sweet potato transplanters (the finger-clip and the clamping-plate ones) for different planting modes (raised beds mulched with plastic film or the bare raised bed systems with varying placement), and it investigates their effects on planting quality, crop growth, and subsequent yield in 2021 in the Huang-Huai-Hai region of China.

2. Materials and Methods

2.1. Equipment Description

2.1.1. Finger-Clip Compound Transplanter

The finger-clip compound transplanter, designed by the Shandong Academy of Agricultural Machinery Sciences and Shandong Huorong Agricultural Technology Development Co., Ltd. (Qingzhou, China), was used for sweet potato slips cultivation of bare raised beds and mulched raised bed systems. It mainly comprises a transmission box, a rotary component, a ridging board, a film pressing wheel, a height adjustment mechanism, a slip taking-planting mechanism, a drive system, a slip delivery mechanism, etc. (Figure 2a). It can accomplish land preparation, ridging, film mulching, drip-irrigation belt laying, and transplanting on two ridges at the same time. During transplanting, the rotary component completes the soil crushing and the soil preparation operations at 300~350 r/min, driven by power from the transmission box, which is connected to the tractor's power take off (PTO). The ridge board squeezes the crushed soil to form two rows of trapezoidal ridges with a height of 30 cm at 85 cm spacing under the traction of the tractor and the pressure of the hydraulic cylinder simultaneously. The drip irrigation laying device and the plastic-film frame mulches the ridge and lays the drip irrigation belt, respectively, and then the slip transplanting apparatus transplants the sweet potato slips by using the slip taking planting mechanism and the slip delivery mechanism at a rotary speed of less than 60 r/min, driven by the ground wheel. Slips are manually placed in the seedling delivery mechanism by the operators sitting on the seats.

(a) (b)

Figure 2. Two kinds of transplanters used for the experiment: (**a**) finger-clip compound transplanter with its slip transplanting apparatus; (**b**) clamping-plate compound transplanter with its seedling delivery mechanism.

2.1.2. Clamping-Plate Compound Transplanter

The clamping-plate compound transplanter (Shandong Jinshuwang Agricultural Machinery Manufacturing Co., Ltd., located in Tengzhou, China) is made up of a suspension frame, rotary blades, a ridge plough, a driving shaft, a slip conveying clamping-plate, a gear box, a soil loader, a slip fixing wheel, etc. (Figure 2b). The transplanter mounts with the tractor by the suspension frame. During the operation, the rotary blades smash soil at 340~360 r/min, driven by the tractor's PTO shaft. The soil is raised and enclosed by the ridge plough to form the raised beds of 30 cm height. Operators place the sweet potato slips in the seedling clips, which are installed on the slip conveying the clamping plate. The slips are then put horizontally vertical on the raised beds with the rotation of the conveying clamping-plate at 30~40 r/min. After this, the sweet potato slips remain covered with soil delivered by the soil loader. The fixing wheel presses the soil over the slips to finish the transplanting in the bare raised bed system. The key parameters of these two transplanters are presented below in Table 1.

Table 1. The key parameters of the two compound transplanters for sweet potato slips.

Parameter	Finger-Clip Compound Transplanter	Clamping-Plate Compound Transplanter
Matched power	120–180 hp	120–180 hp
Working width	1.7 m	1.7 m
Number of ridges	2	2
Transplanting part	Finger-clip type slip taking-planting mechanism	Clamping-plate type slip taking-placing mechanism
Transplant spacing	20–30 cm	20–30 cm
Transplanting depth	4–10 cm	4–10 cm
Slips placement	Boat-shape placement	Horizontal vertical placement
Suitable system	Mulched raised beds system and bare raised beds system	Bare raised beds system
Productivity	0.08–0.13 ha h^{-1}	0.1–0.2 ha h^{-1}

2.2. *Site Description*

Field trials were conducted at Zhangqiu (36°41′ N, 117°32′ E), located in the southeast of the Huang-Huai-Hai region of China, with three crop rotation treatments. In the five years before the experiment, this area had a monsoon climate with an annual average temperature of 10~20 °C, a frost-free period of 167~218 days, and annual rainfall of 450~1100 mm. The accumulated temperature of ≥0 °C is about 5401 °C [31]. In this double cropping area, winter wheat to summer maize is the main crop rotation. When the sweet potato was planted, the winter wheat (end of September to the middle of June) to summer maize (middle of June to end of September) to spring sweet potato (end of April or early May to end of September) rotation is used. According to the USDA texture classification system, the soil in the experiment plots is silt loam, clay (12.3%), silt (74.8%), and sand (12.9%), on average. In the top 30 cm soil layer, soil bulk density, soil moisture, and pH were 1.35 g/cm^3, 12.8%, and 8.3, respectively.

2.3. *Experimental Design*

In the experiment, three treatments were compared: the finger-clip compound sweet potato transplanter for the mulched raised beds system (RB1) (Figure 3a), the finger-clip compound sweet potato transplanter for the bare raised beds system (RB2) (Figure 3b), and the clamping-plate compound sweet potato transplanter for the bare raised bed system (RB3) (Figure 3c). The three treatments were designed in a randomized block with 3 replications. Each plot was 3.5 m wide and 30 m long with an access pathway and guard strip between each. The spring sweet potato slips (variety Jishu 26, and a length of 30 cm~35 cm) with five top nodes were transplanted on 6–7 May and harvested on 8–9 October. Drip irrigation was immediately applied after the transplanting. In the RB1 system, a high-density black polyethylene film (0.02 mm thick, 1.0 m wide) was used as the mulching plastic. In the treatments RB1 and RB2, the sweet potato slips were transplanted as a boat-shape along the ridge direction by using the finger-clip compound sweet potato transplanter, while the sweet potato slips were transplanted as a horizontal vertical placement in the treatment RB3 by using the clamping-plate compound sweet potato transplanter.

(a) (b) (c)

Figure 3. Three designed treatments in this experiment: (**a**) RB1, finger-clip compound transplanter working under mulched raised beds system; (**b**) RB2, finger-clip compound transplanter working under bare raised beds system; (c) RB3, clamping-plate compound transplanter working under bare raised beds system.

The sweet potato slips were planted with the district-recommended plant density of about 49,000 plants/ha with 24 cm × 85 cm plant spacing and planting depth of 5~10 cm. The compound fertilizer [N-P_2O_5-K_2O 10-8-24] (containing total nutrients ≥ 42%, humic acid ≥ 3%, controlled-release K fertilize ≥ 4%) was applied as the basal fertilizer at the rate of 375 kg/ha at transplanting, while 33% pendimethalin EC herbicide (JiangSu Longdeng Chemical Company, Kunshan, China) was sprayed onto the soil surface according to the manufacturer's protocol during the transplanting. About 1.5 months after the transplanting, 80% flumetsulam WG herbicide (Jiangsu Ruibang Pesticide Factory Co., Ltd., Changzhou, China) was carefully used in the three treatments.

2.4. Measurements

2.4.1. Missing Seedling Rate and Qualified Rate of Transplanting Population

The missing seedling rate and the qualified rate of the transplanting population, representing the transplanting quality, were counted—120 theoretical sweet potato slips that should be planted at the 12 split-plots in 3 complete randomized blocks [32,33]. They were calculated using the following equations:

$$Q_M = \frac{N_{LZ}}{N'} \times 100\% \quad (1)$$

$$Q_z = \frac{N_T - (N_{LM} + N_{MM} + N_{CZ} + N_{SM})}{N'} \times 100\% \quad (2)$$

where Q_M is the missing seedling rate (%), Q_Z is the qualified rate of the transplanting (%), N_T is the total planted counts of the sweet potato slips, N' is the theoretically planted counts, N_{LZ} is the missed planted counts of the sweet potato slips, N_{LM} is the exposed planted counts of the slips, N_{MM} is the buried counts of the planted slips, N_{CZ} is the replanted counts of the slips, and N_{SM} is the injured counts of the planted slips.

2.4.2. Precision of Seedling Placement

To calculate the plant spacing of the sweet potato slips, 60 successively planted sweet potato slips were measured of the randomly selected planting row in each plot. To calculate the seeding or the planting depth of crops, the chlorophyll-free stem and coleoptile length (from seed remnants to the onset of green stem) was usually measured as effective depth [34]. For sweet potato slips, the chlorophyll-free stem lengths were not obvious. After 10 days of planting, a mark was made on the five seedlings at the ridge level in each plot. The vertical distance from the lowest position to the marked point was taken as the effective planting depth, and then the sweet potato slips were dug out and the entire stem length below the mark was taken as the effective planting length. The mean planting length was easily obtained. The plant spacing coefficient of variation and the qualified rate of transplanting depth were calculated to assess the transplanting accuracy in each plot using the following equations [32]:

$$CV_X = \frac{\sqrt{\frac{1}{n-1}\sum_{i=1}^{n}\left(X_i - \frac{\sum_{i=1}^{n} X_i}{n}\right)^2}}{\frac{\sum_{i=1}^{n} X_i}{n}} \times 100\% \quad (3)$$

$$V_H = \frac{N_h}{N_T} \times 100\% \quad (4)$$

$$CV_H = \frac{\sqrt{\frac{1}{n-1}\sum_{i=1}^{n}\left(H_i - \frac{\sum_{i=1}^{n} H_i}{n}\right)^2}}{\frac{\sum_{i=1}^{n} H_i}{n}} \times 100\% \quad (5)$$

where CV_X is the plant spacing coefficient of variation; n is the measured number of the planted slips; X_i is the measured plant spacing, cm; V_H is the qualified rate of transplanting depth; N_h is the sweet potato counts of qualified depth; CV_H is the plant depth coefficient of variation; and H_i is the measured plant depth, cm. As the designed transplanting depth was 60 mm, we assumed that the qualified depth was 60 ± 10 mm.

2.4.3. Soil Temperature and Plant Growth

In different treatments, soil temperature was measured at 5 and 10 cm soil depths at 08:00 ($T_{8:00}$), 14:00 ($T_{14:00}$), and 20:00 ($T_{20:00}$). A high precision soil temperature and humidity sensor (JXBS-3001-TR), connected with the weather station, was used. The mean daily soil temperature (T) for 10 days during the period from 10 days to 1 month after transplanting was calculated as follows [35]:

$$T = (2 \times T_{8:00} + T_{14:00} + T_{20:00})/4 \quad (6)$$

The leaf number, the plant height, and the leaf area were all measured to estimate the growth of the spring sweet potato. The samples were measured and obtained within randomly selected areas of 1 m × 1 m from three areas in each plot 1 month after planting. Plant height was calculated from the stem tip to the soil surface. To obtain the leaf area, the leaves were cut and analysed by the LA-S series plant image analysis system (Hangzhou Wanshen Testing Technology Co., Ltd., Hangzhou, China) in a laboratory. After that, the Leaf Area Index (LAI) was calculated as follows [36]:

$$LAI = LA/GA \tag{7}$$

where LAI is the leaf area index, LA is the leaf area in the selected area (m^2), and GA is the ground area of the selected area (m^2).

2.4.4. Weight of Fresh Vines with Leaves and Tuber Yield

At harvest time (i.e., the beginning of October), the weight of the vines with leaves, the number of vines, and the length of the longest vine per plant, which were removed manually in the experiment, were all measured [37]. In each plot, we chose 10 plants randomly.

The tuber yield that was observed in this study included the number of tubers (per plant), the fresh weight of tuber (g $plant^{-1}$), and the yield (t ha^{-1}). During manual harvesting, we collected 10 plants, with an area of the harvest bed that was 170 cm wide and 120 cm long (sampling size), which was taken randomly from each plot. The average number of tubers per plant was measured and categorized as large marketable tubers (≥500 g), medium marketable tubers (≥200 g), and non-marketable tubers (<200 g, or else damaged by insects and diseased tubers) [16]. The total yield per hectare was then calculated using the following equation [38,39]:

$$\text{Yield (t ha}^{-1}) = (10{,}000/\text{scale of sampling plot}) \times \text{yield of sampling plot} \tag{8}$$

2.4.5. Economic Benefit

Input (sweet potato slips, fertiliser, labour, etc.) quantities and the direct cost of all mechanical operations was recorded throughout the field trial, together with the value of outputs (crop yield value), on a common basis (US$ ha^{-1}) [40].

2.5. Data Analysis

The SPSS analytical software package was used for all of the statistical analyses. Mean values were calculated for each of the measurements, and ANOVA was used to assess the effects of the two sweet potato transplanters on both the planting properties and the crop performance of the measures. When the ANOVA indicated a significant F-value, multiple comparisons of annual mean values were performed by the least significant difference (LSD) method. In all analyses, a probability of error smaller than 5% ($p = 0.05$) was considered statistically significant.

3. Results

3.1. Missing Seedling Rate and Qualified Rate of Transplanting Population

Table 2 shows that the mean missing seedling rate Q_M under RB3 treatment of 0.6% appeared to be 59.7% and 77.6% lower ($p > 0.05$) than that under RB1 treatment of 1.4% and RB2 treatment of 2.5%, respectively. This difference was only relevant to the missed counts of the sweet potato slips, while the theoretical planted counts were the same according to Formula (1). To evaluate transplanting quality, the replanted count number N_{CZ} in RB3 of 0.7 was significantly ($p < 0.05$) greater than that in both RB1 and RB2 treatments. However, the difference of exposed counts N_{LM}, buried counts N_{MM}, and injured counts N_{SM} of the planted sweet potato slips were all non-significant ($p = 0.05$) under the three treatments. The qualified rates were also similar in the three treatments.

Table 2. The transplanting quality under the three treatments. Means within a column followed the same letters are not significantly different ($p = 0.05$).

Treatment	Mean Value							Transplanting Quality	
	Theoretical Planted Counts N'	Total Planted Counts N_T	Missed Counts N_{LZ}	Exposed Counts N_{LM}	Buried Counts N_{MM}	Replanted Counts N_{CZ}	Injured Counts N_{SM}	Missing Seedling Rate Q_M (%)	Qualifiec Rate Q_Z (%)
RB1	120.0 a	118.3 a	1.7 a	2.7 a	0 a	0 a	0.3 a	1.4 a	96.1 a
RB2	120.0 a	117.0 a	3.0 a	2.0 a	0.3 a	0 a	0 a	2.5 a	95.6 a
RB3	120.0 a	119.7 a	0.7 a	3.3 a	0 a	0.7 b	0 a	0.6 a	96.1 a

3.2. Precision of Seedling Placement

The planting spacing in the RB1 and RB3 treatments were marginally higher ($p > 0.$(than that in the RB2 treatment (Table 3). However, the plant spacing coefficient of variati in RB1 of 5.1% was 75.2%, significantly smaller ($p < 0.05$) than that in RB3 treatment of 8.9 The mean planting length in the three treatments were all around 200 mm. The plantir depth in the RB1 treatment was 4.0% ($p > 0.05$) and 32.2% ($p < 0.05$) deeper than that in t RB2 and RB3, respectively, and the relative coefficient of variation was slightly lower th in the other treatments. Meanwhile, the qualified rate of the planting depth in the thr treatments was nearly the same, and all were above 95%.

Table 3. Precision of seedling placement under the three treatments. Means within a column follow by the same letters are not significantly different ($p = 0.05$).

Treatment	Plant Spacing		Mean Planting Length (mm)	Planting Depth		
	Spacing Value (cm)	Coefficient of Variation (%)		Depth Value (mm)	Qualified Rate (%)	Coefficient of Variation (%)
RB1	24.3 a	5.1 a	201.8 a	78.1 a	97.1 a	8.7 a
RB2	23.9 a	6.1 ab	198.2 a	75.4 a	96.9 a	9.2 a
RB3	24.2 a	8.9 b	202.9 a	59.2 b	97.5 a	10.6 a

3.3. Soil Temperature and Plant Growth

In general, a soil temperature at 5 cm depth was marginally higher than that at 10 c depth in the three treatments (Table 4). At 5 cm depth, the RB1 increased soil temperatu by 0.3–1.5 °C and 0.1–1.5 °C, respectively, as compared to the RB2 and the RB3 treatmen within the month after the transplanting day. The soil temperature was 5.2%, significant higher in the RB1 than in the RB2 treatment on the 30th day after transplanting. Tl difference between the RB1 and the RB3 on the 10th day and the 30th day was significa at $p = 0.05$ level, independently. Similar results were found in the 10 cm soil depth whe RB1 increased the temperature by 6.4% and 6.0% as compared to the RB2 and the RE respectively, on the 30th day after transplanting.

As shown in Table 5, the difference of leaf number, plant height, and leaf area were a not significant ($p > 0.05$) in the RB1, RB2, and RB3 treatments 1 month after transplantir The leaf number in the RB1 was 10.8% and 26.3% higher than that in the RB2 and the RE relatively, and the RB1 increased the plant height by 8.7% and 6.4% as compared with th RB2 and the RB3, respectively. Meanwhile, the leaf area index in the RB1 was increased b 5.6% and 6.4% compared to the RB2 and RB3 treatments.

Table 4. Soil temperature at 5 cm and 10 cm depth soil layer in three treatments. Means within same transplanting days in the same soil layer followed by the same letter are not significantly different ($p = 0.05$).

Treatment	Soil Layer Depth (cm)	Mean Daily Soil Temperatures (°C)		
		10 Days after Transplanting	20 Days after Transplanting	30 Days after Transplanting
RB1		20.0 a	23.4 ab	29.4 b
RB2	5	19.0 a	23.1 a	28.0 a
RB3		18.6 a	23.3 b	28.0 b
RB1		18.3 a	21.5 a	27.5 a
RB2	10	17.2 a	21.1 a	25.8 a
RB3		17.2 a	21.4 b	25.9 b

Table 5. Plant growth of the three treatments one month after transplanting. Means within a column followed by the same letters are not significantly different ($p = 0.05$).

Treatment	Leaf Number	Plant Height (mm)	Leaf Area Index
RB1	7.2 a	83.6 a	0.125 a
RB2	6.5 a	76.9 a	0.118 a
RB3	5.7 a	78.6 a	0.117 a

3.4. Weight of Fresh Vines with Leaves and Tuber Yield

As shown in Table 6, the RB1 treatment had 7.7% ($p > 0.05$) and 30.2% ($p < 0.05$) more branches in the growth period of nearly five months as compared with the RB2 and RB3, respectively. Meanwhile, the relative weight of the fresh vines with leaves in the RB1 was significantly ($p < 0.05$) increased by 57.5% and 70.8% compared to that in the RB2 and the RB3, respectively. However, the length of the longest vine of each plant was similar (1.5–1.7 m), which may be determined by the growth characteristics of the same sweet potato variety.

Table 6. Weight of fresh vines and tuber yield in three treatments during the experiment. Means within a column by the same letters are not significantly different ($p = 0.05$).

Treatment	Vines (/Plant)			Tubers (/Plant)					
	Total Number	Length of Longest Vine (m)	Weight of Fresh Vines (g)	Total Number	Large Tubers No.	Medium Tubers No.	Fresh Weight (g)	Standard Deviation	Yield (t ha^{-1})
RB1	5.6 a	1.7 a	949.7 a	4.0 a	1.0 a	3.0 a	875.2 a	27.0%	42.9 a
RB2	5.2 ab	1.5 a	602.8 b	4.2 a	1.0 a	3.0 a	653.8 a	24.5%	32.1 a
RB3	4.3 b	1.7 a	556.1 b	5.2 a	0.9 a	4.2 a	706.8 a	26.6%	34.6 a

In this research, the mean tuber number per plant in each treatment was 4–5, while the number of large tubers was about 1 and the number of nedium tubers was about 3–4. The weight of single tubers in the RB3 was slightly more uniform than that in the RB1 treatment, even when it had higher variation of plant spacing and planting depth. The tuber yield per plant was 875.2 g $plant^{-1}$ in the RB1 compared to 653.8 g $plant^{-1}$ in the RB2 and 706.8 g $plant^{-1}$ in the RB3, which indicated that the tubers yield in RB1 was 23.8–33.8% higher than that in the RB2 and the RB3. As a result, the fresh tuber yield was 32.1–42.9 t ha^{-1} in the three treatments.

3.5. Economic Benefit

As shown in Table 7, mean annual input costs for the three treatments varied from 3203.0 US$ ha^{-1} in RB2 to 3337.6 US$ ha^{-1} in the RB1. The RB1 cost the most due to using plastic mulch, even though it used less herbicide and water. Meanwhile, the RB3 cost the least in terms of the labour use of transplanting due to the higher productivity of th machine. However, the difference of input costs among the three treatments was margina Since the RB1 had a greater fresh tuber yield, the farmer profit for the RB1 was 43.8% an 30.2% greater than that for the RB2 and the RB3, respectively.

Table 7. Economic benefit analysis for three treatments.

Treatment	RB1	RB2	RB3
Inputs			
Sweet potato slips (US\$ ha^{-1})	765.6	765.6	765.6
Fertilizer (US\$ ha^{-1})	210.9	210.9	210.9
Herbicide (US\$ ha^{-1})	81.8	93.8	93.8
Plastic mulch and drip irrigation pipe (US\$ ha^{-1})	632.7	485.1	485.1
Mechanical operation cost in transplanting (US\$ ha^{-1})	234.4	234.4	234.4
Labour in transplanting (US\$ ha^{-1})	125.0	117.2	113.3
Irrigation (US\$ ha^{-1})	21.6	30.4	37.8
Mechanical operation cost in other process (US\$ ha^{-1})	703.1	703.1	703.1
Labour use in other process (US\$ ha^{-1})	562.5	562.5	562.5
Total (US\$ ha^{-1})	3337.6	3203.0	3206.5
Outputs			
Yield (US\$ ha^{-1})	42.9	32.1	34.6
Price (US\$ kg^{-1})	0.39	0.39	0.39
Income (US\$ ha^{-1})	16,731.0	12,519.0	13,494.0
Farmer income (US\$ ha^{-1})	13,393.4	9316.0	10,287.5

4. Discussion

The clamping-plate compound sweet potato transplanter had the least missed trans planting counts and the greatest exposed transplanting counts (Table 2). This was due t the reduced action of taking-planting the sweet potato slips, which was one of the typica differences between the clamping-plate type and finger-clip type compound sweet potat transplanters. In the RB3, the planting depth (59.2 mm) was the shallowest and its variatio was the greatest, as shown in Table 3. The reason for this is that sweet potato slips wer placed on the ridge through lifting and through covering the soil on the slips by using th clamping-plate compound sweet potato transplanter. The plant spacing variation (8.9% of the clamping-plate compound sweet potato transplanter by using the soil-coverin method was in agreement with that of the other horizontal transplanter [41], which has similar method of placing the slips. The transplanting depth qualified rate (96.9–97.1% and planting length (198.2–201.8 mm) in the RB1 and the RB2 of the finger-clip compoun sweet potato transplanter were in accordance with Murakami et al. [27]. Available wate for the plant is necessary for rapid early slips growth [10]. The shallower the slips wer planted in the RB3 treatment, the more irrigation was needed. All the transplanting qualit and precision in the RB1, RB2, and RB3 treatments satisfied the sweet potato transplantin requirements [17].

Soil temperature is an important environmental factor for plant growth and devel opment [42,43]. The Huang-Huai-Hai region of China was usually suffering a sudde temperature drop from the end of April to the beginning of May. The soil temperature i the 5–10 cm soil layer of the RB1 treatment was 0.3–1.7 °C and 0.1–1.7 °C higher durin the first month after transplanting than that of the RB2 and the RB3, respectively, wit the help of the plastic mulch. Rao et al. [44] also pointed out that mean soil temperature (19.9 °C) were significantly higher under mulched plots compared to non-mulched soi (19 °C) during their three-year experiment.

The proper soil temperature tended to promote the sweet potato growing processe as shown previously by Bandara et al. [45]. The higher temperature in the RB1 treatmen in the first transplanting month could help to produce better growing conditions, and th plant height and leaf area index were both improved in the RB1 treatment in the initia growing period in this study. The improvements may also be caused by the fact that mor moisture was retained and by enhanced mineral N (29–87%) in the mulched soil for the dry season, as previously indicated by Kundu et al. [46].

Mulched soil enhanced mineral N, P, and K availability is applied for sweet potato [46], while all of those chemical properties are critical for the yield increasing. The mulched raised beds in the RB1 contributed to the significant ($p < 0.05$) increase of 346.9 g plant^{-1} of the aboveground growth and the marginal increase of 221.4 g plant^{-1} of the fresh tubers weight compared with those of the RB2. These results are consistent with those found by Rao et al. [44]. Moreover, using plastic mulch in cool climates seems to increase the aboveground growth of sweet potato significantly, while the storage root yield was less affected [47]. In the RB3, the total tubers number was higher than the other treatments under the horizontal vertical placement by using the clamping-plate compound sweet potato transplanter. The increased number of the fresh tubers in the RB3 was offset by the decrease in the weight of each fresh tuber, and the size of the tubers was more consistent and more popular for fresh sweet potatoes. The yield in the RB3 was slightly higher than that of the RB2 while its weight of the fresh vines of each plant was marginally lower than the RB2, possibly due to the horizontal vertical placement with the varying slips orientation. In RB3, the slips were grown above the ridge furrow by being placed horizontally and vertically to the ridge, while in the RB2, the slips were grown above the ridge by being placed along the ridge during the first few growing months of the growing period. As a result, the distribution of the solar energy in the RB3 was much greater on the ridge areas than that of the RB2, which is crucial for tuber growth as they are planted in the ridge.

The positive effects of mulching and horizontal vertical placement on crop growth and yield were probably responsible for the increased economic benefits in the RB1 and the RB3 treatments. The results agree with those of Hou et al. [21] and Rao et al. [44]. The proportion of labor costs in the mechanized sweet potato production process of this study was 20.1%, which dropped significantly compared with the study of Kassali, in which no machine was used, and in which the labor cost accounted for 68% of the total cost [48]. The use of mechanization in sweet potato production increased the economic benefits. Tang et al. [49] also found that the labor cost was 46.5% of the total cost during sweet potato production in which the transplanting process was accomplished manually. It seems that the use of the mechanized transplanting reduced the labor cost by 26.5%. Yan et al. also pointed out that the labor volume of sweet potato transplanting accounts for about 23% of the whole production process [41], but the labor cost only accounted for 3.7% in this study because of the use of mechanical transplantation. The replacement of labor transplanting with mechanized transplanting thus contributed significantly to the improvement of the economics of sweet potato production.

5. Conclusions

In this study, considerable changes in crop performances and yield due to mulched raised beds and horizontal vertical transplanting placement were observed. The finger-clip compound sweet potato transplanter and the clamping-plate compound sweet potato transplanter satisfies the requirement of sweet potato transplanting among three raised bed cropping systems. With the lowest coefficient variation of plant spacing and planting depth, the finger-clip compound sweet potato transplanter produced the raised beds with a higher temperature by using the plastic mulch for the growth of rapid early slips in the RB1 treatment, thereby improving 57.5–70.8% of the weight of fresh vines and 23.8–33.8% of the yield of tubers compared to both the RB2 and RB3 treatments. However, when the plastic mulch was not used, the clamping-plate compound sweet potato transplanter provided a 7.8% higher yield than the finger-clip compound sweet potato transplanter by placing the slips horizontally vertical to the raised beds. In general, in the areas of the mulched soil planting system, the finger-clip compound sweet potato transplanter could be a suitable option. In the areas without mulch, though, the clamping-plate compound sweet potato transplanter has the potential to increase production.

Author Contributions: Conceptualization, H.L. (Hui Li) and B.W.; methodology, T.H.; machi preparation, J.Z. and H.L. (Hu Liu); field experiment, S.S., Y.S. and X.L.; validation, T.H.; da curation, H.L. (Hu Liu); writing—original draft preparation, H.L. (Hui Li); writing—review a editing, B.W.; project administration, H.L. (Hui Li); funding acquisition, H.L. (Hui Li). All autho have read and agreed to the published version of the manuscript.

Funding: This research was funded by the National Natural Science Foundation of China (Grant N 32201683) and Agricultural Science and Technology Innovation Project of Shandong Academy Agricultural Sciences (Grant No. CXGC2023B04).

Data Availability Statement: Not applicable.

Acknowledgments: Thanks to all of our workmates and the postgraduate students working in t Huang Huai Hai Key laboratory of Modern Agricultural Equipment, who provided their input this research.

Conflicts of Interest: The authors declare no conflict of interest.

References

1. Foley, J.A.; Ramankutty, N.; Bennett, E.M.; Brauman, K.A.; Carpenter, S.R.; Cassidy, E.; Gerber, J.; Hill, J.; Johnston, M.; Monfre C. Solutions for a cultivated planet: Addressing our global food production and environmental sustainability challenges. *Natu* **2011**, *478*, 337–342. [CrossRef] [PubMed]
2. Pardey, P.G.; Beddow, J.M.; Hurley, T.M.; Beatty, T.K.M.; Eidman, V.R. A bounds analysis of world food futures: Global agricultu through to 2050. *Aust. J. Agric. Resour. Econ.* **2014**, *58*, 571–589. [CrossRef]
3. Oladejo, A.O.; Ma, H.; Qu, W.; Zhou, C.; Wu, B.; Yang, X. Influence of ultrasound pretreatments on diffusion coefficients, textu and colour of osmodehydrated sweet potato (*Ipomea batatas*). *Int. J. Food Sci. Technol.* **2017**, *52*, 888–896. [CrossRef]
4. Low, J.; Ball, A.; Magezi, S.; Njoku, J.; Mwanga, R.; Andrade, M.; Tomlins, K.; Dove, R.; Mourik, T.V. Sweet potato developme and delivery in sub-Saharan Africa. *Afr. J. Food Agric. Nutr. Dev.* **2017**, *17*, 11955–11972. [CrossRef]
5. Gao, F.; Gong, Y.; Zhang, P. Production and deployment of virus-free sweetpotato in China. *Crop Prot.* **2000**, *19*, 105–111.
6. Iese, V.; Holland, E.; Wairiu, M.; Havea, R.; Patolo, S.; Nishi, M.; Hoponoa, T.; Bourke, R.M.; Dean, A.; Waqainabete, L. Faci food security risks: The rise and rise of the sweet potato in the Pacific Islands. *Glob. Food Secur.* **2018**, *18*, 48–56. [CrossRef]
7. Getahun, G.T.; Katterer, T.; Munkholm, L.J.; Rychel, K.; Kirchmann, H. Effects of loosening combined with straw incorporati into the upper subsoil on soil properties and crop yield in a three-year field experiment. *Soil Tillage Res.* **2022**, *223*, 1054 [CrossRef]
8. Brandenberger, L.; Shrefler, J.; Rebek, E.; Damicone, J. Sweet Potato Production. Division of Agricultural Sciences and Natur Resources, Oklahoma State University. *Tech. Rep.* **2015**, *8*, HLA-6022.
9. Abrham, T.; Beshir, H.M.; Haile, A. Sweetpotato production practices, constraints, and variety evaluation under different stora types. *Food Energy Secur.* **2020**, *10*, e263. [CrossRef]
10. Parwada, C.; Gadzirai, C.T.; Sithole, A.B. Effect of Ridge Height and Planting Orientation on Ipomoea Batatas (sweet potat Production. *J. Agric. Biotechnol. Sustain.* **2011**, *3*, 72–76.
11. Chagonda, I.; Mapfeka, R.F.; Chitata, T. Effect of Tillage Systems and Vine Orientation on Yield of Sweet Potato (*Ipomoea batat* L.). *Am. J. Plant. Sci.* **2014**, *5*, 3159–3165. [CrossRef]
12. Abdallah, B.; Githiri, S.M.; Kariuki, W.; Saha, H.M. Evaluation of the Performance of Sweet Potato (*Ipomoea batatas*) Clones und Water Stress in the Coastal Lowlands of Kenya. *World J. Agric. Res.* **2020**, *8*, 114–120.
13. Ribeiro, N.P.; Fernandes, A.M.; Silva, R.M.D.; Pelvine, R.A.; Assuncao, N.S. Growth and yield of sweet potato in response to t application of nitrogen rates and paclobutrazol. *Soil Plant Nutr.* **2021**, *80*, e3821. [CrossRef]
14. PEPÓ, P. The effect of different planting methods on the yield and spad readings of sweet potato (*Ipomoea batatas* L.). *J. Agr Environ. Sci.* **2018**, *5*, 7–12. [CrossRef]
15. FAO (Food and Agriculture Organization). FAOSTAT Online Statistical Database: Food Balance 2019. FAOSTAT. Available onlir http://www.fao.org/faostat/zh/#data/QC (accessed on 17 February 2022).
16. Agili, S.; Nyende, B.; Ngamau, K.; Masinde, P. Selection, yield, drought tolerance indices of orange-fleshed sweet potato (Ipomo batatas Lam) hybrid clone. *J. Nutr. Food Sci.* **2012**, *2*, 1–8.
17. Wang, B. Overview of sweet potato production in China. In *Main Cultivation Modes of Sweet Potato in China*, 1st ed.; Zhang, L., M D., Wang, Q., Wang, B., Eds.; China Agricultural Science and Technology Press: Beijing, China, 2012; pp. 1–8.
18. Xu, F.L.; Guo, B.; Ye, B.; Ye, Q.; Chen, H.; Ju, X.; Guo, J.; Wang, Z. Systematical evaluation of GPMIMERG and TRMM3B42V precipitation products in the Huang-Huai-Hai Plain, China. *Remote Sens.* **2019**, *11*, 697. [CrossRef]
19. National Bureau of Statistics of China. *China Statistical Yearbook*; China Statistics Press: Beijing, China, 2021.
20. Ma, D.; E, W. Chapter three- Key technology of green light simplification for fresh edible sweet potato. In *Handbook of Green Lig Simplified Cultivation Techniques for Sweet Potato*; China Agricultural Press: Beijing, China, 2021.
21. Hou, F.Y.; Dong, S.X.; Xie, B.T.; Zhang, H.Y.; Li, A.X.; Wang, Q.M. Mulching with plastic film improved the root quality summer-sown sweet potato (*Ipomoea batatas* (L). Lam.) in northern China. *J. Integr. Agric.* **2019**, *18*, 982–991. [CrossRef]

2. Hu, L.; Hu, Z.; Xie, Y.; Tian, L.; Ji, F.; Wang, B. Study on the route of mechanization of sweet potato (*Ipomoea batatas* lam.) production technology in China. *Chin. Agric. Mech.* **2011**, *5*, 20–25.
3. Chen, L.H.; Younis, T.S.; Allison, M. Horizontal Transplanting of Sweet Potatoes. *Trans. ASAE* **1982**, *25*, 1524–1528. [CrossRef]
4. Hu, L.; Wang, B.; Wang, G.; Yu, Z.; You, Z.; Hu, Z.; Wang, B.; Gao, X. Design and experiment of type 2ZGF-2 duplex sweet potato transplanter. *Trans. Chin. Soc. Agric. Eng.* **2016**, *32*, 8–16.
5. Mukhopadhyay, S.K.; Chattopadhyay, A.; Chakraborty, I.; Bhattacharya, I. Crops that feed the world 5: Sweet potato. Sweet potatoes for income and food security. *Food Secur.* **2011**, *3*, 283–305. [CrossRef]
6. Leighton, C.S. Nutrient and Sensory Quality of Orange-Fleshed Sweet Potato. Master's Thesis, University of Pretoria, Pretoria, South Africa, 2007.
7. Murakami, K.; Takahiko, T.; Yamaguchi, H.; Tanazaki, A. Seedling Transplanter. CN Patent 103988621B, 20 August 2014.
8. Li, H.; He, T.; Liu, H.; Shi, S.; Wang, B.; Zhou, J.; Liu, X.; Wei, G.; Zhang, R. Efficacy of Sweet Potato Transplanting Machine for Different Cultivation Systems in Northern China. *Agriculture* **2022**, *12*, 1184. [CrossRef]
9. Pan, Z.; Li, L.; Chen, D.; Zha, X.; Yang, R. Design and Optimization of a Boat-BottomShaped Transplanting Device for Sweet Potato (Ipomoea batatas) with Low Seedling Damage Rate. *Appl. Sci.* **2022**, *12*, 2817. [CrossRef]
0. Liu, Z.; Wang, X.; Zheng, W.; Lv, Z.; Zhang, W. Design of a Sweet Potato Transplanter Based on a Robot Arm. *Appl. Sci.* **2022**, *11*, 9349. [CrossRef]
1. China Meteorological Administration. China Meteorological Data Network: Global Historical Statistics 2016–2021. China Meteorological Data Service Center. Available online: http://data.cma.cn/ (accessed on 7 March 2023).
2. *JB/T 10291-2013*; Transplanter of Dry Land Plant. Machinery Industry Press: Beijing, China, 2013.
3. *NY/T 3486-2019*; Operating Quality for Vegetable Transplanter. China Agricultural Press: Beijing, China, 2019.
4. Tessier, S.; Saxton, K.E.; Papendick, R.I.; Hyde, G.M. Zero-tillage furrow opener effects on seed environment and wheat emergence. *Soil Tillage Res.* **1991**, *21*, 347–360. [CrossRef]
5. Mao, S.C.; Song, M.Z.; Zhang, C.J.; Han, Y.C.; Xing, J.S.; Zhuang, J.N. Studies on the effects of soil temperature in cotton fields in the wheat and cotton co-growing period under a double cropping system in the Huanghuaihai Plains. *Sci. Agric. Sin.* **1998**, *31*, 5.
6. Cristofori, V.; Rouphael, Y.; Gyves, E.M.; Bignami, C. A simple model for estimating leaf area of hazelnut from linear measurements. *Sci. Hortic.* **2007**, *113*, 221–225. [CrossRef]
7. Mulungu, L.S.; Mwailana, D.J.; Reuben, S.S.; Tarimo, J.P.; Massawe, A.W.; Makundi, R.H. Evaluation on the Effect of Topping Frequency on Yield of Two Contrasting Sweet Potato (*Ipomoea batatas* L.) Genotypes. *J. Appl. Sci.* **2006**, *6*, 1132–1137. [CrossRef]
8. Sugito, Y. *Research Methods: Experiment Method and Scientific Writing*; UB Press, Universitas Brawijaya: Malang, Indonesia, 2013.
9. Widaryanto, E.; Saitama, A. Analysis of plant growth of ten varieties of sweet potato (*Ipomoea batatas* L.) cultivated in rainy season. *Asian J. Plant Sci.* **2017**, *16*, 193–199. [CrossRef]
0. Zhang, X.; Li, H.; He, J.; Wang, Q.; Golabi, M.H. Influence of conservation tillage practices on soil properties and crop yields for maize and wheat cultivation in Beijing, China. *Aust. J. Soil Res.* **2009**, *47*, 362–371. [CrossRef]
1. Yan, W.; Hu, M.; Li, K.; Wang, J.; Zhang, W. Design and Experiment of Horizontal Transplanter for Sweet Potato Seedlings. *Agriculture* **2022**, *12*, 675. [CrossRef]
2. Haque, M.A.; Jahiruddin, M.; Clarke, D. Effect of plastic mulch on crop yield and land degradation in south coastal saline soils of Bangladesh. *Int. Soil Water Conserv. Res.* **2018**, *6*, 317–324. [CrossRef]
3. Shehata, S.A.; Abouziena, H.F.; Abdelgawad, K.F.; Elkhawaga, F.A. Weed Control efficacy, growth and yield of potato (*Solanum tuberosum* L.) as affected by alternative weed control methods. *Potato Res.* **2019**, *62*, 139–155. [CrossRef]
4. Rao, S.A.; Singh, P.; Gonsalves, T. Black plastic mulch affects soil temperature and yield of sweet potato under short season temperate climates. *Int. J. Veg. Sci.* **2023**, *29*, 72–83. [CrossRef]
5. Bandara, G.; Raja, R.K.; Shankle, M.W. Quantifying Growth and Developmental Responses of Sweetpotato to Mid- and Late-Season Temperature. *Agron. J.* **2015**, *107*, 1854–1862.
6. Kundu, D.K.; Singh, R.; Chowdhury, S.R. Effect of rice straw mulch and irrigation on nutrient availability in soil and tuber yield of sweet potato (*Ipomoea batatas* L.) in coastal Orissa. Root and Tuber Crops. In *Nutrition, Food Security and Sustainable Environment*; Naskar, S.K., Nedunchezhiyan, M., Rajasekhara, R.K., Sivakumar, P.S., Ray, R.C., Misra, R.S., Mukherjee, A., Eds.; Regional Entre of Central Tuber Crops Research Institute: Bhubaneswar, India, 2006; pp. 117–122.
7. Dumbuya, G.; Alemayehu, H.A.; Hasan, M.M.; Matsunamid, M.; Shimono, H. Effect of soil temperature on growth and yield of sweet potato (*Ipomoea batatas* L.) under cool climate. *J. Agric. Meteorol.* **2021**, *77*, 118–127. [CrossRef]
8. Kassali, R. Economics of Sweet Potato Production. *Int. J. Veg. Sci.* **2011**, *17*, 313–321. [CrossRef]
9. Tang, C.; Lu, Y.; Jiang, B.; Chen, J.; Mo, X.; Yang, Y.; Wang, Z. Energy, Economic, and Environmental Assessment of Sweet Potato Production on Plantations of Various Sizes in South China. *Agronomy* **2022**, *12*, 1290. [CrossRef]

agronomy

MDPI

Article

Plasma Treated Cattle Slurry Moderately Increases Cereal Yield

Thomas Cottis, Hesam Mousavi * and Svein Øivind Solberg

Faculty of Applied Ecology, Agricultural Sciences, and Biotechnology, Inland Norway University of Applied Science
2418 Elverum, Norway; thomas.cottis@inn.no (T.C.); svein.solberg@inn.no (S.Ø.S.)
* Correspondence: hesam.mousavi@inn.no

Abstract: Plasma treatment offers an approach to enhance the nitrogen (N) content of livestock slurr and biogas digestate, thereby increasing the efficacy of organic fertilizers. This innovative metho is used to produce nitrogen-enriched organic fertilizer (NEO) containing a double concentration o plant-available N. Over three years, we conducted a comprehensive study in 14 spring wheat an barley field trials in Norway. The primary objective was to assess and compare the cereal grain yiel achieved by applying NEO to other conventional fertilizers. The NEO utilized in our research wa derived from the unit developed by the Norwegian company N2 Applied. The results indicated tha 120 kg N ha^{-1} in NEO yielded in the same range of cereal grains as 95 kg N ha^{-1} in mineral fertilize Moreover, the combination of untreated slurry and 55 kg N ha^{-1} in mineral fertilizer Opti-NS yielde the same as 120 kg N ha^{-1} in NEO. Surprisingly a combination of 12 kg N ha^{-1} in mineral fertilize at sowing day and 108 kg N ha^{-1} in NEO at the three-leaf stage led to a higher yield in spring whea than 120 kg N ha^{-1} NEO spread at sowing day in two out of three experimental years. Moreove applying NEO directly to plants has shown no visible signs of harm. Lastly, filtering the slurr resulted in higher cereal grain yields than the untreated slurry. In conclusion, despite possessin the same N content, utilizing NEO yielded a 15–20% lower cereal grain yield than mineral fertilize Nonetheless, 20–30% more yield than the native amount of cattle slurry it derived. However, w have observed an unexplained loss of approximately 17% of the nitrogen in NEO, which does no translate into increased grain yield or nitrogen productivity.

Keywords: agronomy; field crops; fertilization; innovation; wheat; barley; nitrogen

Citation: Cottis, T.; Mousavi, H.; Solberg, S.Ø. Plasma Treated Cattle Slurry Moderately Increases Cereal Yields. *Agronomy* **2023**, *13*, 1549. https://doi.org/10.3390/agronomy13061549

Academic Editors: Umberto Anastasi and Aurelio Scavo

Received: 4 May 2023
Revised: 30 May 2023
Accepted: 1 June 2023
Published: 2 June 2023

1. Introduction

Global food production systems encounter numerous challenges due to rising foo demand, which coincides with population growth [1]. Simultaneously, the detrimenta effects of global warming and soil degradation are progressively diminishing productio capacity [2]. In this context, agroecosystems face substantial societal pressure to foste sustainable food production [3–6].

The beneficial impact of nitrogen (N) fertilization on plant productivity has bee extensively studied and widely acknowledged [7,8]. Moreover, the availability of nitroge (N) is a fundamental necessity in plant production [9,10], and as such, the utilization o N in agroecosystems has undergone a significant transformation in recent decades [11 Nevertheless, the excessive application of nitrogen fertilizers can give rise to significan drawbacks and unfavorable outcomes, despite their initial positive effects [12]. Beyond tha the production process of mineral fertilizers results in environmental pollution, disturbanc of natural processes, and substantially adverse effects on biodiversity and the climate [13 This clarifies the necessity of developing sustainable agricultural amendments based o organic principles.

Over the last twelve years, the Norwegian company N2 Applied has developed a uni to enhance the nitrogen content of slurry or digestate, with electricity and air as the onl inputs [14–16]. The process uses electrical energy to generate an air plasma, where oxyge and nitrogen combine to form a reactive nitrogen gas. The NOx is subsequently absorbe in the slurry as nitrate and nitrite, enriching the slurry with plant-available nitrogen and reducing the pH. The plasma-treated slurry is termed Nitrogen Enriched Organic fertilizers (NEO). The unit is currently accessible for scientific use and testing by producers, with plans for potential commercial availability in Europe in 2023.

The company reports that their units require 50 kWh of electricity per kg N added to the slurry. Consequently, based on an average addition of 1.65 kg N ton^{-1} slurry, the unit would require 82.5 kWh of electricity per ton of treated slurry. Additionally, the company reports that the N2 Applied unit has a daily 5–8 tons capacity.

Since NEO is a novel product, assessing its impact on plant yield and its effectiveness compared to conventional fertilizers, e.g., mineral fertilizers, cattle slurry, etc., is essential before considering its commercialization. Thus, to accomplish this, we at Inland Norway University of Applied Sciences (INN) conducted comprehensive trials to document and compare the effects of NEO on soil health [17,18] and plant yields in the growing chamber [19] and cereal and grass fields at different locations in Norway over three years (2020 to 2022).

The current study investigates and compares the effects of NEO made from cattle slurry on grain yields of spring barley (*Hordeum vulgare* L.) and spring wheat (*Triticum aestivum* L.), hereafter termed barley and wheat, to other conventional fertilizers used in agriculture. Additionally, we anticipate publishing the results from our grass trials in a forthcoming paper.

Since NEO is a novel product with potentially beneficial gains, its effects on plant yields should be elucidated before introducing it into the global markets. Therefore, this study aims to determine the effects of NEO made from cattle slurry on cereal yields in Norway compared to farmers' alternatives, such as mineral fertilizers and untreated cattle slurry. Therefore, the research questions were: (1) What is the fertilization effect of NEO compared to the other alternatives; and (2) Can NEO be spread at three leaf stage without harming the plants, and if so, would such an application produce a higher yield than NEO spread at the sowing day.

We hypothesized that: (1) NEO could produce the same grain yields as mineral fertilizer, and (2) spreading NEO at the three-leaf stage does not harm the plants but instead boost growth and yield compared to spreading NEO on sowing day.

2. Materials and Methods

2.1. Experimental Design

The experimental design consisted of a randomized complete block design with four replicates, encompassing Series 1 and Series 2. In Series 1, the fertilizer plots measured 10 × 3 m, with a harvested area of 1.5 × 8.5 m within each plot. The larger plot size in Series 1 was necessary to facilitate extensive soil sampling for analyzing soil organisms and overall soil health. In Series 2, the fertilizer plots were smaller, measuring 2.5 × 8 m, with harvest plots at 1.5 × 6.5 m.

The results obtained in 2020 provided indicative evidence supporting the notion that filtered slurry yields positive effects. As a result, the plots receiving the filtered slurry treatment in 2020 were transformed into control treatments with no fertilizers in Series 1 for the 2021 and 2022 trials. Additionally, another finding from 2020 indicated that applying NEO at the three-leaf stage of grain plants resulted in lower yields than NEO applied on sowing day. Consequently, a separate series of trials in Series 2 was designed, omitting the NEO application at the three-leaf stage treatment. Furthermore, the N-level treatments in mineral fertilizers were increased to assess the nitrogen effect of NEO better.

2.2. Trials Location

The field trials were located at four representative areas for cereal production in Norway: 1. Tønsberg; 2. Årnes; 3. Hamar; and 4. Stjørdal (Figure 1). Details on the locations and soil types are provided in Table 1.

Figure 1. The map showing the locations in the southern part of Norway, where the field trials we conducted in 2020, 2021, and 2022: Tønsberg 1, Årnes 2, Hamar 3, and Stjørdal 4.

Table 1. The trial numbers, location coordinates, and soil quality information at the trial sites. T barley and wheat were all spring-sown types.

Series	Trial	Location	Crops, Varieties, and Years	Detailed Location and Coordinates	Soil Type and Key Soil Parameters
1	1	3	Barley' Salome' 2020, Wheat 'Betong' 2021, Barley' Bente' 2022	3 km east of Hamar (60.81830° N. 011.17968° E)	Loam. 4.5% organic. pH 7.4
1	2	3	Wheat 'Mirakel' 2020, Barley' Anita' 2021, Wheat 'Betong' 2022	3 km east of Hamar (60.81830° N. 011.17968° E)	Loam. 4.5% organic. pH 7.4
1	3	2	Wheat 'Helmi' 2021	3 km west of Årnes (60.12604° N. 11.39471° E)	Silt loam. 4.0% organic. pH 6.0
1	4	2	Barley 'Brage' 2021	3 km west of Årnes (60.12604° N. 11.39471° E)	Silt loam. 4.0% organic. pH 6.0
2	5	1	Wheat 'Betong' 2021	5 km west of Tønsberg (59.294937° N. 10.318813° E)	Silt loam. 6.5% organic. pH 6.2
2	5	1	Wheat 'Betong' 2022	15 km north of Tønsberg (59.384537° N. 10.232651° E)	Silt loam. 4.8% organic. pH 6.9
2	6	4	Barley 'Thermus' 2021	4 km north of Stjørdal (70.41109° N. 59.3647° E)	Loam. 2.7% organic. pH 6.1
2	6	4	Barley 'Thermus' 2022	4 km north of Stjørdal (70.37496° N. 59.7733° E)	Loam. 2.7% organic. pH 6.1

2.3. Fertilizers

In the trials, we used the following fertilizers:

- Untreated slurry: Cattle Slurry from the Norwegian University of Life Sciences far
- NEO (Nitrogen Enriched organic fertilizer): This is the same slurry as «Untreate slurry» processed through the N2 Applied unit. The available nitrogen in NEO around 50% ammonia, 30% nitrate, and 20% nitrite, and the acidity is down to aroun pH 5.2. The relative levels of nitrate and nitrite vary quite a lot. See Table 2.
- Mineral fertilizer 18-3-15: A commercially available mineral fertilizer produced b Yara [20] with 18% nitrogen (N), 3% phosphorus (P), and 15% potassium (K). The 18 N consists of slightly more ammonia than nitrate. This fertilizer was chosen due to th similarities in plant available nutrients to NEO.

- Mineral fertilizer Opti-NS (27-0-0) [21]: This is an N fertilizer combined with sulfur (S) (3.6%), where the N consists of equal amounts of ammonia and nitrate.

Table 2. Amounts of mineral N, ammonia N, nitrate N, nitrite N, total N (kg N ton^{-1}), and pH in the NEO and untreated cattle slurry used over the three years (average over several analyses per year).

Fertilizer and Year	N-Min (kg ton^{-1})	NH^{4+} (kg ton^{-1})	NO^{3-} (kg ton^{-1})	NO^{2-} (kg ton^{-1})	Total N (kg ton^{-1})	pH
NEO 2020	3.4	1.68	1.24	0.52	4.7	5.3
Untreated 2020	1.7	1.7	0	0	2.8	7.1
NEO 2021	3.2	1.5	0.92	0.8	4.38	5.59
Untreated 2021	1.5	1.5	0	0	2.68	7.17
NEO 2022	3.55	1.66	1.19	0.69	Not analyzed	5.15
Untreated 2022	1.75	1.7	0	0	Not analyzed	7.35

During the production process in N2-Applied's plasma reactor, the untreated slurry undergoes filtering to remove solid particles larger than 5 mm using a screw press, which reduces the original volume by 10%. As a result, the filtered material has a consistency similar to soft coarse peat. This filtering, combined with the plasma treatment, transforms the liquid fertilizer into NEO, which exhibits enhanced soil permeation compared to the untreated liquid slurry.

To determine the appropriate quantities of NEO and other fertilizers for the different trial plots, the company sent samples to AnalyTech Environmental Laboratory in Denmark. In 2021 and 2022, the analysis was conducted on the untreated manure and the pre-produced NEO two weeks before their application in the experimental sites. Table 2 provides the nitrogen and pH values of NEO and untreated slurry for 2020, 2021, and 2022.

In 2020, N2 Applied conducted a test production of NEO in March and sent samples for nitrogen content testing to the Danish lab. The fertilizer amounts for the 2020 trials were calculated based on the results. Unfortunately, an error occurred during the production of NEO intended for the field trials, resulting in lower nitrogen content than initially calculated. As a result, the results from the 2020 trials remain valid but cannot be directly compared. Instead, they serve as supporting material for the results obtained in 2021 and 2022.

The primary objective of our studies was to assess the impact of NEO on crop yield in comparison to other farmer alternatives. Therefore, we established a baseline of 120 kg N ha^{-1} for both wheat and barley, considering it as a typical level for barley in Norway's grain regions, albeit slightly lower than what is commonly used for spring wheat. To achieve the desired nitrogen level of 120 kg per hectare, approximately 40 tons of cattle slurry were processed through the N2 Applied's plasma process unit after filtering. The process converted 40 tons of cattle slurry into 37 tons of NEO containing 120 kg of nitrogen.

In 2020, we conducted two trials in series 1. Unfortunately, due to the abovementioned production error, the nitrogen content in the cattle slurry-based treatments differed in 2020 compared to 2021 and 2022. Table 3 presents the treatments labeled (bold) in series 1 and 2 over three years.

Filtered slurry and NEO have different N contents from year to year. Considering this, adjustments were made to keep the nitrogen content per hectare constant from year to year as the most decisive factor.

In 2021 the N-content in NEO was 3.2 kg N ton^{-1}, and we aimed for fertilization with 120 kg N ha^{-1} in NEO, accordingly 37.5 tons ha^{-1}. As mentioned, 10% of cattle slurry is filtered through NEO production. Thus, the farmers' alternative is to spread 41 tons ha^{-1} of untreated slurry. In 2021 the N-content in the untreated slurry was 1.5 kg N ton^{-1}. This year, applying 41 tons ha^{-1} of untreated slurry to the trial plots provided 61.5 kg N/ha^{-1}. In 2022 the NEO had 3.55 kg N ton^{-1}, resulting in 34 tons ha^{-1}

of NEO reaching 120 kg N ha^{-1}. The untreated had 1.75 kg N ton^{-1}, and with a 10% highe volume than NEO, we applied 37 tons ha^{-1} of untreated slurry—resulting in 65 kg N ha^{-} With this clarification, we used 65 kg N ha^{-1} in the graph labels for the untreated slurry fo 2021 and 2022.

Table 3. The treatments and their labels (bold text) used in our trials.

Treatments in Series 1 in 2020:	Treatments in Series 1 in 2021 and 2022:	Treatments in Series 2 in 2021 and 2022:
MaF51: 51 kg N ha^{-1} in Filtered untreated slurry.	NoF: No fertilizer	NoF: No fertilizer
Ma56: 56 kg N ha^{-1} in untreated slurry	Ma65: 65 kg N ha^{-1} in untreated slurry (manure).	NEO120: 120 kg N ha^{-1} in NEO
NEO102: 102 kg N ha^{-1} in NEO	NEO120: 120 kg N ha^{-1} in NEO	Ma65: 65 kg N ha^{-1} in untreated slurry.
MiNEO 104: 12 kg N ha^{-1} in mineral fertilizer 18-3-15 applied to the trial plots before sowing and 92 kg N/ha^{-1} in NEO at Zadoks GS13 three leaves stage.	MiNEO 120: 12 kg N ha^{-1} in mineral fertilizer 18-3-15 applied to the trial plots before sowing and 108 kg N/ha^{-1} in NEO at Zadoks GS13 three leaves stage.	MaMi120: 65 kg N ha^{-1} in untreated slurry and 55 kg N ha^{-1} in mineral fertilizer Opti-NS.
Mi51: 51 kg N ha^{-1} in mineral fertilizer 18-3-15	Mi65: 65 kg N ha^{-1} in mineral fertilizer 18-3-15	Mi30: 30 kg N ha^{-1} in mineral fertilizer 18-3-15
Mi91: 91 kg N ha^{-1} in mineral fertilizer 18-3-15	Mi91: 91 kg N ha^{-1} in mineral fertilizer 18-3-15	Mi55: 55 kg N ha^{-1} in mineral fertilizer 18-3-15
Mi123: 123 kg N ha^{-1} in mineral fertilizer 18-3-15	Mi120: 120 kg N ha^{-1} in mineral fertilizer 18-3-15	Mi80: 80 kg N ha^{-1} in mineral fertilizer 18-3-15
MaMi123: 56 kg N ha^{-1} in untreated slurry combined with 6.7 kg N ha^{-1} in mineral fertilizer Opti-NS	MaMi120: 65 kg N ha^{-1} in untreated slurry combined with 55 kg N ha^{-1} in mineral fertilizer Opti-NS	Mi105: 105 kg N ha^{-1} in mineral fertilizer 18-3-15
		Mi120: 120 kg N ha^{-1} in mineral fertilizer 18-3-15

It is also necessary to clarify a point regarding the mineral fertilizer plus NEO treatmen (MiNEO120). In earlier testing of NEO, the N2 Applied company had experienced tha NEO could be applied to cereals after germination. Therefore, we agreed to test this in serie 1 by forming the MiNEO120 treatment, where we applied 12 kg N ha^{-1} in mineral fertilize Yara Mila complete fertilizer 18-3-15 (Yara, Oslo, Norway) to the trial plots before sowing combined with 108 kg N ha^{-1} in NEO applied at three leaves stage Zadoks GS13 [22]. Al the other treatments were applied on sowing day by spreading the fertilizers on the tria plots and mixing them into the soil using a disc harrow. The grain was sown a few hour after fertilization.

2.4. Weather Conditions

Table 4 presents May's average temperature, precipitation, and corresponding averag values in all trial locations over 2020–2022. Series 1 had trials in Hamar and Årnes. In 202 Hamar was 1.4 °C colder than average and had less than half of the normal precipitatior In 2021, Hamar and Årnes had a normal average temperature but about 20% more precipi tation than normal. In 2022, Hamar had a normal average temperature but a dry month o May with 23.2 mm less precipitation than the normal 55 mm. The trials in Series 2 were ir Tønsberg and Stjørdal. Tønsberg had 24 mm more rain than average in 2021 and about hal the normal precipitation in 2022. Stjørdal had a dry month in May, with half the norma precipitation in 2021 and average rainfall in 2022 [23].

Table 4. Average temperatures, normal temperatures, precipitation, and normal precipitation for the grain trial locations over the years 2020–2022.

Year	Location	Average Temperature (°C)	Normal Temperature (°C)	Average Precipitation (mm)	Normal Precipitation (mm)
2020	Hamar	8.5	9.9	23	55
2021	Tønsberg	9.9	10.8	95.1	71
2021	Årnes	9.3	10.2	88.4	59
2021	Hamar	9.5	9.9	77.9	55
2021	Stjørdal	9.6	9.0	33.1	63
2022	Tønsberg	11.4	10.8	36.5	71
2022	Hamar	9.8	9.9	31.8	55
2022	Stjørdal	9.6	9.0	72.6	63

2.5. Data Handling, Statistics, and Analysis

Field trial data were first analyzed using ANOVA and Duncan's multiple-range tests of the means. Then, the N effect of NEO was calculated against the nitrogen effect of mineral fertilizer. This was possible as we included a mineral fertilization ladder ranging from 0 kg N-min ha^{-1} to 120 kg N-min ha^{-1}. Next, a linear regression model expressed the relationship between N provided in mineral fertilization (x axis) and grain yield (y axis). The same was done for the N yield data. The regression equations were then used to calculate the N effect of 120 kg N-min ha^{-1} provided in NEO based on yield and N yield data, respectively. This procedure was repeated for each of the trials and finally across all trials, with 95% confidence intervals. Statistical analyses were done in SPSS 28 software (© 2023 IBM (New York, NY, USA). Excel (© 2023 Microsoft (Seattle, WA, USA), and Minitab 21 (2023 Minitab, LLC (State College, PA, USA)), were used for the graphics. Finally, we analyzed samples from all the trial plots for N percentage with the Dumas method to find the Nitrogen yields.

3. Results

3.1. Barley and Wheat Grain and Nitrogen Yield—Series 1 2020

In 2020, when examining barley grain yield (Figure 2A), it was found that NEO102 produced a yield equivalent to that of MiNEO104, Mi91, and MaMi123. However, Mi123 exhibited a significantly higher yield than all other treatments. Additionally, MaF51 demonstrated a yield of 448 kg ha^{-1}, which was significantly higher than that of Ma56.

When considering wheat grain yield (Figure 2A), it was observed that NEO102 yielded significantly more (469 kg ha^{-1}) compared to MiNEO104 while producing a yield similar to that of Mi91 and MaMi123. However, once again, Mi123 displayed a significantly higher yield than the rest of the treatments. Notably, MaF51 demonstrated a significantly higher yield of 756 kg ha^{-1}, surpassing that of Ma56.

The trend in nitrogen yield for barley and wheat (Figure 2B) followed a similar pattern to grain yield; however, the differences between the treatments became more pronounced.

3.2. Barley and Wheat Grain and Nitrogen Yield—Series One, 2021 and 2022

Here, we present the results from three separate trials conducted in barley and wheat as part of Series one in 2021 and 2022. We have analyzed the data separately for grain yield and nitrogen yield.

Regarding barley grain yield (Figure 3A) among the treatments, Mi120 demonstrated the highest yield, surpassing MiNEO120, NEO120, and MaMi120 by 586 kg ha^{-1}, 610 kg ha^{-1}, and 793 kg ha^{-1}, respectively. Notably, MiNEO120 and NEO120 yielded alike, with both treatments significantly outperforming Ma65 and falling within the range of Mi91.

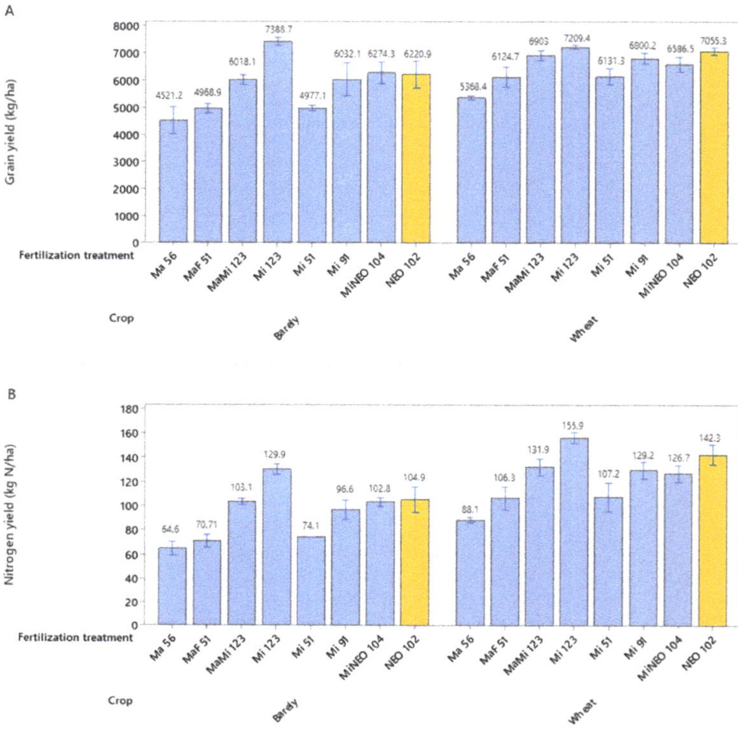

Figure 2. (**A**) Grain yield (15% water content) in kg ha^{-1} and (**B**) nitrogen yield kg N ha^{-1} $\pm$ standard error of the means. Results from the initial trials in 2020 with one trial barley (**left**) and one in spring wheat (**right**). The NEO102 kg N ha^{-1} treatment stands out from oth treatments with its distinct yellow coloration.

Furthermore, regarding wheat grain yield (Figure 3A), unlike the barley results, M NEO120 yielded higher than other treatments; the difference was insignificant to Mi12 NEO120, and MaMi120. NEO120 yielded in the same range as Mi91 but significant surpassed Ma65.

Regarding nitrogen barley yield (Figure 3B), the pattern observed mirrored that grain yield, displaying similar trends across the treatments. However, considering th nitrogen wheat yield (Figure 3B), MiNEO120 exhibited the highest nitrogen yield, reachir 106.1 kg N ha^{-1}. This result was significantly higher than both NEO120 and MaMi12 and it exceeded Mi120 by an additional 8.7 kg N ha^{-1}, although the latter differen was insignificant.

3.3. *Barley and Wheat Grain and Nitrogen Yield—Series Two, 2021–2022*

In series two of the experiments conducted in 2021 and 2022, the barley grain yiel (Figure 4A) of MaMi120 was 5064 kg ha^{-1}, which was similar to the yield of Mi12 However, Mi120 yielded only 170 kg ha^{-1} higher than NEO120, and the difference wa not statistically significant. On the other hand, NEO120 produced a grain yield that fe between the yields of Mi105 and Mi120, with a significantly higher yield of 646 kg ha compared to Ma65.

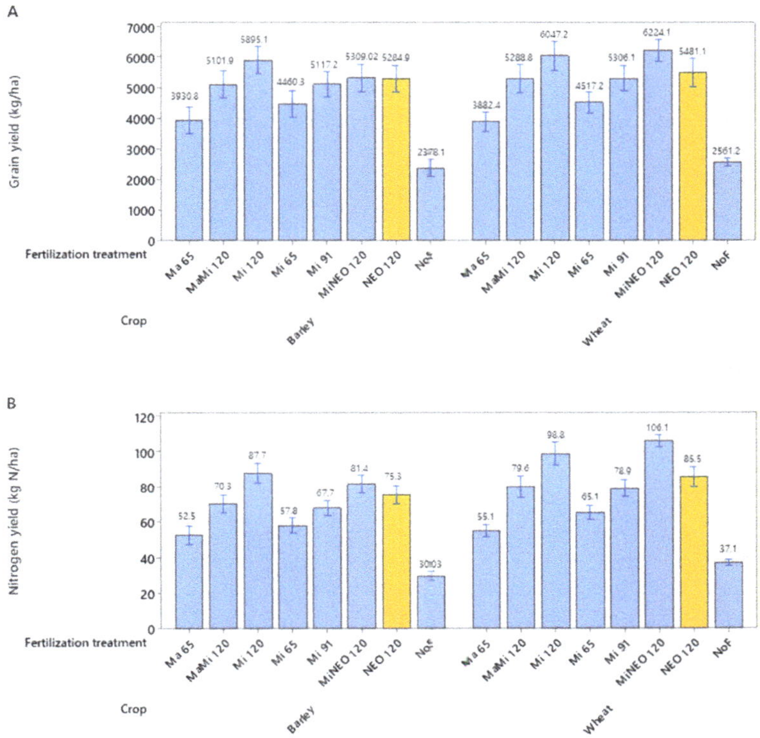

Figure 3. (**A**): Grain yield (15% water content), and (**B**) nitrogen yield, in kg N ha^{-1} ± standard error of the means. Results from Series 1 in 2021 and 2022, with three trials in barley (**left**) and three trials in spring wheat (**right**). The NEO 120 kg N ha^{-1} treatment stands out from other treatments with its distinct yellow coloration.

Regarding wheat grain yield (Figure 4A), MaMi120 yielded significantly higher (451 kg ha^{-1}) than NEO120. On the other hand, NEO120 yielded similar to Mi80 but significantly higher (695 kg ha^{-1}) than ma65.

A similar trend in barley and wheat grain yield was observed for nitrogen yield (Figure 4B). However, the differences between treatments were more pronounced, indicating increased variation in nitrogen yield.

3.4. Nitrogen Effects: Results from All 10 Trials in Series One and Two in 2021 and 2022

The Y axis in Figure 5 represents the nitrogen effect obtained from the range of mineral fertilizers. NEO120 exhibited an equivalent effect on grain yield as 95 kg N ha^{-1} in mineral fertilizer and the same effect on nitrogen yield as 100 kg N ha^{-1} in mineral fertilizer. In simpler terms, 95 kg N ha^{-1} in mineral fertilizer can be substituted with 120 kg N ha^{-1} in NEO when considering wheat and barley grain yield. Similarly, when assessing nitrogen yield, 100 kg N ha^{-1} in mineral fertilizer can be replaced by 120 kg N ha^{-1} in NEO.

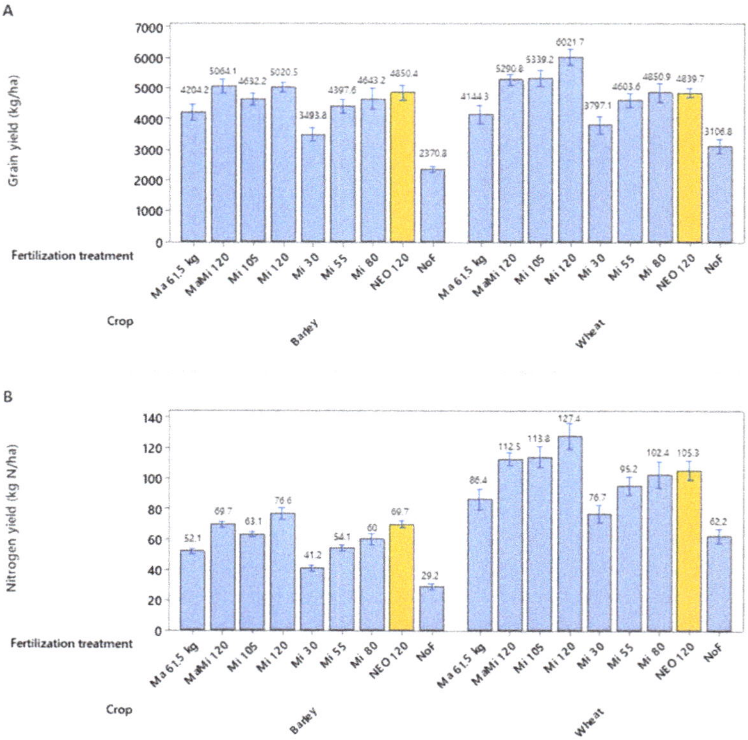

Figure 4. (**A**): Grain yield (15% water content), and (**B**) nitrogen yield, in kg N ha^{-1} ± standard error of the means. Results from Series 2 in 2021 and 2022, with two trials in barley (**left**) and three trials in spring wheat (**right**). The NEO 120 kg N ha^{-1} treatment stands out from other treatments with its distinct yellow coloration.

Ma65 demonstrated a grain yield effect comparable to 50 kg N ha^{-1} in mineral fertilizer but the nitrogen yield from Ma65 was slightly lower. MaMi120 also exhibited the same grain yield nitrogen effect as 95 kg N ha^{-1} in mineral fertilizer. However, the nitrogen yield from MaMi120 was slightly lower in comparison.

3.5. Sum up All Average Yields

The following table, Table 5, presents the average barley and wheat grain yields from the most noteworthy treatments in both series one and two in 2021 and 2022 (columns two and three) and series one only (columns four and five).

When considering the combined results from all trials in barley and wheat (columns two and three), MaMi120 yielded 5083 kg ha^{-1} barley grain and 5290 kg ha^{-1} wheat grain. These results were equivalent to the barley yield obtained from NEO120, while in wheat MaMi120 outperformed NEO120 by 135 kg ha^{-1}. MaMi120 and NEO120 yielded more than 1000 kg ha^{-1} compared to Ma65, with the largest increase observed in wheat. Mi120 yielded 425 kg ha^{-1} higher barley grain and 968 kg ha^{-1} higher wheat grain than NEO120.

Focusing on Series 1 alone (columns three and four), similar yield differences were observed as in the combined results. Additionally, we included the MiNEO120 treatment in this analysis. Regarding barley, MiNEO120 yielded similar to NEO120, while in wheat it outperformed NEO120 by 743 kg ha^{-1} and MaMi120 by 935 kg ha^{-1}. Remarkably, MiNEO120 even surpassed the wheat grain yield of Mi120.

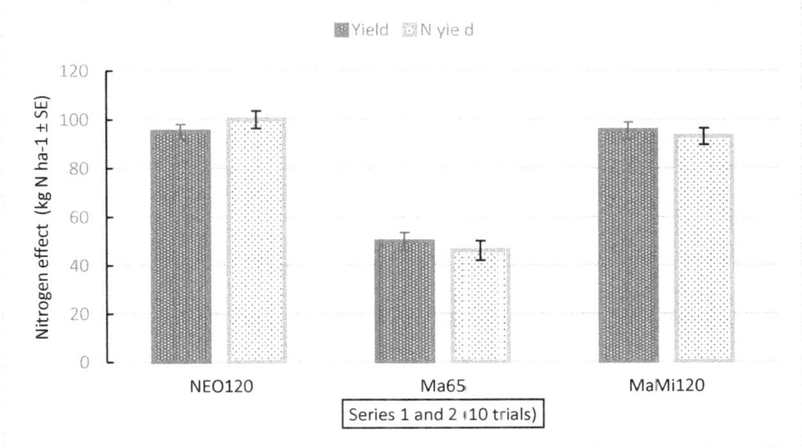

Figure 5. Nitrogen effect of 120 kg N ha^{-1} in NEO (NEO120) compared to the nitrogen effect of 65 kg N ha^{-1} in manure (Ma65) and 120 kg N ha^{-1} in manure with mineral fertilizer (MaMi120). Results are based on data from ten trials in Series 1 and 2 combined. Results are provided for grain yield data and N yield (bars with light pattern).

Table 5. The average yields per ha of the Ma65, MaMi120, NEO120, and Mi120 treatments in all 5 trials in barley and 5 trials in wheat in 2021 and 2022 (Series 1 and 2 combined). Columns four and five give the same only from the trials in Series 1 from the same years, also containing the average yield effects from the MiNEO120 treatment.

Fertilization Treatment	Average Yield (kg ha^{-1}) All Trials 2021 and 2022 Barley (5 Trials)	Average Yield (kg ha^{-1}) All Trials 2021 and 2022 Wheat (5 Trials)	Average Yield (kg ha^{-1}) Series 1 2021 and 2022 Barley (3 Trials)	Average Yield (kg ha^{-1}) Series 1 2021 and 2022 Wheat (3 Trials)
Ma65	4068	4013	3931	3882
MaMi120	5083	5290	5102	5289
NEO120	5068	5155	5285	5481
MiNEO120	-	-	5309	6224
Mi120	5443	6123	5895	6047

4. Discussion

4.1. Nitrogen Fertilization Effect of NEO

The present studies aimed to determine the comparative variances in yield and fertilizer efficacy among NEO, mineral fertilizers, cattle slurry, and other farmers' alternatives. The experiments were located in representative parts of Norway's most crucial grain production areas. The intended nitrogen (N) application level for the study was set at 120 kg N ha^{-1}, which aligns with the average N level commonly used for barley and is slightly lower (around 1–2 kg) than the average N level employed for wheat in practical farming within the region. The yields obtained from applying 120 kg N ha^{-1} using mineral fertilizer fell within the same range as those observed in the official Norwegian variety trials. [24].

During the initial two trials conducted in 2020, we observed a substantial increase in wheat yield when utilizing filtered slurry compared to the untreated slurry, with a difference of 756 kg ha^{-1}. Similarly, in the case of barley, the filtered slurry resulted in a yield increase of 447 kg ha^{-1} compared to the untreated slurry. These findings indicate that by simply filtering the slurry, the fertilizer effect was enhanced by approximately 10–14 percent barley and wheat, respectively. It is worth noting that despite the reduced volume ar applied nitrogen amount due to filtration, the positive impact on yield was still significa This phenomenon can be attributed to the reduced carbon-to-nitrogen (C/N) ratio observ in the filtered slurry, resulting in an enhanced nitrogen mineralization process [25,26]. Th can elucidate a portion of the positive yield effect achieved by implementing the N Applied technology.

Our study focused on assessing the potential of NEO as a substitute for nitrogen (from mineral fertilizers. The results revealed that applying 120 kg N ha^{-1} in NEO result in a similar grain yield as using 95 kg N ha^{-1} in mineral fertilizer, which is a 20% reducti in yield compared to the 120 kg N ha^{-1} in mineral fertilizer. However, NEO exhibit slightly better performance when considering nitrogen yield, with 120 kg N-min ha^{-1} NEO yielding comparable results to 100 kg N ha^{-1} in mineral fertilizer (a 16.7% redu tion) [27–29].

We also examined the impact of combining untreated manure with mineral fertilize on grain yield. Notably, applying 65 kg N ha^{-1} in untreated manure supplemented wi 55 kg N ha^{-1} in mineral fertilizer yielded the same as using 95 kg N ha^{-1} in miner fertilizer and 120 kg N ha^{-1} in NEO for both barley and wheat grain yield. Interesting within 120 kg N ha^{-1} in NEO, 60 kg N is added to the manure through plasma treatme These findings indicate that NEO may suitably replace the combination of untreated manu with mineral fertilizers, which can be an effective strategy to reduce reliance on miner fertilizers while maintaining comparable yields.

To further analyze the results, we examined the average wheat and barley yields, presented in Figures 2–4 and Table 5. Our grain trials set the targeted nitrogen value 120 kg N ha^{-1}. To achieve this level using NEO, we applied 37.5 tons ha^{-1} and 34 tons ha in 2021 and 2022, respectively, considering the varying nitrogen content in NEO. Corr spondingly, the amounts of untreated slurry applied were 41 tons ha^{-1} and 37 tons ha resulting in an average of 39 tons ha^{-1} of untreated slurry. Using this information, w calculated the yields obtained from different combinations of 39 tons ha^{-1} of cattle slur and mineral fertilizer:

When applying 39 tons ha^{-1} of untreated cattle slurry (Ma65), we harvested a barl yield of 4068 kg ha^{-1} and a wheat yield of 4013 kg ha^{-1}. By combining 39 tons ha of untreated slurry with mineral fertilizers (Opti-NS) up to 120 kg N ha^{-1} (MaMi12 we observed improved yields, with barley reaching 5083 kg ha^{-1} and wheat reachir 5290 kg ha^{-1}.

To explore the potential of alternative fertilization techniques, we filtered 39 tons untreated slurry. Then, we processed it through the N2 Applied unit, resulting in 35 tons NEO with a nitrogen content of 120 kg N. Applying 35 tons ha^{-1} of NEO (NEO120) yielde a barley yield of 5068 kg ha^{-1} and a wheat yield of 5155 kg ha^{-1}, the same level as th combination of cattle slurry and mineral fertilizer up to 120 kg N ha^{-1}.

Furthermore, we examined the effects of solely using mineral fertilizers with a nitroge content of 120 kg N ha^{-1}, specifically Yara 18-3-15 (Mi120). This approach resulted in eve higher yields, with barley reaching 5443 kg ha^{-1} and wheat reaching 6123 kg ha^{-1}.

4.2. *NEO at Three Leaf Stage*

Interestingly, throughout our trials, we consistently observed no evidence of damag to barley and wheat plants when NEO was applied during the three-leaf stage. Th finding indicates that applying NEO at this particular growth stage does not result in ar discernible harm to the crops.

The treatment known as MiNEO120 involved the application of NEO (108 kg N ha at the three-leaf stage, while a small quantity (12 kg N ha^{-1}) of mineral fertilizer w applied on the sowing day. This approach yielded a crop production increase of 743 kg ha compared to the sole application of NEO (NEO120) at sowing, based on six trials conducte in 2021 and 2022. Additionally, the MiNEO120 treatment demonstrated slightly high wheat yields than the application of 120 kg N ha^{-1} as mineral fertilizer on the sowing day (Mi120). However, when it came to barley, no significant yield improvement was observed with the MiNEO120 treatment compared to the NEO120 treatment. In fact, the MiNEO120 treatment decreased yield by 586 kg ha^{-1} when compared to the Mi120 treatment.

In contrast, our trials conducted in 2020 using the MiNEO treatment with an N application rate of 104 kg N ha^{-1} resulted in significantly lower yields. Specifically, there was a decrease of 469 kg ha^{-1} compared to using the NEO102 treatment in wheat. However, in the case of barley, both treatments yielded approximately the same results in 2020. These findings highlight the diverse effects that different nitrogen application methods and rates can have on wheat and barley yields.

However, the high yield observed in the MiNEO120 treatment can be partially attributed to trial number three out of the six trials conducted in series one during 2021 and 2022. MiNEO120 yielded a notably higher yield in this particular trial than MiNEO120 in the remaining five trials. We have thoroughly analyzed the weather conditions in the weeks following sowing and the application of NEO in the MiNEO treatment. However, our investigation did not yield any definitive explanations for this discrepancy.

On the other hand, it is widely acknowledged that wheat has a later nitrogen uptake during the growing season compared to barley [30]. This difference in nitrogen absorption timing may help explain why MiNEO120 consistently resulted in higher wheat yields than barley in most of our trials. Another contributing factor could be the high nitrification potential of NEO, which leads to greater availability of plant-accessible nitrate over a concentrated period of 3–4 days following application [18]. This rapid release of nitrate may particularly benefit wheat [31,32].

It is important to note that all other fertilizer treatments, apart from MiNEO120, were applied solely on the sowing day. Therefore, our experiments do not provide insights into how these alternative treatments would have performed if they had been applied partially on the sowing day and partially at the three-leaf stage, similar to the MiNEO120 treatment. Consequently, the only valid comparison is between MiNEO120 and NEO120.

4.3. Limitations and Further Research

The ammonia and nitrate levels in NEO and the mineral fertilizer used in our experiments are similar. However, despite this similarity, it is astonishing that the crop yield obtained from 120 kg N ha^{-1} in NEO is equivalent to the yield achieved from just 95 kg N ha^{-1} in the mineral fertilizer. Furthermore, the yield obtained from 120 kg N ha^{-1} in NEO is the same as that obtained from the combination of untreated slurry and mineral fertilizer Opti-NS at the same nitrogen application rate. However, The nitrogen yields obtained from NEO were slightly superior, delivering an equivalent nitrogen effect as that of 100 kg N ha^{-1} in mineral fertilizer. This suggests that approximately 20 kg ha^{-1}, corresponding to 17% of the plant-available nitrogen in NEO, is lost through other means.

Regrettably, we cannot provide a conclusive explanation for these unexpected outcomes. Notably, the low pH in NEO should typically mitigate the majority of ammonia leakage [33]. Nevertheless, one possibility to consider is that the high soil nitrification potential in the initial days following the application of NEO could lead to nitrate losses without adequate plant absorption. Additionally, there is a risk that nitrogen could be lost as nitrous oxide through denitrification, further exacerbating the situation. Therefore, additional research investigating the emissions or leaching potential of NEO is necessary to gain a clearer understanding of this phenomenon.

5. Conclusions

The current study aimed to investigate and compare the effects of Nitrogen Enriched Organic fertilizer (NEO) made from cattle slurry on barley and wheat grain yields to other conventional fertilizers used in agriculture. The results indicated that 120 kg N ha^{-1} in NEO yielded in the same range of cereal grains as 95 kg N ha^{-1} in mineral fertilizer. Moreover, the combination of untreated slurry and 55 kg N ha^{-1} in mineral fertilizer Opti-NS yielded the same as 120 kg N ha^{-1} in NEO. Surprisingly, the combination of 12 kg N ha^{-1} in minera fertilizer applied at sowing, alongside 108 kg N ha^{-1} in NEO administered at the three-lea stage, resulted in higher wheat yields compared to the application of 120 kg N ha^{-1} o NEO solely spread at sowing in two out of three experimental years. Additionally, th direct application of NEO onto the plants exhibited no observable signs of harm. Lastly it is worth noting that filtering the slurry yielded higher cereal grain yields compare to using the untreated slurry. Thus, while NEO and mineral fertilizers have similar N content, utilizing NEO resulted in a cereal grain yield 15–20% lower than that achieve with mineral fertilizer. However, it still yielded 20–30% higher than the native amount o cattle slurry it originated. Nevertheless, it is worth noting that approximately 17% of th nitrogen in NEO appears to be lost through unidentified means without contributing to grain or nitrogen yields.

Author Contributions: Conceptualization, T.C. and S.Ø.S.; methodology, T.C. and S.Ø.S.; valida tion, T.C. and S.Ø.S.; formal analysis, H.M.; investigation, T.C.; data curation, S.Ø.S. and H.M writing—original draft preparation, T.C., S.Ø.S. and H.M.; writing—review and editing, T.C., S.Ø.S and H.M.; visualization, T.C., S.Ø.S. and H.M.; supervision, S.Ø.S.; project administration, T.C funding acquisition, T.C. and S.Ø.S. All authors have read and agreed to the published version o the manuscript.

Funding: This research was funded by the Research Council of Norway (grant number 30964(Plasmabehandlet husdyrgjødsel—gjødselvirkning, miljøpåvirkning og klimagassutslipp) and th Company N2 Applied.

Data Availability Statement: The data is available upon request.

Acknowledgments: The authors would like to thank the staff at NLR Trøndelag, NLR Innlande NLR Østafjells, and NLR Viken for their help in conducting the field trials. We also thank Morte Tofastrud and Elisabeth Røe at Inland Norway University of Applied Sciences for their support ir project administration. In addition, we would like to acknowledge the project's funders.

Conflicts of Interest: The authors declare no conflict of interest.

References

1. FAO. *The Future of Food and Agriculture, Trends and Challenges*; Food and Agriculture Organization of the United Nations: Rome Italy, 2017; ISBN 978-92-5-109551-5.
2. Intergovernmental Panel on Climate Change. *Climate Change and Land: IPCC Special Report on Climate Change, Desertification, Land Degradation, Sustainable Land Management, Food Security, and Greenhouse Gas Fluxes in Terrestrial Ecosystems*; Cambridge University Press: Cambridge, UK, 2022.
3. Vermeir, I.; Verbeke, W. Sustainable Food Consumption: Exploring the Consumer "Attitude–Behavioral Intention" Gap. *J. Agric Environ. Ethics* **2006**, *19*, 169–194. [CrossRef]
4. Augustin, M.A.; Riley, M.; Stockmann, R.; Bennett, L.; Kahl, A.; Lockett, T.; Osmond, M.; Sanguansri, P.; Stonehouse, W.; Zajac, I. et al. Role of food processing in food and nutrition security. *Trends Food Sci. Technol.* **2016**, *56*, 115–125. [CrossRef]
5. Janker, J.; Mann, S.; Rist, S. Social sustainability in agriculture—A system-based framework. *J. Rural Stud.* **2019**, *65*, 32–42 [CrossRef]
6. Rajic, S.; Đorđević, V.; Tomasevic, I.; Djekic, I. The role of food systems in achieving the sustainable development goals Environmental perspective. *Bus. Strategy Environ.* **2022**, *31*, 988–1001. [CrossRef]
7. Dong, N.-Q.; Lin, H.-X. Higher yield with less nitrogen fertilizer. *Nat. Plants* **2020**, *6*, 1078–1079. [CrossRef] [PubMed]
8. Houlton, B.Z.; Almaraz, M.; Aneja, V.; Austin, A.T.; Bai, E.; Cassman, K.G.; Compton, J.E.; Davidson, E.A.; Erisman, J.W. Galloway, J.N.; et al. A World of Cobenefits: Solving the Global Nitrogen Challenge. *Earth's Future* **2019**, *7*, 865–872. [CrossRef] [PubMed]
9. Bondada, B.R.; Oosterhuis, D.M. Canopy photosynthesis, specific leaf weight, and yield components of cotton under varying nitrogen supply. *J. Plant Nutr.* **2001**, *24*, 469–477. [CrossRef]
10. Rütting, T.; Aronsson, H.; Delin, S. Efficient use of nitrogen in agriculture. *Nutr. Cycl. Agroecosyst.* **2018**, *110*, 1–5. [CrossRef]
11. FAO. *World Fertilizer Trends and Outlook to 2022*; FAO: Rome, Italy, 2019.
12. Upendra, M.S.; Rajan, G.; Gautam, P.P. Nitrogen Fertilization I: Impact on Crop, Soil, and Environment. In *Nitrogen Fixation* Everlon Cid, R., Ademar Pereira, S., Eds.; IntechOpen: Rijeka, Croatia, 2019; Chapter 5.
13. European Commission. Sustainable Use of Nutrients: The Common Agricultural Policy Supports Farmers in the Safe and Efficient Use of Nutrients. Available online: https://ec.europa.eu/info/food-farming-fisheries/sustainability/environmental-sustainability/low-input-farming/nutrients_en (accessed on 18 July 2022).

Graves, D.B.; Bakken, L.B.; Jensen, M.B.; Ingels, R. Plasma Activated Organic Fertilizer. *Plasma Chem. Plasma Process.* **2019**, *39*, 1–19. [CrossRef]
Ingels, R.; Graves, D.B. Improving the Efficiency of Organic Fertilizer and Nitrogen Use via Air Plasma and Distributed Renewable Energy. *Plasma Med.* **2015**, *5*, 257–270. [CrossRef]
N2 Applied. Nitrogen Enriched Organic Fertiliser. Available online: https://n2applied.com/ (accessed on 19 July 2022).
Mousavi, H.; Cottis, T.; Pommeresche, R.; Dörsch, P.; Solberg, S.Ø. Plasma-Treated Nitrogen-Enriched Manure Does Not Impose Adverse Effects on Soil Fauna Feeding Activity or Springtails and Earthworms Abundance. *Agronomy* **2022**, *12*, 2314. [CrossRef]
Mousavi, H.; Solberg, S.Ø.; Cottis, T.; Dörsch, P. Nitrogen-Enriched Organic Fertilizer (NEO) Elevates Nitrification Rates Shortly after Application But Has No Lasting Effect on Nitrification in Agricultural Soils. Preprint. 2023. Available online: https://doi.org/10.21203/rs.3.rs-2565156/v1 (accessed on 10 March 2023).
Mousavi, H.; Cottis, T.; Hoff, G.; Solberg, S.Ø. Nitrogen Enriched Organic Fertilizer (NEO) and Its Effect on Ryegrass Yield and Soil Fauna Feeding Activity under Controlled Conditions. *Sustainability* **2022**, *14*, 2005. [CrossRef]
Yara. YaraMila™. Available online: https://www.yara.com/crop-nutrition/products-and-solutions/global-fertilizer-brands/yaramila/ (accessed on 20 July 2022).
Yara International. Yara Opti NS. Available online: https://www.yara.no/gjoedsel/produkter/yarabela/yarabela-opti-ns-27-0-0-4s/ (accessed on 3 May 2023).
Zadoks, J.C.; Chang, T.T.; Konzak, C.F. A decimal code for the growth stages of cereals. *Weed Res.* **1974**, *14*, 415–421. [CrossRef]
yr.no. Norwegian Meteorological Institute and the Norwegian Broadcasting Corporation; Weather Data History. 2023. Available online: www.yr.no (accessed on 11 March 2023).
Strand, E. *Jord-Og Plantekultur 2021 Forsøk i Korn, Olje-Og Belgvekster, Engfrøavl Og Potet 2020*; NIBIO: Amsterdam, The Netherlands, 2021. Available online: https://nibio.brage.unit.no/nibio-xmlui/handle/11250/2740694 (accessed on 11 March 2023).
Watson, C.A.; Atkinson, D.; Gosling, P.; Jackson, L.R.; Rayns, F.W. Managing soil fertility in organic farming systems. *Soil Use Manag.* **2002**, *18*, 239–247. [CrossRef]
Biswas, D.; Micallef, S.A. *Safety and Practice for Organic Food*; Academic Press: Cambridge, MA, USA, 2019.
Sutton, M.; Howard, C.; Mason, K.; Brownlie, W.; Cordovil, C. Nitrogen Opportunities for Agriculture, Food & Environment. UNECE Guidance Document on Integrated Sustainable Nitrogen Management. 2022. Available online: https://nora.nerc.ac.uk/id/eprint/534033/ (accessed on 10 March 2023).
Jackson, D.R.; Smith, K.A. Animal manure slurries as a source of nitrogen for cereals; effect of application time on efficiency. *Soil Use Manag.* **1997**, *13*, 75–81. [CrossRef]
Ladha, J.K.; Pathak, H.J.; Krupnik, T.; Six, J.; van Kessel, C. Efficiency of Fertilizer Nitrogen in Cereal Production: Retrospects and Prospects. In *Advances in Agronomy*; Academic Press: Cambridge, MA, USA, 2005; Volume 87, pp. 85–156.
Bosshard, C.; Sørensen, P.; Frossard, E.; Dubois, D.; Mäder, P.; Nanzer, S.; Oberson, A. Nitrogen use efficiency of 15N-labelled sheep manure and mineral fertiliser applied to microplots in long-term organic and conventional cropping systems. *Nutr. Cycl. Agroecosyst.* **2009**, *83*, 271–287. [CrossRef]
Alcoz, M.M.; Hons, F.M.; Haby, V.A. Nitrogen Fertilization Timing Effect on Wheat Production, Nitrogen Uptake Efficiency, and Residual Soil Nitrogen. *Agron. J.* **1993**, *85*, 1198–1203. [CrossRef]
Cox, W.J.; Reisenauer, H.M. Growth and ion uptake by wheat supplied nitrogen as nitrate, or ammonium, or both. *Plant Soil* **1973**, *38*, 363–380. [CrossRef]
Fangueiro, D.; Hjorth, M.; Gioelli, F. Acidification of animal slurry—A review. *J. Environ. Manag.* **2015**, *149*, 46–56. [CrossRef] [PubMed]

agronomy

MDP

Article

Optimizing Planting Density in Alpine Mountain Strawberry Cultivation in Martell Valley, Italy

Sebastian Soppelsa *, Michael Gasser and Massimo Zago

Berries and Stonefruits Research Group, Laimburg Research Centre, 39040 Ora-Auer, Italy;
michael.gasser@laimburg.it (M.G.); massimo.zago@laimburg.it (M.Z.)
* Correspondence: sebastian.soppelsa@laimburg.it

Abstract: Optimizing profitability is a challenge that strawberry farmers must face in order to rema competitive. Within this framework, plant density can play a central role. The aim of this two-ye study was to investigate how planting density can induce variations in plant growth and yie performances in an alpine mountain strawberry cultivation (Martell Valley, South Tyrol, Italy), a consequently quantify the farm profit. Frigo strawberry plants cv. Elsanta were planted in soil raised beds and subjected to five different planting density levels (30,000 and 45,000 as large spacir 60,000 as middle spacing; 90,000 and 100,000 plants ha^{-1} as narrow spacing, corresponding to a pla spacing of 28, 19, 14, 9, and 8.5 cm, respectively). Our findings indicate that the aboveground bioma in plants subjected to low planting density was significantly increased by +50% (end of first year) a even doubled in the second year in comparison with plants in high planting density. Those resu were related to higher leaf photosynthetic rate (+12%), and the number of crowns and flower truss per plant (+40% both) ($p < 0.05$). The low yield (about 300 g plant^{-1}) observed in the high plantir density regime was attributable to smaller fruit size during the first cropping year and to both reduced number of flowers per plant and fruit size during the second year ($p < 0.05$). Although t highest yield (more than 400 g plant^{-1}) was obtained with wide plant spacing, the greatest yield p hectare was achieved with high planting densities (28 t ha^{-1} in comparison with 17 t ha^{-1} with lo plant density level). However, the farm profit must take into account the costs (especially related the plant material and harvesting costs) that are higher under the high planting density compar with the other density regimes. Indeed, the maximum farm profit was reached with a density 45,000 plants ha^{-1} which corresponded to EUR 22,579 ha^{-1} (over 2 years). Regarding fruit quali fruits coming from the low plant density level showed a significantly higher color index (+15% mo red color) than fruits from high plant density ($p < 0.05$). In conclusion, our results suggest that middle planting density can be a fair compromise in terms of plant growth, yield, and farm profit

Keywords: *Fragaria x ananassa*; plant spacing; altitude; flowering; fruit quality

Citation: Soppelsa, S.; Gasser, M.; Zago, M. Optimizing Planting Density in Alpine Mountain Strawberry Cultivation in Martell Valley, Italy. *Agronomy* **2023**, *13*, 1422. https://doi.org/10.3390/agronomy13051422

Academic Editors: Aurelio Scavo and Umberto Anastasi

Received: 19 April 2023
Revised: 10 May 2023
Accepted: 12 May 2023
Published: 21 May 2023

1. Introduction

Strawberry is a herbaceous perennial plant belonging to genus *Fragaria* of the fami *Rosaceae* [1]. There are around 24 species of Fragaria in the world, mostly concentrate in China, making it the country with the largest genetic resources of wild strawberry [Nevertheless, the strawberry species cultivated today derives from a natural hybridizatio that occurred in European gardens around the mid-1700's, between two species native America (the South American *F. chiloensis* and the North American *F. virginiana*) [3]. Short thereafter, that new hybrid species, *Fragaria* x *ananassa*, was destined to become a popul fruit crop with a significant economic value [3]. Being a plant with great environment adaptability, it is geographically distributed in various parts of the world [1]. Accordin to data from the Food and Agriculture Organization of the United Nations (FAO), th worldwide production of strawberry was around 8.8 million tons in 2020 [4]. China is th largest strawberry producer in the world (3.3 M tons), followed by United States of Americ

(1.1 M tons), Egypt (0.6 M tons), Mexico (0.6 M tons). and Turkey (0.5 M tons). The leading European country in strawberry production is represented by Spain (0.3 M tons) [4].

According to the Italian National Statistics Institute (data updated to 2021), the national production of strawberries is around 121 thousand tons of which 29 thousand in open fields (1871 ha) and 92 thousand in greenhouses (2631 ha) [5]. Although much of the production is concentrated in southern and central Italy, the significant contribution provided by the northern regions, such as Emilia-Romagna, Veneto, and regions located along the Alpine arc (Piedmont and Trentino-South Tyrol), must not be overlooked, and thanks to this variability of Italian environments, a fresh product is guaranteed throughout the year.

Several areas of strawberry cultivation can be identified in the province of Alto Adige/Südtirol/South Tyrol (in Italian, German, and English, respectively), from Martell Valley to Isarco Valley and Pusteria Valley, thus covering an area of about 100 ha [6]. More in detail, the Martell Valley on the southern side of Venosta Valley allows the cultivation of strawberry (and other berries) in an alpine mountain environment. The beginnings of the "heroic strawberry cultivation" in Martell Valley originated in the 60 s, when a group of farmers, with the help of the Department of Agriculture of the Autonomous Province of Bolzano/Bozen, identified the potential of the strawberry to be grown in this microclimate. The production of strawberries extends from 900 to 1800 m a.s.l., hence the name "Martell Strawberry Valley". Due to the altitude of the growing areas, Martell Valley is considered the highest cultivation area for strawberries in Europe. Late spring planting systems with cold-stored plants are currently the most adopted way by the farmers of the valley. These plants guarantee two crops, one in the year of planting, the other in the following year. The "Martell strawberry" ripens very slowly, in this way the fruits take on unique aromas and fragrances, and make themselves available from June to September, a period in which great national and European productions are absent [7].

Although the Italian strawberry acreage has drastically collapsed over the years (11,000 ha in 1989 to the current 4500 ha in 2021), yield per hectare, however, has resulted in an increase of 47% [5,8]. This improvement is attributable to two factors: breeding programs and growing systems. The first case results in the release of new, more productive cultivars with higher fruit quality parameters and plant resistance/tolerance to pathogens. In the second case, a traditional cultivation system (soil cultivation in open field or protected) has often been replaced by an advanced soilless system [9]. Nevertheless, some strawberry areas are linked to the tradition of the past with a soil cultivation in open field and with historical cultivars (e.g., cultivar Elsanta); this is the case of Martell Valley with very low yields per plant due to both the limiting environmental conditions and the lack of information on some correct agronomic practices (e.g., suitable planting density).

Plant density is simply expressed as the number of individuals per unit ground area [10]. According to several studies conducted primarily on herbaceous crops, plant morphology and productivity are influenced by the manipulation of plant density, more specifically synthesis of chlorophyll, photosynthesis, plant growth, floral induction, and flower formation are affected by different crop spacing [11–15]. The right crop density is certainly essential to obtain a maximum yield and income in strawberry cultivation [16]. Both low and high plant densities can reduce yield and total revenue. In other words, individual plants grown with a large spacing perform their best growth in terms of yield per plant but a low productivity per hectare [16]. On the contrary, as the distance between the plants decreases, a competitive relationship intensifies among individuals for limiting factors such as light, water, and nutrients, leading to a worsening of plant performances [17]. Irrigation and fertilization in open-field conditions are consolidated management practices to overcome or avoid abiotic stresses in relation to water-shortage or nutrient deficiency, respectively [18]. High plant density leads to mutual shading and self-shading of the leaves, thus hindering a correct interception of light [19]. Consequently, plants grown in that condition are subjected to morphological and anatomical changes, producing less biomass (i.e., leaves, roots), delaying flowering more than plants in full sunlight [20]. Looking at scientific literature, the interaction between reproductive phenology in strawberry plant and some environmental parameters (e.g., light intensity, light quality/photoperiod, and temperature) is a topic of particular interest as evidenced by some studies [21–25].

An optimal plant density is calculated by identifying a density threshold beyond which the increase in individual plants does not lead to an increase in revenues [26]. In the study conducted by Wamser et al. [27], the fruit yield of tomato plants cultivated in humid subtropical climate in Calmon, State of Santa Catarina, Brazil (1208 m a.s.l.) was optimized with a plant density of 23,000 plants ha^{-1}, while increasing the number of individual plants increases yield but not the profit.

In another study, also conducted in Brazil on tomato cultivation, Carvalho et al. [28] found an optimal plant density of around 30,000 plants ha^{-1} in Ipameri, State of Goiás (altitude 794 m).

Although many studies were conducted to determine an optimal plant density in several vegetable and fruit crops such as strawberry [14,29–34], the results that emerge from those research publications depend on some environmental and cultivation factors and therefore they have a practical significance in the conditions in which the tests were carried out. A geographical climatic factor such as altitude that affects temperature and radiation has a fundamental role in changing plant responses (e.g., photosynthetic behavior, floral induction, fruit quality) [35–38].

As far as we know, no previous research has investigated the interaction between flowering/yield of strawberry plant and high altitude, combined with different plant densities. The present study aimed to investigate the effects of different plant densities on the growth, flowering, yield, fruit quality, and economic aspects of strawberry plants cv. Elsanta cultivated in a unique alpine mountain environment.

2. Materials and Methods

2.1. Field Management and Experimental Design

The experiment was conducted over two growing seasons (years 2020 and 2021) in an experimental strawberry field managed by the Laimburg Research Centre and located in the municipality of Martell (46°33′30.618″ N; 10°46′53.649″ E; 1.312 m a.s.l.) in South Tyrol, Italy. Martell Valley, a side valley of Venosta Valley included in the Stelvio National Park, is famous for berry production, in particular strawberry and a typical alpine mountain climate characterizes the valley. The soil properties of the 0–20 cm soil layer before planting in May 2020 were as follows: humic loamy sand, pH = 5.1, no free carbonate, organic carbon expressed as humus of 7.3%, phosphorus = 5.0 mg 100 g^{-1}, potassium = 8.0 mg 100 g^{-1} and magnesium = 18.0 mg 100 g^{-1}. Meteorological trends during the growing seasons (from May to August 2020 and 2021) were recorded by iMETOS® weather station with the cloud platform "FieldClimate" (Pessl Instruments, Weiz, Austria) and data are reported in Table 1.

Table 1. Climatic conditions (monthly air temperatures, relative humidity, and rainfall) measured from May to August 2020 and 2021 during the first and second cropping year, respectively.

	Air Temperature (°C)			**Relative Humidity (%)**	**Rainfall (mm)**
	Minimum Temperature	**Maximum Temperature**	**Mean Temperature**		
2020					
May	0.1	22.2	10.9	67.9	56.8
June	3.4	27.6	13.8	76.0	97.6
July	5.8	31.0	16.8	74.6	70.5
August	6.0	31.7	16.3	80.1	120.4
2021					
May	0.0	21.4	9.3	64.9	130.8
June	6.6	30.0	17.5	64.0	39.8
July	7.6	27.8	17.0	71.6	135.0
August	6.2	29.2	15.9	71.8	137.4

In our experimental test, frigo strawberry plants cv. Elsanta (heavy waiting bed (HWB) plants from the nursery: Neessen Aardbei and Aspergeplanten, Grashoek, Netherlands) were planted in soil conditions, precisely on raised beds with white plastic mulch films on the 31 May 2020 and subjected to five different plant density levels (30,000 and 45,000 as large spacing; 60,000 as middle spacing; 90,000 and 100,000 plants ha^{-1} as narrow spacing, corresponding to a plant spacing of 28, 19, 14, 9, and 8.5 cm, respectively) (Figure 1). Plants were managed in the same way in terms of watering, fertilization, and pest control. The field received standard horticultural cares in accordance with the regulation governing integrated production. The experiment setup was organized as a completely randomized block design with 4 replicates composed of 40 plants per experimental unit (i.e., 120 plants per plant density level).

Figure 1. Strawberry plants cv. Elsanta planted in a double row on raised beds and subjected to different plant spacing.

2.2. *Evaluated Parameters*

2.2.1. Morphological and Gas Exchange Parameters

Main characteristics related to plant flowering were evaluated by counting the numb of developed flower trusses per plant by destructively sampling ten randomly selecte plants in each replicate after each harvest period. Thus, each flower truss was careful assessed through a lens to determine the flower number (counted flower pedicel scars) p truss and per plant. Plant growth as affected by plant density level was determined b dissecting the same ten plants per plot previously mentioned. Then, each selected pla was separated into roots and aerial parts (leaves, crowns, flower trusses, and runne and weighed fresh (g fresh weight (FW) plant^{-1}). The number of crowns per plant w evaluated by distinguishing between the main crown and branch crown. Afterwards, plant organs were put in an oven (ED 56, Binder GmbH, Tuttlingen, Germany) at 65 ° until they reached a stable weight and the dry mass was recorded (g dry weight (DW plant^{-1}). In the flowering stage, the net assimilation rate (A, μmol m^{-2} s^{-1}) of leav was evaluated using a portable infrared gas exchange analyzer (CIRAS-2, PP-Systems Hitchin, UK), attached to a PLC-6 cuvette having a measuring window of 2.5 cm^2. Th CO_2 concentration (380 mmol mol^{-1}), PPFD (1500 μmol m^{-2} s^{-1}), leaf temperature (25 °C and air humidity (80%), were controlled by the device. Measurements were perform on a sun-exposed (clear and sunny days between 11:00 a.m. and 13:00 p.m.), young, ful expanded single leaf of four randomly selected plants per planting density.

2.2.2. Yield Parameters

Ripe strawberry fruits (uniformly red) were harvested every four days during th period from mid-July to mid-August 2020 (first harvest year) and throughout the mon of July until the 7 August 2022 (second harvest year). From each experimental unit and each picking time, the commercial production (healthy fruit with a diameter ≥22 mm) an the waste, represented by small fruit (diameter <22 mm), deformed and with the presen of rot, were weighed with a digital scale (Valor™ 2000, OHAUS Europe GmbH, Näniko Switzerland). The total production per plant (g fruit^{-1}) was calculated by dividing th harvested total fruit weight by the number of plants (considered 30 plants per experiment unit). The average fruit weight (g fruit^{-1}) was estimated by randomly sampling 10 fruits each picking time.

2.2.3. Fruit Quality

Fruit quality was assessed on ten healthy strawberries per replicate which correspon to 40 fruits analyzed per treatment. The fruits were sampled at two intermediate pickir times for each harvest year. Flesh firmness was expressed with the Durofel index (D which represents the elasticity of the skin of the fruit (Agrosta® Winterwood instrume Agrosta Sàrl, Serqueux, France). The total soluble solids (°Brix) were determined with refractometer (RFM840, Bellingham-Stanley Ltd., Kent, UK), whereas the titratable acidi (g L^{-1} of citric acid) was measured with a titrator (Flash Automatic Titrator, Steroglas Perugia, Italy) by titrating strawberry pulp to pH 8.2 using 0.1 M NaOH. The extern fruit color was assessed with a colorimeter (CR-400, Konica Minolta, Tokyo, Japan) b measuring the same ten fruits at three different positions around the equatorial side each fruit. The colorimetric coordinates (L*, a*, b*) were used to calculate the color inde [CI = (1000 × a)/(L × b)] with higher CI value, indicating a more intense red color i the fruit [39].

2.2.4. Economic Analysis

A cost–benefit analysis was carried out for each plant density system. Profit is calc lated by subtracting all farm's costs (variable and fixed) from the total revenue. Variab costs vary in relation to production volume, and in our case they referred to labor, pla material, mulch film, pesticides, fertilizers, and fuel. Instead, fixed costs remain the sam regardless of production level and we considered the depreciation and maintenance quot of durable capital (tunnel structure, irrigation system, machinery, and buildings), administrative costs, and interest. Revenue is the total income generated from the sale of the strawberry fruits. Data are presented as total revenue, total costs, and farm profit for two consecutive years of cultivation according to planting density. Moreover, the profitability index, calculated by the ratio between gross income and total costs, is also reported in order to provide an indication of which option (i.e., planting density) is more profitable.

2.3. Statistical Analysis

Data normality was examined with the Shapiro–Wilk test, and homogeneity of variance was confirmed using Flinger–Killeen's test. A two-way ANOVA was performed on data collected from both years and mean separation of the dependent variables obtained with the LSD Fisher's test ($p < 0.05$). In case of significant interaction between "treatments" and "years", results were presented separately for the 2 years in dedicated tables or figures. A one-way ANOVA was performed on photosynthetic data coming from a single cropping year (2021). For non-normal data, Kruskal–Wallis test was applied. No statistical analysis was conducted for the economic part. All analyses were carried out in R v. 3.3.1. (R Development Core Team 2022). Values were expressed as mean ± standard deviation (SD).

3. Results

3.1. Morphological and Gas Exchange Parameters

We observed a worsening of individual plant biomass (roots and aerial part) by reducing the space between plants (Figures 2 and 3). Low plant density treatments showed plants with increased plant biomass (around +50%) in 2020 and doubled the biomass per plant in 2021 compared with values obtained in high plant densities. This result is not attributable to root biomass (no significant differences among treatments) but to the development of the aboveground part intended as leaves, crowns, flower trusses, and stolons.

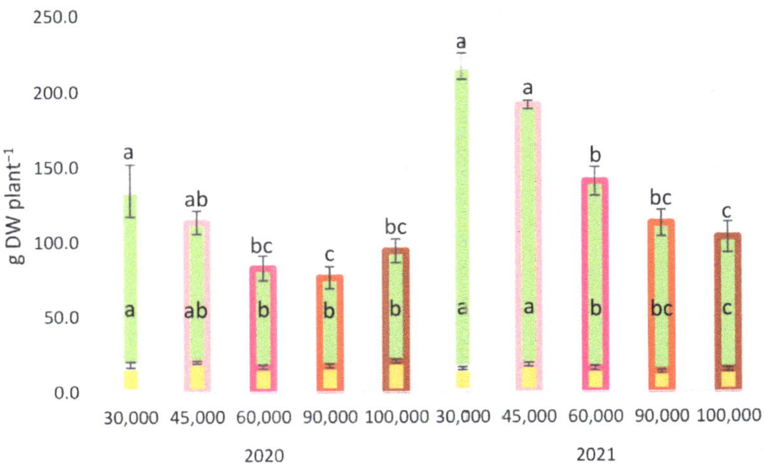

Figure 2. Total biomass (dry weight (DW)) composed of aboveground (green fill) and root biomass (yellow fill) at the end of first (2020) and second (2021) cropping year, as affected by planting densities. Vertical bars indicate means ± SD (n = 4). Within each year, the letters on the top of the bar (total biomass) and the letters on a green background (aboveground biomass) indicate significant differences according to LSD Fisher's test; $p < 0.05$. Root biomass data were not statistically significant.

Figure 3. Strawberry plant biomass at the end of the second cropping year, as affected by planting densities.

The number of primary crowns and formed branch crowns (about three per plant) was statistically not affected by plant spacing treatments in 2020 (Figure 4). In the second year of growth (2021), plants at a larger plant spacing (30,000 or 45,000 plant ha^{-1}) showed more crowns than plants in 90,000 plant ha^{-1} or 100,000 plant ha^{-1} (7.5 and 5.5 crowns, respectively).

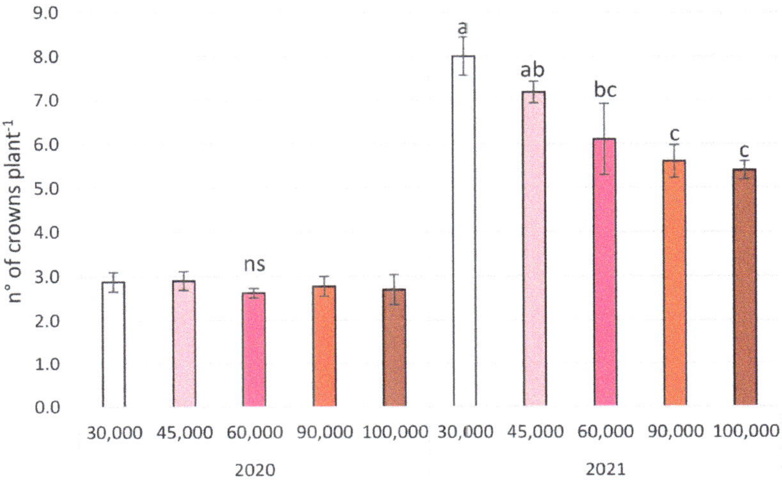

Figure 4. Number of total crowns (main and branch crowns) at the end of first (2020) and second (2021) cropping year, as affected by planting densities. Vertical bars indicate means ± SD (n = 4). Within each year, the letters indicate significant differences according to LSD Fisher's test; $p < 0.05$ (ns: not significant).

The net assimilation rate (A, µmol m^{-2} s^{-1}) was evaluated only in year 2021 (Figure 5). A significantly higher leaf photosynthetic rate (+12%) was measured for leaves in plants subjected to large planting density.

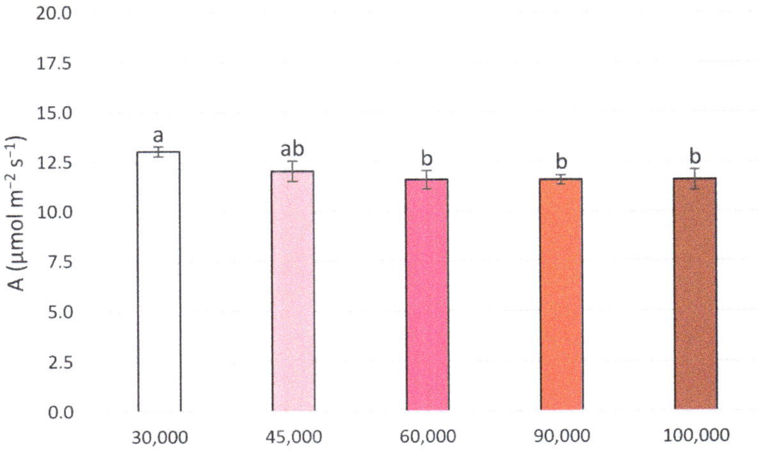

Figure 5. Photosynthetic rate in strawberry leaves during flowering, as affected by planting densities. Vertical bars indicate means ± SD (n = 4). The letters indicate significant differences according to LSD Fisher's test; $p < 0.05$ (ns: not significant).

Floral characteristics were affected by planting density, depending on the croppi year (Figure 6). As floral inductive conditions were the same during the nursery period, significant differences were observed during the first year. A completely different situati in the second year highlighted the influence of plant density on flowering. Indeed, plar subjected to large spacing were characterized by more flower trusses per plant than plar cultivated in high density (8.4 and 5.9, respectively). The highest number of flowers p truss was identified at 45,000 plant ha^{-1}. More flower trusses and flowers per truss larger spacing plants implied that the total number of flowers per plant appeared to significantly greater in those plants compared with plants in high planting densities. T medium plant density (60,000) was significantly similar to narrow plant spacing.

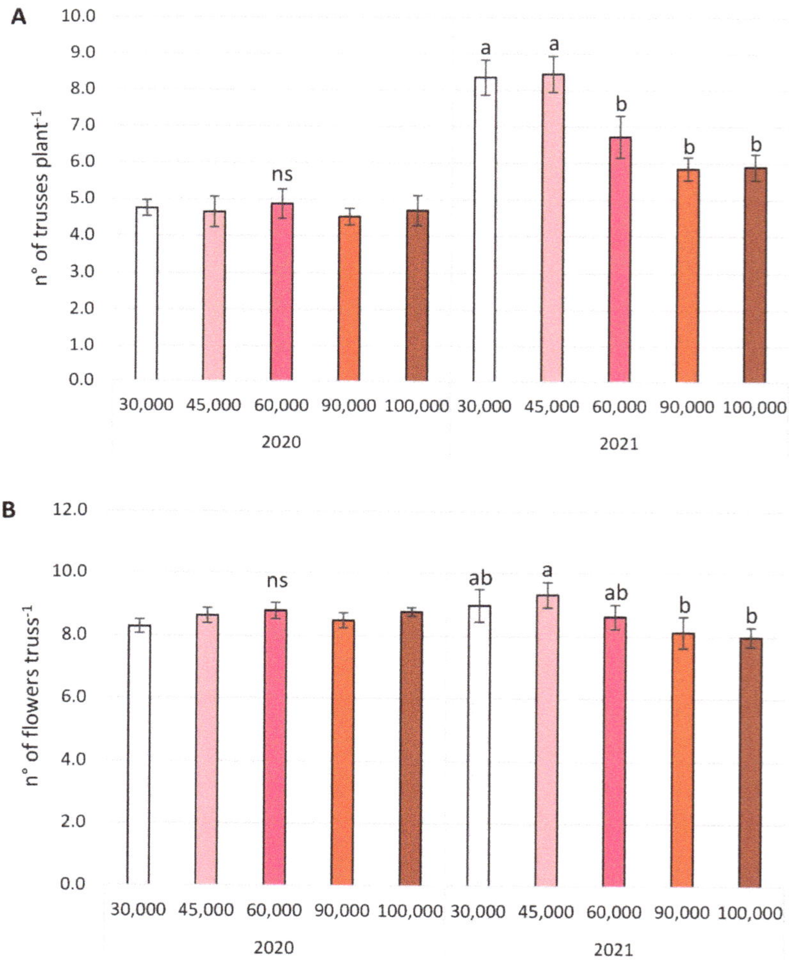

Figure 6. *Cont.*

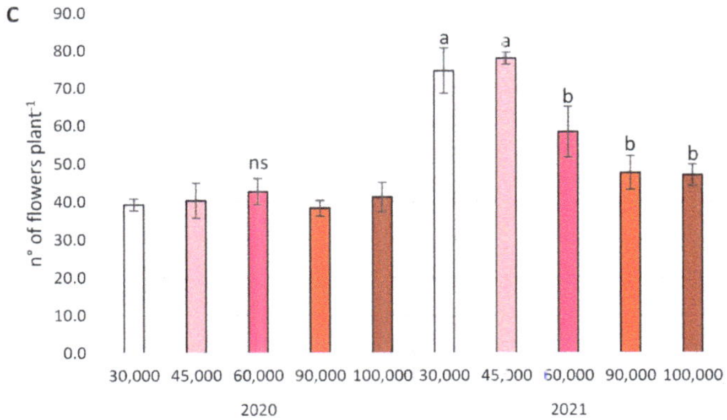

Figure 6. Reproductive characteristics of strawberry plants (flower trusses per plant—(**A**); flowers per truss—(**B**); total flowers per plant—(**C**) at the end of first (2020) and second (2021) cropping year, as affected by planting densities. Vertical bars indicate means ± SD (n = 4). Within each year, the letters indicate significant differences according to LSD Fisher's test; $p < 0.05$ (ns: not significant).

3.2. *Yield Parameters*

The strawberry production and its yield components are reported in Figures 7 and 8 and Table 2. Plants cultivated with wide-middle spacing were characterized by a significantly higher total yield per plant during the first (+36%) and second cropping year (+51%) than those in small spacing (Figure 7). Furthermore, there was a slight advance in fruit ripening in low planting density regimes, highlighting how this agronomic technique can influence the different phenological phases.

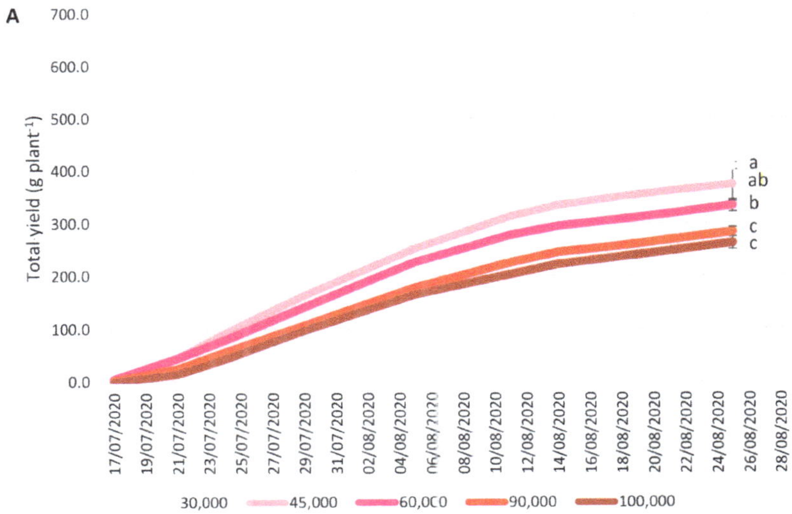

Figure 7. *Cont.*

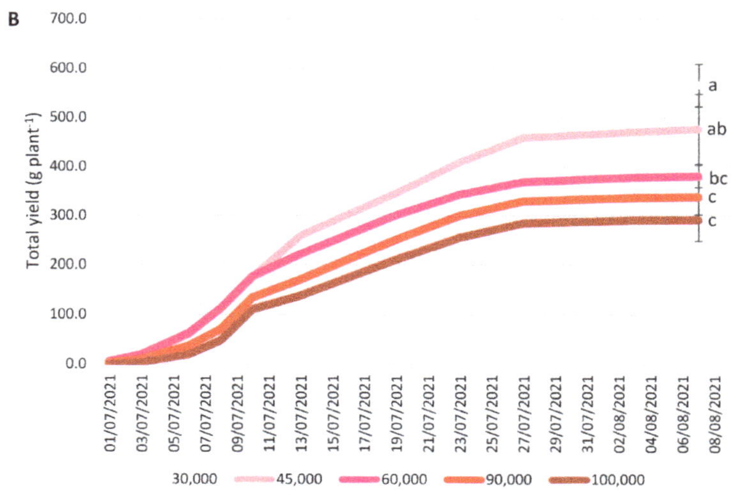

Figure 7. Cumulative total yield of strawberry plants in the first (2020—(**A**) and second (2021—(**B**) cropping year, as affected by planting densities. Vertical bars indicate means ± SD ($n = 4$). The letters indicate significant differences according to LSD Fisher's test; $p < 0.05$ (ns: not significant).

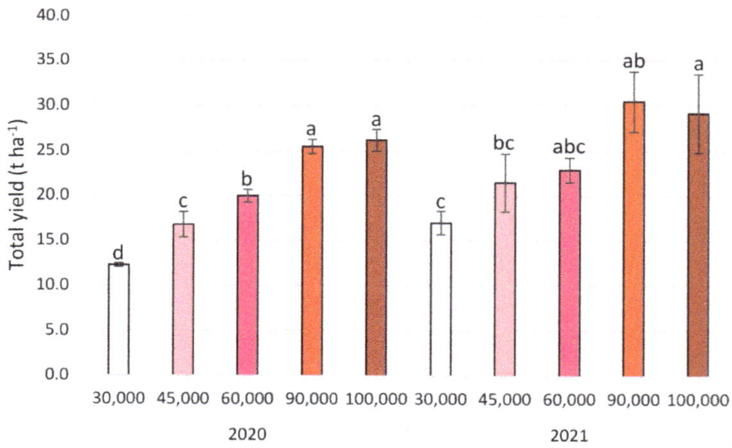

Figure 8. Total yield, expressed as tons per hectare, at the end of first (2020) and second (2021) cropping year, as affected by planting densities. Vertical bars indicate means ± SD ($n = 4$). Within each year the letters indicate significant differences according to LSD Fisher's test; $p < 0.05$ (ns: not significant).

The analysis of yield components shows that the increased production per plant in 30,000 and 45,000 (and partially in 60,000) was due to the significantly highest quantity of first-class commercial berries (+65%) and misshaped fruits (+45%) compared with 90,000 and 100,000. No significant differences emerged for small and rotten fruits (Table 2).

As plant density increased (from 30,000 to 90,000), the productivity, expressed as tons per hectare, increased linearly (Figure 8). No significant difference was found between 90,000 and 100,000. Although the statistical differences were clear during the first cropping year, the productivity values for middle planting densities (i.e., 60,000) were not significantly different from the values displayed in low and high planting densities in the second harvest year (Figure 8).

Table 2. Yield parameters (first-class and second-class marketable yield), as affected by planting densities.

	First-Class Yield (g Plant^{-1})		Small Fruits (g Plant^{-1})		Misshapen Fruits (g Plant^{-1})		Rotten Fruits (g Plant^{-1})	
lant density level (D)								
30,000	394.60 ± 61.18 [1]	A	39.93 ± 4.64	A	50.13 ± 0.02	A	1.94 ± 0.56	A
45,000	336.95 ± 45.23	AB	39.61 ± 7.55	A	45.35 ± 0.73	A	2.71 ± 0.14	A
60,000	266.74 ± 23.06	BC	42.20 ± 2.33	A	45.99 ± 2.97	A	1.89 ± 0.64	A
90,000	232.65 ± 21.67	C	42.83 ± 1.31	A	34.14 ± 0.17	B	1.26 ± 0.55	A
100,000	211.74 ± 17.33	C	32.03 ± 1.18	A	31.09 ± 5.23	B	1.96 ± 0.50	A
Significance	***		ns		***		ns	
Year (Y)								
2020	240.89 ± 24.98		44.13 ± 3.11		43.92 ± 3.58		2.63 ± 0.17	
2021	336.18 ± 51.15		34.51 ± 2.63		38.79 ± 5.14		1.28 ± 0.38	
Significance	***		**		ns		**	
D × Y	ns		ns		ns		ns	

[1] Means ± SD ($n = 4$) followed by the same letter do not significantly differ according to LSD Fisher's test; $p < 0.05$. Two-way ANOVA significant differences: *** $p < 0.001$; ** $p < 0.01$; ns: not significant.

In both cropping seasons (2020 and 2021), the mean fruit weight was significantly higher (around +10%) in plants subjected to low planting densities (30,000 and 45,000) (Figure 9).

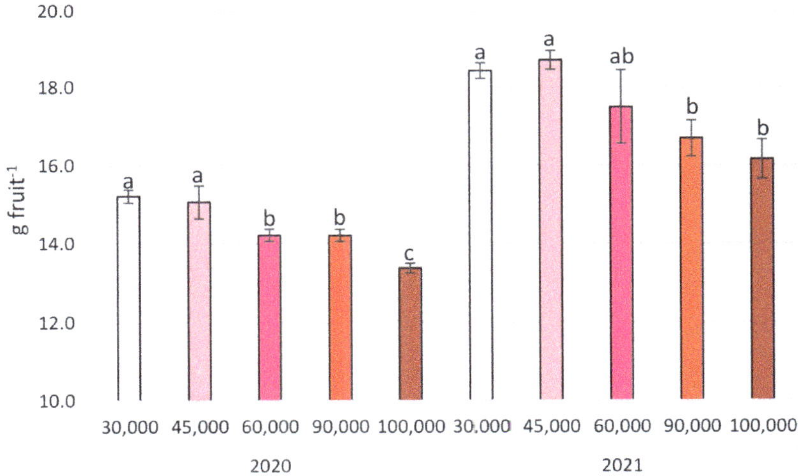

Figure 9. Average fruit weight (g fruit^{-1}) during the first (2020) and second (2021) cropping year, as affected by planting densities. Vertical bars indicate means ± SD ($n = 4$). Within each year, the letters indicate significant differences according to LSD Fisher's test; $p < 0.05$ (ns: not significant).

3.3. Fruit Quality

Strawberry qualitative traits assessed as flesh firmness (FF), total soluble solid (TSS), titratable acidity (TA), and color index (CI) were partially affected by plant density treatments (Table 3). As for FF, its average values were found higher in 2020 than in 2021, whereas the plant density was ineffective on this parameter. No change on TSS and TA was induced by different plant densities, as well as by the factor "year". Plant density had a visible and significant effect on CI, independently from the considered year. CI of fruits presented values ranging from 34 to 40, from light red to red, respectively. The highest CI value was observed in fruits coming from plants cultivated in wide spacing.

Table 3. Fruit quality traits (firmness (FF); total soluble solid (TSS); titratable acidity (TA); and co index (CI)), as affected by planting densities.

	FF (Durofel Index)		TSS (°Brix)		TA (g Acid Citric L^{-1})		CI	
Plant density level (D)								
30,000	36.66 ± 4.87 [1]	A	7.39 ± 0.09	A	6.67 ± 0.03	A	39.64 ± 1.78	A
45,000	36.19 ± 2.48	A	7.23 ± 0.12	A	6.50 ± 0.15	A	38.35 ± 1.10	AB
60,000	35.73 ± 2.63	A	7.04 ± 0.04	A	6.36 ± 0.34	A	36.12 ± 2.20	ABC
90,000	36.23 ± 2.81	A	7.09 ± 0.01	A	6.64 ± 0.04	A	35.86 ± 1.33	BC
100,000	35.87 ± 3.47	A	7.15 ± 0.09	A	6.85 ± 0.07	A	33.85 ± 1.82	C
Significance	ns		ns		ns		***	
Year (Y)								
2020	40.74 ± 0.83		7.15 ± 0.05		6.77 ± 0.06		34.44 ± 1.25	
2021	31.54 ± 0.59		7.21 ± 0.12		6.44 ± 0.18		39.09 ± 1.09	
Significance	***		ns		ns		***	
D × Y	ns		ns		ns		ns	

[1] Means ± SD (n = 4) followed by the same letter do not significantly differ according to LSD Fisher's test; $p < 0.$ Two-way ANOVA significant differences: *** $p < 0.001$; ns: not significant.

3.4. Economic Analysis

A synthesis of cost–benefit analysis is presented in the Figure 10. The total reven (EUR ha^{-1} over 2 cropping years), estimated from multiplying the produced strawberri by the local market price, increased linearly with decreasing the distance between plan (from 30,000 to 90,000). Beyond the planting density of 90,000, there was no increa from an economic point of view. Furthermore, the Figure 10 shows that a narrow pla spacing required a large investment (total costs). Among the tested density treatments, t two extremes (30,000 and 100,000) appeared economically unaffordable with regards to t net farm profit (EUR ha^{-1} over 2 years). A similar farm profit was obtained adopting t planting densities of 60,000 and 90,000, while the maximum farm profit was reached with density of 45,000 which corresponded to EUR 22,579 ha^{-1} (over 2 years).

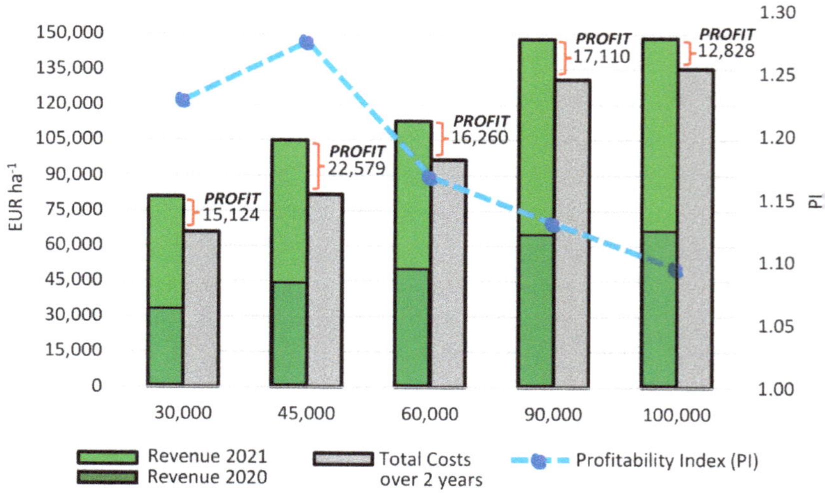

Figure 10. Revenue (green column: dark green = year 2020 and light green = year 2021), total cos (gray column; over 2 years), and net profit (values next to the curly brace; over 2 years), as affected planting densities. Data are expressed as EUR ha^{-1}. The dashed blue line represents the profitabili index (PI), as affected by planting densities.

Considering the profitability index (dashed blue line in the Figure 10), all planting density variants showed a value greater than 1. In particular, the widely spaced variants recorded the highest value (around 1.25).

4. Discussion

A severe intraspecific interaction is caused by an excessive presence of individuals belonging to the same species in a given space, which compete for key input resources [40,41]. Light, nutrients, and water can be considered the three main resource groups [42]. Competition between individual plants occurs primarily in the soil [43]. Root competition intensity increases as resource availability in soil decreases due to resource depletion or mechanisms of allelopathic interactions [44]. Although much research has indicated the correlation between the increase in root biomass accumulation with the decreasing planting density [45], our study did not highlight this relationship, but the increased plant biomass in low density system was due to the aerial part (Figure 2). Poor light exposure for leaves increases the aboveground competition intensity which can diminish with adequate light supply [46]. A large spacing between individual plants allows to maximize the light interception, avoiding negative effects that mutual and self-shading can have [19]. Indeed, our strawberry plants grown with the widest spacing were characterized by leaves with the greatest photosynthetic activity (+12% compared with plants in high planting density) (Figure 5). In addition to plant photosynthesis, light triggers many other essential physiological processes, for example, flower bud initiation and branching, as reported by Yang and Jeong [47] in chrysanthemum plants. Our study confirmed that more flower trusses, flowers, and branch crowns per plants were found in plants more exposed to light (i.e., in low density systems) (Figure 6; Figure 4). However, these findings appeared only during the second cropping year. Auxiliary buds form branch crowns and the apical meristems of crowns develop into terminal inflorescences in autumn, depending on environmental conditions such as daylength and temperature [23,48,49]. Since our plant materials came from the same nursery and the induction/differentiation phase took place under standard conditions for all plants, no significant differences were observed in the reproductive aspects during the first cropping year. On the contrary, subjecting plants to different growing conditions (i.e., planting densities) in autumn 2020 (i.e., at the end of first harvesting year) helped induce morphological changes that would have been evident the following year (2021). Considering this explanation, the significant variations in total and commercial yield in the first harvesting year must be found not in the number of flowers per plant (instead, for the second cropping year) but in the average fruit weight (Table 2; Figure 9). Plant–plant interaction for limited resources can lead to differentiated investment in their growth and reproduction [50]. Strawberry plants under nutrient limitation responded by favoring the vegetative growth (i.e., leaf and root biomass) to the detriment of the plant's reproductive investment, as evidenced by the low yield and very small size of the fruits, observed by Soppelsa et al. [51].

Picking productivity of open-field strawberries (e.g., cultivar Elsanta) for the fresh market is usually at 12–15 kg per hour [52]. In the cultivation area where we carried out the trial, the productivity during harvesting is rather around 10–12 kg per hour (or less), for the same variety. This lower picking productivity can be associated with hostile growing conditions. For instance, the sloping land plots in Martell Valley slow down harvesting operations. Moreover, the exceptional environmental conditions typical of this valley affect pomological traits such as fruit weight. Indeed, Naryal et al. [53] reported that the apricot fruit weight decreases by 0.5 g for every 100 m of increase in altitude. Since in our study a significant difference emerged for the fruit weight parameter (i.e., greater fruit size in low planting density), we took this aspect into account to calculate a differentiated picking productivity in the economic analysis (Figure 10).

As reported in Figures 8 and 10, the increased yield per area with a high plant density (90,000 or 100,000 plant ha^{-1}) led to an increase in total revenue (EUR ha^{-1}) but the total costs (EUR ha^{-1}) also reached a considerably high level. We need an investment cost of

EUR 75,000 per hectare with a low planting density, while more than EUR 130,000 are necessary with a high density, which means an increase of about EUR 60,000 (i.e., +80%). Harvesting costs account for more than 40% of total business costs. Similar percentage among the various planting densities tested. Another important item is represented by the cost of the plants (EUR 0.332 per plant + vat), which varies among the densities tested due to the different number of plants transplanted.

High revenue is not synonymous with high profit, since costs also increase proportionally. Therefore, the profit-maximizing planting spacing was achieved with 45,000 plants ha^{-} The result of our study is consistent with the findings by Matsumoto et al. [54] and Castellanos et al. [55], who state that a middle planting density (or better to say not too high) in upland rice or garlic cultivation is preferable for the highest farm profit. Furthermore the choice of the right planting density has a noticeable influence on opportunity costs, as reported by Jettner et al. [56] comparing different sowing rates of faba bean (*Vicia faba* L.)

Under the described growing conditions, the different plant densities had no significant effect on the main fruit quality traits, except for the color index. With the decrease in the spacing between individual plants, there is less sunlight exposure for fruits, which affects their coloration (Table 3). These findings are confirmed by Martins de Lima et al. [16] Sunlight has a remarkable effect on regulating the biosynthesis of phenolic compounds such as anthocyanins, which is the major pigment in strawberry fruits [57,58]. Having more intense red fruits sometimes reflects the ideal purchase intention of the consumer [59].

As mentioned before, not only can cultural practices affect availability of resources but environmental factors (e.g., elevation) can also play an important role [60]. For example, high-altitudinal levels were observed to change physiological and morphological responses of plants, interfering precisely with resources such as radiation [35]. The positive influence of high elevation on strawberry nursery materials was described by Maroto et al. [61] and Pirlak et al. [62], showing an increased number of leaves, runners, and flowers in that cultivation condition. Although the effects of an agronomic practice such as planting density has been well-investigated on different vegetable and fruit crops including strawberry [10,11,14,32,63], no previous research has been conducted under our imposed experimental conditions (i.e., in alpine mountain environment).

5. Conclusions

Optimizing the planting density is an effective strategy for improving yield and farm profit, especially in alpine mountain environments. Given the results, as summarized in Figure 11, recommendations from this study are: not to exceed the density of 100,000 plants ha^{-1} (economically disadvantageous); adopt medium or low planting densities to have strong plants (e.g., for a third year of continuation); pay attention to correctly manage the picking times with low planting density in order to avoid overripe fruits with a dark intense red color; having fewer plants per hectare reduces the total costs but can increase the business risk (e.g., loss of plants due to crown and root rot, loss of production due to strawberry blossom weevil (*Anthonomus rubi*), etc.); and to encourage producers to adopt wide spacing between plants for more sustainable strawberry growing (we observed a lower incidence of powdery mildew on plants subjected to a low planting density). Further research is needed to examine the agronomic and economic benefits of influencing planting density in a soilless cultivation system for strawberries, a cultivation technique that is increasingly gaining popularity in South Tyrol.

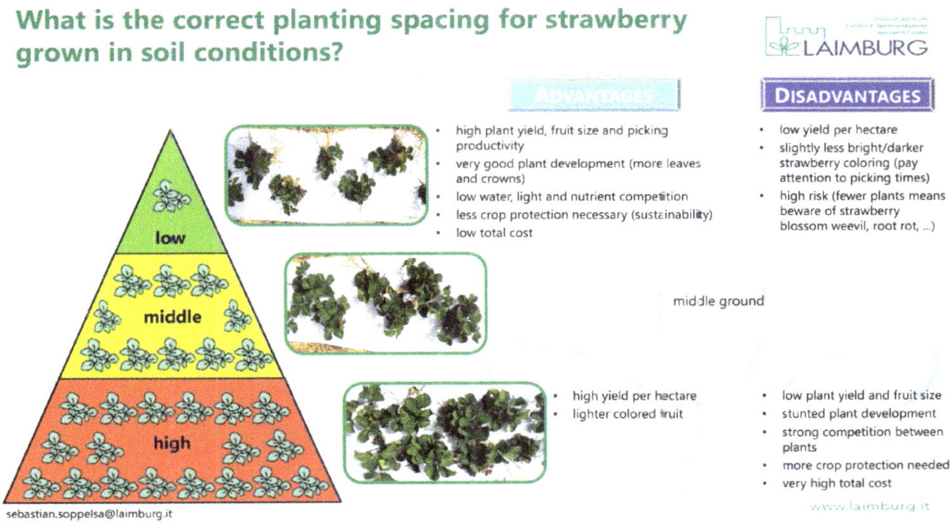

Figure 11. Schematic representation of advantages and disadvantages by adopting different planting densities in strawberry soil cultivation.

Author Contributions: S.S. and M.Z. conceived and designed the experiment; S.S. and M.G. performed the experiment; S.S. analyzed the data; S.S. and M.G. wrote the manuscript. All authors have read and agreed to the published version of the manuscript.

Funding: This research activity was conducted in the framework of the "Action Plan for Mountain Agriculture and Food Sciences 2016–2022, adopted by the Government of South Tyrol (Italy).

Institutional Review Board Statement: Not applicable.

Informed Consent Statement: Not applicable.

Data Availability Statement: Not applicable.

Acknowledgments: Preliminary data from this research study earned the author S.S. the ISHS Young Minds Award during the IX International Strawberry Symposium. S.S. wishes to thank the Symposium Commission for having given him this important recognition. The authors would like to thank Peter Gamper (former mayor of Martello/Martell) for his support during the field activity and for the helpful suggestions regarding the cost–benefit analysis. The authors thank the Department of Innovation, Research University and Museums of the Autonomous Province of Bozen/Bolzano for covering the Open Access publication costs.

Conflicts of Interest: The authors declare no conflict of interest.

eferences

Liston, A.; Cronn, R.; Ashman, T. Fragaria: A Genus with Deep Historical Roots and Ripe for Evolutionary and Ecological Insights. *Am. J. Bot.* **2014**, *101*, 1686–1699. [CrossRef] [PubMed]
Yang, J.; Su, D.; Wei, S.; Chen, S.; Luo, Z.; Shen, X.; Zhang, Z.; Jamil, A.; Tong, J.; Cui, X. Current and Future Potential Distribution of Wild Strawberry Species in the Biodiversity Hotspot of Yunnan Province, China *Agronomy* **2020**, *10*, 959. [CrossRef]
Darrow, G.M. *The Strawberry: History, Breeding, and Physiology*; Holt, Rinehart and Winston: New York, NY, USA, 1966.
FAOSTAT. Available online: http://www.fao.org/faostat/en/#data/QC (accessed on 20 October 2022).
ISTAT. Available online: https://www.istat.it (accessed on 20 October 2022).
ASTAT. Available online: https://astat.provincia.bz.it/it/default.asp (accessed on 20 October 2022).
Cooperative for Regional Development Martello 3B. Available online: https://www.martell.it/it/la-val-martello/ (accessed on 26 October 2022).

8. Bargioni, G. The Italian Strawberry Industry. In Proceedings of the Acta Horticulturae, International Society for Horticultu Science (ISHS), Leuven, Belgium, 1 December 1989; pp. 659–664.
9. Mezzetti, B.; Giampieri, F.; Zhang, Y.; Zhong, C. Status of Strawberry Breeding Programs and Cultivation Systems in Europe a the Rest of the World. *J. Berry Res.* **2018**, *8*, 205–221. [CrossRef]
10. Haque, M.A.; Sakimin, S.Z. Planting Arrangement and Effects of Planting Density on Tropical Fruit Crops—A Review. *Hortic turae* **2022**, *8*, 485. [CrossRef]
11. Djaman, K.; Allen, S.; Djaman, D.S.; Koudahe, K.; Irmak, S.; Puppala, N.; Darapuneni, M.K.; Angadi, S.V. Planting Date and Pla Density Effects on Maize Growth, Yield and Water Use Efficiency. *Environ. Chall.* **2022**, *6*, 100417. [CrossRef]
12. Hou, W.; Khan, M.R.; Zhang, J.; Lu, J.; Ren, T.; Cong, R.; Li, X. Nitrogen Rate and Plant Density Interaction Enhances Radiati Interception, Yield and Nitrogen Use Efficiency of Mechanically Transplanted Rice. *Agric. Ecosyst. Environ.* **2019**, *269*, 183–1 [CrossRef]
13. Rodrigo, V.H.L.; Stirling, C.M.; Teklehaimanot, Z.; Nugawela, A. The Effect of Planting Density on Growth and Development Component Crops in Rubber/Banana Intercropping Systems. *Field Crops Res.* **1997**, *52*, 95–108. [CrossRef]
14. Wright, C.J.; Sandrang, A.K. Density Effects on Vegetative and Reproductive Development in Strawberry Cv. Hapil. *J. Hortic. S* **1993**, *68*, 231–236. [CrossRef]
15. Wu, L.; Deng, Z.; Cao, L.; Meng, L. Effect of Plant Density on Yield and Quality of Perilla Sprouts. *Sci. Rep.* **2020**, *10*, 99 [CrossRef] [PubMed]
16. Martins de Lima, J.; Welter, P.D.; Soares dos Santos, M.F.; Kavcic, W.; Miranda Costa, B.; Fagherazzi, A.F.; Regianini Nerbass, Anneliese Kretzschmar, A.; Rufato, L.; Baruzzi, G. Planting Density Interferes with Strawberry Production Efficiency in Southe Brazil. *Agronomy* **2021**, *11*, 408. [CrossRef]
17. Postma, J.A.; Hecht, V.L.; Hikosaka, K.; Nord, E.A.; Pons, T.L.; Poorter, H. Dividing the Pie: A Quantitative Review on Pla Density Responses. *Plant Cell Environ.* **2021**, *44*, 1072–1094. [CrossRef] [PubMed]
18. Fan, J.; Lu, X.; Gu, S.; Guo, X. Improving Nutrient and Water Use Efficiencies Using Water-Drip Irrigation and Fertilizati Technology in Northeast China. *Agric. Water Manag.* **2020**, *241*, 106352. [CrossRef]
19. Li, T.; Liu, L.-N.; Jiang, C.-D.; Liu, Y.-J.; Shi, L. Effects of Mutual Shading on the Regulation of Photosynthesis in Field-Grov Sorghum. *J. Photochem. Photobiol. B* **2014**, *137*, 31–38. [CrossRef]
20. Pierson, E.A.; Mack, R.N.; Black, R.A. The Effect of Shading on Photosynthesis, Growth, and Regrowth Following Defoliation f Bromus Tectorum. *Oecologia* **1990**, *84*, 534–543. [CrossRef]
21. Samad, S.; Rivero, R.; Kalyandurg, P.B.; Vetukuri, R.R.; Heide, O.M.; Sønsteby, A.; Khalil, S. Characterization of Environmen Effects on Flowering and Plant Architecture in an Everbearing Strawberry F1-Hybrid by Meristem Dissection and Gene Expressi Analysis. *Horticulturae* **2022**, *8*, 626. [CrossRef]
22. Sidhu, V.; Bernier-English, V.; Lamontagne-Drolet, M.; Gravel, V. Effect of Light Quality and Extended Photoperiod on Flower B Induction during Transplant Production of Day-Neutral Strawberry Cultivars. *Can. J. Plant Sci.* **2022**, *102*, 356–367. [CrossRef
23. Sønsteby, A.; Heide, O.M. Dynamics of Dormancy Regulation in 'Sonata' Strawberry and Its Relation to Flowering and Runneri *CABI Agric. Biosci.* **2021**, *2*, 4. [CrossRef]
24. Sønsteby, A.; Hytonen, T. Manipulating Flower Induction through Temperature and Photoperiod Fluctuations. *Int. J. Fruit S* **2005**, *5*, 17–27. [CrossRef]
25. Wang, R.; Eguchi, M.; Gui, Y.; Iwasaki, Y. Evaluating the Effect of Light Intensity on Flower Development Uniformity in Strawber (Fragaria × ananassa) under Early Induction Conditions in Forcing Culture. *HortScience* **2020**, *55*, 670–675. [CrossRef]
26. Dhaliwal, D.S.; Williams, M.M. Understanding Variability in Optimum Plant Density and Recommendation Domains f Crowding Stress Tolerant Processing Sweet Corn. *PLoS ONE* **2020**, *15*, e0228809. [CrossRef]
27. Wamser, A.F.; Valmorbida, J.; Suzuki, A.; Hahn, L.; Mueller, S.; Becker, W.F.; Feltrim, A.L.; Ender, M.M. Planting Density ar Arrangement for the Mechanized Spraying of Vertically Staked Tomatoes. *Hortic. Bras.* **2017**, *35*, 519–526. [CrossRef]
28. Carvalho, F.J.; Carneiro, L.B.; Benett, C.G.S.; Benett, K.S.S.; Martins, A.S.; Silva, A.T.D.; Seleguini, A. Plant Density and Grow Regulator Applications in a Tomato Crop for Industrial Processing. *Rev. Colomb. Cienc. Hortíc.* **2019**, *13*, 397–405. [CrossRef]
29. De Camacaro, M.E.P.; Camacaro, G.J.; Hadley, P.; Dennett, M.D.; Battey, N.H.; Carew, J.G. Effect of Plant Density and Initi Crown Size on Growth, Development and Yield in Strawberry Cultivars Elsanta and Bolero. *J. Hortic. Sci. Biotechnol.* **2004**, 739–746. [CrossRef]
30. Heuvelink, E. Effect of Plant Density on Biomass Allocation to the Fruits in Tomato (*Lycopersicon esculentum* Mill.). *Sci. Hort* **1995**, *64*, 193–201. [CrossRef]
31. Khasmakhi-Sabet, A.; Sedaghathoor, S.; Mohammady, J.; Olfati, J.A. Effect of Plant Density on Bell Pepper Yield and Quality. *I J. Veg. Sci.* **2009**, *15*, 264–271. [CrossRef]
32. Nerson, H. Relationship between Plant Density and Fruit and Seed Production in Muskmelon. *J. Am. Soc. Hortic. Sci.* **2002**, 1 855–859. [CrossRef]
33. Ramos-López, B.I.; Ortiz-Hernández, Y.D.; Morales, I.; Aquino-Bolaños, T. Plant Density on Yield of Husk Tomato (*Physa ixocarpa* Brot.) in Field and Greenhouse. *Ciênc. Rural* **2021**, *51*, e20200992. [CrossRef]
34. Zenobi, S.; Fiorentini, M.; Aquilanti, L.; Foligni, R.; Mannozzi, C.; Mozzon, M.; Zitti, S.; Casavecchia, S.; Al Mohandes Dridi, Orsini, R. Effect of Planting Density in Two Thistle Species Used for Vegetable Rennet Production in Marginal Mediterrane Areas. *Agronomy* **2021**, *11*, 135. [CrossRef]

35. Gale, J. Plants and Altitude—Revisited. *Ann. Bot.* **2004**, *94*, 199. [CrossRef]
36. Guerrero-Chavez, G.; Scampicchio, M.; Andreotti, C. Influence of the Site Altitude on Strawberry Phenolic Composition and Quality. *Sci. Hortic.* **2015**, *192*, 21–28. [CrossRef]
37. Kiełtyk, P. Patterns of Floral Allocation along an Elevation Gradient: Variation in Senecio Subalpinus Growing in the Tatra Mountains. *Alp. Bot.* **2021**, *131*, 117–124. [CrossRef]
38. Wang, H.; Prentice, I.C.; Davis, T.W.; Keenan, T.F.; Wright, I.J.; Peng, C. Photosynthetic Responses to Altitude: An Explanation Based on Optimality Principles. *N. Phytol.* **2017**, *213*, 976–982. [CrossRef] [PubMed]
39. Tessmer, M.A.; Appezzato-da-Glória, B.; Antoniolli, L.R. Influence of Growing Sites and Physicochemical Features on the Incidence of Lenticel Breakdown in 'Gala' and 'Galaxy' Apples. *Sci. Hortic.* **2016**, *205*, 119–126. [CrossRef]
40. Heitholt, J.J.; Sassenrath-Cole, G.F. Inter-Plant Competition: Growth Responses to Plant Density and Row Spacing. In *Physiology of Cotton*; Stewart, J.D., Oosterhuis, D.M., Heitholt, J.J., Mauney, J.R., Eds.; Springer: Dordrecht, The Netherland, 2010; pp. 179–186. ISBN 978-90-481-3194-5.
41. Yang, X.; Zhang, W.; He, Q. Effects of Intraspecific Competition on Growth, Architecture and Biomass Allocation of *Quercus Liaotungensis*. *J. Plant Interact.* **2019**, *14*, 284–294. [CrossRef]
42. Craine, J.M.; Dybzinski, R. Mechanisms of Plant Competition for Nutrients, Water and Light. *Funct. Ecol.* **2013**, *27*, 833–840. [CrossRef]
43. Belcher, J.W.; Keddy, P.A.; Twolan-Strutt, L. Root and Shoot Competition Intensity along a Soil Depth Gradient. *J. Ecol.* **1995**, *83*, 673–682. [CrossRef]
44. Schenk, H.J. Root Competition: Beyond Resource Depletion. *J. Ecol.* **2006**, *94*, 725–739. [CrossRef]
45. Wang, L.-W.; Showalter, A.M.; Ungar, I.A. Effects of Intraspecific Competition on Growth and Photosynthesis of Atriplex Prostrata. *Aquat. Bot.* **2005**, *83*, 187–192. [CrossRef]
46. Wilson, S.D.; Tilman, D. Plant Competition and Resource Availability in Response to Disturbance and Fertilization. *Ecology* **1993**, *74*, 599–611. [CrossRef]
47. Yang, J.; Jeong, B.R. Side Lighting Enhances Morphophysiology by Inducing More Branching and Flowering in Chrysanthemum Grown in Controlled Environment. *Int. J. Mol. Sci.* **2021**, *22*, 12019. [CrossRef]
48. Hytönen, T.; Palonen, P.; Mouhu, K.; Junttila, O. Crown Branching and Cropping Potential in Strawberry (*Fragaria ananassa* Duch.) Can Be Enhanced by Daylength Treatments. *J. Hortic. Sci. Biotechnol.* **2004**, *79*, 466–471. [CrossRef]
49. Koskela, E.A.; Hytönen, T. Control of Flowering in Strawberries. In *The Genomes of Rosaceous Berries and Their Wild Relatives*; Hytönen, T., Graham, J., Harrison, R., Eds.; Compendium of Plant Genomes; Springer: Cham, Switzerland, 2018; pp. 35–48; ISBN 978-3-319-76019-3.
50. Tuller, J.; Marquis, R.J.; Andrade, S.M.M.; Monteiro, A.B.; Faria, L.D.B. Trade-Offs between Growth, Reproduction and Defense in Response to Resource Availability Manipulations. *PLoS ONE* **2018**, *13*, e0201873. [CrossRef] [PubMed]
51. Soppelsa, S.; Kelderer, M.; Casera, C.; Bassi, M.; Robatscher, P.; Matteazzi, A.; Andreotti, C. Foliar Applications of Biostimulants Promote Growth, Yield and Fruit Quality of Strawberry Plants Grown under Nutrient Limitation. *Agronomy* **2019**, *9*, 483. [CrossRef]
52. Commission of the European Communities. *Annex to the Report from the Commission to the Council and the European Parliament on the Situation of the Sector of Soft Fruits and Cherries Intended for Processing*; European Commission: Brussels, Belgium, 2006.
53. Naryal, A.; Dolkar, D.; Bhardwaj, A.K.; Kant, A.; Chaurasia, O.P.; Stobdan, T. Effect of Altitude on the Phenology and Fruit Quality Attributes of Apricot (*Prunus armeniaca* L.) Fruits. *Def. Life Sci. J.* **2020**, *5*, 18–24. [CrossRef]
54. Matsumoto, S.; Tsuboi, T.; Asea, G.; Miyamoto, K.; Maruyama, A.; Takagaki, M.; Kikuchi, M. Effects of Plant Density on the Performance of Selected African Upland Rice Varieties. *Afr. J. Agric. Res.* **2017**, *12*, 2262–2272. [CrossRef]
55. Castellanos, J.Z.; Vargas-Tapia, P.; Ojodeagua, J.L.; Hoyos, G.; Alcantar-Gonzalez, G.; Mendez, F.S.; Alvarez-Sanchez, E.; Gardea, A.A. Garlic Productivity and Profitability as Affected by Seed Clove Size, Planting Density and Planting Method. *HortScience* **2004**, *39*, 1272–1277. [CrossRef]
56. Jettner, R.; Loss, S.P.; Martin, L.D.; Siddique, K.H.M. Responses of Faba Bean (*Vicia faba* L.) to Sowing Rate in South-Western Australia. I. Seed Yield and Economic Optimum Plant Density. *Aust. J. Agric. Res.* **1998**, *49*, 989. [CrossRef]
57. Di Stefano, V.; Scandurra, S.; Pagliaro, A.; Di Martino, V.; Melilli, M.G. Effect of Sunlight Exposure on Anthocyanin and Non-Anthocyanin Phenolic Levels in Pomegranate Juices by High Resolution Mass Spectrometry Approach. *Foods* **2020**, *9*, 1161. [CrossRef] [PubMed]
58. Yoshida, Y.; Koyama, N.; Tamura, H. Color and Anthocyanin Composition of Strawberry Fruit. Changes during Fruit Development and Differences among Cultivars, with Special Reference to the Occurrence of Pelargonidin 3-Malonylglucoside. *J. Jpn. Soc. Hortic.* **2002**, *71*, 355–361. [CrossRef]
59. Bhat, R.; Geppert, J.; Funken, E.; Stamminger, R. Consumers Perceptions and Preference for Strawberries—A Case Study from Germany. *Int. J. Fruit Sci.* **2015**, *15*, 405–424. [CrossRef]
60. Lee, M.A.; Burger, G.; Green, E.R.; Kooij, P.W. Relationships between Resource Availability and Elevation Vary between Metrics Creating Gradients of Nutritional Complexity. *Oecologia* **2021**, *195*, 213–223. [CrossRef] [PubMed]
61. Maroto, J.V.; López-Galarza, S.; San Bautista, A.; Pascual, B. Cold Stored and Fresh Multicrown Strawberry Plants for Autumn-Winter Production in Eastern Spain. In Proceedings of the Acta Horticulturae, International Society for Horticultural Science (ISHS), Leuven, Belgium, 1 September 1997; pp. 545–548.

62. Pirlak, L.; Güleryüz, M.; Bolat, I. The Altitude Affects the Runner Plant Production and Quality in Strawberry Cultivars. I
Proceedings of the Acta Horticulturae, International Society for Horticultural Science (ISHS), Leuven, Belgium, 31 January 200
pp. 305–308.
63. González, J.A.; Mercado, M.I.; Martinez-Calsina, L.; Erazzú, L.E.; Buedo, S.E.; González, D.A.; Ponessa, G.I. Plant Density Effect
on Quinoa Yield, Leaf Anatomy, Ultrastructure and Gas Exchange. *J. Agric. Sci.* **2022**, *160*, 349–359. [CrossRef]

agronomy

MDPI

ticle

ffects of Planting Pre-Germinated Buds on Stand Establishment 1 Sugarcane

ma Varsha Madala [1], Ricardo A. Lesmes-Vesga [1], Calvin D. Odero [1], Lakesh K. Sharma [2] d Hardev S. Sandhu [1,*]

[1] Everglades Research and Education Center, University of Florida, Belle Glade, FL 33430, USA
[2] Department of Soil, Water and Ecosystem Sciences, University of Florida, Gainesville, FL 32603, USA
* Correspondence: hsandhu@ufl.edu; Tel.: +1-561-993-1567

Abstract: Sugarcane (a complex hybrid of *Saccharum* spp.) is propagated vegetatively by using stem pieces of mature cane with healthy buds. Abiotic and biotic stress may cause pre-germination of these buds, which may have an impact on both emergence and plant cane stand establishment. There is very limited information available in the literature. A greenhouse study was conducted with single-budded seed pieces of three levels of bud germination (ungerminated buds, Pop-eyes, and Lalas) from three different cultivars (CP 96-1252, CPCL 05-1201, and CPCL 02-0926) planted in pots and repeated over time. Data on growth parameters (tiller count, primary shoot height, SPAD, and dry biomass of shoots and roots) at early growth showed that Lalas produced more tillers and higher shoot dry biomass than Pop-eyes and ungerminated buds. Both Lalas and Pop-eyes produced higher root dry biomass than ungerminated buds in one of the two experiments. The cultivar had a significant effect on primary shoot height and SPAD. A small plot field experiment was conducted with cultivar CP 96-1252 to validate the results of greenhouse experiments, and similar results were reported for tiller count. The results indicate that pre-germinated buds may have a neutral or positive effect on early sugarcane growth and establishment. Further on-farm research needs to be conducted to confirm these results before using pre-germinated buds as a potential seed source for the late season planting of sugarcane.

Keywords: sugarcane; ungerminated buds; Pop-eyes; Lalas; top visible dewlap leaf (TVD); SPAD (Soil Plant Analysis Development)

ation: Madala, H.V.; smes-Vesga, R.A.; Odero, C.D.; arma, L.K.; Sandhu, H.S. Effects of nting Pre-Germinated Buds on nd Establishment in Sugarcane. *ronomy* **2023**, *13*, 1001. https:// .org/10.3390/agronomy13041001

ademic Editors: Umberto Anastasi d Aurelio Scavo

ceived: 6 March 2023
vised: 24 March 2023
cepted: 27 March 2023
blished: 29 March 2023

1. Introduction

Sugarcane (*Saccharum* spp.) is an important row crop in Florida, with approximately 160,458 ha grown near Lake Okeechobee in southern Florida [1]. Sugarcane is vegetatively propagated using the mature stems (also called stalks) with healthy buds from previous commercial crops or grown specifically for seed cane. Commercial sugarcane in Florida is commonly propagated by laying whole stalks horizontally in the furrow and then chopping each stalk into three to four pieces (billets) with two to three buds on each billet. Buds, also called eyes, sprout when planted in damp soil and, under ideal conditions, primary shoots emerge within 2 to 3 weeks after planting [2]. Hence, seed cane quality (especially healthy buds) is critical to achieving a good crop stand in plant cane. Multiple biotic and abiotic conditions can impact buds and, thus, seed cane quality.

Bud germination before planting (also called pre-germination) occurs especially when the apical growing point dies because of biotic (e.g., insect pest, disease) or abiotic stress factors (e.g., freeze, lodging). This situation ceases the apical dominance and promotes the growth of lateral buds [3]. Based on the level of germination, axillary buds are divided into three categories: ungerminated buds, Pop-eyes, and Lalas (Figure 1). Ungerminated buds do not show any signs of lateral shoot growth or an emergence of green leaf tissue. Some buds only grow bigger and project buds, while others grow 2 to 3 cm, which are known as

Pop-eyes. Pop-eye is a term used to describe the bulging buds on the cane stalk that ha the potential to grow into a full plant when planted in soil. Certain axillary or lateral bu may continue to grow into full-fledged Lalas even if the apical bud continues to operate a very slow rate of growth but has temporarily lost dominance over the lateral buds. At ea node, Lalas are pale green side shoots or branches with wiry adventitious roots growi from the buds. At this point, the buds connected with the oldest green leaves are typical fully mature, and their development is blocked by hormones, such as auxins produc by the meristematic tip, which renders the stem's lateral buds dormant [4]. Sugarca cultivars deal with bud dormancy in different ways. When withered and decaying leav are plucked off growing cane, the buds continue to grow into side shoots called Lalas a are never dormant [5]. If the leaves are not removed until they die naturally, buds becon dormant and likely would not germinate again.

Ungerminated bud | Pop-eye | Lalas

Figure 1. Sugarcane bud germination levels used for the experiment.

There is little understanding of the effects of planting pre-germinated buds on su arcane stand establishment and early growth, and whether it varies among cultiva However, some efforts have been made in other countries, such as Brazil, with promisir results using pre-germinated buds as a novel method to establish sugarcane [6–9]. W hypothesized that pre-germinated buds may have some negative effect on stand establis ment and the effect may vary with genetics. Therefore, a greenhouse study was conducte to evaluate three levels of bud germination for their effect on the early growth of thr sugarcane cultivars.

2. Materials and Methods

2.1. Greenhouse Experiment Setup

Two greenhouse experiments were conducted at the University of Florida, Evergla Research and Education Center (EREC) in Belle Glade, Florida, between January 2021 a June 2021 to determine the effect of different levels of bud germination on early sugarca growth. The soil used for this study came from a field at EREC with a long history (almo 20 years) of sugarcane production. Soil type was Histosol or Dania series muck (Eu hyperthermic, shallow Lithic Haplosaprist) with pH of 7.0, 80% of organic matter conte and 0.66 Mg m^{-3} of bulk density. To eliminate large clods, topsoil (0 to 10 cm soil laye was excavated from a stubble-free and weed-free area and put through a sieve of mesh (4-mm holes). The soil was air dried for 24 h before filling into the pots. A total of 54 plast pots of 6.28-L capacity (20 cm diameter × 20 cm depth) were then filled with 3.85 k of soil. Single-budded seed pieces (with a 2-cm of stalk on either side) were obtained from the middle of mature stalks of three sugarcane cultivars, CPCL 02-0926 [10], CP 96-1252 [11], CPCL 05-1201 [12], and were planted in pots and placed in a greenhouse at EREC. Two seed pieces were planted per pot, widely apart in such a way that the buds on the nodal section of the ungerminated seed piece faced upwards, and the Lalas were kept above the soil surface, and this was conducted similarly for the Pop-eyes. The planted seed pieces were covered with a 3 to 4 cm soil layer. The pots were watered until field capacity (indicated by the drainage of excess water from the pots) and watered at regular intervals (once a week throughout the experiment period). Fertilizers were not applied in this experiment because the soil had enough concentration of all available or mobile nutrients, as reported in lab tests (99.84 mg kg^{-1} P, 470.4 mg kg^{-1} K, 19,036 mg kg^{-1} Ca, 3028.7 mg kg^{-1} Mg, and 113.96 mg kg^{-1} Si). The 54 pots were arranged in groups using a factorial design with sugarcane cultivars and the level of bud germination as two factors. There were six replications of each combination of the two factors. Identical experimental and data-collection procedures were used in both experiments, with the second experiment (March 2021) being a repetition of the first one (January 2021).

2.2. Field Experiment Setup

A field experiment was established in organic soils at EREC (Belle Glade, FL) in March 2022. The experimental design was a randomized complete block design with three treatments (bud germination levels) and four replications. The experimental area comprised of 4 blocks with 3 plots, and each plot was 4.5 m long and 3 rows wide. A 3-row wide alley and 6.09 m long gap was maintained between the blocks and each plot, respectively. With 30.2% of Florida's total sugarcane land under cultivation, CP 96-1252 is the state's most popular commercial sugarcane cultivar and, thus, was used for the field experimental trail [13]. At the time of planting, the sugarcane whole stalks with 3 levels of bud germination (Ungerminated, Pop-eyes, and Lalas) were harvested manually and these stalks were dropped in 15–20 cm deep furrows as pairs. This was followed by chopping the cane stalks into billets (60–90 cm long with 3–4 buds/billet). The furrows were then covered, and other management practices were conducted consistent with the standard commercial sugarcane cultivation in the organic soils of Florida. Data on plant population or tiller count were collected until 95 days after planting to determine the effect of bud germination level on emergence, early growth, and establishment.

2.3. Meteorological Conditions

Whiteman et al. [14] previously documented the environmental impacts on the germination stage in sugarcane. Increases in leaf area were directly associated with the increases in temperature. For sugarcane crop, in the early stages, a temperature of ~30 °C is ideal for plant and stalk growth [15]. Since climatic conditions can influence the crop growth rate, for this experiment, average temperatures and daily precipitation were obtained from the EREC weather station positioned 500 m or less from the experimental location (FAWN weather data). Average temperature and daily precipitation data are provided from January to October 2022 to cover the experimental period (March-June) (Figure 2).

2.4. Growth Measurements

In greenhouse studies, the length of the primary shoot was measured from the base of the plant to the base of the top visible dewlap (TVD) leaf to estimate plant height. Starting at 30 days after planting (DAP), data on primary shoot height was collected every week for 10 weeks. The total number of primary, secondary, and tertiary shoots (known as tillers) was counted at 30, 60, and 96 DAP. The leaf Soil Plant Analysis Development (SPAD) readings were recorded at 70 DAP by using a SPAD-502 chlorophyll meter (Konica Minolta, Tokyo, Japan). SPAD was measured at three spots (at the base of leaf, in the middle, and at the tip of leaf) on the TVD in each pot and the average value was calculated. During harvest at 100 DAP, the plants were uprooted, and the above ground biomass (stems and leaves) was cut from the base using sharp scissors. Similarly, roots were separated from seed pieces and washed thoroughly on a sieve of mesh 10 (2-mm holes) to remove adhering soil particles. The shoots and roots were air dried to remove excessive surface moisture and were weighed on a precision balance (Mettler Toledo Balance XPR204S) to collect data on fresh biomass. To estimate dry biomass, the shoots and roots were maintained in a drying room at 60 °C for 10 days until a constant weight was reached. The dry biomass of shoots and roots was measured using the same precision balance. For the field study, tillers (plant population) were counted at 36, 43, 53, and 95 DAP to determine emergence and early season crop establishment. There were no other data collection in this field experiment.

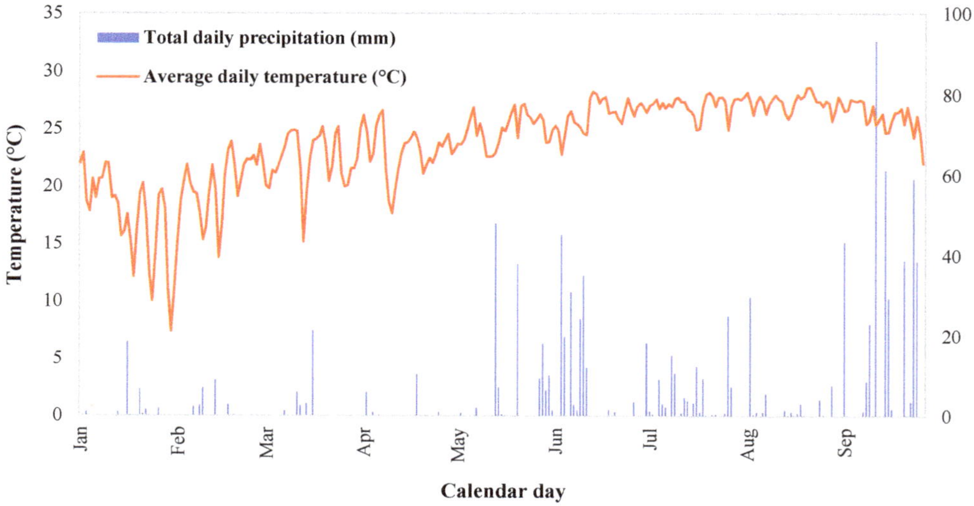

Figure 2. Average temperatures and daily precipitation during the experimental year 2022 (FAWN, 2022).

2.5. Statistical Analysis

Greenhouse data were analyzed using the Proc mixed model in SAS v 9.4 (SAS Institute, Cary, NC, USA). Sugarcane cultivar, bud germination level, and their interactions were considered as fixed effects. Replication and its interaction with other fixed factors were considered as random effects. Means were separated using Tukey's HSD test when treatments and interactions were significant at $p \leq 0.05$. Field data were analyzed using 2-way ANOVA using R programming language (R.4.0.0). Bud germination level treatment was considered as a fixed factor and replication was considered as a random factor. Means were separated using Least Common Difference (LSD) when treatments and interactions were significant at $p \leq 0.05$.

3. Results and Discussion

There was a significant experimental effect on most of the parameters measured in this study. Therefore, data from the first and second greenhouse experiments were analyzed separately. In the first greenhouse experiment, bud germination level showed a significant effect on the number of tillers (primary, secondary, and tertiary shoots) and shoot dry biomass. Cultivar had a significant effect on primary shoot height, SPAD, and root dry biomass, whereas the interaction between the bud germination level and the tested cultivars presented significant effect for SPAD values (Table 1). In the second greenhouse experiment, the bud germination level had a significant effect only on root dry biomass, and cultivar had a significant effect on primary shoot height and root dry biomass. However, there was no significant effect of the interaction cultivar and the bud germination level for any of the measured parameters (Table 1).

Table 1. ANOVA *p*-values for number of tillers and primary shoot height at 96 DAP, SPAD, shoot dry biomass, and root dry biomass in response to sugarcane cultivar and bud germination level.

Fixed Effects	Number of Tillers	Primary Shoot Height (cm)	SPAD	Shoot Dry Biomass (g)	Root Dry Biomass (g)
First Greenhouse Experiment					
Cultivar	0.1378	0.0012 *	0.0002 *	0.0598	0.0452 *
BD level	0.0100 *	0.1325	0.3501	0.0001 *	0.1393
Cultivar × BD level	0.9154	0.5797	0.0584 *	0.1185	0.9582
Second Greenhouse Experiment					
Cultivar	0.1687	<0.0001 *	0.8773	0.1521	<0.0001 *
BD level	0.1012	0.9216	0.9224	0.1942	0.0033 *
Cultivar × BD level	0.9022	0.8500	0.9240	0.3438	0.7147

Asterisk (*) denotes significant differences at $p \leq 0.05$, SPAD: soil plant analysis development, BD level: bud germination level.

3.1. Tiller Count

Lalas produced more tillers per pot (4.61) than Pop-eyes (3.21) and ungerminated buds (2.94) in the first greenhouse experiment, with no significant differences between Pop-eyes and ungerminated buds (Table 2). In the second greenhouse experiment, there was a trend toward increased tiller production with Lalas, but both bud germination level and cultivar had no significant effect on the number of tillers (Table 1). Tillering is primordial for sugarcane as it determines the number of millable canes (NMC). Among the parameters associated with sugarcane yield, the population of stalks present the highest correlation. Therefore, the profitability of the crop depends primarily on the tillers from which the stalks are formed, determining the final number of harvestable stalks [16]. Tillers are functional units, shoots with roots and stem, and leaves that become independent of the mother-shoot and may produce their own tillers, too [17]. Tillering is also influenced by genetic and environmental factors [18]. A higher tillering response in the Lalas and Pop-eye bud germination levels can be attributed to the already existing root and shoot structures. The developed leaves in Lalas and the protruding shoot structures in Pop-eyes can establish faster when planted in soil, and can have a head start with photosynthesis. This might have helped produce more tillers in the given period. In contrast, the ungerminated bud took longer to establish and then produce the primary shoot and tillers within the same time lapse. The genetic makeup of the cultivar also determines its tillering ability. Botanical traits such as bud length, leaf length, and leaf width are higher in CPCL 05-1201, and it has the tendency to produce more stalks (tillers) [12]. Similarly, the sugarcane cultivar CPCL 02-0926 is known to produce more tonnage and is a high yielding cultivar in muck soils [19]. These characteristics support the results where the cultivars CPCL 05-1201 and CPCL 02-0926 reported higher tiller production than CP 96-1252.

Table 2. Mean tiller production and shoot dry biomass in response to different bud germination levels in first greenhouse experiment [a].

Effect	Treatment	Tillers Pot^{-1}	Shoot Dry Biomass (g)
Bud germination level	Lalas	4.61 [a]	27.47 [a]
	Pop-eyes	3.21 [b]	19.80 [b]
	Ungerminated	2.94 [b]	17.74 [b]

[a] Means followed by different letters in a column are significantly different according to Tukey's HSD at $p \leq 0.05$.

In the field study, the bud germination level had a significant effect on the emergence and tiller production at the dates of data collection. Lalas had significantly higher tiller counts compared to Pop-eyes and ungerminated treatments (Table 3). However, there was no significant difference between Pop-eyes and ungerminated buds. Similar results in the field study validate the results of greenhouse studies. In this sense, Lalas present an interesting alternative for sugarcane establishment and production, given that profu tillering is considered as a good crop establishment feature that counteracts eventual hyd stresses and cumulative shooting and tillering failures in the following ratoons [16]. A hi number of tillers present a "physiologic compensatory continuum", which imparts t sugarcane plant an ability to overcome biotic and abiotic stresses [20].

Table 3. Mean number of tillers per hectare (Tillers ha^{-1}) in response to bud germination levels field experiment [a].

Treatment	36 DAP	43 DAP	53 DAP	95 DAP
Lalas	82,075 [a]	103,872 [a]	113,021 [a]	240,304 [a]
Pop-eyes	37,135 [b]	47,630 [b]	52,474 [b]	128,620 [b]
Ungerminated	29,332 [b]	41,172 [b]	47,092 [b]	157,960 [b]

[a] Means followed by different letters in a column are significantly different, DAP: days after planting.

3.2. Primary Shoot Height

Based on the primary shoot height data collected just before harvest, only cultiv had a significant effect, where CPCL 05-1201 reported taller primary shoots than the oth two cultivars in both greenhouse experiments (Figure 3). The temporal variation in t primary shoot height shows that CPCL 05-1201 was similar to the other two cultivars ear in the season (first couple of months), and then CPCL 05-1201 outpaced them (Figure Primary shoot height is an affordable visual observation parameter that could be easi used to characterize the variations in crop growth stages, especially in the grass family [2 Plant height was significantly affected by cultivar. CPCL 05-1201 reported in the secor greenhouse experiment the highest values compared to CPCL 02-0926 and CP 96-12 This could be attributed to genetic differences in plant height and some cultivars m slow in early growth compared to others. Similar findings were reported by Edmé [12] f CPCL 05-1201 cultivar, which reported a higher height in field plantings compared to t reference cultivars CP 78-1628 and CP 89-2143.

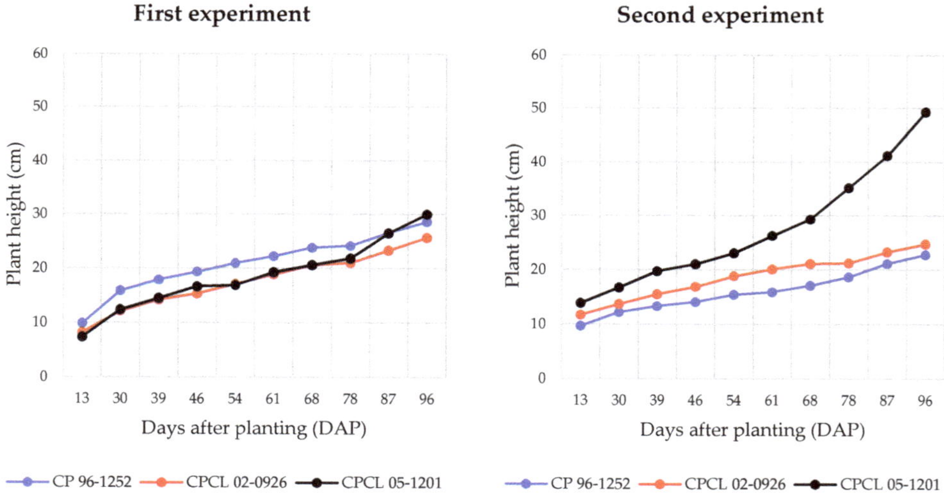

Figure 3. Primary shoot height of three sugarcane cultivars at different days after planting in the fir and second greenhouse experiment.

3.3. SPAD

Cultivar and its interaction with the bud germination level showed a significant effect SPAD values in the first greenhouse experiment (Table 1). SPAD is used to indicate relati chlorophyll content in the leaves of several crops, including sugarcane [22–24]. Among cultivars, CPCL 02-0926 had statistically significant higher SPAD values (38.11) than CPCL 05-1201 (33.36) and CP 96-1252 (31.98) (Table 4), which can be attributed to genetic differences, suggesting that this cultivar might be more tolerant to heat stress [25] and especially to water stress, according to several studies that have found a high correlation between a high SPAD index and drought tolerance in sugarcane [26–29]. Moreover, CPCL 02-0926 showed an adequate Nitrogen (N) leaf concentration compared with the other two cultivars.

Table 4. Mean SPAD in different sugarcane cultivars in first greenhouse experiment [a].

Effect	Treatment	SPAD
Cultivar	CP 96-1252	31.98 [b]
	CPCL 05-1201	33.36 [b]
	CPCL 02-0926	38.11 [a]

[a] Means followed by different letters in a column are significantly different according to Tukey's HSD at $p \leq 0.05$. CP: Canal Point, CPCL: Canal Point and Clewiston, SPAD: soil plant analysis development.

Among the cultivars and bud germination interactions, Lalas had higher SPAD values (36.18) than Pop-eyes (30.43) in CPCL 05-1201, but there was no significant difference between the three bud germination levels in the other two cultivars (Table 5). In the present study, SPAD readings showed a particular trend for the effects of the cultivars and their interaction with the bud germination level. Interestingly, it was found that Lalas presented an enhanced SPAD readings in cultivars CP 96-1252 and CP 05-1201, and did not significantly affect this parameter in CPCL 02-0926 (Table 5), being the last cultivar with the highest N concentration (Table 4). Taking into account that CP 96-1252 and CP 05-1201 reported SPAD readings below 34 (Table 4), Lalas interaction with these cultivars reached SPAD readings above 34, which is considered the suitable N concentration for sugarcane [30]. In other words, Lalas might represent a convenient effect for sugarcane establishment in terms of more efficient N uptake and therefore better tolerance to water stress.

Table 5. ANOVA p-values for SPAD in response to the interaction between cultivar and bud germination level in first greenhouse experiment [a].

Cultivar	Bud Germination Level	SPAD
CP 96-1252	Lalas	34.67 [bcd]
	Pop-eyes	30.23 [d]
	Ungerminated	31.04 [d]
CPCL 05-1201	Lalas	36.18 [abc]
	Pop-eyes	30.43 [d]
	Ungerminated	33.48 [cd]
CPCL 02-0926	Lalas	35.86 [abc]
	Pop-eyes	39.90 [a]
	Ungerminated	38.57 [ab]

[a] Means followed by different letters in a column are significantly different according to Tukey's HSD at $p \leq 0.05$. CP: Canal Point, SPAD: soil plant analysis development.

3.4. Shoot Dry Biomass

Shoot dry biomass was higher in Lalas than Pop-eyes and ungerminated buds in the first greenhouse experiment with no significant difference in the second greenhouse experiment. All cultivars produced similar shoot dry biomass in both experiments (Table 1). Higher shoot dry biomass in Lalas (27.47 g) is attributed to higher tiller count and plant height compared to the other bud germination levels (Table 6). Higher shoot dry biomass may eventually result in higher sugarcane yield at maturity [31]. Sugarcane physiology is poorly understood, but root–shoot relationships have the ultimate effect on the yield [32]. Milligan et al. [33] reported a positive correlation between cane yields with stalk characteristics (stalk number and stalk weight). In the present study, the positive response of shoot dry biomass for the Lalas bud germination level was observed. This statement can be supported by the fact that initial shoot growth of Lalas had a positive response on the total biomass production and accumulation compared to Pop-eye and ungerminated bud levels. Much of the energy produced from photosynthetic activity in Lalas is also consumed in developing new structures, such as new leaves, which contribute for the increased plant height and more biomass in each period compared to ungerminated bud level [17].

Table 6. Shoot dry biomass in response to different bud germination levels in the 2021 first greenhouse experiment [a].

Effect	Treatment	Shoot Dry Biomass (g)
Bud germination level	Lalas	27.47 [a]
	Pop-eyes	19.80 [b]
	Ungerminated	17.74 [b]

[a] Means followed by different letters in a column are significantly different according to Tukey's HSD at $p \leq 0.05$.

3.5. Root Dry Biomass

Root dry biomass was not significantly affected by the bud germination level in the first greenhouse experiment (Table 1), but in the second greenhouse experiment, both Lalas and Pop-eyes had a higher biomass than ungerminated buds (Table 7). The higher root dry biomass values for the bud germination level of Lalas over the other two germination levels (Table 7) were mostly attributed to the presence of pre-existing sett roots at the time of planting. From the discussion above, it is understood that Lalas's germination level is highly efficient in developing new shoot structures within a given period over the other two germination levels. This was similar in developing root structures. The developed sett roots in the first weeks of germination supply water and nutrients to the growing shoot and lead to the production of shoot roots in the later stages [34].

Table 7. Root dry biomass in different sugarcane bud germination levels in second greenhouse experiment [a].

Effect	Treatment	Root Dry Biomass (g)
Bud germination level	Lalas	13.40 [a]
	Pop-eyes	11.78 [a]
	Ungerminated	8.78 [b]

[a] Means followed by different letters in a column are significantly different according to Tukey's HSD at $p \leq 0.05$.

Among cultivars, CPCL 02-0926 had a higher root dry biomass than CP 96-1252 in the first greenhouse experiment. In the second greenhouse experiment, the trend was similar but CPCL 02-0926 had a significantly higher root dry biomass (14.85 g) compared to the other two cultivars, and even CPCL 05-1201 (12.03 g) had a significantly higher root dry biomass than CP 96-1252 (7.06 g) (Table 8). The data collected from both the trails indicated that sugarcane cultivars had a significant effect on root biomass. In both the trials, CPCL 02-0926 had higher root biomass, which can be attributed to the efficiency of the cultivar to establish itself in the muck soils.

Table 8. Root dry biomass in different sugarcane cultivars in first and second greenhouse experiments [a].

Effect	Treatment	Root Dry Biomass (g)	
		1st Experiment	2nd Experiment
Cultivar	CP 96-1252	15.74 [b]	17.06 [c]
	CPCL 05-1201	19.63 [ab]	12.03 [b]
	CPCL 02-0926	10.51 [a]	14.85 [a]

[a] Means followed by different letters in a column are significantly different according to Tukey's HSD at $p \leq 0.05$.

4. Conclusions

Sugarcane planting late in the season often encounters the challenge of poor seedcane quality, which may have been caused by environmental factors such as freeze, lodging due to a hurricane, etc. In the case of freeze damage to the growing point, lateral buds may start germinating into Pop-eyes and Lalas. In the current study, planting of pre-germinated buds (Pop-eyes and Lalas) represent an interesting alternative for efficient sugarcane propagation, as they may provide a head start in early sugarcane growth due to the already existing root and shoot structures. Planting seed cane with Lalas and Pop-eyes may not have any negative effect on yield. Moreover, Lalas show promising performance for a better sugarcane establishment given their higher tillering and shoot dry biomass at early growth (at 3 to 4 months after planting). However, it is important to consider that seedcane used in current greenhouse and field studies was cut manually and handled carefully to avoid any damage to pre-germinated buds. This may not be the case in commercial cane planting in which seedcane is cut mechanically, and mechanical cutting may cause greater damage to pre-germinated buds than ungerminated buds. In greenhouse studies, the Lalas were also kept out of the soil at planting in the pots. Therefore, further on-farm research needs to be conducted to confirm these results before using pre-germinated buds as potential seed source for late season planting of sugarcane.

Author Contributions: Conceptualization, H.V.M. and H.S.S.; Methodology, H.V.M., H.S.S., L.K.S. and C.D.O.; Validation, H.V.M., H.S.S., L.K.S. and C.D.O.; Formal analysis, H.V.M., H.S.S. and R.A.L.-V.; Investigation, H.V.M. and H.S.S.; Data curation, H.V.M., H.S.S. and R.A.L.-V.; Writing—original draft preparation, H.V.M.; Writing—review and editing, H.V.M., H.S.S., R.A.L.-V., L.K.S. and C.D.O.; Visualization, H.V.M.; Supervision, H.S.S.; Project Administration, H.V.M. and H.S.S.; Funding acquisition, H.S.S. All authors have read and agreed to the published version of the manuscript.

Funding: This project was partially funded by USDA-NIFA Hatch project (FLA-ERC-006028) and Florida Sugar Cane League.

Data Availability Statement: Data will be made available at request.

Acknowledgments: We would like to acknowledge Everglades Research and Education sugarcane agronomy lab (Jairo Sánchez, Daniela Jiménez, Anuj Chiluwal, Saroop Sandhu) for assisting in the experimental greenhouse and field setup.

Conflicts of Interest: The authors declare no conflict of interest.

eferences

USDA-NASS. *Crop Production*; National Agricultural Statistics Service (NASS), Agricultural Statistics Board, United States Department of Agriculture (USDA): Washington, DC, USA, February 2022. Available online: https://www.nass.usda.gov/Publications/Todays_Reports/reports/crop0922.pdf (accessed on 10 September 2022).

Sandhu, H.S.; Miller, J.D.; Gilbert, R.A.; Odero, D.C. *Sugarcane Botany: A Brief View*; University of Florida IFAS Extension: Gainesville, FL, USA, 2019; SS-AGR-234. Available online: https://edis.ifas.ufl.edu/publication/SC034 (accessed on 10 September 2022).

Zeiger, E.; Taiz, L.; Moller, I.M.; Murphy, A. *Plant Physiology*, 6th ed.; Sinauer: Sunderland, MA, USA, 2014.

Kochhar, S.L.; Gujral, S.K. *Plant Physiology: Theory and Applications*, 2nd ed.; Cambridge University Press: Cambridge, UK, 2020.

Clements, H.F. *Flowering of Sugarcane: Mechanics and Control*; Hawaii Agricultural Experiment Station, University of Hawaii: Honolulu, HI, USA, 1975; Volume 92.

Otto, R.; Machado, B.A.; da Silva, A.C.M.; de Castro, S.G.Q.; Lisboa, I.P. Sugarcane pre-sprouted seedlings: A novel method for sugarcane establishment. *Field Crop. Res.* **2022**, *275*, 108336. [CrossRef]

dos Santos, S.G.; Chaves, V.A.; Ribeiro, F.D.S.; Alves, G.C.; Reis, V.M. Rooting and growth of pre-germinated sugarcane seedlings inoculated with diazotrophic bacteria. *Appl. Soil Ecol.* **2019**, *133*, 12–23. [CrossRef]

De Oliveira, H.P.; De Melo, R.O.; Baldotto, M.A.; Andrade, M.A.; Baldotto, L.E.B. Performance of pre-sprouted sugarcane seedlings in response to the application of humic acid and plant growth-promoting bacteria. *Semin. Ciênc. Agrár.* **2018**, *39*, 1365–1370. [CrossRef]

Gírio, L.A.D.S.; da Silva, R.P.; de Menezes, P.C.; Carneiro, F.M.; Zerbato, C.; Ormond, A.T.S. Quality of multi-row harvesting in sugarcane plantations established from pre-sprouted seedlings and billets. *Ind. Crop. Prod.* **2019**, *142*, 111831. [CrossRef]

. Glynn, N.C.; Zhao, D.; Comstock, J.C.; Milligan, S.B.; Glaz, B.; Edmé, S.J.; Davidson, R.W.; Gilbert, R.A.; Sood, S.; D'Hont, A.; et al. Registration of 'CPCL 02-0926′ Sugarcane. *J. Plant Regist.* **2013**, *7*, 164–171. [CrossRef]

11. Edme, S.; Tai, P.; Glaz, B.; Gilbert, R.; Miller, J.; Davidson, J.; Dunckelman, J.; Comstock, J. Registration of 'CP 96-1252' Sugarca *Crop. Sci.* **2005**, *45*, 423–424. [CrossRef]
12. Edmé, S.J.; Davidson, R.W.; Zhao, D.; Comstock, J.C.; Sandhu, H.S.; Glaz, B.; Milligan, S.; Hu, C.-J.; Sood, S.; McCorkle, K.; et Registration of 'CPCL 05-1201' Sugarcane. *J. Plant Regist.* **2016**, *10*, 14–21. [CrossRef]
13. Sandhu, H.; Davidson, W. Sugarcane Cultivars Descriptive Fact Sheet: CP 96-1252, CP 01-1372 and CP 00-1101. *EDIS* **20** Available online: https://edis.ifas.ufl.edu/publication/SC102 (accessed on 10 September 2022). [CrossRef]
14. Whiteman, P.; Bull, T.; Glasziou, K. The Physiology of Sugar-Cane VI. Effects of Temperature, Light, and Water on Set Germinati and Early Growth of Saccharum Spp. *Aust. J. Biol. Sci.* **1963**, *16*, 416–428. [CrossRef]
15. Glasziou, K.; Bull, T.; Hatch; Whiteman, P. Physiology of Sugar-Cane VII. Effects of Temperature, Photoperiod Duration, a Diurnal and Seasonal Temperature Changes on Growth and Ripening. *Aust. J. Biol. Sci.* **1965**, *18*, 53–66. [CrossRef]
16. Matsuoka, S. Sugarcane tillering and ratooning: Key factors for a profitable cropping. *Sugarcane Prod. Cultiv. Uses* **2012**, *5*, 137–1
17. Shrivastava, A.K.; Solomon, S.; Rai, R.K.; Singh, P.; Chandra, A.; Jain, R.; Shukla, S.P. Physiological Interventions for Enhanci Sugarcane and Sugar Productivity. *Sugar Tech* **2015**, *17*, 215–226. [CrossRef]
18. Chattha, M.; Ali, A.; Chattha, M. Influence of planting techniques on growth and yield of spring planted sugarcane (*Sacchar officinarum* L.). *Pak. J. Agric. Sci.* **2007**, *44*, 452–455.
19. Sandhu, H.; Davidson, W. Sugarcane Cultivar Descriptive Fact Sheet: CPCL 02-0926 and CP 05-1526. *EDIS* **2019**, *2019*, [CrossRef]
20. Shrivastava, A.K. Yield decline and its management in sugarcane. In *Management and Agro-Industrial Imperatives*; Solomon, Grewal, S.S., Li, Y., Magarey, R.C., Rao, G.P., Eds.; International Book Distributing Co.: Lucknow, India, 2004; pp. 321–345.
21. Bendig, J.; Bolten, A.; Bennertz, S.; Broscheit, J.; Eichfuss, S.; Bareth, G. Estimating Biomass of Barley Using Crop Surface Mod (CSMs) Derived from UAV-Based RGB Imaging. *Remote Sens.* **2014**, *6*, 10395–10412. [CrossRef]
22. Zhao, D.; Irey, M.; LaBorde, C.; Hu, C. Identifying Physiological and Yield-Related Traits in Sugarcane and Energy Cane. *Agron.* **2017**, *109*, 927–937. [CrossRef]
23. Mearaji, H.S.; Ansari, A.; Igdelou, N.K.M.; Lajayer, B.A.; Mohammad, P. Phytohormones and Abiotic Stresses Roles of Phytoh mones in Plants under Abiotic Stresses. Nader Khadem Moghadam Igdelou, Behnam Asgari Lajayer, and Mohammad Pessara In *Handbook of Plant and Crop Physiology*, 4th ed.; Mohammad, P., Ed.; CRC Press: Boca Raton, FL, USA, 2021; pp. 175–213.
24. Clark, A.; Landolt, W.; Bucher, J.; Strasser, R. Beech (*Fagus sylvatica*) response to ozone exposure assessed with a chlorophyl fluorescence performance index. *Environ. Pollut.* **2000**, *109*, 501–507. [CrossRef]
25. Kohila, S.; Gomathi, R. Adaptive physiological and biochemical response of sugarcane genotypes to high-temperature stre *Indian J. Plant Physiol.* **2018**, *23*, 245–260. [CrossRef]
26. Silva, M.D.A.; Jifon, J.L.; Sharma, V.; Da Silva, J.A.G.; Caputo, M.M.; Damaj, M.B.; Guimarães, E.R.; Ferro, M.I.T. Use Physiological Parameters in Screening Drought Tolerance in Sugarcane Genotypes. *Sugar Tech* **2011**, *13*, 191–197. [CrossRef]
27. Silva, M.D.A.; Jifon, J.L.; Da Silva, J.A.G.; Dos Santos, C.M.; Sharma, V. Relationships between physiological traits and productiv of sugarcane in response to water deficit. *J. Agric. Sci.* **2014**, *152*, 104–118. [CrossRef]
28. Silva, M.D.A.; Jifon, J.L.; Dos Santos, C.M.; Jadoski, C.J.; Da Silva, J.A.G. Photosynthetic capacity and water use efficiency sugarcane genotypes subject to water deficit during early growth phase. *Braz. Arch. Biol. Technol.* **2013**, *56*, 735–748. [CrossRef]
29. Dinh, T.H.; Takaragawa, H.; Watanabe, K.; Nakabaru, M.; Kawamitsu, Y. Leaf Photosynthesis Response to Change of Soil Moistu Content in Sugarcane. *Sugar Tech* **2019**, *21*, 949–958. [CrossRef]
30. Jaroenseng, S.; Wongmaneeroj, A.; Amkhab, S. Results of studies the possible correlations between SPAD value and total nitrog contents in the leaves of sugarcane (*Saccharum officinarum* L.). *J. ISSAAS [Int. Society Southeast Asian Agric. Sci.]* **2013**, *16*, 136.
31. Bell, M.; Garside, A. Shoot and stalk dynamics and the yield of sugarcane crops in tropical and subtropical Queensland, Austral *Field Crop. Res.* **2005**, *92*, 231–248. [CrossRef]
32. Otto, R.; Franco, H.C.J.; Faroni, C.E.; Vitti, A.C.; de Oliveira, E.C.A.; Sermarini, R.A.; Trivelin, P.C.O. The Role of Nitrog Fertilizers in Sugarcane Root Biomass under Field Conditions. *Agric. Sci.* **2014**, *05*, 1527–1538. [CrossRef]
33. Milligan, S.B.; Gravois, K.A.; Bischoff, K.P.; Martin, F.A. Crop Effects on Genetic Relationships Among Sugarcane Traits. *Crop. S* **1990**, *30*, 927–931. [CrossRef]
34. Smith, D.; Inman-Bamber, N.; Thorburn, P. Growth and function of the sugarcane root system. *Field Crop. Res.* **2005**, *92*, 169–1 [CrossRef]

agronomy

MDPI

Article

Physiological Quality of Soybean Seeds as a Function of Soil Management Systems and Pre-Harvest Desiccation

Gustavo Ferreira da Silva *, Larissa Chamma, Bruno Cesar Ottoboni Luperini, Prínscilla Pâmela Nunes Chaves, Juliano Carlos Calonego, João Nakagawa and Edvaldo Aparecido Amaral da Silva

Department of Crop Science, School of Agriculture, São Paulo State University (UNESP), Botucatu 18610-034, Brazil

* Correspondence: gustavo.ferreira@unesp.br

Abstract: Soil management systems directly interfere in the soil–plant relationship. However, there are still few studies evaluating the influence of long-term management systems on the physiological quality of soybean seeds. Another little-known topic is the influence of pre-harvest desiccation on the physiological quality of soybean seeds, especially on seed longevity. Thus, the aim of this research was to evaluate the physiological quality of soybean seeds cultivated under conventional tillage and no-tillage systems with and without desiccant use. The experiment was carried out in design is a split plot in a randomized complete block design. The treatments consisted of soil management systems (conventional tillage and no-tillage), with and without pre-harvest desiccation. In the treatment with desiccation, the herbicide Paraquat was applied, when the plants were at the R7.3 phenological stage (most of the seeds had a yellowish coat, with a shiny surface and were already detached from the pod). Seed germination, vigor (first germination count, seedling dry mass, seedling length, time to reach 50% germination (T50), seedling emergence and emergence speed index) and longevity (P50) were evaluated. Seeds cultivated under conventional tillage showed greater vigor for most traits evaluated, with values of T50 and seedling length higher by 24.39% and 24.77%, respectively, compared to NT. In addition, non-desiccation increased the seedling length and dry mass, in 15.45% and 21.59%, respectively. The use of desiccant aiming at seed vigor is dependent on the soil management system. Soybean seed longevity was superior in the no-tillage system, but desiccant application reduced seed longevity.

Keywords: conventional tillage; desiccation; *Glycine max*; longevity; no-tillage; soybean seed

Citation: Silva, G.F.d.; Chamma, L.; Luperini, B.C.O.; Chaves, P.P.N.; Calonego, J.C.; Nakagawa, J.; Silva, E.A.A.d. Physiological Quality of Soybean Seeds as a Function of Soil Management Systems and Pre-Harvest Desiccation. *Agronomy* **2023**, *13*, 847. https://doi.org/10.3390/agronomy13030847

Academic Editors: Umberto Anastasi and Aurelio Scavo

Received: 14 February 2023
Revised: 7 March 2023
Accepted: 11 March 2023
Published: 14 March 2023

1. Introduction

Soybean [*Glycine max* (L.) Merill] has a great impact on the world economy, so any factor that interferes with its development and yield has become relevant. Thus, soil management systems are important to soybean farming, as they aim to promote ideal soil conditions for crop development [1]. There are many different management systems; however, conventional tillage (CT) and no-tillage (NT) are the most common [2].

CT is characterized by soil turning, usually to a depth of 0.20–0.30 m, promoting weed control and satisfactory conditions for crop development. However, due to soil mobilization, plant residues are incorporated, which accelerates the straw decomposition process. In addition, the soil becomes more susceptible to water erosion [3–5].

NT represents a milestone in water and soil conservation in agricultural systems, due to its precepts, such as no soil turning, crop rotation and soil maintenance through straw covering [6]. This system promotes positive alterations in the physical, chemical and biological soil quality, directly interfering in soybean productivity [7–9].

Although there are many reports about the influence of soil management systems on their properties and in soybean yield [10,11], little is known about the influence of NT and CT on soybean physiological quality, mainly seed longevity.

Physiological quality is acquired during the phases of seed development and comprises the acquisition of germination, vigor and storage capacity (longevity) [12], during embryogenesis, grain filling (maturation) and late maturation [13].

Physiological maturity is characterized by the moment when the seed no longer receives nutrients from the mother plant [14]. However, research has revealed that for soybean, the process of acquiring physiological quality continues after the seed is disconnected from the plant [15], making evident the importance of the late maturation phase for the acquisition of seed longevity [16]. During the late maturation phase, longevity increases two-fold until the R9 stage, which corresponds to a mature seed [12,17].

Despite the importance of the late maturation process in soybean seeds for the complete acquisition of seed physiological quality, the use of desiccants in seed production is still common. Desiccation is usually carried out at stages R7.2 (plants with 51% to 75% of leaves and pods yellow) or R7.3 (plants with more than 76% of leaves and pods yellow), with the intent of anticipating the harvest, in order to reduce the seeds exposure to climatic conditions and attacks of pests and diseases at the end of the crop cycle [18–20].

The desiccation process provides advantages related to seed moisture reduction, maturation uniformity and, mainly, and preservation of seed physiological quality, due to the shorter period of exposure in the field, minimizing the irreversible damages of deterioration by moisture [20–22]. However, current research shows that the use of desiccants can reduce the quality of soybean seeds [23–25]. In addition to this factor, it is still unknown whether the soil management system interferes with the soybean seed response to the desiccation process. Thus, the hypothesis of this research is that lower soil thermal amplitude, greater soil moisture conservation and nutrient cycling provided by straw and crop rotation in the no-tillage system, are less stressful for plants and favor greater nutrient absorption, thus contributing to better acquisition of seed physiological quality even under desiccation use. Therefore, the aim of this research was to evaluate the physiological quality of soybean seeds cultivated under a conventional tillage and a no-tillage system with and without desiccant use.

2. Materials and Methods

2.1. Site Description and Experimental Area

The field experiment was carried out at an experimental farm located at Botucatu, SP, Brazil (22°48′57″ S, 48°25′41″ W; 786 m a.s.l.), on a typical Rhodudalf soil, classified as clayey-textured, with chemical and textural characteristics shown in Table 1.

Table 1. Chemical, granulometric and physic soil analysis of seed production experimental area in conventional soil tillage system (CT) and no-tillage (NT), at 0.00–0.20 m depth.

Management System	pH	OM	P	S	H + Al	Ca	Mg	K	Sand	Silt	Clay
	$CaCl_2$	mg dm^{-3}				mmol$_c$ dm^{-3}				g kg^{-1}	
CT	5.0	27.1	61.2	3.6	36.3	39.5	12.7	4.7	147	239	614
NT	5.4	30.9	84.4	4.4	29.6	43.5	14.8	3.3			
	Mac	Mic	TP	Bd		PR					
	cm^3 cm^{-3}			g cm^{-3}		MPa					
CT	0.09	0.44	0.53	1.15		1.62					
NT	0.07	0.44	0.51	1.33		3.33					

pH: active acidity; OM: organic matter; P: exchangeable phosphorus; S: sulphur; H + Al: potential acidity; Ca: exchangeable calcium; Mg: exchangeable magnesium; K: exchangeable potassium; Mac: macroporosity; Mic: microporosity; TP: total porosity; Bd: bulk density; PR: penetration resistance; $CaCl_2$: 0.01 M calcium chloride solution; mg dm^{-3}: milligram per cubic decimeter; mmol$_c$ dm^{-3}: millimol charge per cubic decimeter; g kg^{-1}: gram per kilogram; cm^3 cm^{-3}: cubic centimeter per cubic centimeter; g cm^{-3}: gram per cubic centimeter; MPa: megapascal.

The study used seeds produced in a long-term experimental field, which has been used since 1985 under conventional tillage (CT) and no-tillage (NT) systems. In plots managed under CT, soil preparation is carried out with plowing and harrowing, as shown in Table 2. AIn a NT system, the soil has not been disturbed since 1985, as shown in Table 2.

Table 2. Soil management systems and crop succession used since 1985, highlighting management and species cultivated in fall-winter and spring-summer seasons of each agricultural year.

Year	Management System: Conventional Tillage Fall	Conventional Tillage Spring	No-Tillage Fall	No-Tillage Spring	Season Fall-Winter/Spring-Summer
1985/86	Plowing + harrowing	Plowing + harrowing	Plowing + harrowing	No-tillage	Wheat/soybean
1986/87 to 1994/95	Plowing + harrowing	Plowing + harrowing	No-tillage	No-tillage	Wheat/soybean
1995/96 to 1998/99	Without soil mobilization	Without soil mobilization	No-tillage	No-tillage	Fallow/fallow
1999/00	Plowing + harrowing	Plowing + harrowing	No-tillage	No-tillage	Black oat/maize
2000/01 and 2001/02	Without soil mobilization	Without soil mobilization	No-tillage	No-tillage	Fallow/fallow
2002/03 and 2003/04	Plowing + harrowing	Plowing + harrowing	No-tillage	No-tillage	Black oat/millet-bean
2004/05 and 2005/06	Plowing + harrowing	Plowing + harrowing	No-tillage	No-tillage	Black oat/maize
2006/07	Without soil mobilization	Without soil mobilization	No-tillage	No-tillage	Fallow/soybean
2007/08	Plowing + harrowing	Without soil mobilization	No-tillage	No-tillage	Yellow oat/bean
2008/09	Plowing + harrowing	Without soil mobilization	No-tillage	No-tillage	Yellow oat/bean
2009/10 to 2011/12	Plowing + harrowing	Without soil mobilization	No-tillage	No-tillage	Black oat/maize + brachiaria
2012/13	Without soil mobilization	Plowing + harrowing	No-tillage	No-tillage	Brachiaria/soybean
2013/14	Without soil mobilization	Plowing + harrowing	No-tillage	No-tillage	Wheat/soybean
2014/15	Without soil mobilization	Plowing + harrowing	No-tillage	No-tillage	Safflower/soybean
2015/16	Without soil mobilization	Plowing + harrowing	No-tillage	No-tillage	Safflower/maize
2016/17	Plowing + harrowing	Without soil mobilization	No-tillage	No-tillage	Black oat/maize
2017/18	Plowing + harrowing	Without soil mobilization	No-tillage	No-tillage	Black oat/soybean
2018/19	Plowing + harrowing	Without soil mobilization	No-tillage	No-tillage	Sorghum/soybean
2019/20	Plowing + harrowing	Without soil mobilization	No-tillage	No-tillage	Sorghum/soybean

The region climate, according to the Köppen classification, is CWa type, mesothermic climate with dry winter. The data on maximum, average and minimum temperatures and rainfall during the period of conducting the experiments in the crop seasons 2018/19 and 2019/2020 are shown in Figure 1.

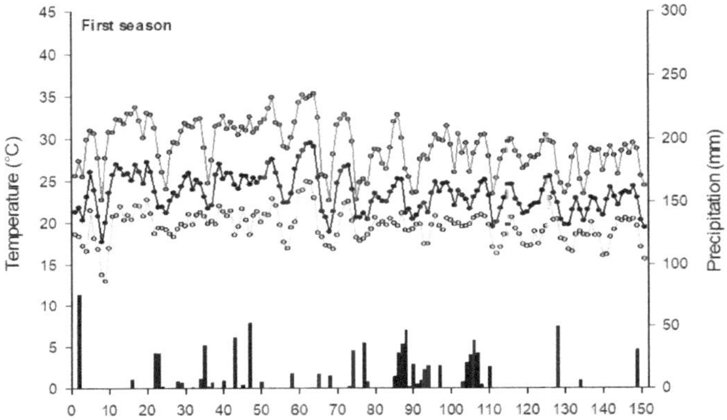

Figure 1. *Cont.*

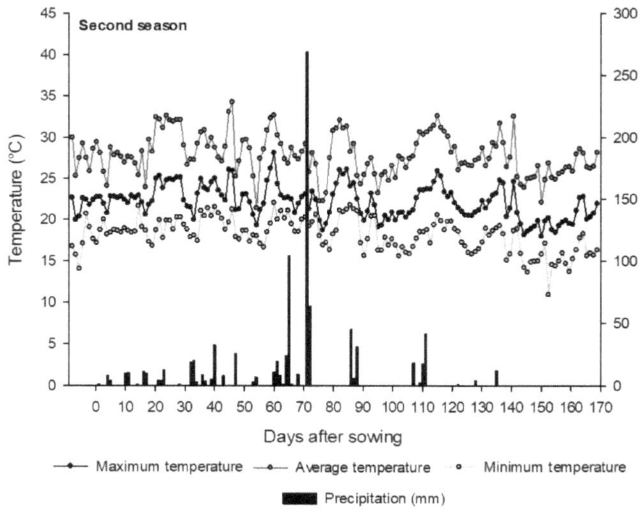

Figure 1. Maximum, average and minimum temperature and rainfall during the two soybea crop seasons.

2.2. Experimental Design and Management

The experiment was carried out in a randomized block design, in a split-plot schem with four replications. The plots (50 m × 9 m) were constituted of two subplots (25 m × 9 m each with a different soil management system, conventional tillage (CT) and no-tillag (NT). These subplots were further divided into two, one with desiccant application (D) an one without (ND).

The experiment was performed during the 2018/19 (first season) and 2019/20 (secon season) seasons, using soybean cultivar TMG 7062. The first season sowing was carrie out on 17 December 2018 and the second on 6 December 2019, using a spacing of 0.45 r between rows, aiming for a density of 300 thousand plants ha^{-1}, using seeds treated wit fungicide Carboxin + Thiram, insecticide Thiamethoxam, inoculant *Bradyrhizobium* sp., an micronutrients Co and Mo. Sowing fertilization was conducted with 60 kg ha^{-1} of K_2O an 60 kg ha^{-1} of P_2O_5, using KCl and simple superphosphate, respectively. In both season soybean was cultivated in succession to sorghum cultivated during fall-winter (Table 2 which supplied the straw for the NT system (NT: 4500 and 4150 kg ha^{-1} of straw in the fir and second season, respectively; CT: 3800 and 3200 kg ha^{-1} of straw in the first and secon season, respectively). In CT, the soil was turned over only in April, before the sorghur sowing, with a harrow at a depth of 0.00–0.20 m (Table 2).

Soybean phytosanitary management involved weed control with the application of he bicide Glyphosate (1.8 kg a.i. ha^{-1}) associated with herbicide Sethoxidim (1.25 kg a.i. ha^{-1} The fungicides Pyraclostrobin + Epoxiconazole (0.08 + 0.03 kg a.i. ha^{-1}, respectively) an Azoxystrobin + Cyproconazole (0.06 + 0.024 kg a.i. ha^{-1}, respectively) and the insecticide Thiamethoxam + Lambda-Cialotrin (0.028 + 0.21 kg a.i. ha^{-1}) were applied preventively

Pre-harvest desiccation was carried out using herbicide paraquat (0.4 kg a.i. ha^{-} 200 L ha^{-1} of spray volume) at the R7.3 stage, when most of the seeds had a yellowis coat, with a shiny surface and were already detached from the pod [11,25,26]. The see water content was of 55 ± 1% and more than 76% of the leaves and pods of the plant in the field were yellow. The desiccant application was carried out with a Jacto Falco AM14/Vortex sprayer, with flat jet tips (fan) model ADI 11002, without wind and with a air temperature of 20 °C.

In the ND treatment, seeds were harvested when they reached the R9 stage, known a the harvest point, in which the seeds have a dry appearance and a water content belo

15% [12,26,27]. Phenological stage characterization at the time of harvest considered the visual characteristics of the plants and seeds.

In the first season, harvest of treatments with desiccation (D) was carried out on 13 March 2019 and the treatments without desiccation (ND) on 16 March 2019; in the second season the desiccated plants (D) were harvested on 15 March 2020 and the non-desiccated (ND) on 19 March 2020. It should be noted that, during this period that the plants remained in the field, after the plant desiccation, there was no rain, and the climatic conditions were similar to the day of harvesting the seeds of the treatment with desiccation (Figure 1).

The pods were harvested and threshed manually and, later, the seeds were stored in a cold chamber (10 °C and 40% relative humidity) for 15 days to stabilize the water content.

2.3. Seed Quality Assessment

After the storage period in a cold chamber, the seed water content was evaluated by the oven method, at 105 ± 3 °C for a period of 24 h [28], with three replications of 15 seeds. Results were expressed as percentage of water on a wet basis.

The germination test was performed with four replications of 25 seeds, using a roll of paper moistened with distilled water equivalent to 2.5 times the dry mass of the paper. The rolls were placed in a germinator at 25 °C. The germination percentage was scored by counting normal seedlings at five (first germination count) and eight days (total germination) [28].

To evaluate radicle protrusion, four replicates of 25 seeds were used, arranged in Petri dishes, using three sheets of filter paper as substrate, which were moistened with distilled water equivalent to 2.5 times the paper dry mass. The Petri dishes were placed in a germinator at 25 °C. The evaluations were carried out every 6 h, counting the number of seeds that presented a radicle with two millimeters of length. The time required to reach 50% germination of viable seeds (T50) was calculated by analyzing the cumulative germination data using the curve-fitting model of the Germinator software [29] and the results were expressed in hours.

For the length and dry mass of seedlings, four replications of 10 seeds from each batch were used, arranged in a roll of paper moistened with distilled water equivalent to 2.5 times the paper dry mass. The seeds were arranged on a line drawn longitudinally in the upper third of the paper, with the seed hilum facing the lower portion of the paper, in order to guide the seedling growth in a straighter line [30]. Paper rolls were conditioned and tilted at 90° in a germinator at 25 °C in the dark. The seedling average length was measured on the seventh day after the beginning of the test. After analyzing the length, the seedlings were kept in an air circulation oven at a temperature of 60 °C to obtain the seed dry weight and the result were expressed in grams.

To evaluate the emergence speed index (ESI), daily counts were performed, considering as emerged seedlings those whose cotyledons were above ground level, at an angle greater than or equal to 90° in relation to the seedling stem, until there were no more emergences. The ESI test was performed on sand under field conditions. Data were submitted to the formula proposed by Maguire [31], in which:

$$ESI = E1/N1 + E2/N2 + \ldots\ En/Nn \tag{1}$$

where: E1, E2, ... En, refers to the number of emerged seedlings computed in the first, second and last counts; N1, N2, ... Nn refers to the number of days from sowing to the first, second and last count. At the end of the test, the total number of emerged seedlings was determined.

For longevity assessment, seeds of each lot were kept at 75% RH (using saturated NaCl solution) and 20 °C for 24 h and then stored in airtight boxes with a saturated NaCl solution and stored at 35 °C and 75% RH [32]. Germination of the seeds was evaluated from the fifth day after beginning of storage until the loss of protrusion capacity. For this, at each moment of evaluation, 25 seeds were removed from each treatment to mount the germination test. Since the intervals between evaluations were shorter at the beginning of storage, and from the moment that a marked loss of germination capacity of the seeds was verified, the tests were carried out at intervals of longer days. From the results obtained, the sigmoidal curve was analyzed for each batch and the longevity was expressed in P50 (time in days to loss 50% of viability) [27].

2.4. Determination of Nutrient in the Seeds

Sulfuric digestion was used to obtain an extract in order to determine the content of N, while P, K, Ca, Mg, S, Cu, Fe, Zn and Mn content in seeds were extracted by nitroperchloric acid digestion and determined by atomic absorption spectrophotometry, as described by AOAC [33].

2.5. Statistical Analysis

Data were tested for normality using the Anderson–Darling test, and homoscedasticity was checked using the Hartley test. Data were submitted to the analysis of variance and the mean differences were discriminated by *t*'s test at 5% probability.

The variables of physiological quality and nutrient content of the seeds were submitted to Pearson's correlation analysis, at the level of 5% of probability. All statistical analyzes were performed in the R software version 3.6.1.

3. Results

The seed water content ranged from 9.0 to 9.8%, without significant differences. The soil management system and desiccant application, in both seasons, did not influence soybean seed germination. However, there was significance for the isolated factors in the vigor and longevity of the seeds (Table 3), but there was no significant interaction between the tested factors.

Table 3. Mean values of germination (%), first germination count (FGC—%), time to reach 50% germination (T50—hours), seedling length (cm), seedling dry mass (g), emergence speed index (ESI) and mean time to loss of 50% viability during storage (P50—days) of soybean seeds as a function of the soil management system and desiccant application.

	First Season		Second Season	
	Conventional Tillage	**No-Tillage**	**Conventional Tillage**	**No-Tillage**
Germination	91.7 a	89.0 a	92.0 a	89.0 a
FGC	91.0 a	80.5 b	83.0 a	78.0 a
T50	48.0 a	54.4 a	57.0 a	43.1 b
Length	21.76 a	23.6 a	21.8 a	16.4 b
Dry mass	0.3405 a	0.3609 a	*	*
ESI	*	*	9.09 a	7.84 b
P50	40.84 a	41.03 a	31.21 b	44.30 a
	With desiccant	**Without desiccant**	**With desiccant**	**Without desiccant**
Germination	90.0 a	90.8 a	91.5 a	89.5 a
FGC	86.8 a	84.8 a	84 a	77 a
T50	49.7 a	52.8 a	49.2 a	50.8 a
Length	20.8 b	24.6 a	19.0 a	19.1 a
Dry mass	0.3083 b	0.3932 a	*	*
ESI	*	*	8.17 a	8.76 a
P50	37.20 b	44.67 a	33.72 b	41.80 a

Means followed by the same lowercase letter on the row do not differ from each other by the *t* test at the 5% probability level. *: there was no isolated significance of the factors.

Regarding seed vigor, the soil management systems affected the variables of first germination count (FGC), time for 50% of seeds to germinate (T50), seedling length and emergence speed index (ESI).

In the first season, the results of FGC in CT were higher than those obtained in NT. In the second season, there was no difference between soil management systems ($p > 0.05$), even with 6.02% more seeds germinated in FCG (Table 3) in CT.

The T50 and seedling length results were influenced by the soil management system only in the second season, with higher values of 24.39% and 24.77%, respectively, in seeds produced in CT (Table 3). Seeds produced in CT also showed higher ESI values in the second season.

Regarding the desiccation factor, the effect on seed vigor was verified in the variables of length and dry mass of seedlings in the first season. In this season, the ND treatment favored seedling growth by 15.45% and dry mass accumulation by 21.59% (Table 3), when compared to seedlings from seeds produced with pre-harvest desiccation.

Dry mass (second season), emergence (first and second seasons) and ESI (first season) showed a significant interaction between soil management and desiccant application factors (Table 4).

Table 4. Means values of dry mass of seedling (g), emergence of seedlings (%) and emergence speed index of soybean seedlings from seeds produced under different soil management systems and desiccant application.

	Dry Mass (g)			
	First Season		Second Season	
	With Desiccant	Without Desiccant	With Desiccant	Without Desiccant
Conventional tillage	ns	ns	0.3058 bB	0.4143 aA
No-tillage	ns	ns	0.4312 aA	0.4493 aA
	Emergence (%)			
Conventional tillage	56.5 bB	87.0 aA	51.0 bB	78.5 aA
No-tillage	72.5 aA	57.5 bB	65.5 aA	51.5 bB
	Emergence speed index			
Conventional tillage	7.02 bB	9.88 aA	ns	ns
No-tillage	6.99 aA	5.67 bB	ns	ns

Means followed by the same lowercase letter in the column and uppercase in the row do not differ from each other by the t test at the 5% probability level. ns: there was no significant interaction between the factors.

In the second season, desiccation reduced the dry mass of seedlings by 26% in CT, but did not affect this variable in NT. In the comparison between management systems, significant differences ($p < 0.05$) were obtained only in treatments with desiccation, where the dry mass of seedlings in chemical management in CT was 29% lower than that obtained in NT (Table 4).

A similar behavior was observed for emergence (first and second seasons) and ESI (first season), where in CT, the desiccant application reduced emergence and ESI by approximately 35% and 29%, respectively. In NT with desiccant application, these variables increased by 21% and 19%, respectively (Table 4).

In general, conventional soil preparation provides more vigorous seed production. However, the response to desiccant application varies depending on the soil management system. In CT there is better seed quality without application, while in NT the desiccant application favors the soybean seed vigor (Tables 3 and 4).

However, the same behavior was not observed in seed longevity (P50). For this trait, the soil management system influenced the response only in the second season, with NT providing a 30% increase in seed storage time without affecting their quality, compared to conventional soil preparation (Table 3). The desiccant factor showed a difference in both seasons, and the absence of desiccant increased seed longevity by 16.72% in the first season and 19.33% in the second season (Table 3).

The physiological quality results of soybean seeds as a function of soil management and desiccant application can be explained by the correlation between the physiological quality variables of the seeds and the nutrient contents exported to them, as can be see in Table 5.

Table 5. Correlation between variables of seed physiological quality and macro and micronutrie content in soybean seeds produced in different soil management systems and desiccant applicatio

	G	FGC	L	DM	E	ESI	T50	P50
				First Season				
Ca	0.070	0.093	0.084	0.397	0.408	0.541 *	−0.338	0.659 **
Mg	−0.108	−0.419	0.123	−0.071	−0.666 **	−0.772 **	0.697 **	0.179
K	0.258	0.263	−0.220	−0.453	−0.316	−0.291	0.110	−0.259
Cu	−0.047	−0.465	0.484	0.705 **	−0.061	−0.238	0.334	0.713 **
Fe	−0.006	0.128	0.151	0.554 *	0.713 **	0.743 **	−0.308	0.559 *
Zn	−0.171	−0.535 *	0.149	0.067	−0.516	−0.734 **	0.644 **	0.313
Mn	−0.156	−0.351	−0.237	−0.523 *	−0.438	−0.665 **	0.317	−0.585 *
P	0.006	−0.391	0.194	−0.026	−0.802 **	−0.929 **	0.633 **	0.081
S	−0.079	−0.249	−0.236	−0.335	−0.461	−0.714 **	0.423	−0.383
N	−0.007	0.071	−0.265	−0.697 **	−0.548 *	−0.512 *	0.126	−0.694 **
				Second season				
Ca	0.319	0.229	0.393	0.229	0.255	0.348	0.443	−0.356
Mg	−0.38	−0.4	−0.359	−0.532 *	0.303	−0.225	−0.625 **	0.347
K	0.417	0.063	−0.367	0.405	−0.543 *	−0.16	0.08	0.048
Cu	−0.106	−0.315	−0.450	0.617 *	−0.48	−0.155	−0.25	0.631 **
Fe	0.106	0.108	0.653 **	0.217	0.542 *	0.661 **	0.877 **	−0.697 **
Zn	−0.117	0.108	0.346	0.565 *	−0.27	0.379	0.429	−0.211
Mn	0.105	0.359	0.535 *	0.044	0.085	0.334	0.717 **	−0.854 **
P	−0.211	−0.560 *	−0.469	0.326	0.001	−0.176	−0.435	0.837 **
S	−0.175	0.271	0.217	0.058	−0.623 **	−0.189	0.162	−0.462
N	−0.245	−0.283	−0.681 **	−0.198	−0.436	−0.573 *	−0.932 **	0.780 **

G: germination; FGC: first germination count; L: seedling lenght; DM: seedling dry mass; E: emergence; E emergence speed index; T50: time to 50% germination; P50: longevity; *: significant at 5% probability; **: significa at 1% probability.

Soybean seed germination showed no significant correlation with nutrients, in bot seasons. FGC was negatively correlated with Zn (first season) and with P (second seaso For seedling length (L), there were significant correlations only in the second season, bein positive for Fe and Mn and negative for N. The seedling dry mass variable, in the fir season, was positively correlated with Cu and Fe and negatively with Mn and N, and in th second season there was a positive correlation with Cu and Zn, and a negative correlatio with Mg (Table 5).

In seedling emergence (E) there was a negative correlation for Mg, P and N and positive correlation for Fe (first season). In the second season, there was a positive correlatio with Fe and a negative correlation with K and S. For the ESI, in the first season, there was positive correlation with Ca and Fe and a negative correlation with Mg, Zn, Mn, P, S and N in the second season a correlation was observed only with Fe, which was positive.

For T50, in the first season, Mg, Zn and P were positively correlated. In the secon season, positive correlations were observed with Fe and Mn, and negative correlations wit Mg and N. As for seed longevity (P50), in the first season there was a positive correlatio with Ca, Cu and Fe, and a negative correlation with Mn and N; in the second season th positive correlations were with Cu, N and P, and the negative ones with Fe and Mn (Table 5

Despite not being the focus of this research, the results of the correlation analysi showed that the micronutrient Fe was positively correlated with the variables of vigo and seed longevity in both seasons, making its importance evident in the physiologic quality of soybean seeds. Thus, new studies aiming to explain the participation of thi nutrient in the seed can contribute to a better understanding of the process of soybean see physiological quality acquisition.

4. Discussion

In this work, we evaluated the physiological quality of soybean seeds cultivated under conventional tillage and no-tillage systems with and without desiccant use. Our results showed that in the no-tillage system, it is common to have a higher soil compaction index in the surface layers, when compared to the conventional system. Thus, it can negatively affect the root development of crops and, consequently, the water and nutrient absorption [34–36]. However, as this is an experiment performed in an area with long-term no-tillage, in which the soil properties are already consolidated [8,9], these factors probably did not harm the soybean plant development, to the point of affecting the seed germination capacity (Table 3).

The lack of desiccation management effect on soybean seed germination is due to the fact that the germination capacity is acquired at the R7.1 stage [12,17], and in this research the pre-harvest soybean desiccation was carried out in R7.3, when the acquisition of this capacity had already taken place. Studies on desiccant application at the physiological maturity stage also revealed that there were no significant differences in germination potential [20,37].

Although the soil management system did not affect seed germination, there was a difference in vigor, through changes in the responses of FGC, T50, length and ESI of soybean seedlings (Table 3). The soybean seed vigor evaluation characteristics were more sensitive for detecting alterations, when compared to the laboratory germination test [28,38]. So, despite the superficial physical and chemical characteristics of the soil managed under no-tillage not having affected germination, these were able to negatively influence seed vigor.

The physiological quality is acquired throughout the plant development stage [12,13]. Thus, the negative results of desiccant application on the vigor (length and dry mass of seedlings) and longevity of the seeds (Table 3), is due to this factor, since, possibly, the desiccation impaired the process of acquiring physiological quality. It should be noted that the desiccant used, paraquat, is a fast-acting contact herbicide, a factor that limits the translocation of the product of photosynthesis into the seeds [39]. So, the negative effect of the desiccant application on the vigor and longevity of the seeds, in this study, is due to physiological processes.

The variables DM, E and ESI (vigor) showed a significant interaction between soil management and desiccant application. So, it was possible to observe that in CT there was better seed quality without desiccant application, while in NT the desiccant application favored soybean seed vigor (Tables 3 and 4). Such results are probably related to the amount of nutrients in the soil available for absorption by the plant (Table 1). Previous works show that in the NT system there is greater nutrient accumulation in the surface layer of the soil, where there is a greater amount of soybean roots [40,41]. Thus, a compensatory effect may have occurred, since the greater supply of nutrients to the plant in NT may have promoted greater accumulation in the seed, and even with the desiccant application, there was no reduction in the nutritional content. In CT, despite having a lower nutrient content in the soil, the non-application of desiccant allowed more time for the absorption of nutrients by the plant and accumulation in the seed, resulting in seeds with greater vigor (DM, E and ESI).

It is worth mentioning that seed vigor is the set of properties that determine the activity and performance of seed lots with acceptable germination, under a wide range of environmental conditions [42]. Thus, from these test results, it was found that the need to apply a desiccant on the soybean seeds is dependent on the soil management system.

Seed longevity is a characteristic that impacts the commercialization of lots, considering that low longevity is associated with loss of vigor and viability [12,32]. In addition to the desiccant factor, soil management systems also affected seed longevity, with NT being the one that promoted the longest storage period (Table 3). Such results may be associated with the seed nutrient content, since in the second season there was a positive correlation with N, P and Cu (Table 5).

N acts on plant growth, on the formation of amino acids, proteins, enzymes and on the chlorophyll molecule [43,44], and its deficiency in the seed can negatively affect the protein content, resulting in a loss of quality [45,46], since the proteins in the seeds act as reserve substances and as chemical reactions catalysts [47]. Thus, during the seed deterioration process, there is a decrease in the content and synthesis of proteins, an increase in the amino acid content, a decrease in the content of soluble proteins and denaturation caused by high temperatures, causing the loss of performance of vital functions [47].

P acts in the formation of the seeds and during the germination process, where it performs fundamental functions, as it is a constituent of the membrane (phospholipids), of nucleic acids and of energy-storing compounds, such as ATP, which is the most important of these compounds [48–50]. Seeds with higher P content present higher initial energy for seedling metabolic activities; consequently, it has a higher physiological quality, contributing to the better performance of the plant in the field, since it makes it less dependent on the existing levels of this element in the soil [51].

Cu, on the other hand, participates in important physiological processes as a structural and metabolic component, as it acts in the composition of proteins involved in oxidation-reduction reactions, in carbohydrate synthesis and as a cofactor of enzymes, such as polyphenol oxidase and superoxide dismutase, which act in the lignin synthesis for cell wall formation and protection from oxidative stress due to the presence of reactive oxygen species [49,52,53]. In this way, the higher the Cu content in the seed, without generating toxicity, the greater the seed longevity.

As in NT there is greater nutrient accumulation [40,41], the greater content of N, P and Cu in the seed, provided by this management system, contributed to greater longevity of the seeds. Through greater protein synthesis, a constituent of DNA and RNA, the seed has a greater energy source for metabolic processes and a greater reduction of oxidative stress during the storage process.

The results of this research showed that the use of desiccant influenced the process of acquiring vigor and longevity of soybean seeds. In addition, it was possible to observe that factors such as the soil management system may be associated with the moment and process of seed quality acquisition. For longevity, a parameter that continues to be acquired until the last stage of seed maturation, this research showed that the desiccant, by accelerating the seed maturation process, impairs the complete acquisition of this characteristic.

Future research involving protein content, enzymatic activity, reactive oxygen species and accumulation of residues may contribute to the understanding of the physiological and biochemical mechanisms of soybean seeds produced in different soil management systems and with or without desiccant application.

5. Conclusions

Our results showed that soil management systems and desiccant use do not influence soybean seed germination. However, conventional soil preparation increases seed vigor when evaluated by the traits of first germination count, dry mass and length of seedling. In addition, the absence of desiccant application promotes the formation of seedlings with greater length and dry mass. In summary, the use of desiccant aiming at seed vigor is dependent on the soil management system. The soybean seed longevity is superior in the no-tillage system, but the desiccant application reduces seed longevity.

Author Contributions: Conceptualization, G.F.d.S. and L.C.; methodology, G.F.d.S., L.C., P.P.N.C and B.C.O.L.; formal analysis, G.F.d.S. and L.C.; investigation, G.F.d.S., L.C., P.P.N.C. and B.C.O.L.; writing—original draft preparation, G.F.d.S. and L.C.; writing—review and editing, J.C.C., J.N. and E.A.A.d.S.; supervision, J.C.C. and E.A.A.d.S.; funding acquisition, G.F.d.S., L.C., J.C.C. and E.A.A.d.S All authors have read and agreed to the published version of the manuscript.

Funding: This research was funded by São Paulo Research Foundation (FAPESP), grant number 2017/50211-9 and by Coordenação de Aperfeiçoamento de Pessoal de Nível Superior (CAPES), grant number 001.

Institutional Review Board Statement: Not applicable.

Informed Consent Statement: Not applicable.

Data Availability Statement: Not applicable.

Acknowledgments: To Conselho Nacional de Desenvolvimento Científico e Tecnológico (CNPq) for the productivity fellowship to E.A.A.d.S. (code 311526/2021) and for granting an award for excellence in research to J.C.C.

Conflicts of Interest: The authors declare no conflict of interest.

References

Cecagno, D.; de Andrade Costa, S.E.V.G.; Anghinoni, I.; Kunrath, T.R.; Martins, A.P.; Reichert, J.M.; Gubiani, P.I.; Balerini, F.; Fink, J.R.; de Faccio Carvalho, P.C. Least limiting water range and soybean yield in a long-term, no-till, integrated crop-livestock system under different grazing intensities. *Soil Tillage Res.* **2016**, *156*, 54–62. [CrossRef]

Torabian, S.; Farhangi-Abriz, S.; Denton, M.D. Do tillage systems influence nitrogen fixation in legumes? A review. *Soil Tillage Res.* **2019**, *185*, 113–121. [CrossRef]

Stefanoski, D.C.; Santos, G.G.; Marchão, R.L.; Petter, F.A.; Pacheco, L.P. Soil use and management and its impact on physical quality. *Rev. Bras. Eng. Agrícola E Ambient.* **2013**, *17*, 1301–1309. [CrossRef]

Sivarajan, S.; Maharlooei, M.; Bajwa, S.G.; Nowatzki, J. Impact of soil compaction due to wheel traffic on corn and soybean growth, development and yield. *Soil Tillage Res.* **2018**, *175*, 234–243. [CrossRef]

Blanco-Canqui, H.; Wortmann, C.S. Does occasional tillage undo the ecosystem services gained with no-till? A review. *Soil Tillage Res.* **2020**, *198*, 104534. [CrossRef]

Blanco-Canqui, H.; Ruis, S.J. No-tillage and soil physical environment. *Geoderma* **2018**, *326*, 164–200. [CrossRef]

Ferreira, C.J.B.; Tormena, C.A.; Severiano, E.D.C.; Zotarelli, L.; Betioli Júnior, E. Soil compaction influences soil physical quality and soybean yield under long-term no-tillage. *Arch. Agron. Soil Sci.* **2021**, *67*, 383–396. [CrossRef]

Silva, G.F.; Calonego, J.C.; Luperini, B.C.O.; Chamma, L.; Alves, E.R.; Rodrigues, S.A.; Putti, F.F.; da Silva, V.M.; de Almeida Silva, M. Soil—Plant relationships in soybean cultivated under conventional tillage and long-term no-Tillage. *Agronomy* **2022**, *12*, 697. [CrossRef]

Silva, G.F.; Matusevicius, A.P.O.; Calonego, J.C.; Chamma, L.; Luperini, B.C.O.; Alves, M.d.S.; Leite, H.M.F.; Pinto, E.d.J.; Silva, M.d.A.; Putti, F.F. Soil–plant relationships in soybean cultivated under crop rotation after 17 years of no-tillage and occasional chiseling. *Plants* **2022**, *11*, 2657. [CrossRef]

10. Calonego, J.C.; Raphael, J.P.A.; Rigon, J.P.G.; de Oliveira Neto, L.; Rosolem, C.A. Soil compaction management and soybean yields with cover crops under no-till and occasional chiseling. *Eur. J. Agron.* **2017**, *85*, 31–37. [CrossRef]

11. Daryanto, S.; Wang, L.; Jacinthe, P.A. Impacts of no-tillage management on nitrate loss from corn, soybean and wheat cultivation: A meta-analysis. *Sci. Rep.* **2017**, *7*, 12117. [CrossRef] [PubMed]

12. Lima, J.J.P.; Buitink, J.; Lalanne, D.; Rossi, R.F.; Pelletier, S.; da Silva, E.A.A.; Leprince, O. Molecular characterization of the acquisition of longevity during seed maturation in soybean. *PLoS ONE* **2017**, *12*, e0180282. [CrossRef]

13. Bewley, J.D.; Bradford, K.J.; Hilhorst, H.W.M.; Nonogaki, H. *Seeds: Physiology of Development, Germination and Dormancy*; Springer: New York, NY, USA, 2013; ISBN 978-1-4614-4692-7.

14. Demir, I.; Ellis, R.H. Changes in seed quality during seed development and maturation in tomato. *Seed Sci. Res.* **1992**, *2*, 81–87. [CrossRef]

15. Finch-Savage, W.E.; Bassel, G.W. Seed vigour and crop establishment: Extending performance beyond adaptation. *J. Exp. Bot.* **2016**, *67*, 567–591. [CrossRef]

16. Vidigal, D.S.; Willems, L.; van Arkel, J.; Dekkers, B.J.W.; Hilhorst, H.W.M.; Bentsink, L. Galactinol as marker for seed longevity. *Plant Sci.* **2016**, *246*, 112–118. [CrossRef]

17. Zanakis, G.N.; Ellis, R.H.; Summerfield, R.J. Seed quality in relation to seed development and maturation in three genotypes of soybean (*Glycine max*). *Exp. Agric.* **1994**, *30*, 139. [CrossRef]

18. Zagonel, J. Herbicide application timing in preharvest desiccation of soybean cultivars with different growth habits. *J. Environ. Sci. Health Part B Pestic. Food Contam. Agric. Wastes* **2005**, *40*, 13–20. [CrossRef]

19. Griffin, J.L.; Boudreaux, J.M.; Miller, D.K. Herbicides as harvest aids. *Weed Sci.* **2010**, *58*, 355–358. [CrossRef]

20. Pereira, T.; Coelho, C.M.M.; Sobiecki, M.; Souza, C.A. Physiological quality of soybean seeds depending on the preharvest desiccation. *Planta Daninha* **2015**, *33*, 441–450. [CrossRef]

21. Delgado, C.M.L.; de Coelho, C.M.M.; Buba, G.P. Mobilization of reserves and vigor of soybean seeds under desiccation with glufosinate ammonium. *J Seed Sci.* **2015**, *37*, 154–161. [CrossRef]

22. Fipke, G.M.; Deak, E.A.; Stecca, J.D.L.; Bernardy, D.; Berger, M.; Tabaldi, L.A.; Martin, T.N. Morphology and enzymatic activity of seedlings from wheat desiccated in pre-harvest. *Acta Sci. Agron.* **2020**, *43*, e44974. [CrossRef]

23. Botelho, F.J.E.; Oliveira, J.A.; Von Pinho, É.V.d.R.; Carvalho, E.R.; Figueiredo, Í.B.D.; Andrade, V. Quality of seeds of soybean obtained from different cultivars subjected to desiccation with different herbicides and application times. *Rev. Agro@Mbiente* **2016**, *10*, 137. [CrossRef]

24. Zuffo, A.M.; Santos, M.D.A.; de Oliveira, I.C.; Alves, C.Z.; Aguilera, J.G.; Teodoro, P.E. Does chemical desiccation and harvest time affect the physiological and sanitary quality of soybean seeds? *Rev. Caatinga* **2019**, *32*, 934–942. [CrossRef]

25. Zuffo, A.M.; Aguilera, J.G.; Carvalho, E.R.; Teodoro, P.E. Harvest times with chemical desiccation and the effects on the enzymat expression and physiological quality of soybean seeds. *Rev. Caatinga* **2020**, *33*, 361–370. [CrossRef]
26. Fehr, W.R.; Caviness, C. *Stages of Soybean Development*; AMES-Iowa State University of Science and Technology: Ames, IA, USA, 197
27. Basso, D.P.; Hoshino-Bezerra, A.A.; Sartori, M.M.P.; Buitink, J.; Leprince, O.; Silva, E.A.A. da Late seed maturation improves th preservation of seedling emergence during storage in soybean. *J. Seed Sci.* **2018**, *40*, 185–192. [CrossRef]
28. Ministério da Agricultura, Pecuária e Abastecimento. *Brasil Regras Para Análise de Sementes*; Ministério da Agricultura, Pecuária Abastecimento: Brasília, Brasil, 2009.
29. Joosen, R.V.L.; Kodde, J.; Willems, L.A.J.; Ligterink, W.; van der Plas, L.H.W.; Hilhorst, H.W.M. Germinator: A software packag for high-throughput scoring and curve fitting of Arabidopsis seed germination. *Plant J.* **2010**, *62*, 148–159. [CrossRef]
30. Nakagawa, J. Testes de vigor baseados no desempenho das plântulas. In *Vigor de Sementes: Conceitos e Testes*; ABRATES: Londrin Brasil, 1999.
31. Maguire, J.D. Speed of germination—Aid in selection and evaluation for seedling emergence and vigor 1. *Crop Sci.* **1962**, 176–177. [CrossRef]
32. Zinsmeister, J.; Lalanne, D.; Terrasson, E.; Chatelain, E.; Vandecasteele, C.; Vu, B.L.; Dubois-Laurent, C.; Geoffriau, E.; Le Signe C.; Dalmais, M.; et al. ABI5 is a regulator of seed maturation and longevity in legumes. *Plant Cell* **2016**, *28*, 2735–2754. [CrossRe [PubMed]
33. AOAC International. *AOAC Official Methods of Analysis of AOAC International*; AOAC International: Rockville, MD, USA, 2016
34. Haghighi, F.; Gorji, M.; Shorafa, M. A study of the effects of land use changes on soil physical properties and organic matter. *Lan Degrad. Dev.* **2010**, *21*, 496–502. [CrossRef]
35. Mangalassery, S.; Sjögersten, S.; Sparkes, D.L.; Sturrock, C.J.; Craigon, J.; Mooney, S.J. To what extent can zero tillage lead to reduction in greenhouse gas emissions from temperate soils? *Sci. Rep.* **2014**, *4*, 4586. [CrossRef] [PubMed]
36. Galdos, M.V.; Pires, L.F.; Cooper, H.V.; Calonego, J.C.; Rosolem, C.A.; Mooney, S.J. Assessing the long-term effects of zero-tillag on the macroporosity of Brazilian soils using X-ray Computed Tomography. *Geoderma* **2019**, *337*, 1126–1135. [CrossRef]
37. Santos, F.L.; Bertacine, F.; Souza, J.S.; Simões, I.; Bossolani, J.W.; Sá, M.E. Influence of the application of desiccation in th physiological quality of soybean seeds. *Rev. Bras. Eng. Biossistemas* **2018**, *12*, 68–76. [CrossRef]
38. Vanzolini, S.; Araki, C.A.d.S.; da Silva, A.C.T.M.; Nakagawa, J. Seedling length test in the evaluation of the physiological quali of soybean seeds. *J. Seed Sci.* **2007**, *29*, 90–96. [CrossRef]
39. Ratnayake, S.; Shaw, D.R. Effects of harvest-aid herbicides on soybean (*Glycine max*) seed yield and quality. *Weed Technol.* **1992**, 339–344. [CrossRef]
40. Deiss, L.; Kleina, G.B.; Moraes, A.; Franzluebbers, A.J.; Motta, A.C.V.; Dieckow, J.; Sandini, I.E.; Anghinoni, I.; Carvalho, P.C.F. Sc chemical properties under no-tillage as affected by agricultural trophic complexity. *Eur. J. Soil Sci.* **2019**, *71*, 1090–1105. [CrossRe
41. Sokolowski, A.C.; Prack McCormick, B.; De Grazia, J.; Wolski, J.E.; Rodríguez, H.A.; Rodríguez-Frers, E.P.; Gagey, M.C.; Debeli S.P.; Paladino, I.R.; Barrios, M.B. Tillage and no-tillage effects on physical and chemical properties of an Argiaquoll soil unde long-term crop rotation in Buenos Aires, Argentina. *Int. Soil Water Conserv. Res.* **2020**, *8*, 185–194. [CrossRef]
42. ISTA; Internacional Seed Testing Association. *International Rules for Seed Testing*; ISTA: Bassersdorf, Switzerland, 2014.
43. Fernie, A.R.; Bachem, C.W.B.; Helariutta, Y.; Neuhaus, H.E.; Prat, S.; Ruan, Y.-L.; Stitt, M.; Sweetlove, L.J.; Tegeder, M.; Wah V.; et al. Synchronization of developmental, molecular and metabolic aspects of source–sink interactions. *Nat. Plants* **202** *6*, 55–66. [CrossRef]
44. Grewal, S.K.; Gill, R.K. Insights into carbon and nitrogen metabolism and antioxidant potential during vegetative phase in quinc (*Chenopodium quinoa* Willd.). *Protoplasma* **2022**, *259*, 1301–1319. [CrossRef]
45. Masclaux-Daubresse, C.; Daniel-Vedele, F.; Dechorgnat, J.; Chardon, F.; Gaufichon, L.; Suzuki, A. Nitrogen uptake, assimilatio and remobilization in plants: Challenges for sustainable and productive agriculture. *Ann. Bot.* **2010**, *105*, 1141–1157. [CrossRe
46. Sandhu, N.; Sethi, M.; Kumar, A.; Dang, D.; Singh, J.; Chhuneja, P. Biochemical and genetic approaches improving nitrogen us efficiency in cereal crops: A review. *Front. Plant Sci.* **2021**, *12*. [CrossRef]
47. Marcos-Filho, J. *Fisiologia de Sementes de Plantas Cultivadas*, 2nd ed.; Abrates: Londrina, Brasil, 2015.
48. Tanaka, R.T.; Mascarenhas, H.A.A.; Borkert, C.M. Nutrição mineral da soja. In *Cultura da Soja Nos Cerrados*; Associação Brasilei para Pesquisa da Potassa e do Fosfato: Piracicaba, Brasil, 1993; pp. 105–135.
49. Marschner, P. *Mineral Nutrition of Higher Plants*; Academic Press: London, UK, 2012.
50. Taiz, L.; Zeiger, E. *Fisiologia e Desenvolvimento Vegetal*, 6th ed.; Artmed: Porto Alegre, Brasil, 2017.
51. Guerra, C.A.; Marchetti, M.E.; Robaina, A.D.; de Souza, L.C.F.; Gonçalves, M.C.; Novelino, J.O. Soybean seed physiologica quality in function of phosphorus, molybdenum and cobalt fertilization. *Acta Sci. Agron.* **2006**, *28*, 91–97. [CrossRef]
52. Yruela, I. Copper in plants: Acquisition, transport and interactions. *Funct. Plant Biol.* **2009**, *36*, 409. [CrossRef] [PubMed]
53. Kabata-Pendias, A. *Trace Elements in Soils and Plants*, 4th ed.; CRC Press: Boca Raton, FL, USA, 2011; ISBN 9780429192036.

agronomy

MDPI

Article

The Border Effects of Dry Matter, Photosynthetic Characteristics, and Yield Components of Wheat under Hole Sowing Condition

Yitao Sun †, Chao Yang †, Huajun Liang, Yuyan Yang, Kangmin Bu, Yongli Dong and Jiangbo Hai *

College of Agronomy, Northwest A&F University, Yangling, Xianyang 712100, China
* Correspondence: haijiangbo@126.com; Tel.: +86-13389221092
† These authors contributed equally to this work.

Abstract: Wheat can be cultivated by hole sowing, but its border effect has not yet been studied. Therefore, we carried out a field experiment from 2021 to 2022 at the Doukou Crop Experimental Demonstration Station (108°52′ E, 34°37′ N) of Northwest A&F University in Jingyang County, Xianyang City, Shaanxi Province, China. The response of dry matter, photosynthetic characteristics, and yield components of wheat to the border effects under the hole sowing method was studied. The results showed specific border effects on each index of five wheat varieties (XN136, XN175, XN527, XN536, and XN765), among which the border effects of XN175 and XN765 were the most significant, with the highest yield. Subsequent correlation analysis revealed that only grain per spike and intercellular carbon dioxide concentration responded negatively to the border effects, and the rest were positively correlated. Finally, we conducted a random forest model analysis of different indicators of wheat varieties with significant border effects. We found that net photosynthetic rate and aboveground dry matter per plant had the most significant impact and contribution to the border effects. In contrast, grain per spike had the most negligible impact on the border effects. Our results fill a gap in the study of the border effects of wheat under hole sowing cultivation for future researchers.

Keywords: border effect; hole sowing; wheat; *Triticum aestivum* L.

Citation: Sun, Y.; Yang, C.; Liang, H.; Yang, Y.; Bu, K.; Dong, Y.; Hai, J. The Border Effects of Dry Matter, Photosynthetic Characteristics, and Yield Components of Wheat under Hole Sowing Condition. *Agronomy* **2023**, *13*, 766. https://doi.org/10.3390/agronomy13030766

Academic Editors: Aurelio Scavo and Umberto Anastasi

Received: 9 February 2023
Revised: 2 March 2023
Accepted: 3 March 2023
Published: 7 March 2023

1. Introduction

Due to the extreme changes in global climate and the rapid growth of population, achieving food supply security under limited arable land conditions is a significant challenge in the 21st century [1–4]. Future food security, therefore, requires further increases in crop yields. According to statistics, to meet global food demand, food production will need to increase by 70–100% by 2050, with an annual increase of more than 4 million tons [5–8], and wheat production needs to grow at 1.7% per year [9–12]. Wheat is one of the most important foods for human beings, which is essential and beneficial to human health. How to maximize the benefits of wheat is of great significance to food production and agricultural income [13–15].

Individuals in the border row usually enjoy better conditions to obtain a higher yield, defined as the border (marginal) effect [16,17]. This effect is usually caused by uncultivated space, which is left between adjacent plots for crop management and differentiation of different varieties [18]. Due to more solar energy, better ventilation, and less nutrient competition, the crop growth and yield of the border rows are better than those of the middle rows [19,20]. Therefore, maximizing the border advantage is essential for improving productivity [21].

The cultivation techniques of wheat can regulate wheat tillering, form a reasonable population, enhance the utilization rate of light energy, and have a great impact on coordinating the relationship between source, sink and flow, increasing yield, and improving quality [22–24]. The sowing method is an important element of cultivation techniques to regulate the growth and development of wheat. Different sowing methods will lead to changes in the structure of wheat population, and therefore the physiological and metabolic processes of plants will change accordingly, affecting the overall growth and development of wheat, and in turn, affecting the yield and quality [25]. The hole sowing cultivation technology of wheat is a high-efficiency agricultural technology integrating rainfall, drought resistance, and efficient utilization of light and heat resources. As a new cultivation technique, it has many excellent characteristics and a good development prospect.

Based on previous research on the effects of different sowing methods and seeding rates on wheat yield and quality, this study further explored the response of different wheat varieties to the border effect of hole sowing. The main purposes are: (1) To explore the response of wheat border effects under the cultivation mode of hole sowing. (2) To explore which wheat varieties are more suitable for hole sowing cultivation. (3) To explore which indicators have significant border effects and the size of the contribution of each indicator to the border effects.

2. Materials and Methods

2.1. Test Designs and Determination Methods

This experiment was carried out at Doukou Crop Experimental Demonstration Station of Northwest A&F University from October 2021 to June 2022. The experimental demonstration station is located in Xinglong Village, Yunyang Town, Jingyang County, Xianyang City, Shaanxi Province, China, 108°52′ E, 34°37′ N. The average temperature and precipitation in 2021–2022 were 10.89 °C and 17.33 mm, respectively (Figure 1). The soil in the experimental field was loam. The soil organic matter content in the 0–20 cm soil layer of the experimental field was 18.03 g·kg^{-1}, the total nitrogen content was 1.31 g·kg^{-1}, the available nitrogen content was 86.3 mg·kg^{-1}, and the available potassium was 227.48 mg·kg^{-1}.

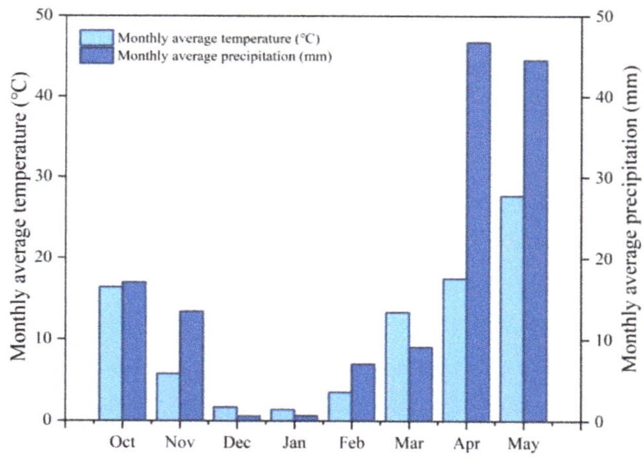

Figure 1. Total precipitation and monthly mean temperature during wheat growth stage from October 2021 to June 2022.

Different wheat varieties, 'XN136', 'XN175', 'XN527', 'XN536', and 'XN765', were selected as experimental materials. These five wheat varieties were provided by the College of Agriculture, Northwest A&F University. The main common characteristics were as follows: they were all semi-winter and semi-dwarf varieties, suitable for planting in the Guanzhong irrigation area of Shaanxi Province, had medium tillering ability, high earning rate, fast filling speed, and medium grain plumpness. The main difference was the plant heights. The average plant heights of each variety were: 77.3 cm for 'XN136', 84.1 cm for

'XN175', 77.1 cm for 'XN527', 76.1 cm for 'XN536', and 79.1 cm for 'XN765'. The sowing density was 168.5 kg·ha^{-1}, and the sowing amount per hole was 12. In order to ensure the accuracy of the experiment, the sowing method of wheat hole sowing used in this experiment was artificial sowing. Firstly, the furrow opener was used to furrow each plot, where 12 furrows (12 rows) were opened in each plot, and the benchmark was used to mark the points of each row. There were 35 mark points in each row, and 12 grains were sown manually at each mark point. In this experiment, different wheat varieties were used as different treatments, with a total of 5 treatments, three replicates, each plot area of 15 m^2, each plot of 12 rows, each row of 35 holes, hole spacing (S1) of 14 cm, and row spacing (S2) of 25 cm (Figure 2). The compound fertilizer (N-P_2O_5-K_2O: 24-15-5) was uniformly applied in the form of base fertilizer at 375 kg·ha^{-1} before tillage. This experiment was sown on 24 October 2021, and harvested on 5 June 2022. Other measures in the experimental field were the same as the requirements of high-yield field cultivation techniques.

(**A**)

(**B**)

Figure 2. *Cont.*

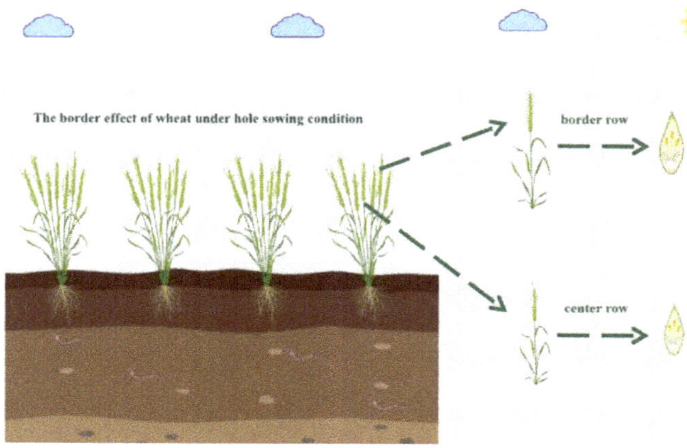

(**C**)

Figure 2. Hole sowing planting map of one plot (**A**), hole sowing wheat growth map (**B**), and th border effect map of wheat (**C**).

For our sampling method, to avoid the influence of side rows on the border effect four holes were randomly sampled in the middle four rows of each plot. Each hole's oute and inner sides were sampled separately, and the average value was used to compare th border effects.

The aboveground dry matter per plant outside and inside each hole of the wheat plar was determined. In the booting stage, heading stage, flowering stage, filling stage, an maturing stage of wheat, four 20 cm plant samples with uniform growth were selecte on the outer and inner sides of each hole. After being brought back to the room on th same day, each part was de-enzymed in a 105 °C oven for 30 min, then the temperatu was reduced to 60–80 °C, and continued to be dried for about eight hours to make it dr quickly, and then removed. The samples then continued to dry for four hours, weighe again, until the weight was constant at which the final weight was measured.

Wheat plants outside and inside each hole were measured for physiological indexe of photosynthetic characteristics (net photosynthetic rate, stomatal conductance, and i tercellular carbon dioxide concentration). In the booting stage, heading stage, flowerin stage, and filling stage of wheat, clear and cloudless weather was selected, and the Li-640 portable photosynthesis system was used to measure the photosynthetic characteristic Four holes were randomly selected from each plot, and four uniform plant samples wer selected on the outside and inside of each hole. The net photosynthetic rate of the middl part of the flag leaf of wheat was measured at 9:00–11:00 a.m. stomatal conductance, an intercellular carbon dioxide concentration. The use of the LI-6400 portable photosynthesi measurement system roughly includes six steps: instrument connection, program loadin instrument correction, data measurement, data transmission, and instrument closing. B fore the measurement, we must first see whether the instrument is connected, and the enter the following steps after the instrument is connected. After the power switch is turne on, the instrument begins to install the OPEN program, which takes about ten minute The configuration file here must be correctly selected and should be consistent with th type of leaf chamber installed on the head of the IRGA analyzer. Because of the change i the surrounding environmental conditions, the zero point of the instrument changes, an therefore, it must be corrected before use as the data will not be reliable otherwise. Whe calibrating the instrument, f3 needs to be selected under the OPEN main program interfac to enter 'Calib Menu'. After entering the calibration menu, seven secondary menus ar displayed on the display screen, among which the first item 'FLOW Meter Zero' (zer adjustment of flowmeter) and the second item 'IRGA Zero' (zero adjustment of infrare gas analyzer, namely correction of CO_2 and H_2O zero points) are necessary operations after each boot. Data measurement is a key step in the use of LI-6400. The data is measured under f4 (New Msmnts, new measurement menu) of the OPEN main interface. Before the experimental data measurement, the H_2O and CO_2 control knobs should be adjusted to BYPASS (if the CO_2 injection system is used, the CO_2 control knob should be adjusted to SCRUB). The measured data is then transmitted to the computer in time.

For yield composition statistics, the number of grains per spike inside each hole was counted. After harvest, the grains were sun-dried to remove impurities. The number of plates used to take each hole outside and inside of a total of 1000 grains were weighed, and repeated three times to calculate the average value of thousand grains. After the wheat matured, the number of effective ears in each plot's 1 m double-row sample section was counted. Each plot was sampled for 1 m^2 of wheat, then threshed with a thresher, dried, weighed with an electronic balance, and calculated for grain yield ($kg \cdot ha^{-1}$).

2.2. Statistical Analysis of Data

The border effect (BE%) was calculated as follows according to Wang et al. [15].

$$BE = \frac{\text{Parameter of border row} - \text{Parameter of center row}}{\text{Parameter of center row}} \times 100 \quad (1)$$

Correlation analysis refers to the analysis of two or more correlated variable elements to measure the degree of correlation between the two variable factors.

Random forest regression is a machine learning technique that can create a set of multiple decision trees, aggregate on the set, and rank the predictors according to the correlation between the predictors and the predictions. It is well known that random forest regression techniques can produce highly accurate predictions and handle many input variables without overfitting.

In this study, the outer side of each hole of wheat was used as the border line, and the middle was used as the center line. Correlation analysis and random forest regression analysis were performed according to each index.

Microsoft Office Excel 2021 and SPSS 26.0 were used for data statistical analysis, and RStudio was used for significant difference analysis and picture drawing. The significance level ($p < 0.05$) was used to determine the average difference using the least significant difference test.

3. Results

3.1. Border Effects of Yield Components

XN136, XN175, and XN765 have significant border effects (Table 1). The border effects of thousand-grain weight and grain per spike of XN175 were the highest, being 15.1% and 14.2%, respectively, followed by XN765 (11.5%, 12%) and XN136 (6.8%, 5.9%). The effective spikes of XN175 were the largest, at 643 × $10^4 \cdot ha^{-1}$, and there was a significant difference between XN175 and XN136, XN536, and XN765. The number of effective spikes per hole of XN175 was the largest, at 23, and the number of effective spikes per plant was 2. The highest yield of XN175 and XN765 was 8587.1 $kg \cdot ha^{-1}$ and 8558.6 $kg \cdot ha^{-1}$, respectively.

3.2. Border Effects of Dry Matter

The aboveground dry matter of wheat at different stages (booting stage, heading stage, flowering stage, filling stage, and maturing stage) was measured. The border effect was analyzed (Figure 3). It can be seen from the figure that the maximum dry matter mass of the five varieties in different stages was at the outer row of wheat, which is the maturing stage of XN175, with a value of 13.17 g/plant. The minimum value of dry matter was found in wheat inline, also wheat XN175, which appeared at booting stage and was 3.37 g/plant. It can be seen that the dry matter of the aboveground plants of the five varieties showed a particular border effect, among which XN136 only had significant differences in the dry matter border effect of the aboveground plants at the heading stage and flowering stage. The dry matter border effect of XN175 in the booting stage and filling stage was significantly different, and the dry matter border effect in the heading stage and the maturing stage was significantly different. The dry matter border effect of XN527 in the five stages was insignificant. XN536 only significantly differed in the dry matter border effect of aboveground dry matter per plant in the flowering stage. The border effect of dry matter per plant above ground of XN765 was significantly different in each stage. It can be seen that the aboveground dry matter of the two wheat varieties, XN175 and XN765, had a significant border effect under hole sowing conditions.

Table 1. Border effects of yield and yield components of wheat at maturity stage in 2021–2022 ($p \leq 0.05$, significant difference when the outline and inline characters of the same variety are completely different; $p > 0.05$, no significant difference when the same or more letters are used).

Variety	Location	Thousand-Grain Weight (g)	Grain Per Spike	Effective Spikes Per Hole	Effective Spikes ($\times 10^4 \cdot ha^{-1}$)	Yield ($kg \cdot ha^{-1}$)	Thousand-Grain Weight (BE%)	Grain Per Spike (BE%)
XN136	outer	56.41 ab	65.67 b	18 a	506.7 c	8358.3 ab	6.8%	5.9%
	inner	52.83 d	62 b					
XN175	outer	57.66 a	75 a	23 a	643 a	8587.1 a	15.1%	14.2%
	inner	50.1 e	65.67 b					
XN527	outer	53.83 b	53 c	22 a	604 ab	7474.9 b	−2.8%	1.3%
	inner	55.39 abcd	52.33 c					
XN536	outer	55.57 abc	42 e	19 a	539 c	8085.3 ab	3.1%	3.3%
	inner	53.9 b	40.67 e					
XN765	outer	53.22 cd	56 c	20 a	570 bc	8558.6 a	11.5%	12%
	inner	47.71 e	50 d					

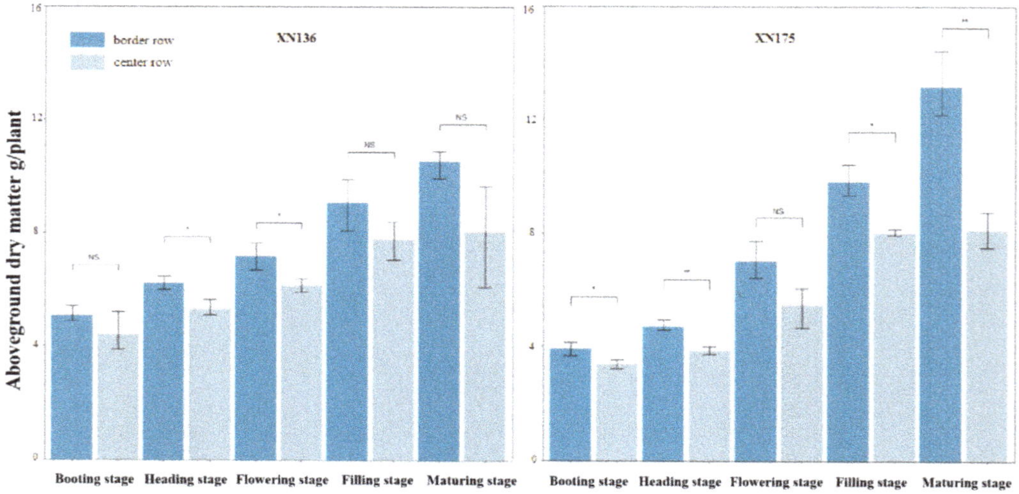

Figure 3. *Cont.*

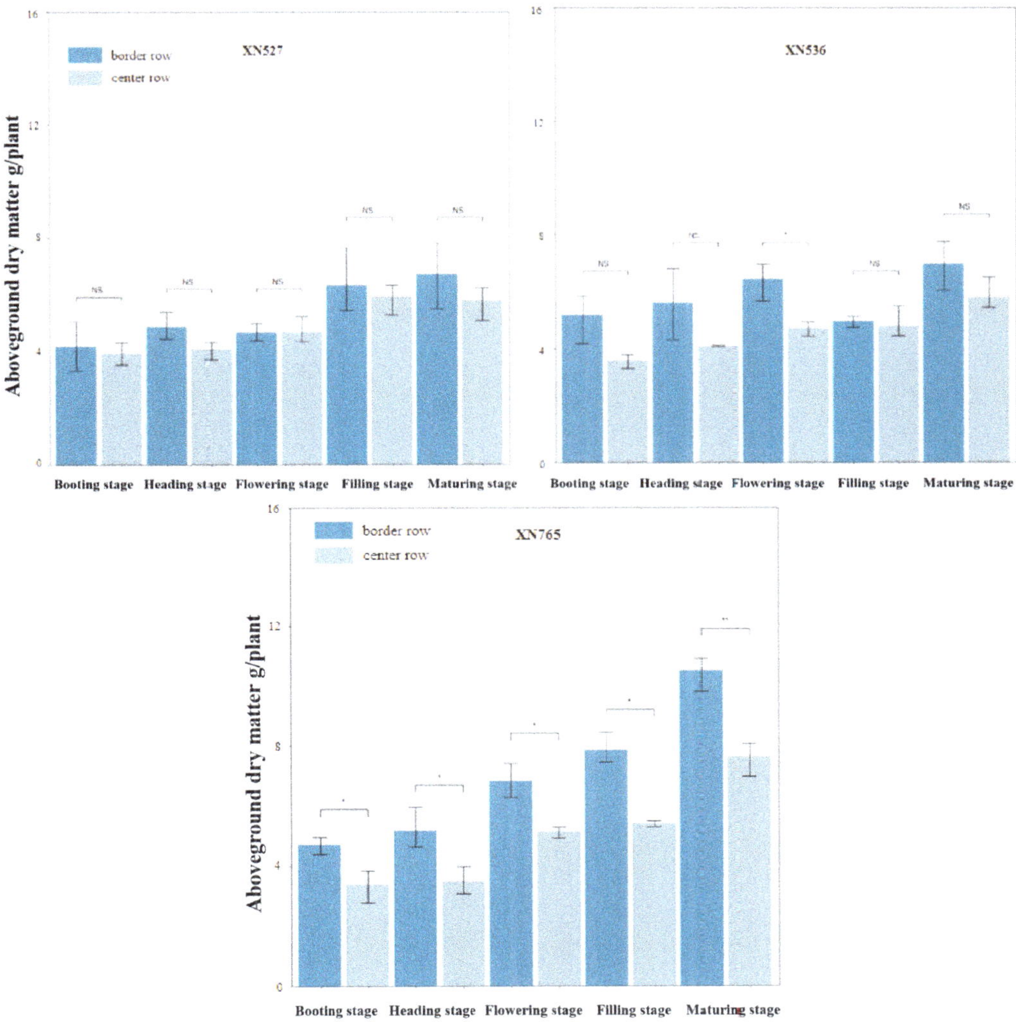

Figure 3. Border effect of aboveground dry matter of wheat in different stages. (*: $p \leq 0.05$, **: $p \leq 0.01$, and NS: non-significant (ANOVA)).

3.3. *Border Effects of Photosynthetic Characteristics*

The photosynthetic characteristics of wheat at different stages (booting stage, heading stage, flowering stage, and filling stage) were measured and the border effects were analyzed.

3.3.1. Border Effects of Net Photosynthetic Rate

The border effect of the net photosynthetic rate in different stages of wheat was analyzed (Figure 4). It can be seen that the net photosynthetic rate of the five varieties in different stages had a specific border advantage. The net photosynthetic rate of the five wheat varieties reached the peak at the heading stage, and reached the lowest value at the filling stage. Among them, only the border effects of XN175 and XN765 were significantly different in particular stages, indicating that these two varieties could exert obvious border effect advantages under hole sowing conditions. Although the remaining three varietie have a certain border effect, the difference was not significant.

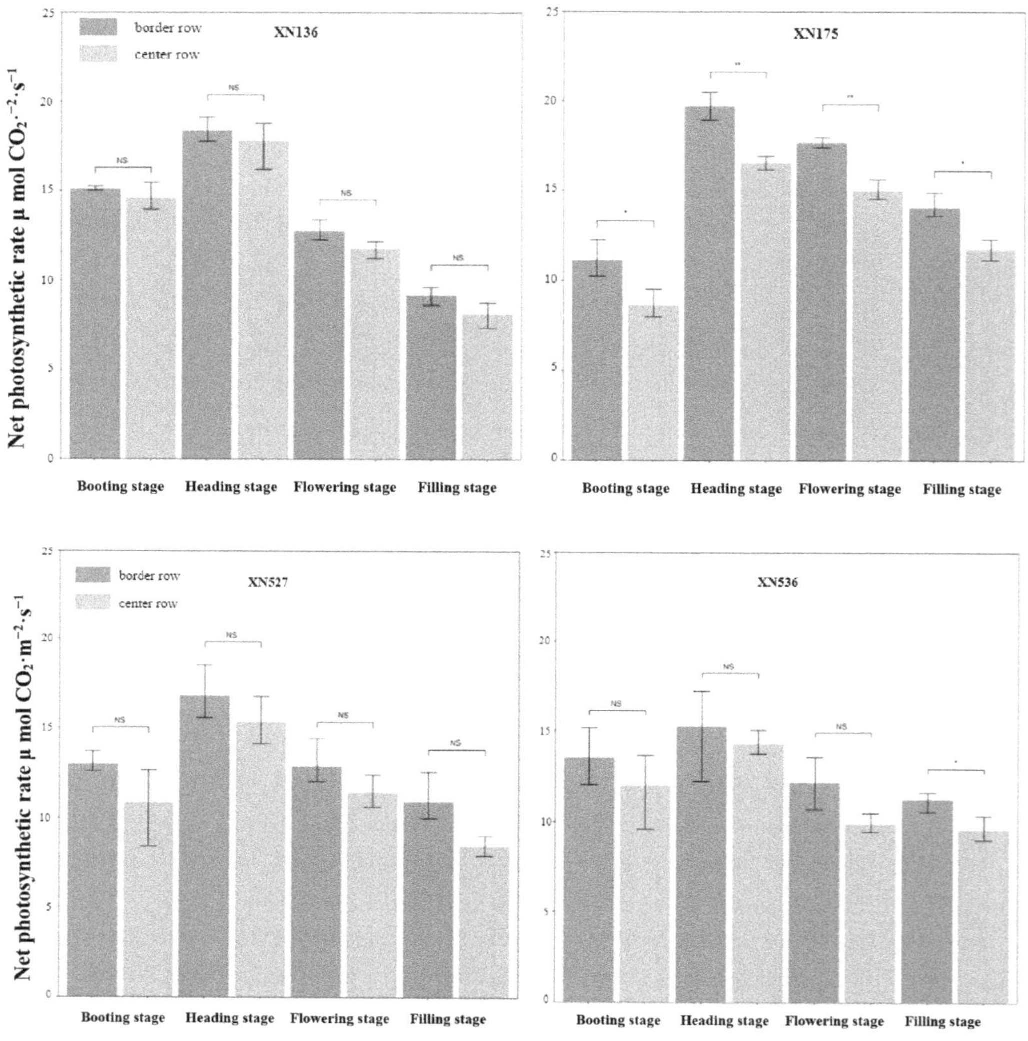

Figure 4. *Cont.*

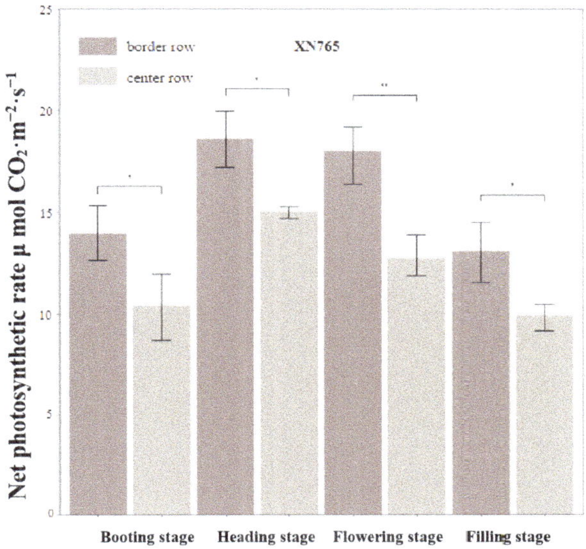

Figure 4. Border effect of net photosynthetic rate of wheat in different stages. (*: $p \leq 0.05$, **: $p \leq 0.01$, and NS: non-significant (ANOVA)).

3.3.2. Border Effects of Stomatal Conductance

The border effect of stomatal conductance in different stages of wheat was analyzed (Figure 5). The stomatal conductance of the three wheat varieties, XN136, XN527, and XN536, had a certain border effect in each stage, but the difference was not significant and all three varieties reached the maximum at the booting stage. This showed an overall downward trend. On the contrary, XN175 and XN765 were significantly different in different stages, where both showed an upward trend from the booting stage to the filling stage, and reached the maximum at the filling stage It can be seen that XN175 and XN765 can play a greater advantage than the other three varieties under hole sowing conditions.

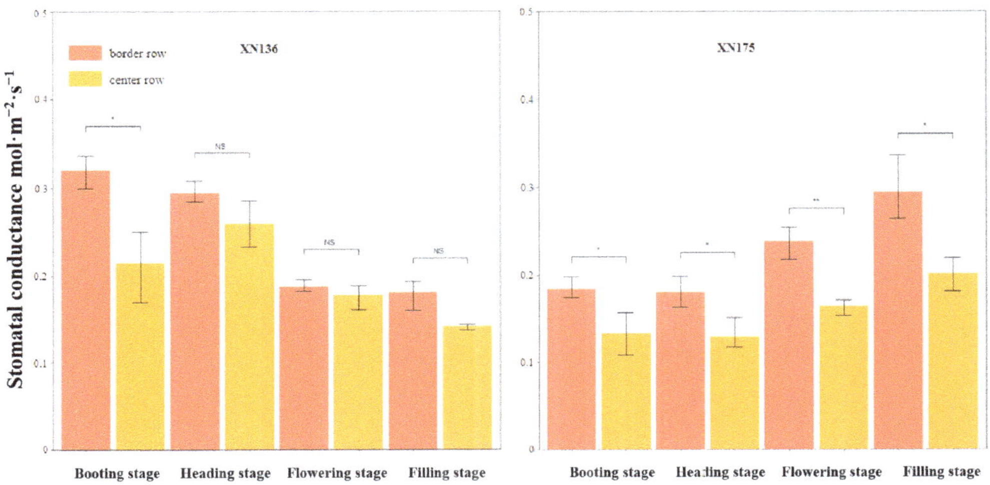

Figure 5. *Cont.*

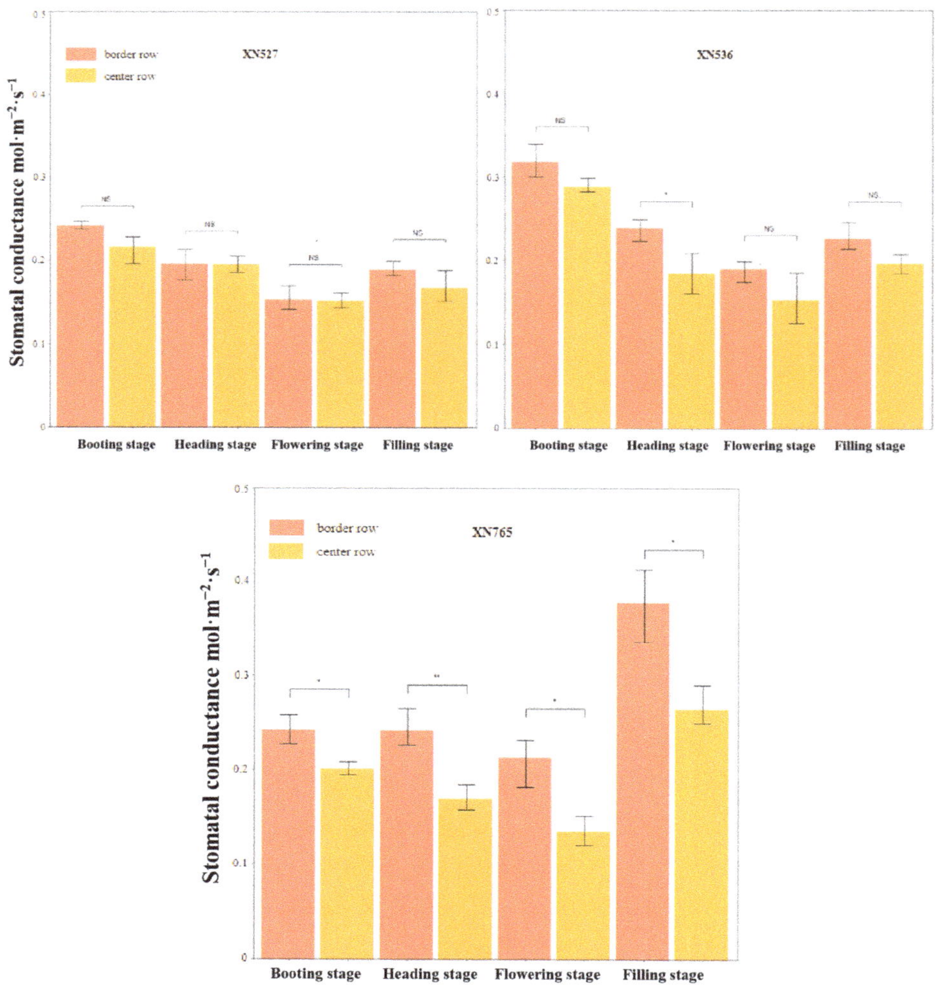

Figure 5. Border effects of stomatal conductance of wheat in different stages. (*: $p \leq 0.05$, **: $p \leq 0.01$, and NS: non-significant (ANOVA)).

3.3.3. Border Effects of Intercellular Carbon Dioxide Concentration

The border effects of wheat intercellular carbon dioxide concentration at different stages were analyzed (Figure 6). The intercellular carbon dioxide concentration of XN136, XN527, and XN536 was relatively stable in different stages. In contrast, the intercellular carbon dioxide concentration of XN175 and XN765 fluctuated wildly and peaked at the filling stage. Overall, the intercellular carbon dioxide concentration of XN175 and XN765 under hole sowing conditions had a more significant border effect.

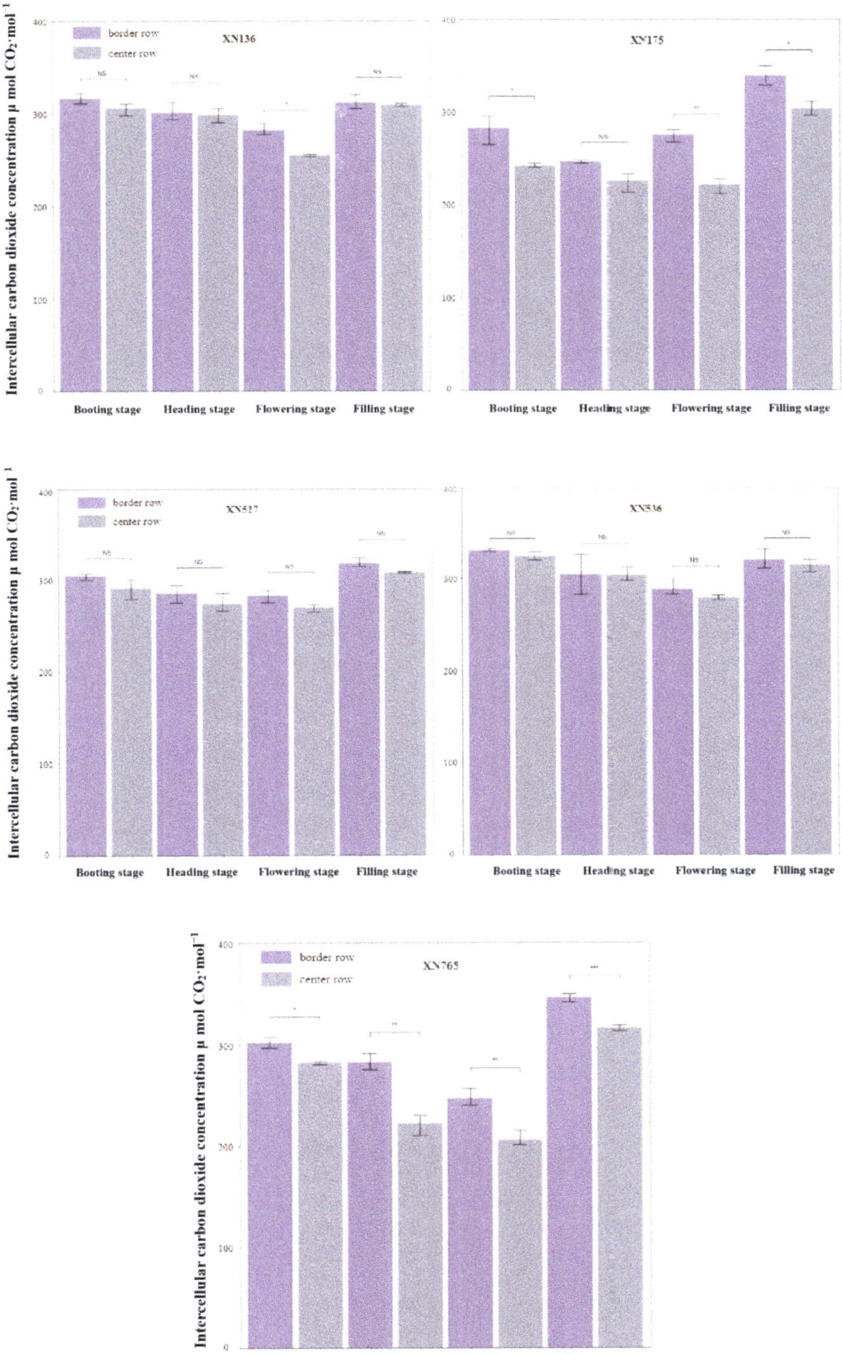

Figure 6. Border effects of intercellular carbon dioxide concentration of wheat in different stages. (*: $p \leq 0.05$, **: $p \leq 0.01$, ***: $p \leq 0.001$, and NS: non-significant (ANOVA)).

3.4. Correlation Analysis of Different Indexes of Wheat

Correlation analysis was performed on net photosynthetic rate, stomatal conductanc intercellular carbon dioxide concentration, dry matter per plant, thousand-grain weigh and grain per spike of the five wheat varieties (Figure 7). In the figure, different colo represent positive and negative correlations, and the color depth represents the correlatio size. The bluer the color, the greater the positive correlation coefficient; the redder the colc the greater the negative correlation coefficient. It was found that only grain per spike and i tercellular carbon dioxide concentration responded negatively to the border effect, and th rest were positively correlated. Among them, grain per spike and aboveground dry matte per plant, stomatal conductance and intercellular carbon dioxide concentration, thousan grain weight, and intercellular carbon dioxide concentration had significant positive co relations with the border effect. There was a significant positive correlation between ne photosynthetic rate and aboveground dry matter per plant, which was the most importar factor affecting the maximum border effect of wheat under hole sowing conditions.

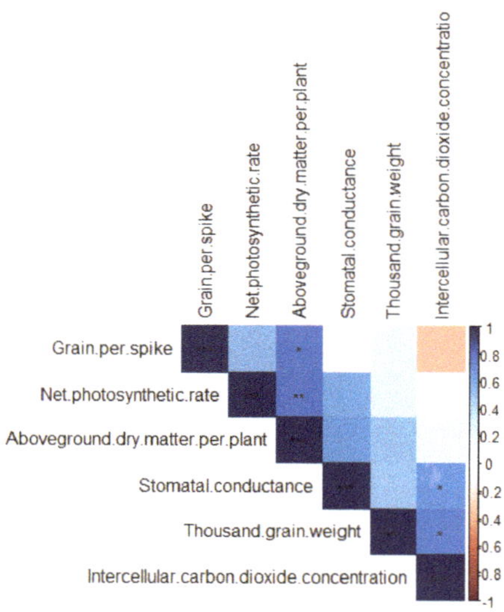

Figure 7. Correlation analysis of different wheat indexes (X axis and Y axis represent different indexe r values in the Figure in different colors, *: $p \leq 0.05$, **: $p \leq 0.01$).

3.5. The Contribution of Different Indicators to Its Significant Border Effects

'Mean decrease Gini' is used to calculate the influence of each variable on the he erogeneity of observations at each node of the classification tree, and to compare th importance of variables. The larger the value, the greater the variable's importance is.

Only XN175 and XN765 showed significant differences in the border effects of differer indicators. Therefore, random forest model analysis was performed on the state, ne photosynthetic rate, stomatal conductance, intercellular carbon dioxide concentratio aboveground dry matter per plant, thousand-grain weight, and grain per spike of these tw varieties (Figure 8). As can be seen from Figure 8, for 'mean decrease Gini', abovegroun dry matter per plant had the most significant response to the border effect, and grain pe spike was the smallest.

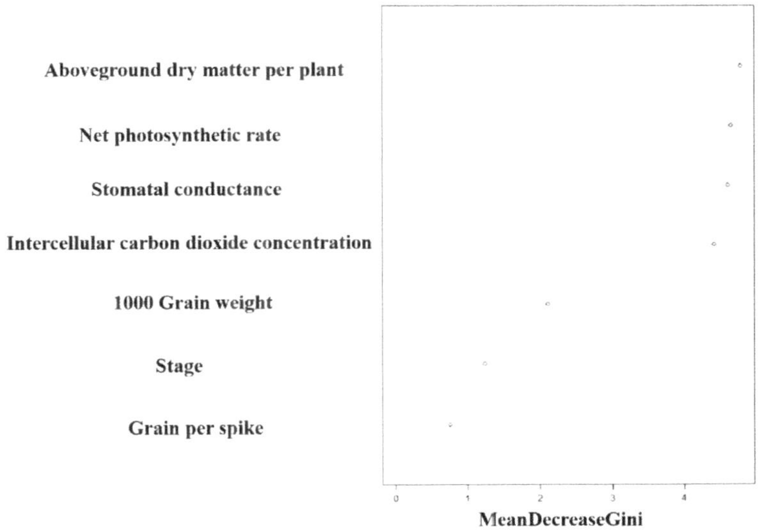

Figure 8. Response of different indexes of wheat cultivars, XN175 and XN765, to border effect (Multi-class area under the curve: 0.9722).

3.6. Difference Analysis of Each Index between Different Wheat Varieties

The photosynthetic characteristics of wheat during the filling stage is the key stage to determine the yield, and about 80% of the nutrients are transported to the wheat grain for accumulation in the middle filling stage. In order to select the most suitable wheat variety for hole sowing among the five varieties, the indexes of wheat filling stage and final yield of the five varieties were compared (Table 2). Among the dry matter per plant, XN175 had the highest value of 8.92 g. Among the photosynthetic characteristics, XN175 had the highest net photosynthetic rate of 12.87 μ mol $CO_2 \cdot m^{-2} \cdot s^{-1}$. The stomatal conductance of XN765 was the largest, which was 0.37 $mol \cdot m^{-2} \cdot s^{-1}$. In the final yield, XN175 was also the highest, which was 8587.1 $kg \cdot ha^{-1}$. Combined with previous results, XN175 and XN765 has been demonstrated to play a greater advantage under hole sowing conditions.

Table 2. Differences in dry matter, photosynthetic characteristics, and final yield of different wheat varieties during grain filling stage from 2021 to 2022. ($p \leq 0.05$, significant difference when the outline and inline characters of the same variety are completely different; $p > 0.05$, no significant difference when the same or more letters are used).

Variety	Dry Matter Per Plant (g)	Net Photosynthetic Rate (μ mol $CO_2 \cdot m^{-2} \cdot s^{-1}$)	Stomatal Conductance ($mol \cdot m^{-2} \cdot s^{-1}$)	Intercellular Carbon Dioxide Concentration (μ mol $CO_2 \cdot mol^{-1}$)	Yield ($kg \cdot ha^{-1}$)
XN136	8.42 a	8.64 c	0.17 c	337.18 a	8358.3 ab
XN175	8.92 a	12.87 a	0.36 ab	336.4 a	8587.1 a
XN527	6.15 bc	9.65 bc	0.26 bc	344.22 a	7474.9 b
XN536	4.90 c	10.4 bc	0.29 ab	349.9 a	8085.3 ab
XN765	6.63 b	11.51 ab	0.37 a	342.88 a	8558.6 a

4. Discussion

The study of the benefits of crop borders usually has two purposes: (1) to avoid the overestimation of crop yields in field trials, and (2) to increase crop productivity by using skip row and rectangular planting patterns. Increasing the dry weight of stubble and non-structural carbohydrate accumulation at harvest of main crops may be an essential strategy for developing high-yield planting practices in rice regeneration systems, by studying the border effects of principal crops and regenerated crops in rice regeneration systems [15]. By measuring the border effect of the rectangular geometry transplanted with wide and narrow hill spacing, by quantifying the size and shape of the hybrid rice planting plot, it was found that for plots with a larger rectangular shape and smaller plot size, the yield estimation will be higher [17]. Maize hybrids have the potential to increase yield through the intercropping system, and through the study of the border effect of maize hybrid intercropping, it was shown that the land equivalent ratio is affected by the use of intercropping hybrids and seasonal climate change [18]. The border effect on the yield of regenerated crops in a mechanized rice regeneration system have also been studied [22]. They proved that the rolling of main crops during mechanical harvesting had a border effect on the yield of the non-rolling zone, thereby reducing the yield loss of regenerated crops.

Our research group has proved that the hole sowing method has an excellent effect on the growth characteristics of wheat, through the comparative test of wheat hole sowing and traditional sowing methods. For example, (1) the effects of different sowing methods on wheat yield and quality was studied by Wu et al. [26], who explored the effects of different sowing methods (drill sowing, wide sowing, and hole sowing) on wheat yield and quality by applying nitrogen fertilizer to wheat 'Xinong 805'. The results showed that the hole sowing treatment increased the flag leaf area of wheat. The application of nitrogen fertilizer increased the dry matter quality of the above-ground part of the wheat in the hole sowing treatment, and the actual yield of the wheat in the hole sowing treatment was the highest, of up to 7430 $kg \cdot ha^{-1}$. The basic seedlings, biomass, and harvest index of wheat under different sowing methods were significantly different. Under the application of nitrogen fertilizer, the storage material transfer amount and contribution rate of each vegetative organ in the hole sowing treatment were the highest. In addition, the hole sowing treatment under topdressing nitrogen fertilizer increased the volume mass, sedimentation value, protein mass fraction, hardness, stability time, tensile area, elongation, and maximum tensile resistance of the grain. For the effects of different sowing methods and sowing rates on wheat yield and quality (2), Qi et al. [27] studied the effects of different sowing methods and sowing rates on grain yield, yield components, protein content, component content, and processing quality of winter wheat. Using high-quality and high-yield winter wheat 'Xinong20' as material, three different sowing methods (drill sowing, wide sowing, and hole sowing) and four different sowing rates (112.5, 150, 187.5, and 225 $kg \cdot ha^{-1}$) were set up for the experiment. Hole sowing is beneficial to the improvement of protein and its components content and processing quality. Increasing the appropriate sowing rate can increase the content of protein and its components.

Previously, no scholars have studied the border effect of wheat under hole-sowing conditions and the main factors affecting its border effect. In this study, the traits of five wheat varieties showed different border effects under the hole-sowing cultivation method. However, only the different indicators of XN175 and XN765 have significant differences. In dry matter, XN175 had significant difference in the boundary effect of dry matter per plant above ground at booting stage and filling stage, while XN765 had significant difference in the boundary effect of dry matter per plant above ground at maturing stage, and the other four stages had significant difference. In the photosynthetic characteristics, the net photosynthetic rate boundary effect of XN175 and XN765 in each stage was significantly different. The stomatal conductance of XN175 and XN765 increased with the growth stage, and had significant boundary effects at different growth stages of wheat. The intercellular carbon dioxide concentration of XN175 was significantly different at booting stage, flowering stage, and filling stage, and the intercellular carbon dioxide concentration of XN765 was significantly different at each stage. It can be seen that these two varieties play an advantage over the other three varieties under the cultivation method of hole sowing, and have higher wheat yields. XN175 and XN765 may be more suitable for bunch planting than the other three varieties, and have significant border effects. This study only studied the border effect of wheat under the condition of hole sowing from the same sowing density. In the future, it is necessary to further study the boundary effect response of sowing density to wheat. At present, there are many cultivation methods, but the traits of different wheat varieties under various cultivation methods should be different. Therefore, it is necessary to establish a model to match the best cultivation methods for wheat in the future.

Through the correlation analysis of different indexes of wheat, it can be found that only grain per spike and intercellular carbon dioxide concentration were negatively correlated with the border effect of wheat under hole sowing conditions, while the rest were positively correlated. Through further random forest model analysis of XN175 and XN765 wheat varieties with significant border effects of each index, it can be found that net photosynthetic rate and aboveground dry matter per plant have the greatest influence on the significant border effect. In contrast, grain per spike has a minor influence on the significant border effect.

5. Conclusions

Under the cultivation mode of hole sowing, different wheat varieties have specific border effects. The varieties with the most significant border effect may be more suitable for hole sowing than other varieties. Under the warm temperate continental monsoon climate conditions, such as those found in the Guanzhong irrigation area in Shaanxi Province, wheat suitable for hole sowing, as a sowing method, can maximize its performance and obtain higher yield. According to our experiment, 'XN175' and 'XN765' had more significant border effects than other varieties under hole sowing conditions. Therefore, 'XN175' and 'XN765' were more suitable for sowing under hole sowing conditions than the other three varieties, and should be fully considered in the popularization and application of hole sowing. Our results fill the gap in the study of the border effect of wheat under the hole-sowing cultivation method. Readers can obtain exciting information from the data analysis of this study, which provides a valuable reference and help for future researchers.

Author Contributions: Y.S.: Investigation, Methodology, Writing—original draft; C.Y.: Writing-original draft; H.L.: Methodology; Y.Y.: Writing—review & editing; K.B.: Writing—review & editing; Y.D.: Methodology; and J.H.: Project administration, Funding acquisition. All authors have read and agreed to the published version of the manuscript.

Funding: The research was financially supported by the Ecological Security and Bioremediation Mechanism of Saline-alkali Soil Improvement in the Middle Yellow River (No. DL2021172002L).

Institutional Review Board Statement: Not applicable.

Data Availability Statement: Not applicable.

Acknowledgments: The authors express their special gratitude to the funding source for the financial assistance and are also thankful to the anonymous reviewers for their constructive and valuable comments on earlier versions of this research article.

Conflicts of Interest: The authors declare no conflict of interest.

eferences

Gong, F.; Wu, X.; Zhang, H.; Chen, Y.; Wang, W. Making better maize plants for sustainable grain production in a changing climate. *Front. Plant Sci.* **2015**, *6*, 835. [CrossRef] [PubMed]
Shewry, P.R.; Hey, S. Do "ancient" wheat species differ from modern bread wheat in their contents of bioactive components? *J. Cereal Sci.* **2015**, *65*, 236–243. [CrossRef]
Singh, M.; Singh, K. Agronomic zinc biofortification of wheat. *Agrochimica* **2019**, *63*, 307–317. [CrossRef]
Mohammed, A.A.B.A.; Omran, A.A.B.; Hasan, Z.; Ilyas, R.A.; Sapuan, S.M. Wheat Biocomposite Extraction, Structure, Properties and Characterization: A Review. *Polymers* **2021**, *13*, 3624. [CrossRef]
Xu, H.; Twine, T.E.; Girvetz, E. Climate Change and Maize Yield in Iowa. *PLoS ONE* **2016**, *11*, e0156083. [CrossRef]
Hegsted, D.M.; Trulson, M.F.; Stare, F.J. Role of Wheat and Wheat Products in Human Nutrition. *Physiol. Rev.* **1954**, *34*, 221–258. [CrossRef] [PubMed]
Jaradat, A.A. Wheat Landraces: A mini review. *Emir. J. Food Agric.* **2013**, *25*, 20–29. [CrossRef]
Li, J.; Yang, J.; Li, Y.; Ma, L. Current strategies and advances in wheat biology. *Crop J.* **2020**, *8*, 879–891. [CrossRef]

9. Yang, F.; Zhang, J.; Liu, Q.; Liu, H.; Zhou, Y.; Yang, W.; Ma, W. Improvement and Re-Evolution of Tetraploid Wheat for Glob Environmental Challenge and Diversity Consumption Demand. *Int. J. Mol. Sci.* **2022**, *23*, 2206. [CrossRef]
10. Leegood, R.C.; Evans, J.R.; Furbank, R.T. Food security requires genetic advances to increase farm yields. *Nature* **2010**, *464*, 83 [CrossRef]
11. Ali, S.; Xu, Y.; Jia, Q.; Ahmad, I.; Wei, T.; Ren, X.; Zhang, P.; Din, R.; Cai, T.; Jia, Z. Cultivation techniques combined wit deficit irrigation improves winter wheat photosynthetic characteristics, dry matter translocation and water use efficiency und simulated rainfall conditions. *Agric. Water Manag.* **2018**, *201*, 207–218. [CrossRef]
12. Martínez-Moreno, F.; Ammar, K.; Solís, I. Global Changes in Cultivated Area and Breeding Activities of Durum Wheat from 18C to Date: A Historical Review. *Agronomy* **2022**, *12*, 1135. [CrossRef]
13. Hu, X.; Ma, J.; Qian, W.; Cao, Y.; Zhang, Y.; Liu, B.; Tang, L.; Cao, W.; Zhu, Y.; Liu, L. Effects of Low Temperature on the Amin Acid Composition of Wheat Grains. *Agronomy* **2022**, *12*, 1171. [CrossRef]
14. de Souza Gonçalves, P.A.; Sousa e Silva, C.R. Efeito de espécies vegetais em bordadura em cebola sobre a densidade populacion de tripes e sirfídeos predadores. *Hortic. Bras.* **2003**, *21*, 731–733. [CrossRef]
15. Zheng, C.; Wang, Y.-C.; Xu, W.-B.; Yang, D.-S.; Yang, G.-D.; Yang, C.; Huang, J.-L.; Peng, S.-B. Border effects of the main an ratoon crops in the rice ratooning system. *J. Integr. Agric.* **2023**, *22*, 80–91. [CrossRef]
16. Liu, G.; Zhang, G.; Hou, P.; Liu, Y.; Li, J.; Ming, B.; Xie, R.; Wang, K.; Li, S. Weak border effects and great uniformity increase yiel of maize (*Zea mays*) under dense population. *Crop Pasture Sci.* **2020**, *71*, 653–659. [CrossRef]
17. Wang, K.; Zhou, H.; Wang, B.; Jian, Z.; Wang, F.; Huang, J.; Nie, L.; Cui, K.; Peng, S. Quantification of border effect on grain yiel measurement of hybrid rice. *Field Crop Res.* **2013**, *141*, 47–54. [CrossRef]
18. Wang, Y.; Zhao, Z.; Li, J.; Zhang, M.; Zhou, S.; Wang, Z.; Zhang, Y. Does maize hybrid intercropping increase yield due to bord effects? *Field Crop Res.* **2017**, *214*, 283–290. [CrossRef]
19. Wang, Y.; Hu, W.; Zhang, X.; Li, L.; Kang, G.; Feng, W.; Zhu, Y.; Wang, C.; Guo, T. Effects of cultivation patterns on winter whe root growth parameters and grain yield. *Field Crop Res.* **2014**, *156*, 208–218. [CrossRef]
20. Ahmed, O. Assessing the Current Situation of the World Wheat Market Leadership: Using the Semi-Parametric Approac *Mathematics* **2021**, *9*, 115. [CrossRef]
21. Hu, Q.; Jiang, W.-Q.; Qiu, S.; Xing, Z.-P.; Hu, Y.-J.; Guo, B.-W.; Liu, G.-D.; Gao, H.; Zhang, H.-C.; Wei, H.-Y. Effect of wide-narro row arrangement in mechanical pot-seedling transplanting and plant density on yield formation and grain quality of japonic rice. *J. Integr. Agric.* **2020**, *19*, 1197–1214. [CrossRef]
22. Zhou, Y.; Zheng, C.; Chen, G.; Hu, R.; Ji, Y.; Xu, Y.; Wu, W. Border Effect on Ratoon Crop Yield in a Mechanized Rice Ratoonin System. *Agronomy* **2022**, *12*, 262. [CrossRef]
23. Cappelli, A.; Cini, E. Challenges and Opportunities in Wheat Flour, Pasta, Bread, and Bakery Product Production Chains: Systematic Review of Innovations and Improvement Strategies to Increase Sustainability, Productivity, and Product Qualit *Sustainability* **2021**, *13*, 2608. [CrossRef]
24. Oda, S. Recent Improvement of Japanese Wheat. *J. Jpn. Soc. Food Sci. Technol.* **2017**, *64*, 165–170. [CrossRef]
25. Sarker, M.; Itohara, Y.; Hoque, M.; Uddin, S.N. Scope and challenges of organic wheat cultivation in Bangladesh. *Aust. J. Crop Sc* **2011**, *5*, 1114–1119.
26. Wu, B.; Dong, Y.; Hai, J. Effect of Hole Seeding Way on Yield and Quality of Winter Wheat. *Acta Agric. Boreali-Occident. Sin.* **201** *28*, 906–913. (In Chinese)
27. Qi, H.; Dong, Y.; Li, C.; Angelique Twizerimana Ren, H.; Hai, J. Effects of Sowing Methods and Seeding Rates on Yield an Quality of Winter Wheat Variety 'Xinong 20'. *Acta Agric. Boreali-Occident. Sin.* **2021**, *30*, 32–40. (In Chinese)

agronomy

Article

Establishing Optimal Planting Windows for Contrasting Sorghum Cultivars across Diverse Agro-Ecologies of North-Eastern Nigeria: A Modelling Approach

Folorunso M. Akinseye [1,2,*], Hakeem A. Ajeigbe [1], Alpha Y. Kamara [3], Akinrotimi O. Omotayo [1], Abdullahi I. Tofa [3] and Anthony M. Whitbread [4]

1 International Crop Research Institute for Semi-Arid Tropics (ICRISAT), Kano P.M.B. 349, Nigeria
2 Centre d'étude régional pour l'amélioration de l'adaptation à la sécheresse (CERAAS), Thies Escale BP 3320, Senegal
3 International Institute of Tropical Agriculture (IITA), Ibadan P.M.B. 5320, Nigeria
4 International Livestock Research Institute (ILRI), Dar es Salaam P.O. Box 34441, Tanzania
* Correspondence: f.akinseye@icrisat.org

Abstract: In the context of climate change, the sowing date and cultivar choice can influence the productivity of sorghum, especially where production is constrained by low soil fertility and early terminal drought across the challenging agro-ecologies of north-eastern Nigeria. Planting within an optimal sowing window to fit the cultivar's maturity length is critical for maximizing/increasing the crop yield following the appropriate climate-smart management practices. In this study, the APSIM crop model was calibrated and validated to simulate the growth and yield of sorghum cultivars with differing maturing periods sown within varying planting time windows under improved agricultural practices. The model was run to simulate long-term crop performance from 1985 to 2010 to determine the optimal planting windows (PWs) and most suitable cultivars across different agro-ecological zones (AEZs). The performance of the model, validated with the observed farm-level grain yield, was satisfactory across all planting dates and cropping systems. The model predicted a lower mean bias error (MBE), either positive or negative, under the sole cropping system in the July sowing month compared to in the June and August sowing months. The seasonal climate simulations across sites and AEZs suggested increased yields when using adapted sorghum cultivars based on the average grain yield threshold of ≥1500 kgha^{-1} against the national average of 1160 kgha^{-1}. In the Sudan Savanna (SS), the predicted optimum PWs ranged from 25 May to 30 June for CSR01 and Samsorg-44, while the PWs could be extended to 10 July for ICSV400 and Improved Deko. In the Northern Guinea Savanna (NGS) and Southern Guinea Savanna (SGS), the optimal PWs ranged from 25 May to 10 July for all cultivars except for SK5912, for which predicted optimal PWs ranged from 25 May to 30 June. In the NGS zone, all cultivars were found to be suitable for cultivation with exception of SK5912. Meanwhile, in the SGS zone, the simulated yield below the threshold (1500 kgha^{-1}) could be explained by the sandy soil and the very low soil fertility observed there. It was concluded that farm decisions to plant within the predicted optimal PWs alongside the use of adapted sorghum cultivars would serve as key adaptation strategies for increasing the sorghum productivity in the three AEZs.

Keywords: adaptation; agro-ecological zones (AEZs); APSIM; adapted sorghums; optimal planting window

Citation: Akinseye, F.M.; Ajeigbe, H.A.; Kamara, A.Y.; Omotayo, A.O.; Tofa, A.I.; Whitbread, A.M. Establishing Optimal Planting Windows for Contrasting Sorghum Cultivars across Diverse Agro-Ecologies of North-Eastern Nigeria: A Modelling Approach. *Agronomy* **2023**, *13*, 727. https://doi.org/10.3390/agronomy13030727

Academic Editors: Umberto Anastasi and Aurelio Scavo

Received: 22 December 2022
Revised: 24 February 2023
Accepted: 25 February 2023
Published: 28 February 2023

1. Introduction

Nigeria is the largest producer of sorghum in West Africa, accounting for about 65–70% of the total sorghum production in the region [1]. Its sorghum production in 2018 was 6.9 million tonnes, accounting for 50% of the total cereal production and occupying about 45% of the total land area devoted to cereal crop production in Nigeria [2]. The production of sorghum in Nigeria, where it is predominantly cultivated in the northern region, has increased overall [3], reaching some seven million tons in 2021, with an average yield of 1160 kgha^{-1} meaning that it is one of the main crops for the country. The increase in production is associated with the dissemination of improved sorghum cultivars that are tolerant to drought and Striga [4]. These cultivars have been promoted through several initiatives by the Federal Government of Nigeria and other development partners. Landraces have long been recognized as a source of traits for local adaptation, stress tolerance, yield stability, and seed nutrition [5]. The long-term selection under variable and low-input environments has resulted in high crop diversity in landraces. The environmental factors contributing to production constraints and low yields include low fertility soils, the length of the growing periods, drought, and water-logging, as well as biotic stresses such as *Striga* parasitism and diseases attacking the foliage, stems, and/or grain [6]. Photoperiod sensitivity is an important trait of West African sorghum germplasm that allows farmers to cope with variations in the planting date (PD) and adapt to environmental constraints [7,8]. The triggering of flowering by day length effectively serves to synchronize the final developmental stages with the end of the rainy season [9]. A major problem in rainfed agriculture in semi-arid regions characterized by short rainy season, occasionally accompanied by in-season drought, is how to determine the optimum sowing date [10]. The delays in the onset of the rainfall, drought, unpredictable periodic dry spells, and shortened rainfall seasons have led to a slight shift in the traditionally recommended sorghum planting dates [11].

Crop management must not only adapt to changing climatic conditions to maintain sufficient production but must do so in a way that reduces greenhouse gas emissions as much as possible—i.e., cropping systems must be climate smart [12]. Transformative changes for climate-smart agriculture must include changes to crops, management, and systems that build resilience to climate change impacts and emit relatively low emissions [13]. Although limited data exist, the available studies have shown that the cultivation of sorghum is relatively low in agricultural emissions compared to other crops [14]. Despite the importance of understanding the potential of sorghum to contribute to a climate-smart future and to food security in Nigeria, as well as in the dryland West Africa region, the promotion of productivity-enhancing technologies (climate-smart strategies) among the farmers is becoming imperative for increasing productivity. Therefore, the choice of a sorghum cultivar with an appropriate planting date should be combined such that the productivity of the sorghum would be optimal when the flowering occurs at least 20 days before the terminal drought in the cropping season [7,15,16]. Thus, matching the phenology to the given biotic and abiotic conditions is a prerequisite for good varietal adaptation to a given environment [7]. Crops adapt to diverse environments through considerable plasticity of phenology, the main determinant of which is rainfall [17] in the semi-arid region; meanwhile, the temperature has a stronger effect in the temperate region. "Manipulating this climatic factor would require adequate knowledge of planting dates so as to accurately synchronize rainfall incidences with crop development" [18].

In north-eastern Nigeria, as applies to other semi-arid regions, the length of the growing period (LGP) is mainly a function of the date of the first rains [19,20], which is delayed as we moved northward and varies widely from year to year. The region is prone to climatic risk, and a good knowledge of the cultivar development cycles relative to the planting date is required for improved productivity. However, with the variable onset and distribution of rainfall as well as the frequent occurrence of drought within the growing season, the farmers' choice of cultivars would depend mainly on their knowledge of the crop's phenology and yield potential in relation to the local characteristics of the wet season [21,22].

In West Africa's semi-arid agro-ecology, favourable conditions for sorghum cultivation usually extend from May to November [20]. Thus, floral initiation takes place under decreasing day length, and the growth duration of photoperiod-sensitive cultivars will be shortened when sowing is delayed [23]. Although photoperiod sensitivity benefits sorghum, in that flowering takes place at a relatively fixed calendar date and allows it to mature after the rains end, despite highly variable sowing dates [24], a high degree of poor grain filling is encountered among the late-planted and late-maturing varieties that run out of water if the sorghum is planted too late in the season [25,26]. In this situation, matching suitable cultivars with their optimal planting windows becomes an important management option. In addition, knowing the extent to which planting can be delayed and the likely yield penalty due to later than the optimal planting [27] is important for increasing the productivity of sorghum in a semi-arid environment.

In semi-arid environments, the planting date decision is important not only because of its effect on yield [28], but also because of the need to minimize the risk of establishment failures and ensure the availability of water for unrestricted plant growth and transpiration [17]. Recommendations concerning the planting dates of crops are usually based on agronomic field experiments that are specific to the fields and regions [29]. The majority of such trials cannot be temporally and spatially replicated across diverse agro-ecologies because of seasonal variations. The determination of the optimum sowing dates for sorghum by field experimentation entails repetition over long periods in order to capture the seasonal variability in the rainfall with the varying photoperiod sensitivity cultivars available. Thus, cropping system models (CSMs) have been a proven methodology for understanding the interactions between climate, soils, farming systems, and management [30,31]. These models, therefore, remain important diagnostic tools for decision-making, not only to capture the effects of variability of the rainfall and edaphic factors on crop productivity, but also to suggest sowing date rules and other crop management strategies for better and more sustainable agriculture [31,32]. Cropping system models such as Agricultural Production Systems sIMulator, APSIM [33,34], describe the dynamics of crop growth, soil water, soil nutrients, and plant residues as a function of climate, cropping history, and soil/crop management in a daily time step. Through the linking of crop growth with soil processes, APSIM is particularly suited for the evaluation of the likely impacts of alternative management practices such as varying planting dates on soil resources and crop productivity. The model has been used intensively in the search for strategies for more efficient production, improved risk management, crop adaptation, and sustainable production [33,35,36]. This work, therefore, seeks to establish the response of diverse sorghum cultivars to different planting windows in the three major agro-ecologies of north-eastern Nigeria. To achieve this, the following objectives were set: (i) evaluate the performance of the APSIM model for simulating the contrasting sorghum cultivars under different management systems, soils, and rainfall patterns; (ii) apply the model to determine the optimal PWs and adapted sorghum cultivars for higher grain yield and resilience in order to minimize crop failure across sites and AEZs.

2. Materials and Methods

2.1. Model Calibration (Experiments, Data Collection, Procedure for Model Calibration and Evaluation)

The experimental data used for the calibration were principally generated from on-station field experiments conducted between 2016 and 2018 under optimal conditions (i.e., no water and nitrogen stress) in two AEZs (Abuja, Southern Guinea savannah, and Kano, Sudan savannah) in northern Nigeria. The experiment was designed to evaluate the effects of sowing dates and nutrient responses on contrasted sorghum cultivars. In Abuja, the experiment was established at the International Institute of Tropical Agriculture (IITA) field station (Latitude 9.16° N, and Longitude 7.35° E), while, in Kano, the experiment was established in two locations: (i) the Bayero University Kano (BUK) Teaching and Research Farm (Latitude 12.98° N and Longitude 9.75° E) and (ii) the ICRISAT research field situated within the Institute for Agricultural Research (IAR) station, Wasai Village, Minjibir (Latitude 12.17° N and Longitude 8.65° E). The details of the experiment and the agronomic data collected have been reported [37,38]. Among the 20+ sorghum cultivars commercially available in Nigeria, five contrasting sorghum cultivars that were considered to be widely cultivated were tested based on their breeding selection history for phenology, photoperiod sensitivity, and grain yield productivity. According to a national cultivar r port [39,40], ICSV-400 is an early maturing cultivar (85–90 days), is photoperiod-insensitiv and has a yield potential from 2.5 to 3.5 t/ha; Improved Deko is medium maturin (90–110 days) and has a low photoperiod sensitivity and a yield potential from 3.5 t 4.0 t/ha; Samsorg-44 and CSR01 are medium maturing and medium photoperiod-sensitiv and have yield potential from 2.0 to 2.5 t/ha; and SK5912 is late maturing (165–175 day and highly photoperiod-sensitive, with a potential yield of 2.5–3.5 t/ha when grown und optimum conditions.

The daily weather was obtained from an automatic weather station (AWS) installe within a 2 km radius of the experiment for the corresponding years of the experimer and was used for calibration. The parameters include the daily maximum and minimur temperature, the solar radiation, and the rainfall. Management practices such as plan ing dates, sowing depth, plant density, type and amount of fertilizer applied in form c NPK, and tillage (type, depth, and fraction of above-ground materials incorporated) we recorded and used for the model setup and simulation. The soil samples were taken b fore planting at each experimental site and were analysed for their physical and chemic properties. The agronomic data, such as dates of flowering and maturity, leaf number pe plant, leaf area index (LAI), yield, and final biomass collected [4], were used to determin the cultivar-specific parameters.

The calibration of the APSIM-sorghum module was implemented within the APSIM 7.10 framework based on the phenology, morphology, yield, and aboveground biomas data described earlier. The model APSIM requires a number of inputs, which include th cultivar type, crop management practices/information, soil properties, and daily weath records (rainfall, minimum temperature, maximum temperature, and solar radiation). Cro development follows a thermal time approach with a reported base (T_b) and optimal (T_{op} and maximum (T_m) temperatures of 11, 32, and 42 °C [41,42]. The thermal time target fc the phase between emergence and panicle initiation is also a function of the day lengt and its duration, when divided by the plastochron (°C degrees per leaf), determines th total leaf number. The total leaf number multiplied by the phyllochron (°C d per lea determines the thermal time to reach the flag leaf stage, which is thus an emergent propert of the model. For parameterizing the genetic coefficients of previously undefined sorghur cultivars, the phenological and morphological stages were based on a combination c observed data and simulation to obtain a yield and above-ground biomass (AGB) that fe within the predefined error limits for each cultivar. Following this method, all coefficien were optimized for further simulation as defined in Table 1. Thereafter, the performanc of the model in simulating the phenology (days to flowering and maturity), morpholog (leaf number per plant and maximum leaf area index (Max_LAI)), grain yield, and AG were compared with the observed values and assessed using mean bias error (MBE), roc mean square error (RMSE), normalized root mean square error ($RMSE_n$) and the tradition R^2 regression statistic (least-squares coefficient of determination) [43]. $RMSE_n$ gives measure (%) of the relative difference between the simulated versus observed data. Th simulation was considered excellent with $RMSE_n$ < 10%, good if 10–20%, acceptable or fa if 20–30%, and poor >30% [44].

Table 1. Genetic coefficients of sorghum cultivars calibrated in the APSIM-sorghum model.

Description of Parameter	Unit	ICSV400	Impr. Deko	CSR01	Samsorg-44	SK5912	Calibration Method (A/B)
Thermal time from emergence to end of juvenile	°C days	180	210	100	100	100	A
Thermal time from end of juvenile to floral initiation	°C days	160	100	100	100	120	A
Photoperiod slope	°C/hour	150	200	500	550	600	A

Table 1. *Cont.*

Description of Parameter	Unit	ICSV400	Impr. Deko	CSR01	Samsorg-44	SK5912	Calibration Method (A/B)
Thermal time from flag leaf to flowering	°C days	170	170	100	100	150	A
Thermal time from flowering to start of grain filling	°C days	80	80	80	80	80	B
Thermal time from flowering to maturity	°C days	560	560	460	500	450	A
Leaf appearance rate (leaf app rate 1)	°C d/leaf	41	56	56	56	56	[31]
Leaf appearance rate (leaf app rate 2)	°C d/leaf	20	28	28	28	28	[31]
Radiation use efficiency(RUE)	g/MJ	1.25	1.25	1.35	1.35	1.65	A
Head grain number determination	g/grain	0.00083	0.0088	0.00083	0.00083	0.0088	A
Maximum grain filling (MaxGFrate)	mg/grain/day	0.09	0.03	0.05	0.05	0.09	A

A: Manual tuning of parameter values; B: Model defaults values: [31] means the parameter calibrated based on the value reported.

$$MBE = 1 - \frac{(\sum_{i=1}^{n} Oi - \sum_{i=1}^{n} Pi)}{\sum_{i=1}^{n} Oi} \tag{1}$$

$$RMSE = \left[\frac{\sum_{i=1}^{n} (Pi - Oi)^2}{n}\right]^{0.5} \tag{2}$$

$$RMSEn\,\% = \left[\frac{\sum_{i=1}^{n} (Pi - Ci)^2}{mean\ of\ observed\ data}\right]^{0.5} \times 100 \tag{3}$$

where n is the number of observations, Pi is the predicted value for the ith measurement and Oi is the observed value for the ith measurement, and O and P represent the mean of the observed and predicted values for all of the parameters studied.

2.2. Model Validation (Experiments, Data Collection, Procedure for Model Validation, and Evaluation)

An independent dataset used for model validation was generated from multi-locational on-farm trials for improved sorghum production technology conducted through the farmers' participatory program between 2013 and 2017. The dataset revealed three distinct cropping systems (intercropping, mixed cropping, and sole cropping) comprising a range of production technologies, including improved sorghum varietal demonstration, seed dressing techniques, conservation agriculture (minimum tillage and conventional tillage), and fertilization strategies aimed at increasing sorghum productivity at the farm level. The additional datasets were obtained from the ICRISAT breeding program from on-farm varietal experiments tested across northern Nigeria spanning four agroecological zones (Sahelian, Sudan Savanna, Northern Guinea, and Southern Guinea Savanna). All the data used are well-documented and include information about basic agronomic management practices such as the sowing date, fertilizer application rate, time of application, planting density, reference geographical coordinates of each farm plot/community, final grain yield, and stalk yield for the five (5) selected and calibrated sorghum cultivars. In addition, variations in the planting date across farms and cultivars were grouped under three months (referred to as "sowing month"), which revealed that 92% of farmers planted in the months of June and July, and only 8% of the farmers sowed in the month of August. Weather data were generated using the downscaled Climate Hazards Group InfraRed Precipitation with Station (CHIRPS) rainfall at a 5.5 km resolution and merged with NASA Power data (temperatures and solar radiation) from the database for Climatology Resource for Agroclimatology, National Aeronautics and Space Administration (NASA) (http://power.larc.nasa.gov, assessed on 25 April 2019) for the corresponding farm's reference coordinates.

Two sources of soil information were obtained for soil parametrization. The first included field-measured soil characteristics and combined the reconnaissance soil survey of Nigeria reported in 1990 and the soil analysis by the Taking Maize Agronomy to Scale in Africa (TAMASA) project in Kano, Kaduna, and Katsina States, respectively. The second soil data source was downscaled ISRIC (International Soil Reference and Information Centre) soil data in 10 × 10 km grids, with the profile layers (in cm) being 5, 15, 30, 60, 100, and 200, used for the corresponding farm's references coordinates. After bias correction of the gridded dataset using the available soil measurement, the soil information was extracted from the ISRIC database [45] for each farm's reference coordinates (the nearest grid point) to run the simulation across the locations. Furthermore, R scripts were developed to (i) append the CHIRPS and NASA power data together and convert each location into a format readily ingestible by APSIM; and (ii) remap the ISRICS gridded soil from 5 cm to 15 cm for the top soil layer as required by APSIM, and then convert these soils into an APSIM SOIL readable format. Following the calibrated cultivar-specific coefficients, an excel executable file was developed that incorporated the management practices, cultivar name, soil, and weather records for the corresponding farm/plot alongside the reported observed grain yield. From the spreadsheet executable file, we created a 3266 APSIM simulation setup that defined different sowing dates, planting densities, and fertilizer applications as reported for the five sorghum cultivars. The model's simulated and observed value was evaluated only for grain yield across the sowing and cropping system using the mean bias error (MBE) and root means square error (RMSE).

2.3. Bias Correction Methods: Daily Observed Rainfall Versus Gridded Rainfall Data (CHIRPS)

Data from nine (9) rainfall observation stations in northern Nigeria with long-term records (1983–2006) were obtained from the climatological unit of the Nigerian Meteorological Agency (NIMET). The Climate Hazards Group InfraRed Precipitation with Station (CHIRPS) data are satellite-based rainfall products with relatively high resolutions (0.05°) and quasi-global coverage (50° S-50° N) for their daily, pentadal, and monthly precipitation datasets [46]. The data were downscaled over the Nigeria grids and extracted for the reference coordinates of the 9 daily observed rainfall stations and 288 different farms coordinates used in the simulations. The bias correction of the gridded data using station-observed data has been shown to increase its applicability to daily time-step agricultural modelling [47]. Two techniques (linear scaling (LS) and empirical quantile mapping (EQM)) were applied to correct the biases in the dataset during validation process. The LS technique shows better accuracy than EQM and replicated the daily observed rainfall data following the study by [48,49].

2.4. Long-Term Simulations of the Contrasted Sorghum Cultivars under Varying Sowing Windows

The simulations were performed across 33 selected sites in Adamawa and Borno States in north-eastern Nigeria for the five calibrated sorghum cultivars. The sites represent the three agroecological zones of the SS, NGS, and SGS (Table 2). The SS has a long dry season followed by a mono-modal rainfall pattern with a distinct rainy season (May–October) and characterized by a high mean temperature (28–32 °C), short growing season (90–110 days), and low rainfall ranging from 600 to 800 mm [50]. Soils in the SS of Nigeria are highly weathered and fragile with low clay content [51]. The dominant soil class of the site is

Alfisol, according to the USDA soil taxonomy [52]. In the NGS, the length of the growing period is between 151 and 180 days [53]. It has a mono-modal rainfall distribution ranging from 900 to 1000 mm annually, and its mean temperatures vary from 28 to 40 °C [54]. According to the world reference baseline, its soils are classified as leached ferruginous tropical soils with high clay content and overlying drift materials [55]. The dominant soil types found in the zone are Alfisols and Entisols, according to the FAO classification. In the SGS, the average maximum temperature in the growing season ranges from 26 to 28 °C, whereas the minimum temperature ranges between 18 and 22 °C [56,57]. The rainfall pattern is mono-modal, with an annual rainfall between 1000 mm and 1524 mm and spread over the 181–210 days that define the growing season [52,56]. The soils in this zone have been identified mainly as Lithosols, Ferralic combisols, Feric acrisols, Oxic haplustalfs and Luvisols [58].

Table 2. Summary of the selected sites for model application of sorghum cultivars under varying planting windows.

S/No	State	LGA	Site	AEZ	Longitude (N)	Latitude (E)
1	Adamawa	Hong	Dulmava	SS	12.9824	10.3014
2		Gombi	Guyaku	SS	12.6634	10.3459
3		Demsa	Mbula Kuli	NGS	12.3016	9.45745
4		Girei	Wuroshi	NGS	12.6164	9.46866
5		Girei	Daneyel	NGS	12.514	9.54761
6		Gombi	Tawa	NGS	12.6856	10.1691
7		Guyuk	Chikila	NGS	11.9719	9.77237
8		Guyuk	Lakumna	NGS	11.9897	9.92083
9		Hong	Hushere Zum	NGS	13.0807	10.1038
10		Numan	Bare	NGS	12.1108	9.5843
11		Numan	Kikan_Kodomti	NGS	11.9878	9.46081
12		Shelleng	Jonkolo-Lama	NGS	12.178	9.89965
13		Shelleng	Lakati-Libbo/	NGS	12.2502	9.69541
14		Song	Sabon Gari	NGS	12.5935	9.84049
15		Song	Suktu	NGS	12.4248	9.63746
16		Demsa	Nassarawo Demsa	SGS	12.1501	9.29625
17		Yola North	Yelwa-Jambore	SGS	12.5046	9.26165
18		Yola South	Fufure	SGS	12.6504	9.1736
19	Borno	Bayo	Balbaya	SS	11.7648	10.5848
20		Bayo	Briyel	SS	11.6497	10.371
21		Bayo	Jara-Dali	SS	11.7316	10.2759
22		Biu	Buratai	SS	12.4158	10.7675
23		Biu	Kabura	SS	12.2653	10.7392
24		Biu	Mathau	SS	12.1097	10.7214
25		Biu	Tum	SS	12.4881	10.8228
26		Hawul	Kwajaffa	SS	12.4831	10.5167
27		Hawul	Puba Vidau	SS	12.1879	10.5224
28		Hawul	Sakwa Hema	SS	12.3894	10.3867
29		Kwayakusar	Kurbo Gayi	SS	11.9575	10.384
30		Shani	Lakundum	SS	12.0506	10.0556
31		Shani	Gwaskara	NGS	12.158	10.2271
32		Kwayakusar	Bila Gusi	NGS	12.0476	10.5192
33		Shani	Kubo	NGS	12.0853	10.14

LGA—Local Government Area, AEZ—Agro-ecological zone; SS—Sudan Savannah, NGS—Northern Guinea savannah, SGS—Southern Guinea savannah.

The soil parameters used were obtained from on-site soil characterization using geospatial buffering points at a 20 km radius using an ArcGIS map of the reference indicating the sites/LGAs. For soil characterization and soil sampling, profile pits were dug in the 33 selected sites in Adamawa and Borno States. The profiles and soil types were classified using the FAO guidelines [59]. All laboratory analyses were carried out at the Analytical Services Laboratory of IITA. The total soil organic carbon (total C) was measured using a modified Walkley and Black chromic acid wet chemical oxidation and spectrophotometri method [60]. The total nitrogen (total N) was determined using a micro-Kjeldahl digestio method [61]. The soil pH in water (S/W ratio of 1:2.5) was measured using a glass electrod pH meter and the particle size distribution, following the hydrometer method [62]. Th available phosphorus was extracted using the Bray-1 method [63]. The phosphorus in th extract was determined calorimetrically according to the molydo-phosphoric blue metho using ascorbic acid as a reducing agent. K was analysed based on the Mehlich 3 extractio procedure [64]. In Adamawa State, most of the topsoils were coarse-textured with highe sand content. In all, 72% had sandy loam, 17% had clay, and 11% had a sandy clay loar texture (Table 3). The soil pH for the selected communities in Adamawa ranged fror 5.9 (Jonkolo-Lama in Shelleng) to 8.0 (Fufure). More than 55% of the soils had pH value for ideal plant growth, indicating neutral (6.1–6.5) to alkaline (8.1–8.3) soil reactions. Th soil organic carbon (OC) content in ranged from 0.22% in Daneyel and Suktu to 0.90% i the Guyuk area. The distribution of soil in the study areas revealed that most of the soi had low (0.4–1.0%) OC levels. The total soil N content in the soils ranged from very lo (<0.05%) to low (0.06–0.1%), with 67% of the study locations falling within the very low N class and 33% of the study sites indicating low N classes. The soil available P varied acros the locations, with very low P (<3.0 mg kg^{-1}) at Woroshi, Tawa, Chikila, Lakumna, Du mava, Hushere-Zum, Jonkolo-Lama, Sabon-Gari, and Yelwa-Jambore. Low soil availabl P (3–7 mg kg^{-1}) was found in Demsa-Nassarawa, Bare, Lakati-Libbo, and Suktu, whil high P (11–32.1 mg kg^{-1}) content was found in Mbula Kuli, Kikan_Kodomti and Fufur The results showed that 50% of the study sites fell within the very low P fertility class, 28 of the sites fell within the low P fertility class, and 22% of the sites fell within the high fertility class. The exchangeable K level across the sites ranged from low to high value with 22% low (<0.15 cmol^{+} kg^{-1}), 44% moderate (0.16–0.3 cmol^{+} kg^{-1}), and 33% hig (>0.3 cmol^{+} kg^{-1}).

Table 3. Physical and chemical properties used for model applications in Adamawa State.

Site	Profile Depth (cm)	BD (g/cm^3)	OC (%)	Sand (%)	Silt (%)	Clay (%)	pH (in H_2O)	N (%)	Meh. P (ppm)	K cmol/kg
Mbula-Kuli	0–200	1.76	0.84	59	23	18	7.8	0.06	32.1	0.5
Demsa-Nassarawo	24–180	2.18	0.66	65	15	20	8.3	0.06	3.8	0.89
Daneyel	31–200	1.76	0.22	81	7	12	7.0	0.01	10.9	0.3
Woroshi	14–94	2.16	0.54	65	19	16	6.4	0.04	1.17	0.36
Guyaku	19–120	1.7	0.35	79	9	12	6.6	0.03	2.14	0.22
Tawa	15–127	1.79	0.62	75	13	12	6.7	0.05	3.38	0.21
Chikila	30–180	2.18	0.90	15	19	66	8.5	0.08	2.55	0.13
Lakumna	20–200	1.77	0.90	25	23	52	7.3	0.10	1.59	0.65
Dulmava	27–201	1.82	0.51	67	15	18	7.5	0.06	1.03	0.17
Hushere-Zum	41–205	1.93	0.46	80	8	12	6.3	0.03	2.41	0.40
Bare	25–200	1.62	0.35	74	9	17	6.6	0.02	4.07	0.20
Kikan_Kodomti	22–200	1.76	0.66	71	9	20	7.3	0.04	13.7	0.20
Lakati-Libbo	27–200	1.83	0.30	78	9	13	7.4	0.01	5.04	0.20
Jonkolo-Lama	15–200	2.06	0.33	78	10	12	5.9	0.02	0.89	0.14
Sabon-Gari	31–200	1.73	0.66	25	33	42	6.2	0.04	1.45	0.4
Suktu	35–210	2.08	0.22	71	11	18	6.3	0.03	6.56	0.20
Yelwa-Jambore	24–155	2.19	0.4	77	11	12	6.5	0.03	1.8	0.09
Fufure	20–145	1.98	0.54	65	17	18	8.0	0.02	32.1	0.10

BD = bulk density, OC = organic carbon content, N = percent Nitrogen, Meh P = Available Phosphorus, an K = potassium.

Similarly, in Borno state, the majority of the soils were coarse-textured with highe sand content. Out of the 15 sites, 47% had sandy loam, 27% had clay, and 26% had a si loamy sand texture (Table 4). The soil pH of water for the communities in Borno Stat ranged from 6.1 to 8.4. More than 70% of the soils had neutral reactions (6.6–7.8), which i the ideal condition for plant growth. The soil OC content in the state ranged from 0.12% t

0.78%. Eight (8) communities equivalent to 53% of the study area had very low OC (<0.4%) levels. The total soil N content in the soils ranged from very low to low, with a very low (<0.05%) status found in the Balbaya, Bila Gusi, Briyel, Buratai, Gwaskara, Jara-Dali, Kubo, Kurba, Mathau, Puba Vidau, Sakwa-Shema, and Tum communities, while the Kwaya Bura, Kwajaffa, and Lakundum communities fell within the low (0.06–0.1%) N fertility class. With the exception of Gwaskara and Lakundum, the top soil available P at all the locations fell within very low (<3.0 mg kg^{-1}) fertility class. The exchangeable K levels were 7% low (<0.15 $cmol^{+}$ kg^{-1}), 33% moderate (0.16–0.3 $cmol^{+}$ kg^{-1}), and 60% high (>0.3 $cmol^{+}$ kg^{-1}) across the sites.

Table 4. Physical and chemical properties used for model applications in Borno State.

Site	Profile Depth (cm)	BD (g/cm^3)	OC (%)	Sand (%)	Silt (%)	Clay (%)	pH (H_2O)	N (%)	Meh. P (ppm)	K cmol/kg
Balbaya	9–200	1.59	0.29	83	7	10	6.1	0.01	1.03	0.0
Briyel	15–200	1.32	0.39	19	29	52	8.4	0.02	2.69	0.4
Jara-Dali	8–200	1.55	0.33	51	13	36	6.6	0.02	1.72	0.3
Buratai	29–150	1.63	0.17	74	8	18	7.6	0.02	2.69	0.6
Kwaya Bura	22–101	1.36	0.78	36	38	26	7.1	0.06	0.89	9.0
Mathau	12.0–94	1.62	0.12	90	0	10	7.4	0.01	2.83	0.8
Tum	12–200	1.40	0.19	28	24	48	7.4	0.01	1.17	0.6
Kwajaffa	30–110	1.31	0.54	16	27	57	7.4	0.06	2.28	0.7
Puba Vidau	10–200	1.32	0.4	18	19	63	8.3	0.02	0.89	0.6
Sakwa Hema	15–170	1.57	0.52	74	9	17	7.0	0.04	0.76	0.1
Bila Gusi	80–200	1.59	0.48	67	15	18	6.5	0.02	2.14	0.1
Kurba Gayi	10–200	1.60	0.32	75	9	16	7.2	0.01	1.03	0.1
Gwaskara	19–200	1.57	0.34	72	13	15	7.1	0.01	11.5	0.1
Kubo	33–200	1.54	o.46	64	13	23	7.3	0.02	1.31	0.8
Lakundum	16–200	1.52	0.73	72	10	18	7.3	0.07	13.6	9.0

BD = bulk density, OC = organic carbon content, N = percent Nitrogen, Meh P = Available Phosphorus and K = potassium.

The long-term (1985–2010) weather data used in the model application was a combination of downscaled CHIRPS (for daily rainfall) and the NASA database for Climatology Resource for Agroclimatology (for minimum and maximum air temperature and solar radiation respectively). The simulations were set up to run at different planting windows using the fertilizer N at the national fertilizer rate of recommendation (NPK 60:30:30 $kghakgha^{-1}$) for sorghum. In the model, 30 kg N were applied at sowing (DAS), with Urea (46% N) top dressed at 30 kg of N ha^{-1} at 30 DAS. The simulation considered an optimum population to be at a 75 cm inter-row by 30 cm intra-row spacing given 44,444 hills/ha against the farmer's lower rate of 22,222 hills/ha. Based on expert knowledge and a previous study [22] that found that the sowing period for sorghum across the three agro ecologies stretches over 60 days, we divided the entire sowing period into four equal planting windows to capture the photoperiod sensitivity of the cultivars. The model was set to consider four (4) planting windows as follows: 16–31 May (PW1), 1–15 Jun (PW2), 16–30 Jun (PW3), and 1–15 Jul (PW4), respectively. In addition, rule-based sowing within the sowing window was applied (cumulative rainfall of 20 mm in 3 rainy events) and implemented at the 33 sites. The sowing depth was set to 5 cm, with a sowing density of 4.5 plant m^2. Considering the farmers' practices in the region, a non-successive simulation (single season, non-rotation mode) was adopted, which implies that the water, organic matter, nitrogen, and phosphorus were reset a few weeks before the start of the growing season.

The optimal window for the sowing dates of the sorghum cultivar was based on the average simulated grain yield over the 26-year period and across the sites in each AEZ. Also, the coefficient of variation (CV%), as the ratio of the standard deviation to the mean simulated grain yield, was used to assess the suitable cultivar for each site and AEZ. The level of variability (high or low percentage) determined whether the cultivar had a high or low suitability for the site based on a mean grain yield of $\geq$1500 kgha^{-1} as the threshold. The threshold was determined as a break-even yield that farmers can produce for marginal economic benefit as described by [22]. The potential evapotranspiration based on the Penman-Monteith equation [37] in the APSIM model was computed as the addition of the simulated soil evaporation and crop transpiration, and, from that, the water use efficiency for the grain yield (WUE_{grain}) was calculated.

3. Results

3.1. Model Performance

As depicted in Table 1, there were differences in the cultivar-specific coefficients across the new sorghum cultivars, particularly in the thermal time that defined the crop vegetative and growth. ICSV400 and Improved Deko had a shorter thermal time requirement (in degree days) to attain the end of the juvenile stage compared to CSR01, Samsorg-44, and SK5912, respectively. Both cultivars (ICSV400 and Improved Deko) were originally bred for drought conditions, which could allow them to serve as a drought escaping mechanism compared to the other cultivars. Also, the calibrated photoperiod slope varied from 11.5 °C/h to 600 °C/H, indicating a shorter degree/hour for low photoperiod sensitivity cultivars such as ICSV400 and improved Deko, while a longer degree/hour was calibrated for the medium and high photoperiod sensitivity cultivars. The thermal time from flowering to physiological maturity above a base temperature of 10 °C was 560 °C days for ICSV400 and improved Deko, indicating a higher value than the degree days of CSR01 (460 °C days), Samsorg-44 (500 °C days), and SK5912 (450 °C days), respectively. The cultivar genetics coefficients for leaf appearance rate followed two steps, i.e., leaf appearance to the development of most leaf ligules (leaf_app_rate 1) and to the last leaf ligule (leaf_app_rate 2). The calibrated values (56 °C d/leaf and 28 °C d/leaf) were the same for all of the varieties except for ICSV400. These values justified the increase in the leaf number (>20) per plant for most West African sorghum cultivars that are photoperiod sensitive.

The performance of the model, presented in Table 5, shows that the simulated days to 50% flowering and to physiological maturity were good and reproduced the observed values with a mean bias error (MBE) ranging from −4 to 4 days (50% flowering) and from 1 to 2 days (physiological maturity). The RMSE of the mean observed estimate of $\leq$10% for all the cultivars confirmed the robustness of the predictions. The model's adjustment of the leaf appearance rate for leaf ligules helps to get an accurate total leaf number (TLN) per plant close to the observed. The estimates of the MBE varied from one to five leaves, and RMSE (of the mean observed) ranged from a high model accuracy (6.4% for improved Deko) to a fairly low accuracy (26.2% for Samsorg-44) for TLN.

Table 5. Statistical evaluation of simulated phenology and morphological traits (LAI and total leaf number/plant) of contrasted sorghum cultivars calibrated from experiment conducted under optimum conditions in Southern Guinea and Sudan Savannah AEZs.

Parameters/ Cultivar	Unit	N	MBE	RMSE		Observed Range	Observed Mean
				Absolute Value	% of Mean Observed		
ICSV-400							
50% Flowering	DAP	11	−1	4	5.4	62–75	68
Physiological Maturity	DAP	11	2	5	4.6	90–106	97
LAI-max	m^2/m^2	11	−0.2	0.8	32.4	1.8–3.0	2.3
Leaf number		4	3.4	3.5	20.5	16–18	17
Improved Deko							
50% Flowering	DAP	7	−4	6	7.9	75 95	84
Physiological Maturity	DAP	7	1	6	5.0	101–122	110
LAI-max	m^2/m^2	7	0.6	0.8	27.0	2.0–3.3	2.5
Leaf number		4	0.4	1.2	6.4	16–19	18
Samsorg-44							
50% Flowering	DAP	4	1	3	3.0	85–114	99
Physiological Maturity	DAP	4	2	4	3.2	112–140	126
LAI-max	m^2/m^2	4	0.2	0.7	26.6	2.2–3.4	3.0
Leaf number		4	5.1	5.2	26.2	19–23	20
CSR01							
50% Flowering	DAP	8	2	8	8.4	84–112	95
Physiological Maturity	DAP	8	1	7	6.1	111–139	123
LAI-max	m^2/m^2	8	0.3	0.4	14.7	2.3–3.7	3.0
Leaf number		8	4	4.1	19.5	19–24	21
SK5912							
50% Flowering	DAP	4	4	5	4.4	95–122	108
Physiological Maturity	DAP	4	2	4	3.0	122–149	135
LAI-max	m^2/m^2	4	0.3	0.6	20.7	2.0–3.3	2.5
Leaf number		4	3.8	4.0	17.6	20.4–25.4	23

Table 5. *Cont.*

N—Number of observations; LAI-max: maximum leaf area index measured during growth; MBE = positive implies over-simulated mean observed; negative implies under-simulated the mean observed value.

The simulated and observed maximum Leaf Area Index (Max_LAI) for all cultivars agrees well with RMSE (% of mean observed), indicating high accuracy for CSR01 and SK5912, low accuracy for improved Deko and Samsorg-44, and very low accuracy for ICSV400. The grain yield and total biomass were acceptably simulated for the contrasted sorghum cultivars within the bounds of statistical errors (Figure 1). For grain yield (Figure 1a), CSR01 had the lowest MBE of −48 kghakgha^{-1}, which under-predicted the observed mean, followed by ICSV-400 (103 kghakgha^{-1}) and improved Deko (114 kghakgha^{-1}), while the highest yield (279 kgha^{-1}) was shown by the cultivar Samsorg-44. The relative RMSE ranged from high accuracy for SK5912 (9.2%) to very low accuracy for ICSV-400 (28.7%). For total biomass (Figure 1b), the relative RMSE ranged from high accuracy for SK5912 (6.9%) to very low accuracy for improved Deko (36.8%).

3.2. Model Validation: Performance with Farm-Level Grain Yield

The performance of the model in simulating grain yield was compared to the observe values under varying planting dates and cropping systems for each sorghum cultiva (Table 6). The planting dates across farms and cultivars were grouped under three month (referred to as "sowing month"), and the number of observations/farms revealed that 92% of farmers planted in the months of June and July, and only 8% of the farmers sowed i the month of August. For ICSV-400, the model under-predicted the mean observed yiel for intercropping and mixed cropping systems, but the model over-predicted the mea observed yield for the sole cropping system across the sowing months. The lowest MB of −977 $kgha^{-1}$ was estimated in theJuly sowing month under the intercropping systen followed by the mixed cropping system, while the highest MBE (781 $kgha^{-1}$) was estimate under the sole cropping system in the month of June. The results showed that the mod over-predicted the mean observed grain yield for Improved Deko across sowing month under the sole cropping system, with the lowest MBE (66 $kgha^{-1}$) estimated for the Jul sowing month, while the highest MBE (548 $kgha^{-1}$) was estimated for August sowin The model over-predicted the grain yield across the sowing months and cropping system except for the June sowing month under sole cropping system, for which lowest MB of −234 $kgha^{-1}$ was estimated. The highest MBE of 624 $kgha^{-1}$ was estimated for Jul sowing under sole cropping. For CSR01, the model under-predicted the mean observe grain yield across sowing months and cropping systems except for the August sowin month under a mixed cropping system. Similarly, for SK5912, the model over-predicted th mean observed grain yield under the sole cropping system across sowing months, whil the model under-predicted across sowing months for the mixed cropping system.

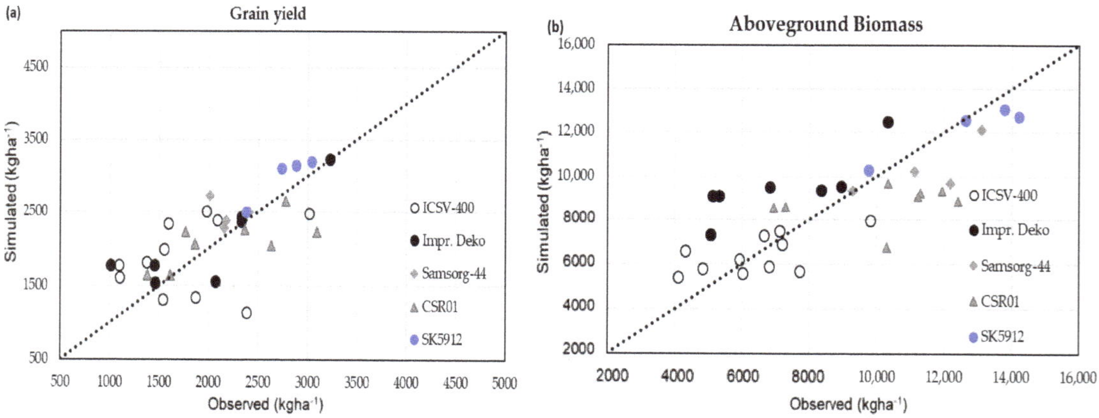

Figure 1. (**a**) Observed vs. simulated grain yield using experiment conducted in 2016–201 growing seasons for cultivar ranges from early to late maturing. ICSV-400 (MBE = 103 kgha− RMSE = 617 $kgha^{-1}$, $RMSE_n$ = 28.7%); Improved Deko (MBE = 114 $kgha^{-1}$, RMSE = 370 kgha− $RMSE_n$ = 18.7%); Samsorg-44 (MBE = 279 $kgha^{-1}$; RMSE = 377 $kgha^{-1}$, $RMSE_n$ = 17.2%); CSR0 (MBE = −48 $kgha^{-1}$, RMSE = 301 $kgha^{-1}$, $RMSE_n$ = 13.8%); SK5912 (MBE = 234 kgha− RMSE = 254 $kgha^{-1}$, $RMSE_n$ = 9.2%). (**b**) Observed vs. simulated total biomass using experimer conducted in 2016–2018 growing seasons for cultivar ranges from early to late maturing. ICSV-40 (MBE = 28 $kgha^{-1}$, RMSE = 1249 $kgha^{-1}$, $RMSE_n$ = 19.5%); Improved Deko (MBE = 2344 kgha− RMSE = 2621 $kgha^{-1}$, $RMSE_n$ = 36.8%); Samsorg-44 (MBE = −1100 $kgha^{-1}$; RMSE = 1432 kgha− $RMSE_n$ = 12.5%); CSR01 (MBE = −976 $kgha^{-1}$, RMSE = 1687 $kgha^{-1}$, RMSEn = 16.5%); SK591 (MBE = −429 $kgha^{-1}$; RMSE = 868 $kgha^{-1}$, $RMSE_n$ = 6.9%).

Table 6. Statistical indices for model validation of contrasted sorghum cultivars across planting date and cropping system from on-farm production technology between 2013 and 2017.

Sowing Month/Cultivar	Cropping System	N	Simulated	Observed	MBE	RMSE
ICSV400				kgha^{-1}		
June	Sole	535	2201	1420	781	1038
July	Intercropping	37	2007	2084	−77	700
	Mixed cropping	27	1698	2646	−948	1229
	Sole	461	2052	1488	564	942
Aug	Intercropping	13	1778	2754	−977	1029
	Mixed cropping	13	1850	2663	−814	959
	Sole	108	1897	1537	360	936
Improved Deko						
June	Sole	178	1656	1426	231	712
July	Sole	111	1554	1488	66	617
Aug	Sole	11	1492	943	548	598
SamSorg-44						
June	Sole	22	1463	1697	−234	808
July	Sole	50	1586	962	624	915
	Intercropping	11	910	750	160	161
Aug	Sole	12	1623	1380	244	738
CSR01						
June	Intercropping	13	1573	2188	−615	624
	Mixed cropping	18	1524	1729	−206	640
	Sole	452	1335	1366	−31	726
July	Intercropping	23	1517	1700	−183	700
	Mixed cropping	13	1297	1973	−676	1203
	Sole	356	1566	1886	−320	952
Aug	Mixed cropping	15	1588	1388	200	258
	Sole	55	1474	1932	−458	940
SK5912						
June	Intercropping	26	1433	1305	128	873
	Mixed cropping	17	1157	1184	−26	834
	Sole	263	1437	1424	13	848
July	Intercropping	11	1147	1576	−429	873
	Mixed cropping	22	1169	2225	−1056	1285
	Sole	320	1587	1408	179	764
Aug	Intercropping	10	1323	1858	−535	824
	Mixed cropping	8	1135	1603	−744	809
	Sole	55	1786	1483	303	800

N—Number of observations/farms.

Figure 2 shows the model performance and the differences between the observed and simulated yield pooled together irrespective of the cropping systems and management practices for each cultivar. The mean observed grain yield for ICSV-400, CSR01, Improved Deko, Samsorg-44, and SK5912 are 1479, 1613, 1431, 1197, and 1446 kgha^{-1}, respectively. Further statistical indices showed that the grain yield of the ICSV-400, Improved Deko, and Samsorg-44 cultivars, respectively, were over-predicted against the mean observed grain yield; meanwhile, the yields of the CSR01, and SK5912 cultivars were slightly under-predicted compared to the mean observed yield. The results revealed low MBEs for CSR01 (−228 kgha^{1}), SK5912 (−241 kgha^{-1}), and Samsorg-44 (102 kgha^{-1}), respectively, with an RMSE of 642 kgha^{-1} estimated for improved Deko, and an RMSE of 655 kgha^{-1} estimated for Samsorg-44. The CV (%) described the level of variability for each cultivar simulated, which shows the lowest value of 8.9% for Samsorg-44, followed by Improved Deko (CV = 12.3%), while the highest variability was observed for CSR01 and SK5912 (CV = 25.5 and 18.4%).

3.3. Seasonal Rainfall and Temperature Trends across the Simulated Sites

The long-term (1985–2010) rainfall indicated that the rainy season starts in May and ends in October, with the highest peak observed in the month of August (Tables 7 and 8) The tables further revealed that about 50–60% of the seasonal rainfall was observed in the months of July and August, with a high inter-seasonal variability indicated by the coefficients of variation (CV), ranging from 18 to 23%. All of the study sites showed a distinct mono-modal rainfall pattern and warming temperature throughout the year Figures 3 and 4 show the average monthly variations in the maximum and minimum temperatures across the selected sites in the Adamawa and Borno States. The maximum temperature uniformly decreases faster than the minimum temperature during the growing season (May–October). In addition, the estimated CV% values for the maximum temperature, ranging from 3.0 to 3.7%, are higher than those of minimum temperature, which range from 2.0 to 2.3% in both states, suggesting that no significant inter-annual variability was observed at the sites for either temperature.

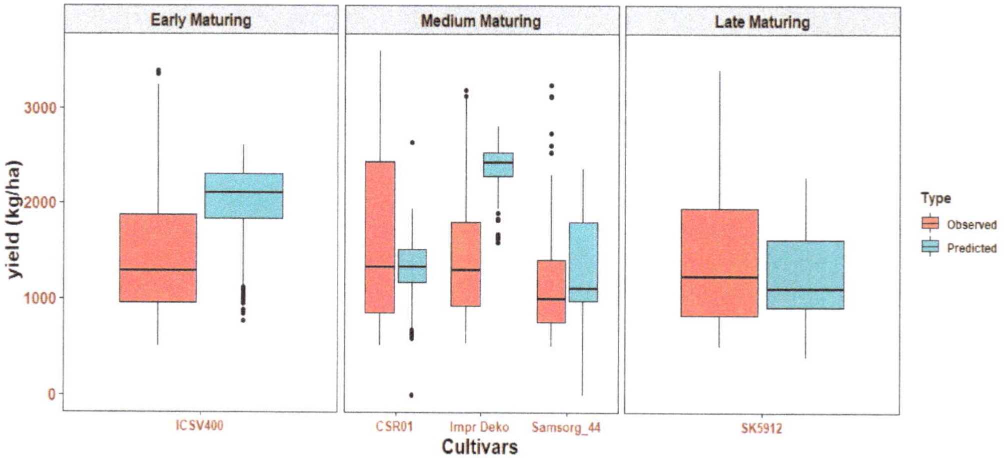

Figure 2. Yield (observed and simulated) using on-farm datasets from the 2013–2017 growing seasons from contrasting environments for five (5) sorghum cultivars ranged from early to late maturing. ICSV-400 (N = 1192; MBE = 535 kgha^{-1}; RMSE = 971 kgha^{-1}, CV = 13.8%); Improved Deko (N = 300; MBE = 960 kgha^{-1}, RMSE = 1169 kgha^{-1}, CV = 12.3%); Samsorg-44 (N = 100; MBE = 102 kgha^{-1}; RMSE = 655 kgha^{-1}, CV = 8.9%); CSR01 (N = 944; MBE = −228 kgha^{-1}, RMSE = 755 kgha^{-1}, CV = 25.5%); SK5912 (N = 731; MBE = −241 kgha^{-1}; RMSE = 879 kgha^{-1}, CV = 18.4%). Coefficient of variations (CV), N = number of observations.

Table 7. Analysis of mean monthly, seasonal rainfall (mm) and level of variability across the simulation sites in Adamawa State (1985–2010).

Site	May	Jun.	Jul.	Aug.	Sep.	Oct.	Seasonal	Stdev	C.V (%)
Demsa-Nassarawo	102.1	121.2	189.3	234.3	172.7	73.5	893	188	21
Mbula Kuli	95.9	115.7	186.5	225.8	168.1	53.6	851	181	21
Daneyel	99.8	118.1	202.9	240.5	156.9	54.4	873	191	22
Woroshi	103.3	126.4	216.5	244.0	156.4	53.7	902	191	21
Guyaku	117.9	155.9	228.9	308.8	176.6	99.1	1087	230	21
Tawa	134.2	149.6	237.1	293.3	192.4	97.2	1104	239	22
Lakumna	91.8	110.3	167.5	258.2	174.9	63.9	872	185	21
Chikila	98.5	106.5	178.4	249.7	165.2	67.8	866	186	21
Hushere Zum	120	133.8	211.7	266.5	196.7	113	1042	241	23
Dulmava	109.9	150.6	225.5	302.8	202.2	113.1	1104	247	22
Bare	91.9	107.4	176.9	244.2	162.9	80.6	864	194	22
Kodomti	91.1	109.5	176.8	243.2	170.2	75.0	866	194	22
Lakati-Libbo	95.2	109.6	186.8	250.2	155.2	74.9	872	191	22
Jonkolo-Lama	97.6	115.0	182.4	268.6	166.2	73.1	903	197	22
Sabon-Gari	99.8	119.5	211.3	269.7	181.8	82.1	964	212	22
Suktu	99.6	116.3	211.4	256.5	157.8	61.5	903	199	22
Yelwa-Jambore	102.1	125.4	206.6	218	163.5	52.2	868	189	22
Fufure	103.8	140.6	220.6	218.5	160.5	51.4	895	190	21

Seasonal—average total seasonal rainfall from May to Oct.; Stdev—Standard deviation from mean; CV—coefficient of variations (in percentage).

Table 8. Analysis of mean monthly, seasonal rainfall (mm) and level of variability across the simulation sites in Borno State from 1985 to 2010.

Site	May	Jun.	Jul.	Aug.	Sep.	Oct.	Seasonal	Stdev	C.V (%)
Balbaya	87.9	141.3	202.9	287.9	167.4	67.4	955	206	22
Briyel	93.2	129.0	174.2	242.7	182.7	61.1	883	182	21
Jara-Dali	78.4	136.8	202.8	289.0	204.4	80.3	992	217	21
Kabura	72.5	142.4	209.7	316.1	149.3	43.4	939	188	20
Mathau	78.3	144.4	204.4	312.1	165.6	51.9	957	174	18
Tum	86.2	149.8	218.1	317.4	170.0	55.9	998	204	20
Buratai	77.4	144.3	210.9	318.4	148.5	45.6	945	191	20
Kwajaffa	99.7	142.3	204.3	306.7	179.3	51.2	983	186	19
Puba Vidau	96.6	144.2	199.6	299.8	188.3	60.3	989	191	19
Sakwa Hema	93.3	144.2	206.9	307.4	176.8	60.2	989	186	19
Bila-Gusi	98.9	124.5	190.6	268.6	183.4	75.7	942	189	20
Kurba Gayi	85.5	145.9	213.1	303.1	166.2	61.1	975	199	20
Gwaskara	83.5	142.1	198.5	295.4	201.6	74.9	996	192	19
Kubo	97.3	121.6	181.9	262.2	192.2	72.3	927	186	20
Lakundum	85.2	146.0	220.2	307.1	158.2	77.9	995	213	20

Seasonal—average total seasonal rainfall from May to Oct.; Stdev—Standard deviation from mean; CV—coefficient of variations (in percentage).

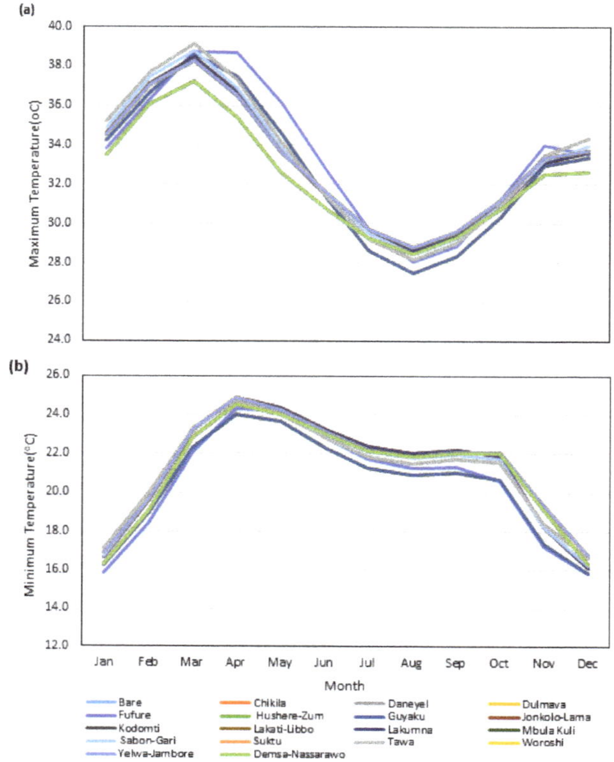

Figure 3. Average monthly variation of (**a**) maximum temperatures and (**b**) minimum temperatures between 1985 and 2010 across the simulation sites in Adamawa State. The coefficients of variation (CV) ranged from 3.0 to 3.7% for maximum temperature and 2.0 to 2.3% for minimum temperature.

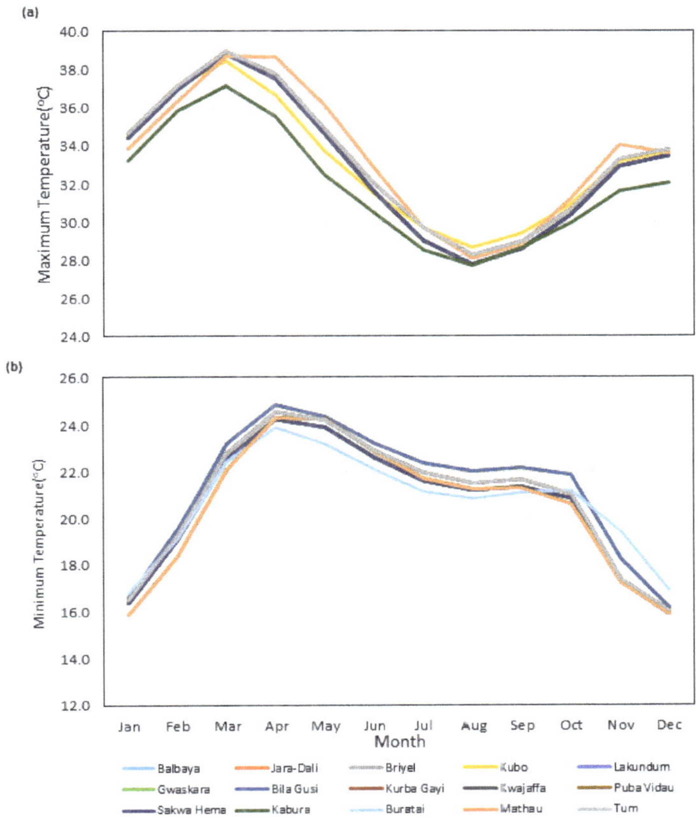

Figure 4. Average monthly variations of (**a**) maximum temperatures and (**b**) minimum temperatures between 1985 and 2010 across the simulation sites in Borno State. The coefficients of variation (CV) ranged from 3.0 to 3.7% for maximum temperature and 2.0 to 2.3% for minimum temperature.

In Adamawa State (Table 7), the seasonal rainfall (May-Oct.) for all of the sites over the 31-year period (1985–2010) ranged from 851 to 1104 mm. It was observed that the rainfall in Dulmava, Hushere Zum, and Guyaku and Tawa was slightly higher (>1000 mm) than in the other locations. The average monthly maximum temperature across the sites over the climatic period ranged from 27.5 to 39.1 °C (Figure 3a), while the average monthly minimum temperature ranged from 15.8 to 24.9 °C (Figure 3b). In Borno State (Table 8), the seasonal rainfall over the 31-year period (1985–2010) across the sites ranged from 883–998 mm with high inter-seasonal variability, varying from 18 to 22%. The average monthly maximum temperature across the sites over the climatic period ranged from 27.8 to 38.9 °C (Figure 4a), while the average monthly minimum temperature ranged from 15.5 to 24.7 °C (Figure 4b).

3.4. Seasonal Analysis of Planting Windows and Sorghum Cultivars on Simulated Grain Yield and Water Use Efficiency (WUE_{grain})

Table 9 shows the mean simulated grain yield (GY) and the water use efficiency for grain yield (WUE_{grain}) of the sorghum cultivars across four different planting windows (PW1, PW2, PW3, and PW4) in the three agro-ecological zones (AEZs) between 1985 and 2010. The mean simulated grain yield and WUE_{grain} showed a decrease with delayed planting (PW1 to PW4) for all five sorghum cultivars. Following the sowing rule strategies implemented for the simulation, the model outputs indicate approximately 45 days of PW, from 25 May to 10 July across AEZs, for all sorghum cultivars except for SK5912, which has approximately 35 days of planting window varying from 25 May and 30 June in the NGS and SGS. A higher mean GY and WUE_{grain} were simulated in the NGS than in the SS and SGS zones. Additionally, the early and medium-maturing sorghum cultivars (ICSV400, Improved Deko, CSR01, and Samsorg44) had higher simulated GY and WUE_{grain} values than those of the late-maturing cultivar (SK5912).

Table 9. Mean simulated grain yield and Water Use Efficiency for grain yield (WUE_{grain}) of sorghum cultivars across different planting windows (PWs) and agro ecological zones.

PW/C	NO	Grain Yield					WUE_{grain}				
		ICSV400	Impr. Deko	CSR01	Samsorg-44	SK5912	ICSV400	Impr. Deko	CSR01	Samsorg-44	SK5912
Sudan Savanna (SS)		$kgha^{-1}$					$kgha^{-1}\ mm^{-1}$				
PW1	420	2321	2211	2340	2097	1703	7.3	6.6	5.0	4.6	3.2
PW2	420	2309	2170	2205	1981	1580	7.3	6.4	4.6	4.2	3.0
PW3	420	2255	2148	1895	1760	1252	6.9	6.2	4.0	3.7	2.4
PW4	420	2228	2145	1778	1613	1128	6.8	6.2	3.8	3.5	2.3
Mean		**2278**	**2168**	**2054**	**1863**	**1416**	**7.1**	**6.3**	**4.4**	**4.0**	**2.7**
Northern Guinea Savanna(NGS)											
PW1	480	2323	2234	2750	2536	2358	7.7	7.0	6.6	6.1	5.0
PW2	480	2315	2188	2677	2447	2128	7.8	6.8	6.0	5.5	4.2
PW3	480	2236	2171	2657	2386	1856	7.2	6.3	5.8	5.3	3.7
PW4	480	2223	2138	2644	2375	1654	7.1	6.5	6.0	5.4	3.5
Mean		**2274**	**2182**	**2682**	**2436**	**1999**	**7.5**	**6.7**	**6.1**	**5.6**	**4.1**
Southern Guinea Savanna(SGS)											
PW1	90	1959	1865	2192	1967	1733	6.7	6.0	5.5	5.1	3.9
PW2	90	1939	1815	2091	1878	1655	6.7	5.9	5.1	4.7	3.6
PW3	90	1920	1841	2106	1898	1488	6.4	5.8	5.1	4.7	3.3
PW4	90	1903	1814	2059	1850	1530	6.4	5.8	4.9	4.5	3.3
Mean	**90**	**1930**	**1834**	**2112**	**1898**	**1602**	**6.6**	**5.9**	**5.2**	**4.7**	**3.5**

Impr.—improved; PW—planting windows [16–31 May (PW1), 1–15 Jun (PW2), 16–30 Jun (PW3), 1–15 Jul (PW4)]; C—Cultivar; NO—Number of observations.

For the SS zone, the optimal sowing window simulated ranged from 25 May to 30 June (PW 1 to PW3) for CSR01 and Samsorg-44 and from 25 May to 15 June (PW1 and PW2) for SK5912, while, for the ICSV400 and Improved Deko cultivars, sowing can extend to 10 July. In the NGS and SGS zones, the optimal planting window ranged from 25 May to 10 July for all sorghum cultivars except for SK5912, for which 25 May to 30 June was simulated to be the optimal planting window. The highest mean WUE_{grain} of 6.4–7.8 $kgha^{-1}\ mm^{-1}$ was simulated for ICSV400. Next to it was improved Deko with a WUE_{grain} of 5.8–6.8 $kgha^{-1}\ mm^{-1}$, and SK5912 was simulated to have the lowest WUE_{grain} (2.3–4.2 $kgha^{-1}\ mm^{-1}$) across the three AEZs

Table 10 shows the mean simulated grain yield for evaluating the adapted sorghum cultivars across sites based on an increased yield threshold of $\geq$1500 $kgha^{-1}$ and against the national average grain yield of 1160 $kgha^{-1}$. In the SS zone, the simulated mean grain yield across the selected sites ranged from 2023 to 2673 $kgha^{-1}$ for ICSV400, 1886–2509 $kgha^{-1}$ for Improved Deko, 1022–3707 $kgha^{-1}$ for CSR01, 939–3324 $kgha^{-1}$ for Samsorg-44, and 730–2847 $kgha^{-1}$ for SK5912, respectively. The CV shows the variability of the simulated

GY across sites, with lower values estimated by ICSV400 (10%) and Improved Deko (9%) compared to higher values estimated by CSR01 (46%), Samsorg-44 (44%), and SK5912 (56%).

Table 10. Mean simulated grain yield (kgha^{-1}) for evaluating adapted sorghum cultivars across sites and AEZs based on increased yield threshold.

AEZ	N-Site	ICSV400	Impr. Deko	CSR01	Samsorg-44	SK5912
Sudan Savanna (SS)	Balbaya	2236	2132	1467	1368	1106
	Briyel	2208	2118	1983	1784	1450
	Buratai	2290	2130	2555	2426	1926
	Dulmava	2363	2201	2249	2001	1439
	Guyaku	2172	2069	1371	1296	1110
	Jara-Dali	2092	2052	1708	1545	1242
	Kabura	2636	2480	3707	3324	2847
	Kurbo Gayi	2673	2509	3645	3247	2445
	Kwajaffa	2328	2221	1624	1448	919
	Lakundum	2276	2151	2612	2289	1621
	Mathau	2013	1886	1067	1029	730
	Puba Vidau	2157	2026	1022	939	783
	Sakwa Hema	2292	2180	1902	1723	1112
	Tum	2157	2098	1847	1653	1096
	Mean	**2278**	**2168**	**2054**	**1863**	**1420**
	CV(%)	**10**	**9**	**46**	**44**	**56**
Northern Guinea Savanna (NGS)	Bare	1926	1803	1269	1174	940
	Bila Gusi	2126	2044	2325	2054	1580
	Chikila	2402	2331	3152	2914	2201
	Daneyel	2028	1945	1807	1615	1247
	Gwaskara	2498	2372	2957	2592	1789
	Hushere Zum	2180	2079	1777	1599	1213
	Jonkolo—Lama	2208	2141	2657	2564	2149
	Kikan_Kodomti	2060	1992	2301	2024	1557
	Kubo	2406	2336	3761	3432	3140
	Lakati-Libbo	2123	2057	1954	1743	1411
	Lakumna	2414	2344	3199	3298	2790
	Mbula Kuli	2237	2176	2685	2366	2120
	Sabon Gari	2495	2360	3387	2999	2680
	Suktu	2314	2221	2874	2538	2052
	Tawa	2458	2328	3127	2753	2294
	Wuroshi	2512	2390	3676	3310	2824
	Mean	**2274**	**2182**	**2682**	**2436**	**1999**
	CV(%)	**11**	**11**	**30**	**30**	**41**
Southern Guinea Savanna (SGS)	Fufure	1306	1155	1010	971	891
	Nassarawo Demsa	2330	2220	3049	2707	2325
	Yelwa-Jambore	2154	2117	2276	2017	1589
	Mean	**1930**	**1834**	**2112**	**1898**	**1602**
	CV(%)	**24**	**27**	**42**	**40**	**43**

Impr.—improved; CV(%)—Coefficients of variations in the percentage.

In NGS zone, the simulated mean grain yield across the sites ranged from 1926 to 2512 kgha^{-1} for ICSV400, 1841–2390 kgha^{-1} for Improved Deko, 1269–3761 kgha^{-1} for CSR01), 1174–3432 kgha^{-1} for Samsorg-44, and 940–3140 kgha^{-1} for SK5912. The variability of the GY across sites indicated low CV% for ICSV400 and Improved Deko (11%) compared to high CV% estimates for CSR01 (30%), Samsorg-44 (30%), and SK5912 (41%). In the SGS zone, the simulated mean grain yield across the sites ranged from 1306 to 2330 kgha^{-1} for ICSV400, 1165–2220 kgha^{-1} for Improved Deko, 1010–3049 kgha^{-1} for CSR01, 971–2707 kgha^{-1} for Samsorg-44, and 891–2325 kgha^{-1} for SK5912. The CV% was generally high for all cultivars, ranging from 24 to 43%. At the mean grain yield threshold of ≥1500 kgha^{-1}, all cultivars simulated were found to be adapted for cultivation except a the Fufure site.

4. Discussion

This study contributes to efforts to develop climate risk strategies for the sorghum based mixed farming systems in northern Nigeria. The evaluation of the model calibratio and its validation with an independent dataset (farm-level yield) under different manage ment, soils, and climatic conditions allow the APSIM-sorghum model to be applied t understanding the dynamics of this heterogeneous farming system. The application c crop modelling to develop adaptation strategies to changing climatic conditions was earlie demonstrated for sorghum by [22,31] and for maize by [65]. The predicted LAI-max an total leaf number (TLN) indicated a low accuracy (RMSE varied from 20 to 30%) due t the relatively higher values simulated for July sowing dates resulting in a higher mea grain yield simulated under calibration. However, the difficulty in predicting TLN coul be linked to the fixed thermal time targets for each of the phases before flowering in th APSIM-sorghum module. These thermal time targets are not directly linked to leaf initia tion and appearance [66]. The predictions of the grain yield (GY) and total biomass (TE ranged from high accuracy $RMSE_n$ (SK5912: 9.2% for GY; 6.9% for TB) to very low accurac $RMSE_n$ (ICSV400: 28.7% for GY; Improved Deko: 36.8% for TB) when evaluated against th observed mean. The low accuracy for GY and TB could be associated with the simulatio of leaf initiation and leaf appearance, which are important for the accurate prediction c morphological traits [31,67].

The use of model evaluation using simple on-station trial datasets is the commo procedure for developing new cultivar parameterizations. However, evaluating model with multi-locational, on-farm trial datasets has proven difficult, with many uncertaintie especially across the different soil, climate, and cropping systems considered [66]. The stud presented here utilized comprehensive data from on-farm trials using different plantin dates, cropping systems, fertilization strategies, soil types, and management regime representing the heterogeneous farming system of northern Nigeria. The performance c the model was satisfactory under varying planting dates (referred to as "sowing month" cropping systems, and sorghum cultivars as described in Table 6. With exception of th CSR01 and SK5912 cultivars, the model's predictions had a lower MBE, either positiv or negative, for the sole cropping system in the July sowing month compared to the Jun and August sowing months. These results could be explained by the pattern of rainfa that serves as a means of crop water utilization, which in turn determines the biomas accumulation for the grain yield. The high rainfall variability across the study sites suggest the importance of matching crop duration to the length of the growing period in the regio because sorghum is a short-day crop and most West African cultivars are photoperio sensitive that could only be produced under rainfed conditions [23,31]. These condition place limits on the use of long-season sorghum cultivars in some locations even within th same AEZ, which permits the choice of early-medium maturing cultivars. Although th soil fertility composition across the sites suggested low values for organic carbon (OC) an nitrogen N, the pH values indicated ideal soils (neutral to alkaline conditions) suitable fo plant growth of sorghum [51].

Our simulations revealed that the optimal PWs and suitable sorghum cultivars wer influenced by the dates of sowing, soil types, rainfall amount, and pattern across site and AEZs. In addition, this has to do with the cultivar's sensitivity or insensitivity t photoperiod and inherently early/late flowering traits [68]. These results corroborate th findings by [23], who reported that inherent soil fertility and rainfall patterns can greatl influence the yield when sowing is delayed. Both early and medium-maturing sorghur cultivars (ICSV400, Improved Deko, CSR01, and Samsorg-44) produced higher GY an WUE_{grain} than those of the late-maturing cultivar (SK5912) at varying PWs and were foun suitable to most sites across the AEZs. The optimal PWs slightly varied among the cultivar and AEZs. Our simulation results suggest an optimal sowing window for the ICSV40 and Improved Deko cultivars from 25 May to 10 July (45 days) and an optimal window for CSR01 and Samsorg-44 from 25 May to 30 June (35 days) in the SS zone. The results further revealed that the planting of CSR01, Samsorg-44, and SK5912 beyond these dates will significantly reduce the mean grain yield by 7%, 9%, and 11%, with no significant yield change estimated for the ICSV400 and Improved Deko cultivars. In the NGS and SGS zones, the optimal PWs ranged from 25 May to 10 July (45 days), except for SK5912, for which 25 May to 30 June (35 days) was simulated.

These results showed the use of early and medium maturing sorghum cultivars with higher yield and the most suitable cultivars to varying soil types simulated across the AEZs. In the SS zone, the level of variability suggests that ICSV400 and Improved Deko were highly suitable for cultivation across the sites; CSR01 and Samsorg-44 were suitable for cultivation in almost all the sites with exception of Guyaku, Balbaya, Mathau Puba Vidau, and Kwajaffa, while the late maturing cultivar (SK5912) adapted for cultivation only in 4 (Buratai, Kabura, Kurbo-Gayi, Lakundum) out of 14 sites. These results suggest only 4 out of the 5 sorghum cultivars may be suitable for cultivation under the current climatic conditions. In NGS, at a mean grain yield threshold of $\geq$1500 kgha^{-1} and the level of variability across the sites, all the cultivars were found to be adapted and suitable for cultivation in most sites, except for CSR01 and Samsorg-44 at Bare, and SK5912 at Bare, Daneyel, Hushere Zum, and Lakati-Libbo, respectively. The simulated yields of all the sorghum cultivars at the Fufure site in the SGS zone were found to be below the yield threshold of $\geq$1500 kgha^{-1}, and these results could be associated with sandy soil in the area and the very low soil fertility resulting in low water retention for crop growth. Also, a late PW reduced the grain yield due to early terminal drought towards the cessation of the growing period, resulting in a high temperature that affects the grain filling period, i.e., slows the rate of grain filling and accelerates senescence, thereby decreasing the photosynthetic activities per unit leaf area [69]. In addition, the increased temperature and water deficit experienced in the late planting window, particularly in PW4, can reduce the crop canopy (leaves and tillers) and decrease the biomass production, which in turn reduces the grain yield.

5. Conclusions

The validation of the model with farm-level grain yield enhanced the predictive capacity of the model for simulating diverse climatically driven yields under different fertilization strategies, sowing dates, and planting densities for the contrasting sorghum cultivars. However, our model application used different PWs based on climate-smart management practices that include the recommended fertilizer application rate and optimal hill population against the farmer practices for sorghum production in Northern Nigeria, geared towards disseminating and increasing the adoption of climate-smart technology, which is the basis for higher productivity. The optimum PWs were simulated as being between 25 May and 30 June for CSR01 and Samsorg-44 but were extended to 10 July for ICSV400 and Improved Deko, while low yield was simulated for SK5912 for all planting windows in the SS zone. In the NGS and SGS zones, the optimal PWs ranged from 25 May to 10 July (45 days) for all cultivars except for SK5912, for which predicted optimal PWs ranged from 25 May to 30 June (35 days). The mean simulated GY for SK5912 fell below the threshold of $\geq$1500 kgha^{-1} in Bare, Daneyel, Hushere Zum, and Lakati-Libbo'. In addition, at the Fufure site in SGS, all of the sorghum cultivars were simulated to be below the yield threshold $\geq$1500 kg ha^{-1} due to sandy soil texture found in the area, with the very low soil fertility resulting in a low water retention capacity for growth. Under climate change, the adoption of appropriate climate-smart technology sorghum will improve food security and reduce greenhouse gas emissions. It may therefore be concluded that the predicted optimal PWs for sorghum would substantially assist the smallholder farmers and seed producers in the region in their choice of cultivars to promote for high yields relative to growing sites and agro-ecologies.

Author Contributions: F.M.A. and H.A.A.: experimental design. F.M.A. and H.A.A.: experimentation. F.M.A., H.A.A., A.M.W. and A.Y.K.: methods. F.M.A., A.O.O. and A.I.T.: statistical analysis. F.M.A., A.O.O. and A.I.T.: manuscript draft. F.M.A., H.A.A., A.M.W. and A.Y.K.: final manuscript. All authors have read and agreed to the published version of the manuscript.

Funding: From 2017–2020, financial support came from the CGIAR Research Programs Grain Legumes and Dryland Cereals (GLDC), Climate Change, Agriculture and Food Security (CCAFS) carried out with support from the CGIAR Trust Fund and through bilateral funding agreements (for details, visit https://www.cgiar.org/funders/) is also acknowledged. The World Bank-funded AICCRA project (Accelerating Impacts of CGIAR Climate Research for Africa) Project ID 173398 is acknowledged for funding the involvement of F.M.A. and A.M.W. in this study. The Integrated Agriculture Activity project funded through the USAID is gratefully acknowledged.

Data Availability Statement: The raw data supporting the conclusions of this manuscript will be made available by the authors, without undue reservation.

Acknowledgments: The authors would like to thank the International Crops Research Institute for the Semi-Arid Tropics (ICRISAT) for their support of the study. We are grateful for the help of the field and laboratory staff of ICRISAT Kano for managing the field trials for this study.

Conflicts of Interest: The authors declare no conflict of interest. The funders had no role in the collection, analysis, or interpretation of data; in the writing of the manuscript, or in the decision to publish the results.

References

1. Ogbonna, A.C. Current Developments in Malting and Brewing Trials with Sorghum in Nigeria: A review. *J. Inst. Brew.* **2011**, *117*, 394–400. [CrossRef]
2. FAOSTAT. FAO Statistical Database (online). Food and Agricultural Organization of the United Nations. Rome. 2019. Available online: http://www.fao.org/faostat/en/#data/ (accessed on 25 July 2020).
3. FAOSTAT. Food and Agricultural Organization of the United Nations (FAO) FAO Statistical Database. 2021. Available online: http://faostat.fao.org (accessed on 19 March 2022).
4. Ajeigbe, H.A.; Akinseye, F.M.; Ayuba, K.; Jonah, J. Productivity and Water Use Efficiency of Sorghum [*Sorghum bicolor* (L.) Moench] Grown under Different Nitrogen Applications in Sudan Savanna Zone, Nigeria. *Int. J. Agron.* **2018**, *2018*, 7676058. [CrossRef]
5. Dwivedi, S.L.; Ceccarelli, S.; Blair, M.W.; Upadhyaya, H.D.; Are, A.K.; Ortiz, R. Landrace Germplasm for Improving Yield and Abiotic Stress Adaptation. *Trends Plant Sci.* **2016**, *21*, 31–42. [CrossRef]
6. Ajeigbe, H.A.; Singh, B.; Emechebe, A. Field evaluation of improved cowpea lines for resistance to bacterial blight, virus and Striga under natural infestation in the West African Savannas. *Afr. J. Biotechnol.* **2008**, *7*, 3563–3568.
7. Kouressy, M.; Dingkuhn, M.; Vaksmann, M.; Heinemann, A.B. Adaptation to diverse semi-arid environments of sorghum genotypes having different plant type and sensitivity to photoperiod. *Agric. For. Meteorol.* **2008**, *148*, 357–371. [CrossRef]
8. Abdulai, A.L.; Kouressy, M.; Vaksmann, M.; Asch, F.; Giese, M.; Holger, B. Latitude and Date of Sowing Influences Phenology of Photoperiod-Sensitive Sorghums. *J. Agron. Crop Sci.* **2012**, *198*, 340–348. [CrossRef]
9. Craufurd, P.Q.; Wheeler, T.R. Climate change and the flowering time of annual crops. *J. Exp. Bot.* **2009**, *60*, 2529–2539. [CrossRef] [PubMed]
10. Marteau, R.; Sultan, B.; Moron, V.; Alhassane, A.; Baron, C.; Traoré, S.B. The onset of the rainy season and farmers' sowing strategy for pearl millet cultivation in Southwest Niger. *Agric. For. Meteorol.* **2011**, *151*, 1356–1369. [CrossRef]
11. Yamusa, A.M.; Abu, S.T.; Yahaya, R.A.; Musa, I.J. Assessing the planting dates of sorghum in a changing climate at Samaru, Northern Nigeria. *Int. J. Biol. Environ. Sci. Trop.* **2013**, *10*, 21–26.
12. FAO. *Climate Smart Agriculture: Policies, Practices and Financing for Food Security, Adaptation and Mitigation*; Food and Agriculture Organization: Rome, Italy, 2010.
13. Lipper, L.; Thornton, P.; Campbell, B.M.; Baedeker, T.; Braimoh, A.; Bwalya, M.; Caron, P.; Cattaneo, A.; Garrity, D.; Henry, K.; et al. Climate-smart agriculture for food security. *Nat. Clim. Chang.* **2014**, *4*, 1068–1072. [CrossRef]
14. Ayinde, T.B.; Ahmed, B.; Nicholson, C.F. Farm-Level Impacts of Greenhouse Gas Reductions for the Predominant Production Systems in Northern Nigeria. In *African Handbook of Climate Change Adaptation*; Oguge, N., Ayal, D., Adeleke, L., da Silva, I., Eds.; Springer: Cham, Switzerland, 2021. [CrossRef]
15. Traore, S.B.; Reyniers, F.-N.; Vaksmann, M.; Kone, B.; Sidibe, A.; Yorote, A.; Yattara, K.; Kouressy, M. Adaptation a 'la se'cheresse des e´cotypes locaux de sorghos du Mali. *Se´cheresse* **2000**, *11*, 227–237.
16. Traoré, P.C.S.; Kouressy, M.; Vaksmann, M.; Tabo, R.; Maikano, I.; Traoré, S.B.; Cooper, P. Climate Prediction and Agriculture: What Is Different about Sudano-Sahelian West Africa? In *Climate Prediction and Agriculture*; Sivakumar, M.V.K., Hansen, J., Eds.; Springer: Berlin/Heidelberg, Germany, 2007. [CrossRef]

Hadebe, S.T.; Mabhaudhi, T.; Modi, A.T. Water use of sorghum (Sorghum bicolor L. Moench) in response to varying planting dates evaluated under rainfed conditions. *Water SA* **2017**, *43*, 91. [CrossRef]

Adetayo, A.O.; Dauda, T.O.; Adetayo, O.B.; Asiribo, O.E.; Issa, F. Rainfall instability difference in the effects of planting dates on growth and yield of maize (Zea mays) in forest savannah eco-climatic zone of Nigeria. *Afr. J. Agric. Res.* **2008**, *3*, 700–707.

Sivakumar, M. Predicting rainy season potential from the onset of rains in Southern Sahelian and Sudanian climatic zones of West Africa. *Agric. For. Meteorol.* **1988**, *42*, 295–305. [CrossRef]

Akinseye, F.M.; Agele, S.O.; Traore, P.C.S.; Adam, M.; Whitbread, A.M. Evaluation of the onset and length of growing season to define planting date—'A case study for Mali (West Africa)'. *Theor. Appl. Clim.* **2015**, *124*, 973–983. [CrossRef]

Lacy, S.M.; Cleveland, D.A.; Soleri, D. Farmer choice of sorghum cultivars in Southern Mali. *Hum. Ecol.* **2006**, *34*, 331–353. [CrossRef]

Akinseye, F.M.; Ajeigbe, H.A.; Traore, P.C.; Agele, S.O.; Zemadim, B.; Whitbread, A. Improving sorghum productivity under changing climatic conditions: A modelling approach. *Field Crop. Res.* **2019**, *246*, 107585. [CrossRef]

Folliard, A.; Traore, P.C.S.; Vaksmann, M.; Kouressy, M. Modelling of sorghum response to photoperiod: A threshold-hyperbolic approach. *Field Crops Res.* **2004**, *89*, 59–70. [CrossRef]

Clerget, B.; Dingkuhn, M.; Goze, E.; Rattunde, H.F.W.; Ney, B. Variability of Phyllochron, Plastochron and Rate of Increase in Height in Photoperiod-sensitive Sorghum Varieties. *Ann. Bot.* **2008**, *101*, 579–594. [CrossRef]

Haussmann, B.I.G.; Rattunde, H.F.; Weltzien-Rattunde, E.; Traoré, P.S.C.; Brocke, K.V.; Parzies, H.K. Breeding Strategies for Adaptation of Pearl Millet and Sorghum to Climate Variability and Change in West Africa. *J. Agron. Crop. Sci.* **2012**, *198*, 327–339. [CrossRef]

Weltzien, E.; Rattunde, H.F.W.; van Mourik, T.A.; Ajeigbe, H.A. Sorghum cultivation and improvement in West and Central Africa. In *Achieving Sustainable Cultivation of Sorghum Volume 2: Sorghum Utilization around the World*; Rooney, W., Ed.; Burleigh Dodds Science Publishing: Cambridge, UK, 2018.

Staggenborg, S.A.; Fjell, D.L.; Devlin, D.L.; Gordon, W.B.; Maddux, L.D.; Marsh, B.H. Selecting Optimum Planting Dates and Plant Populations for Dryland Corn in Kansas. *J. Prod. Agric.* **1999**, *12*, 85–90. [CrossRef]

Santos, R.D.; Boote, K.J.; Sollenberger, L.E.; Neves AL, A.; Pereira LG, R.; Scherer, C.B.; Gonçalves, L.C. Simulated optimum sowing date for forage pearl millet cultivars in multi-location trials in Brazilian semi-arid region. *Front. Plant Sci.* **2017**, *8*, 02074. [CrossRef]

Tofa, A.I.; Chiezey, U.F.; Babaji, B.A.; Kamara, A.Y.; Adnan, A.A.; Beah, A.; Adam, A.M. Modeling Planting-Date Effects on Intermediate-Maturing Maize in Contrasting Environments in the Nigerian Savanna: An Application of DSSAT Model. *Agronomy* **2020**, *10*, 871. [CrossRef]

Bassu, S.; Brisson, N.; Durand, J.L.; Boote, K.J.; Lizaso, J.; Jones, J.W.; Rosenzweig, C.; Ruane, A.C.; Adam, M.; Baron, C.; et al. Do various maize crop models give the same responses to climate change factors? *Glob. Change Biol.* **2014**, *20*, 2301–2320. [CrossRef]

Akinseye, F.M.; Adam, M.; Hoffmann, M.; Traore, P.; Agele, S.; Whitbread, A. Assessing crop model improvements through comparison of sorghum (*Sorghum bicolor* L. moench) simulation models: A case study for West African cultivars. *Field Crop Res.* **2017**, *201*, 19–31. [CrossRef]

Amanullah, M.; Mohamed, C.K.; Safiullah, A.; Selvam, S.; Sivakumar, K. Crop simulation growth model in Cassava. *Res. J. Agric. Biol. Sci.* **2007**, *3*, 255–259.

Keating, B.A.; Carberry, P.S.; Hammer, G.L.; Probert, M.E.; Robertson, M.J.; Holzworth, D.; Huth, N.I.; Hargreaves, J.N.; Meinke, H.; Hochman, Z.; et al. An overview of APSIM, a model designed for farming systems simulation. *Eur. J. Agron.* **2003**, *18*, 267–288. [CrossRef]

Holzworth, D.P.; Huth, N.I.; Devoil, P.G.; Zurcher, E.J.; Herrmann, N.I.; McLean, G.; Chenu, K.; van Oosterom, E.J.; Snow, V.; Murphy, C.; et al. APSIM—Evolution towards a new generation of agricultural systems simulation. *Environ. Model. Softw.* **2014**, *62*, 327–350. [CrossRef]

Van Ittersum, M.K.; Leffelaar, P.A.; Van Keulen, H.; Kropff, M.J.; Bastiaans, L.; Goudriaan, J. On approaches and applications of the Wageningen crop models. *Eur J Agron.* **2003**, *18*, 201–234. [CrossRef]

Whitbread, A.; Robertson, M.; Carberry, P.; Dimes, J. How farming systems simulation can aid the development of more sustainable smallholder farming systems in southern Africa. *Eur. J. Agron.* **2010**, *32*, 51–58. [CrossRef]

Allen, R.G.; Pereira, L.S.; Raes, D.; Smith, M. Crop evapotranspiration—Guidelines for computing crop water requirements. In *FAO Irrigation and Drainage*; Food and Agriculture Organization: Rome, Italy, 1998; Paper 56.

Ajeigbe, H.A.; Akinseye, F.M.; Kunihya, A.; Jonah, J. Sorghum yield and water use under Phosphorus applications in Sudan Savanna zone of Nigeria. *Glob. Adv. Res. J. Agric. Sci.* **2018**, *7*, 245–257.

NACGRAB, FMST. *Crop Varieties Released and Registered in Nigeria*; National Centre for Genetic Resources and Biotechnology, produced by Federal Ministry of Science and Technology: Ibadan, Nigeria, 2009; p. 42.

NACGRAB, FMST. *Guidelines for Registration and Release of New Crop Varieties in Nigeria*; National Centre for Genetic Resources and Biotechnology, produced by Federal Ministry of Science and Technology: Ibadan, Nigeria, 2016; p. 23.

Carberry, P.S.; Hammer, G.L.; Muchow, R.C. Modelling genotypic and environmental control of leaf area dynamics in grain sorghum. III. Senescence and prediction of green leaf area. *Field Crops Res.* **1993**, *33*, 329–351. [CrossRef]

42. Carberry, P.S.; Muchow, R.C.; Hammer, G.L. Modelling genotypic and environmental control of leaf area dynamics in grai sorghum. II. Individual leaf level. *Field Crops Res.* **1993**, *33*, 311–328. [CrossRef]
43. Loague, K.; Green, R.E. Statistical and graphical methods for evaluating solute transport models: Overview and application. *Contam. Hydrol.* **1991**, *7*, 51–73. [CrossRef]
44. Jamieson, P.D.; Porter, J.R.; Wilson, D.R. A test of the computer simulation model ARCWHEAT1 on wheat crops grown in Ne Zealand. *Field Crop. Res.* **1991**, *27*, 337–350. [CrossRef]
45. Funk, C.; Peterson, P.; Landsfeld, M.; Pedreros, D.; Verdin, J.; Shukla, S.; Husak, G.; Rowland, J.; Harrison, L.; Hoell, A.; et a The climate hazards infrared precipitation with stations—A new environmental record for monitoring extremes. *Sci. Data* **201** *2*, 150066. [CrossRef]
46. Batjes, N.H. Harmonized soil profile data for applications at global and continental scales: Updates to the WISE database. *So Use Manag.* **2009**, *25*, 124–127. [CrossRef]
47. Gummadi, S.; Dinku, T.; Shirsath, P.B.; Kadiyala, M.D.M. Evaluation of multiple satellite precipitation products for rainfed maiz production systems over Vietnam. *Sci. Rep.* **2022**, *12*, 485. [CrossRef]
48. Teutschbein, C.; Seibert, J. Bias correction of regional climate model simulations for hydrological climate-change impact studie Review and evaluation of different methods. *J. Hydrol.* **2012**, *456–457*, 12–29. [CrossRef]
49. Crochemore, L.; Ramos, M.-H.; Pappenberger, F. Bias correcting precipitation forecasts to improve the skill of seasonal streamflo forecasts. *Hydrol. Earth Syst. Sci.* **2016**, *11*, 1145–1159. [CrossRef]
50. Adnan, A.A.; Jibrin, J.M.; Kamara, A.Y.; Abdulrahman, B.L.; Shaibu, A.S.; Garba, I.I. CERES–Maize Model for Determining th Optimum Planting Dates of Early Maturing Maize Varieties in Northern Nigeria. *Front. Plant Sci.* **2017**, *8*, 1118. [CrossRef]
51. Shehu, B.M.; Jibrin, J.M.; Samndi, A.M. Fertility Status of Selected Soils in the Sudan Savanna Biome of Northern Nigeria. *Int. Soil Sci.* **2015**, *10*, 74–83. [CrossRef]
52. Dawaki, U.; Dikko, A.; Noma, S.; Aliyu, U. Heavy Metals and Physicochemical Properties of Soils in Kano Urban Agricultur Lands. *Niger. J. Basic Appl. Sci.* **2014**, *21*, 239. [CrossRef]
53. Jagtap, S. Changes in annual, seasonal, and monthly rainfall in Nigeria during 1961-90 and consequences to agriculture. *Disco Innov.* **1995**, *7*, 337–348.
54. Atehnkeng, J.; Ojiambo, P.S.; Donner, M.; Ikotun, T.; Sikora, R.A.; Cotty, P.J.; Bandyopadhyay, R. Distribution and toxigenicit of Aspergillus species isolated from maize kernels from three agro-ecological zones in Nigeria. *Int. J. Food Microbiol.* **200** *122*, 74–84. [CrossRef]
55. Aminu, Z.; Jaiyeoba, I.A. An assessment of soil degradation in Zaria area, Kaduna State, Nigeria. *Ife Res. Publ. Geogr.* **201** *13*, 27–37.
56. Omotosho, J.B.; Agele, S.O.; Balogun, I.A.; Adefisan, E.A. Climate variability, crop-climate modelling and water ecophysiolog research: Implications for plant's capacities for stress acclimation, yield production and food security. *Glob. J. Plant Ecophysic* **2013**, *3*, 56–69.
57. Ayanlade, A.; Adeoye, N.O.; Babatimehin, O. Intra-annual climate variability and malaria transmission in Nigeria. *Bull. Geog Socio-Econ. Ser.* **2013**, *21*, 7–19. [CrossRef]
58. FAO/UNESCO. *FAO–UNESCO Soil Map of the World*; UNESCO: Paris, France, 1974; Volume 1.
59. FAO. *Base Reference Base for Soil Resources. A Framework for International Classification, Correlation and Communication*; World So Resources Reports; FAO: Rome, Grace, 2006; Volume 103, p. 145.
60. Heanes, D. Determination of total organic-C in soils by an improved chromic acid digestion and spectrophotometric procedur *Commun. Soil Sci. Plant Anal.* **1984**, *15*, 1191–1213. [CrossRef]
61. Bremmer, J.M. Nitrogen-Total. In *Method of Soil Analysis Part 3—Chemical Methods*; Sparks, D.L., Ed.; SSSA Book Series 5; SSS Madison, WI, USA, 1996; pp. 1085–1122.
62. Gee, G.W.; Or, D. 2.4 Particle-Size Analysis. In *SSSA Book Series*; Dane, J.H., Topp, C.G., Eds.; Soil Science Society of Americ Madison, WI, USA, 2002; pp. 255–293. [CrossRef]
63. Bray, R.H.; Kurtz, L.T. Determination of total, organic, and available forms of phosphorus in soils. *Soil Sci.* **1945**, *59*, 39–4 [CrossRef]
64. Mehlich, A. Mehlich 3 soil test extractant: A modification of Mehlich 2 extractant. *Commun. Soil Sci. Plant Anal.* **198** *15*, 1409–1416. [CrossRef]
65. Beah, A.; Kamara, A.Y.; Jibrin, J.M.; Akinseye, F.M.; Tofa, A.I.; Adam, A.M. Simulating the Response of Drought–Tolerant Maiz Varieties to Nitrogen Application in Contrasting Environments in the Nigeria Savannas Using the APSIM Model. *Agronomy* **202** *11*, 76. [CrossRef]
66. Hammer, G.L.; Muchow, R.C. Quantifying climatic risk to sorghum in Australia's semi-arid tropics and subtropics: Mod development and simulation. In *Climatic Risk in Crop Production: Models and Management for the Semi-arid Tropics and Subtropic* eds Muchow, R.C., Mellamy, J.A., Eds.; CAB International: Nosworthy Way, Wallingford OX10 8DE, UK, 1991; Chapter 1 pp. 205–232.
67. Zhang, Y.; Feng, L.; Wang, E.; Wang, J.; Li, B. Evaluation of the APSIM-Wheat model in terms of different cultivars, manageme regimes and environmental conditions. *Can. J. Plant Sci.* **2012**, *92*, 937–949. [CrossRef]

68. Craufurd, P.Q.; Mahalakshmi, V.; Bidinger, F.; Mukuru, S.; Chantereau, J.; Omanga, P.; Qi, A.; Roberts, E.; Ellis, R.; Summerfield, R.; et al. Adaptation of sorghum: Characterization of genotypic flowering responses to temperature and photoperiod. *Theor. Appl. Genet.* **1999**, *99*, 900–911. [CrossRef]
69. Slafer, G.A.; Abeledo, L.G.; Miralles, D.J.; Gonzalez, F.G.; Whitechurch, E.M. Photoperiod sensitivity during stem elongation phase as an avenue to rise potential yield in wheat. *Euphytica* **2001**, *119*, 191–197. [CrossRef]

agronomy

MDPI

Article

Winter Oilseed Rape: Agronomic Management in Different Tillage Systems and Seed Quality

Mateusz Sokólski [1], Dariusz Załuski [2], Artur Szatkowski [1] and Krzysztof Józef Jankowski [1,*]

[1] Department of Agrotechnology and Agribusiness, University of Warmia and Mazury in Olsztyn, Oczapowskiego 8, 10-719 Olsztyn, Poland
[2] Department of Genetics, Plant Breeding and Bioresource Engineering, University of Warmia and Mazury in Olsztyn, Plac Łódzki 3, 10-724 Olsztyn, Poland
* Correspondence: krzysztof.jankowski@uwm.edu.pl

Abstract: A three-year study was conducted to analyze agronomic management in the production of winter oilseed rape (WOSR) under different tillage systems. A field experiment was conducted at the University's Agricultural Experiment Station in Bałcyny (north-eastern Poland), in three growing seasons (2016/2017, 2017/2018, and 2018/2019). The experiment had a 3^{5-2} resolution III fractional factorial design with five fixed factors that were tested at three levels of intensity. The experimental factors were: A—tillage: (A0) strip-till, (A1) low-till, (A2) conventional tillage; B—weed control: (B0) pre-emergent, (B1) foliar, (B2) sequential; C—growth regulation: (C0) none, (C1) in fall, (C2)—in fall and spring; D—rate of nitrogen (N) fertilizer applied in spring: (D0) 160, (D1) 200, (D2) 240 kg ha^{-1}; and E—rate of sulfur (S) fertilizer applied in spring: (E0) 0, (E1) 40, (E2) 80 kg ha^{-1}. The crude fat (CF) content of WOSR seeds was highest in the strip-till system (498 g kg^{-1} dry matter, DM), and the total protein (TP) content of seeds was highest (196 g kg^{-1} DM) in low-till and conventional tillage systems. The content of neutral detergent fiber (NDF) was higher in seeds harvested from strip-till and low-till systems than from the conventional tillage system. The seeds of WOSR plants grown in the conventional tillage system accumulated more (by 0.4%) polyunsaturated fatty acids (PUFAs) and less (by 0.5–0.6%) monounsaturated fatty acids (MUFAs). An increase in the N rate from 160–200 to 240 kg ha^{-1} decreased the CF content (495 vs. 484 g kg^{-1} DM) and increased the TP content of seeds (191 vs. 199 g kg^{-1} DM). Sulfur fertilization induced a 34% increase in glucosinolate (GLS) concentrations in WOSR seeds, mainly by enhancing the biosynthesis of alkenyl GLS (by 39%).

Keywords: *Brassica napus* (L.); tillage; weed control; growth regulators; nitrogen and sulfur fertilization; seed quality

Citation: Sokólski, M.; Załuski, D.; Szatkowski, A.; Jankowski, K.J. Winter Oilseed Rape: Agronomic Management in Different Tillage Systems and Seed Quality. *Agronomy* **2023**, *13*, 524. https://doi.org/10.3390/agronomy13020524

Academic Editors: Umberto Anastasi and Aurelio Scavo

Received: 19 January 2023
Revised: 7 February 2023
Accepted: 9 February 2023
Published: 11 February 2023

1. Introduction

Food crops that store carbohydrates as energy reserves (cereals, potatoes, sweet potatoes, etc.) are considered the most economically important around the world. In developing countries, cereals account for as much as 60% of the caloric intake. In developed countries, cereal consumption is relatively high (35% of the caloric intake), but the consumption of fat, including vegetable fats, has also increased [1]. The role of vegetable oils has increased with rapid economic and population growth in the industrial age, and breeding progress (modification of the fatty acid composition of rapeseed—the most important oilseed crop in the temperate zone) [2,3]. In 2010–2020, the global production of four major vegetable oils increased by 5.7 mln Mg per year on average, to reach around 201 mln Mg in the 2020/2021 season, where palm oil (from the fruit of oil palm trees) accounted for 38%, followed by soybean oil—30%, canola oil—14%, and sunflower oil—9% [4].

In terms of nutritional value, oil extracted from the seeds of double-low (canola-quality) rapeseed cultivars is characterized by the most favorable fatty acid profile due to a high content of oleic acid (which decreases blood cholesterol levels), a desirable n-6/n-3

PUFA ratio, a very low content (approx. 6–7%) of nutritionally undesirable saturated fatty acids (SFAs), and an optimal proportion of linolenic acid (which improves neurological function) [5]. Rapeseed oil is used for both culinary and industrial purposes: it is suitable for short-term deep frying, margarine production, production of liquid biofuel components (biodiesel), dyes, varnishes, solvents, etc. [2].

Fat-free residues of rapeseeds are the second most commonly used protein source for livestock after soybean meal [6]. Rapeseed meal is characterized by a favorable amino acid profile, including a relatively high content of exogenous amino acids (methionine and cystine) and minerals (phosphorus and calcium) [7,8]. Even though the concentrations of anti-nutritional factors in seeds of canola cultivars have been considerably reduced, the presence of GLS still decreases the feed value of fat-free seed residues. These compounds impart a bitter flavor to rapeseed press-cake and meal, decrease protein availability, and inhibit the synthesis of thyroid hormones; they may also cause liver damage. Alkenyl GLS exert the most detrimental effects [9]. Fiber is the main factor responsible for decreasing the digestibility and energy value of rapeseed press-cake and meal, and fiber content is nearly twice as high in rapeseed meal than in soybean meal [10].

The quality of rapeseeds is determined by genetic factors (cultivar) and environmental conditions, but it can be considerably modified by agronomic factors and production technology [11]. Tillage and seeding methods have a minor influence on the content of crude fat (CF) and total protein (TP) in the seeds of winter oilseed rape (WOSR) [12–16]. Muśnicki et al. [12,13] demonstrated that shallow pre-sowing plowing increased CF concentration in WOSR seeds by approximately 1%. In turn, Jankowski [16] found that WOSR seeds had lower CF content in the conventional tillage system, compared with simplified tillage. Different tillage systems have no significant effect on TP concentration in WOSR seeds [12–16].

Weeds can considerably compromise the quality and market value of WOSR seeds [17]. Some weed species (e.g., *Sinapis arvensis* L. and *Thlaspi arvense* L.) decrease seed quality by increasing the content of erucic acid and GLS. In canola cultivars, effective weed management contributes to an increase (by up to 10%) in the CF content of seeds [17]. In turn, the absence of chemical weed control significantly affects the fatty acid profile of seeds [17,18].

Growth regulators applied in autumn exert a minor influence on nutrient synthesis in WOSR seeds [16,19]. In a study by Ijaz and Honermeier [19], the CF content of WOSR seeds peaked (454–455 g kg^{-1} dry matter, DM) in the control treatment and after the autumn application of tebuconazole and trinexapac-ethyl. The accumulation of CF in seeds decreased significantly (to 450 g kg^{-1} DM) following the application of metconazole. In contrast, Jankowski [16] observed no significant differences in the concentrations of CF or TP in WOSR seeds in response to growth regulation in autumn.

Chemical growth regulation in spring exerted varied effects on nutrient synthesis in WOSR seeds [20–22]. Ijaz and Honermeier [19], and Matysiak and Kaczmarek [21] found no relationships between the spring application of metconazole, tebuconazole, trinexapac-ethyl [19], and chlorocholine chloride [21] vs. the content of CF and TP in WOSR seeds. Ijaz and Honermeier [19] demonstrated that metconazole induced a minor increase (2%) in the CF content of WOSR seeds, whereas Ijaz et al [22] observed a slight decrease (2%) in the CF content of WOSR seeds when mixtures of tebuconazole and prothioconazole, and difenoconazole and paclobutrazole, were applied in spring. The application of azoxystrobin in combination with triazole fungicides stimulated the synthesis of CF in WOSR seeds, most likely due to prolonged seed formation [22]. Matysiak et al. [23] reported that the timing of growth regulator application (BBCH 30 vs. 50) may affect the quality of WOSR seeds. The CF content of seeds peaked after the application of metconazole (regardless of application timing) and tebuconazole (at BBCH 30). Growth regulators also exert varied effects on the fatty acid profile. The spring application of trinexapac-ethyl increased the concentration of linolenic acid in rapeseed oil by 3% [20], whereas the autumn and spring application of metconazole decreased the concentration of oleic acid by 1% [19].

Nitrogen fertilization is a key determinant of seed quality in WOSR cultivation [16,24,25]. Many researchers have found that increasing N rates induce a decrease in CF concentration [16,24–32] and enhance TP synthesis in WOSR seeds [16,24,27–30]. Nitrogen applic tion in later growth stages has a particularly adverse effect on CF accumulation in WOS seeds. Nitrogen applied at the beginning of flowering contributes to an increase in the T content and a decrease in the CF content of WOSR seeds [1,24].

Sulfur fertilization may have different effects on the nutrient content of WOSR seed According to Sienkiewicz-Cholewa and Kieloch [33] and Fazili et al. [34], S promotes C accumulation in seeds. In turn, Wielebski [35] demonstrated that CF synthesis decreased, an TP concentration increased in WOSR seeds in response to S fertilizer. Jankowski et al. [36,3 and Groth et al. [38] found that S fertilization had no significant influence on the content of C or TP in WOSR seeds. Sulfur considerably modifies GLS concentrations in the biomass of *Bra sicaceae* oilseed crops [37]. In experiments conducted by Jankowski [16] and Groth et al. [38], application increased GLS accumulation in WOSR seeds by up to 24–29%. It should be note that S fertilization increases the content of alkenyl GLS, by 13–15% [35,39] and up to 40% [38 The synthesis of indole GLS is less stimulated by S application, and the S-induced increase i their concentration has been estimated at 5–15% [35–37,39].

The aim of this study was to determine the effects of weed control, growth regulatio and N and S fertilization on the quality of seeds harvested from WOSR plants grow in different tillage systems (conventional tillage, low-till, and strip-till). Five agronomi factors were evaluated at three levels of intensity in a small-area field experiment with 3^{5-2} fractional factorial design. The main effects and two-factor interaction effects wer evaluated with the use of modified fractional design generators [1,40,41].

2. Materials and Methods

2.1. Field Experiment

The presented results were obtained during a small-area field experiment carrie out in 2016–2019 at the Agricultural Experiment Station (AES) in Bałcyny (NE Polan 53°35′46.4″ N, 19°51′19.5″ E). The experiment had a 3^{5-2} resolution III fractional factori design with two replications, where five agronomic factors (A, B, C, D, and E) were teste at three intensity levels (0, 1, and 2) (Table 1).

Table 1. Experimental factors.

Symbol	Agricultural Operation	Level		
		0	1	2
A	Tillage †	strip-till	low-till	conventional tillage
B	Weed control	pre-emergent (0–2 days after sowing) 500 g ha^{-1} metazachlor, 500 g ha^{-1} dimethenamid-P, 250 g ha^{-1} quinmerac	Foliar (BBCH 12–14 ††) 72 g ha^{-1} clopyralid, 24 g ha^{-1} picloram, 12 g ha^{-1} aminopyralid, 750 g ha^{-1} metazachlor	(0–2 days after sowing) 72 g ha^{-1} clomazone (BBCH 12–14) 72 g ha^{-1} clopyralid, 24 g ha^{-1} picloram, 12 g ha^{-1} aminopyralid
C	Growth regulation	none	fall treatment (BBCH 14–15) 210 g ha^{-1} mepiquat chloride, 30 g ha^{-1} metconazole	(BBCH 14–15) 210 g ha^{-1} mepiquat chloride, 30 g ha^{-1} metconazole (BBCH 30–31) 125 g ha^{-1} difenoconazole, 62.5 g ha^{-1} paclobutrazol
D	Spring N rate (kg ha^{-1}) †††	160 (120 + 40) (BBCH 20–30 + 50)	200 (120 + 80) (BBCH 20–30 + 50)	240 (120 + 120) (BBCH 20–30 + 50)
E	Spring S rate (kg ha^{-1}) ††††	0	40 (BBCH 20–30)	80 (BBCH 20–30)

† strip-till with sowing and application of NPK fertilizers to a depth of 10 and 20 cm (50:50%); low-till to depth of 25–30 cm one day before sowing with a seed drill cultivator; conventional tillage—disking to a depth 5–8 cm after harvesting the previous crop, pre-sowing plowing to a depth of 18–20 cm 7–10 days before sowi seedbed preparation, and sowing with a seed drill cultivator. †† Biologische Bundesanstalt, Bundessortena und Chemische Industrie [42]: BBCH 12–14—2–4 true leaves unfolded; BBCH 14–15—4–5 true leaves unfold BBCH 20–30—beginning of the spring growing season; BBCH 30–31—rosette regrowth after winter; BBC 50—beginning of the budding stage. ††† N fertilizer was applied before the spring growing season (BBCH 20–3 at 120 kg N ha^{-1} in the form of: (i) ammonium nitrate (treatments D0E0, D1E0, D2E0) or (ii) ammonium sulfa and ammonium nitrate (treatments D0E1, D1E1, D2E1, D0E2, D1E2, D2E2). The second N rate was applied ammonium sulfate in BBCH stage 50. †††† Sulfur was applied as ammonium sulfate.

The harvested plot areas were 15 m^2 each (10 m by 1.5 m). Winter wheat was the preceding crop in each year of the study. To prepare the plots for WOSR cultivation, winter wheat was cut to a height of 12–15 cm, and the entire straw was collected. The experiment was established on Haplic Luvisol originating from boulder clay [43] with a slightly acidic pH (pH KCl 5.6–6.2) and C_{org} content of 1.04–1.28%. The macronutrient and micronutrient content of the arable layer was determined at: 91.0–221.0 P_2O_5 mg kg^{-1}, 145.0–195.0 K_2O mg kg^{-1}, 89.0–129.3 Mg mg kg^{-1}, 4.4–13.3 SO_4^{2-} mg kg^{-1}, 0.14–0.44 B mg kg^{-1}, 128.0–218.0 Mn mg kg^{-1}, 2.1–4.5 Cu mg kg^{-1}, 5.6–11.8 Zn mg kg^{-1}, and 1680–2000 Fe mg kg^{-1}. The content of C_{org} in soil was determined by the modified Kurmies method (UV-1201V spectrophotometer, Shimadzu Corporation, Kyoto, Japan). Soil pH was measured with a digital pH meter with temperature compensation (20 °C) in deionized water and 1 mol KCl (5:1). Plant-available P and K were extracted with calcium lactate (Egner–Riehm method). Phosphorus was measured by vanadium-molybdenum yellow spectrophotometry (UV-1201V spectrophotometer, Shimadzu Corporation, Kyoto, Japan), and K was measured by atomic emission spectrometry (AES) (Flame Photometers, BWB Technologies Ltd., Newbury, UK). Magnesium was extracted with 0.01 mol of calcium chloride and quantified by atomic absorption spectrophotometry (AAS) (AAS1N, Carl Zeiss, Jena, Germany). Boron concentration was determined colorimetrically (UV-1201V spectrophotometer, Shimadzu Corporation, Kyoto, Japan), and the concentrations of Cu, Zn, Mn, and Fe were determined by AAS after extraction in 1 mol dm^{-3} HCl. The content of sulfide sulfur was analyzed by nephelometry after extraction in an acetate buffer (UV-1201V spectrophotometer, Shimadzu Corporation, Kyoto, Japan).

The seeds of the hybrid WOSR cultivar Kuga were sown between 13 and 23 August at 50 germinating seeds per 1 m^2. Agronomic factors that did not constitute the experimental variables were applied according to good agricultural practice. The following fertilizers were applied before sowing: 40 kg N ha^{-1} (urea), 60 kg P_2O_5 ha^{-1} (enriched superphosphate), and 120 kg K_2O ha^{-1} (potash salt). In the strip-till system (A0), fertilizers were applied together with seeds to a depth of 10 and 20 cm at 50:50%. In low-till (A1) and conventional tillage (A2) systems, fertilizers were distributed on the surface of previous crop residues (stubble). In all treatments, monocotyledonous weeds were controlled with 60 g ha^{-1} propaquizafop (BBCH 12–14). Foliar B fertilizer (2 × 175 g B ha^{-1}) was applied twice in spring (BBCH 20–30 and 50). Pests were controlled chemically in BBCH stages 35–37 (300 g ha^{-1} chlorpiryfos + 30 g ha^{-1} cipermetrin) and BBCH stages 63–67 (60 g ha^{-1} thiacloprid + 6 g ha^{-1} deltamethrin) when the action threshold was exceeded. Fungal diseases were managed with 100 g ha^{-1} dimoxystrobin and 100 g ha^{-1} boscalid (BBCH 63). Winter oilseed rape was harvested at physiological maturity (8–24 July) with a small-plot harvester.

2.2. Seed Quality

The quality of WOSR seeds was evaluated based on the following parameters: the content of CF and TP (in g kg^{-1} DM seeds), proportions of neutral detergent fiber (NDF) and acid detergent fiber (ADF) (in %), fatty acid profile (in %), and the content of total GLS and alkenyl GLS (in μM g^{-1} DM seeds).

Winter oilseed rape seeds were scanned with the use of a NIR Systems 650 near-infrared reflectance spectrometer (FOSS NIR Systems Inc., Silver Spring, USA). A calibration equation derived in the WINISI program was used in the measurements based on the reference data for Kjeldahl (TP), Soxhlet (CF), and van Soest (NDF and ADF) methods. Glucosinolate content was quantified by gas chromatography of trimethylsilyl derivatives of desulfated GLS in the Agilent 7890 gas chromatograph with an HP-5 column (Agilent Technologies Inc., Santa Clara, USA), using the method described by Raney and modified by Michalski et al. [44]. Fatty acid composition was determined by methylation of oil extracted from 0.1 g of ground seeds. Fatty acid methyl esters (FAMEs) were analyzed by gas chromatography (HP 3390A integrator, Hewlett-Packard, Avondale, USA) with a DB-23 capillary column (length: 30 m); operating temperature—200°C (injector and detector temperature—220°C); carrier gas—hydrogen.

2.3. Statistical Analysis

The original 3^{5-2} fractional factorial design with resolution III and 27 treatments per replication was initially generated. However, the experiment involved three growing seasons, which generated large quantities of data for analysis; therefore, in the statistical analysis, the original 3^{5-2} ANOVA model was reduced to 21 treatments, but all attributes of the 3^{5-2} design (III) were retained. Based on this analytical model, ANOVA was used to evaluate the main effects of all fixed factors (A, B, C, D, and E) and their interactions with the random effects of the experimental years (Y × A, Y × B, Y × C, Y × D, and Y × E). The significance of differences between mean values was determined in Tukey's honest significant difference (HSD) test with $p < 0.05$. All analyses were performed in the Statistica 13.3 program [45]. The *F*-values in ANOVA are presented in Table 2.

2.4. Weather Conditions

The weather conditions during the three growing seasons are presented in Figure 1. In all years of the study, the mean daily temperature during the autumn growing season (August-November) was similar to the long-term average (10.3–11.3 vs. 11.4 °C). Different temperatures were noted during winter dormancy (December-March) and the spring-summer growing season (April–July). In the 2018/2019 season, the mean daily temperature during winter dormancy was 1.3 °C higher than the long-term average (0.0 °C). In the remaining years, the mean daily temperature during winter dormancy was comparable to the long-term average. During the spring-summer growing season, the highest mean daily temperature was recorded in 2017/2018, and it exceeded the long-term average by 2.3 °C (15.6 vs. 13.3 °C). In the remaining years, the mean daily temperature during the spring-summer growing season approximated the long-term average (12.9–13.8 vs. 13.3 °C). In the study area, the mean rainfall during the growing season over the last 35 years was 514 mm. In the first and second years of the study, total rainfall during the growing season exceeded the long-term average by 27 and 38%, respectively (655–707 vs. 514 mm). In the third year, precipitation was 9% lower than the long-term average (Figure 1).

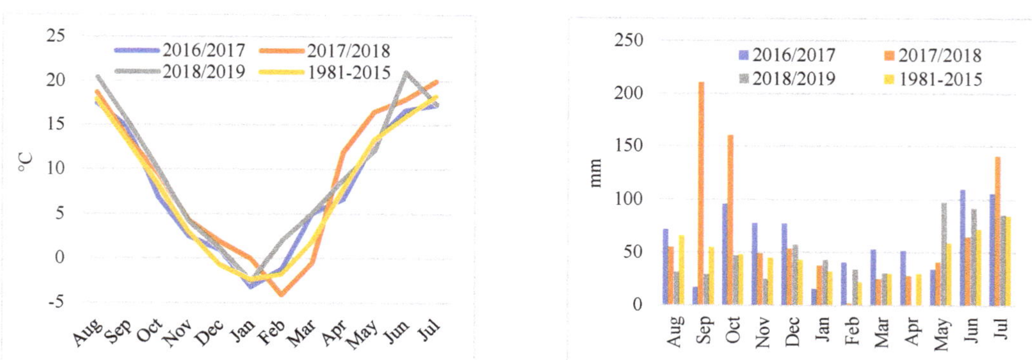

Figure 1. Mean monthly temperature (°C) and total monthly precipitation (mm) during the growing seasons of 2016–2019 vs. the long-term average (1981–2015).

Table 2. *F*-test statistics in ANOVA of seed quality in WOSR production.

Source of Variation	Effect	Crude Fat ($g\ kg^{-1}$ DM)	Total Protein ($g\ kg^{-1}$ DM)	NDF (%)	ADF (%)	GLS ($\mu M\ g^{-1}$)		Fatty Acids (%)								
						Alkenyl	Total	C16:0	C18:0	C18:1	C18:2	C18:3	C20:1	SFAs	MUFAs	PUFAs
Growing season (Y)	random	114.76 **	67.16 **	79.49 **	20.88 **	8.28 **	40.84 **	59.50 **	182.33 **	146.15 **	99.25 **	303.69 **	24.20 **	34.16 **	197.69 **	213.71 **
Tillage system (A)	fixed	7.64 **	4.61 **	5.21 **	1.20 *ns*	0.86 *ns*	2.24 *ns*	2.00 *ns*	1.57 *ns*	5.75 **	1.49 *ns*	9.06 **	0.46 *ns*	2.04 *ns*	7.09 **	5.13 *
Weed control (B)	fixed	0.36 *ns*	0.21 *ns*	0.20 *ns*	0.66 *ns*	0.05 *ns*	0.36 *ns*	1.07 *ns*	0.16 *ns*	0.48 *ns*	0.10 *ns*	0.87 *ns*	2.55 *ns*	1.31 *ns*	0.34 *ns*	0.43 *ns*
Growth regulation €	fixed	0.11 *ns*	0.09 *ns*	0.22 *ns*	0.04 *ns*	0.04 *ns*	0.00 *ns*	2.66 *ns*	0.42 *ns*	0.64 *ns*	4.13 *ns*	0.31 *ns*	18.80 **	2.60 *ns*	2.39 *ns*	2.80 *ns*
Nitrogen fertilization (D)	fixed	10.10 **	6.32 **	11.06 **	0.79 *ns*	0.18 *ns*	0.14 *ns*	0.08 *ns*	0.48 *ns*	0.43 *ns*	0.85 *ns*	0.86 *ns*	0.51 *ns*	0.08 *ns*	0.24 *ns*	0.50 *ns*
Sulfur fertilizati€(E)	fixed	3.47 *ns*	0.53 *ns*	3.62 *ns*	0.05 *ns*	5.15 **	6.02 **	2.04 *ns*	0.09 *ns*	0.70 *ns*	0.02 *ns*	0.92 *ns*	1.60 *ns*	2.16 *ns*	0.11 *ns*	0.13 *ns*
Y × A	random	0.14 *ns*	1.38 *ns*	2.07 *ns*	0.79 *ns*	0.61 *ns*	0.96 *ns*	1.27 *ns*	1.33 *ns*	0.42 *ns*	0.53 *ns*	1.36 *ns*	1.63 *ns*	1.24 *ns*	0.66 *ns*	0.80 *ns*
Y × B	random	0.26 *ns*	0.77 *ns*	0.93 *ns*	1.26 *ns*	0.37 *ns*	0.53 *ns*	1.09 *ns*	1.32 *ns*	0.12 *ns*	0.70 *ns*	0.30 *ns*	0.23 *ns*	1.15 *ns*	0.20 *ns*	0.52 *ns*
Y × C	random	3.29 **	0.91 *ns*	1.73 *ns*	0.45 *ns*	1.12 *ns*	1.26 *ns*	2.68 *ns*	1.28 *ns*	0.25 *ns*	0.34 *ns*	0.09 *ns*	1.62 *ns*	3.09 *ns*	1.30 *ns*	0.21 *ns*
Y × D	random	0.98 *ns*	1.62 *ns*	1.64 *ns*	0.97 *ns*	0.17 *ns*	0.51 *ns*	0.83 *ns*	0.91 *ns*	0.28 *ns*	0.78 *ns*	0.60 *ns*	0.49 *ns*	0.61 *ns*	0.44 *ns*	0.61 *ns*
Y × E	random	2.46 *ns*	0.40 *ns*	2.44 *ns*	0.29 *ns*	2.02 *ns*	2.39 *ns*	2.37 *ns*	2.01 *ns*	2.05 *ns*	5.42 **	1.24 *ns*	13.32 **	2.74 *ns*	6.32 **	4.55 *

ADF—acid detergent fiber; NDF—neutral detergent fiber; GLS—glucosinolates; C16:0—palmitic acid; C18:0—stearic acid; C18:1—oleic acid; C18:2—linoleic acid; C18:3—linolenic acid; C20:1—eicosenoic acid; SFAs—saturated fatty acids; MUFAs—monounsaturated fatty acids; PUFAs—polyunsaturated fatty acids. * significant $p < 0.05$; ** significant $p < 0.01$; *ns*—not significant.

3. Results

3.1. Years

In the first and second years of the study, which were characterized by above-averag precipitation in spring and summer, WOSR seeds accumulated the highest quantities of C The proportions of both crude fiber fractions (NDF and ADF) in WOSR seeds also increase in years with abundant rainfall between April and July. Lower precipitation levels durin the spring-summer growing season (2018/2019) contributed to the accumulation of TP an GLS in WOSR seeds (Table 3). Above-average rainfall in spring and summer (years 1 and supported the synthesis of SFAs (mostly palmitic acid) and PUFAs (mostly linoleic aci and linolenic acid). In the third year of the study, the proportions of SFAs and PUFAs wer 0.73% and 3.69% lower, respectively. Lower precipitation levels during the spring-summe growing season (year 3) promoted the synthesis of oleic acid and eicosenoic acid (MUFA (Table 4).

Table 3. Nutritional value of WOSR seeds (main factors).

Agronomic Factor	Level	Crude Fat (g kg^{-1} DM)	Total Protein (g kg^{-1} DM)	NDF (%)	ADF (%)	Σ alkenyl GLS (μM g^{-1} DM)	Σ GLS (μM g^{-1} DM)
Years	2016/2017	519.1 ^{a}	175.3 ^{c}	28.5 ^{a}	23.6 ^{a}	6.07 ^{b}	8.84 ^{b}
	2017/2018	496.4 ^{b}	189.1 ^{b}	27.6 ^{b}	21.7 ^{b}	5.16 ^{b}	6.47 ^{c}
	2018/2019	457.1 ^{c}	216.3 ^{a}	26.1 ^{c}	20.5 ^{c}	7.47 ^{a}	11.64 ^{a}
Tillage	strip-till	497.6 ^{a}	188.6 ^{b}	27.6 ^{a}	22.2	5.85	8.38
	low-till	489.4 ^{ab}	196.4 ^{a}	27.4 ^{ab}	21.5	6.41	9.12
	conventional tillage	485.6 ^{b}	195.8 ^{a}	27.1 ^{b}	22.0	6.45	9.45
Weed control	pre-emergent	489.8	191.9	27.5	22.0	5.93	8.42
	foliar	492.3	193.7	27.4	22.1	6.28	9.20
	sequential	490.2	195.5	27.4	21.6	6.54	9.35
Growth regulation	none	484.6	195.1	27.1	21.6	5.16	7.54
	BBCH 14–15	492.1	194.4	27.5	21.9	6.98	9.94
	BBCH 14–15 and 30–31	493.6	191.9	27.4	22.1	6.16	8.93
Spring nitrogen rate (kg N ha^{-1})	160	494.5 ^{a}	190.5 ^{b}	27.6 ^{a}	22.0	6.18	8.81
	200	494.1 ^{a}	192.5 ^{ab}	27.5 ^{a}	22.1	6.24	9.11
	240	483.8 ^{b}	198.7 ^{a}	27.0 ^{b}	21.7	6.31	9.16
Spring sulfur rate (kg S ha^{-1})	0	483.7	195.2	27.1	21.7	5.18 ^{c}	7.60 ^{c}
	40	493.6	191.9	27.4	22.1	6.16 ^{b}	8.93 ^{b}
	80	493.9	194.2	27.6	21.9	7.23 ^{a}	10.23 ^{a}

ADF—acid detergent fiber; NDF—neutral detergent fiber; GLS—glucosinolates. BBCH 14–15—4–5 true leave unfolded; BBCH 30–31—rosette regrowth after winter. Means sharing a common letter are not significantl different at $p \leq 0.05$ in Tukey's test. The absence of common letters denotes non-significant differences (*ns*) $p \leq 0.05$ in Tukey's test.

Table 4. Fatty acids (%) (main factors).

Agronomic Factor	Level	C16:0	C18:0	C18:1	C18:2	C18:3	C20:1	SFAs	MUFAs	PUFAs
Years	2016/2017	4.27 ^{a}	1.32 ^{c}	62.40 ^{c}	20.92 ^{a}	10.10 ^{a}	1.00 ^{b}	5.59 ^{a}	63.39 ^{c}	31.02 ^{a}
	2017/2018	4.20 ^{a}	1.52 ^{b}	65.23 ^{b}	19.23 ^{b}	8.80 ^{b}	1.01 ^{b}	5.72 ^{a}	66.24 ^{b}	28.04 ^{b}
	2018/2019	3.12 ^{b}	1.74 ^{a}	66.31 ^{a}	19.48 ^{b}	7.86 ^{c}	1.49 ^{a}	4.86 ^{b}	67.80 ^{a}	27.33 ^{c}
Tillage	strip-till	3.81	1.53	64.85 ^{a}	19.85	8.81 ^{b}	1.15	5.34	66.00 ^{a}	28.66 ^{b}
	low-till	3.85	1.51	64.84 ^{a}	19.81	8.84 ^{b}	1.15	5.36	65.99 ^{a}	28.66 ^{b}
	conventional tillage	3.93	1.54	64.25 ^{b}	19.97	9.10 ^{a}	1.20	5.48	65.45 ^{b}	29.07 ^{a}
Weed control	pre-emergent	3.88	1.53	64.58	19.89	8.93	1.19	5.41	65.77	28.82
	foliar	3.82	1.53	64.61	19.89	8.95	1.20	5.34	65.81	28.85
	sequential	3.91	1.53	64.77	19.85	8.86	1.09	5.43	65.86	28.71

Table 4. *Cont.*

Agronomic Factor	Level	C16:0	C18:0	C18:1	C18:2	C18:3	C20:1	SFAs	MUFAs	PUFAs
Growth regulation	none	3.88	1.52	64.31	20.17	9.01	1.11	5.40	65.42	29.18
	BBCH 14–15	3.84	1.54	64.92	19.69	8.83	1.19	5.38	66.11	28.52
	BBCH 14–15 and 30–31	3.88	1.52	64.59	19.89	8.95	1.18	5.40	65.76	28.84
Spring nitrogen rate (kg N ha^{-1})	160	3.86	1.54	64.60	19.93	8.91	1.17	5.40	65.77	28.84
	200	3.88	1.52	64.83	19.71	8.92	1.16	5.39	65.98	28.63
	240	3.86	1.52	64.61	19.90	8.94	1.17	5.38	65.78	28.84
Spring sulfur rate (kg S ha^{-1})	0	3.86	1.53	64.28	20.11	9.03	1.19	5.38	65.48	29.14
	40	3.88	1.52	64.59	19.89	8.95	1.18	5.40	65.76	28.84
	80	3.85	1.53	65.02	19.67	8.79	1.13	5.39	66.16	28.46

C16:0—palmitic acid; C18:0—stearic acid; C18:1—oleic acid; C18:2—linoleic acid; C18:3—linolenic acid C20:1—eicosenoic acid; SFAs—saturated fatty acids; MUFAs—monounsaturated fatty acids; PUFAs—polyunsaturated fatty acids. BBCH 14–15—4–5 true leaves unfolded; BBCH 30–31—rosette regrowth after winter. Means sharing a common letter are not significantly different at $p \leq 0.05$ in Tukey's test. The absence of common letters denotes non-significant differences (*ns*) at $p \leq 0.05$ in Tukey's test.

3.2. Tillage

Winter oilseed rape grown in low-till and conventional tillage systems accumulated 2–3% less CF and 4% more TP in the seeds than WOSR grown in the strip-till system (Table 3). The content of NDF was higher in seeds harvested from strip-till (27.6%) and low-till (27.4%) systems than those harvested from the conventional tillage system (27.1%). The tillage system did not significantly affect ADF and GLS levels in the seeds (Table 3). The seeds of WOSR plants grown in the conventional tillage system accumulated 0.4% more PUFAs (mostly due to enhanced synthesis of linolenic acid) and 0.5–0.6% less MUFAs (mostly due to a lower proportion of oleic acid) (Table 4). The effects of different tillage systems on the analyzed seed quality parameters were not modified by weather conditions (Y × A) (Table 2).

3.3. Weed Control

None of the tested weed control methods induced significant differences in seed quality parameters, regardless of weather conditions across the years (Y × B) (Table 2).

3.4. Growth Regulation

In years 1 and 2, growth regulators increased the CF content of seeds by 11–61 (fall treatment) to 7–25 g kg^{-1} DM (fall and spring treatments). A reverse relationship was observed in year 3, when growth regulators decreased the CF content of seeds by around 5 g kg^{-1} DM (Table 5). No relationships were found between the application of growth regulators vs. fatty acid profile, and the content of protein, ADF, NDF, and GLS in seeds, regardless of weather conditions (Y × C) (Table 2).

Table 5. The effect of growth regulation on the CF content (g kg^{-1} DM) of WOSR seeds (Y × C).

Growing Season	Growth Regulation at		
	None	BBCH 14–15	BBCH 14–15 and 30–31
2016/2017	512.5 ^{ab}	523.2 ^{a}	519.1 ^{a}
2017/2018	480.5 ^{b}	496.9 ^{b}	505.9 ^{ab}
2018/2019	460.9 ^{bc}	456.0 ^{c}	455.8 ^{c}

BBCH 14–15—4–5 true leaves unfolded; BBCH 30–31—rosette regrowth after winter. Means sharing a common letter are not significantly different at $p \leq 0.05$ in Tukey's test. The absence of common letters denotes non-significant differences (*ns*) at $p \leq 0.05$ in Tukey's test.

3.5. Spring Nitrogen Fertilization

An increase in the N rate from 160 to 200 kg ha^{-1} did not induce significant changes in the content of CF and TP in WOSR seeds (Table 3). In turn, the application of 240 kg N ha^{-1} decreased the CF content (by 10.5 g kg^{-1} DM on average) and increased the TP content (by 7.2 g kg^{-1} DM on average) of WOSR seeds. In addition, the highest N rate (240 kg ha^{-1}) induced a significant decrease in the proportion of NDF (by 0.4–0.5%). The quality of oil (fatty acids) and fat-free seed residues (determined based on the concentrations of total GLS and alkenyl GLS) was not significantly differentiated by spring-applied N rates (Table 2). The effect of N fertilization on the analyzed seed quality parameters was not influenced by weather conditions across the years of the study (Y × D) (Table 2).

3.6. Spring Sulfur Fertilization

Sulfur fertilization (0, 40, and 80 kg ha^{-1}) had no influence on the content of CF, TP, ADF, or NDF in WOSR seeds. A relationship was found between S fertilization and the fatty acid profile of WOSR oil depending on weather conditions (Table 2). An increase in S rate to 80 kg ha^{-1} induced a significant decrease (by 0.49–0.69%) in the concentration of linoleic acid in years when spring and summer precipitation approximated the long-term average (2016/2017 and 2017/2018) (Table 6). The concentration of eicosenoic acid increased (by 0.20%) in response to S fertilizer applied in the year characterized by below-average precipitation in April–July (2018/2019). A rise in the S rate in 2018/2019 (below-average precipitation) contributed to a significant increase in the proportion of PUFAs (by 0.48%) and a decrease in the proportion of MUFAs (0.58%) in WOSR oil (Table 6). Sulfur exerted a strong influence on the content and structure of GLS in seeds (Table 2). The application of 40 and 80 kg S ha^{-1} increased the content of total GLS (by 18% and 35%, respectively) and alkenyl GLS (by 19% and 40%, respectively) in WOSR seeds (Table 3). The proportion of alkenyl GLS in total GLS increased from 68% to 71% under the influence of S fertilization.

Table 6. The effect of spring sulfur rate on the proportions of fatty acids (%) in WOSR oil (Y × E).

Growing Season	Spring Sulfur Rate (kg ha^{-1})		
	0	40	80
		C18:2	
2016/2017	21.23 ^{a}	21.03 ^{a}	20.54 ^{ab}
2017/2018	19.90 ^{b}	19.03 ^{cd}	18.90 ^{d}
2018/2019	19.18 ^{bc}	19.61 ^{bc}	19.57 ^{bc}
		C20:1	
2016/2017	1.00 ^{c}	1.01 ^{c}	0.97 ^{c}
2017/2018	1.02 ^{c}	0.98 ^{c}	1.06 ^{c}
2018/2019	1.57 ^{a}	1.57 ^{a}	1.37 ^{b}
		MUFAs	
2016/2017	62.93 ^{d}	63.19 ^{d}	64.01 ^{d}
2017/2018	65.28 ^{c}	66.46 ^{b}	66.81 ^{ab}
2018/2019	68.22 ^{a}	67.64 ^{ab}	67.64 ^{ab}
		PUFAs	
2016/2017	31.48 ^{a}	31.19 ^{ab}	30.43 ^{b}
2017/2018	28.95 ^{c}	27.84 ^{d}	27.49 ^{d}
2018/2019	26.98 ^{d}	27.49 ^{d}	27.46 ^{d}

C18:2—linoleic acid; C20:1—eicosenoic acid; MUFAs—monounsaturated fatty acids; PUFAs—polyunsaturated fatty acids. Means sharing a common letter are not significantly different at $p \leq 0.05$ in Tukey's test.

4. Discussion

4.1. Tillage

In a study by Jankowski and Budzyński [15], a reduction in the depth of pre-sowing tillage from 30 cm to 10 cm decreased the CF and TP content of seeds by 15 and 3 g kg^{-1}

DM, respectively. In turn, Muśnicki et al. [12,14] demonstrated that a decrease in plowing depth (from 30–32 cm to 10–12 cm) increased the accumulation of CF and TP in seeds by 10 and 18 g kg^{-1} DM, respectively. According to Jankowski [16], a reduction in tillage depth (from 22 cm to 12 cm) increased the CF content of seeds by 8 g kg^{-1} DM. In comparison with conventional tillage, low-till farming and direct sowing decreased the CF content of seeds by around 6–8 g kg^{-1} DM. In the work of Jankowski [16], the TP content of WOSR seeds was not differentiated by the applied tillage implements or plowing depth. In the present study, simplified tillage promoted the accumulation of CF in WOSR seeds. Crude fat content was highest in seeds produced in the strip-till system (497.6 g kg^{-1} DM), followed by seeds grown in the conventional tillage system (lower by 12 g kg^{-1} DM). The TP content of WOSR seeds was lowest in the strip-till system (189 kg^{-1} DM).

4.2. Weed Control

Weed infestation in WOSR stands not only decreases seed yields but also compromises their quality and commercial value [1]. The timing of herbicide application exerts an ambiguous effect on the quality of WOSR seeds. In a study by Adomas [18], the CF content of spring oilseed rape was higher in treatments protected with pre-emergent herbicides (456 g kg^{-1} DM) than those protected with foliar herbicides (451 g kg^{-1} DM). In turn, Hamzei et al. [46] found no correlation between the weed control method and the CF content of WOSR seeds. In the experiment carried out by Gołębiowska and Badowski [47], the application of herbicides with different modes of action (clomazone, metazachlor + quinmerac/clomazone) did not induce significant changes in the CF and TP content of WOSR seeds. Similar observations were made in the current study, where the tested weed control methods did not influence the content of basic nutrients in WOSR seeds. Adomas [18] demonstrated that the proportion of PUFAs in spring rapeseed oil was higher in treatments protected with foliar herbicides, compared with pre-emergent herbicides (328 vs. 311 g kg^{-1} DM), but no changes were found in the concentrations of SFAs or MUFAs. In the work of Mekki et al. [17], chemical weed control contributed to an increase in the content of palmitic acid (by 5%) and oleic acid (by 7%), and a decrease in the concentrations of linoleic acid (by 13%), linolenic acid (by 7%) and erucic acid (by 46%). In the current study, the method of herbicide application had no significant effect on the fatty acid profile of WOSR oil.

4.3. Growth Regulation

An analysis of the literature revealed that the application of growth regulators in autumn was weakly correlated with the nutrient content of WOSR seeds [16,19]. In a study by Jankowski [16], the application of tebuconazole or chlormequat chloride in autumn did not cause significant changes in the CF (431–437 vs. 441 g kg^{-1} DM) or TP (352 vs. 346–351 g kg^{-1} DM) content of WOSR seeds relative to control. Growth regulation in autumn also failed to modify the nutritional value of WOSR seeds in the work of Ijaz and Honermeier [19]. Similar observations were made in the present study, where the CF and TP content of WOSR seeds was not influenced by the application of growth regulators in autumn.

In most of the analyzed studies, growth regulation in spring did not affect the content of basic nutrients in WOSR seeds [16,19–23]. This is consistent with the results of the present study, where the application of growth regulators in spring did not induce changes in CF or TP levels in WOSR seeds. In contrast, Ijaz et al. [22] reported a significant increase in the TP content of WOSR seeds (approx. 3%) and no changes in CF levels in response to growth regulation in spring. The cited authors also found that the effect of this treatment on the fatty acid profile of oil varied across locations. The spring application of growth regulators significantly increased the concentrations of linoleic acid and linolenic acid in oil at Giessen but decreased the proportions of these two acids at Rauischholzhausen [22]. In the present study, growth regulation in spring did not modify the fatty acid profile of WOSR oil, irrespective of environmental or weather conditions.

4.4. Spring Nitrogen Fertilization

Spring N fertilization is a key agronomic factor that influences the quality of WOS seeds. The CF content of seeds generally decreases with a rise in the N rate, and this redu tion is exacerbated by delayed N fertilization. In turn, the TP content of seeds is negativel correlated with CF levels [1]. In studies conducted by Butkutė et al. [28], Jankowski [16 and Sieling et al. [32], WOSR seeds supplied with 160 kg N ha^{-1} accumulated 1–2% mo CF than seeds fertilized with 240 kg N ha^{-1}. Dresbøll et al. [30] observed a clear dro (7%) in the CF content of seeds when the N rate was increased from 120 to 280 kg ha^{-} In turn, Varényiová and Ducsay [25] found no significant differences in the CF conter of seeds between treatments fertilized with 160, 200, and 240 kg N ha^{-1}. According t Jankowski [1], higher N rates do not decrease the CF content of seeds only when they exe no yield-forming effects. In this study, WOSR seeds from treatments supplied with 160 an 200 kg N ha^{-1} in spring accumulated 2% more CF than seeds from treatments fertilize with the highest N rate (240 kg ha^{-1}). In the work of Butkutė et al. [28] and Jankowski [16 the TP content of seeds increased by 2–3% in response to the spring-applied N rate c 240 kg ha^{-1}. Dresbøll et al. [30] found that a very high N rate (280 kg ha^{-1}) induced a 26 increase in the TP content of WOSR seeds in comparison with the treatment fertilized wit 120 kg N ha^{-1}. In turn, Ferguson et al. [31] did not report a significant increase in the T content of seeds under the influence of rising N rates. In the present study, the TP conter of WOSR seeds was higher by 8 g kg^{-1} DM (4%) in treatments supplied with the highest N rate (240 kg ha^{-1}) than in those fertilized with 160 kg N ha^{-1}.

4.5. Spring Sulfur Fertilization

Sulfur fertilization has different effects on the content of CF and TP in *Brassica* seed In a study by Sienkiewicz-Cholewa and Kieloch [33], the S rate of 60 kg ha^{-1} increased C accumulation in WOSR seeds by 2%. According to Fazili et al. [34], S induced a significar (15%) increase in the CF content of oilseed crops (*Brassica campestris* L. and *Eruca sativ* Mill). In studies conducted in north-eastern Poland, S fertilization had a minor effect on C and TP levels in *Brassicaceae* oilseeds [16,36–38,48]. In turn, Wielebski [35] found that a increase in the S rate from 0 to 60 kg ha^{-1} induced a minor, but significant decrease in C accumulation (from 426 to 423 g kg^{-1} DM) and increased the TP content of WOSR seed (from 207 to 210 g kg^{-1} DM). In the present study, CF and TP levels in WOSR seeds we not significantly modified by any of the tested S rates, which is consistent with the finding of Jankowski [16], Jankowski et al. [36], Jankowski et al. [37], and Groth et al. [38].

Sulfur fertilization compromises the quality of WOSR seeds by increasing the GLS content by 13–18% [35,37,39,49] to even 24–29% [16,38]. In the present study, applied in spring increased GLS levels by 34%. The increase in GLS concentrations u der the influence of S fertilization is attributed mainly to the intensified synthesis c alkenyl GLS [1,16,35,37–39,49]. This unfavorable process can compromise the feed valu of fat-free seed residues used in the production of animal feed [37]. The S-induced i crease in the concentration of alkenyl GLS ranged from 21–29% [16,37–39] to even 39 [present study, Table 3]. Alkenyl GLS levels increase under the influence of S fertilizatio mainly due to a rise in the concentrations of glucobrassicanapin (by 28%), gluconapin, an progoitrin (by 16–17%) [35]. In this study and experiments conducted by Wielebski [35 Malarz et al. [39], Ijaz and Honermeier [20], and Groth et al. [38], S fertilization had n effect on the fatty acid profile of WOSR oil.

5. Conclusions

Winter oilseed plants produced in the strip-till system accumulated the largest amoun of CF (498 g kg^{-1} DM). In turn, TP concentration in seeds was highest (196 g kg^{-1} DM in conventional tillage and low-till systems. The seeds of WOSR plants grown in the co ventional tillage system accumulated 0.5% less NDF, 0.4% more PUFAs (due to an increas in the proportion of linolenic acid), and 0.5–0.6% less MUFAs (due to a decrease in th proportion of oleic acid), compared with strip-till systems. The highest N rate (240 kg ha^{-}

decreased the CF content (by 2%) and increased the TP content of seeds (by 4%). Sulfur fertilization increased GLS concentrations (mainly alkenyl GLS) by 34% without differentiating the content of basic nutrients. Sulfur fertilization induced an increase in the proportion of PUFAs (by 0.48%) and a decrease in the proportion of MUFAs (by 0.58%) in WOSR oil only in the season characterized by lower-than-average precipitation in spring and summer. The effect exerted by agronomic management on the quality of WOSR seeds was not influenced by the tillage system.

Author Contributions: Conceptualization, M.S. and K.J.J.; methodology, M.S. and K.J.J.; software, M.S. and D.Z.; validation, M.S., D.Z., A.S. and K.J.J.; formal analysis, M.S. and D.Z.; investigation, M.S., A.S. and K.J.J.; resources, M.S., A.S. and K.J.J.; data curation, M.S., A.S. and K.J.J.; writing—original draft preparation, M.S., D.Z., A.S. and K.J.J.; writing—review and editing, M.S., D.Z., A.S. and K.J.J.; visualization, M.S. and K.J.J.; supervision, M.S. and K.J.J.; project administration, M.S. and K.J.J.; funding acquisition, M.S., D.Z. and K.J.J. All authors have read and agreed to the published version of the manuscript.

Funding: The results presented in this paper were obtained as part of a comprehensive study financed by the National Science Center under the research project entitled "Multi-criteria evaluation of the effectiveness of winter oilseed rape production in various cultivation systems" (grant No. 2018/31/N/NZ9/00536).

Institutional Review Board Statement: Not applicable.

Informed Consent Statement: Not applicable.

Data Availability Statement: Data are contained within the article.

Acknowledgments: We would like to thank the staff of the AES in Bałcyny for technical support during the experiment and Anna Hłasko-Nasalska for technical support during manuscript editing.

Conflicts of Interest: The authors declare no conflict of interest.

References

1. Jankowski, K.J. Winter and spring oilseed rape. In *Crop Production Volume III*; Kotecki, A., Ed.; Wrocław University of Environmental and Life Sciences: Wrocław, Poland, 2020; pp. 305–383. (In Polish)
2. Przybylski, R. Canola/rapeseed oil. In *Vegetable Oils in Food Technology: Composition, Properties and Uses*; Gunstone, F.D., Ed.; Blackwell Publishing: Oxford, UK, 2011; pp. 107–136.
3. Dacko, M.; Oleksy, A.; Synowiec, A.; Klimek-Kopyra, A.; Kulig, B.; Zając, T. Plant-architectural and environmental predictors of seed mass of winter oilseed rape in southern Poland based on the CART trees regression model. *Ind. Crops Prod.* **2023**, *192*, 116109. [CrossRef]
4. Faostat, Food and Agriculture Organization Corporate Statistical Database. Available online: http://www.apps.fao.org (accessed on 14 January 2023).
5. Ghobadi, S.; Hassanzadeh-Rostami, Z.; Mohammadian, F.; Zare, M.; Faghih, S. Effects of canola oil consumption on lipid profile: A systematic review and meta-analysis of randomized controlled clinical trials. *J. Am. Coll. Nutr.* **2019**, *2*, 185–196. [CrossRef]
6. Cheng, H.; Liu, X.; Xiao, Q.; Zhang, F.; Liu, N.; Tang, L.; Wang, J.; Ma, X.; Tan, B.; Chen, J.; et al. Rapeseed meal and its application in pig diet: A review. *Agric.* **2022**, *6*, 849. [CrossRef]
7. Perttilä, S.; Jalava, T.; Rinne, M.; Viana, G.D.S.; Valaja, J. Apparent and standardised ileal digestibility coefficients of amino acids in wheat, soybean meal and rapeseed meal for broilers. *Agric. Food Sci.* **2021**, *2*, 63–73. [CrossRef]
8. Siegert, W.; Zuber, T.; Sommerfeld, V.; Krieg, J.; Feuerstein, D.; Kurrle, U.; Rodehutscord, M. Prececal amino acid digestibility and phytate degradation in broiler chickens when using different oilseed meals, phytase and protease supplements in the feed. *Poult. Sci.* **2019**, *11*, 5700–5713. [CrossRef]
9. Tripathi, M.K.; Mishra, A.S. Glucosinolates in animal nutrition: A review. *Animal Feed Sci. Technol.* **2007**, *1–2*, 1–27. [CrossRef]
10. Li, Z.; Li, Y.; Lv, Z.; Liu, H.; Zhao, J.; Noblet, J.; Wang, F.; Lai, C.; Li, D. Net energy of corn, soybean meal and rapeseed meal in growing pigs. *J. Animal Sci. Biotechnol.* **2017**, *1*, 1–10. [CrossRef]
11. Sikorska, A.; Gugała, M.; Zarzecka, K. Anti-nutritional substances (GSL) in winter oilseed rape depending on the biostimulators used. *Acta Agric. Scand. B Soil Plant Sci.* **2021**, *71*, 472–477. [CrossRef]
12. Muśnicki, C.; Toboła, P.; Muśnicka, B. The influence of various tillage methods and agricultural treatments on winter oilseed rape yields and seed quality. *Zesz. Probl. Post. Nauk Roln.* **1993**, *6*, 7–14. (In Polish)
13. Muśnicki, C.; Toboła, P.; Muśnicka, B. Yield and value of winter oilseed rape (*Brassica napus* L.) seeds produced under a simplified tillage system. *Poznańskie Towarzystwo Przyjaciół Nauk* **1995**, *79*, 81–89. (In Polish)

14. Muśnicki, C.; Toboła, P.; Muśnicka, B. Effect of agronomical and environmental factors on the yield quality of winter oilseed rape. *Rośliny Oleiste Oilseed Crops* **1999**, *2*, 459–469. (In Polish)
15. Jankowski, K.J.; Budzyński, W.S. Effect of the ploughing depth and the weed control method on the winter oilseed rape yield. *Acta Sci. Pol. Agric.* **2003**, *2*, 133–146.
16. Jankowski, K.J. Habitat, Agrotechnological and Economic Conditions of Winter Rapeseed Production for Consumption and Energy Purposes. Disseratations and Monographs. Publisher University Warmia and Mazury in Olsztyn: Olsztyn, Poland, 2007; Volume 131, pp. 1–174. (In Polish)
17. Mekki, B.B.; Sharara, F.A.; El-Rokiek, K.G. Effect of weed control treatments on yield and seed quality of some canola cultivars and associated weeds in newly reclaimed sandy soils. *Am. Eurasian J. Agric. Environ. Sci.* **2010**, *2*, 202–209.
18. Adomas, B. Fatty acids composition in oil of spring rapeseeds in relation to applied herbicides. *Rośliny Oleiste Oilseed Crops* **2003**, *1*, 209–222. (In Polish)
19. Ijaz, M.; Honermeier, B. Effect of different fungicides on seed yield and grain quality of rapeseed under two levels of nitrogen fertilization. In Proceedings of the 13th International Rapeseed Congress, Prague, Czech Republic, 9–11 June 2011; pp. 11–15.
20. Ijaz, M.; Honermeier, B. Effect of triazole and strobilurin fungicides on seed yield formation and grain quality of winter rapeseed (*Brassica napus* L.). *Field Crops Res.* **2012**, *130*, 80–86. [CrossRef]
21. Matysiak, K.; Kaczmarek, S. Effect of chlorocholine chloride and triazoles–tebuconazole and flusilazole on winter oilseed rape (*Brassica napus* var. *oleifera* L.) in response to the application term and sowing density. *J. Plant Prot. Res.* **2013**, *1*, 78–88.
22. Ijaz, M.; Mahmood, K.; Honermeier, B. Interactive role of fungicides and plant growth regulator (Trinexapac) on seed yield and oil quality of winter rapeseed. *Agronomy* **2015**, *3*, 435–446. [CrossRef]
23. Matysiak, K.; Kaczmarek, S.; Adamczewski, K. The influence of trinexapac-ethyl, chlorocholine chloride, metconazole and tebuconazole on plant morphology and yield of winter oilseed rape depending on time of application. *Rośliny Oleiste–Oilseed Crops* **2010**, *2*, 361–372. (In Polish)
24. Narits, L. Effect of nitrogen rate and application time to yield and quality of winter oilseed rape (*Brassica napus* L. var. *oleifera subvar. biennis*). *Agron. Res.* **2010**, *3*, 671–686.
25. Varényiová, M.; Ducsay, L. Effect of increasing spring doses of nitrogen on yield and oil content in seeds of oilseed rape (*Brassica napus* L.). *Acta Fytotech. Zootech.* **2016**, *2*, 29–34. [CrossRef]
26. Barłóg, P.; Grzebisz, W. Effect of timing and nitrogen fertilizer application on winter oilseed rape (*Brassica napus* L.). I. Growth dynamics and seed yield. *J. Agron. Crop Sci.* **2004**, *5*, 305–313. [CrossRef]
27. Rathke, G.W.; Christen, O.; Diepenbrock, W. Effects of nitrogen source and rate on productivity and quality of winter oilseed rape (*Brassica napus* L.) grown in different crop rotations. *Field Crops Res.* **2005**, *2*, 103–113. [CrossRef]
28. Butkutė, B.; Šidlauskas, G.; Brazauskienė, I. Seed yield and quality of winter oilseed rape as affected by nitrogen rates, sowing time, and fungicide application. *Commun. Soil Sci. Plant Anal.* **2006**, *37*, 2725–2744. [CrossRef]
29. Öztürk, Ö. Effects of source and rate of nitrogen fertilizer on yield, yield components and quality of winter rapeseed (*Brassica napus* L.). *Chil. J. Agric. Res.* **2010**, *1*, 132–141. [CrossRef]
30. Dresbøll, D.B.; Rasmussen, I.S.; Thorup-Kristensen, K. The significance of litter loss and root growth on nitrogen efficiency in normal and semi-dwarf winter oilseed rape genotypes. *Field Crops Res.* **2016**, *186*, 166–178. [CrossRef]
31. Ferguson, B.T.; Chastain, T.G.; Garbacik, C.J.; Chastain, B.T.; Wysocki, D.J. Spring nitrogen and cultivar affect seed production in winter canola. *Agronomy J.* **2016**, *3*, 1124–1131. [CrossRef]
32. Sieling, K.; Böttcher, U.; Kage, H. Effect of sowing method and N application on seed yield and N use efficiency of winter oilseed rape. *Agronomy* **2017**, *21*, 21. [CrossRef]
33. Sienkiewicz-Cholewa, U.; Kieloch, R. Effect of sulphur and micronutrients fertilization on yield and fat content in winter rape seeds (*Brassica napus* L.). *Plant Soil Environ.* **2015**, *4*, 164–170. [CrossRef]
34. Fazili, I.S.; Masoodi, M.; Ahmad, S.; Jamal, A.; Khan, J.S.; Abdin, M.Z. Interactive effect of sulfur and nitrogen on growth and yield attributes of oilseed crops (*Brassica campestris* L. and *Eruca sativa* Mill.) differing in yield potential. *J. Plant Nutr.* **2010**, *8*, 1216–1228. [CrossRef]
35. Wielebski, F. The effect of sulphur fertilization on chemical composition of seeds of different breeding forms of winter oilseed rape in the conditions of diverse nitrogen rates. *Rośliny Oleiste Oilseed Crops* **2011**, *1*, 79–95. (In Polish)
36. Jankowski, K.J.; Budzyński, W.S.; Szymanowski, A. Effect of sulfur on the quality of winter rape seeds. *J. Elem.* **2008**, *4*, 521–534.
37. Jankowski, K.J.; Budzyński, W.S.; Kijewski, Ł.; Zając, T. Biomass quality of *Brassica* oilseed crops in response to sulfur fertilization. *Agronomy J.* **2015**, *4*, 1377–1391. [CrossRef]
38. Groth, D.A.; Sokólski, M.M.; Jankowski, K.J. A multi-criteria evaluation of the effectiveness of nitrogen and sulfur fertilization in different cultivars of winter rapeseed—Productivity, economic and energy balance. *Energies* **2020**, *18*, 4654. [CrossRef]
39. Malarz, W.; Kozak, M.; Kotecki, A. The effect of spring fertilization with different sulphur fertilizers on the quality and quantity of seed yield of winter oilseed rape ES Saphir cultivar. *Rośliny Oleiste Oilseed Crops* **2011**, *1*, 107–115. (In Polish)
40. Jankowski, K.J.; Budzyński, W.S.; Załuski, D.; Hulanicki, P.S.; Dubis, B. Using a fractional factorial design to evaluate the effect of the intensity of agronomic practices on the yield of different winter oilseed rape morphotypes. *Field Crops Res.* **2016**, *188*, 50–61. [CrossRef]
41. Załuski, D.; Dubis, B.; Budzyński, W.S.; Jankowski, K.J. Applicability of the 3^{5-2} fractional factorial design in determining the effect of cultivation factors on hulless oat. *Agron. J.* **2016**, *1*, 205–218. [CrossRef]

2. Meier, U.; Growth Stages of Mono- and Dicotyledonous Plants: BBCH Monograph. Julius Kühn-Institut, Quedlinburg, Germany 2018. Available online: https://www.julius-kuehn.de/media/Veroeffentlichungen/bbch%20epaper%20en/page.pdf (accessed on 14 January 2023).
3. IUSS Working Group WRB. *World Reference Base for Soil Resources*, 2nd ed.; World Soil Resources Reports No. 106; FAO: Rome, Italy, 2015.
4. Michalski, K.; Kołodziej, K.; Krzymański, J. Quantitative analysis of glucosinolates in seeds of oilseed rape—effect of sample preparation on analytical results. In Proceedings of the 9th International Rapeseed Congress, Cambridge, UK, 4–7 July 1995; Volume 1, pp. 911–913.
5. StatSoft Inc. Statistica (Data Analysis Software System), 13th ed.; 2017. Available online: http:/www.statsoft.com (accessed on 15 December 2022).
6. Hamzei, J.; Nasab, A.D.M.; Khoie, F.R.; Javanshir, A.; Moghaddam, M. Critical period of weed control in three winter oilseed rape (*Brassica napus* L.) cultivars. *Turk. J. Agric. For.* **2007**, *2*, 83–90.
7. Gołębiowska, H.; Badowski, M. The effect of metazachlor used in mixtures with clomazone and chinomerac on morphology of plants and seed quality of some cultivars of winter oilseed rape. *Acta Sci. Pol. Agric.* **2015**, *3*, 25–38.
8. Jankowski, K.J.; Załuski, D.; Sokólski, M.M. Canola-quality white mustard: Agronomic management and seed yield. *Ind. Crops Prod.* **2020**, *145*, 112138. [CrossRef]
9. Egesel, C.Ö.; Gül, M.K.; Kahriman, F. Changes in yield and seed quality traits in rapeseed genotypes by sulphur fertilization. *Eur. Food Res. Technol.* **2009**, *3*, 505–513. [CrossRef]

agronomy

MDPI

Article

Performance of Winter Wheat (*Triticum aestivum*) Depending on Fungicide Application and Nitrogen Top-Dressing Rate

Zinta Gaile [1,*], Biruta Bankina [1], Ieva Pluduma-Paunina [1], Linda Sterna [1], Gunita Bimsteine [1], Agrita Svarta [1], Janis Kaneps [1], Irina Arhipova [2] and Aigars Sutka [3]

1 Faculty of Agriculture, Latvia University of Life Sciences and Technologies, 2 Liela Str., LV-3001 Jelgava, Latvi
2 Faculty of Information Technologies, Latvia University of Life Sciences and Technologies, 2 Liela Str., LV-3001 Jelgava, Latvia
3 AKPC, Ltd., 109 Vienibas Gatve, LV-1058 Riga, Latvia
* Correspondence: zinta.gaile@lbtu.lv

Abstract: Winter wheat (*Triticum aestivum*) is a crop of which production is associated with rathe large investments for nitrogen fertilization and disease control. The aim of this study was to estimat the effect of five variants of fungicide application and four levels of N (nitrogen) top-dressing ra on the yield and grain quality of winter wheat. Field trials were carried out in Latvia (56° 31′ N 23° 42′ E) for four seasons. Grain yield and quality depended significantly on the conditions c the trial year, as three of them were characterized by drought in varying degrees. Although th average four-year grain yield increased significantly in all fungicide application variants, the effe of this factor was different in individual years. The application of fungicides increased the yiel significantly in one year, decreased significantly in another year, while it had no significant effect o the yield in remaining two seasons. The enhancement of N top-dressing rate increased the grain yiel significantly every year. The interaction between both examined factors was significant; however, th use of higher N rates not always means that also spraying with fungicides has to be more intensiv A clear effect of fungicide application was observed on 1000 grain weight and volume weight, whi the effect of N top-dressing rate was observed on the crude protein, wet gluten and starch conten and Zeleny index.

Keywords: winter wheat; yield; grain quality; leaf diseases; fungicides; nitrogen top-dressing

Citation: Gaile, Z.; Bankina, B.; Pluduma-Paunina, I.; Sterna, L.; Bimsteine, G.; Svarta, A.; Kaneps, J.; Arhipova, I.; Sutka, A. Performance of Winter Wheat (*Triticum aestivum*) Depending on Fungicide Application and Nitrogen Top-Dressing Rate. *Agronomy* **2023**, *13*, 318. https://doi.org/10.3390/agronomy13020318

Academic Editors: Umberto Anastasi and Aurelio Scavo

Received: 13 December 2022
Revised: 10 January 2023
Accepted: 18 January 2023
Published: 20 January 2023

1. Introduction

Mainly bread wheat (*Triticum aestivum*) is grown all over the world, including Europ and Latvia, and the largest bread wheat-growing area is under winter wheat (20 656.1 thousand ha in Europe in 2021, https://ec.europa.eu/ (accessed on 18 November 2022 448.7 thousand ha in Latvia in 2022, https://stat.gov.lv/ (accessed on 18 November 2022) The winter wheat yield is variable depending on the country (climate effect) and annua meteorological conditions, e.g., in the European Union in 2022, the average winter whea yield (bread wheat and spelt wheat (*T. spelta*) yields are reported both together by th Eurostat database) varied from 1.83 t ha^{-1} to 10.87 t ha^{-1} depending on the country. I Latvia in 2022, the average winter wheat yield was 4.36 t ha^{-1}, and exceeded 5.0 t ha^{-1} i some very favourable years, reaching even 7.0–9.0 t ha^{-1} in the best farms. The provisio of all agrometeorological factors (water, photosynthetically active radiation, nutrient soil conditions, suitable cultivar, etc.) as well as disease control is equally importan for obtaining high winter wheat yields with good grain quality. Several leaf and ea diseases have been recognized as an important wheat-yield-limiting factor globally: ta spot (caused by *Pyrenophora tritici-repentis*), e.g., [1–4]; Septoria leaf blotch (caused b *Zymoseptoria tritici*), e.g., [2,4–6]; Septoria nodorum blotch (caused by *Parastagonospor nodorum*), e.g., [7]; powdery mildew (caused by *Blumeria graminis*), e.g., [4,8]; leaf rus (caused by *Puccinia recondita*; previously *Puccinia triticina*), e.g., [8]; stripe rust (caused b

P. striiformis), e.g., [4]; and Fusarium head blight (caused by *Fusarium* spp.), e.g., [6]. Which disease will spread more widely and will be more harmful depends on the specific site and year conditions. In Latvia, tan spot and Septoria leaf blotch have been recognized as the most important wheat leaf diseases [9]. In addition to yield loss, diseases can also cause grain quality decrease [10]. The pressing target among farmers in Latvia is not whether to use a fungicide for winter wheat, but real time application (when and how often) has been of great importance. Nevertheless, after the data analysis of 350 field trials in Sweden, it was found that the use of fungicide was profitable in 188 cases, but was not profitable in 162 cases [11]. It is fairly often concluded that the application of fungicides might not be recommended at all in dry years [5,8], that it is more effective in years with sufficient water provision [4], and that one spraying can provide similar efficacy compared with two or three applications [6,8,12]. These different effects have been obtained in diverse climatic conditions when different diseases prevailed, different fungicides were applied, different cultivars were used, and also the rates of fertilization with nitrogen (N) were different. Several research studies show a strong effect of the fungicide application (F) × nitrogen fertilization rate (N) interaction on grain yield, when higher yields were obtained using the most intensive fungicide application strategy together with the highest N rates [1–3,13]. However, there are also studies in which N × F interaction effect on grain yield has not been established [14], and even in Brinkman et al's [13] study, it did not appear in one out of the nine locations. The literature also shows that the chemical indicators of wheat grain quality were not affected by the F × N interaction [15].

It is observed that producers rarely make fungicide application decisions based solely on expected yield or yield quality losses due to disease outbreaks. Often, the decision is related to the farmer's attitude towards risk, the use of pesticides in general, the farm's financial situation, and other reasons. In addition, no farm has an unsprayed control variant, and if production results show a profit, the unnecessary fungicide application goes unnoticed even financially. However, the fact that fungicides should have been sprayed, but were not, is often obvious [11]. In the most important wheat growing region of Latvia, farmers grow winter wheat comparatively intensively, using high N top-dressing rates and fungicides. For a fungicide application or N top-dressing to be cost-effective, it must pay for itself in increased yield and/or quality. Despite the fact that the impact of nitrogen top-dressing and the application of different fungicides at different timings has been studied thoroughly in the world, the results obtained are contradictory and more research in particular conditions is required. As tan spot and Septoria leaf blotch are the most common wheat diseases in the humid and cool climate [9] of Northern Europe, the triazole (DeMethylation Inhibitor (DMI)) with high efficacy against both diseases was selected in our study—prothioconazole, which was supplemented with spiroxamine (amine, Sterol Biosynthesis Inhibitor (SBI)) in T1 (growth stage (GS) 32–33) treatment and with two active substances from carboxamide group (Succinate Dehydrogenase Inhibitors (SDHI)) in T2 (GS 55–59) treatment; metconazole (DMI) was used in T3 (ear) (GS 63–65) treatment. Therefore, the main objective of this study was to determine the effect of different intensities of fungicide application and N top-dressing rates on winter wheat yield and grain quality. Another objective was to test the hypothesis that an increase in N fertilization rates requires an intensified winter wheat disease control.

Since three trial-years were characterised by shorter or longer periods of drought and heat and a sufficient supply of water was observed only in one year, the hypothesis failed to be proven, but the average four-year yield was significantly affected by both investigated factors. On the other hand, the physical indicators of grain quality were significantly affected by fungicide application, while chemical indicators were significantly affected by increased N top-dressing rates.

2. Materials and Methods

2.1. Trial Site and Studied Factors

Two-factor field experiments with winter wheat (*Triticum aestivum*) were carried out at the Research and Study farm "Peterlauki" (56° 31′ N; 23° 42′ E) of the Latvia University of Life Sciences and Technologies for four years (2017/2018–2020/2021). Soil at the site was Epiabruptic Endostagnic Endoprotocalcic Luvisol in the 2017/2018 and 2019/2020, and Cambic Calcisol in the 2018/2019 and 2020/2021 [16]. Soil reaction (pH_{KCL}) was 6.4–7.0, the content of P_2O_5 was 118–181 mg kg^{-1}, the content of K_2O was 122–262 mg kg^{-1}, and soil organic matter content varied from 29 to 42 g kg^{-1} depending on the year (agrochemical data refer to the 0–20 cm soil depth). Winter wheat cultivar 'Skagen' (DE) was used. 'Skagen' is widely grown in Latvia and is characterised by a high baking quality (elite group) and a comparatively low susceptibility to common leaf diseases (https://www.bundessortenamt de/bsa/sorten/beschreibende-sortenlisten/ (accessed on 30 December 2022)). Plot size was 20 m^2 (2 m × 10 m), and treatments were arranged randomly in four replications.

Studied factors were as follows: (A) fungicide application (F, five treatments), and (B) nitrogen top-dressing rate (N, four treatments). In total, 20 variants were studied.

Fungicide application variants:

F0—control, without fungicide application;

F1—half of a full fungicide dose sprayed at GS 55–59 (T2);

F2—a full fungicide dose sprayed at GS 55–59 (T2);

F3—a full fungicide dose split in two treatments: at GS 32–33 (T1), and at GS 55–59 (T2);

F4—two full fungicide doses split in three treatments: half—at GS 32–33 (T1), half—at GS 55–59 (T2), and full—at GS 63–65.

In this study, the full fungicide dose (100%) was taken as the fungicide dose according to the highest recorded dose of the triazole active substance prothioconazole (DMI) per hectare (200 g ha^{-1}) in one treatment. Prothioconazole was selected due to its efficacy against the most spread wheat leaf diseases in Latvia. In the first treatment (T1), protioconazole was applied at 50% of the full dose (100 g ha^{-1}) in combination with the active ingredient spiroxamine (SBI) (187.5 g ha^{-1}), which is intended to control powdery mildew in cereals. In the second treatment (T2), a fungicide, which, in addition to prothioconazole (a half or a full dose according to the scheme), contains the active substances of the carboxamide group (SDHI)—bixafen and fluopyram (both—48.75 g ha^{-1} in F1, and 97.5 g ha^{-1} in F2). In the third treatment (T3), a full dose of a fungicide containing the active substance metconazole (DMI) (90 g ha^{-1}) was used against Fusarium head blight.

Nitrogen top-dressing variants which were applied:

N120 kg ha^{-1}, divided into two portions 80 + 40 kg ha^{-1};

N150 kg ha^{-1}, divided into two portions 80 + 70 kg ha^{-1};

N180 kg ha^{-1}, divided into three portions 80 + 70 + 30 kg ha^{-1};

N210 kg ha^{-1}, divided into three portions 80 + 80 + 50 kg ha^{-1} (further in the text N120, N150, N180, and N210).

The first portion was given at the time of vegetation renewal in spring, the second portion—at GS 31–32, and the third portion—at GS 49–51. Ammonium nitrate (N 34.4%) was used for the first and third portion of top-dressing, and ammonium sulphate (N 21% and S 24% to provide 28.8 kg ha^{-1} of S) and ammonium nitrate were used for the second portion of top-dressing.

2.2. Crop Management

The agrotechnology used in the trial was typical for the region in production conditions. The pre-crop was always wheat. Traditional soil tillage including ploughing at the depth of 22 cm was used. The rate of basic fertilizer was calculated with the aim to obtain an 8 t ha^{-1} grain yield, and it was given before sowing: 11–25 kg ha^{-1} N, 33–66 kg ha^{-1} P_2O_5, and K_2O depending on the year. Sowing was performed at the optimal time for local conditions (13–27 September depending on the year), and seeds treated with fungicides were used at the rate of 450 (in 2018–2020) to 500 (in 2017) germinable grain m^{-2}. For crop care, herbicides for weed control and plant growth regulators (twice) were applied every year, and insecticides were applied according to the need in 2020 and 2021. Used plant growth regulators were as follows: Cycocel 750 SC (chlormequat chloride, 750 g L^{-1}) 1 L ha^{-1} at the GS 28–29 and Medax Top SC (calcium prohexadione, 50 g L^{-1}, mepiquat chloride, 300 g L^{-1}) 0.75 L ha^{-1} at the GS 33–34. The foliar fertilizer YaraVita Gramitrel (Yara International ASA, Oslo, Norway) (N 3.9%, MgO 15.2%, Cu 3.0%, Mn 9.1%, and Zn 4.1%) was applied in spring together with the plant growth regulator. Grain yield was harvested at GS 89–90, using direct combining (25–26 July 2018, 2019, and 2021, and 2 August 2020), and the yield was recalculated at the 100% purity and 14% moisture.

2.3. Observations and Records Made in the Trial

Crop growth stages (GS 11, 21, 31, 32, 33, 37–39, 49, 51, 55, 59, 61–63, 69, 71, 81, 89–90) according to BBCH scale [17] were recorded every year.

Disease severity (%) was recorded visually five times per season: the whole plant was evaluated at the end of tillering–early stem elongation; three upper leaves were evaluated at flag leaf stage and during heading; two upper leaves were evaluated during the milk stage of maturity. The AUDPC (area under the disease progress curve) was calculated to assess the disease impact during the whole vegetation season [18]. Symptoms for every disease were evaluated separately. Fifty plants/leaves/ears were taken per every plot for evaluation, and leaves were taken to keep the proportion of all levels (flag leaf, 2nd leaf, and 3rd leaf). Wheat ears were evaluated at the early milk maturity stage (data not shown). In addition, the leaf green area (LGA) was recorded visually (in %) at the late milk maturity (GS 77) on the upper two leaves.

In 2020, lodging was observed, which was evaluated using a point scale (9–1, where 9 means no lodging, and 1 means very strong lodging and all stems are bent down at a 90° angle).

Before harvesting, two sample sheafs (each from 0.1 m^2) were taken from every plot to determine the grain/straw ratio, which was later used to calculate the straw yield from the grain yield.

A grain sample of 1 kg was taken from each plot during harvesting to detect grain quality parameters. The Near Infrared Spectroscopy (NIRS) method (analyser Infratec™ NOVA (FOSS, Hillerød, Denmark)) was used to determine the content of crude protein (CP, % in dry matter), wet gluten (WG, for grain at 14% moisture), and starch (SC, % in dry matter), as well as Zeleny index (ZI) and volume weight (VW, kg hL^{-1}). The thousand grain weight (TGW, g) was determined according to the standard ISO 520:2010.

2.4. Data Statistical Processing

The analysis of variance (ANOVA) was used for data analysis both in every specific year and taking into account the data of all four years. Differences between treatments were considered as significant at $p \leq 0.05$ and were detected using Bonferroni test. Significantly different values in tables and figures are labeled with different letters in superscript. Correlation and regression analyses were used for discovering the relations between studied parameters. Data were analysed using IBM Statistics for Windows, Version 23.0.

2.5. Meteorological Conditions during the Study Period

Meteorological conditions were diverse during trial years (Table 1). In autumn (September, October), the best temperature and moisture conditions for good stand establishment were observed in 2019. The autumn of 2017 was overly wet, but those of 2018 and 2020—overly dry. In general, conditions were good for wintering in all four winters, except early spring of 2021, when snow mold (caused by *Microdochium nivale* and/or *Typula* spp.) was observed. The whole spring and summer period of 2018 was extremely hot and dry, but that of 2019 and 2021 was characterised with several drought periods. Only the year 2020 was suitable for the formation of high winter wheat yields (Figure 1).

Table 1. Data of mean air temperatures and precipitation in trial period (2017/2018–2020/202 and in comparison with the long-term-average data, Research and Study farm "Peterlauki", Latvi University of Life Sciences and Technologies.

Month	Average Air Temperature, °C				Norm, °C	Precipitation, mm				Norm, mm
	2017/2018	2018/2019	2019/2020	2020/2021		2017/2018	2018/2019	2019/2020	2020/2021	
Sept.	13.0	14.9	12.7	14.9	12.3	79.8	25.5	53.6	16.2	59.9
Oct.	6.9	8.5	9.0	9.8	6.9	80.0	10.6	36.4	58.4	68.2
Nov.	3.9	3.0	4.4	5.7	2.5	45.4	6.8	48.4	12.6	50.4
Dec.	1.3	−0.9	2.6	0.5	−0.9	×	×	×	×	47.1
Jan.	−1.2	−4.2	3.2	−3.4	−2.7	×	×	×	×	43.6
Febr.	−6.8	1.2	2.5	−5.7	−2.7	×	×	×	×	34.8
Mar.	−2.0	3.0	3.1	1.9	0.7	×	29.6	27.0	13.6	33.8
Apr.	9.0	8.1	6.1	5.9	6.7	69.5	0.0	9.2	4.7	36.0
May	16.1	12.4	9.9	11.1	12	12.0	20.4	30.2	50.6	52.4
June	16.8	19.4	18.7	19.2	15.5	16.0	8.6	139.6	14.8	73.4
July	20.7	16.8	17.0	22.0	17.9	56.5	101.0	47.7	3.2	82.1
Aug.	19.4	17.6	17.7	No data	17.0	34.0	37.8	65.0	No data	69.4

Meteorological data were registered at the trial site by an automatic weather observation station; the norm mear long-term-average data (last 30 years), which were taken from the closest meteorological station (in Jelgava of the Latvian Environment, Geology and Meteorology Centre (https://videscentrs.lvgmc.lv/ (accessed o 5 September 2022)); ×—precipitation data during winter are not shown, because they can be imprecise due precipitation in the form of snow in those months.

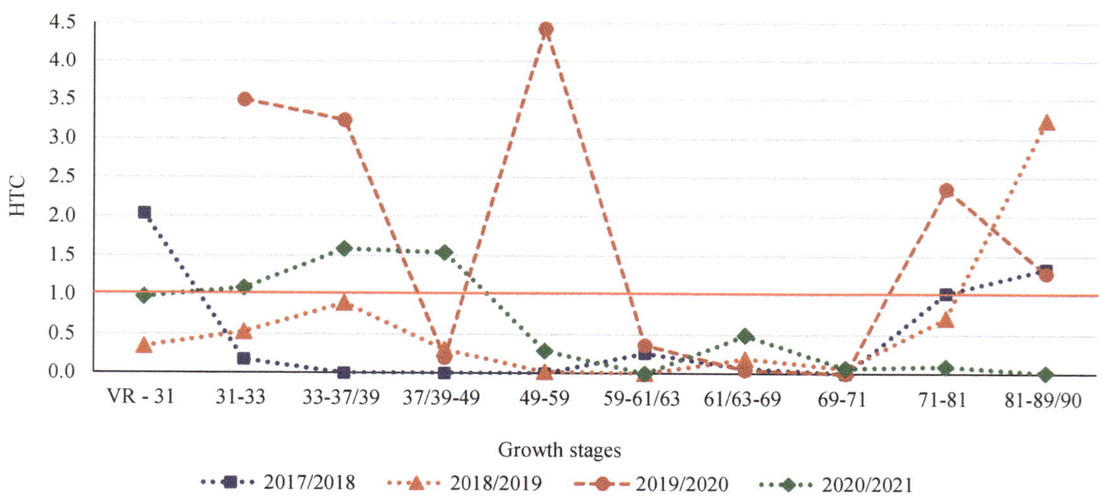

Figure 1. Hydrothermal coefficient (HTC) during various growth stages of winter wheat in 2017/201 2020/2021 at the Research and Study farm "Peterlauki, Latvia University of Life Sciences an Technologies (VR—vegetation renewal; GS—growth stage; HTC was not calculated for the stage VR–G 31 in 2019/2020, when the average day-and-night temperature only once per 21 days exceeded 10 °C

In order to characterise the temperature and moisture conditions more accurately i different wheat development stages in the spring-summer vegetation period, the Selyan nov's hydrothermal coefficient (HTC) [19] was calculated according to the Formula (1

$$\text{HTC} = \frac{\sum R \times 10}{\sum t} \quad (1$$

ΣR—sum of precipitation per period, which has to be characterised;
Σt—sum of temperatures above 10 °C per the same period.

We used the following criteria for the interpretation of HTC: >2—overly wet; 1–2—sufficient moisture provision; <1—insufficient moisture provision; 1.0–0.7—dry conditions; 0.7–0.4—very dry conditions.

Temperature and precipitation in different growth stages caused a critical lack of water, which was observed in the whole season of 2017/2018 and 2018/2019, as well as from the GS 49 to the harvesting maturity in 2020/2021 (HTC below 1 in Figure 1).

A sufficient supply of water during ear formation (up to GS 32) was noted in 2020 and 2021, and up to GS 31—also in 2018. Grain filling stage (see GS 71–81 in Figure 1) was best provided with moisture in the year 2020, when also temperature was more moderate, while the worst situation was observed in 2021.

3. Results

3.1. Crop and Disease Development

The lengths of the spring–summer vegetation period, from the vegetation renewal to GS 89–90 (harvesting maturity), differed depending on the trial year: the longest—in 2019/2020 (120 days); the shortest—in 2017/2018 (104 days). The lengths of specific growth stages (GS) were also diverse depending mainly on the temperature and moisture conditions in a specific year, but did not depend on the studied factors—fungicide application or N top-dressing rate.

Severe lodging (2.2–3.1 points) was observed only once over all four years—in 2020 (data are not shown), when a storm with heavy rain was noted on 29–30 June at the early milk stage (GS 71 was noted on 27 June) and lodging remained until harvesting. The rating did not depend on fungicide treatment but was significantly affected by N top-dressing rate; although, a significantly ($p = 0.02$) lower average rating was noted only for the variant N210. Lodging affected the values of TGW and VW (see below the Section 3.3. Winter wheat grain quality).

The development of diseases differed significantly depending on the year ($p < 0.001$). Tan spot (caused by *P. tritici-repentis*) dominated in three years out of the four, achieving the highest level in 2019. Septoria leaf blotch (caused by *Z. tritici*) proved to be the most important wheat disease only in 2020. All other leaf and ear diseases were observed only occasionally, and their severity did not reach 1% in the untreated variant. Fungicide treatment significantly decreased the severity of tan spot and Septoria leaf blotch ($p < 0.001$), but the efficacy of application schemes depended on the year (Table 2). The influence of nitrogen top-dressing rate on the development of diseases was not significant ($p > 0.05$) on average per four years and in each particular trial year as well.

Table 2. Development of winter wheat leaf diseases depending on fungicide treatment and year (2018–2021) at the Research and Study Farm "Peterlauki", Latvia University of Life Sciences and Technologies (data in AUDPC units).

Fungicide Treatment	Tan Spot					Septoria Leaf Blotch				
	2018	2019	2020	2021	Average	2018	2019	2020	2021	Average
F 0	12.9 a	141.9 a	45.4 a	61.5 a	65.4 A	1.1 a	2.9 a	57.3 a	26.4 a	21.9 A
F 1	6.7 b	90.5 b	17.8 b	37.6 b	38.1 B	0.9 a	1.6 b	21.8 b	13.2 b	9.4 B
F 2	4.1 c	71.3 c	16.4 b	33.6 bc	31.3 B	1.3 a	1.3 b	21.4 b	13.9 b	9.5 B
F 3	3.9 c	60.7 cd	10.5 b	34.5 bc	27.4 B	0.7 a	0.8 a	12.9 c	10.2 b	6.2 B
F 4	3.2 c	45.6 d	12.7 b	27.1 c	22.1 B	1.3 a	0.8 b	16.9 bc	9.2 b	7.1 B

F0—control without fungicide application; F1—half dose applied as T2; F2—full dose applied as T2; F3—full dose applied as split spraying: T1 and T2; F4—two full doses applied as split spraying: T1, T2, and T3. T1—spraying at GS 32–33; T2—spraying at GS 55–59; T3—spraying at GS 63–65. Different letters mean significant differences between disease level (expressed as AUDPC units) for each year (a, b, c, d) and on average for a treatment (A, B).

The efficacy of fungicide application schemes was influenced by disease pressure, and in 2019 (the highest severity of tan spot), a more intensive application of fungicides gave better disease control results. Although the half dose of fungicides gave a lower efficacy in the majority of cases, on average during the four years the differences in disease level between fungicide application variants were not significant.

Leaf green area (LGA) during the late milk ripening (GS 77) fluctuated on average between 50% and 87% in untreated variants and between 62% and 91% in variants with fungicide application. Leaf green area was significantly influenced by the trial year ($p < 0.001$). Fungicide treatment increased the LGA ($p < 0.001$) on average, although the differences between variants with different fungicide application intensities were significant but small. The effect of nitrogen top-dressing rate on LGA was not clearly established in particular years, but on average per trial period a higher dose of nitrogen slightly and significantly increased the LGA (Figure 2).

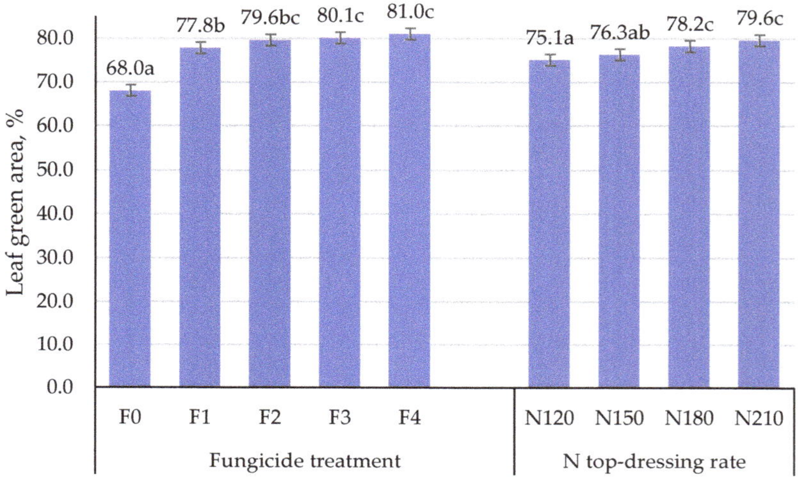

Figure 2. Mean four-year (2018–2021) leaf green area (LGA) of winter wheat at late milk stage (GS 77) at the Research and Study farm "Peterlauki", Latvia University of Life Sciences and Technologies depending on fungicide treatment and N top-dressing rate (F0—control without fungicide application; F1—half dose applied as T2; F2—full dose applied as T2; F3—full dose applied as split spraying: T1 and T2; F4—two full doses applied as split spraying: T1, T2, and T3. T1—spraying at GS 32–33; T2—spraying at GS 55–59; T3—spraying at GS 63–65. N120–210—N top-dressing rate in kg ha^{-1} pure N. Different letters mean significant differences between LGA depending on the specific factor).

3.2. Winter Wheat Grain and Straw Yield

The average four-year winter wheat grain yield was 6.79 t ha^{-1}, which did not reach the planned 8 t ha^{-1}. The average yield was affected significantly by fungicide treatment, N top-dressing, and the interaction of both, but the factor influencing the yield most was the trial year (Table 3).

Any fungicide treatment produced a small but statistically significant increase in the average four-year grain yield compared to the control. However, differences between sprayed variants were inconclusive, and, e.g., the average grain yield in variant F1, where half of the fungicide dose was applied, was equivalent to the yield in variant F4, where two full doses of the fungicide in three sprayings were applied (Table 4).

Moreover, the analysis of variance showed a small but significant effect of F × N interaction on the mean four-year wheat yield (Table 3). Although the effect of F × N interaction was significant, clear regularities that the use of higher N top-dressing rates requires more intensive spraying with fungicides, failed to be proved by our results on average per trial period as well as in separate years. For example, the highest mean yield was obtained in variant F4 (two full fungicide doses split in three sprays), when N210 (the highest rate) was used. Moreover, the rate N180 is high, but its use provided the best yield in fungicide treatment F1 (half dose, one spraying) (Table 4). Such results could be connected with the high influence of the trial year (Table 3, Figure 3) due to the diverse meteorological conditions, especially the differing water supply (Table 1, Figure 1). Fungicide application resulted in a significant wheat grain yield increase only in one trial year (2020); in two years (2019 and 2021) it did not cause significant yield changes, whereas in 2018, the application of fungicides caused a significant negative effect, i.e., the yield decreased (Table 3, Figure 3).

Table 3. Mean squares of winter wheat yield under five fungicide treatments, four N top-dressing rates in four years (2018–2021), and five fungicide treatments and four N rates in every specific year, Research and Study farm "Peterlauki", Latvia University of Life Sciences and Technologies.

Source of Variation	Four-Year Data		Separate Year Data				
	df	Mean Squares	df	2017/2018	2018/2019	2019/2020	2020/2021
Fungicide (F)	4	0.579 ***	4	0.896 ***	0.137	1.955 ***	0.038
N top-dressing rate (N)	3	1.130 ***	3	0.249 ***	0.532 *	0.236 ***	0.434 ***
Year (Y)	3	136.149 ***	–	–	–	–	–
F × N	12	0.145 **	12	0.170 ***	0.082	0.079 ***	0.116 ***
F × Y	12	0.816 ***	–	–	–	–	–
Error	285	0.067	60	0.017	0.175	0.031	0.019
Total	320		80				

Significant at: * $p = 0.05$; ** $p = 0.01$; *** $p < 0.01$.

Table 4. Average four-year (2018–2021) winter wheat grain yield (t ha^{-1}) depending on fungicide treatment and N top-dressing rate at the Research and Study farm "Peterlauki", Latvia University of Life Sciences and Technologies.

Fungicide Treatment	N Top-Dressing Rate				Average for Fungicide Treatment
	N120	N150	N180	N210	
F0	6.51	6.64	6.70	6.75	6.65 a
F1	6.79	6.92	7.01	6.84	6.89 b
F2	6.66	6.81	6.83	6.71	6.75 c
F3	6.54	6.75	6.97	6.87	6.78 cd
F4	6.57	6.89	6.90	7.10	6.86 bd
Average for N top-dressing	6.62 a	6.80 b	6.88 c	6.85 bc	×

F0—control without fungicide application; F1—half dose applied as T2; F2—full dose applied as T2; F3—full dose applied as split spraying: T1 and T2; F4—two full doses applied as split spraying: T1, T2, and T3. T1—spraying at GS 32–33; T2—spraying at GS 55–59; T3—spraying at GS 63–65. N120–210—N top-dressing rate in kg ha^{-1} pure N. Different letters mean significant differences between average yields depending on the studied factor.

Despite the fact that fertilization in general and N top-dressing did not ensure the planned winter wheat yields in three (2018, 2019, and 2021) out of the four years, the increase in N rate always caused a significant increase in grain yield (Tables 3 and 4, Figure 3). Grain yield increased up to the rate of N150–N180 depending on the year.

The mean four-year wheat straw yield was significantly affected only by the trial year ($p < 0.01$), varying from 5.98 to 12.08 t ha^{-1} depending on the year (Figure 4). The exception was 2018, when the yield of straw was significantly ($p = 0.008$) affected by fungicide application, which caused a decrease in straw yields in variants F3 and F4 (data not shown). The straw yield is important in the formation of total wheat biomass, which also requires water and nutrition elements, including part of N top-dressing rate.

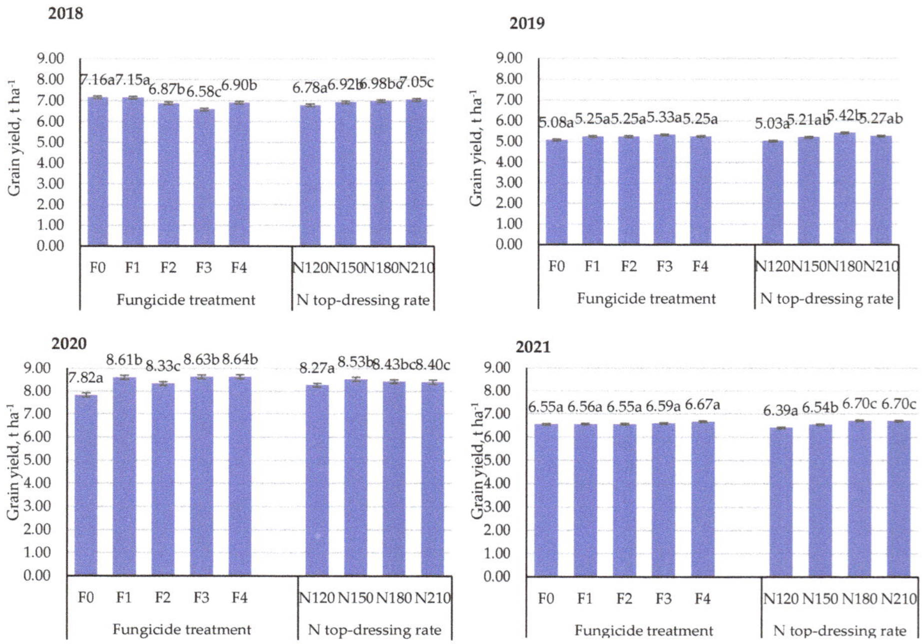

Figure 3. Winter wheat grain yield depending on fungicide treatment, N top-dressing rate, and tria year (2018–2021) at the Research and Study farm "Peterlauki", Latvia University of Life Science and Technologies (F0—control without fungicide application; F1—half dose applied as T2; F2—fu dose applied as T2; F3—full dose applied as split spraying: T1 and T2; F4—two full doses applied a split spraying: T1, T2, and T3. T1—spraying at GS 32–33; T2—spraying at GS 55–59; T3—sprayin at GS 63–65. N120–210—N top-dressing rate in kg ha^{-1} pure N. Different letters mean significar differences between yields depending on the specific factor in the specific year).

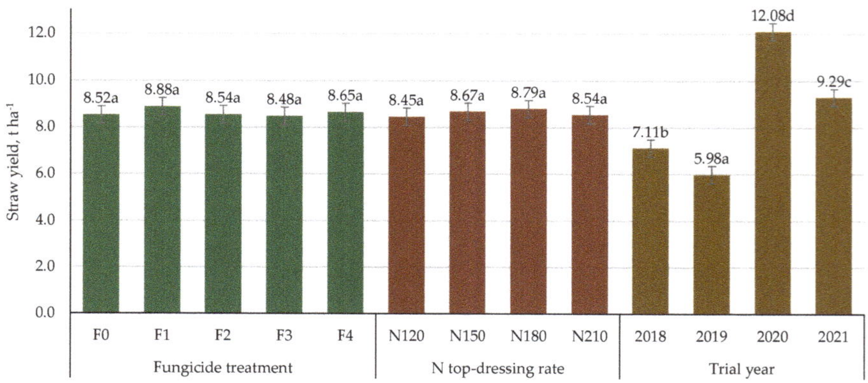

Figure 4. Winter wheat straw yield depending on fungicide treatment, N top-dressing rate, and tria year (2018–2021) at the Research and Study farm "Peterlauki", Latvia University of Life Science and Technologies (F0—control without fungicide application; F1—half dose applied as T2; F2—fu dose applied as T2; F3—full dose applied as split spraying: T1 and T2; F4—two full doses applied a split spraying: T1, T2, and T3. T1—spraying at GS 32–33; T2—spraying at GS 55–59; T3—sprayin at GS 63–65. N120–210—N top-dressing rate in kg ha^{-1} pure N. Different letters mean significar differences between yields depending on the specific factor).

3.3. Winter Wheat Grain Quality

3.3.1. Physical Grain Quality Indicators—1000 Grain Weight (TGW) and Volume Weight (VW)

Four-year average values of TGW and VW significantly depended on fungicide treatment ($p < 0.01$ and $p = 0.046$, respectively) and especially strongly on the trial year ($p < 0.01$ for both indicators; see Table 5). Although TGW increased in three years of fungicide use (2019, 2020, and 2021), this increase was significant only in 2019/2020 (Table 5). VW increased significantly in three trial years except 2020/2021. Lower TGW and VW values were noted in 2020 as a result of heavy lodging, and in 2021—as a result of drought during grain filling (Figure 1). In years when a significant TGW and VW increase was observed after fungicide application, differences between treated variants were noted only once: for VW when treatment F4 did not cause an increase in VW if compared with the control (F0).

Table 5. Winter wheat TGW and VW in trial years (2017/2018–2020/2021) at the Research and Study farm "Peterlauki", Latvia University of Life Sciences and Technologies, and the effect of studied factors on their values.

Trial Year	Average (Min–Max)	Effect of Studied Factors: p-Value		
		F	N	F × N
1000 grain weight (TGW), g				
2017/2018	46.29 b (44.81–47.58)	0.289	0.160	0.366
2018/2019	49.20 a (47.74–50.35)	0.158	0.204	0.988
2019/2020	43.99 c (42.16–45.99)	0.001	0.420	0.653
2020/2021	36.06 d (34.04–37.31)	0.266	0.290	0.724
Volume weight (VW), kg hL^{-1}				
2017/2018	82.73 a (81.77–83.23)	0.006	0.061	0.738
2018/2019	79.72 b (79.27–80.22)	0.001	0.687	0.013
2019/2020	77.92 c (76.72–78.58)	0.001	0.213	0.900
2020/2021	72.56 d (70.08–73.39)	0.880	0.509	0.875

F—fungicide treatment; N—nitrogen top-dressing. Different letters mean significant differences between TGW and VW values depending on trial year.

The relationship between TGW and the important indicators leaf green area at the late milk stage (LGA), grain yield, VW, and crude protein (CP) content in grain was evaluated. The preservation of LGA during milk ripeness stage is important because of grain filling and TGW formation at that time. Frequently, a higher TGW is connected with a higher VW, which was also observed in our study in 2018 and 2019, when a higher TGW and a significantly higher VW were noted. At the same time, TGW is not only a physical grain quality indicator but also a yield-forming component. In all trial years, no significant TGW relationship with any of mentioned indicators was established. The correlation of TGW with LGA and grain yield was significant in three years, and with VW and CP—in two years (Table 6).

Table 6. Correlation of TGW with LGA at the end of milk ripeness, grain yield, VW, and CP depending on trial year (2017/2018–2020/2021), Research and Study farm "Peterlauki", Latvia University of Life Sciences and Technologies, ($n = 20$).

Trial Year	Correlation of TGW with Other Indicators: Correlation Coefficients			
	LGA	Grain Yield	VW	CP
2017/2018	NS	0.688 **	0.639 **	0.639 **
2018/2019	0.592 **	0.478 *	NS	0.655 **
2019/2020	0.765 **	0.576 **	0.899 **	NS
2020/2021	0.445 *	NS	NS	NS

TGW—1000 grain weight; LGA—leaf green area at the end of milk ripeness; VW—volume weight; CP—crude protein. NS means $p > 0.05$; * $p = 0.05$; ** $p = 0.01$.

A significant correlation between LGA and both TGW and VW, and between TGW and VW was also found when data of all four years (n = 80) were included in correlation analysis (p = 0.01; data are not shown).

3.3.2. Chemical Grain Quality Indicators—Crude Protein (CP) and Wet Gluten (WG) Content, Zeleny Index (ZI), and Starch Content (SC)

Unlike the physical quality indicators, the mean four-year values of CP, WG, ZI, and SC were not affected significantly (p > 0.05) by fungicide treatment variants, but they depended significantly on N top-dressing rate and trial year (Table 7).

Table 7. Winter wheat CP, WG, ZI, and SC (measured by NIRS) in trial years (2017/2018–2020/2021), and the effect of studied factors on their values, Research and Studies farm "Peterlauki", Latvia University of Life Sciences and Technologies.

Trial Year	Average (Min–Max)	Effect of Studied Factors: *p*-Value		
		F	N	F × N
	Crude protein (CP), % in dry matter			
2017/2018	11.4 a (10.8–11.9)	0.002	0.001	0.889
2018/2019	13.7 b (13.0–14.2)	0.730	0.001	0.622
2019/2020	13.9 c (13.3–14.5)	0.666	0.001	0.855
2020/2021	13.8 bc (12.0–15.3)	0.886	0.001	0.955
	Wet gluten content (WG), % in grain with 14% moisture			
2017/2018	22.8 a (20.3–24.3)	<0.001	<0.001	0.836
2018/2019	29.1 b (26.9–30.6)	0.481	<0.001	0.626
2019/2020	29.7 b (28.3–31.0)	0.829	<0.001	0.843
2020/2021	28.4 b (22.7–32.9)	0.879	<0.001	0.960
	Zeleny index (ZI) for grain with 14% moisture			
2017/2018	31.9 a (26.0–35.2)	0.001	<0.001	0.961
2018/2019	50.5 b (45.1–54.8)	0.309	<0.001	0.617
2019/2020	52.8 d (48.7–56.7)	0.826	<0.001	0.871
2020/2021	51.3 c (36.2–62.7)	0.820	<0.001	0.921
	Starch content (SC), % in dry matter			
2017/2018	69.7 a (69.2–70.2)	<0.001	<0.001	0.059
2018/2019	68.1 b (67.5–69.1)	0.999	<0.001	0.765
2019/2020	67.1 c (66.1–68.0)	0.165	<0.001	0.943
2020/2021	65.6 d (63.7–67.9)	0.966	<0.001	0.984

F—fungicide treatment; N—nitrogen top-dressing. Different letters mean significant differences between CP, WG, ZI, and SC values depending on trial year.

The year 2018 was the only trial year when fungicide treatment significantly affected all measured parameters (Table 7) but in the direction not desired by the grower. The use of a full fungicide dose (F2 and F3) and two full doses (F4) caused a significant decrease in the CP and WG content, and in ZI. At the same time, SC increased significantly in variants F1, F2, and F3, but in F4 it was equivalent to that of the variant F0. In all other trial years (2019–2021), the use of fungicide did not significantly affect the chemical wheat grain quality indicators. The interaction between both studied factors (F × N) never affected the CP, WG, ZI, and SC values significantly. The increase in nitrogen top-dressing rate caused a significant increase in CP, WG, and ZI up to the rate of N210 in 2018–2020, and up to the rate of N180 in 2021. The rate of N150 always resulted in an increase in CP, WG, and ZI compared with the variant where N120 was used. According to research findings, CP content often correlates with SC. Also in our trial, we observed a significant negative correlation of CP with SC (correlation coefficients depending on the year in 2018–2021, respectively: −0.763, −0.971, −0.972, and −0.996; n = 20; p = 0.01). Thus, changes in SC depending on N top-dressing rate were opposite to the changes in CP content—the gradual decrease in SC was significant up to N210 in 2018, 2019, and 2021; whereas in 2020, although SC decreased gradually, the decrease was significant and of equal value only in variants N180 and N210 compared with the variant N120.

4. Discussion

Tan spot and Septoria leaf blotch were the main diseases during the trial period, which supports our previous conclusions related to the situation in the Baltic region [9,12]. As it has been found before, wheat as a pre-crop for wheat increases the level of wheat diseases, but ploughing can mitigate this impact significantly. The influence of meteorological conditions on the development of diseases has been recognized as more significant than the impact of agronomic practices [9]. The disease progress curves of leaf blotches in the present study differed compared with other findings—a rapid development of diseases started only at the time of flowering or even later in our trial. The results showed that, for example, in 2019, when tan spot achieved the highest level, its severity was 2% at GS 71, and two weeks later (GS 77) achieved 19% in the untreated variant. Similarly, in 2020, when Septoria leaf blotch dominated, the disease severity increased from 1.5% to 11.1% at the same growth stages (from GS 71 to 77). Those peculiarities of leaf disease development explain the comparatively low efficacy of T1, because that time of spraying did not coincide with the development of leaf diseases. The efficacy of fungicide treatment depends not only on the total pressure of diseases during the season, but also on crucial periods for disease development during the season.

More detailed data related to winter wheat leaf disease development depending on fungicide application and N top-dressing rate have been presented in other articles, e.g., by Švarta et al. [20].

The results revealed that the influence of the meteorological conditions on the yield of winter wheat in the research years was greater than the influence of the studied factors (Table 3, Figures 3 and 4), despite the fact that both of them—fungicide application and N top-dressing rate—significantly affected the average four-year yield. According to the obtained data (Table 4), any variant of fungicide application gave a significant increase in the average grain yield, but the differences in yield among the sprayed variants were inconclusive, e.g., the grain yield in the F1 variant (half of the fungicide dose was applied) was equivalent to the yield in the F4 variant (two full fungicide doses were applied in three sprayings). Obtained data are supported by similar results regarding the development of leaf diseases—fungicide application decreased the severity of diseases, but the increase in fungicide treatment intensity was not effective. Similar results have been also obtained in the previous studies in Latvia, where both the pre-developed fungicide application schemes and the two decision support systems were used in one trial, and the applied fungicides included also strobilurins. The results showed that any variant of fungicide application ensured an increase in the yield, but significant differences between yields in the variants differently treated with fungicides were not established [12]. This conclusion is supported by several studies conducted in different conditions and in different regions. A study in Luxembourg [6] found that a single spraying according to the recommendation of a decision support system can provide a yield equivalent to that provided by two or three sprayings according to a previously developed scheme. In Canada it was concluded that a single spraying at GS 39 provided the yields equivalent to those obtained in the variants where a split spraying was applied twice (at GS 30 and 39) [21]. Previous findings in Latvia [12] as well as other studies demonstrate that the yield increase as a result of fungicide application depends on meteorological conditions [5,21–24] and also on other agrotechnical factors (pre-crop, tillage system, etc.) [25]. A study of Byamukama et al. [26] proved that the positive effect of fungicide application for yield increase is better manifested if there are sufficient moisture conditions at the time most favourable for disease development during the vegetation season. During our research, sufficient water provision in the vegetation period was observed only in 2020, when any variant of fungicide application provided a significant yield increase (Figure 3); the other trial years were marked by drought in the stages important for the formation of winter wheat yield (Figure 1). In two years (2019 and 2021), fungicide treatment did not significantly affect the wheat grain yield (Figure 3 and the yield in the variants with fungicide application (F1–F4) was equivalent to that o the control variant (F0). Similar were the findings of Fernandez et al. [27] who carrie out a study of *Triticum durum* in Canada. The researchers established that for diseas control, one fungicide spraying (at GS 62–65) was of equal value to two sprayings (a GS 31 or 49 and at GS 62–65), but the yield in all treated variants was numerically bu not significantly higher than in the control variant. In 2018, when the shortest vegetatio period was observed and the season was extremely hot and dry (Table 1, Figure 1), th effect of fungicide application on the yield was significant but negative, i.e., a significan yield reduction was observed in three variants (F2–F4) (Figure 3). In that year, the diseas severity was low and only tan spot was spread more pronouncedly. We hypothesize that the drought stress together with the stress caused by fungicide spraying resulted in yield reduction in variants F2–F4. Such hypothesis about a combined effect of both stresse was also expressed by Rodrigo et al. [5], who observed that in dry years, all variants c fungicide spraying led to a decrease in grain yield. If decision support systems were use it would have been possible to avoid such a situation, because in conditions that do no promote the spread of diseases, the system would likely not recommend spraying [8]. Th studies comprising simultaneously several cultivars report that cultivars with a lowe yield potential are more responsive to the application of fungicide and provide a greate increase in yield (e.g., in Bhatta et al. [23]—fungicide applied at GS 39; in Byamukam et al. [26]—fungicide applied at GS 60), or—the effect of fungicide application depend on the cultivar's genetic characteristics, which is related to the year of the registratio of the cultivar [28]. The cultivar 'Skagen' used in our study is characterized with a least medium yield potential and has a relatively good field resistance against the mai leaf diseases (3 to 4 points in a 9-point scale, where 9 means the highest susceptibility https://www.bundessortenamt.de/bsa/sorten/beschreibende-sortenlisten/ (accessed o 30 December 2022)). On the other hand, Morgunov et al. [29], while studying the use c fungicides against leaf rust (caused by *Puccinia recondita*), found an increase in yield as result of fungicide application for cultivars with different levels of resistance to this diseas In our study, *Puccinia recondita* was observed only once—at milk ripeness in 2021, when th severity of the disease was low.

Although drought did not contribute to N-use efficiency in three of the four stud years, our results suggest that in each study year separately (Table 3, Figure 3) and o average over the entire period (Table 4), N top-dressing had a significant effect on whea grain yield. On average during the study period, a significant increase in yield was note by increasing the N top-dressing rate to N180; however, the results depended on the specifi year, e.g., in 2020, a significant yield increase was noted up to the rate N150 (Figure 3). Th results were in line with the previous studies, when wheat yield increased significantl with the increase in N rate up to 120 kg ha^{-1} [30], 153 kg ha^{-1} [31], or 180 kg ha^{-1} [32,33 In addition, the N-rate up to which the yield increased significantly, was influenced by th agrometeorological conditions of the specific study.

At the start of the study, it was hypothesized that higher N rates would probabl produce a denser wheat biomass, which would stimulate more diseases in the crop an therefore require a more intensive use of fungicides. However, in our case, the above ground biomass-forming component, straw yield, did not increase by increasing the N top-dressing rate, nor did it depend on either F or F × N; it was influenced only by th conditions of the trial year (Figure 4). In other studies, e.g., [13], it was found that fungicid application strategy at GS 39 or GS 60–65 ensured the highest and a consistent grain yiel increase, especially if it was supplemented with high N rates. Our results also reveale a small significant impact of F × N interaction on the average wheat grain yield over th entire research period and in three separate years (2018, 2020, and 2021; Table 3). Howeve it was not possible to establish any regularity that in more intensive fertilizing options, als a more intensive spraying should be used (Table 4). The inconsistency of the F × N impac on yield increase was even more expressed in specific years (data not shown) compare with the average four-year result. In Argentina, where a significant effect of the F × N interaction on yield was also found in the wheat artificially inoculated with *P. tritici-repentis*, it was established that increasing N rates in a fungicide-untreated control variant did not result in a significant increase in yield, while the yield increase in fungicide-treated (all variants included the use of strobilurins) variants was significant [3]. We observed a significant yield increase by increasing the N top-dressing rate also in untreated control. Our research was performed in a natural background of infection, and the infection level was not high. The results of different studies can differ depending also on cultivar and agrometeorological conditions of a specific study.

The effect of the studied factors on both groups of grain quality indicators—physical (TGW and VW) and chemical (CP, WG, ZI, and SC)—were found to be different. During the four-year trial period, the average physical grain quality indicators were significantly affected by fungicide treatment, but chemical quality indicators—by N top-dressing rate. Both groups were affected significantly by meteorological conditions of the study year, but were not affected by F × N interaction.

The increase in TGW due to fungicide application is established also by other researchers [23,27,34], but Landolfi et al. [34] have also pointed to the importance of the meteorological conditions of the trial year. We found a significant correlation of LGA at the end of milk ripeness stage with TGW, which is in line with MacLean et al. [24], who pointed to the importance of LGA increase for the formation of TGW in the result of fungicide application. The fact that the increase in N top-dressing rate does not significantly increase TGW is consistent with the results of Landolfi et al. [34] but differs from other findings [32,35], where also lower N rates (N60 and N90) and unfertilized control were included. The lowest N top-dressing rate used in our study was N120 kg ha^{-1}, which is not that small at all. Some researchers indicate an increase in TGW up to a certain N top-dressing rate which, when exceeded, decreases the TGW value [36]. For producers, VW is even a more important indicator than TGW; therefore, it is used by grain buyers for price determination. VW depended mostly on the meteorological conditions of the year (Table 5): strong lodging in 2020 and the lack of water during part of the 2021 season (Table 1, Figure 1) caused lower values of VW. As already mentioned, fungicide application increased VW on average per trial period and in two separate years (2019 and 2020), decreased the VW in 2018, and did not cause any significant changes in its value in 2021 In other studies, the observations of the dependence of VW on the use of fungicide were also various. Some results indicate a VW increase [15,26,27], and some results show that the values of VW in control and sprayed variants did not significantly differ. At the same time, it is noted that the nature of VW is related to the climate (meteorological conditions) and the disease severity [5], and that in years with a low disease spread, VW did not increase significantly due to fungicide application [24]. Moreover, the genetic characteristics of the cultivar can affect changes in VW, depending on fungicide application [23]. A study in Sweden revealed that a single fungicide spray (at GS 45–61) resulted in an increase in VW in 12 out of 25 research years [37].

The effect of fungicide treatment on changes in CP content and ZI compared with WG content and SC is described in more detail in the literature. In Latvia, consumers prefer grain with high protein content (for bread baking at least 12%); however, none of four fungicide application variants increased it. Similar results have been obtained in several studies confirming that CP content does not depend on fungicide sprays but depends more on climatic (meteorological) conditions at the study site (including year) [5,15,21,38]. Byamakama et al. [26] found that CP content was slightly but significantly increased by fungicide application; however, the increase depended on the conditions of the study site. The comparison of different timings (GS 60 was compared to GS 39) of fungicide sprays demonstrated that later spraying did not negatively affect the CP content [24]. Some authors link the increase in CP content in the result of fungicide application to both the increase in LGA and the control of Fusarium head blight (caused by *Fusarium* spp.) [23]. In our study, only in 2018, when CP content did not reach 12% in any of the variants (which was connected with untypically hot and dry conditions in the vegetation period), a slight (by 0.3%) but significant decrease in CP content was observed in variants F2–F4. However, the relationship between CP and SC did not change in 2018—it was negative as it is often observed. A small decrease in CP content in wheat grain as a result of fungicide application was also found in Italy, where only two variants were compared (untreated control with fungicide application at GS 55) [34], and in Sweden in separate years [37]. Landolfi et al. [34] wrote that CP decrease affected by fungicide application is probably connected with the higher grain yield. The effect of nitrogen on CP content is well-known, and in our trial, high N rates (N180–210) ensured the highest values of CP content.

The effect of fungicide application on WG content has not been widely studied. Findings in Croatia, similarly to our results, revealed that gluten content is affected by N fertilization and the meteorological conditions in the trial year but not by fungicide (tebuconazole at GS 55) application and F × N interaction [15]. ZI is an indicator characterising the CP quality. Changes in ZI were similar to changes in the values of CP and WG over the trial period and in separate years: ZI was not affected by F and F × N but was significantly affected by N top-dressing rate and meteorological conditions of the year. The effect of fungicide application on the value of ZI has not been found in previous research [5,15] either.

Starch content is not among the traditionally evaluated indicators for food wheat grain quality. SC should be evaluated in cases when wheat grain is intended for the production of ethanol [39] or feed. Similarly to CP and WG content and ZI, SC was also significantly affected by N top-dressing rate; however, the effect was contrary to that of CP, WG, and ZI, i.e., SC decreased with the increase in N rate. Moreover, the effect of the conditions in the trial year was found significant, but fungicide application mainly did not significantly affect the SC (except in 2018).

5. Conclusions

Based on the present findings, it can be concluded that in all studied variants of fungicide application, the average four-year grain yield increased significantly. On the other hand, more intensive spraying did not cause a greater yield increase compared with a single spraying. A strong significant year (Y) effect was noted, and only in one year characterized by normal water supply, a significant yield increase was observed. A yield decrease was observed in one hot and dry year, but fungicide-treated variants and unsprayed control variant provided equivalent grain yields in two years. Our hypothesis that a more intensive fertilization with nitrogen also requires a more intensive fungicide application was not proved, because the mathematically significant F × N interaction was small, its effect did not reveal any regularities, and the yield increase was irregular. Since a significant F × Y interaction effect was also found, it can be concluded that the choice of fungicide application should be related to the spread of diseases, which are dependent also on the year's meteorological conditions. On average, the increase in the rate of N top-dressing up to N180 kg ha^{-1} increased the grain yield significantly. However, the results obtained revealed the importance of meteorological conditions. The wheat straw yield was not affected by F application and N top-dressing rate but mostly depended only on the year's conditions.

The values of TGW and VW of winter wheat on average per four-year research period were significantly affected by fungicide application and especially by the meteorological conditions of the research year, as well as by the interaction between both factors (F × Y). The N top-dressing rate and F × N interaction did not affect significantly ($p > 0.05$) both TGW and VW. Contrary was the effect of the studied factors on grain quality chemical indicators CP, WG, ZI, and SC—they were affected significantly by N top-dressing rate and the conditions of the research year. The highest N rates provided the best values of CP, WG, and ZI and the smallest values of SC. The F × N interaction did not affect significantly any of wheat grain chemical quality indicators.

Author Contributions: Conceptualization, A.S. (Aigars Sutka), B.B. and Z.G.; methodology, Z.G., B.B. and A.S. (Aigars Sutka); formal analyses, I.A.; investigation, Z.G., B.B., I.P.-P., L.S., J.K., G.B. and A.S. (Agrita Svarta); data curation, Z.G., B.B., A.S. (Aigars Sutka) and I.A.; writing—original draft preparation, Z.G.; writing—review and editing, Z.G., B.B., I.A. and A.S. (Aigars Sutka). All authors have read and agreed to the published version of the manuscript.

Funding: This research was funded by the EIP-AGRI project "The development of the decision-making support system for restriction of the diseases, affecting leaves and ears of winter wheat", project No. 18-00-A01612-000003.

Institutional Review Board Statement: Not applicable.

Informed Consent Statement: Not applicable.

Data Availability Statement: The data presented in this study are available on request from the corresponding author. The data will be publicly available after the project is fully completed.

Conflicts of Interest: The authors declare no conflict of interest. The funders had no role in the design of the study; in the collection, analyses, or interpretation of data; in the writing of the manuscript; or in the decision to publish the results.

References

1. Fleitas, M.C.; Schierenbeck, M.; Gerard, G.S.; Dietz, J.I.; Golik, S.I.; Simon, M.R. Breadmaking quality and yield response to the green leaf duration caused by fluxapyroxad under three nitrogen rates in wheat affected by tan spot. *Crop Prot.* **2018**, *106*, 201–209. [CrossRef]
2. Castro, A.C.; Fleitas, C.M.; Schierenbeck, M.; Gerard, G.S.; Simon, M.R. Evaluation of different fungicides and nitrogen rates on grain yield and bread making quality in wheat affected by Septoria tritici blotch and yellow spot. *J. Ceral Sci.* **2018**, *83*, 49–57. [CrossRef]
3. Schierenbeck, M.; Fleitas, M.C.; Gerard, G.S.; Dietz, J.I.; Simon, M.R. Combinations of fungicide molecules and nitrogen fertilization revert nitrogen yield reductions generated by *Pyrenophora tritici-repentis* infections in bread wheat. *Crop Prot.* **2019**, *121*, 173–181. [CrossRef]
4. Wegulo, S.N.; Zwingman, M.V.; Breathnach, J.A.; Baezinger, P.S. Economic returns from fungicide application to control foliar fungal diseases in winter wheat. *Crop Prot.* **2011**, *30*, 685–692. [CrossRef]
5. Rodrigo, S.; Cuello-Hormigo, B.; Gomes, C.; Santamaria, O.; Costa, R.; Poblaciones, M.J. Influence of fungicide treatment on disease severity caused by *Zymoseptoria tritici*, and on grain yield and quality parameters of bread-making wheat under Mediterranean conditions. *Eur. J. Plant Pathol.* **2015**, *141*, 99–109. [CrossRef]
6. El Jarroudi, M.; Kouadio, L.; Junk, J.; Beyer, M.; Paquali, M.; Bock, C.H.; Delfosse, P. Do single, double or triple fungicide sprays differentially affect the grain quality in winter wheat? *Field Crops Res.* **2015**, *183*, 257–266. [CrossRef]
7. Lin, M.; Ficke, A.; Cockram, J.; Lillemo, M. Genetic structure of the Norwegian *Parastagonospora nodorum* population. *Front. Microbiol.* **2020**, *11*, 1280. [CrossRef]
8. El Jarroudi, M.; Kouadio, L.; Beyer, M.; Junk, J.; Hoffmann, L.; Tychon, B.; Maraite, H.; Bock, C.H.; Delfosse, P. Economics of a decision-support system for managing the main fungal diseases of winter wheat in the Grand-Duchy of Luxembourg. *Field Crops Res.* **2015**, *172*, 32–41. [CrossRef]
9. Bankina, B.; Bimšteine, G.; Arhipova, I.; Kaņeps, J.; Darguža, M. Impact of crop rotation and soil tillage on the severity of winter wheat leaf blotches. *Rural Sustain. Res.* **2021**, *45*, 21–27. [CrossRef]
10. Simon, M.R.; Fleitas, M.C.; Castro, A.C.; Schierenbeck, M. How foliar fungal diseases affect nitrogen dynamics, milling, and end-use quality of wheat. *Front. Plant Sci.* **2020**, *11*, 569401. [CrossRef]
11. Djurle, A.; Twengström, E.; Andersson, B. Fungicide treatment in winter wheat The probability of profitability. *Crop Prot.* **2018**, *106*, 182–189. [CrossRef]
12. Bankina, B.; Gaile, Z.; Balodis, O.; Bimšteine, G.; Katamadze, M.; Kreita, D.; Paura, L.; Priekule, I. Harmful winter wheat diseases and possibilities for their integrated control in Latvia. *Acta Agric. Scand. B Soil Plant Sci.* **2014**, *64*, 615–622. [CrossRef]
13. Brinkman, J.M.P.; Deen, W.; Lauzon, J.D.; Hooker, D.C. Synergism of nitrogen rate and foliar fungicides in soft red winter wheat. *Agron. J.* **2014**, *106*, 491–510. [CrossRef]
14. Olesen, J.E.; Jørgensen, L.N.; Petersen, J.; Mortensen, J.V. Effect of rate and timing of nitrogen fertilizer on disease control by fungicides in winter wheat. 1. Grain yield and foliar disease control. *J. Agric. Sci.* **2003**, *140*, 1–13. [CrossRef]
15. Varga, B.; Svečnjak, Z.; Jurkovic, Z.; Pospišil, M. Quality responses of winter wheat cultivars to nitrogen and fungicide applications in Croatia. *Acta Agron. Hung.* **2007**, *55*, 37–48. [CrossRef]
16. IUSS Working Group WRB. *World Reference Base for Soil Resources 2014, International Soil Classification System for Naming Soils and Creating Legends for Soil Maps, Update 2015*; World Soil Resources Reports No. 106; FAO: Rome, Italy, 2015; Available online: https://www.fao.org/3/i3794en/I3794en.pdf (accessed on 2 January 2023).

17. Julius Kühn-Institut. *Growth Stages of Mono- and Dicotyledonous Plants. BBCH Monograph*; Meier, U., Ed.; Julius Kühn-Institu Quedlinburg, Germany, 2018; pp. 18–22. Available online: https://www.julius-kuehn.de/media/Veroeffentlichungen/bbch%2 epaper%20en/page.pdf (accessed on 22 November 2022).
18. Simko, I.; Piepho, H.P. The area under disease progress stairs: Calculation, advantage, an application. *Phytopathology* **2012**, *10* 381–389. [CrossRef] [PubMed]
19. Selyaninov, G.T. *Metody Selskokhozyaistvennoi Kharakteristiki Klimata (Methods of Agricultural Characteristic of Climate)*; Mirovc Agroklimaticheskii Spravochnik (World Agroclimatic Reference Book L); Gidrometeoizdat: Leningrad, Russia, 1937; pp. 5–2 (In Russian)
20. Švarta, A.; Bimšteine, G.; Gaile, Z.; Kaņeps, J.; Plūduma-Pauniņa, I. Winter wheat leaf blotches development depending o fungicide treatment and nitrogen level in two contrasting years. *Agron. Res.* **2022**, *20*, 414–423. [CrossRef]
21. Kutcher, H.R.; Turkington, T.K.; McLaren, D.L.; Irvine, R.B.; Brar, G.S. Fungicide and cultivar management of leaf spot diseases c winter wheat in Western Canada. *Plant Dis.* **2018**, *102*, 1828–1833. [CrossRef]
22. Wegulo, S.; Stevens, J.; Zwingman, M.; Baezinger, P.S. Yield response to foliar fungicide application in winter wheat. I *Fungicides for Plant and Animal Diseases*; Dhanasekaran, D., Thajuddin, N., Paneerselvam, A., Eds.; IntechOpen: London, UK, 201 pp. 227–244. [CrossRef]
23. Bhatta, M.; Regassa, T.; Wegulo, S.N.; Baenzinger, P.S. Foliar fungicide effect on disease severity, yield, and agronomic characteri tics of modern winter wheat genotypes. *Agron. J.* **2018**, *110*, 602–610. [CrossRef]
24. MacLean, D.E.; Lobo, J.M.; Coles, K.; Harding, M.W.; May, W.E.; Peng, G.; Turkington, T.K.; Kutcher, H.R. Fungicide application a anthesis of wheat provides effective control of leaf spotting diseases in western Canada. *Crop Prot.* **2018**, *112*, 343–349. [CrossRef
25. Vrtilek, P.; Smutny, V.; Dryšlova, T.; Neudert, L.; Kren, J. The effect of agronomic measures on grain yield of winter wheat in drie conditions. *Plant Soil Environ.* **2019**, *65*, 63–70. [CrossRef]
26. Byamukama, E.; Ali, S.; Kleinjan, J.; Yabwalo, D.N.; Graham, C.; Caffe-Treml, M.; Mueller, N.D.; Rickertsen, J.; Berzonsky, W.A Winter wheat grain yield response to fungicide application is influenced by cultivar and rainfall. *Plant Pathol. J.* **2019**, *35*, 63–7 [CrossRef] [PubMed]
27. Fernandez, M.R.; May, W.E.; Chalmers, S.; Savard, M.E.; Singh, A.K. Are early foliar fungicide application on durum whea grown in southeast Saskatchewan beneficial in increasing grain productivity? *Can. J. Plant Sci.* **2013**, *94*, 891–903. [CrossRef]
28. Motta-Romero, H.; Niyongira, F.; Boehm, J.D.; Rose, D.J. Effects of foliar fungicide on yield, micronutrients, and cadmium i grain from historical and modern wheat genotypes. *PLoS ONE* **2021**, *16*, e0247809. [CrossRef]
29. Morgunov, A.; Akin, B.; Demir, L.; Keser, M.; Kokhmetova, A.; Martynov, S.; Orhan, S.; Özdemir, I.; Sapakhova, Z.; Yessimbekov M. Yield gain due to fungicide application in varieties of winter wheat (*Triticum aestivum*) resistant and susceptible to leaf rus *Crop Pasture Sci.* **2015**, *66*, 649–659. [CrossRef]
30. Karklins, A.; Lipenite, I.; Ruza, A. Slāpekļa mēslojums un kultūraugu ražas Saldus pētījumu poligonā (N fertilizer use and grai yield in Saldus experimental field). In Proceedings of the Scientific and Practical Conference "Harmonious Agriculture", Jelgav Latvia, 23 February 2017; LLU: Jelgava, Latvia, 2017; pp. 42–49. (In Latvian)
31. Skudra, I.; Ruža, A. Ziemas kviešu slāpekļa papildmēslošanas veidu salīdzinājums integrētā audzēšanas sistēmā (Compariso of winter wheat nitrogen additional fertilizer form within the integrated plant nutrient management). In Proceedings of th Scientific and Practical Conference "Harmonious Agriculture", Jelgava, Latvia, 20–21 February 2014; LLU: Jelgava, Latvia, 201 pp. 30–34. (In Latvian)
32. Litke, L.; Gaile, Z.; Ruža, A. Effect of nitrogen fertilization on winter wheat yield and yield quality. *Agron. Res.* **2018**, *16*, 500–50 [CrossRef]
33. Litke, L.; Gaile, Z.; Ruža, A. Effect of nitrogen rate and forecrop on nitrogen use efficiency in winter wheat (*Triticum aestivum Agron. Res.* **2019**, *17*, 582–592. [CrossRef]
34. Landolfi, V.; Visioli, G.; Blandino, M. Effect of nitrogen fertilization and fungicide application at heading on the gluten protei composition and rheological quality of wheat. *Agronomy* **2021**, *11*, 1687. [CrossRef]
35. Protic, R.; Jovin, P.; Protic, N.; Janvovic, S.; Jovanavic, Ž. Mass of 1000 grains in several winter wheat genotypes at different date of sowing and rates of nitrogen fertilizer. *Rom. Agric. Res.* **2007**, *24*, 39–42.
36. Liniņa, A.; Ruža, A. Slāpekļa mēslojuma un meteoroloģisko apstākļu ietekme uz ziemas kviešu graudu fizikālajiem rādītājien (Influence of nitrogen fertilizer and meteorological conditions on winter wheat grain physical indices). In Proceedings of th Scientific and Practical Conference "Harmonious Agriculture", Jelgava, Latvia, 19–20 February 2015; LLU: Jelgava, Latvia, 201 pp. 70–73. (In Latvian)
37. Wiik, L.; Rosenqvist, H. The economics of fungicide use in winter wheat in Southern Sweden. *Crop Prot.* **2010**, *29*, 11–1 [CrossRef]

38. Blandino, M.; Reyneri, A. Effect of fungicide and foliar fertilizer application to winter wheat at anthesis on flag leaf senescence, grain yield, flour bread-making quality and DON contamination. *Eur. J. Agron.* **2009**, *30*, 275–282. [CrossRef]
39. Jansone, I.; Gaile, Z. Production of bioethanol from starch based agriculture raw material. In Proceedings of the Annual 19th International Scientific Conference "Research for Rural Development 2013", Jelgava, Latvia, 15–17 May 2013; LLU: Jelgava, Latvia, 2013; Volume 1, pp. 35–42.

agronomy

MDPI

Article

A Trade-Off between the Growing Performance and Sowing Density of Sunflower (*Helianthus annuus* L.) under Fertigation in an Arid Saline Area

Shide Dong [1,2,3], Guangmei Wang [1,2,3], Xiaobin Li [4] and Yaohu Kang [4,*]

1 CAS Key Laboratory of Coastal Environmental Processes and Ecological Remediation, Yantai Institute of Coastal Zone Research, Chinese Academy of Sciences, Yantai 264003, China
2 Shandong Key Laboratory of Coastal Environmental Processes, Yantai 264003, China
3 Shandong Saline-Alkali Land Modern Agriculture Company, Dongying 257300, China
4 Key Laboratory of Water Cycle and Related Land Surface Process, Institute of Geographic Sciences and Natural Resources Research, Chinese Academy of Sciences, Beijing 100101, China
* Correspondence: kangyh@igsnrr.ac.cn; Tel.: +86-10-64889061

Abstract: Sunflower is one of the pioneer crops cultivated in salt-affected arid areas. The influences of sowing density on the growth performance of this crop under fertigation conditions have not been well studied. This study arranged four sowing density treatments, 41,667, 35,714, 31,250, and 27,778 plants ha^{-1}, marked as D30, D35, D40, and D45, respectively, to reveal the relationships between soil salinity, growth performance, and sowing density under drip fertigation conditions. The results showed that the electrical conductivity of saturated paste extracts (ECe) decreased during the growing seasons but increased on the topsoil during the non-growing seasons in all of the treatments. The sowing densities had remarkable influences on the ECe in the 0–40 cm soil layer (ECe-40). The average ECe-40 during the two seasons for treatments D30–D45 correspondingly decreased by 7.0%, 33.9%, 11.1%, and 15.8% when compared to the original value. The soil pH in the 0–40 cm soil layer during the two seasons for treatments D30–D45 correspondingly decreased by −0.03, 0.20, 0.20, and 0.27 when compared to the original value. Increasing the spacing in the rows could promote the stem diameter, plant biomass, and proportion of biomass allocated underground. The yield and related yield components in this experiment under fertigation were significantly higher than those under surface irrigation. A sowing density between 31,250 and 35,714 plants ha^{-1} could ensure both the high yield and high morphological quality of the seeds, which could be recommended for sunflower cultivation under drip fertigation conditions.

Keywords: sunflower; sowing density; fertigation; saline soil; yield components

Citation: Dong, S.; Wang, G.; Li, X.; Kang, Y. A Trade-Off between the Growing Performance and Sowing Density of Sunflower (*Helianthus annuus* L.) under Fertigation in an Arid Saline Area. *Agronomy* **2023**, *13*, 179. https://doi.org/10.3390/agronomy13010179

Academic Editors: Umberto Anastasi and Aurelio Scavo

Received: 13 December 2022
Revised: 3 January 2023
Accepted: 4 January 2023
Published: 5 January 2023

1. Introduction

Sunflower is an annual oilseed crop globally cultivated on 24.77 million hectares, and it has an 8% share in the world oilseed market [1]. This crop is a pioneer crop cultivated in salt-affected arid areas [2,3]. On most occasions, sunflower is cultivated under rainfed conditions, and surface irrigation is conducted before sowing in arid or semiarid areas [4]. The Hetao Irrigation District, located in northwest China, is a representative arid area that has approximately half of the irrigated land salt affected [5]. Thus, another surface irrigation event, aiming to leach salt, occurs after the sunflower harvest. More than 600 mm of applied water is needed for sunflower cultivation in this area [6]. However, water competition among different users, caused by water shortages, will predominantly change irrigated lands to rain-fed systems and ultimately increase salinization and decrease sunflower yield. Therefore, the efficient utilization of limited water resources is needed for agricultural production in these arid or semiarid regions.

Drip fertigation has the ability to apply small but frequent irrigation with soluble nutrients and chemicals, which has been found to be superior to the flood method in terms of the potential to save water, increase yield, improve quality, and enhance water and fertilizer use efficiency [7,8]. It has been reported that drip fertigation can improve crop yield by enhancing individual performance, water use efficiency, and seed quality and can support higher plant densities [7,9,10]. The optimal sowing density and yield of sunflower under drip fertigation will be quite different from those under conventional surface irrigation. Further studies on sunflower sowing density under fertigation conditions are needed to increase yields.

Regulating sowing density is an important practice to improve crop yield. Individual shoot biomass and yield decreased with density, while total biomass per area and yield increased with sowing density for grain crops, which was reported by several publications [11,12]. However, the total biomass in a given area was linearly proportional to the sowing density up to a critical density beyond which the total yield did not increase. Eventually, biomass allocation to reproduction may be reduced as well, causing a lower harvest index at very high sowing densities [11,13]. Sowing density also altered plant root distribution, biomass allocation, nutrient uptake, and cell morphology [14–16].

A wide range of sowing densities for the achievement of optimum yields in sunflower can be found in the literature [17,18]. The optimum sowing density for sunflower is influenced by several factors, such as temperature, soil fertility, water availability, and genotype [19,20]. The optimal sowing density to achieve high grain quality and high total yield under drip fertigation conditions remains unclear. Moreover, soil salinity is another important factor that affects agronomic practices. It was reported that an optimal sowing density could form full cover on the ground; on the one hand, it could inhibit weeds [21,22] and, on the other hand, it could reduce soil surface evaporation and prevent salt accumulation in the topsoil (0–20 cm) [3]. The evapotranspiration (ET) in the crop land was correlated with the sowing density, which influenced the soil water content, leaching fraction, and crop water productivity [23]. However, very little attention has been given to the interactions among planting density, soil salinity, yield, and yield components in the literature. Therefore, the objective of this research was to investigate the influence of sowing density on the soil salinity in saline soil and reveal the interactions among sowing density, soil salinity, yield, and yield components under drip fertigation conditions.

2. Materials and Methods

2.1. Experimental Site

Field experiments were conducted in a saline area (41°3′ N, 108°20′ E) from 2018 to 2019 in Wuyuan County, Inner Mongolia Autonomous Region, China. This area has a temperate continental arid climate with annual rainfall and potential evaporation of approximately 170 mm and 2500 mm, respectively. The EC of groundwater was greater than 7.8 dS m^{-1}, and its depth was generally less than 1.5 m. The physical and chemical properties of the tested soil are shown in Table 1. The 0–40 cm soil was silt loam, and the average electrical conductivity of saturated paste extracts (ECe) and the pH in 0–40 cm were 5.7 dS m^{-1} and 8.69, respectively, and the soil was classified as moderately alkaline saline soil [24].

Table 1. The initial physical and chemical properties of the experimental soil.

Soil Depth (cm)	mm for Soil Mechanical Composition (%)			Soil Texture	ECe (dS m^{-1})	pH	Bulk Density (g cm^{-3})
	<0.002	0.002–0.05	0.05–2				
0–10	7.75	73.16	19.07	Silt loam	6.7	8.61	1.51
10–20	7.79	73.03	19.17	Silt loam	5.5	8.68	1.61
20–30	7.84	73.34	18.82	Silt loam	5.2	8.68	1.46
30–40	7.67	73.21	19.12	Silt loam	5.3	8.77	1.54

2.2. Experimental Arrangement

The field experiment consisted of four sowing density treatments, of which the spacing between the sunflower plants in the same row was 30, 35, 40, and 45 cm represented as D30, D35, D40, and D45, respectively, and the distance between the rows was 80 cm for all of the treatments. Thus, the sowing densities corresponding to the above treatments were 41,667, 35,714, 31,250, and 27,778 plants ha^{-1}, respectively. The four treatments, each consisting of three replicated plots, were in a random block arrangement. Each plot had an area of 28 m × 8 m, and there was a one-meter-wide isolation belt between the two adjacent plots. The sunflower cultivar was hybrid sunflower SH363 in this experiment, and the seeds were sown on 28 May 2018 and on 10 June 2019.

Before the experiment, the soil was ploughed and levelled first. Then, the soil was ridged with drip tape buried under the plastic mulches by a multifunction machine (Figure 1A). The top width and height of a ridge were 0.4 m and 0.15 m, respectively, the same as in former studies [25,26]. The intervals between the adjacent ridges were 0.8 m, and the sunflower seeds were sown manually in a single row on the top of each ridge at an interval of the setting spacing according to each treatment. A tensiometer was buried at exactly 0.2 m under the drip emitter, which was nearest to a robust sunflower, in the second replicate plot to schedule drip irrigation (Figure 1A). Thus, there were four tensiometers in this experiment.

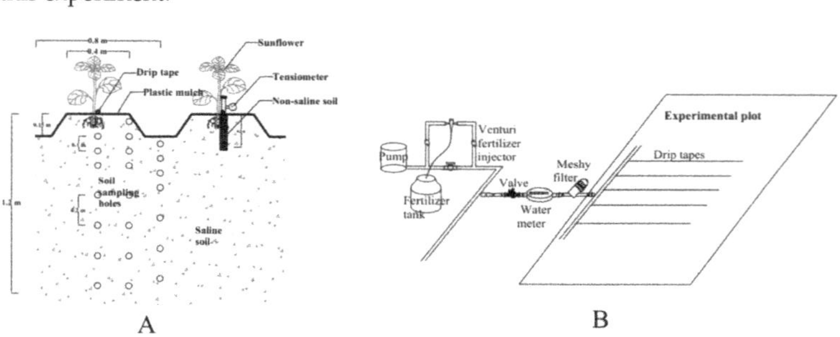

Figure 1. Schematic of the experimental design. (**A**) fertigation system in the experiment and (**B**) ridge planting pattern under drip irrigation.

2.3. Fertigation Scheduling

The irrigation water was pumped from a well with an EC of 1.3 dS m^{-1}. Based on previous studies, 30 mm water was immediately applied after sowing [26]. After the emergence of seedlings, irrigation was scheduled by a soil matric potential (SMP) monitored by tensiometers. The SMP thresholds were set at −10 kPa in the first year and −20 kPa in the second year for salt leaching [27,28]. When more than two of the four monitored SMP values fell below the set threshold, 7 mm water was applied in all of the treatments through a drip irrigation system (Figure 1B). The fertilizer amounts consisted of 180 kg ha^{-1} total N, 100 kg ha^{-1} total P, and 160 kg ha^{-1} total K, which were the same in all of the treatments according to the local conditions. These soluble fertilizers were applied through a venturi fertilizer injector during each irrigation event, according to daily usage [25,29].

2.4. Measurements

2.4.1. SMP

The SMP values monitored by tensiometers at a 0.2 m depth were recorded daily at 15:00 to initiate drip irrigation.

2.4.2. Soil Sampling and ECe

Soil cores were obtained from each plot in all of the treatments using an auger (4.0 cm diameter, 20 cm high) at the beginning and at the end of each growing season (May and September in 2018 and in 2019). Soil samples were obtained at lateral distances of 0, 20, and

40 cm from the emitters. The soil sampling locations and distribution in a profile are shown in Figure 1A. All of the soil samples were air-dried and sieved through a 2 mm sieve. Then, 25 g of sieved soil from each sample was mixed with about 20–25 mL of distilled water to make saturated soil paste according to the standard method [30,31]. To be noted, the amount of the distilled water for the different soil samples might be different as the soil texture affected the water amount that was used to make the saturated soil paste. After 8 h, the extract solution was obtained by centrifuging the saturated soil paste. The electrical conductivity of the saturated soil extracts (ECe) and the pH of the saturated soil extracts were determined by a conductivity meter (DDS-11A, Yulong, Shanghai, China) and a pH meter (pH-3C, Yulong, Shanghai, China), respectively.

2.4.3. Growth Performance

During the flowering stage in the 2018 growing season, whole sunflower bodies were collected to investigate growth performance. Three plants were collected randomly, with both aboveground bodies and whole roots taken into the laboratory, in each replicate plot. Then, the height, ground diameter, and area of each leaf were measured manually. The leaf area was measured based on 1 cm × 1 cm grid paper with coordinates, and the leaf area index (LAI) was calculated according to the measured leaf area and sowing density. The roots separated from the plants were placed into a net bag, soaked in water, and cleaned carefully until all the soil was washed away. Finally, the plant bodies and roots were dried at 65 °C in an oven to estimate the aboveground dry matter weight (AW) and the underground dry matter weight (UW) [32].

2.4.4. Yield and Its Components

Two quadrats (1.6 m × 4 m) in each plot of all of the treatments were selected randomly to estimate the sunflower yield. When the sunflower seeds were ripe, all the sunflower heads in the quadrats were collected manually. The diameter of each head was measured. Seeds were peeled from the heads and naturally wind-dried. The market yield (with immature seeds and impurities extracted), the thousand seed weight (TSW), and the seed setting percentage were measured based on the selected quadrats. Specifically, all of the seeds collected from a repeating plot were mixed in the 2018 season, and the seed length and width were determined based on 100 random seeds in each plot. The irrigation water productivity (IWP) was defined as the ratio of market yield and irrigation water amount [33].

2.5. Data analyses and Statistics

All data gathered in the research were recorded and classified in Microsoft Office Excel 2016. Analyses of variance (ANOVA) were carried out by SPSS 19.0 statistical software (IBM SPSS Inc., Armonk NY, USA). The significant differences in ECe, pH, growth performance, yield, and yield components between the treatments were compared by Tukey's test ($p < 0.05$). Figures were drawn by Surfer 14 (Golden Software Inc., Golden, CO, USA) and SigmaPlot 12.5 (Systat Software Inc., San Jose, CA, USA). The equations adopted in this study are as follows [33–35]:

$$\text{Soil matric potential (SMP, kPa)} = \Psi_{\text{tensiometer}} - \Psi_{\text{gravity}} \quad (1)$$

where $\Psi_{\text{tensiometer}}$ is the reading of a tensiometer dial and Ψ_{gravity} is the gravitational potential between a dial and porous ceramic cup.

$$\text{Leaf area index (LAI)} = A_{\text{leaves}} \times P \quad (2)$$

where A_{leaves} is the total leaf area of one plant and P is the sowing density of the sunflower.

$$\text{Irrigation water productivity (IWP, kg ha}^{-1}\text{ mm}^{-1}) = \frac{Y}{W} \quad (3)$$

where Y is the grain yield of the sunflower and W is the total quantity of water irrigated in the sunflower life circle.

3. Results

3.1. Rainfall, Irrigation, and Soil Matric Potentials

The rainfall amount in the 2018 growing season was 180 mm, which was almost twice that in the 2019 growing season (Table 2). The average annual rainfall in this area was 170 mm, indicating that the climate in 2018 was wetter, while it was drier in 2019 when compared with normal years. Rainfall influenced the amount of applied irrigation water. Basically, the applied irrigation water amount increased as the SMP threshold increased [36]. However, the applied irrigation water amount in 2018, scheduled at −10 kPa, was less than that in 2019, which was scheduled at −20 kPa. More rainfall in 2018 resulted in less applied irrigation water.

Table 2. Amounts of rainfall and irrigation in the 2018 and 2019 growing seasons.

Growing Season	Rainfall (mm)	Applied Irrigation Water Amount (mm)	Total Water Amount (mm)
2018	180	287	467
2019	69	322	491

Interestingly, the total water amounts in these two seasons were between 450 and 500 mm. Daily SMP dynamics (Figure 2) showed that the SMPs in the two seasons all fluctuated around their thresholds, −10 and −20 kPa, respectively, and the vibration amplitude in 2019 was larger than that in the 2018 season, indicating that the SMPs in these two seasons were well controlled through SMP threshold scheduling as anticipated.

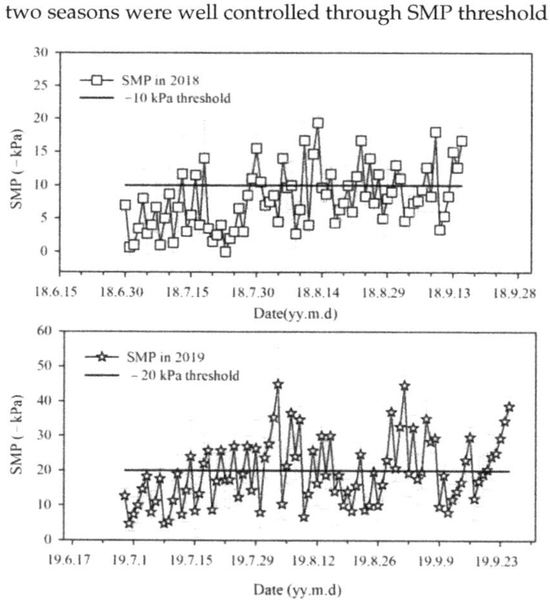

Figure 2. Daily soil matric potential (SMP) dynamics in the 2018 and 2019 growing seasons.

3.2. ECe and pH

The original soil ECe distributions before this experiment showed that the soil salinity decreased with the soil depth (Figure 3A1), and the average ECe within the whole depth indicated that the soil was moderately alkaline saline soil [24]. At the end of the first growing season (Figure 3B1 and Table 3), the ECe in the whole soil profile had firm decreases, and the average ECe showed that the soil changed to mildly alkaline saline soil (Table 3). Significant differences were found between the different treatments for the ECe in the 0–40 cm soil layer (ECe-40) and the ECe in the 0–120 cm soil layer (ECe-120). The D30 treatment had the largest ECe-40 and ECe-120, while the D45 treatment had the lowest ECe-40 and ECe-120.

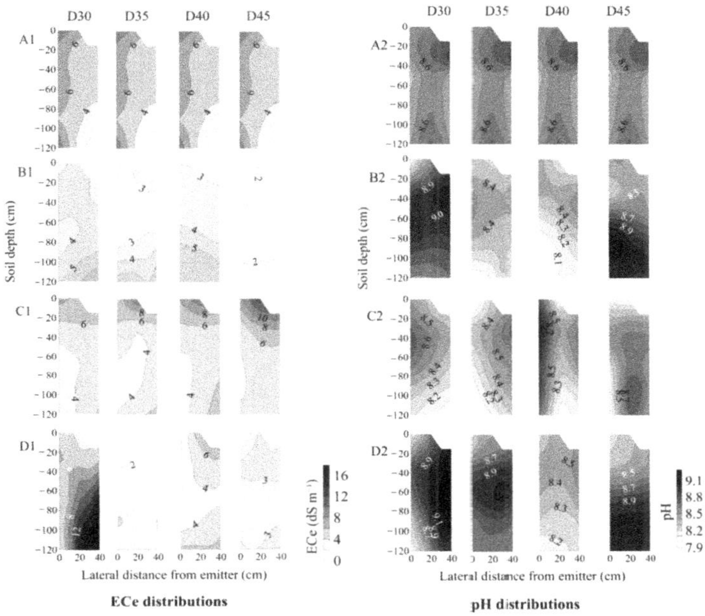

Figure 3. Spatial and temporal distributions of the electrical conductivity of the saturated soil extracts (ECe) and the pH of the saturated soil extracts (pH) in the different treatments. (**A1,A2**) May in 2018; (**B1,B2**) September in 2018; (**C1,C2**) May in 2019; (**D1,D2**) September in 2019. **D30**, **D35**, **D40**, and **D45** indicated treatments with sowing densities of 41,667, 35,714, 31,250, and 27,778 plants ha^{-1}, respectively.

Table 3. The average ECe in the 0–40 cm and 0–120 cm soil layers during the different sampling periods.

Treatments	0–40 cm ECe (dS m^{-1})				0–120 cm ECe (dS m^{-1})			
	May. 2018	Sep. 2018	Apr. 2019	Sep. 2019	May. 2018	Sep. 2018	Apr. 2019	Sep. 2019
D30	5.7 ^{Aab}	4.3 ^{Ab}	6.6 ^{Aa}	5.0 ^{Aab}	5.2 ^{Aa}	4.4 ^{Aa}	4.9 ^{Aa}	8.8 ^{Aa}
D35	5.7 ^{Aa}	2.9 ^{Bb}	6.5 ^{Aa}	1.9 ^{Bb}	5.2 ^{Aa}	3.2 ^{Bb}	5.0 ^{Aa}	2.9 ^{Bb}
D40	5.7 ^{Aa}	3.3 ^{Bb}	7.0 ^{Aa}	4.9 ^{Aab}	5.2 ^{Aa}	4.1 ^{Aa}	5.4 ^{Aa}	4.3 ^{Ba}
D45	5.7 ^{Ab}	2.0 ^{Cc}	9.1 ^{Aa}	3.3 ^{ABc}	5.2 ^{Aa}	1.8 ^{Cb}	6.3 ^{Aa}	2.9 ^{Bb}

Note: Different lowercase letters following the data in the same row in the same soil layer indicate significant differences at $p < 0.05$ among the different sampling periods; different capital letters following the data in the same column indicate significant differences at $p < 0.05$ among different treatments according to Tukey's test.

After one non-growing season (Figure 3C1), salt accumulated in the topsoil. The ECe-40 even became larger than the original value in all of the treatments, and the ECe-120 also increased to its original level. The ECe in the soil profiles in the D35, D40, and D45 treatments all decreased, while that in the D30 treatment increased when the second growing season terminated.

Overall, the average ECe-40 values during the two seasons for treatments D30–D45 were 5.3, 3.8, 5.1, and 4.8 dS m^{-1}, which were correspondingly decreased by 7.0%, 33.9%,

11.1%, and 15.8% when compared with the original value. The average ECe-120 value for treatments D30–D45 were 6.0, 3.7, 4.6, and 3.7 dS m^{-1}, with corresponding decreasin ratios of −16.0%, 28.8%, 11.5%, and 29.5%, respectively. In terms of spatial and tempora changes, the D35, D40, and D45 treatments had more advantages in controlling soil salinit than the D30 treatments.

The pH dynamics varied from the ECe. During the first growing season, the pH values in the 0–40 cm soil layer (pH-40) and in the 0–120 cm soil layer (pH-120) in th D35 and D40 treatments decreased firmly, but they increased in D30 and D45 (Figure 3B right and Table 4). In contrast to the ECe dynamics, the pH decreased or remained stabl instead of increasing after one non-growing season (Figure 3C2). At the end of the secon growing season, noticeable pH increments occurred in the D35 and D45 treatments alon with the salt leaching process (Figure 3D2). Overall, the average pH-40 during the tw seasons for treatments D30–D45 correspondingly decreased by −0.03, 0.20, 0.20, an 0.27 when compared to the original value. The average pH-120 for treatments D30–D4 correspondingly decreased by −0.17, 0.05, 0.16, and −0.11.

Table 4. The average pH in the 0–40 cm and 0–120 cm soil layers during the different sampling period

Treatments	0–40 cm pH				0–120 cm pH			
	May. 2018	**Sep. 2018**	**Apr. 2019**	**Sep. 2019**	**May. 2018**	**Sep. 2018**	**Apr. 2019**	**Sep. 2019**
D30	8.69 ^{Aab}	8.79 ^{Aab}	8.48 ^{Ab}	8.88 ^{Aa}	8.55 ^{Abc}	8.84 ^{Aab}	8.41 ^{Ac}	8.90 ^{Aa}
D35	8.69 ^{Aa}	8.39 ^{Bb}	8.43 ^{Ab}	8.65 ^{Ba}	8.55 ^{Ab}	8.36 ^{Cc}	8.36 ^{Ac}	8.78 ^{Aa}
D40	8.69 ^{Aa}	8.38 ^{Bb}	8.62 ^{Aa}	8.47 ^{BCa}	8.55 ^{Aa}	8.27 ^{Db}	8.53 ^{Aa}	8.37 ^{Ba}
D45	8.69 ^{Aa}	8.43 ^{Bb}	8.44 ^{Ab}	8.39 ^{Cb}	8.55 ^{Aab}	8.74 ^{Ba}	8.51 ^{Ab}	8.74 ^{Aa}

Note: Different lowercase letters following the data in the same row in the same soil layer indicate significar differences at $p < 0.05$ among the different sampling periods; different capital letters following the data in th same column indicate significant differences at $p < 0.05$ among the different treatments.

3.3. Growth Performance

The sunflower growing parameters during the flowering stage in the 2018 season ar shown in Table 5. The plant height (H) first increased and then decreased with increasin spacing in the rows, but non-significant differences were found between the treatment The ground diameter (GD) increased as the spacing in rows increased, and those in the D35 D45 treatments were remarkably larger than those in the D30 treatment. The H/GD rati decreased as the spacing in the rows increased, and the ratios in the D35–D45 treatment were significantly lower than those in the D30 treatment. The LAI values in the D30 an D35 treatments were statistically larger than those in the D40 treatment but not significantl different from those in the D45 treatment. The aboveground dry matter weight (AW) an underground dry matter weight (UW) for an individual plant both increased as the spacin in the rows increased, and the UW/AW ratio exhibited the same trend.

Table 5. The main growing parameters during the flowering stage among treatments in the 201 growing season. H, GD, LAI, AW, and UW are abbreviations for height, ground diameter, leaf are index, aboveground dry matter weight, and underground dry matter weight for an individual plan respectively.

Treatments	H (cm)	GD (mm)	H/GD	LAI	AW (g)	UW (g)	UW/AW
D30	177.3 ^{a}	26.4 ^{b}	67.5 ^{a}	2.8 ^{a}	170.8 ^{c}	49.4 ^{b}	0.29 ^{b}
D35	180.0 ^{a}	33.7 ^{a}	54.3 ^{b}	2.6 ^{a}	264.7 ^{b}	83.8 ^{b}	0.32 ^{ab}
D40	182.0 ^{a}	33.5 ^{a}	54.4 ^{b}	2.1 ^{b}	269.6 ^{b}	85.2 ^{b}	0.32 ^{ab}
D45	177.3 ^{a}	35.6 ^{a}	49.9 ^{b}	2.4 ^{ab}	324.8 ^{a}	135.5 ^{a}	0.41 ^{a}

Note: Different lowercase letters following the data in the same column indicate significant differences at $p < 0.0$ among the different treatments.

The dry biomass parameters in the D45 treatment were significantly larger than those in the rest of the treatments, and the D30 treatment achieved the lowest dry biomass among all of the treatments. The comprehensive growing parameters indicated that increasing the space in the rows could increase the plant ground diameter and the whole plant biomass, especially promoting root growth, which might enhance the ability of plants to absorb water and nutrients and their ability to resist lodging.

3.4. Yield and Its Components

The sunflower market yields in the two seasons both increased first and then decreased as the spacing in the rows increased (Figure 4). The trend curves suggested that the yields in all of the treatments in the 2019 season were higher than those in the 2018 season, and the yields ascended to their peaks, larger than 4000 kg ha^{-1}, when the spacing in the rows was between 35 and 40 cm. The statistical analysis showed that the yield in the D40 treatment in 2019 was higher than those in the D30 and D45 treatments in 2018 and those in the D30 treatment in 2019 and the yield did not vary from the rest of the treatments (Table 6). Notably, the average yields in the two seasons of the D30–D45 treatments firmly increased by 18.0%, 28.9%, 29.7%, and 18.0%, respectively, when compared with the 5-year average yield under surface irrigation.

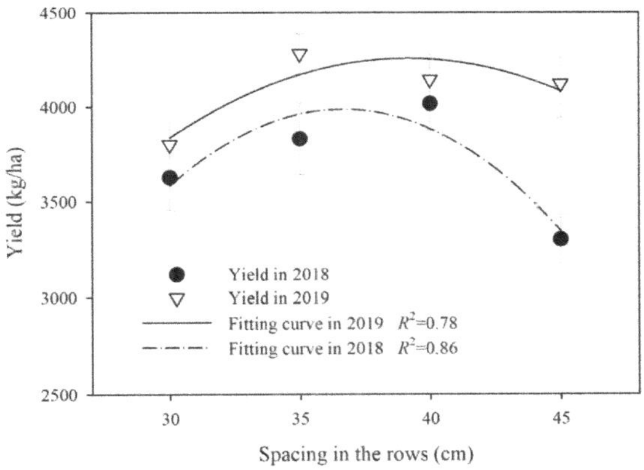

Figure 4. Sunflower yields in the two growing seasons under different treatments (different spacing in the rows). (Error bars indicate the standard deviation).

The irrigation water productivity (IWP) in the D40 treatment in 2018 was 14.0 kg ha^{-1} mm^{-1}, which was the largest and significantly larger than those in the D45 treatment in 2018 and the D30 treatment in 2019, with corresponding IWPs of 11.5 and 11.8 kg ha^{-1} mm^{-1}, respectively. The IWP in each treatment was more than twice that under surface irrigation.

The head diameter and the thousand seed weight (TSW) in the D40 treatment in both of the seasons were the highest, while the lowest values in the D30 treatment were even higher than those under surface irrigation. The setting percentages of seeds in all of the treatments remained comparable with each other, except that in the D30 treatment in the 2018 season the percentage was significantly lower than that in the rest of the treatments. The seed length and width indicated that the seed sizes in the D40 and D45 treatments were apparently larger than those in the D30 and D35 treatments. Overall, it was easy to conclude that the spacing in the rows between 35–40 cm could ensure both the high yield and the high morphological quality of the seeds.

Table 6. Sunflower yield and its components in different treatments in the two growing seasons. IWP and TSW are abbreviations for irrigation water productivity and thousand seed weight, respectively.

Growing Seasons	Treatments	Yield and Its Components						
		Market Yield (kg ha^{-1})	IWP (kg ha^{-1} mm^{-1})	Head Diameter (cm)	TSW (g)	Setting Percentage (%)	* Seed Length (mm)	* Seed Width (mm)
2018	D30	3626.1 ^{cd}	12.6 ^{abc}	19.7 ^{c}	212.3 ^{d}	61.9 ^{c}	22.0 ^{b}	9.2 ^{b}
	D35	3830.9 ^{abc}	13.3 ^{ab}	20.8 ^{bc}	210.5 ^{d}	74.6 ^{ab}	22.6 ^{b}	9.3 ^{b}
	D40	4015.8 ^{abc}	14.0 ^{a}	23.4 ^{a}	237.4 ^{c}	76.2 ^{ab}	23.4 ^{a}	9.9 ^{a}
	D45	3302.5 ^{de}	11.5 ^{c}	21.8 ^{b}	211.5 ^{d}	74.2 ^{b}	23.4 ^{a}	10.2 ^{a}
2019	D30	3800.4 ^{bc}	11.8 ^{bc}	19.9 ^{c}	209.9 ^{d}	76.5 ^{ab}	—	—
	D35	4279.0 ^{a}	13.3 ^{ab}	23.6 ^{a}	243.0 ^{bc}	80.8 ^{a}	—	—
	D40	4141.7 ^{ab}	12.9 ^{abc}	24.6 ^{a}	260.2 ^{ab}	75.7 ^{ab}	—	—
	D45	4118.8 ^{ab}	12.8 ^{abc}	25.0 ^{a}	270.9 ^{a}	78.3 ^{ab}	—	—
† 2016–2020	Surface irrigation	3145.7 ^{e}	5.1 ^{d}	20.6 ^{bc}	165.1 ^{e}	—	—	—

Note: Different lowercase letters following the data in the same column indicate significant differences among the treatments at $p < 0.05$; * the seed length and seed width were not recorded in the 2019 season; † the data of traditional surface irrigation during 2016–2020 were collected from the available publications [5,6,37–40].

4. Discussion

4.1. Influence of Sowing Density on Soil Salinity

This study found that sowing density had a significant influence on soil salinity, which was in accordance with several available publications. Li et al. [41] found that different sowing densities led to differences in soil evaporation and crop transpiration, which caused differences in soil salt accumulation. The crop leaf area index (LAI) increased as the sowing density increased [41,42], and the leaf area duration increased as the sowing density increased. However, high sowing density changed the canopy structure (leaf area distribution) and increased water consumption [13,43]. Our results were consistent with these findings. The sunflower LAI under the D30 treatment (the highest sowing density) was the largest, which might encourage the water consumed by sunflower leaves (known as transpiration) and finally increase the sum of transpiration and soil surface evaporation (known as evapotranspiration, ET) [41]. Since all of the treatments received the same amounts of rainfall and irrigation water, the water for salt leaching decreased as the ET increased. This explained the remarkable salt accumulation phenomenon in the D30 treatment in the 2019 season. However, the salt accumulation phenomenon in the D30 treatment in the 2018 season was not apparent. The SMP threshold for scheduled fertigation in the 2018 season (−10 kPa) was higher than that in the 2019 season (−20 kPa), which changed the vertical water potential gradient. It was calculated that the salt leaching fraction under the condition that irrigation was scheduled by an SMP threshold of −10 kPa was 240% higher than that under −20 kPa [36]. Thus, the soil salinity in the D30 treatment increased when the SMP threshold decreased from −10 kPa in the 2018 season to −20 kPa in the 2019 season. The sowing density had remarkable influences on soil salinity dynamics, the mechanism of which was likely conducted by regulating crop LAI and ET.

4.2. Interactions between Sowing Density and Growth Performance under Fertigation

Individual crop shoot biomass and yield decreased with density, while total biomass per area and yield were linearly proportional to sowing density up to a critical density beyond which the harvest index did not increase [11,12]. High sowing density had a negative influence on individual crop performance when the sowing density was larger than the sowing density threshold. It was reported that poor individual morphological features, such as fewer productive tillers, shorter plants, thinner stalks, reduced hundred-grain weight, and fewer seeds per ear for grain crops would be achieved under a high sowing density [44–47]. A significant decrease in the leaf dry mass, leaf total chlorophyll content, and leaf nitrogen content per leaf area occurred with increasing planting density in each growth stage of maize [16]. In comparison, the total growing performances for the population, such as the LAI, total ear, and biomass, had far greater increasing potentials with increasing planting density [44,48], which meant that a larger sowing density was needed to achieve its peak. The total growth performance for the population was far more sensitive to increasing planting density than individual performance. This was because plants could adjust their growth to their environment through means ranging from whole-plant morphological changes to alterations in the stoichiometry of the photosynthetic apparatus [45]. Nevertheless, these advantages of population growth performance did not translate into yield advantages. Competition for water, nutrients, light, and expanding space altered the plant nutrient allocation, which encouraged more energy to flow to the vegetative organs rather than the reproductive organs [49,50]. Thus, the optimal sowing density for peak yield was normally between the sowing density for the best individual performance and the sowing density for the best population performance. However, peak yield did not achieve the highest economic benefits for crops such as sunflower because seed quality (seed size and hundred-seed weight) had remarkable influences on market price. Thus, a trade-off between seed quality and total yield should be made to determine the optimum sowing density.

The limitations of crop competition for resources under high-density cultivation may be overcome through efficient fertigation practices [51]. Li et al. [41] conducted a five-sowing density experiment for sunflower cultivation under drip fertigation conditions and found that the individual performances decreased less with increasing density, but population performance and yield increased with increasing density. The density corresponding to peak yield and the best population growing performances reached 55,556 plants ha^{-1}, which was much higher than the population density under surface irrigation in this area (normally less than 30,000 plants ha^{-1}), indicating that fertigation could increase yield by supporting more crop density. This result was consistent with our study and together revealed the interactions among sowing density, individual crop performance, total population performance, and yield under drip fertigation conditions.

5. Conclusions

Sowing density had a significant influence on soil salinity, and high-density planting caused salt accumulation in the vertical soil profile during the growing season. The individual growth performance and yield components of sunflower, such as stem diameter, plant biomass, head diameter, thousand-seed weight, and the proportion of biomass allocated underground, decreased remarkably as the sowing density increased. However, the total population growth performance and yield increased with increasing sowing density before a critical density threshold, beyond which the total yield did not increase. The height yield was a dynamic optimal solution that was determined by the sowing density, individual crop performance, total population performance, and irrigation practice. A sowing density between 31,250 and 35,714 plants ha^{-1} is recommended for sunflower cultivation under drip fertigation in arid saline areas with similar conditions.

Author Contributions: Conceptualization, S.D. and Y.K.; methodology, S.D. and X.L.; software, G.W.; writing—original draft preparation, S.D. All authors have read and agreed to the published version of the manuscript.

Funding: This research was funded by the Key Research and Development Program of Shandong Province (Grant No. 2021SFGC0301), the Key Research Program of Frontier Sciences of the Chinese Academy of Sciences (Grant No. QYZDJ-SSW-DQC028), the National Key Research and Development Program of China (Grant No. 2019YFD1002702), and the Shandong Provincial Natural Science Foundation (Grant No. ZR2022QE208).

Data Availability Statement: All relevant data are contained within the article.

Acknowledgments: We would like to thank the Yellow River Delta Saline–Alkali Farmland Ecosys-tem Observation and Research Station of the Yantai Institute of Coastal Zone Research, Chinese Academy of Sciences, for providing the experimental site.

Conflicts of Interest: The authors declare no conflict of interest.

References

1. USDA. *National Agricultural Statistics Service*; USDA-NASS: Washington, DC, USA, 2016.
2. Zeng, W.; Xu, C.; Wu, J.; Huang, J.; Zhao, Q.; Wu, M. Impacts of salinity and nitrogen on the photosynthetic rate and growth o sunflowers (*Helianthus annuus* L.). *Pedosphere* **2014**, *24*, 635–644. [CrossRef]
3. Zhao, Y.; Pang, H.; Wang, J.; Huo, L.; Li, Y. Effects of straw mulch and buried straw on soil moisture and salinity in relation t sunflower growth and yield. *Field Crops Res.* **2014**, *161*, 16–25. [CrossRef]
4. Giannini, V.; Mula, L.; Carta, M.; Patteri, G.; Roggero, P. Interplay of irrigation strategies and sowing dates on sunflower yield i semi-arid Mediterranean areas. *Agric. Water Manag.* **2022**, *260*, 107287. [CrossRef]
5. Zhao, Y.; Li, Y.; Wang, J.; Pang, H.; Li, Y. Buried straw layer plus plastic mulching reduces soil salinity and increases sunflowe yield in saline soils. *Soil Tillage Res.* **2016**, *155*, 363–370. [CrossRef]
6. BWRB. Water Resources Bulletin of Bayannur in 2016–2020. Available online: http://slj.bynr.gov.cn (accessed on 15 September 2022
7. Hu, J.; Gettel, G.; Fan, Z.; Lv, H.; Zhao, Y.; Yu, Y.; Wang, J.; Butterbach-Bahl, K.; Li, G.; Lin, S. Drip fertigation promotes water an nitrogen use efficiency and yield stability through improved root growth for tomatoes in plastic greenhouse production. *Ag Ecosyst. Environ.* **2021**, *313*, 107379. [CrossRef]
8. Kapoor, R.; Kumar, A.; Sandal, S.; Sharma, A.; Raina, R.; Thakur, K. Water and nutrient economy in vegetable crops through dri fertigation and mulching techniques: A review. *J. Plant Nutr.* **2022**, *45*, 2389–2403. [CrossRef]
9. Ibrahim, M.; El-Baroudy, A.; Taha, A. Irrigation and fertigation scheduling under drip irrigation for maize crop in sandy soil. *In Agroph.* **2016**, *30*, 47–55. [CrossRef]
10. Ningaraju, G.; Joseph, P. Effect of drip fertigation on growth and yield of oriental pickling melon (*Cucumis melo* var. *conomon* (L Makino) under high density planting. *Int. J. Sci. Res. Publ.* **2014**, *4*, 1–5.
11. Farnia, A.; Aminbaygi, A.; Fard, A. The effect of plant density on comparison forage yield of spring barley cultivar. *Int. J. Farm All. Sci.* **2014**, *3*, 244–250.
12. Dueri, S.; Brown, H.; Asseng, S.; Ewert, F.; Webber, H.; George, M.; Craigie, R.; Guarin, J.R.; Pequeno, D.N.L.; Stella, T.; et a Simulation of winter wheat response to variable sowing dates and densities in a high-yielding environment. *J. Exp. Bot.* **2022**, *7* 5715–5729. [CrossRef]
13. Zhang, Y.; Xu, Z.; Li, J.; Wang, R. Optimum planting density improves resource use efficiency and yield stability of rainfed maiz in semiarid climate. *Front. Plant Sci.* **2021**, *12*, 752606. [CrossRef]
14. Hecht, V.; Temperton, V.; Nagel, K.; Uwe, R.; Postma, J. Sowing density: A neglected factor fundamentally affecting roo distribution and biomass allocation of field grown spring barley (*Hordeum Vulgare* L.). *Front. Plant Sci.* **2016**, *7*, 944. [CrossRef]
15. Khan, A.; Wang, L.; Ali, S.; Tung, S.; Hafeez, A.; Yang, G. Optimal planting density and sowing date can improve cotton yield b maintaining reproductive organ biomass and enhancing potassium uptake. *Field Crops Res.* **2017**, *214*, 164–174. [CrossRef]
16. Zheng, B.; Zhao, W.; Ren, T.; Zhang, X.; Ning, T.; Liu, P.; Li, G. Low light increases the abundance of light reaction protein Proteomics analysis of maize (*Zea mays* L.) grown at high planting density. *Int. J. Mol. Sci.* **2022**, *23*, 3015. [CrossRef]
17. Borysenko, V.; Karnauh, O.; Nakleka, Y.; Novak, A.; Usik, S.; Koval, G. The influence of plant height and diameter of inflorescence on sunflower productivity depending on sowing density and row spacing. *Taur. Scient. Her.* **2020**, *113*, 28–34. [CrossRef]
18. Pinkovskyi, H.; Tanchyk, S. Management of productivity of sunflower plants depending on terms of sowing and density o standing in arid conditions of the Right-bank Steppe of Ukraine. *Agron. Sci.* **2021**, *76*, 21–38. [CrossRef]
19. Mladenovic, E.; Cvejic, S.; Jocic, S.; Cuk, N.; Cukanovic, J.; Jockovic, M.; Jeromela, A. Effect of plant density on stem and flowe quality of single-stem ornamental sunflower genotypes. *Hortic. Sci.* **2020**, *47*, 45–52. [CrossRef]
20. Soleymani, A. Light response of sunflower and canola as affected by plant density, plant genotype and N fertilization. *J. Photocher Photobiol. B-Biol.* **2017**, *173*, 580–588. [CrossRef]
21. Mouillon, P.; Caldwell, B.A.; Cordeau, S.; Pelzer, C.; Wayman, S.; Ryan, M. Crop density affects weed suppression in organicall managed Sunflower. *Agron. J.* **2020**, *112*, 450–457. [CrossRef]
22. Marin, C.; Weiner, J. Effects of density and sowing pattern on weed suppression and grain yield in three varieties of maize unde high weed pressure. *Weed Res.* **2014**, *54*, 467–474. [CrossRef]
23. Echarte, L.; Echarte, M.; Cerrudo, D.; Gonzalez, V.; Alfonso, C.; Cambareri, M.; Hernandez, M.; Nagore, M.; Maggiora, A Sunflower evapotranspiration and water use efficiency in response to plant density. *Crop Sci.* **2020**, *60*, 357–366. [CrossRef]
24. Bao, S. *Soil and Agricultural Chemistry Analysis*; China Agriculture Press: Beijing, China, 2000. (In Chinese)
25. Dong, S.; Kang, Y.; Wan, S.; Li, X.; Miao, J. Drip-irrigation using highly saline groundwater increases sunflower yield in heavil saline soil. *Agron. J.* **2021**, *113*, 2950–2959. [CrossRef]
26. Kang, Y.; Chen, M.; Wan, S. Effects of drip irrigation with saline water on waxy maize (*Zea mays* L. var. *ceratina* Kulesh) in Nort China Plain. *Agric. Water Manag.* **2010**, *97*, 1303–1309. [CrossRef]

27. Dong, S.; Wan, S.; Kang, Y.; Li, X. Establishing an ecological forest system of salt-tolerant plants in heavily saline wasteland using the drip-irrigation reclamation method. *Agric. Water Manag.* **2021**, *245*, 106587. [CrossRef]
28. Wang, R.; Wan, S.; Kang, Y.; Dou, C. Assessment of secondary soil salinity prevention and economic benefit under different drip line placement and irrigation regime in northwest China. *Agric. Water Manag.* **2014**, *131*, 41–49. [CrossRef]
29. Feng, Z.; Wan, S.; Kang, Y.; Liu, S. Drip fertigation regime for potato on sandy soil. *Emir. J. Food Agr.* **2017**, *29*, 476–484. [CrossRef]
30. Corwin, D.; Lesch, S.; Lobell, D. Laboratory and field measurements. In *Agricultural Salinity Assessment and Management*, 2nd ed.; American Society of Civil Engineers, Environmental and Water Resources Institute: Reston, VA, USA, 2012; pp. 265–341.
31. Visconti, F.; Paz, J. Electrical conductivity measurements in agriculture: The assessment of soil salinity. In *New Trends and Developments in Metrology*; Cocco, L., Ed.; IntechOpen: Rijeka, Croatia, 2016; pp. 99–126.
32. Lu, R. *Methods for Agricultural Chemical Analysis of Soil*; China Agricultural Science and Technology Press: Beijing, China, 2000; pp. 299–301.
33. Chen, X.; Kang, Y.; Wan, S.; Guo, L. Mulches improve ridge-tillage tomato production under drip irrigation with saline water. *Agron. J.* **2019**, *111*, 2116–2127. [CrossRef]
34. Or, D. Who invented the tensiometer? *Soil Sci. Soc. Amer. J.* **2001**, *65*, 1–3. [CrossRef]
35. Ju, Z.; Liu, K.; Zhao, G.; Ma, X.; Jia, Z. Nitrogen fertilizer and sowing density affect flag leaf photosynthetic characteristics, grain yield, and yield components of oat in a semiarid region of Northwest China. *Agronomy* **2022**, *12*, 2108. [CrossRef]
36. Dong, S.; Wan, S.; Kang, Y.; Li, X. Prospects of using drip irrigation for ecological conservation and reclaiming highly saline soils at the edge of Yinchuan Plain. *Agric. Water Manag.* **2020**, *239*, 106255. [CrossRef]
37. Fan, X.; Zhai, Y.; Zhang, Y.; Mi, Z.; Sun, X.; Su, H.; Sun, X.; Gao, H. Fertilizer comparison test on edible sunflower in Hetao Irrigation District. *J. Anh. Agric. Sci.* **2017**, *45*, 24–25. (In Chinese)
38. Yan, H.; Zhong, S.; Wen, X.; Wang, R.; Li, B.; Li, E.; Liu, J.; Li, B.; Duan, Y.; Zhang, T. Effects of different amendments on agronomic characteristics and yield of sunflower in saline-alkali soil. *J. North. Agr.* **2021**, *49*, 35–40. (In Chinese)
39. Yan, S.; Mi, Z.; Sun, X.; Fan, X.; Zhai, Y.; Liu, Z.; Sun, X.; Qin, X.; Du, R.; Liu, X. Study on the improvement effect of new soil amendments for saline-alkali land in sunflower fields in Linhe district. *J. North. Agr.* **2018**, *46*, 54–57. (In Chinese)
40. SBB. Statistical Bulletin of National Economic and Social Development of Bayannur in 2016–2020. Available online: http://tjj.bynr.gov.cn/tjjztzl/tjgb/202104/t20210412_429123.html (accessed on 12 April 2021).
41. Li, J.; Qu, Z.; Chen, J.; Yang, B.; Huang, Y. Effect of planting density on the growth and yield of sunflower under mulched drip irrigation. *Water* **2019**, *11*, 752. [CrossRef]
42. Ren, Y.; Liu, J.; Wang, Z.; Zhang, S. Planting density and sowing proportions of maize–soybean intercrops affected competitive interactions and water-use efficiencies on the Loess Plateau, China. *Eur. J. Agron.* **2016**, *72*, 70–79. [CrossRef]
43. Keipp, K.; Hutsch, B.W.; Schubert, S. How does the harvest index affect water-use efficiency and nutrient-utilization efficiency of sunflowers (*Helianthus annuus* L.)? *J. Agron. Crop Sci.* **2019**, *205*, 519–532. [CrossRef]
44. Tahir, S.; Ahmad, A.; Khaliq, T.; Cheema, M. Evaluating the impact of seed rate and sowing dates on wheat productivity in semi-arid environment. *Int. J. Agr. Biol.* **2019**, *22*, 57–64.
45. Szczepanek, M.; Lemanczyk, G.; Nowak, R.; Graczyk, R. Response of indian dwarf wheat and persian wheat to sowing density and hydrothermal conditions of the growing seasons. *Agriculture* **2022**, *12*, 205. [CrossRef]
46. Yang, G.; Luo, X.; Nie, Y.; Zhang, X. Effects of plant density on yield and canopy micro environment in hybrid cotton. *J. Integr. Agr.* **2014**, *13*, 2154–2163. [CrossRef]
47. Guzman, M.; Vilain, L.; Rondon, T.; Sanchez, J. Sowing density effects in cotton yields and its components. *Agronomy* **2019**, *9*, 349. [CrossRef]
48. Liang, S.; Yoshihira, T.; Sato, C. Grain yield responses to planting density in twin and narrow row cultivation of early cultivars in maize. *Grassl. Sci.* **2020**, *66*, 183–193. [CrossRef]
49. Guo, X.; Zhu, G.; Jiao, X.; Zhou, G. Effects of nitrogen application and planting density on growth and yield of Sesbania pea grown in saline soil. *Curr. Sci.* **2019**, *116*, 758–764. [CrossRef]
50. Rossini, M.; Maddonni, G.; Otegui, M. Inter-plant competition for resources in maize crops grown under contrasting nitrogen supply and density: Variability in plant and ear growth. *Field Crops Res.* **2011**, *121*, 373–380. [CrossRef]
51. Tang, K.; Wang, J.; Yang, Y.; Deng, G.; Yu, J.; Hu, W.; Guo, L.; Du, G.; Liu, F. Fiber hemp (*Cannabis sativa* L.) yield and its response to fertilization and planting density in China. *Ind. Crops Prod.* **2022**, *177*, 114542. [CrossRef]

agronomy

MDPI

Article

Microdosing of Compost for Sustainable Production of Improved Sorghum in Southern Mali

Moumini Guindo [1,*], Bouba Traoré [1,2], Birhanu Zemadim Birhanu [2], Alou Coulibaly [3] and Ramadjita Tabo [2]

1 Institut d'Economie Rurale (IER), Centre Régional de la Recherche Agronomique de Sikasso, Bamako BP 16, Mali; b.traore@cgiar.org

2 International Crops Research Institute for the Semi-Arid Tropics ICRISAT, Niamey BP 12404, Niger; z.birhanu@cgiar.org (B.Z.B.); r.tabo@cgiar.org (R.T.)

3 Institut Polytechnique Rurale de Formation et de Recherche Appliquée (IPR/IFRA) de Katibougou, Koulikoro BP 06, Mali; coulibalynalou@gmail.com

* Correspondence: moumini.guindo1@gmail.com; Tel.: 00223-21260107

Abstract: The depletion of soil organic matter is one of the major challenges constraining agricultural production in the southern zone of Mali. This study evaluated the effects of compost types, methods, and dose applications on the productivity and sustainability of sorghum. Two types of compost (farmer practice and cotton stems) were applied to sorghum at two rates (microdosing at 2.5 t ha^{-1} and broadcasting at 5 t ha^{-1}) and evaluated on 30 farmer fields in 2019 and 2020. The treatments used included CPA (cotton stem compost at 5 t ha^{-1} + 100 kg ha^{-1} DAP), CPA (cotton stem compost at 2.5 t ha^{-1} + 100 kg ha^{-1} DAP), CP (farmer compost at 5 t ha^{-1} + 100 kg ha^{-1} DAP), CP (famer compost at 2.5 t ha^{-1} + 100 kg ha^{-1} DAP), control (100 kg ha^{-1} DAP), and control. The results showed that regardless of the compost type, applying a microdose of 2.5 t ha^{-1} improved the growth rate, plant height, grain yield, and biomass yield by 15%, 18%, 47%, and 27%, respectively, when compared to the control. No statistical difference was observed in the yield of 2061 kg ha^{-1} between applying compost by microdosing at 2.5 t ha^{-1} and broadcasting at 5 t ha^{-1}. It can be inferred that the application of compost by microdosing makes it possible to achieve a 100% fertilized surface compared to broadcasting, with a nitrogen use efficiency of more than 55%. The application of compost by microdosing at 2.5 t ha^{-1} resulted in an economic gain of 334,800 XOF ha^{-1}, which was 27% higher than that obtained with the application of compost by broadcasting at 5 t ha^{-1}. Conversely, the contribution to the improvement of soil nitrogen stock varied from 12–20% with a microdose of 2.5 t ha^{-1} compared to 100% for broadcasting compost at 5 t ha^{-1} per application. Therefore, the availability of cotton stems in the southern zone of Mali presents an opportunity for farmers to implement compost microdose technology to double the fertilized area and improve sorghum productivity.

Keywords: organic manure; broadcasting; yield; fertilizer; Sahel

Citation: Guindo, M.; Traoré, B.; Birhanu, B.Z.; Coulibaly, A.; Tabo, R. Microdosing of Compost for Sustainable Production of Improved Sorghum in Southern Mali. *Agronomy* **2022**, *12*, 1480. https://doi.org/10.3390/agronomy12061480

Academic Editors: Umberto Anastasi and Aurelio Scavo

Received: 12 May 2022
Accepted: 14 June 2022
Published: 20 June 2022

1. Introduction

One of the main constraints of agriculture in Mali remains the depletion of soil organic matter [1]. This situation is aggravated by the practice of continuous land cultivation without sufficient nutrient additions due to low access to mineral fertilizers that are costly [2] but also demographic pressures on land [3] that lead to low crop yields [4]. The bi-annual recommendation of 5 t ha^{-1} of manure [5] is not available to all farmers to cover farm needs [6]. About 68% of Malian farmers are poor smallholders who lack the financial resources to purchase mineral fertilizers [7]. Thus, composting using cereal residues in combination with cattle dung was developed to fill the nutrient gaps and ensure sustainable agricultural production [8]. This practice contributed to the development of cotton sectors in Mali and the Sahel countries [5,9].

Unfortunately, composting is being limited by several factors. For instance, cereal biomass is becoming increasingly insufficient to be used for both animal feed and compost production following increases in human populations that demand plant and animal-based food [10,11].

The southern zone of Mali is characterized by a cotton-based cropping system that occupies 32–41% of cropland [6]. With an average annual production of about 760,000 t of seed cotton [12], there is a significant amount of biomass produced, which remains an opportunity for compost production.

The farmer practice of manure use in the southern zone of Mali is characterized by targeted application, taking into account soil fertility status, access to manure, and the rotation system. Manure is usually applied to poor fields in order to restore fertility lost by crop exports. Thus, the quantities contributed per hectare vary according to the crop and soil types [1]. Faced with the need to remain on the same fields, farmers who are resource-endowed prefer to fertilize cotton and maize and hope that the millet and sorghum crops that follow will benefit from the residual effects of fertilization [11].

Owing to the low availability of manure and limited access to agricultural inputs, including mineral fertilizers, resulting in low crop yield [13], fertilization technology by microdosing or localized plant-hole fertilization with low doses of fertilizer has been developed as an alternative [14,15]. This technology has been shown to improve the productivity of different soils and crops [15]. However, most research on microdosing has largely focused on the application of mineral fertilizers [16]. A previous study conducted in Burkina Faso showed that applying cotton stem compost at a dose of 6 t ha^{-1} combined with mineral fertilizers significantly improves maize yields [8].

In this study, we hypothesized that cotton stem compost applied in microdoses at a rate of 2.5 t ha^{-1} can achieve significantly greater agronomic performance compared to the commonly used broadcasting of 5 t ha^{-1}. Compost application in microdoses can also promote the sustainability of the sorghum production system. Therefore, the objective of this study was to evaluate the effectiveness of compost combined with mineral fertilizer applied in microdoses to sorghum crop. Specifically, the study aimed to evaluate the effectiveness of the half-dose (2.5 t ha^{-1}) of compost by microdosing in comparison with the recommended dose of 5 t ha^{-1} by the broadcasting application method and assess the effects on the indicators of sustainable intensification.

2. Materials and Methods

2.1. Study Sites

The study was carried out in the region of Koutiala (Figure 1) and specifically in the villages of N'Golonianasso (12°43′07″ N 5°69′42″ W), Sirakélé (12°30′50″ N 5°28′40″ W), and Zansoni (12°36′33″ N 5°34′3″ W), which are all located in the most long-standing cotton production area in southern Mali (Figure 1). Cotton is grown on around 30% of the land. The study was carried out during the 2019 and 2020 growing seasons. The highest rainfall amounts were recorded between July and September. From the beginning of July to the second half of September, the rainfall rarely stopped for more than five days. The annual cumulative rainfall recorded in 2019 was 808 mm in N'golonianasso compared to 650 mm in Sirakélé and 831 mm in Zansoni, while in 2020, it was 1019 mm in N'Golonianasso, 890 mm in Sirakélé, and 960 mm in Zansoni. Comparatively, 2020 was the wettest year.

The temperatures ranged from 22 °C to 35 °C. Agricultural production is mainly focused on cotton, maize, sorghum, and small millet cultivation. Plains and rocky highlands dominate the study area. The soils are of the tropical ferruginous type with sandy-loam to sandy-loamy textures, high acidity (pH < 5.6), and low organic matter content.

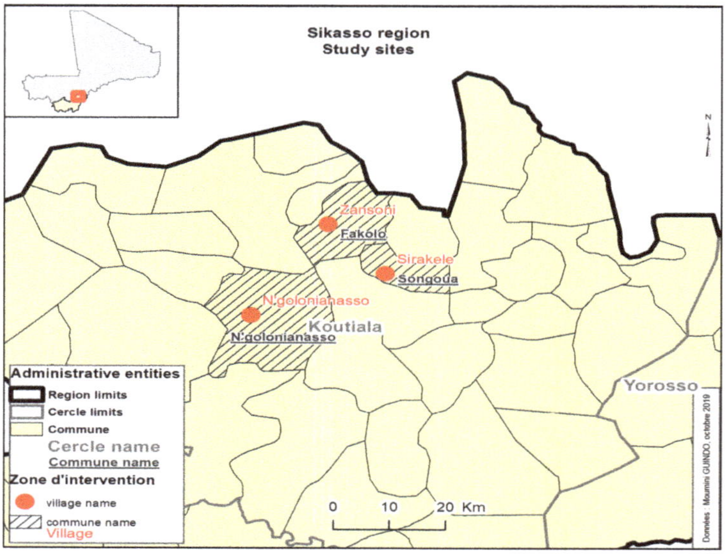

Figure 1. Study sites.

To determine the soil fertility status, composite soil samples were collected prior t the implementation of the trials. The soil samples were packaged and sent to the soi laboratory of Sadoré (ICRISAT-Niamey) for physico-chemical analyses. In each treatmen soil sampling was carried out diagonally towards both ends and in the middle, i.e., a 5 sampling points and a depth of 0–20 cm. The parameters analyzed included soil pH [17 % P (Bray-I) [18], total % nitrogen, and % soil organic carbon (SOC) [19]. The soil % K wa extracted with 1 M NH4OACc solution [20] and determined by flame photometry, whil the soil granulometry was determined by the sedimentation method. Soil physico-chemica properties are shown in Table 1.

Table 1. Soil physico-chemical properties of the study sites.

Soil Characteristics	Study Sites		
	Sirakélé	**N'golonianasso**	**Zansoni**
pH	4.53 ± 1.04	4.55 ± 0.87	4.33 ± 1.00
Total nitrogen (% N)	0.19 ± 0.03	0.28 ± 0.04	0.18 ± 0.03
Assimilable P. (mg/kg)	10.09 ± 3.87	6.84 ± 7.82	13.71 ± 15.86
Exchangeable K. (Cmol+/kg)	0.14 ± 0.05	0.12 ± 0.02	0.18 ± 0.04
Organic matter (% OM)	0.35 ± 0.06	0.61 ± 0.15	0.34 ± 0.06
Clay (% <0.002 mm)	13 ± 7.65	11.83 ± 5.49	7.60 ± 3.65
Fine silt (% 0.05–0.002 mm)	13.40 ± 5.98	41.17 ± 10.19	21.40 ± 12.58
Sand (% >0.05 mm)	73.60 ± 9.13	47.00 ± 13.04	71.00 ± 12.45

The soils used in this investigation had low pH with values ranging from 4.33–4.5 indicating high acidity (Table 1). The organic matter content (% OM) of these soils was als low (<1.5%), as was the total nitrogen content (0.18–0.28% N). The phosphorus conten of soils was low (<15 mg kg^{-1}), as was exchangeable potassium (<0.2 Cmol+/kg^{-1}). Th texture of the soils varied from one village to another and ranged from sandy-silty t loamy-sandy.

2.2. Compost Production

The material used for composting consisted of cotton stems, cattle manure, wood ash, millet glumes/glumelles, and dead leaves of *Pennesetum pediselatum*. In the study area, this grass is known as "*N'Golo*" in the Bambara language. For the compost microdose experiment, the plant material was sorghum (*Sorghum bicolor* [L.] Moench) with the improved variety "Soubatimi" [21]. This variety is dual-purpose with high yield potential ranging from 2.5–3 t ha^{-1} for grain and 10 t ha^{-1} for fodder.

A total of 30 farmers from the three study villages were trained in composting techniques using cotton stems. Each farmer made a compost pile consisting of 500 kg of cotton stems combined with 100 kg of cattle manure and 25 kg each of millet glumes/glumelles, dead leaves of *Pennesetum pediselatum*, and wood ash. The dimensions of the piles were 2 m long, 2 m wide, and 1 m high, with 10 layers of 10 cm. The cotton stems were manually cut into small pieces of about 5–10 cm in length using a cutter. After constituting the compost, the piles were covered with black plastic and tarpaulin. A shed was erected to shelter the compost from the sun. Interventions after composting consisted of watering per week, i.e., 11 waterings in total, and turning of the compost pile on the 45th day. During the turning, large stems and the longer ones were further chopped. On the 90th day, which was the expected date of maturity, the compost was harvested and quantified in the fresh state before being dried in the shade.

2.3. Experimental Set-Up and Compost Application on Farmer Fields

The compost was applied in blocks of 4 treatments in each of the 30 farmer fields and 2 controls. Each farmer field was considered a repetition. The seeding density used was 0.30 m between plants and 0.75 m between rows. A total of 5 sorghum seeds were sown per hole, and 3 seedlings were left after thinning. The agronomic parameters measured included the plant growth rate, measured on the same plants throughout the cycle (seeding-spruce); the height of the plants at harvest; planting density; and grain and biomass yields.

The compost produced was used in the experiment as an organic microdose at a rate of 2.5 t ha^{-1} by placing it at a depth of 7–10 cm, and top dressing by broadcasting was applied at a dose of 5 t ha^{-1} for farmer practice (Table 2). The experiment was carried out in 2019 and 2020, and the carryover effect was evaluated in 2020 from the trial of the previous year.

Table 2. Nutrient application per treatment.

Treatments	N (kg ha^{-1})	P (kg ha^{-1})	K (kg ha^{-1})
Control	0	0	0
Control (DAP 100 kg ha^{-1})	18	46	0
CP (farmer compost at 5 t ha^{-1} + 100 kg ha^{-1} DAP)	55.5	53	29
CP (farmer compost at 2.5 t ha^{-1} + 100 kg ha^{-1} DAP)	36.75	49.5	145
CPA (cotton stem compost at 2. 5 t ha^{-1} +100 kg ha^{-1} DAP)	51.75	51.25	35
CPA (cotton stem compost at 5 t ha^{-1} + 100 kg ha^{-1} DAP)	85.5	56.5	70

DAP = Diammonium Phosphate (18-46-0).

2.4. Nitrogen Use Efficiency and Sustainable Intensification

To assess the increase in grain yield due to nitrogen application (kg grain/kg N), we calculated nitrogen use efficiency using the following formula [22]:

$$\text{NitrogenUseEfficiency} = \frac{\text{Yieldwithnitrogen}\left(\text{kgha}^{-1}\right) - \text{Yieldwithoutnitrogen}\left(\text{kgha}^{-1}\right)}{\text{Amountofnitrogenapplied}\left(\text{kgha}^{-1}\right)}$$

The performance of treatments in terms of system sustainability was assessed through the analysis of indicators covering 5 domains, namely, productivity, profitability, the environment, social, and human conditions [23]. Productivity was determined from grain sorghum yields, while the gross margin (XOF ha^{-1}) was assessed by the difference in loads (compost, DAP fertilizer, seed, and plowing) and products (grains and biomass). As for the environment, the assessment of the partial balance ($\sum$Inputs − $\sum$Outputs, kg ha^{-1}) made it possible to determine the level of nitrogen deficiency in the soil. With regard to dietary energy intake, the indicator chosen was the amount of protein generated per treatment in kilocalories per hectare. A survey of 30 households in the three villages determined the labor needs for the implementation of each treatment or even the need for collective action.

Compost samples were also taken and analyzed to determine the nitrogen composition, C/N ratio, moisture content, and organic matter (% OM). The compost samples were collected from the matured compost heaps at three levels (on the surface, in the middle, and at the bottom) to constitute a composite sample per heap. A total of 500 g of fresh compost for each pile was dried in the oven at 105 °C for 24 h before being submitted for analysis at the Sotuba Sol-Eau-Plante laboratory (Bamako). Organic matter was determined by the loss-on-ignition method, and nitrogen was measured by the Kjeldahl digestion method.

Analysis of variance (ANOVA) was conducted to test the effect of the compost type and application method as fixed effects on the measured sorghum productivity indicators. The Student–Newman–Keuls test was used for means comparison when significant at 5% ($p < 0.05$). Statistical analyses were conducted using Genstat software, 18th edition.

3. Results

3.1. Quantity and Characteristics of Compost

Compost quantities ranged from 595–623 kg across villages, with an average of 613.7 kg per farmer (Table 3). In terms of quality, the nitrogen composition was 1.3–1.5%, and that of organic matter varied from 39–48%.

Table 3. Quantity and characteristics of compost.

Compost	Study Sites		
	Sirakélé	N'golonianasso	Zansoni
Quantity (kg)	616 ± 54	623 ± 27	595 ± 49
Total nitrogen (% N)	1.46 ± 0.18	1.30 ± 0.13	1.37 ± 0.04
C/N ratio	18.68 ± 6.16	20.40 ± 1.45	16.53 ± 0.99
Organic matter (% OM)	45.92 ± 9.75	47.83 ± 10.26	38.93 ± 3.34

3.2. Effect of Compost Application on Sorghum Growth Rate

In 2019, except for the control treatment at 60 days after planting, the sorghum growth rate was similar for all treatments (Figure 2), while in 2020 and in the carryover effect experiment, there was a significant difference between treatments. For instance, the treatment that received improved compost through microdose application (CPA_2.5 t ha^{-1} + 100 kg ha^{-1} DAP) or by broadcasting (CPA_5 t ha^{-1} + 100 kg ha^{-1} DAP) obtained similar daily growth rates of 0.69 cm and 0.96 cm, respectively, from 0–15 days after sowing (DAS) and 15–30 DAS. This growth rate was significantly increased by 8–10% compared to farmer compost treatments applied in microdoses (CP_2.5 t ha^{-1} + 100 kg ha^{-1} DAP) or by broadcasting (CP_5 t ha^{-1} + 100 kg ha^{-1} DAP) and more than 22% for the treatment with 100 kg ha^{-1} DAP and the control.

A daily growth rate of 1.55 cm was obtained between 30–45 DAS for all compost treatments (improved or farmer practice) and application methods (microdosing or broadcasting), while treatment with 100 kg ha^{-1} DAP and the control showed the lowest daily growth rates, which were less than 12% and 20%, respectively. This trend was similar to that observed for the period between 60–75 DAS, with daily growth rates of 3.72 cm and 2.71 cm for the period between 75–90 DAS. The growth rates for the DAP treatment (100 kg ha^{-1}) and control were less than 13% and 33%, respectively, during 60–75 DAS and less than 11% and 27% for the period of 75–90 DAS. A general decrease in the growth rate was observed from 75 DAS for all treatments.

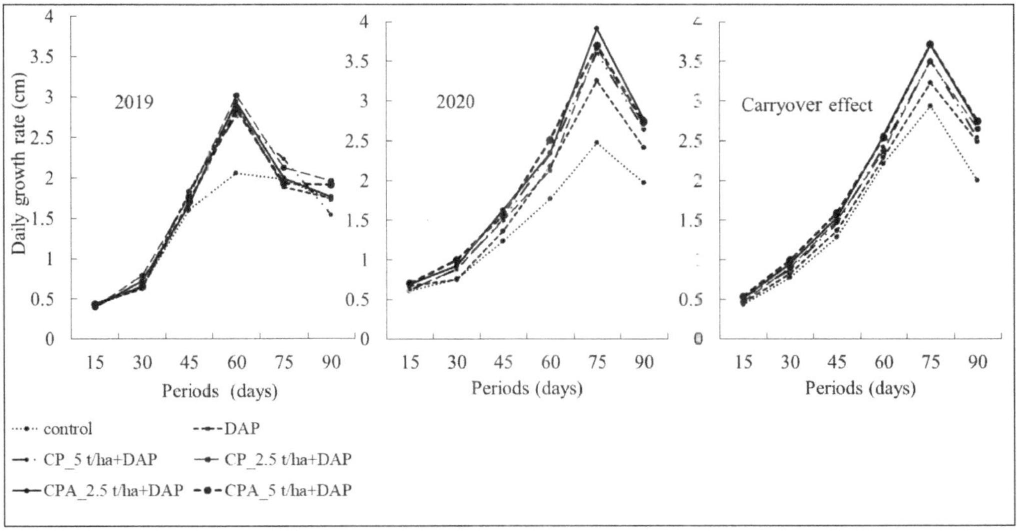

Figure 2. Growth rate for sorghum plant per treatment.

3.3. Effects of Compost on Plant Height, Planting Density, Grain, and Biomass

3.3.1. Direct Effect of Compost Application

The analysis of variance showed highly significant differences ($p < 0.001$) between treatments for plant height, grain yield, and biomass yield (Table 4). Except for the control, all treatments obtained an average height of 245.4 cm at harvest, with an average biomass yield of 11,168 kg ha^{-1}. The control treatment had the shortest plants with a height of less than 16% compared to the tallest plants and also produced the lowest biomass of less than 24% compared to other treatments.

Table 4. Effect of compost application on agronomic parameters of sorghum.

Treatment Effect (Mean of 2019 and 2020)	Height (cm)	Planting Density	Grain Yield (kg ha^{-1})	Biomass Yield (kg ha^{-1})
Control	205.2	80,346	1160	8509
DAP 100 kg ha^{-1}	233.7	84,769	1716	10,408
CP (farmer compost at 5 t ha^{-1} + 100 kg ha^{-1} DAP)	239.1	87,860	1916	11,069
CP (farmer compost at 2.5 t ha^{-1} + 100 kg ha^{-1} DAP)	250.7	87 287	2043	11,787
CPA (improved compost with cotton stems at 2.5 t ha^{-1} + 100 kg ha^{-1} DAP)	248.8	90 168	2320	11,649
CPA (improved compost with cotton stems at 5 t ha^{-1} + 100 kg ha^{-1} DAP)	254.6	82 299	1963	10,926
Mean	238.7	85 453	1853	10,725
p-value	<0.001	0.409	<0.001	<0.001
S.E.D	7.66	5157.9	169	628
% CV	14.5	27.4	41.3	26.5
Year effect				
2019	242.5	87.087	1845	10,384
2020	226.9	80.389	1879	11,780
Mean	238.7	85.453	1853	10,725
p-value	0.006	0.354	0.783	0.002
S.E.D	5.58	3455.1	124.7	439.9
% CV	15.8	27.2	45.3	27.6

For planting density, no statistically significant differences were observed between treatments ($p = 0.409$). On average, there were 85,453 plants per hectare with a coefficient of variation of 27.4 (Table 4). The results on the grain yield showed significant differences ($p < 0.001$) between treatments. The highest yield of 2061 kg ha^{-1} was obtained from improved compost treatments (improved or farmer practice) with a microdose application of 2.5 t ha^{-1} + 100 kg ha^{-1} DAP or with the broadcasting of 5 t ha^{-1} + 100 kg ha^{-1} DAP

(Table 4). The 100 kg ha^{-1} DAP treatment yielded less than 17% compared to the highes yield. The lowest yield of 1160 kg ha^{-1} was obtained with the control.

Over the years, the tallest (242.5 cm) sorghum plants obtained in 2019 significantl decreased by 7% in 2020 ($p < 0.001$) (Table 4). The average grain yield of 1853 kg ha^{-1} wa not statistically different ($p = 0.783$) between the two years. However, the biomass yield c 10,384 kg ha^{-1} in 2019 and 11,780 kg ha^{-1} in 2020 differed significantly ($p < 0.001$).

3.3.2. Carryover Effect of Compost Application

In the carryover effect experiment, the results showed significant differences ($p < 0.001$) be tween treatments regarding plant height at harvest, grain yield, and biomass yield. The applica tion of improved compost by the microdose technique (CPA_2.5 t ha^{-1} + 100 kg ha^{-1} DAP) c broadcasting (CPA_5 t ha^{-1} + 100 kg ha^{-1} DAP) resulted in taller plants (238.5 cm) (Table 5 The plants were significantly higher (by 3%, 6%, and 9.5%, respectively) than those receivin farmer compost treatments in microdoses (CP_2.5 t ha^{-1} + 100 kg ha^{-1} DAP), broadcas ing (CP_5 t ha^{-1} + 100 kg ha^{-1} DAP), and 100 kg ha^{-1} DAP application. The highest grai yield of 2210 kg ha^{-1} was obtained with the improved compost microdose applicatio (CPA_2.5 t ha^{-1} + 100 kg ha^{-1} DAP), which was 27% higher than that obtained by broadcas ing (CPA_5 t ha^{-1} + 100 kg ha^{-1} DAP). The lowest yield of 1131 kg ha^{-1} was obtained wit the control treatment, corresponding to less than 49% compared to the highest yield with th microdose of the improved compost. The highest biomass amount of 12,415 kg ha^{-1} wa also obtained with the improved compost microdose (CPA_2.5 t ha^{-1} + 100 kg ha^{-1} DAP farmer compost (CP_2.5 t ha^{-1} +100 kg ha^{-1} DAP), and treatment with DAP (100 kg ha^{-1} However, the application of these composts (improved or farmer practice) at a dose c 5 t ha^{-1} + 100 kg ha^{-1} DAP produced 8% less biomass than that of the microdose. Th lowest biomass yield (10,227 kg ha^{-1}) was obtained with the control.

Table 5. Carryover effect of compost application on sorghum.

Carryover Effect of Treatments in 2020	Height (cm)	Planting Density	Grain Yield (kg ha^{-1})	Biomass Yield (kg ha^{-1})
Control	198.62	73,259	1131	10,227
DAP 100 kg ha^{-1}	215.7	74,074	1645	11,829
CP (farmer compost at 5 t ha^{-1} + 100 kg ha^{-1} DAP)	*225.07*	*72,926*	*1603*	11,361
CP (farmer compost at 2.5 t ha^{-1} + 100 kg ha^{-1} DAP)	232.26	*73,037*	1941	12,510
CPA (improved compost with cotton stems at 2.5 t ha^{-1} + 100 kg ha^{-1} DAP)	239.85	74,593	2209	12,905
CPA (improved compost with cotton stems at 5 t ha^{-1} + 100 kg ha^{-1} DAP)	237.11	76,444	1640	11,506
Mean	224.77	74,056	1695	11,723
p-value	<0.001	0.882	<0.001	<0.001
S.E.D	1.67	3196.40	199.7	519
% CV	2.9	16.7	23.7	17.1

3.4. *Nitrogen Use Efficiency*

Under experimental conditions, sorghum cultivation without nitrogen applicatio (control) resulted in a mean yield of 1160 kg ha^{-1}. Nitrogen use efficiency varied sig nificantly ($p < 0.001$) depending on the treatment. The highest nitrogen use efficienc was obtained with treatment with DAP (18-46-0) at a dose of 100 kg ha^{-1} and with th improved compost microdose (CPA 2.5 t ha^{-1} + 100 kg ha^{-1} DAP) (Figure 3). One kilogran of nitrogen resulted in 31 kg and 22 kg of sorghum grain with DAP treatment and compos with microdose application (CPA 2.5 t ha^{-1} + 100 kg ha^{-1} DAP), respectively. Broadcastin treatments with CPA_5 t ha^{-1} + 100 kg ha^{-1} DAP and CP_5 t ha^{-1} + 100 kg ha^{-1} DA resulted in a nitrogen use efficiency of 8 kg of sorghum grain per unit of nitrogen.

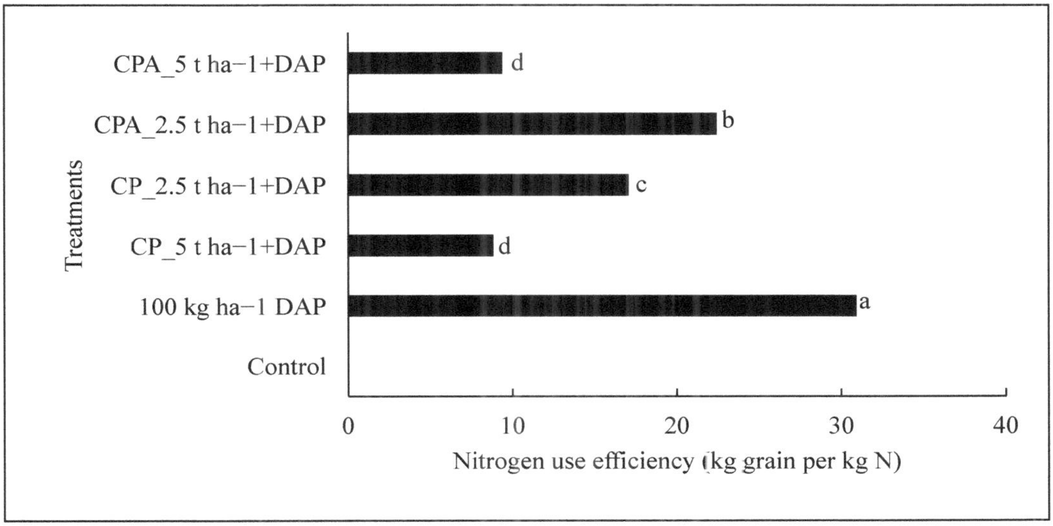

Figure 3. Nitrogen use efficiency (sorghum grain per kilogram of nitrogen used (kg per kg N).

3.5. Indicators of Sustainable Intensification

3.5.1. Compost Contribution to Productivity

The use of the improved compost with the microdose application technique (CPA_2.5 t ha^{-1} + 100 kg ha^{-1} DAP) resulted in the largest yield of 2320 kg ha^{-1}, corresponding to a maximum contribution of 100% to the productivity requirement of the farm (Figure 4), while with the farmer compost (CP_2.5 t ha^{-1} + 100 kg ha^{-1} DAP), the contribution was 90%, which decreased to 80% with the application of farmer compost at 5 t ha^{-1} + 100 kg ha^{-1} DAP. For DAP treatment at 100 kg ha^{-1} and the control, the contribution decreased to 75% and 50%, respectively, compared to that of the improved compost microdose (Figure 4).

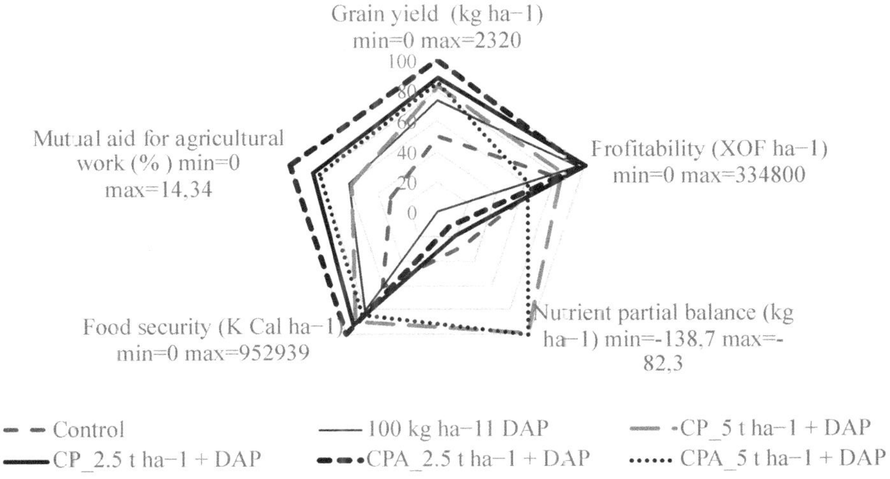

Figure 4. Sustainable intensification for compost application under sorghum production system in southern Mali.

3.5.2. Cost-Effectiveness of the Compost

With 334,800 XOF/ha, microdosing with farmer compost (CP 2.5 t ha^{-1}) or improved compost (CPA 2.5 t ha^{-1}) generated the highest gross margin. This profitability is comparable to that of the 100 kg ha^{-1} DAP treatment (Figure 4). These technologies contributed 95–100% to the economic profitability of the production system. However, this profitability dropped to 80% with the application of farmer compost by the broadcasting system (CP_5 t ha^{-1} + 100 kg ha^{-1} DAP) and the control. The contribution of improved compost in broadcasting (CPA_5 t ha^{-1} + 100 kg ha^{-1} DAP) was the least profitable, with a contribution of about 60% of economic profitability.

3.5.3. Effect of Compost Application on the Environment

Despite the existence of significant differences between treatments ($p < 0.001$), the partial nitrogen balance was negative in all treatments, with values ranging from −82.3 to −138.7 kg ha^{-1}. The broadcasting of compost (improved and farmer practice) at 5 t ha^{-1} + 100 kg ha^{-1} DAP had the largest contribution (100%) to the improvement of the nitrogen balance of soil (Figure 4). The unfertilized control plots and the microdosing treatments of improved compost (CPA_2.5 t ha^{-1} + 100 kg ha^{-1} DAP) and farmer practice (CP_2.5 t ha^{-1} + 100 kg ha^{-1} DAP) still contributed to the improvement of the balance to 30%, 20%, and 10%, respectively. As for the sole use of DAP fertilizer at 100 kg ha^{-1}, it had absolutely no contribution to the improvement of the nitrogen balance but rather drew on the soil reserve.

3.5.4. Labor Assistance for Compost Application

For field activity, the mutual labor assistance received by families is of the order of 14.34% maximum, particularly for agricultural work. This contribution corresponds to 100% of the labor assistance needs and would be required for work on improved compost microdose plots (CPA_2.5 t ha^{-1} + 100 kg ha^{-1} DAP), which showed the highest grain and biomass yields. For the DAP treatment at 100 kg ha^{-1} and farmer compost in broadcasting (CP_5 t ha^{-1} + 100 kg ha^{-1} DAP), labor demand significantly decreased ($p = 0.0147$) to 60%.

3.5.5. Compost Contribution to Food Security

The largest amount of food energy obtained was 952,939 Kcal ha^{-1} (Figure 4) and was obtained with compost application by microdosing (CPA_2.5 t ha^{-1} + 100 kg ha^{-1} DAP). Except for the control, whose contribution to the food energy requirement was 60%, that of other treatments varied from 80% to 90% (Figure 4).

4. Discussion

4.1. Effect of Compost Application on Sorghum Growth

The findings of this study showed that the best daily growth rates of sorghum were obtained with compost application by microdosing (2.5 t ha^{-1} + 100 kg ha^{-1} DAP) and broadcasting (5 t ha^{-1} + 100 kg ha^{-1} DAP) treatment applications. This was observed from 15–90 DAS. The variability of the daily growth rate of the plants between treatments can be explained by the response rate of the sorghum variety to the nutrient contribution from compost. This situation has led to a good biomass yield [24,25]. The observed general decrease in the growth rate observed beyond 75 DAS can be linked to the slow growth rate of the stems due to the translocation of nutrients for the construction of the panicle [26]. At harvest, plants that received compost treatments (improved or farmer practice) in microdoses of 2.5 t ha^{-1} + 100 kg ha^{-1} DAP were significantly taller than those of the control. This difference in size could be due to the effect of the availability of nutrients provided by compost, owing to its mineralization, which was also favored by sufficient rainfall [27]. However, environmental factors, such as the rainfall or drought sequence, can determine the growth of internodes and the height of sorghum, thus limiting the effects of mineral inputs.

4.2. Effect of Compost Application on Sorghum Yield

The application of compost by microdosing (improved or farmer practice) at 2.5 t ha^{-1} + 100 kg ha^{-1} DAP and by broadcasting at 5 t ha^{-1} + 100 kg ha^{-1} DAP both produced an average grain sorghum yield of 2061 kg ha^{-1}. This shows that compost application by microdosing at 2.5 t ha^{-1} can produce an equal grain yield to that of broadcasting at a dose of 5 t ha^{-1}. This performance of microdose technology would be related to the concentration of nutrients in the surface part of the soil and at the level of the active surface of the root system, allowing better absorption of nutrients and water [28,29]. The lowest grain yield of 1160 kg ha^{-1} obtained with the control treatment highlights the importance of fertilizer input and soil poverty in organic matter (0.34–0.61%). Since farmers have difficulties in disposing manure [30], it would be advantageous to adopt organic input based on the application of the compost by microdosing at 2.5 t ha^{-1} to a larger area of cultivated land, compared to the broadcasting method, which uses the double dose of 5 t ha^{-1}.

Our results show that all treatments, except the control, showed the same biomass yield of 11,168 kg ha^{-1}, 24% higher than that of the control. This difference is linked, on the one hand, to the effect of the compost (improved and farmer practice) and, on the other hand, to the dual-use feature of the sorghum variety "Soubatimi", which values the best manure [31]. In retrospect, only microdose treatments (improved and farmer practice) showed the best biomass yield of 12,708 kg ha^{-1}, 10% higher than that of broadcasting. This observation could be linked to the significant effect of the mineralization of the stock of organic matter that was placed basally in the seeding holes [28,32]. Compost with microdosing at 2.5 t ha^{-1} is also critical for the development of above-ground biomass under sufficient rainfall conditions, which would be beneficial to agro-pastoralists, especially for animal fodder.

4.3. Sustainability of the Production System

In the present study, the greatest economic profitability of 334,800 XOF ha^{-1} was achieved with the application of a microdose of 2.5 t ha^{-1} + 100 kg ha^{-1} DAP of improved compost with cotton stems and farmer compost. The profitability was significantly increased by 27% compared to that obtained with the broadcasting of 5 t ha^{-1} + 100 kg ha^{-1} DAP of farmer compost. These results can be attributed to the nitrogen use efficiency of 22 kg of sorghum grain per kg of nitrogen with the application of the compost microdose at 2.5 t ha^{-1} + 100 kg ha^{-1} DAP, compared to only 8 kg of grain per kg of nitrogen with the application of 5 t ha^{-1} + 100 kg ha^{-1} DAP. The application of 56.25 g of compost per plant hole ensures a concentrated supply of nutrients to the plants, reduces losses [33,34], and improves the chemical, physical, and biological properties of soils for subsequent crops in the rotation [35,36]. Although the benefits of microdosing have been demonstrated by several studies [29,37], we found that its contribution to the soil fertility restoration, especially nitrogen (N), was less than that of the application of 5 t ha^{-1} by broadcasting. This can be explained by the absolute amount of organic matter provided per unit area. Comparatively, applying the same dose of organic matter by microdosing and broadcasting would allow the microdose to gain an additional 100% of fertilized area. In the present study, there was no excessive application of nitrogen and hence no negative impacts on the environment [38–40]. When compost is applied as top dressing, followed by a slight soil cover, it mixes with the soil and contributes more to the improvement of the nitrogen balance and the sustainability of the system. The minimal addition of 100 kg ha^{-1} DAP did not improve the nitrogen stock but replenished the nitrogen deficit in the soil since the soil was initially low in nitrogen. Under favorable rainfall conditions, this is explained by the high solubility and leaching losses, as well as volatilization losses [41–43].

One of the pillars of the sustainability of production systems is the social field, generally marked by cohesion through collective aid in labor for agricultural work. In much more productive plots, such as those receiving the compost microdose at 2.5 t ha^{-1} + 100 kg ha^{-1} DAP, the demand for labor assistance is much higher due to the intensity of work and the number of family workers subject to rural exodus [44]. To overcome this constraint of agricultural labor and working time, access to agricultural mechanization (plows, seeders, tractors, an tillers) facilitates the realization of many operations, such as plowing, sowing, weedin and transport [45]. This involvement of agricultural mechanization makes it possible t increase the cultivated area and enhance the value of the workforce for the diversificatio of other activities, generating less painful incomes.

4.4. Challenges Related to the Production and Use of Compost

In southern Mali, the extension of cropland is no longer possible, and, faced with th need for the sustainable intensification of agricultural production systems, farmers ar trying to produce manure through several alternatives (animal pile manure, garbage pile composting, etc.) to meet the needs of organic input. Given the low production of thi organic manure, the burning of crop residue in areas varies from 32% to 62%, dependin on the type of farm [6,46]. As a result, composting, which appears to be one of the potentia options thanks to its fertilizing quality, is becoming increasingly useful and represents major challenge on farms with crop residues. However, the collection and transportation o residues, as well as the availability of family labor and water, are some constraints [47].

For field fertilization, the contribution of compost at 5 t ha^{-1} every two years, a recommended by research [5], is generally out of reach for farmers to cover 100% o cultivated areas [6]. During application, the little manure applied to the field is generally no distributed evenly over all areas, thus promoting irregularity in soil fertility managemen Although the application of compost by microdosing results in the enhanced extraction o nutrients, with harvests thus depleting the soil, it nevertheless has good performance i grain and biomass yield. However, the implementation of microdosing can take time an requires much more labor, causing an additional burden that may hinder the adoption o the technology on a large scale [14].

5. Conclusions

This study evaluated the performance of the technology for applying improved an farmer compost by microdosing at 2.5 t ha^{-1} and by broadcasting at 5 t ha^{-1} in a sorghun production system. It appears that, regardless of the type of compost (farmer practic or improved), the application of compost by microdosing at 2.5 t ha^{-1} can significantl improve the sorghum growth rate, plant height, grain yield, and biomass yield compare to other treatments. The grain yield of 2061 kg ha^{-1} obtained with compost application b microdosing at 2.5 t ha^{-1} or compost broadcasting at 5 t ha^{-1} suggests the possibility o treating 100% of the fertilized field with compost microdosing at 2.5 t ha^{-1}. This gain i fertilized surface area may make it possible to overcome the problem of insufficient organi manure in the Sahel. Depending on rainfall, the biomass yield varied significantly, fron 10,384 kg ha^{-1} in 2019 to 11,780 kg ha^{-1} in 2020. Compost application with microdos technology showed a higher nitrogen use efficiency of more than 55% and an economi gain of more than 27% compared to broadcasting compost. However, the contribution o compost microdose technology to the improvement of the nitrogen stock in the soil pe unit area was less compared to the application of 5 t ha^{-1} of compost by broadcasting. I view of these results, the availability of cotton stems presents an opportunity to intensif compost production to meet the nutrient demands of crops in the Sahel.

Author Contributions: Conceptualization, M.G. and B.T.; methodology, M.G. and B.T.; softwar M.G.; validation, M.G. and B.Z.B.; formal analysis, M.G. and B.T.; investigation, M.G.; resources, B.Z. and R.T.; data curation, M.G. and B.T.; writing—original draft preparation, M.G.; writing—revie and editing, R.T and A.C.; visualization, B.T.; supervision, B.Z.B. and A.C.; project administratio B.Z.B.; funding acquisition, R.T. All authors have read and agreed to the published version o the manuscript.

Funding: This research was funded by USAID, grant number AID-BFS-G-11-00002, and the AP was funded by ICRISAT.

Informed Consent Statement: Informed consent was obtained from all subjects involved in the stud

Acknowledgments: This work was supported by the Africa Research in Sustainable Intensification for the Next Generation (Africa RISING) program in Mali. The authors are very grateful for the financial support provided by the US Agency for Development (USAID) through the International Institute of Tropical Agriculture (IITA). We thank Abdourazak Alio Moussa, colleagues from the N'Tarla research station, and NGO AMEDD for their support. We are also thankful to the anonymous reviewers, whose valuable comments improved the quality of the manuscript.

Conflicts of Interest: The authors declare no conflict of interest. The funders had no role in the design of the study; in the collection, analyses, or interpretation of data; in the writing of the manuscript; or in the decision to publish the results.

References

1. Kouyate, A.B.; Serme, I. Evaluation de l'efficacité du Phosphate Naturel de Tilemsi (PNT) sous différentes pratiques de travail du sol en zone Sahélienne du Mali. *Innov. Space Sci. Res. J.* **2021**, *34*, 845–857.
2. Kintché, K.; Guibert, H.; Sogbedji, J.; Levêque, J.; Bonfoh, B.; Tittonell, P. Long-term mineral fertiliser use and maize residue incorporation do not compensate for carbon and nutrient losses from a Ferralsol under continuous maize–cotton cropping. *Field Crop. Res.* **2015**, *184*, 192–200. [CrossRef]
3. Leyronas, S.; Boche, M.; Baudet, E. L'Afrique Redécouvre les Communs: Une Lecture des Enjeux Fonciers Ruraux Au Mali. In *Agence Française de Développement (AFD). L'Économie Africaine 2020*; La Découverte: Paris, France, 2020; ISBN 9782348057465.
4. Rufyikiri, G. De l'agriculture de Subsistance à Une Agriculture de Marché En Afrique: Une Mutation Impérative Pour Le Développement Du Continent. In *Afrique Durable 2030*; Association Africa 21: Geneva, Switzerland, 2017; Volume 2, pp. 50–68.
5. Bationo, A.; Fairhurst, T.; Giller, K.; Kelly, V.; Lunduka, R.; Mando, A.; Mapfumo, P.; Oduor, G.; Romney, D.; Vanlauwe, B.; et al. *Manuel de Gestion Intégrée de la Fertilité des Sols. Consortium Africain Pour La Santé des Sols*; Fairhurst, T., Ed.; CAB International: Oxfordshire, UK, 2015; 179p, ISBN 9781780645278.
6. Guindo, M.; Traoré, B.; Birhanu, Z.B.; Fred, K.; Coulibaly, A. Biomass and Nutrient Flow Dynamics from Households to Farm Fields: An Experimental Study in Southern Mali. *Nutr. Cycl. Agroecosyst.* **2022**, *under review*.
7. Duan, F. *Aperçu Politique pour le Développement de l'Agroforesterie au Mali*; OXFAM—Research Backgrounder: Boston, MA, USA, 2020; 88p.
8. Ouattara, A. Effets des Techniques de Compostage En Tas Sur La Qualité des Composts et Évaluation de l'Efficacité des Composts Sur La Productivité du Maïs Au Burkina Faso. Mémoire de Diplôme d'Etudes Approfondies, Science Du Sol. MSc thesis, Université Polytechnique de Bobo-Dioulasso, Bobo-Dioulasso, Burkina Faso, 2015; 62p.
9. Diarra, B.; Gian, N.; Konaté, F. Analyse de la Gestion de la Fertilité Organo- Minérale des Sols Dans les Cercles de Koulikoro et de Sikasso (Mali). *Rev. Géographie Trop. D'Environ.* **2019**, *1*, 35–44.
10. Ecofin, A.; Production de Coton du Mali. Rapport Provisoire, la Récolte de Coton a Atteint le Niveau Historique de 760,000 Tonnes en 2021/2022 (Exécutif). 2021. Available online: https://www.agenceecofin.com/Coton/2103-95887-Mali (accessed on 21 March 2022).
11. Blanchard, M.; Coulibaly, K.; Bognini, S.; Dugué, P.; Vall, É. Diversité de la Qualité des Engrais Organiques Produits par les Paysans d'Afrique de l'Ouest: Quelles Conséquences sur les Recommandations de Fumure? *Biotechnol. Agron. Soc. Environ.* **2014**, *18*, 512–523.
12. Cisse, D. Effets des Amendements à Base de Biochar sur les Paramètres Agro-Pédologiques dans une Rotation Coton-Maïs à l'Ouest du Burkina Faso. Ph.D. Thesis, Université de Liège Gembloux, Agro-Bio Tech, Gembloux, Belgium, 2022; 126p.
13. Diarisso, T.; Corbeels, M.; Andrieu, N.; Djamen, P.; Douzet, J.-M.; Tittonell, P. Soil variability and crop yield gaps in two village landscapes of Burkina Faso. *Nutr. Cycl. Agroecosyst.* **2016**, *105*, 199–216. [CrossRef]
14. Sigue, H.; Labiyi, I.; Yabi, J.; Biaou, G. Effet des composantes de la technologie microdose sur la performance économique et financière des exploitations agricoles du Kouritenga et du Zondoma au Burkina Faso. *Afr. Crop Sci. J.* **2019**, *27*, 331–349. [CrossRef]
15. Ibrahim, A.; Pasternak, D.; Fatondji, D. Impact of depth of placement of mineral fertilizer micro-dosing on growth, yield and partial nutrient balance in pearl millet cropping system in the Sahel. *J. Agric. Sci.* **2015**, *153*, 1412–1421. [CrossRef]
16. Adams, A.M.; Gillespie, A.W.; Kar, G.; Koala, S.; Ouattara, B.; Kimaro, A.A.; Bationo, A.; Akponikpe, P.B.I.; Schoenau, J.J.; Peak, D. Long term effects of reduced fertilizer rates on millet yields and soil properties in the West-African Sahel. *Nutr. Cycl. Agroecosyst.* **2016**, *106*, 17–29. [CrossRef]
17. McLean, E. Soil Ph and Lime Requirement. In *Methods of Soil Analysis*; Page, A.L., Ed.; John Wiley & Sons, Ltd.: Madison, WI, USA, 1983; pp. 199–224.
18. Bray, R.H.; Kurtz, L.T. Determination of total, organic, and available forms of phosphorus in soils. *Soil Sci.* **1945**, *59*, 39–46. [CrossRef]
19. Nelson, D.; Sommers, L. Total Carbon, Organic Carbon, and Organic Matter. In *Methods of Soil Analysis: Part 3 Chemical Methods*; Sparks, D.L., Ed.; Soil Science Society of America, American Society of Agronomy: Madison, WI, USA, 1996; pp. 961–1010.
20. Helmke, P.; Sparks, D. Lithium, Sodium, Potassium, Rubidium, and Cesium. In *Methods of Soil Analysis: Part 3 Chemical Methods*; Sparks, D.L., Ed.; Soil Science Society of America, American Society of Agronomy: Madison, WI, USA, 1996; pp. 551–574.

21. Togo, A. Effets de la Fertilisation et de la Date de Semis sur des Variétés de Sorgho à Double Usage. Mémoire, Diplôme d'Ingénieur Agronome. Master's Thesis, Institut Polytechnique Rural de Formation et de Recherche Appliquée, Koulikoroba, Mali, 2019; 50p.
22. Godinot, O. Proposition de Nouveaux Indicateurs d'Efficience d'Utilisation de l'Azote à l'Échelle du Système de Production Agricoles et du Territoire. Ph.D. Thesis, Agro Campus Ouest, Université Européenne de Bretagne, Spécialité Biologie et Agronomie, Rennes, France, 2014; 134p.
23. Hammond, J.; van Wijk, M.; Teufel, N.; Mekonnen, K.; Thorne, P. Assessing smallholder sustainable intensification in the Ethiopian highlands. *Agric. Syst.* **2021**, *194*, 103266. [CrossRef]
24. Chantereau, J.; Cruz, J.-F.; Ratnadass, A.; Trouche, G. *Le Sorgho*, 1st ed.; Agriculture Tropicale En Poche; CTA, Presses Agronomiques de Gembloux: Passage des Déportés, Belgium, 2013; ISBN 9782870161296. Available online: www.pressesagegro.be (accessed on 5 December 2013).
25. Bazongo, P.; Traore, K.; Traore, O.; Yelemou, B.; Sanon, K.B.; Kabore, S.; Hien, V.; Nacro, B.H. Influence des haies de Jatropha sur le rendement d'une culture de sorgho (*Sorghum vulgare*) dans la zone Ouest du Burkina Faso: Cas du terroir de Torokoro. *Int. J. Biol. Chem. Sci.* **2016**, *9*, 2595. [CrossRef]
26. Traore, B.; Serme, I.; Ouattara, B.; Bougma, A.; Ouedraogo, J.; Pouya, B.M.; Ouattara, K. Effet de micronutriments et de nutriments secondaires sur la production du mil et du sorgho en Zone nord soudanienne du Burkina Faso. Effect of micronutrients and secondary nutrients on millet and sorghum production in the North Sudan Zone of Burkina Faso. *Int. J. Biol. Chem. Sci.* **2021**, *15*, 1902–1913. [CrossRef]
27. Garba, O.; Mella, M.T.; Kiari, S.A.; Grema, M.H.I.; Zanguina, A. Valorisation de Glume de Mil et Balle de Riz Par Compostage: Caractérisations Physico-Chimiques des Composts. *Afr. Sci.* **2020**, *17*, 29–38.
28. Tovihoudji, P.G.; Akponikpè, P.I.; Agbossou, E.K.; Bertin, P.; Bielders, C.L. Fertilizer microdosing enhances maize yields but may exacerbate nutrient mining in maize cropping systems in northern Benin. *Field Crop. Res.* **2017**, *213*, 130–142. [CrossRef]
29. Coulibaly, A.; Woumou, K.; Aune, J.B. Sustainable Intensification of Sorghum and Pearl Millet Production by Seed Priming, Seed Treatment and Fertilizer Microdosing under Different Rainfall Regimes in Mali. *Agronomy* **2019**, *9*, 664. [CrossRef]
30. Traore, B.; Guindo, M.; Zemadim, B.; Kizito, F. *Farm Nutrient Dynamics in Southern Mali*; International Institute for Tropical Agriculture (IITA): Ibadan, Nigeria, 2022; 15p.
31. Diancoumba, M.; Worou, O.N.; Nebie, B.; Tangara, A.; Sidibe, M.; Togo, A.; Bendigngar, M.; Desmae, H.; Zemadim, B. Response of Dual-Purpose Sorghum Varieties to Fertiliser and Sowing Dates in Mali's Sudanian Zone. In Proceedings of the Tropentag 2021, Prague, Czech Republic, 14–16 September 2022; 5p.
32. Kouyate, A.B.; Ibrahim, A.; Serme, I.; Dembele, S.G. Sorghum responses to different forms of Tilemsi rock phosphate combined with soluble fertilizers in a low-input production system in Mali. *Int. J. Biol. Chem. Sci.* **2021**, *14*, 3285–3296. [CrossRef]
33. Sissoko, P.; Berti, F.; Gry, S.; Lebailly, P. Effets de l'adoption de la technique du microdosage d'engrais sur la disponibilite et l'accessibilite céréalière des exploitations agricoles à base de mil et de sorgho au Mali. *Agron. Afr.* **2018**, *30*, 193–204.
34. Rosy, C. Identification des Facteurs Environnementaux et Agronomiques Affectant la Réponse à la Fertilisation Microdose dans le Sud du Burkina Faso. Ph.D. Thesis, Faculté des Bioingénieurs, Université Catholique de Louvain, Ottignies-Louvain-la-Neuve, Belgium, 2019; 104p.
35. Doumbia, S.; Dembele, S.G.; Sissoko, F.; Samake, O.; Sousa, F.; Cicek, H.; Adamtey, N.; Fliessbach, A. Evaluation de la fertilité des sols et les rendements de cotonnier, maïs et sorgho à *Gliricidia sepium* (Jacq.) Kunth ex. Walp. *Int. J. Biol. Chem. Sci.* **2020**, *14*, 2583–2598. [CrossRef]
36. Traore, A.; Yameogo, L.P.; Olivier, D.; Traore, K.; Bazongo, P.; Traore, O. Effets Comparés Du Tourteau de Neem [*Azadirachta Indica* (A. Juss)] et du Compost sur le Rendement du Sorgho [*Sorghum Bicolor* (L. Moench)] en Zone Sud-Soudanienne du Burkina Faso. *J. Appl. Biosci.* **2021**, *163*, 16834–16845. [CrossRef]
37. Camara, B.S.; Camara, F.; Berthe, A.; Oswald, A. Micro-Dosing of Fertilizer—A Technology for Farmers' Needs and Resources. *Int. J. AgriSci.* **2013**, *3*, 387–399.
38. Audouin, E.; Vayssières, J.; Odru, M.; Masse, D.; Dorégo, S.; Delaunay, V.; Lecomte, P. Réintroduire l'elevage pour accroitre la durabilite des terroirs villageois d'Afrique de l'ouest: Le cas du bassin arachidier au Sénégal. In *Les Sociétés Rurales Face aux Changements Climatiques et Environnementaux en Afrique de L'Ouest*; IRD: Marseille, France, 2015; pp. 403–427.
39. Wu, P.; Liu, F.; Li, H.; Cai, T.; Zhang, P.; Jia, Z. Suitable fertilizer application depth can increase nitrogen use efficiency and maize yield by reducing gaseous nitrogen losses. *Sci. Total Environ.* **2021**, *781*, 146787. [CrossRef]
40. Godinot, O.; Foray, S.; Lemosquet, S.; Delaby, L.; Édouard, N. De l'animal au territoire, regards sur l'efficience de l'azote dans les systèmes bovins laitiers. *INRAE Prod. Anim.* **2022**, *35*, 43–60. [CrossRef]
41. Maniriho, A. Productivité, Efficience et Profitabilité des Petites Exploitations Agricoles Dans La Région des Sols de Laves au Rwanda. Ph.D. Thesis, Université de Liège, Liège, Belgium, 2021; 241p.
42. Traore, A.; Bandaogo, A.; Yameogo, P.L.; Anadi, K.R.; Traore, K.; Bazongo, P.; Traore, O. Effects of Zaï Hole Sizing on Maize Yield in the South Sudan Zone of Burkina Faso. *Int. J. Innov. Appl. Stud.* **2021**, *33*, 435–442.
43. Carton, S.; Levavasseur, F. *Performances Agronomiques et Environnementales de la Méthanisation Agricole dans un Contexte de Grandes Cultures Céréalières (sans Élevage) et Recommandations de Bonnes Pratiques*; Rapport Étude Commandée par le Ministère de l'Agriculture et de l'Alimentation; AgroParisTech; INRAE: Paris, France, 2022; 87p.
44. Rangé, C.C. *Structures Agraires et Accès des Jeunes à la Terre—Revue de Littérature*; Hal-01870605; HAL: Lyon, France, 2018; 29p.

5. Side, C.S.; Havard, M. Développer Durablement la Mécanisation pour Améliorer la Productivité de l'agriculture Familiale en Afrique Subsaharienne. *Int. J. Adv. Stud. Res. Afr.* **2015**, *6*, 34–43.
6. Sissoko, F.; Coulibaly, D.; Cisse, O.; Dugue, P. Évaluation de l'arrière effet de la culture du coton sur la production cerealiere en zone cotonniere du Mali. In Proceedings of the AGRAR-2013, 1st Conférence de la Recherche Africaine sur l'Agriculture, l'Alimentation et la Nutrition, Yamoussoukro, Ivory Coast, 4–6 June 2013; pp. 149–160.
7. Dagbenonbakin, G.D.; Djenontin, A.J.; Adjovi, N.A.; Igué, M.A.; Mensah, G.A. Production et utilisation de compost et gestion des residus de recolte. In *Fiche Technique*; Dépôt Légal No. 6529 Du 18/01/2013; Bibliothèque National (BN) du Bénin: Porto-Novo, Benin, 2013; 10p, Available online: http://www.slire.net/download/2155/ft_n_04_dagb_nonbakin_et_al_compost.pdf (accessed on 18 January 2013).

agronomy

MDPI

Article

Analysis on Heat Characteristics for Summer Maize Cropping in a Semi-Arid Region

Zhiwei Wang [1,†], Weiwei Sun [1,†], Xiaoli Liu [2], Yangyang Li [1], Brian Collins [3], Najeeb Ullah [4] and Youhong Song [1,5,

1 School of Agronomy, Anhui Agricultural University, Hefei 230036, China; zwwang@stu.ahau.edu.cn (Z.W. swwahau@163.com (W.S.); liyangyang@ahau.edu.cn (Y.L.)
2 School of Engineering, Anhui Agricultural University, Hefei 230036, China; xl.liu@ahau.edu.cn
3 College of Science and Engineering, James Cook University, Townsville, QLD 4811, Australia; brian.collins@jcu.edu.au
4 Faculty of Science, Universiti Brunei Darussalam, Jalan Tungku Link, Gadong BE1410, Brunei; n.ullah@uq.edu.au
5 Centre for Crop Science, Queensland Alliance for Agriculture and Food Innovation, The University of Queensland, Brisbane, QLD 4350, Australia
* Correspondence: y.song@ahau.edu.cn
† These authors contributed equally to this work.

Abstract: Heat stress during flowering is a critical limitation for summer maize production. Howeve the incidence of heat varies with years and locations, and it poses a great risk to successful maiz reproduction and kernel setting. Therefore, it is essential to provide a sound quantification of hea occurrence in relation to maize growth and development. Here, we analyzed the characteristic of heat occurrence based on climate data for over 60 years on Huaibei Plain, China. The effectiv accumulated temperature showed a slight interannual variation. The average maximum temperatur ($\overline{T}_{max}$) during flowering was 32 °C–33 °C, which was approximately 2 °C higher than that ove the whole growing season. The probability (P) for the daily T_{max} > 33 °C during flowering wa closer to 50% and this maximum temperature ranged between 33 °C and 37 °C. The five level from normal to extreme heat for T_{max} were defined. Across the six studied sites, the mild level hea stress accounted for most of incidents (P, 25–50%), followed by moderate (P, 13–25%) and sever (P, 0.5–13%), and the minimum for extreme heat stress (P, 0.5%). Four phases bracketing flowerin during maize development were given, i.e., 1 week prior to anthesis, 1 week during anthesis, 1 wee for anthesis-silking, and 1week post silking. There was a greater probability for heat stress inciden from anthesis to silking compared to the other developmental stages. Additionally, maize grain yiel slightly increased with the increase in T_{max} to 33 °C, but it declined as T_{max} surpassed 33 °C. I conclusion, the pattern and characteristics of heat stress were quantified bracketing maize flowerin These findings assist to advise summer maize cropping strategies on the semi-arid and semi-humi Huaibei Plain, China or similar climate and cropping regions.

Keywords: *Zea mays* L.; bracketing flowering; heat stress incidents; cropping risk

Citation: Wang, Z.; Sun, W.; Liu, X.; Li, Y.; Collins, B.; Ullah, N.; Song, Y. Analysis on Heat Characteristics for Summer Maize Cropping in a Semi-Arid Region. *Agronomy* **2022**, *12*, 1435. https://doi.org/10.3390/agronomy12061435

Academic Editors: Umberto Anastasi and Aurelio Scavo

Received: 9 May 2022
Accepted: 11 June 2022
Published: 15 June 2022

1. Introduction

Maize (*Zea mays* L.), as a staple food, plays an important role in securing food securit and is also one of the most important cereal crops as a source of feed and industrial rav material for humans and animals [1–3]. However, maize production is subject to increasin risk of heat stress because of climate change [4]. The average temperature has risen b 0.46 °C from the preindustrial period to the 1971−2000 period [5], and the increasing tren of temperature may continue during the 21st century [6,7]. In addition to the increase global surface temperature, the occurrence frequency and intensity of heat stress are als increasing significantly [8,9]. It is reported that a short duration of heat stress during th critical flowering stage can lead to a huge yield loss caused by heat-induced reproductiv failure [10]. Thus, heat stress during flowering has become one of the major meteorological disasters for maize cropping [6,11].

Heat stress during the flowering interferes in a series of events such as tasseling, pollen shedding, silking, pollination, pollen germination, pollen tube growth, and fertilization [12–14]. It is reported that in the pre-anthesis phase (7 days before pollination) heat accelerates tasseling and heat during anthesis phase reduces pollen shedding time, impairs pollen morphology, and induces pollen sterility [15–17]. In addition, high temperature at the silking phase inhibits anther dehiscence [18,19], pollen germination, and pollen tube extension [20]. Furthermore, post-silking (15 days) heat hinders fertilization and kernel development [20,21], which finally leads to a decrease in the seed-setting rate and grain yield [4,14]. Heat stress during different developmental phases variably affects maize reproductive growth; therefore, it is critical to quantify the occurrence characteristics of heat stress during these different phases, i.e., pre-anthesis, anthesis, silking, and post-silking.

Huaibei Plain in North Anhui, located at the south of Huang-Huai-Hai Plain in China, is one of the major areas under maize production in Anhui Province [22]. A rotation system with summer maize sown in early June and winter wheat sown in early October was commonly adopted on the Huaibei Plain over last many decades and is likely to continue for the next decade [18,23]. As such, maize flowering normally occurs in late July or early August in this region, which often coincides with heat incidence [24,25]. It is reported that the frequency of heat stress was approximately once in every 1.7 years on Huaibei Plain [24]. The frequency of moderate and severe heat stress during maize flowering was higher than 15% and 20%, respectively [24]. In addition, the number of days, timing, duration, and severity of heat events are becoming more frequent under global warming [26]. Therefore, it is necessary to quantify the characteristics of heat stress with reference to maize flowering, to provide guidelines for maize-cropping systems.

However, the characteristics of heat stress occurrence on Huaibei Plain or nearby regions are rarely analyzed. In particular, heat occurrence analysis in relation to the specific maize flowering stage is not reported. Therefore, the objective of this study is to quantify the spatiotemporal characteristics of heat stress occurrence with reference to the critical period bracketing maize flowering based on the historical meteorological data from six sites on Huaibei Plain, China.

2. Materials and Methods

2.1. Study Area and Data Sources

Six sites, i.e., Shouxian (116.78° E, 32.55° N, south on the map), Bengbu (117.38° E, 32.92° N, east on the map), Fuyang (115.82° E, 32.90° N, west on the map), Suzhou (116.98° E, 33.63° N, northeast on the map), Bozhou (115.78° E, 33.85° N, northwest on the map), and Dangshan (116.35° E, 34.42° N, north on the map) were chosen on Huaibei Plain, Anhui Province, China (Figure 1). Daily maximum temperature (T_{max}), minimum temperature (T_{min}), and mean temperature (T_{mean}) data around maize growing season and flowering stage were obtained from the China Meteorological Data Service Centre (CMDC, http://data.cma.cn/, accessed on 7 January 2016). Maize grain yield data are only available from 1998 to 2017 at the four sites, i.e., Fuyang, Bozhou, Suzhou, and Bengbu, since in earlier years no such detailed online records were available in the local agricultural bureau.

Maize-cropping characteristics are based on typical local farming practices, i.e., sowing around mid–early June, anthesis to silking around late July to early August, and harvested by the end of September or early October [23,27]. Critical developmental stages bracketing flowering usually occurred from 15 July to 15 August (32 days) in this region or nearby.

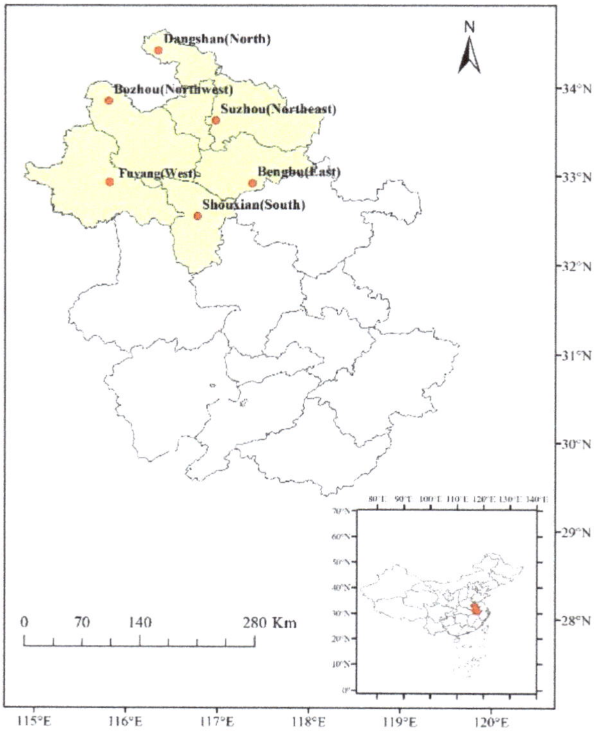

Figure 1. The locations of Shouxian (south, 116.78° E, 32.55° N), Bengbu (east, 117.38° E, 32.92° N), Fuyang (west, 115.82° E, 32.90° N), Suzhou (northeast, 116.98° E, 33.63° N), Bozhou (northwest, 115.78° E, 33.85° N), and Dangshan (north, 116.35° E, 34.42° N) on Huaibei Plain, North Anhui Province, China.

2.2. Data Analysis during Maize Growing Season and Flowering Stage over 61 Years (1957–2017)

The effective accumulated temperature referred to the sum of daily temperatures with a base temperature of 8 °C was quantified during the summer maize growing season across the six sites. The characteristics of average annual maximum temperature ($\overline{T}_{max}$), minimum temperature ($\overline{T}_{min}$), and mean temperature ($\overline{T}_{mean}$), and its coefficient of variation (CV) value were analyzed for both the entire maize growing season and for the flowering stage for the six sites.

A threshold temperature of 33 °C is reported to induce heat stress during maize flowering [28,29] and is selected for this study. First, the number of days with daily T_{max} lower than, greater than, or equal to 33 °C were counted over the 32-day period around flowering, and the probability of the number of days for T_{max} < or ≥ 33 °C were determined. Furthermore, from 33 °C to 40 °C, the occurrence of days with T_{max} higher than specific thresholds were calculated (hereafter, heat days, HD).

According to classification standards issued by the China Meteorological Administration (CMA, http://www.cma.gov.cn/en2014/, accessed on 20 March 2012), the heat stress warnings fall into three levels, i.e., (i) yellow warning signal, when $T_{max} \geq 35$ °C for two consecutive days; (ii) orange warning signal, when $T_{max} \geq 37$ °C for one day; (iii) red warning signal, when $T_{max} \geq 40$ °C for one day. Then the occurrence probability of heat stress duration days (DDs, form ≥1 d to ≥7 consecutive days) and heat stress warning were calculated. The five levels for daily maximum temperature i.e., I (normal), II (mild), III (moderate), IV (severe) and V (extreme) were classified based on T_{max}, duration days

(DDs), occurrence probability (OP, %), and warning color (WC) (Table 1). Moreover, the 32-day period (from 15 July to 15 August) around flowering was divided into four phases: pre-anthesis (A, from 15 to 24 July), anthesis (B, from 25 to 31 July), silking (C, from 1 to 7 August), and post-silking (D, from 8 to 15 August). The probability of heat stress occurrence was analyzed in each phase with different temperature thresholds. Finally, the daily T_{max} average in each phase were extracted to investigate the distribution of the heat events over years. A heat map was generated to represent the detailed characteristics of the heat events over time across various sites.

Table 1. Classification and criteria for heat stress on Huaibei Plain, China.

Level	Classification	Maximum Temperature (T_{max})	Duration Days (DDs)	Occurrence Probability (OP, %)	Warning Colour (WC)
I	Normal	T_{max} < 33 °C	DDs ≥ 1	P > 50	Green
II	Mild	33 °C ≤ T_{max} < 35 °C	DDs ≥ 1	25 < P ≤ 50	Reseda
III	Moderate	35 °C ≤ T_{max} < 37 °C	DDs ≥ 2	13 < P ≤ 25	Yellow
IV	Severe	37 °C ≤ T_{max} < 40 °C	DDs ≥ 1	0.5 < P ≤ 13	Orange
V	Extreme	T_{max} ≥ 40 °C	DDs ≥ 1	P ≤ 0.5	Red

2.3. Inverse Distance Weighting Method Quantifies the Spatial Distribution of Heat Stress Occurrence Probability

The inverse distance weighting (IDW) [30] method is one of the most widely used deterministic methods in spatial interpolation, which is characterized by high speed, convenient computation, and interpretation. IDW sums the values of nearby points multiplied by a weighting factor that is a decreasing function of distance. For this operation, ArcGIS 10.2 software (Environmental Systems Research Institute (ESRI) Inc., Redlands, CA, USA) was used to quantify spatial distribution of heat stress occurrence probability on Huaibei Plain.

2.4. Relationship between Maize Grain Yield and Average Daily T_{max} around Maize Flowering in 1998–2017

Climate change has had bidirectional effects on maize production over the past 20 years on Huang-Huai-Hai Plain (including Huaibei Plain) [31]. The optimal temperature for maize plants in daytime ranges from 22 to 32 °C [32], while the temperature above 33 °C does harm to maize production during flowering [28,29]. The relationship of maize grain yield and the average daily T_{max} in each phase, i.e., pre-anthesis from 15 to 24 July, anthesis from 25 to 31 July, silking from 1 to 7 August, and post-silking from 8 to 15 August, was conducted using the recent 20 years of data (from 1998 to 2017, except for 2003, as the record is missing because of extreme meteorological disasters at four sites, i.e., Bengbu, Fuyang, Suzhou, and Bozhou. Then, the data were analyzed with the relationship between maize grain yield and T_{max} < 33 °C and T_{max} ≥ 33 °C embedded in Microsoft Office Excel (Microsoft Excel 2016, Microsoft, Redmond, WA, USA) software.

3. Results

3.1. Interannual Variation of Effective Accumulated Temperature during Maize Growing Season

The interannual variation of effective accumulated temperature during the summer maize growing season over years across the six sites ranged from 1800 °Cd to 2200 °Cd (Figure 2), with the highest of 2215.8 °Cd at Bengbu, and the lowest of 1800.8 °Cd at Dangshan. In addition, CV values, as an index reflecting the fluctuation degree of effective accumulated temperature between years, were relatively small for the six sites and ranged from 3.85% in Shouxian to 4.16% in Suzhou, indicating that the interannual variation of effective accumulated temperature during maize growing season was relatively stable.

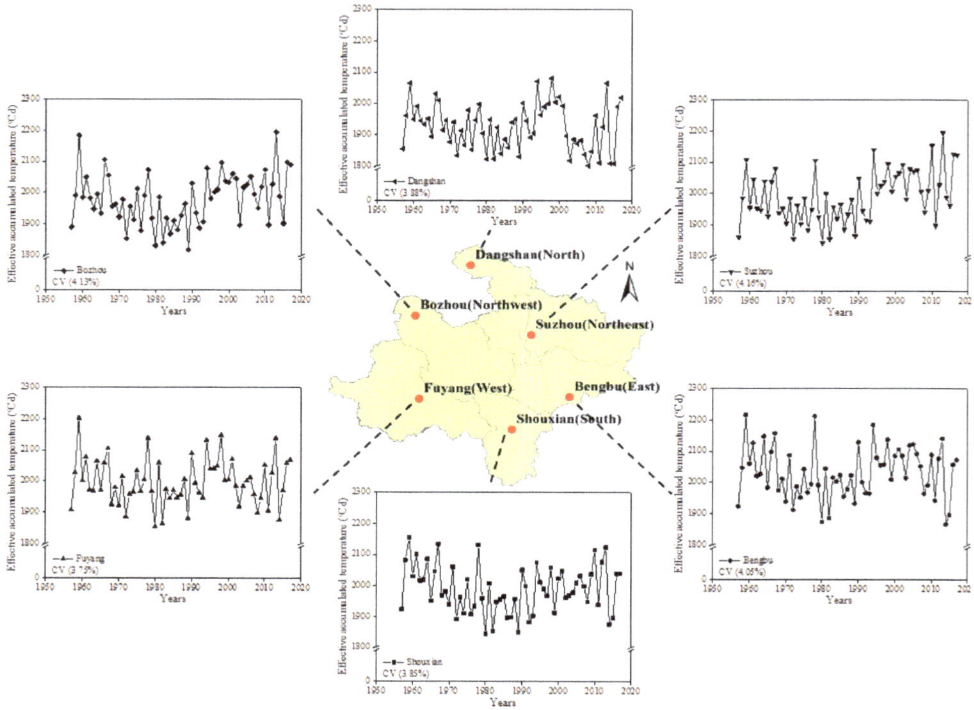

Figure 2. Interannual variation of effective accumulated temperature from 10 June to 30 Septembe corresponding to the summer maize growing season in the past 61 years (1957–2017) for six sites, i.e Shouxian, Bengbu, Fuyang, Suzhou, Bozhou, and Dangshan, on Huaibei Plain, China.

3.2. *The Characteristics of* $\overline{T}_{max}$, $\overline{T}_{min}$, $\overline{T}_{mean}$ *and CV Value during Maize Growing Season and Flowering Stage*

The characteristics of $\overline{T}_{max}$, $\overline{T}_{min}$, $\overline{T}_{mean}$ are shown during maize growing seaso and the flowering stage by working out the average of data of 61 years of the study site (Table 2). Approximately $\overline{T}_{max}$ of 33 °C was calculated for the entire growing seaso and $\overline{T}_{max}$ around the flowering stage ranged from 32 °C to 33 °C. The $\overline{T}_{min}$ and $\overline{T}_{mea}$ also showed similar characteristics for these periods, that is, the average temperature a flowering stage was approximately 2 °C higher than that of the entire growing season. Th CV for different developmental phases and sites varied between 2.6% and 4.5% (Table 2 The results showed that average temperature during the flowering stage was relativel higher than that of the growing season, and the stable interannual temperature variatio may be the reason for the minor fluctuation of effective accumulated temperature durin the maize growing season.

Table 2. Quantitative of characteristics of $\overline{T}_{max}$ (°C), $\overline{T}_{min}$ (°C) and $\overline{T}_{mean}$ (°C) and its CV (%) value during the maize growing season and flowering stage over 61 years in six sites, i.e., Shouxian, Bengbu, Fuyang, Suzhou, Bozhou, and Dangshan.

Period	Sites	$\overline{T}_{max}$, °C (CV, %)	$\overline{T}_{min}$, °C (CV, %)	$\overline{T}_{mean}$, °C (CV, %)
Growing season	Shouxian	29.97 (3.23)	22.01 (2.88)	25.61 (2.65)
	Bengbu	30.48 (3.23)	22.30 (3.08)	25.98 (2.78)
	Fuyang	30.46 (3.06)	21.79 (2.78)	25.66 (2.61)
	Suzhou	30.41 (2.81)	21.73 (3.81)	25.62 (2.86)
	Bozhou	30.51 (2.82)	21.42 (3.83)	25.52 (2.84)
	Dangshan	30.09 (2.65)	20.99 (3.27)	25.04 (2.64)
Flowering stage	Shouxian	32.31 (4.44)	24.73 (3.15)	28.15 (3.59)
	Bengbu	32.94 (4.52)	25.08 (3.24)	28.59 (3.78)
	Fuyang	32.67 (4.23)	24.52 (3.20)	28.16 (3.61)
	Suzhou	32.48 (3.99)	24.47 (3.72)	28.02 (3.59)
	Bozhou	32.44 (3.89)	24.17 (3.75)	27.87 (3.62)
	Dangshan	31.96 (3.64)	23.86 (3.68)	27.42 (3.51)

3.3. The Spatial Distribution of Probability for Days in 32-Day Period with $T_{max} \geq 33$ °C

Figure 3 shows the spatial distribution of probability with $T_{max} \geq 33$ °C over 61 years on Huaibei Plain. The probability for $T_{max} \geq 33$ °C was estimated as 44.6% for Shouxian, 52.2% for Bengbu, 49.6% for Fuyang, 45.1% for Suzhou, 45% for Bozhou, and 38.3% for Dangshan (Figure 3). This suggested that $T_{max} \geq 33$ °C occurred for nearly half of the time during the flowering phase of the crop (32-day period). Consequently, it causes a high risk for a maize crop, exposing it to heat stress events in this region.

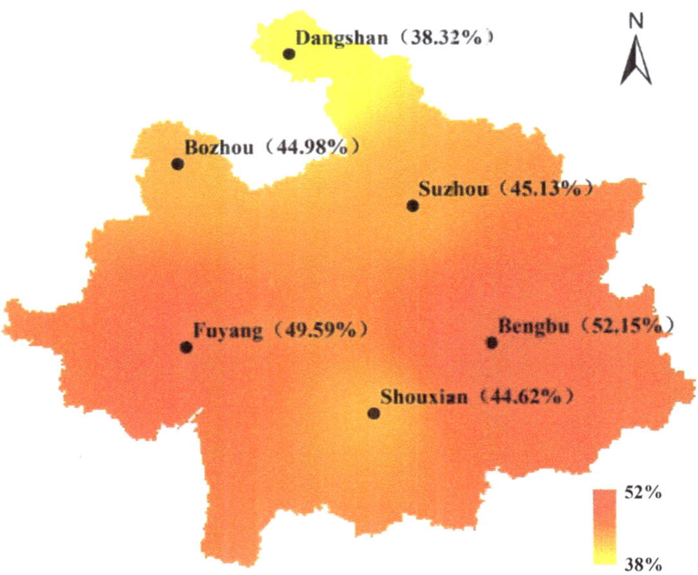

Figure 3. The Spatial distribution of probability for days with $T_{max} \geq 33$ °C during the flowering stage in six sites, i.e., Shouxian, Bengbu, Fuyang, Suzhou, Bozhou, and Dangshan, on Huaibei Plain, China.

3.4. The Occurrence Frequency of HDs

Within the 32 days bracketing flowering, the occurrence of days with T_{max} between 33 and 40 °C are shown in Figure 4. As T_{max} increased, the days corresponding to T_{max} gradually decreased across all the sites. When the temperature threshold reached 39–40 °C, the number of days were close to zero. The fitted T_{max}–day curve showed that the HDs decline slowly, followed by dropping quickly, and lastly plateaued as the temperature increased. Most of the HDs occurred between 33–37 °C (Figure 4), but there were considerable spatial variations across the sites.

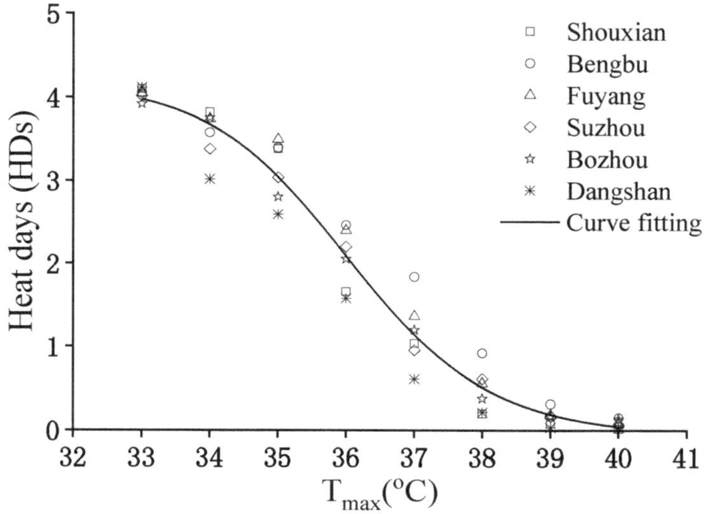

Figure 4. The occurrence frequency and curve fitting of days with T_{max} greater than the specified temperature thresholds during the flowering stage for six sites, i.e., Shouxian (square), Bengbu (circle), Fuyang (triangle), Suzhou (rhombus), Bozhou (pentacle), and Dangshan (asterisk), on Huaibei Plain, China.

3.5. The Occurrence Probability and Classification of Heat Events

The continuous sustained high temperature, particularly during the maize flowering stage, significantly damages the crop yield. Thus, we quantified the occurrence probability of DDs at different temperature thresholds. There was a 50% chance for $T_{max} \geq 33$ °C occurring for at least one day during flowering. With higher thresholds and longer DDs, the probability of heat stress decreased, but the probability of heat stress from ≥33 °C to ≥35 °C was still very high (Table 3). In addition, the occurrence probability of moderate (yellow warning) and severe (orange warming) heat stress was 25% and 10%, respectively. For the extreme heat of the red warning, the probability was estimated to be less than 0.5%, suggesting that heat events of ≥40 °C have rarely occurred in the region (Table 3).

Heat events around maize flowering were classified into four phases (A, B, C, and D) based on T_{max} and OP over the years across the six studied sites, as described in the previous section in detail. At $T_{max} \geq 33$ °C, the OP of heat stress in four phases was about 30% (28.5% in D phase in Dangshan) to 60% (60.9% in B phase in Bengbu), which gradually decreased as the temperature threshold increased. The OP was always highest during phase B between $T_{max} \geq 33$ °C to 36 °C and lowest during phase D compared with all other phases (Table 3). The OP of heat stress in each phase was highest for Bengbu and the lowest for Dangshan. The results showed that heat stress was more likely to occur at anthesis–silking compared to the other phases during flowering. However, HDs with T_{max} ≥ 37 °C have been more likely to occur in phase A (Table 4).

Table 3. The occurrence probability (%) of duration days (DDs) from 1 day to over 7 consecutive days and heat stress warnings during the flowering stage in six selected sites, i.e., Shouxian, Bengbu, Fuyang, Suzhou, Bozhou, and Dangshan, on Huaibei Plain, China.

T_{max}	Shouxian							Bengbu							Fuyang						
	≥1 d	≥2 d	≥3 d	≥4 d	≥5 d	≥6 d	≥7 d	≥1 d	≥2 d	≥3 d	≥4 d	≥5 d	≥6 d	≥7 d	≥1 d	≥2 d	≥3 d	≥4 d	≥5 d	≥6 d	≥7 d
≥33 °C	44.6	41.3	37.6	32.4	29.5	25.9	24.7	52.2	48.9	45.1	40.2	35.9	31.8	30.0	49.6	45.7	42.7	39.2	34.5	31.7	27.0
≥34 °C	31.9	29.0	25.5	22.1	18.8	15.0	13.1	39.7	36.3	33.0	28.0	24.5	20.4	18.9	37.0	33.8	30.9	26.4	23.2	18.8	16.3
≥35 °C	20.0	17.4	15.3	13.0	11.0	8.1	6.9	28.4	25.7	22.7	18.5	14.9	11.8	11.2	25.3	22.7	19.9	15.8	12.7	9.6	8.7
≥36 °C	9.4	7.2	5.9	4.3	3.8	3.1	3.1	17.9	16.0	13.7	10.6	8.1	7.6	6.4	14.0	11.7	9.5	7.1	6.3	5.0	3.4
≥37 °C	4.2	3.1	2.5	1.9	1.9	1.6	1.0	10.1	8.9	7.2	5.6	5.2	4.5	2.9	6.9	5.6	4.5	3.2	3.0	2.3	1.9
≥38 °C	0.9	0.5	0.5	0.3	0.3	0.3	0.0	4.3	3.3	2.2	1.9	1.5	1.2	0.9	2.7	1.7	1.3	1.3	0.9	0.4	0.4
≥39 °C	0.3	0.2	0.2	0.2	0.0	0.0	0.0	1.4	1.1	1.0	0.7	0.3	0.0	0.0	0.8	0.4	0.3	0.3	0.3	0.0	0.0
≥40 °C	0.1	0.0	0.0	0.0	0.0	0.0	0.0	0.5	0.3	0.3	0.3	0.3	0.0	0.0	0.3	0.2	0.2	0.2	0.0	0.0	0.0

T_{max}	Suzhou							Bozhou							Dangshan						
	≥1 d	≥2 d	≥3 d	≥4 d	≥5 d	≥6 d	≥7 d	≥1 d	≥2 d	≥3 d	≥4 d	≥5 d	≥6 d	≥7 d	≥1 d	≥2 d	≥3 d	≥4 d	≥5 d	≥6 d	≥7 d
≥33 °C	45.1	41.5	37.8	34.0	29.7	25.8	24.0	45.0	41.0	36.8	32.1	27.8	23.9	21.8	38.3	34.4	29.7	24.5	20.8	18.8	16.0
≥34 °C	32.6	29.8	26.8	22.4	17.1	15.0	12.9	32.8	30.0	25.9	21.6	17.9	15.9	13.7	25.4	22.2	17.6	14.5	10.9	8.6	7.3
≥35 °C	22.2	19.7	16.0	12.9	9.4	8.7	7.1	21.1	18.5	15.0	12.3	9.3	8.0	6.8	15.9	13.1	10.8	8.6	5.5	4.8	3.8
≥36 °C	12.7	10.0	8.4	7.1	5.5	4.7	4.4	12.4	9.8	7.6	6.4	5.3	4.8	3.3	7.9	5.6	4.0	2.6	2.2	1.4	1.1
≥37 °C	5.7	4.5	3.3	2.9	2.3	2.0	1.1	5.9	4.5	3.5	2.6	2.2	1.4	0.8	2.8	1.9	1.2	0.9	0.5	0.0	0.0
≥38 °C	2.6	2.3	1.7	1.3	1.1	0.3	0.0	2.2	1.6	1.0	0.7	0.5	0.0	0.0	0.9	0.5	0.5	0.2	0.0	0.0	0.0
≥39 °C	0.7	0.5	0.2	0.2	0.0	0.0	0.0	1.0	0.9	0.6	0.4	0.0	0.0	0.0	0.2	0.2	0.2	0.0	0.0	0.0	0.0
≥40 °C	0.2	0.0	0.0	0.0	0.0	0.0	0.0	0.4	0.3	0.3	0.0	0.0	0.0	0.0	0.1	0.1	0.0	0.0	0.0	0.0	0.0

Table 4. The occurrence probability (%) of heat stress in four phases, i.e., pre-anthesis (A, 15–24 July), anthesis (B, 25–31 July), silking (C, 1–7 August), and post-silking (D, 8–15 August), around maize flowering in six selected sites, i.e., Shouxian, Bengbu, Fuyang, Suzhou, Bozhou, and Dangshan, on Huaibei Plain, China.

T_{max}	Shouxian				Bengbu				Fuyang				Suzhou				Bozhou				Dangshan			
	A	B	C	D	A	B	C	D	A	B	C	D	A	B	C	D	A	B	C	D	A	B	C	D
≥33 °C	43.6	54.1	46.8	35.7	52.1	60.9	54.3	42.6	49.5	59.5	48.7	41.8	45.1	53.2	45.2	38.1	46.6	51.1	44.7	37.9	39.8	45.4	40.3	28.5
≥34 °C	33.1	38.9	34.2	21.9	40.5	47.3	42.2	29.3	39.5	42.2	39.1	27.0	35.6	39.3	33.3	22.1	36.4	38.9	32.6	23.0	28.7	31.4	26.0	15.2
≥35 °C	21.5	24.6	21.3	12.5	30.5	35.4	29.5	18.6	27.9	30.4	27.2	15.6	24.3	27.6	24.6	12.5	24.3	26.7	22.0	11.3	18.7	21.5	17.3	5.7
≥36 °C	10.5	12.2	8.2	6.4	21.8	22.2	18.0	9.0	18.2	16.6	15.2	6.4	15.2	16.2	13.3	5.9	15.2	16.4	12.6	4.9	10.5	10.1	6.0	2.9
≥37 °C	4.6	5.4	3.5	3.1	13.1	11.9	10.3	4.5	9.5	7.5	6.6	3.3	7.4	7.5	4.2	3.3	8.4	6.6	5.4	2.5	4.3	2.1	2.8	1.4
≥38 °C	1.0	1.6	0.5	0.6	5.6	4.4	4.4	2.5	3.3	2.8	2.1	1.0	4.3	3.3	1.2	1.2	3.8	1.6	2.1	0.8	1.3	0.7	0.9	0.4
≥39 °C	0.2	1.2	0.0	0.0	2.0	1.2	1.6	1.2	1.1	0.7	0.5	0.4	1.5	0.7	0.5	0.0	1.6	0.7	0.9	0.4	0.5	0.0	0.2	0.0
≥40 °C	0.0	0.2	0.0	0.0	0.5	0.7	0.7	0.4	0.2	0.2	0.2	0.0	0.5	0.2	0.0	0.0	0.8	0.7	0.0	0.0	0.3	0.0	0.0	0.0

3.6. *The Occurrence, Intensity, and Distribution of Heat Events over Four Phases, i.e., Pre-Anthesis (A), Anthesis (B), Silking (C), and Post-Silking (D)*

The heat map shows the occurrence, intensity, and distribution of heat events a pre-anthesis (A), anthesis (B), silking (C), and post-silking (D) over the years across th studied sites (Figure 5). The mild heat events occurred frequently across the four phase during the last 61 years, while extreme heat events rarely occurred across the six location Furthermore, the studied 61-year period was divided into three successive periods: first 2 years (1957–1976), second 20 years (1977–1996), and last 21 years (1997–2017). During th first 20 years, mild and moderate heat stress occurred frequently during the four phase with a low frequency of occurrence of severe and extreme heat stress (Figure 5I), but, in th second period, the normal situation (no heat stress), and mild and moderate heat stres occurred occasionally (Figure 5II). In the last 21-year period, the mild to moderate hea stress occurred often, and moderate to severe heat stress occurred at times, especiall between 2010 and 2017, particularly during anthesis (Figure 5III). It should be noted tha extreme heat stress could possibly occur at any of the four phases in different years an sites. This suggested that climate change could lead to more uncertain extreme heat event

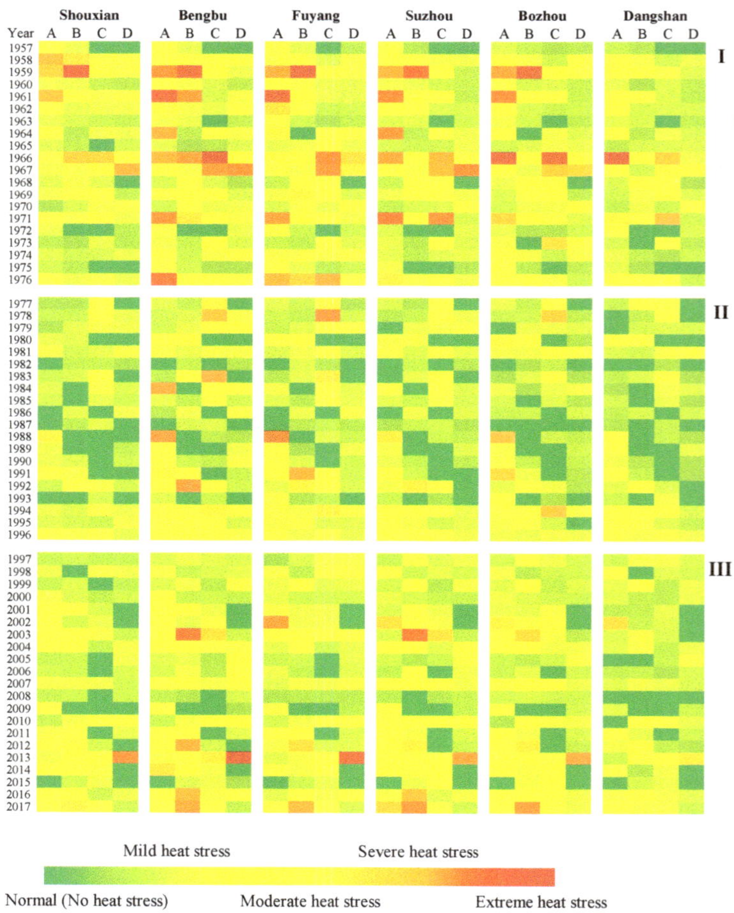

Figure 5. The occurrence, intensity, and distribution of heat events in three successive phases of th

first 20 years (**I**, 1957–1976), second 20 years (**II**, 1977–1996), and last 21 years (**III**, 1997–2017) in four phases (A, pre-anthesis; B, anthesis; C, silking; and D, post-silking) at six sites, i.e., Shouxian, Bengbu, Fuyang, Suzhou, Bozhou, and Dangshan, on Huaibei Plain, China. Note: green color indicates no heat stress; reseda color indicates mild heat stress; yellow color indicates moderate heat stress; orange color indicates severe heat stress; and red color indicates extreme heat stress.

3.7. Relationship between Maize Grain Yield and Average Daily T_{max} around Maize Flowering

The relationship between maize grain yield and average daily T_{max} around flowering for the last 20 years was calculated (Figure 6). The average maize grain yield ranged between 4–6 t ha^{-1} at different average daily T_{max}. Specifically, grain yield slightly increased as the T_{max} increased up to 33 °C (R^2 from 0.0005 to 0.4528), while it declined as the T_{max} surpassed 33 °C (R^2 from 0.001 to 0.3957), except for in Fuyang and Suzhou at the silking phase, indicating that with the bidirectional impact of increasing temperature on maize production over the past 20 years on Huaibei Plain, the trends below and above 33 °C are not significant. Therefore, T_{max} around 33 °C may be a temperature threshold for maize production. Maize grain yield negatively responds to the increase in T_{max} during flowering, and the higher the T_{max} threshold was, the more significantly did the grain yield decline.

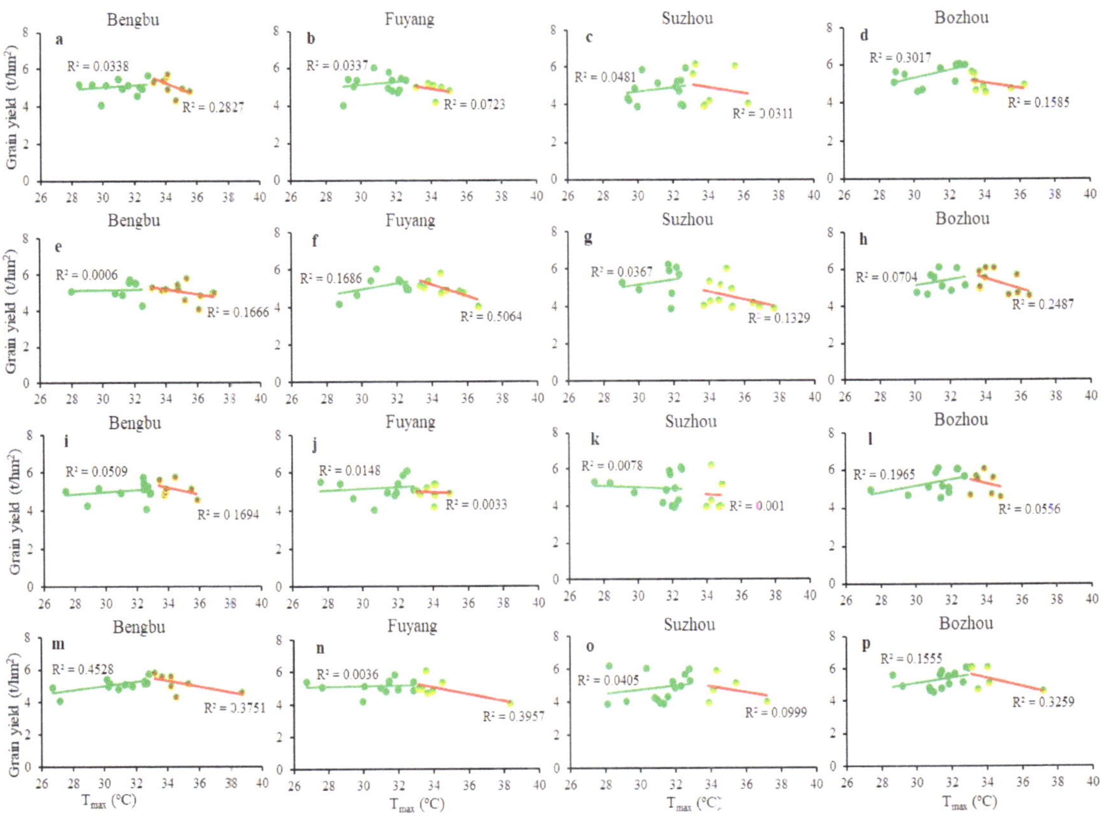

Figure 6. The relationship and linear fitting between maize grain yield and T_{max} during flowering stage, including four phases, i.e., pre-anthesis (from 15 to 24 July, top subfigure layer), anthesis (from

25 to 31 July, upper-middle subfigure layer), silking (from 1 to 7 August, lower-middle subfigure layer), and post-silking (from 8 to 15 August, bottom subfigure layer) at four sites, i.e., Bengbu (**a**,**e**,**i**,**m**), Fuyang (**b**,**f**,**j**,**n**), Suzhou (**c**,**g**,**k**,**o**), and Bozhou (**d**,**h**,**l**,**p**), on Huaibei Plain, China. Note: green and red lines represent fitting of the relationship between maize grain yield and average daily T_{max} < and ≥ 33 °C, respectively. Green and light green dots represent maize grain yield under average daily T_{max} < and ≥ 33 °C, respectively.

4. Discussion

4.1. The Characteristics of Heat Risk Occurrence over Years in Huaibei Plain, China

With a warming climate, heat events occur more frequently, especially during maize flowering [26], and the average maximum temperature (T_{maxx}) during the flowering stage was consistently higher than that of the entire growing season. The comprehensive occurrence frequency of heat stress of summer maize on Huaibei Plain was relatively high, and the occurrence frequency of extreme heat stress was about 8% [24]. In this study, the probability of heat stress is estimated to be close to 50%, and the occurrence frequency of mild, moderate, severe, and extreme heat events was about 50%, 25% to 50%, 0.5% to 13%, and less than 0.5%, respectively, indicating that heat stress occurred once every 2 years, with a frequent occurrence of $T_{max} \geq 33$ °C and $T_{max} \geq 35$ °C and occasional occurrence of $T_{max} \geq 37$ °C. This difference in the occurrence frequency and probability of heat stress is due to varying temperature threshold classification criteria. In further study, remote sensing data can be used as the data source, which can better reflect the heat stress of the study area compared with the calculation of spatial distribution of high temperature based on meteorological point data [33].

Globally, the occurrence intensity and frequency of heat stress events are increasing significantly [9,34], but there was little information on how the number of heat days (HDs) varies with temperature. In this study, the number of HDs showed a slow decline, followed by a quick drop, and lastly it plateaued with increases in temperature, and the most HDs occurred between 33 °C and 37 °C. In addition, heat stress during the maize flowering stage has been estimated using different methods and indicators such as extreme degree days, daily relative humidity, and seasonal rainfall [8,24,25]. The grades of heat stress were divided into slight, moderate, and server levels in terms of the occurrence time and duration of heat stress [24,34]. The heat stress classification criteria proposed in this paper comprehensively considered the T_{max}, DDs, OP, and WC, and it is more suitable for the quantification of heat stress characteristics. Furthermore, our study suggested the high probability of heat stress occurrence during anthesis to silking compared to other stages, indicating heat stress has a potential adverse effect on maize anther dehiscence and pollen vigor. The findings showed the onset and duration of heat stress, as well as the level and frequency of its occurrence, which can provide critical information to local farmers to accurately manage their crops for heat risk occurrence.

4.2. Effect of Heat Risk on Rainfed Maize Cropping in Huaibei Plain, China

The effect of heat stress on maize flowering varies with thresholds. In particular, prolonged exposure to temperatures above 32 °C can reduce pollen viability and germination rate to levels down to zero in many genotypes [15]. It is reported that temperatures above 35 °C can greatly reduce ovule fertilization in maize and T_{max} above 36 °C greatly reduced pollen viability [35] because of tapetum layer disintegration [36]. Furthermore, no pollen grain germination was recorded at T_{max} above 38 °C [32], as this temperature inhibits anther dehiscence [18] by reducing anther apical pore width [19]. Maize cropping is extremely sensitive to heat stress during flowering. However, the probability of T_{max} above 33 °C, 35 °C, and 37 °C are about 50%, 20%, and 6% in study area, respectively, so that reproductive processes such as anther dehiscence, pollen number, viability, and germination rate could be affected [4]. In addition, this study shows that heat stress occurred for several consecutive days on Huaibei Plain, such as the probability of seven hot days at above 35 °C was between 3.8% (Dangshan) and 11.2% (Bengbu). This continuous heat stress can further elevate the maize canopy temperature to accelerate tasseling and anthesis, and results in a prolonged anthesis–silking interval (ASI) [14,37,38], which caused great risk to maize production during flowering.

Pre-anthesis heat affects the pollination process by advancing tasseling, accelerating tassel inflorescences, and shifting the duration of pollen shedding [4,14,39], although heat stress has little effect on silk elongation and silking date [4,21] and grain yield remained stable with the T_{max} increases during this phase. When the heat stress coincides with post-silking, the pollen tube growth and pollination and fertilization process were disrupted [17,20,21]. Meanwhile, the frequency of heat events will increase the risk of drought in that they produce positive feedbacks together that intensify their effects [40,41]. Higher temperature may increase potential evapotranspiration, causing more rapid soil drying and greater severity of drought by increasing a vapor pressure deficit [42]. Farmers usually take some additional agronomic measures, i.e., selection of varieties, optimum sowing data, adequate water, fertilizer management, and application of exogenous substances, to adapt to the warmer temperature for reducing the effect of climate change during maize flowering.

4.3. Maize Cropping Strategy for Heat Risk

Climate change is likely to increase the number of hot days and the probability of heat stress around maize flowering on Huaibei Plain [24]. Thus, farmers have to adjust their cropping strategies in response to climate change, particularly via the selection of heat-resistant hybrids and management, such as optimizing the sowing date [43,44], applying a plant growth regulator [45], and improving irrigation systems [46,47]. For example, maize grain yield can be increased by hybrid choices and optimizing sowing dates and cultivar selection under warming of 1.5 °C and 2 °C [11,43]. However, the traditional maize–wheat rotation system dictates a fixed maize sowing time, which is more likely to suffer from heat stress prior to and post flowering, thereby reducing maize yield [48]. Thus, early or late sowing may be possible under warming scenarios [11]. In addition, applying urea in combination with nitrapyrin and plant growth regulator (gibberellic acid) in the pre-flowering stage has the potential to mitigate N_2O emission, improve N response efficiency, and increase maize yield under hot climatic conditions [45]. Moreover, improving irrigation is also an effective means of adaptation for compromising the sensitivity to heat stress by lowering the maize canopy temperature [49]. These agronomic management strategies can effectively alleviate or avoid increased heat stress incidents during maize production.

Crop models have been widely used to provide support, management, and decision-making for maize production [50,51]. For example, the World Food Studies (WOFOST) model and agricultural production systems simulator (APSIM)-maize model were used to simulate sowing dates, cultivar shifts, and climate adaptation for maize cropping, and to predict the potential maize yield based on climate and crop conditions around flowering [52,53]. Therefore, it is feasible to use crop models to predict the occurrence characteristics and countermeasures of heat events under climate change. Notably, the relationship between grain yield and threshold temperature point (33 °C) should be taken into account in model prediction, so as to further analyze the specific climatic suitability and optimize sowing dates and response strategies for heat stress during the maize flowering stage. A schematic diagram of heat stress occurrence patterns, characteristics, and coping strategies of heat risk is provided as Figure 7.

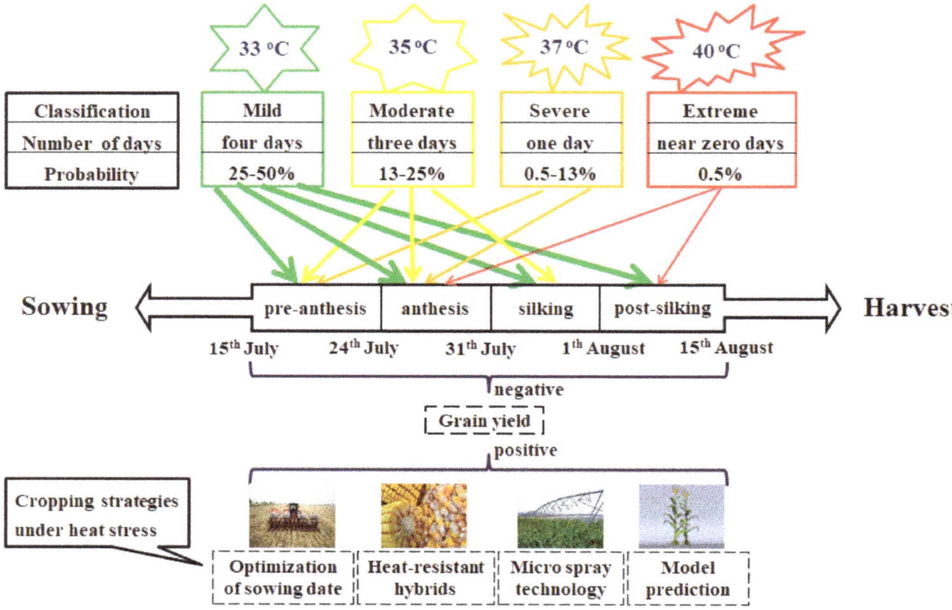

Figure 7. Schematic diagram of occurrence patterns, characteristics, and coping strategies of hea stress in relation to maize cropping on Huaibei Plain, China or similar climate and cropping region Note: green, yellow, orange, and red arrow symbols represent mild, moderate, severe, and extrem heat stress, respectively. The thickness and direction of the arrow symbols indicate the probabilit and phases of the occurrence of heat stress.

5. Conclusions

Heat stress during the maize flowering stage is detrimental for successful maiz reproduction and final grain yield formation. In this study, the characteristics of hea stress occurrence in relation to reproductive development were described. The effectiv accumulated temperature was about 2000 °Cd with little interannual variation. The averag maximum temperature (T_{max}) in the flowering stage was about 33 °C, which was highe than that for the entire growing season. The probability of days with $T_{max} \geq 33$ °C i estimated to be 50%, and the most HDs occurred between 33 °C and 37 °C. The five-leve classifications of heat stress were identified based on T_{max}, DDs, OP, and WC. In additio the heat stress for $T_{max} \geq 35$ °C occurred more frequently while the stress for $T_{max} \geq 37$ ° occurred occasionally and for $T_{max} \geq 40$ °C occurred rarely across the studied years an sites. Our study suggested that compared with other phases, heat stress was more likely t occur from anthesis to silking. In addition, the maize grain yield slightly increased as th T_{max} increased to 33 °C, but it declined as the T_{max} surpassed 33 °C, especially in anthesi These findings will help to guide summer maize cropping under climate change.

Author Contributions: Data curation, formal analysis, investigation, and writing—original draf Z.W.; formal analysis, resources, and software, W.S.; methodology and resources, X.L.; formal analysi and methodology, Y.L.; formal analysis and methodology, B.C.; formal analysis and writing—revie and editing, N.U.; methodology, funding acquisition, supervision, and writing—review and editin Y.S. All authors have read and agreed to the published version of the manuscript.

Funding: The work was supported by a grant from the National Key R&D Program of Chin (No. 2017YFD0301307), the Belt and Road Special Foundation of the State Key Laboratory of Hydro ogy-Water Resources and Hydraulic Engineering (No. 2021491011), and the Natural Science Found tion of Anhui Province of China (No. 1808085ME158).

Institutional Review Board Statement: Not applicable.

Informed Consent Statement: Not applicable.

Data Availability Statement: Not applicable.

Acknowledgments: The authors thank Yulei Zhu for meteorological data analysis, Jinpeng Li and Huihui Liu for advice.

Conflicts of Interest: The authors declare no conflict of interest.

References

1. Long, S.P.; Zhu, X.; Naidu, S.L.; Ort, D.R. Can improvement in photosynthesis increase crop yields? *Plant Cell Environ.* **2006**, *29*, 315–330. [CrossRef] [PubMed]
2. Rufino, C.D.; Fernandes-Vieira, J.; Martín-Gil, J.; Abreu, J.D.; Tavares, L.C.; Fernandes-Correa, M.; Martin-Ramos, P. Water stress influence on the vegetative period yield components of different maize genotypes. *Agronomy* **2018**, *8*, 151. [CrossRef]
3. Chen, G.; Cao, H.; Chen, D.; Zhang, L.; Zhao, W.; Zhang, Y.; Wen-qi, M.A.; Jiang, R.; Zhang, H.; Zhang, F. Developing sustainable summer maize production for smallholder farmers in the North China Plain: An agronomic diagnosis method. *J. Integr. Agric.* **2019**, *18*, 1667–1679. [CrossRef]
4. Lizaso, J.I.; Ruiz-Ramos, M.; Rodríguez, L.; Gabaldon-Leal, C.; Oliveira, J.A.; Lorite, I.J.; Sánchez, D.; García, E.; Rodríguez, A. Impact of high temperatures in maize: Phenology and yield components. *Field Crop. Res.* **2018**, *216*, 129–140. [CrossRef]
5. Vautard, R.; Gobiet, A.; Sobolowski, S.; Kjellström, E.; Stegehuis, A.; Watkiss, P.; Mendlik, T.; Landgren, O.; Nikulin, G.; Teichmann, C.; et al. The European climate under a 2 °C global warming. *Environ. Res. Lett.* **2014**, *9*, 034006. [CrossRef]
6. Lobell, D.B.; Gourdji, S.M. The influence of climate change on global crop productivity. *Plant Physiol.* **2012**, *160*, 1686–1697. [CrossRef]
7. Lizaso, J.I.; Ruiz-Ramos, M.; Rodríguez, L.; Gabaldon-Leal, C.; Oliveira, J.A.; Lorite, I.J.; Rodríguez, A.; Maddonni, G.A.; Otegui, M.E. Modeling the response of maize phenology, kernel set, and yield components to heat stress and heat shock with CSM-IXIM. *Field Crop. Res.* **2017**, *214*, 239–252. [CrossRef]
8. Lobell, D.B.; Hammer, G.L.; Mclean, G.; Messina, C.; Roberts, M.J.; Schlenker, W. The critical role of extreme heat for maize production in the United States. *Nat. Clim. Change* **2013**, *3*, 497–501. [CrossRef]
9. Huang, M.; Wang, J.; Wang, B.; Liu, D.L.; Feng, P.; Yu, Q.; Pan, X.; Waters, C. Assessing maize potential to mitigate the adverse effects of future rising temperature and heat stress in China. *Agric. For. Meteorol.* **2021**, *311*, 108673. [CrossRef]
10. Jagadish, S.V.K.; Craufurd, P.Q.; Wheeler, T.R. High temperature stress and spikelet fertility in rice (*Oryza sativa* L.). *J. Exp. Bot.* **2007**, *58*, 1627–1635. [CrossRef]
11. Huang, M.; Wang, J.; Wang, B.; Liu, D.; Yu, Q.; He, D.; Wang, N.; Pan, X. Optimizing sowing window and cultivar choice can boost China's maize yield under 1.5 °C and 2 °C global warming. *Environ. Res. Lett.* **2020**, *15*, 024015. [CrossRef]
12. Xiong, W.; Matthews, R.B.; Holman, I.; Lin, E.; Xu, Y. Modelling China's potential maize production at regional scale under climate change. *Clim. Change* **2007**, *85*, 433–451. [CrossRef]
13. Lin, Y.; Feng, Z.; Wu, W.; Yang, Y.; Zhou, Y.; Xu, C. Potential impacts of climate change and adaptation on maize in Northeast China. *Agron. J.* **2017**, *109*, 1476–1490. [CrossRef]
14. Wang, Y.; Tao, H.; Tian, B.; Sheng, D.; Xu, C.; Zhou, H.; Huang, H.; Wang, P. Flowering dynamics, pollen, and pistil contribution to grain yield in response to high temperature during maize flowering. *Environ. Exp. Bot.* **2019**, *158*, 80–88. [CrossRef]
15. Herrero, M.P.; Johnson, R.R. High temperature stress and pollen viability of maize1. *Crop Sci.* **1980**, *20*, 796–800. [CrossRef]
16. Begcy, K.; Nosenko, T.; Zhou, L.; Fragner, L.; Weckwerth, W.; Dresselhaus, T. Male sterility in maize after transient heat stress during the tetrad stage of pollen development. *Plant Physiol.* **2019**, *181*, 683–700. [CrossRef]
17. Jiang, Y.; Lahlali, R.; Karunakaran, C.; Warkentin, T.D.; Davis, A.R.; Bueckert, R.A. Pollen, ovules, and pollination in pea: Success, failure, and resilience in heat. *Plant Cell Environ.* **2019**, *42*, 354–372. [CrossRef]
18. Shao, R.; Yu, K.; Li, H.; Jia, S.; Yang, Q.; Zhao, X.; Zhao, Y.; Liu, T. The effect of elevating temperature on the growth and development of reproductive organs and yield of summer maize. *J. Integr. Agric.* **2021**, *20*, 1783–1795. [CrossRef]
19. Wang, Y.; Liu, X.; Hou, X.; Sheng, D.; Dong, X.; Gao, Y.; Wang, P.; Huang, S. Maximum lethal temperature for flowering and seed set in maize with contrasting male and female flower sensitivities. *J. Agron. Crop Sci.* **2021**, *207*, 679–689. [CrossRef]
20. Rattalino Edreira, J.I.; Mayer, L.I.; Otegui, M.E. Heat stress in temperate and tropical maize hybrids: Kernel growth, water relations and assimilate availability for grain filling. *Field Crop. Res.* **2014**, *166*, 162–172. [CrossRef]
21. Rattalino Edreira, J.I.; Carpici, E.B.; Sammarro, D.; Otegui, M.E. Heat stress effects around flowering on kernel set of temperate and tropical maize hybrids. *Field Crop. Res.* **2011**, *123*, 62–73. [CrossRef]
22. Ahmad, M.I.; Shah, A.N.; Sun, J.; Song, Y. Comparative Study on Leaf Gas Exchange, Growth, Grain Yield, and Water Use Efficiency under Irrigation Regimes for Two Maize Hybrids. *Agriculture* **2020**, *10*, 369. [CrossRef]
23. Li, D.; Qi, H.; Ma, X. The climate index and assessment about drought and flood in maize's key growth stage in Huaibei plain in Anhui province. *Chin. Agric. Sci. Bull.* **2013**, *29*, 208–216. (in Chinese).
24. Li, D.; Sun, Y.; Sun, Y. Use of integrated climatic index to determine high temperature damage to summer maize at florescence in the Huaibei Plain. *Chin. J. Eco-Agric.* **2015**, *23*, 1035–1044. (in Chinese).

25. Chen, H.; Li, S. Prediction of high temperature disasters risk during summer maize flowering period under future climate warming background in Henan Province. *Chin. J. Eco-Agric.* **2020**, *28*, 1–13. (in Chinese).
26. Wang, Y.; Sheng, D.; Zhang, P.; Dong, X.; Yan, Y.; Hou, X.; Wang, P.; Huang, S. High temperature sensitivity of kernel formation in different short periods around silking in maize. *Environ. Exp. Bot.* **2021**, *183*, 104343. [CrossRef]
27. Yuan, H.; Cui, Y.; Ning, S.; Jiang, S.; Yuan, X.; Tang, G. Estimation of maize evapotraspiration under drought stress-A case study of Huaibei Plain, China. *PLoS ONE.* **2019**, *14*, e0223756. [CrossRef]
28. Tao, Z.; Sui, P.; Chen, Y.; Li, C.; Nie, Z.; Yuan, S.; Shi, J.; Gao, W. Subsoiling and ridge tillage alleviate the high temperature stress in spring maize in the North China Plain. *J. Integr. Agric.* **2013**, *12*, 2179–2188. [CrossRef]
29. Gao, Z.; Feng, H.-Y.; Liang, X.-G.; Lin, S.; Zhao, X.; Shen, S.; Du, X.; Cui, Y.-H.; Zhou, S.-L. Adjusting the sowing date of spring maize did not mitigate against heat stress in the North China Plain. *Agric. For. Meteorol.* **2021**, *298–299*, 108274. [CrossRef]
30. Shepard, D. A two-dimensional interpolation function for irregularly-spaced data. In *Proceedings of the 1968 ACM National Conference*; Blue, R.B.S., Rosenberg, A.M., Eds.; ACM Press: New York, NY, USA, 1968; pp. 517–524.
31. Lu, W.; Yu, H.; Cao, S.; Chen, C. Effects of climate warming on growth process and yield of summer maize in Huang-Huai-Hai Plain in last 20 years. *Sci. Agric. Sin.* **2015**, *48*, 3132–3145. (in Chinese).
32. Sánchez, B.; Rasmussen, A.; Porter, J.R. Temperatures and the growth and development of maize and rice: A review. *Glob. Change Biol.* **2014**, *20*, 408–417. [CrossRef] [PubMed]
33. Hu, X.; Zhao, Z.; Zhang, L.; Liu, Z.; Li, S.; Zhang, X. A high-temperature risk assessment model for maize based on MODIS LST. *Sustainability* **2019**, *11*, 6601. [CrossRef]
34. Guan, Y.; Liu, J.; He, Q.; Li, R.; Mi, X.; Qin, X. Risk probability of heat injury during summer maize flowering period in North China Plain based on information diffusion theory. *Chin. J. Agrometeorol.* **2021**, *42*, 606–615. (in Chinese).
35. Siebers, M.H.; Slattery, R.A.; Yendrek, C.R.; Locke, A.M.; Drag, D.; Ainsworth, E.A. Simulated heat waves during maize reproductive stages alter reproductive growth but have no lasting effect when applied during vegetative stages. *Agric. Ecosyst. Environ.* **2017**, *240*, 162–170. [CrossRef]
36. Djanaguiraman, M.; Prasad, P.V.; Boyle, D.L.; Schapaugh, W.T. Soybean pollen anatomy, viability and pod set under high temperature stress. *J. Agron. Crop Sci.* **2013**, *199*, 171–177. [CrossRef]
37. Bassetti, P.; Westgate, M.E. Emergence, elongation, and senescence of maize silks. *Crop Sci.* **1993**, *33*, 271–275. [CrossRef]
38. Vanaja, M.; Sathish, P.; Kumar, G.V.; Razzaq, A.; Vagheera, P.; Lakshmi, N.J.; Yadav, S.K.; Sarkar, B.; Maheswari, M. Elevated temperature and moisture deficit stress impact on phenology, physiology and yield responses of hybrid maize. *J. Agrometeorol.* **2017**, *19*, 295–300. [CrossRef]
39. Hatfield, J.L.; Prueger, J.H. Temperature extremes: Effect on plant growth and development. *Weather Clim. Extrem.* **2015**, *10*, 4–10. [CrossRef]
40. Teixeira, E.I.; Fischer, G.; van Velthuizen, H.; Walter, C.; Ewert, F. Global hot-spots of heat stress on agricultural crops due to climate change. *Agric. For. Meteorol.* **2013**, *170*, 206–215. [CrossRef]
41. Suzuki, N.; Rivero, R.M.; Shulaev, V.; Blumwald, E.; Mittler, R. Abiotic and biotic stress combinations. *New Phytol.* **2014**, *203*, 32–43. [CrossRef]
42. Teskey, R.; Wertin, T.; Bauweraerts, I.; Ameye, M.; Mcguire, M.A.; Steppe, K. Responses of tree species to heat waves and extreme heat events. *Plant Cell Environ.* **2015**, *38*, 1699–1712. [CrossRef] [PubMed]
43. Liu, Z.; Hubbard, K.G.; Lin, X.; Yang, X. Negative effects of climate warming on maize yield are reversed by the changing of sowing date and cultivar selection in Northeast China. *Glob. Change Biol.* **2013**, *19*, 3481–3492. [CrossRef] [PubMed]
44. Zhou, B.; Yue, Y.; Sun, X.; Wang, X.; Wang, Z.; Ma, W.; Zhao, M. Maize grain yield and dry matter production responses to variations in weather conditions. *Agron. J.* **2016**, *108*, 196–204. [CrossRef]
45. Dawar, K.; Sardar, K.; Zaman, M.; Müller, C.; Sanz-Cobena, A.; Khan, A.; Borzouei, A.A.; Pérez-Castillo, A.G. Effects of the nitrification inhibitor nitrapyrin and the plant growth regulator gibberellic acid on yield-scale nitrous oxide emission in maize fields under hot climatic conditions. *Pedosphere* **2021**, *31*, 323–331. [CrossRef]
46. Kang, S.; Hao, X.; Du, T.; Tong, L.; Su, X.; Lu, H.; Li, X.; Huo, Z.; Li, S.; Ding, R. Improving agricultural water productivity to ensure food security in China under changing environment: From research to practice. *Agric. Water Manag.* **2017**, *179*, 5–17. [CrossRef]
47. Wu, J.; Zhang, J.; Ge, Z.; Xing, L.; Han, S.; Shen, C.; Kong, F. Impact of climate change on maize yield in China from 1979 to 2016. *J. Integr. Agric.* **2021**, *20*, 289–299. [CrossRef]
48. Tian, B.; Zhu, J.; Nie, Y.; Xu, C.; Meng, Q.; Wang, P. Mitigating heat and chilling stress by adjusting the sowing date of maize in the North China Plain. *J. Agron. Crop Sci.* **2019**, *205*, 77–87. [CrossRef]
49. Zhou, L.; Turvey, C.G. Climate change, adaptation and China's grain production. *China Econ. Rev.* **2014**, *28*, 72–89. [CrossRef]
50. Elliott, J.; Müller, C.; Deryng, D.; Chryssanthacopoulos, J.; Boote, K.J.; Büchner, M.; Foster, I.; Glotter, M.; Heinke, J.; Iizumi, T.; et al. The global gridded crop model intercomparison: Data and modeling protocols for phase 1 (v1.0). *Geosci. Model Dev.* **2015**, *8*, 261–277. [CrossRef]
51. Yin, X.; Leng, G. Modelling global impacts of climate variability and trend on maize yield during 1980–2010. *Int. J. Clim.* **2021**, *41*, E1583–E1596. [CrossRef]

2. Xiao, D.; Qi, Y.; Shen, Y.; Tao, F.; Moiwo, J.P.; Liu, J.; Wang, R.; Zhang, H.; Liu, F. Impact of warming climate and cultivar change on maize phenology in the last three decades in North China Plain. *Theor. Appl. Clim.* **2016**, *124*, 653–661. [CrossRef]
3. Bassu, S.; Fumagalli, D.; Toreti, A.; Ceglar, A.; Giunta, F.; Motzo, R.; Zajac, Z.; Niemeyer, S. Modelling potential maize yield with climate and crop conditions around flowering. *Field Crop. Res.* **2021**, *271*, 108226. [CrossRef]

agronomy

MDPI

Article

Genetic Divergence and Spatial Configuration Influence the Weed Spectrum, Herbage Yield and Nutritive Quality of Temperate Cowpea

Muhammad Aamir Iqbal [1,*], Arslan Ahmed [1], Muhammad Imran [2], Hanaa E. Ahmed [3], Rehab M. Hafez [3] and Asmaa A. Hamad [3,4]

1 Department of Agronomy, Faculty of Agriculture, University of Poonch Rawalakot, Rawalakot 12350, Pakistan; arslanahmedupr1@gmail.com
2 Department of Entomology, Faculty of Agriculture, University of Poonch Rawalakot, Rawalakot 12350, Pakistan; muhammadimran@upr.edu.pk
3 Botany and Microbiology Department, Faculty of Science, Cairo University, Giza 12613, Egypt; helbadawy@sci.cu.edu.eg (H.E.A.); rehabhafez@sci.cu.edu.eg (R.M.H.); asmahamad@yahoo.com (A.A.H.)
4 Department of Biology, College of Science, Taif University, P.O. Box 11099, Taif 21944, Saudi Arabia
* Correspondence: aamir1801@yahoo.com

Abstract: Under a changing climate, the biologically viable management of weeds and the exploratio of the genetic divergence of spreading and towering cultivars of forage cowpea in different rov configuration systems hold the potential to boost sustainable feed supply for dairy animals. field study was undertaken to sort out the most nutritive and high-biomass-producing cultiva (Cowpea−,2007 and Rawan−,2010) of cowpea and optimize the row configuration (R × R of 15, 3 45 and 60 cm) to manage the weed spectrum. The results revealed that Rawan-2010 remained superic in the 15 cm row configuration by recording 39% lesser weed density (WD) than the correspondin value recorded by the same cultivar sown in the 60 cm row configuration. The same treatmer combination recorded a 20% lesser fresh weed weight than Cowpea−,2007 sown in the same rov configuration, while it exhibited a 5.6 g m^{-2} lesser corresponding value of dry weed weight. I contrast, Cowpea-2010 sown in the 45 cm row configuration recorded the maximum yield attribute (stem girth, leaf and branch numbers, leaf area, fresh and dry weights per plant), except plant heigh (PH), which resulted in 7% and 13% higher green herbage yield (GH) and dry matter biomass (DM respectively, than the same cultivar sown in the 30 cm row configuration. Pertaining to nutritiona value, Rawan-2010 in the 45 cm row configuration yielded the maximum crude protein and minimur crude fiber content, while the same cultivar gave the greatest ash content in the wider row spacin With GH, the correlation analyses indicated an antagonistic association for PH, a moderately linea relationship between stem girth and branch numbers and a strong direct association between lea area and fresh plant weight.

Keywords: planting geometry; biomass; crude protein; correlation analysis; leguminous forages

Citation: Iqbal, M.A.; Ahmed, A.; Imran, M.; Ahmed, H.E.; Hafez, R.M.; Hamad, A.A. Genetic Divergence and Spatial Configuration Influence the Weed Spectrum, Herbage Yield and Nutritive Quality of Temperate Cowpea. *Agronomy* **2022**, *12*, 1323. https://doi.org/10.3390/agronomy12061323

Academic Editors: Umberto Anastasi and Aurelio Scavo

Received: 15 March 2022
Accepted: 25 May 2022
Published: 30 May 2022

1. Introduction

Globally, the skyrocketing population necessitates the proportional enhancemen of milk and meat production, which require a sustainable supply of quality forages i abundant quantities throughout the year [1]. Among forage crops, cereals such as sorghun maize, oat, barley, and millets yield copious quantities of green herbage; however, thes cereals have low protein and digestibility. Thus, in order to maintain milk productio dairy animals need to be fed expensive protein additives that lead to a significant hike i the cost of production and a depletion of net returns [2]. Legumes hold bright perspective to overcome the nutritional quality concerns related to forages due to having superic nutritional quality along with the potential to gain nitrogen (N) through the biologica

N fixation process. Among legume forages, cowpea (*Vigna unguiculata* (L.) Walp) is of pivotal pertinence in the USA, many African countries, China and especially in South Asian countries (Pakistan, India and Bangladesh) [3]. It encompasses numerous advantageous characteristics, including unmatched drought tolerance, superior adaptability to harsh climatic conditions, and the unique option to be grown as a dual-purpose crop (forage and grain). Additionally, cowpea has the potential to be adjusted in a variety of farming systems (irrigated or rainfed, arid or semi-arid, tropical or temperate) owing to its established tolerance against abiotic stresses such as drought, heat, salinity, soil erosion and, more importantly, its unmatched ability to thrive well in toxic soils [4–7].

Nevertheless, in order to become a viable alternative to traditional multi-cut leguminous forages (clover species, alfalfa, etc.), cowpea cultivation must demonstrate a competitive yield advantage in terms of higher herbage yield and nutritive quality. In addition, the changing climate, which has been characterized by global warming and the shifting of rainfall patterns and seasonal distributions, necessitates the cultivation of drought- and heat-resilient crops like cowpea in rainfed farming systems of temperate regions [8–10]. However, cowpea forage yield in temperate areas has remained suboptimal compared to other single-cut forages, primarily owing to low-yielding cultivars and outdated agronomic production technology packages. Presently, towering and spreading types of cowpea cultivars are available in Pakistan, having moderate biomass yield potential in the range of 14–30 tons per hectare with an appropriate level of nutritional quality [11,12]. However, to the best of our information, research findings on the comparative performance of forage cowpea cultivars in terms of herbage yield and nutritional quality under temperate conditions are very scant. Besides its genetic potential, agronomic production technology, especially the row configuration of towering and spreading types of cultivars, imparts significant influence on weed composition, crop plants' growth and herbage yield [13–16]. Row configuration alterations have been effectively used to manage soil moisture deficiency through skipping one or multiple rows of the crop in what is commonly referred to as a skip row configuration, while on the other hand, adding one or more rows by reducing inter-row spacing is called an additive row configuration. There is a serious lack of field-trial-based evidence regarding row configuration optimization for boosting herbage yield and nutritional quality traits of cowpea cultivars.

As a result of the changing climate, infestations of indigenous and exotic weeds have become robust in temperate regions, necessitating the evaluation of biologically viable ways to keep them below a threshold level. Previous studies demonstrated that altering row configurations, especially closer ones, offered weed suppression in a biologically viable way and also resulted in the robust growth of plants on either side of the rows [17,18]. Likewise, wider row spacing for grain crops recorded 21–43% higher weed density owing to the agro-botanical superiority of weeds in acquiring growth resources like moisture, nutrients and solar radiation [19]. Additionally, it was revealed that cowpea-sorghum intercropping in 30 cm spatial arrangements resulted in significantly lesser (57%) density and biomass production (29% lesser fresh weight and 37% lower dry weight compared to wider row configurations) of weeds [19–21]. In addition to weed management, narrow row configurations remained effective in controlling wind erosion, aided in conserving moisture and reduced crusting of the soil surface, which led to improved soil health and structure. Additionally, a narrow planting configuration served as a viable way to use the available moisture efficiently in dry farming [22,23]. Moreover, it was also observed that changes in plant configuration imparted a significant influence on leaf area index and canopy development through the alteration of evapotranspiration partitioning between the soil surface and crop plants. These findings were supported by another study, whereby Staggenborg et al. [24] attained higher biomass yields from narrow rows in comparison to wider rows under optimum fertility and moisture conditions. It was also inferred that under moisture-deficient conditions, no significant differences in biomass yield by different row configuration treatments was evident. In contrast, Bandaru et al. [25] found superior herbage and grain yields from narrow rows of crops coupled with wide intra-row spacing among seedlings. However, it was suggested that the clumping of narrow and wide row configuration remained superior only under sub-optimal conditions, while this effect lost superiority as growth conditions were optimized. M'Khaitir and Vanderlip [26] reported that increasing the plant population remained effective in boosting plant growth and biomass production, especially when soil moisture remained sufficient. Again, conclusive findings are lacking pertaining to row configuration optimization and its impact on weed density, herbage yield and the nutritional quality of cowpea in temperate regions.

In light of the changing climatic scenario and emerging market opportunities in the dairy industry, we set out to reinvestigate the potential fit of forage cowpea in the temperate Himalayan region of Pakistan. However, the prime need of the time is to bridge research and knowledge gaps pertaining to the interactive effect of genetic potential and row configuration on weed density, growth attributes, biomass yield and the nutritional quality of cowpea. To achieve this goal, we hypothesized that harmonizing row configuration with genetic divergence (towering- or spreading-type growth habit) might boost vegetative growth, herbage yield, quality attributes and weed suppression through the effective use of farm inputs and environmental resources in spatio-temporal dimensions. In contrast, temperate climatic conditions could potentially restrict the expression of genetic potential and neutralize the influence of row arrangement on the growth, yield and quality traits of cowpea. Thus, this field trial was undertaken with the objective of harmonizing the row configuration of cowpea cultivars having varying growth habits in rainfed farming under a temperate climate.

2. Materials and Methods

2.1. Meteorological Features and Physico-Chemical Description of Experimental Locality

The field trial was executed at the research area (main campus) of the University of Poonch Rawalakot, Azad Jammu and Kashmir, Pakistan, during 2018–2019. The geographical coordinates [27] of the study site are presented in Figure 1. The test crop (cowpea) was sown after the harvesting of winter wheat on June 22 and 25 of 2018 and 2019, respectively. The meteorological features regarding temperature and rainfall of the study site during the crop growth season (mean values of both years) are presented in Figure 2. The study locality entails rainfed farming systems receiving sufficient precipitation during the crop growth cycle to support the economical production of crops like maize, soybean, sorghum, and a variety of vegetables that are primarily grown at the subsistence level.

Prior to the cultivation of the test crop (cowpea), the physicochemical analyses of the experimental block were performed by collecting soil samples from two depths (0–15 cm and 15–30 cm) from four corners and the middle of the experimental block. Subsequently, the soil samples (belonging to both soil depths) were homogenized thoroughly by hand mixing, and thereafter, the samples were shade dried, grounded and sieved (using a sieve having a 2 mm pore size). For the estimation of pH, the soil was mixed with water (1:2.5 ratios) to prepare the paste that was subjected to the glass electrode for determining the pH [28]. The electrical conductivity (EC) of the soil samples was estimated using a conductivity meter [29–33]. In addition, organic carbon (OC) content was evaluated using the wet oxidation method, while the Walkley–Black protocol was followed for assessing the organic matter (OM) content [29]. Moreover, total nitrogen (N) was estimated with the help of the Kjeldahl apparatus for distillation and H_2SO_4 (concentrated acid) titration [30]. Phosphorous (P) was determined by using Olsen's method, which entails the reaction of 0.5 N $NaHNO_3$ at 8.5 pH with a soil:extractant paste (1:10 ratio) and the subsequent use of spectrophotometer (882 nm) in a system containing H_2SO_4 [31]. Finally, potassium (K) availability was estimated by an ammonium acetate extraction (shaking soil samples with an ammonium acetate solution of 0.5 M for 30 min) method that caused K^+ ion displacement, and a flame photometer was used for their detection. For recording micronutrient concentrations in soil samples, an extraction method encompassing ammonium acetate solution (CH_3COONH_4) was reacted with soil paste (pH = 3.0) for iron (Fe) estimation. Thereafter, a colorimetric method along with a spectrophotometer (510 nm wavelength) was put into practice to determine Fe content. Moreover, the concentration of micronutrients, including boron (B), zinc (Zn), copper (Cu) and manganese (Mn), were estimated by following the extraction method that involved diethyle-netriaminepentaacetic acid [32–34]. The experimental soil's texture was loam, having a pH and OM of 7.8 and 1.05%, respectively, indicating the dire need to appropriately fertilize the soil to achieve the potential yield. The bulk density of the experimental soil was 1.24 cm^{-3}, while the EC was 0.45 dS m^{-1}, indicating the soil was normal without salinity. As far as the macronutrients of the soil samples were concerned, the NPK concentrations remained at 87, 5.8 and 183 mg kg^{-1}, respectively. The micronutrients were in an appropriate range, such as B (1.18 mg kg^{-1}), Mn (19.2 mg kg^{-1}), Fe (14.2 mg kg^{-1}), Cu (1.82 mg kg^{-1}) and Zn (1.29 mg kg^{-1}).

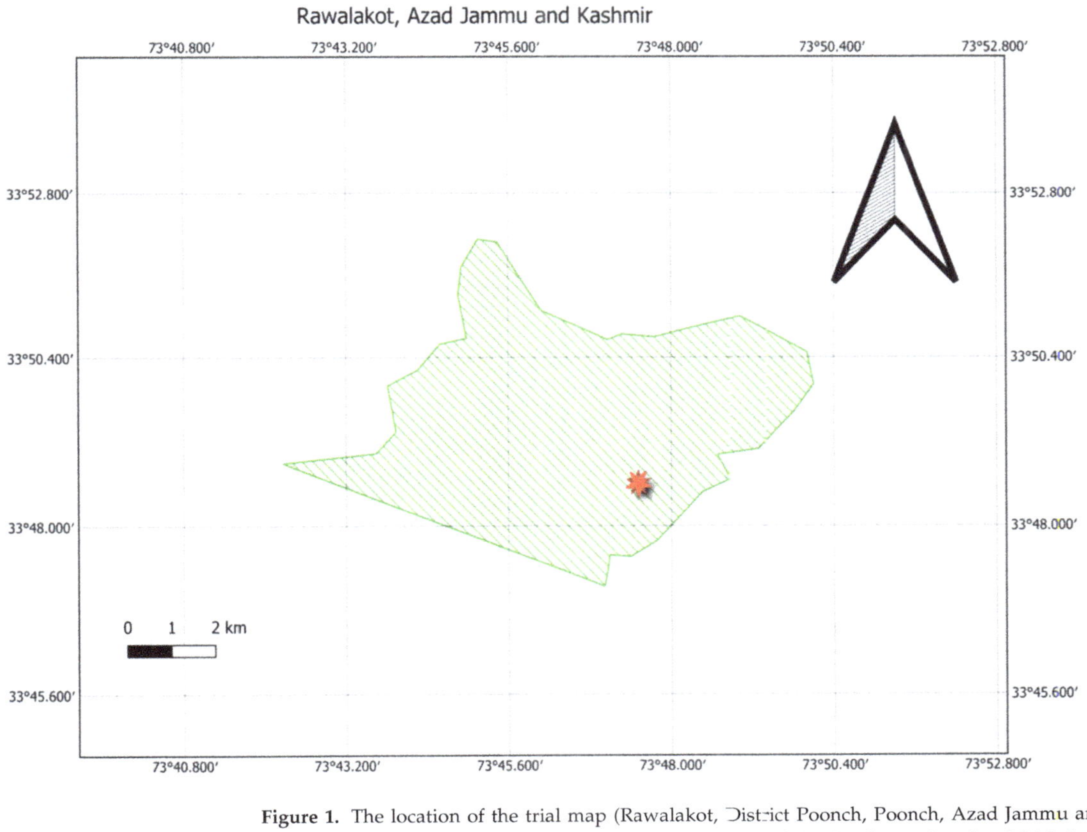

Figure 1. The location of the trial map (Rawalakot, District Poonch, Poonch, Azad Jammu and Kashmir, Pakistan) prepared for this study with the help of QGIS software (version 3.24.3, Bern, Switzerland), whereby the red star indicates the approximate location of the trial and the half-arrow depicts the North direction.

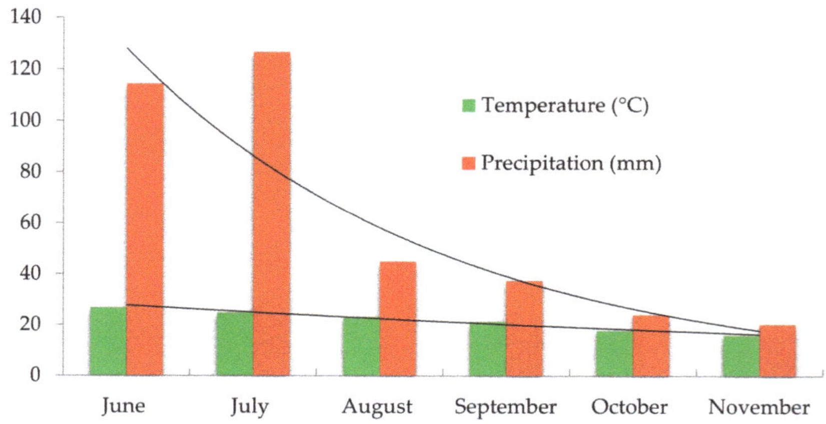

Figure 2. The study area's (Rawalakot, District Poonch, Azad Jammu and Kashmir, Pakistar meteorological features (temperature and rainfall) during the course of field trials.

2.2. *Details of Treatments and Experiment's Execution*

The field experiment was constituted of cowpea cultivars (Cowpea−,2007 an Rawan−,2013) of varying growth habits (spreading and towering ones) and different ro configurations including R × R = 15 cm, R × R = 30 cm, R × R = 45 cm and R × R = 60 cn In this way, there were eight treatment combinations in total. The execution of the field tri was performed as per randomized complete block design (RCBD) in a regular arrangemen The replications of all experimental units were maintained in triplicate. The experiment units had a net plot size of 3.6 m × 5 m, maintained after excluding the land area occupie by walking paths of 0.60 m width and 0.45 m wide bunds surrounding the experimer tal plots. In addition, fellow areas of 0.60 m were maintained among the experiment units, while a 5 m fellow area was kept among replications. Regarding fertilization, th co-application of organic (chicken manure at the rate of 5 tons ha^{-1}) and mineral fertilize (DAP at the rate of 60 kg ha^{-1}) was done as a basal dose owing to rainfed conditions. Th seeds of cowpea cultivars were hydro-primed (by pre-soaking seeds dipping in sterilize water for 12 h) in order to achieve rapid and vigorous seed germination as recommende by Iqbal et al. [35]. Thereafter, seeds were shade dried on clean muslin cloth sheets an subsequently stored at 10 °C.

Regarding seed-bed preparation, a tractor-driven common cultivator was used t plough the field thrice, while planking (wooden plank) followed each ploughing. A fin seedbed was prepared, having been thoroughly pulverized. Cowpea cultivars were sow as per treatment using a 30 kg seed rate ha^{-1} in the last week of June during both year using a single-row hand drill, and plant-to-plant spacing was maintained at 15 cm.

2.3. *Response Variables Recordings*

For recording the data of the response variables, ten plants were randomly selecte from the central rows of experimental units, and their averages were then computed fo further analyses. Plant height was determined with the help of a tailor's tap from the plan base to the tip of the uppermost leaf, and leaf area was estimated using a portable digit leaf area meter. The stem girth of cowpea plants was recorded using a vernier calipe The green herbage yield was estimated after harvesting all plants in every unit that wer separately bundled and weighed using a spring balance in the field. Thereafter, the biomas yields of experimental units were converted into a hectare basis by following Equatio (1). For the estimation of crude protein content, a macro-KJeldahl apparatus was used fc nitrogen measurement, which was multiplied by a constant of 6.25. Likewise, the H_2SO

and NaOH digestion method was followed for assessing crude fiber contents. In addition, the soxhlet extraction methodology was followed in order to determine the extractable ether percentage of forage samples of all treatments preserved in triplicate. Finally, ashing (at 600 °C) of forage samples was performed as per the muffle furnace technique for the calculation of total ash contents [35,36].

$$\text{Herbage yield of cowpea} = \frac{\text{Yield per plot} \times 10{,}000\ \text{m}^2}{\text{Plot area (m}^2\text{)}} \quad (1)$$

2.4. Statistical Analyses

The data were recorded for all response variables under study. These data were thoroughly arranged and subsequently subjected to statistical analyses using Bartlett's test, which exhibited a non-significant impact of the year, and thus, yearly data transformation into the mean values was performed for sorting out the statistical significance among employed treatments. After that, Fisher's analysis of variance (ANOVA) was put into use for estimating overall significance, while a comparison for treatment means was made using Tukey's honest significant difference (HSD) test at the level of probability of 5% using the SAS statistical package (9.2 Version, SAS Institute, Cary, NC, USA) [37,38].

3. Results and Discussion

3.1. Weeds Density and Biomass

During the course of field investigation, different types of weeds were identified in experimental units, including *Conyza bonariensis*, *Parthenium hysterophorus*, *C. canadensis*, *Cannabis sativa*, *Tagetes minuta*, *C. japonica*, *Xanthium strumarium*, *Centaurea cyanus*, *Lamium album*, *Leonurus cardiac* and *Strobilanthes urticifolia*. The findings revealed that weed density (WD) and biomass varied significantly among experimental units encompassing treatment combinations of different cowpea cultivars and row configurations (Figure 3). It was observed that Cowpea−,2007 in all row configurations recorded comparatively higher WD, especially in the row configuration of 60 cm (C_1P_4), while narrow row spacing (15 cm) (C_1P_1) allowed much lesser (39%) WD (Figure 3A). In a similar fashion, the row configuration of 45 cm (C_1P_3) recorded comparatively greater WD than 30 cm, while it exhibited a significantly lower corresponding value in comparison to the row configuration of 60 cm. In contrast, cowpea cultivar Rawan-2010 sown in the 15 cm row configuration (C_2P_1) outperformed Cowpea-2007 by recording a more significantly meager WD than the same cultivar sown in the 60 cm row spacing (C_2P_4). Following the trend, this cultivar also recorded a higher number of weeds in a wider row configuration of 60 cm compared to the 45 cm row configuration (C_2P_3). Pertaining to the fresh weight of weeds (WFW), cowpea cultivar Cowpea−,2007 sown in the row configuration of 60 cm resulted in the maximum values of weed fresh weight, while the same cultivar recorded lesser WFW in the 30 cm row configuration (Figure 3B). However, Rawan−,2010 remained superior by exhibiting a comparatively lesser WFW, especially in the row configuration of 30 cm (C_2P_1), which was 20% lesser than Cowpea−,2007 sown in the same row configuration. In addition, Rawan−,2010 sown in the 45 cm row configuration outperformed the 60 cm row spacing, but it remained inferior to the 30 cm spacing as far as WFW in forage cowpea was concerned (Figure 3C). Moreover, C_1P_4 (Cowpea−,2007 sown in the 60 cm row configuration) recorded the maximum weed dry weight (WDW), which was 7.8 g m^{-2} greater than the WDW produced by the same cultivar in the 30 cm row configuration. However, Rawan-2010 remained superior in the 15 cm row configuration (C_2P_1) by recording the minimum WDW, which was 5.6 g m^{-2} lesser than the same cultivar sown in a wider row spacing of 60 cm. Overall, Rawan−,2010 in the 30 cm row configuration recorded a 17% less WDW in comparison to Cowpea−,2007 sown in the corresponding row spacing. These findings corroborate with those of Abbas et al. [39], who opined that wider row spacing significantly enhanced the weed density owing to the greater space available for weed seeds to germinate and thrive vigorously, which led to greater intra-species competition and, ultimately, cowpea plants suffered adversely. It was also inferred that although closer row spacing reduced weed density, it also resulted in higher plant-plant competition for growth resources, which led to stunted plant height and lower vegetative growth traits of crop plants. Similarly, in our trial, the spreading type of cultivar in a closer row spacing of 30 cm provided lesser space for weed growth, which reduced weed density. Additionally, intense competition for limited growth resources, especially moisture and nutrients in closer spaced row configurations, might also be attributed to lower weed fresh weight [14–17]. In agreement with our findings, it was concluded that sub-optimal wider planting arrangements (60 cm and higher) resulted in significantly higher weed density (23–29%) and dry weight (34–41%) owing to superior agro-botanical traits of weeds, which promoted the vigorous growth of weeds by virtue of their higher nutrients and moisture uptake compared to crop plants [15,21].

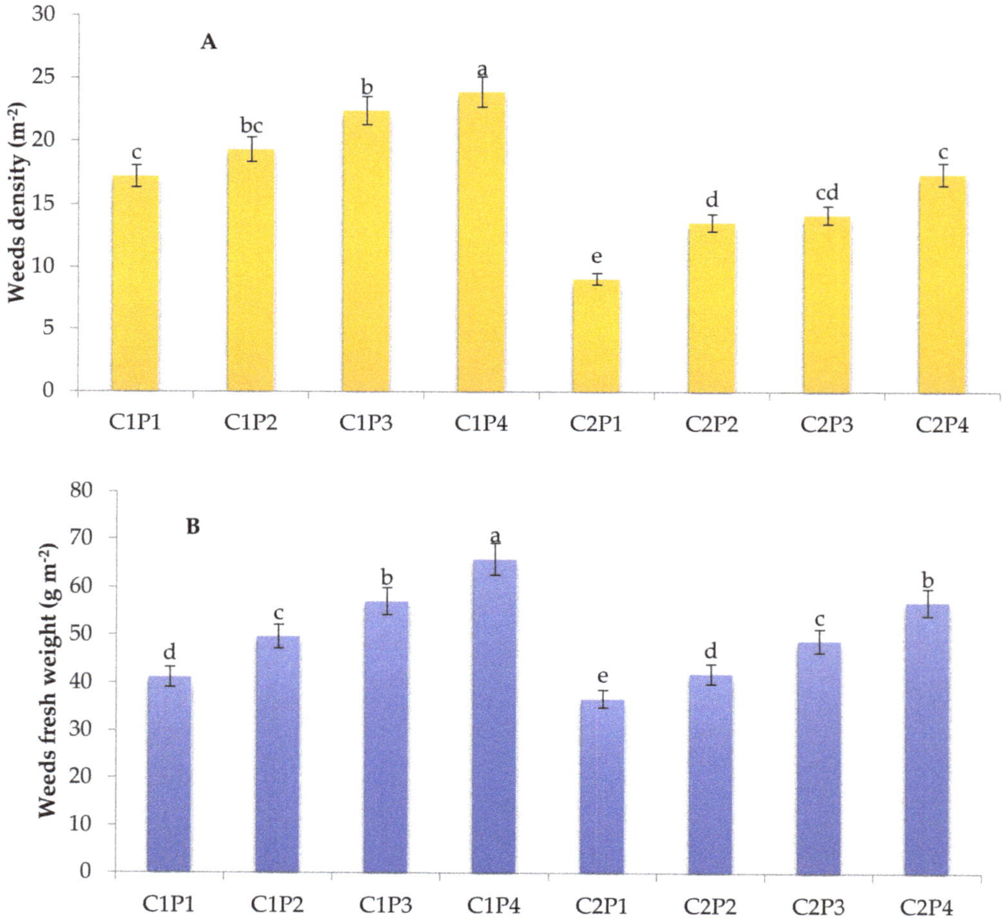

Figure 3. *Cont.*

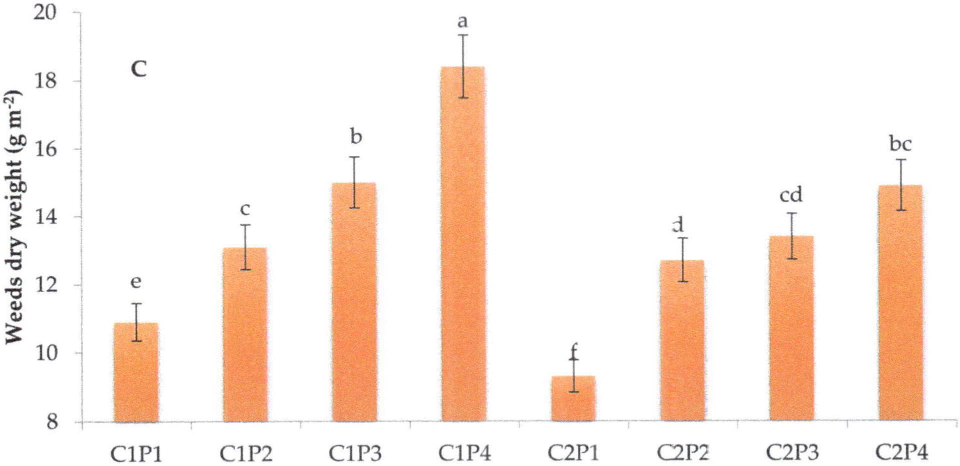

Figure 3. Weed infestation as indicated by (**A**) the density of weeds, (**B**) fresh weight of weeds and (**C**) dry weight as influenced by the genetic divergence of cowpea cultivars and row configuration under a temperate climate. C_1 = cowpea cultivar Cowpea−2007, C_2 = cowpea cultivar Rawan−2010, P_1 = R × R of 15 cm, P_2 = R × R of 30 cm, P_3 = R × R of 45 cm, P_4 = R × R of 60 cm. Column bars having different letters significantly vary at $p = 0.05$.

3.2. Yield Attributes

The research findings showed the significant influence of cowpea cultivars' genetic divergence and row configuration on the vegetative growth traits of cowpea plants under temperate climatic conditions (Table 1). As far as plant height (PH) was concerned, Cowpea−,2007 recorded the tallest plants, especially in the row configuration of 30 cm (C_1P_2), and it was followed by the same cultivar sown in the 45 cm row spacing, which recorded a 4% lesser PH than C_1P_2. The Rawan-2010 exhibited a significantly lesser PH, as C_2P_3 gave 40% less PH compared to C_1P_2; however, it remained superior to C_1P_4. In contrast to PH, Rawan−,2010 outmatched Cowpea−,2007 sown in all row configurations in terms of stem girth (Sg). In particular, the 45 cm row configuration recorded the maximum Sg. It was followed by the same cultivar sown in the 30 cm row spacing, while Cowpea−,2007 exhibited the most thin stemmed plants, particularly in the row configuration of 60 cm. Pertaining to the number of branches (BN) and leaves (LN) per plant of cowpea, Rawan−,2010 in the 45 cm row configuration (C_2P_3) recorded the maximum values, which were 8% and 6%, respectively, higher than the following treatment combination of C_2P_2. The minimum values of BN and LN were exhibited by C_1P_1, which were 79% and 87% lower than the best performing treatment combination of C_2P_3. Regarding the leaf area (La) per plant at 56 DAS (La_1) and 75 DAS (La_2), the maximum values of La_1 and La_2 were exhibited by the cowpea cultivar Rawan-2010 planted in the row configuration of R × R of 45 cm (C_2P_3), while the following treatment combination of C_2P_2 produced 84% and 79% lesser values of La_1 and La_2 in comparison to C_2P_3. The minimum La at both recordings was demonstrated by Cowpea−,2007 sown in a narrow row configuration of 15 cm (C_1P_1) by recording a 53% and 51% fewer La_1 and La_2 than the most well-performing treatment combination of C_2P_3. Interestingly, the narrowest row configuration of Cowpea-2007 remained statistically at par with the wider row spacing (C_1P_4) as far as La_1 and La_2 were concerned (Table 1). The research findings of this field trial corroborate with those of previously reported conclusions [40–42], whereby plant height was reported to be a genetically controlled trait, and appropriate agronomic management, especially optimal planting density, also remained significantly effective in producing taller plants. However, in our study, Cowpea-2007 remained taller primarily owing to its geneti traits compared to the spreading type of Rawan-2007, as both cultivars were subjected t similar row configurations in the field. Besides genetics, the 45 cm row configuration migh have resulted in the better attainment of growth resources like solar radiation, macro- an micro-nutrients, and moisture, which promoted the stem diameter and number of branche per plant [43–47]. Moreover, the significantly lower leaf area in wider row configuration might be attributed to higher weed densities, which could have restricted nutrient suppl to crop plants, and ultimately, a reduced leaf area was recorded for both cultivars of forag cowpea [14,15,48].

Table 1. Yield traits of cowpea under the interactive effect of genetic divergence and row configuratio in a temperate climate.

Treatments	Plant Height (cm)	Stem Girth (cm)	Branches Number per Plant	Leaf Number per Plant	Leaf Area per Plant (cm^2) at 56 DAS	Leaf Area per Plant (cm^2) at 70 DAS
C_1P_1	68.3 ± 0.28^{d}	5.16 ± 0.13^{e}	9.3 ± 0.04^{g}	24.3 ± 0.09^{f}	75.9 ± 0.24^{g}	91.1 ± 0.51^{h}
C_1P_2	77.7 ± 0.19^{a}	6.01 ± 0.55^{d}	12.6 ± 0.51^{d}	27.6 ± 0.50^{d}	88.6 ± 0.39^{e}	109.4 ± 0.32^{e}
C_1P_3	74.8 ± 0.81^{b}	5.13 ± 0.92^{e}	11.1 ± 0.63^{e}	26.3 ± 0.15^{e}	79.3 ± 0.63^{f}	100.9 ± 0.18^{f}
C_1P_4	58.3 ± 0.64^{c}	4.66 ± 0.28^{f}	10.7 ± 0.45^{f}	24.3 ± 0.45^{f}	76.3 ± 0.45^{g}	94.7 ± 0.29^{g}
C_2P_1	53.3 ± 0.37^{g}	6.03 ± 0.35^{d}	14.0 ± 0.07^{c}	40.3 ± 0.27^{c}	93.3 ± 0.17^{d}	115.0 ± 0.11^{d}
C_2P_2	54.7 ± 0.19^{f}	6.65 ± 0.41^{b}	15.5 ± 1.01^{b}	44.6 ± 0.11^{b}	107.6 ± 0.53^{b}	126.6 ± 0.65^{b}
C_2P_3	55.8 ± 0.41^{e}	7.07 ± 0.22^{a}	16.7 ± 0.50^{a}	46.1 ± 0.50^{a}	116.7 ± 0.21^{a}	137.1 ± 0.29^{a}
C_2P_4	53.1 ± 0.34^{g}	6.10 ± 0.19^{c}	15.3 ± 0.83^{b}	41.3 ± 0.33^{c}	100.3 ± 0.32^{c}	121.5 ± 0.32^{c}

C_1 = cowpea cultivar Cowpea−2007, C_2 = cowpea cultivar Rawan-2010, P_1 = R × R of 15 cm, P_2 = R × R of 30 cr P_3 = R × R of 45 cm, P_4 = R × R of 60 cm, DAS = days after sowing. Values having different letters within sam column vary significantly at $p = 0.05$.

3.3. Plant Fresh and Dry Weights, Green Herbage Yield and Dry Matter Biomass

The results of this field trial demonstrated that genetic divergence and row configur tion significantly influenced fresh plant weight at 60 DAS (WF_1) and 80 DAS (WF_2), dr weight at 60 DAS (WD_1) and 80 DAS (WD_2), green herbage (GH) and dry matter (DM yields (Table 2). As far as WF_1 and WF_2 were concerned, Rawan-2010 sown in 45 cm rov spacing (C_2P_3) remained superior to the rest of the row configurations and Cowpea−200 under all row spacings. This treatment combination was followed by C_2P_2; however, it pro duced 6% and 5% lesser WF_1 and WF_2, respectively, compared to C_2P_3. The correspondin minimum values of WF_1 and WF_2 were recorded for Cowpea−2007 sown in the closest rov configuration of R × R of 15 cm (C_1P_1), which were 24% and 22%, respectively, less tha C_2P_3. Regarding WD, the maximum WF_1 and WF_2 were exhibited by the cowpea cultiva of Rawan-2010 sown in the row configuration of 45 cm, while the lowest corresponding va ues were given by Cowpea−2007 planted in 15 cm R × R. In a similar fashion, Rawan-201 remained outmatched, especially in the 45 cm row configuration (C_2P_3) by recording th maximum GH and DM, while it was followed by C_2P_2, which produced 7% and 13% lesse GH and DM, respectively. Cowpea−,2007 sown in 15 cm row spacing could not perforr on par with the rest of the treatments and recorded the minimum GH and DM, which wer 35% and 68% lesser than C_2P_3. These findings are in contrast to Bange et al. [49], wh reported no significant influence of skip or additive row configuration on plant growt and weights; rather, it was opined that row spacing was usually governed by farmin needs like machinery use considerations. However, in our field trial, row configuratio influenced solar radiation interception owing to the towering and spreading nature of th cultivars, which promoted plant fresh and dry weights in 45 cm spacing, while too clos and more wide row configurations reduced plant weights owing to the intense competitio for growth resources and weed interference, respectively, which led to significantly lowe herbage yield and dry matter production. These findings support previously reported results [50–54] whereby changing the row spacing modified the water reserves available in the soil along with the pattern by which moisture became available and was uptaken by the crop plants. It was also suggested that narrow row spacing could boost plant growth in soils having poor soil structure that restrict root exploration into the skip area, and thus, narrower row spacing could record higher biomass yield owing to the superior plant population [55–57]. Conversely, narrower row spacing might also be practiced on good-structured soils having higher moisture availability because it might compensate for closer row spacing, and ultimately, plants with higher fresh and dry weights could be produced, as reported by Kerby et al. [58]. However, it was demonstrated that plant population impact in varying row configurations (replacement and additive) showed no consistent association between plant growth and plant population [56,59]; however, we may assume that these insignificant results were owing to the testing of towering type cultivars, as otherwise, an amalgamation of spreading- and towering-type cultivars could have responded positively to row configurations in terms of plant fresh and dry weights along with green herbage production [54,57,60]. Moreover, better yield attributes by Rawan-2010 in the row configuration of 45 cm might be attributed to the significantly higher green herbage yield and dry matter production, especially regarding higher plant fresh and dry weights. The greater genetic potential, higher capacity to intercept photosynthetically active radiation, improved root architecture for the uptake of moisture and nutrients along with the greater leaf area that triggered biosynthesis of carbohydrates were reported to be prime factors in boosting the biomass production of forage legumes like cowpea, soybean, cluster bean and ricebean [47,50,54,61–65].

Table 2. Fresh and dry weight per plant, green herbage yield and dry matter yield under the interactive effect of genetic divergence and row configuration in a temperate climate.

Treatments	Plant Fresh Weight (g) at 60 DAS	Plant Fresh Weight (g) at 80 DAS	Plant Dry Weight (g) at 60 DAS	Plant Dry Weight (g) at 80 DAS	Green Herbage Yield (t ha^{-1})	Dry Matter Yield (t ha^{-1})
C_1P_1	86.2 ± 0.35 ^{f}	100.6 ± 0.61 ^{f}	28.1 ± 0.11 ^{f}	34.0 ± 0.28 ^{e}	12.3 ± 0.24 ^{f}	2.9. ± 0.23 ^{g}
C_1P_2	92.1 ± 0.17 ^{d}	104.2 ± 0.28 ^{d}	30.9 ± 0.42 ^{d}	37.7 ± 0.07 ^{d}	15.2 ± 0.51 ^{c}	3.4 ± 0.50 ^{e}
C_1P_3	88.6 ± 0.63 ^{e}	102.9 ± 0.19 ^{e}	29.8 ± 0.29 ^{e}	37.2 ± 0.23 ^{d}	14.1 ± 0.16 ^{d}	3.1 ± 0.83 ^{f}
C_1P_4	88.1 ± 0.55 ^{ef}	102.1 ± 0.26 ^{e}	29.1 ± 0.09 ^{e}	35.1 ± 0.41 ^{de}	13.0 ± 0.45 ^{e}	3.0 ± 0.45 ^{d}
C_2P_1	101.2 ± 0.13 ^{c}	116.9 ± 0.41 ^{bc}	34.7 ± 0.47 ^{c}	41.9 ± 0.38 ^{cd}	14.3 ± 0.37 ^{d}	3.9 ± 0.37 ^{c}
C_2P_2	102.1 ± 0.42 ^{b}	117.9 ± 0.55 ^{b}	35.1 ± 0.461 ^{b}	43.1 ± 0.23 ^{b}	15.6 ± 0.11 ^{b}	4.4 ± 0.16 ^{b}
C_2P_3	107.9 ± 0.69 ^{a}	123.8 ± 0.12 ^{a}	38.6 ± 0.25 ^{a}	47.9 ± 0.30 ^{a}	16.7 ± 0.50 ^{a}	4.9 ± 0.51 ^{a}
C_2P_4	101.4 ± 0.29 ^{c}	114.0 ± 0.19 ^{c}	34.6 ± 0.33 ^{c}	42.0 ± 0.14 ^{c}	15.3 ± 0.63 ^{c}	4.0 ± 0.63 ^{c}

C_1 = cowpea cultivar Cowpea−2007, C_2 = cowpea cultivar Rawan−2010, P_1 = R × R of 15 cm, P_2 = R × R of 30 cm, P_3 = R × R of 45 cm, P_4 = R × R of 60 cm, DAS = days after sowing. Values having different letters with same column vary significantly at $p = 0.05$.

3.4. Nutritional Quality Attributes

The research findings depicted the significant influence of genetic divergence and row configuration on the nutritional value of forage cowpea (Figure 4). The protein content (Cp) enhances the nutritional value of forages, and the maximum Cp content was produced by Rawan-2010 sown in a row configuration of 60 cm (C_2P_4), while the following treatment combination of C_2P_3 recorded fewer Cp (Figure 4a. The minimum Cp content was recorded for Cowpea−2007 sown in the closer row configuration of 15 cm (C_1P_1). Overall, Rawan-2010 outperformed Cowpea−2007, and a row configuration of 60 cm remained unmatched as far as the Cp of forage cowpea was concerned. The increased crude fiber (Cf) content deteriorates the nutritive quality of forages, and Cowpea−2007 remained inferior, especially in a row configuration of 15 cm, by producing the maximized Cf (Figure 4b). Interestingly,

C_2P_3 (Rawan-2010 in 45 cm row spacing) gave the minimum Cf, which was 18% less in comparison to C_1P_1. This treatment combination was followed by C_2P_2, which exhibited 11% less Cf content than C_1P_1. In a similar fashion to Cp, Cowpea−2007 remained inferior to Rawan-2007 under all spatial arrangements by demonstrating increased Cf content. Pertaining to total ash (Ta) content, Rawan-2010 sown in the widest row configuration (C_2P_4) resulted in the maximum Ta content, and it was followed by the same cultivar sown in the spatial arrangement of 30 cm (C_2P_3) (Figure 4c). Cowpea-2007 could not perform on par with Rawan-2010, and the minimum value of Ta was exhibited by the closest row spacing of 15 cm, which was 15% less than the most well-performing treatment combination of C_2P_4. The research findings of this field trial remained in line with those of Iqbal et al. [22], who opined that leaves are a rich source of crude protein, and a higher leaf number along with a greater leaf area resulted in a significantly higher Cp content of forage soybean. It was also suggested that optimized row arrangement favored nutrient uptake, especially of nitrogen, which boosted amino acid biosynthesis, and ultimately, greater Cp content was recorded, which improved the nutritive value of the forage for dairy animals. Additionally, genetic divergence was also reported to be one of the prime reasons for differences in Cp in forage legume crops, and it was inferred that genetic potential and agronomic package determine the Cp content of forage crops [40,47]. Moreover, it was opined that high planting densities delayed crop switching to reproductive growth, which might be presumably attributed to a higher accumulation of amino acids, which improved Cp content [53]. There exists an inverse relationship between crude fiber content and nutritive quality of forage, as most fractions of fiber are non-digestible by ruminants, and thus, their higher concentration deteriorates feed quality. The lowest quality forage with the maximum fiber content was recorded for the towering type cultivar sown in the narrow row spacing, which was presumably attributed to restricted nutrient supply owing to intense competition for growth resources. Ultimately, higher fiber content was produced. Additionally, the same cultivar had a higher plant height, and owing to its taller stem enriched with fiber, greater fiber contents were recorded. These findings are in agreement with previous studies [47,64,65], whereby taller plants resulted in higher fiber content, and there existed an antagonistic association between crude protein and fiber contents. Similarly, total ash presents the mineral constituents of forages, which are required by animals to maintain the normal functioning of the body. Higher mineral contents improve the nutritive value of feeds [2]. The spreading type cultivar in the widest row configuration resulted in the maximum ash, presumably owing to higher nutrient uptake. These findings corroborate with those of [60,61,63], who opined that row configuration imparted significant influence on mineral constituents of forage crops (both cereals and leguminous forages) through optimization of solar radiation interception and uptake of macro- and micro-nutrients from the soil solution.

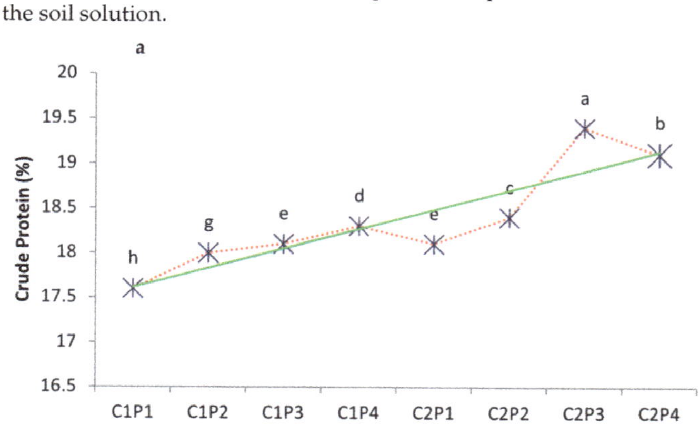

Figure 4. *Cont.*

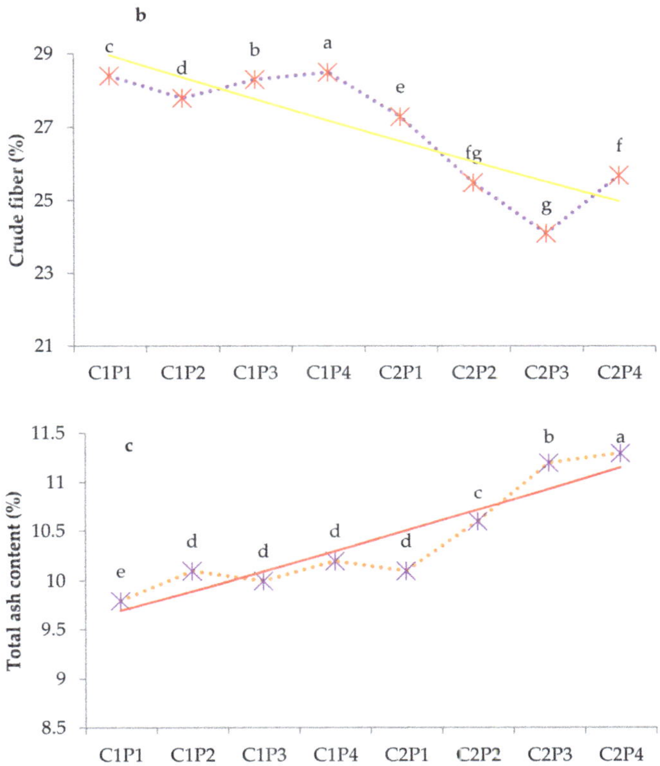

Figure 4. Nutritive quality: (**a**) protein content, (**b**) crude fiber content and (**c**) total ash content of forage cowpea cultivars sown under varying row configurations (details of treatments are presented in the footnote of Tables 1 and 2). Different letters show significant difference at $p = 0.05$.

3.5. Correlation among Yield Attributes, Seed Yield and Biological Yield

As per correlation analyses, green herbage yield had significant linear relationships with vegetative yield attributes except for the plant height of temperate cowpea cultivars sown under varying configurations. The results indicated a significantly negative association of green herbage yield with plant height ($R^2 = -0.84^*$) (Figure 5a) and a moderately significantly direct relationship with stem girth ($R^2 = 0.82^*$) (Figure 5b) and the number of leaves per plant ($R^2 = 0.64^*$) (Figure 5c). In contrast, leaf area per plant ($R^2 = 0.89^{**}$) of cowpea cultivars sown under different row configurations had a significantly stronger direct relationship with herbage yield (Figure 5d). In a similar fashion, fresh weight per plant of cowpea cultivars ($R^2 = 0.87^{**}$) (Figure 5e) was strongly associated with green herbage yield of cowpea compared to other growth attributes like stem girth and number of leaves per plant. Lastly, correlation analysis of dry weight per plant with dry matter yield of cowpea cultivars sown in different row configurations ($R^2 = 0.87^{**}$) also exhibited a stronger linear association (Figure 4f). These findings pertaining to the negative correlation of plant height with green herbage yield of cowpea cultivars are in contradiction with those of Iqbal et al. [46], who opined that plant height was linearly associated with soybean yield as taller plants assisted in dominating weed populations. Additionally, greater plant height imparted an upper edge to crop plants for up-taking more nutrients from soil solution along with efficient interception of photosynthetically active radiation (PAR) [61,62] compared to many types of indigenous and exotic weeds. Previously, it was reported that stem girth and the number of leaves per plant along with leaf area exhibited stronger direct associations with yield because the maximum stem girth, being the heaviest pa of the plant, contributed significantly towards herbage yield, while greater leaf numbe and leaf area triggered the photosynthetic rate, which led to maximum biosynthesis c carbohydrates and, ultimately, enhanced yield [54,55,61]. It was also suggested that amon vegetative growth traits, leaf area per plant might be used as a reliable indicator for th estimation of crop yield. Likewise, fresh plant weight was directly associated with th green herbage yield of cowpea, indicating higher growth supported by optimized rov configuration. Presumably, an improved light environment and use of radiation serv as vital factors enabling plants to attain weight as per genetic potential, which leads t maximized biomass production. It has been suggested that better canopy architecture an greater leaf pigments enhanced intercepted PAR which promoted biomass accumulatio by crop plants and thus contributed significantly to boosting crop biomass yields [54 Moreover, optimized row configurations facilitated positive changes in canopy structur along with photosynthetic capacity, which improved above-ground biomass accumulatio as reported by Xue et al. [54] and Zhang et al. [60]. Furthermore, greater fresh weight pe plant resulted in higher dry weight per plant, which was linearly associated with the dr matter yield of cowpea. This might be attributed to better growth and biomass accumula tion which increased dry weight per plant and, in turn, increased the dry matter yield c forage legumes as reported by Lithourgidis et al. [63], Ismail and Hall [64], Iqbal et al. [55 and Basaran et al. [53].

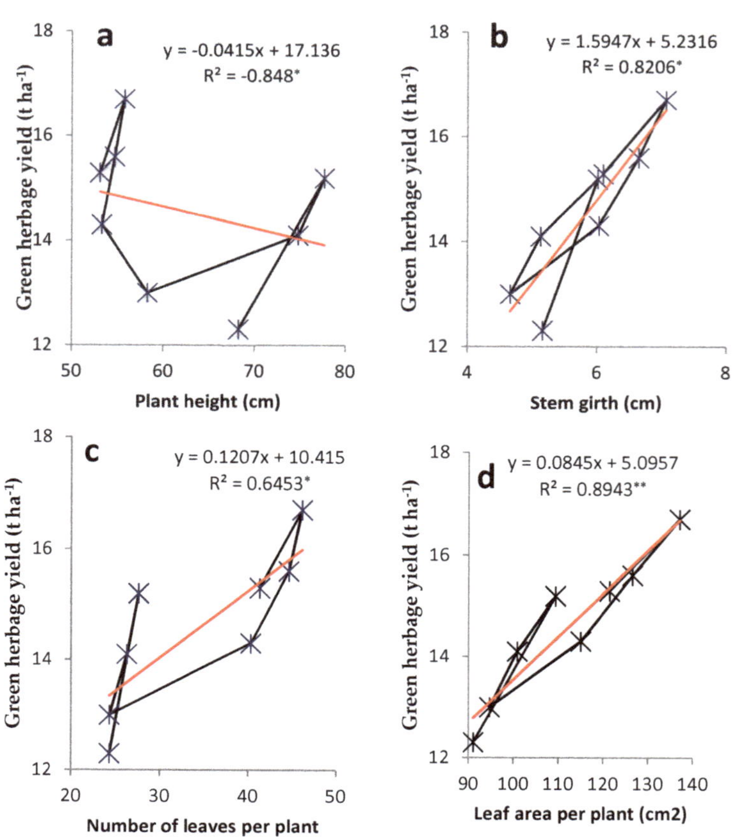

Figure 5. *Cont.*

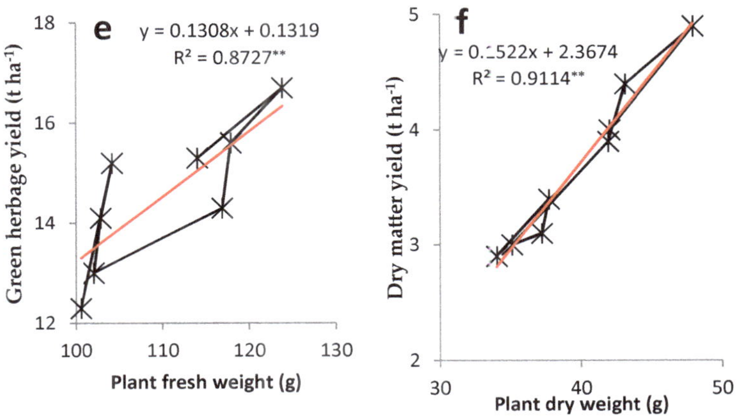

Figure 5. Correlation analyses depicting the relationship between herbage yield and yield attributes such as (**a**) plant height (**a**), stem girth (**b**), leaf area per plant (**c**), number of leaves per plant, (**d**) leaf area per plant, (**e**) fresh weight per plant and (**f**) dry weight per plant. ** and * show highly significant and moderate significant differences at $p = 0.05$ respectively.

4. Conclusions

Here, we have explored the production potential of cowpea to be promoted as an alternative leguminous forage crop in the summer-rainfall environment of the temperate Himalayan region of Pakistan. In this region, farmers and agronomists have been consistently pointing out the dire need for a new competitive forage legume crop in order to provide sustainable supplies of quality forage to dairy animals. From the research findings of this field trial, it might be inferred that cowpea (Rawan-2010 sown in 45 cm row configuration) could potentially serve as a resilient short-duration forage crop having the potential to provide abundant green forage (around 17 t ha^{-1}) of superior quality (higher crude protein and total ash content and lower crude fiber content). Additionally, the spreading type of cowpea in a narrow row configuration (30 cm) remained effective in suppressing weed density and their fresh and dry weights, which holds a bright perspective for regions having intensive weed infestations. Moreover, the cultivation package, low requisite mechanization, and profitable farming of numerous food legumes have already been established in the Azad Jammu and Kashmir region of Pakistan; therefore, the expansion of forage cowpea would not pose a challenge, and farmers might conveniently incorporate cowpea in prevalent farming systems of the region. Moreover, the same scenario pertaining to cultivar growth habit and row configuration might be applicable to other cropping areas having similar pedo-climatic conditions globally Furthermore, the capability of cowpea to withstand the intermittent drought spells and inconsistent rainfall under rainfed conditions in temperate areas needs further field evaluation. Meanwhile, future studies must encompass an evaluation of the economic viability and profitability of forage cowpea compared to other food legumes.

Author Contributions: Conceptualization, M.A.I.; methodology, M.A.I., A.A. and M.I.; validation, M.A.I. and M.I.; formal analysis, A.A.H., R.M.H. and M.A.I.; writing—original draft preparation, A.A.H., R.M.H., M.A.I. and H.E.A.; writing—review and editing, A.A.H., R.M.H., M.A.I. and H.E.A. All authors have read and agreed to the published version of the manuscript.

Funding: This research received no external funding

Institutional Review Board Statement: Not applicable.

Informed Consent Statement: Not applicable.

Conflicts of Interest: The authors declare no conflict of interest.

Abbreviations

WD	weeds density
PH	plant height
GH	green herbage yield
DM	dry matter biomass
R × R	row to row spacing
RCBD	randomized complete block design
DAP	di-ammonium phosphate
DAS	days after sowing
ANOVA	analysis of variance
CP	crude protein

References

1. Faville, M.J.; Cao, M.; Schmidt, J.; Ryan, D.L.; Ganesh, S.; Jahufer, M.Z.Z.; Hong, S.W.; George, R.; Barrett, B.A. Divergent Genomic Selection for Herbage Accumulation and Days-To-Heading in Perennial Ryegrass. *Agronomy* **2020**, *10*, 340. [CrossRef]
2. Iqbal, M.A.; Asif, I.; Akbar, N.; Khan, H.Z.; Abbas, R.N. A study on feed stuffs role in enhancing the productivity of milch animals in Pakistan-Existing scenario and future prospect. *Glob. Vet.* **2015**, *14*, 23–33.
3. Iqbal, M.A.; Hamid, A.; Siddiqui, M.H.; Hussain, I.; Ahmad, T.; Ishaq, T.; Ali, A. A meta-analysis of the impact of foliar feeding of micronutrients on productivity and revenue generation of forage crops. *Planta Daninha* **2019**, *37*, e019189237. [CrossRef]
4. El-Taher, A.M.; Abd El-Raouf, H.S.; Osman, N.A.; Azoz, S.N.; Omar, M.A.; Elkelish, A.; Abd El-Hady, M.A.M. Effect of Salt Stress and Foliar Application of Salicylic Acid on Morphological, Biochemical, Anatomical, and Productivity Characteristics of Cowpea (*Vigna unguiculata* L.) Plants. *Plants* **2022**, *11*, 115. [CrossRef] [PubMed]
5. Iqbal, M.A.; Imtiaz, H.; Abdul, H.; Bilal, A.; Saira, I.; Ayman, S.; Celaleddin, B.; Rana, D.K.; Imran, M. Soybean herbage yield, nutritional value and profitability under integrated manures management. *Anais Acad. Brasil. Cienc.* **2021**, *93*, e20181384. [CrossRef]
6. Ehlers, J.D.; Hall, A.E. Cowpea (*Vigna unguiculata* L. Walp.). *Field Crop. Res.* **1997**, *53*, 187–204. [CrossRef]
7. Huynh, B.L.; Ehlers, J.D.; Huang, B.E.; Muñoz-Amatriaín, M.; Lonardi, S.; Santos, J.R.; Roberts, P.A. A Multi-Parent Advanced Generation Inter-Cross (MAGIC) population for genetic analysis and improvement of cowpea (*Vigna unguiculata* L. Walp.). *Plant J.* **2018**, *93*, 1129–1142. [CrossRef]
8. Raina, A.; Laskar, R.A.; Tantray, Y.R.; Khursheed, S.; Wani, M.R.; Khan, S. Characterization of Induced High Yielding Cowpea Mutant Lines Using Physiological, Biochemical and Molecular Markers. *Sci. Rep.* **2020**, *10*, 3687. [CrossRef]
9. Jayathilake, C.; Visvanathan, R.; Deen, A.; Bangamuwage, R.; Jayawardana, B.C.; Nammi, S.; Liyanage, R. Cowpea: An overview on its nutritional facts and health benefits. *J. Sci. Food Agric.* **2018**, *98*, 4793–4806. [CrossRef]
10. Baker, C.; Modi, A.T.; Nciizah, A.D. Weeding Frequency Effects on Growth and Yield of Dry Bean Intercropped with Sweet Sorghum and Cowpea under a Dryland Area. *Sustainability* **2021**, *13*, 12328. [CrossRef]
11. Iqbal, M.A. Improving germination and seedling vigour of cowpea (*Vigna unguiculata* L.) with different priming techniques. *Am.-Eur. J. Agric. Environ. Sci.* **2015**, *15*, 265–270.
12. Iqbal, M.A. Evaluation of forage cowpea and hey as a feed resource for ruminant production: A mini-review. *Glob. Vet.* **2015**, *14*, 747–751.
13. Choudhary, V.K.; Kumar, P.S. Weed suppression, nutrient leaching, water use and yield of turmeric (*Curcuma longa* L.) under different land configurations and mulches. *J. Clean. Prod.* **2018**, *210*, 795–803. [CrossRef]
14. Arce, G.D.; Pedersen, P.; Hartzler, R.G. Soybean seeding rate effects on weed management. *Weed Technol.* **2009**, *23*, 17–22. [CrossRef]
15. Kolb, L.N.; Gallandt, E.R.; Mallory, E.B. Impact of spring wheat planting density, row spacing, and mechanical weed control on yield, grain protein, and economic return in Maine. *Weed Sci.* **2012**, *60*, 244–253. [CrossRef]
16. Mennan, H.; Ngouajio, M.; Isık, D.; Kaya, E. Effects of alternative management systems on weed populations in hazelnut (*Corylus avellana* L.). *Crop Prot.* **2006**, *25*, 835–841. [CrossRef]
17. Rajcan, I.; Swanton, C.J. Understanding maize-weed competition: Resource competition, light quality and the whole plant. *Field Crop Res.* **2001**, *71*, 139–150. [CrossRef]
18. Ryan, M.R.; Mirsky, S.B.; Mortensen, D.A.; Teasdale, J.R.; Curran, W.S. Potential synergistic effects of cereal rye biomass and soybean planting density on weed suppression. *Weed Sci.* **2011**, *59*, 238–246. [CrossRef]
19. Stephenson, D.O.; Barry, J.; Brecke, B.J. Weed management in single- vs. twin-row cotton (*Gossypium hirsutum*). *Weed Technol.* **2010**, *24*, 275–280. [CrossRef]
20. Abudulai, M.; Kusi, F.; Seini, S.S.; Seidu, A.; Nboyine, J.A.; Larbi, A. Effects of planting date, cultivar and insecticide spray application for the management of insect pests of cowpea in northern Ghana. *Crop Prot.* **2017**, *100*, 168–176. [CrossRef]
21. Adigun, J.; Osipitan, A.O.; Lagoke, S.T.; Adeyemi, R.O.; Afolami, S.O. Growth and yield performance of cowpea (*Vigna unguiculata* (L.) Walp) as influenced by row-spacing and period of weed interference in South-West Nigeria. *J. Agric. Sci.* **2014**, *6*, 188. [CrossRef]

2. Iqbal, M.A.; Iqbal, A.; Abbas, R.N. Spatio-temporal reconciliation to lessen losses in yield and quality of forage soybean (*Glycine max* L.) in soybean-sorghum intercropping systems. *Bragantia* **2018**, *77*, 283–291 [CrossRef]
3. Iqbal, M.A. Comparative performance of forage cluster bean accessions as companion crops with sorghum under varied harvesting times. *Bragantia* **2018**, *77*, 476–484. [CrossRef]
4. Staggenborg, S.A.; Fjell, D.L.; Devlin, D.L.; Gordon, W.B.; Marsh, B.H. Grain sorghum response to row spacings and seeding rates in Kansas. *J. Prod. Agric.* **1999**, *12*, 390–395. [CrossRef]
5. Bandaru, V.; Stewart, B.A.; Baumhardt, R.L.; Ambati, S.; Robinson, C.A.; Schlegel, A. Growing dryland grain sorghum in clumps to reduce vegetative growth and increase yield. *Agron. J.* **2006**, *98*, 1109–1120. [CrossRef]
6. M'Khaitir, Y.O.; Vanderlip, R.L. Grain sorghum and pearl millet response to date and rate of planting. *Agron. J.* **1992**, *84*, 579–582. [CrossRef]
7. Iqbal, M.A.; Rahim, J.; Naeem, W.; Hassan, S.; Khattab, Y.; Sabagh, A. Rainfed winter wheat (*Triticum aestivum* L.) cultivars respond differently to integrated fertilization in Pakistan. *Fresen. Environ. Bull.* **2021**, *30*, 3115–3121.
8. Black, C.A. *Methods of Soil Analysis, Part II*; American Society of Agronomy: Madison, WI, USA, 1965.
9. Sparks, D.L.; Page, A.L.; Helmke, P.A.; Leoppert, R.H.; Soltanpour, P.N.; Tabatabai, M.A.; Johnston, G.T.; Sumner, M.E. *Methods of Soil Analysis*; Soil Science Society of America: Madison, WI, USA, 1996.
0. Naresh, R.K.; Tomar, S.S.; Kumar, D.; Samsher, P.; Singh, S.; Dwivedi, P.; Kumar, A. Experiences with rice grown on permanent raised beds: Effect of crop establishment techniques on water use, productivity, profitability and soil physical properties. *Rice Sci.* **2014**, *21*, 170–180. [CrossRef]
1. AOAC. *Official Methods of Analysis*, 17th ed.; Association of Official Analytical Chemists: Arlington, VI, USA, 2003.
2. Jackson, M.L. *Soil Chemical Analysis*; Prentice Hall: Englewood Cliffs, NJ, USA, 1962.
3. Piper, C.S. *Soil and Plant Analysis*; Press Adelaide: Adelaide, Australia, 1950.
4. Ryan, J.; Estefan, G.; Rashid, A. *Soil and Plant Analysis Laboratory Manual*, 2nd ed.; International Center for Agricultural Research in the Dry Areas (ICARDA): Aleppo, Syria; National Agriculture Research Center (NARC): Islamabad, Pakistan, 2001.
5. Iqbal, M.A.; Raza, R.Z.; Zafar, M.; Ali, O.M.; Ahmed, R.; Rahim, J.; Ijaz, R.; Ahmad, Z.; Bethune, B.J. Integrated Fertilizers Synergistically Bolster Temperate Soybean Growth, Yield, and Oil Content. *Sustainability* **2022**, *14*, 2433. [CrossRef]
6. Jabbar, A.; Iqbal, A.; Iqbal, M.A.; Sheikh, U.A.A.; Rahim, J.; Khalid, S.; Hafez, R.M.; Shah, A.-u.-H.; Khan, A.A.; Bazmi, M.S.A.; et al. Egyptian Clover Genotypic Divergence and Last Cutting Management Augment Nutritive Quality, Seed Yield and Milk Productivity. *Sustainability* **2022**, *14*, 5833. [CrossRef]
7. Steel, R.G.D.; Torrie, J.H., Dickey, D. *Principles and Procedures of Statistics: A Biometrical Approach*; McGraw Hill Book Co., Inc.: New York, NY, USA, 1997; pp. 172–177.
8. SAS Institute. *The SAS System for Windows: Version 8.02*; SAS Institute Inc.: Cary, NC, USA, 2003.
9. Abbas, R.N.; Arshad, M.A.; Iqbal, A.; Iqbal, M.A.; Imran, M.; Raza, A.; Chen, J.-T.; Alyemeni, M.N.; Hefft, D.I. Weeds Spectrum, Productivity and Land-Use Efficiency in Maize-Gram Intercropping Systems under Semi-Arid Environment. *Agronomy* **2021**, *11*, 1615. [CrossRef]
0. Shilpashree, N.; Devi, S.N.; Manjunathagowda, D.C.; Muddappa, A.; Abdelmohsen, S.A.M.; Tamam, N.; Elansary, H.O.; El-Abedin, T.K.Z.; Abdelbacki, A.M.M.; Janhavi, V. Morphological Characterization, Variability and Diversity among Vegetable Soybean (*Glycine max* L.) Genotypes. *Plants* **2021**, *10*, 671. [CrossRef] [PubMed]
1. Manoj, K.N.; Shekara, B.G.; Sridhara, S.; Jha, P.K.; Prasad, P.V.V. Biomass Quantity and Quality from Different Year-Round Cereal–Legume Cropping Systems as Forage or Fodder for Livestock. *Sustainability* **2021**, *13*, 9414. [CrossRef]
2. Osipitan, O.A.; Fields, J.S.; Lo, S.; Cuvaca, I. Production Systems and Prospects of Cowpea (*Vigna unguiculata* (L.) Walp.) in the United States. *Agronomy* **2021**, *11*, 2312. [CrossRef]
3. Jayawardhane, J.; Goyali, J.C.; Zafari, S.; Igamberdiev, A.U. The Response of Cowpea (*Vigna unguiculata*) Plants to Three Abiotic Stresses Applied with Increasing Intensity: Hypoxia, Salinity, and Water Deficit. *Metabolites* **2022**, *12*, 38. [CrossRef]
4. Salama, H.S.A.; Nawar, A.I.; Khalil, H.E. Intercropping Pattern and N Fertilizer Schedule Affect the Performance of Additively Intercropped Maize and Forage Cowpea in the Mediterranean Region. *Agronomy* **2022**, *12*, 107. [CrossRef]
5. Amaral, J.B.C.; Lopes, F.B.; Magalhães, A.C.M.d.; Kujawa, S.; Taniguchi, C.A.K.; Teixeira, A.d.S.; Lacerda, C.F.d.; Queiroz, T.R.G.; Andrade, E.M.d.; Araújo, I.C.d.S.; et al. Quantifying Nutrient Content in the Leaves of Cowpea Using Remote Sensing. *Appl. Sci.* **2022**, *12*, 458. [CrossRef]
6. Iqbal, M.A.; Bethune, B.J.; Iqbal, A.; Abbas, R.N.; Aslam, Z.; Khan, H.Z.; Ahmad, B. Agro-botanical response of forage sorghum-soybean intercropping systems under atypical spatio-temporal patterns. *Pak. J. Bot* **2017**, *49*, 987–994.
7. Iqbal, M.A.; Hamid, A.; Hussain, I.; Siddiqui, M.H.; Ahmad, T.; Khaliq, A.; Ahmad, Z. Competitive indices in cereal and legume mixtures in a South Asian environment. *Agron. J.* **2019**, *111*, 242–249. [CrossRef]
8. Rudra, S.G.; Shruti, S.; Jha, S.K.; Rajeev, K. Physico-chemical and functional properties of cowpea protein isolateas affected by the dehydration technique. *Leg. Res.* **2016**, *39*, 370–378.
9. Bange, M.P.; Carberry, B.S.; Marshall, J.; Milroy, S.P. Row configuration as a tool for managing rain-fed cotton systems: Review and simulation analysis. *Aust. J. Exp. Agric.* **2005**, *45*, 65–77. [CrossRef]
0. Hakan, G.; Avcioglu, R.; Soya, H.; Kir, B. Intercropping of corn with cowpea and bean Biomass yield and silage quality. *Afr. J. Biotech.* **2008**, *22*, 4100–4104.

51. Rahman, A.N.; Larbi, A.; Kotu, B.; Marthy Tetteh, F.; Hoeschle-Zeledon, I. Does Nitrogen Matter for Legumes? Starter Nitroge Effects on Biological and Economic Benefits of Cowpea (*Vigna unguiculata* L.) in Guinea and Sudan Savanna of West Afric *Agronomy* **2018**, *8*, 120. [CrossRef]
52. Zaheer, S.; Arif, M.; Akhtar, K.; Khan, A.; Khan, A.; Bibi, S.; Muhammad Aslam, M.; Ali, S.; Munsif, F.; Jalal, F.; et al. Grazing an Cutting under Different Nitrogen Rates, Application Methods and Planting Density Strongly Influence Qualitative Traits an Yield of Canola Crop. *Agronomy* **2020**, *10*, 404. [CrossRef]
53. Basaran, U.; Ayan, I.; Acar, Z.; Mut, H.; Asci, O.O. Seed yield and agronomic parameters of cowpea (*Vigna unguiculata* L genotypes grown in the Black Sea region of Turkey. *Afr. J. Biotechnol.* **2011**, *10*, 13461. [CrossRef]
54. Iqbal, M.A.; Iqbal, A.; Ahmad, Z.; Raza, A.; Rahim, J.; Imran, M.; Sheikh, U.A.A.; Maqsood, Q.; Soufan, W.; Sahloul, N.M.A.; et a Cowpea [*Vigna unguiculata* (L.) Walp] herbage yield and nutritional quality in cowpea-sorghum mixed strip intercroppin systems. *Rev. Mex. Cienc. Pec.* **2021**, *12*, 402–418. [CrossRef]
55. Iqbal, M.A.; Siddiqui, M.H.; Afzal, S.; Ahmad, Z.; Maqsood, Q.; Khan, R.D. Forage productivity of cowpea [*Vigna unguiculata* (L Walp] cultivars improves by optimization of spatial arrangements. *Rev. Mex. Cien. Pec.* **2018**, *9*, 203–219. [CrossRef]
56. Adams, P.D.; Weaver, D.B. Brachytic stem trait, row spacing, and plant population effects on soybean yield. *Crop Sci.* **1998**, *3* 750–755. [CrossRef]
57. Kwapata, M.B.; Hall, A.E. Response of contrasting vegetable-cowpea cultivars to plant density and harvesting of young gree pods. I. Pod production. *Field Crops Res.* **1990**, *24*, 1–10. [CrossRef]
58. Kerby, T.A.; Cassman, K.G.; Keeley, M. Genotypes and plant densities for narrow-row cotton systems. 1. Height, nodes, earlines and location of yield. *Crop Sci.* **1990**, *30*, 644–649. [CrossRef]
59. Xue, W.; Lindner, S.; Nay-Htoon, B.; Dubbert, M.; Otieno, D.; Ko, J.; Muraoka, H.; Werner, C.; Tenhunen, J.; Harley, P. Nutritiona and developmental influences on components of rice crop light use efficiency. *Agric. For. Meteorol.* **2016**, *223*, 1–16. [CrossRef]
60. Zhang, L.; van der Werf, W.; Bastiaans, L.; Zhang, S.; Li, B.; Spiertz, J.H.J. Light interception and utilization in relay intercrops c wheat and cotton. *Field Crops Res.* **2008**, *107*, 29–42. [CrossRef]
61. Liu, X.; Rahman, T.; Yang, F.; Song, C.; Yong, T.; Liu, J.; Zhang, C.; Yang, W. PAR interception and utilization in different maiz and soybean intercropping patterns. *PLoS ONE* **2017**, *12*, e0169218. [CrossRef]
62. Mattera, J.; Romero, L.A.; Cuatrín, A.L.; Cornaglia, P.S.; Grimoldi, A.A. Yield components, light interception and radiation us efficiency of lucerne (*Medicago sativa* L.) in response to row spacing. *Eur. J. Agron.* **2013**, *45*, 87–95. [CrossRef]
63. Raza, M.A.; Feng, L.Y.; van der Werf, W.; Cai, G.R.; Khalid, M.H.B.; Iqbal, N.; Hassan, M.J.; Meraj, T.A.; Naeem, M.; Khan, I.; et a Narrow-wide-row planting pattern increases the radiation use efficiency and seed yield of intercrop species in relay- intercroppin system. *Food Energy Secur.* **2019**, *8*, e170. [CrossRef]
64. Lithourgidis, A.S.; Vasilakoglou, I.B.; Dhima, K.V.; Dordas, C.A.; Yiakoulaki, M.D. Forage yield and quality of common vetc mixtures with oat and triticale in two seeding ratios. *Field Crops Res.* **2006**, *99*, 106–113. [CrossRef]
65. Ismail, A.M.; Hall, A.E. Semi-dwarf and standard-height cowpea responses to row spacing in different environments. *Crop Sc* **2001**, *40*, 1618–1623. [CrossRef]

agronomy

MDPI

Article

Seed Priming and Foliar Application of Nutrients Influence the Productivity of Relay Grass Pea (*Lathyrus sativus* L.) through Accelerating the Photosynthetically Active Radiation (PAR) Use Efficiency

Purabi Banerjee [1], Visha Kumari Venugopalan [2], Rajib Nath [1], Prodip Kumar Chakraborty [3], Ahmed Gaber [4,*], Walaa F. Alsanie [5], Bassem M. Raafat [6] and Akbar Hossain [7,*]

1 Department of Agronomy, Bidhan Chandra Krishi Viswavidyalaya, Mohanpur 741252, West Bengal, India; itsmepurabi1@gmail.com (P.B.); rajibbckv@yahoo.com (R.N.)
2 Division of Crop Sciences, ICAR—Central Research Institute for Dryland Agriculture, Hyderabad 500059, Telengana, India; visha.venugopal@gmail.com
3 Department of Agricultural Meteorology and Physics, Bidhan Chandra Krishi Viswavidyalaya, Mohanpur 741252, West Bengal, India; pkc_bckv@rediffmail.com
4 Department of Biology, College of Science, Taif University, P.C. Box 11099, Taif 21944, Saudi Arabia
5 Department of Clinical Laboratories Sciences, The Faculty of Applied Medical Sciences, Taif University, P.O. Box 11099, Taif 21944, Saudi Arabia; w.alsanie@tu.edu.sa
6 Department of Radiological Sciences, College of Applied Medical Sciences, Taif University, P.O. Box 11099, Taif 21944, Saudi Arabia; bassemraafat@tu.edu.sa
7 Department of Agronomy, Bangladesh Wheat and Maize Research Institute (BWMRI), Nashipur, Dinajpur 5200, Bangladesh
* Correspondence: a.gaber@tu.edu.sa (A.G.); akbar.hossain@bwmri.gov.bd (A.H.)

Citation: Banerjee, P.; Venugopalan, V.K.; Nath, R.; Chakraborty, P.K.; Gaber, A.; Alsanie, W.F.; Raafat, B.M.; Hossain, A. Seed Priming and Foliar Application of Nutrients Influence the Productivity of Relay Grass Pea (*Lathyrus sativus* L.) through Accelerating the Photosynthetically Active Radiation (PAR) Use Efficiency. *Agronomy* **2022**, *12*, 1125. https://doi.org/10.3390/agronomy12051125

Academic Editors: Umberto Anastasi, Aurelio Scavo and Adrian C. Newton

Received: 14 November 2021
Accepted: 5 May 2022
Published: 6 May 2022

Abstract: The efficiency of a crop to intercept and utilize solar radiation for photosynthates production serves as one of the deciding factors of the productive potential of the crop stand. Interception and use efficiency of photosynthetically active radiation (PAR) were estimated in relay grass pea under different nutrient management schedules in consecutive two crop seasons of 2017–2018 and 2018–2019. Treatments were two levels of seed priming (i.e., 1. S_1: Without seed priming and 2. S_2: Seed priming with ammonium molybdate at 0.5 g kg^{-1} seed) and five levels of foliar-applied nutritions with various combinations of 2% Urea and 0.5% NPK (19:19:19) shuffling their times of application, replicated thrice laying out in a factorial randomized block design. Seed priming along with twice sprays of NPK (19:19:19) at pre-flowering followed by a second one after 15 days recorded maximum leaf area index (LAI) and total chlorophyll content augmenting greater interception and use efficiency of PAR with highest biomass accumulation, crop growth rate (CGR) and leaf nutrient contents leading to a significant increase in seed yield over control (1596.70 and 1182.00 kg ha^{-1}, respectively) in a pooled analysis. LAI and total chlorophyll content established linear relationships with PAR interception explaining about 94 and 88% variations in intercepted PAR at 90 DAS. Intercepted PAR during different phenophases was positively correlated to dry matter accumulation and net photosynthetic rate with polynomial relationships. Seed yield of grass pea varied about 95 and 96% respectively during 2017–2018 and 2018–2019 with the variations in PAR interception at the pod developmental stage.

Keywords: foliar spray; grass pea; intercepted PAR; PAR use efficiency; seed priming

1. Introduction

Light interception and its direct impact on crop growth have been important concepts with respect to field crops [1]. Like many other crops, the amounts of incoming intercepted photosynthetically active radiation (I PAR) and radiation use efficiency (RUE) of the canopy for biomass production have been highlighted as the most important determinants of the productive potentiality of the leguminous crop stands like mungbean [2], pigeon pea [3], lentil [4], etc. Basu et al. [5] recorded up to 97% variation in intercepted PAR, which could be explained by the biomass accumulation in case of transplanted rice. On the other hand, Oluwasemire and Odugbenro [6] noted the maximum increment in plant biomass for groundnut with a PAR interception to the tune of 55–60%. Further studies indicated that the incoming PAR intercepted by crop canopy is largely governed by the leaf area index (LAI) and canopy architecture [7]. Basically, leaf area is one of the major determinants of PAR interception and its utilization for biomass accumulation and net photosynthesis [8]. Expanding leaf area is a commendable attribute to the overall growth rate of any crop leading to extensive interception of solar radiation and eventually contributing to better economic harvests [9]. On the other hand, the radiation conversion efficiency of a crop into plant biomass equally depends upon the physiological characteristics of the crop [10] as well as on environmental conditions [11]. In this context, the leaf chlorophyll content of a plant is one of the fundamental attributing physiological characteristics related to photosynthetic capacity. Accelerated chlorophyll biosynthesis invariably leads to capturing more incoming solar radiation and a greater rate of net photosynthesis [12]. Notably, RUE is also enhanced with the increase in PAR interception [13]. However, improvement in RUE clearly indicates a higher rate of photosynthesis, which in turn contributes to better yield and nutrient use efficiency. In this context, Worku and Demisie [3] observed around 88% correlation between dry matter production and RUE regarding pigeon pea. In addition, Jena et al. [7] registered up to 4.12 g MJ^{-1} RUE in mustard with increasing biomass production.

Grass pea (*Lathyrus sativus* L.) is generally relay-cropped using the residual soil moisture in rice-fallow during *rabi* season in India [14]. Basically, it is a protein-rich pulse crop (28%) containing considerable proportions of several minerals like calcium, phosphorus and iron [15]. It is considered as an 'insurance crop' as it produces reliable yields when all other crops fail due to a harsh environment. Compared to the other pulse crops, grass pea is a remarkable drought-tolerant crop that thrives with minimal external inputs and consequently is an ideal legume for resource-poor farmers [16].

Seed priming is a recent technology to magnify the rate and synchrony of crop seeds germination, vigour and establishment of seedlings and subsequent attainments of biomass, yield attributing characters and yield of pulse crops [17]. Nutrient seed priming can serve as a simple but effective agronomic practice to meet the nutrient demand of the crop in the early growth stages and eventually increase the final yield in case of relay sowing of pulse crops. In rice fallows, seed priming with KH_2PO_4 [18], sodium molybdate [19] has been earlier reported to increase grass pea production owing to accelerated crop growth and better uptake of nutrients from soil. Basically, molybdenum (Mo) is a vital micronutrient regulating different physiological and biochemical mechanisms in grain legumes [20]. In particular, its direct involvement in the synthesis and activity of nitrogenase and nitrate reductase enzymes, regulating symbiotic N fixation and N assimilation by triggering rhizobial activity has been cited by earlier literature [21]. Application of ammonium molybdate at a dose of 0.5 g kg^{-1} seed has been observed to increase root nodulation of grass pea up to 80–90% along with up-gradation of economic yield to the tune of 30% [14].

The foliar fertilization technique provides the crops plants with a quick supply of nutrients reaching directly to the site of photosynthesis without any wastage [22]. Especially in indeterminate legumes, foliar application of nutrients is very much proficient as it provides sufficient time for conversion of late formed flowers into pods in addition to stimulation of balanced partitioning of photoassimilates from source to sink [23]. Foliar feeding of urea and NPK (19:19:19) was found to be beneficial in the case of green gram, black gram, lentil, grass pea and chickpea [24,25] by delaying senescence and thereby facilitating photosynthesis. The positive influences of NPK foliar nutrition and their interactions are inevitably attributed to the indispensable role of nitrogen (N), phosphorus (P) and potassium (K) in the physiological development of plants [15]. Application of N helps to expand leaf area as N is considered as the primary constituent of leaf chlorophyll maximizing the photosynthetic capacity and overall growth of crop plants [26]. Generally, fertilization with N increases the vegetative growth, total carbohydrate, soluble sugars and NPK content of plants [27]. Modulation of dry matter and protein contents in grain legume crops in terms of both qualitative and quantitative points of view through N application is a very well-known fact. Legume crops go through gradual leaf senescence well before their maturity, which obstructs the yield by breaking the normal source–sink relationship [28]. This specific setback can be overcome through the foliar spray of nitrogen [29], whereas P stimulates root, seed and fruit development along with aiding in vital metabolic functions of plants [30]. In addition, P also departs energy in the form of ATP for nitrogen metabolism and hence enhances BNF, increasing rhizobial colonization, leaf area, photosynthesis, carbon partitioning and biomass accumulation [31]. Phosphorus has a stimulating effect on the growth parameters, total carbohydrate, soluble sugars and minerals contents and influences the productivity by affecting the processes of energy storage and transfer [32]. Potassium addition significantly stimulates root and shoot growth, and enhances the BNF and protein content of pulse grains [33], besides regulating the water economy in the plant body through osmoregulation and maintenance of leaf water potential [34]. Notably, Randhawa et al. [35] reported an interception of PAR of around 460 MJ m^{-2} along with maximum total dry matter and RUE using a nutrient management schedule consisting of NPK in terms of maize.

Indeed, there is a paucity of information regarding the impact of PAR interception and PAR use efficiency on grass pea production in the lower Gangetic plains of Eastern India. This study had been undertaken with the specific objectives of quantifying the amount of intercepted PAR and PAR use efficiency of winter grown grass pea as well as evaluating their interaction with the growth, physiology and seed yield of relay grass pea as influenced by seed priming with Mo and foliar nutrition with urea and NPK.

2. Materials and Methods

2.1. Location of the Study

The field experiment was pursued at the 'A–B' block, District Seed Farm (22°93′ N, 88°53′ E, 9.75 m above the mean sea level) of Bidhan Chandra Krishi Viswavidyalaya, Nadia, West Bengal, India during two subsequent *rabi* seasons (October–March) of 2017–2018 and 2018–2019.

2.2. Soil and Weather Conditions

The soil of the study site was well-drained Gangetic alluvium (order: Inceptisol, suborder: Aqepts, great group: Haplaquepts) with moderate fertility and nearly neutral in reaction, categorised under the textural class of sandy loam with a neutral soil reaction. The detailed physicochemical properties of the soil of the research plots have been depicted in Table 1. Meteorological features of the experimental site in both years have been presented graphically in Figure 1.

Table 1. Details of the experimental soil before experimentation.

Soil Property	Value		Procedures Followed
	2017–2018	2018–2019	
pH	7.3	7.4	Glass electrode pH meter [36]
Electrical conductivity (dS m^{-1})	0.18	0.17	EC meter [37]
Organic carbon (%)	0.56	0.54	Wet oxidation method [38]
Available nitrogen (kg ha^{-1})	231.28	227.17	Modified Kjeldahl method [39]
Available phosphate (kg ha^{-1})	34.51	35.73	0.5 M $NaHCO_3$ extract [40]
Available potassium (kg ha^{-1})	188.83	190.75	Neutral N NH_4OAc extract [39]
Available molybdenum (ppm)	0.03	0.04	Ammonium oxalate extract [41]
Available boron (ppm)	0.51	0.53	Azomethine H [42]
Available zinc (ppm)	0.26	0.21	DTPA-TEA extract [43]
Available manganese (ppm)	0.85	0.94	DTPA-TEA extract [43]
Available iron (ppm)	0.59	0.56	DTPA-TEA extract [43]

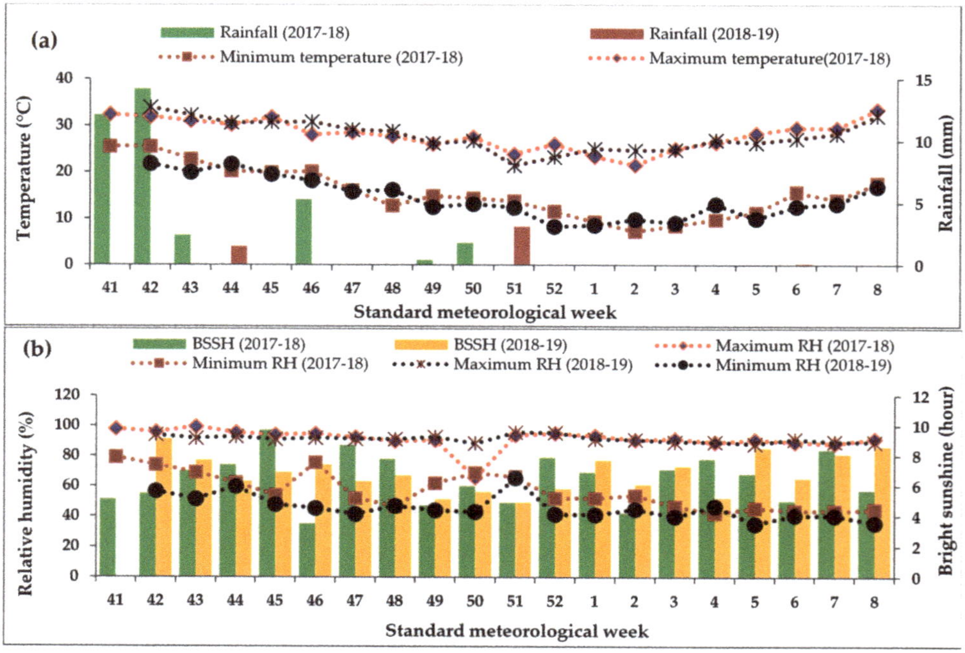

Figure 1. Meteorological features (**a**) rainfall and temperature; (**b**) relative humidity (RH) and brigh sunshine hours (BSSH)) of the experimental site during 2017–2018 and 2018–2019.

2.3. Treatments and Design

The field experiment was arranged in a factorial randomized block design consistin of 2 levels of the 1st factor (seed priming) and 5 levels of the 2nd factor (foliar sprays) i various combinations with a total of 10 treatments replicated thrice. Grass pea seeds c the variety Ratan (Bio L-212) were used for the whole experiment. Detailed treatments ar presented in Table 2.

Table 2. Treatment details of the experiment.

	Treatments
	Seed priming (S)
S_1	No seed priming
S_2	Seed priming with ammonium molybdate at 0.5 g kg^{-1} seed
	Foliar sprays of nutrient (F)
F_1	No foliar spray
F_2	Foliar spray of 2% Urea at the pre-flowering stage
F_3	Foliar spray of 2% Urea at the pre-flowering stage and 15 days after 1st spray
F_4	Foliar spray of 0.5% NPK (19:19:19) at the pre-flowering stage
F_5	Foliar spray of 0.5% NPK (19:19:19) at the pre-flowering stage and 15 days after 1st spray

2.4. Experimental Procedures

The event of land preparation was completely excluded for relay grass pea crop in thi experiment. Generally, grass pea crop requires a seed rate of 40 kg ha^{-1} for line sowin However, the seeds were sown at the rate of 80 kg ha^{-1} in individual experimental plots c 5 m × 3 m through broadcasting on a standing rice crop as per the recommended practice of relay cropping. Half of the seeds were primed with ammonium molybdate at the rat of 0.5 g kg^{-1} of seed for 8 h followed by shade dry and the rest were kept dry on the day before sowing. Before an hour of sowing, all the seeds were treated with *Rhizobium* biofertilizer at the rate of 20 g kg^{-1} of seed for better nodulation. Basal dose of fertilizers application as well as irrigation were completely excluded in case of cultivation of relay grass pea.

One manual weeding was done at 25–30 days after sowing for proper stand establishment of the crop. Foliar sprays with 2% Urea and 0.5% NPK (19:19:19) were done as per the treatment wise allotments in the morning hours spraying with the help of a knapsack sprayer by one labourer simply walking along with the individual plots. The exact amounts per plot requirements of fertilizers were calculated as per the treatment schedule and the same was mixed with the tap water (at the rate of 500 lit ha^{-1}) inside the spray tank for better accuracy of the dose. Spraying of fungicide including SAAF (Mancozeb + Carbendazim) @ 2.5 g lit^{-1} of water was done at 60 DAS as a plant protection measure.

2.5. Data and Their Estimation Procedures

The observations of PAR were measured starting from vegetative (15–45 DAS) up to the pod filling stage (75–105 DAS) at 11.30 h at 30 days intervals using Line quantum sensor (APOGEE Logan UT). The instrument was placed 25 cm above the crop across the rows to estimate incident radiation. Then, it was kept horizontally under the canopy and placed likewise 25 cm higher the soil surface to measure the transmitted radiation from the bottom of the canopy. The reflected PAR was measured from the same position by simply inverting the sensor. Intercepted PAR (I PAR) and PAR use efficiency (PARUE) were calculated following Equations (1) and (2) [44]:

$$\text{I PAR } (\%) = \frac{\text{PAR}_{(O)} - \text{T PAR} - \text{R PAR}_{(C)}}{\text{PAR}_{(O)}} \times 100 \quad (1)$$

where $\text{PAR}_{(O)}$ = incident PAR above the canopy, T PAR = transmitted PAR through the canopy towards the soil surface, and $\text{R PAR}_{(C)}$ = reflected PAR from the canopy

$$\text{PARUE } (\text{g/Mega mole}) = \frac{\text{Dry matter accumulation } (\text{g/m}^2)}{\text{I PAR } (\text{Mega mole/m}^2)} \quad (2)$$

For taking observations of growth attributes of grass pea, 20 plants were tagged through random selection excluding the border rows from each plot. For growth analysis, dry matter accumulation, crop growth rate (CGR), and leaf area index (LAI) of grass pea crop was worked out at vegetative (30 DAS), flowering (60 DAS) and pod filling stage (90 DAS) from 10 randomly selected plants.

LAI was computed following the expression [45]:

$$\text{LAI} = \frac{\text{Leaf area per plant } (\text{m}^2) \times \text{ Number of plants}}{\text{Ground area } (\text{m}^2)} \quad (3)$$

CGR was estimated using the following formula of Watson [45] and expressed in g m^{-2} day^{-1}:

$$\text{CGR} = \frac{1}{\text{G}} \times \frac{W_2 - W_1}{t_2 - t_1} \quad (4)$$

where W_1 = total dry weight of plant at time t_1, W_2 = total dry weight of plant at time t_2 and G = ground area.

The leaf chlorophyll contents were estimated at 30, 60 and 90 DAS. It was measured by taking absorbance readings at 480, 510, 645 and 663 nm wavelengths against a blank one with only 80% acetone in a Systronics-105 spectrophotometer. The chlorophyll a and b, total chlorophyll and carotenoid were estimated with the following formula given by Arnon [46], all expressed in mg g^{-1} of fresh leaf weight:

$$\text{Chlorophyll a} = (12.7 \times A_{663}) - (2.69 \times A_{645}) \times V/W \times 1000 \quad (5)$$

$$\text{Chlorophyll b} = (22.9 \times A_{665}) - (4.68 \times A_{663}) \times V/W \times 1000 \quad (6)$$

$$\text{Total chlorophyll} = (20.2 \times A_{645}) + (8.02 \times A_{663}) \times V/W \times 1000 \quad (7)$$

$$\text{Carotenoid} = (7.6 \times A_{480}) - (1.49 \times A_{510}) \times V/W \times 1000 \quad (8)$$

where V = Extract volume (mL), W = Fresh weight of leaf tissue (g), and A = Absorbance.

The net photosynthetic rate of grass pea leaves was measured with a portable handheld photosynthesis system (CI-340 Handheld Photosynthesis system, CID Bio-Science, Inc Camas, WA, USA) and expressed in µmol m^{-2} s^{-1}. The measurements were obtained on clear sunny days from the fully developed upper leaves of five selected plants from 11:30 a.m. to 12.30 p.m. at 30, 60 and 90 DAS.

The available nitrogen, phosphorous and potassium content in grass pea leaves were determined respectively by the modified Kjeldahl method [39], Olsen's method [40] and flame photometer method [39].

2.6. Statistical Analysis

Data were statistically analysed by implementing the analysis of variance (ANOVA) techniques proposed by Gomez and Gomez [47] for factorial randomized block design. Pooled analysis was exercised in case of similar data from both years. Treatment means were compared by employing the F-test. The significant differences between the treatments were compared by a critical difference at a 5% level of significance. The regression analysis was carried out by SPSS 7.5 software, (SPSS 7.5 copyright, 1997 by SPSS Inc., USA Base 7.5 Application guide). Tukey's posthoc test was performed to compare the differences between mean values.

3. Results

3.1. Prevailing Weather Conditions during Grass Pea Growth

The details of the meteorological parameters pertaining to the period of experimentation are presented in Figure 1a,b. The temperature throughout the months of the cropping period during *rabi* seasons (October 2017 to February 2018 and October 2018 to February 2019) ranged between 8.8 to 32.1 °C and 10.1 to 32.4 °C, respectively. During both of the years under experimentation, the average maximum and minimum temperature showed a decreasing trend from November to January. However, the average mean temperature tended to increase thereafter up to February. The crop experienced a very scanty rainfall during its growing seasons during both the experimental years. The maximum relative humidity varied between 90.0 to 97.5% and 89.8 to 92.9% while minimum relative humidity ranged from 44.5 to 75.2% and 32.8 to 59.6% during the experimentation period of 2017–2018 and 2018–2019. There was a variation in the bright sunshine hour being maximum in 2017–2018 and 2018–2019 in November (7.6 h) and February (7.9 h), respectively, while minimum sunshine hours were recorded in October (5.6 h) and December (5.9 h) during the consecutive seasons of the experiment. Maximum rainfall during the cropping period of 2017–2018 and 2018–2019 was 7.7 mm (October) and 0.7 mm (February), respectively.

3.2. Interception of Photosynthetic Active Radiation (PAR) by Grass Pea Canopy

The percent interception of PAR has gradually escalated accordingly with the advancement of phenophases of the crop up to 90 DAS in the pooled estimation of the experimental years (Figure 2). Maximum interceptions were recorded with seed priming with ammonium molybdate (84.18 and 87.72%) and sprays of 0.5% foliar NPK (19:19:19) (88.87 and 91.79%) twice during 60 and 90 DAS, respectively, which were significantly higher compared to their corresponding treatments.

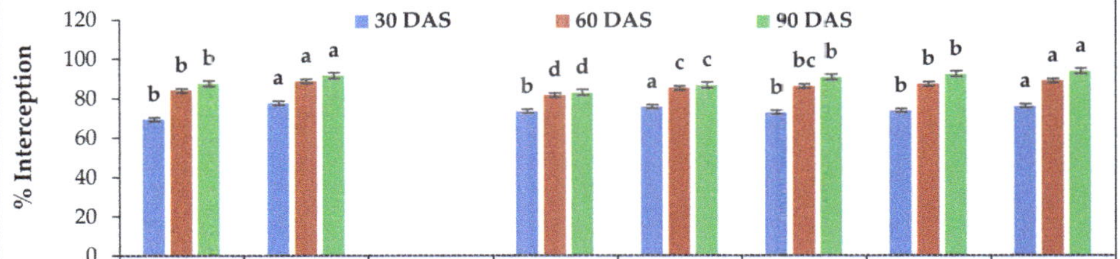

	No priming	Seed priming with A. molybdate		No spray	2% urea (once)	2 % urea (twice)	NPK 19:19:19 (once)	NPK 19:19:19 (twice)
30 DAS	69.51	77.84		73.53	75.92	73.01	73.94	76.37
60 DAS	84.18	88.87		81.83	85.53	86.44	87.61	89.32
90 DAS	87.72	91.79		83.11	86.91	91.01	92.63	94.15

Figure 2. Percent interception of PAR during at different growth stages of grass pea (pooled means of 2 years) (Different letters in all bars indicate the significant differences between means.)

3.3. *Effect of Seed Primimg and Foliar Spray of Nutrients on Growth Characters of Grass Pea*

Dry matter accumulation of relay grass pea progressively advanced with the development of the crop up to the pod development stage, i.e., 90 DAS (Figure 3). Interestingly, LAI and CGR also exhibited similar increasing trends till 90 DAS but with a decreasing rate from flowering (60 DAS) towards pod development.

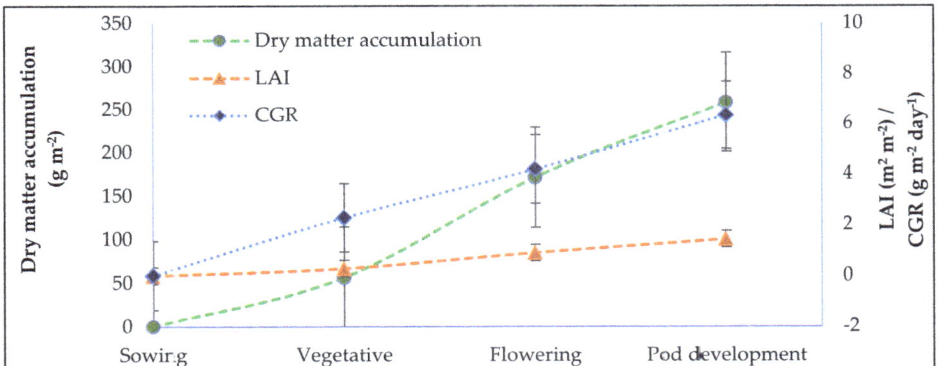

Figure 3. Growth characters at different phenophases of grass pea (pooled means of 2 years).

Significant variation was observed regarding growth traits of grass pea in terms of dry matter accumulation, LAI and CGR among the seed priming and foliar sprayed plots under pooled estimation (Tables 3–5, respectively). Molybdenum seed priming recorded greater dry biomass accumulation (58.84, 174.55 and 264.45 g m^{-2}) and crop growth rate (2.68, 4.23 and 6.95 g m^{-2} day^{-1}) at 30, 60 and 90 DAS, respectively, which were statistically significant over control. Accordingly, seed priming also attained enlarged LAI of about 19.23, 4.59 and 4.28%, respectively, at 30, 60 and 90 DAS according to the pooled over data. During 60 and 90 DAS, higher dry biomass accumulation (176.97 and 269.40 g m^{-2}) and CGR (5.55 and 7.74 g m^{-2} day^{-1}) were attained with the treatments where 0.5% NPK (19:19:19) spray was applied twice irrespective of seed priming. In case of foliar sprays, the lowest LAI was found without sprays. At the pod developmental stage (90 DAS), folia sprays of 2% urea two times recorded a 9.09% increase, whereas a 16.67% increase wa achieved with 0.5% NPK (19:19:19) foliar spray at pre-flowering and pod developmenta stages. Interaction effects among the two factors of the experiment were found to b statistically significant in the later stages of growth of grass pea.

Table 3. Dry matter accumulation (g m^{-2}) in grass pea at different growth stages (pooled means c 2 years).

Treatment	30 DAS	60 DAS	90 DAS
	Seed priming (S)		
No priming	50.75 ± 0.38 b	167.19 ± 0.25 b	251.54 ± 0.25 b
Mo seed priming	58.84 ± 0.19 a	174.55 ± 0.48 a	264.45 ± 0.54 a
CD ($p \leq 0.05$)	3.36	3.48	4.47
	Foliar sprays of nutrient (F)		
No spray	55.82 ± 0.76 b	163.93 ± 0.25 e	246.63 ± 0.28 e
2% Urea (once)	56.38 ± 0.60 a	168.41 ± 0.54 d	252.21 ± 0.38 d
2% Urea (twice)	56.23 ± 0.50 a	170.77 ± 0.14 c	258.04 ± 0.42 c
0.5% NPK 19:19:19 (once)	56.18 ± 0.20 a	174.27 ± 0.27 b	263.69 ± 0.25 b
0.5% NPK 19:19:19 (twice)	56.86 ± 0.29 a	176.97 ± 0.47 a	269.40 ± 0.37 a
CD ($p \leq 0.05$)	0.03	2.13	3.65
	Interaction		
S F	NS	3.82	5.86

NS—Non-significant. Different letters denote significant differences between means.

Table 4. LAI of grass pea at different growth stages (pooled means of 2 years).

Treatment	30 DAS	60 DAS	90 DAS
	Seed priming (S)		
No priming	0.26 ± 0.01 b	0.87 ± 0.01 b	1.40 ± 0.01 b
Mo seed priming	0.31 ± 0.02 a	0.91 ± 0.02 a	1.46 ± 0.02 a
CD ($p \leq 0.05$)	0.02	0.02	0.03
	Foliar sprays of nutrient (F)		
No spray	0.27 ± 0.01 d	0.85 ± 0.01 e	1.32 ± 0.01 e
2% Urea (once)	0.28 ± 0.01 c	0.88 ± 0.01 d	1.39 ± 0.02 d
2% Urea (twice)	0.29 ± 0.02 b	0.89 ± 0.01 c	1.44 ± 0.02 c
0.5% NPK 19:19:19 (once)	0.30 ± 0.01 a	0.92 ± 0.01 b	1.48 ± 0.01 b
0.5% NPK 19:19:19 (twice)	0.28 ± 0.01 c	0.92 ± 0.01 a	1.54 ± 0.02 a
CD ($p \leq 0.05$)	NS	0.02	0.04
	Interaction		
S × F	NS	0.02	0.03

NS—Non-significant. Different letters designate significant differences between means.

3.4. Effect of Seed Priming and Foliar Spray of Nutrients on Physiology of Grass Pea

Relatively higher total chlorophyll contents in grass pea leaves were observed with M seed priming as compared to no priming (1.09 vs. 1.15, 1.40 vs. 1.50, 0.93 vs. 1.02 mg g^{-1} c fresh weight) at 30, 60 and 90 DAS, respectively (Figure 4). Foliar spray of nutrients took significant positive role in improving the total chlorophyll content. This varied in the rang of 1.09–1.27 mg g^{-1} of fresh weight (30 DAS), 1.41–1.62 mg g^{-1} of fresh weight (60 DAS and 0.92–1.15 mg g^{-1} of fresh weight in the pooled estimation. However, the twice folia spray of 0.5% NPK (19:19:19) attained the highest values followed by twice 2% urea spra which were statistically significant over control.

Table 5. CGR (g m^{-2} day^{-1}) of grass pea at different growth stages (pooled means of 2 years).

Treatment	30 DAS	60 DAS	90 DAS
	Seed priming (S)		
No priming	1.89 ± 0.03 b	3.84 ± 0.02 b	5.49 ± 0.02 b
Mo seed priming	2.68 ± 0.04 a	4.23 ± 0.02 a	6.95 ± 0.03 a
CD ($p \leq 0.05$)	0.11	0.12	0.19
	Foliar sprays of nutrient (F)		
No spray	2.21 ± 0.01 b	2.42 ± 0.02 d	5.02 ± 0.01 e
2% Urea (once)	1.58 ± 0.04 b	3.56 ± 0.02 c	6.36 ± 0.02 d
2% Urea (twice)	2.22 ± 0.03 b	4.49 ± 0.03 b	7.18 ± 0.02 c
0.5% NPK 19:19:19 (once)	2.81 ± 0.02 a	4.90 ± 0.01 b	7.32 ± 0.03 b
0.5% NPK 19:19:19 (twice)	2.85 ± 0.03 a	5.55 ± 0.03 a	7.74 ± 0.04 a
CD ($p \leq 0.05$)	0.20	0.23	0.30
	Interaction		
S × F	NS	0.32	0.43

NS—Non-significant. Different letters indicate significant differences between means.

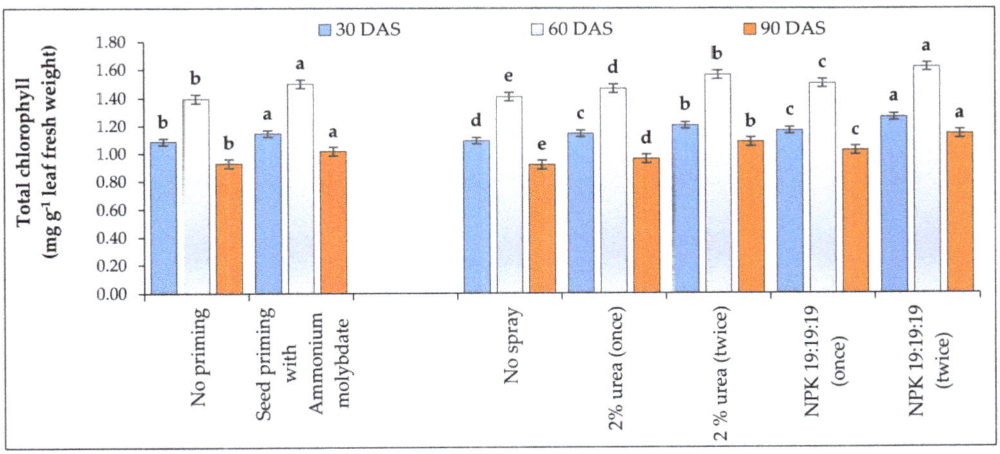

Figure 4. Effect of seed priming and foliar sprays on total leaf chlorophyll content of grass pea at different growth stages (pooled means of 2 years) (Different letters in all bars denote significant differences between means.)

The rate of photosynthesis in the above-ground parts of relay grass pea grown during *rabi* seasons of 2017–2018 and 2018–2019 progressively increased up to 60 DAS and afterwards a gradual decrease was observed (Figure 5). In accordance with leaf chlorophyll content, a significantly higher rate of net photosynthesis was observed under the treatment with seed priming irrespective of foliar nutrients application throughout the growing period as compared to control. Pooled results showed that Mo seed priming attained a higher rate of photosynthesis (7.98, 16.27 and 6.13 µmol m^{-2} s^{-1}) at 30, 60 and 90 DAS, respectively, which were statistically significant over control. Among the different foliar sprayed treatments, 0.5% NPK (19:19:19) spray at pre-flowering and 15 days after 1st spray reached the maximum rate of net photosynthesis (18.25 µmol m^{-2} s^{-1}) followed by 2% urea spray at pre-flowering and 15 days after 1st spray (16.82 µmol m^{-2} s^{-1}) at the flowering stage concerning the pooled over means.

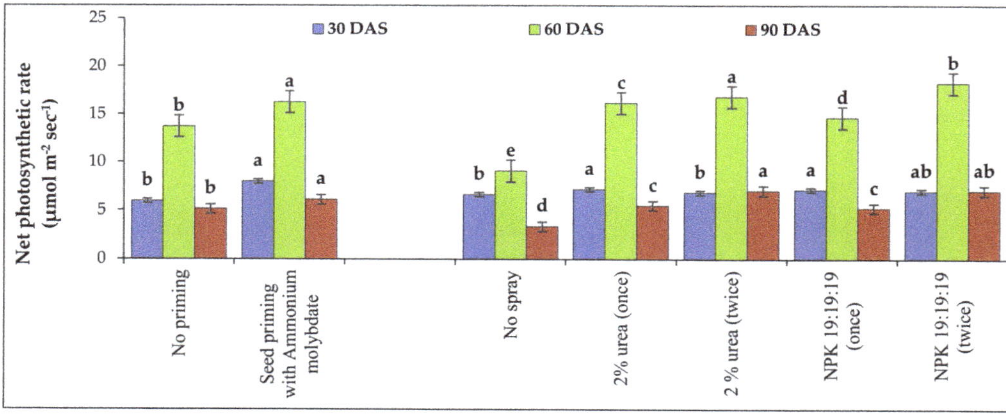

Figure 5. Effect of seed priming and foliar sprays on net photosynthetic rate of grass pea at different growth stages (pooled means of 2 years) (Different letters in all bars indicate significant differences between means.)

3.5. Growth and Physiology of Grass Pea with Respect to Intercepted PAR

Intercepted PAR established linear relationships with respect to both LAI and total leaf chlorophyll contents of grass pea throughout its growing period under this study (Table 6). Pooled estimation revealed that about 94% and 88% variations in intercepted PAR could be explained by the variations, respectively, in LAI and total chlorophyll content at 90 DAS.

Table 6. Impact of leaf area index (x) and total chlorophyll (z) on cumulative intercepted PAR (y).

Growth Stages	Impact of Leaf Area Index (x)			Impact of Total Chlorophyll (z)		
	Regression Equation	R^2	Relation	Regression Equation	R^2	Relation
30 DAS	y = 2.2063x + 0.0368	0.74	Linear	y = 0.5551z + 0.0173	0.67	Linear
60 DAS	y = 0.7066x − 0.0256	0.83	Linear	y = 0.2237z + 0.2717	0.85	Linear
90 DAS	y = 0.3957x + 0.1251	0.94	Linear	y = 0.3772z + 0.2965	0.88	Linear

The efficiency in PAR interception among the various treatments was verified with the trend in dry biomass accumulation as well as with the pattern of net photosynthetic rate. Both the dry matter accumulation and net photosynthetic rate were estimated to be polynomial functions of intercepted PAR throughout the growth stages of grass pea. The magnitude of R^2 values showed its significance in those relationships (Figure 6). R^2 values indicated that about 83.15, 93.76 and 96.69% variations in dry matter accumulation at 30, 60 and 90 DAS, respectively, could be explained by the differentiation in cumulative intercepted PAR, whereas these variations reached the tune of 76.74, 78.64 and 83.33% at the respective intervals with respect to the rate of net photosynthesis.

3.6. Photosynthetic Active Radiation Use Efficiency (PARUE) of Grass Pea

The accumulation rate of dry biomass per unit interception of PAR i.e., the PARUE were found to be significantly higher in case of seed priming (0.09, 0.22 and 0.43 g Mega mole^{-1}) compared to without priming (0.07, 0.19 and 0.41 g Mega mole^{-1}) at 30, 60 and 90 DAS. However, the application of 0.5% NPK (19:19:19) spray at pre-flowering following the second one at 15 days intervals recorded the highest PARUE (0.25 and 0.50 g Mega mole^{-1}) at the respective intervals among all the foliar-applied treatments (Figure 7).

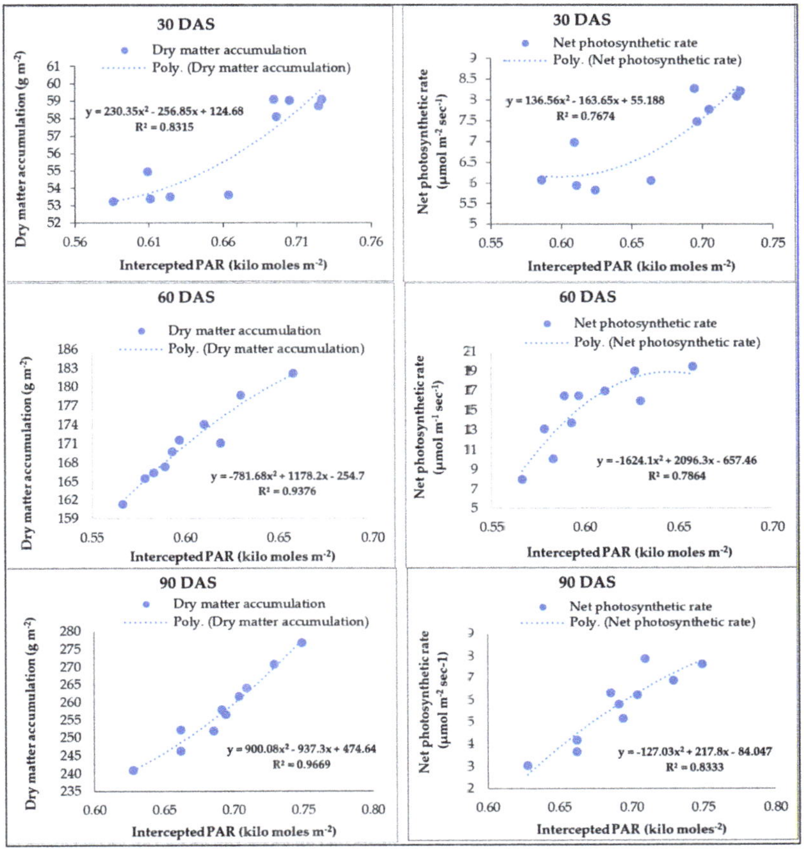

Figure 6. Impact of intercepted PAR on dry matter accumulation and net photosynthetic rate of grass pea.

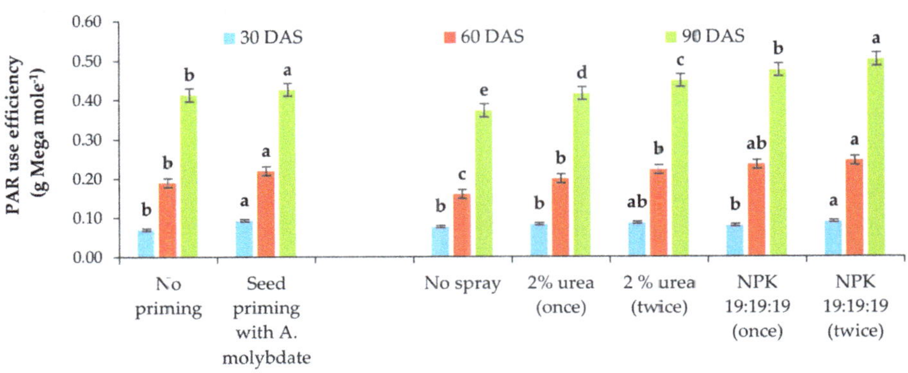

Figure 7. PAR use efficiency at different growth stages of grass pea (pooled means of 2 years). (Different letters in all bars indicate significant differences between means.)

3.7. Seed Yield of Grass Pea

Seed yield of grass pea was magnified with the treatments efficiently enhancing cro growth and net photosynthetic rate, eventually intercepting a greater amount of PA in both years. Seed priming with ammonium molybdate recorded significantly highe seed yield compared to control (1509.99 and 1350.40 kg ha^{-1}) under pooled estimation c 2017–2018 and 2018–2019. Among the foliar sprayed plots, foliar 0.5% NPK (19:19:19) a pre-flowering and 15 days after 1st spray registered to the tune of 1589.39 kg ha^{-1} see yield, which was statistically significant over the others (Figure 8).

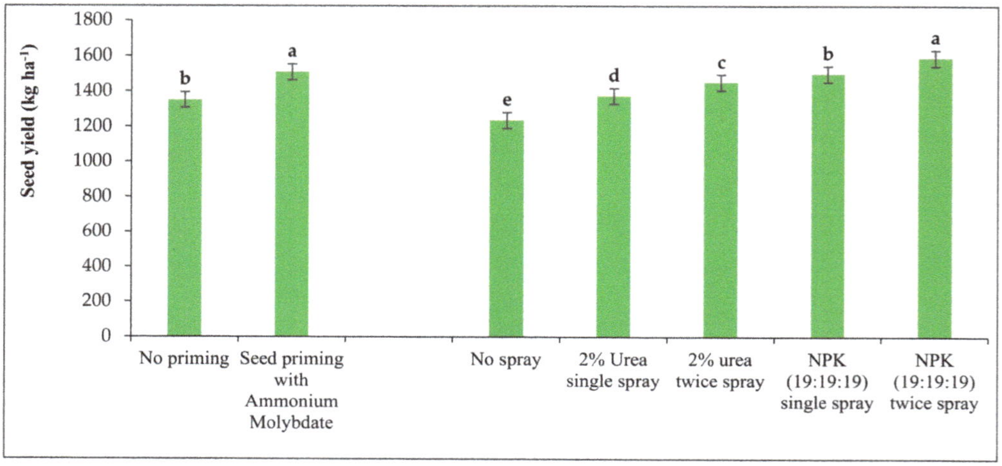

Figure 8. Seed yield of grass pea as influenced by seed priming and foliar nutrition (pooled means c 2 years) (Different letters in all bars denote significant differences between means.)

For the season 2017–2018, the variations obtained in yield was 96.2% governed by th variations in PAR at 30 DAS and PAR at 90 DAS (Table 7). Moreover, variations in PAR a 90 DAS alone dictates 98.3% of the variations observed in yield. The variations in PAR a 90 DAS govern 95.1% of the variations obtained in yield in 2018–2019.

Table 7. Effect of intercepted PAR on seed yield of grass pea.

Regression Equations	R^2	Adj. R^2	Significance
2017–2018			
$Y = -1090.263 + 3483.124\ PAR_{90}{}^{**} + 317.668\ PAR_{30}{}^{*}$	0.966	0.962	30.530
$Y = -1013.469 + 3676.854\ PAR_{90}$	0.987	0.983	20.637
2018–2019			
$Y = -2225.833 + 5096.12\ PAR_{90}{}^{**}$	0.957	0.951	33.626

* Significant at 5%, ** significant at 1% level of probability.

3.8. Impact of Seed Primimg and Foliar Nutrition on Nutrients Content in Grass Pea Leaves

Pooled analysis presented in Table 8 revealed that seed priming with ammoniun molybdate facilitated maximum leaf N, P and K contents (0.86, 0.25 and 1.11%, respectivel which were statistically significant over control. Twice foliar spray of 0.5% NPK (19:19:1 attained maximum nutrients in leaf estimation among the foliar sprayed plots. Next t this, the treatment with twice sprays of 2% urea recorded higher values of leaf N conten However, a single spray of NPK (19:19:19) achieved more P and K contents as compared t twice sprays of 2% urea.

Table 8. Effect of seed priming and foliar sprays on leaf nutrients (N, P and K) content (%) in grass pea (pooled means of 2 years).

Treatment	N (%)	P (%)	K (%)
	Seed priming (S)		
No priming	0.78 ± 0.01 b	0.22 ± 0.01 b	1.07 ± 0.01 b
Mo seed priming	0.86 ± 0.02 a	0.25 ± 0.01 a	1.11 ± 0.02 a
CD ($p \leq 0.05$)	0.01	0.01	0.01
	Foliar sprays of nutrient (F)		
No spray	0.65 ± 0.01 e	0.17 ± 0.01 e	0.99 ± 0.01 e
2% Urea (once)	0.84 ± 0.01 d	0.21 ± 0.01 d	1.06 ± 0.01 d
2% Urea (twice)	0.88 ± 0.01 b	0.24 ± 0.01 c	1.11 ± 0.02 c
0.5% NPK 19:19:19 (once)	0.79 ± 0.01 c	0.27 ± 0.02 b	1.13 ± 0.25 b
0.5% NPK 19:19:19 (twice)	0.93 ± 0.01 a	0.30 ± 0.02 a	1.17 ± 0.37 a
CD ($p \leq 0.05$)	0.02	0.01	0.02
	Interaction		
S × F	0.02	NS	NS

NS—Non-significant. Different letters designate significant differences between means.

4. Discussion

4.1. Impact of Seed Primimg and Foliar Spray of Nutrients on Growth Traits and Physiology

Initial Mo application was found to be strongly associated with extension of canopy coverage, which maintained a progressive increment in LAI and CGR even after the reproductive growth set in. Nevertheless, an increasing rate of LAI and CGR with a declining pattern after flowering (60 DAS) might be due to a simultaneous onset of the reproductive stage with leaf senescence and a reduced rate of newer leaf emergence of grass pea owing to terminal heat and moisture stress [48]. In fact, the crop was exposed to a constant rise in ambient temperature coupled with deficit atmospheric humidity and soil moisture particularly at the time of seed filling due to lack of rainfall and exclusion of irrigation and a decline in soil moisture storage due to and irrigation. As a consequence, the crop might have survived with lower water consumption hampering the normal rate of net photosynthesis. Probably, this phenomenon was more prevalent in case of avoidance of any kind of nutrient use, which drastically brought down the overall growth rate in those treatments. Enhancement in plant growth with Mo application was cited with respect to several winter pulse crops including lentil [49], chickpea [50], garden pea [51], grass pea [16], etc. No specific pattern in crop growth was found among the foliar sprayed treatments at 30 DAS as the spraying schedule started from 45 DAS onwards. Additionally, foliar spray of NPK at the pre-flowering stage followed by an additional one after 15 days with special reference to grass pea happened to be a fantastic way out to flourish with extended leaf area throughout the reproductive phase of this crop.

4.2. Growth and Physiology of Grass Pea in Connection with Intercepted PAR

From Table 2 and Figure 2, it was evident that grass pea crop intercepted a greater amount of PAR with successive enlargement in leaf area throughout the growing period. This finding was in agreement with Worku and Demisie [3]. The introduction of the exclusive combination of micronutrient Mo and macronutrients (NPK) might have helped in profuse branching and leaf production resulting in higher final biomass production. Due to lesser canopy coverage, the treatment without priming or foliar spray always intercepted least amount of PAR. Availability of Mo in the form of seed priming might have facilitated better nitrogen metabolism. In addition, Mo is associated with the absorption and translocation of iron (Fe) in plants [52]. In this connection, Fe plays a pivotal role in chloroplast development, chlorophyll biosynthesis and energy transfer in plants [53]. Thus, the physiological efficiency in terms of photosynthetic activity of grass pea was probably boosted with the active participation of Mo in this regard [54]. In addition, application of NPK might be attributed to amplifying the expansion of leaf area, chlorophyll content and nutrients assimilation capacity of the crop [55]. The efficiency of foliar NPK was clearly portrayed by the study of leaf photosynthesis. Maximum photosynthesis was positively correlated with leaf nitrogen, phosphorus and potassium content [56]. Longstreth and Nobel [57] reported that plant mineral status could markedly influence the photosynthesis owing to modified leaf chlorophyll content. These improved features related to leaf area expansion and enhanced production of photosynthetic pigments augmented better PAR interception and photosynthetic efficiency, ultimately magnifying the productivity of crops [35]. Positive interaction between leaf area extension and PAR interception have already been recorded earlier [8]. Interception of PAR and its impact on growth and physiology has been recorded by a number of authors in terms of different legumes. In some of the cases, the relationships were linear [9] and, in other instances, these were found to be polynomial [58].

4.3. PAR Use Efficiency (PARUE)

Higher use efficiency of I PAR with the application of Mo seed prining and 0.5% NPK (19:19:19) spray at pre-flowering following the second one at a 15 day interval recorded implied better efficiency in terms of conversion of energy to dry matter in the particular treatments. In other words, this treatment with seed priming along with ammonium molybdate at 0.5 g kg^{-1} seed combined with twice foliar sprays of 0.5% NPK (19:19:19) utilized maximum energy to produce the greater volume of biomass with better LAI and improved rate of crop growth. Foliar nutrition might have triggered the grass pea crop growth and aided in flourishing profuse canopy coverage, which in turn led to greater interception and use efficiency of solar radiation [4]. Rosati and Dejong [59] suggested that PARUE was improved with N fertilization. Randhawa et al. [35] observed a positive impact of supplemental NPK on plant growth by modification of the shape and size of the crop canopy, thereby obtaining higher use efficiency of intercepted solar radiation. Notably, biomass accumulation per unit energy use was at a maximum during the later phases of grass pea growth under the present experiment. Similar trends were found under mungbean [2] and lentil [60]. This might occur in the pulse crops because of late emerging vegetative flushes in these crops with the intercepted solar radiation.

4.4. Yield and Leaf Nutrients Content of Grass Pea in Relation to I PAR

In the present experiment, seed priming with ammonium molybdate at the rate of 0.5 g kg^{-1} seed and foliar 0.5% NPK (19:19:19) at pre-flowering and 15 days after the 1st spray established a remarkable influence regarding augmentation of seed yield. Similar positive outcomes in response to seed priming with Mo in economic yield of chickpea [61], cowpea [62] and grass pea [63] and that of lentil [64] and grass pea [54] with respect to foliar spraying of 0.5% NPK (19:19:19) was reported earlier. Increment in leaf nutrient contents through Mo seed priming were cited by a number of literature works regarding chickpea [65], lentil [66], mungbean [67], peanut [68], etc. Involvement of Mo in vital physiological and biochemical functions, especially regarding the functioning of leghemoglobin protein and nitrogenase enzyme required for rhizobial activity in legumes for N fixation and its subsequent assimilation related to nitrate reductase activity has already been reported to manifest momentous impact on legume growth and productivity [20]. Navaz et al. [19] revealed the synergistic effect of Mo on escalating the N, P and K contents in grass pea stover. However, foliar NPK induced enhancement in nutrient content in pigeon pea leaves was reported by Gowda et al. [69]. In a nutshell, nutrient application in the form of seed priming with Mo and foliar NPK remarkably contributed to improved photosynthesizing capacity and better source to sink partitioning through considerable capture of solar radiation eventually brought about a spectacular increase in biomass and seed yield. In particular, foliar nutrition with NPK might have fostered the cell division and enzymatic activity through regulation of water economy inside the grass pea plants. This eventually accelerated the flower production, photosynthetic rate, translocation of photosynthates to the seed, pod formation and seed development and turning up with higher seed yield [16].

Basically, the optimum temperature range for grass pea growth ranges from 10–25 °C. However, it requires around 15 °C temperature for healthier seedling growth during the vegetative stage [70]. In fact, mean daily maximum temperature above 25 °C has been considered as the upper threshold limit for heat stress in cool season crops [71]. The higher mean daily maximum temperature coupled with lower mean relative humidity that the crop experienced during the pod developmental stage were visibly beyond the optimum range (Figure 1). Hence, the crop had definitely been exposed to heat stress during this stages, which is critical from the production point of view of grass pea. On the other hand, higher temperatures combined with lower relative humidity have a specific role in increasing the evapotranspiration loss from soil as well as crop canopy, which can imply apparent moisture stress at the reproductive stage of this crop. Decline in relative humidity in the air owing to the higher atmospheric temperature and rainfall scarcity might have substantially attributed to intensifying the impacts of heat and moisture stress inside the crop by means of depleting the soil moisture storage [63,72]. In this context, the crop faced adverse impacts of these abiotic stresses on overall growth and physiological development without the external supply of plant nutrients, consequently acquiring lesser photosynthetic area and harvesting lower amounts of photosynthetically active portion of solar radiation biomass production, ultimately hampering seed set and yield potential [73]. Optimum supply of plant nutrients might have successfully endeavoured for mitigation of the terminal heat and moisture stress with simultaneous increment in PAR interception in the crop of the corresponding treatments. Apart from this, the greater sunshine hours during the growing period of grass pea in both years might have contributed to better interception of solar radiation and corresponding upgradation of photosynthetic activity [54].

5. Conclusions

Characteristics of radiation interception is one of the fundamental contributing unique features with respect to field crops production. On the other hand, LAI and CGR could be considered as vital indices to influence light interception in grass pea crop through expansion of canopy coverage. Limitations in production owing to restricted PAR capture and photosynthetic activity were evident from the reduced growth rate, depleted chlorophyll, and nutrients content in leaves. Considering the findings of the present experiment, it may be concluded that integration of seed priming with ammonium molybdate at 0.5 g kg^{-1} seed along with exogenous application of 0.5% NPK (19:19:19) spray at pre-flowering and 15 days after 1st spray may be adopted by the grass pea farmers in case of its relay sowing for immense potential of this combination with respect to interception and use efficiency of PAR sustaining growth and production potential under Lower Gangetic plains of Eastern India.

Author Contributions: Conceptualization, P.B., V.K.V., R.N. and P.K.C.; methodology and visualization, P.B., V.K.V., R.N. and P.K.C.; software, P.B., V.K.V. and R.N.; validation, P.B., V.K.V. and R.N.; formal analysis, P.B., V.K.V., R.N. and A.H.; investigation, P.B., V.K.V. and R.N.; resources, P.B., V.K.V., R.N. and P.K.C.; data curation, P.B., V.K.V., R.N. and P.K.C. and A.H.; writing—original draft preparation, P.B., V.K.V., R.N. and P.K.C.; writing—review and editing, A.G., W.F.A., B.M.R. and A.H.; supervision and project administration, A.G., W.F.A., B.M.R. and A.H.; funding acquisition, A.G., W.F.A., B.M.R. and A.H. All authors have read and agreed to the published version of the manuscript.

Funding: This research work was financially supported by the Department of Agronomy, Bidhan Chandra Krishi Viswavidyalaya, Mohanpur-741252, Nadia, West Bengal, India. The current study was also partially funded by the Taif University Researchers for funding this research with a Supporting Project number (TURSP-2020/53), Taif University, Taif, Saudi Arabia.

Institutional Review Board Statement: Not applicable.

Informed Consent Statement: Not applicable.

Data Availability Statement: Most of the data are available in all tables and figures of the manuscript

Conflicts of Interest: The authors would hereby like to declare that there is no conflict of interest fc the article.

References

1. Pradhan, S.; Sehgal, V.K.; Bandyopadhyay, K.K.; Panigrahi, P.; Parihar, C.M.; Jat, S.L. Radiation interception, extinction coefficier and use efficiency of wheat crop at various irrigation and nitrogen levels in a semiarid location. *Ind. J. Plant Physiol.* **2018**, 2. 416–425. [CrossRef] [PubMed]
2. Tzudir, L.; Basu, S.; Maji, S.; Bera, P.S.; Nath, R.; Mazumdar, D.; Chakraborty, P.K. Impact of weather variables on dry matte accumulation and yield of mungbean [*Vigna radiata* (L.) Wilczek] varieties under different dates of sowing. *Legume Res.* **2016**, 3: 427–434.
3. Worku, W.; Demisie, W. Growth, light interception and light use efficiency of pigeon pea (*Cajanas cajan*) to planting density i Southern Ethiopia. *J. Agron.* **2012**, *11*, 85–93. [CrossRef]
4. Venugopalan, V.K.; Nath, R.; Sengupta, K.; Nalia, A.; Banerjee, S.; Chandran, M.A.S.; Ibrahimova, U.; Dessoky, E.S.; Attia, A.C Hassan, M.M.; et al. The response of lentil (*Lens culinaris* Medik.) to soil moisture and heat stress under different dates of sowin and foliar application of micronutrients. *Front. Plant Sci.* **2021**, *12*, 679469. [CrossRef]
5. Basu, S.; Maji, S.; Dutta, S.K.; Jena, S.; Nath, R.; Chakraborty, P.K. Impact of PAR interception at different time points on total dr matter production in rice (*Oryza sativa* L.) crop transplanted on different dates. *J. Food Agric. Environ.* **2014**, *12*, 285–291.
6. Oluwasemire, K.O.; Odugbenro, G.O. Solar Radiation Interception, Dry Matter Production and Yield among Different Plan Densities of *Arachis* spp. in Ibadan, Nigeria. *Agric. Sci.* **2014**, *5*, 864–874. [CrossRef]
7. Jena, S.; Basu, S.; Maji, S.; Bandyopadhyay, P.; Nath, R.; Chakraborty, P.K.; Chakraborty, P.K. Variation in absorption c photosynthetic active radiation (PAR) and PAR use efficiency of wheat and mustard grown under intercropping system. *Biosca* **2015**, *10*, 107–112.
8. Jain, G.; Sandhu, S.K. Radiation interception and growth dynamics in mustard under different dates of sowing. *J. Pharn Phytochem.* **2019**, *8*, 499–504.
9. Azam, M.; Hussain, A.; Wajid, S.A.; Maqsood, M. Effect of sowing date, irrigation and plant densities on radiation interceptio and its utilization efficiency in lentils. *Int. J. Agric. Biol.* **2002**, *4*, 217–219.
10. Zhang, Y.; Tang, Q.; Zou, Y.; Li, D.; Qin, J.; Yang, S.; Chen, L.; Xia, B.; Peng, S. Yield potential and radiation use efficiency of supe hybrid rice grown under subtropical conditions. *Field Crop. Res.* **2009**, *114*, 91–98. [CrossRef]
11. Hundal, S.S.; Kaur, P.; Malikpuri, S.D.S. Radiation use efficiency of mustard cultivars under different sowing dates. *J. Agrome* **2004**, *6*, 70–75.
12. Krause, G.H.; Weis, E. Chlorophyll Fluorescence and Photosynthesis: The Basics. *Ann. Rev. Plant Biol.* **2003**, *42*, 313–34 [CrossRef]
13. Gautam, P.; Lal, B.; Nayak, A.K.; Raja, R.; Panda, B.B.; Tripathi, R.; Shahid, M.; UKumar, U.; Baig, M.J.; Chatterjee, D.; et a Inter-relationship between intercepted radiation and rice yield influenced by transplanting time, method, and variety. *Int. Biometeorol.* **2019**, *63*, 337–349. [CrossRef]
14. Banerjee, P.; Mukherjee, B.; Nath, R. Chapter 5: Prospects of Grass pea in Rice Fallow in Eastern India. Sustainable Production c Pulses in Diverse Agro-ecosystems. In *Stress Management and Livelihood Security*; Kumar, N., Nath, C.P., Singh, N.P., Eds.; Scientifi Publishers: Jodhpur, India, 2021; Volume 2, pp. 63–77.
15. Banerjee, P.; Mukherjee, B.; Ghosh, A.; Pramanik, M.; Nath, R. Influence of seed priming and foliar nutrition on quality an nutrient uptake of relay grass pea (*Lathyrus sativus* L.) in Gangetic plains of West Bengal. *Int. J. Curr. Microbiol. App. Sci.* **2020**, 2864–2872. [CrossRef]
16. Banerjee, P.; Visha Kumari, V.; Nath, R.; Bandopadhyay, P. Seed priming and foliar nutrition studies on relay grass pea afte winter rice in lower Gangetic plain. *J. Crop. Weed* **2019**, *15*, 72–78. [CrossRef]
17. Ghasemi-Golezani, K.; Jabbarpour-Bonyadi, Z.; Shafagh-Kolvanagh, J.; Nikpour-Rashidabad, N. Effects of Water Stress an hydro-priming duration on field performance of lentil. *Intl. J. Farm Alli. Sci.* **2013**, *2*, 922–925.
18. Bhowmick, M.K.; Dhara, M.C.; Duary, B.; Biswas, P.K.; Bhattacharyya, P. Improvement of lathyrus productivity through see priming and foliar nutrition under rice-*utera* system. *J. Crop. Weed.* **2014**, *10*, 277–280.
19. Navaz, M.; Kumar, S.; Shrivastava, G.K.; Mandavi, M.; Salam, P.K.; Pandey, N. Impact of foliar spray of nutrients and see treatment on uptake of phosphorus of plant and seed of lathyrus (*Lathyrus sativus* L.) under relay cropping system. *Int. J. Faun Biol. Stud.* **2018**, *5*, 1–2.
20. Banerjee, P.; Nath, R. Prospects of molybdenum fertilization in grain legumes—A review. *J. Plant Nutr.* **2022**, *45*, 1425–144 [CrossRef]
21. Banerjee, P.; Das, P.; Sinha, S. Importance of molybdenum for the production of pulse crops in India. *J. Plant Nutr.* **2022**, 4. 300–310. [CrossRef]
22. Bhowmick, M.K. Effect of foliar nutrition and basal fertilization in lentil under rainfed conditions. *J. Food Legume* **2008**, 2 115–116.

23. Das, S.K.; Jana, K. Effect of seed hydro-priming and urea spray on yield parameters, yield and quality of lentil (*Lens culinaris* Medikus). *Legume Res.* **2016**, *39*, 830–833. [CrossRef]
24. Das, S.K.; Jana, K. Effect of foliar spray of water soluble fertilizer at pre flowering stage on yield of pulses. *Agric. Sci. Digest* **2015**, *35*, 275–279. [CrossRef]
25. Mandre, B.K.; Singh, R.P.; Dubey, M.; Waskle, U.; Birla, V. Effect of Foliar Application of Nutrients on Growth and Yield Attributing Characters of Black Gram. *Int. J. Curr. Microbiol. App. Sci.* **2020**, *9*, 419–428. [CrossRef]
26. Zahoor, A.; Riaz, M.; Ahmad, S.; Ali, H.; Khan, M.B.; Javed, K.; Anjum, M.A.; Zia-ul-Haq, M.; Khan, M.A. Ontogeny growth and radiation use efficiency of *Helianthus annuus* L., as affected by hybrids, nitrogenous regimes and planting geometry under irrigated arid conditions. *Pak. J. Bot.* **2010**, *42*, 3197–3207.
27. Khalid, A.K. Effect of nitrogen fertilization on morphological and biochemical traits of some Apiaceae crops under arid region conditions in Egypt. *Bioscience* **2013**, *5*, 15–21. [CrossRef]
28. Bhagariya, S.; Shah, K.A.; Chaudhry, K.M. Effect of Different Nutrient Management Practices on Yield Potential of Summer Cowpea [*Vigna unguiculata* L.] under South Gujarat Condition. *Int. J. Curr. Microbiol. App. Sci.* **2020**, *9*, 3267–3276. [CrossRef]
29. Palta, J.A.; Nandwal, A.S.; Kumari, S.; Turner, N.C. Foliar nitrogen applications increase the seed yield and protein content in chickpea (*Cicer arietinum* L.) subject to terminal drought. *Aust. J. Agric. Res.* **2005**, *56*, 105–112. [CrossRef]
30. Mmbaga, G.W.; Mtei, K.M.; Ndakidemi, P.A. Extrapolations on the use of Rhizobium inoculants supplemented with phosphorus (P) and potassium (K) on growth and nutrition of legumes. *J. Agric. Sci.* **2014**, *5*, 1207–1226. [CrossRef]
31. Mitran, T.; Meena, R.S.; Lal, R.; Layek, J.; Kumar, S.; Datta, R. Role of Soil Phosphorus on Legume Production. In *Legumes for Soil Health and Sustainable Management*; Meena, R., Das, A., Yadav, G., Lal, R., Eds.; Springer: Singapore, 2018; pp. 487–510.
32. Srivastava, T.K.; Ahlawat, L.P.S.; Panwar, J.D.S. Effect of phosphorus, molybdenum and biofertilizers on productivity of pea (*Pisum sativum* L.). *Ind. J. Plant Physiol.* **1998**, *3*, 237–239.
33. Srinivasarao, C.; Ali, M.; Ganeshamurthy, A.N.; Singh, K.K. Potassium requirements of pulse crops. *Better Crops Int.* **2003**, *17*, 9–11.
34. Wang, M.; Zheng, Q.; Shen, Q.; Guo, S. The critical role of potassium in plant stress response. *Int. J. Mol. Sci.* **2013**, *14*, 7370–7390. [CrossRef] [PubMed]
35. Randhawa, M.S.; Maqsood, M.; Shehzad, M.A.; Chattha, M.U.; Chattha, M.B.; Nawaz, F.; Yasin, S.; Abbas, T.; Nawaz, M.M.; Khan, R.D.; et al. Light interception, radiation use efficiency and biomass accumulation response of maize to integrated nutrient management under drought stress conditions. *Turk. J. Field Crop.* **2017**, *22*, 134–142. [CrossRef]
36. Jackson, M.L. *Soil Chemical Analysis*; Prentice Hall of India Private Ltd.: New Delhi, India, 1963.
37. Piper, C.S. *Soil and Plant Analyses*; Prentice Hall of India: New Delhi, India, 1942; p. 498.
38. Walkley, A.; Black, C.A. Estimation of organic carbon by the chromic acid titration method. *Soil Sci.* **1934**, *47*, 29–38. [CrossRef]
39. Jackson, M.L. *Soil and Plant Analysis*; Asia Publishing House: Bombay, India; New Delhi, India, 1967; pp. 30–38.
40. Olsen, S.R.; Cole, C.V.; Watanable, F.S.; Dean, L.A. *Estimation of Available Phosphorus in Soils by Extraction with Sodium Carbonate*; United State Department of Agriculture Circular: Washington, DC, USA, 1954; Volume 939, pp. 1–9.
41. Grigg, J.L. Determination of the available molybdenum of soils. *N. Z. J. Sci. Tech. Sect.* **1953**, *A-34*, 404–414.
42. John, M.K.; Chuah, H.H.; Neufeld, J.H. Application of improved azomethine-H method to the determination of boron in soils and plants. *Anal. Lett.* **1975**, *8*, 559–568. [CrossRef]
43. Lindsay, W.L.; Norvell, W.A. Development of a DTPA soil test for zinc, iron, manganese, and copper. *Soil Sci. Soc. Am. J.* **1978**, *42*, 421–428. [CrossRef]
44. Dhaliwal, L.K.; Hundal, S.S.; Chahal, S.K. Agroclimatic indices of Indian mustard (*Brassica juncea*) under Punjab conditions. *Ind. J. Agric. Sci.* **2007**, *77*, 82–91.
45. Watson, D.J. The physiological basis of variation in yield. *Adv. Agron.* **1952**, *6*, 103–109.
46. Arnon, D.I. Copper enzymes in isolated chloroplasts. Polyphenoloxidase in Beta vulgaris. *Plant Physiol.* **1949**, *24*, 1. [CrossRef]
47. Gomez, K.A.; Gomez, A.A. *Statistical Procedure for Agricultural Research*; John Wiley and Sons Inc.: New York, NY, USA, 1984; p. 180.
48. Sinclair, T.R.; Muchow, R.C. Radiation use efficiency. *Adv. Agric.* **1999**, *65*, 215–265. [CrossRef]
49. Singh, D.; Khare, A.; Singh, S. Effect of phosphorus and molybdenum nutrition on yield and nutrient uptake in lentil (*Lens culinaris* L.). *Ann. Plant Soil Res.* **2017**, *19*, 37–41.
50. Gupta, S.C.; Gangwar, S. Effect of molybdenum, iron and microbial inoculants on symbiotic traits, nutrient uptake and yield of chickpea. *J. Food Legume* **2012**, *25*, 45–49.
51. Rabby, A.K.M.Z.; Paul, A.K.; Sarker, J.R. Effects of nitrogen and molybdenum on the growth and yield of garden pea (*Pisum sativum* L.). *Int. J. Bio-Resour. Stress Manage.* **2011**, *2*, 230–235.
52. Kim, S.A.; Guerinot, M.L. Mining iron: Iron uptake and transport in plants. *FEBS Lett.* **2007**, *581*, 2273–2280. [CrossRef]
53. Khan, N.; Tariq, M.; Ullah, K.; Muhammad, D.; Khan, I.; Rahatullah, K.; Ahmed, N.; Ahmed, S. The effect of molybdenum and iron on nodulation, nitrogen fixation and yield of chickpe genotypes (*Cicer arietinum* L.). *IOSR J. Agric. Vet. Sci.* **2014**, *7*, 63–79. [CrossRef]
54. Banerjee, P.; Ghosh, A.; Visha Kumari, V.; Nath, R. Effect of canopy temperature on physiological processes of grass pea as influenced by seed priming and foliar fertilization. *J. Agromet.* **2021**, *23*, 340–343. [CrossRef]

55. Walker, A.P.; Beckerman, A.P.; Gu, L.; Kattge, J.; Cernusak, L.A.; Domingues, T.F.; Scales, J.C.; Wohlfahrt, G.; Wullschleger, S.D.; Woodward, F.I. The relationship of leaf photosynthetic traits—*V* and *J*—to leaf nitrogen, leaf phosphorus, and specific leaf area: A meta-analysis and modeling study. *Ecol. Evol.* **2014**, *4*, 3218–3235. [CrossRef]
56. Gago, J.; Daloso, D.M.; Figueroa, C.M.; Flexas, J.; Fernie, A.R.; Nikoloski, Z. Relationships of leaf net photosynthesis, stomatal conductance, and mesophyll conductance to primary metabolism: A multispecies meta-analysis approach. *Plant Physiol.* **2016**, *171*, 1–29. [CrossRef]
57. Longstreth, D.J.; Nobel, P.S. Nutrient influences on leaf photosynthesis: Effects of nitrogen, phosphorus and potassium for *Gossypium hirsutum* L. *Plant Physiol.* **1980**, *65*, 541–543. [CrossRef]
58. Basu, S.; Dutta, S.K.; Fangzauva, D.; Jena, S.; Maji, S.; Nath, R.; Chakraborty, P.K. PAR interception and dry matter accumulation in groundnut (*Arachis hypogaea* L.) cultivars sown at different time periods in the Gangetic Plains of West Bengal. *J. Agromet.* **2013**, *15*, 201–204.
59. Rosati, A.; Dejong, T.M. Estimating photosynthetic radiation use efficiency using incident light and photosynthesis of individual leaves. *Ann. Bot.* **2003**, *91*, 869–877. [CrossRef]
60. Chakraborty, A. Growth and yield of lentil (*Lens culinaris* L.) as affected by Boron and Molybdenum application in lateritic soil *J. Crop. Weed.* **2009**, *5*, 88–91.
61. Datta, J.K.; Kundu, A.; Hossein, S.D.; Banerjee, A.; Mondal, N.K. Studies on the impact of micronutrient (molybdenum) on germination, seedling growth and physiology of bengal gram (*Cicer arietinum*) under laboratory condition. *Asian J. Crop Sci.* **2011**, *3*, 56–67. [CrossRef]
62. Chatterjee, R.; Bandyopadhyay, S. Effect of boron, molybdenum and biofertilizers on growth and yield of cowpea (*Vigna unguiculata* L. Walp.) in acid soil of eastern Himalayan region. *J. Saudi Soc. Agric. Sci.* **2015**, *16*, 332–336. [CrossRef]
63. Banerjee, P.; Venugopalan, V.K.; Nath, R. Response of soil moisture regime of relay grass pea (*Lathyrus sativus* L.) to seed priming and foliar fertilization in new alluvial zone of West Bengal. *Agric. Sci. Digest.* **2022**. [CrossRef]
64. Maji, S.; Das, S.; Reza, H.; Mandi, S.; Banerjee, A.; Nath, R.; Mondal, S.; Roy, S.; Bandopadhyay, P. Best management practices for lentil yield intensification as relay crop in rice fallows of the lower Gangetic plains of West Bengal. In Proceedings of the National Conference on Innovative Farming for Food and Livelihood security in Changing Climate, Kalyani, India, 12–13 January 2018; p. 102.
65. Kumar, J.; Sharma, M. Effect of phosphorus and molybdenum on yield and nutrient uptake by chickpea (*Cicer arietinum* L.). *Adv. Plant Sci.* **2005**, *18*, 869–873.
66. Togay, Y.; Togay, N.; Dogan, Y. Research on the effect of phosphorus and molybdenum applications on the yield and yield parameters in lentil (*Lens culinaris* Medic.). *Afr. J. Biotechnol.* **2008**, *7*, 1256–1260.
67. Samant, T.K. Effect of *Rhizobium* and molybdenum inoculation on yield, economics, nodulation and nitrogen uptake in mungbean (*Vigna radiata* L.). *Int. J. Chem. Stud.* **2017**, *5*, 1376–1379.
68. Mandou, M.S.; Chotangui, A.H.; Nkot, L.N.; Nwaga, D. Effect of Rhizobia inoculation phosphorus and molybdenum application on nodulation, N uptake and yield of peanut (*Arachis hypogaea* L.). *Int. J. Agron. Agric. Res.* **2017**, *11*, 103–113.
69. Gowda, K.M.; Halepyati, A.S.; Koppalkar, B.G.; Rao, S. Yield, nutrient uptake and economics of pigeonpea (*Cajanus cajan* L. Millsp.) as influenced by soil application of micronutrients and foliar spray of macronutrients. *Karnataka J. Agric. Sci.* **2015**, *28*, 266–268.
70. Tenikecier, H.S.; Orak, A.; Çubuk, M.G.; Deveci, S. Effect of Different Temperatures on Germination and Early Seedling Growth of Grass Pea (*Lathyrus Sativus* L.). *Türk Tarım Doğa Bilimleri Derg.* **2021**, *8*, 17–22. [CrossRef]
71. Sita, K.; Sehgal, A.; Hanumantha Rao, B.; Nair, R.M.; Vara Prasad, P.V.; Kumar, S.; Gaur, P.M.; Farooq, M.; Siddique, K.H.M.; Varshney, R.K.; et al. Food Legumes and Rising Temperatures: Effects, Adaptive Functional Mechanisms Specific to Reproductive Growth Stage and Strategies to Improve Heat Tolerance. *Front. Plant Sci.* **2017**, *8*, 1658. [CrossRef]
72. Banerjee, P.; Mukherjee, B.; Venugopalan, V.K.; Nath, R.; Chandran, M.A.S.; Dessoky, E.S.; Ismail, I.A.; El-Hallous, E.I.; Hossain, A. Thermal Response of Spring–Summer-Grown Black Gram (*Vigna mungo* L. Hepper) in Indian Subtropics. *Atmosphere* **2021**, *12*, 1489. [CrossRef]
73. Banerjee, P.; Venugopalan, V.K.; Nath, R.; Althobaiti, Y.S.; Gaber, A.; Al-Yasi, H.; Hossain, A. Physiology, Growth, and Productivity of Spring–Summer Black Gram (*Vigna mungo* L. Hepper) as Influenced by Heat and Moisture Stresses in Different Dates of Sowing and Nutrient Management Conditions. *Agronomy* **2021**, *11*, 2329. [CrossRef]

agronomy

MDPI

rticle

;reenhouse Gas Emissions and Yield Production from an)rganic and Conventional Fertilization on Quinoa

›rge Alvar-Beltrán, Anna Dalla Marta, Roberto Vivoli, Leonardo Verdi * and Simone Orlandini

Department of Agriculture, Food, Environment and Forestry (DAGRI), University of Florence, 50144 Florence, Italy; jorge.alvar@unifi.it (J.A.-B.); anna.dallamarta@unifi.it (A.D.M.); roberto.vivoli@unifi.it (R.V.); simone.orlandini@unifi.it (S.O.)
* Correspondence: leonardo.verdi@unifi.it; Tel.: +39-0552755741

Abstract: The high nutritional properties of quinoa have resulted in a production increase worldwide. The resistance to environmental stresses renders this crop suitable for sustainable farming systems. Few studies have examined the impact of different agricultural management strategies and its contribution to climate change. In this work, we quantify soil greenhouse gas (GHG) emissions, in terms of carbon dioxide (CO_2), methane (CH_4) and nitrous oxide (N_2O), and crop productivity (yields and biomass) under conventional (urea) and organic (digestate) fertilization. Significant differences ($p < 0.05$) in N_2O cumulative emissions are reported between digestate (50–100 kg N ha^{-1}), urea (50–100 kg N ha^{-1}) and the control (0 kg N ha^{-1}). Higher cumulative GHG emissions are observed under 100 kg N ha^{-1} of digestate (337.8 kg C ha^{-1} CO_2 and 0.23 kg N ha^{-1} for N_2O) compared to treatments with lower nitrogen (N) inputs. However, yield and biomass production do not show significant differences ($p > 0.05$) with increasing nutrient application. Hence, this study opens the discussion about the pros and cons of increasing fertilization to improve yields besides providing agricultural extension workers with additional information to promote sustainable quinoa production worldwide.

Keywords: digestate; urea; nitrogen; carbon dioxide; methane: nitrous oxide; global warming potential

itation: Alvar-Beltrán, J.; Dalla arta, A.; Vivoli, R.; Verdi, L.; rlandini, S. Greenhouse Gas nissions and Yield Production from Organic and Conventional rtilization on Quinoa. *Agronomy* 22, *12*, 1015. https://doi.org/ .3390/agronomy12051015

cademic Editors: Umberto Anastasi d Aurelio Scavo

ceived: 14 March 2022
ccepted: 20 April 2022
iblished: 23 April 2022

1. Introduction

The use of fertilizers in agriculture are responsible for a large share of anthropogenic emissions [1]. According to the International Panel on Climate Change (IPCC), at global level, 1% of the applied nitrogen (N) in agriculture is lost as nitrous oxide (N_2O) emissions [2]. Other studies suggest that N_2O losses can range between 0.03 and 14% of the applied N depending on the regions, soil texture, and crops [1]. It is also estimated that N_2O emissions from agriculture represent approximately 80% of the global anthropogenic emissions. The magnitude of the environmental issue increases when considering organic fertilizers, which are important sources of nutrients sometimes missing in synthetic fertilizers, and when misused, can become large sources of greenhouse gases (GHG) [3,4]. Generally, due to their own composition, liquid-organic fertilizers provide higher N_2O emissions compared to solid-organic fertilizers [1,5]. Organic fertilizers can improve nutrient availability, biodiversity, and microbial activity in the soil. Conversely, as soil organic carbon (C) increases, so does the microbial activity which consequently accelerates carbon dioxide (CO_2) emissions through soil respiration [6]. Recent literature reports a direct effect of organic fertilization, with compost and digestate, on methane (CH_4) emission increase [7,8].

Due to its high tolerance to biotic and abiotic stresses, and nutritional properties, the cultivation of quinoa is gaining scientific attention [9]. Much of the scientific efforts have been devoted to the study of quinoa's resilience to water, salinity, temperature, and nutrient stress, but little is known about the linkages between GHG emissions and increasing fertilization. Hence, the compounded effects of agricultural strategies on improving crop yields represents an area of major interest in agriculture [10]. For quinoa, discrepancie emerge between the actual effect of fertilization on biomass production and seed yield For example, study [11] observes a positive relationship between yields and N fertilizatio rates. Similarly, study [12] shows a yield increase of up to 75–100 kg N ha^{-1}. With regard to the type of fertilizer, most of the research on quinoa has assessed the effect of syntheti fertilizers, particularly urea ($CO(NH_2)_2$) and ammonium nitrate (NH_4NO_3), on quino yields. Ref. [11] describes an increase in plant height and seed yield with rising NH_4NO rates. The same study reports an increase in yield from 0.9 to 9.2 g $plant^{-1}$ under 0 an 122 kg N ha^{-1}, respectively. Similar substantial yield enhancements are displayed whe N-fertilization rates are increased from 40 to 120 kg N ha^{-1} [13], as well as with calciur ammonium nitrate (NH_4NO_3), with yields increasing from 1.8 to 3.5 Mg ha^{-1} under 0 an 120 kg N ha^{-1}, respectively [14].

On the contrary, the authors of [15,16] report little to no differences in terms c yield and biomass production up to 25 kg N ha^{-1}. In addition, studies [15–17] sugges that N-fertilization has a minor effect on quinoa performance, requiring approximatel 25 kg N ha^{-1} per ton of quinoa seeds produced. Ref. [10] affirm that N splitting is not a advantage for fast-developing species such as quinoa, particularly in terms of seed yiel and protein content.

Despite the previous studies, there is still limited information on quinoa's perfo mance under organic crop-systems. Existing studies have assessed different organi fertilizers, such as compost and cow manure, but do not show notable differences i yields [18]. Instead, Ref. [10] report a yield increase with organic N-fertilization (60, 12 and 180 kg N ha^{-1} of slurry), observing a yield of 2.20 Mg ha^{-1} under 180 kg N ha^{-1} c slurry. Similarly, study [19] report 1.20 Mg ha^{-1} (Ayacucho) and 1.70 Mg ha^{-1} (Huancave ica) with 35 Mg ha^{-1} and 54 Mg ha^{-1} of poultry fertilizer (guano), respectively. From th former literature review, we conclude that countless factors are responsible on improvin yields and, consequently, require additional scientific attention; in particular, on the linl ages between fertilization and yields, and the environmental costs associated with highe agricultural inputs [20]. Lastly, recent studies highlight the differences in GHG emission because of the type of fertilizer (organic and synthetic), pedoclimatic conditions, and agr cultural management strategies [1]. This emphasizes the need to conduct case-specifi observations to obtain information about emission dynamics from agricultural systems.

In the present study, digestate and urea are assessed at different rates (0, 50, an 100 kg N ha^{-1}) to better understand their effect on quinoa yields, biomass production, an environmental impacts.

2. Materials and Methods

2.1. Experimental Design and Agricultural Management Strategies

This study was run between May and August 2019 at the experimental field of Istitut Tecnico Agrario Statale (ITAS) (43° 47′ 06″ N and 11° 13′ 06″ E; 40 m.a.s.l.) in Tuscany, Ital A randomized complete block design (RCBD) was used, divided in two types of fertilize (urea (U) and digestate (D)) with two different N-levels (50 and 100 kg N ha^{-1}) and control (0 kg N ha^{-1}); hereafter referred to as Control, 50D, 100D, 50U, and 100U. Eac treatment included three replicates, for a total of 16 plots (Figure 1). Each plot sized ±4 m (2.1 m width and 1.8 m length), with four rows spacing 70 cm and plants distancing 10 cr from each other (15–20 plants m^{-2} and ±75 plants $plot^{-1}$). Thus, experimental design wa Control (0 kg N ha^{-1}); 50 D (50 kg N ha^{-1} as digestate); 100 D (100 kg N ha^{-1} as digestate 50 U (50 kg N ha^{-1} as urea); 100 U (100 kg N ha^{-1} as urea).

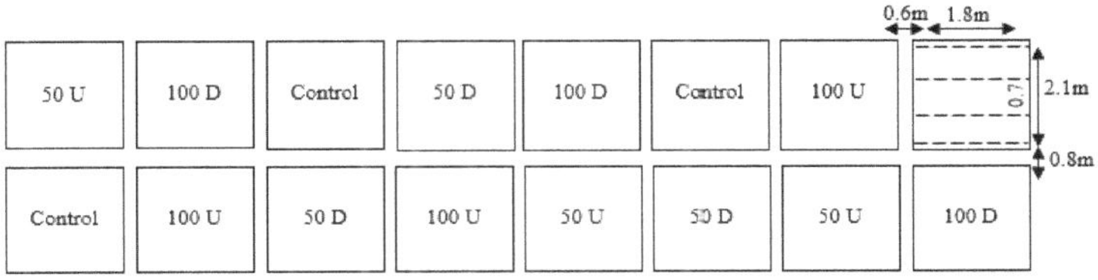

Figure 1. Experimental design with different N-fertilization levels: Control (0 kg N ha^{-1}); 50 D (50 kg N ha^{-1} as digestate); 100 D (100 kg N ha^{-1} as digestate); 50 U (50 kg N ha^{-1} as urea); 100 U (100 kg N ha^{-1} as urea).

The genotype Titicaca was sown on 8 May and harvested on 26 August, with a total cycle duration of 110 days. Urea ($CO(NH_2)_2$) and the liquid fraction of digestate from pig slurries (Table 1) were manually spread in two equal doses, and immediately incorporated into the soil (at 20 cm depth) using a tractor (BCS 710 ACTION). Fertilization was carried out 28 and 49 days after sowing (DAS), coinciding with the stem elongation and flowering phase, respectively. Mechanical weeding was done at the same time as fertilizer application using the same walking tractor. The soil texture in the study field was characterized for being sandy-loam (56% sand, 15% clay, and 29% silt) with a total N content of 0.15 g kg $soil^{-1}$ at 0–20 cm depth.

Table 1. Elemental characterization of fertilizers (urea and digestate) used in this experiment.

	Units	Urea	Digestate
Organic C	%	-	3.02
N content total	%	46	0.39
$N\text{-}NH_4^+$	%	-	0.30
$N\text{-}NO_3^-$	%	-	<0.01
P content total	mL l^{-1}	-	452
K content total	mL l^{-1}	-	2457
Dry matter	%	100	1.89

2.2. *Crop and Agrometeorological Measurements*

The monitoring of plant's phenology was performed during the entire growing season. Measurements were taken from the two middle rows to avoid any side effects. For the evaluation of crop performance (yield and biomass production), all the plants in the middle rows (38 plants $plot^{-1}$) were sampled at physiological maturity and manually harvested using a sickle. Dry weight of grains and biomass (stems and leaves) were determined by drying samples in an oven at 80 °C for 48 h. To standardize the results, yield (grains) and biomass production (stems and leaves) were converted to kg ha^{-1}. Harvest index (%) was calculated as the ratio between yield (grains) and biomass (Table 2). To complete the agrometeorological analysis, weather information (daily average, maximum and minimum temperature, and precipitation) was obtained from both the Regional Hydrological Services of Tuscany (SIR) and the Functional Centre of the Tuscany Region (CFR).

Table 2. ANOVA test (±Standard Deviation) for seed yield (kg ha^{-1}), aboveground biomass (kg ha^{-1}), and harvest index.

Fertilizer Type	N-Treatment (kg ha^{-1})	Yield (kg ha^{-1})	Biomass (kg ha^{-1})	Harvest Index (%)
Control	0	844 ± 126 a	3188 ± 382 a	22.4 ± 3.27 a
Digestate	50	750 ± 154 a	3188 ± 382 a	23.6 ± 3.46 a
	100	792 ± 138 a	3832 ± 856 a	20.9 ± 2.42 a
Urea	50	852 ± 106 a	3410 ± 338 a	25.0 ± 3.07 a
	100	894 ± 422 a	4064 ± 2244 a	22.7 ± 2.22 a
Control		844 ± 126 a	3188 ± 318 a	22.4 ± 3.27 a
Digestate		772 ± 132 a	3510 ± 690 a	22.3 ± 3.40 a
Urea		874 ± 276 a	3738 ± 1480 a	23.9 ± 2.72 a
	0	844 ± 126 a	3188 ± 318 a	22.4 ± 3.27 a
	50	800 ± 130 a	3298 ± 344 a	24.3 ± 3.40 a
	100	844 ± 286 a	3948 ± 1524 a	21.8 ± 2.30 a

Means that do not share a letter are significantly different from each other ($p < 0.05$).

2.3. Greenhouse Gas Measurements

GHG measurements were recorded biweekly from the middle rows during the entire growing season: at 7 DAS (hereafter, M1), 16 DAS (M2), 28 DAS (M3), 33 DAS (M4), 49 DAS (M5), 63 DAS (M6), 77 DAS (M7), 93 DAS (M8), and 110 DAS (M9), respectively. GHG monitoring was performed using 16 static chambers (one per plot) and a portable gas analyzer (XCGM-400, Madur Sensonic). Soil emissions measurements were carried out between the rows avoiding the inclusion of plants inside static chambers. Chambers were assembled following the USDA-ARS GRACEnet Project Protocols [21], as described by [8]. Chambers were made by two parts: chamber lids [8] and chamber collars that were inserted into the soil at 5 cm depth to avoid any mechanical damages to the plant's root system. The portable gas analyzer used non-dispersive infrared (NDIR) technology for the analysis of CO_2 (±0 ppm), CH_4 (±10 ppm), and N_2O (±1 ppm). Gas concentration inside the chambers (ppm), area (314 cm^2), volume (9420 cm^3), closing time (one hour), and molecular weight of each gas were considered for computing soil GHG fluxes. Cumulative soil GHG emissions at the end of growing season were expressed as kg C ha^{-1} for CO_2 and CH_4, and as kg N ha^{-1} for N_2O.

For the observations on emissions fluxes and yields analysis, a yield-scaled emission calculation was used for each gas. Yield-scaled emissions were expressed as the net contributes of 1 kg of quinoa grains (kg CO_2 eq.) The kg CO_2 eq. were obtained from gas-specific Global Warming Potentials (GWP) (CH_4 and N_2O with values of 28 and 298, respectively) [22]. The total impact of each treatment was determined as kg CO_2 eq. and computed separately for the cumulative sum of each GHG (Table 3).

Table 3. Total cumulative GHG emissions during the growing season and daily GHG fluxes.

		GHGs during the Growing Season			GHGs Average Day^{-1}		
Fertilizer Type	N-Treatment (kg ha^{-1})	CO_2 (kg C ha^{-1})	CH_4 (kg C ha^{-1})	N_2O (kg N ha^{-1})	CO_2 (kg C ha^{-1} day^{-1})	CH_4 (kg C ha^{-1} day^{-1})	N_2O (kg N ha^{-1} day^{-1})
Control	0	223.1 ± 16.4 b	1.73 ± 0.27	0.04 ± 0.02 c	23.3 ± 1.6 b	0.21 ± 0.03	0.003 ± 0.001 c
Digestate	50	293.6 ± 20.9 ab	2.40 ± 0.34	0.15 ± 0.01 b	30.6 ± 2.9 ab	0.30 ± 0.04	0.015 ± 0.001 b
	100	337.8 ± 19.7 a	2.29 ± 0.26	0.23 ± 0.01 a	36.1 ± 2.3 a	0.28 ± 0.03	0.025 ± 0.003 a
Urea	50	220.7 ± 32.4 b	2.23 ± 0.94	0.13 ± 0.04 b	23.4 ± 3.7 b	0.27 ± 0.10	0.012 ± 0.005 b
	100	229.1 ± 45.5 b	2.22 ± 0.66	0.11 ± 0.03 b	23.5 ± 5.1 b	0.26 ± 0.08	0.010 ± 0.003 b
	Control	223.1 ± 16.4 b	1.73 ± 0.27	0.04 ± 0.02 c	23.3 ± 1.6 b	0.21 ± 0.03	0.003 ± 0.001 c
	Digestate	315.7 ± 30.3 a	2.34 ± 0.28	0.19 ± 0.05 a	33.3 ± 3.8 a	0.29 ± 0.03	0.020 ± 0.006 a
	Urea	224.9 ± 35.6 b	2.22 ± 0.72	0.12 ± 0.03 b	23.5 ± 4.0 b	0.26 ± 0.08	0.011 ± 0.004 b
	0	223.1 ± 16.4	1.73 ± 0.27	0.04 ± 0.02 b	23.3 ± 1.6	0.21 ± 0.03	0.003 ± 0.001 b
	50	257.1 ± 46.8	2.31 ± 0.64	0.14 ± 0.03 a	27.0 ± 4.9	0.28 ± 0.07	0.013 ± 0.004 a
	100	283.5 ± 67.3	2.25 ± 0.45	0.17 ± 0.07 a	29.8 ± 7.8	0.27 ± 0.06	0.017 ± 0.008 a

Means that do not share a letter are statistically significantly different from each other ($p < 0.05$).

2.4. Statistical Analysis

The GHG emissions (CO_2, CH_4, and N_2O), yields, and biomass were examined by analysis of variance (ANOVA). The Tukey HSD test with a critical p value lower than 0.05 was used as pairwise comparison to test the significance between different N-fertilization rates (0, 50, 100 kg N ha^{-1}) and types of fertilizer (digestate and urea). The statistical package used to run the ANOVA analysis was Minitab 19.

3. Results

3.1. Meteorological Information

The average temperature recorded during the growing season was 23.0 °C (Figure 2). July and August were the warmest months, with an average temperature of 26.1 °C. Average maximum and minimum temperatures were of 29.3 °C and 17.0 °C, with maximum and minimum absolute values of 39.5 °C and 4.2 °C, respectively. The total amount of precipitation was 197.2 mm, distributed over 28 precipitation events. Most of the precipitation was recorded in May (135.2 mm), with erratic rainfall from June to September.

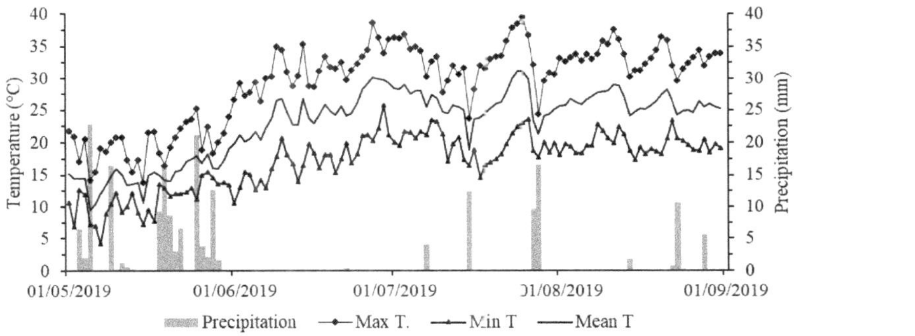

Figure 2. Average, maximum and minimum temperatures (°C), and precipitation (mm) observed during the growing season (May–August 2019).

3.2. Carbon Dioxide Emissions

Despite a high variability in the results, soil CO_2 emissions during the growing season showed similar trends in all treatments (Figure 3). Emissions increased following crop growth, peaking twice at each fertilization event (at 28 and 49 DAS, corresponding to stem elongation and flowering). Regardless of the N rate (50 and 100 kg N ha^{-1}), treatments with digestate produced higher emissions than synthetic fertilizers and the control (Table 3). Between the two fertilization events, and from M6 to the end of the experiment, soil CO_2 emissions experienced a notable decrease with time and, therefore, at harvesting (M9), soil CO_2 emissions were close to initial levels. Significantly higher ($p < 0.05$) cumulative CO_2 emissions were observed under 50D and 100D treatments (293.6 and 337.8 kg C ha^{-1}, respectively) than for the other treatments (Table 3). However, no significant emission differences ($p > 0.05$) were reported between the control, 50U and 100U (Table 3). Similar cumulative CO_2 emissions ($p < 0.05$) were recorded for the control and urea treatments, and these were approximately 30% lower to those of digestate (Table 3). Overall, changes in N fertilization rate (0, 50, and 100 kg N ha^{-1}) did not affect the CO_2 emission rate ($p > 0.05$) (Table 3).

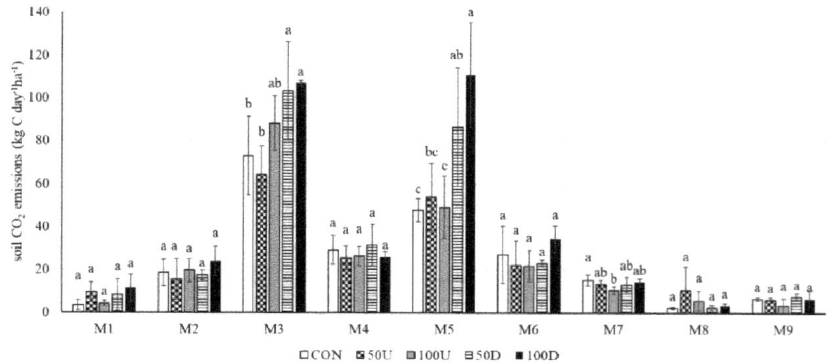

Figure 3. CO_2 emission fluxes (kg C day^{-1} ha^{-1}) during the growing cycle for the control (CON urea 50 kg N ha^{-1} (50U), urea 100 kg N ha^{-1} (100U), digestate 50 kg N ha^{-1} (50D), and digestat 100 kg N ha^{-1} (100D). Results that do not share letters are significantly different from each othe ($p < 0.05$).

3.3. *Methane Emissions*

As for CO_2, two CH_4 emission peaks were observed in all treatments (M3 and M5 coinciding with each fertilization event (Figure 4). Nevertheless, during the second fe tilization event (M5), emissions were considerably higher (0.9 to 1.2 kg C day ha^{-1}) t those observed during the first fertilization event (0.1 to 0.3 kg C day ha^{-1}). No significan CH_4 emission differences ($p > 0.05$) were displayed between treatments. However, durin the second fertilization treatment, higher CH_4 emissions were observed in both digestat treatments, suggesting a marginal effect of water and methanogenic bacteria content o digestate. This was also confirmed by the cumulative CH_4 emissions, with higher value detected under 50D and 100D, though without displaying significant differences ($p > 0.05$ amongst treatments (Table 3). In M2 and M9, the soil acted as a CH_4 sink with low level of CH_4 oxidation. Overall, during the growing season, no statistical differences ($p > 0.05$ were recorded for soil CH_4 emissions in all treatments, suggesting that CH_4 emission were neither related to different N rates (0, 50, and 100 kg N ha^{-1}) nor to different types c fertilizer (digestate and urea) (Table 3).

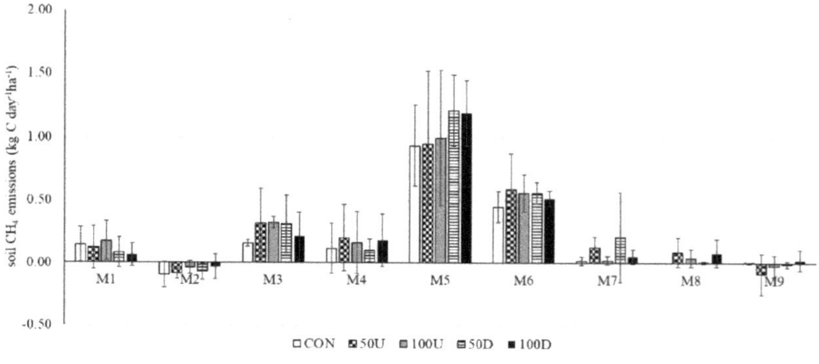

Figure 4. CH_4 emission fluxes (kg C day^{-1} ha^{-1}) during the growing cycle for the control (CON urea 50 kg N ha^{-1} (50 U), urea 100 kg N ha^{-1} (100 U), digestate 50 kg N ha^{-1} (50 D), and digestat 100 kg N ha^{-1} (100 D). Letters are not displayed since no statistical differences were observed betwee treatments ($p < 0.05$)).

3.4. Nitrous Oxide Emissions

For the N_2O emissions fluxes from the soil, similar trends were detected to those of CO_2 and CH_4, with an N_2O increase at each fertilization event (M3 and M5) and following measurement (M6) (Figure 5). Nevertheless, during the second fertilization event (M5), N_2O emissions showed a different trend to that of CO_2 and CH_4, based on the type of fertilizer (digestate and urea) and N rate (50 and 100 kg N ha^{-1}). For the control, N_2O emissions were not reported during the entire growing cycle, except for M3. Regardless of the amount of N (50 and 100 kg N ha^{-1}), urea had a longer emission lag-time than digestate and, therefore, higher N_2O emissions were observed at M6 than at M5. In fact, digestate was more reactive than urea, showing an emission peak under 50D and 100D immediately after each fertilization event. Nevertheless, only significantly higher ($p < 0.05$) emissions were observed under 100D. As for CO_2 and CH_4, N_2O emissions decreased between the two fertilization events, with emissions returning to initial levels at the end of the experiment. From the analysis of cumulative N_2O emissions (Table 3), results showed higher N_2O emissions ($p < 0.05$) at 100D compared to other treatments (CON and 50D). Significant differences ($p < 0.05$) were reported between 100D and 100U, 50U and 50D, as well as when comparing 100D, 100U, 50U, and 50D to the control. Regardless of the N supply, N_2O emissions were significantly affected by the type of fertilizer ($p < 0.05$), with digestate producing the highest N_2O emissions compared to urea. N_2O emissions were affected by different N rates, being 100 kg N ha^{-1} the treatment (100D and 100U) with highest N_2O emissions.

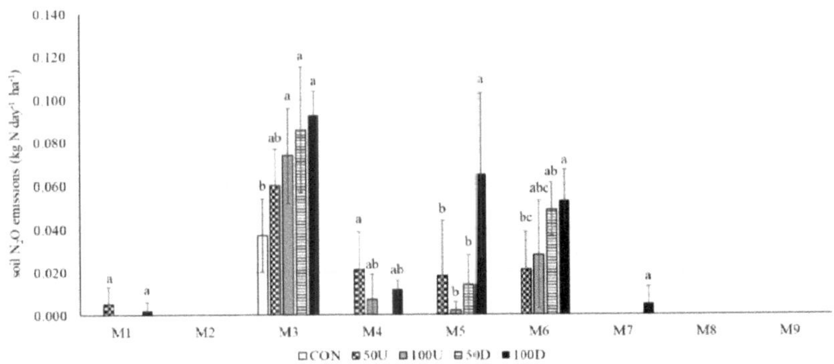

Figure 5. N_2O emission fluxes (kg N day^{-1} ha^{-1}) during the growing cycle for the control (CON), urea 50 kg N ha^{-1} (50U), urea 100 kg N ha^{-1} (100U), digestate 50 kg N ha^{-1} (50D), and digestate 100 kg N ha^{-1} (100D). Results that do not share letters are significantly different from each other ($p < 0.05$).

3.5. Quinoa Yields and Biomass Production

From the analysis of results, we observed no differences in yield and biomass neither between different N rates (0, 50, and 100 kg N ha^{-1}) nor types of fertilizers (urea and digestate) (Table 2). Even though the highest seed yields were obtained under 100U and 50U (894 and 852 kg ha^{-1}, respectively), no significant yield differences ($p > 0.05$) were found when comparing 100D, 50D, and the control. A similar behavior was observed with biomass production, with no significant differences between different N rates nor fertilizers. While the highest dried-biomass production was reported under 100U and 100D with 4064 and 3832 kg ha^{-1}, respectively, the lowest was found both under 50U and the control (3188 kg ha^{-1}). The average harvested index for all the treatments was 23.7%.

3.6. Yield-Scaled Emissions

The ratio between cumulative GHG emissions (Table 3) and observed yields (Table 2) were crucial for determining the environmental impacts of quinoa. Relative CO_2, CH_4, and N_2O emissions, expressed as kg and g of C and N kg seed^{-1} produced (depending on the gas), showed alike behavior in all treatments (CON, 50D, 100D, 50U, and 100U). For all three gases (CO_2, CH_4, and N_2O), values were highest at 100D followed by 50D, 100U, 50U, and, finally, the control. For N_2O, significant differences ($p < 0.05$) for relative emissions were observed under 100D when compared to the other treatments (Table 3). The gas-specific GWP for each GHG was used to calculate the yield-scaled emission (Table 4) for each treatment. Yield-scaled emissions were significantly higher for digestate (0.075 kg CO_2eq kg seed^{-1}) than for the control (0.013 kg CO_2 eq. kg seed^{-1}). Despite the different levels of N fertilization, no significant differences were reported between treatments. In addition, the gas-specific emissions per kg of seeds produced for different types of fertilizers showed that digestate had higher N_2O emissions than urea.

Table 4. Yield-scaled emissions (CO_2, CH_4, and N_2O) per kg of quinoa seed produced and Global Warming Potential (GWP) (kg CO_2eq kg seed^{-1}) of each treatment.

Fertilizer Type	N-Treatment (kg ha^{-1})	CO_2 (kg C kg Seed^{-1})	CH_4 (mg C kg Seed^{-1})	N_2O (mg N kg Seed^{-1})	CH_4 (kg CO_2 eq kg Seed^{-1})	N_2O (kg CO_2 eq kg Seed^{-1})	Total (kg CO_2 eq kg Seed^{-1})
Control	0	0.266 ± 0.024	2.1 ± 0.3	0.04 ± 0.02 b	0.057 ± 0.007	0.013 ± 0.006 b	0.337 ± 0.024
Digestate	50	0.402 ± 0.082	3.2 ± 0.2	0.21 ± 0.05 a	0.090 ± 0.006	0.061 ± 0.016 a	0.553 ± 0.100
	100	0.438 ± 0.097	3.0 ± 0.7	0.30 ± 0.06 a	0.083 ± 0.021	0.089 ± 0.018 a	0.609 ± 0.134
Urea	50	0.265 ± 0.076	2.7 ± 0.1.5	0.15 ± 0.05 ab	0.076 ± 0.042	0.045 ± 0.016 ab	0.386 ± 0.128
	100	0.301 ± 0.145	3.0 ± 1.9	0.15 ± 0.08 ab	0.084 ± 0.052	0.044 ± 0.025 ab	0.429 ± 0.213
	Control	0.266 ± 0.024 b	2.1 ± 0.3	0.04 ± 0.02 b	0.057 ± 0.007	0.013 ± 0.006 b	0.337 ± 0.024 b
	Digestate	0.420 ± 0.083 a	3.1 ± 0.5	0.25 ± 0.07 a	0.087 ± 0.014	0.075 ± 0.021 a	0.581 ± 0.110 a
	Urea	0.283 ± 0.105 ab	2.9 ± 1.5	0.15 ± 0.06 b	0.080 ± 0.043	0.045 ± 0.019 b	0.408 ± 0.159 ab
	0	0.266 ± 0.024	2.1 ± 0.3	0.04 ± 0.02 b	0.057 ± 0.007	0.013 ± 0.006 b	0.337 ± 0.024
	50	0.334 ± 0.103	3.0 ± 1.0	0.18 ± 0.06 ab	0.083 ± 0.028	0.053 ± 0.017 ab	0.470 ± 0.137
	100	0.370 ± 0.133	3.0 ± 1.3	0.22± 0.10 a	0.084 ± 0.035	0.067 ± 0.031 a	0.519 ± 0.187

Means that do not share a letter are statistically significantly different from each other ($p < 0.05$).

4. Discussion

4.1. Carbon Dioxide Emissions

Our results showed that CO_2 emissions peak at each fertilization event, afterwards, emissions declined in a similar way to that reported by [8,23]. The CO_2 emissions peaked in spite of the amount of N (50 and 100 kg N ha^{-1}) and type of fertilizer (urea and digestate), with similar findings as those observed in the control (0 kg N ha^{-1}). Hence, the increase in CO_2 emissions was not just attributed to fertilization treatments, but to agricultural practices, in particular, mechanical weeding. Since the control was subjected to the same agricultural practices as that of the other treatments, the peak in CO_2 displayed a similar behavior among treatments. Soil CO_2 emissions were produced during harrowing, releasing CO_2 trapped in the soil pores [24,25]. The destruction of soil aggregates, through harrowing, improved soil aeration and the incorporation of organic matter into the soil As a result, microorganism activity was enhanced which led to a rapid decomposition of soil organic matter [25]. In addition, CO_2 emissions from agricultural soils were affected by a wide range of factors such as soil moisture, temperature, soil organic matter, and pH, among others, thereby rendering emission dynamics highly variable [8,26,27]. The compounded effect of all these factors resulted in a high variability in CO_2 emissions.

Furthermore, in this study, we observed higher CO_2 emissions under higher N rates of digestate (100D) than under lower N rates of both digestate (50D) and/or urea (100U and 50U). This was probably due to the combined effect of high-water and organic matter content in digestate, which uniformly reached the rhizosphere. Moreover, water played a key role in the development of soil microbial activity. For example, the action of water in digestate, together with high atmospheric temperatures (Figure 1), created optimal microbial growing conditions, thereby increasing metabolic activity and enhancing microbial respiration, which consequently resulted in the release of CO_2. Similarly, organic matter in digestate also enhanced the soil microbial growth. Previous studies demonstrated that organic matter represented a fundamental source of carbon (C) for the metabolism of soil bacteria and, thus, further increased CO_2 emissions [28,29].

With regards to soil CO_2 emissions from urea, our results did not show significant differences on cumulative soil CO_2 emissions between the control, 50U, and 100U. Therefore, given that the soil CO_2 emissions from urea were triggered by hydrolysis (occurring under the presence of water and urease enzyme), and that the C contained in urea was lost by volatilization, no significant differences were reported among treatments.

4.2. Methane Emissions

The soil CH_4 emissions observed in this experiment were negligible compared to those of CO_2. Nevertheless, CH_4 emission trends were similar to those of CO_2, increasing after each fertilization event. Dried conditions during the growing cycle, intercalated by heavy rainfall events, were responsible for low CH_4 cumulative emissions (28.9 kg C ha^{-1}, average of all treatments). These values were consistent with the findings of [30,31], affirming that soil water and anaerobic conditions were the main drivers of CH_4 emissions. Despite the discrepancies found within the literature, study [32] concluded that N fertilization, and in particular NH_4^+, inhibit CH_4 oxidation due to its similar molecular size and the low specificity of the monooxygenase enzymes of methanotrophic bacteria. This competition was first described by study [33], demonstrating the inhibitory effect of NH_4^+ on CH_4 oxidation. As a result, the CH_4 emissions from digestate can be ascribed to the intrinsic content of methanogenic bacteria within digestate [34]. In addition, due to the combined effect of physical disturbances within the soil from harrowing, and the use of different types of fertilizers, CH_4 emissions from digestate were similar to those observed in urea and the control. The authors of [24,30] affirmed that land use changes, from natural conditions to agricultural lands, strongly reduced the CH_4 oxidation potential of soils, which switched from being a sink to become a CH_4 source. In this experiment, CH_4 absorption from the soil was observed at M2 and M9, when harrowing did not take place (Figure 4). Hence, tillage was the main factor responsible for CH_4 emissions. Tillage was previously shown to have a greater effect on CH_4 emissions than the type of fertilizers and N fertilization rates by directly disturbing the methane-oxidizing community at the soil level [35].

4.3. Nitrous Oxide Emissions

The present study revealed the effect of N rate on N_2O emissions, with a positive relationship between increasing N rates and N_2O emissions, mostly from digestate. Due to the shallowed root system of the selected quinoa genotype, less N was taken up by the root system and, therefore, higher residual N was left in the soil prone to environmental losses (soil leaching and volatilization). As for CO_2 and CH_4, N_2O emissions followed a similar behavior during the growing season, with two emission peaks at each fertilization event. According to studies [29,36], N_2O emissions were strongly related to soil moisture conditions and available N-compounds, mostly NH_4^+. During the first fertilization event, all treatments showed alike trend, though higher emissions under 100D due to a higher water (>98%), NH_4^+ (>75% of total N), and organic C content present in digestate. The compounded effect of all these factors accelerated soil N_2O emissions. Relevant N_2O emissions were also reported for 50D as well as for urea treatments. N_2O emissions from the latter treatments were associated to heavy precipitation observed before applying fertilizers (Figure 2), which improved soil water content and enhanced urea decomposition. Urea had a high NH_4^+ content, but as a granular fertilizer, required more water to degrade [30]. Harrowing performed at M5 (43 DAS) disaggregated the soil and ensured homogenous infiltration of water from precipitation. For example, the 4 mm of rain observed at 68 DAS resulted in urea degradation. Thus, the late production of N_2O emissions were, to some extent, explained by the delayed response of urea during the second fertilization even Similarly, other studies showed a strong correlation between soil water content and N_2(emissions following the urea fertilization event, even if precipitation occurred several day after fertilization [37,38]. Moreover, a decrease in N_2O emissions was observed after the las fertilization event, from M7 to the end of the experiment, and reaching zero at M8 (Figure 5 In agreement with the latter observations were the findings made by [4,36]. N_2O emission from the control were only observed at M3, probably due to the simultaneous effect o weather conditions, harrowing, and the intrinsic soil N content. In this line, study [29 showed a slight increase in N_2O emissions when soil pores were filled with more than 60 of water. Study [4] indicated that the physical conditions of fertilizers were responsible fo increasing N_2O emissions by 23% from digestate when compared to urea. Moreover, th dried climatic conditions occurring after M6, resulted in N_2O emission differences betwee digestate and urea. Hence, the type of fertilizer had an impact on the amount of N_2(emissions, which was related to the soil water content, N compounds, and C concentratio present in fertilizers. This was consistent with studies [39,40], who reported an increas of NH_4^+-compounds in the soil with increasing N_2O emissions. Overall, the higher wate content found on digestate played a key role on N_2O emission dynamics and, therefor showed a significantly higher N_2O emissions than in urea.

From the analysis on cumulative N_2O emissions, 100D produced the highest N_2(emissions because of the higher NH_4^+, organic C, and water content found in digestat (Table 1). The key role of water was also observed in the cumulative N_2O emissions fron 100U, showing very little differences when compared to 50U and 50D (Table 3). Due t its intrinsic characteristics of high-water content, N-compounds, and C content, digestat generated higher N_2O emissions than urea.

4.4. *Quinoa Yields and Environmental Impacts*

The herein study did not show statistically significant differences in terms of biomas and yield production, neither for different types of fertilizers (digestate and urea) and N fertilization levels (0, 50, and 100 kg N ha^{-1}). The soil N content was adequate to satisfy th N requirements of quinoa and, consequently, the additional N from fertilizers did not affec the final seed yield. Similar observations were made by studies [15–17,41], concludin that N requirements of quinoa were relatively low. However, the present results wer discrepant to those of studies [10,19], which reported significant yield differences whe applying higher amounts of poultry 1.7 Mg seed ha^{-1} with 54 Mg ha^{-1} poultry fertilize and slurry fertilizer (2.2 Mg ha^{-1} with 180 kg N ha^{-1}). The findings of studies [12–14 were also in accordance with the latter research, showing a steady yield increase of up t 75–100 kg N ha^{-1} and 120 kg N ha^{-1}. In addition, study [42] suggested that yields in sout Italy were considerably lower (1.4 Mg ha^{-1}) if sowing in May rather than in April. In ou experiment, the sowing was delayed to May due to adverse weather conditions of Apri 2020 and, therefore, lower yields (800 kg ha^{-1}) to those of study [42] were reported. Thi was due to high temperatures during the flowering stage (end of June) (Figure 2), whic had a detrimental effect on seed pollination [43].

From the assessment of quinoa's environmental impacts, this study reported lower va ues (0.46 kg CO_2eq kg^{-1} of quinoa seed, average of all treatments) to those of studies [19,44 in Peru (1.03 and 0.88 kg CO_2eq kg^{-1} of quinoa harvested, respectively). While the latte study used a Life Cycle Assessment (LCA) software to estimate GHG emissions, our stud was based on field observations. Lastly, the daily CO_2 emissions observed in our wor (27.4 kg CO_2-C ha^{-1} day^{-1} kg $seed^{-1}$, average of all treatments) halved those of maiz (40–60 kg CO_2-C ha^{-1} day^{-1}), whereas those of N_2O (0.132 kg N_2O-N ha^{-1} average value of all treatments) were notably lower to those of maize (2–7 kg N_2O-N ha^{-1} during th growing season) [4,25,27]. This was invariably the result of higher agricultural input (fertilizers, pesticides, water, etc.) of maize compared to those of quinoa.

5. Conclusions

The expansion of crops with a low environmental impact and high nutritional properties, coupled with low-impact agricultural strategies, are increasingly drawing scientific and public attention. In this study, we evaluated GHG emissions of quinoa cultivation using different types of fertilizer and N rates in central Italy, Tuscany. We observed that the direct soil GHG emissions of digestate are higher (30% for CO_2 and 40% for N_2O) than those of urea. Although we only evaluated the direct emissions from the soil, further impact assessments should consider the indirect emissions from the production and subsequent spreading of fertilizers. However, as a by-product of a renewable resource, digestate is assumed to be a zero-impact fertilizer during the production phase, besides being accepted as an effective strategy for reusing resources within a farm. The former is key when comparing to conventional fertilizers such as urea, which requires less harrowing but has higher production emissions. For this reason, a more in-depth analysis is recommended both in the direct and indirect emissions deriving from different types of fertilizers. To conclude, based on our observations, the low N requirements of quinoa masked the effect of fertilization. Therefore, the actual convenience of fertilization must be regularly evaluated according to the agroclimatic conditions of each site. This study represents a starting point for the definition of low-impact quinoa production with the prospects of developing more sustainable farming systems around the world.

Author Contributions: Conceptualization, L.V. and J.A.-B.; Methodology, L.V., J.A.-B. and R.V.; Formal Analysis, J.A.-B.; Writing, L.V. and J.A.-B.; Writing—Review and Editing, A.D.M. and S.O.; Supervision, A.D.M. and S.O. All authors have read and agreed to the published version of the manuscript.

Funding: This work was partially funded by the project SYSTEMIC "An integrated approach to the challenge of sustainable food systems adaptive and mitigatory strategies to address climate change and malnutrition", Knowledge hub on Nutrition and Food Security, in a joint action of JPI HDHL, JPI OCEANS and FACCE JPI launched in 2019 under the ERA NET ERA HDHL (n 696295).

Institutional Review Board Statement: Not applicable.

Informed Consent Statement: Not applicable.

Data Availability Statement: Not applicable.

Acknowledgments: The authors acknowledge Euro Panracci from the Department of Agricultural, Food and Environmental Sciences (University of Perugia) for the provision of quinoa seed. Special recognitions to Azienda Agricola Marchese De' Frescobaldi—Fattoria di Corte for providing this research with digestate.

Conflicts of Interest: The authors declare no conflict of interest.

References

1. Walling, E.; Vaneeckhaute, C. Greenhouse gas emissions from inorganic and organic fertilizer production and use: A review of emission factors and their variability. *J. Environ. Manag.* **2020**, *276*, 111211. [CrossRef] [PubMed]
2. De Klein, C.; Novoa, R.S.A.; Ogle, S.; Smith, K.A.; Rochette, P.; Wirth, T.C. N_2O emissions from managed soils, and CO_2 emis-sions from lime and urea application. In *2006 IPCC Guidelines for National Greenhouse Gas Inventories*; Eggleston, S., Buendia, L., Miwa, K., Ngara, T., Tanabe, K., Eds.; Institute for Global Environmental Strategies, Tanabe: Hyama, Japan, 2006.
3. Nkoa, R. Agricultural benefits and environmental risks of soil fertilization with anaerobic digestates: A review. *Agron. Sustain. Dev.* **2014**, *34*, 473–492. [CrossRef]
4. Verdi, L.; Kuikman, P.J.; Orlandini, S.; Mancini, M.; Napoli, M.; Dalla Marta, A. Does the use of digestate to re-place mineral fertilizers have less emissions of N_2O and NH_3? *Agric. For. Meteorol.* **2019**, *269*, 112–118. [CrossRef]
5. Cayuela, M.L.; Aguilera, E.; Sanz-Cobena, A.; Adams, D.C.; Abalos, D.; Barton, L.; Ryals, R.; Silver, W.L.; Alfaro, M.A.; Pappa, V.A. Direct nitrous oxide emissions in Mediterranean climate cropping systems: Emission factors based on a meta-analysis of available measurement data. *Agric. Ecosyst. Environ.* **2017**, *238*, 25–35. [CrossRef]
6. Oertel, C.; Matschullat, J.; Zurba, K.; Zimmermann, F.; Erasmi, S. Greenhouse gas emissions from soils—A review. *Geochemistry* **2016**, *76*, 327–352. [CrossRef]
7. Severin, M.; Fuss, R.; Well, R.; Garlipp, F.; Van den Weghe, H. Soil, slurry and application effects on greenhouse gas emissions. *Plant Soil Environ.* **2015**, *61*, 344–351. [CrossRef]

8. Verdi, L.; Mancini, M.; Ljubojević, M.; Orlandini, S.; Dalla Marta, A. Greenhouse gas and ammonia emissions from soil: The effect of organic matter and fertilisation method. *Ital. J. Agron.* **2018**, *13*, 260–266. [CrossRef]
9. Vega-Gálvez, A.; Miranda, M.; Vergara, J.; Uribe, E.; Puente, L.; Martínez, E.A. Nutrition facts and functional potential of quinoa (*Chenopodium quinoa* willd.), an ancient Andean grain: A review. *J. Sci. Food Agric.* **2010**, *90*, 2541–2547. [CrossRef]
10. Jacobsen, S.-E.; Christiansen, J.L. Some Agronomic Strategies for Organic Quinoa (*Chenopodium quinoa* Willd.). *J. Agron. Crop Sci.* **2016**, *202*, 454–463. [CrossRef]
11. Alandia, G.; Jacobsen, S.-E.; Kyvsgaard, N.C.; Condori, B.; Liu, F. Nitrogen Sustains Seed Yield of Quinoa under Intermediate Drought. *J. Agron. Crop Sci.* **2016**, *202*, 281–291. [CrossRef]
12. Iqbal, S.M.B.S.; Afzal, I. Evaluating the Response of Nitrogen Application on Growth Development and Yield of Quinoa Genotypes. *Int. J. Agric. Biol.* **2014**, *16*, 886–892.
13. Jacobsen, S.E.; Jørgensen, I.; Stølen, O. Cultivation of quinoa (*Chenopodium quinoa*) under temperate climatic conditions in Denmark. *J. Agric. Sci.* **1994**, *122*, 47–52. [CrossRef]
14. Kaul, H.-P.; Kruse, M.; Aufhammer, W. Yield and nitrogen utilization efficiency of the pseudocereals amaranth, quinoa, and buckwheat under differing nitrogen fertilization. *Eur. J. Agron.* **2005**, *22*, 95–100. [CrossRef]
15. Benhalbib, O. Chapter 6.1.5: Status of quinoa production and research in Morocco. In *State of the Art Report on Quinoa around the World in 2013*; Bazile, D., Bertero, H.D., Nieto, C., Eds.; FAO: Rome, Italy, 2015; pp. 478–491.
16. Alvar-Beltrán, J.; Saturnin, C.; Dao, A.; Dalla Marta, A.; Sanou, J.; Orlandini, S. Effect of drought and nitrogen fertilisation on quinoa (*Chenopodium quinoa* Willd.) under field conditions in Burkina Faso. *Ital. J. Agrometeorol.* **2019**, *1*, 33–43.
17. Moreale, A. In the Quinoa Project: Wageningen University. 1993. Available online: http://edepot.wur.nl/354101 (accessed on 27 August 2018).
18. Bilalis, D.; Kakabouki, I.; Karkanis, A.; Travlos, I.; Triantafyllidis, V.; Hela, D. Seed and Saponin Production of Organic Quinoa (*Chenopodium quinoa* Willd.) for different Tillage and Fertilization. *Not. Bot. Horti Agrobot. Cluj-Napoca* **2012**, *40*, 42–46. [CrossRef]
19. Cancino-Espinoza, E.; Vázquez-Rowe, I.; Quispe, I. Organic quinoa (*Chenopodium quinoa* L.) production in Peru: Environmental hotspots and food security considerations using Life Cycle Assessment. *Sci. Total Environ.* **2018**, *637–638*, 221–232. [CrossRef]
20. Sellami, M.H.; Pulvento, C.; Lavini, A. Agronomic Practices and Performances of Quinoa under Field Condi-tions: A Systematic Review. *Plants* **2021**, *10*, 72. [CrossRef]
21. Parkin, T.B.; Venterea, R.T. USDA-ARS GRACEnet project protocols, chapter 3. In *Chamber-Based Trace Gas Flux Measurements*; Impedans: Dublin, Ireland, 2010; (Replace Original Version of April 2003).
22. IPCC. Climate Change. The Physical Science Basis. In *Contribution of Working Group I to the Fifth Assessment Report of the Intergovernmental Panel on Climate Change*; Cambridge University Press: Cambridge, UK; New York, NY, USA, 2013.
23. Maucieri, C.; Barbera, A.C.; Borin, M. Effect of injection depth of digestate liquid fraction on soil carbon dioxide emission and maize biomass production. *Ital. J. Agron.* **2016**, *11*, 6–11. [CrossRef]
24. Omonode, R.A.; Vyn, T.J.; Smith, D.R.; Hegymegi, P.; Gál, A. Soil carbon dioxide and methane fluxes from long-term tillage systems in continuous corn and corn–soybean rotations. *Soil Tillage Res.* **2007**, *95*, 182–195. [CrossRef]
25. Ussiri, D.A.N.; Lal, R. Long-term tillage effects on soil carbon storage and carbon dioxide emissions in continuous corn cropping system from an alfisol in Ohio. *Soil Till. Res.* **2009**, *104*, 39–47. [CrossRef]
26. Paustian, K.; Six, J.; Elliott, E.T.; Hunt, H.W. Management options for reducing CO_2 emissions from agricul-tural soils. *Biogeochemistry* **2002**, *48*, 147–163. [CrossRef]
27. Zhai, L.-M.; Liu, H.-B.; Zhang, J.-Z.; Huang, J.; Wang, B.-R. Long-Term Application of Organic Manure and Mineral Fertilizer on N_2O and CO_2 Emissions in a Red Soil from Cultivated Maize-Wheat Rotation in China. *Agric. Sci. China* **2011**, *10*, 1748–1757. [CrossRef]
28. Sänger, A.; Geisseler, D.; Ludwig, B. Effects of moisture and temperature on greenhouse gas emissions and C and N leaching losses in soil treated with biogas slurry. *Biol. Fertil. Soils* **2011**, *47*, 249–259. [CrossRef]
29. Pezzolla, D.; Bol, R.; Gigliotti, G.; Sawamoto, T.; López, A.L.; Cardenas, L.; Chadwick, D. Greenhouse gas (GHG) emissions from soils amended with digestate derived from anaerobic treatment of food waste. *Rapid Commun. Mass Spectrom.* **2012**, *26*, 2422–2430. [CrossRef]
30. Snyder, C.S.; Bruulsema, T.W.; Jensen, T.L.; Fixen, P.E. Review of greenhouse gas emissions from crop production systems and fertilizer management effects. *Agric. Ecosyst. Environ.* **2009**, *133*, 247–266. [CrossRef]
31. Le Mer, J.; Roger, P. Production, oxidation, emission and consumption of methane by soils: A review. *Eur. J. Soil Biol.* **2001**, *37*, 25–50. [CrossRef]
32. Tiwari, S.; Singh, J.S.; Singh, D.P. Methanotrophs and CH_4 sink: Effect of human activity and ecological perturbations. *Clim. Chang. Environ. Sustain.* **2015**, *3*, 35–50. [CrossRef]
33. Schimel, J.P.; Holland, E.A.; Valentine, D. Controls on methane flux from terrestrial ecosystems. In *Agricultural Ecosystem Effects on Trace Gases and Global Climate Change*; Harper, L.A., Mosier, A.R., Duxbury, J.M., Rolston, D.E., Eds.; American Society of Agronomy: Madison, WI, USA, 1993; pp. 167–182.
34. Verdi, L.; Mancini, M.; Napoli, M.; Orlandini, S.; Dalla Marta, A. Soil carbon emissions from maize under different fertilization methods in an extremely dry summer in Italy. *Ital. J. Agrometeorol.* **2019**, *2*, 3–10.
35. Ball, B.C.; Scott, A.; Parker, J.P. Field N_2O, CO_2 and CH_4 fluxes in relation to tillage, compaction and soil quality in Scotland. *Soil Tillage Res.* **1999**, *53*, 29–39. [CrossRef]

6. Wulf, S.; Maeting, M.; Clemens, J. Application technique and slurry co-fermentation effects on ammonia, nitrous oxide and methane emissions after spreading: II. Greenhouse gas emissions. *J. Environ. Qual.* **2002**, *31*, 1795–1801. [CrossRef]
7. Luo, J.; Ledgard, S.; Lindsey, S. Nitrous oxide emissions from application of urea on New Zealand pasture. *N. Z. J. Agric. Res.* **2007**, *50*, 1–11. [CrossRef]
8. Soares, J.R.; Cassman, N.A.; Kielak, A.M.; Pijl, A.; Carmo, J.B.; Lourenco, K.S.; Laanbroek, H.J.; Cantarella, H.; Kuramae, E.E. Nitrous oxide emissions related to ammonia-oxidizing bacteria and mitigation options from N fertilization in a tropical soil. *Sci. Rep.* **2016**, *6*, 30349. [CrossRef] [PubMed]
9. Mosier, A.R. Exchange of gaseous nitrogen compounds between agricultural systems and the atmosphere. *Plant Soil* **2001**, *228*, 17–27. [CrossRef]
0. Signor, D.; Cerri, C.E.P.; Conant, R. N_2O emissions due to nitrogen fertilizer applications in two regions of sugarcane cultivation in Brazil. *Environ. Res. Lett.* **2013**, *8*, 015013. [CrossRef]
1. Alvar-Beltrán, J.; Dao, A.; Marta, A.D.; Saturnin, C.; Casini, P.; Sanou, J.; Orlandini, S. Effect of Drought, Nitrogen Fertilization, Temperature, and Photoperiodicity on Quinoa Plant Growth and Development in the Sahel. *Agronomy* **2019**, *9*, 607. [CrossRef]
2. Pulvento, C.; Riccardi, M.; Lavini, A.; d'Andria, R.; Iafelice, G.; Marconi, E. Field trial evaluation of two chenopodium quinoa genotypes grown under rain-fed conditions in a typical Mediterranean environment in South Italy. *J. Agron. Crop Sci.* **2010**, *196*, 407–411. [CrossRef]
3. Alvar-Beltrán, J.; Verdi, L.; Marta, A.D.; Dao, A.; Vivoli, R.; Sanou, J.; Orlandini, S. The effect of heat stress on quinoa (cv. Titicaca) under controlled climatic conditions. *J. Agric. Sci.* **2020**, *158*, 255–261. [CrossRef]
4. Vázquez-Rowe, I.; Larrea-Gallegos, G.; Villanueva-Rey, P.; Gilardino, A. Climate change mitigation opportunities based on carbon footprint estimates of dietary patterns in Peru. *PLoS ONE* **2017**, *12*, e0188182. [CrossRef]

agronomy

MDPI

Article

Comparative Analysis of Rice and Weeds and Their Nutrient Partitioning under Various Establishment Methods and Weed Management Practices in Temperate Environment

Aijaz Nazir [1], Mohammad Anwar Bhat [2], Tauseef Ahmad Bhat [2], Suhail Fayaz [2], Mohammad Salim Mir [2], Umar Basu [3], Shafat Ahmad Ahanger [4], Suhail Altaf [4], Bisma Jan [2], Bilal Ahmad Lone [1], Muntazir Mushtaq [5,*], Mohamed El-Sharnouby [6], Milan Skalicky [7], Marian Brestic [7,8] and Ayman El Sabagh [9,*]

1 Agromet Unit, Faculty of Horticulture, Sher-e-Kashmir University of Agricultural Sciences and Technology of Kashmir, Shalimar, Srinagar 190025, India; magreyaijaz357@gmail.com (A.N.); alonebilal127@gmail.com (B.A.L.)
2 Division of Agronomy, Faculty of Agriculture, Sher-e-Kashmir University of Agricultural Sciences and Technology of Kashmir, Wadura, Sopore 193201, India; anwaragri@gmail.com (M.A.B.); tauseekk@gmail.com (T.A.B.); bhatsuhailm@gmail.com (S.F.); mirsalim@skuastkashmir.ac.in (M.S.M.); bismajan6@gmail.com (B.J.)
3 Division of Plant Pathology, Sher-e-Kashmir University of Agricultural Sciences and Technology, Jammu 180009, India; basuumar1608@gmail.com
4 Division of Plant Pathology, Faculty of Agriculture, Sher-e-Kashmir University of Agricultural Sciences and Technology of Kashmir, Wadura, Sopore 193201, India; shafatahanger99@gmail.com (S.A.A.); suhailaltaf474@gmail.com (S.A.)
5 Division of Germplasm Evaluation, ICAR-National Bureau of Plant Genetic Resources, Pusa Campus, New Delhi 110012, India
6 Department of Biotechnology, College of Science, Taif University, P.O. Box 11099, Taif 21944, Saudi Arabia; m.sharnouby@tu.edu.sa
7 Department of Botany and Plant Physiology, Faculty of Agrobiology, Food, and Natural Resources, Czech University of Life Sciences Prague, Kamycka 129, 16500 Prague, Czech Republic; skalicky@af.czu.cz (M.S.); marian.brestic@uniag.sk (M.B.)
8 Laboratory Slovak University of Agriculture in Nitra, Tr. A. Hlinku 2, 94901 Nitra, Slovakia
9 Department of Agronomy, Faculty of Agriculture, Kafrelsheikh University, Kafr El-Shaikh 33516, Egypt
* Correspondence: muntazirbt@gmail.com (M.M.); ayman.elsabagh@agr.kfs.edu.eg (A.E.S.)

Citation: Nazir, A.; Bhat, M.A.; Bhat, T.A.; Fayaz, S.; Mir, M.S.; Basu, U.; Ahanger, S.A.; Altaf, S.; Jan, B.; Lone, B.A.; et al. Comparative Analysis of Rice and Weeds and Their Nutrient Partitioning under Various Establishment Methods and Weed Management Practices in Temperate Environment. *Agronomy* **2022**, *12*, 816. https://doi.org/10.3390/agronomy12040816

Academic Editors: Umberto Anastasi and Aurelio Scavo

Received: 9 February 2022
Accepted: 23 March 2022
Published: 27 March 2022

Abstract: A research trial was conducted at Agronomy Farm (SKUAST-K, Wadura, Jammu & Kashmir), during *kharif* 2017 and 2018 to evaluate nutrient removal in rice under various rice establishment methods and weed control measures. The study comprised of two factors: rice establishment techniques {Transplanting (TPR); Direct seeding (DSR) and System of rice intensification (SRI)} as main plot treatments and weed control measures {Butachlor @ 1500 g a.i ha^{-1} (B); Penoxsulam @ 22.5 g a.i ha^{-1} (P); Pyrazosulfuron ethyl + Pretilachlor @ 15 and 600 g a.i ha^{-1} (PP); Bensulfuron methyl + Pretilachlor @ 60 and 600 g a.i ha^{-1} (BP); 2 Conoweeding/Hand Weeding (CW/HW); Weed free (WF) and weedy check (WC)} as sub-plot treatments meant to evaluate the best establishment method and weed management practice for rice. Over DSR and transplanted rice, the SRI technique yielded a significant increase in dry biomass accumulation (17.04 and 17.20 t ha^{-1}) and grain (7.9 and 8.17 t ha^{-1}) and straw (9.60 and 10.17 t ha^{-1}) yields. Penoxsulam herbicide significantly showed higher grain and straw yield of 8.19 and 8.28 t ha^{-1} and 10.13 and 10.44 t ha^{-1}, respectively, than other weed management measures by comparing the means using critical difference. TPR excelled in reducing dry weed biomass more than other established methods. All herbicides considerably reduced dry weed biomass, but Penoxsulam herbicide showed the greatest reduction in dry weed biomass and proved superior against complex weed flora. Weeds showed maximum contribution towards total Biomass under DSR, among rice establishment techniques. In contrast, among different weed control measures, it was maximum in weedy check treatment (Untreated Control) and minimum in penoxsulam treatment. SRI significantly excelled in crop (grain and straw) nutrient uptake compared to the DSR and TPR method, although different crop establishment techniques non-significantly influenced nutrient concentrations. Furthermore, penoxsulam treatment demonstrated higher crop (grain and straw) nutrient uptake among the various weed management measures. However, available soil nutrients were observed among establishment techniques, highest in DSR and lowest in SRI. Moreover, direct-seeded rice excelled SRI and transplanted rice in weed nutrient uptake, and among the different herbicidal treatments, penoxsulam recorded the lowest uptake in weeds. Nutrient budgeting demonstrated that DSR showed the maximum percentage of nutrient removal by weeds, and the minimum ratio was in TPR. In contrast, the lowest rate of nutrients removed via weeds were seen in penoxsulam application under various weed management measures.

Keywords: biomass; herbicide; Himalayas; nutrient uptake; penoxsulam; rice; SRI; weed; yield

1. Introduction

Rice production is the agricultural sector's backbone in sub-tropical and tropical countries, including India, the world's 2nd-largest producer and consumer [1]. By the end of 2025, demand for rice, the world's most important staple food, is predicted to reach 800 metric tons [2,3]. As the world's population expands, developing countries must focus on producing sufficient rice on limited arable land through crop genetic improvement, managerial optimization, and socioeconomic factors. Doubling of production on a sustainable basis is required to feed more than 9 billion people by 2050. Although more than 75 percent of rice production comes from 79 m ha of irrigated lowland area, it is assumed that 17 m ha of Asia's irrigated rice crop will face water scarcity. In contrast, 22 m ha will face economic water scarcity [4,5], raising concerns about the sustainability of rice production under flooded conditions. Furthermore, transplanted rice has higher greenhouse gas emissions (CH_4 and N_2O emissions), contributing to global warming [6]. As a result, alternative rice cultivation techniques should be executed to reduce rice-related hazardous gas emissions. The direct-seeded rice (DSR) is gaining importance in recent years to minimize water and labour scarcity issues while retaining a sustainable output. As a result, adopting DSR has several benefits, including minimal use of irrigation water, time consumption, reduced labour scarcity, mitigation of climate change, increased output of succeeding crops, and so on [6]. Many variables and weeds that limit DSR productivity are the most severe biological constraints causing economic losses. Due to maximum plant density and biomass production during the vegetative phase, DSR has a higher nutritional requirement than a transplanted crop; hence, senescence occurs earlier as a nutrient deficiency is developed at the reproductive phase [7]. To make more efficient water use, rice-growing practises transitioning from traditional rice to other alternative rice techniques. The system of rice intensification (SRI) invloves laternate wetting and drying rice fields [5,8]. More than half of the nitrogen used in flooded rice production is lost to the environment through volatilization, leaching, surface runoff, and denitrification, polluting freshwater and marine habitats. Changes in soil physical, chemical, and biological features caused by puddling and submergence, on the other hand, have some positive effects on soil quality, which influences the availability of some macro- and micronutrients, nutrient content, and crop uptake pattern. The transition from traditional rice production to aerobic rice has increased micronutrient deficiency, particularly in Fe, posing a new challenge to iron availability. Micronutrient deficiencies, such as Fe and Zn, are a severe hazard to the world's population's health. The SRI is a water-conserving production system that increases fame and interest as it improves rice productivity and improves nutrient uptake and nutrient usage efficiency [9,10]. For water-saving rice techniques, nutrient management tools are critical because changes in soil redox potential significantly impact the availability of soil nutrients, transport in soil, and removal by crop plants. Flooded condition in rice results in an anoxic condition as the soil is submerged, while aerobic rice soil develops a completely different soil environment than traditional flooded rice soil. Because of the early weed emergence under favourable soil conditions, crops and weeds compete for different factors viz. nutrients, space, water, and light [11]. Weeds cause various losses, fluctuating from 50% to complete failure of crops [7], and weeds can absorb up to nine times more nutrients in the unweeded plot. Fertilizer use and consumption in rice have risen dramatically in recent decades [12]. It has been reported that weeds absorb more than 60% of applied fertilizers, resulting in poorer nutrient availability for crops [13]. Weed nutrient removal is dependent on the period of their growth, but due to labour scarcity and increased wages, controlling weeds in transplanted rice at critical stages by manual weeding alone is very difficult. Herbicides with a single mechanism of action will not be effective against a wide range of weeds. So, to control these broad-spectrum weeds, herbicide formulations with various modes of action combined with hand weeding will result in effective weed control, lesser nutrient loss via weeds, accompanied by more crop nutrient uptake. Many herbicides, successfully used to control weeds in rice crops, with diverse compositions are recommended to avoid residue buildup, weed flora shifts, and the increase of herbicide-resistant weeds [14]. The current herbicide use trend is to identify an efficient weed control measure by using low dose high-efficiency herbicides, which will reduce overall herbicide use and make application easier and more cost-effective. Given the above, this research experiment was conducted to find suitable crop establishment techniques and weed control strategies for rice to minimize nutrient losses, reducing the cost of cultivation and making rice cultivation as profitable as possible.

2. Materials and Methods

2.1. Experimental Site Description

A research trial was studied at Agronomy Farm, SKUAST-K, Wadora-Sopore, J&K, in the 2 consecutive *kharif* seasons (2017 and 2018). Wadora is situated at 34°34′ North latitude, 74°40′ East longitude, and at 1587 m altitude from mean sea level. In the mid-latitude temperate zone, the research farm has 812 mm mean annual precipitation. During the experimental years, the total rainfall of 339.4 and 352.5 mm was received in 2017 and 2018, respectively, with minimum temperatures fluctuating from 5.25 to 19.30 °C and 5.41 to 18.6 °C. Maximum temperatures fluctuated from 24.5 to 32.7 °C and 22.0 to 31.8 °C, respectively (Figures 1 and 2), and mean relative humidity (R.H) ranged from 67.3 to 90.8% and 60.1 to 92.1% (maximum R.H) and 32.1 to 64.4 percent and 33.3 to 70.4 percent (minimum R.H), respectively. This location has a cold temperate climate, with minus winter temperatures and moderate temperatures in the summer. The rice crop in this area has a growth period of 140–150 days. Samples from the top 20 cm of the soil profile were taken to assess initial nutrients. The texture of soil was silty-clay-loam, with a neutral pH and medium range of organic carbon (O.C), medium in available nitrogen (N), phosphorus (P) and potassium (K) (Table 1).

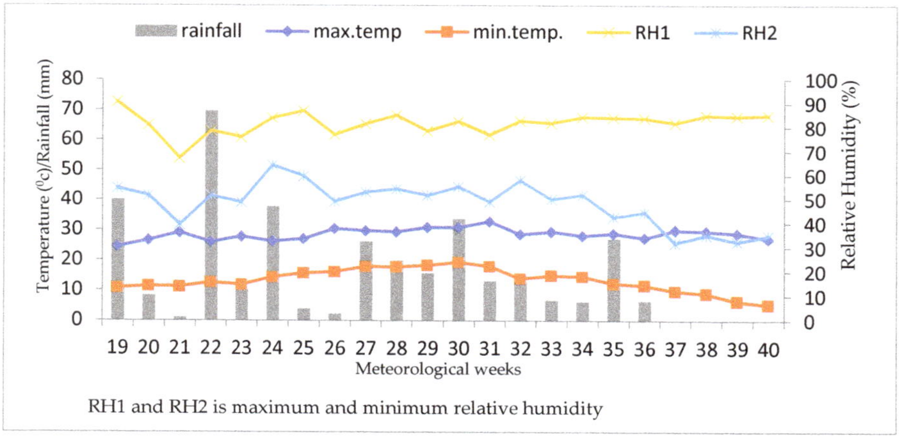

Figure 1. Average weekly meteorological data (2017).

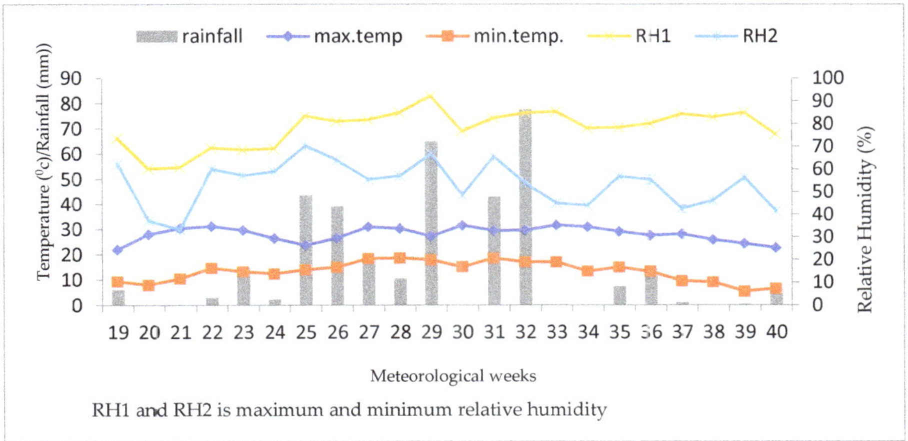

Figure 2. Average weekly meteorological data (2018).

Table 1. Initial soil status of experimental field.

Characteristics	Status	Range	Method Used
A. Physical			
Texture		Silty-clay-loam	International Pipette Method [15]
B. Chemical			
pH	6.9	Neutral	Blackman's glass electrode pH meter [16]
O.C (%)	0.99%	Medium	Walkely and Black rapid titration method [16]
Available N	380 (kg ha^{-1})	Medium	Alkaline Potassium permanganate method [17]
Available P	19.7 (kg ha^{-1})	Medium	Extraction with 0.5 M NaHCO3 [18]
Available K	280 (kg ha^{-1})	Medium	Flame photometer method [16]

2.2. Design of Experiment

A split-plot design with three replications was used to carry out the experiment with three rice establishment techniques in main plots and seven weed control measures as a sub-factor. The treatments of the main plot consisted of (i) TPR (Transplanting), (ii) DSR (Direct Seeding), (iii) SRI (System of Rice Intensification) and sub-plot treatments comprised of (i) Butachlor @ 1500 g a.i. ha^{-1} (B), (ii) Penoxsulam @ 22.5 g a.i. ha^{-1} (P), (iii) Pyrazosulfuron ethyl + pretilachlor @ 15 and 600 g a.i. ha^{-1} (PP), (iv) Bensulfuron methyl + pretilachlor @ 60 and 600 g a.i. ha^{-1} (EP), (v) 2 Cono-weeding/Handweeding (CW/HW), (vi) Weed free (WF) and (vii) Weedy check (WC). No manual weeding was followed in the weedy check treatment, but in the weed-free treatment, multiple manual weedings were received.

2.3. Crop Management Practices

Pre-germinated rice seeds were sown with a row spacing of 20 cm in DSR on 17 May and were also used for growing nursery in case of transplanting (TPR) method. 25 days old seedlings were used for transplanting, with 20 cm × 10 cm spacing. In the last week of May, 12-day-old seedlings (SR-4) were transplanted for the SRI method with 25 cm × 25 cm spacing. In experiment plots, well-decomposed 10 t ha^{-1} of FYM was mixed during field preparation. N, P_2O_5, and K_2O were applied in the proportion of 120:60:30 kg ha^{-1}. At transplanting, full quantity of P_2O_5 and K_2O, as well as 1/2 N, were involved; however, at active tillering and panicle initiation phase the rest 1/2 N was given in two splits. Th herbicides butachlor, pyrazosulfuron-ethyl + pretilachlor, and bensulfuron methyl + pret lachlor were applied as pre-emergent at three DAS/DAT penoxsulam as post-emergent a 10 DAS/DAT in each establishment method as per treatment. In SRI and DSR, no continu ous standing water was maintained throughout the vegetative stage, but a thin water laye was continued from flowering to soft dough stage; however, up to the dough stage, 2–3 cr of water was maintained under TPR.

2.4. Collection and Processing

2.4.1. Rice (Grain + Straw)

Plants in a 0.5 m × 0.5 m quadrant from each plot of treatment were collected a 15-day intervals and sun-dried (3–4 days), oven-dried (60–65 °C) to attain a constant weigh The crop's dry biomass noted in grams was transformed into t ha^{-1}. The crop was harveste in the second week of October grain, and straw yields were measured in kg ha^{-1} and wer later converted into t ha^{-1}. For analysis purposes, rice grain and straw samples wer collected from different treatments separately, oven-dried (60–65 °C) until the attainmen of constant weight, and grinded with Yarco grinder in the laboratory.

2.4.2. Weeds

Weeds uprooted from each plot at harvest were sundried, followed by oven dryin (60–65 °C) until they reached a constant weight. For analysis, the Yarco grinder was used t grind the oven-dried plant samples. The weight of weed biomass was measured in t ha−

2.5. Nitrogen (N), Phosphorus (P) and Potassium (K) Estimation

Chemical analysis was performed on the ground samples placed in labelled bag A 0.5 g sample was digested with concentrated H_2SO_4 @ 10 mL plus digestion-mixtur (H_2SO_4 + $HClO_4$ + HNO_3) to assess the nitrogen content. The micro Kjeldahls metho was used to determine total nitrogen (N). Phosphorus content was evaluated with a spec trophotometer by the Vanads-Molybdo-phosphoric yellow method by digestion in a tr acid-mixture (HNO_3:$HClO_4$:H_2SO_4 = 10:4:1). A flame photometer was used to quantify th plant sample's potassium concentration (percent) [16]. The uptake of N, P and K by weed straw, and crop grain, estimated by multiplying dry matter production with their respectiv content values, was expressed as kg ha^{-1}. After the crop was harvested, soil samples wer obtained from each plot up to a depth of 15 cm and were shade dried and labelled. Afte drying, the soil samples were crushed, then sieved through a 2 mm screen, and for lal analysis, the composite sample was collected. For each soil sample, the available so nitrogen was estimated in kg ha^{-1} using the alkaline potassium permanganate method [17 The available P of soil samples noted in kg ha^{-1} was assessed using 0.5 N NaHCO3 at a pH of 8.5 [18]. The K content of samples was measured in kg ha^{-1} using an extraction metho with 1N ammonium acetate at a pH of 7.0 [16].

2.6. Statistical Analysis

The recorded data were analyzed statistically using analysis of variance subjected t split-plot design with the help of R software. At a significance level of 0.05, the treatmen averages were compared using the critical difference (C.D) test. Regression analysis of yiel with crop and weed dry matter and nutrient uptake by crop and weeds was compute and regression equations were fitted to estimate the response of yield explained by dr matter and nutrient uptake.

3. Results

3.1. Dry Matter Accumulation

Rice establishment techniques and weed control measures illustrated significant influ ence on dry Biomass production in rice. Dry matter accumulation improved exponentiall up to 100 days after sowing (DAS), then decreased at a declining rate until maturit

(Table 2). SRI performed better than TPR and DSR, with dry matter accumulation of 17.04 and 17.20 t ha^{-1} at harvest, respectively. In contrast, direct-seeded rice had the lowest dry matter accumulation of 12.31 and 12.65 t ha^{-1} at 2017 and 2018 (Table 1). SRI showed a 13.48 and 27.75 percent increase in dry matter accumulation than transplanted and direct-seeded rice at harvest in 2017, and the comparable figures for 2018 were 12.83 and 26.43 percent, respectively. However, as compared to other herbicides, the plots under Penoxsulam herbicide significantly recorded the higher dry matter accumulation of 15.73 and 116.30 t ha^{-1}, with a percentage increase of 18.59 and 17.94 percent when compared to weedy check during 2017 and 2018, respectively.

Table 2. Dry biomass (qha^{-1}) of rice crop as influenced by rice establishment techniques and weed control measures.

Treatments	40 (DAS)		55 (DAS)		70 (DAS)		85 (DAS)		100 (DAS)		115 (DAS)		130 (DAS)		Maturity	
	2017	2018	2017	2018	2017	2018	2017	2018	2017	2018	2017	2018	2017	2018	2017	2018
Rice establishment techniques																
TPR	1.09	1.42	7.7	8.2	53.4	54.4	110.3	112.8	124.7	127.3	140.4	144.0	145.1	148.5	147.4	149.9
DSR	1.00	1.26	6.3	6.7	45.8	45.5	98.5	100.7	108.8	110.5	117.8	118.4	120.6	122.1	123.1	126.5
SRI	1.17	1.57	8.5	9.2	55.8	57.9	118.1	120.8	136.3	139.3	151.3	154.8	166.1	169.3	170.4	172.0
SE (m) ±	0.003	0.01	0.06	0.04	0.19	0.19	0.34	0.31	1.12	0.95	0 80	0.74	1.39	1.41	1.48	1.49
C.D. ($p \leq 0.05$)	0.01	0.04	0.26	0.19	0.79	0.79	1.37	1.25	4.53	3.85	3 22	3.01	5.60	5.69	5.98	6.01
Weed control measures																
B	1.01	1.34	6.9	7.5	46.5	47.4	103.9	105.4	117.4	121.0	124.2	129.1	131.1	136.8	134.1	139.8
P	1.17	1.5	8.1	8.6	56.5	57.4	116.5	118.0	134.6	138.2	144.7	150.6	154.3	160.0	157.3	163.0
PP	1.14	1 47	7.8	8.3	54.1	55.0	112.5	115.0	129.8	133.4	138.2	144.1	148.6	154.3	151.6	157.3
BP	1.09	1 42	7.5	8.0	51.7	52.6	109.4	111.9	127.4	131.1	135.2	140.1	141.7	147.4	144.7	150.4
2 CW/HW	1.04	1.37	7.2	7.8	49.4	50.3	106.1	108.5	122.2	125.8	129.7	134.6	137.2	142.9	140.2	145.9
WF	1.22	1.55	8.4	8.9	60.0	60.9	118.9	121.4	139.0	143.6	152.0	157.9	160.3	166.0	163.3	169.0
WC	0.94	1.28	6.7	7.2	43.8	44.7	96.6	102.1	110.6	112.3	117.2	123.1	125.1	130.8	128.1	133.8
SE (m) ±	0.01	0.01	0.01	0.02	0.10	0.12	0.13	0.16	1.55	1.53	1.92	1.85	1.36	1.53	1.30	1.47
C.D. ($p \leq 0.05$)	0.05	0.05	0.05	0.06	0.31	0.36	0.40	0.47	4.67	4.43	5.76	5.35	4.10	4.43	3.90	4.25

AS = Days after sowing, TPR = Transplanting, DSR = Direct seeding, SRI = System of rice intensification, B = Butachlor (1500 g a.i. ha^{-1}), P = Penoxsulam (22.5 g a.i. ha^{-1}), PP = Pyrazosulfuron ethyl + pretilachlor (15 and 600 g a.i. ha^{-1}), BP = Bensulfuron methyl + pretilachlor (60 and 600 g a.i. ha^{-1}), 2CW/HW = 2 Cono-weeding/Hand-weeding, WF = Weed free and WC = Weedy check.

3.2. Yield of Rice (Grain + Straw)

Rice yield (grain + straw) was affected significantly under different rice establishment approaches and weed control measures (Table 3). SRI produced a maximum grain yield of 7.92 and 8.17 t ha^{-1} and straw yield of 9.60 and 10.17 t ha^{-1}, followed by TPR, whereas DSR produced significantly lower grain yield (6.01 and 6.24 t ha^{-1}) and straw yield (8.00 and 8.32 t ha^{-1}). SRI method excelled transplanting and DSR method in grain yield by 24.11 and 10.47 percent and 23.62 and 12.23 percent, respectively. During 2017 and 2018, Penoxsulam herbicide produced significantly more grain yield (8.19 and 8.28 t ha^{-1}) and straw yield (10.13 and 10.44 t ha^{-1}) than WC and other weed control techniques.

3.3. Weed Dry Matter

Significant differences were noticed during the study period in weed biomass and its contribution to total Biomass. Rice establishment techniques and weed control approaches influenced significantly dry weed biomass (Table 3). Compared to SRI and DSR, dry weed biomass was found less under transplanted rice at harvest, with reductions of 9.22 and 43.10 percent in 2017 and 10.04 and 45.06 percent in 2018. On contrary, DSR had recorded maximum dry weed biomass. Among the different herbicidal treatments, Penoxsulam herbicide significantly lowered dry weed biomass (0.25 and 0.23 t ha^{-1}), reducing 67.08 and 69.38 percent in dry weed biomass at harvest compared to weedy check, respectively.

Table 3. Yield (t ha^{-1}) of rice, weed biomass (t ha^{-1}) and biomass share of weeds as influenced by rice establishment techniques and weed control measures.

Treatments	Grain—Yield (t ha^{-1})		Straw—Yield (t ha^{-1})		Weed—Biomass (t ha^{-1})		Total—Biomass (Rice + Weeds) (t ha^{-1})		Weeds Share of Total Biomass (%)	
	2017	2018	2017	2018	2017	2018	2017	2018	2017	2018
Rice establishment techniques										
TPR	7.09	7.17	8.88	9.21	0.22	0.20	16.19	16.58	1.36	1.21
DSR	6.01	6.24	8.00	8.32	0.69	0.68	14.70	15.24	4.69	4.46
SRI	7.92	8.17	9.60	10.17	0.27	0.25	17.79	18.59	1.52	1.34
SE (m) ±	0.25	0.28	0.26	0.29	0.01	0.01	0.52	0.49	-	-
C. D. ($p \leq 0.05$)	0.75	0.84	0.79	0.87	0.03	0.04	1.51	1.48	-	-
Weed control measures										
B	6.04	6.43	8.23	8.67	0.47	0.45	14.74	15.55	3.19	2.89
P	8.19	8.28	10.13	10.44	0.25	0.23	18.57	18.95	1.35	1.21
PP	7.59	7.77	9.31	9.51	0.27	0.25	17.17	17.53	1.57	1.43
BP	7.27	7.45	9.03	9.37	0.29	0.27	16.59	17.09	1.75	1.58
2 CW/HW	7.06	7.14	8.72	8.96	0.34	0.32	16.12	16.42	2.11	1.95
WF	9.00	9.29	10.75	11.19	0.00	0.00	19.75	20.48	0.00	0.00
WC	4.10	4.39	6.06	6.50	1.15	1.12	11.31	12.01	10.17	9.33
SE (m) ±	0.24	0.28	0.23	0.21	0.01	0.01	0.55	0.52	-	-
C. D. ($p \leq 0.05$)	0.72	0.84	0.69	0.61	0.04	0.04	1.66	1.61	-	-

3.4. *Weeds Share of Total Biomass*

The findings of our experiment revealed that weeds contributed significantly to total biomass production under various establishment methods and weed control measures. As a result, the maximum share (%) of weeds (4.69 and 4.46%) towards total biomass production was observed under DSR treatment among different methods of crop establishment during 2017 and 2018, respectively (Table 3). However, under various weed management approaches the maximum contribution of weeds to total biomass production was 10.17 and 9.33%, noted under weedy check treatment.

3.5. *Uptake of Nitrogen (N), Phosphorus (P) and Potassium (K) in Rice*

Crop establishment techniques and weed control measures showed non-significant influence on NPK concentration in rice. In contrast, establishment methods and weed management practices affected the uptake of nitrogen, phosphorus, and potassium. During both years, SRI significantly improved nitrogen, phosphorus, and potassium uptake by rice over transplanting and DSR. SRI achieved the maximum N-uptake of 83.95 and 89.05 kg ha^{-1} in grain, followed by TPR and DSR, among the methods of rice establishment (Table 4). Uptake of phosphorus and potassium in grain followed the same trend. The higher phosphorus uptake (Table 5) and potassium (Table 6) noted 19.01 and 20.42 kg ha^{-1} and 23.76 and 24.51 kg ha^{-1}, respectively in grain under SRI. In straw, the highest nitrogen, phosphorus and potassium uptake were 47.04 and 50.85 kg ha^{-1}, 6.72 and 7.12 kg ha^{-1} and 150.77 and 164.80 kg ha^{-1}, respectively, under SRI, being significantly superior to transplanted rice and DSR. During 2017 and 2018, SRI's advantages in increasing total nitrogen, phosphorus and potassium uptake than transplanted rice were 18.37 and 20.72 percent N, 15.89 and 16.70 percent P, and 9.88 and 11.78 percent K, respectively, while its superiority over direct seeded rice was 28.6 and 29.07 percent N, 29.96 and 32.06 percent P, 21.70 and 24.06 percent K, respectively.

During 2017 and 2018, among the various weed control measures, penoxsulam noted the maximum N uptake by grain (83.54 and 86.11 kg ha^{-1}), and straw (48.62 and 51.16 kg ha^{-1}) (Table 4). The highest phosphorus uptake achieved under SRI in rice was 18.84 and 19.86 kg ha^{-1} (grain) and 6.08 and 7.30 kg ha^{-1} (straw), respectively (Table 5). Similarly, a significantly higher uptake of K i.e., 24.57 and 24.83 kg ha^{-1} by grain and 157.95

and 166.96 kg ha^{-1} by straw, was observed under SRI (Table 6). During 2017, penoxsulam @ 22.5 g a.i. excelled WC and B treatments in increasing total nitrogen, phosphorus and potassium uptake by 50.66 and 27.61 percent N, 49.23 and 21.99 percent P, and 44.38 and 22.73 percent K. At the same time, the corresponding values for 2018 were 47.99 and 24.57 percent N, 50.11 and 24.07 percent P, and 41.88 and 20.63 percent K, respectively.

Table 4. Influence of rice establishment techniques and weed control measures on nitrogen content (%) and uptake (kg ha^{-1}) of rice.

Treatment	Grain N Content (%)		Straw N Content (%)		Grain N Uptake (kg ha^{-1})		Straw N Uptake (kg ha^{-1})		Total N Uptake (kg ha^{-1})	
	2017	2018	2017	2018	2017	2018	2017	2018	2017	2018
Rice establishment techniques										
TPR	0.93	0.95	0.47	0.47	65.18	67.61	41.74	43.29	106.92	110.9
DSR	0.98	0.99	0.44	0.45	58.90	61.78	35.20	37.44	94.1	99.22
SRI	1.06	1.09	0.49	0.5	83.95	89.05	47.04	50.85	130.99	139.9
SE (m) ±	0.001	0.007	0.002	0.001	0.90	1.30	1.87	1.81	3.73	3.63
C. D. ($p \leq 0.05$)	N.S	N.S	N.S	N.S	3.54	5.08	5.42	5.45	11.20	10.90
Weed control measures										
B	0.97	0.99	0.47	0.45	58.59	63.66	37.08	39.88	95.67	103.54
P	1.02	1.04	0.50	0.48	83.54	86.11	48.62	51.16	132.16	137.27
PP	1.0	1.02	0.50	0.48	75.90	79.25	44.69	45.65	120.59	124.9
BP	0.99	1.01	0.49	0.47	71.97	75.25	42.31	44.04	114.28	119.29
2 CW/HW	0.97	0.99	0.48	0.46	68.48	70.69	40.11	42.11	108.59	112.8
WF	1.03	1.05	0.51	0.49	92.70	97.55	49.37	55.95	142.07	153.5
WC	0.94	0.96	0.46	0.44	38.54	42.14	26.66	29.25	65.2	71.39
SE (m) ±	0.007	0.007	0.002	0.002	0.93	0.95	1.34	1.67	3.53	4.03
C. D. ($p \leq 0.05$)	* N.S	N.S	N.S	N.S	2.66	2.74	4.03	5.02	10.60	12.10

* N.S means non-significant.

Table 5. Influence of rice establishment techniques and weed control measures on phosphorus content (%) and uptake (kg ha^{-1}) of rice.

Treatment	Grain P Content (%)		Straw P Content (%)		Grain P Uptake (kg ha^{-1})		Straw P Uptake (kg ha^{-1})		Total P Uptake (kg ha^{-1})	
	2017	2018	2017	2018	2017	2018	2017	2018	2017	2018
Rice establishment techniques										
TPR	0.23	0.23	0.06	0.07	16.31	16.49	5.33	6.45	21.64	22.94
DSR	0.22	0.22	0.06	0.06	13.22	13.72	4.80	4.99	18.02	18.71
SRI	0.24	0.25	0.07	0.07	19.01	20.42	6.72	7.12	25.73	27.54
SE(m) ±	0.003	0.003	0.0001	0.0003	0.25	0.27	0.22	0.25	0.93	1.06
C.D. ($p \leq 0.05$)	N.S	N.S	N.S	N.S	0.99	1.13	0.87	0.77	2.80	3.20
Weed control measures										
B	0.24	0.24	0.46	0.06	14.50	15.42	4.94	5.20	19.44	20.62
P	0.23	0.24	0.49	0.07	18.84	19.86	6.08	7.30	24.92	27.16
PP	0.23	0.23	0.48	0.07	17.45	17.88	5.59	6.66	23.04	24.54
BP	0.23	0.23	0.47	0.07	16.71	17.14	5.42	6.56	22.13	23.7
2 CW/HW	0.23	0.23	0.47	0.06	16.23	16.43	5.23	5.38	21.46	21.81
WF	0.23	0.24	0.5	0.07	20.70	22.28	7.52	7.83	28.22	30.11
WC	0.22	0.22	0.45	0.06	9.02	9.65	3.63	3.90	12.65	13.55
SE (m) ±	0.004	0.004	0.0003	0.0003	0.32	0.33	0.15	0.16	1.16	1.20
C. D. ($p \leq 0.05$)	* N.S.	N.S.	N.S	N.S	0.93	0.96	0.45	0.49	3.50	3.62

* N.S means non-significant.

Table 6. Influence of rice establishment techniques and weed control measures on potassium conte (%) and uptake (kg ha^{-1}) of rice.

Treatments	Grain K Content (%)		Straw K Content (%)		Grain K Uptake (kg ha^{-1})		Straw K Uptake (kg ha^{-1})		Total K Uptake (kg ha^{-1})	
	2017	2018	2017	2018	2017	2018	2017	2018	2017	2018
Rice establishment techniques										
TPR	0.29	0.30	1.54	1.58	20.56	21.51	136.72	145.49	157.28	167.00
DSR	0.29	0.29	1.49	1.51	17.42	18.09	119.23	125.66	136.65	143.75
SRI	0.30	0.30	1.57	1.62	23.76	24.51	150.77	164.80	174.53	189.31
SE(m) ±	0.002	0.002	0.004	0.005	0.26	0.16	4.43	3.63	4.77	4.60
C.D. ($p \leq 0.05$)	N.S.	N.S.	N.S	N.S	1.05	1.10	12.13	11.01	14.33	13.90
Weed control measures										
B	0.29	0.29	1.50	1.54	17.52	18.64	123.51	133.58	141.03	152.22
P	0.30	0.30	1.56	1.60	24.57	24.83	157.95	166.96	182.52	191.79
PP	0.31	0.31	1.55	1.59	23.51	24.09	144.37	151.27	167.88	175.36
BP	0.29	0.30	1.54	1.58	21.07	22.36	139.09	148.08	160.16	170.44
2 CW/HW	0.29	0.29	1.52	1.56	20.47	20.72	132.56	139.79	153.03	160.51
WF	0.30	0.30	1.58	1.61	26.99	27.86	169.77	180.08	196.76	207.94
WC	0.29	0.29	1.48	1.52	11.88	12.72	89.63	98.74	101.51	111.46
SE(m) ±	0.004	0.004	0.005	0.005	0.35	0.24	3.45	3.26	4.26	4.88
C.D. ($p \leq 0.05$)	N.S	* N.S.	N.S	N.S	1.03	0.72	10.21	9.73	12.80	14.66

* N.S means non-significant.

3.6. Nitrogen (N), Phosphorus (P) and Potassium (K) Concentration and Uptake in Weeds

During both years of the experiment, the results demonstrated that weeds in rice cult vated using the direct seeding approach had greater nutritional N, P and K concentration As compared to other herbicidal treatments, Weedy check increased the concentration o nitrogen, phosphorus, and potassium in weeds in both years (Table 7), yet all herbicida treatments remained at par in terms of N, P and K concentrations. Nitrogen, phospho rus, and potassium uptake of weeds under different crop establishment approaches wer affected significantly.

Table 7. Influence of rice establishment techniques and weed control measures on nutrient co centration (%) of weeds.

Treatments	N		P		K	
	2017	2018	2017	2018	2017	2018
Rice establishment techniques						
TPR	1.5 (1.32)	1.50 (1.31)	1.37 (0.91)	1.36 (0.89)	1.52 (1.38)	1.52 (1.38)
DSR	1.51 (1.33)	1.51 (1.32)	1.37 (0.91)	1.37 (0.90)	1.53 (1.40)	1.53 (1.40)
SRI	1.51 (1.32)	1.50 (1.32)	1.37 (0.91)	1.37 (0.90)	1.53 (1.40)	1.53 (1.40)
SE (m) ±	0.0003	0.0003	0.0005	0.0005	0.002	0.002
C. D. ($p \leq 0.05$)	0.001	0.001	0.0018	0.0018	0.006	0.006
Weed control measures						
B	1.59 (1.55)	1.59 (1.54)	1.44 (1.07)	1.43 (1.07)	1.62 (1.64)	1.62 (1.64)
P	1.59 (1.54)	1.59 (1.53)	1.43 (1.06)	1.43 (1.04)	1.62 (1.63)	1.61 (1.60)
PP	1.59 (1.54)	1.59 (1.53)	1.43 (1.06)	1.43 (1.04)	1.61 (1.59)	1.62(1.63)
BP	1.59 (1.54)	1.59 (1.53)	1.43 (1.07)	1.43 (1.05)	1.62 (1.63)	1.62 (1.64)
2 CW/HW	1.59 (1.54)	1.59 (1.53)	1.43 (1.06)	1.42 (1.04)	1.62 (1.62)	1.62 (1.62)
WF	1.00 (0.00)	1.00 (0.00)	1.00 (0.00)	1.00 (0.00)	1.00 (0.00)	1.00 (0.00)
WC	1.60 (1.57)	1.60 (1.56)	1.44 (1.07)	1.43 (1.05)	1.62 (1.64)	1.62 (1.64)
SE (m) ±	0.0006	0.0006	0.0005	0.0005	0.004	0.004
C. D. ($p \leq 0.05$)	0.0017	0.0017	0.0015	0.0015	0.012	0.012

Note: Data in parenthesis are square root transformed values.

During both years, DSR excelled SRI and TPR in weed nutrient uptake. The highest uptake of N by weeds recorded under DSR (Table 8) was 10.85 and 10.50 kg ha^{-1}. When related to direct seeding, the transplanting method resulted in 38.12 percent less depletion of N uptake by weeds in 2017 and 39.68 percent in 2018. P uptake by weeds demonstrated a similar pattern as it was noticed under DSR, the uptake of P was higher among the different treatments of rice establishment techniques. Likewise, significantly the uptake of K remained higher under DSR as compared to SRI and TPR.

Table 8. Influence of rice establishment techniques and weed control measures on nutrient uptake of weeds (kg ha^{-1}).

Treatments	N		P		K	
	2017	**2018**	**2017**	**2018**	**2017**	**2018**
Rice establishment techniques						
TPR	1.98 (3.39)	1.90 (3.09)	1.74 (2.32)	1.67 (2.09)	2.01 (3.55)	1.94 (3.26)
DSR	3.20 (10.85)	3.15 (10.50)	2.72 (7.47)	2.67 (7.14)	3.27 (11.42)	3.23 (11.14)
SRI	2.14 (4.17)	2.07 (3.87)	1.87 (2.87)	1.81 (2.63)	2.18 (4.39)	2.11 (4.10)
SE (m) ±	0.01	0.01	0.02	0.01	0.01	0.01
C. D. ($p \leq 0.05$)	0.04	0.03	0.06	0.03	0.04	0.03
Weed control measures						
B	2.80 (7.28)	2.73 (6.91)	2.39 (5.02)	2.33 (4.71)	2.87 (7.69)	2.81 (7.35)
P	2.12 (3.80)	2.03 (3.46)	1.85 (2.61)	1.77 (2.35)	2.15 (3.97)	2.07 (3.64)
PP	2.20 (4.15)	2.11 (3.80)	1.91 (2.85)	1.83 (2.57)	2.24 (4.38)	2.16 (4.04)
BP	2.26 (4.44)	2.18 (4.08)	1.97 (3.08)	1.90 (2.80)	2.32 (4.71)	2.24 (4.37)
2 CW/HW	2.44 (5.26)	2.37 (4.91)	2.11 (3.62)	2.04 (3.34)	2.50 (5.56)	2.43 (5.22)
WF	1.00 (0.00)	1.00 (0.00)	1.00 (0.00)	1.00 (0.00)	1.00 (0.00)	1.00 (0.00)
WC	4.25 (18.04)	4.19 (17.60)	3.56 (12.35)	3.50 (11.90)	4.34 (18.87)	4.30 (18.54)
SE (m) ±	0.02	0.02	0.03	0.02	0.03	0.02
C. D. ($p \leq 0.05$)	0.07	0.06	0.10	0.06	0.08	0.07

Note: Data in parenthesis are square root transformed values.

During the study, weedy check significantly enhanced weed nitrogen, phosphorus, and potassium uptake from the rest of weed control plots, followed by butachlor, illustrating superiority significantly among herbicidal treatments (Table 8). Uptake of nitrogen i.e., 3.80 and 3.46 kg ha^{-1} by weeds, was lower in penoxsulam treated plots. In 2017, 24.28 and 50.11 percent of weeds N uptake was less depleted under penoxsulam treatment, and the corresponding values for 2018 were 25.64 and 51.55 percent. The uptake of P by weeds showed a similar pattern as it was found that P uptake was lowest under penoxsulam treatment that recorded 22.59 and 48.03 percent less depletion in P uptake by weeds in 2017. The corresponding figures for 2018 were 24.03 and 49.42 percent. Similarly, the lowest K uptake by weeds was realized under penoxsulam that, demonstrated 22 08 and 50.46 percent less depletion in potassium uptake via weeds in 2017. The corresponding values for 2018 were 26.33 and 51.86% than butachlor and weedy check, respectively.

3.7. Nutrients Budgeting (Rice + Weeds) for Removal of Nitrogen (N), Phosphorus (P) and Potassium (K)

Total nitrogen, phosphorus, and potassium uptake (Table 9) in rice was significantly affected under different crop establishment techniques, and the maximum uptake was found under SRI (331.25 and 356.75 kg ha^{-1}) followed by TPR (285.84 and 300.84 kg ha^{-1}) and DSR (248.77 and 261.68 kg ha^{-1}). Weed control practices also significantly influenced the total nitrogen, phosphorus, and potassium uptake by crop. The maximum values of 339.6 and 356.22 kg ha^{-1} under weed control measures were recorded under penoxsulam treated plots, apart from weed-free plots.

Table 9. Influence of rice establishment techniques and weed control measures on total nutrient uptake of the crop (kg-ha^{-1}).

Treatment	Total N + P + K Crop Uptake (kg ha^{-1})		Total N + P + K Weed Uptake (kg ha^{-1})		N + P + K Uptake Weeds Share (%)	
	2017	2018	2017	2018	2017	2018
Rice establishment techniques						
TPR	285.84	300.84	9.26	2.09	3.14	0.69
DSR	248.77	261.68	29.74	7.14	10.68	2.66
SRI	331.25	356.75	11.43	2.63	3.34	0.73
Weed control measures						
B	256.14	276.38	19.99	4.71	7.24	1.68
P	339.6	356.22	10.38	2.35	2.97	0.66
PP	311.51	324.8	11.38	2.57	3.52	0.79
BP	296.57	313.43	12.23	2.8	3.96	0.89
2 CW/HW	283.08	295.12	14.44	3.34	4.85	1.12
WF	367.05	391.55	0.00	0.00	0.00	0.00
WC	179.36	196.4	49.26	11.9	21.55	5.71

Weeds also compete for nutrients and recorded significant uptake of total nitrogen, phosphorus and potassium under various establishment techniques and weed control practices. Maximum uptake by weeds under different crop establishment treatments was observed under DSR (29.74 and 7.14 kg ha^{-1}), then was followed by SRI (11.43 and 2.63 kg ha^{-1}) and TPR (9.26 and 2.09 kg ha^{-1}). Under different weed control measures, the maximum NPK total uptake was noticed in WC treatment by weeds (Table 9). Percent of nutrients removed via weeds indicated that significantly more significant amounts of nitrogen, phosphorus, and potassium were removed during the study period. The maximum contribution of 10.68 and 2.66% by weeds was noticed under DSR among different establishment techniques (Figure 3; Table 9). However, under various weed management measures, a weedy check showed a maximum contribution of 21.55 and 5.71% towards removing nutrients by weeds during our research (Figure 4).

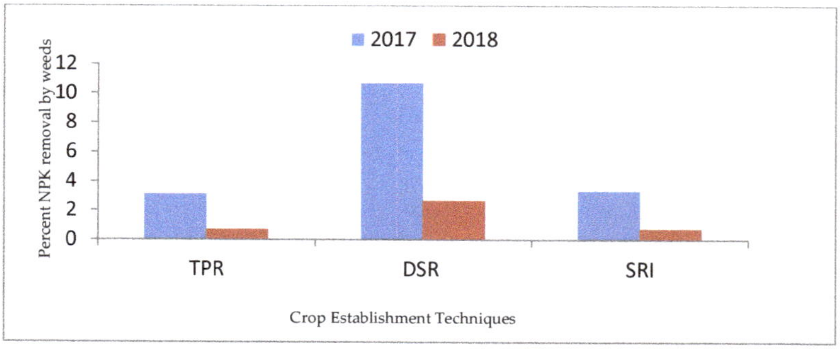

Figure 3. Proportion of nutrient uptake by weeds from total uptake (plants + weeds NPK uptake) under crop establishment techniques.

3.8. Soil Nutrient Status

3.8.1. Available Nitrogen (N)

During the period of study, the available nitrogen status of 235.76 and 219.76 kg ha^{-1}, respectively, under SRI was seen lowest, which significantly differed compared to TP and DSR, demonstrating that SRI removed more N from the soil than transplanted and direct-seeded rice (Table 10). However, the highest values of available N status of soil were registered highest under DSR (266.41 and 250.41 kg ha^{-1}). All the weed control measures significantly affected the soil available N status, but the highest status of available N (273.93 and 257.94 kg ha^{-1}) was recorded under weedy check. The lowest available N status of 239.55 and 223.55 kg ha^{-1} in soil was observed under penoxsulam treatment, revealing that penoxsulam treatment removed more N from the soil among the different weed control measures which possibly is due to higher Biomass resulting in more extraction of nitrogen from soil.

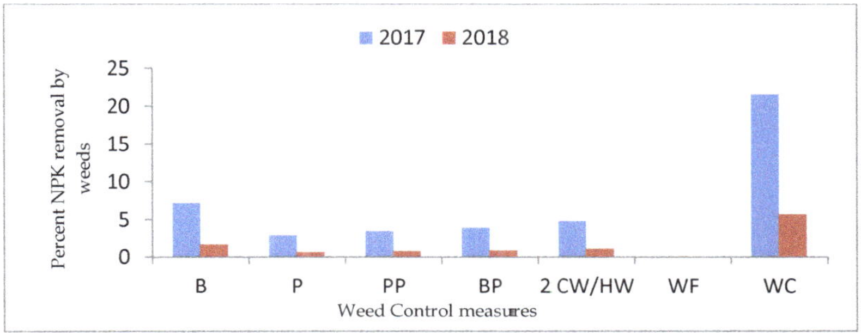

Figure 4. Proportion of nutrient uptake by weeds from total uptake (plants + weeds NPK uptake) under different weed control measures.

Table 10. Influence of rice establishment techniques and weed control measures on available nutrient status of soil (kg ha^{-1}).

Treatments	Available N		Available P		Available K	
	2017	2018	2017	2018	2017	2018
Rice establishment techniques						
TPR	255.7	239.7	20.08	19.85	143.8	135.5
DSR	266.4	250.4	23.31	26.56	159.2	153.4
SRI	235.7	219.7	17.82	15.60	134.2	124.5
SE(m) ±	1.71	2.07	0.34	0.35	1.64	1.91
C. D. ($p \leq 0.05$)	5.21	6.21	1.01	1.05	4.94	5.72
Weed control measures						
B	263.8	247.8	22.16	24.57	157.9	150.2
P	239.5	223.5	18.18	16.03	132.0	123.7
PP	246.24	230.2	19.25	18.18	138.2	130.1
BP	253.0	237.0	20.35	20.42	145.6	137.6
2 CW/HW	259.3	243.3	21.52	22.78	152.9	145.1
WF	232.5	216.5	16.99	14.19	124.5	116.1
WC	273.9	257.9	24.39	28.54	169.2	161.7
SE(m) ±	2.63	2.21	0.30	0.29	1.93	2.28
C. D. ($p \leq 0.05$)	7.90	6.73	0.90	0.88	5.80	6.84

3.8.2. Available Phosphorus (P)

The study revealed that among different crop establishment methods, the available P status of soil recorded under SRI (17.82 and 15.60 kg ha^{-1}) was lowest than transplanted and direct seeded rice, respectively illustrating that more P from the soil was removed under SRI than transplanted and DSR (Table 10). Under DSR, the highest status of available P (23.31 and 26.56 kg ha^{-1}) in soil was demonstrated during the two consecutive years of study. Concerning herbicidal applications, the available phosphorus in the soil, recorded lowest under penoxsulam was 18.18 and 16.03 kg ha^{-1}, respectively, than remaining herbicidal treatments, while highest values (24.39 and 28.54 kg ha^{-1}) in soil were observe under weedy check treatment, indicating that more P from the soil has been removed i Penoxsulam treatment.

3.8.3. Available Potassium (K)

Among different crop establishment methods, the available soil K recorded under SR (134.27 and 124.54 kg ha^{-1}) was lowest among other rice establishment techniques, whil DSR excelled in demonstrating the higher values of available K (159.27 and 153.40 kg ha^{-1} in soil (Table 10), confirming more removal of K from the soil in SRI than transplanted an direct-seeded rice. For weed management practices, the lowest available K status in soi recorded under penoxsulam was 132.03 and 123.75 kg ha^{-1}, indicating that more K has bee removed in penoxsulam treatment from the soil. Among weed control treatments. Howeve weedy check treatment registered higher availability of K (169.22 and 161.79 kg ha^{-1}) i the soil during research.

3.9. Regression and Correlation Studies

Among rice establishment techniques and weed control measures, crop attributes vi crop dry matter, and total N uptake by crop illustrated a positive correlation with rice grai yield. The coefficient of determination was highly significant for rice grain yield with dr crop matter (0.18 and 0.26) (Figure 5) and total N uptake (0.87 and 0.91) (Figure 6) in ric establishment techniques. The variations in dry crop matter and total N uptake could b attributed to 18 and 26% and 87 and 91% during 2017 and 2018, respectively. Among th different weed management measures, the coefficient of determination for grain yield wit crop dry matter (0.05 and 0.50) (Figure 7) and total N uptake (0.99 and 0.99) (Figure 8 was noted significant. The deviations in dry crop biomass and total N-uptake could b attributed to 5 and 50% and 99 and 99% during 2017 and 2018, respectively. Among th various crop establishment methods, rice grain yield and weed parameters viz. weed dr matter and NPK uptake by weeds demonstrated a negative correlation. Coefficient o determination among rice grain yield and dry matter of weeds (0.92 and 0.96) (Figure 9), N uptake by weeds (0.70 and 0.61) (Figure 10), P uptake by weeds (0.70 and 0.61) (Figure 11 and K uptake by weeds (0.70 and 0.61) (Figure 12) was found significant in various ric establishment techniques. In conclusion, variations in weed dry matter and NPK uptake b weeds could be explained to 92 and 96%, 70 and 61%, 70 and 61%, and 70 and 61%, durin 2017 and 2018, respectively. Furthermore, there was a negative correlation between wee biomass with weed NPK uptake and grain yield among the different weed managemen approaches. The coefficient of determination for grain yield with weed dry matter (0.9 and 0.96) (Figure 13), N uptake by weeds (0.95 and 0.97) (Figure 14), P uptake by weed (0.95 and 0.97) (Figure 15) and K uptake by weeds (0.95 and 0.97) (Figure 16) was recorde significant. Conclusion: variations in weed dry matter and NPK uptake by weeds coul be explained to 92 and 96%, 95 and 97%, 95 and 97%, and 95 and 97% during 2017 an 2018, respectively.

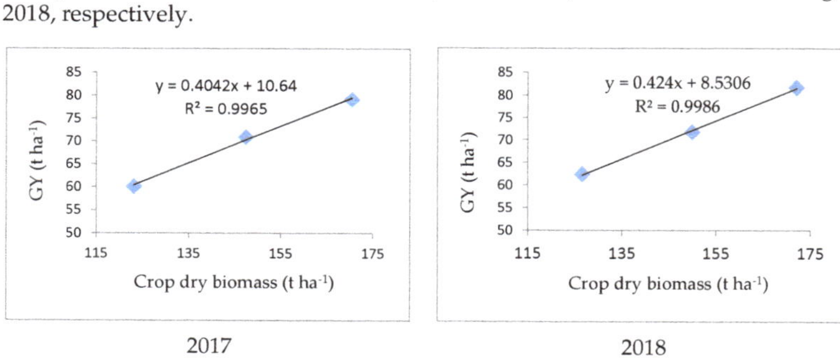

Figure 5. Linear regression line among crop establishment methods between dry matter accumulatio vs. Grain yield (GY).

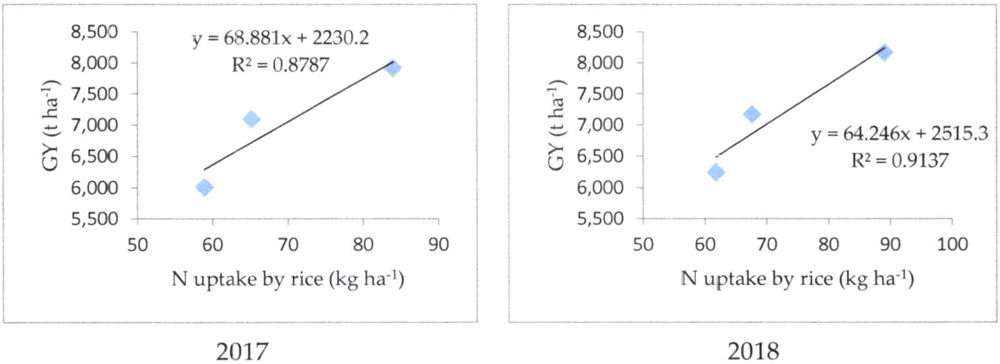

Figure 6. Linear regression line among crop establishment methods between N uptake vs. Grain yield (GY).

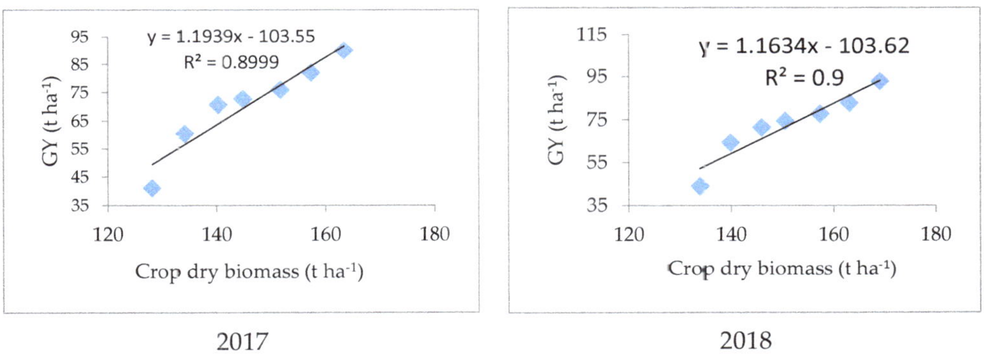

Figure 7. Linear regression line among weed management practices between dry matter accumulation vs. Grain yield (GY).

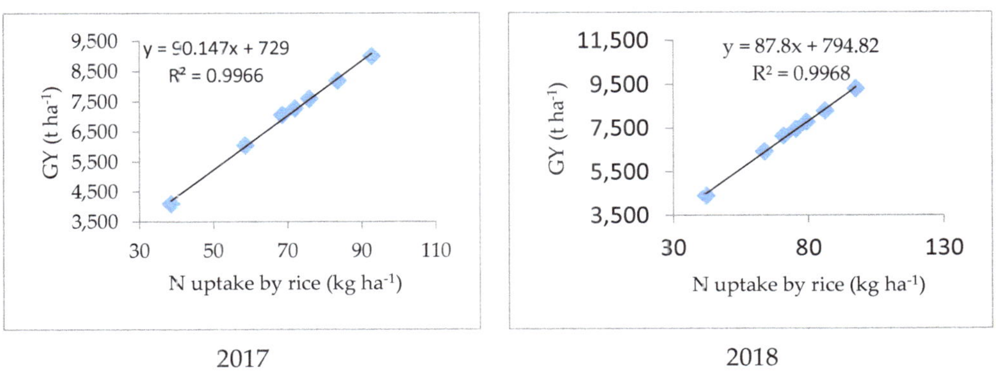

Figure 8. Linear regression line among weed management practices between N uptake vs. Grain yield (GY).

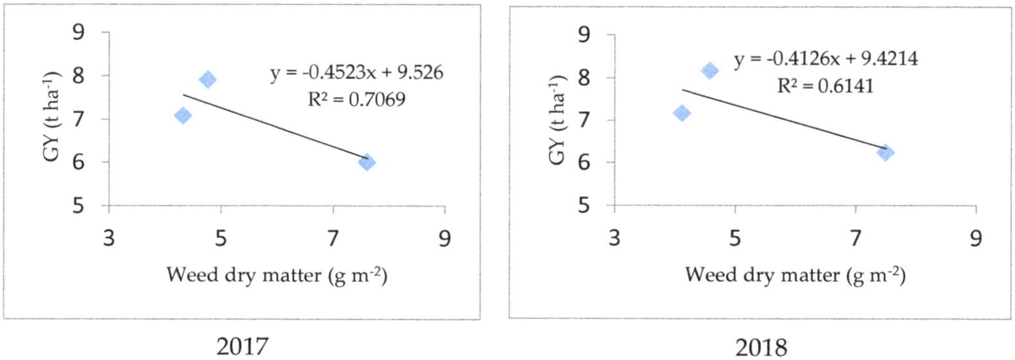

Figure 9. Linear regression line among crop establishment methods between weed dry matters vs. Grain yield (GY).

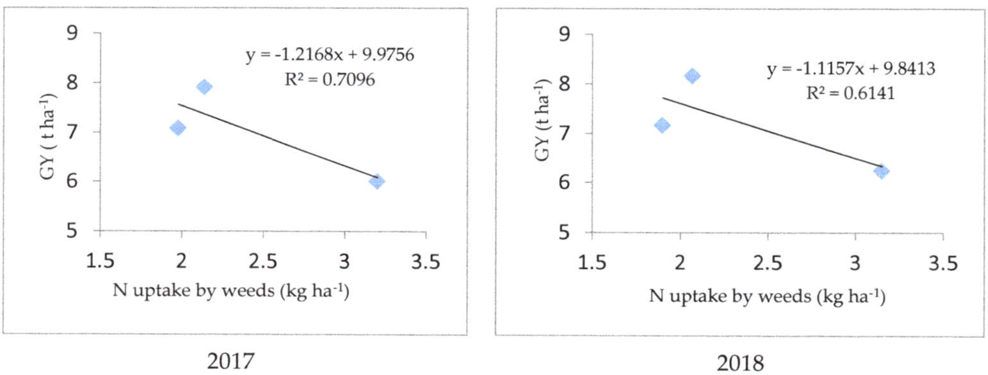

Figure 10. Linear regression line among crop establishment methods between N uptake by weeds vs. Grain yield (GY).

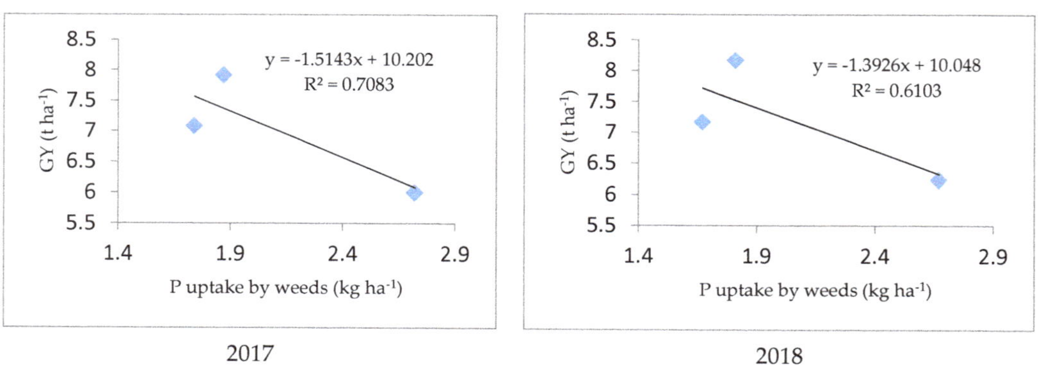

Figure 11. Linear regression line among crop establishment methods between p uptake by weeds vs. Grain yield (GY).

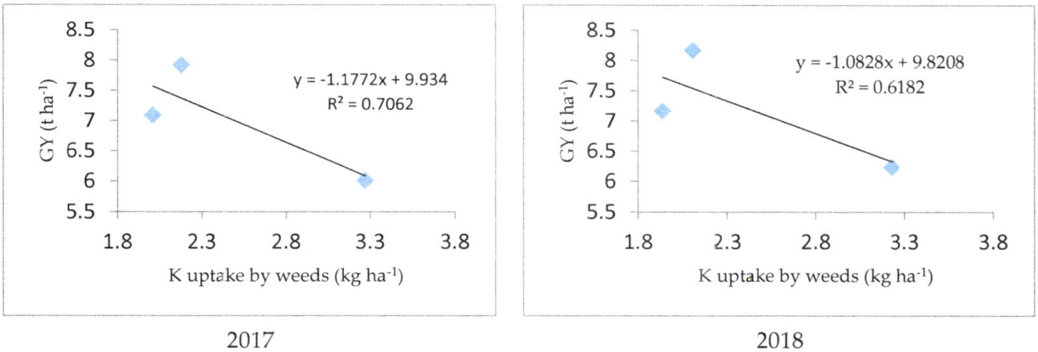

Figure 12. Linear regression line among crop establishment methods between K uptake by weeds vs. Grain yield (GY).

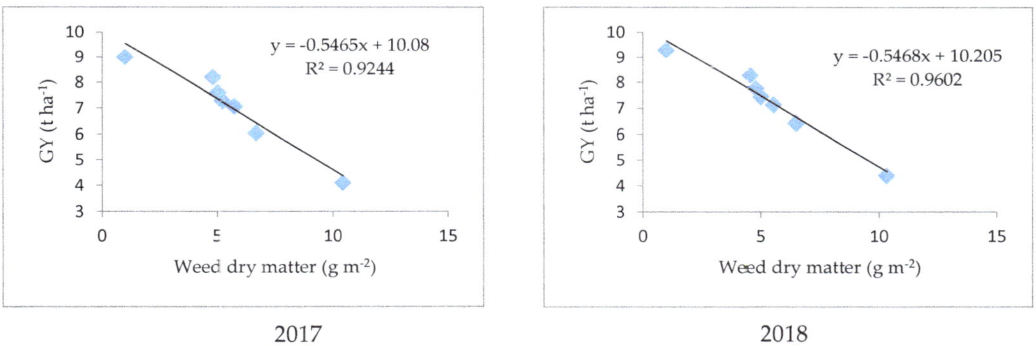

Figure 13. Linear regression line among weed management practices between weed dry matter vs. Grain yield (GY).

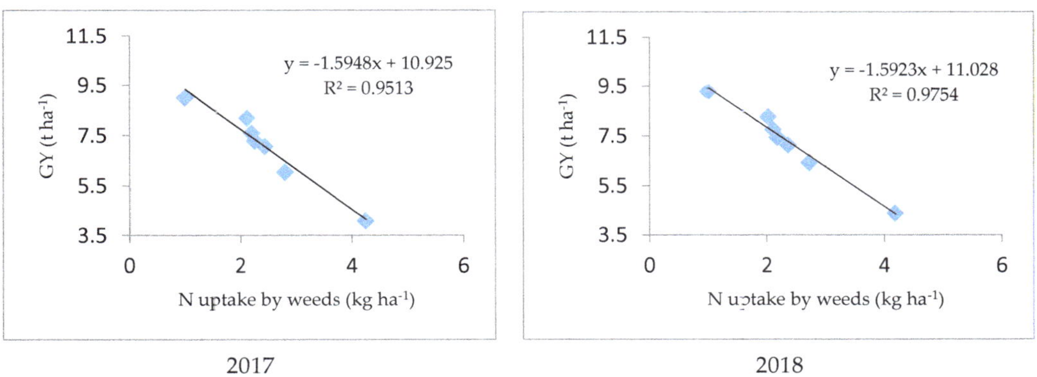

Figure 14. Linear regression line among weed control measures between N-uptake by weeds vs. Grain yield (GY).

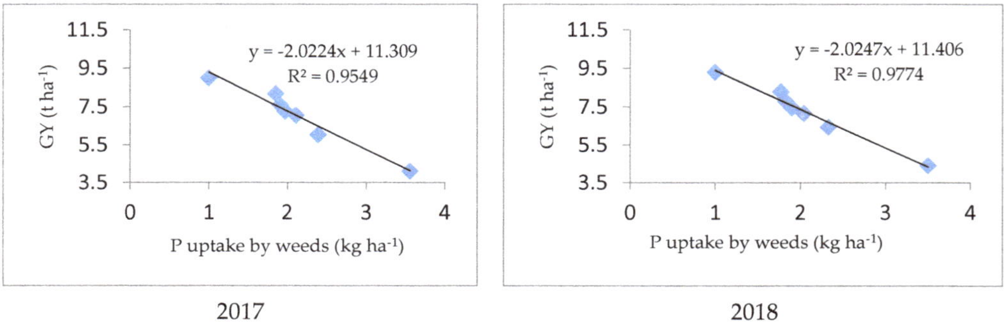

Figure 15. Linear regression line among weed control measures between P-uptake by weeds vs. Grain yield (GY).

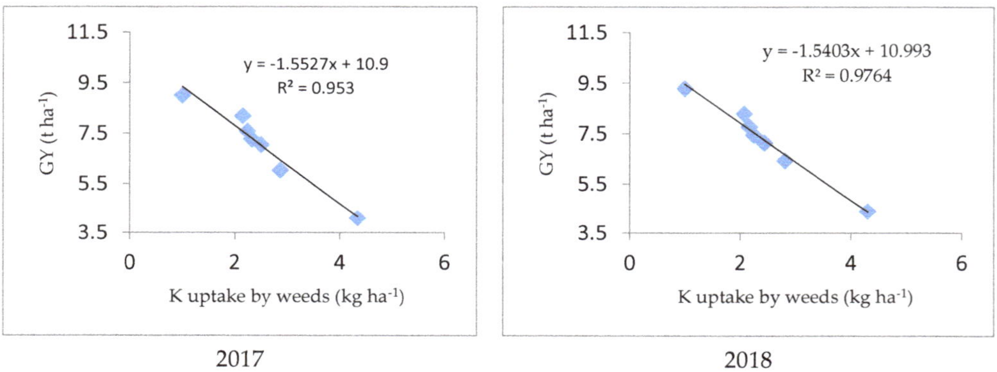

Figure 16. Linear regression line among weed control measures between K-uptake by weeds vs. Grain yield (GY).

4. Discussion

Under SRI, the dry matter was observed highest compared to other crop establishment techniques. This could be described as soils becoming more robust due to broader spacing, which improves soil's organic matter retention, nutrient cycling, and biological fertility, boosting crop development such as dry matter output. Increased plant height and other growth characteristics such as LAI may be responsible for higher dry matter output under the SRI approach [19–21]. Under different weed control measures, the maximum dry matter was achieved in penoxsulam treated plots. Less weed competition led to better crop growth during the initial growth period, which boosted nutrient availability and light, favouring better photosynthate buildup. Herbicide use reduced weed growth and allowed the crop to reach its full potential, resulting in higher dry matter accumulation [20–22].

The highest yield (grain + straw) observed in SRI is described due to the enhanced expression of yield attributes. Under SRI, transplanting young seedlings promotes better tillering and rooting, resulting in greater root volume, improved tillers, more filled spikelets and maximum grain weight. In addition, wider spacing (25 cm × 25 cm) encourages canopy and root growth, enhancing grain filling [23–25]. Several scientists support our research findings [20,21,26,27]. Penoxsulam treatment demonstrated higher values of grain and straw yield. Herbicide application inhibited weed growth and permitted the rice crop to receive adequate nutrient supply. Production of more photosynthates via more effective tillers per metre^{-2} and proper dry matter partitioning (source to sink) resulted in higher grain and straw yields [28,29]. Several studies demonstrated similar findings [20,21,30].

Maximum weed biomass noticed in DSR is attributed to hospitable environment received in DSR by weeds, compared to SRI and TPR rice methodologies. Puddling under TPR provided favourable environments at an initial stage for short crop growth, responsible for weed smothering. Preparation of land in standing water destroys the existing weed flora. However, puddling promotes rice growth, resulting in the suppression of weeds [24,31,32]. Under DSR, dry tillage, aerobic environment due to absence of flooding [33] resulted in higher weed biomass. Penoxsulam application resulted in a maximum reduction in dry weed biomass. Penoxsulam is an excellent herbicide that effectively controls weeds and has low toxicity to rice seedlings [34–36].

Under different crop establishment techniques, the maximum share (%) of weeds towards total Biomass observed in DSR demonstrated that weeds received favourable environmental conditions due to dry tillage and aerobic conditions since flooding was absent, resulting in a maximum share of weeds to total Biomass [7]. However, under different weed control measures, the percentage of weeds to total biomass production obtained in weedy check could be illustrated due to vast weed flora and higher dry weed biomass [7].

SRI proved to be most efficient technology of rice production for better nutrient content and uptake than DSR and TPR as SRI has excellent potential for better use of resources. Further, the findings of our study revealed that removal of total nitrogen, phosphorus and potassium by rice were noted highest in SRI, under rice establishment methods. This could be attributed to the square geometry of the hills, wider spacing, and the planting of a single seedling under SRI that improved the interception of photosynthetically active radiation and uptake of nutrients. Plants with increased root growth under SRI have access to utilize subsoil nutrients. It's also possible that SRI soil and water management strategies, such as alternate wetting and drying, can boost microbial P solubilization [37,38]. The crop's nutrient uptake is determined by nutrient status and yield. Rice establishment techniques illustrated a significant impact on yield. Hence uptake of nutrients varied significantly despite the non-significant nutrient content variation. Percentage of nutrients removed by weeds illustrated that lower amounts of nitrogen, phosphorus, and potassium by weeds were accumulated in TPR due to the smothering of weeds due to continuous submergence. However, dry tillage and aerobic environment due to lack of flooding conditions contributed to maximum percent removal of nitrogen, phosphorus, and potassium by weeds under DSR [7].

Among the different herbicidal treatments, penoxsulam resulted in maximum nutrient content and uptake by the crop. Since weeds compete for macronutrient uptake thereby demonstrated a significant effect on total uptake. The nutrient uptake by the crop is determined by its nutrient content and above-ground biomass, which indicates the crop's better growth due to less weed interaction [38,39]. Further, the percent (%) removal of N, P and K due to weeds demonstrated that significantly lower amounts of N, P and K were accumulated in penoxsulam treated plots due to maximum reduction in weed biomass since it effectively controlled complex weed flora. However, N, P and K were removed maximum in weedy check due to weeds under various weed control measures, illustrating complex weed flora and higher dry weed biomass [7].

Available soil nitrogen, phosphorus and potassium observed lowest in SRI under establishment methods might be attributed to more removal of N, P and K from soil than TPR and DSR, since SRI recorded higher yield and maximum uptake of nutrients [7]. Under weed control treatments, the available N, P and K status of soil was found lowest under penoxsulam treatment, illustrating that penoxsulam, because of higher biomass production, removed more nutrients from the soil among other weed control treatments and could be ascribed to effective weed control provided by penoxsulam herbicide, favoured the crop growth and resulted in higher crop biomass followed by maximum crop nitrogen, phosphorus and potassium uptake [7,40].

5. Conclusions

At the outset of our detailed experiment, we can conclude that among the crop establishment methods, SRI demonstrated the higher accumulation of dry matter, yield (grain + straw) and maximum nutrient (N, P and K) uptake by rice. In contrast, the dry weed biomass and removal of nitrogen, phosphorus and potassium by weeds were illustrated higher in DSR. Under weed control measures, penoxsulam treatment showed the highest accumulation of dry matter, rice yield and maximum nutrient (N, P and K) removal by rice, however, weed biomass and nutrient removal by weeds were observed highest in weedy check plots. The contribution of weeds towards nitrogen, phosphorus and potassium removal was illustrated highest under DSR. In contrast, penoxsulam application contributed the lowest in nutrient removal by weeds as far as weed control treatments were studied. Among crop establishment techniques and weed control measures, the treatment with penoxsulam application under SRI resulted in minimum uptake of nitrogen, phosphorous and potassium by weeds as compared to other treatment combinations, which indicate that the above-mentioned treatment combination can be employed in the Northwestern region of India for optimum resource utilization to boost rice productivity.

Author Contributions: Conceptualization, A.N., M.A.B. and T.A.B.; methodology, A.N., S.F., M.S.M., S.A.A. and U.B.; software, T.A.B., M.M., B.A.L. and M.S.; validation, M.A.B., S.A. and M.B.; formal analysis, B.A.L., T.A.B., M.M., S.A., M.E.-S. and M.B.; investigation, M.A.B., B.A.L., S.F., M.S.M. and A.E.S.; resources, B.J., S.A.A. and U.B.; data curation, A.E.S., M.S. and M.E.-S.; writing—original draft preparation, A.N., M.A.B., T.A.B., M.S.M., B.J. and S.F.; writing—review and editing, M.S., M.E.-S., M.B., A.E.S., S.A.A., U.B., S.A. and M.M. All authors have read and agreed to the published version of the manuscript.

Funding: This Research was funded by the Researchers Supporting Project number (TURSP-2020/139), Taif University, Taif, Saudi Arabia.

Institutional Review Board Statement: Not applicable.

Informed Consent Statement: Not applicable.

Data Availability Statement: Not applicable.

Acknowledgments: The authors extend their appreciation to Taif University Researchers Supporting Project number (TURSP-2020/139), Taif University, Taif, Saudi Arabia. The authors express the deepest appreciation to the teachers and technical staff of Division of Agronomy; Faculty of Agriculture, Sher-e-Kashmir University of Agricultural Sciences and Technology of Kashmir, Wadura, Sopore-193201, India for providing all the necessary facilities, suggestions, help, cooperation, and praise to complete the research.

Conflicts of Interest: The authors declare no conflict of interest.

References

1. IFPRI. International Food Policy Research Institute. 2010. Available online: http://www.ifpri.org (accessed on 20 August 2021).
2. Kubo, M.; Purevdorj, M. The future of rice production and consumption. *J. Food Distrib. Res.* **2004**, *35*, 128–142.
3. Rana, N.; Rahim, M.S.; Kaur, G.; Bansal, R.; Kumawat, S.; Roy, J.; Deshmukh, R.; Sonah, H.; Sharma, T.R. Applications and challenges for efficient exploration of omics interventions for the enhancement of nutritional quality in rice (*Oryza sativa* L.). *Crit. Rev. Food Sci. Nutri.* **2020**, *60*, 3304–3320. [CrossRef] [PubMed]
4. Lal, B.; Gautam, P.; Joshi, E. Different rice establishment methods for producing more rice per drop of water: A review. *Int. J. Res. Biosci.* **2013**, *2*, 1–12.
5. Midya, A.; Saren, B.K.; Pramanik, K. Aerobic rice culture, System of Rice Intensification (SRI) and System of Assured Rice Production (SARP): Emerging water savings production technologies for rice yield stability in tropics under shrinking water resource base. In Proceedings of the Souvenir Paper of 1st International Conference on Bio-Resource, Environment and Agricultural Sciences, Visva-Bharati, West Bengal, India, 4–6 February 2017; pp. 61–66.
6. Kaur, J.; Singh, A. Direct seeded rice: Prospects, problems/constraints and researchable issues in India. *Curr. Agric. Res. J.* **2017**, *5*, 13–32. [CrossRef]
7. Dhaliwal, S.S.; Sharma, S.; Shukla, A.K.; Sharma, V.; Bhullar, M.S.; Dhaliwal, T.K.; Alorabi, M.; Alotaibi, S.S.; Gaber, A.; Hossain, A. Removal of Biomass and Nutrients by Weeds and Direct-Seeded Rice under Conservation Agriculture in Light-Textured Soils of North-Western India. *Plants* **2021**, *10*, 2431. [CrossRef] [PubMed]

Yadav, G.S.; Shivay, Y.S.; Kumar, D.; Babu, S. Enhancing iron density and uptake in grain and straw of aerobic rice through mulching and rhizo-foliar fertilization of iron. *Afr. J. Agric. Res.* **2013**, *8*, 5447–5454.

Kassam, A.; Stoop, W.; Uphoff, N. Review of SRI modifications in rice crop and water management and research issues for making further improvements in agricultural and water productivity. *Rice Water Environ.* **2011**, *9*, 163–180. [CrossRef]

0. Shahane, A.A.; Shivay, Y.S.; Prasanna, R.; Kumar, D. Nutrient removal by rice–wheat cropping system as influenced by crop establishment techniques and fertilization options in conjunction with microbial inoculation. *Sci. Rep.* **2020**, *10*, 21944. [CrossRef]

1. Zimdahl, R.L. *Fundamentals of Weed Science*, 4th ed.; Academic Press: Cambridge, MA, USA, 2013.

2. Hemalatha, K.; Singh, Y.; Kumar, S. Leaf colour chart-based nitrogen and weed management impacts on weeds, yield and nutrient uptake in dry direct-seeded rice. *Indian J. Weed Sci.* **2020**, *52*, 318–321. [CrossRef]

3. Puniya, R.; Pandey, P.C.; Bisht, P.S.; Singh, D.K. Nutrient uptake by crop and weeds as influenced by trisulfuron, trisulfuron + pretilachlor and bensulfuron methyl in transplanted rice. *Indian J. Weed Sci.* **2007**, *39*, 239–240.

4. Saha, M.; Banerjee, H.; Pal, S. Relative efficacy of herbicides in wheat. *Indian J. Weed Sci.* **2006**, *38*, 127–128.

5. Piper, C.S. *Soil and Plant analysis, Asian ed.*; Hans Publishers: Bombay, UK, 1996.

6. Prasad, T.N.; Pandey, R.D ; Sahni, R.P. Response of early paddy to water regimes and nitrogen in calcareous soil. *Indian J. Agron.* **1990**, *35*, 364–370.

7. Jackson, M.L. *Soil Chemical Analysis*; Prentice Hall of India Pvt. Ltd.: New Delhi, India, 1973; Volume 15, pp. 13–22.

8. Subbiah, B.V.; Asija, G.L. A rapid procedure for the estimation of available nitrogen in soil. *Curr. Sci.* **1956**, *25*, 259–260.

9. Olsen, S.R.; Cole, C.V.; Watanabe, F.S.; Dean, L.A. *Estimation of Available Phosphorus in Soils by Extraction with Sodium Bicarbonate*; United States Department of Agriculture Circular: Washington, DC, USA, 1939; Volume 1954, pp. 1–19.

0. Saha, A.; Bharti, V. Effect of different crop establishment methods on growth, yield and economics of rice (*Oryza sativa* L.). *Environ. Ecol.* **2010**, *28*, 519–522.

1. Nazir, A.; Bhat, M.A.; Bhat, T.A.; Rashid, Z.; Mohi-ud-din, R.; Fayaz, S.; WAni, S.A. Crop establishment and weed management effect on weed parameters and rice yield under temperate zone of Kashmir. *Indian J. Weed Sci.* **2020**, *52*, 217–221.

2. Nazir, A.; Bhat, M.A.; Bhat, T.A.; Fayaz, S.; Bahar, F.A.; Mohi-ud-din, R. Effect of crop establishment methods and weed management practices on yield and yield attributes of rice (*Oryza sativa* L.) under temperate conditions of Kashmir. *Appl. Biol. Res.* **2021**, *23*, 171–178. [CrossRef]

3. Bhat, M.A.; Hussain, A.; Ganai, M.A.; Mushki, G.M. Effect of herbicides used alone and in combination on weeds and transplanted rice under temperate conditions of Kashmir. *Appl. Biol. Res.* **2011**, *13*, 75–78.

4. Miller, B.C.; Hill, J.E.; Roberts, S.R. Plant Population Effects on Growth and Yield in Water-Seeded Rice. *Agron. J.* **1991**, *83*, 291–297. [CrossRef]

5. Chauhan, B.S.; Ahmed, S.; Awan, T.H. Performance of sequential herbicides in dry-seeded rice in the Philippines. *Crop Prot.* **2015**, *74*, 124–130. [CrossRef]

6. Xu, L.; Li, X.; Wang, X.; Xiong, D.; Wang, F. Comparing the Grain Yields of Direct-Seeded and Transplanted Rice: A Meta-Analysis. *Agronomy* **2019**, *9*, 767. [CrossRef]

7. Jehangir, I.A.; Hussain, A.; Sofi, N.R.; Wani, S.H.; Ali, O.M.; Abdel Latef, A.A.H.; Raja, W.; Bhat, M.A. Crop Establishment Methods and Weed Management Practices Affect Grain Yield and Weed Dynamics in Temperate Rice. *Agronomy* **2021**, *11*, 2137. [CrossRef]

8. Mahajan, G.; Chauhan, B.S.; Gill, M.S. Optimal nitrogen fertilization timing and rate in dry-direct seeded rice in north-west India. *Agron. J.* **2012**, *103*, 1676–1682. [CrossRef]

9. Singh, R.; Singh, S.P.; Singh, V.P.; Sirazuddin; Verma, H.; Shukla, D.K. Weed control in dry seeded rice with penoxsulam. *Int. J. Basic Appl. Agric. Res.* **2016**, *14*, 379–382.

0. Martini, L.F.D.; Burgos, N.R.; Noldin, J.A.; de Avila, L.A.; A Salas, R. Absorption, translocation and metabolism of bispyribac-sodium on rice seedlings under cold stress. *Pest Manag. Sci.* **2014**, *71*, 1021–1029. [CrossRef]

1. Shan, F.A.; Bhat, M.A.; Ganai, M.A.; Hussain, A.; Bhat, T.A. Effect of crop establishment and weed control practices on the performance of rice (*Oryza sativa* L.). *Appl. Biol. Res.* **2012**, *14*, 79–85.

2. Chauhan, B.; Awan, T.H.; Abugho, S.B.; Evengelista, G.; Yadav, S. Effect of crop establishment methods and weed control treatments on weed management, and rice yield. *Field Crop. Res.* **2015**, *172*, 72–84. [CrossRef]

3. Chauhan, B.S.; Johnson, D.E. The Role of Seed Ecology in Improving Weed Management Strategies in the Tropics. *Adv. Agron.* **2010**, *105*, 221–262.

4. Chauhan, B. Weed Ecology and Weed Management Strategies for Dry-Seeded Rice in Asia. *Weed Technol.* **2012**, *26*, 1–13. [CrossRef]

5. Timsina, J.; Haque, A.; Chauhan, B.S.; Johnson, D.E. Impact of tillage and rice establishment methods on rice and weed growth in the rice-maize-mungbean rotation in northern Bangladesh. In Proceedings of the 28th International Rice Research Conference, Hanoi, Vietnam, 8–12 November 2010.

6. Subramanyam, D.; Raghava, R.C.; Srinivasulu, D. Effect of puddling, water and weed management practices on weed dynamics and yield of transplanted rice (*Oryza sativa* L.). *Indian J. Weed Sci.* **2006**, *38*, 37–41.

7. Ehsanullah, A.; Jabran, N.K.; Habib, T. Comparison of different planting methods for optimization of plant population of fine rice (*Oryza sativa* L.) in Punjab (Pakistan). *Pak. J. Agric. Sci.* **2007**, *44*, 597–599.

8. Rani, S.; Sukumari, P. Root growth, nutrient uptake and yield of medicinal rice njavara under different establishment techniques and nutrient sources. *Am. J. Plant Sci.* **2013**, *4*, 1568–1573. [CrossRef]

39. Naz, S.; Nandan, R.; Roy, D.K. Effect of crop establishment methods and weed management practices on productivity, economic and nutrient uptake in direct seeded rice (*Oryza sativa* L.). *Int. J. Curr. Microbiol. Appl. Sci.* **2020**, *9*, 3002–3009. [CrossRef]
40. El-Desoki, E.R. Effect of some weed control treatments on transplanting rice and nutrients uptake by rice and weeds. *J. Agric. Sc Mansoura Univ.* **2003**, *28*, 23–35.

agronomy

MDPI

Article

How Cover Crop Sowing Date Impacts upon Their Growth, Nutrient Assimilation and the Yield of the Subsequent Commercial Crop

Paul Cottney [1,2,*], Lisa Black [1], Paul Williams [2] and Ethel White [1,†]

[1] Agri-Food and Biosciences Institute, Crossnacreevy BT6 9SH, UK; Lisa.Black@afbini.gov.uk (L.B.); Ethel.whiteni@gmail.com (E.W.)
[2] Institute for Global Food Security, School of Biological Sciences, Queen's University Belfast, Belfast BT9 5DL, UK; p.williams@qub.ac.uk
* Correspondence: pcottney02@qub.ac.uk; Tel.: +44-028-9054-8007
† The author has retired.

Abstract: Cover crops are typically sown post-harvest of commercial crops, prior to winter, which means that as sowing date is delayed, so will biomass production potential. The wide range of benefits associated with cover crops relies on them to produce sufficient biomass. Therefore, it must be identified how late certain species of cover crops can be sown. In the climatic conditions of Northern Ireland, not only has no research been conducted on how cover crops perform at various sowing dates but also their effect on the subsequent commercial crop yield has not been investigated. Addressing these issue will in turn help provide recommendations to maximise and encourage later sowing of cover crops. Consequently, five species of cover crops were chosen, from a range of families, then sown on 14 August, 7 September and 27 September. This is to mimic when land becomes fallow post-harvest of typical crops/rotations to this region. It was found that tillage radish (*Raphanus sativus* L.), when sown on the earliest date, could accumulate a maximum of 261 kg/ha of nitrogen (N), whereas, when sown on the last date, phacelia (*Phacelia tanacetifolia* L.) significantly outperformed all other species and assimilated 70 kg/ha of N. The cover crops were then incorporated into the soil and over-sown with spring barley (*Hordeum vulgare* L.). However, the spring barley yield was unaffected by any treatments. This trial shows that the non-leguminous species chosen are highly effective in assimilating nutrients when sown mid-August until early-September.

Keywords: nitrogen assimilation; spring barley yield; weed management; carbon assimilation; biofertiliser; light interception

Citation: Cottney, P.; Black, L.; Williams, P.; White, E. How Cover Crop Sowing Date Impacts upon Their Growth, Nutrient Assimilation and the Yield of the Subsequent Commercial Crop. *Agronomy* **2022**, *12*, 369. https://doi.org/10.3390/agronomy12020369

Academic Editors: Umberto Anastasi and Aurelio Scavo

Received: 8 December 2021
Accepted: 24 January 2022
Published: 1 February 2022

1. Introduction

When the sowing of cover crops is delayed later into autumn, their exposure to conditions critical for growth is reduced. This is detrimental to cover crop growth rate due to a decrease in the average ambient air and soil temperature as well as a continuously diminishing quality and quantity of light. In addition, soil moisture will typically increase and can negatively impact on soil trafficability which is critical for machinery to operate and sow the cover crop. Sowing early post-harvest of the previous commercial crop is important for biomass growth and also for nutrient accumulation [1] and grazing potential [2,3]. Early sowing of cover crops is the optimum practice to maximise their benefits but this is not always possible as their sowing date depends on the harvest date of the prior crop [1]. Early harvested crops (mid-July to early-August), e.g., winter barley (*Hordeum vulgare* L.), are more conducive to sowing early but these crops do not account for all rotations that have fallow land over winter prior to a spring crop. Therefore, to maximise the area of land sown with cover crops, they must also be planted after crops with harvests that are later than winter barley, such as winter wheat (*Triticum aestivum* L.), spring barley (*Hordeum vulgare* L.) and vegetables. This raises the need for investigation of how late in the autumn, in the climatic conditions of Northern Ireland (NI), can cover crops be sown and grow successfully, and which species are better suited to later sowing dates.

Different species of cover crops exhibit variances in how they perform under delayed sowing, due to their vigour and competitiveness [4]. Producing high biomass increases nutrient assimilation [5]. Therefore, delayed sowing will reduce effectiveness in mitigating against N leaching [6]. This is because the potential to scavenge N has been found to relate to leaf expansion rate along with radiation interception [6] and the depth and density of the roots [5]. Delayed sowing of cover crops can also affect weed suppression ability [7]. This is important as an integrated management tool for growers, especially those that are organic, where identification of the best species and management practices to suppress weeds is essential [8]. Brust et al. [9] found that forage radish (*Raphanus sativus var. oleiformis* L.) was consistent in reducing weed biomass at all test sites (3 across 2 years in Germany) and that phacelia reduced weed dry matter by 77% relative to fallow. These authors noted that fast competitive growth boosted light interception, enhanced the plant canopy and, therefore, increased ability to shade out weeds.

Cover crops add carbon (C) to soils, but this benefit will also be reduced with delayed sowing, due to the decline in biomass which may also reduce the C concentration within the plant, as more mature early-sown crops may have more C-rich structural compounds. Delayed sowing, therefore, could affect the carbon:nitrogen (C:N) ratio [8,10], which influences the rate of N mineralisation [11,12]. The speculated effect would be a decrease in C:N due to a lower proportion of structural C compounds within the plants. This decrease in C:N ratio could result in a higher mineralisation of nutrients from the cover crop biomass which could better supply the commercial crop with N. However, less N accumulated by the cover crop could result in more N leached over winter, which could reduce the quantity of N available to the commercial crop, in comparison to earlier sowing. This in turn could affect spring barley grain yields, if N fertiliser rate is insufficient.

Objective

From a list of sixteen species of cover crops investigated by Cottney et al. [13], five species have been chosen from a range of families which showed potential to increase nutrient cycling and grain yield in that greenhouse experiment. The chosen species include forage rape (*Brassica napus* L.), tillage radish (*Raphanus sativus* L.), vetch (*Vicia villosa* L.), westerwolds (*Lolium multiflorum* L.) and phacelia (*Phacelia tanacetifolia* L.) to be investigated under field conditions. The species will be sown at three different dates to represent sowing of cover crops after harvest of winter barley, winter wheat/spring barley, and to represent a delayed commercial crop harvest. The objective is to investigate the effect on:

(1) Cover crop biomass,
(2) Cover crop nutrient assimilation, and
(3) Consequence on spring barley yield supplemented with a reduced rate N program (70 kg/ha of inorganic N).

2. Materials and Methods

2.1. Experimental Design

A split-plot experimental design was generated using the procedure AGHIERARCHICAL [14] with two treatments replicated across 8 blocks during cover crop growth. The experiment was reduced in replication to 4 blocks during spring barley growth due to resource limitations. Treatment factors included planting date (whole plot) and species (sub-plot), each of which had 3 and 6 levels, respectively. The whole plot was completely randomised, with the sub-plots randomised within the whole plots. An unplanted control of bare fallow and the 5 species of cover crops (vetch, forage rape, tillage radish, westerwolds and phacelia) were sown, as shown in Appendix (Table A1). At each sowing date, the control was cultivated with the disc and left unsown as a bare fallow. This mirrors farmer practice of stale seedbed creation whereby fallow land may be cultivated to both destroy and encourage more weeds to grow. Hence, each control at each sowing date is different to each other where each control will be presented separately in the data.

The experiment was planned to be repeated for two years but the second year replication was not possible due excessive rainfall.

2.2. Location

The trial site was located in Hillsborough, Co. Down, NI (latitude 54.445117, and longitude 6.096430). The soil is a clay loam to 30 cm with a particle distribution analysis of 44.7% sand, 33.5% silt and 22.9% clay in a 15 cm profile, with a deep clay after 35 cm in the profile. The soil analysis is shown in Table 1, with soil temperatures, ambient air and average site rainfall shown in Figures A1–A3, respectively. Soil and ambient air temperatures were logged using a Tinytag Plus 2 TGP-4510 datalogger and a soil probe (PB-5001) (West Sussex, UK) measuring to a soil depth of 15 cm.

Table 1. Standard soil test and soil mineral nitrogen (SMN) prior to sowing the cover crops.

Parameter	Unit	Value
pH		5.87
Phosphorous (P)	mg/L	58.2
Potassium (K)	mg/L	202.5
Magnesium (Mg)	mg/L	180.3
Sulphur (S)	mg/L	18.3
Total soil N	%	0.34
Total soil C	%	3.68
Nitrite ($NO^-{}_2$) + nitrate ($NO^-{}_3$)	mg/kg	14.9
Ammonium ($NH^+{}_4$)	mg/kg	14.9
Total SMN *	mg/kg	29.9

* Bulk density of 1.2 g/cm^3.

2.3. Crop Management

The previous crop was whole-cropped spring barley, with the prior rotation having been all cereals for 8 previous years. Additional nutrients were supplied in the form of slurry to remove this as a limiting factor to cover crop growth, using 35 m^3/ha of pig slurry (finisher) (Table A2). The slurry was applied on 3 August by tanker with a dribble bar attached and metered using a flow sensor. A Lemkin disc Heliodor disc (Alpen, Germany) was used to cultivate prior to each sowing date to a depth of 10 cm. Cover crops were then sown using a Wintersteiger plot sower (Essex, England) following the recommended sowing rates advised by RAGT Seeds (Wilson, 2018, personal communication) (A 4.1). Sowing dates (SD) were 14 August 2018 (SD 1), 7 September 2018 (SD 2), and 27 September 2018 (SD 3).

A Bomford flail (Worcestershire, UK) was used to mechanically mulch, and thus terminate, the cover crops on 28 February 2019. The plots were then mouldboard ploughed (Kverneland, Merseyside, UK) to 20 cm on 11 April 2019. A seedbed was created by using a power-harrow (parallel to the plots) (Kverneland, Merseyside, UK) and the spring barley variety KWS Irina was sown to establish 325 seeds/m^2 accounting for the TGW (thousand grain weight), its% germination and field losses. using a Wintersteiger plot sower on 19 April 2019. Plot size during cover crop growth was 1.68 m × 16 m and 1.68 m × 12 m yielded from the subsequent plots of spring barley. The spring barley received a spray programme to control weeds, pests and diseases, with 70 kg/ha of inorganic N applied (Table A3) using a Sissis high accuracy fertiliser applicator on 23 May 2019. Plots were yielded using a Sampo plot harvester combine (Pori, Finland) on 18 September 2019. Prior to harvest, plots were visually assessed and scored for percentage crop damage of lodging, leaning, necking and brackling, as well as chickweed growth through the spring barley, which was scored on a 1–9 scale (1 = the highest, 9 = the lowest).

2.4. Data Collection

2.4.1. Soil Mineral Nitrogen (SMN)

Initial soil samples from each block were sampled to a depth of 15 cm on 2 Augus 2018. On 25 February 2019, the control plots with nothing planted were sampled to 15 cn depth for soil mineral nitrogen (SMN). Fifty grams of soil sieved to 4 mm was mixed witl 100 mL of 2 M KCl, using an additional, 10 g of the soil sample to obtain the dry matte The mix of soil and KCl was shaken in an orbital shaker for 1 h at 200 RPM, centrifuged a 2970 g for 4 min and the liquid fraction filtered through No. 40 Whatman filter paper. Tw blanks were run with each set of extractions to determine and adjust for any contaminatio Soil N was transformed into kg/ha using a bulk density of 1.2 g/cm^3 and multiplying b the sampled depth of 15 cm.

2.4.2. Leaf Area Index (LAI)

Ceptometer readings were taken using an AccuPAR LP-80 (METER Group, Inc. Pul man, Washington, DC, USA) which calculated the leaf area index (LAI) using a mode (documented in the manual). Measurements were taken monthly, on the same sub-plot when weather conditions allowed (readings required dry and bright conditions resultin in a variation in measurement date). The LAI is an effective measurement which is quic and non-destructive to demonstrate crop growth, senescence and damage by frost ove time. The ceptometer measures light levels above and below crop and builds a LAI whicl is the amount of green cover (metres squared) per metre of ground area. Differences in th model's accuracy will exist between species as each cannot be assumed as homogenou in building this estimation of plant cover. However, it does provide a low-cost method c comparing crop growth over time using non-invasive techniques.

2.4.3. Biomass Sampling

Cover crop biomass was determined on 4 February 2019 using a 0.71 × 0.71 m quadra The brassicas (tillage radish and forage rape) had the roots extracted and washed under tap as this biomass was deemed to be a large proportion of the total biomass. The weed were separated from cover crop biomass. The roots were washed under a tap to remov soil. All biomass fractions were weighed, chopped to 4 cm using a stainless steel knife an 100 g subsamples taken. The subsamples were washed with deionised water, then dried a 60 °C for 48 h until a constant weight.

2.4.4. N and C Determination

Dried samples of the cover crop above-ground biomass, roots and weeds were mille to 1.0 mm using a Cyclotec 293 mill (FOSS, Cheshire, UK). N and C were analysed us ing the Dumas dry combustion method with a Trumac CN analyser (Leco Corporatior Michigan, USA) furnace temperature 1350 °C, with quality controls of an in-house verifie reference material run every 20 samples. Nutrient accumulation was calculated throug multiplication of the relative nutrient% by biomass. Nutrient accumulation of the weed was generated by multiplying the weed biomass of each plot by the average nutrien concentration of the weed biomass from the relative controls of SD 1, SD 2 and SD 3.

2.4.5. Energy Dispersive X-Ray Fluorescence (EDXRF)

EDXRF was used to measure a broad-spectrum nutrient profile in a select list of th best performing cover crops. Samples were milled to 1 mm (as described in Section 2.4.4 with 2.5–3.5 g (depending on species) loaded into sample cups to a depth >4 mm. T create a pellet, 300 PSI was applied for 20 s. A certified reference material (mixed Polis herbs INCT-MPH-2) was used in each batch of sample, allowing recoveries to be detecte and coefficient of variation (CV) to be gauged. Only recoveries of 100 +/− 20% with maximum CV of 10% were used as parameters to accept the specific nutrients from th profile measured. Nutrient uptakes were calculated through multiplication of cover cro biomass by its relative nutrient concentration. The nutrient accumulation of the weeds was not added on to the results shown.

2.5. Statistical Analysis

Genstat Version 18 [14] was used to analyse parameters of cover crop growth. Restricted maximum likelihood (REML) was used to analyse the cover crop nutrient accumulation and spring barley yield due to the unequal number of observations as REML produces predicted means.

The ceptometer measurements were analysed using REML as the monthly data were correlated. To account for unequally spaced measurements, the power model of order 1 was applied to the random component (Day Number). The fixed model included was Day Number + Sowing date + Species + Day Number × Sowing date + Day Number × Species + Sowing date × Species + Day Number × Sowing date × Species, and random components were Rep + Rep × Whole plot + Rep × Whole plot × Sub-plots + Plot × Day Number.

Fisher's unprotected post hoc analysis was applied to discriminate differences between species. Grain yield had covariates included in the REML analysis to adjust yield due to crop damage from chickweed, lodging and brackling which were scored on a plot basis prior to harvest. REML analysis of repeated measures was used to analyse the ceptometer readings of LAI with power model of order 1 applied to the random component of Day Number to take account of the correlation structure of the unequally spaced repeat measurements.

Results are deemed significant if probability due to random chance is under 5% ($p < 0.05$) and tendencies are regarded under 10% ($p < 0.10$).

3. Results

3.1. Leaf Area Index (LAI)

LAI was significantly affected by both sowing date ($p < 0.001$) and species ($p < 0.001$) and exhibited significant interaction ($p < 0.001$) (Table 2). The LAI indirectly shows the natural senescence and the degree the species were affected by frost. Species most affected by the frost/winter temperatures experienced include tillage radish and the phacelia as seen by the large LAI declines in SD 1 and SD 2 between November and December (Figures 1 and A2). Westerwolds exhibited the largest LAI, but westerwolds sown at SD 2 exhibited the second lowest LAI.

LAI on 16 October was higher in all species sown at SD 1, than from the two later sowings. By mid-November, all species, except westerwolds, sown on SD 2 had increased LAI/interception. By mid-December, LAI had decreased in all species sown on SD 1 and were similar to those of the species sown on SD 2. Changes in LAI between mid-December and mid-February of the first two sowing dates were relatively small, decreasing for most species. All species, when sown later (SD 3, 27 September 2018) had low LAIs.

Table 2. REML analysis of leaf area index (LAI).

Treatment	N.D.F *	Chi Probability	LSD
Day Number	3	<0.001	0.248
Sowing Date	1	<0.001	0.41
Species	5	<0.001	0.308
Day Number × Sowing Date	3	<0.001	0.494
Day Number × Species	15	<0.001	0.615
Sowing Date × Species	5	<0.001	0.576
Day Number × Sowing Date × Species	15	<0.001	1.061

* Number of degrees freedom.

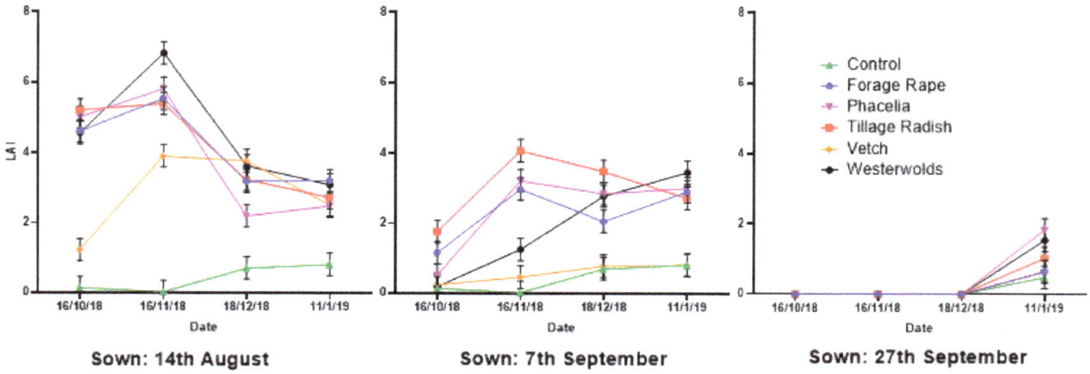

Figure 1. Leaf area index (LAI) of the three different sowing dates for each cover crop measured at the monthly dates. Error bars represent the standard error of the mean (SEM). Early-sown crops were more affected by frosts than later sowing. N = 8 for each mean.

3.2. Biomass Production

Figure 2 shows the fractions of biomass produced by the cover crops at the different sowing dates. The letters show Fisher's unprotected post hoc LSD (0.05) for total production inclusive of cover crop, roots and weeds. Total biomass was affected by sowing date ($p < 0.001$) and species ($p < 0.001$), and exhibited a significant interaction ($p < 0.001$) (Table A4). Biomass production in all species decreased with later sowing. The extent of the decrease varied with species, hence the significant interaction. Tillage radish and forage rape produced the greatest overall biomass at SD 1 ($p < 0.05$) (Figure 2), with tillage radish producing 6447 kg/ha DM including roots and the forage rape producing 6026 kg/ha including roots.

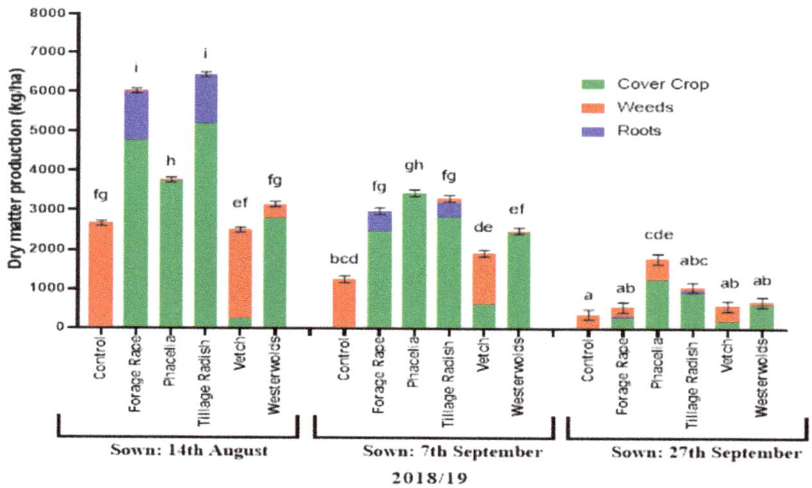

Figure 2. A stacked bar chart to show the biomass produced from above ground cover crops, their roots and also the weeds which grew in each treatment. Error bars represent the standard error of the mean (SEM). Means followed by a common letter are not significantly different by Fisher's unprotected post hoc least significant difference (LSD) at the 0.05% level of significance which equals 884.7. N = 8 for each mean.

Between SD 1 and SD 2, phacelia did not display a significant reduction in biomass—3732 and 3464 kg/ha (excluding weeds), respectively. At SD 2 and SD 3, of all the species, phacelia produced the greatest levels of biomass. At SD 3, phacelia produced 1830 kg/ha, almost two-fold more biomass than tillage radish (967 kg/ha). The vetch was the lowest biomass-producing cover crop at any date. Delaying planting from SD 1 to SD 2, numerically increased the vetch biomass from 302 to 728 kg/ha. Vetch and westerwolds produced similar levels of biomass at both SD 1 and SD 2, but the vetch produced significantly less cover crop biomass and significantly more weeds. It must also be noted that late-sown forage rape was the only species that exhibited crop damage caused by pigeons, which reduced biomass.

3.3. Roots

As would be expected, root biomass was substantial in tillage radish and forage rape, hence its measurement in these species. Total roots biomass in these species was affected by sowing date. SD 1 led to the largest accumulation of over 1200 kg/ha of each species, which declined to 429 kg/ha for tillage radish and 587 kg/ha for forage rape at SD 2 to less than 100 kg/ha of root biomass recorded at SD 3, for either species.

3.4. Weeds

Overall weed growth (measured from control plots) was 2744, 2271 and 717 kg/ha at SD 1, SD 2 and SD 3, respectively. Weed growth was affected by sowing date ($p < 0.05$), where later sowing reduced weed biomass found in the cover crops (Table A4). This declined on average from 534 to 285 kg/ha and finally to 211 kg/ha at SD 1, SD 2 and SD 3, respectively. Vetch was the only species that did not suppress weeds ($p < 0.05$). A significant interaction between sowing date and species was found ($p < 0.001$) (Table A4).

3.5. Cover Crop% N

N concentration (%) was not affected by sowing date, but species exhibited a significant difference ($p < 0.001$) and a significant interaction ($p < 0.01$) (Table A4). Although sowing date was not significant, a significant interaction ($p < 0.01$) with species was exhibited (Table A4.) This suggests that there must have been cross over of means, caused by treatments, as outlined by Grace-Martin [15]. The control exhibited a significantly lower ($p < 0.05$) N concentration (2.6%) in comparison to all other species (means not presented). The vetch had the highest% N of 5.15% when planted on SD 1 but when planted at the latest sowing date it did not have any significantly higher% N than any of the controls of just weeds.

3.6. N Accumulation (Biomass x% N)

SD 1 resulted in the largest N accumulation ($p < 0.001$) for all species, including the control (Table A4). A maximum accumulation of 261 kg N/ha occurred in tillage radish, with forage rape accumulating a similar amount of 255 kg N/ha. At SD 2, tillage radish accumulated 162 kg N/ha, which was the largest at that date compared to the control which accumulated the least N (25 kg/ha). At SD 3, phacelia outperformed all other cover crops and accumulated 70 kg N/ha, whilst N uptake in tillage radish and forage rape decreased further than the other species. When weed N accumulation was subtracted from the vetch N accumulation, this species only accumulated 18 kg N/ha at SD 1, 25 kg N/ha at SD 2 and 6 kg N/ha at SD 3, further reflecting the low biomass production seen in Figure 3.

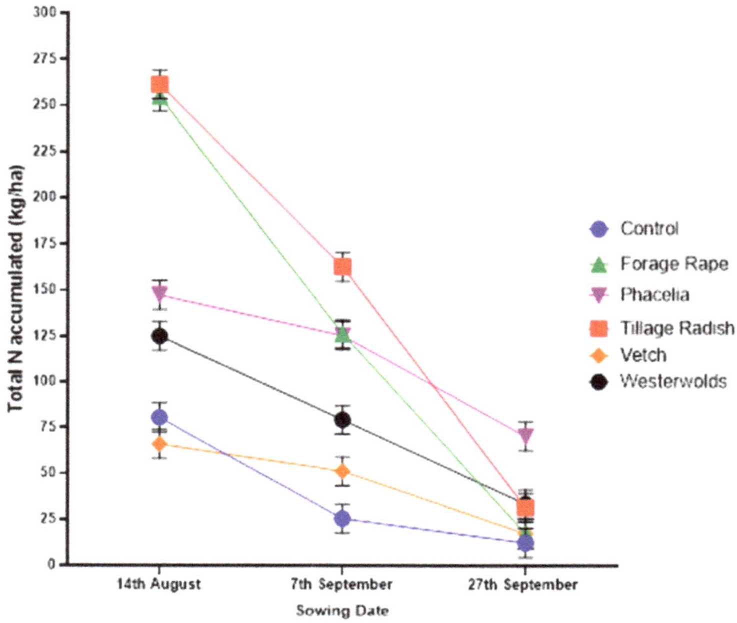

Figure 3. N accumulation for each treatment of cover crop which includes the above ground biomas cover crop roots and weeds produced for each of the three sowing dates. Error bars represent th standard error of the mean (SEM). Fisher's unprotected $LSD_{0.05}$ = 41.0. N = 8 for each mean.

The control accumulated over 80 kg/ha of N at SD 1, 25 kg/ha on SD 2 and 12 kg/h on SD 3. This reflects a clear reduction in weed growth with delayed sowing. The amoun of N accumulated in the control at SD 1 is atypical, as normal practice would not allo these weeds to produce viable seeds. Alternatively, a herbicide or additional cultivatio would be used to destroy them. However, at later sowing, the weeds did not mature an the amount of weed N was higher in the phacelia sown at date 3 than the total weed N o the relative SD 3 control (14.6 versus 12.3 kg/ha). This shows a potential complementar competitive effect.

Sowing date significantly affected root N uptake ($p < 0.001$). At SD 1, tillage radis accumulated 48 kg N/ha compared to forage rape at 37 kg/ha. At SD 2, this declined t 15 kg/ha and 18 kg N/ha, respectively, and at SD 3 a maximum of 5 kg N/ha was detecte in tillage radish.

3.7. Carbon (C) Accumulation

C accumulation exhibited significant ($p < 0.001$) differences in sowing date, species an a significant interaction ($p < 0.001$). Tillage radish and forage rape accumulated the greates total C of 2359 and 2361 kg/ha, respectively (Figure 4). Phacelia exhibited the greates growth at SD 3, which resulted in the largest C accumulation (Figure 4). C accumulation i the roots was only affected by sowing date ($p < 0.001$) due to no significant difference in th species average concentration. The controls produced the least C at each sowing date.

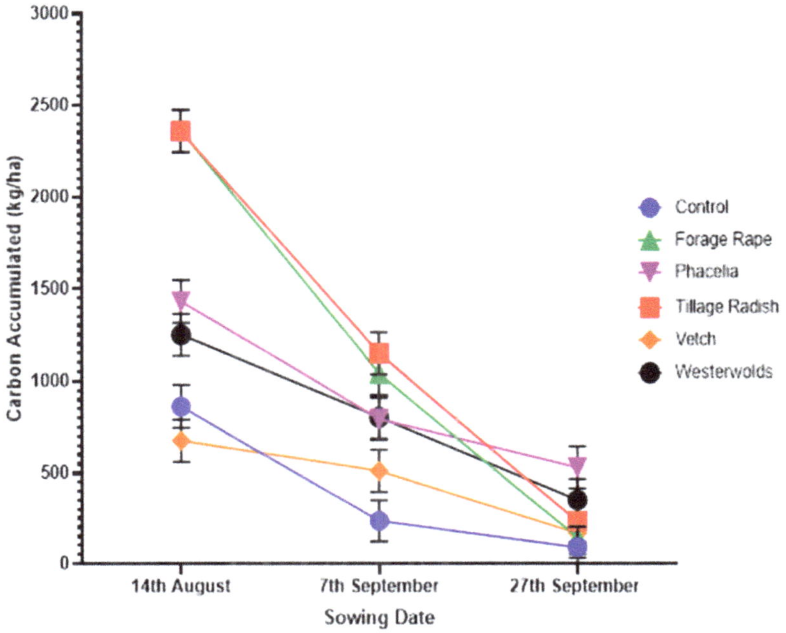

Figure 4. Carbon (C) accumulation for each treatment of cover crop which includes the above ground biomass, cover crop roots and weeds produced for each of the three sowing dates. Error bars represent the standard error of the mean (SEM). Fisher's unprotected $LSD_{0.05} = 327.7$. N = 6 for each mean.

3.8. C:N Ratio

SD 1 had a significantly higher C:N ratio than SD 2 or SD 3 ($p < 0.05$) (Table 3). C:N ratio was significantly different ($p < 0.01$) between species, with the tillage radish having the lowest C:N ratio of all species ($p < 0.05$). The C:N ratio exhibited a low range varying between 8.0 to 10.7.

Table 3. C:N ratio of cover crop biomass at the different sowing dates (excludes roots).

Sowing Date	Control	Forage Rape	Phacelia	Tillage Radish	Vetch	Westerwolds	Sowing Date Average
1	10.6	8.9	10.8	9	8	10.4	9.6 [b]
2	10.5	8	6.4	6.9	8.4	8	8.0 [a]
3	11	9.1	7	8.1	8.5	9.3	8.8 [a]
Species Average	10.7 [c]	8.7 [ab]	8.1 [ab]	8.0 [a]	8.3 [ab]	9.2 [bc]	8.8
			Statistical analysis				
	Treatment		*p*-value	SEM *	LSD #		
	Sowing Date		<0.001	0.494	0.912		
	Species		<0.01	0.946	1.285		
	Sowing Date × Species		0.06	0.489	2.205		

* Standard error of the mean. # Least significant difference (0.05) Means which do not share the same letters are significantly ($p < 0.05$) different to each other. N = 6 for each mean.

3.9. Macro-Nutrient Assimilations (P, K, and S) of Cover Crops

Macro-nutrient assimilation values are presented in Figure 5, and whilst Table 4 shows the significance of treatments applied. Effects of sowing date and species exhibited significant ($p < 0.001$) differences for P, K, and S, and all interactions were significant ($p < 0.001$).

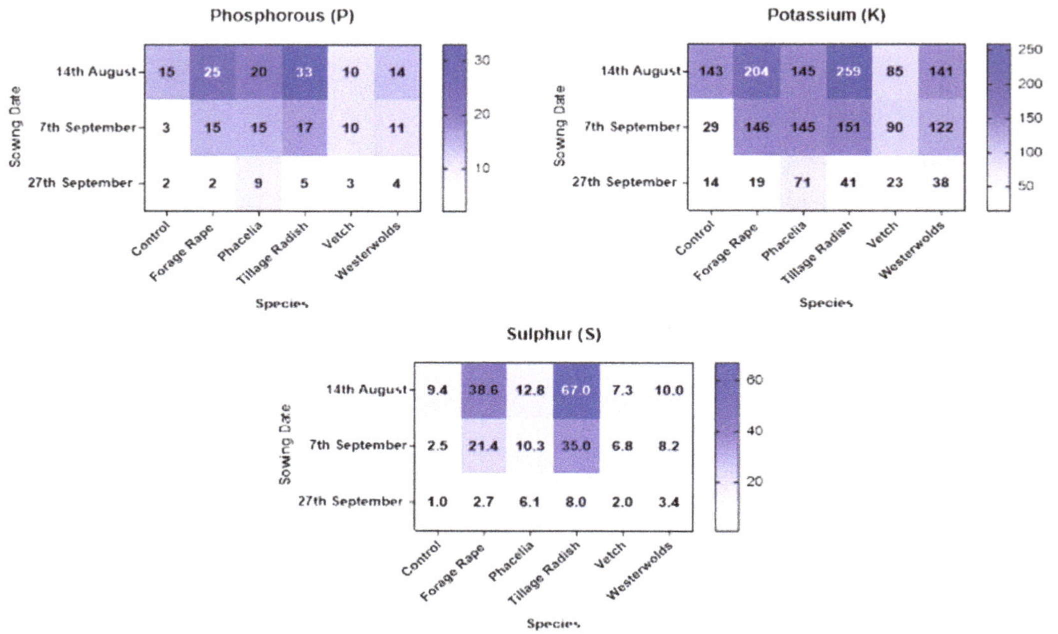

Figure 5. Nutrient assimilation of P, K and S of the cover crops in response to sowing dates (kg/ha).

Table 4. *p*-values for the nutrient assimilations for the treatments of sowing date, species and their interaction.

	P Uptake (kg/ha)			K Uptake (kg/ha)			S Uptake (kg/ha)		
	Chi-Prob	SEM *	LSD #	Chi-Prob	SEM	LSD	Chi-Prob	SEM	LSD
Sowing Date	$p < 0.001$	0.9	2.7	$p < 0.001$	9.3	22.6	$p < 0.001$	1.12	2.97
Species	$p < 0.001$	1.0	3.8	$p < 0.001$	9.3	32	$p < 0.001$	1.19	4.14
Sowing Date × Species	$p < 0.001$	0.9	6.6	$p < 0.001$	9.0	55.4	$p < 0.001$	1.07	7.16

* Standard error of the mean. # Least significant difference.

Tillage radish at SD 1 accumulated the greatest P (33 kg/ha), K (259 kg/ha), and S (67 kg/ha), demonstrating that this species has the largest potential to accumulate nutrients when sown early (Figure 5). The slurry added 50 kg/ha of P. In comparison to the control, the tillage radish accumulated almost two-fold more P, with the tillage radish accumulating 25 kg/ha P, which is half of what the slurry added. At SD 2, tillage radish, forage rape and phacelia accumulated 94%, 91% and 91%, respectively of the 160 kg/ha of K added by the slurry. However, at SD 1, tillage radish and forage rape accumulated 99 and 44 kg/ha more K than supplied by the slurry. The high concentration of S in the tillage radish and its large biomass resulted in an uptake of 67 kg/ha S being accumulated. This is three-fold greater S than added by the slurry. At SD 2, which represents sowing after a normal harvest of cereal crops, there is less variation in the uptake between tillage radish, forage rape and phacelia. However, at SD 3, phacelia accumulated the greatest levels of P, and K but not S.

3.10. Grain Yield

At harvest of the spring barley, many plots exhibited excessive chickweed growth due to partial resistance to the herbicides used. This could have affected grain yield. Therefore, the chickweed was visually scored prior to harvest on a plot basis, along with lodging, leaning and brackling. Grain yield was analysed using REML with the covariates of lodging, leaning and brackling to produce predicted mean grain yields. Chickweed was the only covariate with a significant effect ($p < 0.001$) on grain yield. The REML analysis using the covariates found that grain yield was unaffected by any sowing date or species (Table 5) and yields varied between 6.9 and 7.9 t/ha (85% DM). Numerically, barley, following phacelia, exhibited the highest species average yield and the highest individual yield when sown at the latest sowing date. Westerwolds sown at SD 3 exhibited a 1 t/ha lower grain yield than phacelia. All sowing dates had similar mean grain yields.

Table 5. 2018/19 spring barley yield (t/ha) and REML analysis.

Sowing Date	Control	Forage Rape	Phacelia	Tillage Radish	Vetch	Westerwolds	Sowing Date Average
1	7.52	7.28	7.41	7.5	7.51	7.67	**7.48**
2	7.43	7.34	7.45	7.31	7.44	7.54	**7.42**
3	7.41	7.61	7.94	7.32	7.72	6.88	**7.48**
Species average	**7.45**	**7.41**	**7.6**	**7.38**	**7.56**	**7.36**	**7.46**
			REML Analysis + covariates				
	Parameter			***p*-value**	**SEM #**	**LSD ***	
Covariate	*Chickweed*			**<0.001**			
Covariate	*Leaning*			0.94			
Covariate	*Lodging*			0.42			
Covariate	*Brackling*			0.11			
Variate	Sowing Date			0.85	0.106	0.403	
Variate	Species			0.81	0.108	0.412	
Variate	Sowing Date × Species			0.47	0.104	0.778	

Standard error of the mean. * Least significant difference. N = 4 for each mean.

4. Discussion

4.1. Grain Yield

The success of cover crops can be judged on the grain yields produced, due to its economic importance. However, the grain yield was not significantly affected by any of the treatments or combinations in this single year study, which would question the value of using cover crops. The grain yields exhibited are all regarded as very high for spring barley [16,17]. A second year's replication of the trial was attempted but adverse weather of high rainfall during the planned sowing dates meant that it was impossible to initiate because the soil was not trafficable.

The lack of difference in grain yield may be due to an oversupply of N which masked effects between treatments. Prior to the planting of the cover crops (August 2018), an average of 60 kg/ha of inorganic N was found in the 15 cm soil profile and 263 kg N/ha (total N) supplied from slurry and the additional 70 kg N/ha in the form of inorganic fertiliser. This meant that N could have been oversupplied causing the lack of differences. In this trial, winter rainfall was low (Figure A3) and could have retained more N in the soil and from the slurry. Table A5 shows the SMN of control plots (fallow) sampled on 25 February 2019 post-winter and prior to flailing. This ranges from 9.5 to 24.4 kg N/ha (15 cm profile), which shows a considerable decline between the initial (August) SMN and the post-winter (February) measurement. This decline could be a combination of leaching/loss of N or immobilisation into the soil. Based on the February assessment of SMN and using a typical estimation for N availability from slurry, was why an extra an additional 70 kg N/ha was applied to the spring barley. A study by White et al. [18] investigating a range of applications of N applied in the autumn found that, in NI, SMN tested in the spring was not a good predictor of soil N supply. Retrospectively, this should not have been applied (in this year). This may have been due to sufficient rainfall and soil temperatures that promoted a high rate of N mineralisation in the soil which, in turn, supported the N requirement of the spring barley. However, White et al. [18] found that application of 800 kg N/ha in the autumn in NI still required additional N to be applied in the spring/summer growing season to maximise the yield of wheat as a result of the N losses and immobilisation.

The different cover crop residue quantities, and qualities, can add variability to nutrient supply of the commercial crop [10] as increasing C supply to soil microorganisms can immobilise nutrients in the microbial biomass and compete with roots for N [19,20]. This can reduce commercial crop yield [21]. This trial suggests two reasons why cover crop are not immobilising N in the conditions applied, and that the residue is mineralising at sufficient rate to supply the spring barley with adequate nutrients.

1. No significant increase in yield was found, in any treatment, in response to th 70 kg/ha of inorganic N applied.

2. The sowing dates produced various quantities of biomass, whereby increase biomass could have increased immobilisation.

Slurry was applied prior to sowing to ensure that nutrients were not a limiting facto in order to maximise cover crop biomass, and thus identify nutrient uptake. The rational of the treatment design implemented, e.g., applying slurry, was that it was thought tha the cover crops were going to have low N mineralisation rates with potential N immobi isation, particularly the brassica species as found by Couëdel et al. [11]. Findings fror this experiment and Cottney et al. [13] show a considerable breakdown of nutrients fror the cover crops, with the ability to replace all inorganic N required by spring crops. I this trial, N uptake in the cover crops varied from 20 to 261 kg/ha, whereas the contrc had 17 kg N/ha on average in February. Assuming grain N of 1.5% and that the grai accounts for 80% of N accumulated, a spring barley crop with a grain yield of 7 t/ha an 8 t/ha (15% moisture content) would require 111 and 127 kg N/ha, respectively. If SMN from the February control (only 15 cm) plots' average is subtracted, this means the barle requires an estimated 94 and 110 kg N/ha. This suggests that the spring barley acquired N from sources other than the fertiliser and residue, especially in the control plots of SD which accumulated 12 kg N/ha in the weed residue. Therefore, the 70 kg of inorganic N should not have been applied to the spring barley. However, verification of this decisio would have required a treatment of plots with a zero rate N fertiliser regime as well a the 70 kg/ha of inorganic N. Increasing the reliability of these results would require mor replication across different years and particularly at different sites, as this was a site c high fertility.

4.2. Biomass Yield

To maximise the area sown to cover crops requires species which produce adequat growth when sown late. This gives farmers the assurance that they are not wastin their resources. This research demonstrates that, at SD 3, phacelia produced the greates biomass and was the best-suited species in that slot. At SD 1, a maximum biomass growt of 6447 kg/ha DM was recorded from tillage radish, which reduced to 307 kg/ha DM following the final SD. When cover crops are grown as a livestock feed, the cost of forage i proportional to the level of growth and utilisation of that crop, thus favoring high biomas to dilute costs and make the crop profitable. The quantity of biomass produced fror the forage rape was 6062, 3346 and 307 kg/ha following SD 1, 2 and 3, which is hig in comparison to a trial conducted in Ireland by Keogh et al. [3]. They found that th maximum dry matter yield of forage rape (cv. Stego) when sown on the 1 August wa 4548 kg DM (shoot + root DM), declining to 3047 kg when planted on 15 August an 1091 kg DM on 31 August. All had been supplemented with 120 kg N/ha. The othe species investigated in that experiment was stubble turnips (cv. Delilah) which exhibite larger biomass yields at the latter two sowing dates. Furthermore, only the late-sown (SD 3 forage rape was damaged considerably by pigeons. It is, therefore, unsuitable when sow late. This phenomenon was only observed in late-sown plots and is presumed to be due t a combination of production of a dense canopy when sown at date 1 and 2. Consequentl pigeons could not land to graze.

4.3. N Accumulation

Tillage radish accumulated 261 kg N/ha at SD 1, which is almost twice the N re quirement for a 6 t/ha spring barley crop [22]. This large accumulation was driven b high biomass and high% N. Phacelia, forage rape and tillage radish contained over 4% N, whereas a study by Wendling et al. [23] found that phacelia, tillage radish (daikor and turnip rape (forage rape) had a N concentration of 2.13, 2.22 and 2.04%, respectively. The higher% N in this trial reflects the high levels of N contained in the soil this trial was conducted on. Moreover, these cover crops can increase concentration of N in response to additional nutrients (slurry) [13]. This is a mechanism to increase N uptake and reduce the C:N ratio. Wendling et al. [23] reported biomass yields of 6.3, 6.3 and 4.4 t/ha and N uptakes of 120, 139 and 132 kg/ha for phacelia, tillage radish and forage rape, respectively. In this trial, tillage radish, phacelia and forage rape all accumulated considerably higher quantities of nutrients despite having similar biomass. This is due to the considerably greater concentration of N. This study found that vetch had the greatest N concentration when planted early (5.16%), which declined in response to delaying sowing to 2.94 and 3.09% at SD 2 and SD 3, respectively. In comparison, Lawson et al. [8] found that delayed sowing did not affect N concentration in the hairy (winter) vetch (variety not stated) and in comparison,% N averaged 4.1% and the monoculture was 31% weeds with a biomass of 1.4 t/ha. The species' average% N was considerably higher than many other studies [8,24]. This may be due to high residual N with the species increasing their relative N concentration in response to the additional nutrients in slurry, thus creating a luxury uptake.

The vetch, a legume, should have fixed additional N into the rhizosphere. The process and productivity of N fixation is not only temperature-dependent but also relies on bacterial infection of the roots. This occurs 3–4 days post-germination and takes 3–5 weeks to produce visible and active root nodules [25], which means that delayed sowing would highly effect the vetch's ability to biologically fix N. Li et al. [26] estimated that legumes fix 24 kg of N per tonne of biomass produced. Extrapolating those findings means that in this study only moderate levels of N could have been fixed, as only 242, 638 and 189 kg/ha of biomass was produced from SD 1, SD 2 and SD 3, respectively. However, this study underestimates the N contained in the roots as they were not considered. This has been found to equate to 30–50% of plant N [26].

The N in the taproots of the brassicas was evaluated but, again, it does not account for minor roots as well as N rhizodeposits through sloughed-off root hair cells, N in root exudates and root fragments which have been found to account for an additional 4.6–10.3% of total plant N for brassicas (tillage radish, winter turnip rape and oilseed radish [27]. This means that the N accumulations have been underestimated due to difficulty in accurately extracting these N rhizodeposits under field conditions. Furthermore, at SD 1, there will be a greater proportion of N rhizodeposits because the biomass in the roots declined with delaying sowing, as seen in Figure 2.

Tillage radish accumulated 261 kg N/ha, which is considerably more N than any spring commercial crop requires. Furthermore, it also contained 260 kg of K and 67 kg of S and is thus a considerable bio-fertiliser. However, this uptake of macro-nutrients could diminish soil nutrient availability in the spring and affect commercial crop yields, but would depend on nutrient mineralisation rate. Therefore, subsequent trials must investigate immediate impact on soil fertility. The cultivated control accumulated 80 kg N/ha in the weeds at SD 1 and is arguably not representative of farmer practice as it would have been destroyed with herbicides to avoid weeds producing viable seeds. Therefore, it is an overestimate of what typically fallow land would have accumulated. The Nutrient Management Guide RB209 [22] estimates soil residual N following harvest of various crops and rotations, where a SNS of 70 kg/ha would be relatively high. This means that phacelia sown late (accumulating 70 kg/ha at SD 3) has the potential to deplete these SNS reserves and could be beneficial to mitigate against N leaching. Therefore, the functions of cover crops at later sowing dates would change to more predominantly environmental considerations of N accumulation to reduce leaching and physical soil protection, due to numerous other benefits being linked to biomass production, such as effect on soil biology [28].

The decline in both biomass and N uptake of the roots at later sowing of the tillage radish and forage rape suggests that this might reduce their ability to "biodrill"—a term referring to the ability of roots to grow through a plough pan (layer of compacted soil) to enhance soil structure [29]. This is due to the number of roots being similar, since the same number of seeds were planted, although individual root size has been reduced. This could be detrimental to growing deep or exerting a positive effect on the soil profile, as thicker roots can exert higher penetrative pressure [30,31].

4.4. C:N Ratio

The% N of species in this experiment was high. This led to a low C:N ratio, especially in comparison to other studies. For example, forage rape had a C:N ratio of 22 when sown as a sole crop by Couëdel et al. [11], where tillage radish (same variety) and phacelia had a C:N ratio of 18.6 and 20.6 reported by Wendling et al. [23]. In this trial, the highest C:N ratio was 10.0 in the control, with all cover crop species having lower C:N ratios. This suggests that the plant nutrients will break down quickly, leading to a net mineralisation [12]. This may have been observed, as the different sowing dates producing various quantities of biomass and thus N uptake did not negatively influence spring barley yield. In subsequent trials, the N offtake of spring barley must be evaluated to help identify effects. Couëdel et al. [11] found that N mineralisation to the commercial crop ranged from −6 kg/ha for mustard to 20 kg/ha in the forage rape (same variety). Silgram and Harrison [10] estimated that cover crops with a C:N ratio below 25–30 are required for net mineralisation in year one. However, Couëdel et al. [11] concluded that a threshold of below 15 was required for net mineralisation within 6 months.

The cover crops in this research were destroyed as early as weather permitted and, if delayed to a later date, C:N ratio would have increased due to greater amounts of structural compounds within the plant. This could reduce the rate of release of nutrients from the residue, and decrease the transfer of nutrients to the following commercial crop, and thus increase the N requirement for that crop [32]. The C:N can be modified by species choice and supply of supplementary nutrients in the form of slurry, as found by Cottney et al. [13], meaning that nutrient mineralisation can be manipulated. Jensen et al. [33] found that N released from plant residues after 217 days reached a maximum of 40% for those with a C:N below 10. When applied to this study, 104 kg of N would have been released from the tillage radish on SD 1.

4.5. C Accumulation

Increasing soil organic matter levels is beneficial not only for soil functions including porosity, biological activity, nutrient retention and soil structure but can also to help offset anthropogenic carbon dioxide emissions (CO2) [34]. This study shows that early sowing is paramount for returning C. Forage rape returned over 2361 kg C on SD 1 equivalent to 8.66 t/ha CO^2 [35] despite forage rape producing less biomass than the tillage radish. Forage rape had a higher% C than tillage radish which resulted in the larger C accumulation. In comparison, the control accumulated the lowest C in biomass. This demonstrates the environmental benefit of cover crops compared to fallow. Alternative ways to return organic material to soil is to incorporate the commercial crop straw but this comes at a cost when this material is a commodity. Average straw yields for spring barley in 2017 and 2018 was 3.5 t/ha and 5 t/ha at 46% C (determined in a prior experiment) which means a return of 1.6 t/ha and 2.3 t/ha of C. This organic material will have a higher C:N ratio and will be slower to decompose, which may be better to improve long-term organic matters [36,37]. However, the typically high straw prices in regions such as NI contribute considerably to gross margins [38], whereby incorporating straw as a method to return organic matter is not justified financially/economically and that cover crops shown are just as effective at returning similar levels of C to the soil. This study does not account for additional sources of C that the cover crops are returning which originate from root exudates. They can be a significant proportion of photosynthetic C transformed into plant and microbiome usable products as Swinnen et al. [39] found in wheat and barley that rhizodeposited C represented 7–15% of total C assimilated.

4.6. P, K, and S Uptake of Cover Crops

The slurry applied 50 kg/ha P (90% availability) [22], and tillage radish accumulated 33 kg/ha of P at SD 1. This is a substantial uptake of the added nutrients which could have considerable environmental benefits. In comparison, phacelia at SD 3 accumulated 9 kg/ha P, and had the greatest accumulation at this date. Whilst nutrient loss from groundwater was not measured, trapping nutrients in biomass is a mechanism of protection against loss from the system [6]. Furthermore, on low nutrient index soils, the cover crops could sequester nutrients and cause a negative effect on the growth of the commercial crop due to nutrient competition. Couëdel et al. [40] found a maximum S uptake of 23 kg/ha for tillage radish whilst forage rape (mosa) had an uptake of 17 kg/ha, whereas this study found that the same species exhibited almost a three-fold greater uptake. Slurry supplied 24 kg/ha of S, which suggests that tillage radish can effectively sequester the S applied. Early-sown tillage radish and forage rape accumulated large amounts of K which could either enhance nutrient cycling if this mineralises quickly, or could cause competition for K in the spring barley crop. Slurry added 160 kg/ha K, whereas tillage radish accumulated 260 kg/ha thereby facilitating the fulfilment of the cover crop requirement and replenishing the soil. Soil tests post-cover crop or commercial crop were not conducted but investigating this could be important to determine whether cover crops facilitate nutrient cycling.

4.7. Weed Suppression

Weed pressure decreased naturally with later sowing and was observed through diminished biomass in the control with later planting. At SD 1 and SD 2, tillage, forage rape and brassicas exhibited almost total weed suppression. This demonstrates that these species, and in particular, forage rape could be a viable alternative to the chemical control of weeds. Forage rape at SD 1 and SD 2 was unaffected by the frost, retaining its canopy and was thus able to shade weeds and provide almost total weed suppression. This is observed in the final ceptometer reading, whereby forage rape had the highest LAI in comparison to the other species. The weed suppression in phacelia and tillage radish was high due to their profuse growth, also found by Brust et al. [9]. LAI was only measured on one date for SD 3 for two reasons. Firstly, the ceptometer uses two different sensors to measure incoming radiation. This can create an error which increases in plots with sparse biomass. Secondly, it was not possible to measure very low biomass as the wand could not get under its canopy, creating a null measurement.

If planting for weed suppression, the best cover crops are tillage radish, phacelia, forage rape and westerwolds. This is of particular importance in organic systems. However, weeds at SD 3 could be regarded as beneficial to trap more nutrients and add to soil structure protection due to growing roots anchoring the soil.

4.8. Recommendations to Maximise and Encourage Later Sowing of Cover Crops

Cottney et al. [41] found that a lower proportion of farmers considered planting cover crops after commercial crops harvested in September. This reduces the amount of land sown to cover crops. By planting later, through using better suited species as identified in this study, would reduce the amount of land left fallow over winter and could mitigate against loss of nutrients such as N, and also provide many more benefits. This study has found that species choice when sowing later is critical. Another strategy to improve species competitiveness when late sown, is to increase seed rate. This was not investigated in this trial but could be implemented on farm to enhance growth. Other strategies include using a mixture of species to encourage competition (also referred to as over-yielding) which was demonstrated in a study by Wendling et al. [42]. It was also found in this trial that the N fractions in the late-sown phacelia contained more weed N than the relative control covered in weeds. However, where supplementary N is applied, increased competition from a mixture is diminished in comparison to the sole species alone [42]. Mixtures were not investigated in this trial due to the exponential number of combinations and seeding rates, whereas testing the individual species is of greater importance. This study recommends, that if using a mixture of species or sole crop (one species) in planned rotations wher harvest dates are after the start of September, to ensure that phacelia makes up a larg proportion of that mix, or is the sole crop to be used.

Another strategy to encourage later sowing is to adopt subsidies. In the Republic o Ireland, with similar climatic conditions, which subsidises the practice of cover croppin growers were more likely to plant after later-sown cereal rotations such as winter whea and spring barley compared to those in NI [41]. The aims of the farmers, and species usec changed from being focused on grazing by livestock to a primary focus on soil structure soil health and capture leachable nutrients. Therefore, subsidies could be adopted in N especially as this trial has demonstrated that cover crops can capture considerable nutrient in these climatic conditions.

5. Conclusions

Spring barley yield was unaffected by later sowing of cover crop species. Phaceli competed considerably better than all other species when late sown, accumulating 70 kg/h of N with a considerable biomass. This is a large amount of N that is protected over winte in comparison to fallow. Furthermore, the N accumulation by phacelia is almost hal the requirement of a subsequent spring barley crop, if the N in the biomass mineralise sufficiently. Early-sown tillage radish and forage rape accumulated over 250 kg N/h which is almost twice what spring crops require. This further highlights how unproductiv fallow land is and that modern agriculture cannot close the nutrient cycle status quo.

Author Contributions: Conceptualization, P.C., P.W., L.B. and E.W.; Investigation, P.C.; Writing–Original Draft, P.C.; Writing—Review and Editing, P.C., P.W., E.W. and L.B.; Visualization, P.C Resources, L.B. and E.W.; Supervision, P.W., L.B. and E.W. All authors have read and agreed to th published version of the manuscript.

Funding: This research was funded by DAERA grant number (17/1/01).

Institutional Review Board Statement: Not applicable.

Informed Consent Statement: Not applicable.

Data Availability Statement: Not applicable.

Acknowledgments: The authors would like to thank all AFBI staff involved in helping undertak this trial, in particular, Colin Garrett Ashley Cathcart and Gary Weir as well as all the farm sta who helped with setting up the trial. Seed was kindly provided by RAGT Seeds. The authors ar also grateful to the AFBI statistical analysis branch and acknowledge the guidance and analysi conducted by Michelle Allen (AFBI, Newforge), and likewise, Hugh McKeating and Alan Wrigh (AFBI, Newforge) of the soils branch for analysing soil and plant samples.

Conflicts of Interest: The authors declare no conflict of interest. The funders had no role in the desig of the study; in the collection, analyses, or interpretation of data; in the writing of the manuscript, o in the decision to publish the results.

Appendix A

Table A1. Cover crop, varieties, sowing rates and plants sown/m^2.

Species	Variety	Recommended Sowing Rate (kg/ha)	Chosen Sowing Rate (kg/ha)	Plants Sown/m^2
Tillage Radish	Daikon	25–30	25	177
Forage rape	Mosa	10	10	278
Phacelia	Natra	10	8	462
Vetch (Hairy)	Villana	25–30	25	85
Westerwolds	Magnum	40–46	40	763
Control/fallow	-	-	-	-

Table A2. Nutrient concentration of slurry on a fresh basis and quantity (kg/ha) of nutrients applied.

Parameter	K	P	S	Mg	NH_4^+	Total N	DM
	mg/kg	mg/kg	mg/kg	mg/kg	%	%	%
	4583.5	1438.5	671	874	0.5	0.8	7.8
			Nutrient application (kg/ha)				
Rate 35 m^3/ha	160.4	50.4	23.5	30.6	176.2	263.7	

Nutrient concentrations are reported on a fresh basis.

Table A3. 2018/19 Spring barley agronomy and crop protection.

Date	Active Ingredient	Chemical	Reason	Rate	Manufacturer
	Lambda-cyhalothrin	Warrior	Aphids	50 mL/ha	Syngenta
28 May 2019	Manganese 500 g/L	Mantrac	Manganese Trace Elements	1 lt/ha	YARA
	Chlorothalonil	Bravo	Disease	1 lt/ha	Syngenta
	Mecoprop-P	Headland Charge	Weeds	2 lt/ha	Headland Agrochemicals Limited
	Metsulfuron-methy + tribenuron-methy	Ally Max ® SX	Weeds	42 g/ha	Dupont
	Prothioconazole + bixafen	Siltra Xpro	Disease	0.6 lt/ha	Bayer
6 June 2019	Trinexapac-ethyl	Moddus	Growth—Regulator	0.2 lt/ha	Syngenta
	Chlormequat chloride	3C Chlormequat 750	Growth—Regulator	1 lt/ha	O-BASF
	Manganese 500 g/L	Mantrac	Manganese Trace Eements	0.6 lt/ha	YARA
2 July 2019	Prothioconazole + bixafen	Siltra Xpro	Disease	0.6 lt/ha	Bayer
	Chlorothalonil	Bravo	Disease	1 lt/ha	Syngenta
		Fertiliser application			
Date	**Rate**	**Product**	**Manufacturer**	**Nitrate—N**	**Ammoniacal—N**
23 May 2019	70 Kg N/ha	Yara Can (27%)	YARA	13.5%	13.5%

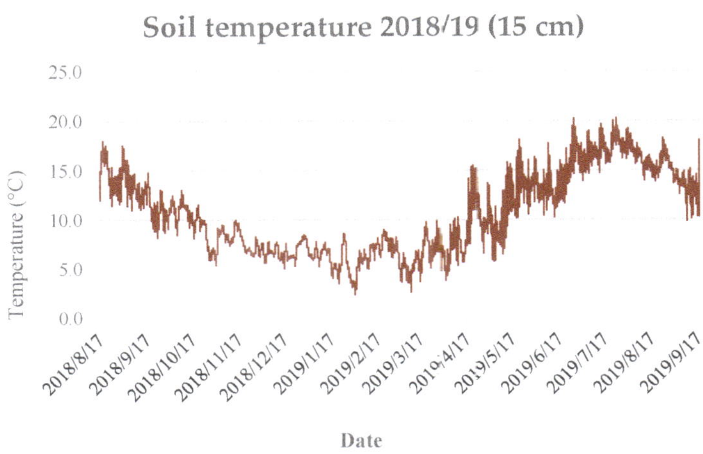

Figure A1. Recorded soil temperatures during both the cover crop and spring barley growth (°C).

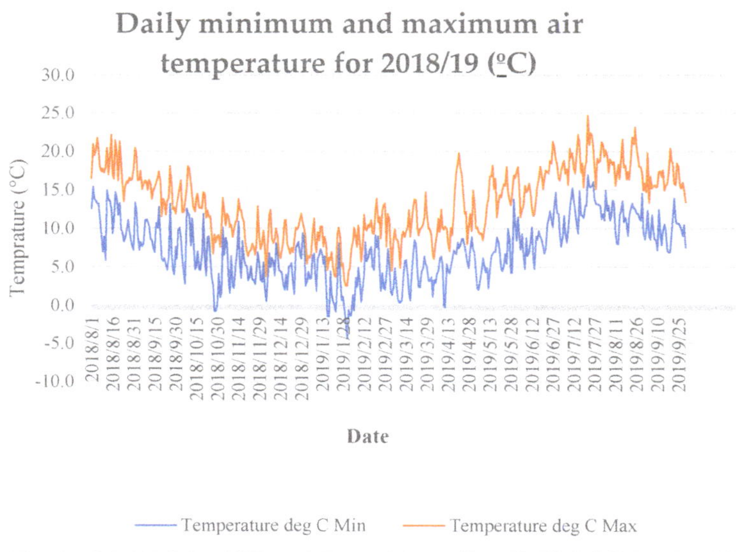

Figure A2. Recorded daily maximum and minimum air temperatures during both the cover crop and spring barley growth (°C).

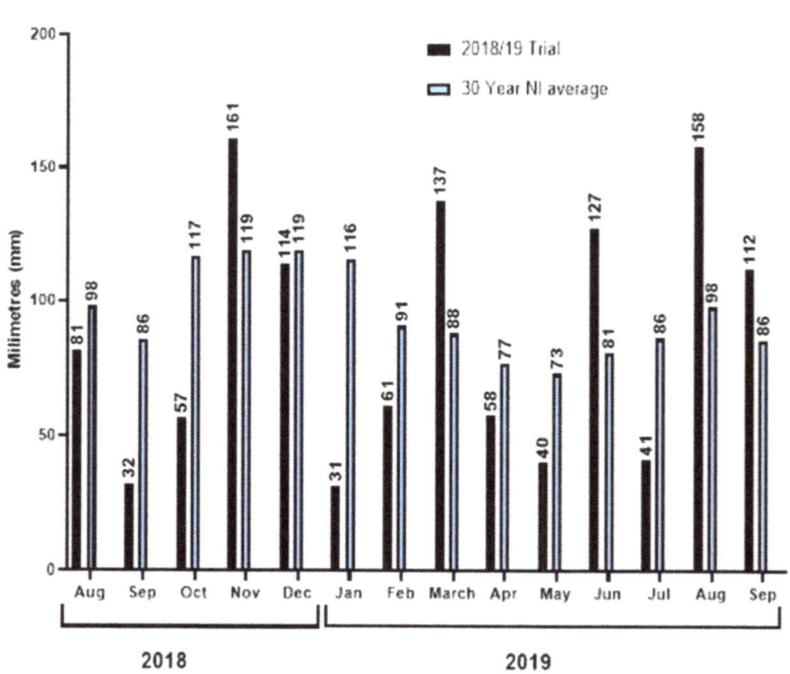

Figure A3. Recorded rainfall during both the cover crop and spring barley growth (mm). Recorded from an official weather station located Crossnacreevy, Belfast. The 30 year NI average was obtained from the Met Office.

Table A4. REML analysis p-values for parameters of cover crop growth.

Parameter	Total CC * BIOMASS (kg/ha)	Total Root (R) Biomass (kg/ha)	Total Weed (W) Biomass (kg/ha)	Above Ground Biomass (CC + W) (kg/ha)	Total Biomass (CC + R+ W) (kg/ha)	% N CC (%)	% Carbon CC (%)	C:N Ratio CC	% N Roots (%)
Sowing date (SD)	<0.001	<0.001	<0.05	<0.001	<0.001	0.12	<0.001	<0.001	<0.001
Species	<0.001	0.64	<0.001	<0.001	<0.001	<0.001	<0.001	<0.01	<0.05
SD × Species	<0.001	0.52	<0.001	<0.001	<0.001	<0.01	0.08	0.06	<0.05
Parameter	% Carbon roots (%) *	Root C N ratio	CC N uptake (kg/ha)	Root N uptake (kg/ha)	Weed N uptake (kg/ha)	Total N uptake (CC + root + weed) (kg/ha)	CC carbon accumulation (kg/ha)	Root carbon accumulation(kg/ha)	Total carbon (CC + root + weed) accumulation (kg/ha)
Sowing date (SD)	<0.05	<0.001	<0.001	<0.001	0.35	<0.001	<0.001	<0.001	<0.001
Species	<0.001	<0.001	<0.001	0.18	<0.001	<0.001	<0.001	0.58	<0.001
SD × Species	0.87	0.20	<0.001	0.35	0.07	<0.001	<0.001	0.46	<0.001

* CC—cover crop.

Table A5. Soil mineral nitrogen (SMN) of control plots measured on 29 February 2019.

Sowing Date	Average SMN in kg/ha (15 cm)
1	24.4
2	9.5
3	16.1
Mean	16.7

N = 4 for each mean.

References

1. Toom, M.; Tamm, S.; Talgre, L.; Tamm, I.; Tamm, Ü.; Narits, L.; Hiiesalu, I.; Mae, A.; Lauringson, E.; Andres, M.E. The Effect of Sowing Date on Biomass and Nitrogen Accumulation of Five Winter Cover Crop Specie. Agricultural Research & Technology. *Open Access J.* **2021**, *25*, 133–140.
2. Finney, D.; Kaye, J. Biomass Production and Carbon/Nitrogen Ratio Influence Ecosystem Services from Cover Crop Mixtures. *Agron. J.* **2016**, *108*, 39–52. [CrossRef]
3. Keogh, B.; McGrath, T.; Grant, J. The effect of sowing date and nitrogen on the dry-matter yield and nitrogen content of forage rape (*Brassica napus* L.) and stubble turnips (*Brassica rapa* L.) in Ireland. *Grass Forage Sci.* **2012**, *67*, 2–12. [CrossRef]
4. Akbari, P.; Herbert, S.; Hashemi, M.; Barker, A.; Omid, R.; Zandvakili, O.; Emami Bistgani, Z. Winter Annual Rye Seeding Date Influence on Nitrogen Recovery and Ammonia Volatilization from Late Fall Surface-Applied Manure. *Agronomy* **2020**, *10*, 931. [CrossRef]
5. Hashemi, M.; Farsad, A.; Sadeghpour, A.; Weis, S.; Herbert, S. Cover-crop seeding-date influence on fall nitrogen recovery. *J. Plant Nutr. Soil Sci.* **2013**, *176*, 69–75. [CrossRef]
6. Teixeira, E.I.; Johnstone, P.; Chakwizira, E.; Ruiter Jd Malcolm, B.; Shaw, N.; Zyskowski, R.; Khaembah, E.; Sharp, J.; Meenken, E.; Fraser, P.; et al. Sources of variability in the effectiveness of winter cover crops for mitigating N leaching. *Agric. Ecosyst. Environ.* **2016**, *220*, 226–235. [CrossRef]
7. Sturm, D.; Kunz, C.; Peteinatos, G.; Gerhards, R. Do cover crop sowing date and fertilization affect field weed suppression? *Plant Soil Environ.* **2017**, *63*, 82–88.
8. Lawson, A.; Cogger, C.; Bary, A.; Fortuna, A.-M. Influence of Seeding Ratio, Planting Date, and Termination Date on Rye-Hairy Vetch Cover Crop Mixture Performance under Organic Management. *PLoS ONE* **2015**, *10*, e0129597. [CrossRef]
9. Brust, J.; Weber, J.; Gerhards, R. Do cover crop mixtures have the same ability to suppress weeds as competitive monoculture cover crops? *Jul. -Kühn-Arch.* **2014**, *443*, 422–430.
10. Silgram, M.; Harrison, R. *The Mineralisation of Nitrogen in Cover Crops: A review*; Department of Environment and Rural Affairs: London, UK, 1998.
11. Couëdel, A.; Alletto, L.; Tribouillois, H.; Justes, É. Cover crop crucifer-legume mixtures provide effective nitrate catch crop and nitrogen green manure ecosystem services. *Agric. Ecosyst. Environ.* **2018**, *254* (Suppl. C), 50–59. [CrossRef]
12. Justes, E.; Mary, B.; Nicolardot, B. Quantifying and modelling C and N mineralization kinetics of catch crop residues in soil: Parameterization of the residue decomposition module of STICS model for mature and non mature residues. *Plant Soil* **2009**, *325*, 171–185. [CrossRef]
13. Cottney, P.; Black, L.; White, E.; Williams, P.N. The Correct Cover Crop Species Integrated with Slurry Can Increase Biomass, Quality and Nitrogen Cycling to Positively Affect Yields in a Subsequent Spring Barley Rotation. *Agronomy* **2020**, *10*, 1760. [CrossRef]
14. *VSN International Genstat for Windows 18th*; International V: Hemel Hempstead, UK, 2017.
15. Grace-Martin, K. Interpreting Interactions when Main Effects are Not Significant. Making Statistics Make Sense 2020. Available online: https://www.theanalysisfactor.com/interactions-main-effects-not-significant (accessed on 4 February 2020).
16. DAERA. Final Results of the June Agricultural Census. In *Statistical News Release*; Department of Agriculture, Environment and Rural Affairs: Belfast, UK, 2018.

17. DAERA. *Crop Yield and Production Estimates in DAERA*; Department of Agriculture, Environment and Rural Affairs: Belfast, UK 2019.
18. White, E.M.; Wilson, F.E.A.; Kettlewell, P.S.; Sklyvester-Bradley, R.; Watson, C. Nitrogen nutrition of wheat at three contrastin sites in the UK. *Asp. Appl. Biol.* **1997**, *50*, 241–247.
19. Kuzyakov, Y.; Xu, X. Competition between roots and microorganisms for nitrogen: Mechanisms and ecological relevance. *New Phytol.* **2013**, *198*, 656–669. [CrossRef]
20. Sainju, U.; Whitehead, W.; Singh, A. Cover crops and nitrogen fertilization effects on soil aggregation and carbon and nitroge pools. *Can. J. Soil Sci.* **2003**, *83*, 155–165. [CrossRef]
21. Kaye, J.; Finney, D.; White, C.; Bradley, B.; Schipanski, M.; Alonso-Ayuso, M.; Hunter, M.; Burgess, M.; Mejia, C. Managin nitrogen through cover crop species selection in the U.S. mid-Atlantic. *PLoS ONE* **2019**, *14*, e0215448. [CrossRef]
22. AHDB. Nutrient Management Guide (RB209). In *Agriculture & Horticulture Development Board*; AHDB, Ed.; AHDB: Warwickshire UK, 2017.
23. Wendling, M.; Büchi, L.; Amosse, C.; Sinaj, S.; Walter, A.; Charles, R. Nutrient accumulation by cover crops with different roc systems. *Asp. Appl. Biol.* **2015**, *129*, 91–95.
24. Wendling, M.; Büchi, L.; Amosse, C.; Sinaj, S.; Walter, A.; Charles, R. Influence of root and leaf traits on the uptake of nutrients i cover crops. *Plant Soil* **2016**, *409*, 419–434. [CrossRef]
25. White, C.A.; Holmes, H.F.; Morris, N.L.; Stobart, R.M. A review of the benefits, optimal crop management practices an knowledge gaps associated with different cover crop species. In *AHDB*; Agriculture and Horticulture Development Board Cambridge, UK, 2016.
26. Li, X.; Sorensen, P.; Li, F.; Petersen, S.O.; Olesen, J.E. Quantifying biological nitrogen fixation of different catch crops, and residua effects of roots and tops on nitrogen uptake in barley using in-situ 15N labelling. *Plant Soil* **2015**, *395*, 273–287. [CrossRef]
27. Kanders, M.J.; Berendonk, C.; Fritz, C.; Watson, C.; Wichern, F. Catch crops store more nitrogen below-ground when considerin Rhizodeposits. *Plant Soil* **2017**, *417*, 287–299. [CrossRef]
28. Finney, D.M.; Buyer, J.S.; Kaye, J.P. Living cover crops have immediate impacts on soil microbial community structure an function. *J. Soil Water Conserv.* **2017**, *72*, 361. [CrossRef]
29. Chen, G. Alleviation of Soil Compaction by Brassica Cover Crops. In *Geography*; University of Maryland, College Park: Marylan MD, USA, 2009.
30. Materechera, S.A.; Dexter, A.R.; Alston, A.M. Penetration of very strong soils by seedling roots of different plant species. *Plant Soil* **1991**, *135*, 31–41. [CrossRef]
31. Rosolem, C.A.; Foloni, J.S.S.; Tiritan, C.S. Root growth and nutrient accumulation in cover crops as affected by soil compactio *Soil Tillage Res.* **2002**, *85*, 109–115. [CrossRef]
32. Hunter, M.C.; Schipanski, M.E.; Burgess, M.H.; LaChance, J.C.; Bradley, B.A.; Barbercheck, M.E.; Kaye, J.P.; Mortensen, D.A Cover Crop Mixture Effects on Maize, Soybean, and Wheat Yield in Rotation. *Agric. Environ. Lett.* **2019**, *4*, 180051. [CrossRef]
33. Jensen, L.S.; Salo, T.; Palmason, F.; Breland, T.A.; Henriksen, T.M.; Stenberg, B.; Pedersen, A.; Lundström, C.; Esala, M. Influenc of biochemical quality on C and N mineralisation from a broad variety of plant materials in soil. *Plant Soil* **2005**, *273*, 307–32 [CrossRef]
34. FAO. Soil carbon sequestration. In *SOLAW Background Thematic Report-TR04B*; FAO: Rome, Italy, 2006.
35. Paustian, K.; Larson, E.; Kent, J.; Marx, E.; Swan, A. Soil C Sequestration as a Biological Negative Emission Strategy. *Front. Clim* **2019**, *1*. [CrossRef]
36. Ghimire, B.; Ghimire, R.; VanLeeuwen, D.; Mesbah, A. Cover Crop Residue Amount and Quality Effects on Soil Organic Carbo Mineralization. *Sustainability* **2017**, *9*, 2316. [CrossRef]
37. Puttaso, A.; Vityakon, P.; Saenjan, P.; Trelo-ges, V.; Cadisch, G. Relationship between residue quality, decomposition patterns, an soil organic matter accumulation in a tropical sandy soil after 13 years. *Nutr. Cycl. Agroecosystems* **2011**, *89*, 159–174. [CrossRef]
38. Department of Agriculture, Environment and Rural Affairs. Statistical Review of Northern Ireland Agriculture. In *DAERA*; Department of Agriculture, Environment and Rural Affairs: Belfast, UK, 2017.
39. Swinnen, J.; Veen, J.A.V.; Merckx, R. Carbon fluxes in the rhizosphere of winter wheat and spring barley with conventional v integrated farming. *Soil Biol. Biochem.* **1995**, *27*, 811–820. [CrossRef]
40. Couëdel, A.; Alletto, L.; Justes, E. Crucifer-legume cover crop mixtures provide effective sulphate catch crop and sulphur gree manure services. *Plant Soil* **2018**, *426*, 61–76. [CrossRef]
41. Cottney, P.; Williams, P.N.; Black, L.; White, E. The Perception and Use of Cover Crops within the Island of Ireland Annals c Applied Biology (accepted December 2020). *Ann. Appl. Biol.* **2021**, *179*, 34–47. [CrossRef]
42. Wendling, M.; Büchi, L.; Amossé, C.; Jeangros, B.; Walter, A.; Charles, R. Specific interactions leading to transgressive overyielding in cover crop mixtures. *Agric. Ecosyst. Environ.* **2017**, *241*, 88–99. [CrossRef]

agronomy

Article

Leaching of Glyphosate and AMPA from Field Lysimeters

Marco Milan *, Francesco Vidotto and Silvia Fogliatto

Dipartimento di Scienze Agrarie, Forestali e Alimentari (DISAFA), Sezione di Agronomia, Università di Torino, Largo Paolo Braccini 2, 10095 Grugliasco (Torino), Italy; francesco.vidotto@unito.it (F.V.); silvia.fogliatto@unito.it (S.F.)
* Correspondence: marco.milan@unito.it

Abstract: Leaching of glyphosate and AMPA as affected by the time elapsed between the spraying and first leaching event was studied on large-scale lysimeters in a two-year study. The leaching events were induced by irrigation interventions able to deliver 336 L, equivalent to a rainfall of 40 mm. Four groups of three lysimeters were randomly selected between the 12 lysimeters available. They were irrigated on either one day after herbicide treatment (1 DAT), 7 DAT, 14 DAT or 28 DAT. The same group of lysimeters were irrigated a second time 14 days after the first irrigation, corresponding to a period of time of 15 DAT (1 + 14), 21 DAT (7 + 14), 28 DAT (14 + 14) and 42 DAT (28 + 14). In both years, lysimeters were sprayed with glyphosate (360 L ha^{-1}) at a rate of 12 L ha^{-1}, the maximum field rate allowed on the label. Our results pointed out that the leaching of glyphosate and AMPA is effectively event-driven and highlighted the importance of the first rainfall event in moving glyphosate through the soil, increasing the potential risk of water contamination. Overall, both chemicals showed a risk of water contamination. Glyphosate may persist more than usually considered, and its residues were found in leached waters from lysimeters treated 30 days before the leaching event. Other factors may affect the movement of these two compounds through the soil profile after spraying: temperature pattern and soil moisture. Finally, the results of this study refer to a very high application rate of glyphosate. Hence, at lower field rates, observed concentrations can likely be minor.

Keywords: leaching; water pollution; degradation; herbicide; metabolite; fate

Citation: Milan, M.; Vidotto, F.; Fogliatto, S. Leaching of Glyphosate and AMPA from Field Lysimeters. *Agronomy* **2022**, *12*, 328. https://doi.org/10.3390/agronomy12020328

Academic Editors: Umberto Anastasi and Aurelio Scavo

Received: 20 December 2021
Accepted: 24 January 2022
Published: 27 January 2022

1. Introduction

The concern about the environmental and health effects of glyphosate and AMPA arising in the last years is likely related to the increased use of the parent compound due to the introduction of genetically modified glyphosate-resistant crops and to the possibility to use the herbicide in pre-harvest [1,2].

Glyphosate is considered a non-mobile herbicide, being well-retained by clay particles, organic matter and iron hydroxides, save in light soil, where preferential flows may occur. The transportation of glyphosate is also facilitated by a water soluble colloidal fraction [1]. AMPA is extremely more mobile than the parent compound [1,2]. Starting from spraying, glyphosate undergoes microbial degradation through two pathways: the formation of sarcosine and glycine and the formation of AMPA [3]. The mobility of both chemicals in the soil depends on their persistence and their rate of adsorption [4].

Glyphosate average soil half-life is generally less than 30 days [5,6]; however, different soil characteristics may greatly affect its soil residence time [1,7,8]. Chemical or physical degradation is considered negligible [1,9]. The movement of glyphosate and AMPA in the soil is also affected by the presence of iron-oxides, which may retain both chemicals in the soil matrix diminishing their transport through the soil profile [10]. The risk of leaching of both glyphosate and AMPA is affected by soil characteristics. Soil characteristics affect the residence time of the chemicals in the soil, hence their proneness to be degraded or transported along the soil profile [11]. For instance, on non-structured soils without macropores, leaching is limited. However, glyphosate leaching might be expected in coarse and oxide-poor soils, as well as on soils with a low glyphosate sorption capacity [1,3] Furthermore, the adsorption of glyphosate on soil particles can be affected by the presence of phosphate, which competes for sorption sites [9], even though this phenomenon may be limited to a few soils [12]. Both chemicals have a low leaching potential according to the GUS index, but AMPA, compared to glyphosate, has a higher water solubility, being 140-fold more soluble in water than glyphosate [5].

Glyphosate and AMPA, as other commonly used pesticides, may move into the water through many different phenomena such as runoff, leaching and drift [13]. Despite residues of glyphosate and AMPA in surface waters being widely reported worldwide [14–16], their presence in groundwaters is significantly lower. For instance, in Italy, AMPA and glyphosate residues were detected in 43% and 66% of the surface-water monitoring points, respectively (22% and 52% of them exceeded the law limit), but only in 5% and 8% in groundwaters, respectively (about 2% of them exceeded the law limit). Similar data are reported in France [14]. Other studies reported the presence of glyphosate and AMPA in groundwaters. Rendon-Von Osten and Dzul-Caamal [17] monitored the presence of glyphosate in groundwaters in the agricultural areas of Yucatan, Mexico. They found glyphosate in all monitored wells, with concentrations up to 1.41 $\mu g\ L^{-1}$. However, there is no indication about the depth of monitored wells.

The presence of glyphosate and AMPA in groundwaters has also been reported by Scribner et al. [18] in a monitoring study carried out in the USA. Van Stempvoort et al. [3], based on previous studies and their findings, postulated that glyphosate migration to deep groundwaters is limited due to microbial degradation and sorption processes. In an agricultural area of India characterized by years of glyphosate application, shallow groundwaters showed contamination, with concentrations ranging from 1 to 4 $\mu g\ L^{-1}$ for glyphosate and up to 11 $\mu g\ L^{-1}$ for AMPA. Similar levels of concentrations were found in Sri Lanka (from 0.7 to 3.5 $\mu g\ L^{-1}$) in abandoned wells close to fields [19]. Leaching of glyphosate and AMPA from 1 m depth lysimeters was monitored in a 3-year study in Central Italy. Both chemicals were found in leachates, AMPA with a more pronounced frequency compared to the parental compound. The concentrations ranged from 0.5 to 13.5 $\mu g\ L^{-1}$ for glyphosate, while AMPA from 1 to 24.9 $\mu g\ L^{-1}$ (maximum peak) [20]. Previous studies had highlighted the importance of the time interval elapsed from the spraying and first leaching event as well as the intensity of the first rainfall [3,21,22]. Similar findings were also obtained by Kjær et al. and Norgaard et al. [11,23].

The risk of water contamination by pesticides can be derived considering the physical and chemical properties of the chemicals (e.g., GUS index, solubility) and the properties of the soil (texture) [24,25]. Field studies can be carried out using specific devices such as lysimeters. Lysimeters are devices commonly used to estimate the leaching of pesticides in agricultural soils [26,27]. While several studies focused on the presence of glyphosate and AMPA in surface waters, less information is available regarding their risk of leaching, particularly at significant soil depths.

The aim of this study was to evaluate the leaching potential of glyphosate and AMPA on agricultural soil as affected by the time elapsed between spraying and the first leaching event. Moreover, the relationship between the time of herbicide application, the first event of leaching and the entity of residues transported in leached waters was assessed. The study was carried out on large-scale lysimeters under field-like conditions.

2. Materials and Methods

The study was hosted at the experimental station of the Dipartimento di Scienze Agrarie, Forestali e Alimentari of the University of Torino, Italy, in 2013 and 2014. The experimental station is located in the municipality of Carmagnola (NW Italy, 44°53′08.99″ N, 7°41′11.33″ E; WGS84), about 40 km far from the city of Turin.

A group of 12 lysimeters built in 1991 was used during the two-year trial. Lysimeters are disposed of two adjacent rows (Figure 1). Each lysimeter has a rectangular shape made of high-density polyethylene with a surface area of 8.4 m^2 (2.8 × 3 m) and a depth of 1.8 m. To facilitate water discharge, a series of polyethylene tubes arranged horizontally were laid at the bottom part of the lysimeters. Soil was separated from the tubes by three layers of gravel (30 cm), sand (30 cm), and non-woven polypropylene fiber, respectively. These layers constituted the drain component for each lysimeter. At the time of installation, lysimeters were filled with disturbed soil taken from the surrounding experimental station soil.

Figure 1. The lysimeters used during the two-year trial are included in the red box.

The main characteristics of the soil, which is classified as typic udifluvent, are reported in Table 1.

Table 1. Main physical characteristics of the soil.

Soil Component	%
Sand	34.8
Silt	59.8
Clay	5.4
Organic matter *	0.44

* average in the 0–2.2 m depth.

In order to ensure field-like conditions, the lysimeters were buried, and the adjacent area followed the same agronomic practices adopted in the lysimeters. At the experimental fields, the groundwater level was about 6 m deep with negligible seasonal variation. In the 0–0.5 m soil depth, the soil bulk density was 1.30 mg m^{-3}, and the water content at saturation averaged 0.56 mm^3 mm^{-1} [28,29].

Zavattaro et al. [30] performed a physical and hydrological characterization of the lysimeters soil eight years after their installation. The main soil hydrological parameters and the soil bulk density, evaluated at 0–20 cm and 20–50 cm soil depth, resulted in very similar between lysimeters and the undisturbed soil. The bulk density and water tension were measured at 0, 33, and 1500 kPa. The soil of lysimeters did not suffer from compaction due to machinery transit as the lysimeter soil was spade-tilled. Bulk density differences were encountered only in the plowed layer of the soil surrounding the lysimeters. No significant differences between the lysimeter soil and the undisturbed one were found in the deeper layers. The water storing capacity of the soil was evaluated at field capacity and at permanent wilting point. Only at field capacity, few differences were observed for the volumetric water content [30,31]. According to a study carried out in 2007, the soi infiltration rate was 70 mm/h [32].

2.1. Agronomic Practices Adopted in the Previous Years and Lysimeter Preparation

Until 2012 lysimeters were cultivated with maize following the local agronomic pra tices in terms of fertilization and crop protection. In that period, tillage operations withi the lysimeters were manually performed by using a spade until to a depth of 30 cm, whil weed control in maize was carried out in pre-emergence with mixtures of different herb cides (see [21]). At that time, the crop was cultivated within both the lysimeter and all th surrounding surfaces to mimic field-like conditions. About a week prior to the start of eac growing season trial, the lysimeters were left to fully discharge the percolating water, i present, to ensure the absence of gravitational water flows. Lysimeters were sprayed onl after the end of percolation.

In both years (2013 and 2014), lysimeters were sprayed with Taifun MK® (Adama Itali srl, Grassobbio, Italy), a herbicide containing 360 g L^{-1} of glyphosate, at rate of 12 L ha^{-} The chosen application rate was the maximum allowed on the label. Treatment was applie using a backpack sprayer (Bellspray Inc dba R&D Sprayers, Opelousas, LA, USA; mode D-201-S), with 2-L bottle header, with an aluminum CO_2 cylinder, equipped with 4 nozzl spray boom. In the previous years, glyphosate was never applied on lysimeters to contro weed infestation. In 2013, herbicide application occurred on 17 June 17, while in 201 on 5 June. In case of unfavorable weather forecast, during lysimeter preparation an after herbicide application, temporary covers were set up on the lysimeters and remove immediately after rainfall.

Four groups of three lysimeters were randomly selected between the 12 lysimeter available. Each group was irrigated (first irrigation) at different times after treatment: a one day after herbicide treatment (1 DAT), 7 DAT, 14 DAT and 28 DAT. The same lysimeter were irrigated a second time (second irrigation) 14 days after the first irrigation, corre sponding to a period of time from treatment of 15 DAT (1 + 14), 21 DAT (7 + 14), 28 DA (14 + 14) and 42 DAT (28 + 14). Each lysimeter was irrigated using water withdraw from a 30 m-deep well located 150 m far from the lysimeter facility. Three samples o irrigation water were analyzed to verify the absence of glyphosate and AMPA in the wate In addition, the experimental site hosted an official monitoring point of the groundwate monitoring network of the Piedmont region (Monitoring point: 00105910002—TF2 Tett Frati—Carmagnola—GWB-S5a). The analysis carried out each year by the regional au thority for environmental protection in the 13 m-deep aquifers did not find residues o both chemicals (<0.01 µg L^{-1}). The amount of water distributed on each lysimeter durin a single irrigation was 336 L, a quantity corresponding to a 40 mm rainfall. This amount o rainfall per event was calculated considering the last ten years' meteorological pattern o the zone in the period of potential herbicide application. In a previous trial carried out o the same lysimeters in 2011 and 2012, this amount of water was able to produce importan leaching. At the time of irrigation, each lysimeter was irrigated separately. The irrigatio required about 30 min to deliver the selected quantity of water. Irrigation was carried ou by means of a hose with a dispersion device attached to its end.

A 200 L collection tank placed 2.5 m deep into an inspection chamber was used t collect the percolated water drained by gravity into each lysimeter. The percolated wate flowed from the lysimeter into the tank by means of a valve, with a manual regulatio After each percolation event, the water was withdrawn from the collection tank by electri pump. The total percolated volume was measured with an in-line flow meter (K24 Turbin meter, Piusi Instruments, Suzzara, Italy).

The presence of percolated water was monitored in the irrigated lysimeters startin from one day after irrigation. About a week after the irrigation, the water drained at th bottom of each lysimeter was collected and the full volume was measured. Three sample per lysimeter were collected from the entire volume of leached water. The total leache volume was collected with a submersible drainage pump (Calpeda, Montorso Vicentin

Italy). The water samples were then put into 1 L graduated square polyethylene bottles (Kartell®, Noviglio, Italy) and immediately stored in a −25 °C cold room until analysis. In order to assess the presence of background residues from the previous season, in 2014, before the starting of the new trial, three samples of leached water were collected from each lysimeter previously used in 2013 trial (blank samples).

2.2. Soil Moisture Measurements

Soil moisture was measured just before each planned irrigation in the upper soil layers (0–5 cm soil depth), taking the soil sample by means of a trowel. Soil samples were not collected using core samplers to avoid the creation of preferential ways. After collection, soil samples were immediately weighted, then let dry into laboratory stoves at 105 °C for 24 h., hence re-weighted.

2.3. Glyphosate and AMPA Analysis

The analyses were performed using high-performance liquid chromatography (HPLC) using a Varian instrument (Agilent Technologies Italia, Cernusco sul Naviglio, Italy) equipped with a ternary pump (Pro Star mod. 230) and a fluorescence detector used at excitation and emission wavelengths of 266 nm and 305 nm, respectively. The column was a Supelco-sil® (Sigma Aldrich, Saint Louis, MO, USA) LC-NH2 (25 mm × 4.6 µm, 5 µm) with a mobile phase (75/25 v/v) composed by KH_2PO_4 0.1 N: acetonitrile (Carlo Erba® reagents, Cornaredo, Italy). The mobile phase was pumped at an isocratic rate flow rate of 1 mL min^{-1}. The retention time for glyphosate was 9.5 min, for AMPA 4.6 min. Before the analysis, each water sample was prepared to evaporate 100 mL of the initial sample until a final volume of 4 mL using a Rotavapor (Rotavapor® R-100, Buchi, Cornaredo, Italy) set at 50 °C. The evaporation process was facilitated by adding small aliquots of acetone (Carlo Erba® reagents, Cornaredo, Italy). The reduced sample was introduced in a 5 mL flask and filled to the total volume by adding deionized water. The sample was finally filtered using a 0.45 µm nylon filter. Once filtered, each sample underwent derivatization. The derivatization was performed preparing a solution with 400 µL of borate buffer (0.05 M, pH 10), 200 µL of 9 fluorenyl-methyl chloroformates (FMOC-Cl) (Sigma-Aldrich®, Merck Group, Darmstadt, Germany) dissolved in acetonitrile, 200 µL of the sample within a test tube with screw cap. The solution was vortexed for one minute, then left undisturbed for 60 min. After this period, 600 µL of 2% concentrated H_3PO_4 was added to the solution that was vigorously shacked. After this step, 2 mL of ethyl ether (Carlo Erba® reagents, Cornaredo, Italy) were added, and the solution was shaken another time. Once the separation of the phases occurred, 1.6 mL of the aqueous phase were withdrawn and transferred in a vial for analytical determination. The limit of quantification (LOQ) was 0.1 µg L^{-1} for both chemicals.

2.4. AMPA/Glyphosate Ratio (AMPA/GLY Ratio)

The metabolite/parent compound ratio (MPR) has been considered a way to discriminate between diffuse and point pollution sources [33–35]. An MPR above 1 is an indicator of diffuse pollution: parent compounds are transported slowly along the soil profile, and they have time to be partially or totally degraded. By contrast, a low MPR value indicates rapid leaching of the parent compound. We use the AMPA/GLY ratio to explain the differences in AMPA and glyphosate concentrations observed at different time intervals from spraying and first leaching event. The AMPA/GLY ratio was calculated for all sampling dates when concentrations data were available. In case of concentrations below the quantification limit, the value of the correspondent limit of quantification was considered for the calculation of the ratio.

2.5. Statistical Analysis

The t test was used to individuate statistical differences ($\alpha \leq 0.05$) in glyphosate and AMPA concentrations between years at the same temporal interval from irrigation. A bi-variate correlation analysis was performed to verify the existence of a linear relationship between the percolated volumes measured at the first and the second irrigation and corresponding glyphosate and AMPA concentrations. The software SPSS, version 27.00 (SPSS released 2020, IBM Corporation, Armonk, NY, USA), was used to perform the analysis.

2.6. Weather Conditions

In Figure 2, the meteorological data observed during the period of the study are reported. All the weather data were collected from the meteorological station located 150 m far from the lysimeters. The meteorological station is part of the regional weather network system. In 2013 two periods of high temperatures were recorded, during the days that preceded the spraying and in the last ten days of July (which corresponded to the end of the trial). In 2014, the average temperature was lower than in the previous year (more than 1 °C). During 2013, the good weather conditions during the weeks after herbicide application did not require the covering of lysimeters for rainfall events. In the night between 30 and 31 July 2013, just before the irrigation at 28 DAT (Days After Treatment) and 14 + 14 DAT, a sudden storm discharged 24 mm of rain on the uncovered lysimeters. The irrigation planned that day has considered the rainfall fallen during the night, and thus the amount of water distributed was 130 L. In 2014, we recorded two heatwaves, the former just in the days following herbicide application, the latter in the middle of July. Both periods, of about 7 days each, were characterized by max and min mean temperatures significantly higher than the average. For instance, during the first heatwave from 7 June to 13 the maximum average temperature was 33.5 °C and the minimum 16.5 °C, compared to the average temperature of the period June–July of 28.7 °C and 15 °C, respectively. These periods of high daily temperatures likely amplified the evaporation processes, influencing the total amount of water available for percolation.

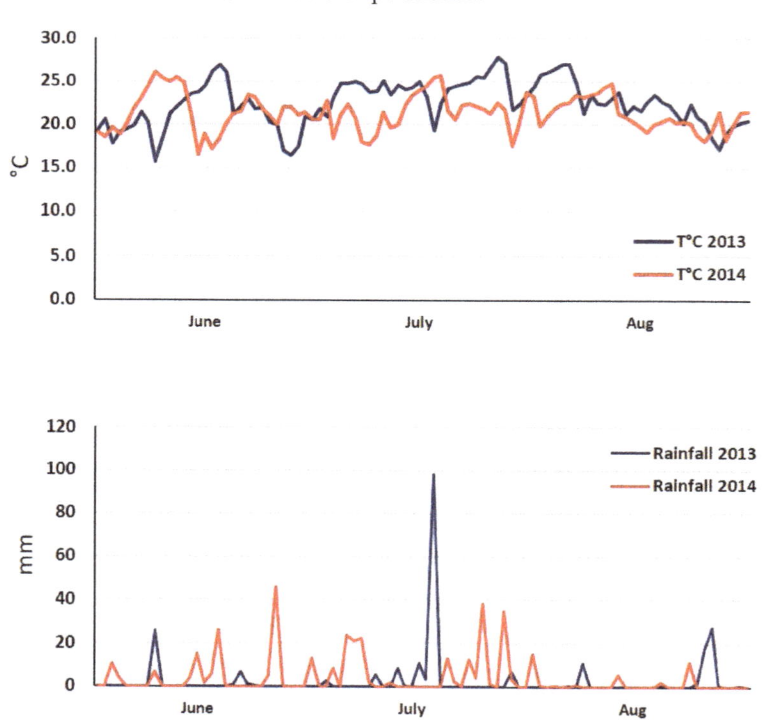

Figure 2. Meteorological trend observed in 2013 and 2014.

3. Results

3.1. Percolation Volumes

In both seasons, the total volumes percolated through the lysimeters after each planned leaching event were measured (Table 2). The amount of percolation water leached through the soil profile was likely affected by the soil conditions before and after the leaching event. In general, the highest volumes were observed after the second irrigation. In 2013 the lowest amount of leached waters was observed at the leaching event planned at 7 and 28 DAT and at 42 DAT. The repetition of irrigation caused a more pronounced movement of water along the soil profile. This was evident by the observation of the leached volumes. The percolated volumes monitored in 2014 resulted lower compared to those observed in 2013 due to the different meteorological conditions that occurred during the trial. The highest temperatures recorded during the 2014 trial have likely affected the evaporation rates of the soil. This is confirmed by the moisture data measured before each planned irrigation (Table 3). In the first four irrigation times (1, 7, 14, 28 DAT), soil moisture percentage was always about one point below the level observed in the previous season before the execution of the irrigation (Table 3). The average amount of percolated water resulting from the three lysimeters at 28 DAT (38 l) was unexpected considering the null or negligible percolation volumes observed in the previous sampling times. However, only two days before the irrigation (on 1 July), a severe storm occurred during the afternoon. For a while, part of the covering structure installed to protect the lysimeters was removed by the strength of the wind. The technicians were able to re-establish the impermeable cover quite soon, but an imprecise amount of rain has certainly reached the soil surface. To confirm this, the moisture level measured on the soil surface just before the irrigation was higher than that observed at the two previous sampling times.

Table 2. Volumes of percolated waters (±SE) during the two-year study. Values are the arithmetic mean of three replications.

Days Elapsed between the Treatment and the First Irrigation	2013	2014
DAT (1° irrigation)	Percolated water (L)	
1	57 (±13)	NL
7	12 (±2)	7 (±2)
14	14 (±8)	NL
28	11 (±5)	38 (±23)
Days Elapsed between the treatment and the Second Irrigation	Percolated water (L)	
DAT (2° irrigation)		
(15) 1 + 14	53 (±10)	72 (±23)
(21) 7 + 14	21 (±5)	18 (±9)
(28) 14 + 14	146 (±45)	103 (±3)
(42) 28 + 14	3 (±2)	2 (±1)

Note: NL: no leaching.

Table 3. Soil moisture at the first 5 cm depth measured before the irrigation of lysimeters. Arithmetic mean of three replications ± SE.

DAT	2013	2014
	Soil moisture (%) ± SE	
1	19.6 (±0.29) a	18.9 (±0.09) b
7	18.4 (±0.82) a	17.3 (±0.25) b
14	17.9 (±0.79) a	17.0 (±0.16) b
28	20.8 (±0.53) a	18.3 (±0.14) b
(15) 1 + 14	18.1 (±0.38)	20.5 (±0.27)
(21) 7 + 14	20.2 (±1.15)	20.6 (±0.18)
(28) 14 + 14	20.2 (±0.28)	20.5 (±0.30)
(42) 28 + 14	20.2 (±0.35) a	19.5 (±0.03) b

Data values with different letters are statistically different (Students's *t*-test; $\alpha = 0.05$).

3.2. Glyphosate and AMPA Concentrations

3.2.1. Season 2013

During 2013 the highest concentration of glyphosate was recorded at the first leachin event (1 DAT) with a value of 1.39 µg L^{-1} (Table 4). A higher percolation volume, compare to the following leaching events, characterized this sampling. In the following week glyphosate concentration remained quite stable, never exceeding 0.60 µg L^{-1}. In wate samples collected after the second round of irrigation, glyphosate residues still remain te times higher than the LOQ (0.1 µg L^{-1}) at 15 and 21 DAT. The leached water collecte from lysimeters irrigated a month after spraying still showed relevant traces of glyphosat (0.27 µg L^{-1}). AMPA had a similar trend in the first set of irrigation, even though wit lower concentration values (Table 4). The concentration peak (0.97 µg L^{-1}) was measure in percolated waters collected from lysimeters irrigated at 1 DAT. On percolated water collected after the repetition of irrigation, the highest AMPA concentration was reache at 15 DAT (0.84 µg L^{-1}). At 42 DAT, residues of AMPA were above 1 µg L^{-1}. At th first leaching event, there was a significant correlation between percolated volumes an concentration values for both chemicals (Table 5). The analysis carried out on the blan samples collected in 2014, just before the beginning of the new experimental season, showe residues of glyphosate and AMPA below 0.1µg L^{-1} in all the analyzed samples.

Table 4. Glyphosate and AMPA concentrations detected in percolated waters in 2013 and 201 Arithmetic mean of three replications ±SE.

	2013	2014	2013	2014	2013	2014
DAT 1° Irrigation	**GLY µg/L**	**GLY µg/L**	**AMPA µg/L**	**AMPA µg/L**	**AMPA/GLY**	**AMPA/GLY**
1	1.39 (±0.60)	-	0.97 (±0.35)	-	0.7	-
7	0.59 (±0.28) a	0.19 (±0.09) b	0.22 (±0.10)	<0.1	0.4	≤0.5
14	0.57 (±0.22)	-	0.18 (±0.05)	-	0.3	-
28	0.27(±0.13)	0.52 (±0.40)	0.11 (±0.01)	<0.1	0.4	≤0.2
DAT 2° irrigation						
(15) 1 + 14	1.04 (±0.45) a	0.13 (±0.04) b	0.84 (±0.28) A	0.22 (±0.09) B	0.8	1.7
(21) 7 + 14	1.19 (±0.70)	<0.1	0.11 (±0.01)	<0.1	0.1	≤1
(28) 14 + 14	0.28 (±0.10)	0.12 (±0.02)	0.08 (±0.01) B	0.52 (±0.30) A	0.3	4.3
(42) 28 + 14	<0.1	<0.1	1.07 (±0.25)	<0.1	≥1	≥1

GLY: glyphosate; Data values with different letters are statistically different (Students's *t*-test; $\alpha = 0.05$). Lowercas letters: differences in glyphosate concentrations between years; Capital letters: differences in AMPA concentration between years.

Table 5. Correlation between percolated volumes and glyphosate and AMPA concentrations in 201 and 2014 according to the two irrigations period.

	2013		2014	
	GLY	AMPA	GLY	AMPA
1° irrigation	0.725 **	0.665 *	0.850 *	-
2° irrigation	−0.462	−0.375	0.337	0.875 *

** ($\alpha \leq 0.01$); * ($\alpha \leq 0.05$).

3.2.2. Season 2014

In 2014, due to the unfavorable weather conditions, it was not possible to collec samples after the first (1 DAT) and the third (14 DAT) planned irrigation. Overall, th percolated volumes were greatly lower than in the previous season. Samples collecte at 7 and 28 DAT both showed residues of glyphosate above the detection limit. On th contrary, AMPA residues resulted below the limit of quantification. The repetition o irrigation determined a flux of water along the soil profile, allowing the collection of wate samples after each event. Glyphosate residues never exceeded 0.13 µg L^{-1}, while in th case of AMPA, a concentration peak of 0.52 μg L^{-1} was reached at 28 DAT, likely due to the important percolation flux (see Table 2). The high concentrations of AMPA recorded at 15 and 28 DAT are likely justified by the fact that after the first irrigations at 1 and 14 DAT, no percolation was detected. The repetition of the irrigations induced the leaching of AMPA residues formed until that moment. A positive correlation between percolated volumes and AMPA concentrations was found (Table 5).

3.2.3. AMPA/GLY Ratio

In 2013, the AMPA/GLY ratio was below 1 (on average 0.4) in almost all the sampling dates, with the exception of the last sampling date (Table 4). These values may reflect the rapid transport, due to the irrigations, of the parent compound through the soil, which may have delayed the formation of the metabolite. During 2014, in a few cases, it was not possible to calculate the MPR ratio due to the absence of leaching. The high AMPA/GLY values observed at 15 and 28 DAT (1.7 and 4.3 at 15 and 28 DAT in 2014), derived from the high concentrations of AMPA observed. The high AMPA/GLY ratio may be explained by both the rapid degradation of glyphosate due to the high temperature that occurred the day following herbicide application and by the high interaction of the parent compound residues with soil, which may bring faster degradation or higher absorption. The fact that, in the last sampling dates, the MPR ratio reached high values was quite unexpected considering that from the spraying to first irrigation, enough time elapsed to allow the parent compound to be degraded by microorganisms. During the permanence in the soil of the parent compound, biodegradation processes are favored, and metabolite formation occurs. At the moment of the first irrigations, no leached water was observed in the lysimeters due to the particular weather conditions as described in the previous paragraphs.

4. Discussion

The present study dealt with the mobility of glyphosate and AMPA in long-established field lysimeters. The information obtained may help to explain the behavior of these chemicals in field-like conditions under different scenarios in terms of the occurrence of percolation events. The lysimeters used in this study allowed the collection of percolated water to a high depth (1.8 m) compared to other studies carried out worldwide both on lysimeter and at field scale [1], giving the possibility to understand the mobility of these chemicals at uncommon depths.

According to our results, the mobility of glyphosate and AMPA seems to be related to the amount of percolation water involved and to the time elapsed from the spraying to the leaching event. Similar findings are reported by Giuliano et al. [36]. At a higher amount of percolation, volumes generally correspond to greater chemical residues in leached waters. In addition, the results demonstrated that glyphosate is more susceptible to leaching in case of important rainfall very close to spraying time; this behavior was previously seen by other authors [1,3,12,20]. According to Napoli et al., rainfall occurring within two weeks after spraying may lead to a leaching of glyphosate until a depth of at least 1 m [20]. Similarly, Al-Rajab et al. [37] found glyphosate residues only 18 days after the first percolation. Other authors found that in tile drains posed at 1 m depth, glyphosate and AMPA concentrations frequently exceeded 0.1 μg L^{-1}. In this study, the concentration found at 1 m depth in one of the experimental sites were on average of 0.54 μg L^{-1} for glyphosate and 0.17 μg L^{-1} for AMPA, but they refer to an application field rate 3-fold lower (4 L ha^{-1}) than that used in the current study (12 L ha^{-1}). In the same study, it is reported that heavy rains fallen soon after herbicide application may carry to marked leaching with concentrations of up to 11 μg L^{-1} for glyphosate and 0.6 μg L^{-1} for AMPA at 19 days after application [15]. Giuliano et al. [36] found high percolation peaks of mesotrione and glyphosate in water samples collected during the season from tension plates lysimeters. All these data fit with our findings.

While the highest concentrations of glyphosate were detected at leaching events close to herbicide spraying, the presence of AMPA residues did not follow a regular pattern.

This is because AMPA is a metabolite and its formation depends on the availability and degradation of the parent compounds as well as by other concomitant factors. In addition, AMPA can be adsorbed by phosphonate groups [3], and its biodegradation is considered slower than that of its parent compound [38]. Furthermore, its release in the soil occurred over a longer period of time compared to glyphosate [11]. Our results pointed out that the leaching of glyphosate and AMPA is effectively event-driven and highlighted the importance of the first rainfall event in moving glyphosate through the soil, increasing the potential risk of water contamination. This was observed also by Kjær et al. and Rasmussen et al. [11,39].

However, our results showed that other factors could affect the movement of these two compounds through the soil profile after spraying: temperature pattern and soil moisture. Similar findings were observed by Al-Rajab et al. (2008) [37]. Even though high temperatures may increase glyphosate degradation, the rapid transport of the chemical in the deeper soil layers due to close heavy rainfall may diminish the formation of AMPA, favoring the sorption of glyphosate on the soil matrix. The dryer the soil, the higher the risk of easy transport of pesticides through the soil profile by rainfall events by means of macropore flow [39,40] and reduced microbial activity. In 2014, glyphosate and AMPA fates were likely influenced by the climatic conditions that occurred during the trial, and in particular by the two recorded heatwaves, the former only a few days after spraying. The high temperatures boosted the microbial activity and significantly increased the evaporation processes. The importance of the weather conditions before and after the pesticide application, as well as the initial moisture conditions of the soil, are highlighted by Rasmussen et al. [39]. Our results showed that residues of glyphosate could be found in leached waters from lysimeters treated 30 days before the leaching event. That means that glyphosate may persist more than usually considered [5,41]. A possible explanation derives from the results of Bento et al. [42], which reported a reduced degradation of glyphosate and AMPA at dryer conditions.

In our study, we used the AMPA/GLY ratio to assess the percolation dynamics of the two compounds in relation to the occurrence of leaching events. In our study, in both years, AMPA/GLY ratio was generally below one, indicating a common degradation of the parent compound. After herbicide application, the microbial activity starts to degrade the parent compound, generating the metabolite. However, there are certain conditions that can alter this natural trend; in high permeable soils, the transfer of the parent compound can be very rapid, and microbial activity has insufficient time to degrade the chemical. The high MPR values observed in 2014 are probably related to the specific weather conditions that occurred after herbicide spraying and the irrigation events. The weather conditions affected the percolation dynamics limiting or, in two cases, annulling the leaching phenomena. In particular, the highest AMPA/GLY ratio observed at 15 and 28 DAT indicates a significant prevalence of the metabolite over the parent compound. As a possible explanation of these findings, we may consider that even when a leaching event does not produce leached waters, it determines the movement of the chemical through the soil profile. When no significant leaching occurred, chemicals remained confined in the soil matrix; hence they are more available to microbial degradation.

5. Conclusions

Glyphosate and AMPA can be transferred to deeper soil layers at concentrations above the law limits (0.1 $\mu g\ L^{-1}$ is the maximum allowable concentration in the European Union for a generic pesticide in groundwater), even in case of leaching events far from spraying. Our results showed that AMPA might pose a risk of contamination of groundwaters as well as its parent compound. The weather and soil conditions can affect the dynamic of glyphosate movement and likely its degradation pattern. Both chemicals showed a potential risk of water contamination. Finally, we may consider that these results refer to a very high application rate of glyphosate. Hence, at lower field rates, concentrations can likely be minors.

Author Contributions: Conceptualization, M.M. and F.V.; data curation, M.M. and S.F.; formal analysis, M.M. and S.F.; funding acquisition, F.V., investigation, M.M.; methodology, M.M. and F.V.; supervision, F.V.; writing—original draft preparation, M.M.; writing—review and editing, M.M, F.V. and S.F. All authors have read and agreed to the published version of the manuscript.

Funding: This research received no external funding.

Institutional Review Board Statement: Not applicable.

Informed Consent Statement: Not applicable.

Data Availability Statement: Data are available from the authors upon request.

Acknowledgments: The authors thank the field technicians who made a valuable contribution during the study.

Conflicts of Interest: The authors declare no conflict of interest.

References

1. VeVereecken, H. Mobility and Leaching of Glyphosate: A Review. *Pest Manag. Sci. Former. Pestic. Sci.* **2005**, *61*, 1139–1151. [CrossRef] [PubMed]
2. Van Bruggen, A.H.; He, M.M.; Shin, K.; Mai, V.; Jeong, K.; Finckh, M.; Morris, J., Jr. Environmental and Health Effects of the Herbicide Glyphosate. *Sci. Total Environ.* **2018**, *616*, 255–268. [CrossRef] [PubMed]
3. Borggaard, O.K.; Gimsing, A.L. Fate of Glyphosate in Soil and the Possibility of Leaching to Ground and Surface Waters: A Review. *Pest Manag. Sci. Former. Pestic. Sci.* **2008**, *64*, 441–456. [CrossRef] [PubMed]
4. Okada, E.; Costa, J.L.; Bedmar, F. Adsorption and Mobility of Glyphosate in Different Soils under No-till and Conventional Tillage. *Geoderma* **2016**, *263*, 78–85. [CrossRef]
5. Lewis, K.A.; Tzilivakis, J.: Warner, D.J.; Green, A. An International Database for Pesticide Risk Assessments and Management. *Hum. Ecol. Risk Assess.* **2016**, *22*, 1050–1064. [CrossRef]
6. Simonsen, L.; Fomsgaard, I.S.; Svensmark, B.; Spliid, N.H. Fate and Availability of Glyphosate and AMPA in Agricultural Soil. *J. Environ. Sci. Health Part B* **2008**, *43*, 365–375. [CrossRef]
7. Gimsing, A.L.; Borggaard, O.K.; Jacobsen, O.S.; Aamand, J.; Sørensen, J. Chemical and Microbiological Soil Characteristics Controlling Glyphosate Mineralisation in Danish Surface Soils. *Appl. Soil Ecol.* **2004**, *27*, 233–242. [CrossRef]
8. Nguyen, N.K.; Dörfler, U.; Welzl, G.; Munch, J.C.; Schroll, R.; Suhadolc, M. Large Variation in Glyphosate Mineralization in 21 Different Agricultural Soils Explained by Soil Properties. *Sci. Total Environ.* **2018**, *627*, 544–552. [CrossRef]
9. Grandcoin, A.; Piel, S.; Baures, E. AminoMethylPhosphonic Acid (AMPA) in Natural Waters: Its Sources, Behavior and Environmental Fate. *Water Res.* **2017**, *117*, 187–197. [CrossRef]
10. Rampazzo, N.; Rampazzo Todorovic, G.; Mentler, A.; Blum, W.E. Adsorption of Glyphosate and Aminomethylphosphonic Acid in Soils. *Int. Agrophys.* **2013**, *27*, 203–209. [CrossRef]
11. Kjær, J.; Olsen, P.; Ullum, M.; Grant, R. Leaching of Glyphosate and Amino-methylphosphonic Acid from Danish Agricultural Field Sites. *J. Environ. Qual.* **2005**, *34*, 608–620. [CrossRef] [PubMed]
12. Borggaard, O. Does Phosphate Affect Soil Sorption and Degradation of Glyphosate? A Review. *Trends Soil Plant Sci. J.* **2011**, *2*, 11.
13. Carter, A. How Pesticides Get into Water-and Proposed Reduction Measures. *Pestic. Outlook* **2000**, *11*, 149–156. [CrossRef]
14. ANSES. Glyphosate: ANSES Reviews the Monitoring Data. The French Agency for Food, Environmental and Occupational Health & Safety. 2019. Available online: https://www.anses.fr/en/content/glyphosate-anses-reviews-monitoring-data (accessed on 20 December 2021).
15. Paris, P.; Pace, E.; Maschio, G.; Ursino, G. Rapporto Nazionale Pesticidi Nelle Acque. Dati 2017–2018. Edizione 2020. 2020. Available online: https://sinacloud.isprambiente.it/portal/apps/sites/?fromEdit=true#/portalepesticidi/pages/rapporti (accessed on 20 December 2021).
16. Poiger, T.; Buerge, I.J.; Bächli, A.; Müller, M.D.; Balmer, M.E. Occurrence of the Herbicide Glyphosate and Its Metabolite AMPA in Surface Waters in Switzerland Determined with On-Line Solid Phase Extraction LC-MS/MS. *Environ. Sci. Pollut. Res.* **2017**, *24*, 1588–1596. [CrossRef] [PubMed]
17. Rendon-von Osten, J.; Dzul-Caamal, R. Glyphosate Residues in Groundwater, Drinking Water and Urine of Subsistence Farmers from Intensive Agriculture Localities: A Survey in Hopelchén, Campeche, Mexico. *Int. J. Environ. Res. Public Health* **2017**, *14*, 595. [CrossRef]
18. Scribner, E.A.; Battaglin, W.A.; Gilliom, R.J.; Meyer, M.T. *Concentrations of Glyphosate, Its Degradation Product, Aminomethylphosphonic Acid, and Glufosinate in Ground- and Surface-Water, Rainfall, and Soil Samples Collected in the United States, 2001–2006*; Scientific Investigation Report 2007-5122. U.S.; Department of the Interior, U.S. Geological Survey: Reston, VA, USA, 2007.
19. Jayasumana, C.; Paranagama, P.; Agampodi, S.; Wijewardane, C.; Gunatilake, S.; Siribaddana, S. Drinking Well Water and Occupational Exposure to Herbicides Is Associated with Chronic Kidney Disease, in Padavi-Sripura, Sri Lanka. *Environ. Health* **2015**, *14*, 6. [CrossRef]

20. Napoli, M.; Cecchi, S.; Zanchi, C.A.; Orlandini, S. Leaching of Glyphosate and Aminomethylphosphonic Acid through Silty Cla Soil Columns under Outdoor Conditions. *J. Environ. Qual.* **2015**, *44*, 1667–1673. [CrossRef]
21. Milan, M.; Ferrero, A.; Fogliatto, S.; Piano, S.; Vidotto, F. Leaching of S-Metolachlor, Terbuthylazine, Desethyl-Terbuthylazin Mesotrione, Flufenacet, Isoxaflutole, and Diketonitrile in Field Lysimeters as Affected by the Time Elapsed between Spraying an First Leaching Event. *J. Environ. Sci. Health Part B* **2015**, *50*, 851–861. [CrossRef]
22. Norgaard, T.; Moldrup, P.; Ferré, T.P.; Olsen, P.; Rosenbom, A.E.; de Jonge, L.W. Leaching of Glyphosate and Aminomethylpho phonic Acid from an Agricultural Field over a Twelve-year Period. *Vadose Zone J.* **2014**, *13*, vzj2014-05. [CrossRef]
23. Kladivko, E.J.; Brown, L.C.; Baker, J.L. Pesticide Transport to Subsurface Tile Drains in Humid Regions of North America. *Cri Rev. Environ. Sci. Technol.* **2001**, *31*, 1–62. [CrossRef]
24. Flury, M. Experimental Evidence of Transport of Pesticides through Field Soils—A Review. *J. Environ. Qual.* **1996**, *25*, 25–4 [CrossRef]
25. Nicholls, P.H. Factors Influencing Entry of Pesticides into Soil Water. *Pestic. Sci.* **1988**, *22*, 123–137. [CrossRef]
26. Bergström, L. Use of Lysimeters to Estimate Leaching of Pesticides in Agricultural Soils. *Environ. Pollut.* **1990**, *67*, 325–34 [CrossRef]
27. Winton, K.; Weber, J.B. A Review of Field Lysimeter Studies to Describe the Environmental Fate of Pesticides. *Weed Technol.* **199** *10*, 202–209. [CrossRef]
28. Grignani, C.; Zavattaro, L.; Sacco, D.; Gilardi, M. Misure Di Evapotraspirazione Da Sistemi Colturali Foraggeri Utilizzand Lisimetri a Percolazione. In Proceedings of the Congress Agro-Meteorology to Monitor Water Consumption, Sassari, Ital November 1999; pp. 165–174.
29. Zavattaro, L.; Grignani, C. Deriving Hydrological Parameters for Modeling Water Flow under Field Conditions. *Soil Sci. Soc. An J.* **2001**, *65*, 655–667. [CrossRef]
30. Zavattaro, L.; Grignani, C.; Sacco, D.; Ferraris, S. Confronto Tra Metodi per La Misura Dell'evapotraspirazione Utilizzand Lisimetri a Percolazione. In Proceedings of the Congress Agro-Meteorology to Monitor Water Consumption, Sassari, Ital November 1999; pp. 357–365.
31. Zavattaro, L.; Grignani, C. Rappresentatività Di Lisimetri a Percolazione Di Grandi Dimensioni per Lo Studio Dei Consumi Idri Delle Colture. In Proceedings of the Congress Agro-Meteorology to Monitor Water Consumption, Sassari, Italy, November 199 pp. 349–356.
32. Rivoira, C. *L'efficienza di Diversi Metodi Irrigui in Risicoltura*; Università di Torino: Torino, Italy, 2007.
33. Thurman, E.M.; Goolsby, D.A.; Meyer, M.T.; Mills, M.S.; Pomes, M.L.; Kolpin, D.W. A Reconnaissance Study of Herbicide and Their Metabolites in Surface Water of the Midwestern United States Using Immunoassay and Gas Chromatography/Mas Spectrometry. *Environ. Sci. Technol.* **1992**, *26*, 2440–2447. [CrossRef]
34. Bozzo, S.; Azimonti, G.; Villa, S.; Di Guardo, A.; Finizio, A. Spatial and Temporal Trend of Groundwater Contamination fron Terbuthylazine and Desethyl-Terbuthylazine in the Lombardy Region (Italy). *Environ. Sci. Processes Impacts* **2013**, *15*, 366–37 [CrossRef]
35. Adams, C.D.; Thurman, E.M. Formation and Transport of Deethylatrazine in the Soil and Vadose Zone. *J. Environ. Qual.* **1991**, *2* 540–547. [CrossRef]
36. Giuliano, S.; Alletto, L.; Deswarte, C.; Perdrieux, F.; Daydé, J.; Debaeke, P. Reducing Herbicide Use and Leaching in Agronomicall Performant Maize-Based Cropping Systems: An 8-Year Study. *Sci. Total Environ.* **2021**, *788*, 147695. [CrossRef]
37. Al-Rajab, A.J.; Amellal, S.; Schiavon, M. Sorption and Leaching of 14 C-Glyphosate in Agricultural Soils. *Agron. Sustain. De* **2008**, *28*, 419–428. [CrossRef]
38. Sassman, J.; Pienta, R.; Jacobs, M.; Cioffi, J. *Pesticide Background Statements: Herbicides*; Forest Service, U.S. Department c Agriculture: Washington, DC, USA, 1984; Volume 1.
39. Rasmussen, S.B.; Abrahamsen, P.; Nielsen, M.H.; Holm, P.E.; Hansen, S. Effects of Single Rainfall Events on Leaching c Glyphosate and Bentazone on Two Different Soil Types, Using the DAISY Model. *Vadose Zone J.* **2015**, *14*, vzj2014-11. [CrossRe
40. Lewan, E.; Kreuger, J.; Jarvis, N. Implications of Precipitation Patterns and Antecedent Soil Water Content for Leaching c Pesticides from Arable Land. *Agric. Water Manag.* **2009**, *96*, 1633–1640. [CrossRef]
41. Al-Rajab, A.J.; Schiavon, M. Degradation of 14C-Glyphosate and Aminomethylphosphonic Acid (AMPA) in Three Agricultura Soils. *J. Environ. Sci.* **2010**, *22*, 1374–1380. [CrossRef]
42. Bento, C.P.; Goossens, D.; Rezaei, M.; Riksen, M.; Mol, H.G.; Ritsema, C.J.; Geissen, V. Glyphosate and AMPA Distribution i Wind-Eroded Sediment Derived from Loess Soil. *Environ. Pollut.* **2017**, *220*, 1079–1089. [CrossRef] [PubMed]

agronomy

MDPI

Article

Soil Macroinvertebrate Response to Paddy Rice Farming Pathways in Mpologoma Catchment, Uganda

Jackson Bunyangha [1,2,*], Agnes W. Muthumbi [2], Nathan N. Gichuki [2], Mwanjalolo J. G. Majaliwa [3] and Anthony Egeru [4]

1 Department of Science, Technical and Vocational Education, Makerere University, Kampala P.O. Box 7062, Uganda
2 Department of Biology, Faculty of Science and Technology, University of Nairobi, Nairobi P.O. Box 30197, Kenya; amuthumbi@uonbi.ac.ke (A.W.M.); ngichuki@uonbi.ac.ke (N.N.G.)
3 Department of Geography, Geo-Informatics and Climatic Sciences, Makerere University, Kampala P.O. Box 7062, Uganda; majaliwam@gmail.com
4 Department of Environmental Management, Makerere University, Kampala P.O. Box 7062, Uganda; egeru81@gmail.com
* Correspondence: jbjackson2011@gmail.com; Tel.: +256-703-140853

Abstract: Agricultural practices play a major role in influencing soil fauna abundance and diversity. Interest in these practices has increased owing to the growing need for sustainable agricultural systems in this era of increasing agricultural intensification. In this study, two paddy rice farming pathways (smallholder and large-scale commercial) and an adjacent natural wetland in Mpologoma catchment were studied to determine the response of soil macroinvertebrates to paddy rice farming pathways. Eighteen macroinvertebrate taxa were observed, some of which were not the usual soil taxa (Hirudinea, Decapoda, Ephemeroptera, Trichoptera, and Odonata). SIMPER analysis showed that Oligochaeta, Gastropoda, and Coleoptera were the major taxa responsible for dissimilarity among sites. Macroinvertebrate richness and diversity also varied among sites. Some taxa showed habitat exclusivity: Diptera, Odonata, and Trichoptera were exclusive to both rice paddies; Decapoda, Chilopoda, Diplopoda, and Blattodea to natural wetland; Diplura and Ephemeroptera were exclusive to large-scale commercial paddies. NMDS ordination showed that macroinvertebrate distribution among sites was strongly correlated with soil pH and calcium and moderately correlated with phosphorus. These results indicate that wetland conversion to rice paddies could affect macroinvertebrate richness and diversity and underscore the importance of soil environment in influencing the macroinvertebrate community in rice paddies.

Keywords: soil fauna; soil quality; agricultural systems; macroinvertebrates; rice paddies; wetlands

Citation: Bunyangha, J.; Muthumbi, A.W.; Gichuki, N.N.; Majaliwa, M.J.G.; Egeru, A. Soil Macroinvertebrate Response to Paddy Rice Farming Pathways in Mpologoma Catchment, Uganda. *Agronomy* **2022**, *12*, 312. https://doi.org/10.3390/agronomy12020312

Academic Editors: Umberto Anastasi, Aurelio Scavo and Jiaen Zhang

Received: 23 November 2021
Accepted: 13 January 2022
Published: 26 January 2022

1. Introduction

Establishing sustainable food production systems is a major global concern, and the need to balance the social, economic, and environmental aspects of crop production has attracted vast research in the area of agricultural sustainability [1,2]. In the rice production sector, however, much of the attention has been focused on yield maximization to meet the high rice demand for nutrition and poverty reduction [3–9]. Research on the environmental impacts of rice production has been limited [10]. In particular, the response of soil invertebrates to paddy rice farming pathways has received negligible focus, yet these invertebrates play a key role in leveraging sustainable paddy rice production [11,12].

Soil invertebrates range in size from the small microfauna (average size < 0.2 mm), such as nematodes, through the medium-sized mesofauna (0.2–2 mm), such as microarthropods and enchytraeids, to the largest macrofauna (>2 mm), such as arthropods, molluscs, annelids, and crustaceans [13,14]. By decomposing organic matter, modifying soil structure, and mediating nutrient cycling, among other functions, soil invertebrates enhance soil quality for sustainable crop production [13–19]. However, conservation of soil invertebrates continues to be a challenge in agricultural landscapes. This is largely attributed to their dwelling in the surface litter or in nests and burrows, such that they create in the top 20 cm of the soil profile, where they are prone to physical, chemical, or biological disturbance [13,20,21]. The sensitivity to disturbance is of utmost ecological significance in the function of soil invertebrates as bioindicators [22–25], and changes in their community characteristics can provide valuable feedback on prevailing soil management practices.

Many authors have highlighted the effects of conventional farm management practices on soil invertebrates [26–29]. They have shown how agricultural practices, such as tillage and field traffic (machine compaction), can affect soil invertebrate populations and reduce crop yields [30–32]. Irrigation has also been reported to reduce soil arthropod abundance: Menta et al. [20] observed that Acari, Collembola, and Hymenopteran numbers were higher in corn and wheat fields which had conservation (non-irrigated) soil management practices than in fields with conventional (irrigated) soil management practices. John et al. [33] also observed that microarthropod (mite and collembola) preferred a non-flooded crop rotation environment, unlike the enchytraeids. Similarly, agrochemical use is also reported to interfere with the soil invertebrate environment. For instance, Förster et al. [34] observed that the use of fungicide carbendazim and insecticide lambda-cyhalothrin reduces millipede *Trigoniulus corallinus* and earthworm *Pontoscolex corethrurus* abundance in terrestrial systems. However, it should be noted that, although the majority of the studies have reported detrimental effects, not all conventional farm management practices cause negative effects on soil biodiversity. For example, nitrogen (N) fertilizer addition has been demonstrated to increase mesostigmata mite richness and collembolan abundance, whereas reduced soil disturbance increased the species richness, abundance, and diversity of oribatid mites and collembola [35].

Most studies on the effect of management practice on soil biodiversity in croplands have been carried out in terrestrial areas. Studies targeting wetland environments and rice paddies in particular are few and have mainly focused on the effect of crop rotation [33,36]. No study has dealt with the response of soil macroinvertebrates to paddy rice farming pathways. A farming pathway is defined, in this study, as the course of agricultural management practices followed on a particular farm during a cropping cycle. Accordingly, two categories of paddy rice farming pathways are recognized in the paddy rice farming system: subsistence smallholder pathway and large-scale commercial farming pathway. These pathways differ in many respects including extent of mechanization, sown crop varieties, irrigation, technical agronomic services (fertilizer, pesticide, extension services), and farm size [2,37,38]. Subsistence smallholder farmers cultivate a threshold of 2 ha of land mainly for home consumption and are generally more resource constrained than their large-scale commercial counterparts [38–40]. We assumed that the management practices along each of these paddy rice farming pathways could have unique implications on the community characteristics of the soil macroinvertebrates.

Therefore, in this study, we investigated the response of soil macroinvertebrate taxa in terms of occurrence frequency, richness, abundance, evenness, and diversity in the different paddy rice farming pathways. Macroinvertebrate densities were further correlated with soil characteristics, such as pH, organic matter, nitrogen, phosphorus, potassium, and texture. We hypothesized that the frequency, abundance, richness, evenness, and diversity of macroinvertebrates would reduce along land use intensification gradient from natural wetland, through smallholder paddies, to large-scale commercial paddies. Furthermore, since paddies have unique soil and hydrological conditions, their macroinvertebrate community would be expected to at least encompass some terrestrial or aquatic taxa, such as the arthropod larvae and pupae.

2. Materials and Methods

2.1. Study Sites

This study was carried out in Mpologoma catchment, a renowned paddy rice growing watershed, in eastern Uganda (Figure 1). The catchment derives its name from River Mpologoma, which forms the major drainage network in the region. In the Mpologoma River's drainage system are several wetlands that the local communities have largely converted for rice growing. Majority of the rice paddies are of the smallholder type but there is also one large-scale commercial farm known as the Kibimba Rice Scheme. The Kibimba Rice Scheme is located at latitude 0°32′2462 N and longitude 33°52′9350 E and covers an area of 4350 hectares. It was established in 1973 by the government of Uganda to increase food production [37,41]. However, it is currently a private venture. The farmed area covering 3900 ha is divided into 18 blocks and each block is further subdivided into 4–6 strips of 1–4 plots. This area receives an annual rainfall of between 900–1400 mm and has 2 rainy seasons, late February–May and August–November, with a peak in April [41].

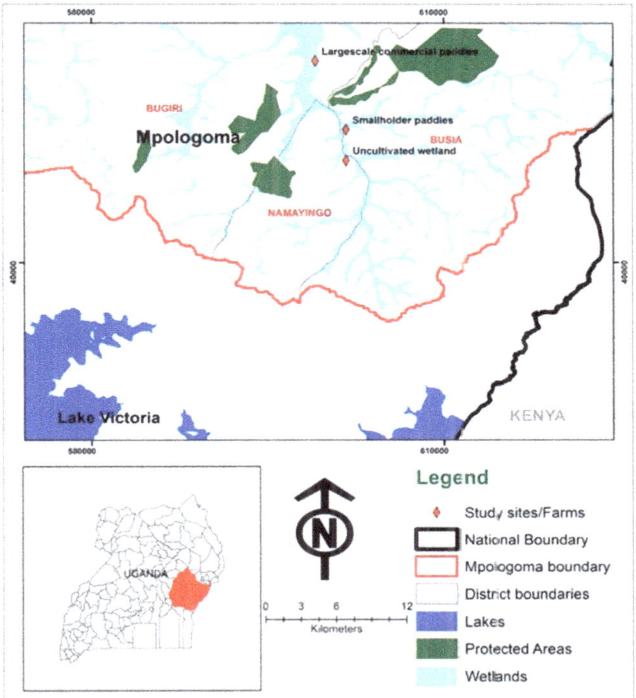

Figure 1. Location of the study sites in Mpologoma catchment.

Rice farming at the large-scale commercial farm (COM) is highly mechanized from land preparation to harvesting. The soil is clayish (Table S3). Ploughing of the land and seeding are mainly carried out twice a year in September–October and March–April. Rice growing is possible throughout the year due to the readily available irrigation water supplied from the reservoir (dam) constructed across River Kibimba. Flooding and drainage of rice fields is regulated through sluice gates constructed across the water canals. Herbicides, fertilizers, and pesticides are regularly used on the farm during the production cycle. The most commonly used herbicides include Glyphosate, which is mainly applied to control volunteer plants that emerge after primary cultivation, and D-Amine for post-emergence weeds in the first weeks of transplanting. The 1st fertilizer application to seedlings is carried out 11 days after sowing seeds in the nursery, while transplanting occurs at 25–28 days. Further fertilizer applications are carried out 5–7 days after transplanting the seedlings, then 30 days later, and the last dose is applied at 60 days after transplanting. The fields are flooded with water from the 65th day and this can go on till the 100th day. Afterwards, the plots are drained in preparation for harvesting about 20 days later. Rice varieties at the commercial farm are the short-duration type (mainly K 23, K 85, and Pusa) that mature between 105 and 120 days. Dimethoate and Beam can be applied when required to control stem borers and leaf fungi, respectively. Harvesting mainly occurs during the June–July and December–January dry seasons and it is carried out by combine harvesters. Fallowing is rare but if any, it may be between 1 and 2 growing seasons.

The smallholder rice paddies (SHD), on the other hand, are rain-fed and largely tilled using hand hoes or ox-ploughs. The soil texture is clay loam (Table S3). The plots are small, some measuring about 50 ft by 100 ft, and a farmer may use one or more plots for rice farming. Slash-and-burn is commonly practiced for field preparation. Planting usually occurs during the rainy seasons of late February–early March and late August–early September and the crop is harvested during the June–July and December–January dry seasons after about $3^{1}/_{2}$–4 months. Land preparation, planting, weeding, and rice harvesting rely on family or hired human labor. Agrochemicals, such as fertilizers, herbicides, and pesticides are not used on these farms. After harvest, some farmers immediately hand-till their fields in preparation for the next season, although others may choose to fallow their rice fields for a season or two.

A third sampling site, an uncultivated natural wetland (Uncul), was adopted as a control. The Uncul soil texture is loam (Table S3). The natural wetland is fed by river water flowing into the reservoir that supplies the large-scale commercial farm. It had pristine wetland vegetation cover and was hardly influenced by human activities, like that in the rice paddies. The vegetation mainly comprised *Cyperus papyrus* L., *Cyperus latifolius* Poir. Cycads, *Impatiens tinctoria* A.Rich, *Phragmites australis* (Cav.) Trin and *Echinocloa pyramidalis* (Lam.) Hitch., and Chase.

All the three sites (COM, SHD, and Uncul) were similar by being wetland areas located close to each other within a 10 km radius and had been under the particular land use/cover type and climatic conditions for at least 30 years [42].

2.2. Soil Sampling and Macroinvertebrate Extraction

Soil sampling for macroinvertebrate and chemical characteristics was carried out between September 2018 and December 2019. For macroinvertebrate analysis, four replicate plots were randomly selected in the large-scale and small-scale rice paddies for each crop stage: young rice (about one month after transplanting), mature rice (at harvest), and fallow (unplowed rice fields after harvest for at least one season). This was aimed at capturing the variability in macroinvertebrate community within and between farming pathways. In each plot, 3 monoliths measuring 25 × 25 × 15 cm deep were randomly dug out along a diagonal transect. The exact position of the monoliths was marked out using a wooden quadrat of 25 × 25 cm outside dimensions. In the first 2 days of invertebrate collection in September 2018, monoliths were initially dug to a depth of 30 cm [43,44]. However, because deep layers (15–30 cm) had barely any macroinvertebrates, sampling depth was adjusted to 0–15 cm in the later sampling. Similarly, soil replicates were collected from four randomly selected plots located in the natural wetland. However, since there were not crop stages in the natural wetland, 12 soil samples were collected from the 4 plots. Therefore, in total, 84 soil samples were collected. At each sampling point, litter within the 25 cm quadrat was initially hand-sorted for any surface macroinvertebrates before the monolith was dug out. Then, the monolith was carried to the sorting area, crumbled and all large invertebrates (>5 mm long) and visible under a magnifying glass were carefully hand-sorted and removed with forceps. The soil sample was subsequently mixed with water in a bucket and sieved over a 2 mm sieve to maximize extraction of macroinvertebrates. The collected macroinvertebrates were preserved in 10% formalin and carried to the laboratory where they were stained with Rose Bengal to ease sorting and counting under a binocular dissecting microscope.

Soil samples for chemical analysis were also collected from the top 0–15 cm adjacent to the sampling locations for soil samples for fauna analysis. The samples were then mixed homogeneously, air-dried, their gravel, roots, and large organic residues were removed, and then they were passed through a 2 mm sieve. Generally, analysis of soil samples for chemical parameters was performed following procedures described in [45]. The parameters measured included pH, texture (%), soil organic matter (%), available phosphorus (ppm), available nitrogen (%), available potassium (cmol/Kg of soil), available sodium (cmol/Kg of soil), and available calcium (cmol/Kg of soil). Soil texture was classified following the USDA soil texture triangle classification [46].

2.3. Data Analysis

Macroinvertebrate data were collected in terms of abundance of invertebrate taxa per soil sample. Frequency of occurrence of each taxon was then derived by considering the number of soil samples in which a particular taxon was present against the total number of replicates over the entire study. Community indices, such as richness, evenness, and diversity were also obtained. Diversity and evenness of soil macroinvertebrates were, respectively, calculated using Equations (1) and (2) [47–50], as follows:

$$H' = -\sum_{i=1}^{s} p_i * In(p_i) \quad (1)$$

$$J = H'/In(s) \quad (2)$$

where H' is the Shannon–Wiener diversity index, s is the total number of taxa collected, p_i is the proportion of individuals of taxon i (n_i) relative to all individuals (N) from all the collected taxa for a particular land use ($p_i = n_i/N$), and J is Pielou's measure of evenness.

Rank–abundance curves were then produced to display the relative abundance of macroinvertebrate taxa for each farming pathway and natural wetland. Macroinvertebrate abundances, richness, evenness, and diversity were further compared across sites and crop stages using one-way ANOVA in R 4.0.4 [51]. This aimed to establish whether or not the differences in means varied among the categories. Count data were standardized to densities (ind./m^2) [43], normality, and homoscedasticity evaluated using the Shapiro–Wilk test and Levene's test, respectively.

To visualize how farming pathway related with the macroinvertebrate community, non-metric multidimensional scaling (NMDS) was performed based on Bray–Curtis dissimilarity index [52,53]. The results were plotted in an NMDS ordination plane, in a two-dimensional space. Differences in the macroinvertebrate community among the different patterns visualized with NMDS were analyzed using a permutational multivariate analysis of variance (PERMANOVA) test. After a significant PERMANOVA test ($p \leq 0.05$), SIMPER (similarity percentages) analysis was performed to examine which invertebrate groups were driving the differences in community among the sites. The significance of contribution of each macroinvertebrate group to the differences and which of the soil environmental factors (pH, SOM, N, P, K, Na, Ca, and texture) were significantly correlated to the first two axes of the NMDS ordination plane were tested using the envfit function. NMDS ordination, PERMANOVA, SIMPER, and envfit function were performed with the vegan package (2.5–7) in R 4.0.4 [51,54].

3. Results

3.1. Macroinvertebrate Community Composition

A total of 18 macroinvertebrate taxa were observed in the soil samples. They belonged to 3 phyla Annelida, Mollusca, and Arthropoda at 61.3%, 17.9%, and 20.8%, respectively (Table S1). The arthropods were mainly larvae and nymphs. Oligochaeta (57.3%), Gastropoda (17.9%), Coleoptera (12.4%), Hirudinea (4.0%), Hymenoptera (2.1%), Diptera (1.6%), and Trichoptera (1.1%) were the most abundant groups and constituted about 96% of the macroinvertebrates. Arachnida, Isopoda, Dermaptera, and Orthoptera comprised about 3% while the rest of the taxa Decapoda, Diplopoda, Chilopoda, Diplura, Ephemeroptera, Blattodea, and Odonata totaled less than 1%. In addition, Oligochaeta was the most ubiquitous taxon followed by Coleoptera, Gastropoda, and Hirudinea (Table S2). Macroinvertebrate occurrence frequencies across sites and crop stages did not follow a specific pattern. For instance, oligochaetes occurred in 100%, 97%, and 72% of the soil samples collected from the natural wetland, smallholder paddies, and large-scale commercial paddies, respectively; meanwhile, Coleoptera were mainly found in soil samples from the large-scale commercial paddies (66.7%), followed by natural wetland (58.3%), and least in the smallholder paddies (33.3%). Similarly, the majority of the Gastropods and Hirudinea were encountered in the large-scale commercial paddies (72.2% and 50.0%, respectively), followed by smallholder paddies (33.3% and 38.9%, respectively), and natural wetland (25.0% and 8.3%, respectively). Overall, eight of the taxa were common to all the three sites. One taxon (Isopoda) was common to both large-scale commercial paddies and natural wetland, while three taxa (Diptera, Odonata, and Trichoptera) were common to both large-scale commercial and smallholder paddies. Four taxa (Decapoda, Chilopoda, Diplopoda, and Blattodea) were exclusive to the natural wetland, while two taxa (Diplura and Ephemeroptera) were exclusive to the large-scale commercial paddies. No taxa were exclusive to smallholder paddies and none to both smallholder and natural wetland. Taxa such as Hirudinea, Decapoda, Trichoptera, Ephemeroptera, and Odonata which are not regular soil candidates, especially in terrestrial soils, were also encountered in the samples.

The overall macroinvertebrate density was 304.8 ± 327.6 ind./m^2 and ranged between 284.0 ± 210.3 ind./m^2 (in the natural wetland), 307.6 ± 357.2 ind./m^2 (in the large-scale commercial paddies), and 308.9 ± 336.0 ind./m^2 (in the smallholder farms) (Table 1). Density per replicate ranged between 16 and 1584 ind./m^2 in the large-scale rice paddies, 32 and 1872 ind./m^2 in the smallholder paddies, and 32 and 736 ind./m^2 in the natural wetland. Across the crop stages, macroinvertebrate density was highest in the smallholder harvest (353.3 ± 493 ind./m^2) and lowest in the commercial fallow (250.7 ± 247.5 ind./m^2). There was high variability in density among replicates and it muffled any significant differences in total density between crop stages and sites. Nevertheless, some taxa showed significant differences in their density among sites. For instance, Oligochaeta, the most abundant taxon overall, had highest density (253.8 ± 345.2 ind./m^2) in the smallholder paddies compared with the large-scale paddies (86.2 ± 193.7 ind./m^2) and the natural wetland (202.7 ± 200.7 ind./m^2) and this variation among sites was significant ($p < 0.05$). Post hoc comparison indicated that the difference was mainly between the smallholder and large-scale pair ($p < 0.05$).

The second most abundant taxon was Gastropoda. It recorded significantly higher densities ($p < 0.05$) in the large-scale paddies (111.1 ± 212.7 ind./m^2) compared with smallholder rice paddies (12.4 ± 31.6 ind./m^2) and natural wetland (10.7 ± 21.9 ind./m^2). A post hoc test showed that the difference was also mainly between the smallholder and large-scale paddies ($p < 0.05$). The other important taxon in terms of density was Coleoptera. Coleoptera had a relatively higher density in the large-scale commercial paddies (72.0 ± 214.4 ind./m^2) compared with smallholder rice paddies (9.8 ± 24.9 ind./m^2) and natural wetland (18.7 ± 19.1 ind./m^2), but the difference among sites was not significant ($p > 0.05$). The rest of the taxa recorded relatively low densities and the differences were also not significant among sites and crop stages ($p > 0.05$).

Table 1. Mean density ± SD (ind.m^{-2}) of soil macroinvertebrate taxa by site and crop stage.

Taxa	Crop Stages						Sites			Overall
	Com-You	Com-Har	Com-Fal	Shd-You	Shd-Har	Shd-Fal	Com	Shd	Natural Wetland	
Oligochaeta	176 ± 313.8	58.7 ± 77.3	24.0 ± 33.1	186.7 ± 260.4	313.3 ± 503.4	261.3 ± 221.8	86.2 ± 193.7	253.8 ± 345.2	202.7 ± 200.7	174.7 ± 278.7
Hirudinea	10.7 ± 19.7	13.3 ± 19.1	18.7 ± 16.5	14.7 ± 27.7	4.0 ± 9.9	22.7 ± 22.1	14.2 ± 18.3	13.8 ± 22.0	1.3 ± 4.6	12.2 ± 19.2
Gastropoda	140 ± 300	49.3 ± 74.9	144.0 ± 205.1	22.7 ± 48.9	13.3 ± 22.5	1.3 ± 4.6	111.1 ± 212.7	12.4 ± 31.6	10.7 ± 21.9	54.5 ± 148.3
Arachnida	-	2.7 ± 9.2	6.7 ± 12.7	2.7 ± 6.2	1.3 ± 4.6	-	3.1 ± 9.2	1.3 ± 4.5	5.3 ± 14.2	2.7 ± 8.5
Isopoda	-	2.7 ± 9.2	-	-	-	-	0.9 ± 5.3	-	9.3 ± 27.7	1.7 ± 11.1
Decapoda	-	-	-	-	-	-	-	-	1.3 ± 4.6	0.2 ± 1.7
Hymenoptera	1.3 ± 4.6	1.3 ± 4.6	-	8.0 ± 19.9	12.0 ± 32.8	5.3 ± 18.5	0.9 ± 3.7	8.4 ± 24.0	17.3 ± 50.3	6.5 ± 24.9
Coleoptera	6.7 ± 8.2	164 ± 356	45.3 ± 69.2	9.3 ± 12.7	2.7 ± 6.2	17.3 ± 40.6	72.0 ± 214.4	9.8 ± 24.9	18.7 ± 19.1	37.7 ± 143.5
Diptera	5.3 ± 14.2	9.3 ± 18.6	1.3 ± 4.6	13.3 ± 22.5	4.0 ± 7.2	1.3 ± 4.6	5.3 ± 13.8	6.2 ± 14.5	-	5.0 ± 13.1
Dermaptera	-	6.7 ± 10.7	-	-	1.3 ± 4.6	-	2.2 ± 6.8	0.4 ± 2.7	5.3 ± 7.9	1.9 ± 5.5
Diplura	1.3 ± 4.6	1.3 ± 4.6	-	-	-	-	0.9 ± 3.7	-	-	0.4 ± 2.5
Chilopoda	-	-	-	-	-	-	-	-	2.7 ± 9.2	0.4 ± 3.5
Diplopoda	-	-	-	-	-	-	-	-	1.3 ± 4.6	0.2 ± 1.7
Ephemeroptera	-	-	1.3 ± 4.6	-	-	-	0.4 ± 2.7	-	-	0.2 ± 1.7
Orthoptera	1.3 ± 4.6	-	5.3 ± 14.2	1.3 ± 4.6	-	4.0 ± 9.9	2.2 ± 8.7	1.8 ± 6.4	6.7 ± 12.7	2.7 ± 8.5
Blattodea	-	-	-	-	-	-	-	-	1.3 ± 4.6	0.2 ± 1.7
Odonata	-	1.3 ± 4.6	1.3 ± 4.6	-	1.3 ± 4.6	-	0.9 ± 3.7	0.4 ± 2.7	-	0.6 ± 3.0
Trichoptera	-	18.7 ± 35.3	2.7 ± 6.2	1.3 ± 4.6	-	-	7.1 ± 21.8	0.4 ± 2.7	-	3.2 ± 14.6
Overall	342.7 ± 401.7	329.3 ± 404	250.7 ± 275.5	260.0 ± 247.5	353.3 ± 493	313.3 ± 223.7	307.6 ± 357.2	308.9 ± 336.0	284.0 ± 210.3	304.8 ± 327.6

Abbreviations: Com—large-scale commercial paddies; Shd—smallholder paddies; You—young; Har—harvest/mature; Fal—fallow.

3.1.1. Richness, Evenness, and Diversity

Macroinvertebrate richness ranged between 3.47 ± 1.38 in the large-scale paddies, 3.42 ± 2.07 in the natural wetland, and 2.64 ± 1.22 in the smallholder paddy soil replicates (Table 2). The difference in richness among sites was significant ($p < 0.05$) and a post hoc test indicated that it was significant between the smallholder and large-scale paddies pair ($p < 0.05$). Similarly, the diversity of macroinvertebrates was 0.85 ± 0.48, 0.72 ± 0.65 and 0.56 ± 0.44 in the large-scale commercial paddies, natural wetland, and smallholder paddies, respectively. It significantly varied among sites ($p < 0.05$) and was also between the smallholder and large-scale commercial paddies pair ($p < 0.05$). On the other hand, variation in evenness among sites was not significant ($p > 0.05$).

Table 2. Mean ± SD richness, evenness and diversity of macroinvertebrates across crop stages and sites.

Index	Crop Stages						Sites		
	Com-You	Com-Har	Com-Fal	Shd-You	Shd-Har	Shd-Fal	Com	Shd	Uncul
Richness per replicate	2.83 ± 1.27	4.08 ± 0.90	3.50 ± 1.68	3.17 ± 1.47	2.25 ± 1.29	2.50 ± 0.67	3.47 ± 1.38	2.64 ± 1.22	3.42 ± 2.07
Pileou's Evenness (J′)	0.53 ± 0.36	0.80 ± 0.16	0.65 ± 0.35	0.61 ± 0.30	0.35 ± 0.37	0.54 ± 0.31	0.66 ± 0.31	0.50 ± 0.34	0.53 ± 0.35
Shannon's Diversity (H′)	0.59 ± 0.46	1.10 ± 0.32	0.87 ± 0.53	0.77 ± 0.46	0.39 ± 0.48	0.52 ± 0.31	0.85 ± 0.48	0.56 ± 0.44	0.72 ± 0.65

Abbreviations: Com—large-scale commercial paddies; Shd—smallholder paddies; You—young; Har—harvest/mature; Fal—fallow.

3.1.2. Rank Abundance Relationship

The macroinvertebrate community of the sampled sites was characterized by a few abundant taxa (Figure 2). The steepest slope was for smallholder rice paddies with one dominant taxon, followed by natural wetland with a less steep slope than for smallholder paddies, and also dominated by one taxon, and then the largescale commercial paddies with a shallower gradient but with three relatively more evenly abundant taxa. The majority of the taxa fell in the rare category, arbitrarily defined in this study as a relative abundance of 0.5% or less in a given farming pathway or natural wetland over the entire study.

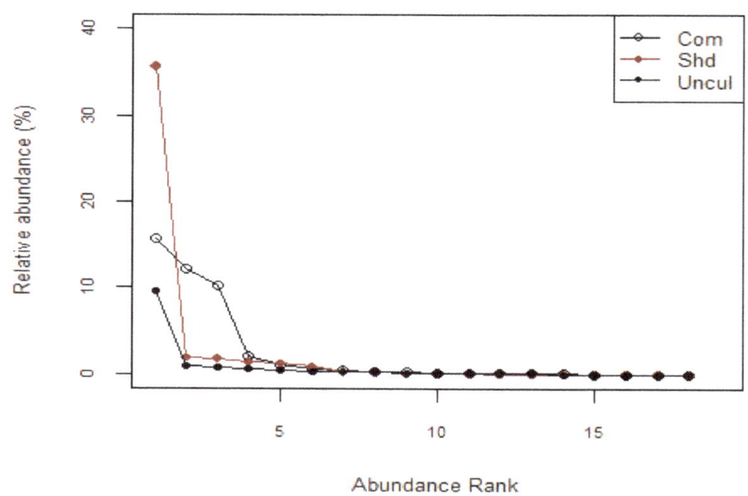

Figure 2. Rank abundance curves for the soil macroinvertebrates observed in the paddy rice farming pathways in Mpologoma catchment. Abbreviations: Com—large-scale commercial paddies; Shd—smallholder paddies; Uncul—natural wetland.

3.1.3. Community Ordination

The macroinvertebrate community in the large-scale commercial rice paddies (COM), smallholder rice paddies (SHD), and natural wetland (Uncul) showed some overlap (Figure 3), especially between SHD and Uncul, and 15% of the variation in communities was due to management differences as confirmed by PERMANOVA ($p \leq 0.001$, $R^2 = 0.151$). A subsequent pairwise SIMPER analysis to examine the average contribution of the different macroinvertebrate groups to the overall dissimilarity between site pairs revealed that Oligochaeta, Gastropoda, and Coleoptera were the most influential taxa (Table 3). Oligochaetes contributed largest to the variation in each site pair. In addition, the first 2 invertebrate taxa accounted for over 60% of the dissimilarity in each pairwise site comparison, while with the third taxon, Coleoptera, over 75% of the dissimilarity between sites could be explained. Further analysis of the intrinsic taxa indicated that Oligochaeta, Gastropoda, Coleoptera, and Odonata were the only taxa that had significant contribution to the overall dissimilarity among sites (Table 4).

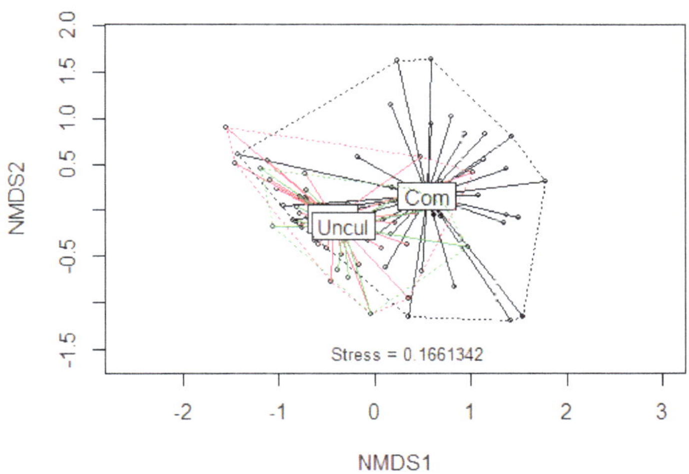

Figure 3. Bray-Curtis based NMDS plot of soil macroinvertebrate community composition. Points represent macroinvertebrate samples condensed on the two axes of NMDS plot. Spider diagrams connect the points to the respective land use type: Com (black), Uncul (green), and Shd (red).

Table 3. Cumulative percentage contribution of most influential taxa to the overall similarity between site pairs.

Macroinvertebrate Community Pair	Oligochaeta	Oligochaeta and Gastropoda	Oligochaeta, Gastropoda, and Coleoptera
Com–Shd	47	69	82
Com–Uncul	43	65	78
Shd–Uncul	59	67	74

Abbreviations: Com—large-scale commercial paddies; Shd—smallholder paddies; Uncul—natural wetland.

Table 4. Taxa driving the overall distribution pattern across sites.

Taxa	Squared Correlation Coefficient, R^2	*p*-Value
Oligochaeta	0.5245	0.001 ***
Gastropoda	0.3677	0.001 ***
Coleoptera	0.1871	0.001 ***
Odonata	0.1477	0.002 **

Significance codes: ** $p < 0.01$, *** $p < 0.001$.

3.2. Soil Environment and Its Relationship with Macroinverterbate Community

Generally, soil characteristics varied across the sites (Table S3). The overall average pH was 5.73 ± 0.06 and ranged between 5.54 ± 0.11 in the large-scale commercial paddies, 5.81 ± 0.08 in smallholder paddies, and 6.03 ± 0.11 in natural wetland. The difference in pH among sites was significant ($p < 0.05$) and the post hoc test showed that it was mainly from the natural wetland and large-scale paddies pair ($p < 0.05$). Similarly, soil organic matter (SOM), nitrogen (N), phosphorus (p), and calcium (Ca) contents were, respectively, 37.53 ± 3.74%, 0.46 ± 0.06%, 0.24 ± 0.02 ppm, and 51.64 ± 11.48 cmol/Kg in the natural wetland, 26.36 ± 2.5%, 0.46 ± 0.02%, 0.18 ± 0.01 ppm, and 30.63 ± 5.21 cmol/Kg in the smallholder paddies, and 17.7 ± 1.38%, 0.43 ± 0.02%, 0.13 ± 0.01 ppm, and 8.17 ± 1.44 cmol/Kg in the large-scale commercial rice paddies. There were significant differences ($p < 0.001$) in the variation of phosphorus and calcium among sites. Post hoc test indicated that the differences in p and Ca were between the pairs: natural wetland and large-scale paddies ($p < 0.001$ and $p < 0.001$, respectively), smallholder and large-scale paddies ($p < 0.01$ and $p < 0.01$, respectively), and natural wetland and smallholder paddies ($p < 0.01$ and $p < 0.1$, respectively). Similarly, the percentage of clay varied ($p < 0.001$) such that it was highest in the large-scale paddies (41.39 ± 1.57%), followed by smallholder paddies (32.56 ± 1.73%), and lowest in the natural wetland (24.50 ± 3.42%). On the other hand, sand also varied ($p < 0.01$) being highest in the natural wetland (44.25 ± 1.56%), followed by smallholder paddies (40.64 ± 2.31%), and large-scale paddies (33.89 ± 1.56%). Post hoc test showed that the differences in sand and clay composition were between the pairs: smallholder and large-scale commercial paddies ($p < 0.05$ and $p < 0.01$, respectively) and natural wetland and large-scale commercial paddies ($p < 0.05$ and $p < 0.001$, respectively). No significant differences were observed for SOM, sodium (Na), nitrogen (N), potassium (K), and silt.

Soil environment variables were further correlated with the NMDS ordination scores of the macroinvertebrate community. Soil pH, Ca, P, and sand had a negative correlation while clay content had a positive correlation (Figure 4). Basing on R^2 and the p-values to discriminate the strength of the relationship, it was shown that there was a strong correlation for pH and calcium ($p < 0.001$), moderate intensity correlation for phosphorus ($p < 0.01$), and a low intensity correlation for clay and sand ($p < 0.05$) (Table 5). SOM, N, Na, K, and silt were not significantly correlated with the NMDS ordination pattern of the macroinvertebrates in this study.

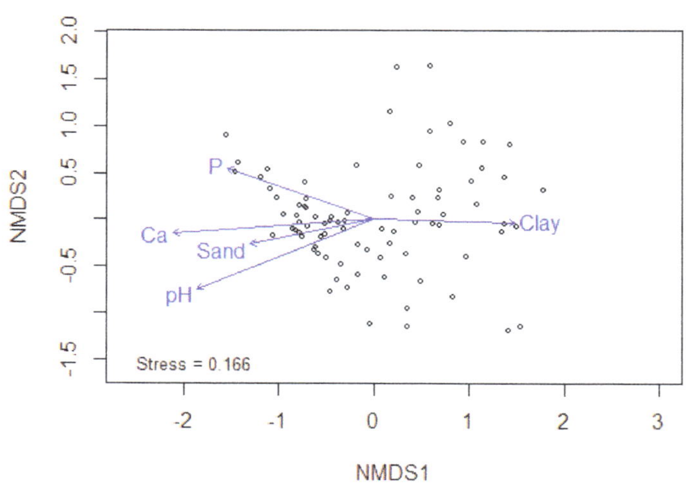

Figure 4. Soil environment variables correlated with macroinvertebrate samples in an ordination system. Points represent samples. Arrows show the significant vectors while their direction represents the gradient direction of the environmental driver.

Table 5. Correlation analysis of soil characteristics with NMDS scores.

Variable	R-Squared	p-Value
pH	0.1954	0.001 ***
SOM [1]	0.0231	0.394
N	0.0291	0.305
P	0.1256	0.007 **
K	0.0398	0.193
Na	0.0397	0.181
Ca	0.2156	0.001 ***
Sand	0.0855	0.027 *
Clay	0.1121	0.012 *
Silt	0.0094	0.694

[1] SOM—soil organic matter. R-squared—squared correlation coefficient. Significance codes: * p-value < 0.05, ** p-value < 0.01, *** p-value < 0.001.

4. Discussion

The study has shown that macroinvertebrate taxa responded differently to the paddy rice farming pathways in Mpologoma catchment during the 2018–2019 period. In particular, the most ubiquitous oligochaetes were encountered in 100% of the samples from the natural wetland but the frequency reduced in the large-scale commercial and smallholder rice paddies. Similar studies have observed higher or lower frequencies of some oligochaete species in the natural grassland and/or hardwood forest than in the cultivated areas [55]. In the current study, however, oligochaetes were not identified to the lower levels; hence, differences that could arise at species levels were not captured. Nevertheless, a similar trend would be obtained with some oligochaete species showing higher or lower frequencies in the natural wetland compared with the cultivated paddies. In the rice paddies, oligochaete density also varied: it was significantly higher in the smallholder than large-scale commercial paddies. Variation in oligochaete frequency and density among sites could be attributed to the differences in soil conditions. Generally, the magnitude of physical soil disturbance through tillage and chemical disturbance through agrochemical use increases with farming intensity from the pristine natural wetland condition where the soil is largely at rest to the cultivated condition where the soil is regularly worked for farming. Such disturbance has been reported to affect the distribution of soil invertebrates within the soil column and across habitats [21,32,35]. At the large-scale commercial farm, pesticides and inorganic fertilizers are heavily used coupled with high mechanization and associated soil compaction. Conversely, the smallholder paddy rice farmers in this area do not use inorganic fertilizers; they rely on the natural fertility of the alluvial soil deposited by moving water. Soil compaction is also minimal. Therefore, these factors could have made the soil environment at the large-scale commercial farm less conducive to oligochaetes compared with that in the smallholder rice paddies and natural wetland [32,34,56–58]. In addition, the lowest oligochaete density observed in the commercial fallow stage was attributed to the farm management dynamics whereby plots that were not in active agricultural use were not supplied with irrigation water. Consequently, this could have increased the concentration of agrochemical residues in those plots which jeopardized oligochaete survival [59].

Gastropod density also varied among sites, being highest in the large-scale commercial rice paddies and lowest in the natural wetland. Gastropods are largely herbivorous organisms and some are avid feeders on young rice stems [60–62]. Therefore, the stable food supply at the large-scale farm where rice is grown throughout the year coupled with reduced predation from diplopods and chilopods [63] could have been major drivers for the high density of gastropods in the large-scale rice paddies. Additionally, the scaring away of birds by the bird chasers who are employed by the commercial farm could also have significantly contributed to the reduced predation on snails by the predatory birds such as the African Jacana (Actophilornis africanus) and ducks, among others [41].

Densities for individual arthropod taxa were low except coleoptera and did not significantly vary among sites. Nevertheless, some arthropods showed habitat exclusiveness. In this category were the detritivorous diplopods which were exclusive to the natural wetland. Their exclusivity could be attributed to litter availability because the short fallow periods (usually one season or none) in the cultivated large-scale and smallholder rice paddies coupled with the routine soil disturbance may not have allowed sufficient time for litter to accumulate in these fields compared with the natural wetland where the soil had negligible perturbations. Similarly, Diplurans were only found in the large-scale commercial rice paddies, as previously documented [64]. This preference could be related to high prey abundances, such as insect larvae (e.g., Coleopteran nymphs and Trichopteran larvae), in the large-scale paddies compared with the natural wetland and smallholder rice paddies. Furthermore, some studies have documented high Acari and Collembola abundance in vineyards [20,65] relative to other arthropods, but in our study these taxa had low densities and insignificantly contributed to the dissimilarity among sites. This implies that it is difficult to make generalizations about macroinvertebrate abundances and diversity in varying agricultural systems because the agricultural practices vary and are complex, yet the response of macroinvertebrates might be a habitat-specific occurrence [66].

Land use conversion from natural wetland to cultivated rice fields, especially conversion to the large-scale commercial paddy rice farming, was also associated with increased richness and diversity of macroinvertebrates. Indeed, the rank–abundance relationship confirmed a higher macroinvertebrate evenness in the large-scale rice paddies than other sites. This contrasts with similar studies where diversity of macroinvertebrates decreased with increasing land use intensification [55,67]. The high diversity at the large-scale commercial farm could be attributed to the near-mosaic farm setup, where the bunds of natural vegetation that separate plots, strips, and blocks, and the road network which provides access to the plots, could have created refugia that enhanced macroinvertebrate diversity [68].

From the ordination system, macroinvertebrate communities of smallholder paddies and natural wetland seemed to be highly similar. This was attributed to the high proximity of the smallholder paddies to the natural wetland which could have easily allowed fauna mobility between the two adjacent sites in the landscape [20,69]. Furthermore, the NMDS regression showed that soil pH, calcium, phosphorus, clay, and sand were the major paddy soil variables that had significant relationships with the ordination pattern. Similar studies [65] have reported factors such as soil pH, SOM, and temperature to be important soil environment variables that drive invertebrate distribution, especially arthropods, in vineyards. They found texture to be insignificant; however, we found that texture, especially clay and sand content, were significantly correlated with macroinvertebrate distribution in rice paddies. This implies that the soil environment of rice paddies could be influenced by some unique factors unlike terrestrial croplands, and these would need further exploration. Nevertheless, this study has highlighted that changes in soil properties could play a role in driving the community characteristics of macroinvertebrates in rice paddies.

This study recorded some macroinvertebrate taxa that are not regular soil residents. These included Hirudinea, Decapoda, Ephemeroptera, Trichoptera, and Odonata. Of these, Hirudinea have previously been observed in some studies [70]. What could be driving such taxa to wetlands and paddies in particular is not certain, but could be attributed to the wetland environment itself, as some taxa have to return to water to breed, making flooded wetlands suitable habitats. Additionally, the changing climate could be impacting the breeding behavior of these taxa in ways that are not currently known. Our study used an unbalanced design where some samples especially from the natural wetland were less than those from other sites. We think that this could have somehow affected these results, though some authors (e.g., Milliken and Johnson [71]) argue that lack of balance does not usually affect results in a single-factor ANOVA.

5. Conclusions

This study aimed to evaluate the response of soil macroinvertebrates to paddy rice farming pathways during the 2018–2019 period. A total of 18 macroinvertebrate taxa were observed, dominated by Oligochaeta. Three taxa: Oligochaeta, Gastropoda, and Coleoptera accounted for most of the dissimilarity among sites. Some taxa showed habitat exclusivity: Decapoda, Chilipoda, Diplopoda, and Blattodea were exclusive to the natural wetland, while two taxa (Diplura and Ephemeroptera) were exclusive to the large-scale commercial paddies. No taxa were exclusive to smallholder paddies and none to both smallholder and natural wetland. However, three taxa (Diptera, Odonata, and Trichoptera) were common to both large-scale commercial and smallholder paddies. Considering richness and diversity, large-scale commercial paddies generally had higher richness and diversity than natural wetland and smallholder paddies. Among the soil environmental variables, pH, phosphorus, calcium, clay, and sand seemed to most influence the macroinvertebrate community in the study area. We also observed some taxa that are not usually in soils. These included Hirudinea, Decapoda, Ephemeroptera, Trichoptera, and Odonata. Further studies are needed to extend experiments to other paddy rice farming contexts, monitor macroinvertebrate diversity over a longer term in a balanced design, and explore the potential of using such taxa in monitoring soil quality in rice paddies. The studies would identify indicator macroinvertebrate species that are tolerant or intolerant to chemical and physical soil disturbance in the paddies. It would also be necessary to identify any other apparently new macroinvertebrate taxa in paddies that could be using wetlands during some of their growth stages to respond to the changing global climate. Rice farmers, especially the smallholders, should also mosaic their paddy rice fields with natural areas to create refugia, increasing macroinvertebrate biodiversity.

Supplementary Materials: The following are available online at: https://www.mdpi.com/article/10.3390/agronomy12020312/s1, Table S1: Percentage composition of soil macroinvertebrate taxa by site and crop stage, Table S2: Proportion of soil samples where macroinvertebrate taxa were encountered per site and crop stage, Table S3: Soil characteristics (Mean ± SE) per site and crop stage.

Author Contributions: Conceptualization, J.B., A.W.M., N.N.G. and A.E.; methodology, J.B., A.W.M. and N.N.G.; validation, J.B., A.W.M. and N.N.G.; formal analysis, J.B.; investigation, J.B. and A.W.M.; data curation, J.B., A.W.M., N.N.G., M.J.G.M. and A.E.; writing—original draft preparation, J.B., A.W.M., N.N.G., M.J.G.M. and A.E.; writing—review and editing, J.B., A.W.M., N.N.G., M.J.G.M. and A.E.; visualization, J.B. and A.W.M.; supervision, A.W.M., N.N.G. and M.J.G.M.; project administration, J.B.; funding acquisition, J.B. and A.E. All authors have read and agreed to the published version of the manuscript.

Funding: This research was funded by RUFORUM under the Social and Environmental Trade-offs in African Agriculture (SENTINEL) Project of the Global Challenges Research Fund (GCRF), Research Councils UK (Grant # RU/2018/GTA/DRG/039), and Makerere University (DELP Small Research Grant).

Institutional Review Board Statement: Not applicable. This study did not involve animal or human subjects.

Informed Consent Statement: Not applicable. This study did not involve human subjects.

Data Availability Statement: The data presented in this study are openly available in FigShare at https://doi.org/10.6084/m9.figshare.17040803.v1 (accessed on 9 November 2021).

Acknowledgments: We thank the farm owners who allowed us to carry out research in their fields. We are very grateful to our field assistants Benjamin Simiyu and David from Kibimba farm for assisting with language interpretation and other field-related activities that smoothened our fieldwork in the farmers' paddies. We appreciate Aguttu Gorret from Makerere University for chemical analysis of soil samples and the technical staff of the School of Biological Sciences, University of Nairobi, for lab and materials during macroinvertebrate sample collection and lab analysis.

Conflicts of Interest: The authors declare no conflict of interest.

References

1. FAO. *Sustainable Agriculture: A Tool to Strengthen Food Security and Nutrition in Latin America and the Caribbean: 2014–2015 Highlights*; FAO: Rome, Italy, 2016.
2. Stringer, L.C.; Fraser, E.D.G.; Harris, D.; Lyon, C.; Pereira, L.; Ward, C.F.M.; Simelton, E. Adaptation and Development Pathways for Different Types of Farmers. *Environ. Sci. Policy* **2020**, *104*, 174–189. [CrossRef]
3. Van Nguyen, N.; Ferrero, A. Meeting the Challenges of Global Rice Production. *Paddy Water Environ.* **2006**, *4*, 1–9. [CrossRef]
4. Norman, J.C.; Kebe, B. *African Smallholder Farmers: Rice Production and Sustainable Livelihoods. International Rice Commission Newsletter*; FAO: Rome, Italy, 2006.
5. Chenoune, R.; Belhouchette, H.; Paloma, S.G.Y.; Capillon, A. Assessing the Diversity of Smallholder Rice Farms Production Strategies in Sierra Leone. *NJAS-Wagening J. Life Sci.* **2016**, *76*, 7–19. [CrossRef]
6. Rodenburg, J.; Zwart, S.J.; Kiepe, P.; Narteh, L.T.; Dogbe, W.; Wopereis, M.C.S. Sustainable Rice Production in African Inland Valleys: Seizing Regional Potentials through Local Approaches. *Agric. Syst.* **2014**, *123*, 1–11. [CrossRef]
7. Tomich, T.P.; Lidder, P.; Coley, M.; Gollin, D.; Meinzen-Dick, R.; Webb, P.; Carberry, P. Food and Agricultural Innovation Pathways for Prosperity. *Agric. Syst.* **2019**, *172*, 1–15. [CrossRef]
8. OECD-FAO. *Agricultural Outlook 2018–2027*; OECD Publishing: Paris, France; FAO: Rome, Italy, 2018.
9. Annual Report 2018-International Rice Research Institute (IRRI) by Glenn Concepcion-Issuu. Available online: https://issuu.com/glennconcepcion/docs/irri_annual_report_2018 (accessed on 9 November 2021).
10. Pingali, P.L.; Rosegrant, M.W.; Prabhu, L.; Mark, W. Confronting the Environmental Consequences of the Green Revolution In Asia. *Educ. New Environ. Leadersh. Asia* **1994**, *2*, 61–74.
11. Roger, P.A.; Heong, K.L.; Teng, P.S. Biodiversity and Sustainability of Wetland Rice Production: Role and Potential of Microorganisms and Invertebrates. In *The Biodiversity of Microorganisms and Invertebrates: Its Role in Sustainable Agriculture*; CAB International: Wallingford, UK, 1991.
12. Keesstra, S.D.; Bouma, J.; Wallinga, J.; Tittonell, P.; Smith, P.; Cerdà, A.; Montanarella, L.; Quinton, J.N.; Pachepsky, Y.; Van Der Putten, W.H.; et al. The Significance of Soils and Soil Science towards Realization of the United Nations Sustainable Development Goals. *Soil* **2016**, *2*, 111–128. [CrossRef]
13. Lavelle, P.; Bignell, D.; Lepage, M.; Wolters, V.; Roger, P.; Ineson, P.; Heal, O.W.; Dhillion, S. Soil Function in a Changing World: The Role of Invertebrate Ecosystem Engineers. *Eur. J. Soil Biol.* **1997**, *33*, 159–193.
14. Lavelle, P.; Decaëns, T.; Aubert, M.; Barot, S.; Blouin, M.; Bureau, F.; Margerie, P.; Mora, P.; Rossi, J.P. Soil Invertebrates and Ecosystem Services. *Eur. J. Soil Biol.* **2006**, *42*, S3–S15. [CrossRef]
15. Schmidt, A.; John, K.; Arida, G.; Auge, H.; Brandl, R.; Horgan, F.G.; Hotes, S.; Marquez, L.; Radermacher, N.; Settele, J.; et al. Effects of Residue Management on Decomposition in Irrigated Rice Fields Are Not Related to Changes in the Decomposer Community. *PLoS ONE* **2015**, *10*, e0134402. [CrossRef] [PubMed]
16. Brussaard, L.; de Ruiter, P.C.; Brown, G.G. Soil Biodiversity for Agricultural Sustainability. *Agric. Ecosyst. Environ.* **2007**, *121*, 233–244. [CrossRef]
17. Frouz, J. Effects of Soil Macro- and Mesofauna on Litter Decomposition and Soil Organic Matter Stabilization. *Geoderma* **2018**, *332*, 161–172. [CrossRef]
18. Gardi, C.; Montanarella, L.; Arrouays, D.; Bispo, A.; Lemanceau, P.; Jolivet, C.; Mulder, C.; Ranjard, L.; Römbke, J.; Rutgers, M.; et al. Soil Biodiversity Monitoring in Europe: Ongoing Activities and Challenges. *Eur. J. Soil Sci.* **2009**, *60*, 807–819. [CrossRef]
19. Schmidt, A.; Auge, H.; Brandl, R.; Heong, K.L.; Hotes, S.; Settele, J.; Villareal, S.; Schädler, M. Small-Scale Variability in the Contribution of Invertebrates to Litter Decomposition in Tropical Rice Fields. *Basic Appl. Ecol.* **2015**, *16*, 674–680. [CrossRef]
20. Menta, C.; Conti, F.D.; Fondón, C.L.; Staffilani, F.; Remelli, S. Soil Arthropod Responses in Agroecosystem: Implications of Different Management and Cropping Systems. *Agronomy* **2020**, *10*, 982. [CrossRef]
21. Barbercheck, M.E.; Neher, D.A.; Anas, O.; El-Allaf, S.M.; Weicht, T.R. Response of Soil Invertebrates to Disturbance across Three Resource Regions in North Carolina. *Environ. Monit. Assess.* **2009**, *152*, 283–298. [CrossRef]
22. Kibblewhite, M.G.; Ritz, K.; Swift, M.J. Soil Health in Agricultural Systems. *Philos. Trans. R. Soc. B Biol. Sci.* **2008**, *363*, 685–701. [CrossRef]
23. Herrick, J.E. Soil Quality: An Indicator of Sustainable Land Management? *Appl. Soil Ecol.* **2000**, *15*, 75–83. [CrossRef]
24. Gedoz, M.; Freitas, E.M.; da Silva, V.L.; Johann, L. Edaphic Invertebrates as Indicators of Soil Integrity Quality. *Floresta E Ambiente* **2021**, *28*, e20200069. [CrossRef]
25. Parisi, V.; Menta, C.; Gardi, C.; Jacomini, C.; Mozzanica, E. Microarthropod Communities as a Tool to Assess Soil Quality and Biodiversity: A New Approach in Italy. *Agric. Ecosyst. Environ.* **2005**, *105*, 323–333. [CrossRef]
26. Neave, P.; Fox, C.A. Response of Soil Invertebrates to Reduced Tillage Systems Established on a Clay Loam Soil. *Appl. Soil Ecol.* **1998**, *9*, 423–428. [CrossRef]
27. Errouissi, F.; Moussa-Machraoui, S.B.; Ben-Hammouda, M.; Nouira, S. Soil Invertebrates in Durum Wheat (*Triticum Durum* L.) Cropping System under Mediterranean Semi-Arid Conditions: A Comparison between Conventional and No-Tillage Management. *Soil Tillage Res.* **2011**, *112*, 122–132. [CrossRef]
28. Cates, A.M.; Ruark, M.D.; Hedtcke, J.L.; Posner, J.L. Long-Term Tillage, Rotation and Perennialization Effects on Particulate and Aggregate Soil Organic Matter. *Soil Tillage Res.* **2016**, *155*, 371–380. [CrossRef]

9. Jones, D.T.; Susilo, F.X.; Bignell, D.E.; Hardiwinoto, S.; Gillison, A.N.; Eggleton, P. Termite Assemblage Collapse along a Land-Use Intensification Gradient in Lowland Central Sumatra, Indonesia. *J. Appl. Ecol.* **2003**, *40*, 380–391. [CrossRef]
0. Sharley, D.J.; Hoffmann, A.A.; Thomson, L.J. The Effects of Soil Tillage on Beneficial Invertebrates within the Vineyard. *Agric. For. Entomol.* **2008**, *10*, 233–243. [CrossRef]
1. Karlen, D.L.; Wollenhaupt, N.C.; Erbach, D.C.; Berry, E.C.; Swan, J.B.; Eash, N.S.; Jordahl, J.L. Long-Term Tillage Effects on Soil Quality. *Soil Tillage Res.* **1994**, *32*, 313–327. [CrossRef]
2. Bluett, C.; Tullberg, J.N.; McPhee, J.E.; Antille, D.L. Soil and Tillage Research: Why Still Focus on Soil Compaction? *Soil Tillage Res.* **2019**, *194*, 104282. [CrossRef]
3. John, K.; Zaitsev, A.S.; Wolters, V. Soil Fauna Groups Respond Differentially to Changes in Crop Rotation Cycles in Rice Production Systems. *Pedobiologia* **2021**, *84*, 150703. [CrossRef]
4. Förster, B.; Garcia, M.; Francimari, O.; Römbke, J. Effects of Carbendazim and Lambda-Cyhalothrin on Soil Invertebrates and Leaf Litter Decomposition in Semi-Field and Field Tests under Tropical Conditions (Amazônia, Brazil). *Eur. J. Soil Biol.* **2006**, *42*, S171–S179. [CrossRef]
5. Cole, L.; Buckland, S.M.; Bardgett, R.D. Influence of Disturbance and Nitrogen Addition on Plant and Soil Animal Diversity in Grassland. *Soil Biol. Biochem.* **2008**, *40*, 505–514. [CrossRef]
6. Saifutdinov, R.A.; Sabirov, R.M.; Zaitsev, A.S. Springtail (Hexapoda: Collembola) Functional Group Composition Varies between Different Biotopes in Russian Rice Growing Systems. *Eur. J. Soil Biol.* **2020**, *99*, 103208. [CrossRef]
7. Government of Uganda. *Uganda National Rice Development Strategy (N RDS): 2008–2018*; Ministry of Agriculture, Animal Industry and Fisheries: Kampala, Uganda, 2012.
8. Kolawole, O. *Smallholder Farming Practices: Synthesis, Critical Reflections and Policy Issues*; Nova Science: New York, NY, USA, 2019; pp. 249–260.
9. Index. Available online: https://ipad.fas.usda.gov/highlights/2015/06/Southeast_Asia/Index.htm (accessed on 9 November 2021).
0. Murphy, S. *Changing Perspectives: Small-Scale Farmers, Markets and Globalization*; IIED: London, UK; Hivos: The Hague, The Netherlands, 2010.
1. BirdLife Data Zone. Available online: http://datazone.birdlife.org/index.php/site/factsheet/kibimba-rice-scheme-iba-uganda (accessed on 9 November 2021).
2. Bunyangha, J.; Majaliwa, M.J.G.; Muthumbi, A.W.; Gichuki, N.N.; Egeru, A. Past and Future Land Use/Land Cover Changes from Multi-Temporal Landsat Imagery in Mpologoma Catchment, Eastern Uganda. *Egypt. J. Remote Sens. Space Sci.* **2021**, *24*, 675–685. [CrossRef]
3. Swift, M.; Bignell, D. *Standard Methods for Assessment of Soil Biodiversity and Land Use Practice*; ICRAF: Bogor, Indonesia, 2001; p. 40.
4. Baillie, I.C.; Anderson, J.M.; Ingram, J.S.I. Tropical Soil Biology and Fertility: A Handbook of Methods. *J. Ecol.* **1990**, *78*, 547. [CrossRef]
5. Okalebo, J.R. *Laboratory Methods of Soil and Plant Analysis: A Working Manual*, 2nd ed.; SACRED Africa: Nairobi, Kenya, 2002; pp. 1–131.
6. United States Department of Agriculture. USDA Textural Soil Classification. In *Soil Mechanics Level I Module 3-USDA Textural Soil Classification*; United States Department of Agricultur: Washington, DC, USA, 1987; pp. 1–53.
7. Maurer, B.A.; McGill, B.J. Measurement of Species Diversity. *Biol. Divers. Front. Meas. Assess* **2011**, *345*, 55–65.
8. Help, C.H.R.; Herman, P.M.J.; Soetaert, K. Indices of Diversity and Evenness. *Oceanis* **1998**, *24*, 61–87.
9. Kim, B.R.; Shin, J.; Guevarra, R.B.; Lee, J.H.; Kim, D.W.; Seol, K.H.; Lee, J.H.; Kim, H.B.; Isaacson, R.E. Deciphering Diversity Indices for a Better Understanding of Microbial Communities. *J. Microbiol. Biotechnol.* **2017**, *27*, 2089–2093. [CrossRef]
0. Brower, J.E.; Zar, J.H.; Von Ende, C.N. *Field and Laboratory Methods for General Ecology*; McGraw-Hill Education: New York, NY, USA, 1997; p. 288.
1. R: The R Project for Statistical Computing. Available online: https://www.r-project.org/ (accessed on 9 March 2021).
2. Kenkel, N.C.; Orloci, L. Applying Metric and Nonmetric Multidimensional Scaling to Ecological Studies: Some New Results. *Ecology* **1986**, *67*, 919–928. [CrossRef]
3. Clarke, K.R. Non-Parametric Multivariate Analyses of Changes in Community Structure. *Aust. J. Ecol.* **1993**, *18*, 117–143. [CrossRef]
4. Oksanen, J.; Blanchet, F.G.; Friendly, M.; Kindt, R.; Legendre, P.; McGlinn, D.; Minchin, P.R.; O'Hara, R.B.; Simpson, G.L.; Solymos, P.; et al. Vegan: Community Ecology Package. 2020. Available online: https://cran.r-project.org/web/packages/vegan/vegan.pdf (accessed on 9 March 2021).
5. Callaham, M.A.; Richter, D.D.; Coleman, D.C.; Hofmockel, M. Long-Term Land-Use Effects on Soil Invertebrate Communities in Southern Piedmont Soils, USA. *Eur. J. Soil Biol.* **2006**, *42* (Suppl. 1), 150–156. [CrossRef]
5. França, F.M.; De Paiva, T.C.B.; Marcantônio, A.S.; Teixeira, P.C.; Ferreira, C.M. Acute Toxicity and Ecotoxicological Risk Assessment of Rice Pesticides to Lithobates Catesbeianus Tadpoles. *J. Environ. Sci. Health-Part B Pestic. Food Contam. Agric. Wastes* **2015**, *50*, 406–410. [CrossRef]
7. Gimeno-García, E.; Andreu, V.; Boluda, R. Heavy Metals Incidence in the Application of Inorganic Fertilizers and Pesticides to Rice Farming Soils. *Environ. Pollut.* **1996**, *92*, 19–25. [CrossRef]
3. Gunstone, T.; Cornelisse, T.; Klein, K.; Dubey, A.; Donley, N. Pesticides and Soil Invertebrates: A Hazard Assessment. *Front. Environ. Sci.* **2021**, *9*, 122. [CrossRef]

59. Calderbank, A. The Occurrence and Significance of Bound Pesticide Residues in Soil. *Rev. Environ. Contam. Toxicol.* **1989**, *108*, 71–103. [CrossRef]
60. Fang, L.; Wong, P.K.; Lin, L.; Lan, C.; Qiu, J.W. Impact of Invasive Apple Snails in Hong Kong on Wetland Macrophytes, Nutrients, Phytoplankton and Filamentous Algae. *Freshw. Biol.* **2010**, *55*, 1191–1204. [CrossRef]
61. Sheldon, S.P. The Effects of Herbivorous Snails on Submerged Macrophyte Communities in Minnesota. *Ecology* **2013**, *68*, 1920–1931. [CrossRef]
62. Yusa, Y.; Sugiura, N.; Wada, T. Predatory Potential of Freshwater Animals on an Invasive Agricultural Pest, the Apple Snail Pomacea Canaliculata (Gastropoda: Ampullariidae), in Southern Japan. *Biol. Invasions.* **2006**, *8*, 137–147. [CrossRef]
63. Barker, G.M. Millipedes (Diplopoda) and Centipedes (Chilopoda Myriapoda) as Predators of Terrestrial Gastropods. *Nat. Enemies Terr. Molluscs* **2004**, 405–426.
64. Widyastuti, R. Feeding Rate of Soil Animals in Different Ecosystems in Pati, Indonesia. *Hayati J. Biosci.* **2006**, *13*, 119–123. [CrossRef]
65. Ghiglieno, I.; Simonetto, A.; Orlando, F.; Donna, P.; Tonni, M.; Valenti, L.; Gilioli, G. Response of the Arthropod Community to Soil Characteristics and Management in the Franciacorta Viticultural Area (Lombardy, Italy). *Agronomy* **2020**, *10*, 740. [CrossRef]
66. Bedano, J.C.; Domínguez, A.; Arolfo, R.; Wall, L.G. Effect of Good Agricultural Practices under No-till on Litter and Soil Invertebrates in Areas with Different Soil Types. *Soil Tillage Res.* **2016**, *158*, 100–109. [CrossRef]
67. Barros, E.; Pashanasi, B.; Constantino, R.; Lavelle, P. Effects of Land-Use System on the Soil Macrofauna in Western Brazilian Amazonia. *Biol. Fertil. Soils* **2002**, *35*, 338–347. [CrossRef]
68. Keppel, G.; Van Niel, K.P.; Wardell-Johnson, G.W.; Yates, C.J.; Byrne, M.; Mucina, L.; Schut, A.G.T.; Hopper, S.D.; Franklin, S.E. Refugia: Identifying and Understanding Safe Havens for Biodiversity under Climate Change. *Glob. Ecol. Biogeogr.* **2012**, *21*, 393–404. [CrossRef]
69. Tuck, S.L.; Winqvist, C.; Mota, F.; Ahnström, J.; Turnbull, L.A.; Bengtsson, J. Land-Use Intensity and the Effects of Organic Farming on Biodiversity: A Hierarchical Meta-Analysis. *J. Appl. Ecol.* **2014**, *51*, 746–755. [CrossRef] [PubMed]
70. Li, W.; Dou, Z.; Cui, L.; Zhao, X.; Zhang, M.; Zhang, Y.; Gao, C.; Yang, Z.; Lei, Y.; Pan, X. Soil Fauna Diversity at Different Stages of Reed Restoration in a Lakeshore Wetland at Lake Taihu, China. *Ecosyst. Health Sustain.* **2020**, *6*, 1722034. [CrossRef]
71. Milliken, G.A.; Johnson, D.E. *Analysis of Messy Data. Volume 1: Designed Experiments*, 1st ed.; Van Nostrand Reinhold: New York, NY, USA, 1984; p. 127.

agronomy

MDPI

Article

Effects of Two Varieties and Fertilization Regimes on Growth, Fruit, and Silymarin Yield of Milk Thistle Crop

Vasiliki Liava [1], Anestis Karkanis [1,*], Nicholaos Danalatos [2] and Nikolaos Tsiropoulos [3]

1 Laboratory of Weed Science, Department of Agriculture Crop Production and Rural Environment, University of Thessaly, Fytokou St., 38446 Volos, Greece; vasiliki.liava@gmail.com
2 Laboratory of Agronomy and Applied Crop Physiology, Department of Agriculture Crop Production and Rural Environment, University of Thessaly, Fytokou St., 38446 Volos, Greece; danal@uth.gr
3 Analytical Chemistry and Pesticides Laboratory, Department of Agriculture Crop Production and Rural Environment, University of Thessaly, Fytokou St., 38446 Volos, Greece; ntsirop@uth.gr
* Correspondence: akarkanis@uth.gr; Tel.:+30-2421-093-135

Abstract: Milk thistle is an alternative crop to winter cereals for southern Europe as this species is drought tolerant and its fruits contain silymarin. The aim of this study was to assess the impact of two varieties and fertilization regimes (sheep manure and inorganic fertilizer) on crop productivity. A two-factor experiment was conducted in a randomized split-plot design with three replicates. The varieties were Palaionterveno and Spata, while the fertilization treatments were control, sheep manure, and calcium ammonium nitrate applied at 75 and 125 kg N ha^{-1}. Variety and fertilization significantly affected plants development and productivity, as well as oil and silymarin yield. The use of manure and inorganic nitrogen fertilizer increased rosette diameter, oil and silymarin yield, above-ground biomass, and fruit yield. The influence of inorganic fertilization, regardless of the application dose, was more apparent than organic fertilization. Moreover, variety significantly affected plants growth and silymarin content, as well as silymarin composition. The variety Spata had the greatest silymarin content, reaching 4.40%, and a high silybin B concentration. In conclusion, the selection of a suitable variety is important for achieving high fruit and silymarin yields, while inorganic nitrogen fertilization can maximize the productivity of the milk thistle crop.

Keywords: flavonolignans content; inorganic fertilization; productivity; quality; *Silybum marianum*

Citation: Liava, V.; Karkanis, A.; Danalatos, N.; Tsiropoulos, N. Effects of Two Varieties and Fertilization Regimes on Growth, Fruit, and Silymarin Yield of Milk Thistle Crop. *Agronomy* **2022**, *12*, 105. https://doi.org/10.3390/agronomy12010105

Academic Editors: Carlos Cantero-Martínez and Carmelo Maucieri

Received: 7 December 2021
Accepted: 30 December 2021
Published: 1 January 2022

1. Introduction

Milk thistle (*Silybum marianum* (L.) Gaertn.) is a well-known source of silymarin, which has anticancer [1], hepatoprotective, anti-inflammatory [2], anti-collagenase [3], immunomodulatory [4], and neuroprotective properties [5]. Due to its pharmaceutical properties, this species is an important medicinal plant and it is cultivated in many European countries such as Austria, Bulgaria, Czech Republic, Germany, and Poland [6–8]. However, except for silymarin production, milk thistle can be cultivated for oil production [9,10]. This oil is edible [11] and can be used in cooking [12] or in the pharmaceutical and cosmetic industries [11,13,14].

Milk thistle crop can be included in rotation systems since it is a low input crop, while its fruits have high economic value since the demand for silymarin is high [15,16]. The optimization of cultivation practices is really important to increase crop productivity [17]. Cultivation practices such as plant density [18], irrigation, and fertilization [16] affect both plants development and productivity of this crop. The application of organic or inorganic fertilizers at appropriate doses contributes significantly to the increase in milk thistle crop yield. Nitrogen and phosphorus application led to increased fruit yield [19], while the application of manure enhanced the plants height, silymarin content [20], and fruit and silymarin yield [16]. In contrast, in Bulgaria under different climatic conditions, the use of nitrogen, phosphorus, and potassium negatively affected silymarin accumulation in the fruits; although, silymarin yield was increased [21].

Except for cultivation techniques, genetic material plays an important role in plants development and productivity [15]. For instance, Shokrpour et al. [22], assessing different milk thistle genotypes in Iran, observed that the plant height ranged from 131.87 to 160.61 cm and the fruit yield from 889 to 2416 kg ha^{-1}, revealing the importance of genetic material in increasing crop yields. Moreover, genetic material can affect both the silymarin content and its composition in flavonolignans (silybins A and B, isosilybins A and B, silychristin, and silydianin) and taxifolin [23–25]. In the literature, there is limited information about the productivity and quality (e.g., silymarin content) of genotypes originating from southern Europe, a region that is characterized by semi-arid conditions. For instance, Arampatzis et al. [25] evaluated several native genotypes originating from Greece and observed that silymarin content ranged from 2.3% to 7.7%. These genotypes varied in flavonolignans and taxifolin content and only two genotypes (Spata and Kastoria) exhibited both the highest silymarin and silybin A + B content. In Italy, a country with similar climatic conditions to Greece, twenty-six milk thistle genotypes, originating from Italy and other countries, varied both in silymarin content and composition [23]. Four of these genotypes exhibited high silybin and silychristin content, while fourteen genotypes had high silydianin content [23]. Moreover, genotypes originating from southern Europe can have high productivity since Arampatzis et al. [18] reported that the fruit yield of a genotype originating from Greece ranged from 1444 to 2222 kg ha^{-1} depending on plant density. The assessment of milk thistle genotypes adaptation to semi-arid climate conditions of this region is crucial to maximize both the crop productivity and the commercial value of the final product (fruits or silymarin extracts). Thus, the evaluation of milk thistle genotypes of Greek origin, which exhibit high productivity and quality (e.g., high silymarin content), under low input and high input conditions is important in order to select genotypes that could be included in breeding programs. In this context, the aims of this study were (1) to assess two milk thistle varieties in terms of productivity and quality (e.g., silymarin content), (2) to evaluate the impact of sheep manure and inorganic fertilizer on crop yield and quality, (3) to examine the interaction effects of variety and fertilization on fruit and silymarin yield of milk thistle crop.

2. Materials and Methods

2.1. Study Site, Growing Conditions, and Experimental Design

A two-year experiment was set up at the experimental field of the University of Thessaly in Velestino (Thessaly Region, Greece) during the growing seasons 2019–2020 and 2020–2021. The soil was sandy clay loam with a pH of 7.4. In the first experimental year, the total precipitation from November to May was 368.6 mm, while in the second experimental year was 273.5 mm. Two varieties of milk thistle originating from Greece were sown on 29 October in both years. The row spacing was 50 cm, while the density of the plants in the row was 13 plants m^{-1} (Figure 1).

A two-factor experiment was conducted in a randomized split-plot design with three replicates. Variety and fertilization were the main plot and sub-plot factors, respectively. The tested varieties were Palaionterveno and Spata, while the fertilization treatments were control without fertilization, sheep manure, and calcium ammonium nitrate applied at two doses (Table 1). The selection of the two varieties was based on the content and composition of silymarin. According to Arampatzis et al. (2019b), Spata has high silymarin (5.9–7.7%) and silybin A + B content, while Palaionterveno is characterized by lower silymarin (2.4–3.3%) and silybin A + B content compared with Spata.

Figure 1. Experimental field (plot size: 2 × 3 m, density: 26 plants m^2) of milk thistle in the second experimental period on 20 November 2020.

Table 1. Description of organic and inorganic fertilization treatments.

Fertilizers	Dose	Application Time
Sheep Manure	13 t ha^{-1}	Pre-sowing
Chemical properties: C/N ratio: 10.4, organic matter: 47.5%, pH: 7.3, total nitrogen (TN): 22,695 mg kg^{-1}, phosphorus (P): 773 mg kg^{-1}, potassium (K): 3739 mg kg^{-1}, magnesium (Mg): 3549 mg kg^{-1}, copper (Cu): 3.1 mg kg^{-1}, zinc (Zn): 24.3 mg kg^{-1}, manganese (Mn): 62.1 mg kg^{-1}, iron (Fe): 29.5 mg kg^{-1}, boron (B): 17.8 mg kg^{-1}, and sodium (Na): <100 mg kg^{-1}.		
Calcium ammonium nitrate	75 kg N ha^{-1} applied at two doses (25 and 50 kg N ha^{-1})	1st dose: 15 January 2020 and 13 January 2021
	125 Kg N ha^{-1} applied at two doses (50 and 75 kg N ha^{-1})	2nd dose: 3 March at both seasons

2.2. Measurements

2.2.1. Agronomic Parameters

Within each sub-plot, rosette diameter and height were measured for five plants in the central rows avoiding plants at the edges of the rows. The rosette diameter was measured at 144 and 138 DAS (days after sowing) in 2020 and 2021, respectively, while the maximum height of plants was recorded at 193 and 190 DAS in 2020 and 2021, respectively. For the above-ground dry biomass determination, four consecutive plants from a central row were selected at the growth stage where the plants had the maximum height, and then after drying of samples at 60 °C for four days the dry biomass was estimated. Moreover, the number of inflorescences per plant was measured in five plants per treatment. Harvest was made manually in two central rows (1 m per row) at the end of May. After the harvest, the fruits were separated from the other parts of the inflorescences and the 1000-fruit weight was measured in three samples of 100 fruits.

2.2.2. Chemical Composition Analysis: Oil and Silymarin

Oil was extracted from powdered dry fruit samples with hexane according to the procedure described in the previous study of Arampatzis et al. [25]. After the oil extraction, firstly the defatted fruit samples were extracted with methanol using a Soxhlet extraction apparatus and then, the silymarin determination was made by a HPLC system (HP 1100 Liquid Chromatograph, Hewlett-Packard GmbH, Waldbronn, Germany) with a UV detector and coupled to a ternary-delivery system following the analytical conditions described by Arampatzis et al. [25]. The identification and quantification of silymarin compounds (flavonolignans and taxifolin) were made according to the procedure described in our previous work [17]. Finally, oil and silymarin yield (kg ha^{-1}) was calculated according to Equations (1) and (2).

$$\text{Oil yield} = \text{oil content} \times \text{fruit yield} \tag{1}$$

$$\text{Silymarin yield} = \text{silymarin content} \times \text{fruit yield} \quad (2)$$

2.3. Statistical Analysis

The results of morphological parameters (rosette diameter, height), fruit yield and its components, above-ground biomass, and quality parameters (oil, silymarin, flavonolignans and taxifolin content) were statistically analyzed using the SigmaPlot 12 statistical package (Systat Software, San Jose, CA, USA). A two-way analysis of variance (ANOVA) was conducted to assess the effects of two factors (variety and fertilization) and their interactions on growth, yield, and quality of milk thistle, while the differences between means were separated by Fisher's least significant dereference (LSD) test at $p = 0.05$.

3. Results

3.1. Plants Growth Traits

Variety and fertilization exhibited positive effects on plants growth traits. In 2020 and 2021, the maximum rosette diameter and plant height were recorded in the variety Spata (Table 2). In general, plants growth was greater during the first year. Concerning fertilization regimes, in both growing seasons, the application of organic and inorganic fertilization significantly increased both rosette diameter and plant height. An exception was the application of sheep manure during the second season as there was no difference among this treatment and the untreated control at the plant height.

Table 2. Effects of two milk thistle varieties (Palaionterveno and Spata) and fertilization regimes (sheep manure and inorganic fertilizer) on rosette diameter and plant height of milk thistle.

Treatmentss	Rosette Diameter (cm) 2020	Rosette Diameter (cm) 2021	Plant Height (cm) 2020	Plant Height (cm) 2021
Varieties				
Palaionterveno	69.0 b	52.2 b	170.3 b	159.6 b
Spata	72.9 a	60.4 a	201.3 a	177.6 a
$LSD_{5\%}$	0.74	1.85	4.51	5.22
Fertilization				
Control	65.6 d	46.0 d	169.6 d	153.7 c
Sheep manure	69.0 c	51.5 c	184.1 c	156.4 c
CAN-75 kg N ha^{-1}	72.8 b	59.7 b	190.9 b	176.3 b
CAN-125 kg N ha^{-1}	76.3 a	67.9 a	198.7 a	188.0 a
$LSD_{5\%}$	1.04	2.62	6.37	7.39
F-values and significant differences				
Variety (V)	126.603 ***	86.628 ***	212.446 ***	53.062 ***
Fertilization (F)	179.866 ***	120.464 ***	33.691 ***	44.131 ***
V × F	3.153 ^{ns}	0.366 ^{ns}	0.846 ^{ns}	1.782 ^{ns}

CAN: calcium ammonium nitrate. For each factor, means followed by different letters within the same column show significant differences according to the LSD test. *** significant at $p \leq 0.001$ and ns = not significant.

The application of calcium ammonium nitrate (CAN) regardless of the application dose enhanced plants development compared to organic fertilization. For instance, the application of CAN fertilizer increased rosette diameter and plant height by 5.2–24.2% and 3.6–16.8%, respectively, compared to the use of manure.

The greatest values of dry weight were measured in 2020, specifically in the variety Spata, while there was an interaction effect between the two factors (Table 3). During the two-year experiment, in both varieties, the minimum above-ground dry biomass was recorded in the unfertilized plots. The organic fertilization significantly influenced this trait especially in the first growing period, increasing the above-ground biomass of the variety Palaionterveno by 29.4%. Moreover, CAN fertilizer further enhanced this trait. For instance, in the variety Spata, inorganic nitrogen fertilizer increased the dry biomass by 43.6–56.1% and 44.1–45.7% compared to the control treatment in 2020 and 2021, respectively. The results also revealed that the low inorganic fertilizer dose affected more the variety Spata compared with Palaionterveno, while the high dose similarly affected the two varieties.

Table 3. Interaction effects between variety and fertilization on aboveground dry biomass of milk thistle crop.

Varieties	Fertilization	Above-Ground Biomass (kg ha^{-1}) 2020	2021
Palaionterveno	Control	15,293.7 e	15,097.1 d
	Sheep manure	21,651.3 d	18,405.0 c
	CAN-75 kg N ha^{-1}	26,506.0 bc	22,218.2 b
	CAN-125 kg N ha^{-1}	29,340.7 b	26,808.9 a
Spata	Control	17,282.7 e	15,465.9 d
	Sheep manure	22,952.3 cd	18,221.9 c
	CAN-75 kg N ha^{-1}	30,648.0 b	27,671.5 a
	CAN-125 kg N ha^{-1}	39,347.3 a	28,475.2 a
$LSD_{5\%}$		4236.02	2419.16
F-values and significant differences			
Varieties (V)		19.041 ***	10.254 **
Fertilization (F)		61.013 ***	100.774 ***
V × F		3.916 *	4.951 *

CAN: calcium ammonium nitrate. Means followed by different letters within the same column show significant differences according to the LSD test. *, **, and *** significant at $p \leq 0.05$, $p \leq 0.01$, and $p \leq 0.001$, respectively.

3.2. Productivity

In both years, variety and fertilization had a significant impact on the number of inflorescences (Table 4). In particular, in 2020, the variety Palaionterveno had the greatest value, while in 2021, Spata reached the maximum number. In both years, organic fertilization did not augment the number of inflorescences compared with control. However, the application of CAN fertilizer significantly increased this trait, especially in the second growing period when the high dose of the fertilizer was applied. Moreover, neither variety nor fertilization affected the 1000-fruit weight. In 2021, the 1000-fruit weight was higher as it ranged from 23.7 to 25.0 g, while in 2020, it ranged from 21.7 to 22.4 g.

Table 4. Effects of two milk thistle varieties (Palaionterveno and Spata) and fertilization regimes (sheep manure and inorganic fertilizer) on yield parameters (number of inflorescences, 1000-fruit number, and fruits per central inflorescence) of milk thistle.

Treatments	Number of Inflorescences (No Plant^{-1})		1000-Fruit Weight (g)		Fruits Per Central Inflorescence (No)	
	2020	2021	2020	2021	2020	2021
Varieties						
Palaionterveno	6.7 a	4.9 b	21.7 a	24.3 a	147.5 a	73.6 a
Spata	5.5 b	5.7 a	22.4 a	24.3 a	99.7 b	58.9 b
$LSD_{5\%}$	1.04	0.57	-	-	6.41	4.12
Fertilization						
Control	5.3 b	3.5 c	21.8 a	23.9 a	103.7 d	49.2 d
Sheep manure	4.3 b	3.5 c	22.3 a	23.7 a	113.9 c	57.3 c
CAN-75 kg N ha^{-1}	7.2 a	6.5 b	22.4 a	25.0 a	129.3 b	72.5 b
CAN-125 kg N ha^{-1}	7.5 a	7.8 a	21.7 a	24.6 a	147.4 a	85.8 a
$LSD_{5\%}$	1.48	0.80	-	-	9.06	5.83
F-values and significant differences						
Varieties (V)	6.177 *	11.692 **	3.518 ^{ns}	0.0129 ^{ns}	249.724 ***	57.223 ***
Fertilization (F)	9.826 ***	64.289 ***	0.664 ^{ns}	1.922 ^{ns}	39.871 ***	69.863 ***
V × F	0.757 ^{ns}	0.231 ^{ns}	0.524 ^{ns}	1.532 ^{ns}	0.140 ^{ns}	1.629 ^{ns}

CAN: calcium ammonium nitrate. For each factor, means followed by different letters within the same column show significant differences according to the LSD test. *, **, and *** significant at $p \leq 0.05$, $p \leq 0.01$, and $p \leq 0.001$, respectively. ns = not significant.

Concerning the number of fruits on the central inflorescence, there was a significant difference between the two varieties, and Palaionterveno had the highest value. Fertilization significantly affected this parameter as the lowest number was recorded in the unfertilized control, followed by the use of manure. In 2021, inorganic fertilization increased the number of fruits on the central inflorescence by up to 42.7% and 33.2% compared with control and organic fertilization. Moreover, in 2020, the number of fruits per inflorescence in each treatment was almost doubled compared with the equal treatment in 2021. However, the impact of fertilization was more intense in the second growing season.

In the first growing season, fruit yield was significantly affected only by fertilization, as there was no significant difference among the two varieties (Table 5). Organic and inorganic fertilization enhanced fruit yield, although the use of manure affected to a lesser extent this trait. Moreover, the high rate of CAN fertilizer increased fruit yield by 19.7% (Spata) and 22.7% (Palaionterveno) compared with the application of manure. In 2021, there was an interaction effect between variety and fertilization on fruit yield. The yield ranged from 726.5 kg ha^{-1} to 1504.1 kg ha^{-1} and there were significant differences between the two varieties in the same fertilization regime, as the greatest values were recorded in Palaionterveno.

Table 5. Effects of two milk thistle varieties (Palaionterveno and Spata) and fertilization regimes (sheep manure and inorganic fertilizer) on fruit yield of milk thistle crop.

Fertilization	Fruit Yield (kg ha^{-1})			
	2020		2021	
	Varieties		Varieties	
	Spata	Palaionterveno	Spata	Palaionterveno
Control	963.0 Da	975.8 Da	726.5 Cb	856.7 Da
Sheep manure	1184.7 Ca	1083.5 Ca	781.6 Cb	1018.6 Ca
CAN-75 kg N ha^{-1}	1339.9 Ba	1245.9 Ba	976.7 Bb	1307.7 Ba
CAN-125 kg N ha^{-1}	1475.9 Aa	1401.5 Aa	1128.5 Ab	1504.1 Aa
$LSD_{5\%\ fertilization}$	93.50		76.99	
$LSD_{5\%\ varieties}$	-		54.43	
F-values and significant differences				
Varieties (V)	4.231 ^{ns}		109.256 ***	
Fertilization (F)	42.086 ***		84.297 ***	
V × F	0.710 ^{ns}		4.483 ^{ns}	

CAN: calcium ammonium nitrate. Small letters show significant differences between the two varieties, while capital letters show significant differences among the fertilization treatments. *** Significant at $p \leq 0.001$, respectively. ns = not significant.

Assessing each variety separately, the inorganic nitrogen fertilization led to significantly higher fruit yield compared with organic fertilization and control. The high rate of CAN fertilizer increased the fruit yield by 32.3% in Palaionterveno and 30.7% in Spata compared with the use of manure. Finally, organic fertilization had no impact on Spata as there was no difference between manure and control.

3.3. Oil Content and Yield

In 2020, Palaionterveno had significantly higher oil content compared with Spata (Table 6). However, in 2021, there was no difference between the two varieties. Moreover, in both growing seasons, neither organic nor inorganic fertilization had an impact on oil content. Concerning oil yield, in 2020, the application of manure did not enhance this trait compared with the untreated control (Table 7). However, inorganic fertilization significantly increased oil yield regardless of the fertilization rate. In particular, nitrogen fertilization increased oil yield by up to 21.3% compared with the use of manure. In 2021, there was an interaction effect between variety and fertilization on oil yield. In Palaionterveno, organic and inorganic fertilization significantly increased oil yield, while in Spata, organic fertilization had no impact. In both varieties, the application of inorganic nitrogen fertilization affected more the oil yield compared with the use of manure. Assessing the two varieties in the equal treatment, there were significant differences in all the cases as Palaionterveno had the greatest values. The highest oil yield (377.4 kg ha^{-1}) was recorded in Palaionterveno when the inorganic fertilizer was applied at a high rate. In general, fertilization affected more the variety Palaionterveno than Spata.

Table 6. Effects of two milk thistle varieties (Palaionterveno and Spata) and fertilization regimes (sheep manure and inorganic fertilization) on oil content in milk thistle.

Treatments	Oil Content (%)	
	2020	2021
Variety		
Palaionterveno	23.3 a	24.7 a
Spata	22.1 b	24.9 a
$LSD_{5\%}$	0.72	-
Fertilization		
Control	23.2 a	24.6 a
Sheep manure	22.8 a	24.5 a
CAN-75 kg N ha^{-1}	22.6 a	25.1 a
CAN-125 kg N ha^{-1}	22.1 a	24.9 a
F-values and significant differences		
Variety (V)	10.720 *	2.667 ^{ns}
Fertilization (F)	1.894 ^{ns}	2.822 ^{ns}
V × F	1.661 ^{ns}	1.759 ^{ns}

CAN: calcium ammonium nitrate. For each factor, means followed by different letters within the same column show significant differences according to the LSD test. * Significant at $p \leq 0.05$. ns = not significant.

Table 7. Effects of two milk thistle varieties (Palaionterveno and Spata) and fertilization regimes (sheep manure and inorganic fertilizer) on oil yield of milk thistle.

Fertilization	Oil Yield (kg ha^{-1})			
	2020		2021	
	Varieties		Varieties	
	Spata	Palaionterveno	Spata	Palaionterveno
Control	215.7 Ca	235.4 Ca	181.2 Cb	207.7 Da
Sheep manure	264.5 Ba	253.3 BCa	193.1 Cb	247.9 Ca
CAN-75 kg N ha^{-1}	303.8 Aa	280.9 Ba	246.2 Bb	326.0 Ba
CAN-125 kg N ha^{-1}	314.3 Aa	321.7 Aa	279.5 Ab	377.4 Aa
$LSD_{5\%\ fertilization}$	28.09		19.35	
$LSD_{5\%\ varieties}$	-		13.68	
F-values and significant differences				
Varieties (V)	0.035 ^{ns}		100.740 ***	
Fertilization (F)	18.407 ***		89.666 ***	
V × F	1.024 ^{ns}		5.776 ^{ns}	

CAN: calcium ammonium nitrate. Small letters show significant differences between the two varieties, while capital letters show significant differences among the fertilization treatments. *** Significant at $p \leq 0.001$, respectively. ns = not significant.

3.4. Silymarin Content and Yield

The results of the two-year experiment indicate that silymarin content is primarily influenced by genetic material (Table 8). The variety Spata had significantly higher silymarin content than Palaionterveno. Organic and inorganic fertilization had no impact on this trait. In contrast, in 2020 and 2021, inorganic nitrogen fertilization increased the silymarin yield by 21.5–30.7% and 28.1–37.8%, respectively, compared to the untreated control. The impact of manure on this trait was noticeable only in 2020. Finally, in both varieties, the concentration of silymarin was greater in the second year; although, the maximum silymarin yields were recorded in the first year owing to the higher fruit yields.

Table 8. Effects of two milk thistle varieties (Palaionterveno and Spata) and fertilization regimes (sheep manure and inorganic fertilizer) on silymarin content and yield of milk thistle.

Treatments	Silymarin Content (%)		Silymarin Yield (kg ha^{-1})	
	2020	2021	2020	2021
Varieties				
Palaionterveno	2.40 b	2.51 b	28.3 b	29.4 b
Spata	4.04 a	4.40 a	50.0 a	39.7 a
$LSD_{5\%}$	0.15	0.09	3.56	2.18
Fertilization				
Control	3.31 a	3.50 a	32.1 c	27.1 c
Sheep manure	3.24 a	3.44 a	37.2 b	29.9 c
CAN-75 kg N ha^{-1}	3.14 a	3.43 a	40.9 b	37.7 b
CAN-125 kg N ha^{-1}	3.20 a	3.44 a	46.3 a	43.6 a
$LSD_{5\%}$	-	-	5.04	3.09
F-values and significant differences				
Varieties (V)	536.430 ***	1858.094 ***	167.331 ***	100.694 ***
Fertilization (F)	1.059 ns	0.474 ns	12.699 ***	53.650 ***
V × F	0.664 ns	0.614 ns	0.952 ns	0.358 ns

CAN: calcium ammonium nitrate. For each factor, means followed by different letters within the same column show significant differences according to the LSD test. *** Significant at $p \leq 0.001$. ns = not significant.

3.5. Silymarin Active Constituents

In both growing seasons, variety significantly affected silymarin composition. In 2021 there were differences between the two varieties in all the evaluated compounds, while in the first year, the accumulation of isosilybin B, silydianin, and isosilychristin in the fruits of varieties was not significantly different. In both years, taxifolin, silybin A and B, silychristin and isosilybin A content were higher in the variety Spata (Tables 9 and 10). The silybin A + B content in the variety Spata was 79.3–81.9% higher than that in Palaionterveno. In contrast, the dominant components of Palaionterveno were silydianin and isosilychristin. Finally, the organic and inorganic nitrogen fertilization did not influence the accumulation of silymarin constituents, as there were no significant differences between the fertilization regimes.

Table 9. Effects of two milk thistle varieties (Palaionterveno and Spata) and fertilization regimes (sheep manure and inorganic fertilizer) on the content of silymarin active constituents in fruits of milk thistle in 2020.

Treatments	Silymarin Active Constituents (mg/g)-2020						
	TXF	SCS	SDN + ISCS	SBA	SBB	ISBA	ISBB
Varieties							
Palaionterveno	2.37 b	2.20 b	10.79 a	0.79 b	2.20 b	3.56 b	2.14 a
Spata	4.86 a	5.84 a	9.28 a	4.66 a	9.77 a	4.02 a	1.97 a
$LSD_{5\%}$	0.41	0.66	-	0.46	0.97	0.20	-
Fertilization							
Control	3.87 a	3.66 a	11.29 a	2.64 a	5.69 a	3.88 a	2.10 a
Sheep manure	3.47 a	4.16 a	9.79 a	2.64 a	6.49 a	3.72 a	2.10 a
CAN-75 kg N ha^{-1}	3.42 a	4.21 a	9.48 a	2.62 a	5.86 a	3.79 a	2.00 a
CAN-125 kg N ha^{-1}	3.71 a	4.04 a	9.57 a	3.00 a	5.91 a	3.77 a	2.03 a
F-values and significant differences							
Varieties (V)	169.358 ***	137.984 ***	4.196 ns	316.275 ***	275.831 ***	23.747 ***	3.744 ns
Fertilization (F)	1.188 ns	0.633 ns	1.321 ns	0.731 ns	0.581 ns	0.493 ns	0.339 ns
V × F	0.510 ns	1.357 ns	1.247 ns	0.885 ns	0.735 ns	0.586 ns	0.692 ns

Taxifolin: TXF, silychristin: SCS, silydianin + isosilychristin: SDN + ISCS, silybin A: SBA, silybin B: SBB, isosilybin A: ISBA, isosilybin B: ISBB, and calcium ammonium nitrate: CAN. For each factor, means followed by different letters within the same column show significant differences according to the LSD test. *** Significant at $p \leq 0.001$ respectively. ns = not significant.

Table 10. Effects of two milk thistle varieties (Palaionterveno and Spata) and fertilization regimes (sheep manure and inorganic fertilizer) on the content of silymarin active constituents in fruits of milk thistle in 2021.

Treatments	Silymarin Active Constituents (mg/g)-2021						
	TXF	**SCS**	**SDN + ISCS**	**SBA**	**SBB**	**ISBA**	**ISBB**
Varieties							
Palaionterveno	2.52 b	2.27 b	10.97 a	0.91 b	2.26 b	3.81 b	2.33 a
Spata	5.04 a	6.20 a	9.12 b	6.37 a	11.08 a	4.12 a	2.09 b
$LSD_{5\%}$	0.25	0.18	0.19	0.25	0.34	0.17	0.11
Fertilization							
Control	3.86 a	4.31 a	10.16 a	3.65 a	6.73 a	4.03 a	2.26 a
Sheep manure	3.67 a	4.31 a	10.18 a	3.55 a	6.62 a	3.90 a	2.18 a
CAN-75 kg N ha^{-1}	3.82 a	4.19 a	9.83 a	3.66 a	6.61 a	3.96 a	2.26 a
CAN-125 kg N ha^{-1}	3.76 a	4.13 a	10.01 a	3.70 a	6.72 a	3.97 a	2.16 a
F-values and significant differences							
Varieties (V)	449.584 ***	2136.812 ***	409.389 ***	2163.437 ***	3015.044 ***	15.523 ***	19.612 ***
Fertilization (F)	0.509 ns	1.119 ns	3.186 ns	0.320 ns	0.157 ns	0.498 ns	0.852 ns
V × F	0.244 ns	0.557 ns	1.059 ns	0.912 ns	0.234 ns	0.187 ns	2.205 ns

Taxifolin: TXF, silychristin: SCS, silydianin + isosilychristin: SDN + ISCS, silybin A: SBA, silybin B: SBB, isosilybin A: ISBA, isosilybin B: ISBB, and calcium ammonium nitrate: CAN. For each factor, means followed by different letters within the same column show significant differences according to the LSD test. *** significant at $p \leq 0.001$, respectively. ns = not significant.

4. Discussion

4.1. Milk Thistle Varieties

Variety significantly affected plants growth, since the rosette diameter and height of plants in Spata were higher by up to 13.57% and 15.40%, respectively, compared with Palaionterveno. The impact of genetic material in plants growth parameters was observed in previous studies as Ram et al. [15], Gresta et al. [26], Shokrpour et al. [22], and Sulas et al. [27] recorded plant height from 78 cm to 207 cm in various milk thistle genotypes. Moreover, varieties significantly affected the number of fruits per inflorescence as in Palaionterveno this number was higher by up to 32.38% than that in Spata. The number of fruits per inflorescence in these two varieties of Greek origin ranged from 58.9 to 147.5 and was similar to that observed in previous studies. Stancheva et al. [28] reported values from 119.4 to 185.5, while Shokrpour et al. [22] recorded fewer fruits per inflorescence, ranging from 51.6 to 101.4. In contrast, 1000-fruit weight was not affected by variety. With regard to fruit yield, in 2021, the highest values were recorded in the variety Palaionterveno, while in 2020, there were no differences between the two varieties. In 2020, in Palaionterveno, a reduction in the seed germination was observed and as a consequence, the plant density was reduced by 20% resulting in a lower fruit yield than that in 2021. According to Arampatzis et al. [18], a high plant density can lead to the greatest seed yield. In another study, Shokrpour et al. [22] assessed various ecotypes and recorded yields from 889 to 2416 kg ha^{-1}. The above results show that the selection of a productive variety is important in order to maximize crop yield.

The oil content was ranged from 22.1 to 25.1% in the two varieties. In other studies conducted in Italy and Greece, Martinelli et al. [23] and Arampatzis et al. [25] observed that oil content in fruits of several milk thistle genotypes had higher values that ranged from 26.7 to 31.7% and from 24.7 to 31.1%, respectively. Compared with other species, the oil content of milk thistle is similar to hemp (*Cannabis sativa* L.) in which is ranging from 25.5 to 28.2% [29], but less than that in sesame (*Sesamum indicum* L.) and sunflower (*Helianthus annuus* L.) seeds that oil content is ranging from 34.2 to 36.5% and from 37.9 to 51%, respectively [30]. In 2021, the oil content in both varieties was higher compared with that in 2020, probably due to the wetter weather conditions that prevailed during April and May in 2020. Similarly, in Iran, water stress increased oil accumulation in the fruits of the plants [10].

Silymarin content in the fruits of the two varieties ranged from 2.4 to 4.4%, and variety Spata had significantly higher silymarin content than Palaionterveno. This finding is in agreement with Arampatzis et al. [25] as Spata had the highest silymarin content between thirty genotypes. In previous studies conducted in Greece, Italy, and India, the silymarin content in the fruits of several milk thistle genotypes ranged from 2.0 to 7.72% [15,23,25]. Moreover, the accumulation of silymarin active components was different, and Spata had a high content of silybin B, while silybin A and B constituted 35–39% of silymarin in Spata and 12% in Palaionterveno. In 2021, the silymarin accumulation was higher by 4.38–8.18% compared with 2020, probably due to the dryer weather conditions that prevailed during the second growing season. However, it is well documented in other studies that water stress causes an increase in silymarin accumulation in the fruits [10,18].

4.2. Fertilization Regimes

The application of sheep manure significantly increased rosette diameter and aboveground biomass in both years compared with the untreated control, while plant height was affected only in the first growing season. Similarly, Saad-Allah et al. [20], reported that the use of poultry manure increased plant height and dry biomass of milk thistle, while in rice (*Oryza sativa* L.), the application of manure increased plant height and the effect was more obvious when the manure was combined with urea [31]. However, the plant growth was greater when the calcium ammonium nitrate was applied compared to the use of sheep manure. The application of the sheep manure was not sufficient to fully meet plants nitrogen requirements due to the slow nitrogen mineralization from this organic fertilizer. This result can be explained since milk thistle is characterized by rapid growth in the period of mid-March to early May and as a consequence, the requirements are more intense during this period. Similarly, Popin et al. [32] reported that the use of urea or manure increased the height of maize (*Zea mays* L.) compared with control, and the effect of urea was more intense. It is also important to point out that the application of calcium ammonium nitrate at a high dose led to the maximum rosette diameter, aboveground biomass, and height of plants. Previously, an increase in nitrogen fertilizer dose increased plants height or biomass of different milk thistle genotypes [17,33]. Moreover, the use of sheep manure or calcium ammonium nitrate significantly influenced the number of fruits per inflorescence. Similarly, Afshar et al. [16] observed that poultry manure increased this trait. In contrast, 1000-fruit weight was not affected by fertilization.

The use of sheep manure beneficially affected fruit yield. This finding is in agreement with previous experiments as Saad-Allah et al. [20] observed that the application of chicken manure led to higher fruit yield compared to control, and an augmentation in the rate of the applied manure further increased the fruit yield. In general, the application of manure increased yield in other crops such as winter wheat (*Triticum aestivum* L.), maize, rice, and the effect was more obvious when the manure was combined with chemical fertilizers [31,34,35]. However, milk thistle plants show rapid growth and produce high aboveground biomass, and thus the nitrogen needs cannot be completely covered by sheep manure due to the slow nitrogen mineralization. As a result, the application of calcium ammonium nitrate led to a greater yield than manure. In a previous study, nitrogen fertilization significantly increased plants productivity, especially the application of a high dose [17]. In 2021, there was an interaction effect on fruit yield between the two factors, since sheep manure influenced the fruit yield only in Palaionterveno. Similarly, in rice [36] and bread wheat [37], there was an interaction between genotype and nitrogen fertilization on grain yield.

Moreover, fertilization had no impact on oil content. Similarly, Afshar et al. [16] observed that manure application had no impact on oil content in milk thistle fruits, while Li et al. [38] reported that the application of nitrogen fertilizers did not affect the oil content in sunflower seeds. These results show that the oil content mainly depends on the genetic material. Moreover, oil yield ranged from 181.2 to 377.4 kg ha^{-1} similar to previous studies that recorded 217.5–376.1 kg ha^{-1} [16] and 353–591 kg ha^{-1} [18] in different genotypes.

However, sunflower shows greater oil productivity (669 to 1210 kg ha^{-1}) [30] compared to milk thistle crop. In our study, the application of calcium ammonium nitrate led to higher oil yield compared with manure; consistent with the maximum oil yield in canola crop that was observed in the plots where the manure was applied in [39]. Moreover, the calcium ammonium nitrate increased the oil yield of both varieties, while sheep manure affected this parameter only in the variety Palaionterveno. Similarly, in milk thistle [16], canola (*Brassica napus* L.) [39], and hemp [29] crops, the application of manure increased oil yield compared to control.

Silymarin content was not influenced by sheep manure and calcium ammonium nitrate application. In other studies, Afshar et al. [16] observed that the use of manure did not affect the silymarin content, while Saad-Allah et al. [20] mentioned that manure application led to a higher silymarin accumulation than that in control. Regarding the effects of fertilization on the content of silymarin constituents, our results indicated no significant differences between sheep manure and calcium ammonium nitrate, while Geneva et al. [21] and Stancheva et al. [28] reported that the soil and foliar fertilization (NPK) decreased the content of silybins A and B, silychristin, silydianin, and taxifolin. However, both varieties showed silymarin yield (27.1–50 kg ha^{-1}) comparable to previous studies that recorded 13.3–63.3 kg ha^{-1} [6,18], while fertilization positively affected silymarin yield. Sheep manure induced an increase only in the first year of the experiment, while calcium ammonium nitrate, especially in a high dose, influenced silymarin yield of the two varieties in both years. These findings are in agreement with that of Geneva et al. [21], Afshar et al. [16], and Liava et al. [17] in different genotypes, as they observed that the application of manure or inorganic fertilizers enhanced silymarin yield owing to higher fruit yield.

5. Conclusions

The outcome indicates that variety and fertilization are important factors that can influence fruit and silymarin yield of milk thistle crop. Spata had greater silymarin content and yield, and higher accumulation of silybin A and B compared with Palaionterveno, while fertilization regimes had no impact on flavonolignans and taxifolin content in the fruits. The results of the experiment clearly show that the application of sheep manure and calcium ammonium nitrate promoted plants growth and yield. In general, the application of the inorganic fertilization significantly improved the productivity of this crop compared with sheep manure; although, further studies are needed to evaluate the effects of the combination of inorganic with organic fertilization on this crop. Finally, our results revealed that Spata can be exploited in breeding programs owing to high silymarin productivity and high silybins A and B content. The development of new varieties with the desired traits will help to further improve the productivity of this crop and the quality of the final product (fruits or silymarin extracts).

Author Contributions: Investigation, methodology, writing—original draft preparation, V.L.; writing—review and editing, A.K.; review and editing, N.D. and N.T.; methodology, supervision, A.K. and N.T. All authors have read and agreed to the published version of the manuscript.

Funding: This study was implemented within the framework of the doctoral fellowship of V. Liava by the State Scholarships Foundation (IKY) of Greece.

Institutional Review Board Statement: Not applicable.

Informed Consent Statement: Not applicable.

Data Availability Statement: The data presented in this study are available in this article.

Acknowledgments: The first author V. Liava would like to thank IKY for awarding the scholarship that was funded by the Project "Strengthening Human Research Resources through the Implementation of Doctoral Research" from the resources of the Operational Programme "Human Resources Development, Education and Lifelong Learning", 2014–2020, which is co-financed by Greece and the European Union (European Social Fund-ESF).

Conflicts of Interest: The authors declare no conflict of interest.

References

1. Kim, S.-H.; Choo, G.-S.; Yoo, E.-S.; Woo, J.-S.; Lee, J.-H.; Han, S.-H.; Jung, S.-H.; Kim, H.-J.; Jung, J.-Y. Silymarin inhibit proliferation of human breast cancer cells via regulation of the MAPK signaling pathway and induction of apoptosis. *Oncol. Let* **2021**, *21*, 492. [CrossRef]
2. Ferraz, A.C.; Almeida, L.T.; da Silva Caetano, C.C.; da Silva Menegatto, M.B.; Souza Lima, R.L.; de Senna, J.P.N.; de Oliveir Cardoso, J.M.; Perucci, L.O.; Talvani, A.; Geraldo de Lima, W.; et al. Hepatoprotective, antioxidant, anti-inflammatory, an antiviral activities of silymarin against mayaro virus infection. *Antivir. Res.* **2021**, *194*, 105168. [CrossRef]
3. Vostálová, J.; Tinková, E.; Biedermann, D.; Kosina, P.; Ulrichová, J.; Rajnochová Svobodová, A. Skin protective activity of silymari and its flavonolignans. *Molecules* **2019**, *24*, 1022. [CrossRef]
4. Abbasirad, F.; Shaygannejad, V.; Hosseininasab, F.; Mirmosayyeb, O.; Mahaki, B.; Moayedi, B.; Esmaeil, N. Significant im munomodulatory and hepatoprotective impacts of silymarin in MS patients: A double-blind placebo-controlled clinical trial. *In Immunopharmacol.* **2021**, *97*, 107715. [CrossRef]
5. Rathore, P.; Arora, I.; Rastogi, S.; Akhtar, M.; Singh, S.; Samim, M. Collagen nanoparticle-mediated brain silymarin delivery An approach for treating cerebral ischemia and reperfusion-induced brain injury. *Front. Neurosci.* **2020**, *14*, 538404. [CrossRef [PubMed]
6. Andrzejewska, J.; Sadowska, K.; Mielcarek, S. Effect of sowing date and rate on the yield and flavonolignan content of the fruits o milk thistle (*Silybum marianum* L. Gaertn.) grown on light soil in a moderate climate. *Ind. Crops Prod.* **2011**, *33*, 462–468. [CrossRef
7. Koláčková, P.; Růžičková, G.; Šafránková, I.; Hron, K.; Hrůzová, K. Evaluation of the growth dynamics and morphologica characteristics of genetic sources of *Silybum marianum* (L.) Gaertn. *Acta Univ. Agric. Silvic. Mendel. Brun.* **2015**, *63*, 1141–114 [CrossRef]
8. Denev, P.N.; Ognyanov, M.H.; Georgiev, Y.N.; Teneva, D.G.; Klisurova, D.I.; Yanakieva, I.Z. Chemical composition and antioxidan activity of partially defatted milk thistle (*Silybum marianum* L.) seeds. *Bulg. Chem. Commun.* **2020**, *52*, 182–187.
9. Duran, D.; Ötleş, S.; Karasulu, E. Determination amount of silymarin and pharmaceutical products from milk thistle wast obtained from cold press. *Acta Pharm. Sci.* **2019**, *57*, 85–101. [CrossRef]
10. Majidi, M.M.; Shafiei-Koij, F.; Pirnajmedin, F.; Jami, M.; Radan, Z. Fatty acid profile, silymarin content and production propertie of milk thistle (*Silybum marianum*) germplasm under different water environments. *Crop Pasture Sci.* **2021**, *72*, 302–310. [CrossRef
11. Harrabi, S.; Ferchichi, A.; Sakhri, H.; Feki, M.; Hossaineian, F. Phospholipid and n-alkane composition, anti-α-glucosidase an anti-cyclooxygenase activities of milk thistle oil. *Eur. Food Res. Technol.* **2021**, *247*, 1557–1567. [CrossRef]
12. Fathi-Achachlouei, B.; Azadmard-Damirchi, S. Milk thistle seed oil constituents from different varieties grown in Iran. *J. Am. O Chem. Soc.* **2009**, *86*, 643–649. [CrossRef]
13. Ionescu, N.; Popescu, M.; Bratu, A.; Istrati, D.; Ott, C.; Meghea, A. Valuable Romanian vegetable oils and extracts with hig pharmaco-cosmetic potential. *Rev. Chim.* **2015**, *66*, 1267–1272.
14. Meddeb, W.; Rezig, L.; Abderrabba, M.; Lizard, G.; Mejri, M. Tunisian milk thistle: An investigation of the chemical compositio and the characterization of its cold-pressed seed oils. *Int. J. Mol. Sci.* **2017**, *18*, 2582. [CrossRef] [PubMed]
15. Ram, G.; Bhan, M.; Gupta, K.; Thaker, B.; Jamwal, U.; Pal, S. Variability pattern and correlation studies in *Silybum marianu* Gaertn. *Fitoterapia* **2005**, *76*, 143–147. [CrossRef] [PubMed]
16. Afshar, R.K.; Chaichi, M.; Assareh, M.; Hashemi, M.; Liaghat, A. Interactive effect of deficit irrigation and soil organic amendment on seed yield and flavonolignan production of milk thistle (*Silybum marianum* L. Gaertn.). *Ind. Crops Prod.* **2014**, *58*, 166–17 [CrossRef]
17. Liava, V.; Karkanis, A.; Tsiropoulos, N. Yield and silymarin content in milk thistle (*Silybum marianum* (L.) Gaertn.) fruits affecte by the nitrogen fertilizers. *Ind. Crops Prod.* **2021**, *171*, 113955. [CrossRef]
18. Arampatzis, D.; Karkanis, A.; Tsiropoulos, N. Impact of plant density and mepiquat chloride on growth, yield, and silymari content of Silybum marianum grown under Mediterranean semi-arid conditions. *Agronomy* **2019**, *9*, 669. [CrossRef]
19. Vozhehova, R.A.; Fedorchuk, M.I.; Lavrynenko, Y.O.; Kokovikhin, S.V.; Lykhovyd, P.V.; Biliaieva, I.M.; Nesterchuk, V.V. Effect o agrotechnological elements on milk thistle (*Silybum marianum*) productivity. *Regul. Mech. Biosyst.* **2018**, *9*, 156–160. [CrossRef]
20. Saad-Allah, K.; Fetouh, M.; Elhaak, M. Induction of milk thistle (*Silybum marianum* L. Gaertn) growth and phytochemical production by natural stimulants. *J. Appl. Res. Med. Aromat. Plants.* **2017**, *6*, 101–110. [CrossRef]
21. Geneva, M.; Zehirov, G.; Stancheva, I.; Iliev, L.; Georgiev, G. Effect of soil fertilizer, foliar fertilizer, and growth regulato application on milk thistle development, seed yield, and silymarin content. *Commun. Soil Sci. Plant Anal.* **2008**, *39*, 17–2 [CrossRef]
22. Shokrpour, M.; Giglou, M.T.; Asghari, A.; Bahrampour, S. Study of some agronomic attributes in milk thistle (*Silybum marianu* Gaertn.) ecotypes from Iran. *J. Med. Plant Res.* **2011**, *5*, 2169–2174.
23. Martinelli, T.; Potenza, E.; Moschella, A.; Zaccheria, F.; Benedettelli, S.; Andrzejewska, J. Phenotypic evaluation of a milk thistl germplasm collection: Fruit morphology and chemical composition. *Crop Sci.* **2016**, *56*, 3160–3172. [CrossRef]
24. Poppe, L.; Petersen, M. Variation in the flavonolignan composition of fruits from different *Silybum marianum* chemotypes an suspension cultures derived therefrom. *Phytochemistry* **2016**, *131*, 68–75. [CrossRef] [PubMed]

25. Arampatzis, D.; Karkanis, A.; Tsiropoulos, N. Silymarin content and antioxidant activity of seeds of wild *Silybum marianum* populations growing in Greece. *Ann. Appl. Biol.* **2019**, *174*, 61–73. [CrossRef]
26. Gresta, F.; Avola, G.; Guarnaccia, P. Agronomic characterization of some spontaneous genotypes of milk thistle (*Silybum marianum* L. Gaertn.) in Mediterranean environment. *J. Herbs Spices Med. Plants* **2006**, *12*, 51–60. [CrossRef]
27. Sulas, L.; Re, G.A.; Bullitta, S.; Piluzza, G. Chemical and productive properties of two Sardinian milk thistle (*Silybum marianum* (L.) Gaertn.) populations as sources of nutrients and antioxidants. *Genet. Resour. Crop Evol.* **2016**, *63*, 315–326. [CrossRef]
28. Stancheva, I.; Georgiev, G.; Geneva, M.; Ivanova, A.; Dolezal, M.; Tumova, L. Influence of foliar fertilization and growth regulator on milk thistle seed yield and quality. *J. Plant Nutr.* **2010**, *33*, 818–830. [CrossRef]
29. Laleh, S.; Jami Al-Ahmadi, M.; Parsa, S. Response of hemp (*Cannabis sativa* L.) to integrated application of chemical and manure fertilizers. *Acta Agric. Slov.* **2021**, *117*, 1–15. [CrossRef]
30. Ebrahimian, E.; Seyyedi, S.M.; Bybordi, A.; Damalas, C.A. Seed yield and oil quality of sunflower, safflower, and sesame under different levels of irrigation water availability. *Agric. Water Manag.* **2019**, *218*, 149–157. [CrossRef]
31. Ismael, F.; Ndayiragije, A.; Fangueiro, D. New fertilizer strategies combining manure and urea for improved rice growth in Mozambique. *Agronomy* **2021**, *11*, 783. [CrossRef]
32. Popin, G.V.; dos Santos, A.K.B.; Ferrão, G.E.; Lourenço, D.A.; Siqueira-Neto, M. Effect of organic N-sources on maize yield components. *Aust. J. Crop Sci.* **2019**, *13*, 1215–1222. [CrossRef]
33. Školníková, M.; Škarpa, P.; Vagnerova, L. Response of milk thistle [*Silybum marianum* L. (Gaertn.)] on nitrogen and sulphur fertilization. *MendelNet* **2017**, *24*, 116–120.
34. Holík, L.; Hlisnikovský, L.; Kunzová, E. The effect of mineral fertilizers and farmyard manure on winter wheat grain yield and grain quality. *Plant Soil Environ.* **2018**, *64*, 491–497. [CrossRef]
35. Yang, C.; Du, W.; Zhang, L.; Dong, Z. Effects of sheep manure combined with chemical fertilizers on maize yield and quality and spatial and temporal distribution of soil inorganic nitrogen. *Complexity* **2021**, *2021*, 4330666. [CrossRef]
36. Gewaily, E.E.; Ghoneim, A.M.; Osman, M.M.A. Effects of nitrogen levels on growth, yield and nitrogen use efficiency of some newly released Egyptian rice genotypes. *Open Agric.* **2018**, *3*, 310–318. [CrossRef]
37. Belete, F.; Dechassa, N.; Molla, A.; Tana, T. Effect of nitrogen fertilizer rates on grain yield and nitrogen uptake and use efficiency of bread wheat (*Triticum aestivum* L.) varieties on the Vertisols of central highlands of Ethiopia. *Agric. Food Secur.* **2018**, *7*, 78. [CrossRef]
38. Li, W.P.; Shi, H.B.; Zhu, K.; Zheng, Q.; Xu, Z. The quality of sunflower seed oil changes in response to nitrogen fertilizer. *Agron. J.* **2017**, *109*, 2499–2507. [CrossRef]
39. Gao, J.; Thelen, K.D.; Min, D.H.; Smith, S.; Hao, X.; Gehl, R. Effects of manure and fertilizer applications on canola oil content and fatty acid composition. *Agron. J.* **2010**, *102*, 790–797. [CrossRef]

agronomy

MDPI

Article

Weed-Free Durations and Fertilization Regimes Boost Nutrient Uptake and Paddy Yield of Direct-Seeded Fine Rice (*Oryza sativa* L.)

Rana Nadeem Abbas [1], Asif Iqbal [1], Muhammad Aamir Iqbal [2,*], Omar M. Ali [3], Raees Ahmed [4], Raina Ijaz [5], Adel Hadifa [6] and Brandon J. Bethune [7]

1 Department of Agronomy, Faculty of Agriculture, University of Agriculture Faisalabad, Faisalabad 38040, Pakistan; nadeem4u3@yahoo.com (R.N.A.); asifiqbaldehasi@yahoo.com (A.I.)
2 Department of Agronomy, Faculty of Agriculture, University of Poonch Rawalakot, Rawalakot 12350, Pakistan
3 Department of Chemistry, Turabah University College, Turabah Branch, Taif University, P.O. Box 11099, Taif 21944, Saudi Arabia; om.ali@tu.edu.sa
4 Department of Plant Pathology, Faculty of Agriculture, University of Poonch Rawalakot, Rawalakot 12350, Pakistan; raees@upr.edu.pk
5 Department of Horticulture, Faculty of Agriculture, University of Poonch Rawalakot, Rawalakot 12350, Pakistan; rainaijaz@upr.edu.pk
6 Rice Research and Training Center, Field Crops Research Institute, Agricultural Research Center, Sakha 33717, Egypt; adelhadifarrtc@gmail.com
7 Department of Plant Sciences, University of California Davis, Davis CA 95616, USA; Bbjbethune@outlook.com
* Correspondence: muhammadaamir@upr.edu.pk

Citation: Abbas, R.N.; Iqbal, A.; Iqbal, M.A.; Ali, O.M.; Ahmed, R.; Ijaz, R.; Hadifa, A.; Bethune, B.J. Weed-Free Durations and Fertilization Regimes Boost Nutrient Uptake and Paddy Yield of Direct-Seeded Fine Rice (*Oryza sativa* L.). *Agronomy* **2021**, *11*, 2448. https://doi.org/10.3390/agronomy11122448

Academic Editors: Umberto Anastasi and Aurelio Scavo

Received: 11 November 2021
Accepted: 29 November 2021
Published: 30 November 2021

Abstract: Under the changing climate, fertilization regimes and weed infestation management in aromatic direct-seeded fine rice (DSR) remain vital for curbing environmental hazards and ensuring food security. A multi-year field study was undertaken to appraise the influence of fertilization techniques and weed-free periods on weed dynamics, nutrient uptake and paddy yield in a semi-arid environment. Treatments included two fertilization methods (broadcasting and side placement) and five weed-free durations (20, 30, 40, 50 post-seeding days, DAS) along with a weed-free crop for a whole season. Weed competition for a season-long crop (weedy check) was maintained for each fertilizer application method. Our results revealed that the side placement of fertilizers resulted in a significantly lower weed density and biomass, even under season-long weed competition. The highest paddy yield was recorded for a crop without weeds, while weed-free duration of up to 50 DAS followed it. The uptake of nitrogen (N), phosphorus (P) and potassium (K) for a weed-free duration of up to 50 DAS were only 19%, 9% and 8%, respectively, as compared to the weedy check. The uptake of N, P and K by weeds in the broadcast method was 18%, 30% and 24% higher, compared to side-placed fertilizers. The period of 20–50 DAS remained critical in DSR as far as weed control was concerned. Thus, the side placement of fertilizers and controlling weeds for up to 50 days after rice sowing can be recommended for general adoption in semi-arid agro-ecological conditions.

Keywords: broadcasted fertilization; side-dressing; paddy growth; weeds competition; macro-nutrients

1. Introduction

Rice (*Oryza sativa* L.) constitutes the major staple crop after wheat, which feeds billions of people across the globe and is hence referred to as the global grain [1–3]. It is being grown in all habitable continents of the world, owing to its wide adaptability to a wide range of pedo-climatic conditions [4,5]. In Asia, it is cultivated with irrigation systems, while in Pakistan, rice is ranked the third most prominent crop, covering 10% of the cultivated area and contributing 17% to the total cereal production [6–8]. However, rice cultivation through nursery transplanting in a puddled field is cumbersome, time-consuming and is a prodigal water use method [9]. Under the changing climate, looming water crises and the uncertainty of climatic optima have endangered the sustainability of transplanted rice's production systems, which no longer seems a feasible technique, especially in South Asian countries, such as Pakistan, China, India and Bangladesh. It is estimated that 800 L of water are applied for producing 1 kg paddy globally, whereas the corresponding value in Pakistan stands at 3000–5000 L of water [7]. Another consequence of puddled transplanted rice is the delayed sowing of wheat, which reduces its yield by 33% [10].

Direct seeding of rice (DSR) has emerged as an alternate and pragmatic approach to tackling production constraints [10] and imparting sustainability to the rice-wheat cropping system [11,12]. The DSR offers several advantages, including faster plant growth, avoidance of transplanting shock, ease in cultivation, less labor and water requirements as well as early maturity [13–17]. Additionally, a significant reduction in methane emission was also accomplished with DSR [6]. However, weed infestation has remained the prime biological constraint limiting the productivity of DSR by up to 75–80% [14,18]. High weed infestation in DSR was attributed to recalcitrant weed flora, dry conditions, frequent tillage, alternate wet and dry periods and the lesser competitive ability of 30–35 day old rice seedling against genetically superior weed flora, as these have better plant infrastructure [19]. The critical period of weed competition (CPWC) in DSR is a prerequisite to be determined for devising a strategy to keep weeds below the threshold level [20]. Johnson et al. [21] determined the CPWC in rice as 5–79 DAS (days after sowing), while in contrast, Anwar et al. [22] concluded that the CPWC for DSR was 7–50 DAS. Additionally, it was documented that the presence of weeds in DSR beyond 20 DAS adversely affected yield components which led to reduced growth as well as reduced paddy yield. It was also inferred that the reduction in paddy yield was proportional to CPWC up to 30 DAS [20,23,24].

Under changing climate scenarios, the severity of competition rendered by weeds in DSR can be modified by optimizing cultural practices that simultaneously affect both crops and weeds. Among these cultural practices, fertilizer management remains instrumental in determining the competitive outcomes of the weed-crop association [23,24]. Deep banding or the side placement of nitrogenous fertilizers (whereby fertilizers might be placed in the side of crop rows manually or drilled) reduces the growth of weeds and nitrogen (N) uptake compared to broadcasted fertilizers [25]. Rasmussen et al. [26] inferred that reducing weed density and fertilizer management concurrently increased the paddy yield by 28%. The broadcast fertilizer remained inferior to the side-band application method, owing to an increase in weed emergence and reduction in crop growth which reduced grain yield by 10% [27]. These findings suggested that the manipulation of crop fertilization application techniques may impart significant influence on weed density in DSR and must be evaluated in tandem with respect to CPWC [28,29].

Numerous studies have evaluated weed dynamics and the resultant yield losses in DSR, either by using increasing weed-free or weedy duration approaches [20,22,30–33]. Nevertheless, none of these has assessed weed competition in response to fertilization techniques in DSR. Numerous studies have also elaborated the potential of fertilizer application techniques for maintaining nutrient status, along with boosting fertilizer use efficiency (FUE) [24,25,34]. The fertilization application technique involving fertilizer placement at a depth of 5 cm and 5 cm apart from crop seeds significantly improved the above-ground biomass and paddy yield in comparison to the manual surface broadcast fertilization method [26]. This method (burying of fertilizers below 5 cm) promoted the growth of rice roots which triggered the growth of crop plants during the early vegetative growth stages [18]. Additionally, appropriate fertilization methods, such as band fertilization, effectively reduced the quantities of fertilizers needed for DSR without adversely affecting the rice plants' growth and paddy yield [27]. The broadcasted fertilizers resulted in higher losses by emissions, while deep fertilization increased the nutrient absorption by rice roots by minimizing loses as leaching and gaseous emissions [33]. The side placement of fertilizers significantly enhanced peroxidase and catalase in DSR. In addition, it inferred that mechanically deep-buried fertilization delayed rice leaf senescence by improving the activities of antioxidant enzymes and reducing the malonic dialdehyde in DSR, which led to a higher paddy yield. Under the changing climate, the optimization of fertilization method has become even more important owing to its potential for 40% and 54% reduction in methane (CH4) and nitric oxide (NO) emissions, respectively [6]. Therefore, optimization of the fertilization method constitutes a potent strategy to mitigate environmental pollution which is contributed to by methane emissions from rice fields.

It is hypothesized that fertilization regimes could exert a growth-restricting influence on weed growth by reducing nutrient availability to weeds. In contrast, a specific weed-free duration could potentially boost the rice yield in DSR culture. Therefore, the present study was undertaken to appraise the influence of the fertilization methods and weed-free duration on weed dynamics, nutrient uptake and the paddy yield of DSR.

2. Materials and Methods

2.1. Description of Meteorological and Physico-Chemical Characteristics of Experimental Site

The Agronomic Research Farms of the University of Agriculture Faisalabad, Pakistan (31.4504° N, 73.1350° E, altitude of 186 m) [35], was the location of the experiments undertaken for this study during two consecutive years (2018 and 2019). The sowing of the experiment was done after the harvest of the winter wheat crop. The mean temperature and rainfall of the experimental site during both growing seasons (end of May to mid October), as per the recordings of meteorological observatory located in the close vicinity of our experimental site, are presented in Figure 1. The irrigations (3 inches depth and last irrigation of 4 inches depth) were applied fortnightly through the flood irrigation method, however the interval between irrigations was reduced, keeping in view the high temperature and crop needs. In total, 10–14 irrigations were applied on an average during both years of the study.

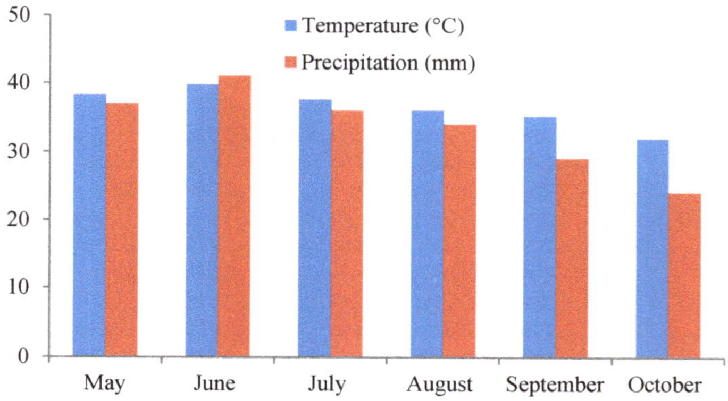

Figure 1. The meteorological features (temperature and precipitation) of experimental site (Faisalabad, Punjab, Pakistan) during crop growing seasons, (2 years mean data).

To conduct pre-experiment soil physico-chemical analyses, soil samples (0–15 cm and 15–30 cm depths) were taken from the four corners and the middle of the experimental area (net plot size was 6 m × 3 m, while there were 12 experimental plots per replication making total experimental area 648 m^2, excluding water channels, walking paths and replication separating fellow area). The collected samples from both depths were thoroughly homogenized by hand for subsequent analyses. Thereafter, the samples were dried under shade, grounded and subsequently sieved with the help of a sieve with a pore size of 2 mm. For measuring the pH, the samples were prepared using 1:2.5 ratios of soil and water while the glass electrode was used to record the pH [36]. The electrical conductivity (EC) was also measured using the conductivity meter [23]. The wet oxidation method was followed for organic carbon (OC) estimation volumetrically. Meanwhile, the Walkley–Black methodology was used to determine the organic matter (OM) content of the soil samples [37]. To estimate the total nitrogen (N) content, a Kjeldahl apparatus was used for distillation which followed titration using concentrated H_2SO_4 [38]. Additionally, the phosphorous (P) content of the soil samples was estimated by performing Olsen's method (0.5 N $NaHNO_3$ at 8.5 pH by maintaining soil: extractant ratio of 1:10) using spectrophotometer (by setting wavelength at 882 nm in a system containing sulfuric acid) [39]. At the same time, potassium (K) was determined by following the standard procedure involving the ammonium acetate extraction of air-dried soil samples through shaking them with an ammonium acetate solution (0.5 M) for 30 min, which led to the displacement of positively charged K ions and a flame photometer was used to detect them. Subsequently, the methodology described in [37] was used for the K calculation. As far as the micronutrients were concerned, the available iron (Fe) content was estimated using an extraction method with an ammonium acetate solution (CH_3COONH_4) by maintaining 3.0 pH. Thereafter, a spectrophotometer (510 nm wavelength) was used by following the colorimetric method to determine the Fe content of the soil extracts. Furthermore, the rest of the micronutrients, including zinc (Zn), boron (B), copper (Cu) and manganese (Mn), were estimated using the extraction method involving diethylenetriaminepentaacetic acid [40–42]. The soil had a loam texture, with a pH 8.2, while OM remained at only 0.55%, indicating exhaustive cultivation of the soil. The soil had a bulk density and EC of 1.39 cm^{-3} and 0.47 dS m^{-1}, respectively. The NPK contents remained 77, 4.1 and 113 mg kg^{-1}, respectively. Among the micronutrients, B, Mn, Fe, Cu and Zn were 1.00, 20.9, 10.17, 1.8 and 1.21 mg kg^{-1} of soil, respectively.

2.2. Experimentation Details

The field trial was comprised of two fertilization methods (FM), including broadcasted fertilizers (BF) and the side placement (SP) method, along with six weed-free periods (WFD), for instance, 20, 30, 40 and 50 days after sowing (DAS), and full crop season weed-free (WF) and a weedy check (WC) for comparison. The fertilizer was broadcasted in each plot separately, while side placement was done along the crop lines using a single row hand drill. The experiment was executed using a randomized complete block design (RCBD) with factorial scheme, while three replications of each experimental treatment were maintained. The unit plot size (excluding the water channel and field path area) was 6 m × 3 m (experimental plots were separated by 2 feet wide bunds and 5 feet fellow area, similarly 5 m fellow area was maintained among the replications). Rice seeds (cv. Super Basmati that is a short stature, early maturing, highly aromatic cultivar that is currently being cultivated over a large area in the rice-belt of Punjab province) were hydro-primed for 24 hr to obtain a higher germination and vigorous seedling growth. The soaked seeds were shade dried and stored at 10 °C. To ensure that the seed-bed obtained fine tilth, three ploughings with a tractor-driven common cultivator were performed, while subsequent planking was also done to pulverize the soil. The crop was sown (50 kg seed rate ha^{-1}) during the last week of May with a single row hand drill in 20 cm apart rows. Chemical fertilizers, such as urea (125 kg N ha^{-1}), di-ammonium phosphate (55 kg P ha^{-1}) and sulphate of potash (40 kg K ha^{-1}), were applied manually in both techniques (broadcast and side placement involving fertilizer dressing at 10 cm away from the seeds and at a depth of 5 cm). At the sowing time, P and K and half of N were applied with the last ploughing. The remaining N was equally divided and subsequently applied at the tillering and panicle formation stages. The crop was kept free of weeds for different duration's: 0, 20, 30, 40, 50 DAS and season long. Weeds were manually removed up to the completion of the respective weed-free period. Once a specific weed-free period was accomplished, weeds were not controlled and competition with rice seedlings was allowed. The first irrigation was applied at 5 DAS, while the subsequent irrigation scheduling depended on the weather conditions and crop needs.

2.3. Data Recordings of Response Variables

Data on weed growth (density WD and dry biomass WB) were recorded at harvesting from two randomly selected quadrats (100 cm × 100 cm = 1 m^2) from each experimental unit plot. The weed density of grasses, sedges and broader leaf weeds was measured.

Weeds were clipped off near the soil surface and were sundried for five days. Weeds' dry biomass was taken separately using oven-dried (for 48 h at a constant temperature of 70 °C) samples. Data pertaining to the response variables were recorded by randomly selecting fifteen plants per experimental plot, and their average was computed. Panicle bearing tiller (m^{-2}) numbers were counted using two random sites of each experimental plot and were subsequently averaged. The random sampling of kernels was done in each plot for recording 1000-kernels weight. The crop in each experimental plot was manually harvested (on October 14 during the first year and October 18 during the subsequent year) and then tied into bundles. The experimental plots were separately threshed for recording the paddy yield and were subsequently converted into tons per hectare. For the nutrient uptake by rice and weeds, the grinding of oven-dried material was completed (Cyclotec 1093 Sample Mill, Sweden) and the material was subsequently passed through a 40-meshscreen. The N-P-K concentrations in rice and weeds' samples (collected at harvest) were determined as described in the laboratory manual of ICARDA [43].

2.4. Statistical Analyses

The data regarding parameters under investigation were arranged and subsequently analyzed statistically by employing Bartlett's test, which showed that the year had a non-significant effect and, as a result, data about the year were transformed into mean values and used for subsequent analyses. Thereafter, Fisher's ANOVA technique was employed, and a comparison of treatment means was made using Duncan's multiple range test (DMRT) at 5% level of probability [44].

3. Results and Discussion

3.1. Weeds Density and Dry Biomass

The results revealed that the fertilization regimes and weed-free duration significantly influenced the weeds (grasses, sedges and broad leaf) density and dry matter (Figure 1). In the experimental plots, weeds of the *Gramineae* family were *Echinochloa crusgalli*, *Echinochloa colonum*, *Dactyloctenium aegyptium*, *Eclipta prostrata*, *Cynodon dactylon*, *Leptochloa chinensis* and *Eleusine indica*. The weeds of the Cyperaceae family were *Cyperus rotundus* and *Cyperus iria* (sedges), while some of the broad-leave weeds were *Trianthema portulacastrum*, *Ipomoea aquatica* and *Portulaca oleracea*. It was observed that weeds emerged profusely between 15–25 DAS, and the pace of weed emergence was at a peak up to 40 DAS in weedy check plots. After 40 DAS, weeds emergence and their growth slowed remarkably, owing to the rice seedling establishment in the season-long weed-free treatment. Weed biomass remained statistically on par at 40 and 50 DAS for broadcast and side placed fertilizers (Figure 2b), indicating 20–30 DAS as the most critical period for attaining biomass by weed flora.

Regarding weed density and dry biomass, the interactive effect of FM with WFD remained significant ($p \leq 0.05$) (Figure 2a,b), and weed biomass remained significantly lesser, especially at 20 and 30 DAS for SP fertilizers compared to BF (Figure 2b). In the WC plot, WD and WB were 301.67 m^2 and 144.89 g m^2, respectively, for BF against 286.33 m^2 and 132.87 g m^2 in SP. For the WF plots up to 50 DAS, WD and WB were reduced to 88.33 m^2 and 45.893 g m^2, respectively, while their corresponding values for SP were 51 m^2 and 40 g m^2. This corresponds to 42 and 13% higher WD and WB for BF, compared to SP. These declined significantly as the WFD was increased up to 50 DAS. However, the WB recorded by BF and SP remained statistically at par to each other, at 40 and 50 DAS (Figure 2b).

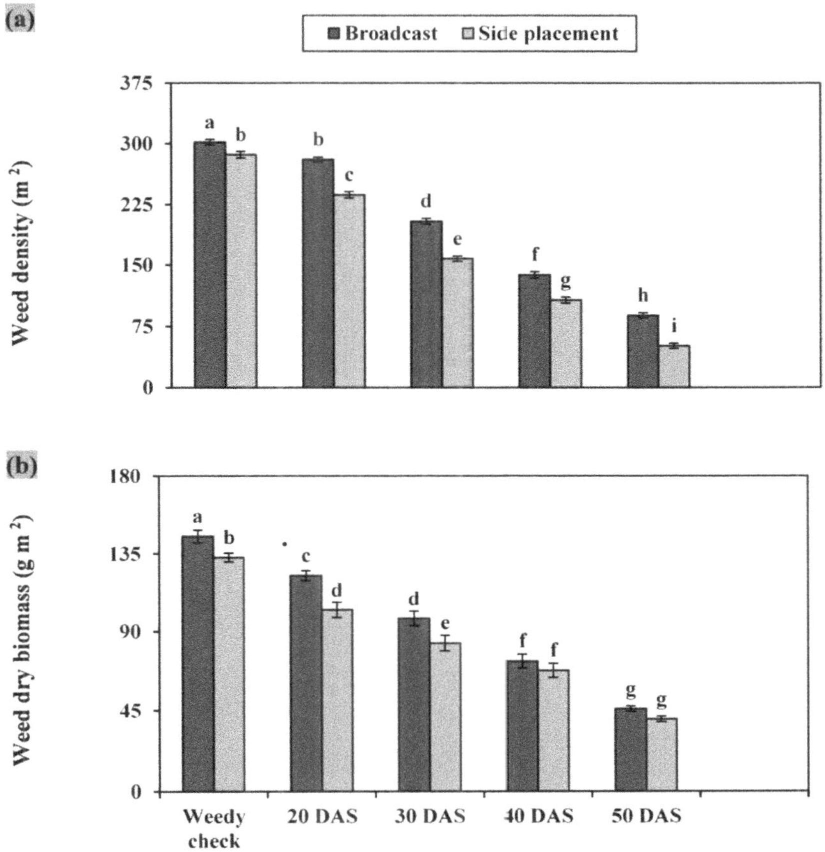

Figure 2. Effects of different weed-free periods on (**a**) weed density and (**b**) weed dry biomass in direct-seeded rice under different fertilizer application methods. Vertical bars above mean denote the standard error (±) of six replicates and letters depict significane at 5% probability level. Units on *x*-axis remain same for both a and b figures, (2 years mean data).

The grass weed proportion remained higher compared to broad leaf weed sedges up till 30 DAS, nevertheless, it manifested a sharp increment thereafter (Figure 3). Although the broad leaf weed proportion was lesser during the first 30 DAS, however, they continued to show up until the end of the season. Later in the season, narrow-leaved weeds gradually replaced the broadleaved weeds. It was observed that FM had no significant effect on the different weed types. Weed density and sedge biomass remained the same for a whole season. Previously, it was concluded that broad-leaf weeds dominated initially for up to 30 DAS, while grasses surpassed broadleaves afterwards. Additionally, it was noted that DSR had a higher density of grasses followed by broad-leaf weeds and sedges [45]. Our results suggested that the weed density and biomass were higher by 20% and 10%, respectively, for BF compared to SP (Figure 2). This might be attributed to the nutrients being readily available to weeds in broadcasted plots, leading to a profound increase in weed infestation in comparison to side-placed fertilizers. Our findings are also in concurrence with those of [46], who opined that the side-placement fertilization method recorded 40% lower weed density, owing to significantly lesser available nutrients in comparison to broadcasted fertilizer. It was also inferred that side-injected fertilizers resulted in a significantly higher nutrient uptake (23%) by crop plants and lesser nutrient losses (over 31%) which led to a noticeable decline in the free nutrients that were available to weeds seedling and, ultimately, their competitive superiority was decreased. Notably, crop plants were able to attain robust earlier growth. Rasmussen et al. [26] also reported a lower density of weeds (10%) and weed biomass (55%) when fertilizer was side-applied as this resulted in a slow release of nutrients compared to broadcasted fertilizers, which were readily absorbed by weeds and lost to the environment through leaching and gaseous emissions. It was concluded broadcasted fertilizers were equally available to weeds and crop plants, however owing to botanical superiority, weeds were able to extract more nutrients than crop plants which caused a significant increase in their fresh and dry biomass [2,14,21,25,26]. Weeds were able to absorb and utilize environmental resources such as solar radiation and CO_2, more efficiently and, resultantly, tended to record more vigorous growth compared to crop plants [46]. More importantly, weeds occupied more space owing to rigorous growth and the resulting lesser space in the rhizosphere remained available to crop plants for nutrient absorption from the soil solution [28,31]. However fertilization did not significantly effect weed biomass when weeds were controlled by herbicides. However, under unrestricted weed growth, fertilizer placement above the soil surface increased weed biomass substantially more than the placement of fertilizer below the soil surface [28].

3.2. Yield Components and Paddy Yield

The yield components contributing to paddy yield were largely influenced by fertilization methods and the duration of the weed-free period, and their interaction also remained significant ($p \leq 0.05$) (Figure 4a–d). The number of panicles (m^{-2}) of DSR in the weed-free plots of up to 20 DAS were more numerous for SP than BF (Figure 4b). However, for the rest of the weed-free periods, both fertilizer application methods remained statistically alike (Figure 4a). Side placement of fertilizer also realized more kernels per panicle in those plots that were kept free of weeds for up to 20 DAS and throughout the growing season than the broadcast method (Figure 3b). The treatment effect on the 1000-kernel weight was relatively less pronounced, with the exception of those plots in which weeds were controlled for up to 40 DAS, where the side placement application of fertilizer produced heavier kernels (Figure 4c). In the plots which had no weeds for 20–50 DAS and, afterwards, were subjected to unrestricted weed growth, side placement of fertilizer produced a distinct yield advantage over the broadcast method (Figure 4d). However, FM remained non-significant when either weeds were controlled or not controlled at all throughout the growing season. The numbers of tillers having panicles, kernels in each panicle and weight of 1000 kernels and paddy yield were increased with a weed-free duration. Moreover, weed-free conditions for the whole season were instrumental in improving (5–10%) the yield attributes, especially under the side-placed fertilization treatment (Figure 4a–c). Interestingly in comparison to WF, weed competition for the whole season reduced the amount of panicles bearing tillers, kernels per panicle and kernel weight, as well as the paddy yield by 40%, 50%, 20% and 75%, respectively (Figure 4). These findings are in agreement with those of [47] who reported that fertilizer placement remained instrumental in boosting grain yield, as compared to broadcasted fertilizers, owing to lesser wastage through leaching and volatilization, along with maximized uptake by crop plants. Likewise, a number of previous studies [26,27,48–52] reported the superiority of side-placed fertilization for increasing the grain yield by 28% compared to broadcasted fertilization. It was also inferred that the rice yield was lowered by 0.75 kg for each kg of weed biomass, as weeds had overtaken the crop plants in terms of acquiring growth resources, such as CO_2 and solar radiation [20]. Chauhan and Johnson [32] revealed a net loss of 24% in paddy yield when weeds kept growing for 28 DAS of rice. Weed competition for the whole crop season resulted in over 80% grain yield loss owing to severe competition for growth resources, especially moisture and nutrients. It was also inferred that, owing to weeds' genetic superiority, as indicated by better root architecture, imparted edge over crop plants in terms of nutrients acquisition and ultimately robust growth of weeds, led to a serious decline in rice growth. Khaliq and Matloob [20] observed that weed infestation up to 200 m^{-2} significantly reduced paddy yield (51–64%) compared to the WF crop. Furthermore, previous studies found that weed interference had a varied impact on rice growth, depending upon the duration of the weed's dominance and the growth stage of the rice seedlings. This implies that if weed-free conditions are maintained for this stipulated period, it equates to providing weed-free conditions for a whole season, which has the potential to significantly increase paddy yield. These findings corroborate with earlier results where weed presence beyond 55 DAS did not remain drastic for rice [20,21]. Hence, it was suggested that subsequent weed control could be of little use and, rather, it could be economically unviable, incurring additional expenditures and leading to a higher cost of production [22–24].

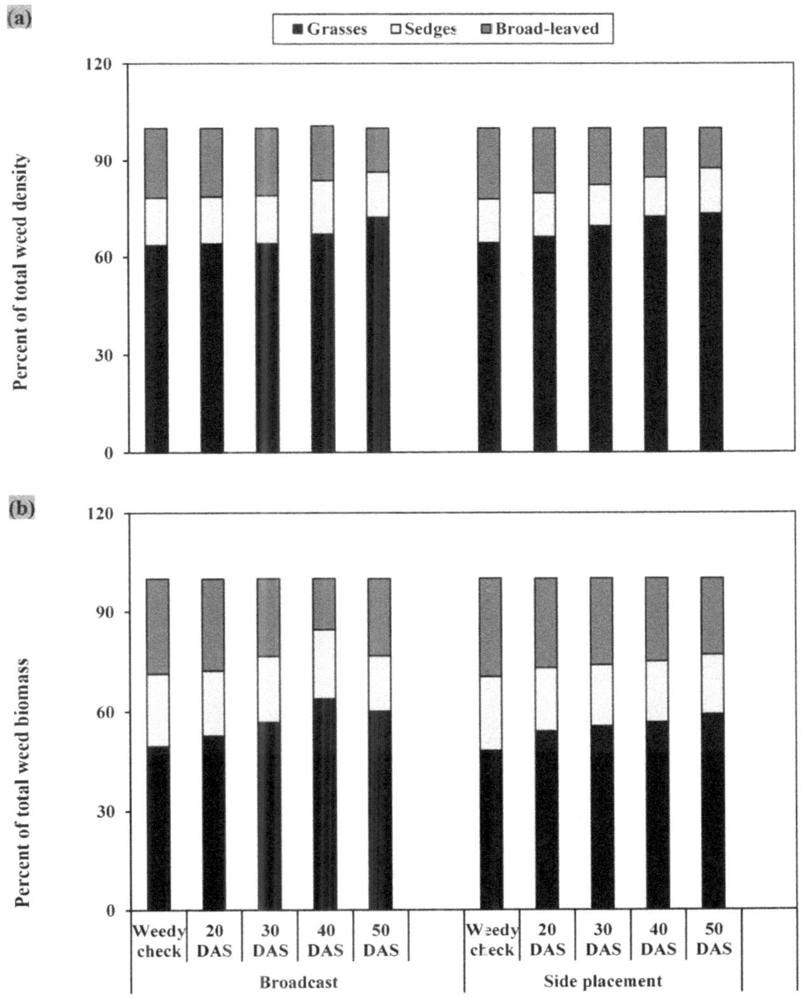

Figure 3. Relative proportion of different weed types (grasses, sedges and broad leaf weeds) on percentage basis in total (**a**) weed density and (**b**) weed dry biomass in direct-seeded rice under different fertilizer application methods. Units on x-axis remain same for both a and b figures, (2 years mean data and six replicates).

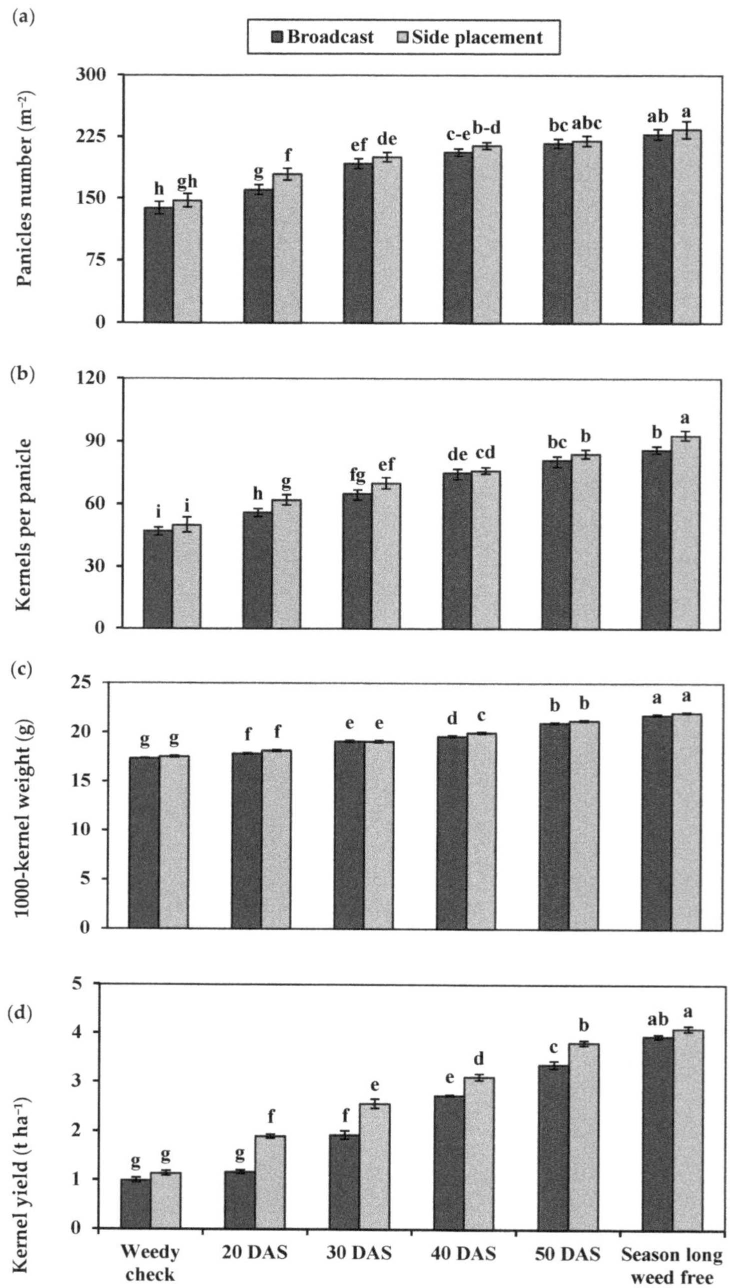

Figure 4. Effects of different weed-free periods on (**a**) number of panicles, (**b**) kernels per panicle, (**c**) 1000 kernels weight and (**d**) kernel yield of direct-seeded rice under different fertilizer application methods. Vertical bars above mean denote the standard error (±) of six replicates and letters depict significane at 5% probability level. Units on *x*-axis remain same for a, b, c and d figures, (2 years mean data).

3.3. Nutrient Uptake by Weeds and Directly Seeded Rice

The amount of absorbed nutrients by weeds and rice seedlings was significantly affected by the fertilization technique (FM) and weed-free durations (WFD) (Figures 5 and 6). The interactive effect of the FM and WFD remained significant for nutrient uptake; however, N absorption was the only exception whereby the two fertilization techniques did not differ (Figure 5a). Weeds extracted more nutrients from BF, suggesting their competitive advantage imparted by better plant architecture and genetic diversity. The nutrient uptake by weeds in plots kept weeds free for up to 50 DAS was only 19% (N), 9% (P) and 8% (K) compared to the weedy check, while NPK uptake remained 18%, 30%, and 24% higher for BF compared to SP (Figure 5). The NPK uptake of rice seedlings increased significantly with the weed-free duration; however, an opposite trend was observed for weeds. It was also inferred that being WF for beyond 50 DAS also produced a higher uptake of NPK by the rice seedlings. The interaction of FM and WFD remained significant, as far as nutrient uptake was concerned (Figure 6). The SP of the fertilizers improved the nutrient uptake in DSR compared to the BF. The maximum decline of N (70%), P (90%) and K (95%) were recorded in DSR, which confronted season-long weed competition, compared to WF. The SP recorded N, P and K uptake of 15%, 26% and 31% higher than BF (Figure 6). These results are also in harmony with those of [25–28], who recorded losses of 37 kg N, 30 kg P_2O_5 and 37 kg K_2O m^{-2} as a result of unchecked weed growth, and it was inferred that weeds dominated crop plants in terms of acquiring nutrients and, resultantly, significantly lesser nutrients could make their way to the targeted crop plants. It was also suggested that the presence of weeds throughout the crop season remained more drastic for rice seedlings in comparison to the fertilization technique, owing to a superior root network which enabled weeds to extract more nutrients, including N, P and K. Additionally, previous field investigations have also revealed that the presence of season-long weeds and type of fertilization placement depleted 21 kg N, 19 kg P_2O_5 and 77 kg K_2O m^{-2}, which led to a serious decline in the growth and development of rice plants and, ultimately, yield attributes, along with the paddy yield, were noticeably declined compared to the weedy check crop [30,31,53–57]. It was inferred that side-injected fertilizers remained effective in boosting yield attributes and grain yield by over 19%, compared to broadcasted fertilizers, as the nutrient uptake was 43% higher. It was concluded that broadcasted fertilizers resulted in the vigorous growth of weeds up to 50 DAS, which imparted adverse effects on the nascent rice seedling and, resultantly, yield attributes were seriously compromised. In comparison, N losses as volatilization were reduced by over 73% [53–57]. Prior studies have also concluded that the broadcast technique of fertilization was responsible for the excessive use of fertilizers and must be replaced with side-dressed fertilization in order to boost crop yield through higher nutrient use efficiency and through overcoming the votalization and leaching challenges in an economically viable way.

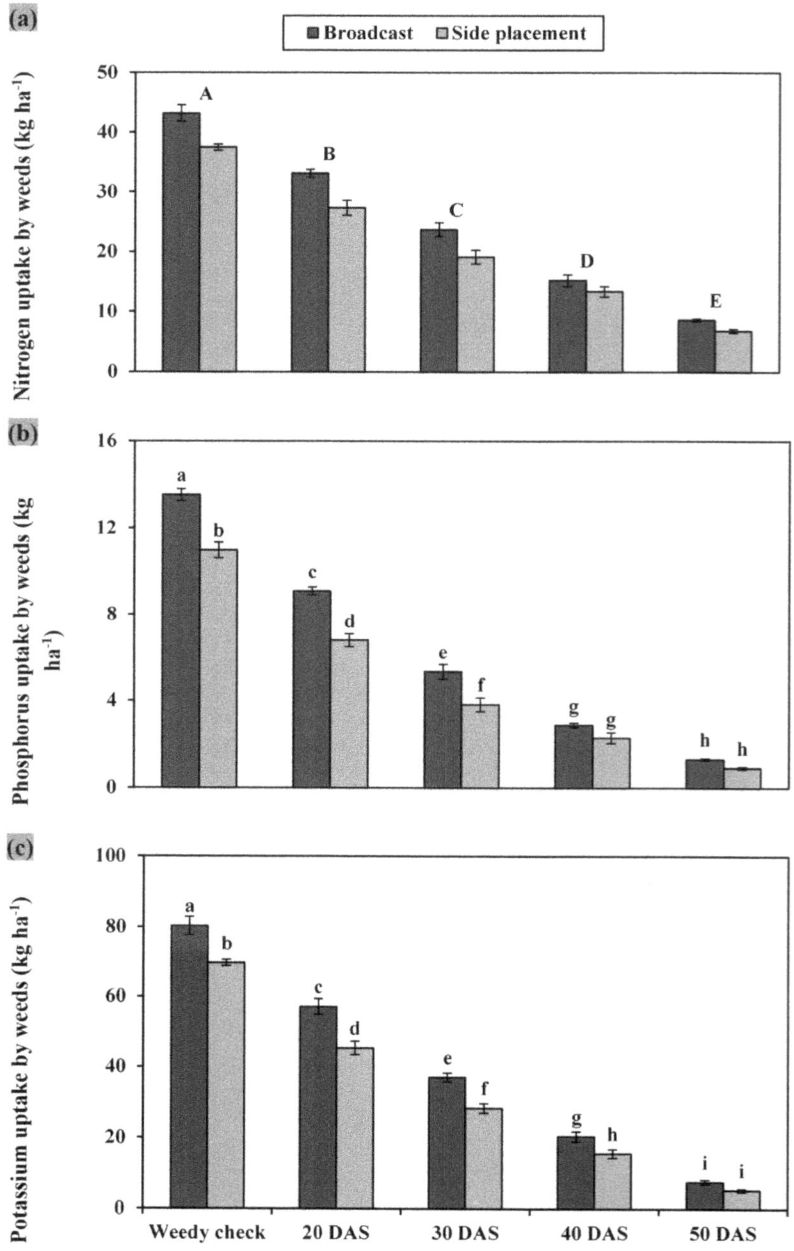

Figure 5. Effects of different weed-free periods on (**a**) nitrogen uptake, (**b**) phosphorous uptake and (**c**) potassium uptake by weeds in direct-seeded rice under different fertilizer application methods. Vertical bars above mean denote the standard error (±) of six replicates small letters depict significance at 5% probability level. Units on x-axis remain same for a, b and c figures. Capital letters represent same letter for both fertilizers methods, (2 years mean data).

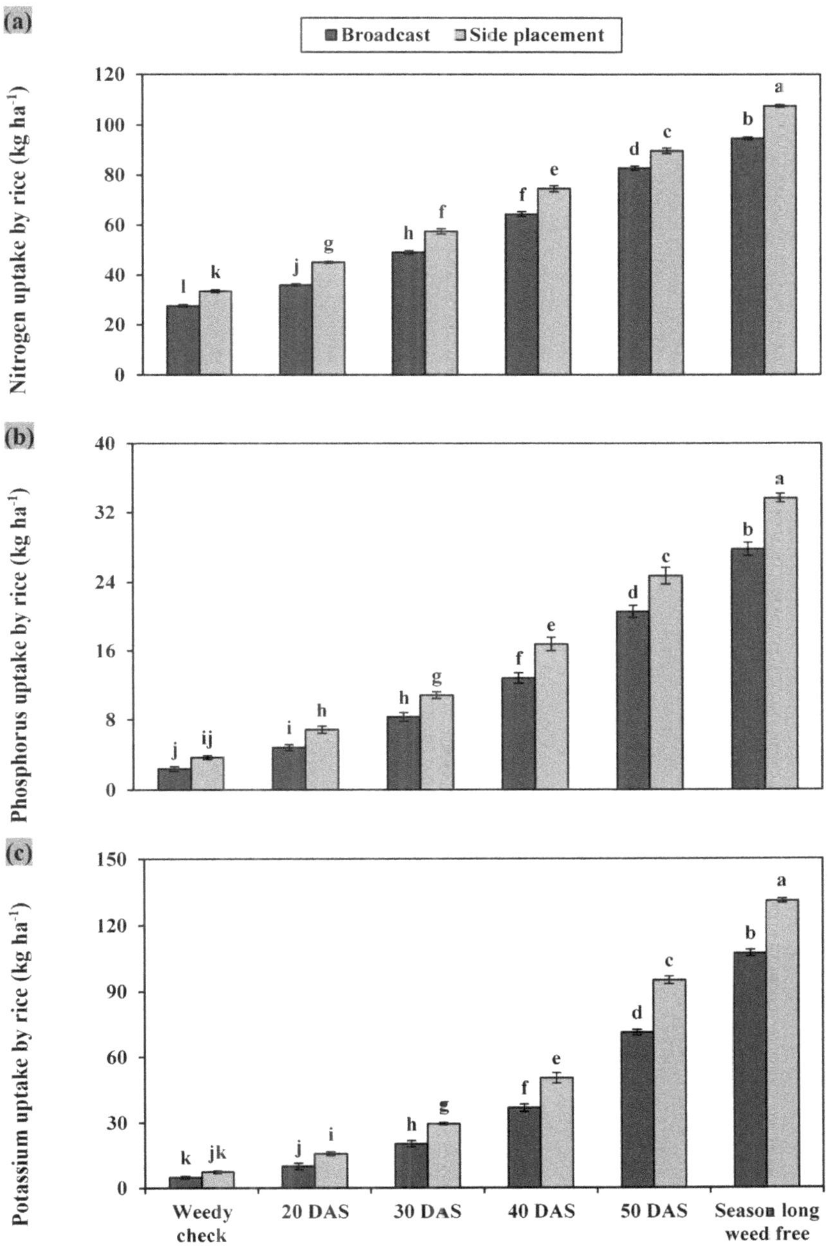

Figure 6. Effects of different weed-free periods on (**a**) nitrogen uptake, (**b**) phosphorous uptake and (**c**) potassium uptake by direct-seeded rice under different fertilizer application methods. Vertical bars above mean denote the standard error (±) of six replicates and letters depict significance at 5% probability level. Units on *x*-axis remain same for a, b and c figures, (2 years mean data).

4. Conclusions

The findings proved to be in line with the postulated hypothesis, as the fertilization technique and weed-free duration effectively influenced the weed infestation, yield attributes, paddy yield and nutrient uptake by crop plants in direct-seeded fine rice (DSR). Our study suggested that maintaining weed-free plots for the whole crop season remained instrumental in yielding the highest paddy yield, and it was also inferred that DSR requires 10 to 50 DAS weed-free days to achieve the optimum growth, maximum yield attributes and the highest paddy yield. Considering the fertilizer application methods, side-placement was better for improving the rice yield and yield components and also recorded a higher nutrient uptake in DSR. Thus, these results might be helpful in boosting the DSR yield in semi-arid and arid regions of the world by using the biologically viable agronomic practices of weeding and fertilization techniques. These findings may serve as a reference to conduct further studies involving other fertilizer application methods, such as band placement and fertigation, aimed at increasing nutrient availability to crop plants and reducing their losses through absorption by weeds and other wastages (leaching, volatilization and other gaseous emissions). At the same time, the impact of the weed-free duration in terms of economic viability also needs to be quantified in order to determine the economic viability of weed-free durations and fertilization application techniques.

Author Contributions: Conceptualization, R.N.A. and M.A.I.; methodology, R.N.A., M.A.I. and A.I.; validation, M.A.I., R.A. and R.I.; formal analysis, M.A.I. and A.I.; writing—original draft preparation, R.N.A., M.A.I. and A.I.; writing—review and editing, R.I., O.M.A., A.H. and B.J.B.; funding acquisition, M.A.I. and O.M.A. All authors have read and agreed to the published version of the manuscript.

Funding: This research was funded by the Taif University Researchers Supporting Project number TURSP-2020/81, Taif University, Taif, Saudi Arabia.

Institutional Review Board Statement: Not applicable.

Informed Consent Statement: Not applicable.

Conflicts of Interest: The authors declare no conflict of interest.

References

1. Gómez de Barreda, D.; Pardo, G.; Osca, J.M.; Catala-Forner, M.; Consola, S.; Garnica, I.; López-Martínez, N.; Palmerín, J.A.; Osuna, M.D. An Overview of Rice Cultivation in Spain and the Management of Herbicide-Resistant Weeds. *Agronomy* **2021**, *11*, 1095. [CrossRef]
2. Vitalini, S.; Orlando, F.; Vaglia, V.; Bocchi, S.; Iriti, M. Potential role of Lolium multiflorum Lam. In the management of rice weeds. *Plants* **2020**, *9*, 324. [CrossRef]
3. Islam, A.K.M.M.; Nasir, M.; Akter Mou, M.; Yeasmin, S.; Islam, M.S.; Ahmed, S.; Anwar, M.P.; Hadifa, A.; Baazeem, A.; Iqbal, M.A.; et al. Preliminary reports on comparative weed competitiveness of Bangladeshi monsoon and winter rice varieties under puddled transplanted conditions. *Sustainability* **2021**, *13*, 5091. [CrossRef]
4. Martins, M.B.; Agostinetto, D.; Fogliatto, S.; Vidotto, F.; Andres, A. Aeschynomene spp. Identification and Weed Management in Rice Fields in Southern Brazil. *Agronomy* **2021**, *11*, 453. [CrossRef]
5. Martin, R.; Chhun, S.; Yous, S.; Rien, R.; Korn, C.; Srean, P. Survey of Weed Management Practices in Direct-Seeded Rice in North-West Cambodia. *Agronomy* **2021**, *11*, 498. [CrossRef]
6. Haque, M.M.; Datta, J.; Ahmed, T.; Ehsanullah, M.; Karim, M.N.; Akter, M.S.; Iqbal, M.A.; Baazeem, A.; Hadifa, A.; Ahmed, S.; et al. Organic amendments boost soil fertility and rice productivity and reduce methane emissions from paddy fields under sub-tropical conditions. *Sustainability* **2021**, *13*, 3103. [CrossRef]
7. Ahmad, M.; Akhtar, M.; Anwar, M. Basmati rice: Progress, issues and prospects for rice crop in Pakistan. *Soil Environ.* **2005**, *30*, 50–57.
8. Akram, M. Aromatic rice of Pakistan-a review. *Pak. J. Agric. Sci.* **2009**, *22*, 154–160.
9. Saha, S.; Munda, S.; Singh, S.; Kumar, V.; Jangde, H.K.; Mahapatra, A.; Chauhan, B.S. Crop Establishment and Weed Control Options for Sustaining Dry Direct Rice Production in Eastern India. *Agronomy* **2021**, *11*, 389. [CrossRef]
10. Ali, S.; Liu, Y.; Ishaq, M.; Shah, T.; Ilyas, A.; Din, I. Climate change and its impact on the yield of major food crops: Evidence from Pakistan. *Foods* **2017**, *6*, 39. [CrossRef] [PubMed]
11. Kumar, V.; Jat, H.S.; Sharma, P.C.; Singh, B.; Gathala, M.K.; Malik, R.K.; Kamboj, B.R.; Yadav, A.K.; Ladhaa, J.K.; Raman, A.; et al. Can productivity and profitability be enhanced in intensively managed cereal systems while reducing the environmental footprint of

production? Assessing sustainable intensification options in the breadbasket of India. *Agric. Ecosyst. Environ.* **2018**, *252*, 132–147. [CrossRef]
2. Rao, A.N.; Brainard, D.C.; Kumar, V.; Ladha, J.K.; Johnson, D.E. Preventive weed management indirect-seeded rice: Targeting the weed seed bank. *Adv. Agron.* **2017**, *144*, 45–142.
3. Liu, H.; Hussain, S.; Zheng, M.; Peng, S.; Huang, J.; Cui, K.; Nie, L. Dry direct-seeded rice as an alternative to transplanted-flooded rice in Central China. *Agron. Sustain. Dev.* **2015**, *35*, 285–294. [CrossRef]
4. Hakim, A.R.; Juraimi, S.M.; Rezaul, K.; Khan, M.S.I.; Islam, M.S.; Choudhury, M.K.; Soufan, W.; Alharby, H.; Bamagoos, A.; Iqbal, M.A.; et al. Effectiveness of herbicides to control rice weeds in diverse saline environments. *Sustainability* **2021**, *13*, 2053. [CrossRef]
5. Ali, S.; Iqbal, M.A.; Afzal, S.; Afzal, M.I. Growth, kernel quality and yield assessment of aerobic aromatic rice (*Oryza sativa* L.) under different mulching types and spatial arrangements. *World J. Agric. Sci.* **2015**, *12*, 84–90.
6. Iqbal, M.A.; Ali, S. Evaluation of yield and yield components of aerobic fine rice (*Oryza sativa* L.) as influenced by different mulches and planting patterns. *Am. Eur. J. Agric. Environ. Sci.* **2015**, *14*, 1089–1094.
7. Iqbal, M.A. Productivity and quality of direct seeded rice under different types of mulches and planting patterns: A review. *Am. Eur. J. Agric. Environ. Sci.* **2015**, *14*, 1240–1247.
8. Marchesi, C.; Saldain, N.E. First report of herbicide-resistant Echinochloa crus-galli in Uruguayan rice fields. *Agronomy* **2019**, *9*, 790. [CrossRef]
9. Rao, A.N.; Johnson, D.E.; Sivaprasad, B.; Ladha, J.K.; Mortimer, A.M. Weed management in direct-seeded rice. *Adv. Agron.* **2007**, *93*, 153–255.
0. Khaliq, A.; Matloob, A. Weed crop competition period in three fine rice cultivars under direct seeded rice culture. *Pak. J. Weed Sci. Res.* **2011**, *17*, 229–243.
1. Johnson, D.E.; Wopereis, M.C.S.; Mbodj, D.; Diallo, S.; Powers, S.; Haefele, S.M. Timing of weed management and yield losses due to weeds in irrigated rice in the Sahel. *Field Crop. Res.* **2004**, *85*, 31–42. [CrossRef]
2. Anwar, M.P.; Juraimi, A.S.; Samedani, B.; Puteh, A.; Man, A. Critical period of weed control in aerobic rice. *Sci. World J.* **2012**, *20*, 1–10. [CrossRef] [PubMed]
3. Di-Tomaso, J.M. Approaches for improving crop competitiveness through manipulation of fertilization strategies. *Weed Sci.* **1995**, *43*, 491–497. [CrossRef]
4. Blackshaw, R.E.; Molnar, L.J.; Larney, F.J. Fertilizer, manure and compost effects on weed growth and competition with winter wheat in western Canada. *Crop Protect.* **2005**, *24*, 971–980. [CrossRef]
5. Iqbal, M.A. Nano-Fertilizers for Sustainable Crop Production under Changing Climate: A Global Perspective. In *Sustainable Crop Production*, 1st ed.; Hasanuzzaman, M., Ed.; Intechopen Ltd.: London, UK, 2019; pp. 1–10.
6. Rasmussen, K.; Rasmussen, J.; Petersen, J. Effects of fertilizer placement on weeds in weed harrowed spring barley. *Acta Agric. Scand. B-Plant Soil Sci.* **1996**, *46*, 192–196. [CrossRef]
7. Kirkl, K.J.; Beckie, H.J. Contribution of nitrogen fertilizer placement to weed management in spring wheat (*Triticum aestivum* L.). *Weed Technol.* **1998**, *12*, 507–514. [CrossRef]
8. Mahajan, G.; Timsina, J. Effect of nitrogen rates and weed control methods on weeds abundance and yield of direct-seeded rice. *Arch. Agron. Soil Sci.* **2011**, *57*, 239–250. [CrossRef]
9. Khaliq, A.; Matloob, A.; Ihsan, Z.A.; Abbas, R.N.; Aslam, Z.; Rasul, F. Supplementing herbicides with manual weeding improves weed control efficiency, growth and yield of dry seeded rice. *Int. J. Agric. Biol.* **2013**, *15*, 191–199.
0. Khaliq, A.; Mahmood, S.; Matloob, A.; Khan, M.B.; Awan, I.U. Optimizing seeding density and tank mixture of herbicides help reduce yield losses in dry seeded fine rice. *Pak. J. Weed Sci. Res.* **2012**, *18*, 167–181.
1. Cathcart, R.J.; Chandler, K.; Swanton, C.J. Fertilizer N rate and the response of weeds to herbicide. *Weed Sci.* **2004**, *52*, 291–296. [CrossRef]
2. Chauhan, B.S.; Johnson, D.E. Row spacing and weed control timing affect yield of aerobic rice. *Field Crops Res.* **2011**, *121*, 226–231. [CrossRef]
3. Chauhan, B.S.; Abugho, S.B. Fertilizer placement affects weed growth and grain yield in dry-seeded rice (*Oryza sativa* L.) systems. *Am. J. Plant Sci.* **2013**, *4*, 1260–1264. [CrossRef]
4. Evans, S.; Knezevic, S.; Lindquist, J.; Shapiro, C.; Blankenship, E. Nitrogen application influences the critical period for weed control in corn. *Weed Sci.* **2003**, *51*, 408–417. [CrossRef]
5. Abbas, R.N.; Arshad, M.A.; Iqbal, A.; Iqbal, M.A.; Imran, M.; Raza, A.; Chen, J.-T.; Alyemeni, M.N.; Hefft, D.I. Weeds spectrum, productivity and land-use efficiency in maize-gram intercropping systems under semi-arid environment. *Agronomy* **2021**, *11*, 1615. [CrossRef]
6. Addis, H.; Andreas, K.; Theib, O.; Stefan, S. Linking selected soil properties to land use and hill slope- A watershed case study in the Ethiopian highlands. *Soil Water Res.* **2016**, *11*, 163–171. [CrossRef]
7. Black, C.A. *Methods of Soil Analysis, Part II*; American Society of Agronomy: Madison, WI, USA, 1965.
8. Sparks, D.L.; Page, A.L.; Helmke, P.A.; Leoppert, R.H.; Soltanpour, P.N.; Tabatabai, M.A.; Johnston, G.T.; Sumner, M.E. *Methods of Soil Analysis*; Soil Science Society of America: Madison, WI, USA, 1996.

39. Naresh, R.K.; Tomar, S.S.; Kumar, D.; Samsher, P.; Singh, S.; Dwivedi, P.; Kumar, A. Experiences with rice grown on permanen raised beds: Effect of crop establishment techniques on water use, productivity, profitability and soil physical properties. *Rice Sc* **2014**, *21*, 170–180. [CrossRef]
40. AOAC. *Official Methods of Analysis*, 17th ed.; Association of Official Analytical Chemists: Arlington, VI, USA, 2003.
41. Jackson, M.L. *Soil Chemical Analysis*; Prentice Hall: Englewood Cliffs, NJ, USA, 1962.
42. Piper, C.S. *Soil and Plant Analysis*; Press Adelaide: Adelaide, Australia, 1950.
43. Ryan, J.; Estefan, G.; Rashid, A. *Soil and Plant Analysis Laboratory Manual*, 2nd ed.; International Center for Agricultural Researcl in the Dry Areas (ICARDA): Aleppo, Syria; National Agriculture Research Center (NARC): Islamabad, Pakistan, 2001.
44. Steel, R.G.D.; Torrie, J.H.; Dickey, D. *Principles and Procedures of Statistics: A biometrical Approach*; McGraw Hill Book Co. Inc.: New York, NY, USA, 1997; pp. 172–177.
45. Ahmed, S.; Alam, M.J.; Hossain, A.; Islam, A.K.M.M.; Awan, T.H.; Soufan, W.; Qahtan, A.A.; Okla, M.K.; El Sabagh, A. Interactiv Effect of Weeding Regimes, Rice Cultivars, and Seeding Rates Influence the Rice-Weed Competition under Dry Direct-Seede Condition. *Sustainability* **2021**, *13*, 317. [CrossRef]
46. Javed, T.; Afzal, I.; Mauro, R.P. Seed Coating in Direct Seeded Rice: An Innovative and Sustainable Approach to Enhance Grai Yield and Weed Management under Submerged Conditions. *Sustainability* **2021**, *13*, 2190. [CrossRef]
47. Mohanta, S.; Banerjee, M.; Malik, G.C.; Shankar, T.; Maitra, S.; Ismail, I.A.; Dessoky, E.S.; Attia, A.O.; Hossain, A. Productivity anc Profitability of Kharif Rice Are Influenced by Crop Establishment Methods and Nitrogen Management in the Lateritic Belt of th Subtropical Region. *Agronomy* **2021**, *11*, 1280. [CrossRef]
48. Guan, S.; Fukami, K.; Matsunaka, H.; Okami, M.; Tanaka, R.; Nakano, H.; Sakai, T.; Nakano, K.; Ohdan, H.; Takahashi, K Assessing Correlation of High-Resolution NDVI with Fertilizer Application Level and Yield of Rice and Wheat Crops Usin Small UAVs. *Remote Sens.* **2019**, *11*, 112. [CrossRef]
49. Xu, L.; Li, X.; Wang, X.; Xiong, D.; Wang, F. Comparing the Grain Yields of Direct-Seeded and Transplanted Rice: A Meta-Analysi *Agronomy* **2019**, *9*, 767. [CrossRef]
50. Basavalingaiah, K.; Ramesha, Y.M.; Paramesh, V.; Rajanna, G.A.; Jat, S.L.; Dhar Misra, S.; Kumar Gaddi, A.; Girisha, H.C Yogesh, G.S.; Raveesha, S.; et al. Energy Budgeting, Data Envelopment Analysis and Greenhouse Gas Emission from Rice Prc duction System: A Case Study from Puddled Transplanted Rice and Direct-Seeded Rice System of Karnataka, India. *Sustainabilit* **2020**, *12*, 6439. [CrossRef]
51. Shekhawat, K.; Rathore, S.S.; Chauhan, B.S. Weed Management in Dry Direct-Seeded Rice: A Review on Challenges anc Opportunities for Sustainable Rice Production. *Agronomy* **2020**, *10*, 1264. [CrossRef]
52. Chen, Y.; Fan, P.; Mo, Z. Deep Placement of Nitrogen Fertilizer Affects Grain Yield, Nitrogen Recovery Efficiency, and Roo Characteristics in Direct-Seeded Rice in South China. *J. Plant Growth Regul.* **2021**, *40*, 379–387. [CrossRef]
53. Yao, Y.L.; Zhang, M.; Tian, Y.H.; Zhang, B.W.; Zhao, M.; Zeng, K.; Yin, B. Urea deep placement for minimizing NH3 loss in ar intensive rice cropping system. *Field Crops Res.* **2018**, *218*, 254–266. [CrossRef]
54. Xu, J.Z.; Liao, L.X.; Tan, J.Y.; Shao, X.H. Ammonia volatilization in gemmiparous and early seedling stages from direct seedin rice fields with different nitrogen management strategies: A pots experiment. *Soil Till Res.* **2013**, *126*, 169–176. [CrossRef]
55. Wang, W.Q.; Peng, S.B.; Liu, H.Y.; Tao, Y.; Huang, J.L.; Cui, K.H.; Nie, L.X. The possibility of replacing puddled transplante flooded rice with dry seeded rice in central China. A review. *Field Crops Res.* **2017**, *214*, 310–320. [CrossRef]
56. Pan, S.G.; Wen, X.C.; Wang, Z.M.; Umair, A.; Tian, H.; Duan, M.Y.; Mo, Z.W.; Fan, P.S.; Tang, X.R. Benefits of mechanized dee placement of nitrogen fertilizer in direct-seeded rice in South China. *Field Crops Res.* **2017**, *203*, 139–149. [CrossRef]
57. Zhu, C.; Xiang, J.; Zhang, Y.; Zhu, D.; Chen, H. Mechanized transplanting with side deep fertilization increases yield and nitroger use efficiency of rice in Eastern China. *Sci. Rep.* **2019**, *9*, 5653. [CrossRef]

agronomy

MDPI

Article

Long-Term Integrated Nutrient Management in the Maize–Wheat Cropping System in Alluvial Soils of North-Western India: Influence on Soil Organic Carbon, Microbial Activity and Nutrient Status

Salwinder Singh Dhaliwal [1], Sandeep Sharma [1], Vivek Sharma [1], Arvind Kumar Shukla [2], Sohan Singh Walia [3], Majid Alhomrani [4], Ahmed Gaber [5,*], Amardeep Singh Toor [1], Vibha Verma [1], Mehakpreet Kaur Randhawa [1], Lovedeep Kaur Pandher [1], Prabhjot Singh [1] and Akbar Hossain [6,*]

1 Department of Soil Science, Punjab Agricultural University, Ludhiana 141004, India; ssdhaliwal@pau.edu (S.S.D.); sandyagro@pau.edu (S.S.); sharmavivek@pau.edu (V.S.); amardeep@pau.edu (A.S.T.); vermavibha@pau.edu (V.V.); mehakpreet-soil@pau.edu (M.K.R.); lovedeeppandher@pau.edu (L.K.P.); prabh@pau.edu (P.S.)
2 Indian Institute of Soil Science, Bhopal 462038, India; arvindshukla2k3@yahoo.co.in
3 Department of Agronomy, Punjab Agricultural University, Ludhiana 141004, India; waliass@pau.edu
4 Department of Clinical Laboratories Sciences, The Faculty of Applied Medical Sciences, Taif University, P.O. Box 11099, Taif 21944, Saudi Arabia; m.alhomrani@tu.edu.sa
5 Department of Biology, College of Science, Taif University, P.O. Box 11099, Taif 21944, Saudi Arabia
6 Department of Agronomy, Bangladesh Wheat and Maize Research Institute, Dinajpur 5200, Bangladesh
* Correspondence: a.gaber@tu.edu.sa (A.G.); akbarhossainwrc@gmail.com (A.H.)

Citation: Dhaliwal, S.S.; Sharma, S.; Sharma, V.; Shukla, A.K.; Walia, S.S.; Alhomrani, M.; Gaber, A.; Toor, A.S.; Verma, V.; Randhawa, M.K.; et al. Long-Term Integrated Nutrient Management in the Maize–Wheat Cropping System in Alluvial Soils of North-Western India: Influence on Soil Organic Carbon, Microbial Activity and Nutrient Status. *Agronomy* **2021**, *11*, 2258. https://doi.org/10.3390/agronomy11112258

Academic Editors: Umberto Anastasi and Aurelio Scavo

Received: 26 September 2021
Accepted: 4 November 2021
Published: 8 November 2021

Abstract: Integrated nutrient management (INM) is a widely recognized tool to ensure sustainable crop productivity while preserving soil fertility. The addition of organic manures in soil has been evidenced to improve soil characteristics, in addition to improving nutrient availability. The soil samples, with five treatment combinations of chemical fertilizers with farmyard manure (FYM), were collected from a 17-year-old field experiment conducted at PAU, Ludhiana to investigate the effect of INM on the buildup of organic carbon (OC), microbial community, soil nutrient status and improvement in soil physical properties under the maize–wheat cropping system. The INM technique enhanced the OC content (0.44 to 0.66%), available N (152.8 to 164.9 kg ha^{-1}), P (22.8 to 31.4 kg ha^{-1}) and K (140.6 to 168.0 kg ha^{-1}) after 17 years. The DTPA-extractable and total micronutrients (Zn, Cu, Fe, and Mn) status also improved significantly with FYM supplementation. The organic source, coupled with inorganic fertilizers, improved the water holding capacity, total porosity, soil respiration, microbial biomass C, microbial biomass N, and potentially mineralizable N. However, pH, EC, and bulk density of soil decreased with the addition of FYM, coupled with chemical fertilizers.

Keywords: organic manure; inorganic fertilizers; cropping pattern; soil physicochemical and biological properties

1. Introduction

In north-western India, the continuous rice–wheat cropping has led to the exhaustion of natural resources and deteriorated soil fertility, producing agricultural outcomes [1]. Thus, a paradigm shift in cropping systems with different crops is required to maintain soil health and sustainable yield. Alternate cropping systems and soil management practices may prove beneficial to improve soil fertility and maintain environmental health. For crop diversification, maize-wheat cropping system has been identified as a suitable alternative to rice–wheat system [2,3]. Moreover, maize accounts for a significant fraction of global food consumption. The acreage under maize has increased in the past few years, as it helps to maintain soil health, in contrast to the rice–wheat cropping system. [4].

Although production, under intensive cultivation, is increasing year after year, it is depleting the huge amount of macro and micronutrients from the soil. Injudicious application of micronutrient fertilizers, declined use of crop residues and organic manures, as well as potential crop harvests in the last few decades, have resulted in micronutrient deficiencies in north-western India [5]. Excessive supplementation of inorganic fertilizers has also deteriorated the soil structure and declined soil organic matter (SOM) and microbial activity. Integrated nutrient management (INM) is the feasible solution for sustaining the crop productivities, as nutrient requirements of both the crops are high and have shown superior response towards higher levels of nutrient application [6]. The balanced use of nutrients is the key to improving the sustainable production of crops [7]. The inorganic fertilizers, through soil or foliar application, have shown tremendous results in terms of agricultural productivity [8,9]. Furthermore, the use of inorganic nutrient sources coupled with organic sources is a feasible approach for higher agricultural productivity and monitoring soil health [10]. The utilization of well-decomposed farmyard manure (FYM) in soil management practices is a well-known practice for enhancing crop yield, enhancing SOM, promoting microbial activities, promoting friendly soil environmental management [11,12], increasing the total organic sources supply, and increasing the plant-available macro and micronutrients in soil. The decomposition of plant residues favors the conversion of unavailable plant nutrients into an available form, increasing their plant absorption [13]. Besides improving the nutrient availability, organic manures also affect the soil physical and biological characteristics, as well as possessing residual effects on the succeeding crops. Previous reports have evidenced the greater residual impact of organic manures on the succeeding wheat crop [14,15].

The organic manures, being low in available nutrients, cannot substitute all nutrients required for yield sustainability [16]. On the other hand, the supplementation of nutrients solely through chemical fertilizers is insufficient to meet the complete nutrient demand of agricultural plants. Hence, INM has been identified as a viable option to improve soil health and sustain agricultural productivity on a long-term basis. For instance, the yield outcomes of the pearl millet–wheat cropping system were improved when nutrients were supplied through both FYM and inorganic fertilizers, over the sole use of inorganic fertilizers [17]. The integrated use of organic and inorganic N fertilizers in MWCS increased the SOM content and microbial activity, and thus improved the soil fertility [18]. The INM system seems to be an environmental-friendly approach that offers an advantage of the least impact on food quality. To date, the in-depth knowledge of build-up of soil carbon status, microbial community, and soil properties with INM under MWCS is scant in alluvial soils of north-western India. Hence, an attempt was made to study the impact of different levels of FYM along with inorganic fertilizers on soil organic carbon status, microbial community, and nutrient status under MWCS.

2. Material and Methods

2.1. Site Specification and Treatment Details

The experiment was planned with the sole objective for the yield sustainability of maize and wheat crops grown in a sequence and maintenance of soil health under the INM technique. The long-term field experiment on MWCS was carried out on permanent plots established since the Kharif 2001 season at the research farm, Department of Agronomy, Punjab Agricultural University (PAU), Ludhiana (30°56′ N, 75°52′ E, and 247 m above mean sea level), India. The experiment comprised five treatment combinations with three replications in a completely randomized block design with plot size 22.5 m × 7.5 m (Table 1).

Different treatment combinations consisted of the addition of nitrogen, phosphorus, and potassic fertilizers, in combination with farmyard manure (FYM), under the conventional tillage system. In brief, the experimental field was subjected to 2 ploughings, followed by planking to get a fine seed bed. Wheat variety PBW 343 was sown in the first week of November, and after harvesting of wheat, the crop maize variety PMH 1 was sown in the last week of June each year. Maize and wheat attained physiological maturity at 90 and 125 days, respectively.

Table 1. Treatment details of the sustainable production system model in the maize–wheat system (2001–2017).

Treatments	Maize				Wheat	
	Practice	N-P_2O_5-K_2O (kg ha^{-1})	FYM (t ha^{-1})	$ZnSO_4$ (kg ha^{-1})	Practice (Row to Row Spacing)	N-P_2O_5-K_2O (kg ha^{-1})
T_1	FP * (55,000 plants ha^{-1})	100-30-0	6	0	FP (22.5 cm)	150-60-0
T_2	RP ** (75,000 plants ha^{-1})	120-60-30	10	25	RP (15 cm)	120-60-30
T_3	RP (75,000 plants ha^{-1})	180-60-30	10	25	RP (15 cm)	150-60-30
T_4	RP (75,000 plants ha^{-1})	Fertilizer on soil test basis (100-0-30)	6	25	RP (15 cm)	120-60-30
T_5	RP (75,000 plants ha^{-1})	120-60-30	10	25	Wheat is replaced with Gobhi sarson followed by mungbean	Gobhi Sarson-100:30:0, mungbean-0:0:0

FP *—farmers practice; RP **—recommended practice.

Treatments differed in terms of nitrogen and FYM levels. The crop residues of previous crops were removed. The well-rotten FYM was obtained from PAU dairy shed, which was decomposed for 6 months in a pit. The FYM was added 15 days prior to sowing of the maize crop. The pH, EC, and OC of FYM were 7.21, 1.52, and 203.81 g kg^{-1}. The nutrient content in FYM was recorded as N = 1.16%, P= 0.48%, and K = 0.56%, on dry weight basis. Farmers add 6 t ha^{-1} FYM, whereas, under RP, 10 t ha^{-1} FYM is added. The urea (46% N), diammonium phosphate (DAP; 18% N, 46% P_2O_5), and muriate of potash (MOP; 60% K_2O) were used as a source of N, P, and K respectively. Nitrogen was applied in three equal splits to both the crops. One-third N was applied at the time of sowing; whereas, the remaining doses were applied with first and second irrigation. Four irrigations were applied to the maize crop; whereas, 5 irrigations were applied to the maize crop. Whole P and K fertilizers were applied at the time of sowing of maize and wheat crops, respectively. The FYM and Zn were applied only during the maize crop.

2.2. Initial Physicochemical Characteristics of the Experimental Soil

The physicochemical and biological properties of initial soil samples in 2001 at 0–15 cm (D1) and 15–30 cm (D2) depth have been given in Table 2. The soil of the experimental field was determined in 2001. The soil was loamy sand in texture (Typic Ustochrept), lying in an Ustic soil moisture regime, with bulk density 1.72 g cm^{-3}, total porosity 30.5%, water holding capacity (WHC) 48.6%, and organic carbon (OC) 0.40%.

2.3. Soil Analysis

In total, 30 composite soil samples from each block (5 treatments × 3 replications × 2 depths) were collected after 17 years with a screw auger after maize crop harvest in October 2017 (experiment terminated). Immediately after collection, the samples were separated into two halves. One half of the sample was immediately stored at 4 °C to assay soil microbiological properties, and the other half was air-dried, sieved through a 2.0 mm plastic sieve, and stored for physicochemical analysis. Among soil characteristics, bulk density and WHC were estimated, employing the weighing bottle method and Keen's box method [19,20]. Total porosity was determined using the procedure given by Prihar and Verma [21]. The pH and EC of soil samples were estimated using pH meter and EC meter [22]. The available N, P, and K were determined using the alkaline $KMnO_4$ method, Olsen extractable P method, and neutral ammonium acetate method, respectively [23–25]. Diethylene triamine-pentaacetic acid (DTPA)-extractable soil micronutrients (Zn, Cu, Fe, and Mn) were determined by using DTPA–TEA buffer in the ratio of 1:2 and then their concentration was estimated in atomic absorption spectrophotometer (AAS) [26]. Total macro and micronutrients were estimated by using the method given by Page et al. [27] Total N in soil was estimated by the micro-Kjeldahl method. The total P, total K, and micronutrients in soil were determined by digesting the soil samples with diacid (i.e. HNO_3 and $HClO_4$ in the ratio of 9:4) and these digests were analyzed for total P, K, and DTPA extractable soil micronutrients after appropriate dilutions. Total P and K content were measured by employing the molybdenum blue method and flame photometric method, respectively. For micronutrients estimation, total Zn, Cu, Fe, and Mn contents were measured using AAS (Varian AAS FS 240 Model).

Table 2. Physicochemical and biological properties of initial soil samples (2001).

Soil Properties	Depth (D1)	Depth (D2)
Bulk density (g cm^{-3})	1.72	1.69
Total porosity (%)	30.5	29.9
Water holding capacity (%)	48.6	48.3
pH (1:2 soil: water suspension)	7.60	7.80
EC dSm^{-1} (1:2 soil: water suspension)	0.30	0.24
Organic carbon (g kg^{-1})	4.0	3.5
Available N (Kg ha^{-1})	119.7	102.3
Available P (Kg ha^{-1})	14.4	12.8
Available K (Kg ha^{-1})	128.6	124.7
Total N (%)	0.12	0.08
Total P (%)	0.25	0.19
Total K (%)	0.27	0.21
DTPA-Extractable Zn (mg kg^{-1})	1.26	0.68
DTPA-Extractable Cu (mg kg^{-1})	0.30	0.22
DTPA-Extractable Fe (mg kg^{-1})	3.83	2.56
DTPA-Extractable Mn (mg kg^{-1})	3.48	2.65
Total Zn (mg kg^{-1})	112.5	86.5
Cu (mg kg^{-1})	13.5	10.4
Fe (%)	2.6	1.8
Mn (mg kg^{-1})	132.8	97.6
PMN (mg kg^{-1} 7 d^{-1})	8.6	6.7
MBC (mg kg^{-1})	82.9	65.4
MBN (mg kg^{-1})	23.4	12.9
CO_2-C (mg kg^{-1} 10 d^{-1})	1.8	0.8

PMN—potentially mineralizable nitrogen; MBC—microbial biomass carbon; MBN—microbial biomass nitrogen; CO_2-C—soil respiration.

2.4. Soil Carbon and Soil Microbiological Analysis

The OC content in soil was estimated by using the wet combustion method [28]. The potentially mineralizable nitrogen (PMN) in the soil was estimated by following the procedure described by Keeney [29]. The microbial biomass nitrogen (MBN) was determined from the soil, as described by Keeney and Nelson [30], and the mineral nitrogen released by the microbial component was measured. The chloroform fumigation and incubation procedure was employed for the estimation of microbial biomass carbon (MBC) in the soil [31]. Soil respiration was measured by the chloroform fumigation and incubation procedure (CFIM) [31]. The amount of CO_2-C produced by soil microorganisms during respiration was measured and CO_2-C (soil respiration) was expressed as mg per kg of soil over a 10 day period [32].

2.5. Statistical Analysis

The data were analyzed by using statistical analysis software (SPSS software, 19.0 SPSS Institution Ltd., Chicago, IL, USA). One-way analysis of variance (ANOVA), followed by Duncan's multiple range test, was performed to determine the treatment effects at 0.05 level of probability [33].

3. Results

3.1. Impact of INM on Soil Carbon and Microbiological Composition

The maximum OC build-up was obtained in T_3 treatment and showed non-significant variation with all other treatments except treatment T_1 (Figure 1).

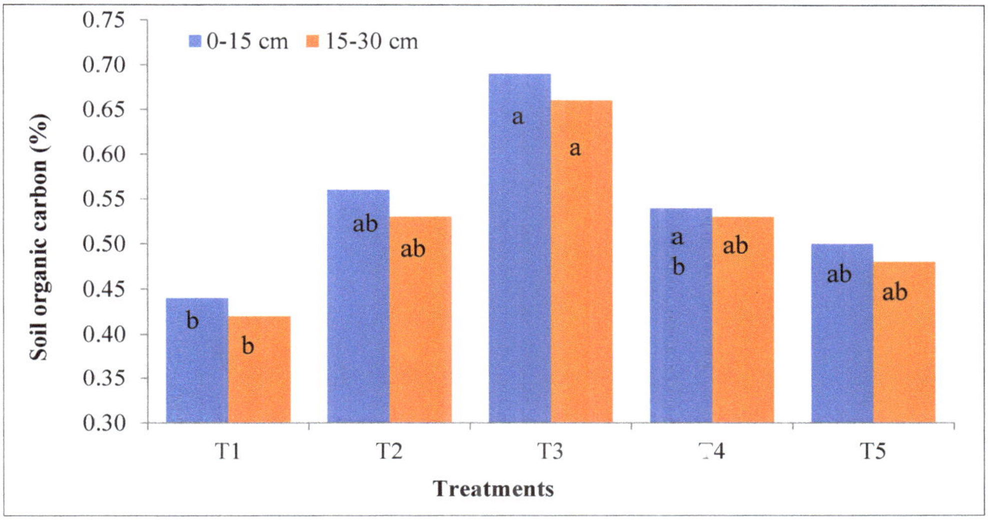

Figure 1. Effect of INM technique on soil OC under the maize–wheat system. In bars, means with similar letter(s) are statistically identical, as per $LSD_{0.05}$.

The buildup of OC content was observed under all treatments over their initial level. The PMN ranged from 10.3 to 13.7 mg kg^{-1} 7 d^{-1} in D1 and from 7.2 to 10.6 mg kg^{-1} 7 d^{-1} in D2. Among different treatments, PMN was significantly greater in treatment T_3 as compared with the other treatments and was lowest in treatment T_5. The soil MBC varied from 116.3 to 132.8 mg kg^{-1} in D1 and from 42.7 to 56.6 mg kg^{-1} in D2 (Figure 2). Application of chemical fertilizers with FYM enhanced the MBC content over their initial levels which were reported to be 82.9 and 65.4 mg kg^{-1} in D1 and D2, respectively. Among different treatments, T_3 treatment resulted in maximum content of MBC followed by treatments T_2, T_4, T_1 and T_5, respectively. The MBN showed a similar trend as MBC and it ranged from 42.7 mg kg^{-1} in T_5 to 56.6 mg kg^{-1} in T_3 in D1 and from 36.8 mg kg^{-1} in T_5 to 44.7 mg kg^{-1} in T_3 in D2 (Figure 2). The addition of chemical fertilizers with FYM improved the CO_2-C content to a significant extent in all treatments over its initial levels, which were reported to be 1.8 and 0.8 mg kg^{-1} 10 d^{-1} in D1 and D2, respectively. It was found maximum in treatment T_3 (4.9 mg kg^{-1} 10 d^{-1}) and showed non-significant variation with treatments T_2 (4.4 mg kg^{-1} 10 d^{-1}) and T_4 (4.1 mg kg^{-1} 10 d^{-1}) and lowest variation in T_1 (3.7 mg kg^{-1} 10 d^{-1}) in D1. In D2, it was highest in treatment T_3 (3.7 mg kg^{-1} 10 d^{-1}) and showed non-significant variation with treatments T_2 (2.9 mg kg^{-1} 10 d^{-1}) and lowest in T_5 (2.1 mg kg^{-1} 10 d^{-1}).

3.2. Impact of INM on Soil Physical Characteristics

Bulk density, total porosity, and WHC ranged from 1.59 to 1.68 g cm^{-3}, 31.4 to 37.6%, and 50.9 to 59.6%, respectively, in D1 (Table 3). In D2, these ranged from 1.52 to 1.62 g cm^{-3}, 29.2 to 36.3%, and 47.7 to 56.9%, respectively. The maximum bulk density was reported in T_5 and was lowest in T_3, However, total porosity and WHC followed the opposite trend, with maximum values in T_3 and the lowest in T_5 in D1, while the lowest values were found in T_1 in D2.

The pH values in soil samples of D1 ranged from 7.33 to 7.48 and from 7.30 to 7.47 in soil samples of D2, under all treatments. The pH values decreased from their initial levels in all treatments. Lower pH values were reported under treatments in which 10 t ha^{-1} FYM had been added (T_2, T_3, and T_5) as compared to treatments in which 6 t ha^{-1} FYM was added (T_1 and T_4). A similar trend was followed in soil samples of depth D2. The soil EC values varied from 0.21 to 0.27 dS m^{-1} and 0.18 to 0.25 dS m^{-1}, respectively. The higher magnitude of EC was recorded in treatment T_3, while lower values were reported in treatments T_1 and T_5.

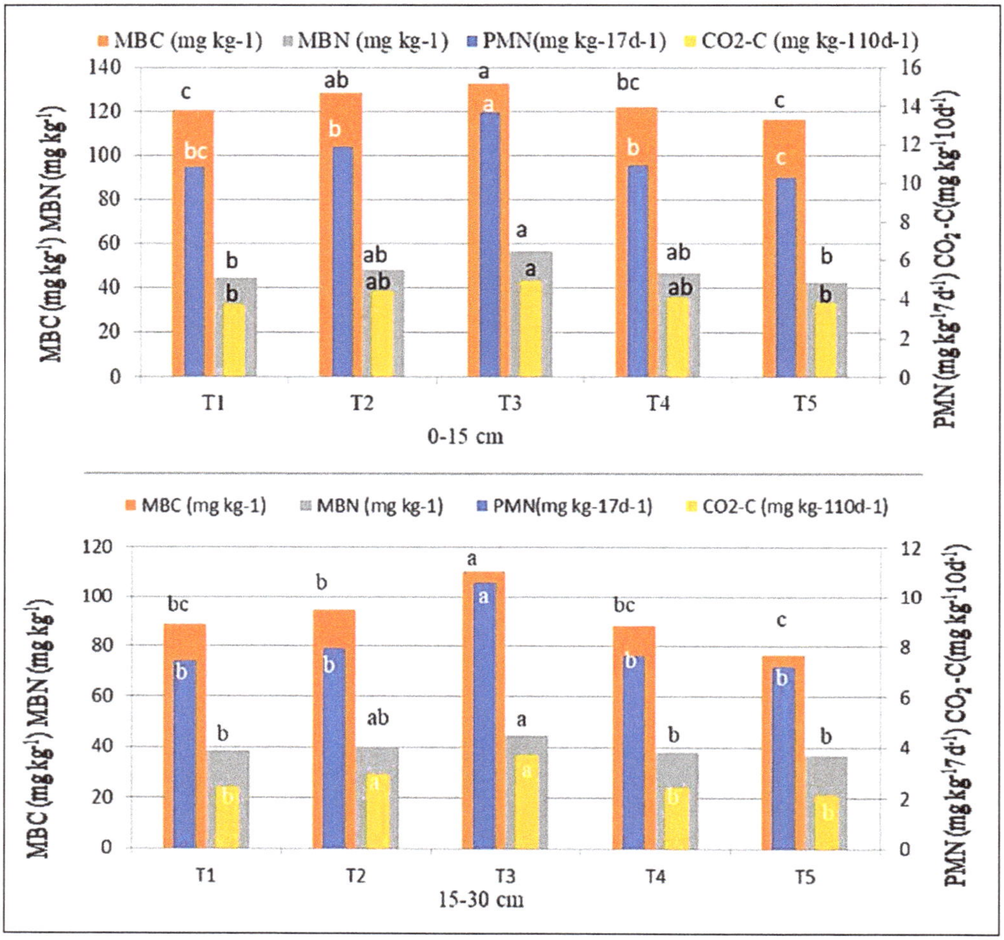

Figure 2. Effect of INM on soil microbiological properties under the maize–wheat system at 0–15 (**top**) and 15–30 cm (**bottom**) depth. PMN—potentially mineralizable nitrogen; MBC—microbial biomass carbon; MBN—microbial biomass nitrogen; CO_2-C—soil respiration. In bars, means with similar letter(s) are statistically identical, as per $LSD_{0.05}$.

Table 3. Effect of INM technique on soil's physicochemical properties under the maize–wheat system.

Treatments	Bulk Density ($g\ cm^{-3}$)	Total Porosity (%)	Water Holding Capacity (%)	pH	EC ($dS\ m^{-1}$)
		D1			
T1	1.64 ^{ab}	31.8 ^{cd}	51.7 ^{c}	7.45 ^{ab}	0.22 ^{b}
T2	1.61 ^{ab}	34.7 ^{b}	56.8 ^{ab}	7.34 ^{c}	0.23 ^{ab}
T3	1.59 ^{b}	37.6 ^{a}	59.6 ^{a}	7.33 ^{c}	0.27 ^{a}
T4	1.65 ^{ab}	33.5 ^{bc}	52.5 ^{bc}	7.48 ^{a}	0.24 ^{ab}
T5	1.68 ^{a}	31.4 ^{d}	50.9 ^{c}	7.37 ^{bc}	0.21 ^{b}
Mean	1.63	33.8	54.3	7.39	0.23
Initial	1.72	30.5	48.6	7.6	0.3
LSD ($p \leq 0.05$)	0.08	1.9	4.8	0.09	0.04
		D2			
T1	1.58 ^{ab}	29.2 ^{d}	47.7 ^{c}	7.44 ^{a}	0.18 ^{b}
T2	1.53 ^{ab}	33.6 ^{b}	53.6 ^{ab}	7.33 ^{b}	0.20 ^{ab}
T3	1.52 ^{b}	36.3 ^{a}	56.9 ^{a}	7.30 ^{b}	0.25 ^{a}
T4	1.57 ^{ab}	32.1 ^{bc}	50.4 ^{bc}	7.47 ^{a}	0.21 ^{ab}
T5	1.62 ^{a}	30.6 ^{cd}	49.2 ^{bc}	7.34 ^{b}	0.18 ^{b}
Mean	1.56	32.4	51. 6	7.38	0.2
Initial	1.69	29.9	48.3	7.8	0.24
LSD ($p \leq 0.05$)	0.09	1.9	5.4	0.08	0.05

Treatments detail in Table 1; two depths, i.e., D1 (0–15 cm) and D2 (15–30 cm). In the column, means with similar letter(s) are statistically identical, as per $LSD_{0.05}$.

3.3. Impact of INM on Available and Total Macronutrients (NPK) in Soil

The observations regarding available N content indicated that the N content was low in both D1 and D2 and the maximum value in soil samples of D1 was 164.9 kg ha^{-1} in treatment T_3. The result of treatment T_3 was statistically different from treatment T_1, in which the least available N (150.8 kg ha^{-1}) was recorded. The concentration of available N in soil reduced with depth (D2), where N contents ranged between 150.6 and 163.4 kg ha^{-1}. The available P levels in soils of depths D1 and D2 enhanced significantly from their initial values of 14.4 and 12.8 kg ha^{-1}, respectively. The available P content in soil improved when nutrients were supplemented through the combined use of chemical fertilizers with organic FYM (T_1, T_2, T_3, and T_5) and also in treatment T_3, which favored the significantly higher buildup of P content (31.4 kg ha^{-1}) more than all other treatments. Soil supplemented with FYM and chemical fertilizers recorded a higher level of available K content over its initial level (Table 4). However, a maximum increase (168.0 kg ha^{-1} in D1 and 166.2 kg ha^{-1} in D2) was observed in treatment T_3. The lowest available K was observed in treatment T_1 in which K was not added through chemical fertilizers but only 6 t ha^{-1} FYM was incorporated in the soil.

Total N content in the present study ranged from 0.16% in T_5 to 0.25% in T_3 treatments in soils of depth D1 and from 0.15% in T_5 to 0.21% in T_3 treatments in D2 (Table 5). Initially, the value of total N in soil was 0.12% in depth D1 and 0.08% in depth D2. All the treatments recorded a decline in total N content with the increase in soil depth. The highest content was observed in treatment T_3, followed by T_2 and T_4, and the lowest content was found in T_5. A similar trend was followed in the soils of depth D2. Total P content of soil ranged from 0.38–0.53% in D1 and 0.34–0.50% in D2 soil samples (Table 5).

A significant buildup of P was observed in all treatments that received chemical fertilizers with FYM. Total P content decreased in soils of depth D2, as compared with the soils of sample D1 under all treatments. In soil samples of D1, treatment T_3 recorded maximum content of total P and showed non-significant variation with all other treatments, except T_1. However, in soil samples of D2, treatment T_3 was significantly superior to all other treatments. The treatments which included the application of FYM coupled with chemical fertilizers recorded a significant buildup of total K in soil, which varied from 0.32

to 0.39% in D1 and from 0.25 to 0.31% in D2 soil samples (Table 5). The results of total K content recorded a higher level in D1 than in D2 soil samples. The maximum total K content was reported in treatment T_3 and lowest in T_2 in both soil layers.

Table 4. Effect of INM technique on available N, P, and K in soil under the maize–wheat system.

Treatments	Available					
	N (kg ha^{-1})		P (kg ha^{-1})		K (kg ha^{-1})	
	D1	D2	D1	D2	D1	D2
T_1	152.8 ^{b}	150.6 ^{c}	22.8 ^{b}	20.4 ^{b}	140.6 ^{b}	138.5 ^{b}
T_2	161.2 ^{a}	158.8 ^{ab}	25.4 ^{b}	22.6 ^{b}	151.2 ^{ab}	148.9 ^{b}
T_3	164.9 ^{a}	163.4 ^{a}	31.4 ^{a}	30.2 ^{a}	168.0 ^{a}	166.2 ^{a}
T_4	159.8 ^{ab}	155.3 ^{bc}	24.2 ^{b}	21.9 ^{b}	148.8 ^{b}	145.6 ^{b}
T_5	158.9 ^{ab}	156.4 ^{bc}	23.2 ^{b}	20.7 ^{b}	145.0 ^{b}	143.7 ^{b}
Mean	159.5	156.9	25.4	23.2	150.7	148.6
Initial	119.7	102.3	14.4	12.8	128.6	124.7
LSD ($p \leq 0.05$)	7.9	6.3	4.7	5.5	18.4	16.8

Treatments detail in Table 1; two depths, i.e., D1 (0–15 cm) and D2 (15–30 cm). In the column, means with similar letter(s) are statistically identical, as per $LSD_{0.05}$.

Table 5. Effect of INM technique on total N, P, and K in soil under the maize–wheat system.

Treatments	% Total					
	N		P		K	
	D1	D2	D1	D2	D1	D2
T_1	0.19 ^{ab}	0.16 ^{b}	0.42 ^{bc}	0.40 ^{bc}	0.33 ^{b}	0.29 ^{ab}
T_2	0.22 ^{ab}	0.19 ^{ab}	0.48 ^{ab}	0.43 ^{b}	0.32 ^{b}	0.25 ^{c}
T_3	0.25 ^{a}	0.21 ^{a}	0.53 ^{a}	0.50 ^{a}	0.39 ^{a}	0.31 ^{a}
T_4	0.21 ^{ab}	0.18 ^{ab}	0.45 ^{abc}	0.42 ^{b}	0.36 ^{ab}	0.28 ^{b}
T_5	0.16 ^{b}	0.15 ^{b}	0.38 ^{c}	0.34 ^{c}	0.35 ^{ab}	0.29 ^{ab}
Mean	0.21	0.18	0.45	0.42	0.35	0.28
Initial	0.12	0.08	0.25	0.19	0.27	0.21
LSD ($p \leq 0.05$)	0.07	0.04	0.09	0.06	0.04	0.02

Treatments detail in Table 1; two depths, i.e., D1 (0–15 cm) and D2 (15–30 cm). In the column, means with similar letter(s) are statistically identical, as per $LSD_{0.05}$.

3.4. *Impact of INM on DTPA-Extractable and Total Micronutrients (Zn, Cu, Fe, and Mn) in Soil*

Among micronutrient cations, DTPA-extractable Zn varied from 2.92–3.88 mg kg^{-1} in the D1 and 2.34–3.48 mg kg^{-1} in D2 soil samples in different treatments. A significant increase in Zn was observed in treatments T_3 (3.88 mg kg^{-1}), T_2 (3.70 mg kg^{-1}), T_4 (3.54 mg kg^{-1}), and T_5 (3.38 mg kg^{-1}) as compared with T_1 (2.92 mg kg^{-1}), in which no additional dose of Zn was added through $ZnSO_4$. The improved Cu content (0.44–0.84 mg kg^{-1} in D1 and 0.34–0.62 mg kg^{-1} in D2 soil samples) was recorded in all treatments over their initial levels (Table 6).

The maximum content of DTPA-extractable Cu was recorded in T_3 treatment showed non-significant variation with T_2 and T_4 treatments in D1 and with T_2, T_4, and T_5 treatments in D2 soil samples. On the contrary, the DTPA-extractable Fe contents in soil recorded a significant improvement in all the treatments over its initial value of 3.88 mg kg^{-1} (Table 6). The DTPA-extractable Fe content varied from 10.12 to 19.66 mg kg^{-1} and 8.48 to 14.58 mg kg^{-1} in D1 and D2 soil samples, respectively, under different treatments. The DTPA-extractable Mn in the current study increased in D1 and D2 soil samples from 11.16 to 18.38 mg kg^{-1} and 9.24 to 15.08 mg kg^{-1}, respectively, as compared with its initial value (3.48 mg kg^{-1} and 2.65 mg kg^{-1}, respectively). The treatments T_2, T_3, and T_4 showed non-significant variation with reason to DTPA extractable Mn in both layers of soil (Table 6).

Table 6. Effect of INM technique on DTPA-extractable micronutrients in soil under the maize–wheat system.

Treatments	Zn (mg kg^{-1})		Cu (mg kg^{-1})		Fe (mg kg^{-1})		Mn (mg kg^{-1})	
	D1	D2	D1	D2	D1	D2	D1	D2
T_1	2.92 ^{b}	2.34 ^{b}	0.44 ^{b}	0.32 ^{b}	11.74 ^{bc}	10.26 ^{ab}	11.16 ^{b}	9.24 ^{b}
T_2	3.70 ^{a}	3.22 ^{a}	0.60 ^{ab}	0.47 ^{ab}	14.02 ^{b}	12.36 ^{ab}	16.34 ^{a}	13.12 ^{ab}
T_3	3.88 ^{a}	3.48 ^{a}	0.84 ^{a}	0.62 ^{a}	19.66 ^{a}	14.58 ^{a}	18.38 ^{a}	15.08 ^{a}
T_4	3.54 ^{a}	3.38 ^{a}	0.58 ^{ab}	0.46 ^{ab}	10.12 ^{c}	9.68 ^{b}	14.94 ^{ab}	12.42 ^{ab}
T_5	3.38 ^{ab}	3.12 ^{a}	0.48 ^{b}	0.36 ^{ab}	10.76 ^{bc}	8.48 ^{b}	11.82 ^{b}	9.64 ^{b}
Mean	3.48	3.11	0.59	0.45	13.26	11.07	14.53	11.90
Initial	1.26	0.68	0.30	0.22	3.83	2.56	3.48	2.65
LSD ($p \leq 0.05$)	0.57	0.60	0.32	0.28	3.27	4.71	3.93	4.11

Treatments detail in Table 1; two depths, i.e., D1 (0–15 cm) and D2 (15–30 cm). In the column, means with similar letter(s) are statistically identical, as per $LSD_{0.05}$.

The results for total Zn content demonstrated the superior level of total Zn in all the treatments over treatment T_1 (Table 7). The total Zn content ranged from 160.0 to 196.7 mg kg^{-1} and 134.8 to 176.9 mg kg^{-1}, respectively, under all treatments. The highest Zn content was recorded in the T_3 treatment and showed non-significant variation with treatments T_2 and T_4. The total Zn content was reduced with soil depth. The variation in Cu content was found from 18.0 mg kg^{-1} in T_1 to 26.8 mg kg^{-1} in T_3 in D1 and from 15.4 mg kg^{-1} in T_1 to 24.3 mg kg^{-1} in T_3 in D2 soil samples. Soil supplemented with FYM and chemical fertilizers recorded an increased total Cu over its initial levels. The total Fe concentration ranged from 2.7 to 3.9% in D1, in which it increased in all treatments over its initial value (2.6%). Its higher content was reported in T_2, T_3, and T_4 treatments. while lower content was found in T_1 and T_5 treatments. Total Mn content of soil varied from 170.3 to 224.3 mg kg^{-1} in D1 and 148.4 to 202.9 mg kg^{-1} in D2 soil samples. Total Mn content in soil showed an appreciable increase over its initial levels.

Table 7. Effect of INM technique on total micronutrients in soil under the maize–wheat system.

Treatments	Zn (mg kg^{-1})		Cu (mg kg^{-1})		Fe (%)		Mn (mg kg^{-1})	
Depth	D1	D2	D1	D2	D1	D2	D1	D2
T_1	160.0 ^{c}	134.8 ^{c}	18.0 ^{d}	15.4 ^{c}	2.7 ^{c}	2.1 ^{b}	170.3 ^{c}	148.4 ^{b}
T_2	182.3 ^{ab}	152.6 ^{bc}	24.3 ^{ab}	21.6 ^{ab}	3.6 ^{ab}	2.9 ^{a}	190.0 ^{b}	166.2 ^{b}
T_3	196.7 ^{a}	176.9 ^{a}	26.8 ^{a}	24.3 ^{a}	3.9 ^{a}	3.1 ^{a}	224.3 ^{a}	202.9 ^{a}
T_4	176.7 ^{abc}	158.5 ^{ab}	22.0 ^{bc}	19.8 ^{abc}	3.4 ^{ab}	2.6 ^{ab}	184.0 ^{bc}	156.6 ^{b}
T_5	163.3 ^{bc}	139.6 ^{c}	20.0 ^{cd}	16.8 ^{bc}	3.1 ^{bc}	2.2 ^{b}	174.0 ^{c}	151.2 ^{b}
Mean	175.8	152.5	22.2	19.6	3.3	2.6	188.5	165.1
Initial	112.5	86.5	13.5	10.4	2.6	1.8	132.8	97.6
LSD ($p \leq 0.05$)	21.2	18.8	3.7	3.4	0.6	0.5	14.1	29.1

Treatments detail in Table 1; two depths, i.e., D1 (0–15 cm) and D2 (15–30 cm). In the column, means with similar letter(s) are statistically identical, as per $LSD_{0.05}$.

3.5. Correlation Analysis among Different Soil Parameters

The correlation analysis of OC and microbiological characteristics with other soil characteristics have been presented in Figure 3. The soil OC content showed a strong positive correlation with soil porosity, water holding capacity, and soil EC; however, it was negatively correlated with soil pH and bulk density. Similarly, the soil microbiological properties suggested a positive correlation with soil porosity, WHC, and soil EC to a greater extent. The soil pH and bulk density showed a non-significant correlation with soil microbiological properties. Among different soil characteristics, soil OC showed the highest correlation (i.e., (r = 0.95, $p \leq 0.05$)) with soil porosity, which was followed by a correlation of CO_2-C with soil pH and soil EC (r = 0.90, $p \leq 0.05$).

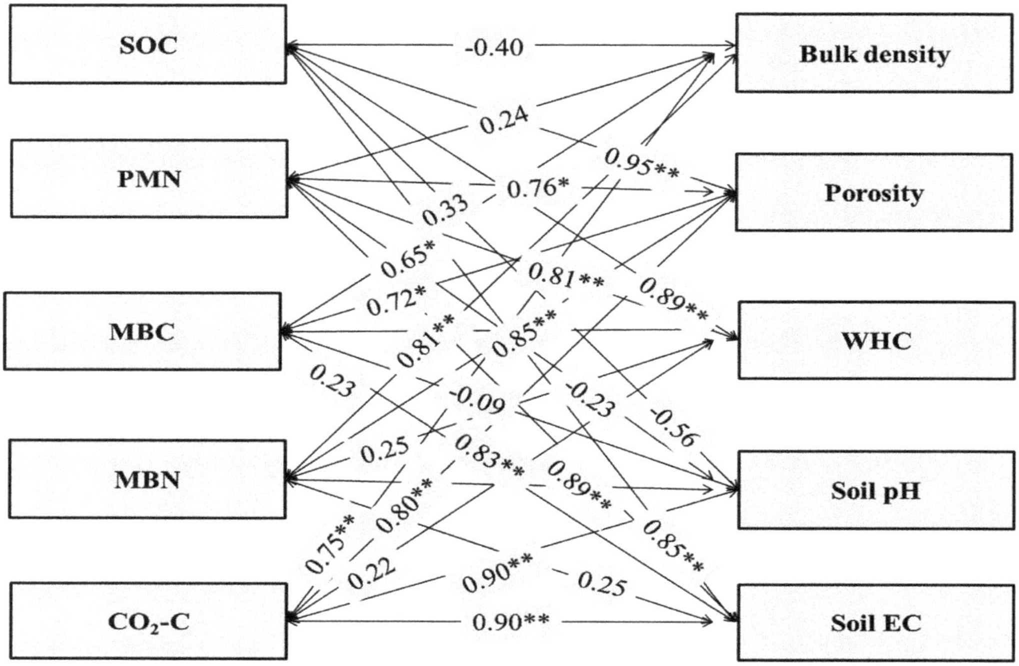

Figure 3. The correlation coefficient of soil OC and microbiological community with soil properties (**-correlation is significant at the 0.01 level; *-correlation is significant at the 0.05 level).

4. Discussion

4.1. Impact of INM on Soil Carbon and Microbiological Composition

Combined supplementation of fertilizers with FYM showed a notable impact on the OC contents of the D2 (15–30 cm). Similar improvement in OC content with combined addition of FYM and chemical fertilizers over inorganic fertilizer alone under MWCS in an Alfisol has also been reported [34]. An improvement in OC content might be associated with the SOM supplementation in the form of FYM, improved root anatomy, and more plant residue addition, with the higher application of nutrients through manure and chemical fertilizers [35].

The PMN reduced with the soil depth in all treatments and increased over its initial levels in D1 and D2 soil samples. The PMN is widely associated with the potential N supplying capability of soil [36]. Higher PMN in all treatments suggests the accumulation of mineralizable N pools in the soil through organic manure addition [37]. The combined addition of FYM and chemical fertilizers enhanced the MBC content over their initial levels in D1 and D2 soil samples, which may be related to improved root growth and crop residues addition after harvesting [38]. Additionally, the addition of organic matter through manure application may provide a favorable environment for enhanced microbial activity and transformations of micronutrients in agricultural soils [39]. The results are concordant with the results reported by Nath et al. [40].

The reduced MBN content with soil depth might be associated with the low OC content in D2 soil samples. The balanced supplementation of organic manure and FYM resulted in the appropriate nutrient availability, which further improved the rhizosphere activity and growth parameters of the plant. The improvement in these parameters resulted in a higher mineralization rate of N and also higher OC content in the soil. The results corroborate the findings of Chang et al. [41]. The increase in CO_2-C (soil respiration) in integrated treatments could have resulted from available carbon substrate through manure, easily mineralizable organic compounds, and other essential nutrients (N and P) for soil microorganisms, available through chemical fertilizers and manure [42]. Higher soil respiration suggested the higher metabolically activity of microbial biomass in soil.

4.2. Impact of INM on Soil Physicochemical Characteristics

Soil bulk density reduced, compared with its initial levels, under all the treatments and total porosity and WHC increased over their initial level. Similar results have already been reported for bulk density and total porosity, with the addition of FYM either alone or integrated use of NPK and FYM in soil samples collected after wheat harvest [43]. This could be ascribed to the produced soil particle binding agents such as polysaccharides and bacterial gums from the microbial breakdown of organic manures. These molecules decrease the soil bulk density by promoting soil aggregation and hence improve the porosity [44]. The improvement in the structural characteristics of soil with FYM supplementation influenced the WHC of soil positively [45].

The soil pH values reduced with an increase in soil depth. Soil pH is also reduced with FYM application, which might be associated with the release of organic acid during microbial decomposition of FYM [46]. The changes in soil pH with FYM supplementation may be owed to oxidation of organic matter and release of carbon dioxide in the soil [47]. The addition of NPK fertilizers resulted in higher EC, which increased the salts accumulation in the soil. This was also due to the decomposition of organic matter added through FYM [48].

4.3. Impact of INM on Available and Total NPK in Soil

The use of INM demonstrated a significant improvement in available N contents as compared with their initial level, which might be related to the N mineralization from the applied fertilizers during decomposition. Higher N availability in the treatments applied with FYM might be due to the slow-release of organically bound nutrients from FYM. It improves the complexation of metal ions, and, thus, increases the bioavailability of nutrient elements to plants [1]. The FYM also provides a favorable environment for the conversion of non-available plant nutrient form to available plant nutrients and slowly release available carbon [49]. The trend for total N followed a similar trend of OC level as the soil-internal cycling is associated with OC; thus, an increase in total N has been recorded with the increase in organic carbon content [37]. Higher content of total N in plots supplemented with organic sources and 50% of recommended NPK fertilizers has been observed in the literature [50].

The addition of FYM to the soil resulted in increased available P content in the soil by mineralization or solubilizing the native P reserves. The elevation in available P content with the application of FYM, along with chemical fertilizers under MWCS, was also reported by Rajneesh et al. [51]. The organic manure increased the nutrient retention capacity of the soil by enhancing the SOM; thus, the available nutrient level of soil required for optimum crop productivity was improved [52]. Mani et al. reported an increase in total P content in soil under treatment in which FYM had been added with NPK, Zn, and phosphate solubilizing bacteria [7]. The application of FYM increased total P in the soil as it acts as P source and also facilitates the retention of P in soil [53]. The increase in available K on FYM addition may be related to the reduced K fixation and release of K, due to the interaction of FYM with clay [54]. Another possible reason for the improvement in total K content might be based on the fact that FYM retains K ions on the exchange sites of its decomposed products, which reduces its leaching loss [55].

4.4. Impact of INM on DTPA-Extractable and Total Micronutrients in Soil

Extractable DTPA increased under all treatments over its initial level as FYM had been added in all treatments at different rates. This could have been due to the fact that FYM supplies an extensive amount of Zn to the soil as well as facilitates the biological and chemical changes that favor the dissolution of non-available Zn [56]. The increase in total Zn content among different treatments might be associated with Zn supplementation through chemical fertilizer and organic manure [18].

The increment in available Cu contents in soil with FYM supplementation might be attributed to its reduced redox potential, which resulted in an increased release of bioavailable micronutrients in the soil over the sole use of synthetic fertilizers. The improved DTPA-extractable Cu content may be due to its complexation with organic molecules released during FYM decomposition, which increased its availability by prohibiting fixation, oxidation, precipitation, and leaching. Nutrient supplementation through FYM in conjugation with chemical fertilizers increased total Cu in soil over its initial level. Addition of FYM to the soil forms organic chelates in soil, which decrease the probability of retaining Cu ions and encourage the increase in microorganism populations, which enhance the plant accessibility of soil micronutrients [38].

The increased availability of Fe with the addition of FYM may be attributed to its increased availability due to the decrease in soil pH by the virtue of organic manure [57]. The enhancement in the soil redox potential with the addition of FYM increased total Fe content [58]. The application of FYM resulted in the buildup of DTPA-extractable Mn in soil which may be attributed to the supply of Mn in the soil through manure. The DTPA-extractable Mn content was greater in the FYM-treated plots, due to Mn release during FYM decomposition. Apart from that, organic acids and humic substances released from FYM decomposition encourage the Mn mobilization from solid phase to soil solution [59]. The micronutrients levels decreased with an increase in soil depth under all treatments. Similar observations were recorded by Sharma and Shweta [60].

5. Conclusions

The long-term study concluded that the integrated use of farmyard manure, coupled with chemical fertilizers in maize–wheat cropping system, had significant improvement in soil organic carbon and soil microbiological community of soil. The data on the build-up of macronutrients (N, P, and K) and DTPA-extractable micronutrients (Zn, Cu, Fe, and Mn) also remarkably improved when the balanced amount of nutrients was supplied through the integrated application of mineral and FYM. Among different treatments, the treatment in which an additional 50% dose of nitrogen was added over its recommended value of soil was found best to sustain the agricultural outcomes of the maize–wheat system in the loamy sand soil of Punjab.

Author Contributions: Conceptualization, S.S.D., A.K.S., A.S.T. and S.S.W.; methodology and visualization, S.S.D., P.S., L.K.P. and S.S.W.; software, S.S.D.; validation, S.S.D., P.S., L.K.P. and S.S.W.; formal analysis, S.S.D. and A.H.; investigation, S.S.D., P.S., L.K.P. and S.S.W.; resources, S.S.D., A.K.S., A.S.T. and S.S.W.; data curation, S.S.D. and A.H.; writing—original draft preparation, S.S.D., S.S., V.S., A.K.S., S.S.W., A.S.T., V.V., M.K.R., L.K.P. and P.S.; writing—review and editing, M.A., A.G. and A.H.; supervision and project administration, M.A., A.G. and A.H.; funding acquisition, M.A., A.G. and A.H. All authors have read and agreed to the published version of the manuscript in the journal.

Funding: The current research was supported by "All India Coordinated Research Project for Micronutrients and Secondary Plant Nutrients in Soil and Plants", the Indian Institute of Soil Science, Bhopal, India. The research was also partially supported by Taif University Researchers Supporting Project number (TURSP 2020/257), Taif University, Taif, Saudi Arabia for funding.

Institutional Review Board Statement: Not applicable.

Informed Consent Statement: Not applicable.

Data Availability Statement: Not applicable.

Acknowledgments: The authors are thankful to the Directorate of Research, PAU, Ludhiana, Project Coordinator, "All India Coordinated Research Project for Micronutrients and Secondary Plant Nutrients in Soil and Plants", Indian Institute of Soil Science, Bhopal India. The authors also extend their appreciation to Taif University Researchers Supporting Project number (TURSP 2020/257), Taif University, Taif, Saudi Arabia for funding.

Conflicts of Interest: The authors declare no competing interest.

References

1. Hashim, M.; Dhar, S.; Vyas, A.K.; Singh, C.B. Yield trends and changes in physico-chemical properties of soil in maize-wheat cropping system under integrated nutrient management. *J. Environ. Biol.* **2017**, *38*, 727–734. [CrossRef]
2. Meena, H.N.; Singh, S.K.; Meena, M.S.; Jorwal, M. *Crop Diversification in Rice-Wheat Cropping System with Maize in Haryana*; Extension Bulletin-2/2021; ICAR-Agricultural Technology Application Research Institute: Jodhpur, India, 2021; pp. 1–12.
3. Brankov, M.; Simic, M.; Dragicevic, M. The influence of maize-winter wheat rotation and pre-emergence herbicides on weeds and maize productivity. *Crop. Prot.* **2021**, *143*, 105558. [CrossRef]
4. Sharma, P.C.; Jat, H.S.; Kumar, V.; Gathala, M.K.; Datta, A.; Yaduvanshi, N.P.S.; Choudhary, M.; Sharma, S.; Singh, L.K.; Saharawat, Y.; et al. Sustainable intensification opportunities under current and future cereal systems of north-west India. In *Technical Bulletin: CSSRI/Karnal/2015/4*; Central Soil Salinity Research Institute: Karnal, India, 2015; p. 46.
5. Bharti, B.; Sharma, R.P. Long term effect of integrated nutrient management on soil properties and availability of nutrients in a Typic Hapludalfs under maize-wheat cropping. *Int. J. Environ. Agric. Res.* **2017**, *3*, 43–48. [CrossRef]
6. Sharma, V.; Singh, M.J.; Khokhar, A.K. Productivity, nutrient uptake and soil properties as influenced by integrated nutrient management in maize-wheat cropping system under rainfed conditions of sub-montane Punjab. *Agric. Res. J.* **2020**, *57*, 839–847. [CrossRef]
7. Mani, D.; Upadhyay, S.K.; Kumar, C.; Balak, S.; Pathak, N. Effect of integrated nutrient management system on nutrient uptake and yield of maize (*Zea mays*). *New Agric.* **2011**, *22*, 5–14.
8. Ferrari, M.; Dal Cortivo, C.; Panozzo, A.; Barion, G.; Visioli, G.; Giannelli, G.; Vamerali, T. Comparing Soil vs. Foliar nitrogen supply of the whole fertilizer dose in common wheat. *Agronomy* **2021**, *11*, 2138. [CrossRef]
9. Brankov, M.; Simić, M.; Dolijanović, Ž.; Rajković, M.; Mandić, V.; Dragičević, V. The Response of maize lines to foliar fertilizing. *Agriculture* **2020**, *10*, 365. [CrossRef]
10. Kumar, B.; Dhar, S.; Paul, S.; Paramesh, V.; Dass, A.; Upadhyay, P.K.; Kumar, A.; Abdelmohsen, S.A.M.; Alkallas, F.H.; El-Abedin, T.K.Z.; et al. Microbial Biomass Carbon, Activity of Soil Enzymes, Nutrient Availability, Root Growth, and Total Biomass Production in Wheat Cultivars under Variable Irrigation and Nutrient Management. *Agronomy* **2021**, *11*, 669. [CrossRef]
11. Blair, N.R.; Faulkner, D.; Till, A.R.; Poulton, P.R. Long-term management impacts on soil C, N and physical fertility. *Soil. Till. Res.* **2005**, *91*, 30–38. [CrossRef]
12. Kundu, S.; Bhattacharyya, R.; Parkash, V.; Ghosh, B.N.; Gupta, H.S. Carbon sequestration and relationship between carbon addition and storage under rainfed soybean-wheat rotation in a sandy loam soil of the India Himalayas. *Soil. Till. Res.* **2006**, *92*, 87–95. [CrossRef]
13. Verma, K.; Bindra, A.; Singh, J.; Negi, S.C.; Datt, N.; Rana, U.; Manuja, S. Effect of integrated nutrient management on growth, yield attributes and yield of maize and wheat in maize-wheat cropping system in mid hills of Himachal Pradesh. *Int. J. Pure Appl. Biosci.* **2018**, *6*, 282–301. [CrossRef]
14. Tiwari, C.; Sharma, K.; Khandelwal, S.K. Effect of green manuring through Sesbania cannabina and Sesbania rostrata and nitrogen application through urea to maize (*Zea mays*)–wheat (*Triticum aestivum*) cropping system. *Indian J. Agron.* **2004**, *49*, 15–21.
15. Yadav, R.L.; Yadav, D.V.; Duttamajumdar, S.K. Rhizospheric environment and crop productivity: A review. *Indian J. Agron.* **2008**, *53*, 1–17.
16. Sheoran, S.; Raj, D.; Antil, R.S.; Mor, V.S.; Dahiya, D.S. Productivity, seed quality, and nutrient use efficiency of wheat (*Triticum aestivum* L.) under organic, inorganic, and INM practices after 20 years of fertilization. *Cereal Res. Commun.* **2017**, *45*, 315–325. [CrossRef]
17. Brar, B.S.; Singh, J.; Singh, G.; Kaur, G. Effect of long-term application of inorganic and organic fertilizers on soil organic carbon and physical properties in maize wheat rotation. *Agronomy* **2015**, *5*, 220–238. [CrossRef]
18. Zhang, S.; Li, Z.; Yang, X. Effects of long-term inorganic and organic fertilization on soil micronutrient status. *Commun. Soil Sci. Plant Anal.* **2015**, *46*, 1778–1790. [CrossRef]
19. Jalota, S.K.; Khera, R.; Ghuman, B.S. *Methods in Soil Physics*; Narosa Publishing House: Delhi, India, 1998; pp. 41–45.
20. Richard, L.A. Diagnosis and improvement of saline and alkali soils. In *Agriculture Hand Book No. 60*; USDA: Washington, DC, USA, 1954; pp. 7–33.
21. Prihar, S.S.; Verma, K.S. A rapid method for direct determination of air porosity of soil. *Soil Sci.* **1969**, *107*, 145–147. [CrossRef]
22. Jackson, M.L. A manual of methods useful for instruction and research in soil chemistry, physical chemistry, soil fertility and soil genesis. In *Soil Chemical Analysis-Advanced Course*, 2nd ed.; Department of Science. University of Wisconsin Madison: Madison, WI, USA, 1973.
23. Subbiah, B.V.; Asija, G.L. A rapid procedure for estimation of available nitrogen in soils. *Curr. Sci.* **1956**, *25*, 259–260.
24. Olsen, S.R.; Cole, C.V.; Watanabe, F.S.; Dean, L.A. *Estimation of Available Phosphorus by Extraction with Sodium Bicarbonate (Circular 39)*; US Department of Agriculture: Washington, DC, USA, 1954.
25. Merwin, H.D.; Peech, M. Exchangeability of soil potassium in sand, silt and clay fractions as influenced by the nature of the complimentary exchangeable cations. *Soil Sci. Soc. Am. Proc.* **1950**, *15*, 125–128. [CrossRef]

26. Lindsay, W.L.; Norvel, W.A. Development of DTPA soil test for zinc, copper, iron and manganese. *Soil Sci. Soc. Am. J.* **1978**, *42*, 421–428. [CrossRef]
27. Page, A.L.; Miller, R.H.; Keeney, D.R. *Methods of Soil Analysis. Part 2. Chemical and Microbiological Properties*, 2nd ed.; American Society of Agronomy and Soil Science Society of America: Madison, WI, USA, 1982.
28. Walkley, A.; Black, C.A. An examination of the Digtjareff method for determination of soil organic matter and a proposed modification of chromic acid titration method. *Soil Sci.* **1934**, *37*, 29–39. [CrossRef]
29. Keeney, D.R. Nitrogen—Availability indices. In *Methods of Soil Analysis Part 2*; Bottomley, P.J., Angle, J.S., Weaver, R.W., Eds.; John Wiley & Sons: Hoboken, NJ, USA, 1982; pp. 711–733.
30. Keeney, D.R.; Nelson, D.W. Nitrogen—Inorganic forms. In *Methods of Soil Analysis Part 2*; Bottomley, P.J., Angle, J.S., Weaver, R.W., Eds.; John Wiley & Sons: Hoboken, NJ, USA, 1982; pp. 643–698.
31. Jenkinson, D.S. Determination of microbial biomass carbon and nitrogen in soil. In *Advances in Nitrogen Cycling in Agricultural Ecosystems*; Wilson, J.R., Ed.; CAB International: Wallingford, UK, 1988; pp. 368–386.
32. Anderson, J.P.E.; Domsch, K.H.A. Physiological method for the quantitative measurement of microbial biomass in soils. *Soil Biol. Biochem.* **1978**, *10*, 215–221. [CrossRef]
33. Panse, V.G.; Sukhatme, P.V. *Statistical Methods for Agricultural Workers*, 4th ed.; ICAR: New Delhi, India, 1985; p. 359.
34. Kumari, G.; Thakur, S.K.; Kumar, N.; Mishra, B. Long term effect of fertilizers, manure and lime on yield sustainability and soil organic carbon status under maize (*Zea mays*)–wheat (*Triticum aestivum*) cropping system in Alfisols. *Indian J. Agron.* **2013**, *58*, 152–158.
35. Dhaliwal, S.S.; Naresh, R.K.; Mandal, A.; Walia, M.K.; Gupta, R.K.; Singh, R.; Dhaliwal, M.K. Effect of manures and fertilizers on soil physical properties, build-up of macro and micronutrients and uptake in soil under different cropping systems, a review. *J. Plant. Nutr.* **2019**, *42*, 2873–2900. [CrossRef]
36. Russell, C.A.; Dunn, B.W.; Batten, G.D.; Williams, R.L.; Angus, J.F. Soil tests to predict optimum fertilizer nitrogen rate for rice. *Field Crop. Res.* **2006**, *97*, 286–301. [CrossRef]
37. Kaur, K.; Kapoor, K.K.; Gupta, A.P. Impact of organic manures with and without mineral fertilizers on soil chemical and biological properties under tropical conditions. *J. Soil Sci. Plant. Nutr.* **2005**, *168*, 117–122. [CrossRef]
38. Dhiman, D.; Sharma, R.; Sankhyan, N.K.; Sepehya, S.; Sharma, S.K.; Kumar, R. Effect of regular application of fertilizers, manure and lime on soil health and productivity of wheat in an acid Alfisol. *J. Plant. Nutr.* **2019**, *42*, 2507–2521. [CrossRef]
39. Dhaliwal, S.S.; Naresh, R.K.; Mandal, A.; Singh, R.; Dhaliwal, M.K. Dynamics and transformations of micronutrients in agricultural soils as influenced by organic matter build-up: A Review. *Environ. Sustain. India* **2019**, *1–2*, 100007. [CrossRef]
40. Pal, J.; Palatty, A.E. Long term effect of nutrient management on biological properties of acid soil under maize-wheat cropping system. *J. Pharmacog. Phytochem.* **2019**, *SP5*, 208–211.
41. Nath, D.J.; Ozah, B.; Baruah, R.; Barooah, R.C.; Borah, D.K.; Gupta, M. Soil enzymes and microbial biomass carbon under rice-toria sequence as influenced by nutrient management. *J. Indian Soc. Soil Sci.* **2012**, *60*, 20–24.
42. Chang, E.H.; Wang, C.H.; Chen, C.L.; Chung, R.S. Effects of long-term treatments of different organic fertilizers complemented with chemical N fertilizer on the chemical and biological properties of soil. *Soil Sci. Plant. Nutr.* **2014**, *60*, 451–499. [CrossRef]
43. Salehi, A.; Fallah, S.; Sourki, A.A. Organic and inorganic fertilizer effect on soil CO_2 flux, microbial biomass, and growth of *Nigella sativa* L. *Int. Agrophys.* **2017**, *31*, 103–116. [CrossRef]
44. Meena, K.B.; Alam, M.S.; Singh, H.; Bhat, M.A.; Singh, A.; Mishra, A.K.; Thomas, T. Influence of farmyard manure and fertilizers on soil properties and yield and nutrient uptake of wheat. *Int. J. Chem. Studies* **2018**, *6*, 386–390.
45. Choudhary, A.K.; Thakur, R.C.; Kumar, N. Effect of integrated nutrient management on soil physical and hydraulic properties in rice-wheat crop sequence in N-W Himalayas. *Indian J. Soil Conser.* **2008**, *36*, 97–104.
46. Katkar, R.N.; Kharche, V.K.; Sonune, B.A.; Wanjari, R.H.; Singh, M. Long-term effect of nutrient management on soil quality and sustainable productivity under sorghum-wheat crop sequences in vertisols of Akola, Maharashtra. *Agropedology* **2012**, *22*, 103–114.
47. Liang, Q.; Chen, H.; Gong, Y.; Fan, M.; Yang, H.; Lal, R.; Kuzyakov, Y. Effects of 15 years of manure and inorganic fertilizers on soil organic carbon fractions in a wheat-maize system in the North China Plain. *Nutr. Cycl. Agroecosyst.* **2012**, *92*, 21–33. [CrossRef]
48. Meena, B.P.; Biswas, A.K.; Singh, M.; Chaudhary, R.S.; Singh, A.B.; Das, H.; Patra, A.K. Long-term sustaining crop productivity and soil health in maize–chickpea system through integrated nutrient management practices in Vertisols of central India. *Field Crop. Res.* **2019**, *232*, 62–76. [CrossRef]
49. Bhatt, M.K.; Raverkar, K.P.; Labanya, R.; Bhatt, C.K. Effects of long-term balanced and imbalanced use of inorganic fertilizers and organic manure (FYM) on soil chemical properties and yield of rice under rice-wheat cropping system. *J. Pharmacog. Phytochem.* **2018**, *7*, 703–708.
50. Wolf, D.C.; Wagner, G.H. Carbon transformations and soil organic matter formation. In *Principles and Applications of Soil Microbiology*; Pearson Education: London, UK, 2005; p. 320.
51. Ahmad, W.; Shah, Z.; Khan, F.; Ali, S.; Malik, W. Maize yield and soil properties as influenced by integrated use of organic, inorganic and bio-fertilizers in a low fertility soil. *Soil Environ.* **2013**, *32*, 121–129.
52. Rajneesh Sharma, R.P.; Snakhyan, N.K.; Kumar, R. Long-term effect of fertilizers and amendments on depth-wise distribution of available NPK, micronutrient cations, productivity and NPK uptake by maize-wheat system in an acid Alfisol of North-Western Himalayas. *Commun. Soil Sci. Plant. Anal.* **2017**, *48*, 2193–2209.

3. Mondal, S.; Mallikarjun, M.; Ghosh, M.; Ghosh, D.C.; Timsina, J. Influence of integrated nutrient management (INM) on nutrient use efficiency, soil fertility and productivity of hybrid rice. *Arch. Agron. Soil Sci.* **2016**, *62*, 1521–1529. [CrossRef]
4. Prashanth, D.V.; Krishnamurthy, R.; Naveen, D.V. Long-term effect of integrated nutrient management on soil nutrient status, content and uptake by finger millet crop in a typic kandiustalf of eastern dry zone of Karnataka. *Commun. Soil Sci. Plant. Anal.* **2019**, *51*, 161–174.
5. Urkurkar, J.S.; Tiwari, A.; Chitale, S.; Bajpai, R.K. Influence of long-term use of inorganic and organic manures on soil fertility and sustainable productivity of rice (*Oryza sativa*) and wheat (*Triticumaestivum*) in Inceptisols. *Indian J. Agric. Sci.* **2010**, *80*, 208–212.
6. Bansal, K.L. Potassium balance in multiple cropping systems in Vertisol at Jabalpur. *J. Potassium Res.* **1992**, *85*, 52–58.
7. Mohrana, P.C.; Sharma, B.M.; Biswas, D.R. Changes in the soil properties and availability of micronutrients after six-year application of organic and chemical fertilizers using stcr-based targeted yield equations under pearl millet-wheat cropping system. *J. Plant. Nutr.* **2016**, *34*, 56–65.
8. Singh, N.J.; Athokpam, H.S.; Devi, K.N.; Chongtham, N.; Singh, N.B.; Sarma, P.T.; Singh, S.D. Effect of farmyard manure and press mud on fertility status of alkaline soil under maize-wheat cropping sequence. *Afr. J. Agric. Res.* **2015**, *10*, 2421–2431.
9. Dhaliwal, S.S.; Manchanda, J.S.; Walia, S.S.; Dhaliwal, M.K. Differential response of manures in transformation of DTPA and total zinc and iron in rice transplanted on light textured soils of Punjab. *Int. J. Sci. Environ. Technol.* **2013**, *2*, 300–312.
0. Sharma, R.P.; Shweta, S. Soil fertility as influenced by vermicompost application in potato under wet temperate conditions of Himachal Pradesh. *Annal. Biol.* **2013**, *29*, 346–352.

agronomy

MDPI

Article

Potassium and Water-Deficient Conditions Influence the Growth, Yield and Quality of Ratoon Sugarcane (*Saccharum officinarum* L.) in a Semi-Arid Agroecosystem

Rajan Bhatt [1], Jagdish Singh [2], Alison M. Laing [3], Ram Swaroop Meena [4], Walaa F. Alsanie [5], Ahmed Gaber [6,*] and Akbar Hossain [7,*]

1 Regional Research Station (RRS), Kapurthala, Panjab Agricultural University (PAU), Ludhiana 144601, Punjab, India; rajansoils@pau.edu
2 Regional Research Station (RRS), Gurdaspur, Panjab Agricultural University (PAU), Ludhiana 144601, Punjab, India; jagdishsingh@pau.edu
3 CSIRO Agriculture & Food, Brisbane 4067, Australia; alison.laing@csiro.au
4 Department of Agronomy, Banaras Hindu University, Varanasi 221005, Uttar Pradesh, India; meenars@bhu.ac.in
5 Department of Clinical Laboratories Sciences, The Faculty of Applied Medical Sciences, Taif University, P.O. Box 11099, Taif 21944, Saudi Arabia; w.alsanie@tu.edu.sa
6 Department of Biology, College of Science, Taif University, P.O. Box 11099, Taif 21944, Saudi Arabia
7 Bangladesh Wheat and Maize Research Institute, Dinajpur 5200, Bangladesh
* Correspondence: a.gaber@tu.edu.sa (A.G.); akbarhossainwrc@gmail.com (A.H.)

Citation: Bhatt, R.; Singh, J.; Laing, A.M.; Meena, R.S.; Alsanie, W.F.; Gaber, A.; Hossain, A. Potassium and Water-Deficient Conditions Influence the Growth, Yield and Quality of Ratoon Sugarcane (*Saccharum officinarum* L.) in a Semi-Arid Agroecosystem. *Agronomy* **2021**, *11*, 2257. https://doi.org/10.3390/agronomy11112257

Academic Editors: Umberto Anastasi and Aurelio Scavo

Received: 11 September 2021
Accepted: 3 November 2021
Published: 8 November 2021

Abstract: Groundwater and soil potassium deficiencies are present in northern India. Sugarcane is a vital crop in the Indian Punjab; it is grown on approximately 91,000 hectares with an average yield of 80 tonnes ha^{-1} and a sugar recovery rate of 9.59%. The role of potassium (K) fertilizer under both sufficient and deficient irrigation in ratoon sugarcane crops is not well documented. We conducted a split-plot ratoon cane experiment during 2020–2021 at the Gurdaspur Regional Research Station of Punjab Agricultural University, India, on K-deficient soils. Main treatments were fully irrigated (I_1) and water stressed (I_0) conditions, with sub-treatments reflecting K fertilizer application rates of 0 (M_1), 67 (M_2), 133 (M_3), and 200 (M_4) kg K ha^{-1}. The ratoon sugarcane performance was assessed in terms of growth, productivity, sugar quality and incidence of key insect pests. At harvest, trends in the growth and yield parameters in I_1 were improved over the I_0 treatment, with cane height (+12.2%), diameter (+3.3%), number of internodes (+5.4%), biomass yield (+7.6%) and cane yield (+5.9%) all higher, although little significant difference was observed between treatments. Ratoon cane yield under irrigation was 57.1 tonnes ha^{-1}; in water-stressed conditions, it was 54.7 tonnes ha^{-1}. In terms of sugarcane quality parameters, measured 12 months after harvesting the initial seed crop, values of Brix (+3.6%), pol (+3.9%), commercial cane sugar percentage (+4.0%) and extractable sugar percentage (+2.8%) were all higher in the irrigated treatments than the water-stressed plots. Irrigated treatments also had a significantly lower incidence of two key insect pests: top borer (*Scirpophaga excerptalis*) was reduced by 18.5% and stalk borer (*Chilo auricilius*) by 21.7%. The M_3 and M_4 treatments resulted in the highest cane yield and lowest incidence of insect pests compared to other K-fertilizer treatments. Economic return on K-fertilizer application increased with increasing fertilizer dosage. Under the potassium-deficient water-stressed conditions of the region of north India, a fertilizer application rate of 133 kg K ha^{-1} is recommended to improve ratoon sugarcane growth, yield, and quality parameters and economic returns for sugarcane farmers.

Keywords: water stress; potassium fertilizer; Brix; sugarcane yield; insect-pest incidence

1. Introduction

The increase in intensive agricultural practices in northern India (i.e., Punjab, Haryana) over recent decades, combined with conventional crop establishment and irrigation meth ods, has resulted in the lowering of the underground water table and an increase in water-deficient conditions in which farmers produce crops [1–3]. Sugarcane (*Saccharum* spp. complex) is a commercially viable crop which is cultivated not only for edible-sugar products but also as a source of biomass for bioelectricity and second-generation bioethanol. Water stress has a negative impact on sugarcane development and productivity. Improving sugarcane survival and growth rates during periods of water stress is important to achieve sustainable agronomic production in northern India. Sugarcane quality and performance under water stress can be measured in terms of crop water use efficiency (WUE) [3]. Sugarcane plants have evolved a variety of molecular processes which limit the use of resources such as water and which regulate plant development in response to environmental conditions [4,5]. Water stress reduces the leaf-water potential and stomatal openings, resulting in down-regulation of photosynthesis-related genes and lower plant-CO_2 availability [6]. Stress responses involve several molecular networks including signal transduction [7–10]. Improved methods must be developed, tested, and recommended to farmers to reduce sugarcane WUE while improving plant quality and productivity.

Sugarcane is cultivated in north India under sub-tropical conditions. It is an important industrial and food crop with high sugar concentration, and it is extensively used commercially, e.g., as a source of ethanol to blend with petrol [11–13]. Average sugarcane production across the whole of India was around 362 M tonnes with productivity of 71.5 tonnes ha^{-1}. In the Indian Punjab, sugarcane is grown on 91,000 hectares with an average yield of 80.35 tonnes ha^{-1}, and a sugar recovery rate of 9.6%, similar to the national average but lower than that achieved in nearby states where more potassium fertilizer is applied [14].

To sustainably cultivate sugarcane, judicious use of nutrients is necessary as under-application may lead to significant yield and quality loss, as well as depleting the soil [13,15]. It is estimated that for every 100 tonnes of sugarcane produced, key nutrient requirements (i.e., those taken up by 100 tonnes of cane) are: nitrogen (N) 208 kg ha^{-1}, phosphorus (P) 53 kg ha^{-1}, potassium (K) 280 kg ha^{-1}, sulphur (S) 30 kg ha^{-1}, iron (Fe) 3.4 kg ha^{-1}, manganese (Mn) 1.2 kg ha^{-1}, and copper (Cu) 0.6 kg ha^{-1} [12]. While sugarcane K requirements are high (above those of N and P), in practice, little K is applied, even in K-deficient soils [16]. Potassium is an essential plant nutrient which improves plant nutrition and metabolism, N- and water-use efficiencies, root growth, and which regulates the opening of leaf stomata, particularly under water-stressed conditions [17,18]. Additionally, K aids in the functioning of plant enzymes, acting as a catalyst for the activation of around 60 [19–21]. K is also involved in seed germination, transport of photosynthate from leaves to rest plant [22–26], maintaining a balance of cations and anions within plant parts, protein synthesis, photosynthesis, energy transfer [16,27,28], and stress resistance [8,17,18]. K also interacts with other plant nutrients such as N to enhance their use efficiencies and reduce overall cultivation costs of sugarcane cultivation [28–32].

In northern India, sugarcane is grown from seed, and the initial harvest is called the "seed crop". Crops are not destroyed at this first harvest; instead, the sugarcane plant is managed to produce a subsequent "ratoon crop," which improves the economics of sugarcane production. Production costs are lower in ratoon crops than in seed crops, as the costs of land preparation and crop establishment are eliminated [12,13]. Furthermore, early tissue drying and nitrogen flushing mean that the ratoon crop is harvested over a longer window, extending the crushing schedule of sugar factories [33]. Yields of the ratoon crop are lower than those of the seed crop; this may be a result of increased bulk density [13,34,35], poor fertilizer use [14,36], and/or increased incidence of pests and diseases. Other factors which contribute to low ratoon-crop yields are a poor choice of cultivar, low air temperatures, poor quality irrigation water, and weed competition [37]. The relatively lower air temperature of northern India reduces the number of shoots that resprout after the harvest of the seed crop. Previous recommendations to increase the yields of ratoon sugarcane crops in northern India have included mulching the bare soil surface between plants with crop residues or intercropping short-duration vegetable or pulse crops between cane rows, but little success has been observed in reducing yield gaps [15,38,39].

Sugarcane plants have relatively high nutritional requirements [40], and a shortage in any one key nutrient can adversely affect plant performance in terms of productivity and cane juice quality [13]. It is important to maintain a balanced application of nutrients to the cane across both seed and ratoon crops [40]. K fertilizer application in K-deficient soils improves plant performance and reduces water, nutrient, and pesticide footprints by improving input use efficiency. Hence, for water- and K-deficient soils, quantifying the appropriate application rate of K fertilizer is important to ensure the sustainability of sugarcane production, and particularly of ratoon sugarcane production.

Given its importance in sugarcane production, applications of between 60 and 120 kg ha^{-1} K are recommended [41–43]. However, at some K-deficient sites, deficits of up to700 kg ha^{-1} have been recorded [38,44]. There are currently no standardized recommendations for K fertilizer application in north India, even on known K-deficient soils [13,15]. As groundwater levels in the region have also been observed to be low [1,2], the role of K fertilizer in K-deficient and water-stressed conditions is worthy of investigation. We conducted an experiment at the Gurdaspur Regional Research Station of Punjab Agricultural University during 2020–2021 on a ratoon sugarcane crop. Our objectives were to (1) identify standardized K-fertilizer recommendations in low K soils under water-stressed conditions to achieve improved ratoon-crop growth, yield and quality; (2) identify the optimal K-fertilizer dosage to reduce the incidence of insect pests; and (3) to calculate the benefit-to-cost ratio of K-fertilizer treatments.

Hypothesis: Judicious use of K fertilizer under I_1 and I_0 plots at deficient sites (<137.5 kg K_2O ha^{-1}) resulted in significantly lesser insect-pest incidence, and higher growth, yield and quality parameters which further add to the livelihoods of the cane farmers of the region.

2. Material and Methods

2.1. Experiment Location and Inherent Soil Fertility

The experiment was conducted between March 2020 and March 2021 at the Gurdaspur Regional Research Station of Punjab Agricultural University (PAU), India, in a split-plot design with irrigation as the main treatment and sub-treatments of different rates of K-fertilizer. The experimental site was located at 32°49.383′ N and 75°42.588′ E, at an elevation of 225 m. The soil was sandy loam in texture, with a neutral (7.3) pH, an EC of 0.045 dS m^{-1}, moderate soil organic carbon (0.65%), and relatively high in available phosphorus (26.5 kg P ha^{-1}) and low in available potassium (97.5 kg K ha^{-1}) using ammonium acetate method (using flame photometer), as previously reported [14,40]. The threshold value is 137.5 kg K_2O ha^{-1}. Further, soil bulk density was 1.62 g per cm^3 at the surface 0–15 cm.

2.2. Weather during the Experiment

A meteorological station at the site recorded daily maximum and minimum temperature, class A pan evaporation, and rainfall. During the experimental period 822.4 mm rain was received, evaporation 1419.3 mm, average maximum air temperature ranged between 17.1 to 35.9 °C, and average minimum air temperature between 7.4 and 25.7 °C (Figure 1).

2.3. Experimental Treatments and Recorded Observations

The experiment was a split-plot design, with irrigation level as the main treatment and applications of muriate of potassium fertilizer in the sub-plots. There were 24 treatment plots: (a) in 12 of these plots were water-stressed and (b) in 12 plots received the standard irrigation for sugarcane in this region. In both the water-stressed and unstressed plots, there were three replicates of four potassium-fertilizer treatments viz., 0, 40, 80, and 120 kg K ha^{-1}. Irrigation was either applied fully (I_1) or to achieve water-stressed plants (I_0). In the fully irrigated plots, sufficient water was applied throughout the experimental period to ensure it was non-limiting to ratoon-crop growth; however, if more than 20 mm rain was received in any 24 hours, irrigation was suspended. In the water-stressed plots, irrigation ceased at the key sugarcane growth stages of germination, tillering, and grand growth after three weeks. Sub-plots treatments were: 0 kg ha^{-1} K fertilizer (M_1), 67 kg ha^{-1} K fertilizer (M_2), 133 kg ha^{-1} (M_3) and 200 kg ha^{-1} K fertilizer (M_4). Excepting water and K fertilizer management, all other sugarcane agronomic practices followed the recommendations of PAU, Ludhiana [14].

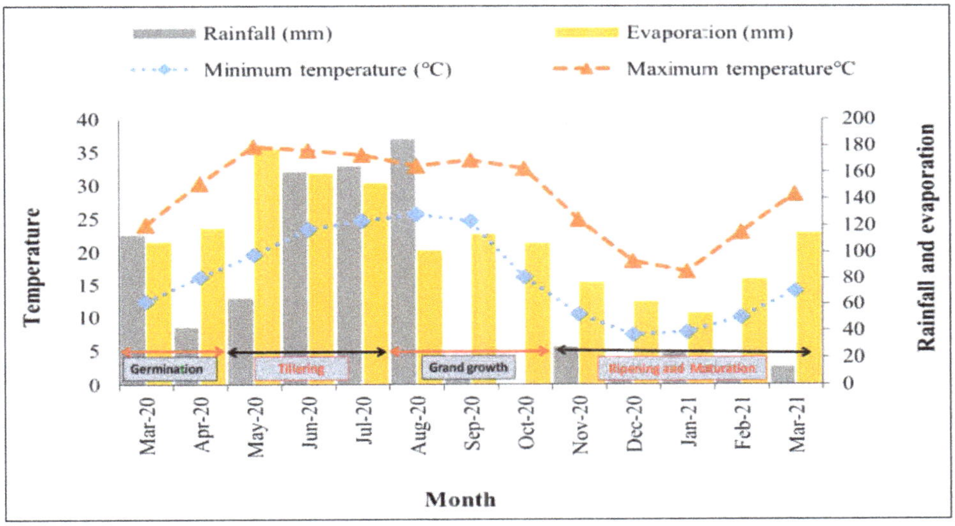

Figure 1. Weather conditions at the Gurdaspur Regional Research Station from March 2020 to March 2021.

The experimental treatments were applied to a ratoon crop of the sugarcane cultivar CoPb 91, which was planted at 75 cm inter-row spacing in 6 m long and 4.5 m broad plots following the harvesting of the seed crop on 15 March 2020. There were three replications of each treatment and sub-treatment plot.

Five canes were tagged in each experimental plot. Measurements of sugarcane growth were taken from these five canes of the number of resprouted canes (at 35 DAH, days after harvesting), average cane height (at 116, 155, 178, 200, 277, and 312 DAH), average cane stalk diameter in the middle of the stalk, the number of nodes per cane. Measurements of sugarcane quality (Brix, pol, percentage purity, extraction percentage, and commercial cane sugar (CCS) as both a percentage and a weight per hectare) were recorded from ten representative, pest- and disease-free, canes from each experimental plot 10 and 12 months after the harvest of the seed crop, on 13 November 2020 and 26 February 2021, respectively, following standard experimental protocols [13,15]. Sugarcane juice was extracted from the harvested canes using a cane crusher to assess Brix and other quality metrics, using standard protocols [41]. At maturity, the number of millable canes in each 27 m^2 plot was manually counted and each plot was manually harvested and processed to record final yield and biomass data in tonnes per hectare.

The presence of early shoot borer (*Chilo infuscatellus*) was manually observed and recorded at 65 DAH. The incidence of two other critical sugarcane pests, stalk borer (*Chilo auricilius*) and top borer (*Scirpophaga excerptalis*), was manually observed and recorded when the ratoon crop was harvested.

2.4. Calculations

The commercial cane sugar (CCS) percentage was calculated using Equation (1):

$$CCS\ (\%) = \{Sucrose\% - (Brix\% - Sucrose\ \%) \times 0.4\} \times 0.74 \quad (1$$

where, 0.4 is the multiplication factor and 0.74 is the crusher factor.

A weight-per-area CCS was determined using Equation (2), as reported in [29,45]:

$$CCS\ (t/ha) = CCS\ (\%) \times \text{sugarcane yield}\ (t\ ha^{-1})/100 \quad (2$$

The benefit-to-cost (B:C) ratio of additional applied K fertilizer in the ratoon cane was calculated using Equation (3), as reported by [15] and [31]:

$$B{:}C = \text{Value of sugarcane yield}\ (Rs\ ha^{-1})/\text{Cost of K fertilizer}\ (Rs\ ha^{-1}) \quad (3$$

where, the cost of muriate of potassium fertilizer was 19,000 INR t^{-1} and the value of the sugarcane yield was the amount of sugarcane produced (tonnes ha^{-1}) multiplied by the sugarcane price, 2950 INR t^{-1}. The B:C ratio is dimensionless.

2.5. Statistical Analysis

Pooled data for the main and sub-plot treatments and their interactions were subjected to analysis of variance (ANOVA) using the STAR (Statistical Tool for Agricultural Research) software package. Statistical significance was inferred at $p \leq 0.05$. The cane growth, yield and quality data were analysed as per the procedure given by Gomez and Gomez for split-plot design using OPSTAT program developed by Chaudhary Charan Singh Haryana Agricultural University, Hisar, India. R software [46] was used to investigate correlations between the different quality attributes.

3. Results

3.1. Ratoon Crop Productivity

The fully irrigated (I_1) plot fertilized with 133 kg K ha^{-1} (M_3) had more resprouted canes, more millable canes, greater cane length and diameter, more leaves, and higher Brix yield and biomass (Table 1).

Table 1. Average sugarcane height under irrigation and potassium treatments.

Treatments	Cane Height (cm) at DAH					
	116	155	178	200	277	312
Irrigation treatment						
I_1	68.9 ^{a}	155.5 ^{a}	206.6 ^{a}	221.7 ^{a}	248.0 ^{a}	261.5 ^{a}
I_0	60.9 ^{b}	137.3 ^{b}	183.1 ^{a}	198.3 ^{a}	228.2 ^{b}	233.1 ^{a}
Significance level ($p \leq 0.05$)	**	**	SS	SS	**	NS
CV (%)	6.7	3.2	9.2	9.3	2.5	10.0
Potassium fertilizer treatment						
M_1	62.0 ^{a}	142.0 ^{a}	186.5 ^{a}	208.8 ^{b}	215.3 ^{c}	217.5 ^{c}
M_2	64.9 ^{a}	145.8 ^{a}	194.3 ^{a}	227.5 ^{ab}	233.3 ^{bc}	234.7 ^{bc}
M_3	65.5 ^{a}	147.8 ^{a}	197.0 ^{a}	239.5 ^{a}	248.0 ^{ab}	254.2 ^{ab}
M_4	66.7 ^{a}	150.0 ^{a}	201.5 ^{a}	246.7 ^{a}	260.3 ^{a}	260.0 $_a$
Significance level ($p \leq 0.05$)	SS	SS	SS	**	**	**
CV (%)	7.9	9.9	7.7	6.6	8.9	7.4
I × M	SS	SS	SS	SS	SS	SS

DAH, days after harvesting the initial seed crop of sugarcane; main plot treatments are I_1 (fully irrigated) and I_0 (water-stressed); subplot treatments are M_1 (0 kg K ha^{-1}), M_2 (67 kg K ha^{-1}), M_3 (133 kg K ha^{-1}), and M_4 (200 kg K ha^{-1}); CV, coefficient of variation; ** denotes significance at $p \leq 0.05$; SS, statistically similar. The superscript similar letter within a continuous column indicates no statistical difference while different letters denote significant differences at $p \leq 0.05$.

Average sugarcane height was significantly higher in this treatment than in the treatment with the same K fertilizer but with water stress (i.e., I_0 M_3); by 9.4% at 116 DAH, by 7.6% at 155 DAH, and by 12.2% at 312 DAH. There was no significant difference in average cane height between the zero-K control treatment (M_1) and plots where K fertilizer was applied at 116, 155 or 178 DAH. From 200 DAH onwards, there were clear differences between treatments, and by harvest (312 DAH) cane height was highest (19.5% above M_1) in M_4, and 7.9% to 16.9% higher in M_2 and M_3, respectively (Table 1). Further, I_1 plots had significantly higher cane height at 116, 155 and 277 DAH as compared to I_0 plots.

Average cane diameter did not differ significantly between irrigation treatments for most of the ratoon crop growing season, although greater measurements were recorded in the fully irrigated I_1 treatment (Table 2). Relative to the M_1 treatment average cane diameter differed from 237 DAH in the M_4 treatment; cane diameter in the M_2 and M_3 treatments was not always significantly different from the control treatment. At harvest, the average cane diameter in the M_4 treatment was 11.3% greater than in the M_1 treatment. There was no statistical difference in the number of leaves per plant between the irrigation treatments, nor between the K-fertilizer treatments, at any time from harvesting the seed crop to harvesting the ratoon crop (Table 2).

Table 2. Average sugarcane diameter and number of leaves under irrigation and potassium treatments.

Treatments	Cane Diameter (cm) at DAH				Leaves per Plant at DAH	
	200	237	277	312	200	237
			Irrigation treatment			
I_1	28.6 ^{a}	28.1 ^{a}	28.9 ^{a}	28.5 ^{a}	9.6 ^{a}	16.7 ^{a}
I_0	28.5 ^{a}	27.9 $_b$	28.2 ^{a}	27.6 ^{a}	9.4 ^{a}	15.6 ^{a}
F-test ($p \leq 0.05$)	SS	**	SS	SS	SS	SS
CV (%)	10.6	5.6	3.1	5.4	9.4	10.4
			Potassium fertilizer treatment			
M_1	27.8 ^{a}	26.6 ^{c}	27.4 -	26.5 ^{b}	9.3 ^{a}	15.3 ^{a}
M_2	28.5 ^{a}	27.5 ^{bc}	28.3 ^{ab}	27.7 ^{ab}	9.4 ^{a}	15.6 ^{a}
M_3	28.8 ^{a}	28.4 ^{ab}	29.2 ^{a}	28.4 ^{ab}	9.6 ^{a}	16.5 ^{a}
M_4	29.3 ^{a}	29.4 ^{a}	29.6 ^{a}	29.5 ^{a}	9.8 ^{a}	16.9 ^{a}
F-test ($p \leq 0.05$)	SS	**	**	**	SS	**
CV (%)	5.5	4.4	4.4	5.5	8.0	5.3
I × M	SS	SS	SS	SS	SS	SS

DAH, days after harvesting the initial seed crop of sugarcane; main plot treatments are I_1 (fully irrigated) and I_0 (water-stressed); subplot treatments are M_1 (0 kg K ha^{-1}), M_2 (67 kg K ha^{-1}), M_3 (133 kg K ha^{-1}), and M_4 (200 kg K ha^{-1}); CV, coefficient of variation; ** denotes significance at $p \leq 0.05$; SS, statistically similar. The superscript similar letter within a continuous column indicates no statistical difference while different letters denote significant differences at $p \leq 0.05$.

There were no statistical differences between irrigation treatments in the average number of internodes per plant or in the average Brix at any time during the ratoon crop growing season, although internodes were lower and Brix higher in the fully irrigated treatment (Table 3). Relative to the M_1 control, the M_4 fertilizer treatment had 12.5% and 12.0% more internodes at 200 and 237 DAH; however, at later samplings (277 and 312 DAH), there was no significant difference in the number of internodes per plant between any K-fertilizer treatments. There were no significant differences in Brix between any K-fertilizer treatments, although trends suggested that higher Brix was associated with greater K-fertilizer application.

Table 3. Average number of internodes per plant and average Brix under irrigation and potassium treatments.

Treatments	Internodes per Plant at DAH				Brix at DAH	
	200	237	277	312	277	312
			Irrigation treatment			
I_1	9.5 ^{a}	12.8 ^{a}	10.7 ^{a}	13.9 ^{a}	20.5 ^{a}	20.8 ^{a}
I_0	10.8 ^{a}	13.5 ^{a}	10.7 ^{a}	14.7 ^{a}	20.3 ^{a}	19.5 ^{a}
F-test ($p \leq 0.05$)	SS	SS	SS	SS	SS	SS
CV (%)	7.2	11.2	8.7	3.9	8.3	7.7
			Potassium fertilizer treatment			
M_1	9.6 ^{b}	12.5 ^{b}	10.1 ^{a}	12.7 ^{a}	19.5 ^{a}	18.1 ^{a}
M_2	9.9 ^{b}	12.7 ^{b}	10.5 ^{a}	14.5 ^{a}	20.0 ^{a}	19.5 ^{a}
M_3	10.3 ^{b}	13.5 ^{ab}	10.9 ^{a}	14.8 ^{a}	20.7 ^{a}	21.0 ^{a}
M_4	10.8 ^{a}	14.0 ^{a}	11.4 ^{a}	15.2 ^{a}	21.2 ^{a}	21.9 ^{a}
F-test ($p \leq 0.05$)	**	**	SS	**	NS	NS
CV (%)	6.6	6.5	8.5	5.6	14.7	11.7
I × M	SS	SS	SS	SS	SS	SS

DAH, days after harvesting the initial seed crop of sugarcane; main plot treatments are I_1 (fully irrigated) and I_0 (water-stressed); subplot treatments are M_1 (0 kg K ha^{-1}), M_2 (67 kg K ha^{-1}), M_3 (133 kg K ha^{-1}), and M_4 (200 kg K ha^{-1}); CV, coefficient of variation; ** denotes significance at $p \leq 0.05$; SS, statistically similar. The superscript similar letter within a continuous column indicates no statistical difference while different letters denote significant differences at $p \leq 0.05$.

There were no significant differences between irrigation treatments in terms of the number of resprouted shoots in the ratoon crop, the number of millable canes, or in the sugarcane biomass and yield at harvest, although in all parameters, observations were more favourable in the fully irrigated treatment (Table 4). Similarly, there were no significant differences in these parameters between any of the K-fertilizer treatments, although trends suggested improved outcomes with increasing K fertilizer application, with the greatest resprouted shoots (52.3%), number of millable canes (60,000 ha^{-1}), biomass yield (10.9 tonnes ha^{-1}) and cane yield (61.0 tonnes ha^{-1}) in the M_4 treatment.

Table 4. Sugarcane resprouting percentage, number of millable canes, and biomass and cane yields under irrigation and potassium treatments.

Treatments	Resprouted Ratoon 35 DAH (%)	NMC (000/ha)	Biomass Yield (t ha^{-1})	Cane Yield (t ha^{-1})
		Irrigation treatment		
I_1	40.1 ^{a}	55.4 ^{a}	10.53 ^{a}	57.1 ^{a}
I_0	37.1 ^{a}	47.2 ^{a}	9.79 ^{a}	54.7 ^{a}
F-test ($p \leq 0.05$)	SS	SS	SS	SS
CV (%)	14.6	7.6	5.0	3.5
		Potassium fertilizer treatment		
M_1	32.7 ^{a}	47.9 ^{a}	9.32 ^{a}	50.8 ^{b}
M_2	36.6 ^{a}	48.4 ^{a}	10.15 ^{a}	53.8 ^{b}
M_3	50.5 ^{a}	59.9 ^{a}	10.28 ^{a}	58.1 ^{a}
M_4	52.3 ^{a}	60.0 ^{a}	10.88 ^{a}	61.0 ^{a}
F-test ($p \leq 0.05$)	SS	SS	SS	**
CV (%)	10.6	8.7	8.8	6.2
I × M	SS	SS	SS	SS

DAH, days after harvesting the initial seed crop of sugarcane; NMC, number of millable canes; main plot treatments are I_1 (fully irrigated) and I_0 (water-stressed); subplot treatments are M_1 (0 kg K ha^{-1}), M_2 (67 kg K ha^{-1}), M_3 (133 kg K ha^{-1}), and M_4 (200 kg K ha^{-1}); CV, coefficient of variation; ** denotes significance at $p \leq 0.05$; SS, statistically similar. The superscript similar letter within a continuous column indicates no statistical difference while different letters denote significant differences at $p \leq 0.05$.

3.2. Insect Pest Occurrence

Under fully irrigated conditions, there was a significantly lower incidence of top borer (−18%) and stalk borer (−29%) than under water-stressed conditions, and no difference in the incidence of shoot borer (Table 5).

Table 5. The average incidence of key insect pests under irrigation and potassium treatments.

Treatments	Shoot Borer (%)	Top Borer (%)	Stalk Borer (%)
	Irrigation treatment		
I_1	6.3 ^{a}	7.1 ^{b}	6.6 ^{b}
I_0	7.7 ^{a}	8.4 ^{a}	8.5 ^{a}
F-test ($p \leq 0.05$)	SS	**	**
CV (%)	5.2	8.6	3.7
	Potassium fertilizer treatment		
M_1	7.7 ^{a}	8.3 ^{a}	8.2 ^{a}
M_2	6.8 ^{a}	7.7 ^{ab}	7.2 ^{a}
M_3	6.3 ^{a}	7.0 ^{b}	7.2 ^{a}
M_4	7.2 ^{a}	8.0 ^{a}	7.6 ^{a}
F-test ($p \leq 0.05$)	SS	**	SS
CV (%)	7.6	9.2	12.3
I × M	SS	SS	SS

Main plot treatments are I_1 (fully irrigated) and I_0 (water-stressed); subplot treatments are M_1 (0 kg K ha^{-1}), M_2 (67 kg K ha^{-1}), M_3 (133 kg K ha^{-1}), and M_4 (200 kg K ha^{-1}); CV, coefficient of variation; ** denotes significance at $p \leq 0.05$; SS, statistically similar. The superscript similar letter within a continuous column indicates no statistical difference while different letters denote significant differences at $p \leq 0.05$.

Under different K-fertilizer treatments, there was no significant difference in the incidence of shoot borer or stalk borer, although the trend was for higher levels of both pests under the M_1 (0 kg K ha^{-1}) treatment, and the M_2 (67 kg K ha^{-1}) and M_3 (133 kg K ha^{-1}) treatments had the lowest incidence of both shoot borer and stalk borer. The M_3 treatment had 15.6% less incidence of top borer than the M_1 treatment, while the M_1, M_2 and M_4 treatments did not differ significantly.

3.3. Ratoon Crop Quality

Irrigation treatment did not affect the Brix, purity, commercial cane sugar, or extractable sugar percentage at either 10 or 12 months after harvesting the seed crop (Tables 6 and 7). Pol was 4.1% higher in the fully irrigated (I_1) treatment 10 months after harvesting the seed crop, but this difference was no longer significant two months later. At 10 months after harvesting the seed crop, there was no significant difference in Brix between any irrigation treatment; however, two months later, the K-fertilized treatments had 4.5% (M_2), 7.5% (M_3) and 9.0% (M_4) higher Brix than the M_1 control treatment (Tables 6 and 7).

Similarly, at 10 months after seed crop harvest, the pol percentage was 5.5% higher in M_3 and M_4 than in M_1; at 12 months after seed crop harvest, the pol percentages were 7.1% and 9.8% higher in M_3 and M_4, respectively, than in M_1. The extractable sugar percentage was higher than M_1 in M_3 (+10.9%) and M_4 (+14.3%) 10 months after seed crop harvest; two months later, there was no significant difference between M_1, M_2, or M_3, while the extractable sugar percentage in M_4 was 11.3% higher than in M_1. The commercial cane sugar percentage was 4.5% to 9.0% higher than the control in all K-fertilizer treatments at 10 months after harvesting the seed crop; two months later there was no significant difference between CCS in M_1 and M_2, while M_3 (+7.0%) and M_4 (+10.0%) were higher than M_1. In the weight-per-area, CCS data, M_3 (+20.7%) and M_4 (+29.3%) were higher than the M_1 control at 10 months after harvesting the seed crop. Two months later, all K-fertilized treatments were higher than the control, by 10.5% (M_2), 23.1% (M_3), and 32.3% (M_4). There were no significant differences in purity between any fertilizer treatments at either sampling interval (Tables 6 and 7).

Table 6. Average sugarcane quality parameters 10 months after harvesting the seed crop unde irrigation and potassium treatments.

Treatments	Average Sugarcane Quality Parameters 10 Months After Harvesting					
	Brix (°)	Pol (%)	Purity (%)	CCS (%)	Extraction (%)	CCS (Tonnes ha^{-1})
			Irrigation treatment			
I_1	18.8 ^{a}	17.2 ^{a}	91.7 ^{a}	12.1 ^{a}	53.5 ^{a}	6.8 ^{a}
I_0	18.0 ^{a}	16.5 ^{b}	89.7 ^{a}	11.6 ^{a}	52.8 ^{a}	6.4 ^{a}
F-test ($p \leq 0.05$)	SS	**	SS	SS	SS	SS
CV (%)	4.6	1.0	4.6	3.4	6.7	7.4
			Potassium fertilizer treatment			
M_1	17.5 ^{a}	16.3 ^{c}	89.2 ^{a}	11.1 ^{c}	49.6 ^{b}	5.8 ^{b}
M_2	18.2 ^{a}	16.6 ^{b}	91.4 ^{a}	11.6 ^{b}	53.2 ^{ab}	6.3 ^{b}
M_3	18.7 ^{a}	17.1 ^{a}	91.3 ^{a}	12.0 ^{ab}	55.0 ^{a}	7.0 ^{a}
M_4	19.1 ^{a}	17.3 ^{a}	90.8 ^{a}	12.1 ^{a}	56.7 ^{a}	7.5 ^{a}
F-test ($p \leq 0.05$)	SS	**	SS	**	**	**
CV (%)	7.7	2.2	5.2	2.4	6.5	7.3
I × M	SS	SS	SS	SS	SS	SS

CCS, commercial cane sugar; extraction, extractable sugar percentage; main plot treatments are I_1 (fully irrigatec and I_0 (water-stressed); subplot treatments are M_1 (0 kg K ha^{-1}), M_2 (67 kg K ha^{-1}), M_3 (133 kg K ha^{-1} and M_4 (200 kg K ha^{-1}); CV, coefficient of variation; ** denotes significance at $p \leq 0.05$; SS, statistically simila The superscript similar letter within a continuous column indicates no statistical difference while different letter denote significant differences at $p \leq 0.05$.

Table 7. Average sugarcane quality parameters 12 months after harvesting the seed crop unde irrigation and potassium treatments.

Treatments	Average Sugarcane Quality Parameters 12 Months After Harvesting					
	Brix(°)	Pol (%)	Purity (%)	CCS (%)	Extraction (%)	CCS (Tonnes ha^{-1})
			Irrigation treatment			
I_1	21.3 ^{a}	19.6 ^{a}	92.1 ^{a}	13.8 ^{a}	58.6 ^{a}	7.9 ^{a}
I_0	20.6 ^{a}	18.9 ^{a}	91.9 ^{a}	13.3 ^{a}	57.0 ^{a}	7.3 ^{a}
F-test ($p \leq 0.05$)	SS	SS	SS	SS	SS	SS
CV (%)	7.3	9.0	1.6	9.6	7.0	12.3
			Potassium fertilizer treatment			
M_1	19.9 ^{c}	18.3 ^{c}	91.9 ^{a}	12.9 ^{b}	54.8 ^{b}	6.5 ^{c}
M_2	20.8 ^{b}	19.1 ^{b}	92.2 ^{a}	13.5 ^{ab}	56.6 ^{b}	7.2 ^{b}
M_3	21.4 ^{ab}	19.6 ^{ab}	91.5 ^{a}	13.8 ^{a}	58.8 ^{ab}	8.0 ^{a}
M_4	21.7 ^{a}	20.1 ^{a}	92.3 ^{a}	14.2 ^{a}	61.0 ^{a}	8.6 ^{a}
F-test ($p \leq 0.05$)	**	**	SS	**	**	**
CV (%)	3.1	2.4	3.4	3.5	6.0	7.3
I × M	SS	SS	SS	1.52	SS	SS

CCS, commercial cane sugar; extraction, extractable sugar percentage; main plot treatments are I_1 (fully irrigatec and I_0 (water-stressed); subplot treatments are M_1 (0 kg K ha^{-1}), M_2 (67 kg K ha^{-1}), M_3 (133 kg K ha^{-1} and M_4 (200 kg K ha^{-1}); CV, coefficient of variation; ** denotes significance at $p \leq 0.05$; SS, statistically simila The superscript similar letter within a continuous column indicates no statistical difference while different letter denote significant differences at $p \leq 0.05$.

3.4. Correlations between Quality Parameters

Ten months after harvesting the seed crop, Brix was moderately positively correlate with pol and the extractable sugar percentage, weakly positively correlated with bot commercial cane sugar values, and moderately negatively correlated with purity (Table 8

Table 8. Correlations between sugarcane quality parameters 10 and 12 months after harvesting the seed crop.

	10 Months after Harvesting the Seed Crop					
	Brix (o)	**Pol (%)**	**Purity (%)**	**CCS (%)**	**Extraction (%)**	**CCS (Tonnes ha^{-1})**
Brix (o)	1	0.6	−0.6	0.2	0.4	0.1
Pol (%)	0.6	1	0.2	0.9	0.5	0.5
Purity (%)	−0.6	0.2	1	0.6	0.1	0.4
CCS (%)	0.2	0.9	0.6	1	0.5	0.6
Extraction (%)	0.4	0.5	0.1	0.5	1	0.5
CCS (tonnes ha^{-1})	0.1	0.5	0.4	0.6	0.5	1
	12 Months after Harvesting the Seed Crop					
	Brix (o)	**Pol (%)**	**Purity (%)**	**CCS (%)**	**Extraction (%)**	**CCS (Tonnes ha^{-1})**
Brix (o)	1	0.8	−0.1	0.7	0.3	0.7
Pol (%)	0.8	1	0.5	1.0	0.2	0.8
Purity (%)	−0.1	0.5	1	0.6	−0.2	0.3
CCS (%)	0.7	1.0	0.6	1	0.1	0.8
Extraction (%)	0.3	0.2	−0.2	0.1	1	0.4
CCS (tonnes ha^{-1})	0.7	0.8	0.3	0.8	0.4	1

CCS, commercial cane sugar; extraction, extractable sugar percentage.

Strong positive correlations were observed between pol and the percentage CCS, with moderate positive correlations between pol and the extractable sugar percentage and the weight-per-area CCS, and a weak positive correlation between pol and purity. Moderate positive correlations were observed between purity and both CCS values and a weak positive correlation between purity and the extractable sugar percentage. The percentage CCS was associated with moderate positive correlations with both the extractable sugar percentage and the weight-per-area CCS, while the extractable sugar percentage was moderately positively correlated with the weight-per-area CCS.

Two months later, correlations between Brix and other parameters had become more positive: strong positive correlations were observed with pol and both CCS values, and a weak positive correlation was observed between pol and the extractable sugar percentage, while the correlation between Brix and purity was weakly negative (Table 8).

Correlations between pol and other parameters had also become more positive, except the correlation between pol and the extractable sugar percentage, which went from moderately to weakly positively correlated. There was no change in the correlation between purity and the percentage CCS, while the correlations between purity and extractable sugar percentage and between purity and weight-per-area CCS went from weakly positive to weakly negative and from moderately positive to moderately negative, respectively. The correlation between the percentage CCS and the extractable sugar percentage changed from moderately to weakly positive, while that between the percentage CCS and the weight-per-area CCS changed from moderately to strongly positive. The correlation between the extractable sugar percentage and the weight-per-area CCS did not change significantly between the sampling intervals.

3.5. Economic Analysis

Higher economic benefits were achieved under the fully irrigated treatments than under those with water stress (Table 9).

As well, yields increased with increasing K-fertilizer application. The highest yields were achieved in the I_1M_4 treatment; these were 25.5% higher than those of the I_1M_1 treatment. Similarly, yields in the I_0M_4 treatment were 14.2% higher than those of the I_0M_1 treatment. Increasing fertilizer resulted in increased income: the income achieved in I_1M_2 and in I_0M_2 was 9145 and 7965 INR ha^{-1} more than in the I_1M_1 and I_0M_1 treatments, respectively; however, at the maximum K-fertilizer rate, additional income was 38,350 INR ha^{-1} in I_1M_4 (above I_1M_1) and 21,240 INR ha^{-1} in I_0M_4 (above I_0M_1). The benefit-to-cost ratios reflected these data, and the highest B:C (10.1) was achieved in the fully irrigated treatment with the highest K-fertilizer applied (I_1M_4). The lowest B:C (6.3) was achieved in the water-stressed treatment with the lowest K-fertilizer applied (I_oM_2).

Table 9. Benefit-to-cost ratio of the ratoon crop under irrigation and potassium treatments.

Treatments	Fertilizer Cost (Rs ha^{-1})	Recorded Yield (Tonnes ha^{-1})	Reported Response	Additional Income due to Applied K (Rs ha^{-1})	Benefit-Cost Ratio	Overall Trend
I_1M_1	0	51.0	–	–	–	
I_1M_2	1273	54.1	3.1	9145	7.18	3.36
I_1M_3	2546	59.4	8.4	24,780	9.73	4.20
I_1M_4	3800	64.0	13.0	38,350	10.09	3.92
I_0M_1	0	50.7	–	–	–	
I_0M_2	1273	53.4	2.7	7965	6.26	
I_0M_3	2546	56.8	6.1	17,995	7.07	
I_0M_4	3800	57.9	7.2	21,240	5.59	

Change in cane yield is the change under different fertilizer treatments with irrigation treatment held constant.; I_1 is the fully irrigated treatment and I_0 the water-stressed treatment; the fertilizer treatments are M_1 (0 kg K ha^{-1}), M_2 (67 kg K ha^{-1}), M_3 (133 kg K ha^{-1}), and M_4 (200 kg K ha^{-1}); the cost of K fertilizer was 19,000 INR t^{-1}; sugarcane price was 2950 INR t^{-1}.

4. Discussion

4.1. Ratoon Sugarcane Performance under Irrigation

I_1 and I_0 treatment plots received a total of 13 and 10 irrigations, respectively, each with a depth of 50 mm. Thus, water stress equivalent to the lack of 150 mm irrigation water was expected in I_0 treatments; however, this stress was reduced due to receipt of 822.4 mm rainfall (Figure 1) during the experimental period, which largely coincided with the skipped irrigations. It is likely that differences between I_1 and I_0 treatments would have been stronger without this unforeseen rainfall.

Under fully irrigated conditions (all I_1 treatments), sugarcane growth parameters were improved, albeit not significantly different from the measurements observed under water-stressed conditions (all I_0 treatments; Tables 1–3). This may be a result of improved moisture availability [47,48], N use efficiency [49], significantly lower incidence of both stalk borer and top borer in I_1 plots (Table 5), all of which contribute to improving cane growth [50–52]. Under mild water stress, ratooned sugarcane has insect-pests incidence jumped while decreases are observed in stomatal conductance, transpiration rate, internal CO_2 concentration, and photosynthetic rate [53,54]. Water shortages result in cane yield reductions of up to 60% [55–57]. Sugarcane is most susceptible to water stress throughout the tillering and stem elongation phases [58,59], with stem and leaf growth being the most affected [55]. The physical responses to water stress in sugarcane are most commonly leaf rolling, stomatal closure, restriction of stalk and leaf growth, leaf senescence, and reduced leaf area [60]. Furthermore, both cell division and cell elongation are disrupted by water stress [59], with stem and leaf elongation being the most severely affected growth processes [61,62].

Irrigation did not affect the incidence of early shoot borer; however, stalk borer and top borer were observed in significantly higher numbers under water stress conditions (Table 5). This may be a consequence of poor nutrient movement from the leaves to other plant parts [13,14,53].

Under the fully irrigated conditions (I_1 treatments), ratoon sugarcane quality metrics at both 10 and 12 months after harvesting the seed crop were all better than metrics under the water-stressed conditions (I_0 treatments); albeit, the differences were not statistically significant (Tables 6 and 7). These trends may be the result of irrigation which improved metabolic and physiological activities, nutrient uptake and movement within the sugarcane plant from leaves, and higher fertilizer use efficiency [13,16,40,54–57,59]. At 12 months of ratoon canes, Brix relations with other quality parameters improved while remaining negative with purity (Table 8).

4.2. Ratoon Sugarcane Performance under Potassium Fertilizer

The M_3 treatment, with 133 kg K ha^{-1} performed better than any other K treatment in terms of shoot resprouting and other sugarcane performance metrics (Tables 1–4). Of the plant growth metrics recorded, treatments with both lower (i.e., M_2, 67 kg K ha^{-1}) and higher (i.e., M_4, 200 kg K ha^{-1}) rates of K fertilizer did not achieve as well as those recorded in the M_3 treatment. This may be a result of improved sugarcane metabolism [17,18,56], recorded significantly lower incidence of insect-pests (Table 5) which are further responsible for poor performance of canes in M_4 plots, better enzyme activation [19,21,58], carbohydrate transport [61], balancing of hormones and auxin levels [54], and sugarcane root growth and development [11,15,56,62]. Of the three insect pests studied, only the incidence of the top borer was significantly lower in M_3 as compared to M4 plots affected by the potassium fertilizer rate. Of the necessary plant nutrients, K is required in higher quantities. The performance of canes growing in K-deficient soils will be adversely affected by little or no K fertilizer [62]. Sugarcane productivity is influenced by the inherent capacity of the soil to supply K in the soil solution [63]. Consequently, K is a crucial element in achieving sustainable ratoon sugarcane production [64], as it activates photosynthesis, protein synthesis, starch production, and protein and sugar translocation [46,65]. The transfer of photosynthates in sugarcane is significantly reduced when K is in deficit [22,27,62]. Sugarcane crops react significantly to K fertilization only in soils with low available K [30].

Potassium deficiency reduces sugarcane growth, yields, and quality, while all are improved by applying sustainable fertilizer K to deficient soils [47]. Sugarcane responds to K fertilizers by increasing cane yield without changing the sucrose concentration in the cane [30]. In ratooned sugarcane, Shukla et al. [12] reported the following effects of K fertigation (66 kg K ha^{-1} administered with irrigation water): (i) enhanced dry matter accumulation at all development stages, (ii) increased the number of sprouted buds in ratoon cane stubble, and (iii) higher numbers of millable canes as a result of robust tillers generated in the ratooned cane. Moisture stress reduced cane yield when K was inadequate, while moisture stress had no effect on yield when sufficient K (above 133 kg K ha^{-1}) was supplied [65].

K-fertilization in K-deficient soils improves the transportation of nutrients from the leaves to the entire plant, resulting in comparatively fewer sweat leaves which are not preferred by sucking insect pests. This may explain why incidences of the major insect pests in stalk borer, early shoot borer, and top borer was reduced in the M_3 treatment (Table 5).

At both 10 and 12 months after harvesting the seed crop, higher K fertilizer application rates improved sugarcane quality parameters relative to the M_1 control (0 kg K ha^{-1} applied: Tables 6 and 7). The highest sugarcane juice quality was observed in the M_3 treatment, with 133 kg K ha^{-1} applied. This may be because the addition of K fertilizer improves sugarcane root growth and development (by improving input use efficiency), which might be due to translocation of photosynthates [22–27], which made the leaves bitter and reduced insect-pest incidence [13,14,62–64]. Further, K plays a key role in regulating stomatal openings through which water transpires from the plant to the atmosphere, thereby regulating transpiration losses under water stress [49].

Overall, the M_3 sub-treatment 133 kg K ha^{-1} performed best in terms of ratoon growth, and sugarcane production and quality, particularly under water stress conditions. The incidence of insect pests was also lowest in the M_3 treatment as compared to the other plots [15]. In general, in northern India, all sugarcane leaves are removed from the field prior to establishing the next crop: little of the K taken up by the plant is available to be returned to the soil after harvest. The importance of sufficient K-fertilizer application in sugarcane production on soils inherently low in K has been demonstrated here.

5. Conclusions

This experimental research has demonstrated that ratoon sugarcane performanc in north India is somewhat affected by irrigation and potassium treatments. Unde water-stress conditions, a trend for reduced ratoon productivity was observed, althoug this was not statistically significant. Relative to control treatments with no K fertilize adding K has elsewhere been reported to improve plant growth; however, in this experi ment, no significant differences in average sugarcane height, diameter, or internodes pe plant were observed in the ratoon crop. Adding K fertilizer improved sugarcane qualit (e.g., measured in terms of Brix, pol, purity, extractable sugar percentage and commercia cane sugar) relative to a baseline with no K fertilizer. Significantly higher sugarcane qualit and reductions in key insect pests were observed in the treatment where 133 kg K ha^- was applied, in both irrigated and water-stressed plots. Further research to extend thes experimental results and to examine, in more detail, relationships between key qualit parameters such as pol and commercial cane sugar variables should be conducted to opti mise ratoon quality and sugar recovery rate. We recommend that in the K-deficient soils o northern India, applications of 133 kg K ha^{-1} should be standard, regardless of irrigatio application, to improve ratoon sugarcane growth, yield and quality, and ultimately t enable smallholder farmers to improve their livelihoods through more sustainable an climate-smart sugarcane production.

Author Contributions: Conceptualization, R.B., J.S. and A.H.; methodology and visualization, R.B R.S.M. and J.S.; software, R.B., R.S.M. and A.H.; validation, R.B. and J.S.; formal analysis, R.B., J.S R.S.M. and A.H.; investigation, R.B. and J.S.; resources, R.B.; data curation, R.B., J.S., R.S.M. and A.H writing—original draft preparation, R.B., J.S., R.S.M., A.M.L. and A.H.; writing—review and editin A.G., W.F.A., A.H., R.S.M. and A.M.L.; supervision and project administration, R.B., A.H., A.G. an W.F.A.; funding acquisition, R.B., A.H., A.G. and W.F.A. All authors have read and agreed to th published version of the manuscript in Agronomy.

Funding: This research work was supported by the Director of the Regional Research Statior Gurdaspur Potassium Research Institute of India, and International Potassium Institute, Switzerlanc for providing necessary support for carrying out these cane ratoon experiments during 2020–2021 an by the Taif University Researchers for funding this research with Project number (TURSP-2020/53 Taif University, Taif, Saudi Arabia.

Institutional Review Board Statement: Not applicable.

Informed Consent Statement: Not applicable.

Data Availability Statement: Most of the data are available in all Tables and Figures of the manuscript

Acknowledgments: The authors want to acknowledge the Director, Regional Research Statior Gurdaspur Potassium Research Institute of India, and International Potassium Institute, Switzerlan for providing necessary support for carrying out these cane ratoon experiments during 2020–202 The authors also appreciate Taif University Researchers Supporting Project number (TURSP-2020/53 Taif University, Taif, Saudi Arabia.

Conflicts of Interest: The authors would hereby declare that there is no conflict of interest for the articl

References

1. Bhatt, R.; Singh, P.; Hussain, A.; Tamsina, J. Rice-wheat system in the north-west Indo-Gangetic Plains of South Asia: Issues an technological interventions for increasing productivity and sustainability. *Paddy Water Environ.* **2021**, 1–21, Epub ahead of prin [CrossRef]
2. Bhatt, R.; Arora, S. Soil matric potential based irrigation using tensiometers for conserving irrigation water. *Curr. Sci.* **2021**, *12* 197–200.
3. Bhatt, R. Resources Management for Sustainable Sugarcane Production. In *Resources Use Efficiency in Agriculture*; Kumar, S Meena, S.R., Jhariya, K.M., Eds.; Springer: Singapore, 2020; pp. 650–685.
4. Nishiyama, R.; Watanabe, Y.; Leyva-Gonzalez, M.A.; VanHa, C.; Fujita, Y.; Tanaka, M.; Sekid, M.; Shinozakie, K.Y.; Shinozakif, K Estrellab, K.H.; et al. Arabidopsis AHP2, AHP3, and AHP5histidinephos-photransfer proteins function as redundant negativ regulators of drought stress response. *Proc. Natl. Acad. Sci. USA* **2013**, *110*, 4840–4845. [CrossRef] [PubMed]

5. Ha, C.V.; Leyva-Gonzalez, M.A.; Osakabe, Y.; Tran, U.T.; Nishiyama, R.; Watanabe, Y.; Tanakae, M.; Sekie, M.; Yamaguchif, S.; Dong, N.V.; et al. Positive regulatory role of strigolactone in plant responses to drought and salt stress. *Proc. Natl. Acad. Sci. USA* **2014**, *111*, 581–856. [CrossRef] [PubMed]
6. Osakabe, K.; Osakabe, Y. Plantlightstress. In *Encyclopaedia of Life Sciences*; Robinson, S.A., Ed.; Nature Publishing Group: London, UK, 2012.
7. Osakabe, Y.; Arinaga, N.; Umezawa, T.; Katsura, S.; Nagamachi, K.; Tanaka, H.; Ohiraki, H.; Yamada, K.; Seo, A.; Abo, M.; et al. Osmotic stress responses and plant growthc ontrolled by potassium transporters in *Arabidopsis*. *Plant Cell* **2013**, *25*, 609–624. [CrossRef]
8. Osakabe, Y.; Shinozaki, Y.; Shinozaki, K.; PhanTran, L.S. Sensing the environment: Key roles of membrane-localized kinases in plant perception and response to abiotic stress. *J. Exp. Bot.* **2013**, *64*, 445–458. [CrossRef]
9. Negi, J.; Matsuda, O.; Nagasawa, T.; Oba, Y.; Takahashi, H.; Kawai-Yamada, M.; Uchimiya, H.; Hashimoto, M.; Iba, K. CO_2 regulator SLAC1 and its homologues are essential for anion homeostasis in plant cells. *Nature* **2008**, *452*, 483–486. [CrossRef]
10. Vahisalu, T.; Kollist, H.; Wang, Y.F.; Nishimura, N.; Chan, W.Y.; Valerio, G. SLAC1 is required for plant guard cell S-typeanionchannel function in stomatal signalling. *Nature* **2008**, *452*, 487–491. [CrossRef]
11. Priya, S.R.K.; Bajpai, P.K.; Suresh, K.K. Stochastic models for sugarcane yield forecasting. *Ind. J. Sugarcane Technol.* **2015**, *30*, 1–5.
12. Shukla, S.K.; Yadav, R.L.; Singh, P.N.; Singh, I. Potassium nutrition for improving stubble bud sprouting, dry matter partitioning, nutrient uptake, and winter initiated sugarcane (Saccharum spp. hybrid complex) ratoon yield. *Eur. J. Agron.* **2009**, *30*, 27–33. [CrossRef]
13. Bhatt, R.; Oliveira, M.W.; da Silva, V.S.G. Sugarcane nutrition for food and environmental security. *Braz. J. Dev.* **2021**, *6*, 64431–64467. [CrossRef]
14. PAU. *Package of Practices for crops of Punjab-Kharif*; Punjab Agricultural University: Ludhiana, India, 2021; pp. 55–66.
15. Bhatt, R.; Singh, P. Sugarcane response to irrigation and potash levels in subtropics. *Agric. Res. J. PAU* **2021**, in press.
16. Bhatt, R.; Sharma, M. Potassium Scenario—A case study in the Kapurthala district of Punjab, India. *Agric. Res. J.* **2011**, *48*, 24–27.
17. Cakmak, I. The role of potassium in alleviating detrimental effects of abiotic stresses in plants. *J. Plant Nutr. Soil Sci.* **2005**, *168*, 521–530. [CrossRef]
18. Wang, M.; Zheng, Q.; Shen, Q.; Guo, S. The critical role of potassium in plant stress response. *Inter. J. Mol. Sci.* **2013**, *14*, 7370–7390. [CrossRef]
19. Hawkesford, M.; Horst, W.; Kichey, T.; Lambers, H.; Schjoerring, J.; Skrumsagermoller, I.; White, P. Function of macronutrients. In *Marschner's Mineral Nutrition of Higher Plants*; Marschner, P., Ed.; Academic Press: London, UK, 2012; pp. 135–189.
20. Mengel, K. Potassium. In *Handbook of Plant Nutrition*; Barker, A.V., Pilbeam, D.J., Eds.; CRC Press: Boca Raton, FL, USA, 2007; pp. 91–120.
21. Haeder, H.E.; Mengel, K. Translocation and respiration of assimilates in tomato plants as influenced by K nutrition. *Z. Pflanz. Bodenkd.* **1972**, *131*, 139–148. [CrossRef]
22. Mengel, K.; Viro, M. Effect of Potassium Supply on the Transport of Photosynthates to the Fruits of Tomatoes (*Lycopersicon esculentum*). *Physiol. Planta* **1974**, *30*, 295–300. [CrossRef]
23. Hartt, C.E. Effect of potassium deficiency upon translocation of ^{14}C in attached blades and entire plants of sugarcane. *Plant Physiol.* **1969**, *44*, 1461–1469. [CrossRef]
24. Hartt, C.E. Effect of potassium deficiency upon translocation of ^{14}C in detached blades of sugarcane. *Plant Physiol.* **1970**, *45*, 183–187. [CrossRef]
25. Brunt, V.; Sultenfuss, J.H. Better crops with plant food. *Potassium Funct. Potassium* **1998**, *82*, 4–5.
26. Berg, W.K.; Cunningham, S.M.; Brouder, S.M.; Joern, B.C.; Johnson, K.D.; Volence, J.J. Influence of phosphorus and potassium on alfalfa yield, taproot C and N pools, and transcript levels of key genes after defoliation. *Crop Sci.* **2009**, *49*, 974–982. [CrossRef]
27. Filho, J.O. Potassium nutrition of sugarcane. In *Potassium in Agriculture*; Munson, R.D., Ed.; American Society of Agronomy, Crop Science Society of America, Soil Science Society of America: Madison, WI, USA, 1985; pp. 1045–1062.
28. Kwong, K.F. The effects of potassium on growth, development, yield, and quality of sugarcane. In *Potassium for Sustainable Crop Production and Food Security, Proccedings of the First National Potash Symposium, Dar es Salaam, Tanzania, 28–29 July 2015*; International Potash Institute: Zug, Switzerland, 2002; pp. 430–444.
29. Kumar, A.; Babar, L.; Mohan, N.; Bansal, S.K. Effect of Potassium Application on Yield, Nutrient Uptake and Quality of Sugarcane and Soil Health. *Ind. J. Fert.* **2019**, *15*, 782–786.
30. Marshner, H. *Mineral Nutrition of Higher Plants*, 2nd ed.; Academic Press: Cambridge, MA, USA, 1995; p. 889.
31. Sidhu, P.S. Mineralogy of potassium in soils of Punjab, Haryana, Himachal Pradesh, and Jammu and Kashmir. In *Mineralogy of Soil Potassium*; Potash Research Institute of India: Gurgaon, India, 1982; pp. 6–14.
32. Grewal, J.S.; Kanwar, J.S. *Potassium and Ammonium Fixation in Indian Soils (A Review)*; Indian Council of Agricultural Research: New Delhi, India, 1973; p. 75.
33. Vasudeo, R.; Naidu, R.; Lakshmikantham, M. Ratooning sugarcane in Madras. *Madras Agric. J.* **1946**, *4*, 3–12.
34. Verma, R.S. *Sugarcane Ratoon Management*; International Book Distributing Company: Lucknow, India, 2002; p. 102.
35. Choudhary, H.R.; Singh, R.K. Effect of sequential application of herbicides on weeds and productivity of spring-planted sugarcane (*Saccharum officinarum* L.). *Bioscan* **2016**, *11*, 687–690.

36. Sundara, B.; Tripathi, B.K. Available N changes and N balance under multi-ratooning of sugarcane varieties in a tropical vertisol. In Proceedings of the ISSCT XXIII (II), New Delhi, India, 22–26 February 1999; pp. 80–88.
37. Dissanayake, N.; Hoy, J.W. Organic material soil amendment effects on root rot and sugarcane growth and characterization of the materials. *Plant Dis.* **1999**, *83*, 1039–1046. [CrossRef]
38. Kanwar, R.; Kaur, S.H. Further studies on improving the productivity of stubble crops in low temperate area of north India. In Proceedings of the National Seminar on Ratoon Management, Lucknow, India, 14–15 March 1981; pp. 27–34.
39. Verma, R.S.; Yadav, R.L. Intercropping in sugarcane for improving stubble sprouting under low temperature conditions in subtropical India. *Bhartiya Sugar* **1988**, *13*, 45–48.
40. Jagtap, S.M.; Jadhav, M.B.; Kulkarm, R.V. Effect of levels of NPK on yield and quality of sugarcane (cv. Co. 7527). *Ind. Sugar* **2006**, *56*, 35–40.
41. Bhatt, R.; Singh, P.; Kumar, R.; Kashyp, L.; Bansal, S.K. Sugarcane growth, yield and quality parameters as affected by different irrigation and potash levels in Punjab, India. In Proceedings of the International Conference Canecon-2021 Entitled, "Sugarcane for Sugar and Beyond", Coimbatore, India, 19–22 June 2021; Palaniswami, C., Hemaprabha, G., Viswanathan, R., Karuppaiyan, R., Mohanraj, K., Mahadeva Swamy, H.K., Ram, B., Eds.; pp. 336–340, ISBN 978-93-85267-30-7.
42. Hunsigi, G. *Sugarcane in Agriculture and Industry*; Prism Books Pvt. Ltd.: Bangalore, India, 2001; pp. 125–138, 207.
43. Wood, A.W.; Schroeder, B.L. Potassium: A critical role in sugarcane production, particularly in drought conditions. In Proceedings of the Australian Society of Sugarcane Technologists, Brisbane, Australia, 4–7 May 2004; Available online: http://www.cabdirect.org/abstracts/20043079912.html (accessed on 7 May 2020).
44. Wood, R.A. The roles of nitrogen, phosphorus and potassium in the production of sugarcane in South Africa. *Fertil. Res.* **1990**, *26*, 87–98. [CrossRef]
45. Cotlear, C.B.G. Sugarcane Juice Extraction and Preservation, and Long-Term Lime Pretreatment of Bagasse. Ph.D. Dissertation, Office of Graduate Studies of Texas A&M University, College Station, TX, USA, 2004.
46. Bhatt, R.; Singh, P.; Kumar, R. *Assessment of Potash in Improving Yield and Quality of Sugarcane under Water Stressed and Unstressed Conditions*; Final project report submitted to Indian Potash Limited (IPL): Gurgaon, India, 2021.
47. Ghaffar, A.; Saleem, M.F.; Fiaz, N.; Nadeem, M.A.; Wains, G.M. Yield and quality of sugarcane as influenced by different doses of potash and its time of application. *Pak. J. Agric. Sci.* **2013**, *50*, 345–350.
48. Bansal, S.K.; Imas, P.; Nachmansohn, J. The Impact of Potassium Fertilization on Sugarcane Yields: A Comprehensive Experiment of Pairwise Demonstration Plots in Uttar Pradesh, India. *e-ifc. Int. Potash Inst.* **2018**, *54*, 13–20. Available online: https://www.ipipotash.org/uploads/udocs/e-ifc-54-sep-2018-india-sugarcane.pdf (accessed on 9 October 2021).
49. Meade, G.P.; Chen, J.C.P. *Can Sugar Handbook*, 10th ed.; Wiley-Inter Science Publication: New York, NY, USA, 1977; p. 405.
50. Waraich, E.A.; Ahmad, R.; Hur, R.G.M.; Ahmad, A.; Mahmood, N. Response of foliar application of KNO3 on yield, yield components and lint quality of cotton (Gossypium hirsutum L.). *Afr. J. Agric. Res.* **2011**, *6*, 5457–5463.
51. Medeiros, D.B.; da Silva, E.C.; Mansur Custodio Nogueira, R.J.; Teixeira, M.M.; Buckeridge, M.S. Physiological limitations in two sugarcane varieties under water suppression and after recovering. *Theor. Exp. Plant Physiol.* **2013**, *25*, 213–222. [CrossRef]
52. Basnayake, J.; Jackson, P.A.N.; Inman-Bamber, G.; Lakshmanan, P. Sugarcane for water-limited environments. Variation in stomatal conductance and its genetic correlation with crop productivity. *J. Exp. Bot.* **2015**, *66*, 3945–3958. [CrossRef] [PubMed]
53. Lakshmanan, P.; Robinson, N. Stress physiology: Abiotic stresses. In *Sugarcane: Physiology, Biochemistry, and Functional Biology*; Moore, P.H., Botha, F.C., Eds.; John Wiley & Sons, Inc.: Chichester, UK, 2014; pp. 411–434.
54. Basnayake, J.; Jackson, P.A.N.; Inman-Bamber, G.; Lakshmanan, P. Sugarcane for water-limited environments. Genetic variation in cane yield and sugar content in response to water stress. *J. Exp. Bot.* **2012**, *63*, 6023–6033. [CrossRef]
55. Gentile, A.; Dias, L.I.; Mattos, R.S.; Ferreira, T.H.; Menossi, M. MicroRNAs and drought responses in sugarcane. *Front. Plant Sci.* **2015**, *6*, 58. [CrossRef]
56. Inman-Bamber, N.; Smith, D. Water relations in sugarcane and response to water deficits. *Field Crops Res.* **2005**, *92*, 185–202. [CrossRef]
57. Machado, R.; Ribeiro, R.; Marchiori, P.; Machado, D.; Machado, E.; Landell, M. Biometric and physiological responses to water deficit in sugarcane at different phenological stages. *Pesqui. Agropecuária Bras.* **2009**, *44*, 1575–1582. [CrossRef]
58. Inman-Bamber, N.; Lakshmanan, P.; Park, S. Sugarcane for water-limited environments: Theoretical assessment of suitable traits. *Field Crops Res.* **2012**, *134*, 95–104. [CrossRef]
59. Inman-Bamber, N.; Bonnett, G.; Spillman, M.; Hewitt, M.; Jackson, J. Increasing sucrose accumulation in sugarcane by manipulating leaf extension and photosynthesis with irrigation. *Aust. J. Agric. Res.* **2008**, *59*, 13–26. [CrossRef]
60. Korndörfer, G.H.; Oliveira, L.A. O potássio na cultura da cana-de-açúcar. In *Potássio na Agricultura Brasileira*; Yamada, T., Roberts, T.L., Eds.; Esalq/USP: Piracicaba, Brazil, 2005.
61. Schultz, N.; Lima, E.; Pereira, M.G.; Zonta, E. Residual effects of nitrogen, potassium and vinasse, fertilization on cane plant and ratoon harvested with and without straw burning. *Rev. Bras. Ciência Solo* **2010**, *34*, 811–820. [CrossRef]
62. Weber, H.; Daros, E.; Zambon, J.L.C.; Ido, O.T.; Barela, J.D. Ratoon sugarcane productivity recovering with NPK fertilization. *Sci. Agrar.* **2002**, *2*, 73–77.
63. Mengel, K.; Seçer, M.; Koch, K. Potassium effect on protein formation and amino acid turnover in developing wheat grain. *Agron. J.* **1981**, *73*, 74–78. [CrossRef]

4. Ashraf, M.Y.; Hussain, F.; Akhter, J.; Gul, A.; Ross, M.; Ebert, G. Effect of different sources and rates of nitrogen and supra optimal level of potassium fertilization on growth, yield and nutrient uptake by sugarcane. Grown under saline conditions. *Pak. J. Bot.* **2008**, *40*, 1521–1531.
5. Olivot, T.; Dal'Col Lúcio, A. Metan: An R package for multi-environment trial analysis. *Methods Ecol. Evol.* **2020**, *11*, 783–789. [CrossRef]

agronomy

MDPI

Article

Assessing the Productivity, Quality and Profitability of Orange Fleshed Sweet Potatoes Grown in Riverbank of the Tista Floodplain Agro-Ecological Zone of Bangladesh

Abdullah-Al Mahmud [1], Md Jahangir Alam [1], Simon Heck [2], Wolfgang J. Grüneberg [3], Debashish Chanda [4], E. H. M. Shofiur Rahaman [4], Md Samim Hossain Molla [5], Md Mazharul Anwar [6], Md Al-Amin Hossain Talukder [5], Md Akkas Ali [7], Md Nurul Amin [8], Majid Alhomrani [9], Ahmed Gaber [10,*] and Akbar Hossain [11,*]

1 On-Farm-Research Division, Bangladesh Agricultural Research Institute, Gaibandha 5700, Bangladesh; mahmud.tcrc@gmail.com (A.-A.M.); jahangir.bari@gmail.com (M.J.A.)
2 International Potato Center (CIP), Africa Regional Office, ILRI Campus, P.O. Box 25171-00603, Nairobi 00100, Kenya; s.heck@cgiar.org
3 International Potato Center (CIP), Avenida La Molina 1895, Distrito de La Molina Lima 12, Lima 15023, Per w.gruneberg@cgiar.org
4 International Potato Center (CIP), Dhaka 1213, Bangladesh; d.chanda@cgiar.org (D.C.); E.RAHAMAN@CGIAR.ORG (E.H.M.S.R.)
5 On-Farm-Research Division, Bangladesh Agricultural Research Institute, Rangpur 5402, Bangladesh; samimmolla@yahoo.com (M.S.H.M.); alamintalukder@yahoo.com (M.A.-A.H.T.)
6 On-Farm-Research Division, Bangladesh Agricultural Research Institute, Rajshahi 6201, Bangladesh; anwar.sci.bari@gmail.com
7 On-Farm-Research Division, Bangladesh Agricultural Research Institute, Gazipur 1701, Bangladesh; akkasbari@gmail.com
8 Breeder Seed Production Center, Bangladesh Agricultural Research Institute, Panchagrah 5020, Banglades nurul01141@yahoo.com
9 Department of Clinical Laboratories Sciences, The Faculty of Applied Medical Sciences, Taif University, P.O. Box 11099, Taif 21944, Saudi Arabia; m.alhomrani@tu.edu.sa
10 Department of Biology, College of Science, Taif University, P.O. Box 11099, Taif 21944, Saudi Arabia
11 Department of Agronomy, Bangladesh Wheat and Maize Research Institute, Dinajpur 5200, Bangladesh
* Correspondence: a.gaber@tu.edu.sa (A.G.); akbarhossainwrc@gmail.com (A.H.)

Citation: Mahmud, A.-A.; Alam, M.J.; Heck, S.; Grüneberg, W.J.; Chanda, D.; Rahaman, E.H.M.S.; Molla, M.S.H.; Anwar, M.M.; Talukder, M.A.-A.H.; Ali, M.A.; et al. Assessing the Productivity, Quality and Profitability of Orange Fleshed Sweet Potatoes Grown in Riverbank of the Tista Floodplain Agro-Ecological Zone of Bangladesh. *Agronomy* **2021**, *11*, 2046. https://doi.org/10.3390/agronomy11102046

Academic Editors: Umberto Anastasi and Aurelio Scavo

Received: 5 September 2021
Accepted: 8 October 2021
Published: 12 October 2021

Abstract: Orange fleshed sweet potatoes (OFSP) are desirable for high productivity and profitabilit and their distribution to improve the nutrition of river bank inhabitants of Gaibandha and Rangpu districts of Bangladesh. In this context, a field trial was conducted in two riverbank-based farmer fields such as Saghata, Gaibandha, and Pirganj, Rangpur, particularly in the Active Tista Floodplai Agro-ecological Zone of Bangladesh. Four OFSP varieties were evaluated, i.e., G1: BARI SP-8; G BARI SP-12; G3: BARI SP-14; G4: BARI SP-15, along with one local cultivar as a control (Red ski with white flesh). Significant variations among the sweet potato genotypes were noted for a numbe of tuberous roots plant^{-1}, length of root diameter of roots, tuberous root weight plant^{-1}, root yiel (fresh), root yield (dry), beta-carotene yield, as well as energy output. Over the locations, BARI SP-1 produced about 73% higher root yield (32.00 t ha^{-1}) and it was like the BARI SP-8 (31.07 t ha^{-1} which produced about 68% higher yield in comparison with local cultivar (18.51 t ha^{-1}). Acros the location, BARI SP-8 performed better in root yield (31.89 t ha^{-1}) in Gaibandha, 69% superio to local cultivar, whereas BARI SP-12 performed better in Rangpur (33.66 t ha^{-1}), which was 86 greater than the local sweet potato cultivar. Considering the root dry yield production, BARI SP- produced the highest in the Gaibandha location after that Rangpur location. Further, BARI SP-14 ha wider adaptability and stability over the year and location depended on the AMMI model. The bet carotene yield (Vitamin-A precursor) ranged 336–2957 kg ha^{-1} among the OFSP varieties, wherea the highest (2957 kg ha^{-1}) carotene was recorded in BARI SP-14, similar to BARI SP-15 (2952 kg ha^{-1} but was much lower in BARI SP-8 and BARI SP-12. Moreover, BARI SP-8 and BARI SP-12 were als economically profitable in terms of gross margin (3233 and 3364 US$ ha^{-1}, respectively), net retur (3039 and 3170 US$ ha^{-1}, respectively) and BCR (3.21 and 3.31, respectively, vs. 1.91) due to highe returns with a similar production cost of the local cultivar. The results suggested that BARI SP-is economically profitable in the riverbank areas of Gaibandha, and BARI SP-12 is suitable for the riverbank areas of Rangpur.

Keywords: sweet potato; beta-carotene; yield; profitability

1. Introduction

Sweet potato, a perennial root crop belonging to the family of Convolvulaceae, has several flesh colors (white, yellow, cream, purple and orange) [1,2]. Among them, the orange, white and cream flesh sweet potatoes are commonly cultivated and consumed. It is the seventh most important crop in the world [3,4]. It is grown globally in more than a hundred countries with an average yield of 12.20 t ha^{-1} from 8.62 million ha of land [5]. The most commonly edible parts of the sweet potatoes are the tuberous roots, while the leaves are also important [1,2] and an important staple food across the Asian, African and Pacific region countries of the world. Sweet potato is also being used as cattle feed [1,6,7]. Sweet potato is considered as a healthy food having a low level of fat and protein, but rich in carbohydrates.

Orange-fleshed sweet potatoes (OFSP) are being considered as resilient crops due to its high carotenoid content (Precursor of Vitamin-A) and good yields, and also rich carbohydrates, vitamins and minerals [3,4] which can improve the nutrition of under-privileged farmers in numerous developing nations. Orange-fleshed sweet potatoes are also high-yielding with the capacity to generate more edible energy than wheat, rice or cassava per unit area [3]. Its root flesh and green leaves are great sources of antioxidants [8], minerals (Zn, K, Na, Mn, Ca, Mg and Fe), fiber and vitamin C [9].

Sweet potato, one of the preferred root crops due to the highest dry matter content for human consumption, out of which 70% of it is composed of starch [1,10]. A good sweet potato variety possesses a great amount of dry matter which is treated as an essential characteristic [11]. It can also be considered as one of the best meals assessed for long-time space travel, due to the fact of their nutritional attributes [12].

OFSP's are also quite excellent sources of vitamin-A [8,13] and its main pigments, especially β-carotene and carotenoids, which are closely associated with the improvement of the immune system of human beings, and reduces the risk of cardiovascular complexities, age-related macular degeneration, and cataract development [14]. So, beta-carotene enriched sweet potatoes could be used successfully in small-scale interventions in the riverbanks and rural areas to improve nutrition status and to combat vitamin-A deficiency-induced diseases.

In Bangladesh, the cultivation of sweet potatoes is concentrated in the riverbanks and riverine islands (called the Char area; formed from sedimentation). Around 6.5 million (around 4%) of the Bangladeshi people live in the riverine islands (Char areas) and most of them are marginalized. Poverty is the common to the riverine island people; some of them are vulnerable and they usually cultivate sweet potato local cultivars in their fallow lands where other crops are not cultivated. Cultivating OFSP in the riverine islands of northern Bangladesh have a reasonable benefit. The soils in the northern riverine islands/Char areas are sandy and sandy loam type where water scarcity is common for rice cultivation and also challenging to cultivate other cereals such as wheat and maize. Sweet potatoes are cultivated on various soils, although good drained medium-textured and light (sandy to sandy loam) soils with a pH of 4.5–7.0 [1,15] are preferred for better vegetative growth and root development. At present, sweet potato cultivation in Bangladesh is about 0.13 million hectares (Mha) with a production of 1.47 million tons (Mt) [5]. OFSP varieties are generally less drought tolerant than the white-fleshed (WF) cultivars [16] and the average yield of OFSP at farm level in the riverbanks and riverine islands is about 10–12 t ha^{-1} [5] where crops grown with less irrigation and minimum inputs compared to other crops grown;

while the potential yield can be as high as 35–40 t ha^{-1} [17]. There is significant potential to increase the yield of sweet potato by bridging the yield gap in these communities.

Genotypes and environmental interactions are associated with the performance of the varieties that show stability when cultivating in different environments and is essential for achieving new and improved genotypes [18]. The Additive Main effect and Multiplicative Interaction (AMMI) biplot is used to explain test location and genotype performance in test environments [19]. The AMMI biplot analysis is an important Genotype Environment Interaction (GEI) assessment strategy that helps plant breeders/agriculturists to identify and select higher performing genotypes in specific environments [20,21]. AMMI Stability Value (ASV) illustrates the distance of origin from the point of adjustment between the IPCA2 (Interaction Principal Component Axes for the environment) value versus the IPCA1 (Interaction Principal Component Axes for genotype) value of the AMMI model [22]. This has led to the need for adaptation and stability testing to obtain high quality and adaptive genotypes in different locations. The new superior varieties selected in the multi-environment test are expected to be as stable and uniform as possible, although they are born in different environments. Sweet potato is one of the main crops of the river island which reflect a significant source of nutrition and an attractive, important role in the upkeep of food security and increasing the profits of sweet potato growers [23,24]. It is commonly cultivated in the country, especially concentrated in the northern riverbanks and give more profits with fewer investments [25].

Life in the riverbanks and riverine islands is both unpredictable and insecure as they are facing major hazards such as flash flooding, riverbank erosion and cost of land. Numerous Char inhabitants fight to make or buy sufficient food to consume, and malnutrition and micronutrient shortcomings are widespread in these areas. Identification of promising varieties for riverbanks and riverine islands from the existing International Potato Center (CIP) bred sweet potato varieties may provide farmers with higher yields and help ensure food security in Bangladesh. Therefore, the International Potato Center (CIP) and Bangladesh Agriculture Research Institute (BARI) are introducing OFSP as a resilient and healthy food crop that can provide both economic opportunities and nutritional benefits to these Char farmers.

At present, OFSP is studied for its versatility and adaptability in diversified climatic conditions. The present study considered the performance of CIP-bred sweet potato varieties in the northern parts of Bangladesh and the impact on the economy of sweet potato cultivation in the Char area with the following objectives: (i) to assess the field performance and stability of OFSP varieties at field level in riverbanks and riverine islands; (ii) to calculate the cost and income of sweet potato cultivation at the farm level.

2. Materials and Methods

2.1. The Site, Season, Climatic Condition and Nutrient Status of the Experimental Field

The trial was carried out in two locations, namely Pirganj (25°23′ N and 89°18′ E) of Rangpur and Saghata (25°10′ N and 89°58′ E) of Gaibandha districts, Bangladesh (representing AEZ 2: Active Tista Floodplain agro-ecological zone of Bangladesh), during November–March (winter time) of 2018–2019 and 2019–2020 crop season.

The experimental sites were in a sub-tropical climate zone and characterized by little rainfall (42–53 mm) during the crop growing season (November–March) in the year. The monthly mean maximum temperature for the period of the sweet potato crop growth and development was mostly lower than the long-term averages in both locations, with some exception of 2018–2019 in Rangpur. On the other hand, the monthly mean minimum temperature was a little bit superior in Gaibandha than the long-term mean, but similar in Rangpur. The second crop year (2019–2020) was cooler in both the sites particularly in January, and the mean maximum temperature was about 1–3 °C lower than the 1st year cropping season (Figure 1a,b). The monthly average maximum temperature was 26–34 °C in Gaibandha and 22.8–30.2 °C in Rangpur and the average monthly minimum temperature was 12–18 °C in Gaibandha and 11.1–17.3 °C in Rangpur, respectively. January was the coolest month (average temperature vary between 11.1–12 °C) and March was the warmest month (average temperature vary between 30.2–34 °C) in both locations (Figure 1). The crop received a total of 29 and 19 mm rainfall in Gaibandha and 120 and 42 mm rainfall in Rangpur (Figure 1c,d). Before conducting the experiment in the field, pre-planted soil samples were collected to a depth of 15 cm in both locations and analyzed in the SRDI (Soil Resource Development Institute) laboratory. Soil properties were presented of the site shown in Table 1.

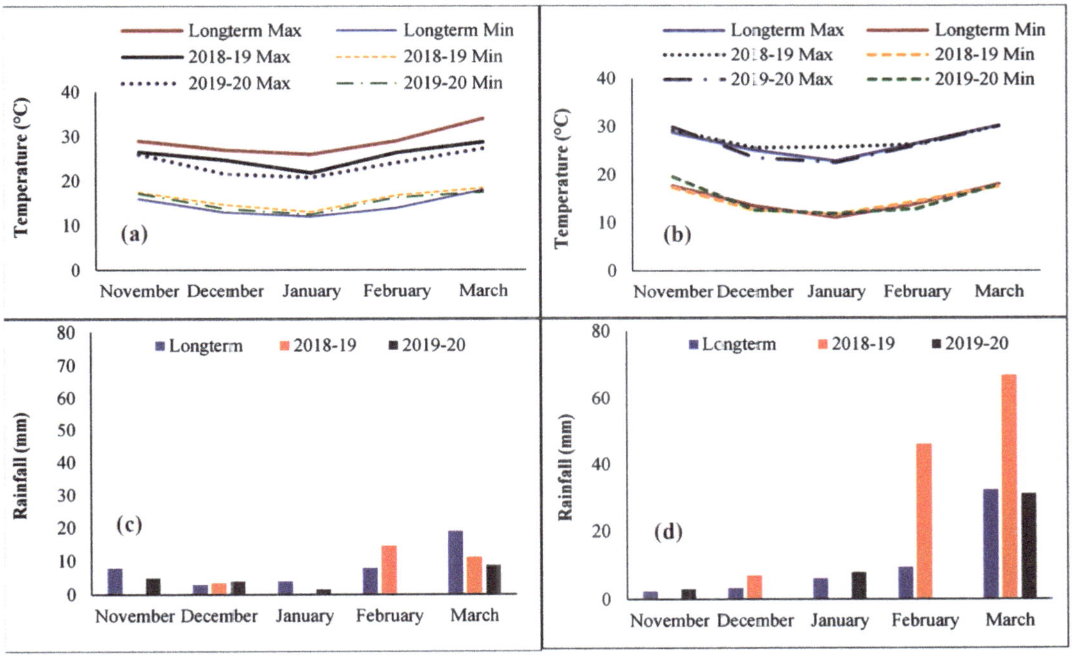

Figure 1. Monthly average maximum and minimum temperature of (**a**) Saghata, Gaibandha, and (**b**) Pirganj, Rangpur; (**c**) monthly mean rainfall (mm) of Saghata, Gaibandha, and (**d**) monthly mean rainfall (mm) of Pirganj, Rangpur, during two years compared to long-term (2000–2019) standards at Bangladesh Meteorological Department.

Table 1. Initial soil condition (0–15 cm) of OFSP trial plots at Saghata of Gaibandha and Pirganj of Rangpur in Bangladesh.

Locations	pH	OM (%)	Total N (%)	K	P	S	Zn	B
				meq/100 g	µg/g Soil			
Saghata, Gaibandha	6.45 Slightly Acidic	0.83 VL	0.05 L	0.14 M	9.68 L	15.2 M	0.24 VL	0.25 L
Pirganj, Rangpur	6.40 Slightly Acidic	0.27 VL	0.02 VL	0.26 M	30.10 VH	2.18 VL	0.45 VL	0.12 VL

Very low = VL; Low = L; Medium = M and Very high = VH.

2.2. Planting Materials, Design of the Experiment, and Crop Management

The trials were laid out at the farm level by following a randomized complete block design with six dispersed replications. Four BARI-released, vitamin-A enriched sweet potato cultivars were used, viz., BARI SP-8, BARI SP-12, BARI SP-14 and BARI SP-15, along with one local cultivar as a check (Table 2).

Table 2. Major Characteristics and year release of OFSP varieties used in the trial during both crop seasons.

Name of the Variety/Cultivar	Pedigree	Year of Release	Characteristics
BARI Mistialu-8 (BARI SP-8)	CIP-440025	2008	Skin color: Red, Flesh color: Yellow, Dry matter: 33.71 ± 1%, Beta-carotene: 1.08 mg/100 g FW *, Fe: 7.86 mg/kg, Zn: 14.76 mg/kg
BARI Mistialu-12 (BARI SP-12)	CIP-440001	2013	Skin color: Yellow, Flesh color: Orange, Dry matter: 22.04 ± 1%, Beta-carotene: 3.60 mg/100 g FW *, Fe: 14.76 mg/kg, Zn: 8.09 mg/kg
BARI Mistialu-14 (BARI SP-14)	CIP-441132	2017	Skin color: Light orange, Flesh color: Orange, Dry matter: 29.46 ± 1%, Beta-carotene: 10.10 mg/100 g FW *, Fe: 5.17 mg/kg, Zn: 6.47 mg/kg
BARI Mistialu-15 (BARI SP-15)	CIP-440267.2	2017	Skin color: Pink, Flesh color: Orange, Dry matter: 28.91 ± 1%, Beta-carotene: 10.39 mg/100 g FW *, Fe: 13.25 mg/kg, Zn: 6.47 mg/kg
Local Variety	Local cultivar	-	Skin color: Pink, Flesh color: White, Dry matter: 36.5 ± 1%, Beta-carotene: Trace/nil

* FW means fresh weight.

Sweet potato vines were planted between 1–10 November in both crop years ove the two locations, with the spacing 60 cm across row and 30 cm within row. Unit plo size ranged from 200 to 400 m^2 across the locations. The trial plot area was fertilize with 120 kg N ha^{-1} as urea, 30 kg P ha^{-1} as triple superphosphate (TSP), 60 kg K ha^{-} as muriate of potash (MoP), 15 kg S ha^{-1} as gypsum 4 kg Zn ha^{-1} as zinc sulfate (an 1 kg B ha^{-1} as boric acid. The 50% of the urea and MoP with the whole amount of TS gypsum, zinc sulfate and boric acid were applied during the final land preparation. Th remaining urea and MoP were applied at 35 DAP (days after planting). For good crop stan and higher root yield, irrigation was applied at 30, 45, 60, 75, 90, 115 DAP, maintainin two-third (6 cm) of the valley in both locations.

Some infestation of weevil occurred in the vines in both locations during vine prepara tion. The weevil infestation in the vines of potatoes was controlled by dipping the vines i the Ripcord (Cypermethrin) solution before planting and also were applied (at 60 DAP a well as earthing up (30, 60 and 90 DAP). The crops were harvested on 25 to 30 March eac of the years. The tuberous root yield was collected from 2 m × 2 m (4 m^2) at the center o the plot at each location and converted into t ha^{-1}. Ten plants were randomly selected and the tuberous roots number plant^{-1} was averaged. Similarly, the length of root (cm) the diameter of root (cm), and per plant root weight were also measured following th same procedure.

2.3. Calculation of Root Dry Yield (t ha^{-1})

For dry matter measurement, about 100 g of sweet potato was collected for eac variety and was oven-dried for about 24 h at 80 °C. Finally, the amount of root dry matte (%) was determined using the following equation:

$$\text{Dry matter (\%)} = \text{Sample dry weight/Total sample weight} \times 100 \quad (1$$

The dry root yield of sweet potato was calculated from the fresh tuberous root yiel and % dry matter content using the formula:

$$\text{Dry tuberous root yield (t ha}^{-1}\text{)} = \text{fresh root yield (t ha}^{-1}\text{)} \times \text{\% dry matter content}/100 \quad (2$$

2.4. β-Carotene Yield Calculation (kg ha^{-1})

The cut roots were collected in a compound stack and five root samples (weighin 100 to 300 g) were taken to determine b-carotene. The cut roots were washed and cleane in tap water and allowed to air dry. Dried roots were peeled, and every root was slashe longitudinal direction in four parts. Two parallel sections of individual roots were taken t prepare for the 100-g compound sample which was placed in a transparent polythene ba and freeze-dried at −31 °C for 72 h. The dried samples were weighed, ground into flou in a stainless-steel mill, and stored in brown paper bags. The amount of root dry matter was computed from flesh and dry weight and expressed as a percentage. About 2 g of ground sweet potato sample was taken with 5 ml acetone and then acetone-petroleum ether (20:80; v/v) was added. After filtration and rotational evaporation process at 35 °C, the remaining solvent was removed to N_2 atmosphere and then dissolved in 2 ml petroleum ether. β-carotene (Sigma-Aldrich Corp., St. Louis, USA) stock and standard solutions and sample solutions measured 450 nm on spectrophotometer [26].

2.5. Economic Performance

The variable costs used in the analysis include land preparation, cutting of vines, planting of vines in the main field, fertilizers, insecticides, irrigation, harvesting and cleaning, etc. (Table 3).

Table 3. The production cost of sweet potato used in the economic analysis.

Items	Amount ha^{-1}	Unit Price (US$)	Total Cost (US$)	% of Total
• Variable costs				
• Vine (No. ha^{-1})	56,000	0.006	336	24.47
• Land preparation	1	84	84	6.12
• Human labor (Man-days)				0.00
• Vine cutting	10	4.8	48	3.50
• Vine plantation	28	4.8	134	9.76
• Fertilizer	2	4.8	10	0.73
• Irrigation	8	4.8	38	2.77
• Weeding	15	4.8	72	5.24
• Insecticide	3	4.8	14	1.02
• Harvesting and cleaning	30	4.8	144	10.49
• Total labor	96		461	33.58
• Fertilizer				0.00
• Urea (kg)	260	0.192	50	3.64
• TSP (kg)	150	0.264	40	2.91
• MoP (kg)	120	0.18	22	1.60
• Gypsum (kg)	83	0.12	10	0.73
• Zinc sulphate (kg)	10	1.8	18	1.31
• Boric acid (kg)	6	1.44	9	0.66
• Irrigation	1	90	90	6.55
• Insecticide	1	60	60	4.37
• Total variable cost			1179	85.87
• Fixed Cost				0.00
• Interest on operating capital (%)	0.09		44	3.20
• Land rental value	1	360	150	10.92
• Total fixed cost			194	14.13
• Total cost			1373	

These variable costs were determined based on information provided by local farmers in the communities surrounding the trials. Fixed costs are costs that do not change with the change in the amount and type of production, for example, the price of land rent, and interest on operating costs. The land rental price includes the rental cost for sweet potato production based on information provided by local farmers. The cost of land rent was determined according to the duration of the crop (5 months). Interest on Operating Capital (IOC) was determined by [27] the following equation:

$$\text{Interest on operating capital (IOC)} = \text{TVC} \times \text{I} \times \text{t}/(100 \times 12) \quad (3)$$

where TVC = total variable cost, I = interest rate per annum (9%, at present the interest rate of the bank in Bangladesh) and t = crop production period in months (like above).

The total cost in sweet potato production was estimated by the sum of total variable cost and fixed costs. The price of inputs and outputs were estimated in local currency (Bangladeshi Taka, BDT) based on the average values in the respective areas (Gaibandha and Rangpur). The above prices were converted to US$ using an exchange rate of 1 US$ = 84.69 BDT. The total (gross) return was estimated from the quantity of harvested tuberous root (t ha^{-1}) and the price of their farm-gate. The prices of sweet potatoes were 142 US$ t^{-1}. The gross margin was estimated from the difference between gross return and total variable cost. The net return was determined from the difference between gross return and total cost. Finally, the benefit-cost ratio (BCR) was estimated from gross return divided by total cost.

Preliminary information was collected by personal interview (PI), key informant interview (KII) and focus group discussion (FGD). It was organized by assigning a sorted, open, and closed-end questionnaire and a checklist. Sweet potato farmers were interviewed directly by enumerators to collect preliminary data on sweet potato growers and yields. Most of the tabular analysis was conducted with mean and percentage calculations

The profit margin of sweet potato growers, traders and retailers were estimated using the following formula:

$$\text{NP} = \text{TR} - \text{TC} \quad (4)$$

where NP = Net Profit (US$), TR = Total Return (US$) and TC = Total cost.

2.6. Statistical Analysis

Data on different attributes recorded for two years were analyzed by ANOVA (using STAR' statistical package developed by Biometrical Division, International Rice Research Institute (IRRI), Manila, Philippines) two evaluate the differences between treatments and the means were separated using LSD (least significant difference) at 5% level of significance. The results of different attributes with their interactions were found statistically significant and have been presented accordingly. Again, the stability parameter was analyzed using "metan" statistical package [28] in R studio version 1.4.110.

3. Results

3.1. Impact of OFSP Variety, Year and Location on Yield-Related Characters, Fresh and Dry Root Yield, and Beta-Carotene Production in the Northern Riverbanks

The main effect and interaction effects of the factors (sweet potato varieties, locations and years) on various yield contributing characters (tuberous roots $plant^{-1}$; Tuberous roots length; Tuberous root diameter; Root yield $plant^{-1}$) are presented in Table 4. The average tuberous root $plant^{-1}$ of sweet potato varieties were ranged from 3.29 to 5.16 $plant^{-1}$ where the lowest tuberous root was observed in the local variety and the highest was from BARI SP-12 which was similar to BARI SP-8 and BARI SP-14 (Table 4). Compared to local varieties, OFSP varieties, BARI SP-8, BARI SP-12, BARI SP-14 and BARI SP-15 produced an average of 54.71%, 56.83%, 46.50% and 37.08% more tuberous roots $plant^{-1}$, respectively. The present findings agree with previous findings [29,30] and the variation in tuberous roots $plant^{-1}$ was genotypic.

Table 4. Effect of Variety, environment, and growing season on no. of tuberous roots plant^{-1}, length of root, the diameter of root and tuberous root yield plant^{-1}.

Treatment	No. of Tuberous Root Plant^{-1}	Length of Root (cm)	Diameter of Root (cm)	Tuberous Root Weight Plant^{-1} (g)
		Variety (V)		
BARI SP-8	5.09 (54.71%)	15.19 (23.10%)	4.34 (52.28%)	463.71(65.33%)
BARI SP-12	5.16 (56.83%)	15.34 (24.31%)	4.21 (47.72%)	491.81 (75.35%)
BARI SP-14	4.82 (46.50%)	14.62 (18.48%)	3.96 (38.95%)	443.61(58.17%)
BARI SP-15	4.51 (37.08%)	15.08 (22.20%)	3.35 (35.09%)	430.51(53.50%)
Local	3.29	12.34	2.85	280.47
$LSD_{0.05}$	0.42	1.29	0.36	27.38
		Environment (E)		
Gaibandha	4.52	14.47	3.82	418.77
Rangpur	4.63	14.55	3.88	425.27
$LSD_{0.05}$	ns	ns	ns	ns
		Year (Y)		
2018–2019	4.64	14.26	3.81	411.26
2019–2020	4.52	14.76	3.88	432.78
$LSD_{0.05}$	ns	ns	ns	27.38

Values within the parenthesis indicate the values increased (%) over check; V, Variety; E, Environment (locations) and Y, Year; ns: non-significant; $LSD_{0.05}$ means significant at 5% level of probability.

Considering the two-year average data, the tuberous root length of sweet potato varieties ranged from 12.42 to 15.42 cm and 12.25 to 14.96 cm in Gaibandha and Rangpur, respectively (Table 5). In Gaibandha, the highest root length (15.42 cm) was attained from BARI SP-8 and BARI SP-12 which was identical to BARI SP-14 (15.16 cm). Again, in Rangpur, the highest root length was attained from BARI SP-15 (16.22 cm) followed by BARI SP-12 (15.26 cm) and BARI SP-8 (14.96 cm). In both locations, the lowest root length was attained from the local variety with a value of 12.42 and 12.25 cm, respectively. Average root length was attained from OFSP varieties, 23.10% from BARI SP-8, 24.31% from BARI SP-12, 18.48% from BARI SP-14, and 22.20%, respectively (Table 5).

Table 5. Interaction effect of variety (V) and environment (E) on root length of sweet potato.

Variety (V)	Environment/Locations		Mean Increased over Local Check (%)
	Gaibandha	Rangpur	
BARI SP-8	15.42	14.96	23.10
BARI SP-12	15.42	15.26	24.36
BARI SP-14	15.16	14.07	18.48
BARI SP-15	13.93	16.22	22.20
Local	12.42	12.25	-
$LSD_{0.05}$		1.29	

$LSD_{0.05}$ means significant at 5% level of probability.

Average of two-year data across locations, the tuberous root diameter of sweet potato varieties ranged from 2.85 to 4.34 cm where BARI SP-8 attained the maximum root diameter trailed by BARI SP-12 and BARI SP-14. Compared to local varieties, OFSP varieties, BARI SP-8, BARI SP-12, BARI SP-14, and BARI SP-15 produced an average of 52.28%, 47.72%, 38.95%, and 35.09% more tuberous root diameter, respectively. All-time low root diameter was attained from the local variety. In Rangpur, the average root yield plant^{-1} among the sweet potato genotypes varied from 274.4 to 509.9 g plant^{-1} and in Gaibandha it was from 286.5 to 473 g plant^{-1} (Table 6). Considering the mean of the two years, BARI SP-12 was the highest yielding followed by BARI SP-8. The local cultivar (check) produced the lowest root yield plant^{-1} in both locations. The average root yield plant^{-1} of typical sweet potato genotypes were varied from 280.4 to 491.8 g plant^{-1} where BARI SP-12 was the highest yielding followed by BARI SP-8 and BARI SP-14 (Table 4). Compared to local varieties BARI SP-8, BARI SP-12, BARI SP-14 and BARI SP-15 produced an average of 65.33%, 75.35%, 58.17% and 53.50% more tuberous roots weight plant^{-1}, respectively. Considering the crop season, average more root was produced in 2019–2020 compared to 2018–2019 crop season and considering the location mean, more root was produced in Rangpur than Gaibandha (Table 6).

Table 6. Interrelation impacts of variety (V), Year (Y) and environment (E) on average root weight plant^{-1} (g) of sweet potato.

Varieties (V)	Gaibandha			Rangpur		
	2018–2019	2019–2020	Mean	2018–2019	2019–2020	Mean
BARI SP-8	464.19	474.29	469.24 (63.77%)	423.68	492.66	458.17 (66.96%)
BARI SP-12	444.97	502.39	473.68 (65.32%	502.01	517.87	509.94 (85.82%)
BARI SP-14	428.67	449.37	439.02 (53.22%)	440.24	456.15	448.20 (63.33%)
BARI SP-15	420.96	429.8	425.38 (48.46%)	432.05	439.22	435.64 (58.75%)
Local	282.73	290.3	286.52	273.13	275.71	274.42
$LSD_{0.05}$				27.38		

$LSD_{0.05}$, significant at 5% level of probability.

The main effect and interaction effects of the factors (sweet potato varieties, locations and years) on various fresh root yields, dry root yield and beta-carotene yield were presented in Tables 7–9. Fresh tuberous root yield of different sweet potato genotypes in Gaibandha was ranged from 18.91 to 31.89 t ha^{-1} where BARI SP-8 was the highest yielding and BARI SP-12 was the 2nd highest yielding.

Table 7. Fresh and dry root yield (t ha^{-1}), beta carotene yield (kg ha^{-1}) of sweet potato varieties.

Treatment	Yield of Fresh Roots (t ha^{-1})	Yield of Dry Roots (t ha^{-1})	Beta Carotene Yield (kg ha^{-1}) (DW Basis)
Variety (V)			
BARI SP-8	31.07 (67.86%)	10.47 (66.99%)	336.19
BARI SP-12	32.00 (72.88%)	7.05 (12.44%)	1152.29
BARI SP-14	29.28 (58.18%)	8.62 (37.48%)	2955.90
BARI SP-15	28.41(53.48%)	8.21 (30.94%)	2951.41
Local	18.51	6.27	Nil
$LSD_{0.05}$	1.27	0.38	59.49
Environment (E)			
Gaibandha	27.64	8.10	1463.97
Rangpur	28.07	8.15	1510.26
$LSD_{0.05}$	ns	ns	37.63
Year (Y)			
2018–2019	27.14	7.91	1460.16
2019–2020	28.56	8.34	1514.07
$LSD_{0.05}$	1.27	0.38	37.63

Values within the parenthesis indicate the values increased (%) over check; V, Variety, E, Environment (locations) and Y, Year; ns: non-significant; $LSD_{0.05}$ means significant at 5% level of probability.

Table 8. Interrelation impacts of variety (V), and environment (E) on fresh yield (t ha^{-1}) and dry yield (t ha^{-1}) of sweet potato.

Variety (V)	Yield of Fresh Roots (t ha^{-1})		% Yield Increase over Local Check	Yield of Dry Roots (t ha^{-1})		% Yield Increase over Local Check
	Gaibandha	Rangpur		Gaibandha	Rangpur	
BARI SP-8	31.89	30.24	67.86	10.75	10.20	66.99
BARI SP-12	30.34	33.66	72.88	6.69	7.42	12.44
BARI SP-14	28.98	29.58	58.18	8.54	8.71	37.48
BARI SP-15	28.08	28.75	53.51	8.12	8.31	30.94
Local	18.91	18.11		6.40	6.13	
$LSD_{0.05}$ (V × E)	1.27			0.38		

V, Variety, E, Environment (locations) and Y, Year; $LSD_{0.05}$ means significant at 5% level of probability.

Table 9. Interrelation impacts of variety (V), and year (Y) on fresh yield (t ha^{-1}) and dry yield (t ha^{-1}) of sweet potato.

Variety (V)	Yield of Fresh Roots (t ha^{-1})		% Yield Increase over Local Check	Yield of Dry Roots (t ha^{-1})		% Yield Increase over Local Check
	2018–2019	2019–2020		2018–2019	2019–2020	
BARI SP-8	29.30	32.84	67.86	9.88	11.07	66.99
BARI SP-12	31.25	32.74	72.88	6.89	7.22	12.44
BARI SP-14	28.67	29.88	58.18	8.45	8.80	37.48
BARI SP-15	28.15	28.68	53.51	8.14	8.29	30.94
Local	18.34	18.68		6.21	6.32	
$LSD_{0.05}$ (V × E)	1.27			0.38		

V, Variety, E, Environment (locations) and Y, Year; $LSD_{0.05}$ means significant at 5% level of probability.

In Rangpur, the fresh root yield among the sweet potato genotypes ranged from 18.11 to 30.24 t ha^{-1}, where BARI SP-12 was the highest yielding and BARI SP-8 was the 2nd highest yielding. On the contrary, the local cultivar (check) generated the lowest root yield in Gaibandha and Rangpur 18.91 and 18.11 t ha^{-1}, respectively. In the 2018–2019 crop season, BARI SP-12 was the highest root yielding (31.25 t ha^{-1}) trailed by BARI SP-8 (29.30) and in 2019–2020 crop season BARI SP-8 was the highest fresh root yielder (32.84 t ha^{-1}) which was identical as BARI SP-12 (31.74 t ha^{-1}) followed by BARI SP-14 and BARI SP-15. The local cultivar was the lowest yielding in a couple of years (18.34 t ha^{-1} and 18.66 t ha^{-1}. Considering the mean of both years results across locations the uppermost fresh tuberous root yield (32.00 t ha^{-1}) was attained from BARI SP-12 and the 2nd uppermost was from BARI SP-8 (31.07 t ha^{-1}) and the 3rd uppermost yield was from BARI SP-14 (29.28 t ha^{-1}) which was 72.88%, 67.86%, and 58.18% higher fresh root yield ha^{-1}, respectively, compared to local cultivar (Tables 7–9). The lowest mean fresh root yield was attained from the local variety (18.51 t ha^{-1}) in Rangpur (Table 8).

Root dry yield of different sweet potato genotypes in Gaibandha was ranged from 6.40 to 11.75 t ha^{-1} where BARI SP-8 produced the maximum dry yield followed by BARI SP-14 and BARI SP-15. Similar dry root yield among the genotypes was also observed in Rangpur and was ranged from 6.13 to 10.19 t ha^{-1}, where BARI SP-8 was the highest dry root yielder followed by BARI SP-14 and BARI SP-15. Local variety (check) produced the lowest dry root yield in Gaibandha and Rangpur 6.40 and 6.13 t ha^{-1}, respectively. In both years, BARI SP-8 remained the highest dry root yielder (9.87 and 11.07 t ha^{-1} in 2018–2019 and 2019–2020 crop season) followed by BARI SP-14 (8.44 and 8.88 t ha^{-1}) and BARI SP-15 (8.13 and 8.29). The local cultivar proved to be the lowest dry root yielder in a couple of years (6.20 and 6.26 t ha^{-1}). Reflecting the average of two years results across locations the BARI SP-8 attained the highest dry root yield (10.47 t ha^{-1}) and BARI SP-14 (8.62 t ha^{-1}) attained the 2nd highest, and BARI SP-15 (8.21 t ha^{-1}) attained the 3rd highest dry yielder which produced 67.86%, 37.48%, and 30.94% higher dry root yield (t ha^{-1}), respectively, in comparison with local cultivar. The lowest mean dry root yield was attained from the local cultivar (6.26 t ha^{-1}).

Yield (Kg ha^{-1}) of β-Carotene (DW Basis)

β-Carotene production of different sweet potato genotypes was ranged from 335.53 to 2957.09 ha^{-1} where BARI SP-14 was the average highest beta-carotene yielder, followed by BARI SP-15. BARI SP-12 and BARI SP-8 was the lowest beat-carotene yielder, and no/trace beta-carotene yield was determined from the local cultivar (Table 7).

3.2. Selection for Stable Sweet Potato Genotype Based on Stability Variance and AMMI Analysis

The tuberous root yield stability for each variable across the location-year was stated by Shukla [31] and reported that varieties along with minimum values of Shukla's stability variance (σ^2i) are more stable. BARI SP-14 demonstrated the lowest Shukla's variance stability (−0.41) that seemed to be more stable over the location-year effect. In addition. Shukla's stability variance (σ^2i) criteria are also suitable for the BARI SP-14. The genotypes BARI SP-8 (32.46 t ha^{-1}) and BARI SP-15 (30.60 t ha^{-1}) produced a higher yield considering all years and locations (Tables 8 and 9).

Furthermore, the AMMI Stability value (ASV) is used to define genotype and location adaptability [32]. The genotypes with the maximum ASV value are taken into account unstable and more responsive to specific environments. On the other hand, the genotypes with the minimum ASV value are envisaged as stable genotypes over the environment (Table 10). The BARI SP-15 and BARI SP-14 genotypes were the best for yield and stability based on ASV value whereas the BARI SP-8 and BARI SP-12 showed the highest yield performance and were the most unstable based on ASV value.

Table 10. AMMI analysis provides stability parameters for tuber yield in sweet potato.

Variety	Mean	Shukla's Stability Variance (σ2i)	AMMI-Stability Value (ASV)	Note
BARI SP-8	31.07	3.08	2.50	Specific adaptation
BARI SP-12	32.00	3.33	2.84	Specific adaptation
BARI SP-14	29.28	−0.41	0.31	Wider adaptation
BARI SP-15	28.42	0.01	0.55	Wider adaptation
Local	18.51	1.04	1.32	Specific adaptation to the marginal environment

The principal component score (IPCA1) for both genotypes and environments were plotted against the mean tuber yield and the AMMI biplot1 was drawn (Figure 2A). The Principal Component interaction (IPCA1) score was plotted against the Principal Component interaction (IPCA2) for assessing the adaptation of genotypes (Figure 2B). The variety BARI SP-14 which is situated within the circle was stable and the other varieties were unstable because of their disperse position (Figure 2B).

3.3. Profitability Analysis of OFSP at Farm Level in the Northern Riverbanks

The summary statistics of survey variables of 80 sweet potato farms are presented in Table 11.

Table 11. Summary statistics for survey variables under different sweet potato varieties.

Variables	Sample Mean	Standard Deviation	Minimum	Maximum
Gross Output (US Dollar)	2133.40	238.05	1599.40	2828.51
Land Size (hectares)	0.43	1.05	0.11	1.09
Labor (man-days)	91.72	85.71	63.42	133.70
Variable Cost (taka)	805.43	525.39	547.51	1295.56
Fixed Cost (taka)	103.48	56.66	86.17	130.02
Age (years)	42.50	11.13	25.00	66.00
Education (years)	8.31	4.40	0.00	14.00

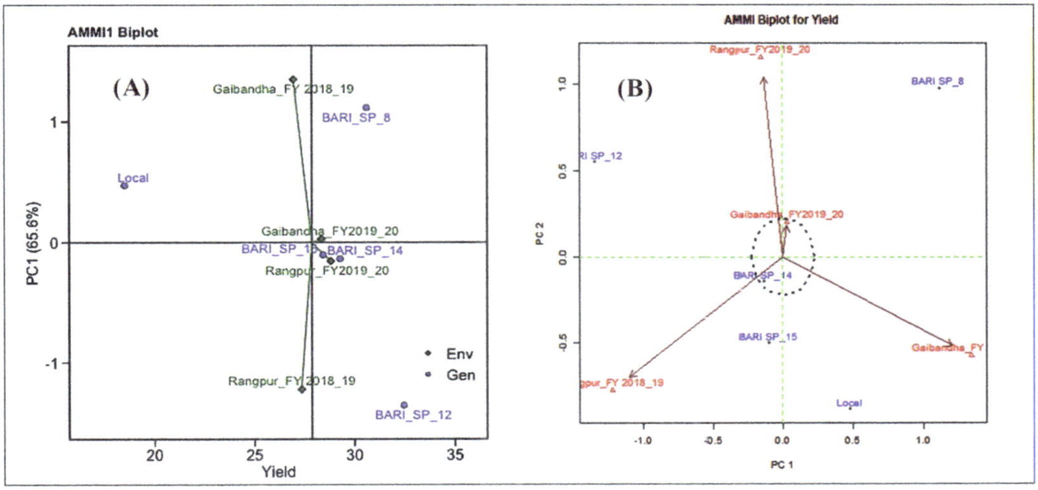

Figure 2. (**A**) AMMI 1 and (**B**) AMMI 2 biplot of IPCA1 (Interaction Principal Component Axes for genotype) axis against the mean yield of five sweet potato varieties evaluated in Gaibandha and Rangpur for tubers yield in sweet potato during 2018–2019 and 2019–2020 in Bangladesh.

The sample average gross output was 2133.40 US$ with a standard deviation of 238.05 that indicated the large variability of output among the sweet potato farms. Variations of standard deviation for the entire variable were large due to variation of land size and the minimum land size of sweet potato farms was 0.11 and the maximum was 1.09. Human labor was employed 91.72 US$ per farm for producing the crop in the crop year (2018–2019). It indicated that the existing production technology among farm households was labor intensive. The variable cost involved among the cost of inputs of the sample farms, e.g., seed cost, hired labor, manure, fertilizer, insecticide, irrigation, etc. Conversely, fixed labor involved the rental value of land and family labor. Age and education are important socio-economic factors that keep a great role in the on-farm operation. The OFSP regarding all the varieties were made at a profit with positive gross margin, net return and BCR > 1, whatever the amount of labor cost in the economic analysis (Table 12). The gross margin was 97–132% higher (1406–1914 US$ ha^{-1}) than that of the local sweet potato variety (1450 US$ ha^{-1}) due to higher tuberous root yield (53–73%) with the same production costs. Net return ranged from 2662–3170 US$ ha^{-1}, about three folds higher (av. 2914 US$ ha^{-1}) than the local sweet potato variety (1256 US$ ha^{-1}). Finally, the BCR in OFSP ranged from 2.94–3.31, while the minimum value was observed in the local variety (1.91).

Table 12. Economic profitability of OFSP. Data are presented mean ± standard error in US$ (1 US$ = 84.69 BDT).

Varieties	Yield (t ha^{-1})	Gross Return (US$ ha^{-1})	Total Variable Cost (US$ ha^{-1})	Total Fixed Cost (US$ ha^{-1})	Total Cost (US$ ha^{-1})	Gross Margin (US$ ha^{-1})	Net Return (US$ ha^{-1})	Benefit-Cost Ratio (BCR)
BARI SP-8	31.07 ± 2.7	4412 ± 383	1179	194	1373	3233 ± 383	3039 ± 383	3.21 ± 0.28
BARI SP-12	32.00 ± 2.6	4543 ± 368	1179	194	1373	3364 ± 368	3170 ± 368	3.31 ± 0.27
BARI SP-14	29.28 ± 1.7	4157 ± 235	1179	194	1373	2978 ± 235	2784 ± 235	3.03 ± 0.17
BARI SP-15	28.41 ± 1.5	4035 ± 212	1179	194	1373	2856 ± 212	2662 ± 212	2.94 ± 0.15
Local	18.51 ± 1.1	2629 ± 159	1179	194	1373	1450 ± 159	1256 ± 159	1.91 ± 0.12

4. Discussion

The tuberous root yield of sweet potato displays several positive associations with the plant and root characteristics [33]. The higher root yield of the tuberous roots is reliant on the higher tuberous root weight resulted in higher tuberous root diameter, more leaf number per plant, increase vine length, more tuberous root plant^{-1}, increase the number of sweet potato vines plant^{-1}, and a minimum influenced by increase tuberous root length [33]. Orange flesh sweet potato genotypes have high beta-carotene [8,29] in comparison with sweet potato cultivars having white, cream or yellow flesh. The color of the original flesh may also indicate the intensity of the pigment included. The more intense of the color of the root flesh, the higher the amount of beta-carotene [34]. The morphological and yield-contributing characters of sweet potatoes will be determined by the influence of the growing environment on the genetic makeup of the crop. These two factors will inter-relate throughout the plant's growing period so that the shape of the tuberous roots look like each other or dissimilar. If the impact of the growing condition is major, rather than the genetic impact, then there may be morphological distinctions of the varieties/cultivars [35]. These growing conditions/environments included climatic and soil conditions as well as water availability [36]. Some of the morphological characters such as leaf size, color, stem, petiole, skin and flesh color of sweet potato are stable and are not influenced by growing conditions, while morphological characters such as the length of vines and leaf stalk, leaf shape and yield of tuberous roots which can be easily changed as affected by the environment [37]. Our current findings on plant roots, root lengths, root diameters and root production agree with the findings [30,38–40].

In Bangladesh, due to high crop competition in the plain ecology and the use of local farming in the production system, the potential for sweet potato production is comparatively low, and the quantity and quality are also reduced. In our field trial, it was proved that the newly developed OFSP varieties produced higher root yield than the local cultivars, which indicates that the new varieties have good genetic characteristics to provide high yields in the various environments, and our statement agrees with [32]. Moreover, climatic conditions and intercultural operations according to cultivars/varieties also affect the productivity of sweet potatoes [41–43]. Root yield varies greatly between sweet potato varieties/cultivars and even individual plants of the same variety/cultivars such as those affected by cultivation, breeding material, and growing environment and edaphic condition [44]. Genetic and environmental factors also affect on the morphological and physiological character, yield and dry matter production [45].

Therefore, it is very important to select suitable sweet potato varieties based on environmental conditions [41]. The overall yield performance of all studied varieties was comparatively low in Gaibandha than Rangpur location considering their potential yield. Stability analysis provides the level of productivity of a genotype to a certain environment [46]. BARI SP-14 had a relatively good yield maintaining stability in unfavorable locations and responding well to favorable locations followed by BARI SP-15. Further, the probable causes for its low yield were sandy to sandy loam soil having less moisture holding capacity, which may also be due to inadequate irrigation application at the time of root growth and development, inadequate intercultural management practices, and considered as neglected crops in the Char/riverbank areas.

Beta-carotene may vary from place to place and from year to year. This beta-carotene yield was initially controlled by the genetic factor. It is affected by the amount of irrigation during crop growth and the amount of fertilizer used for root production. But fertilizer application has a positive effect on beta-carotene content and generally agrees with the study [47–49] which revealed that increased potassium and zinc fertilization increased carotene levels in sweet potato roots. However, the amount of carotene increase fluctuates within the sweet potato genotype. Smoleń and Sady [50] stated that nitrogen fertilization alone has no significant effects on the extent of carotenoid (in carrots). Thus, the difference between carotene components can be estimated by applying specific macro-components. Again, when the roots are stored in the soil until needed, the carotenoid and β-carotene content of the roots may be affected by the sweet potato variety and storage age of the roots [51,52].

The profit may be more or less due to the variation in the yield of sweet potato. The harmful aspects of sweet potato cultivation include lack of sweet potato vines (planting material) at farm level [53], awareness of sweet potato farmers about OFSP, storage capacity [54] and lack of processing facilities, lack of suitability, marketing structure and high marketing costs among producers [53–56]. The cost and returns analysis showed that labor cost was 33.58% followed by cost of acquisition of vines found 24.47% of the total cost of sweet potato cultivation and profitability of OFSP production.

The main problem faced by the sweet potato growers was the unavailability of quality vines. The second problem was indicated by farmers was the lack of storage facilities and lack of knowledge about storage techniques. Most farmers do not choose storage and so the roots were sold almost immediately in the local market and, as a result, people are mistreated by sweet potato farmers by marketing [43,46].

The farmers also stated that they faced yield loss due to sweet potato weevil attacks. In addition to this, unpredicted weather due to uneven rainfall also influenced the sweet potato root yield. Numerous sweet potato growers also reported a shortage of labor at the time of planting and harvesting sweet potatoes.

These limits can be forwarded by on time and sufficient supply of planting materials (vines), generating awareness about OFSP varieties with improved cultivation systems and updating the market approach of sweet potatoes in Bangladesh resulting in the extended better ways for escalating farmers' income along with nutrition.

5. Conclusions

Growing OFSP in the riverbanks/Char areas of northern Bangladesh provides agronomic and economic benefits to producers. Farmers who find it difficult to grow rice, maize or wheat in the riverbanks/Char areas because of sandy and fallow land and shortage of water (especially for rice) may produce sweet potato as an alternative. Farmer's yield of OFSP in the riverbanks/Char area has an average of about 18.51 t ha^{-1} and entails a considerably reduced amount of irrigation and other inputs relative to other crops. However, among the cultivars, BARI SP-12 produced 73% more root yield (32 t ha^{-1}) like BARI SP-8 by 68% higher yield (31.07 t ha^{-1}) compared to the local cultivar over the locations. In terms of feasibility, BARI SP-12 performed the best and yielded 33.66 t ha^{-1} in Rangpur whereas BARI SP-8 performed the best in Gaibandha with an average yield of 31.89 t ha^{-1}, which was 86 and 69% higher than that of the local cultivars. Conversely, the beta carotene (Vitamin-A precursor) was the greatest in BARI SP-14 (2957 kg ha^{-1}) like BARI SP-15 (2952 kg ha^{-1}), whereas it was much lower in BARI SP-12 and BARI SP-8. In addition, BARI SP-8 and BARI SP-12 was found stable in specific locations and found economically profitable due to the highest root yield in the respective areas. Besides these two varieties, BARI SP-14 and BARI SP-15 were found stable in both locations. Finally, it may be concluded that BARI SP-8 and BARI SP-12 are suitable for cultivation in the Char areas of Gaibandha and Rangpur in terms of a good yield. Furthermore, BARI SP-14 and BARI SP-15 are found stable for both locations and can be cultivated in the Char areas. These four OFSP varieties can serve as a good source of beta-carotene (vitamin-A) among the Char dwellers' as well as the whole nation of Bangladesh.

Author Contributions: Conceptualization, A.-A.M., M.A.A., M.S.H.M.; methodology, A.-A.M., M.J.A. and M.N.A.; software, A.-A.M., M.J.A.; validation, A.-A M., S.H., W.J.G. and E.H.M.S.R.; formal analysis, A.-A.M., M.J.A., M.N.A.; investigation, A.-A.M., M.J.A. and M.A.-A.H.T.; resources, A.-A.M.; data curation, A.-A.M., M.J.A. and A.H.; writing—original draft preparation, A.-A.M., W.J.G., S.H. and D.C.; writing—review and editing, S.H., W.J.G., M.M.A., M.A., A.G., M.A.A., A.H. and A.-A.M.; visualization, A.-A.M., A.H. and M.J.A.; supervision, A.-A.M.; project administration, A.-A.M., M.A., A.G. and D.C.; funding acquisition, A.-A.M., M.A., A G., A.H. and D.C. All authors have read and agreed to the published version of the manuscript.

Funding: The authors would like to thank the International Potato Center (CIP, Bangladesh) authority for awarding research expenses and also the Taif University Researchers Supporting Project number (TURSP 2020/257), Taif University, Taif, Saudi Arabia.

Institutional Review Board Statement: Not applicable.

Informed Consent Statement: Not applicable.

Data Availability Statement: Data may be available after request to the corresponding author.

Acknowledgments: We earnestly give thanks to the Director-General of Bangladesh Agricultural Research Institute (BARI) for delivering the services for this study and the authors extend their appreciation to Taif University Researchers Supporting Project number (TURSP 2020/257), Taif University, Taif, Saudi Arabia.

Conflicts of Interest: The author declares no conflict of interest.

References

1. Woolfe, J.A. *Sweet Potato: An Untapped Food Resource*; Cambridge University Press: Cambridge, UK, 1992; p. 643.
2. Bovell-Benjamin, A.C. Sweet potato: A review of its past, present, and future role in human nutrition. *Adv. Food Nutr. Res.* **2007**, *52*, 1–59.
3. Low, J.; Nyongesa, M.; Quinn, S.; Parker, M. (Eds.) *Potato and Sweetpotato in Africa. Transforming the Value Chains for Food and Nutrition Security*; CABI International: Boston, MA, USA, 2015; 632p, ISBN 978-1-78064-420-2. [CrossRef]
4. CIP (International Potato Centre). 2017. Available online: http://cipotato.org/research/sweetpotato-in-Africa (accessed on 11 August 2021).
5. FAO. *The State of Food and Agriculture, Climate Change, Agriculture and Food Security*; FAO: Rome, Italy, 2016; Available online: http://www.fao.org/3/i6030e/i6030e.pdf (accessed on 14 August 2021).
6. Murugan, S.; Paramasivam, S.K.; Nedunchezhiyan, M. Sweet potato as animal feed and fodder. *Fruit Vege. Cere. Sci. Biotech.* **2012**, *6*, 106–114.
7. Chakrabarti, A.; Kumari, R.; Dey, A.; Bhatt, B.P. Sweet Potato—An Excellent Source of Livestock Feed. Krishi Sewa. 2014. Available online: http://www.krishisewa.com/cms/articles/production-technology/391-sweet-potato.html (accessed on 13 August 2021).
8. Teow, C.C.; Truong, V.D.; McFeeters, R.F.; Thompson, R.L.; Pecota, K.V.; Yencho, G.C. Antioxidant activities, phenolic and beta-carotene contents of sweet potato genotypes with varying flesh colours. *Food Chem.* **2007**, *103*, 829–838. [CrossRef]
9. Antia, B.S.; Akpan, E.J.; Okon, P.A.; Umoren, I.U. Nutritive and anti-nutritive evaluation of sweet potatoes (*Ipomoea batatas*) leaves. *Pakistan J. Nutr.* **2006**, *2*, 166–168. [CrossRef]
10. Slafer, G.A.; Savin, R. Source—Sink relationships and grain mass at different positions within the spike in wheat. *Field Crops Res.* **1994**, *37*, 39–49. [CrossRef]
11. Mbwaga, Z.; Mataa, M.; Msabaha, M. Quality and yield stability of orange-fleshed sweet potato (*Ipomoea batatas*) varieties grown in different agro-ecologies. In Proceedings of the 8th African Crop Science Society Conference, Minya, Egypt, 27–31 October 2007; pp. 339–345.
12. Wilson, C.D.; Pace, R.D.; Bromfield, E.; Jones, G.; Lu, J.Y. Sweet potato in a vegetarian menu plan for NASA's Advanced Life Support Program. *Life Support Biosph. Sci.* **1998**, *5*, 347–351.
13. Wu, X.; Sun, C.J.; Yang, L.H.; Zeng, G.; Liu, Z.Y.; Li, Y.M. Beta-carotene content in sweet potato varieties from China and the effect of preparation on beta-carotene retention in the Yanshu. *Innov. Food Sci. Emerg. Technol.* **2008**, *9*, 581–586. [CrossRef]
14. Byers, T.; Perry, G. Dietary carotenes, vitamin C and vitamin E as protective antioxidants in human cancers. *Ann. Rev. Nutr.* **1992**, *12*, 139–159. [CrossRef] [PubMed]
15. Ahn, P.M. *Tropical Soils and Fertilizer Use*; Longman Scientific and Technical: Harlow, UK, 1993.
16. Tumwegamire, S.; Kapinga, R.; Zhang, D.; Crisnam, C.; Agilli, S. Opportunities for promoting orange-fleshed sweet potato among food based approach to combat vitamin A deficiency in sub-Saharan Africa. *Afr. Crop Sci. J.* **2004**, *12*, 241–253.
17. Azad, A.K.; Miaruddin, M.; Oahab, M.A.; Sheikh, M.H.R.; Nag, B.L.; Rahman, M.H.H. *Krishi Projukti Hatboi (Handbook on Agro-Technology)*, 9th ed.; Bangladesh Agricultural Research Institute: Gazipur, Bangladesh, 2020; pp. 35–37.
18. Hongyu, K.; García-Peña, M.; de Araújo, L.B.; Dias, C.T.S. Statistical analysis of yield trials by AMMI analysis of genotype × environment interaction. *Biom. Lett.* **2014**, *51*, 89–102. [CrossRef]
19. Kivuva, B.M.; Githiri, S.M.; Yencho, G.C.; Sibiya, J. Genotype × Environment interaction for storage root yield in sweet potato under managed drought stress conditions. *J. Agril. Sci.* **2014**, *6*, 41–56.
20. Adebola, P.O.; Shegro, A.; Laurie, S.M.; Zulu, L.N.; Pillay, M. Genotype × environment interaction and yield stability estimate of some sweet potato [*Ipomoea batatas* (L.) Lam] breeding lines in South Africa. *J. Plant Breed. Crop Sci.* **2013**, *5*, 182–186. [CrossRef]
21. Amare, B.; Abay, F.; Tsehaye, Y. Evaluation of sweet potato (*Ipomea batatas* L.) varieties for total storage root yield in south. *Am. J. Trade Policy* **2014**, *1*, 27–32. [CrossRef]
22. Purchase, J.L. Parametric Analysis to Describe Genotype × Environment Interaction and Yield Stability in Winter Wheat. Ph.D. Thesis, University of the Orange Free State, Bloemfontein, South Africa, 1997; pp. 19–22.

3. Prakash, P.; Avinash, K.; Devesh, R.; Debdutt, B. Economic analysis of sweet potato farming and marketing in Odisha. *J. Root Crops* **2016**, *42*, 163–167.
4. Prakash, P.; Kishore, A.; Roy, D.; Behura, D.; Immanuel, S. Bio fortification for reducing hidden hunger: A value chain analysis of sweet potato in India. *Agril. Econ. Res. Rev.* **2017**, *30*, 20–30.
5. GAINS Project. Technological Empowerment and the Creation of Sustainable Livelihoods in Odisha State in India. 2013. Available online: http://www.orihort.in/Application/Download/Generating%20Advances%20in%20Incmes%20and%20Nutrition%20through%20Sweet%20Potato%20(GAIN).pdf (accessed on 8 September 2017).
6. Yildirim, Z.; Tokuşoğlu, Ö.; Öztürk, G. Determination of sweet potato [*Ipomoea batatas* (L.) Lam.] genotypes suitable to the Aegean region of Turkey. *Turkish J. Field Crops* **2011**, *16*, 48–53.
7. Alam, M.J.; Humphreys, E.; Sarkar, M.A.R.; Sudhir, Y. Intensification and diversification increase land and water productivity and profitability of rice-based cropping systems on the High Ganges River Floodplain of Bangladesh. *Field Crops Res.* **2017**, *209*, 10–26. [CrossRef]
8. Olivoto, T.; Lúcio, A.D.; Da Silva, J.A.; Marchioro, V.S.; de Souza, V.Q.; Jost, E. Mean performance and stability in multi-environment trials I: Combining features of AMMI and BLUP techniques. *Agron. J.* **2019**, *111*, 2949–2960. [CrossRef]
9. Rahman, M.H.; Patwary, M.M.A.; Barua, H.; Hossain, M.; Nahar, S. Evaluation of Orange Fleshed Sweet potato (*Ipomoea batatas* L.) Genotypes for Higher Yield and Quality. *Agriculturists* **2013**, *11*, 21–27. [CrossRef]
0. Rahaman, E.H.; Hossain, M.M.; Hoque, M.E.; Ali, M.R.; Mahmud, A.A.; Hasan, M.M.; Mohanta, H.C.; Bhuiyan, M.K.; Islam, M.S.; Hossain, M.J. Farmers preference and suitability assessment of BARI released Orange-Fleshed Sweet potato varieties in Charlands of Gaibandha. *Bangladesh Hort* **2015**, *2*, 1–11.
1. Shukla, G.K. Some statistical aspects of partitioning genotype environmental components of variability. *Heredity* **1972**, *29*, 237–245. [CrossRef]
2. Purchase, J.L.; Hatting, H.; Van Deventer, C.S. Genotype × environment interaction of winter wheat (*Triticum aestivum* L.) in South Africa: II. Stability analysis of yield performance. *S. Afr. J. Plant Soil* **2000**, *17*, 101–107. [CrossRef]
3. Reddy, R.; Soibam, H.; Ayam, V.S.; Panja, P.; Mitra, S. Morphological characterization of sweet potato cultivars during growth, development and harvesting. *Indian J. Agric. Res.* **2018**, *52*, 46–50. [CrossRef]
4. Saraswati, P.; Soplanit, A.; Syahputra, A.T.; Kossay, L.; Muid, N.; Ginting, E.; Lyons, G. Yield trial and sensory evaluation of sweet potato cultivars in Highland Papua and West Papua Indonesia. *J. Trop. Agric.* **2013**, *51*, 74–83.
5. Hughes, A.R.; Inouye, B.D.; Johnson, M.T.J.; Underwood, N.; Vellend, M. Ecological Consequences of Genetic Diversity. *Ecol. Lett.* **2008**, *11*, 609–623. [CrossRef] [PubMed]
6. Mwololo, J.K.; Mburu, M.W.K.; Muturi, P.W. Performance of sweet potato varieties across environments in Kenya. *Int. J. Agric. Agri. Res.* **2012**, *2*, 1–11.
7. Purbasari, K.; Sumadji, A.R. Studi variasi ubi jalar (*Ipomoea batatas* L.) berdasarkan karakter morfologi di Kabupaten Ngawi. *Florea J. Biol. Pembelajarannya* **2018**, *5*, 78–84. [CrossRef]
8. Jahan, M.A.; Siddique, M.A.R. Studies on Growth, Yield and Storage of Sweet Potato as Influenced by Variety and Time of Harvest. Master's Thesis, Department of Horticulture, Bangladesh Agricultural University, Mymensingh, Bangladesh, 2001; pp. 1–84.
9. Rafique, F.; Fatema, K.; Rahman, M.H.; Hossain, M.M. Vegetative growth and yield performance of eight sweet potato genotypes. *Bangladesh Hortic.* **2015**, *1–2*, 103–110.
0. Hayati, M.; Zakaria, S.; Efendi; Anhar, A. Morphological characteristics and yields of several sweet potato (*Ipomoea batatas* L.) tubers. In Proceedings of the IOP Conference Series: Earth and Environmental Science 425 (2020) 012055, the 1st International Conference on Agriculture and Bioindustry, Banda Aceh, Indonesia, 24–26 October 2019. Available online: https://iopscience.iop.org/article/10.1088/1755-1315/425/1/012055/pdf (accessed on 13 August 2021).
1. Mahmud, A.A.; Hassan, M.M.; Alam, M.J.; Molla, M.S.H.; Ali, M.A.; Mohanta, H.C.; Alam, M.S.; Islam, M.A.; Talukder, M.A.H.; Ferdous, M.Z.; et al. Farmers' Preference, Yield, and GGE-Biplot Analysis-Based Evaluation of Four Sweet potato (Ipomoea batatas L.) Varieties Grown in Multiple Environments. *Sustainability* **2021**, *13*, 3730. [CrossRef]
2. Antiaobong, E.E.; Bassey, E.E. Constraints and prospects of sweet potato (*Ipomoea batatas* (L.) Lam) production in humid environment of southeastern Nigeria. In Proceedings of the Second African Regional Conference on Sustainable Agriculture, Uyo, Nigeria, 4–7 November 2008; Governor's Office Annex: Uyo, Nigeria; pp. 68–72.
3. Antiaobong, E.E.; Bassey, E.E. Characterization and evaluation of six sweet potato varieties (*Ipomoea batatas* (L.) Lam) for quantitative and qualitative characters and tolerance to Cylas puncticollis, Boh in high humid environment of southeastern Nigeria. *J. Agril. Res. Pol.* **2009**, *4*, 17–21.
4. Ravi, V.; Saravanan, R. Crop Physiology of Sweet potato. Fruit, Vegetable and Cereal Science and Biotechnology. In *Global Science Books*; 6 (Special Issue I); Central Tuber Crops Research Institute: Thiruvananthapuram, India, 2012; pp. 17–29.
5. Ravi, V.; Indira, P. Crop Physiology of Sweet potato. *Hortic. Rev.* **1999**, *23*, 277–316.
6. Bernardo, R. *Breeding for Quantitative Traits in Plants*, 3rd ed.; University of Minnesota–Twin Cities, Stemma Press Woodbury: Woodbury, MN, USA, 2020.
7. Laurie, S.M.; Faber, M.; van Jaarsveld, P.J.; Laurie, R.N.; du Plooya, C.P.; Modisane, P.C. β-Carotene yield and productivity of orange-fleshed sweet potato (*Ipomoea batatas* L. Lam.) as influenced by irrigation and fertilizer application treatments. *Sci. Hortic.* **2012**, *142*, 180–184. [CrossRef]

48. George, M.S.; Lu, G.; Zhou, W. Genotypic variation for potassium uptake and utilization efficiency in sweet potatoes. *Field Crop Res.* **2002**, *77*, 7–15. [CrossRef]
49. Abd El-Baky, M.M.H.; Ahmed, A.A.; El-Nemr, M.A.; Zaki, M.F. Effect of potassium fertilizer and foliar zinc application on yiel and quality of sweet potato. *Res. J. Agric. Biol. Sci.* **2010**, *6*, 386–394.
50. Smoleń, S.; Sady, W. The effect of various nitrogen fertilization and foliar nutrition regimes on the concentrations of sugars carotenoids and phenolic compounds in carrot (*Daucus carota* L.). *Sci. Hortic.* **2009**, *120*, 315–324. [CrossRef]
51. K'osambo, L.M.; Carey, E.E.; Misra, A.K.; Wilkes, J.; Hagenimana, V. Influence of age, farming site, and boiling on pro-vitamin A content in sweet potato (*Ipomoea batatas* (L.) Lam.) storage roots. *J. Food Comp. Anal.* **1998**, *11*, 305–321. [CrossRef]
52. Chattopadhyay, A.; Chakraborty, I.; Mukopadhayay, S.K.; Kumar, P.K.; Sen, H. Compositional changes of sweet potato as influenced by cultivar, harvest date and cooking. *Acta Hortic.* **2006**, *703*, 211–218. [CrossRef]
53. Ahmed, M.R. Yield Performance of Sweet Potato Cultivation at Farmers' Field. Master's Thesis, Department of Agricultural Extension and Information System, Sher-e-Bangla Agricultural University, Dhaka, Bangladesh, 2018; p. 64.
54. Abrham, T.; Beshir, H.M.; Haile, A. Sweet potato production practices, constraints, and variety evaluation under different storage types. *Food Energy Secur.* **2020**, *10*, e263.
55. Sugri, I.; Maalekuu, B.K.; Gaveh, E.; Kusi, F. Sweet potato Value Chain Analysis Reveals Opportunities for Increased Income and Food Security in Northern Ghana. *Adv. Agri.* **2017**, *2017*, 1–14. [CrossRef]
56. Ezin, V.; Quenum, F.; Bodjrenou, R.H.; Kpanougo, C.M.I.; Kochoni, E.M.G.; Chabi, B.I.; Ahanchede, A. Assessment of production and marketing constraints and value chain of sweet potato in the municipalities of Dangbo and Bonou. *Agric. Food Secur.* **2018**, *7*, 15. [CrossRef]

agronomy

MDPI

Article

Soil Test Based Fertilizer Application Improves Productivity, Profitability and Nutrient Use Efficiency of Rice (*Oryza sativa* L.) under Direct Seeded Condition

Vijay Kant Singh [1,2], Poonam Gautam [1], Gangadhar Nanda [3,4], Salwinder Singh Dhaliwal [2], Biswajit Pramanick [3,5,*], Shiv Singh Meena [1], Walaa F. Alsanie [6], Ahmed Gaber [7], Samy Sayed [8] and Akbar Hossain [9,*]

1. Department of Soil Science, G.B. Pant University of Agriculture & Technology, Pantnagar 263145, Uttarakhand, India; vksingh@pau.edu (V.K.S.); drpgautam@rediffmail.com (P.G.); meenashiva20@gmail.com (S.S.M.)
2. Department of Soil Science, Punjab Agricultural University, Ludhiana 141004, Punjab, India; ssdhaliwal@pau.edu
3. Department of Agronomy, G.B. Pant University of Agriculture & Technology, Pantnagar 263145, Uttarakhand, India; gangadhar.nanda4@gmail.com
4. Animal Production Research Institute, Rajendra Prasad Central Agricultural University, Pusa 848125, Bihar, India
5. Department of Agronomy, Rajendra Prasad Central Agricultural University, Pusa 848125, Bihar, India
6. Department of Clinical Laboratories Sciences, The Faculty of Applied Medical Sciences, Taif University, P.O. Box 11099, Taif 21944, Saudi Arabia; w.alsanie@tu.edu.sa
7. Department of Biology, College of Science, Taif University, P.O. Box 11099, Taif 21944, Saudi Arabia; a.gaber@tu.edu.sa
8. Department of Science and Technology, University College-Ranyah, Taif University, P.O. Box 11099, Taif 21944, Saudi Arabia; s.sayed@tu.edu.sa
9. Bangladesh Wheat and Maize Research Institute, Dinajpur 5200, Bangladesh

* Correspondence: biswajit@rpcau.ac.in (B.P.); akbarhossainwrc@gmail.com (A.H.)

Citation: Singh, V.K.; Gautam, P.; Nanda, G.; Dhaliwal, S.S.; Pramanick, B.; Meena, S.S.; Alsanie, W.F.; Gaber, A.; Sayed, S.; Hossain, A. Soil Test Based Fertilizer Application Improves Productivity, Profitability and Nutrient Use Efficiency of Rice (*Oryza sativa* L.) under Direct Seeded Condition. *Agronomy* **2021**, *11*, 1756. https://doi.org/10.3390/agronomy11091756

Academic Editors: Umberto Anastasi and Aurelio Scavo

Received: 10 August 2021
Accepted: 28 August 2021
Published: 31 August 2021

Publisher's Note: MDPI stays neutral with regard to jurisdictional claims in published maps and institutional affiliations.

Abstract: A field investigation on direct seeded rice (DSR) was carried out in the two consecutive rice growing seasons of 2017 and 2018 at Pantnagar, Uttarakhand, India for the development and validation of soil test crop response (STCR) to fertilizer and for assessing the performance of STCR-treatments as compared to the general recommended dose (GRD) in terms of yield, nutrient uptake and use efficiency, and the economics of DSR. For producing 1 Mg of rice-grain, the required nutrients (N, P, and K) were 2.01 kg, 0.44 kg, and 3.06 kg; the contribution from the soil was 22.05%, 37.34%, and 41.48%; from applied farmyard manure 23.25%, 28.34%, and 16.80%, from fertilizer 38.08%, 49.93%, and 252.98%; and from fertilizer with FYM 44.83%, 60.57%, and 278.70%; for N, P, and K, respectively. The STCR approach, with or without FYM, at both the target yields (4.5 Mg ha^{-1} and 5.0 Mg ha^{-1}) markedly enhanced the grain yield (20.2% to 32.3%) and production efficiency over the GRD. It also exhibited a higher NPK uptake and use efficiency, along with better profitability, than the GRD. Therefore, the STCR-targeted yield approach could improve the yield, economics, and efficiency of nutrient use for direct seeded rice.

Keywords: direct seeded rice; soil test crop response; nutrient use efficiency; grain yield; net return

1. Introduction

Rice (*Oryza sativa* L.) is mostly grown using the transplanting establishment method, which requires more time, water, and labor for nursery preparation. The profit margins of transplanted rice (TPR) have been reduced continuously due to the higher water input and labor intensive transplanting, as well as higher labor costs [1]. The water and labor demand may be reduced by growing direct-seeded rice (DSR) instead of TPR [2]. There has been a shift towards DSR in South East Asia during recent times [3]. In India, DSR produces about 2 to 12 per cent higher GY than TPR [4]. Direct seeding helps in reducing the water consumption by up to 30 per cent, as it does not require seedling raising in a nursery, puddling operations in the main field, transplanting, or maintaining 4 to 5 cm of water in the main field after transplanting. Furthermore, DSR matures about 8 to 10 days earlier than transplanted rice.

According to the conventional estimate, food grain demand will be 355 Mt by 2030 in India; while on the other hand, the response ratio (RR) and factor productivity of crops are continuously declining every year, due to the applied fertilizer in intensive cultivation [5]. The use efficiencies of nitrogen, phosphorus, and potassium are 30–50%, 15–20%, and 60–70%, respectively [5], which is certainly low and increases the cost of cultivation. During the past few decades, the use of fertilizers for enhancing food production has increased many fold, and which, if it exceeds the crop requirements, often causes environmental pollution. The imbalanced use of inorganic fertilizers in India has resulted in a net negative balance of nutrients of about 8 to 10 MT y^{-1} [6], and by 2025, the extent of this negative balance may rise up to 15 MT y^{-1}. Resource-poor farmers of the nation used to follow an imbalanced fertilization, which disturbs the nutrient availability, leading to a decrease in soil productivity in the long run [7]. Apart from this, increasing fertilizer prices and their availability is one of the main hurdles to balanced fertilization. Excessive chemical fertilizer application has aggravated the deficiencies of secondary and micro-nutrients in different soils. Furthermore, inadequate nutrition of crops worsens the situation, in terms of declined soil fertility.

Organic manures (OM) are a valuable source of nutrients, but their sole application is not sufficient to meet the nutrient requirements of high yielding varieties and often results in poor crop yields [8]. Furthermore, using the generally recommended dose (GRD) of fertilizer is not able to maintain yields vis-à-vis the economic returns of crops, due to fatigue in soil health, and this requires refinement for balanced crop nutrition [7]. Therefore, the sole use of neither OM nor chemical fertilizer can enhance the sustainability of an intensive production system [9]. The use of an appropriate combination of OM and chemical fertilizers [10], depending on soil fertility status [11], is a step forward for providing balanced fertilization to crops. Such integrated nutrient management (INM) can increase the income of farmers [12]. The continuous application of the GRD of fertilizer along with FYM enhances rice grain yields and their sustainability [7,13,14].

Harnessing the potential yields of high yielding varieties of crops requires the application of optimum doses of nutrients [15]. However, an inadequate and imbalanced fertilizer use for crop production, without proper knowledge of the inherent soil capabilities and crop requirements, is also one of the causes that prevent gaining the full yield potential of crops and the deterioration of soil health, as well as economic losses to farmers [8,16], and often resulting in an adverse impact on crops and the soil, in terms of nutrient toxicity and deficiency [8]. Furthermore, fertilizer use requires knowledge of the expected GY response, which depends upon the crop nutrient requirements, nutrient supply from indigenous sources, and the fate of fertilizers applied to the soil in the short and long term [17]. Therefore, a comprehensive approach, considering soil tests, field research, and profitability, could be employed for fertilizer use. Thus, the soil test crop response (STCR) methodology can be adopted for calculating nitrogen (N), phosphorus (P), and potassium (K) requirements as needed. NPK requirements are linearly correlated with the target yield (TY), depending on the soil test values (STVs). In the STCR approach, the fertilizer doses are prescribed according to the developed fertilizer adjustment equations, after the establishment of a significant relationship between STVs, the added fertilizer nutrients, and the crop response [8] for a particular soil type. Thus, precise fertilizer recommendations can be made using this approach, as it involves data of soil and plant analysis [8]. A higher response ratio is also observed along with a higher benefit–cost ratio as the nutrient application is based on demand and the correction of soil nutrient imbalances [18]. There is a need to develop a balanced nutrient management strategy, involving the STCR methodology for the sole use of chemical fertilizers and the integrated use of chemical fertilizers and

FYM for DSR. Hence, the present study was carried out to (i) develop fertilizer equations for chemical mode and integrated mode using the STCR methodology; (ii) validate these equations for achieving target yields; and (iii) assess the performance of STCR treatments with the various prevailing nutrient management strategies, in terms of grain and straw yield, economics, nutrient uptake, and use efficiency.

2. Material and Methods

2.1. Experimental Site

The study area falls in the Tarai region, with a sub-tropical-humid climate. Three kinds of field studies, i.e., 1. fertility gradient (FG) experiment with wheat in 2016–17, 2. test crop experimentation with DSR in 2017–18 to develop fertilizer equation, and 3. verification experimentation in 2018–19 were conducted at the B2 block of NEBCRC, Pantnagar University, U.S. Nagar, Uttarakhand, India (29° N latitude, 79°29′ E longitude, 243.84 m above MSL). During the rice-growing season, the total rainfall was 1355.40 mm and 1641.30 mm for 2017 and 2018, respectively. The maximum temperature fluctuated from 30.8 to 38.2 °C, while the minimum temperature fluctuated from 14.5 to 26.4 °C during 2017. During 2018, the maximum temperature fluctuated from 29.6 to 37.2 °C, while the minimum temperature fluctuated from 12.0 to 26.9 °C (Figure 1).

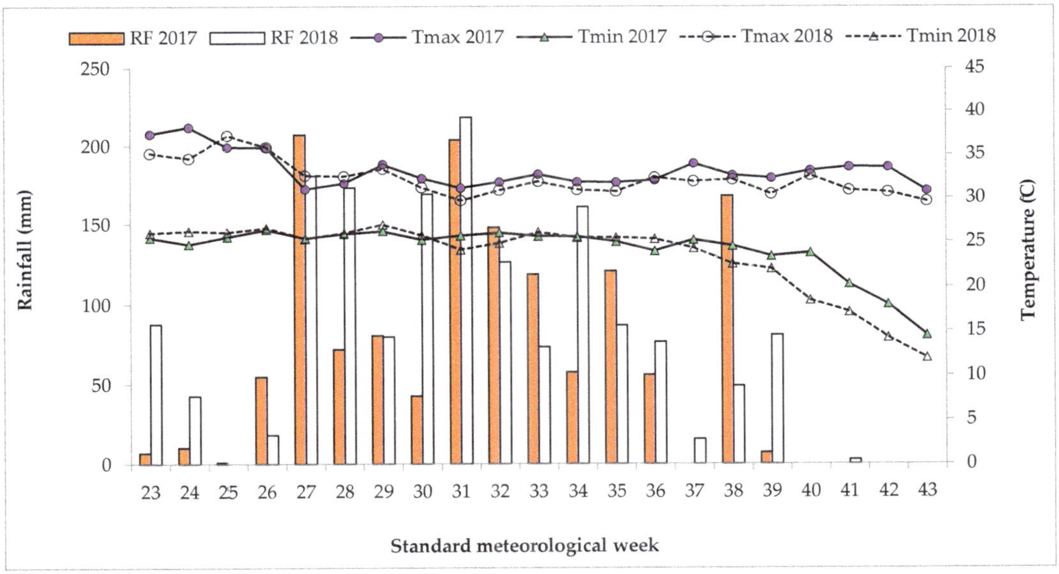

Figure 1. Standard meteorological weekly variation in the weather during the crop growing season.

The soil texture of the study area was clay-loam, under the taxonomically categorized great group named, Hapludoll. Before the start of experimentation, soil samples were randomly collected from different spots at a depth of 0–15 cm from the experimental field, and after making a composite; it was shade dried and processed and analyzed for various chemical properties. The results revealed that the values of pH [19], electrical conductivity [19], WBC (Walkley and Black organic C) [20], available N [21], available P [22], and available K [23] were 7.33, 0.41 dS m^{-1}, 0.57%, 150.53, 15.64, and 141.12 kg ha^{-1}, respectively.

2.2. Fertility Gradient (FG) Experiment

A FG experiment was performed in 2016 to nullify the previous effects on soil fertility and create an artificial FG prior to testing crops, as per Ramamoorthy et al. [24]. Strips of 75.0 m × 7.5 m area were made, and these strips were fertilized with three levels o N, P_2O_5, and K_2O (0-0-0, 100-100-100, and 200-200-200 kg ha^{-1} applied to strip I, II, anc III, respectively). A uniform dose of $ZnSO_4$ at 25 kg ha^{-1} was applied to all strips. Ar exhaust crop of wheat (variety: UP 2526) was grown during *Rabi* 2016–17 to stabilize th soil fertility and create an artificial FG. After the harvest of the wheat crop, 24 sample from the surface soil were collected from each strip and were analyzed for available N, I and K, by the method adopted for analysis of the initial soil sample, in order to assess th development of FG.

2.3. Test Crop (DSR) Study

Three varied fertility gradient strips were again split into twenty-four plots (21 trea ments + 3 controls), resulting in a total of 72 (24 × 3) plots with a size of 5 m × 3 m each. Th treatments were different identified groupings of 4 levels of N (0, 60, 120 and 180 kg ha^{-1} P_2O_5 (0, 30, 60 and 90 kg ha^{-1}), and K_2O (0, 20, 40 and 60 kg ha^{-1}), which were randomizec in each strip. The treatments (0, 5, and 10 Mg ha^{-1} FYM) were superimposed across th strips. Details of treatments are given in Table S1.

After the layout, samples were again taken from the soil surface to determine th available N, P, and K of each experimental plot before sowing, as per the procedur mentioned for analysis of the initial soil sample. The sources of N, P, K, and Zn wer urea, single super phosphate, muriate of potash, and zinc sulphate, respectively, while th organic source was FYM. Half the amount of N, the full amount of P, K, Zn, and FYM wer applied as basal. Healthy rice seeds were sown continuously at a line spacing of 20 cm a 2–3 cm depth on 8 June 2017 and were covered manually. The remaining N was applied ir two equal amounts (30 days after sowing (DAS) and 50 DAS) as a top dressing. At harves (24 October 2017), grain and by-product (straw) samples of DSR were taken from each plo for the estimation of total N, P, and K content. The applied FYM had 25% moisture, 0.63 % total N, 0.13 % total P, and 0.60% total K.

2.4. Plant Analysis

Total N, P, and K contents of the economic plant parts, i.e., grain and by-product were obtained as per the standard procedure [19]. Nutrient uptake by grain and straw was computed by multiplication of GY (kg ha^{-1}) with nutrient content in grain (%) anc SY (kg ha^{-1}) with nutrient content in the by-product (%), respectively. The summation o nutrient uptake in the grain and in straw gives the total nutrient uptake by the crop.

2.5. Basic Parameters (NR, CS, and CF)

Calculation of the basic parameters, i.e., NR, CS, and CF was computed following the formulae illustrated by Ramamoorthy et al. [24]. The fertilizer nutrient requisite fo targeted productivity was calculated as follows:

Chemical mode:

$$F_N = \frac{NR}{CF} \times 100\,T - \frac{CS}{CF} \times SN$$

$$F_{P2O5} = \frac{NR}{CF} \times 2.29 \times 100T - \frac{CS}{CF} \times 2.29 \times SP$$

$$F_{K2O} = \frac{NR}{CF} \times 1.20 \times 100T - \frac{CS}{CF} \times 1.20 \times SK$$

Integrated mode:

$$F_N = \frac{NR}{CF*} \times 100\,T - \frac{CS}{CF*} \times SN - \frac{CFYM}{CF*} \times FYM - N$$

$$F_{P2O5} = \frac{NR}{CF*} \times 2.29 \times 100\,T - \frac{CS}{CF*} \times 2.29 \times SP - \frac{CFYM}{CF*} \times 2.29 \times FYM - P$$

$$F_{K2O} = \frac{NR}{CF*} \times 1.20 \times 100\,T - \frac{CS}{CF*} \times 1.20 \times SK - \frac{CFYM}{CF*} \times 1.20 \times FYM - K$$

Here, F_N, F_{P2O5}, and F_{K2O} stand for fertilizer nitrogen, phosphorus, and potassium (kg ha^{-1}), respectively. NR denotes the nutrient requirement of nitrogen, phosphorus, and potassium (kg ha^{-1}); CF, CS, and CFYM are %-share of corresponding nutrient (N/P/K) of the total nutrient uptake from fertilizer without FYM, soil, and FYM. CF* represents the % contribution of the corresponding nutrients from fertilizers with FYM. T represents targeted yield (Mg ha^{-1}); SN, SP, and SK correspond to the STVs for the available N, P, and K (kg ha^{-1}) in the soil. FYM-N, FYM-P, and FYM-K correspond to the N, P, and K content added through FYM (kg ha^{-1}).

2.6. Verification Experiment

A field experiment was conducted with DSR (variety: ND 359) during *Kharif* 2018–19 to assess the performance of STCR treatments with different nutrient management strategies in terms of GY, SY, economics, nutrient uptake, and use efficiency in a randomized block design (RBD) with three replicates. Table 1 contains the treatment details.

Table 1. Treatment details of the verification experiment.

Treatment	Symbol	$N\text{-}P_2O_5\text{-}K_2O$-FYM*Applied (kg ha^{-1}/Mg ha^{-1} *)
Control	CK	0-0-0-0
General recommended fertilizer dose	GRD	120-60-40-0
GRD + 5 t FYM ha^{-1}	GRDFYM	120-60-40-5
Soil test based fertilizer dose (STB)	STB	200-60-40-0
STB + 5 t FYM ha^{-1}	STBFYM	200-60-40-5
STCR based fertilizer dose for TY_1	STCR TY_1	143-62-36-0
STCR TY_1 INM	STCR TY_1FYM	105-44-31-5
STCR based fertilizer dose for TY_2	STCR TY_2	169-72-43-0
STCR TY_2 INM	STCR TY_2FYM	127-52-37-5
Farmer's practice	FP	130-40-20-0

Target yield level 1= TY_1 = 4.5 Mg ha^{-1}; target yield level 2= TY_2 = 5.0 Mg ha^{-1}; * FYM application rate in terms of Mg ha^{-1} and other fertilizers were applied as kg ha^{-1}.

An initial soil sample from the surface soil was collected prior to the sowing of DSR. Fertilizer prescription equations developed in test crop experiments on DSR crops were used to calculate the amount of fertilizer nutrients for achieving TYs of 4.5 Mg ha^{-1} (TY1) and 5.0 Mg ha^{-1}(TY2). The FYM used in this experiment had 0.58 % total N, 0.15 % total P, and 0.54% total K. The sowing operation for this experiment was performed on 12 June 2018, and the crop was harvested at full maturity stage on 25 October 2018. Standard agronomic practices were followed for the growing of the DSR.

2.7. Yield and Nutrient Uptake

The DSR was harvested and threshed manually. Grain yield (GY) (kg) and straw yield (SY) (kg) was recorded from the net plot leaving the border rows and was later converted to Mg ha^{-1}. Harvest index was determined as follows:

$$\text{Harvest index (HI)} = GY \times 100/(GY + SY)$$

PE (kg ha^{-1} d^{-1}) was calculated by dividing GY (kg ha^{-1}) by the duration of the crop (days), which was constant for all the treatments (136 days). Collected samples of rice grain and by-product (rice-straw) from each treatment were processed and analyzed for total N, P, and K content by adopting the procedure given in the test crop experiment.

2.8. Economic Analysis

The cost of fertilizer (Indian rupee (INR) ha^{-1}) for various treatments in the verification experiment was worked out separately, considering the prevailing prices of fertilizers in INR at the time of their use. Gross return (value of additional yield) was calculated based on the MSP (price for minimum support) of rice set by the Indian government during the year 2018–19 and expressed as INR ha^{-1}. Net return (INR ha^{-1}) was calculated by subtracting the fertilizer cost from the gross return. B:C ratio was worked out as follows:

- B:C ratio = Net return (INR ha^{-1})/Fertilizer cost (INR ha^{-1})
- Economic efficiency was calculated as follows:
- Economic efficiency (INR ha^{-1} d^{-1}) = Net return (INR ha^{-1})/duration (days).

where the duration of DSR was constant for all the treatments (136 days).

2.9. Nutrient Use Efficiency

Nutrient (N/P/K) use efficiency parameters were calculated using the following formulae, as per [25]:

- Agronomic efficiency of nutrient (kg grain (kg nutrient)$^{-1}$)
- Agronomic efficiency (AE) = (GYF − GYC)/AFN
- Recovery efficiency of nutrient (%)
- Recovery efficiency (RE) = (TNUF − TNUC)/AFN × 100
- Partial factor productivity of nutrient (kg grain (kg nutrient)$^{-1}$)
- Partial factor productivity(PFP) of nutrient = (GYF)/AFN
- Reciprocal internal use efficiency of nutrient (kg Mg^{-1} grain yield)
- Reciprocal internal use efficiency (RIUE) of nutrient = GNU/GY

where TNUF is the total nutrient uptake of DSR from the fertilized plot (kg ha^{-1}), TNUC is the total nutrient uptake of DRS from the control plot (kg ha^{-1}), AFN is the amount of applied fertilizer nutrient (kg ha^{-1}), GYF is the grain yield of the fertilized plot (kg ha^{-1}), and GYC is the grain yield of the control plot (kg ha^{-1}). GY is the grain yield (Mg ha^{-1}). GNU is the nutrient uptake by the grain.

2.10. Statistical Analysis

Descriptive statistics was used for the test crop experiment. Data recorded in verification experiments were analyzed using the ANOVA technique [26]. Treatment means were compared with a LSD-test (least significant difference) with a probability level of 0.05.

3. Results and Discussion

3.1. FG Establishment Experiment

To allow maximum deviations in the fertility strips, gradient experiments were conducted to minimize the factors related to soil and other management practices that could affect the crop yields. Table 2 showed the range and average STVs for N, P, and K, indicating significant variation with respect to fertility strips. Strip III is highest in nutrient level, as the maximum fertilizer was applied in comparison to strip I and II. STVs for N, P, and K varied from 125.4 to 200.7 kg N ha^{-1}, 14.4 to 21.7 kg P ha^{-1}, and 122.1 to 173.6 kg K ha^{-1}, with the mean 168.8, 17.8, and 151.7 kg ha^{-1}, respectively. The STVs for N, P, and K increased with increasing fertility levels, from strip I to strip III. Ammal et al. [27] also reported that the average level of N, P, and K STVs increased with increasing fertility level, and the highest STVs were reported in strip III. The highest level STVs of N in strip III might be due to the addition of two folds more NPK fertilizers than the onefold and control [27]. The STVs of P and K in strip III were the highest owing to having a graded fertilization [27]. Dwivedi et al. [28] also explained that the reason behind the higher STVs of P might be due to its fixation, due to its immobile nature in soil.

The multiple linear regression (MLR) study showed that the effect of the strips on the STVs of N, P, and K was highly significant when it was taken as a dependent variable, separately (Table 3). This proved that the experiment created a significant fertility gradient; furthermore, it made the soil suitable for the test crop in the STCR experiment. Similar findings have also been reported by [29–32].

Table 2. Descriptive statistics of available soil nutrients (0–15), nutrients after the soil fertility gradient experiment.

Strip		Soil Available Nutrients (kg ha^{-1})		
		N	P	K
Strip I	Range	125.4–200.7	14.4–19.2	122.1–172.5
	Mean ± SD	165.2 ± 20.2	16.9 ± 1.1	144.3 ± 13.3
	(CV %)	(12.2)	(6.5)	(9.2)
	Median	163.1	17.0	143.4
Strip II	Range	125.4–200.7	15.8–19.5	125.4–173.6
	Mean ± SD	169.3 ± 21.9	17.7 ± 1.0	154.3 ± 14.1
	(CV %)	(12.9)	(5.5)	(9.1)
	Median	175.6	17.7	156.8
Strip III	Range	150.5–188.1	16.6–21.7	133.3–171.4
	Mean ± SD	172.0 ± 13.1	18.9 ± 1.5	156.5 ± 11.0
	(CV %)	(7.6)	(7.8)	(7.0)
	Median	175.6	18.2	157.4
All strips	Range	125.4–200.7	14.4–21.7	122.1–173.6
	Mean ± SD	168.8 ± 18.7	17.8 ± 1.5	151.7 ± 13.8
	(CV %)	(11.1)	(8.1)	(9.1)
	Median	169.3	17.7	153.4

SD, standard deviation; CV (%), co-efficient of variation (%).

Table 3. R^2, CV (%), and SD of whole plots.

Dependent Variable	*p* Level	R^2	Average	SD	CV (%)
SN	<0.01	0.76	168.82	18.71	11.08
SP	<0.01	0.79	17.81	1.42	7.99
SK	<0.01	0.71	151.73	13.77	9.08

SN, SP, and SK denote soil nitrogen, phosphorus, and potassium, correspondingly.

3.2. Yield and Nutrient Uptake

Descriptive statistics for GY, SY, and nutrient uptake for DSR is given in Table 4. The GY and SY of DSR in the whole plot ranged from 2273 to 6705 kg ha^{-1} and 3750 kg ha^{-1} to 12,386 kg ha^{-1}, with mean values of 4291 kg ha^{-1} and 7699 kg ha^{-1}, respectively. The range of GY and SY was 2273–5341 kg ha^{-1}, 3864–11,250 kg ha^{-1} in strip I; 2841–6705 kg ha^{-1}, 3750–12,273 kg ha^{-1} in strip II; and 3636–6477 kg ha^{-1}, 4545–12,386 kg ha^{-1} in strip III. The mean for GY and SY was 3902 kg ha^{-1} and 7014 kg ha^{-1}, 4276 kg ha^{-1} and 7869 kg ha^{-1}, and 4697 kg ha^{-1} and 8215 kg ha^{-1}, respectively, in strips I, II, and III. Likewise, GY and SY, total nitrogen uptake (TUN), phosphorus uptake (TUP), and potassium uptake (TUK) followed the order strip I < strip II < strip III. In the whole plot TUN, TUP, and TUK ranged from 24.0 to 171.0 kg ha^{-1}, 4.3 to 45.7 kg ha^{-1}, and 44.0 to 238.6 kg ha^{-1}, with a mean of 87.6 kg ha^{-1}, 19.3 kg ha^{-1}, and 131.6 kg ha^{-1}, respectively. TUN in strips I, II, and III ranged from 24.0–100.9 kg ha^{-1}, 43.2–152.1 kg ha^{-1}, and 53.9–171.0 kg ha^{-1}, with mean values of 66.2 kg ha^{-1}, 88.2 kg ha^{-1}, and 108.5 kg ha^{-1}, respectively. TUP ranged from 4.3–18.1 kg ha^{-1}, 4.3–25.5 kg ha^{-1}, and 10.1–45.7 kg ha^{-1}, with a mean of 11.4 kg ha^{-1}, 16.0 kg ha^{-1}, and 30.4 kg ha^{-1}, respectively. TUK ranged from 44.0–186.4 kg ha^{-1}, 60.2–238.6 kg ha^{-1}, and 72.0–236.8 kg ha^{-1}, with mean of 106.5 kg ha^{-1}, 136.1 kg ha^{-1}, and 152.1 kg ha^{-1}, respectively.

Table 4. Descriptive statistics of DSR GY and nutrient uptake.

Strip		Grain Yield (kg ha^{-1})	Straw Yield (kg ha^{-1})	N Uptake (kg ha^{-1})	P Uptake (kg ha^{-1})	K Uptake (kg ha^{-1})
Strip I	Range	2273–5341	3864–11,250	24.0–100.9	4.3–18.1	44.0–186.4
	Mean ± SD (CV%)	3902 ± 899 (23.0)	7014 ± 1752 (25.0)	66.2 ± 17.3 (24.4)	11.4 ± 3.5 (31.0)	106.5 ± 32.8 (30.8)
	Median	3864	6875	71.4	11.5	104.7
Strip II	Range	2841–6705	3750–12,273	43.2–152.2	4.3–25.5	60.2–238.6
	Mean ± SD (CV%)	4276 ± 1098 (25.7)	7869 ± 2254 (28.6)	88.2 ± 30.9 (32.7)	16.0 ± 6.0 (37.4)	136.1 ± 46.1 (33.9)
	Median	4034	7727	86.6	17.5	135.2
Strip III	Range	3636–6477	4545–12,386	53.9–171.0	10.1–45.7	72.0–236.8
	Mean ± SD (CV%)	4697 ± 897 (19.1)	8215 ± 2245 (27.3)	108.5 ± 30.7 (26.4)	30.38 ± 9.6 (31.6)	152.07 ± 45.8 (30.1)
	Median	4375	8580	125.2	32.4	155.0
All strips	Range	2273–6705	3750–12,386	24.0–171.0	4.3–45.7	44.0–238.6
	Mean ± SD (CV%)	4291 ± 1010 (23.5)	7699 ± 2129 (27.6)	87.6 ± 32.5 (34.6)	19.3 ± 10.6 (55.0)	131.6 ± 45.6 (34.7)
	Median	4091	7580	84.0	16.6	124.2

SD, standard deviation; CV, coefficient of variation (%).

The grain yield of DSR had a strong correlation ($p < 0.001$) with total N uptake ($r^2 = 0.578$), followed by total K uptake ($r^2 = 0.523$) and total P uptake ($r^2 = 0.376$) (Figure 2). Variability in FG and the application of variable doses of nitrogenous, phosphorus, and potassium fertilizers gave impactful results on the GY and SY of DSR. The highest GY and SY were obtained in strip III due to the overall highest nutrient application. Plots where NPK fertilizer was applied with the FYM obtained higher GY and SY, as compared to sole NPK fertilizer application. Expected outcomes were also reported by [11] in beetroot, [33] in wheat, [34] in rice, and [35] in maize crops under different conditions. TUN was highest in strips III and II compared to I, as also reported by [36], and it might have been the sufficient availability of N fertilizer in the III strip which created a favorable N uptake. An adequate dose of N application, and enhanced absorption and accumulation resulted in higher GY, SY, and uptake (NPK), as also reported by [37]. The possible reason behind the highest TUP being in strip III was attributed to better root proliferation, having a graded P application [31,38–41]. The higher dose of N application stimulated the vegetative and root foraging capacity, meaning the crops require additional P and K, and increased the TUP in the crops [42–44]. The highest total potassium uptake by a crop was recorded with strip I to strip III and might be attributed to the higher application of fertilizer potassium [31,41]. Panaullah et al. [45] also reported that the majority of potassium uptake was in straw, as compared to grain. A similar effect of FYM on phosphorus uptake by crop plants has previously been reported [46,47].

3.3. Evolution of Basic Parameters

The basic parameters NR, CS, CFYM, CF, and CF* were required for the computation of prescription equations for the TY of DSR, with and without FYM, which is necessary for developing fertilizer doses. Basic parameters are developed with the help GY, TUN, TUP, TUK, initial STVs of N, P, and K, applied nutrient rates (N, P, and K) through fertilizers and FYM. These parameters are shown in Table 5.

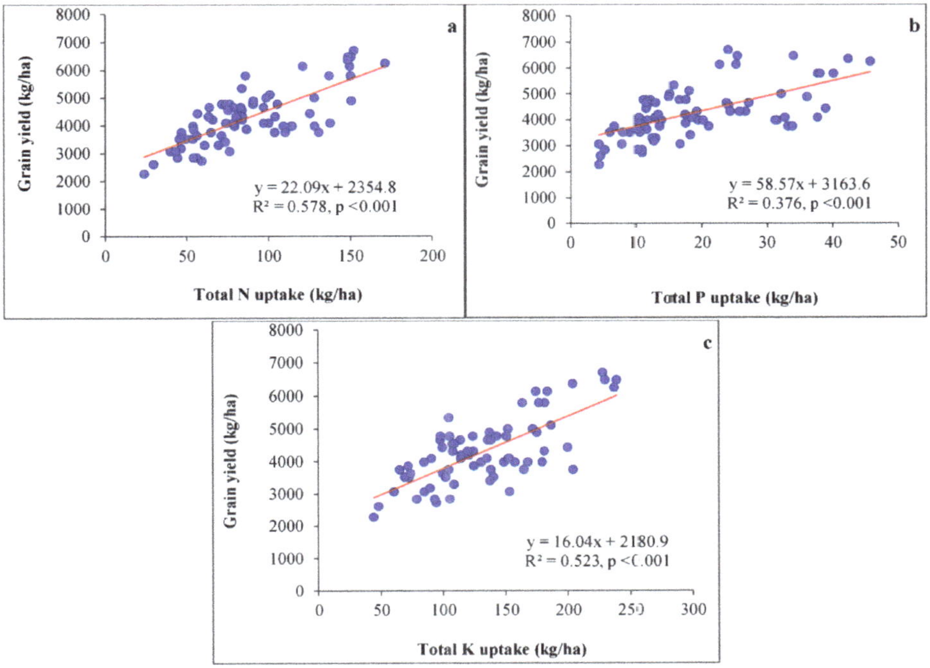

Figure 2. Linear correlations between grain yield and total N (**a**), P (**b**), and K (**c**) uptake in direct-seeded rice in the test crop experiment.

Table 5. Basic parameters for calculating fertilizer requirement, with and without FYM, for the targeted yield of DSR.

Basic Parameter	Nutrient		
	N	P	K
Nutrient requirement (NR) (kg 100 kg^{-1})	2.01	0.44	3.06
Soil nutrient-supply (CS) (%)	22.05	37.34	41.48
Nutrients from fertilizers only (CF) (%)	38.08	49.93	252.98
Nutrientsupply by fertilizer with FYM (CF*) (%)	44.83	60.57	278.70
Nutrients from FYM only (CFYM) (%)	23.25	28.34	16.80

The NR was 2.01 kg, 0.44 kg, and 3.06 kg for N, P, and K, respectively. The NR for K was 1.52 and 6.95 times higher than N and P, respectively. Similar results were also reported by [48], who observed 1.48 kg N, 1.05 kg P_2O_5, and 1.86 K_2O as required per q of rice grain, respectively. In DSR, CS-22.05, 37.34%, and 41.48%; CF-38.08, 49.93%, and 252.98%; CF*-44.83, 60.57%, and 278.70%; and CFYM- 23.25, 28.34, and 16.80, respectively, for N, P, and K was found. The contribution of nutrients from the fertilizer was high compared to the contribution from the soil, owing to having a higher and rapid nutrient availability in the inorganic form from the fertilizers. CFYM was calculated with the help of data using the FYM treated and control plots, and it followed the order P > N > K. Bera et al. [49] also reported a similar trend. The nutrient supply through a native source in all the plots and their interaction effects might be the reason that the addition of CS, CFYM, CF, and CF* was not equal in percentage. The higher value of CF and CF* for potassium for the uptake in native soil source might be due to the interaction effects of the optimum supply of N and P, combined with the priming effects of potassium; this may increase the release of soil exchangeable and non-exchangeable potassium [50]. Similar findings were also reported by [27,51].

3.4. Prescription Equations in Chemical and Integrated Mode in DSR

Fertilization equations based on STCR were made to get the target yield of DSR with and without FYM by taking fundamental crop-production parameters *viz.* nutrient requirement (NR), efficiencies of fertilizers (CF and CF*), soil test (CS), and organic source FYM (CFYM) (Table 6).

Table 6. Fertilizer-prescription equations as per the STCR for the target-yield of DSR.

Nutrient Requirement (kg ha^{-1})	Without FYM	With FYM
Nitrogen (FN)	5.28 T—0.579 SN	4.48 T—0.492 SN—0.519 FYM-N
Phosphorus (FP_2O_5)	2.02 T—1.71 SP	1.66 T—1.41 SP—1.07 FYM-P
Potassium (FK_2O)	1.45 T—0.200 SK	1.32 T—0.179 SK—0.072 FYM-K

T is yield target in q ha^{-1}; 1 quintal (q) = 100 kg; SN, SP, SK are alkaline $KMnO_4$-N, Olsen's-P and NH_4OAc-K in kg ha^{-1}, respectively; FYM-N, FYM-P, and FYM-K are the amounts of N, P, and K in kg ha^{-1} applied through FYM, respectively.

Sharma and Singh [52], Benbi and Benipal [53], and Verma et al. [39] also developed fertilizer prescription equations based on STCR. Prescription equations are simple to use, and by putting values of the TY and STVs of N, P, and K into the equation, one can find the fertilizer requirement precisely for that crop in certain climatic conditions. Tables 2–4 show that the high STVs of NPK require smaller amounts of additional chemical and organic fertilizers. For the range of STVs of N, P, and K and TY of 4000, 4500, and 5000 kg ha^{-1}, ready reckoners were prepared for NPK alone and NPK with 5 Mg ha^{-1} FYM. It is understandable that in the experimental outcomes the fertilizer N, P_2O_5, and K_2O requirements for the desired TY of DSR decreased with increasing STVs. For the TY of 4000, 4500, and 5000 kg ha^{-1} of DSR without FYM with STVs of available N, P, and K as 200:20:120 kg ha^{-1}, the amounts of fertilizer N, P_2O_5, and K_2O to be applied are 95.40, 121.80, and 148.20 kg N ha^{-1}; 46.44, 56.53, and 66.62 kg P_2O_5ha^{-1} 1; and 34.46, 41.72, and 48.98 kg K_2O ha^{-1}, respectively. However, when 5 Mg ha^{-1} FYM was applied along with NPK, the amount of fertilizer N, P_2O_5, and K_2O were reduced to 64.45, 86.85, and 109.25 kg N ha^{-1}; 31.11, 39.42, and 47.73 kg P_2O_5ha^{-1}; and 29.18, 35.78, and 42.38 kg K_2O ha^{-1}, respectively. FYM application in combination with chemical fertilizer resulted in savings of the chemical fertilizer and, ultimately, the cost of cultivation. The requirement of fertilizer when FYM was applied along with chemical fertilizer and when chemical fertilizer was applied alone was worked out separately for a particular STV and TY, and the difference between the two was a fertilizer nutrient equivalent (FNE) to FYM. Applying 5 Mg FYM ha^{-1} along with chemical fertilizers, on average, saved 36.04 kg N ha^{-1}, 16.62 kg P_2O_5 ha^{-1}, and 4.60 kg K_2O ha^{-1} at the range of STVs and varying TY. The effect of FYM varied with crops, soil fertility level, and TY. FNE to FYM decreased with increasing STVs, while it increased with increasing TYs. Moreover, in the short term experiments, fertilizer savings were not up to the mark, but in the long term, they might give better results, as OM addition improves the crop sustainability and soil quality by improving the soil physical, chemical, and biological health. A similar finding for FNE and fertilizer savings with FYM was also reported by [54] for rice.

3.5. Verification Experiment

Verification trials are an important way to ascertain the validity of the results acquired from fertilizer prescription equations, before recommendations to farmers for higher profitability and efficiency than the GRD.

3.6. Yield and Production Efficiency

The production efficiency (PE) was affected by the different nutrient management treatments and are presented in Figure 3. Different nutrient management treatments gave significantly higher GY and SY than the CK (2386 and 3977 kg ha^{-1}). The highest GY and SY were obtained with STCR TY2FYM (4962 kg ha^{-1} and 6174 kg ha^{-1}), which was comparable with that of STCR TY2 (4924 kg ha^{-1}), STCR TY1FYM (4545 kg ha^{-1}), STCRTY1 (4508 kg ha^{-1}), and STBFYM (4432 kg ha^{-1}) for GY and STBFYM (6250 kg ha^{-1}); and STCR TY2 (6174 kg ha^{-1}), STCR TY1FYM (5795 kg ha^{-1}), GRDFYM (5871 kg ha^{-1}), STCRTY1 (5682 kg ha^{-1}), and GRD (5644 kg ha^{-1}) for SY. The STCR-IPNS based fertilizer recommendations resulted in 21.20 % in TY1 and 32.32 % in TY2 increases in the GY of DSR over the GRD.

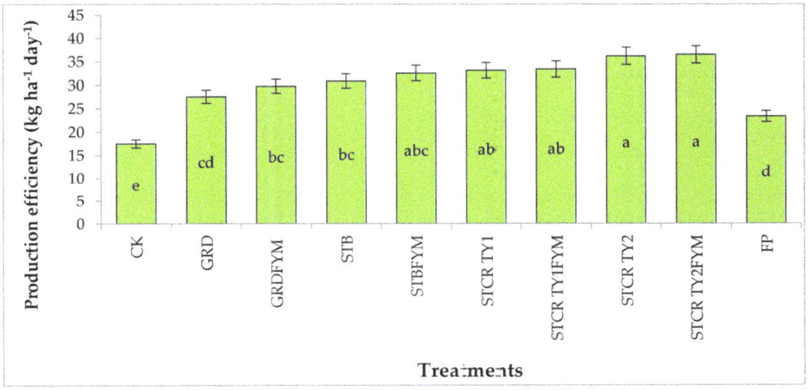

Figure 3. DSR production efficiency under different treatment combinations. Bars with the same letter are not different at a 5% probability level; error bars indicate the standard deviation of the respective means.

The chemical mode of the STCR based fertilizer recommendation also gave a higher GY (20.21% in TY1 and 32.31% in TY2) compared to the GRD. The STCR-based fertilizer recommendation gave markedly higher rice GY than other approaches of fertilizer prescription [55]. In general, organic manures+chemical fertilizers significantly increase the available soil nutrient content and also improve the soil environment [56,57]. The increased yield under the STCR approach, with and without FYM, might have been due to the balanced application of fertilizers, as per the soil test, and the crop demand for growth and development. Furthermore, sufficient nutrient availability might have played a significant role in the rice-physiology for increased dry matter production and resulted in higher SY under the STCR-TY approach. Better vegetative growth, along with high yield attributes, resulted in a higher GY of rice [58]. In a similar experiment with two TYs (4.5 Mg ha^{-1} and 5.0 Mg ha^{-1} for maize), Venkatesh et al. [59] reported that a STCR-target yield based fertilizer recommendation with or without FYM led to 13.14–35.38% and 11.67–26% enhancement in GY and SY, respectively, under maize-lentil rotation. However, the integrated nutrient management based-STCR-TY approach led to a higher GY and SY than the sole use of chemical fertilizers. The results of the verification experiment revealed the achievement of attaining the desired TY of DSR was noted as +1.00, +0.17, −0.76, and −1.52 per cent with STCR TY1FYM, STCR TY1, STCR TY2FYM, and STCR TY2, respectively, which were within a variation of ±10%, proving the validity of the equations [60]. In the present study, the STCR-based INM schedule registered a relatively higher percent achievement than the STCR-chemical fertilizer based approach. Furthermore, a lower targeted yield was more easily achieved than the higher ones. Similar findings were also reported by [61]. The maximum HI was recorded in STCRTY1 (44.30%), which differed significantly from CK (37.55%) and FP (37.51%). Higher dry matter partitioning to grain might be the reason for the higher HI under STCRTY1 treatment. The PE was highest in STCRTY2FYM (36.49 kg ha^{-1}d^{-1}), which was comparable with STCRTY2 (36.21 kg ha^{-1}d^{-1}), STCRTY1FYM (33.42 kg ha^{-1}d^{-1}), STCRTY1 (33.14 kg ha^{-1}d^{-1}), and STBFYM (32.59 kg ha^{-1}d^{-1}). The lowest PE was recorded with CK (17.55 kg ha^{-1}d^{-1}) which was markedly lower than the other treatments. The higher PE with STCRTY2FYM was due to the higher GY under this treatment (Table 7). Furthermore, the adoption of the STCR-TY approach led to a significant increase in PE, ranging from 5.57 kg ha^{-1} d^{-1} to 8.92 kg ha^{-1} d^{-1}, over the GRD.

Table 7. Yield and harvest index of direct-seeded rice under different nutrition.

Treatments	Grain Yield (kg ha^{-1})	Straw Yield (kg ha^{-1})	Harvest Index (%)
CK	2386 ^{e}	3977 ^{c}	37.55 ^{b}
GRD	3750 ^{cd}	5644 ^{ab}	39.93 ^{ab}
GRDFYM	4053 ^{bc}	5871 ^{ab}	40.84 ^{ab}
STB	4205 ^{bc}	5341 ^{b}	44.05 ^{a}
STBFYM	4432 ^{ab}	6250 ^{a}	41.45 ^{ab}
STCRTY$_1$	4508 ^{ab}	5682 ^{ab}	44.30 ^{a}
STCR TY$_1$FYM	4545 ^{ab}	5795 ^{ab}	44.05 ^{a}
STCR TY$_2$	4924 ^{a}	6174 ^{ab}	44.03 ^{a}
STCR TY$_2$FYM	4962 ^{a}	6402 ^{a}	43.91 ^{a}
FP	3182 ^{d}	5303 ^{b}	37.51 ^{b}
Significance level	*	*	*

* Significance at 5% level; values followed by the same letter are not different at a 5% probability level.

3.7. NPK Uptake

Data on NPK uptake (grain, straw, and total) by DSR under different treatments are given in Table 8. The maximum N uptake by grain, straw, and total N uptake was recorded in STCR TY2FYM, which was significantly higher than all other treatments, except STCR TY2 (79.65 kg ha^{-1}) and STCR TY1FYM (75.30 kg ha^{-1}) for grain N uptake; STCR TY1FYM (39.22 kg ha^{-1}), STBFYM (39.14 kg ha^{-1}), and STCR TY2 (37.47 kg ha^{-1}) for straw N uptake; and STCR TY2 (117.12 kg ha^{-1}) and STCR TY1FYM (114.52 kg ha^{-1}) for total N uptake. The lowest N uptake by grain, straw, and total N uptake was recorded in CK (26.73 kg ha^{-1}, 10.23 kg ha^{-1}, and 36.96 kg ha^{-1}). The highest P uptake in grain, as well as total P uptake, was found with treatment STCR TY2FYM (18.44kg ha^{-1} and 29.62 kg ha^{-1}) which was significantly higher than the other treatments. However, the maximum P uptake in straw, found in STCR TY2FYM (11.18 kg ha^{-1}), was on par with STCR TY2 (9.97 kg ha^{-1} and STBFYM (9.56 kg ha^{-1}) and significantly higher than the other treatments.

The highest P uptake in grain, as well as total P uptake, was found with treatment STCR TY2FYM (18.44kg ha^{-1} and 29.62 kg ha^{-1}) which was significantly higher than other treatments. However, maximum P uptake in straw found in STCR TY2FYM (11.18 kg ha^{-1} was at par with STCR TY2 (9.97 kg ha^{-1}) and STBFYM (9.56 kg ha^{-1}) and significantly higher than other treatments. The lowest values for P uptake by grain, straw and total P uptake were recorded with CK (4.86 kg ha^{-1}, 3.82 kg ha^{-1}and 8.67 kg ha^{-1}). Maximum rice-grain K-uptake, rice-straw K-uptake and total K-uptake was found in STCR TY2FYM (31.55 kg ha^{-1}, 106.76 kg ha^{-1}and 138.30 kg ha^{-1}) which was at par with STCR TY2 (29.93 kg ha^{-1}, 98.59 kg ha^{-1}and 128.52 kg ha^{-1}). The lowest values for P uptake by grain, straw, and total P uptake were recorded with CK (4.86 kg ha^{-1}, 3.82 kg ha^{-1}, and 8.67 kg ha^{-1}). The maximum rice-grain K-uptake, rice-straw K-uptake, and total K-uptake was found in STCR TY2FYM (31.55 kg ha^{-1}, 106.76 kg ha^{-1}, and 138.30 kg ha^{-1}), which was at par with STCR TY2 (29.93 kg ha^{-1}, 98.59 kg ha^{-1}, and 128.52 kg ha^{-1}). The lowest grain and straw and total K uptake were recorded with CK (7.83 kg ha^{-1}, 49.72 kg ha^{-1} and 57.55 kg ha^{-1}). The favorable soil conditions with STCR treatments might have paved the way for better absorption and mobilization, in tune with the growth and activity of roots, which may have caused a better production of dry matter and absorption of nutrients and increased grain and straw yield, and N, P, and K contents, which were reflected in their higher uptakes.

Table 8. NPK uptake by DSR under different nutrition.

Treatments	Nitrogen Uptake (kg ha^{-1})			Phosphorus Uptake (kg ha^{-1})			Potassium Uptake (kg ha^{-1})		
	Grain	Straw	Total	Grain	Straw	Total	Grain	Straw	Total
CK	26.73 ^{g}	10.23 ^{e}	36.96 ^{f}	4.86 ^{f}	3.82 ^{g}	8.67 ^{g}	7.83 ^{g}	49.72 ^{e}	57.55 ^{f}
GRD	50.75 ^{ef}	18.38 ^{d}	69.13 ^{e}	8.47 ^{e}	6.55 ^{ef}	15.02 ^{f}	16.50 ^{f}	77.69 ^{cd}	94.19 ^{de}
GRDFYM	58.65 ^{cde}	26.04 ^{bc}	84.69 ^{d}	10.81 ^{d}	7.79 ^{cdef}	18.60 ^{e}	19.26 ^{ef}	82.39 ^{cd}	101.65 ^{cd}
STB	57.86 ^{de}	26.17 ^{bc}	84.03 ^{d}	11.53 ^{cd}	7.40 ^{def}	18.93 ^{de}	21.36 ^{de}	75.15 ^{cd}	96.51 ^{de}
STBFYM	67.27 ^{bcd}	39.14 ^{a}	106.41 ^{bc}	12.31 ^{cd}	9.56 ^{abc}	21.87 ^{cd}	22.99 ^{cd}	89.17 ^{bc}	112.16 ^{c}
STCRTY$_1$	70.48 ^{bc}	30.63 ^{b}	101.10 ^{c}	12.90 ^{cd}	7.96 ^{cde}	20.85 ^{cde}	25.04 ^{cd}	82.87 ^{cd}	107.91 ^{cd}
STCR TY$_1$FYM	75.30 ^{ab}	39.22 ^{a}	114.52 ^{ab}	13.40 ^{bc}	8.95 ^{bcd}	22.35 ^{c}	26.40 ^{bc}	88.24 ^{bc}	114.64 ^{bc}
STCR TY$_2$	79.65 ^{ab}	37.47 ^{a}	117.12 ^{ab}	15.45 ^{b}	9.97 ^{ab}	25.42 ^{b}	29.93 ^{ab}	98.59 ^{ab}	128.52 ^{ab}
STCR TY$_2$FYM	83.50 ^{a}	43.35 ^{a}	126.85 ^{a}	18.44 ^{a}	11.18 ^{a}	29.62 ^{a}	31.55 ^{a}	106.76 ^{a}	138.30 ^{a}
FP	40.83 ^{f}	20.97 ^{cd}	61.81 ^{e}	6.71 ^{e}	5.95 ^{f}	12.66 ^{f}	14.97 ^{f}	70.19 ^{d}	85.16 ^{e}
Significance level	*	*	*	*	*	*	*	*	*

* Significance at 5% level; values followed by the same letter are not different at 5% probability level.

3.8. Fertilizer Economics and Economic Efficiency

The data on the fertilizer economics in DSR, as affected by different treatments (Table 9), revealed that the value of the additional yield of DSR over CK was the highest with STCR TY2FYM (2576 kg ha^{-1}), followed by STCR TY2 (2538 kg ha^{-1}), and was the lowest with FP (795 kg ha^{-1}). Similarly, the highest net return was noted in STCR TY2 (INR 38,738 ha^{-1}), followed by STCR TY2FYM (INR 38,293 ha^{-1}), and was the lowest in FP (INR 10,431 ha^{-1}). Similarly, the highest B:C ratio was noted in STCR TY2 (6.83), followed by STCRTY1 (6.68), and was the lowest in FP (2.99). The STCR-based nutrient recommendation net return and B:C cost ratio ranged from INR 31,696–38,738 ha^{-1} and 5.21–6.83, respectively, higher than the GRD (INR 19,287 ha^{-1} and 4.21, respectively). The highest B:C ratio with STCR TY2 might have been due to the highest net return with this treatment. Similarly, a higher benefit–cost ratio with STCR-based fertilizer treatments than the GRD or farmer practice has been reported [62]. Net returns from the improved practice (STCR technology) were substantially higher than the FP for DSR. Similarly, higher indices of economic analysis, such as gross and net return and benefit:cost ratio, than the GRD in transplanted rice under rainfed Alfisols have been noticed [15]. The highest economic efficiency was recorded with STCR TY2 (INR 285 ha^{-1} d^{-1}), which was due to the highest net return being obtained under this treatment (Table 10). The economic efficiency of the DSR was enhanced by INR 91 ha^{-1} d^{-1} to INR 143 ha^{-1} d^{-1} due to the STCR-target yield based nutrient recommendation, compared to the application of GRD.

3.9. Nutrient Use Efficiency (NUE)

Table 10 shows the NUE as influenced by various nutrition. Nutrient management practices caused a significant variation in NUE under the DSR system of rice. In general, the STCR treatments had a higher AEN compared to the other treatments. The highest AEN occurred with STCR TY2FYM (17.4 kg grain (kg N)$^{-1}$), which was statistically comparable with STCR TY1FYM (17.1 kg grain (kg N)$^{-1}$), STCR TY2 (15.0 kg grain (kg N)$^{-1}$), and STCR TY1 (14.8 kg grain (kg N)$^{-1}$). STCR TY2FYM increased the AEN significantly by 6.0 kg grain (kg N)$^{-1}$ compared with GRD. The lowest AEN occurred in FP (6.1 kg grain (kg N)$^{-1}$). The AEP and AEK were significantly higher in STCR-based nutrient management practices than the application of the GRD alone, except for the AEK in STCR TY2FYM and STCR TY1FYM, where they were at par with the GRD. An AEN of 6.8–34.2 [63] and AEK of 28.4–55.3 [64] in rice crops have been reported. Similarly, the REN, REP, and REK increased significantly with STCR treatments compared to application of the GRD alone, except for REK, where the STCR TY1FYM and GRD were comparable.

Table 9. Fertilizer-economics in DSR under different treatments.

Treatments	Additional Yield (kg ha^{-1})	Value of Additional Yield (INR)	Fertilizer and FYM Cost (INR)	Net Return (INR ha^{-1})	B:C Ratio	Economic Efficiency (INR ha^{-1} day^{-1})
CK	0	–	–	–	–	-
GRD	1364 ^{cd}	23,864 ^{cd}	4576	19,287 ^{de}	4.21 ^{bcd}	142 ^{cd}
GRDFYM	1667 ^{bc}	29,167 ^{bc}	7076	22,090 ^{cd}	3.12 ^{d}	162 ^{bcd}
STB	1818 ^{bc}	31,818 ^{bc}	5504	26,314 ^{bcd}	4.78 ^{abcd}	193 ^{bc}
STBFYM	2045 ^{abc}	35,795 ^{abc}	8004	27,791 ^{abcd}	3.47 ^{cd}	204 ^{bc}
STCRTY$_1$	2121 ^{ab}	37,121 ^{ab}	4836	32,286 ^{abc}	6.68 ^{a}	237 ^{ab}
STCR TY$_1$FYM	2159 ^{ab}	37,784 ^{ab}	6089	31,696 ^{abc}	5.21 ^{abc}	233 ^{abc}
STCR TY$_2$	2538 ^{a}	44,413 ^{a}	5675	38,738 ^{a}	6.83 ^{a}	285 ^{a}
STCR TY$_2$FYM	2576 ^{a}	45,076 ^{a}	6782	38,293 ^{ab}	5.65 ^{ab}	282 ^{a}
FP	795 ^{d}	13,920 ^{d}	3489	10,431 ^{e}	2.99 ^{d}	77 ^{d}
Significance level	*	*	-	*	*	*

* Significance at 5% level; N: 11.6, P_2O_5: 38.89, and K_2O:21.27 INR kg^{-1}, FYM: 0.5 INR kg^{-1}, rice grain: 17.50 INR kg$^{-1.}$ values followed by the same letter are not different at a 5% probability level.

Table 10. Nutrient use efficiency in DSR, as influenced by different treatments.

Treatments	NUE				PUE				KUE			
	AE	RE	PFP	RIUE	AE	RE	PFP	RIUE	AE	RE	PFP	RIUE
CK	-	-	-	11.20 ^{h}	-	-	-	2.03 ^{h}	-	-	-	3.28 ^{h}
GRD	11.4 ^{bc}	26.8 ^{cde}	31.3 ^{bc}	13.53 ^{f}	22.7 ^{de}	10.6 ^{e}	62.5 ^{de}	2.26 ^{g}	34.1 ^{bc}	91.6 ^{d}	93.8 ^{de}	4.40 ^{g}
GRDFYM	11.8 ^{bc}	33.8 ^{c}	28.7 ^{c}	14.47 ^{e}	25.5 ^{cde}	15.2 ^{d}	61.9 ^{e}	2.67 ^{f}	27.9 ^{c}	73.9 ^{d}	67.9 ^{g}	4.75 ^{fg}
STB	9.1 ^{cd}	23.5 ^{de}	21.0 ^{de}	13.77 ^{f}	30.3 ^{cd}	17.1 ^{d}	70.1 ^{cde}	2.74 ^{ef}	45.5 ^{ab}	97.4 ^{cd}	105.1 ^{cd}	5.08 ^{ef}
STBFYM	9.2 ^{cd}	31.4 ^{cd}	20.0 ^{e}	15.17 ^{d}	31.2 ^{cd}	20.2 ^{cd}	67.7 ^{de}	2.78 ^{de}	34.3 ^{bc}	91.5 ^{d}	74.2 ^{fg}	5.18 ^{de}
STCRTY$_1$	14.8 ^{ab}	44.9 ^{b}	31.5 ^{bc}	15.63 ^{cd}	34.2 ^{bc}	19.6 ^{c}	72.7 ^{cd}	2.86 ^{d}	58.9 ^{a}	139.9 ^{ab}	125.2 ^{b}	5.55 ^{cd}
STCR TY$_1$FYM	17.1 ^{a}	61.5 ^{a}	36.0 ^{a}	16.57 ^{ab}	43.6 ^{ab}	27.6 ^{b}	91.9 ^{a}	2.95 ^{c}	42.6 ^{bc}	112.6 ^{bcd}	89.6 ^{def}	5.82 ^{bc}
STCR TY$_2$	15.0 ^{ab}	47.4 ^{b}	29.1 ^{c}	16.10 ^{bc}	35.2 ^{abc}	23.3 ^{bc}	68.4 ^{de}	3.13 ^{b}	59.0 ^{a}	165.1 ^{a}	114.5 ^{bc}	6.12 ^{ab}
STCR TY$_2$FYM	17.4 ^{a}	60.7 ^{a}	33.5 ^{ab}	16.80 ^{a}	44.8 ^{a}	36.5 ^{a}	86.3 ^{ab}	3.71 ^{a}	45.4 ^{ab}	142.4 ^{ab}	87.5 ^{ef}	6.37 ^{a}
FP	6.1 ^{d}	19.1 ^{e}	24.5 ^{d}	12.83 ^{g}	19.9 ^{e}	10.0 ^{e}	79.5 ^{bc}	2.11 ^{h}	39.8 ^{bc}	138.0 ^{abc}	159.1 ^{a}	4.70 ^{g}
Significance level	*	*	*	*	*	*	*	*	*	*	*	*

* Significance at 5% level; Values followed by the same letter are not different at 5% probability level.

The REK in this experiment varied from 73.85–165.05, which is in the range of 65.7–366.2% reported by [65] in a rice crop. Except for STCR TY1FYM, all other STCR-based nutrient management practices were comparable with the GRD in the case of PPFN. Except for STCR TY2FYM and STCR TY1FYM, all other STCR-based nutrient management practices were comparable with the GRD in the case of PPFP. Except for STCR TY2 and STCR TY1, all other STCR-based nutrient management practices were comparable with the GRD in case of PPFK. A PFPK of 62.42–191.33 has been reported [64]. In general, the reciprocal internal use efficiency (RIUE) followed the order RIUEN>RIUEK>RIUEP [25]. The RIUE for nitrogen, phosphorus, and potassium was enhanced markedly with STCR-based nutrient management practices, compared to GRD alone. Significantly lower RIUE for N, P, and K were recorded with CK, which was at par with FP in the case of RIUEP. A balanced application of plant nutrients increases the GY and nutrient use efficiency [66]. The relatively higher AE with a STCR-based application of fertilizers compared to common recommendations might be due to the balanced supply and efficient utilization of nutrients due to synergistic effects of fertilizers and the applied FYM [18].

4. Conclusions

The fertilizer prescription equations and ready reckoners developed in the present experiment, considering the crop NPK demand and soil NPK supply, fertilizer, and FYM, can successfully be used as a guide for implementation of integrated nutrient management in DSR. Fertilizers equations developed for STCR-INM and STCR chemical mode achieved the desired target yields with −1.5 to +1.0 percent, proving the validity of these equations. Nutrient management through the STCR-INM and STCR chemical mode could raise the yield, economic return, and efficiency of applied nutrients in direct-seeded rice over the generally recommended dose and prevailing farmer practices. Therefore, STCR-INM and STCR chemical mode based nutrient management can be recommended as an effective tool for balanced fertilization.

Supplementary Materials: The following are available online at https://www.mdpi.com/article/10.3390/agronomy11091756/s1, Table S1: Detailed N, P and K treatment combinations used for test crop experimentation.

Author Contributions: Conceptualization: V.K.S., P.G., G.N., S.S.D., S.S.M. and B.P.; methodology: V.K.S., P.G. and G.N.; formal analysis: V.K.S., P.G. and G.N.; data curation: V.K.S., P.G. and G.N.; statistical expertise: V.K.S., P.G. and G.N.; writing—original draft preparation: V.K.S., P.G., G.N., S.S.D., S.S.M. and B.P.; writing—review and editing: P.G., G.N., W.F.A., A.G., S.S. and A.H.; visualization: V.K.S., P.G. and G.N.; supervision: P.G. and G.N.; project administration: A.G., S.S. and A.H.; funding acquisition: W.F.A., A.G., S.S. and A.H. All authors have read and agreed to the published version of the manuscript in 'Agronomy'.

Funding: This research was funded by the ICAR-SRF Fellowship programme of the Government of India. The work was also supported by the Taif University Researchers Supporting Project number (TURSP-2020/53), Taif University, Taif, Saudi Arabia.

Institutional Review Board Statement: Not applicable.

Informed Consent Statement: Not applicable.

Data Availability Statement: Data will be available after request.

Acknowledgments: Authors are grateful to the Head, Department of Soil Science, GBPUAT, Pantnagar for providing the necessary facilities and the researchers associated with the ICAR–AICRP on soil test crop response. The first author is also grateful to the ICAR-SRF Fellowship programme of the Government of India for financial support. The authors appreciated Taif University Researchers Supporting Project number (TURSP-2020/53), Taif University, Taif, Saudi Arabia for providing financial support to conduct the research.

Conflicts of Interest: The authors have no conflict of interest to disclose.

Ethical Statement: No living organism (human or animal) was involved in conducting the present experiments.

References

1. Pandey, S.; Velasco, L.E. Economics of alternative rice establishment methods in Asia: A strategic analysis. *Soc. Sci. Div. Discuss. Pap. Int. Rice Res. Inst. Los Banos Philipp.* **1999**, *1*, 12–18.
2. Gill, G.; Humphreys, E.; Kukal, S.S.; Walia, U.S. Effect of water management on dry seeded and puddled transplanted rice. Part 1: Crop performance. *Field Crops Res.* **2011**, *120*, 112–122.
3. Pandey, S.; Velasco, L. Trends in crop establishment methods in Asia and research issues. In *Rice Is Life: Scientific Perspectives for the 21st Century*; Manila IRRI: Manila, Philippines, 2005; pp. 178–181.
4. Hossain, M.; Lewis, D.; Bose MLChowdhury, A. *Rice Research, Technological Progress, and Impacts on the Poor: The Bangladesh Case (Summary Report)*; International Food Policy Research Institute: Washington, DC, USA, 2003.
5. Dey, P. Soil Health Management. *Bull. Indian Soc. Soil Sci.* **2016**, *30*, 79–97.
6. Pramanick, B.; Brahmachari, K.; Ghosh DBera, P.S. Influence of foliar application seaweed (*Kappaphycus* and *Gracilaria*) saps in rice (*Oryza sativa*)–potato (*Solanum tuberosum*)–blackgram (*Vigna mungo*) sequence. *Indian J. Agron.* **2018**, *63*, 7–12.
7. Singh, D.K.; Pandey, P.C.; Nanda, G.; Gupta, S. Long-term effects of inorganic fertilizer and farmyard manure application on productivity, sustainability and profitability of rice-wheat system in Mollisols. *Arch. Agron. Soil Sci.* **2019**, *65*, 139–151. [CrossRef]
8. Sharma, V.K.; Pandey, R.N.; Sarvendra, K.; Chobhe, K.A.; Suresh, C. Soil test crop response based fertilizer recommendations under integrated nutrient management for higher productivity of pearl millet (*Pennisetum glaucum*) and (*Triticum aestivum*) wheat under long term experiment. *Indian J. Agric. Sci.* **2016**, *86*, 1076–1081.
9. Patel, A.; Neupane, M.P.; Nanda, G.; Singh, S.P. Effect of NPK management and bioinoculants on growth and yield of wetland rice (*Oryza sativa* L.). *Environ. Ecol.* **2016**, *34*, 181–185.
10. Mahajan, A.; Gupta, R.D. Role of INM in Sustainable Rice–Wheat Cropping System. In *Integrated Nutrient Management (INM) in a Sustainable Rice-Wheat Cropping System*; Mahajan, A., Gupta, R.D., Eds.; Springer: Dordrecht, The Netherlands, 2009. [CrossRef]
11. Santhi, R.; Bhaskaran, A.; Natesan, R. Integrated fertilizer prescriptions for beetroot through inductive cum targeted yield model on an alfisol. *Commun. Soil Sci. Plant Anal.* **2011**, *42*, 1905–1912. [CrossRef]

12. Nanda, G.; Sravan, U.S.; Singh, A.; Singh, S.P. Effect of NPK Levels and Bio-Organics on Growth, Yield and Economics of Basmat Rice (*Oryza sativa* L.) cv HUBR 10-9. *Environ. Ecol.* **2016**, *34*, 1530–1534.
13. Majhi, P.; Rout, K.K.; Nanda, G.; Singh, M. Soil quality for rice productivity and yield sustainability under long-term fertilizer and manure application. *Commun. Soil Sci. Plant Anal.* **2019**, *50*, 1330–1343. [CrossRef]
14. Majhi, P.; Rout, K.K.; Nanda, G.; Singh, M. Long term effects of fertilizer and manure application on productivity, sustainability and soil properties in a rice-rice system on Inceptisols of Eastern India. *Commun. Soil Sci. Plant Anal.* **2021**, *52*, 1631–1644 [CrossRef]
15. Sharma, G.K.; Mishra, V.N.; Maruti Sankar, G.R.; Patil, S.K.; Srivastava, L.K.; Thakur, D.S.; Rao, C.S. Soil-test-based optimum fertilizer doses for attaining yield targets of rice under midland Alfisols of Eastern India. *Commun. Soil Sci. Plant Anal.* **2015**, *46* 2177–2190. [CrossRef]
16. Singh, S.R.; Maitra, D.N.; Kundu, D.K.; Majumdar, B.; Saha, A.R.; Mahapatra, B.S. Integrated fertilizer prescription equations for recommendations of fertilizers in jute–rice–garden pea sequence on alluvial soil of eastern India. *Commun. Soil Sci. Plant Anal.* **2015**, *46*, 1–15. [CrossRef]
17. Dobermann, A.; Witt, C.; Abdulrachman, S.; Gines, H.C.; Nagarajan, R.; Son, T.T.; Tan, P.S.; Wang, G.H.; Chien, N.V.; Thoa, V.T.K. et al. Soil fertility and indigenous nutrient supply in irrigated rice domains of Asia. *Agron. J.* **2003**, *95*, 913–923. [CrossRef]
18. Rao, A.S.; Srivastava, S. Soil test based fertiliser use: A must for sustainable agriculture. *Fertil. News* **2000**, *45*, 25–38.
19. Jackson, M.L. *Soil Chemical Analysis Prentice*; Hall of India Private Limited: New Delhi, India, 2000; p. 498.
20. Walkley, A.; Black, I.A. An examination of Degtjareff method for determining soil organic matter and a proposed modification of the chromic acid titration method. *Soil Sci.* **1934**, *37*, 29–38. [CrossRef]
21. Subbaiah, B.V.; Asija, G.L. A rapid procedure for the determination of available nitrogen in soil. *Curr. Sci.* **1956**, *25*, 259–260.
22. Olsen, S.R.; Cole, C.V.; Watanabe, F.S.; Dean, L.A. *Estimation of Available Phosphorus in Soils by Extraction with Sodium Bicarbonate* No. 939; US Dept Agric Circular: Washington, DC, USA, 1954.
23. Hanway, J.J.; Heidel, H. Soil analysis methods as used in Iowa state college soil testing laboratory. *Iowa Agric.* **1952**, *57*, 1–31.
24. Ramamoorthy, B.; Narasimham, R.L.; Dinesh, R.S. Fertilizer application for specific yield targets on Sonora 64 (wheat). *Indian Farming* **1967**, *17*, 43–45.
25. Singh, P.; Benbi, D.K.; Verma, G. Nutrient Management Impacts on Nutrient Use Efficiency and Energy, Carbon, and Net Ecosystem Economic Budget of a Rice–Wheat Cropping System in Northwestern India. *J. Soil Sci. Plant Nutr.* **2021**, *21*, 559–577 [CrossRef]
26. Gomez, K.A.; Gomez, A.A. *Statistical Procedures in Agricultural Research*, 2nd ed.; Wiley: New York, NY, USA, 1984.
27. Ammal, U.B.; Coumaravel, K.; Sankar, R.; Dey, P. Fertilizer Prescriptions under STCR-IPNS for Rice-Rice Cropping Sequence on An Inceptisol (Typic Ustropept). *Indian J. Agric. Res.* **2019**, *53*, 698–702. [CrossRef]
28. Dwivedi, B.S.; Shukla, A.K.; Singh, V.K.; Yadav, R.L. Improving nitrogen and phosphorus use efficiencies through inclusion of forage cowpea in the rice–wheat systems in the Indo-Gangetic Plains of India. *Field Crops Res.* **2003**, *80*, 167–193. [CrossRef]
29. Gayathri, A.; Vadivel, A.; Santhi, R.; Boopathi, P.M.; Natesan, R. Soil test based fertilizer recommendation under integrated plant nutrition system for potato (*Solanum tuberosum* L.) in hilly tracts of Nilgiris district. *Indian J. Agric. Res.* **2009**, *43*, 52–56.
30. Mahajan, G.R.; Pandey, R.N.; Datta, S.C.; Kumar, D.; Sahoo, R.N.; Parsad, R. Fertilizer nitrogen, phosphorus and sulphur prescription for aromatic hybrid rice (*Oryza sativa* L.) using targeted yield approach. *Proc. Natl. Acad. Sci. India Sect. B Biol. Sci.* **2014**, *84*, 537–547. [CrossRef]
31. Singh, Y.V. STCR based gradient experiment with sorghum (*Sorghum biocolor* L.) to NPK fertilizers in the alluvial soil. *Indian J. Crop Ecol.* **2014**, *2*, 83–86.
32. Arya, A. Optimization of Fertilizer Doses through STCR Approach for Cauliflower (*Brassica oleracea* L var. Botrytis) Grown in Mollisols. Doctoral Dissertation, GB Pant University of Agriculture and Technology, Pantnagar, India, 2019.
33. Mahajan, G.R.; Pandey, R.N.; Datta, S.C.; Kumar, D.; Sahoo, R.N.; Parsad, R. Soil test based fertilizer recommendation of nitrogen phosphorus and sulphur in wheat (*Triticum aestivum* L.) in an Alluvial soil. *Int. J. Agric. Environ. Biotechnol.* **2013**, *6*, 271.
34. Ahmed, S.; Basumatary, A.; Das, K.N.; Medhi, B.K. Targeted yield based fertilizer prescriptions for autumn rice (*Oryza sativa* L. in inceptisols of Assam, India. *Indian J. Agric. Res.* **2015**, *49*, 437–441. [CrossRef]
35. Luthra, N. STCR Approach for Optimizing Integrated Plant Nutrients Supply to Obtain Better Growth and Yield of Hybrid Maize (*Zea mays* L.). Ph.D. Dissertation, GB Pant University of Agriculture and Technology, Pantnagar, India, 2019.
36. Radha Madhav, M.; Ravikumar, A.; Venkateswaralu, B. Effect of different sources of nitrogen on growth, yield and nutrient uptake of rice. *Agric. J.* **1996**, *43*, 119–122.
37. Sandhu, S.S.; Mahal, S.S. Performance of rice (*Oryza sativa*) under different planting methods, nitrogen levels and irrigation schedules. *Indian J. Agron.* **2014**, *59*, 392–397.
38. Verma, V.S.; Saxena, K.K. Response of french bean to graded doses of nitrogen, phosphorus and potassium in silty-loam soil of central Uttar Pradesh. *Indian J. Agron.* **1995**, *40*, 67–71.
39. Verma, T.S.; Suri, V.K.; Paul, J.; Vishavavidyalaya, H.P.K. Prescription-based fertilizer recommendations for rice, maize and wheat in different Agro-climatic zones of Himachal Pradesh. *J. Indian Soc. Soil Sci.* **2002**, *50*, 272–277.
40. Verma, M.; Singh, Y.V.; Ajay Babu Verma, S.; Meena, R.; Sahi, S.K. Soil test crop response based gradient experiment on rice (*Oryza sativa* L.) to NPK fertilizers in the alluvial soil. *Indian J. Agric. Allied Sci.* **2015**, *1*, 51–53.

41. Verma, M.; Singh, Y.V.; Dey, P.; Babu, A. Soil test based fertilizer recommendation for mustard (*Brassica juncea* L.) in Eastern Plain Zone of Uttar Pradesh, India. *Int. J. Curr. Microbiol. Appl. Sci.* **2017**, *6*, 155–161. [CrossRef]
42. Surekha, K.; Kumar, R.M.; Nagendra, V.; Sailaja, N.; Satyanarayana, T. 4R Nitrogen Management for Sustainable Rice Production. *Better Crop.* **2016**, *10*, 16–19.
43. Pasha, L.M.; Reddy, M.D.; Reddy, M.G.; Devi, M.U. Effect of irrigation schedule, weed management and nitrogen levels on weed growth in rice (*Oryza sativa*) under aerobic conditions. *Indian J. Weed Sci.* **2011**, *43*, 54–60.
44. Singh, A.S.; Pandey, N.; Ahmad, A.; Verma, N. *Nutrient Uptake and Economics of Rice under Aerobic Condition in CPE Based Irrigation Schedules and Nutrient Level in Raipur, India*; Jawaharlal Nehru Krishi Vishwa Vidyalaya: Jabalpur, India, 2013; p. 42.
45. Panaullah, G.M.; Timsina, J.; Saleque, M.A.; Ishaque, M.; Pathan, A.B.M.B.U.; Connor, D.J.; Saha, P.K.; Quayyum, M.A.; Humphreys, E.; Meisner, C.A. Nutrient uptake and apparent balances for rice-wheat sequences. III. Potassium. *J. Plant Nutr.* **2006**, *29*, 173–187. [CrossRef]
46. Pande, J. *Soil Test Crop Response Based Balanced Fertilization of Cabbage (Brassica oleracea var. Capitata) Grown in a Mollisol of Uttarakhand*; G.B. Pant University of Agriculture and Technology: Pantnagar, India, 2010.
47. Dubey, A.; Tiwari, A.; Pandey, R.L.; Dubey, A. Effect of organic manure in seed yield and seed attributes of rajmash (*Phaseolus vulgaris* L.) under *Vertisols* of Chhattisgarh plains, India. *Int. J. Curr. Res.* **2014**, *6*, 9469–9472.
48. Chatterjee, D.; Srivastava, A.; Singh, R.K. Fertilizer recommendations based on targeted yield concept involving integrated nutrient management for potato (Solanum tuberosum) in tarai belt of Uttarakhand. *Indian J. Agric. Sci.* **2010**, *80*, 1048.
49. Bera, R.; Seal, A.; Bhattacharyya, P.; Das, T.H.; Sarkar, D.; Kangjoo, K. Targeted yield concept and a framework of fertilizer recommendation in irrigated rice domains of subtropical India. *J. Zhejiang Univ. Sci. B* **2006**, *7*, 963–968. [CrossRef] [PubMed]
50. Ray, P.; Jana, K.; Maitra, A.K.; Saha, D.N.; Chaudhury, M.N.; Saha, J.S.; Saha, A.R. Fertilizer prescription on soil test basis for jute, rice and wheat in a Typic Ustochrept. *J. Indian Soc. Soil Sci.* **2000**, *48*, 79–84.
51. Ammal, U.B.; Sankar, R.; Coumaravel, K.; Dey, P. Fertilizer Prescriptions for Rice Based on STCR–IPNS. *Int. J. Curr. Microbiol. App. Sci.* **2020**, *9*, 2517–2522. [CrossRef]
52. Sharma, B.M.; Singh, R V. Soil-test-based fertilizer use in wheat for economic yield. *J. Indian Soc. Soil Sci.* **2005**, *53*, 356–359.
53. Benbi, D.K.; Benipal, D.S. Fertilizer recommendations based on soil tests for yield targets of mustard and rapeseed and their validations under farmers' field conditions in Punjab. *J. Indian Soc. Soil Sci.* **2006**, *54*, 316–321.
54. Basavaraja, P.K.; Mohamed Saqeebulla, H.; Dey, P.; Prakash, S.S. Fertilizer prescription equations for targeted yield of rice (*Oryza sativa* L.) and their validation under aerobic condition. *Int. J of Agri. Sci.* **2016**, *8*, 1003–1008.
55. Mittal, S.; Sharma, S.K. Effect of different approaches of fertilizer recommendations on yield, nutrient uptake and economics of rice under SRI. *J. Pharmacogn. Phytochem.* **2018**, *7*, 761–765.
56. Tiwari, A.; Dwivedi, A.K.; Dikshit, P.R. Long-term influence of organic and inorganic fertilization on soil fertility and productivity of soybean–wheat system in a Vertisol. *J. Indian Soc. Soil Sci.* **2002**, *50*, 472–475.
57. Anderson, J.; Blackie, M.; Eilitta, M.; Fernandes, E.; Sanginga, N.; Smaling, E.; Spencer, D. *A Consultative Review of the Rockefeller Foundation's Activities to Improve and Sustain Soil Fertility in East and Southern Africa*; The Rockefeller Foundation: New York, NY, USA, 2002.
58. Sengar, S.S.; Wade, L.J.; Baghel, S.S.; Singh, R.K.; Singh, G. Effect of nutrient management on rice (Oryza sativa) in rainfed lowland of southeast Madhya Pradesh. *Indian J. Agron.* **2000**, *45*, 315–322.
59. Venkatesh, M.S.; Hazra, K.K.; Ghosh, P.K.; Singh, K.K. Improving productivity of maize-lentil rotation in alkaline Fluvisol following soil test crop response (STCR)-targeted yield approach of nutrient management. *Arch. Agron. Soil Sci.* **2021**, 1–15. [CrossRef]
60. Velayutham, M.; Sankar, M.G.R.; Reddy, K.C.K. Soil test-crop response research work in India for fertilizer recommendation. In Proceedings of the International Symposium on Soil Testcrop Response Correlation Studies, Dhaka, Bangladesh, 7–10 February 1984.
61. Katharine, S.P.; Santhi, R.; Maragatham, S.; Natesan, R.; Ravikumar, V.; Dey, P. Soil test based fertilizer prescriptions through inductive cum targeted yield model for transgenic cotton on Inceptisol. *IOSR J. Agric. Vet. Sci.* **2013**, *6*, 36–44. [CrossRef]
62. Arunima, G.; Antaryami, M.; Jena, B. Soil test based fertilizer recommendation for targeted yield of pumpkin (*Cucurbita moschata*) under rice-pumpkin cropping system in an inceptisol of Orissa. *Environ. Ecol.* **2011**, *29*, 574–576.
63. Duan, Y.H.; Shi, X.J.; Li, S.L.; Sun, X.F.; He, X.H. Nitrogen use efficiency as affected by phosphorus and potassium in long-term rice and wheat experiments. *J. Integr. Agric.* **2014**, *13*, 588–596. [CrossRef]
64. Banerjee, H.; Ray, K.; Dutta, S.K.; Majumdar, K.; Satyanarayana, T.; Timsina, J. Optimizing potassium application for hybrid rice (Oryza sativa L.) in coastal saline soils of West Bengal, India. *Agronomy* **2018**, *8*, 292. [CrossRef]
65. Fageria, N.K.; Dos Santos, A.B.; De Moraes, M.F. Yield, potassium uptake, and use efficiency in upland rice genotypes. *Commun. Soil Sci. Plant Anal.* **2010**, *41*, 2676–2684. [CrossRef]
66. Selvi, D.; Santhy, P.; Dhakshinamoorthy, M.; Maheshwari, M. Microbial population and biomass in rhizosphere as influenced by continuous intensive cultivation and fertilization in an Inceptisol. *J. Indian Soc. Soil Sci.* **2004**, *52*, 254–257.

agronomy

MDPI

Article

Improvement of Soil Health and System Productivity through Crop Diversification and Residue Incorporation under Jute-Based Different Cropping Systems

Mukesh Kumar [1,2], Sabyasachi Mitra [1], Sonali Paul Mazumdar [1], Bijan Majumdar [1], Amit Ranjan Saha [1], Shiv Ram Singh [1,3], Biswajit Pramanick [2,*], Ahmed Gaber [4], Walaa F. Alsanie [5] and Akbar Hossain [6,*]

[1] ICAR-Central Research Institute for Jute and Allied Fibres (ICAR-CRIJAF), Barrackpore, Kolkata 700120, West Bengal, India; mukesh.agro@gmail.com (M.K.); mitrasaby@gmail.com (S.M.); sonalipmazumdar110@gmail.com (S.P.M.); bmajumdar65@gmail.com (B.M.); amit.saha60@yahoo.co.in (A.R.S.); shivramsingh22@gmail.com (S.R.S.)

[2] Department of Agronomy, Dr. Rajendra Prasad Agricultural University (RPCAU), Pusa, Samastipur 848125, Bihar, India

[3] ICAR-Indian Institute of Sugarcane Research (ICAR-IISR), Lucknow 226002, Uttar Pradesh, India

[4] Department of Biology, College of Science, Taif University, P.O. Box 11099, Taif 21944, Saudi Arabia; a.gaber@tu.edu.sa

[5] Department of Clinical Laboratories Sciences, The Faculty of Applied Medical Sciences, Taif University, P.O. Box 11099, Taif 21944, Saudi Arabia; w.alsanie@tu.edu.sa

[6] Department of Agronomy, Bangladesh Wheat and Maize Research Institute, Dinajpur 5200, Bangladesh

* Correspondence: biswajit@rpcau.ac.in (B.P.); akbarhossainwrc@gmail.com (A.H.)

Citation: Kumar, M.; Mitra, S.; Mazumdar, S.P.; Majumdar, B.; Saha, A.R.; Singh, S.R.; Pramanick, B.; Gaber, A.; Alsanie, W.F.; Hossain, A. Improvement of Soil Health and System Productivity through Crop Diversification and Residue Incorporation under Jute-Based Different Cropping Systems. *Agronomy* **2021**, *11*, 1622. https://doi.org/10.3390/agronomy11081622

Academic Editors: Umberto Anastasi and Aurelio Scavo

Received: 13 July 2021
Accepted: 12 August 2021
Published: 16 August 2021

Abstract: Crop diversity through residue incorporation is the most important method for sustaining soil health. A field study was conducted over five consecutive years (2012–2017) to see the impact of residue incorporartions in Inceptisol of eastern India. The main plot treatments had five cropping systems (CS), namely, fallow−rice−rice (FRR), jute−rice−wheat (JRW), jute−rice−baby corn (JRBc), jute−rice−vegetable pea (JRGp), jute−rice−mustard−mungbean/green gram (JRMMu), which cinsisted of four sub-plots with varied nutrient and crop residue management (NCRM) levels, namely crops with no residue +75% of the recommended dose of fertilizers (RDF) (F_1R_0), crops with the residue of the previous crops +75% RDF (F_1R_1), crops with no resiude +100% RDF (F_2R_0), and crops with residue +100% RDF (F_2R_1). The highest system productivity was obtained for JRBc (15.3 Mg·ha^{-1}), followed by JRGp (8.81 Mg·ha^{-1}) and JRMMu (7.61 Mg·ha^{-1}); however, the highest sustainability index was found with the JRGp cropping system (0.88), followed by JRMMu (0.82). Among the NCRMs, the highest productivity (8.78 Mg·ha^{-1}) and sustainability index (0.83) were recorded in F_2R_1. Five soil parameters, namely, bulk density, available K, urease activity, dehydrogenase activity, and soil microbial biomass carbon (SMBC), were used in the minimum data-set (MDS) for the calculation of the soil quality index (SQI). The best attainment of SQI was found in the JRGp system (0.63), closely followed by the JRMMu (0.61) cropping system.

Keywords: crop diversification; soil quality; crop residue; rice; jute; legume

1. Introduction

The eastern part of the Indo-Gangetic Plain (IGP) is often considered as the supreme agro-ecological zone of the globe [1]. Cropping systems using rice as a base crop, like the double cropping systems of rice−rice and rice−potato, and intensive triple cropping systems like jute−rice−rice and jute−rice-mustard, have been practiced for a long in this region [2,3]. Crop intensification is needed in order to produce more food per unit area per unit time, so as to assure food security for the burgeoning Indian population, expected to exceed 1.60 billion people in 2050 [4]. Nevertheless, crop intensification poses a major threat regarding sustainability issues for the rice-based system in this region. Excess nutrient removal through crops being harvested in an area year after year have altered the soil fertility status [5]. Moreover, intensive and poor-planned inorganic nutrition in intensive cropping-system has led to nutrient imbalance in the soil, and thus a decline in soil fertility, which is of great concern for production sustainability and soil health in this region [2]. Henceforth, the current need is to sustain soil fertility with some alternative, but easy methods, like returning the crop residue back to the soil through suitable crop diversification, as well as using some promising crops that are believed to boost soil health. Establishing an efficient cropping sequence with the best possible crop management can ensure soil fertility as well as profitability [6–9].

Diversified cropping systems are one of the key options that can benefit to the soil by supplying organic matter, although the amount and quality vary with the type of crops grown in a field. Crop diversification includes crops that return plenty of organic matter (through the roots, root exudates, leaves, and stems) into the soil, as well as being re able to sustain or enhance economic yields and reduce nutrient loss through runoff and leaching. Legume crop inclusion in crop rotation, either through conventional cropping systems or conservation tillage systems, can return the crop residue to the soil and augment the soil organic carbon (SOC), making the overall system sustainable [10]. All positive attributes of soil are increased with residue return. Soil loss, through erosion, soil-nutrient leaching loss, etc., is also reduced through residue retention in the soil [11–14]. All of these factors influence soil quality. Many earlier findings [15–17] have demonstrated that crop-residue incorporation under conservation-tillage practice improves soil attributes such as SOC; available N, P, and K; soil aggregates; water-holding capacity (WHC); soil aeration; soil enzymes, as assessing the soil quality and its trend of change by different crop management practices is key for making agricultural practices more sustainable. According to Karlen et al. [18] soil quality is "the capacity of a specific kind of soil to function within natural or managed ecosystem boundaries, to sustain plant and animal productivity, maintain or enhance water and air quality, and support human health and habitation". As soil is a complex phenomenon, soil quality cannot be measured directly. However, soil quality indicators that are sensitive to change in land use management and conservation practices are measurable attributes. These attributes also influence the ability of soil regarding sustainable crop production. Thus, it is necessary to develop minimum data-sets of varied soil quality indicators under different determining parameters, such as fields [19], landscapes, and regions [20–22]. The jute-based cropping system for the jute growing region of India is mainly a jute—rice—wheat cropping system [3]. This system needs to be diversified by including legumes or oilseeds or high-value crops like baby corn in order to maintain soil health and economic return. In this study, we diversified the jute—rice—wheat system by replacing winter crop wheat with mustard, vegetable pea, or baby corn.

The soil quality index (SQI) was developed by Andrews et al. [23] to quantify the physical, chemical, and biological properties. This soil quality index (SQI) is based on a combination of soil properties to provide a better indication of soil quality than through individual parameters. SQI is a tool for evaluating the sustainability of different soil and crop management practices [21,24,25]. Higher values of SQI indicate a better soil quality, which means the soil will perform better and produce a higher yield in a more sustainable manner than those with a lower SQI. Hence, research on determining soil quality as affected by the various cropping systems and crop management has significance in the global scientific community [22]. Plenty of research has been carried out regarding the long-term nutrition and tillage that influence soil quality [25–28], but little information [21,29,30] is available about the SQI of diverse cropping systems and the integration of different types of crop residues with nutrients.

To bridge the wide research gap, the present investigation was executed with the following objectives: (i) to evaluate the system productivity of diverse cropping systems with crop residue and nutrient management and (ii) to estimate the soil quality index of different cropping systems with various nutrient and crop residue management practices in Inceptisols, in the eastern part of the Indo-Gangetic plain.

2. Materials and Methods

2.1. Experimental Site

The research work was performed at ICAR-CRIJAF, Barrackpore, India (22°45′ N, 88°26′ E, 9.0 MSL) during 2012–2017 in Inceptisols, where the order of soil falls under the new alluvial zone (NAZ). The primary crop at the study site was jute (*Corchorus olitorius* L.) mono-cropping. The soil was 54%, 34%, and 12% sand, silt, and clay, respectively, which falls under the category of loam soil. Prior to commencing the study, soil was sampled from the entire experimental field at a 0–15 cm depth, and was analysed subsequently after making a composite sample. The initial value of the soil pH was 7.1 (neutral), with 1.43 Mg/m^3 and 16.4 cmol/kg bulk density and cation exchange capacity, respectively, while the organic carbon and available N, P, and K were 6.80 g/kg, 126 mg/kg, 17 mg/kg and 103 mg/kg, respectively. The total rainfall/annum over the experimental period, i.e., 2012–2013, 2013–2014, 2014–2015, 2015–2016, and 2016–2017 was 1434 mm, 1810 mm, 1337 mm, 1682 mm, and 1552 mm, respectively (Figure 1). The rainfall was well distributed. June to August is known as monsoon season (rainy season), and had the maximum amount of rainfall during experimentation period (2012–2017). The average rainfall during the five-year study was 1563 mm/annum, with a maximum rainfall of 1810 mm/annum in 2013–2014 and a minimum of 1337 mm/annum in 2014–2015. During the research work the average mean minimum and maximum temperature ranged from 20.7 to 21.1 °C and 30.7 to 34.4 °C, respectively. Considering the entire study during 2012–2017, the average values of the monthly minimum relative humidity (RH) varied from 58 to 63%, while the average monthly maximum RH varied between 92% and 94%.

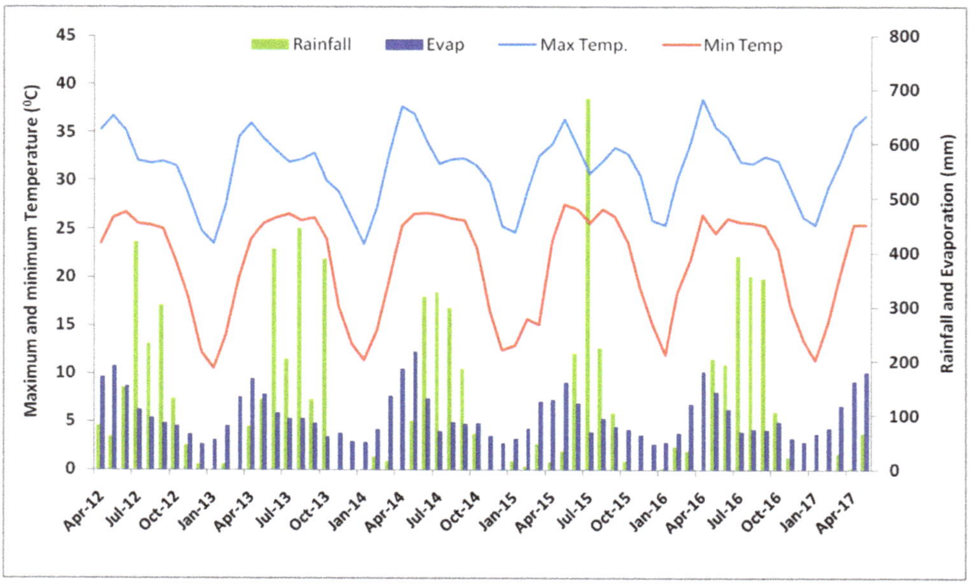

Figure 1. Monthly rainfall. Mean monthly maximum and minimum temperature and evaporation during the year of experimentation 2012–2017.

2.2. Experimental Design and Treatments

A two-factor statistical design (split-plot design) was used in the current study, with one factor being different cropping systems (CSs) and the other being different nutrient and crop residue managements (NCRMs). Five different CSs were designed in the main plots, namely, fallow–rice–rice (FRR), jute–rice–wheat (JRW), jute–rice–baby corn (JRBc), jute–rice–vegetable pea (JRGp), jute–rice–mustard–mungbean (JRMMu), with four different NCRM practices superimposed in the sub-plots, namely, 75% of the recommended dose of fertilizers (RDF) without crop residue (F1R0), 75% RDF with crop residue (F1R1), 100% RDF without crop residue (F2R0), and 100% RDF with crop residue (F2R1). This set of treatment combinations was replicated three times. The same cropping systems (CSs) and nutrient and crop residues managements (NCRMs) were repeated in their respective fixed plots for five years. The RDF for different crops and crop calendar for the sowingand harvesting of each crop are provided in Table A1. The harvested residues of rice, wheat, and corn were weighed, and a fixed amount of 4 Mg·ha^{-1} for these residues was incorporated in the soil, while the amount of residue for vegetable pea and mungbean was 2 Mg·ha^{-1}. These amounts of residues for the respective crops were fixed for the entire study period. All of the residues were incorporated into the soil with the help of tractor-drawn single disc-harrowing, after which they were cultivated with a rotavator in their particular cropping-systems just before sowing of jute (April). Residues were added and incorporated once in a year, i.e., before sowing the jute. Before sowing the other crops, primary and secondary tillage operations were completed. However, the mustard and mungbean in the JRMMu cropping system were sown just one day after harvesting the rice and mustard, respectively, without any tillage operations in order to accommodate four crops in the year, thus saving the time required for tillage operations. A puddling operation was done before transplanting the rice crop.

2.3. Soil Sampling and Analysis

Soil samples were collected from five random places from each plot of 20 m^2, at a 0–15 cm depth, using an auger (5 cm diameter), after the end of the five-year long experiment. These five samples from each plot were mixed together to make a composite sample of 500 g. Thereafter, the fresh soil samples were divided into two parts, in which a part of the soil sample was air-dried, ground, and sieved (through 2 mm mesh) for the chemical analysis. The soil organic carbon (SOC) was determined using 0.167 M $K_2Cr_2O_7$ [31], available N from the $KMnO_4$ extraction methods [32], available phosphorus (Av.P) from the Olsen method [33], and available potassium (Av.K) from the 1N NH_4OAc method, following the method illustrated by Jackson (1973) [34]. As per the chloroform (CCl_4) fumigation extraction methods of Vance et al. [35], estimations of the microbial biomass carbon (SMBC) and microbial biomass N (SMBN) were worked out in the laboratory using the remainder of the fresh soil samples. The viable and cultural microbial population was enumerated by adopting the standard-serial-dilution-plate technique with a selective medium for specified groups of soil microorganisms. Such as enumeration of bacteria was done using nutrient agar containing 50 mg/L cycloheximide [36], *Azotobacter* by Ashby's N free mannitol agar, and phosphate solubilizing bacteria (PSB) by Pikovskaya's agar media [37]. The dehydrogenase activity (DHA) in the soil was estimated as per the method given by Tabatabai (1982) [38], which is based on the reduction of idonitotetrazolium chloride (TTC) into triphenyl formazan (TPF), thereafter examining the sample using a spectrophotometer at a wavelength of 485 nm. The estimation of the fluorescein diacetate hydrolytic activity (FDA) was done following the standard procedure mentioned by Schnurer and Rosswall [39], in which the fluorescein released after the reduction was measured on with a spectrophotometer at a wavelength of 490 nm. Urease activity (UA) was determined by quantifying the NH4+ released during 2 h of soil incubation at 370 °C after distillation with magnesium oxide (MgO) using a boric acid indicator and back titration with 0.005 N sulphuric acid [40]. The acid phosphatase (ACP) and alkaline phosphatase (ALP) activities the in soil were estimated through the detection of p-nitrophenol (PNP) released after incubation for 1 h at 37 °C at pH 6.5 of soil with p-Nitrophenyl phosphate disodium [41].

2.4. System Productivity and Sustainable Yield Index (SYI)

To compute the productivity of the different cropping systems, the crop equivalent yield (CEY) was calculated based on the price of jute fibre using the following formula.

$$\text{CEY} = \text{yield of crop in system } (\text{Mg}\cdot\text{ha}^{-1}) \times \text{price of crop (Rs/t)/price of jute (Rs/t)} \quad (1)$$

System productivity was calculated by adding the jute yield and crop equivalent yield of the other crops for the respective years.

The sustainable yield index (SYI) of the system was calculated based on the data of five years of system productivity, following the formula given by Singh et al. [42].

$$\text{SYI} = (Y_A - \alpha)/Y_h \quad (2)$$

where Y_A is the average system productivity of the particular treatment, α is the standard deviation, and Y_h is the highest system productivity recorded in a set of management practice during a year.

2.5. Soil Quality Index (SQI) Calculation

The procedure to determine the soil quality index had the following four steps: (i) setting the goal, (ii) selection of minimum data set (MDS) of indicators that represent the best soil function, (iii) scoring the MDS indicators, and (iv) the integration of an indicator score into a relative soil quality index [21,25]. In this study, the system productivity of different cropping systems was defined as the goal, because the farmers wanted to get greater productivity from their land area. A total of 16 soil attributes were analyzed during the investigation, of which only 12 soil attributes were included in the MDS for soil quality using the principal component analysis (PCA). Only those soil attributes that differed significantly among treatments were selected as representative of MDS through PCA, as suggested by Andrews et al. [42]. Principal component (PC) variables that had high factor loading and eigen values were assumed to be the best representation of those attribute variables. Therefore, the variables selected in this study explained at least 5% of the data variation and had PCs with eigen values >1 [43]. Moreover, only highly weighted factor loadings were selected within each PC, namely those with absolute values within 10% of the highest factor loading. In the situation when more than one factor fell into a single PC, a multivariate correlation coefficient was applied to exclude the redundant variables from the MDS [23]. Among the significantly correlated variables of one PC, the higher factor loading variable was retained. The variables that were not correlated but were highly weighted in PC were considered important and were kept in the MDS. After the selection of the MDS indicators, every observation of each MDS indicator was assigned a score using a non-linear scoring method [43]. Indicators were arranged based on its value, whether a higher value was considered "good" or "bad", and according to its function soil. In the present study, SMBC, DHA, urease, and available K were retained in the MDS and scored as "more is better", as they were good with a higher value. However, bulk density was considered good with a lower range, and thus was scored as "less is better". For indicators that fell under "more is better", each observation was divided by the highest observed value, so this received a score of 1. In contrast to the "less is better" indicators, each observation was divided by the lowest observed value so that the lowest observed value received a score of 1. Non-linear scoring function (NLSF):

$$\text{NLSF (Y)} = 1/[1 + e\hat{}\{-b(x - A)\}] \quad (3)$$

where x is the soil property value, A is the baseline or value of the soil property where the score is equal to 0.5, and b is the slope.

Once the variables were assigned a score, the MDS variables for each observation were weighted using the PCA results. As in the total data set, each PC explained a certain amount (%) of the variation. This percentage is divided by the total percentage of variation explained by all PCs with eigen vectors >1, only if the weighted factor for the variables was chosen for a given PC. Then, the weighted MDS variable scores were summed up for each observation to calculate the soil quality index (SQI), using the following equation:

$$\text{Soil quality index (SQI)} = \sum_{i=1}^{n} (\text{Wi} \times \text{Si}) \quad (4)$$

where Si is the score indicated for the ith variable and Wi is the weightage of the factor for the ith variable, which is derived from the PCA. Here, the assumption is that a higher SQI indicates a better soil quality or better performance of soil. Furthermore, the present contribution of each final key indicator towards SQI was also calculated. The SQI was calculated by assigning the estimated factors to the soil quality indicators as follows:

$$\text{SQI} = 0.353 \times S_{\text{SMBC}} + 0.131 \times S_{\text{DHA}} + 0.136\, S_{\text{UA}} + 0.117 \times S_{\text{AvK}} + 0.103 \times S_{\text{BD}} \quad (5)$$

where S is the score for the subscripted variable and the coefficients are the weighting factors.

2.6. Statistical Analysis

Analysis of variance (ANOVA) was performed to determine the effects of the treatment of the cropping systems and NCRM practices on the system productivity and soil quality attributes. SPSS Windows version 20.0 (SPSS Inc., Chicago, IL, USA) was used for the statistical analysis of the data (PCA, scoring functions). The least-square difference (LSD) was performed post hoc in order to find the pair-wise comparison of the significant difference of all of the data, with a level of significance at $p \leq 0.05$ [44].

3. Results and Discussion

3.1. System Productivity and Sustainable Yield Index

System productivity differed considerably ($p \leq 0.05$) among the varied cropping systems and NRCM practices during all of the experimental years. System productivity was significantly higher (15.3 Mg·ha^{-1}) in JRBc compared with the other cropping systems. The system productivity of the remaining cropping systems recorded was as follows: JRGp (8.81 Mg·ha^{-1}) > JRMMu (7.61 Mg·ha^{-1}) > JRW (6.65 Mg·ha^{-1}) > FRR (4.41 Mg·ha^{-1}; Table 1). Although the system productivity was equilvalent to the NRCM practices during the first three years of the experiment (2012–2015), a significantly higher system productivity was recorded with F_2R_1 than for the remaining NRCM treatments during the last two years (2015–2017) of experimentation. The pooled data for five years of system productivity indicated significant ($p \leq 0.05$) variation among the different cropping systems. The interaction between CS and NCRM was found to be non-significant ($p \leq 0.05$) for system productivity. The highest system productivity was recorded in JRBc, followed by JRGp. This is mainly due to the higher yield and market price of baby corn compared with what the other crops produced [3]. System productivity is directly related to and functions with the market price and yield of crops, hence their crop yield and its value decide the system productivity. Among the NRCM practices, the system productivity was similar among the treatments, but the highest system productivity (8.78 Mg·ha^{-1}) was recorded for the F_2R_1 treatment. The application of 100% NPK to each test crop included in the system with crop residue enhanced the crop productivity of each test crop, and thereby increased the system productivity. This is because the use of crop residues is likely to improve the soil tilth through a reduced bulk density and by increasing soil aggregation. Moreover, the addition of crop residues also increased the SOC and available N, P, and K content, which resulted in increased crop productivity by improving the nutrient acquisition and reducing wind and water erosion, and prevented nutrient losses from run-off and leaching [11,12,45]. Our results were corroborated with the results of Sharma and Behera [46], who also reported that crop residue incorporation led to increased soil health and crop productivity.

Table 1. System productivity and sustainable yield index (SYI) of different cropping systems under nutrient and crop residue management practices.

Cropping Systems (CS)	System Productivity (Mg·ha^{-1})						SYI
	2012–2013	2013–2014	2014–2015	2015–2016	2016–2017	Pooled	
			Cropping system				
FRR	4.64	4.80	4.74	3.81	3.96	4.41	0.79
JRW	6.29	7.21	7.83	5.89	6.05	6.65	0.75
JRBc	18.3	15.3	13.7	17.8	11.5	15.3	0.71
JRGp	8.94	8.80	9.41	7.87	9.10	8.81	0.88
JRMMu #	8.13	8.51	8.13	7.23	7.10	7.61	0.82
SEm (±)	0.42	0.29	0.34	0.31	0.12	0.29	
LSD ($p \leq 0.05$)	1.16	0.85	0.98	0.91	0.37	0.84	
		Nutrient and residue management practices (NRCM)					
F_1R_0	9.10	8.72	8.41	7.86	7.21	8.26	0.76
F_1R_1	9.12	8.80	8.67	8.30	7.40	8.51	0.80
F_2R_0	9.40	9.91	8.95	8.38	7.54	8.67	0.79
F_2R_1	9.41	9.12	9.05	8.77	8.04	8.78	0.83
SEm (±)	0.25	0.18	0.21	0.15	0.10	0.18	
LSD ($p < 0.05$)	NS	NS	NS	0.42	0.30	0.36	
CS × NRCM	NS	NS	NS	NS	NS	NS	

J—jute; R—rice; w—wheat, Bc—baby corn; Gp—vegetable pea; M—mustard; Mu—mung bean (green gram); F_1R_0—75% NPK/RDF without crop residue; F_1R_1—75% NPK/RDF with crop residue F_2R_0—100% NPK without crop residue; F_2R_1—100% NPK with crop residue; # crop was sown on zero tillage; SEm—standard error of mean; LSD—least significant difference; S—significant; NS—non-significant.

The highest sustainable yield index (SYI) was the JRGp cropping system (0.88), followed by JRMMu (0.82), and the lowest was for JRBc (0.71). Nonetheless, with the highe system productivity of JRBc compared with the other systems, year-wise variations existed which led to a decreased SYI. Although the market value of baby corn was higher than the other products, fluctuation in the market price (varies from INR. 10–20/kg cob), which depend on the availability and demand of baby corn in the local market, led to decreased SYI. Among the NRCM practices, the highest SYI was recorded (0.83) for F_2R_1, followed by F_1R_1 (0.80). The lowest SYI was recorded where lower doses of mineral fertilizer were applied without crop residue. SYI is the function of crop yield and the fluctuation in yield/standard deviation (α). During the course of the study, the fluctuation of crop yield which determines the standard deviation, was less where 100% NPK was applied with crop residue, and led to a higher SYI. The benefit of crop residue incorporation is that could provide proper soil health to the region to help sustain the crop yield. The high content of SOC under F_2R_1 could increase the microbial activity and soil enzymes, as microbes use SOC as a carbon substrate, which helps to increase the availability of N, P, and K The highest SYI was recorded under F_2R_1 because of the greater availability of N, P, and K, and accelerated nutrient acquisition resulting increased crop yields of each of the test crops [45,47].

3.2. Soil pH and Bulk Density (BD)

The soil pH did not exhibit significant differences with different cropping systems and NRCM practices (Table 2). Bulk density differed significantly ($p \leq 0.05$) among the cropping systems and NCRM practices. The JRGp system had a significantly lower bulk density (1.43 Mg/m^3) compared with the JRMMu system (1.57 Mg/m^3), but was not significant compared with the rest of the cropping systems. Among the NCRM practices a significantly lower bulk density was recorded in F_2R_1 (100% NPK with crop residue compared with F_1R_0, but was it was similar to F_1R_1. The differences in BD under the JRGp system are mainly associated with the addition of legume crops, which are responsible for lowering the BD. The effect of legume crops on bulk density has been reported by Lati et al. [48]. The bulk density of the soil decreased with loosening the soil through tillage and through the incorporation of crop residue [7]. The interaction effect of CS and NRCM practices on pH and BD was non-significant.

Table 2. Soil physio-chemical properties influenced by different cropping systems and nutrient and crop residue management practices.

Cropping System/Soil Attributes	pH	BD (Mg/m^3)	SOC (g/kg)	Av N (mg/kg)	Av P (mg/kg)	Av K (mg/kg)
			Cropping system (CS)			
FRR	7.11	1.47	6.94	109.7	22.5	83.8
JRW	7.13	1.46	6.57	108.5	18.9	88.1
JRBc	7.16	1.49	6.48	103.8	16.4	99.3
JRGp	7.11	1.43	7.37	111.4	25.1	101.6
JRMMu #	7.07	1.55	7.27	111.4	22.7	104.5
SEm (±)	0.06	0.04	0.23	6.10	2.18	5.26
LSD ($p \leq 0.05$)	NS	0.10	0.55	NS	5.03	12.35
		Nutrients and crop residue management practices (NCRM)				
F_1R_0	7.07	1.52	6.65	103.2	20.1	92.2
F_1R_1	7.13	1.46	6.90	110.6	21.3	95.1
F_2R_0	7.15	1.51	6.84	106.4	20.9	94.8
F_2R_1	7.11	1.43	7.32	115.7	22.4	99.8
SEm (±)	0.06	0.03	0.22	3.02	0.77	2.53
LSD ($p \leq 0.05$))	NS	0.08	0.50	6.16	1.56	6.50
CS × NCRM	NS	NS	NS	NS	NS	NS

J—jute; R—rice; W—wheat; Bc—baby corn; Gp—vegetable pea; M—mustard; Mu—mung bean (green gram); F_1R_0—75% NPK/RDF without crop residue; F_1R_1—75% NPK/RDF with crop residue; F_2R_0—100% NPK without crop residue; F_2R_1—100% NPK with crop residue; # crop was sown on zero tillage; BD—bulk density; SOC—soil organic carbon; AvN—available N; AvP—available P; AvK—available K, SEm—standard error of mean; LSD—least significant difference; S—significant; NS—non-significant.

3.3. Soil Organic Carbon and Available N, P and K

Soil organic carbon (SOC) was significantly ($p \leq 0.05$) higher (7.3 mg/kg) in the JRGp cropping system compared with the JRW and JRBc cropping systems, but it was similar to the JRMMu system (7.27 g/kg; Table 2). The JRBc cropping system exhibited the lowest amount of SOC. The higher SOC for the cropping system including vegetable pea (Gp) and mungbean (Mu) might be attributed to the higher decomposition of residues due to the lower C:N ratio of leguminous crops. On the contrary, cereal dominated systems like JRW and JRBc, which consisted of nutrient-exhaustive crops like corn, rice, and wheat, had a higher C:N ratio in their residues, hindering the decomposition of SOC, therefore resulting in a lower SOC [9]. Among the NRCM practices, a significantly ($p \leq 0.05$) higher SOC was recorded in F_2R_1 compared with F_1R_0 The short term variability in SOC might be due to the addition of crop residues in soil [45,49], through increasing the cropping intensity [50] by diversifying with leguminous crops, which resulted in increased SOC [47]. Available N (Av.N) did not differ significantly among the cropping systems, but available P and K differed significantly ($p \leq 0.05$). However, the maximum available N in the soil was recorded in the JRGp and JRMMu cropping systems. Significantly higher available P (Av. P) was recorded in the JRGp cropping system compared with the JRW and JRBc systems, but it was similar to the JRMMu and RR cropping systems. Similarly, significantly higher available K (av.K) was recorded in JRMMu compared with other cropping systems, but it was statistically at the same level as the JRGp and JRBc cropping systems. The maximum available N in the cropping system with legume crops like pea and mung bean/green gram might be because legumes obtain atmospheric nitrogen through N-fixation for their own requirements, and subsequently release N into the soil as a result of nodulation, root, leaf, etc., incorporation and decomposition throughout growth, thus adding considerable quantities of N, P, and K to the soil [14,46,51,52]. Among NRCM practices, significantly higher available primary nutrients (av. NPK) were recorded in F_2R_1 compared with F_1R_0 and F_2R_0, and it was similar to F_1R_1. A substantial increase in the N content of soil due to crop residue incorporation has also been reported by Bakht et al. [12] and Shafi et al. [11]. Furthermore, Gupta et al. [53] found a higher available phosphorus and potassium content when practicing return of crop-residue to the soil in a 4-year study compared with a no residue return, i.e., residue removal.

3.4. Soil Microbial Properties

Cropping system and crop residue management had a significant effect on soil microbial activities (Table 3). Soil microbial biomass nitrogen (SMBN) and soil microbial biomass carbon (SMBC) were significantly ($p \leq 0.05$) higher in the JRMMu cropping system compared with the other systems. Among the NRCM practices, higher SMBN and SMBC were recorded in F_2R_1 compared with the other NRCM treatments. The microbial population was differed according to cropping systems and NRCM practices—the bacterial population was the highest in the JRW cropping system, but the lowest count was recorded in the FRR cropping system. Contrary to that, the *Azotobacter* population was the highest in JRBc, followed by the JRMMu system. A comparatively higher PSB was recorded in the JRGp and JRMMu cropping systems than for the other cropping systems. The JRGp and the JRMMu cropping systems, which had pea and mungbean, respectively, required a higher P nutrition for N fixation. The legume crops have the capacity to solubilize P quickly compared with cereals, by secreting root exudates that enhance the growth and activities of PSB [54,55]. Among the NCRM practices, all of the soil microbes were higher in F_2R_1 compared with the other NCRM practices. The bacterial population was higher where more nitrogen and crop residue were applied because the availability of the growth substances for bacteria were higher from the crop residues and applied N, which led to increased bacteria growth that was responsible for the decomposition of crop residues. Crop residues with comparative considerable constituents of cellulose, chitin, lignin, etc., are degraded by these soil bacteria [56,57].

Table 3. Soil microbial properties influenced by different cropping systems and nutrient and crop residue management practices.

Treatments	SMBN (mg/kg Soil)	SMBC (mg/kg Soil)	Bacteria (10^6 cfu/g Soil)	Azotobacter (10^4 cfu/g Soil)	PSB (10^6 cfu/g Soil)
Cropping system (CS)					
FRR	19.1	174.4	34.2	25.6	15.3
JRW	20.3	173.8	43.9	28.3	15.5
JRBc	20.5	171.3	42.4	43.7	17.5
JRGp	22.0	180.8	42.8	35.3	20.7
JRMMu #	23.5	190.6	37.5	38.7	20.8
SEm (±)	0.66	2.77	1.87	3.52	0.88
LSD ($p \leq 0.05$)	1.90	7.99	5.39	10.04	2.55
Nutrients and crop residue management practices (NCRM)					
F_1R_0	17.4	173.2	33.5	33.4	15.5
F_1R_1	20.6	177.5	42.0	35.3	18.9
F_2R_0	21.7	176.7	38.7	33.9	18.1
F_2R_1	24.6	185.2	46.40	40.5	19.0
SEm (±)	0.49	0.91	1.55	2.11	0.86
LSD ($p \leq 0.05$)	1.40	2.62	4.48	NS	2.48
CS × NCRM	NS	*	*	NS	*

J—jute; R—rice; W—wheat; Bc—baby corn; Gp—vegetable pea; M—mustard; Mu—mung bean (green gram); F_1R_0—75% NPK/RDF without crop residue; F_1R_1—75% NPK/RDF with crop residue F_2R_0—100% NPK without crop residue; F_2R_1—100% NPK with crop residue; # crop was sown on zero tillage; SMBC—soil microbial biomass carbon; SMBN—soil microbial biomass N; SEm—standard error of mean; * significant at $p \leq 0.05$ level; LSD—least significant difference; S—significant; NS—non-significant.

3.5. Soil Enzyme Activities

The cropping systems and the residues returned to the soil significantly alter the enzymatic activities in the soil (Table 4). The dehydrogenase (DHA) activity was significantly higher (3.86 μg TPF/g soil/hr) in the JRMMu system compared with the other cropping systems. The residue returned to soil definitely modified the soil microclimate, which manipulated the microbial metabolism [58,59] and led to a significantly enhanced DHA activity on the surface soil. The fluorescein diacetate hydrolytic (FDH) activity was also significantly higher (16.7 μg fluorescein/g/h) in JRMMu; however, it was similr to the JRGp (16.46 μg fluorescein/g/h) and FRR (16.6 μg fluorescein/g/h) cropping systems. Although the acid phosphatase (AcP) activity was non-significant in the cropping systems, the highest value (513.8 μg *p*NP/g/h) was recorded in the JRGp cropping system. In contrast, the alkaline phosphatase (AlP) was significantly higher (810.8 μg *p*NP/g/h) in the JRMMu cropping system compared with JRW and FRR, but it was similar to the JRGp (739.7 μg *p*NP/g/h) cropping system. The urease (UA) activity was significantly higher (59.8 μg NH_4^+/g/h) in the JRGp cropping system compared with the other systems, except for the JRMMu system. Among the NCRM practices, the DHA and FDH activities were significantly higher in F_2R_1 compared with the other NCRM practices. Although the activity of AcP, AlP, and UA was non-significant among the NCRM practices, a higher value was recorded for the F_2R_1 treatment. Returning crop residues to the soil alters the soil enzyme activities by providing a more congenial environment for the growth of microbes; as a result, a soil matrix with more enzyme accumulation was observed. Adding more residues to the soil results in a higher SOC, which is important for forming stable complexes with free enzymes [60]. Moreover, jute added a large amount of shedding leaves (~1 Mg·ha^{-1}) into the soil, which contain a considerable amount of N, P, and K, and improved the rhizosphere, hence promoting the growth and multiplication of the soil microbes, MBC, and soil enzymes [61,62]. The inclusion of legumes and/or their residue in the soil of a cereal-based system has been reported to improve the soil organic matter status, soil enzyme activity, and soil respiratory activity [59,63].

Table 4. Soil enzymes are influenced by different cropping systems and nutrient and crop residue management practices.

Treatments	DHA (μg TPF/g/h)	FDA (μg Fluorescein/g/h)	AcP (μg *p*NP/g/h)	AlP (μg *p*NP/g/h)	Urease (μg NH_4^+/g/h)
		Cropping system (CS)			
FRR	3.51	16.56	367.39	674.18	33.3
JRW	3.51	14.78	402.08	542.35	34.52
JRBc	3.40	15.26	359.38	736.55	37.45
JRGp	3.36	16.46	513.75	793.75	59.79
JRMMu #	3.36	16.67	488.04	810.60	42.02
SEm (±)	0.08	0.39	43.40	37.98	5.69
LSD ($p \leq 0.05$)	0.23	1.14	NS	109.67	16.43
		Nutrients and crop residue management practices (NCRM)			
F_1R_0	3.06	14.52	400.74	653.49	33.88
F_1R_1	3.40	15.73	414.95	730.74	35.51
F_2R_0	3.52	16.42	440.78	726.28	34.88
F_2R_1	4.14	17.12	448.95	735.28	45.45
SEm (±)	0.06	0.17	17.67	28.32	3.079
LSD ($p \leq 0.05$)	0.18	0.49	34.5	32.64	8.89
CS × NCRM	*	*	NS	NS	NS

J—jute; R—rice; W—wheat; Bc—baby corn; Gp—vegetable pea; M—mustard; Mu—mung bean (green gram); F_1R_0—75% NPK/RDF without crop residue; F_1R_1—75% NPK/RDF with crop residue F_2R_0—100% NPK without crop residue; F_2R_1—100% NPK with crop residue; # crop was sown on zero tillage; DHA—dehydrogenase; FDA—fluorescein diacetate hydrolytic; AcP-Acid phosphatase; AlP—alkaline phosphatase; UA—urease. * Significant at $p \leq 0.05$ level; SEm—standard error of mean; LSD—least significant difference; S—significant; NS—non-significant.

3.6. Soil Quality Index (SQI)

The goal of this study was to enhance and sustain the system productivity, which applied the PCA method for selecting MDS, i.e., minimum data set from the varied bio-physicochemical soil attributes [5]. Following Brejda et al. [24], and Wander and Bollero [64], the MDS was composed of principal components (PCs) with an eigen value ≥ 1 and PCs with a minimum of 5% data variations. PCs with a higher eigen value with variability were the best describing parameters for SQI. The cumulative variance of the four principal components (PCs) was 68.6% (Table 5). The amount of variability explained by PC-1, PC-2, PC-3, and PC-4 was 32.18, 16.10, 11.09, and 9.19%, respectively. Within each PC, only the highest weighted factors possessing absolute values within 10% of the highest factor loading were retained for MDS. For those PCs where more than one factor was retained for MDS, the redundant factor was eliminated using multivariate correlation coefficients [23]. The variables that were significantly correlated among the variables in each PC were eliminated, and only the highest eigen value among the correlated variables was retained in MDS.

Table 5. Principle component analysis of the soil variables.

Principle Components	PC-1	PC-2	PC-3	PC-4
Eigenvalue	3.862	1.932	1.332	1.103
Variance (%)	32.18	16.10	11.09	9.19
Cumulative variance (%)	32.18	48.28	59.38	68.57
Eigen vector				
BD	0.236	0.198	0.437	**0.766**
SOC	**0.641**	0.243	0.205	−0.033
AvN	0.569	−0.348	−0.244	0.035
AvP	0.575	0.299	−0.552	0.06
AvK	−0.127	−0.308	**0.722**	−0.258
SMBC	**0.752**	−0.223	0.106	0.313
SMBN	0.630	−0.404	0.325	−0.184
DHA	0.616	**−0.557**	−0.185	−0.18
FDA	**0.698**	−0.432	−0.084	0.111
AcP	0.546	0.442	0.205	−0.414
AlP	0.544	**0.541**	0.071	−0.294
UA	0.548	**0.575**	0.092	0.097

BD—bulk density; SOC—soil organic carbon; AvN—available N; AvP—available P; AvK—available K; SMBC—soil microbial biomass carbon; SMBN—soil microbial biomass N; DHA—dehydrogenase; FDA—fluorescein diacetate hydrolytic; AcP—acid phosphatase; AlP—alkaline phosphatase; UA—urease.

In contrast, the non-significant correlated values were preferably retained in MDS. Consequently, in PC-1, three highly weighted factors namely, SMBC, FDA, and SOC, were selected, but SMBC was retained for MDS, as it had the highest eigen value and significant correlation with SOC (R^2 = 0.418 **) and with FDA (R^2 = 0.355 **; Table 6).

In PC-2, there were three highly loaded factors, namely, UA (0.575), DHA (0.557), and AlP (0.541), but two parameters, i.e., DHA and UA, were selected for the MDS because DHA was not significantly correlated with UA (R^2 = 0.014) and AlP (R^2 = 0.051), whereas UA was retained for the MDS because of the high factor loading, despite having a significant correlation with AlP (R^2 = 0.460 **). In PC-3 and PC-4, only available K and bulk density, respectively, had the highest factor loading, and thus both were retained for MDS. Finally, five soil indicators, i.e., SMBC, DHA, UA, Av. K, and BD, were selected for MDS in Inceptisol. According to Biswas et al. [65], UA and BD changed through soil management practices are of a good quality indicators in Inceptisol.

Table 6. Correlation matrix of higher weighted factor loading under different PC.

	BD	SOC	SMBC	FDA	AvK	DHA	Alphos	Urease
BD	1.00							
SOC	0.18	1.00						
SMBC	0.301 *	0.418 **	1.00					
FDA	0.10	0.355 **	0.611 **	1.00				
AvK	0.012	−0.02	−0.093	−0.018	1.00			
DHA	−0.128	0.211	0.494 **	0.578 **	−0.024	1.00		
Alphos	0.282*	0.390 **	0.313 *	0.063	−0.099	0.051	1.00	
UA	0.072	0.412 **	0.216	0.196	−0.143	0.014	0.460 **	1.00

* Correlation is significant at the 0.05 level (two-tailed); ** Correlation is significant at the 0.01 level (two-tailed). BD—bulk density; SOC—soil organic carbon; AvN—available N; AvP—available P; AvK—available K; SMBC—soil microbial biomass carbon; SMBN—soil microbial biomass; DHA—dehydrogenase; FDA—fluorescein diacetate hydrolytic; AcP-Acid phosphatase; AlP—alkaline phosphatase; UA—urease. .

However, although SOC was selected as a good soil indicator by many researchers [27,66,67], in our study, SOC had a lower factor loading value than SMBC. SMBC also had a strong significant correlation with SOC; hence, SMBC was taken for the MDS. In addition, SMBC is strongly influenced by management factors in the short term, thus it provides an indication of the soil's functions, i.e., ability to store and recycle nutrients and energy. It also serves as a sensitive indicator of change in organic matter levels and its equilibrium [21,62,67]. The set of indicators chosen for MDS should be easily measurable; cost-effective; and vary with crop, soil, and climatic conditions [21,43,67]. Each PC explains a certain amount of the variation (%) in the total dataset. The weighted factors for the indicators chosen under a given PC were calculated from the variation of each PC divided by the cumulative percentage of the variation explained by all PCs with eigen values $\geq$1. The weighted factors (i.e., % variation of each PC/cumulative % variation explained by all PCs) for PC-1, PC-2, PC-3, and PC-4 were 0.47, 0.23, 0.16, and 0.13, respectively.

The MDS indicator values were normalized on a scale of 0–1 using the non-linear scoring function (NLSF), as shown in equation 3, following the approach of Andrews et al. [43] and Masto et al. [66]. After deciding the shape of the predicting response, i.e., more is better, less is better, or optimum is better for soil function, the limits or threshold values were assigned for each indicator (Table 7), and the scoring function curves of all MDS indicators are presented in Figure 2.

Table 7. Scoring functions (SF), threshold values, and weight for the MDS indicators.

Soil Indicator	Lower Threshold	Upper Threshold	A/Basline	Slope	References
SMBC	0	350	150	0.029	Masto et al. [66]
DHA	0	5.5	1.5	1.260	Masto et al. [66]
UA	12	74	24	0.10	Biswas et al. [65]
AvK	20	180	90	0.095	Masto et al. [66]
BD	1.2	2.1	1.5	−13.5	Bhaduri et al. [27]

SMBC—soil microbial biomass carbon; DHA—dehydrogenase; UA—urease; AvK—available K; BD—bulk density.

The score of each MDS was derived from the curve and was used for the SQI calculation for the cropping system and NCRM practices. Among the five cropping systems, SQI varied from 0.51 to 0.63 (Figure 3), and the highest SQI was in JRGp (0.63) and the lowest was in FRR (0.51). However, the SQI of JRGp was the same as for JRMMu (0.61). Among the NCRM practices, the highest SQI was 0.63 in the F_2R_1 (100% NPK with crop residue) treatment, which was significantly higher than the rest of the NCRM treatments. The SQI for F_1R_1 and F_2R_0 was similar. The highest SQI was recorded for the cropping systems with legume crops and for nutrients applied along with the crop residue. Legume cropd and their residues have been well known to enhance soil fertility through soil N_2-fixation, nutrient recycling, falling leaves, and root exudates, thereby sustaining productivity [11,14,46,67], which is evident from the value of SQI, which was higher in the cropping systems with legume crops in their sequence.

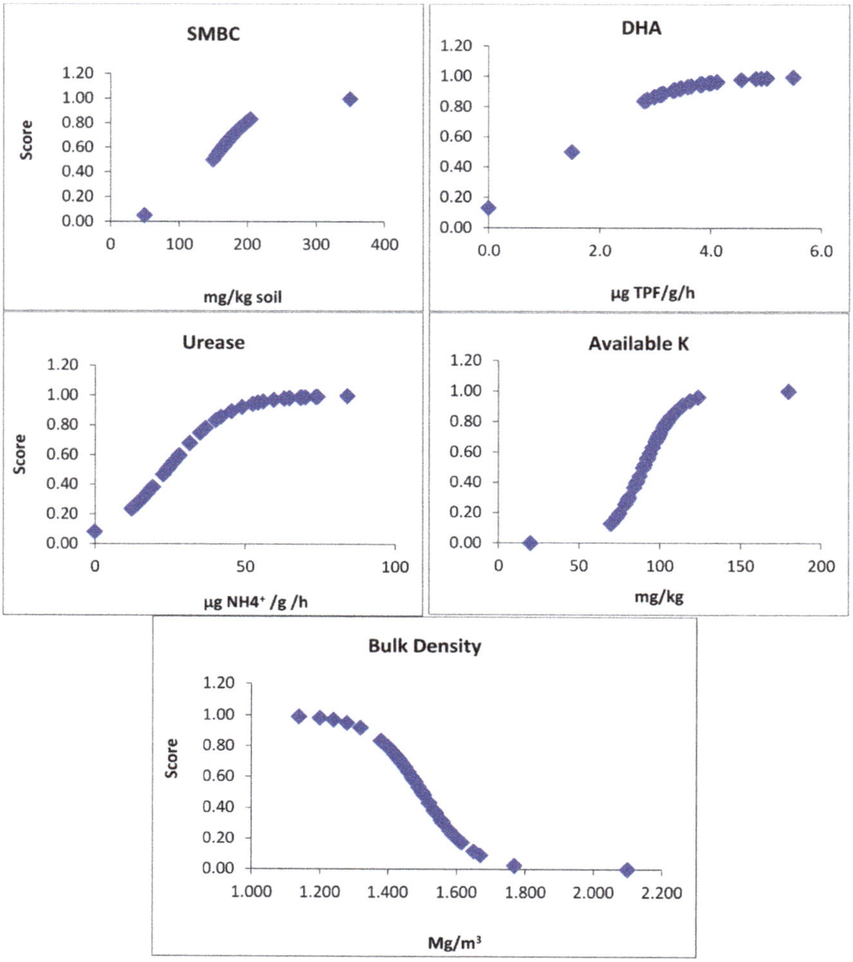

Figure 2. Nonlinear scoring functions for six ninimum data set (MDS)/soil quality indicator. SMBC—soil microbial biomass carbon; DHA—dehydrogenase.

The average contribution of each MDS varied among all of the cropping systems and NCRM practices, but the highest contribution was recorded with SMBC (24.1%), followed by DHA (11.8%), urease (8.76%), available K (7.01%), and BD (5.63%) in all cropping systems and NCRM practices. The contribution of BD, Av K, and UA varied among cropping systems, but not in NCRM practices. The higher contribution was for BD (6.7%) towards SQI development compared with Av. K (4.5%) and urease (4.7%) in the FRR cropping system, but the reverse trend was observed in the other cropping systems. A positive and linear correlation ($y = 1.954x - 0.377$; $R^2 = 0.51$) was recorded between the sustainable yield index and SQI (Figure 4). It indicated that sustainability increased with increasing the soil quality, adding legumes in the cropping system, and residue incorporation in the soil, which also increased SQI.

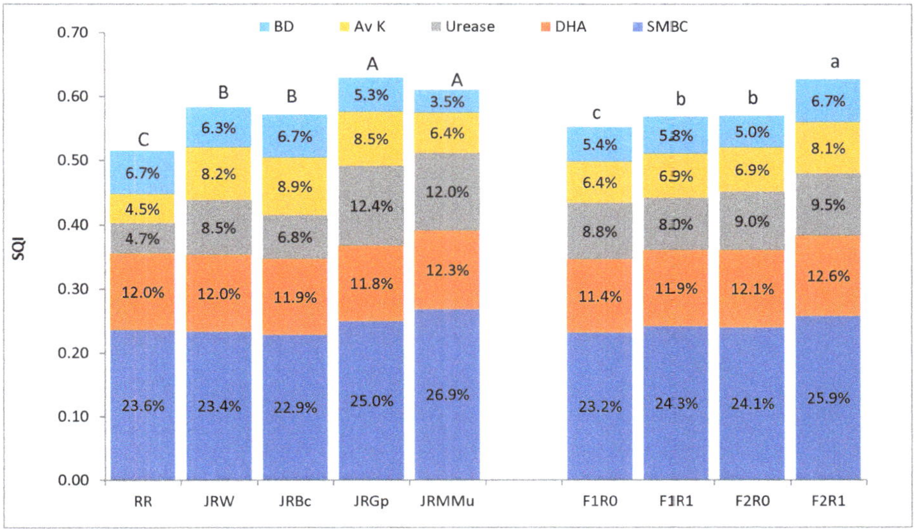

Figure 3. Soil quality index and contribution of soil quality indicator for different cropping system (CS) and nutrient and crop residue management (NRCM) practices. J-Jute, R-Rice, W-Wheat, Bc-Baby corn; Gp: Vegetable pea; M-Mustard, Mu-mung bean (green gram); F_1R_0-75% NPK/RDF without crop residue; F_1R_1-75% NPK/RDF with crop residue F_2R_0-100% NPK without crop residue; F_2R_1-100% NPK with crop residue. BD—bulk density; SOC—soil organic carbon; AvK—available K; SMBC—soil microbial biomass carbon; DH—dehydrogenase. Capital and small similar letters mean there are non-significant difference between CS and NRCM, respectively.

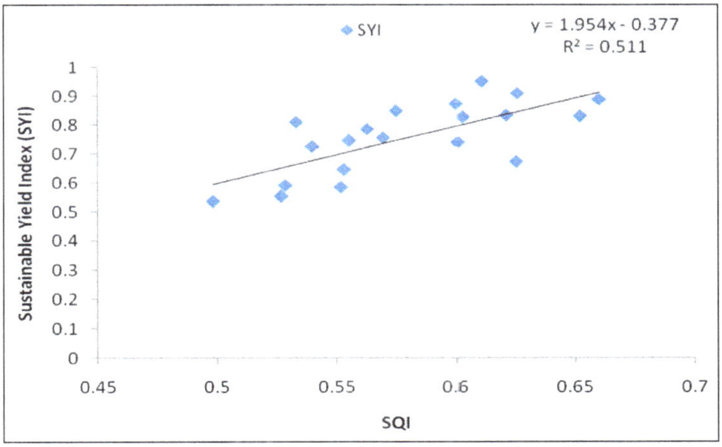

Figure 4. Relationship between sustainable yield index (SYI) and soil quality index (SQI).

4. Conclusions

From the consideration and analysis of the experimental data, it can be concluded that the system productivity of the jute–rice–baby corn (JRBc) system was higher than for the other cropping systems, followed by the jute–rice–pea (JRGp) and jute–rice–mustard–mungbean (JRMMu) cropping systems, while sustainability was higher in the JRGp cropping system. Both system productivity and sustainability were higher when the recommended dosed of fertilizers were applied with crop residue (F_2R_1) in the soil. All the important soil attributes were higher in JRGp and JRMMu for the 100% NPK applied with crop residue incorporation. The highest soil quality index (SQI) was seen for the jute–rice–vegetable pea (JRGp) and jute–rice–mustard–mung bean (JRMMu) when grown with the application of 100%NPK with crop residues. The five soil indicators, i.e., SMBC, DHA, UA, AvK, and BD, were selected for MDS in Inceptisols, in which SMBC contributed the highest towards the soil quality index (SQI) determination out of all of the cropping systems and NRCM practices. Hence, diversification/intensification of rice–rice or jute–rice cropping should be included vegetable pea (JRGp) or mustard–mung bean (JRMMu) provided with 100% NPK/RDF applied with crop residue in order to sustain the cropping system and higher soil quality.

Author Contributions: Conceptualization, M.K., S.M., S.P.M., B.M., A.R.S., S.R.S. and B.P.; methodology, M.K.; software, M.K. and B.P.; validation, M.K., S.P.M. and S.M.; formal analysis, M.K., B.P. and A.H.; investigation, M.K.; resources, S.M. and S.P.M.; data curation, M.K., B.P. and A.H.; writing—original draft preparation, M.K., S.M., S.P.M., B.M., A.R.S., S.R.S. and B.P.; writing—review and editing, A.G., A.H. and W.F.A.; visualization, M.K., S.M., S.P.M. and B.P.; supervision, S.M. and S.P.M.; project administration, S.M., A.G., W.F.A. and S.P.M.; funding, A.G., A.H. and W.F.A. All authors have read and agreed to the published version of the manuscript.

Funding: The current research work was funded by the Director of the Central Research Institute for Jute and Allied Fibres and the Taif University Researchers Supporting Project number (TURSP-2020/53), Taif University, Taif, Saudi Arabia.

Institutional Review Board Statement: Not applicable.

Informed Consent Statement: Not applicable.

Data Availability Statement: Most of the recorded data are available in all Tables in the manuscript.

Acknowledgments: We are highly thankful to the Director of the Central Research Institute for Jute and Allied Fibres for providing facilities and encouragement for this study. The authors also extend their appreciation to Taif University for funding current work by Taif University Researchers Supporting Project number (TURSP-2020/53), Taif University, Taif, Saudi Arabia.

Conflicts of Interest: The authors would hereby like to declare that there is no conflict of interests.

Appendix A

Table A1. Details of cropping systems and recommended dose of fertilizers (RDF).

Particulars	Jute	Rainy Rice	Winter Rice	Wheat	Baby Corn	Vegetable Pea	Mustard	Mungbean
Land preparation	One harrowing and one rotavator to incorporate residues in case of residue application and while no-residue was there then only one rotavator	Puddling	Puddling	One rotavator	One rotavator	One rotavator	No tillage	No tillage
Sowing/transplanting time (same for all the years of study)	15 April	12 August	24 January	25 November	15 November	25 November	11 November	5 February
Variety	JRO-204	Khitish	Khitish	PBW 343	G-5414	Azad P-3	B-54	Pant mung-5
Crop duration (Days)	110	120	140	130	90	100	85	70
Spacing (cm)	25 × 7 cm	20 × 15 cm	20 × 15 cm	20 cm	50 × 15 cm	40 × 10 cm	35 × 5 cm	35 × 10 cm
Harvesting	5 August	12 November	13 June	5 April	15 March	5 Marc	4 February	10 April
100 % Recommended doses of fertilizer (RDF) with crop residue								
Fertilizer–N (kg/ha)	80	120	80	120	100	40	60	25
Fertilizer–P_2O_5 (kg/ha)	40	60	40	60	60	60	30	60
Fertilizer–K_2O (kg/ha)	40	60	40	40	40	40	30	40
75% RDF with or without crop residue								
Fertilizer–N (kg/ha)	60	90	60	90	75	30	45	18.75
Fertilizer–P (kg/ha)	30	45	30	45	45	45	22.5	45
Fertilizer–K (kg/ha)	30	45	30	30	30	30	22.5	30

References

1. Timsina, J.; Connor, D.J. Productivity and management of rice-wheat cropping systems: Issues and challenges. *Field Crops Res.* **2001**, *69*, 93–132. [CrossRef]
2. Biswas, B.; Ghosh, D.C.; Dasgupta, M.K.; Trivedi, N.; Timsina, J.; Dobermann, A. Integrated assessment of cropping systems in the Eastern Indo-Gangetic plain. *Field Crop Res.* **2006**, *99*, 35–47. [CrossRef]
3. Kumar, M.; Singh, S.R.; Jha, S.K.; Shamna, A.; Mazumdar, S.P.; Singh, A.; Kundu, D.K.; Mahapatra, B.S. System productivity, profitability and resource use efficiency of jute based cropping systems in the eastern Indo-Gangetic plain. *Ind. J. Agric. Sci.* **2014**, *84*, 209–213.
4. UN-Pop. World Population Prospects: The 2017 Revision, Online Demographic Profiles. 2017. Available online: http://www.population.un.org/wpp/Graphs/DemographicProfiles/ (accessed on 12 March 2020).
5. Chaudhury, J.; Mandal, U.K.; Sharma, K.L.; Ghosh, H.; Mandal, B. Assessing soil quality under long-term rice-based cropping system. Commun. *Soil Sci. Plant Anal.* **2005**, *36*, 1141–1161. [CrossRef]
6. Hobbs, P.R.; Gupta, R.K. Rice–wheat cropping systems in the Indo-Gangetic plains: Issues of water productivity in relation to new resource-conserving technologies. In *Water Productivity in Agriculture: Limits and Opportunities for Improvement*; Kijne, J.W., Barker, R., Molden, D., Eds.; CAB International: Wallingford, UK, 2003; pp. 239–253.
7. Bhattacharyya, R.; Prakash, V.; Kundu, S.; Srivastva, A.K.; Gupta, H.S. Soil aggregation and organic matter in a sandy clay loam soil of the Indian Himalayas under different tillage and crop regimes. *Agric. Ecosyst. Environ.* **2009**, *132*, 126–134. [CrossRef]
8. Ramesh, P.; Panwar, N.R.; Singh, A.B.; Ramana, S.; Rao, A.S. Impact of organic manure combinations on the productivity and soil quality in different cropping systems in central India. *J. Plant Nutr. Soil Sci.* **2009**, *172*, 577–585. [CrossRef]
9. Kar, S.; Pramanick, B.; Brahmachari, K.; Saha, G.; Mahapatra, B.S.; Saha, A.; Kumar, A. Exploring the best tillage option in rice based diversified cropping systems in alluvial soil of eastern India. *Soil Tillage Res.* **2021**, *205*, 104761. [CrossRef]
10. Yang, X.M.; Kay, B.D. Rotation and tillage effects on soil organic carbon sequestration in a typicHapludalf in Southern Ontario. *Soil Tillage Res.* **2001**, *59*, 107–114. [CrossRef]
11. Shafi, M.; Bakht, J.; Jan, M.T.; Shah, Z. Soil C and N dynamics and maize (*Zea mays*) yield as affected by cropping systems and residue management in North western Pakistan. *Soil and Tillage Res.* **2007**, *94*, 520–529. [CrossRef]
12. Bakht, J.; Shafi, M.; Jan, M.T.; Shah, Z. Influence of crop residue management, cropping system and N fertilizer on soil N and C dynamics and sustainable wheat (*Triticumaestivum*) production. *Soil Tillage Res.* **2009**, *104*, 233–240. [CrossRef]
13. Singh, Y.; Gupta, R.K.; Singh, J.; Singh, G.; Singh, G.; Ladha, J.K. Placement effects on rice residue decomposition and nutrient dynamics on two soil types during wheat cropping in rice–wheat system in north Western India. *Nutr. Cycl. Agroecosyst.* **2010**, *88*, 471–480. [CrossRef]
14. Kumar, T.K.; Rana, D.S.; Nain, L. Legume residue and N management for improving productivity and N economy and soil fertility in wheat (*Triticum aestivum*)-based cropping systems. *Natl. Acad. Sci. Lett* **2019**, *42*, 297–307. [CrossRef]
15. Islam, K.R.; Weil, R.R. Land use effects on soil quality in a tropical forest ecosystem of Bangladesh. *Agric. Ecosys. Environ.* **2000**, *79*, 9–16. [CrossRef]
16. Lobe, I.; Amelung, W.; Du Preez, C.C. Losses of carbon and nitrogen with prolonged arable cropping from sandy soils of the South African Highveld. *Eur. J. Soil Sci.* **2001**, *52*, 93–101. [CrossRef]
17. Jaskulska, I.; Romaneckas, K.; Jaskulski, D.; Galezewski, L.; Breza-Boruta, B.; Debska, B.; Lemanowicz, J. Soil properties after eight years of the use of strip-till one-pass technology. *Agronomy* **2020**, *10*, 1596. [CrossRef]
18. Karlen, D.L.; Mausbach, J.W.; Doran, J.W.; Cline, R.G.; Harris, R.F.; Schuman, G.E. Soil quality: A concept, definition and framework for evaluation. *Soil Sci. Soc. Am. J.* **1997**, *61*, 4–10. [CrossRef]
19. Bastida, F.; Moreno, J.L.; Hernández, T.; García, C. Microbiological degradation index of soils in a semi aridsemi-arid climate. *Soil Biol. Biochem.* **2006**, *38*, 3463–3473. [CrossRef]
20. Cluzeau, D.; Guernion, M.; Chaussod, R.; Martin-Laurent, F.; Villenave, C.; Cortet, J.; Ruiz-Camacho, N.; Pernin, C.; Mateille, T.; Philippot, L.; et al. Integration of biodiversity in soil quality monitoring: Baselines for microbial and soil fauna parameters for different land-use types. *Eur. J. Soil Biol.* **2012**, *49*, 63–72. [CrossRef]
21. Sharma, K.L.; Mandal, U.K.; Srinivas, K.; Vittal, K.P.R.; Mandal, B.; Grace, J.K.; Ramesh. V. Long-term soil management effects on crop yields and soil quality in a drylandAlfisol. *Soil Tillage. Res.* **2005**, *83*, 246–259 [CrossRef]
22. Bhaduri, D.; Purakayastha, T.J. Long-term tillage, water and nutrient management in rice–wheat cropping system: Assessment and response of soil quality. *Soil Tillage Res.* **2014**, *144*, 83–95. [CrossRef]
23. Andrews, S.S.; Mitchel, J.P.; Mancinelli, R.; Larlen, D.I.; Hartz, T.K.; Horwarth, W.R.; Pettygrove, G.S.; Scow, K.M.; Munk, D.S. On farm assessment of soil quality in California's central valley. *Agron. J.* **2002**, *94*, 12–23. [CrossRef]
24. Brejda, J.J.; Moorman, T.B.; Karlen, D.L.; Dao, T.H. Identification of regional soil quality factors and indicators I.Central and Southern High Plains. *Soil Sci. Soc. Am. J.* **2000**, *64*, 2115–2124. [CrossRef]
25. Masto, R.E.; Chhonkar, P.K.; Singh, D.; Patra, A.K. Soil quality response to long-term nutrient and crop management on a semi-arid Inceptisol. *Agric. Ecosyst. Environ.* **2007**, *118*, 130–142. [CrossRef]
26. Manna, M.C.; Swarup, A.; Wanjari, R.H.; Ravankar, H.N.; Mishra, B.; Saha, M.N. Long term effect of fertilizer and manure application on soil organic carbon storage, soil quality and yield sustainability under subhumid and semi-arid tropical India. *Field Crop Res.* **2005**, *93*, 264–280. [CrossRef]

27. Bhaduri, D.; PurakayasthaPurakayastha, T.J.; Patra, A.K.; Chakraborty, D. Evaluating soil quality under a long-term integrated tillage–water–nutrient experiment with intensive rice–wheat rotation in a semi-arid Inceptisol. *India. Environ. Monit. Assess.* **2014**, *186*, 2535–2547. [CrossRef] [PubMed]
28. Choudhary, M.; Jat, H.S.; Datta, A.; Yadav, A.K.; Sapkota, T.B.; Mondal, S.; Meena, R.P.; Sharma, P.C.; Jat, M.L. Sustainable intensification influences soil quality, biota, and productivity in cereal-based agroecosystems. *Appl. Soil Ecol.* **2018**, *126*, 189–198. [CrossRef]
29. Sinha, N.K.; Chopra, U.K.; Singh, A.K. Cropping system effects on soil quality for three agro-ecosystems in india. *Expl Agric.* **2014**, *50*, 321–342. [CrossRef]
30. Sofi, J.A.; Bhat, A.G.; Kirmai, N.A.; Wani, J.A.; Lone, A.H.; Ganie, M.A.; Dar, G.I.H. Soil quality index as affected by different cropping systems in northwestern Himalayas. *Environ. Monit. Assess.* **2016**, *188*, 161–174. [CrossRef]
31. Walkley, A.; Black, I.A. An examination of Degtjareff method for determining organic carbon in soils: Effect of variations in digestion conditions and of inorganic soil constituents. *Soil Sci.* **1934**, *63*, 251–263. [CrossRef]
32. Subbiah, B.V.; Asija, G.L. A rapid procedure for assessment of available nitrogen in soils. *Curr. Sci.* **1956**, *31*, 196–260.
33. Olsen, S.; Cole CWatanabe, F.; Dean, L. Estimation of available phosphorus in soils by extraction with sodium bicarbonate. In *United* States Department of Agriculture Circular No. 939; USDA: Washington, DC, USA, 1954.
34. Jackson, M.L. *Soil Chemical Analysis*; Prentice Hall of India Pvt., Ltd.: New Delhi, India, 1973.
35. Vance, E.D.; Brookes, P.C.; Jenkinson, D.S. An extraction method for measuring soil microbial biomass C. *Soil Biol. Biochem.* **1987**, *19*, 703–707. [CrossRef]
36. Parkinson, D.; Gray, T.R.G.; Williams, S.T. Methods for studying the ecology of soil microorganisms. In *International Biological Programme Handbook 19*; Blackwell Scientist Publications: Oxford, UK, 1971.
37. Pikovskaya, A.I. Mobilization of phosphorus in soil in connection with vital activity of some microbial species. *Microbiology* **1948**, *17*, 362–370.
38. Tabatabai, M.A. Soil enzymes. In *Methods of Soil Analysis. Part 2. Chemical and Microbiological Properties*; Academic Press: New York, NY, USA, 1982; pp. 903–947.
39. Schnurer, J.; Rosswall, T. Fluorescein di-acetate hydrolysis as a measure of total microbial activity in soil and litter. *Appl. Environ. Microb.* **1982**, *43*, 1256–1261. [CrossRef] [PubMed]
40. Tabatabai, M.A.; Bremner, J.M. Assay of Urease Activity in Soils. *Am. J. Soil Sci. Soc.* **1972**, *41*, 350–352. [CrossRef]
41. Tabatabai, M.A.; Bremner, J.M. 1969. Use of p-nitrophenyl phosphate for assay of soil phosphatase activity. *Soil Biol. Biochem.* **1969**, *1*, 301–307. [CrossRef]
42. Singh, R.P.; Das, S.K.; BhaskaraRao, U.M.; Narayana Reddy, M. *Towards Sustainable Dryland Agricultural Practices, Technical Bulletin*; Central Research Institute for Dryland Agriculture: Hyderabad, India, 1990; p. 106.
43. Andrews, S.S.; Karlen, D.I.; Mitchell, J.P. A comparison of soil quality indexing methods for vegetable production systems in Northern California. *Agric. Ecosyst. Environ.* **2002**, *90*, 25–45. [CrossRef]
44. Gomez, K.A.; Gomez, A.A. *Statistical Procedures for Agricultural Research*; John Wiley & Sons Inc.: New York, NY, USA, 1984.
45. Gangwar, K.S.; Singh, K.K.; Sharma, S.K.; Tomar, O.K. Alternative tillage and crop residue management in wheat after rice in sandy loam soils of Indo-Gangetic plains. *Soil Tillage Res.* **2006**, *88*, 242–252. [CrossRef]
46. Sharma, A.R.; Behera, U.K. Recycling of legume residue for nitrogen economy and higher productivity in maize (*Zea mays*)–wheat (*Triticumaestivum*) cropping system. *Nutr. Cycl. Agroecosyst.* **2009**, *83*, 197–210. [CrossRef]
47. Ghosh, P.K.; Venkatesh, M.S.; Hazra, K.K.; Kumar, N. Long term effect of pulses and nutrient management on soil organic carbon dynamics and sustainability on an Inceptisol of Indo-Gangetic Plains of India. *Exp. Agric.* **2012**, *48*, 473–487. [CrossRef]
48. Latif, M.A.; Mehuys, G.R.; Mackenzie, A.F.; Alli, I.; Faris, M.A. Effects of legumes on soil physical quality in maize crops. *Plant Soil* **1992**, *140*, 15–23. [CrossRef]
49. Das, T.K.; Bhattacharyya, R.; Sharma, A.R.; Das, S.; Saad, A.A.; Pathak, H. Impacts of conservation agriculture on total soil organic carbon retention potential under an irrigated agro-ecosystem of the western Indo-Gangetic Plains. *Eur. J. Agron.* **2013**, *51*, 34–42. [CrossRef]
50. Hutchinson, J.J.; Campbell, C.A.; Desjardins, R.L. Some perspectives on carbon sequestration in agriculture. *Agric. For. Meteorol.* **2007**, *142*, 288–302. [CrossRef]
51. Peoples, M.B.; Herridge, D.F.; Ladha, J.K. Biological nitrogen fixation: An efficient source of nitrogen for sustainable agricultural production. *Plant Soil* **1995**, *174*, 3–28. [CrossRef]
52. Gupta, R.K.; Yadvinder-Singh Ladha, J.K.; Singh, J.; Bijay-Singh Singh, G.; Pathak, H. Yield and phosphorus transformations in a rice–wheat system with crop residueand phosphorus management. *Soil Sci. Soc. Am. J.* **2007**, *71*, 1500–1507. [CrossRef]
53. Kundu, B.S.; Nehra KYadav, R.; Tomar, M. Biodiversity of phosphate solubilizing bacteria in rhizosphere of chickpea, mustard and wheat grown in different regions of Haryana. *Ind. J. Microbiol.* **2009**, *49*, 120–127. [CrossRef]
54. Mishra, P.K.; Mishra, S.; Selvakumar, G.; Bisht, J.K.; Kundu, S.; Gupta, H.S. Coinoculation of *Bacillus thuringeinsis*-KR1 with *Rhizobium leguminosarum* enhances plant growth and nodulation of pea (*Pisumsativum* L.) and lentil (*Lens culinaris* L.). *World J. Microbiol. Biotechnol.* **2009**, *25*, 753–761. [CrossRef]
55. Zhao, S.; Li, K.; Zhou, W.; Qiu, S.; Huang, S.; He, P. Changes in soil microbial community, enzyme activities and organic matter fractions under long-term straw return in north-central China. *Agric. Ecosyst. Environ.* **2016**, *216*, 82–88. [CrossRef]

56. Arcand, M.; Helgason, B.; Lemke, R. Microbial crop residue decomposition dynamics in organic and conventionally managed soils. *Appl. Soil Ecol.* **2016**, *107*, 347–359. [CrossRef]
57. Sinsabaugh, R.L.; Follstad Shah, J.J. Ecoenzymatic stoichiometry of recalcitrant organic matter decomposition: The growth rate hypothesis in reverse. *Biogeochemistry* **2011**, *102*, 31–43. [CrossRef]
58. Singh, G.; Bhattacharyyaa, R.; Das, T.K.; Sharma, A.R.; Ghosh, A.; Das, S.; Jha, P. Crop rotation and residue management effects on soil enzyme activities, glomalin and aggregate stability under zero tillage in the Indo-Gangetic Plains. *Soil Tillage. Res.* **2018**, *184*, 291–300. [CrossRef]
59. Gianfreda, L.; Ruggiero, P. Enzyme activities in soil. In *Nucleic Acids and Proteins in Soil*. Nannipieri, P., Smalla, K., Eds.; Springer: Berlin, Germany, 2006; pp. 257–311.
60. Chang, E.H.; Chung, R.S.; Tsai, Y.W. Effect of application rates of organic fertilizers on soil enzymes activities and microbial counts. *Soil Sci. Plant Nutr.* **2007**, *53*, 132–140. [CrossRef]
61. Singh, S.R.; Kundu, D.K.; Tripathi, M.K.; Dey, P.; Saha, A.R.; Kumar, M.; Singh, I.; Mahapatra, B.S. Impact of balanced fertilization on nutrient acquisition, fibre yield of jute and soil quality in New Gangetic alluvial soils of India. *Appl. Soil Ecol.* **2015**, *92*, 24–34. [CrossRef]
62. Chander, K.; Goyal, S.; Mundra, M.C.; Kapoor, K.K. Organic matter, microbial biomass and enzyme activity of soils under different crop rotations in the tropics. *Biol. Fertil. Soils* **1997**, *24*, 306–310. [CrossRef]
63. Wander, M.M.; Bollero, G.A. Soil quality assessment of tillage impacts to Illinois. *Soil Sci. Soc. Am. J.* **1999**, *63*, 961–971. [CrossRef]
64. Biswas, S.; Hazra, G.C.; Purakayastha, T.J.; Saha, N.; Mitran, T.; Roy, S.S.; Basak, N.; Mandal, B. Establishment of critical limits of indicators and indices of soil quality in rice-rice cropping systems under different soil orders. *Geoderma* **2017**, *292*, 34–48. [CrossRef]
65. Masto, R.E.; Chhonkar, P.K.; PurakayasthaTJPatra, A.K.; Singh, D. Soil quality indices for evaluation of long term land use and soil management practices in semi-arid sub-tropical India. *Land Degrad. Dev.* **2008**, *19*, 1–14. [CrossRef]
66. Shahid, M.; Nayak, A.K.; Shukla, A.K.; Tripathi, R.; Kumar, A.; Mohanty, S.; Bhattacharyya, P.; Raja, R.; Panda, B.B. Long-term effects of fertilizer and manure applications on soil quality and yields in a sub-humid tropical rice-rice system. *Soil Use Manag.* **2013**, *29*, 322–332. [CrossRef]
67. Sharma, K.L.; Ramachandrappa, B.K.; Suma Chandrika, D.; Sathish, A.; Dhanpal, G.N.; Rao, C.S.; Shankar, M.A.; Grace, J.K.; Sankar, G.M.; Chary, G.R ; et al. Effect of organic manure and crop residue based long-term nutrient management systems on soil quality changes under sole finger millet (*Eleusinecoracana* (L.) Gaertn.) and Groundnut (*Arachishypogaea* L.)–Finger millet rotation in rainfed Alfisol. *Commun. Soil Sci. Plant Anal.* **2016**, *47*, 899–914.

agronomy

MDPI

Article

Weeds Spectrum, Productivity and Land-Use Efficiency in Maize-Gram Intercropping Systems under Semi-Arid Environment

Rana Nadeem Abbas [1], Muhammad Awais Arshad [1], Asif Iqbal [1], Muhammad Aamir Iqbal [2,*], Muhammad Imran [3], Ali Raza [4,5], Jen-Tsung Chen [6], Mohammed Nasser Alyemeni [7] and Daniel Ingo Hefft [8]

1 Department of Agronomy, Faculty of Agriculture, University of Agriculture Faisalabad, Faisalabad 38040, Pakistan; nadeem4u3@yahoo.com (R.N.A.); awaisarshaduaf@yahoo.com (M.A.A.); asifiqbaldehasi@yahoo.com (A.I.)
2 Department of Agronomy, University of Poonch Rawalakot, Rawalakot 12350, Pakistan
3 Department of Entomology, University of Poonch Rawalakot, Rawalakot 12350, Pakistan; muhammadimran@upr.edu.pk
4 Fujian Provincial Key Laboratory of Crop Molecular and Cell Biology, Fujian Agriculture and Forestry University, Fuzhou 350002, China; alirazamughal143@gmail.com
5 Key Lab of Biology and Genetic Improvement of Oil Crops, Oil Crops Research Institute, Chinese Academy of Agricultural Sciences (CAAS), Wuhan 430062, China
6 Department of Life Sciences, National University of Kaohsiung, Kaohsiung 81148, Taiwan; jentsung@nuk.edu.tw
7 Botany and Microbiology Department, College of Science, King Saud University, Riyadh 11451, Saudi Arabia; mnyemeni@ksu.edu.sa
8 University Centre Reaseheath, Reaseheath College, Nantwich CW5 6DF, UK; daniel.hefft@reaseheath.ac.uk
* Correspondence: aamir1801@yahoo.com

Citation: Abbas, R.N.; Arshad, M.A.; Iqbal, A.; Iqbal, M.A.; Imran, M.; Raza, A.; Chen, J.-T.; Alyemeni, M.N.; Hefft, D.I. Weeds Spectrum, Productivity and Land-Use Efficiency in Maize-Gram Intercropping Systems under Semi-Arid Environment. *Agronomy* **2021**, *11*, 1615. https://doi.org/10.3390/agronomy11081615

Academic Editors: Umberto Anastasi and Aurelio Scavo

Received: 19 July 2021
Accepted: 10 August 2021
Published: 13 August 2021

Abstract: To ensure food security on sustainable basis, reducing weeds interference and boosting land use efficiency are critical. A field study was conducted at research farm of University of Agriculture Faisalabad, Pakistan, to sort out the most productive maize-gram intercropping system under semi-arid environment. Treatments included sole maize in single row (60 cm apart) (T_1) and double rows (90 cm apart) (T_2) strips, sole black (T_3) and green gram (T_4) crops, six single rows (60 cm apart) of maize with twelve double rows (20 cm) of black (T_5) and green gram (T_6), three double rows (90 cm apart) of maize with three sets of quadratic rows (20 cm apart) of black (T_7) and green gram (T_8). The experiment was executed in regular arrangement of randomized complete block design with three replications. The results revealed that T_1 produced the highest grain yield (6.97 t ha^{-1}) of maize and significantly lower weeds infestation compared to wider row spacing (T_2). Among intercropping systems, T_8 significantly decreased weeds density (16.33 plants m^{-2}) and their fresh (20.93 g m^{-2}) and dry weights (5.63 g m^{-2}), while the maximum land use efficiency as indicated by unmatched land equivalent ratio and intercropping advantage were recorded by T_7 and T_8. Interestingly, green gram in intercropping recorded over 58% higher productivity than black gram. We conclude that maize-green gram intercropping hold potential to impart sustainability to maize production by reducing weeds infestation (431% lower than sole maize) and could be a viable option for smallholder farmers in semi-arid environment.

Keywords: sustainable intercropping; companion crops; *Vigna mungo*; *Vigna radiata*; living mulch; land equivalent ratio

1. Introduction

Intensive agriculture is providing substantial yields of cereals but has caused serious environmental degradation, largely owing to excessive use of mineral fertilizers and chemical pesticides [1]. For ensuring nutritional security on sustainable basis under changing climate, developing innovative farming systems for cereals are indispensable.

especially in Asian countries like Pakistan, India, Saudi Arabia, China and Bangladesh. These countries are confronting profound environmental degradation as evident through global warming and unpredictable variation in precipitation regimes which have adversely affected the farming systems across the continent [2]. The development of farming systems that are biologically viable, economically attractive, farmer friendly, technologically adoptable and environmentally sustainable are direly needed. Intercropping of cereals like maize (*Zea mays* L.) with legumes (green and black grams) may improve resources (light, moisture, mineral nutrients etc.) utilization efficiency due to complementary use of inputs in temporal and spatial dimensions [3]. In addition, intercropping systems exploit complementarities of species to attain sustainable intensification by multiplying crops outputs per unit of land area with substantial slicing of anthropogenic inputs. Maize plants hold competitive advantage over legumes by virtue of deeper and rapidly spreading roots system, while legumes fulfill a greater part of their nitrogen requirement from biological nitrogen fixation process [4,5]. Furthermore, it was reported that strip cropped maize with legumes developed deeper and extended roots network into the soil for exploring lower soil horizons owing to competition for moisture uptake [6]. Although overall productivity on intercropping systems remained on higher side, however, maize and legume intercrops witness individual yields reduction in intercropping systems [1,3,4], which constitutes the most pertinent challenge especially in semi-arid environment. Moreover, changing climate requisites evaluating atypical maize production systems that may potentially boost productivity without requiring additional farm inputs.

The choice of legume for intercropping with maize determines the productivity of intercropping systems by ensuring compatibility in utilizing growth resources [1]. Compared to solo crop equivalents, overall intercropping systems productivity and land use efficiency as indicated by land equivalent ratio were significantly (23–47%) increased [7,8]. Similarly, green gram (*Vigna radiata* L.) and black gram (*Vigna mungo* L.) may impart sustainability to maize-legume intercropping system by enhancing land use efficacy attained through higher utilization efficiency of farm applied inputs. However, optimization of intercropping system may potentially reduce the degree of inter and intra species competition and boost the added benefits offered by cereal-legume intercropping systems [9–13], which continues to remain an unexplored aspect under irrigated conditions of semi-arid environment. This is of the utmost importance as numerous types of species-specific mechanisms alter the physiological response of intercrops and directly determine the extent of added advantage offered by intercropping system.

Recently, the changing climate and global warming scenarios have given rise to various types of exotic and indigenous weeds along with causing intensification of their infestation [14,15]. Weeds such as awn-less barnyard grass (*Echinochloa colona*), field-bind weed (*Convulvulus arvensis*) etc. keep on emerging and produce abundant quantities of seeds until they are managed by tillage, weedicides, or employing crop competition through intercropping [1,16]. Herbicides are being used extensively to manage weeds in maize and green or black gram; however, there are very scant post emergence options, especially for perennial weeds. In addition, persistent herbicides usage having similar modes of action may potentially lead to the evolution of resistance in weeds. Many summer weeds including *Sonchus oleraceaus* L. have developed resistance to commonly used herbicides like glyphosate [17–19]. Besides ecosystem disruption, injudicious use of herbicides has serious health consequences due to high shelf-life of their active ingredients. Moreover, the lack of new effective herbicides release on commercial scale has caused shifts in weed population, growing environmental concerns owing to pollution, and skyrocketing prices of herbicides which have necessitated curbing and limiting the use of herbicides. Under these conditions living mulch as an intercrop may prove beneficial in controlling weeds and increase yield per unit of land without damaging the environment. Cereal-legumes intercropping systems reduced yield attributes (plant height, stem girth, leaf area, plants fresh and dry weights etc.) and biomass productivity of intercrops, however overall yield per unit area was increased by over 37% [1]. Additionally, legumes intercropping with cereals intensified the competition for growth resources which reduced intercrops yield by 23–37% and therefore, it was suggested to select legume intercrops having compatibility with cereals for growth resources utilization in temporal and spatial dimensions [10,11]. Moreover in cereal-legume intercropping systems, it has also been reported that cereals like maize and sorghum hold competitive advantage in acquiring growth resources by virtue of superior agro-botanical traits compared to most of legumes companion crops [1,12,20]. Legumes such as black and green gram sown in appropriate intercropping systems with maize might potentially reduce weeds infestation by providing them lesser space for growth.

Moreover, challenges posed by climate change and declining soil fertility have multiplied the risks of crop failure for small land holders in Indo-Pak subcontinent [21]. The quest has peaked to find out the cropping systems which provide yield stability along with being sustainable in long run. Therefore, there has been increasing interest in integrating cultural practices like intercropping to reduce our reliance on herbicides and develop a more effective and biologically viable weed control strategy. It was hypothesized that maize may perform differently in intercropping with legume intercrops owing to variability of growth resources utilization in temporal and spatial dimensions, while optimization of intercropping system could potentially suppress weeds infestation due to inter-species competition and lesser growth space available to weeds flora. So, this multi-year field experiment was performed with dual objectives to optimize intercropping systems of green and black gram with maize for suppressing weeds infestation and to quantify the impact of different intercropping systems on the productivity of intercrops and land use efficiency.

2. Materials and Methods

2.1. Experimental Site and Treatments Details

The experiments were performed at the Agricultural Graduate Research Farm, University of Agriculture, Faisalabad, Pakistan (31.4504° N, 73.1350° E, altitude of 186 m) [20]. The sowing of the experiment was done after the harvest of winter wheat crop. The mean temperature and rainfall of the experimental site during both growing seasons (April-August) remained 27.7 °C and 83 mm respectively, as per meteorological observatory located at the close vicinity of our experimental site. The field trial was executed to study the comparative weed control potential of green and black gram intercropping in maize. The experiment was comprised of treatments including sole maize in single row (60 cm apart) strips (T_1), sole maize in double rows strips (90 cm apart) (T_2), sole black (T_3) and green gram (30 cm apart rows) (T_4), six single rows (60 cm apart) of maize with twelve double rows (20 cm) of black gram (T_5), six single rows (60 cm apart) of maize with twelve double rows (20 cm apart) of green gram (T_6), three double rows (90 cm apart) of maize with three sets of quadratic rows (20 cm apart) of black gram (T_7) and three double rows (90 cm apart) of maize with three quadratic rows (20 cm apart) of green gram (T_8). The schematic presentation of treatments regarding maize intercropping with green and black gram under varying row placements has been given in Figure 1. The field experiments were arranged in randomized complete block design with three replications, while net plot size (excluding field bunds, sub water channels and field pathways) area (length × width) was maintained at 5.0 m × 3.6 m. There were eight experimental plots per replication, while the experiment was comprised of total 24 plots.

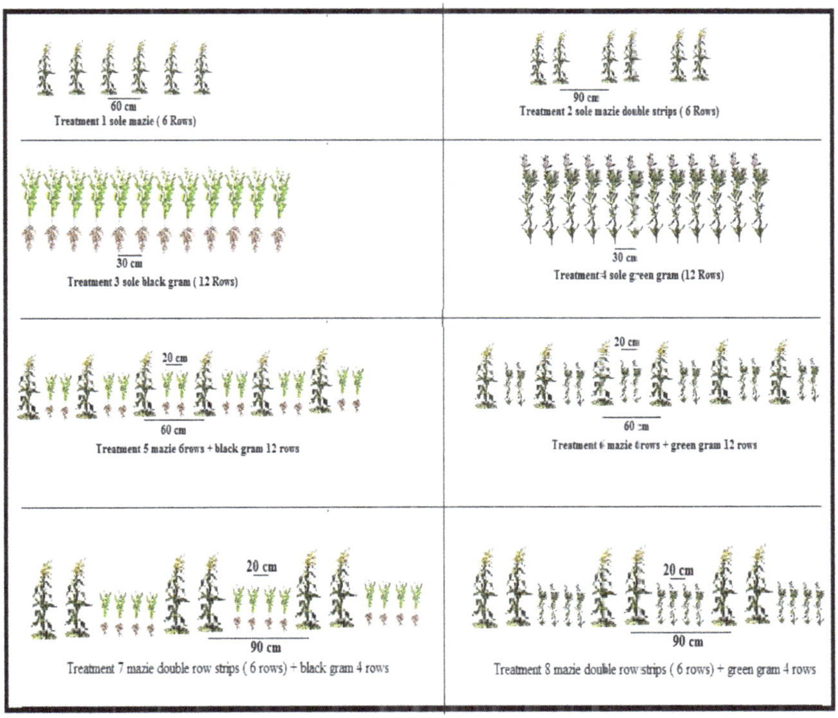

Figure 1. Schematic sketch of row placement of black and green grams in maize at various planting patterns in the field.

2.2. Site Physico-Chemical Properties

Pre-experiment soil analyses were performed by taking the soil samples from experimental site at two different depths (0–15, 0–30) from four corners and center of experimental block that were thoroughly homogenized for subsequent analyses. All the samples were air dried, grounded and sieved using 2 mm sieve. The glass electrode pH meter was used for measuring the pH of soil samples (soil and water in 1:2.5 ratio) [22] while electrical conductivity (EC) was determined with the help of conductivity meter [23]. Wet oxidation method was used for determining the organic carbon (OC) volumetrically. The soil organic matter (OM) was estimated by following Walkley–Black methodology [24]. For estimating total nitrogen (N) content, distillation in Kjeldahl apparatus was performed that was followed by titration with the concentrated H_2SO_4 [25]. Additionally, Olsen's method (0.5 N $NaHNO_3$ at 8.5 pH by maintaining soil: extractant ratio of 1:10) using spectrophotometer at 882 nm wavelength in a sulfuric acid system) was used for determining the available phosphorous (P) content [26], while standard procedure (ammonium acetate extraction involving air dried soil samples shaking with 0.5 M ammonium acetate solution for 30 min which effectively displaced positively charged K ions that were determined using flame photometer) as outlined by [27] was put into use to calculate potassium (K). Among micronutrients, available iron (Fe) was extracted using 1 N NH_4OAC at pH of 3.0. Subsequently, the extract was subjected to analysis using spectrophotometer at 510 nm wavelength by colorimetric method. Moreover, boron, zinc, copper and manganese contents in soil samples were estimated using diethylenetriaminepentaacetic acid extraction method [28–30].

Soil of the experimental site had a loam texture with pH of 8.1, while OM was only 0.51% indicating severely exhaustive utilization of soil. The soil had EC and bulk density of 0.42 dS m^{-1} and 1.40 cm^{-3} respectively. The NPK contents were 71, 4.3 and 110 mg

kg^{-1} respectively. The micronutrient B, Mn, Fe, Cu and Zn were 1.02, 20.4, 10.1, 1.9 and 1.1 mg kg^{-1} of soil, respectively.

2.3. Planting Material and Crop Husbandry

Maize hybrid (DK-919) was sown manually using the recommended kernel rate of 25 kg ha^{-1}, while erect type cultivars of black gram (cv. Arooj-97) and green gram (cv. AZRI Mung-2006) were sown using the recommended seed rate of 30 and 25 kg ha^{-1} respectively. The plant-plant spacing for maize was maintained at 25 cm, while 10 cm was the distance between green and black gram plants. Hoeing was done manually after 20 days of sowing to remove the early weed-crop competition. Fertilizers (urea, di ammonium phosphate and potassium sulphate) were applied at the rate 150, 100 80 kg ha^{-1} N-P-K, respectively. Full doses of P and K, while one-third of N fertilizer were applied at the time of seed bed preparation. The remaining N was applied in two equal splits with irrigations at 15 and 30 days after sowing (DAS). All the other agronomic practices were performed uniformly in all experimental plots.

2.4. Weeds Dynamics

The densities of individual weeds (*Echinochloa colona, Trianthema portuclacastrum, Convulvulus arvensis and Convulvulus esculentus*) and total weeds were counted (from an area of 1 square meter) per experimental unit using a rectangular quadrate at 20, 40, 60 DAS and at the time of crop harvesting. Weeds were cut with the help of sickle and weighed using an electric balance. Subsequently, weeds were sun dried for one week then kept in an oven at 42 °C and weighed repeatedly until constant dry weight was achieved after 24 h. Thereafter, all the samples were weighted individually and collectively using a digital balance. All intercrops after harvesting were left in the field for two weeks for sun drying and thereafter tied into bundles and stocked for four week. Then maize cobs were separated from the stalks and allowed drying in sunshine for five days to achieve 10% grain moisture content before shelling. Randomly, ten plants from each plot were used to record thousand grains weight and their average was worked out. The biological yield (grain yield + stalks yield) and grain yield were recorded on per plot basis to determine the harvest indices of maize, black and green gram using Formula (1);

$$\text{Harvest Index} = \text{Grain Yield}/\text{Biological Yield} \times 100 \quad (1)$$

2.5. Land Use Efficiency

Land use efficiency was measured using land-equivalent ratio which was calculated as described by Formula (1).

$$\text{LER} = \text{LER (Maize)} + \text{LER (green/black bean)} \quad (2)$$

$$\text{LER (Maize)} = \text{Grain yield of intercropped maize}/\text{Grain yield of sole maize} \quad (3)$$

$$\text{LER (black gram)} = \text{Grain yield of intercropped black gram}/\text{Grain yield of sole black gram} \quad (4)$$

$$\text{LER (green gram)} = \text{Grain yield of intercropped green gram}/\text{Grain yield of sole green gram} \quad (5)$$

2.6. Statistical Analysis

The collected data were subjected to analyses of variance (ANOVA) technique and subsequently to assign significance among treatment means, Tukey's Honest significance test was employed at 5% probability level with the help of "SAS" statistical package. The correlation analyses (n = 8) for determining the direct or inverse relationship between weeds density and their fresh and dry weights with grain yield of intercropped green gram and black gram were conducted using Microsoft's Excel program [31].

3. Results

3.1. Weeds Infestation

The results revealed that monocultures of maize (60 cm spaced single row strips and 90 cm spaced double rows strips) differed significantly in terms of weeds infestation as wider row spacing (T_2) recorded higher weeds density along with their fresh and dry biomasses compared to T_1 during both seasons. In addition, it was noted that weeds density was significantly reduced by maize-gram intercropping systems in comparison to T_1 and T_2 treatments (Table 1). Among maize intercropping systems with green gram and black gram, T_8 remained superior by recording the minimum weeds density along with their fresh and dry weights, while the highest corresponding values of weeds density, fresh and dry weights were exhibited by T_5 during both years. Among weed species at final harvest, the highest presence of *Echinochloa colona* and *Trianthema portuclacastrum* were noted for T_2, while T_7 remained effective in suppressing the infestation of these weeds (Figure 2). Contrarily, T_8 remained superiorly unmatched by recording the minimum infestations of *Convulvulus arvensis* and *Convulvulus esculentus*, while their highest infestations were recorded in T_2 (Figure 1).

Table 1. Weeds density (WD), fresh (WFW) and dry (WDR) weights in maize, black gram and green gram sole crops and in maize-gram intercropping systems under semi-arid conditions.

Treatments	WD (m^{-2})		WFW ($g\ m^{-2}$)		WDW ($g\ m^{-2}$)	
	2018	2019	2018	2019	2018	2019
T_1	60 ± 0.93 b	61 ± 1.13 b	112 ± 1.23 b	113 ± 0.54 b	27 ± 0.09 b	27 ± 0.62 b
T_2	85 ± 0.18 a	84 ± 0.94 a	121 ± 0.65 a	120 ± 0.19 a	33 ± 0.84 a	32 ± 1.27 a
T_3	53 ± 1.01 c	53 ± 0.14 bc	68 ± 0.18 c	69 ± 1.01 c	20 ± 0.39 c	21 ± 0.75 bc
T_4	46 ± 0.62 d	45 ± 019 c	64 ± 0.53 c	65 ± 0.34 cd	17 ± 0.91 cd	16 ± 0.91 c
T_5	37 ± 0.81 e	38 ± 1.05 d	49 ± 0.74 c	48 ± 0.66 d	14 ± 0.84 de	14 ± 0.28 d
T_6	31 ± 0.15 f	30 ± 0.24 e	41 ± 0.08 e	40 ± 1.14 e	13 ± 0.22 e	13 ± 1.05 de
T_7	27 ± 1.11 f	28 ± 0.81 e	39 ± 1.27 e	39 ± 0.29 ef	12 ± 1.11 e	12 ± 0.22 e
T_8	16 ± 0.43 g	15 ± 1.17 f	20 ± 0.17 f	20 ± 1.25 f	5 ± 0.35 f	5 ± 1.27 f

In each column, standard deviations followed by unalike letters differ significantly from each other at $p \leq 0.05$. T_1 = sole maize in 60 cm distanced single rows, T_2 = sole maize in 90 cm distanced double row strips, T_3 = sole black gram in 30 cm distanced single rows, T_4 = sole green gram in 30 cm distanced single rows, T_5 = six single rows (60 cm apart) of maize with twelve double rows (20 cm) of black gram, T_6 = six single rows (60 cm apart) of maize with twelve double rows (20 cm) of green gram, T_7 = three double rows (90 cm apart) of maize with three sets of quadratic rows (20 cm apart) of black gram, T_8 = three double rows (90 cm apart) of maize with three sets of quadratic rows (20 cm apart) of green gram).

3.2. Yield Attributes, Grain and Biological Yields and Harvest Index of Maize

Solo maize crops performed differently as T_1 remained the most superior treatment by recording the highest 1000 grains weight along with grain and biological yields during both crop growing seasons (Table 2). Among intercropping systems, green and black gram sown as living mulch significantly reduced 1000 grains weight, grain and biological yields of maize. However, T_8 exhibited the heaviest 1000 grains weight, grain and biological yields. Contrarily, T_7 could not perform at par to other intercropping systems by recording the least 1000 grainS weight, grain and biological yields of maize. Moreover, T_1 gave numerically higher harvest index, however it remained non-significant among solo and intercropping treatments.

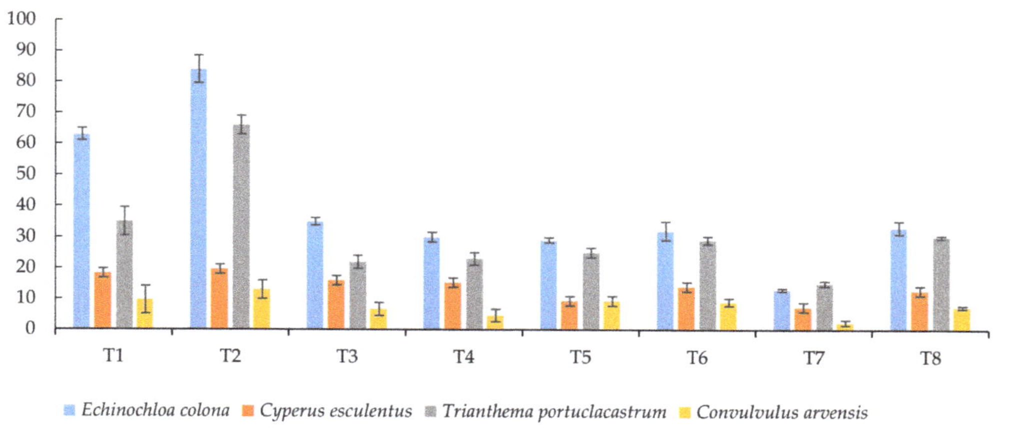

Figure 2. Density (m^{-2}) of different weed species in sole and maize-gram intercropping systems at final harvest. (T_1 = sole maize in 60 cm distanced single rows, T_2 = sole maize in 90 cm distanced double row strips, T_3 = sole black gram in 30 cm distanced single rows, T_4 = sole green gram in 30 cm distanced single rows, T_5 = six single rows (60 cm apart) of maize with twelve double rows (20 cm) of black gram, T_6 = six single rows (60 cm apart) of maize with twelve double rows (20 cm) of green gram, T_7 = three double rows (90 cm apart) of maize with three sets of quadratic rows (20 cm apart) of black gram, T_8 = three double rows (90 cm apart) of maize with three sets of quadratic rows (20 cm apart) of green gram).

Table 2. 1000 grains weight (GW), grain yield (GY), biological yield (BY) and harvest index (HI) of sole maize and in intercropping systems with green gram and black gram under semi-arid conditions.

Treatments	GW (g)		BY (t ha^{-1})		GY (t ha^{-1})		HI (%)	
	2018	2019	2018	2019	2018	2019	2018	2019
T_1	242.67 ± 1.71 a	244.31 ± 0.21 a	16.91 ± 0.37 a	16.51 ± 1.12 a	6.97 ± 1.04 a	6.84 ± 0.24 a	41.02	41.42
T_2	238.67 ± 0.94 ab	239.05 ± 0.31 b	16.19 ± 0.18 b	16.10 ± 1.05 b	6.56 ± 0.99 b	6.49 ± 0.24 b	40.48	40.31
T_3	-	-	-	-	-	-	-	-
T_4	-	-	-	-	-	-	-	-
T_5	227.33 ± 0.19 cd	225.08 ± 1.05 cd	15.19 ± 0.55 cd	15.23 ± 0.98	6.00 ± 1.12 cd	6.10 ± 0.29 cd	39.16	0.40
T_6	221.33 ± 0.84 cd	223.64 ± 0.16 cd	14.82 ± 0.81 cd	14.76 ± 0.43 cd	5.72 ± 0.67 cd	5.66 ± 0.17 cd	38.58	38.32
T_7	216.67 ± 0.71 d	213.991.14 d	14.78 ± 0.52 d	14.61 ± 1.18 d	5.69 ± 0.53 d	5.61 ± 1.15 d	38.49	38.11
T_8	231.00 ± 1.13 c	2.290.34 c	15.32 ± 1.10 c	15.16 ± 0.55	6.06 ± 0.94 c	6.00 ± 0.15 c	39.86	38.76

In each column, standard deviations followed by unalike letters differ significantly from each other at $p \leq 0.05$. T_1 = sole maize in 60 cm distanced single rows, T_2 = sole maize in 90 cm distanced double row strips, T_3 = sole black gram in 30 cm distanced single rows, T_4 = sole green gram in 30 cm distanced single rows, T_5 = six single rows (60 cm apart) of maize with twelve double rows (20 cm) of black gram, T_6 = six single rows (60 cm apart) of maize with twelve double rows (20 cm) of green gram, T_7 = three double rows (90 cm apart) of maize with three sets of quadratic rows (20 cm apart) of black gram, T_8 = three double rows (90 cm apart) of maize with three sets of quadratic rows (20 cm apart) of green gram). T_3 and T_4 did not contain maize crop, so presented as (-) in the table.

3.3. Grain Yield of Sole and Intercropped Black and Green Gram Crops

The intercropping of legumes with maize significantly reduced the grain yield of both green gram and black gram compared to their sole crop equivalents (Tables 3 and 4). The results revealed that solo crops of black gram (T_3) and green gram (T_4) recorded the maximum grain yields than intercrops yields. In intercropping systems with maize, the maximum yields of black and green gram were noted for T_7 and T_8 respectively. Interestingly, T_5 and T_6 remained the most inferior intercropping systems as far as grain yield of both intercrops was concerned as yield reduction of black and green were 37–39% and 38–41% in comparison to their sole crop equivalents during both cropping seasons.

Table 3. Grain yield (GY) of black gram sown as sole crops and in intercropping systems with maize under semi-arid conditions.

Treatments	GY (t ha^{-1})	
	2018	2019
T_3	0.81 ± 1.14 a	0.80 ± 0.34 a
T_5	0.51 ± 0.34 c	0.52 ± 1.01 c
T_7	0.62 ± 0.97 b	0.61 ± 0.18 b

In each columns given means followed by unalike letters are differ significantly from each other at $p \leq 0.05$. T_3 = sole black gram in 30 cm distanced single rows, T_5 = six single rows (60 cm apart) of maize with twelve double rows (20 cm) of black gram, T_7 = three double rows (90 cm apart) of maize with three sets of quadratic rows (20 cm apart) of black gram.

Table 4. Grain yield (GY) of green gram sown as sole crops and in intercropping systems with maize under semi-arid conditions.

Treatments	GY (t ha^{-1})	
	2018	2019
T_4	0.86 ± 0.67 a	0.84 ± 1.18 a
T_6	0.53 ± 0.18 c	0.51 ± 0.93 c
T_8	0.65 ± 0.73 b	0.66 ± 0.23 b

In each columns given means followed by unalike letters are differ significantly from each other at $p \leq 0.05$. T_4 = sole green gram in 30 cm distanced single rows, T_6 = six single rows (60 cm apart) of maize with twelve double rows (20 cm) of green gram, T_8 = three double rows (90 cm apart) of maize with three sets of quadratic rows (20 cm apart) of green gram).

3.4. Land Equivalent Ratios and Intercropping Advantage

The results revealed that maize-green gram and maize-black gram intercropping systems exhibited land equivalent ratio (LER) of over 1, which indicates substantial yield advantage of intercropping over mono cropping system of maize (Table 5). The maximum LER of maize was exhibited by T_8 which was at par to rest of the intercropping treatments. As far as LERs of green and black gram intercrops were concerned, T_7 and T_8 showed the highest LER for black and green gram intercrops respectively. In terms of total LER (LER of maize + LER of intercrop), T_8 and T_7 remained superior by recording the maximum total LER as well as intercropping advantage (IA) of maize-gram intercropping systems. Moreover, T_5 remained inferior to the rest of intercropping systems by recording the minimum total LER along with IA.

Table 5. Land equivalent ratio (LER) of maize, green gram and black gram as affected by maize-pulses intercropping systems under semi-arid conditions. (Means of 2-years data).

Intercropping Systems	Maize LER	Black Gram LER	Green GramLER	Total LER	IA (%)
T_5	0.86	0.63	-	1.49	49
T_6	0.87	-	0.63	1.50	50
T_7	0.87	0.76	-	1.63	63
T_8	0.87	-	0.77	1.64	64

T_5 = six single rows (60 cm apart) of maize with twelve double rows (20 cm) of black gram, T_6 = six single rows (60 cm apart) of maize with twelve double rows (20 cm) of green gram, T_7 = three double rows (90 cm apart) of maize with three sets of quadratic rows (20 cm apart) of black gram, T_8 = three double rows (90 cm apart) of maize with three sets of quadratic rows (20 cm apart) of green gram). T_5 and T_7 treatments did not include green gram, while black gram was not included in T_6 and T_8 treatments, so their absence is presented with (-).

3.5. Correlation of Weeds Infestation and Gram Yield

The correlation analysis was conducted to determine interrelationship (direct or inverse) between weeds infestation and grain yield of intercrops. The variation in weeds density (Figure 3A), fresh weight (Figure 3B) and dry weight (Figure 3C) were inversely proportional to grain yield of intercropped black gram and green gram crops indicating the effectiveness of intercropping systems in suppressing the weeds biomass. Correlation model analysis displayed that enhancement in every 1 g m^{-2} grain yield of intercropped pulses decreased weed density 1.18 m^{-2}, fresh weight 1.5 g m^{-2}, and dry weight 0.52 g m^{-2} of weed infestation.

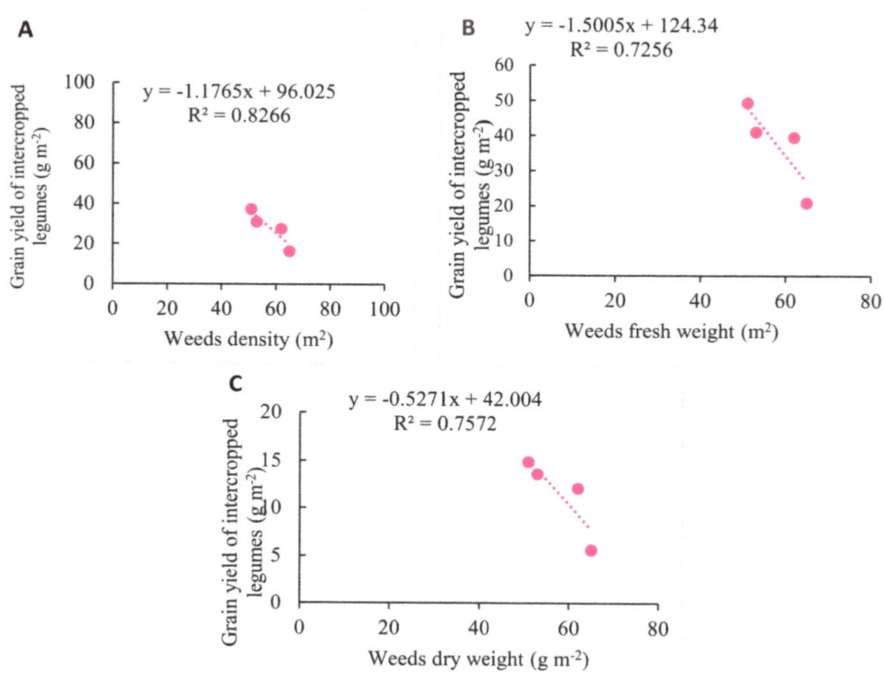

Figure 3. Interrelationship of weeds density and their fresh and dry weights with grain yield of intercropped pulses. Sole crop were excluded and mean values of following four intercropping systems have been used for correlation; T_5 = six single rows (60 cm apart) of maize with twelve double rows (20 cm) of black gram, T_6 = six single rows (60 cm apart) of maize with twelve double rows (20 cm) of green gram, T_7 = three double rows (90 cm apart) of maize with three sets of quadratic rows (20 cm apart) of black gram, T_8 = three double rows (90 cm apart) of maize with three sets of quadratic rows (20 cm apart) of green gram).

4. Discussion

The research findings were in line with postulated hypothesis as maize in intercropping with black or green gram suppressed weeds infestation. Our results exhibited that weeds density and biomass were significantly suppressed by intercropping systems particularly three double rows (90 cm apart) of maize sown with three quadratic rows (20 cm apart) of green and gram (T_7 and T_8) remained superior compared to maize monocultures (Table 1). Less weeds interference in intercropping systems might be attributed to severe competition offered by intercropped legumes for vital resources like space, light, nutrients and moisture which put most of the weeds out of competition [32]. Additionally, shading effects rendered by intercrops (green and black gram) canopies have been previously inferred to impart adverse impacts on weeds germination, growth and biomass production which led to reduced fresh and dry weights of weeds flora [33,34]. Contrastingly, monocultures recorded significantly higher weeds density and biomass probably owing to lesser competition for growth resources and availability of abundant sunlight for photosynthesis in the absence of spreading canopies of intercrops especially in maize monoculture having 90 cm apart rows. These findings corroborate with those of [35], who inferred that in comparison to cereals monocultures, cereal-legumes intercropping effectively suppressed weeds growth by restricting space and mineral nutrients availability which boosted growth and grain yields of companion crops. Similar findings were also reported by [33], whereby intercropping resulted in a lower weed biomass and maximized the yield in a biologically viable way. Weeds suppression effect owing to lesser available space available in cereal-legumes intercropping system was increased by closely spaced row strips of companion crops [35]. In another study, weeds suppression up to 65% was reported in cereal-legumes intercropping under semi-arid conditions [2]. Moreover, intercropping of cereals with spreading types of legumes (cowpea, cluster bean etc.) remained effective in reducing weeds incidence by reducing weed-seeds bank in the upper soil horizons [36]. Contrastingly, it was inferred that although legumes as intercrops enhanced weed control but also caused significant reduction in crops yield [13,15], therefore exploring compatibility among intercrops needs further studies.

The yield attributes especially 1000 grains weight is one of the vital indicator of maize grain yield which may be utilized as a reliable indicator to project grain yield (GY) of cereals including maize. The monoculture of maize (T_1) outperformed T_2 treatment by recording the maximum 1000 grains weight along with GY and biological yield (BY) (Table 2). This might be attributed to lesser weeds infestation and fragile interspecies competition for soil and environmental growth resources which assisted in higher partitioning and translocation of more assimilates towards reproductive plant parts. However, 1000 grains weight along with GY and BY of maize were significantly reduced in intercropping systems especially with green gram compared to sole maize. This might be due to less plant competition in monoculture for soil derived growth resources especially moisture and nutrients along with environmental resources (light and CO2) in contrast to intercropping systems [37–40]. The reduction in intercropped maize BY might be attributed to allocation of resources in different direction than uni-directional movement in sole cropping system [41,42]. More inter-row and inter-crop competition for resource utilization tended to disturb the source to sink relationship [43,44] and ultimately GY of maize was reduced in intercropping with green and black gram. Intercropping of maize with black and green gram non-significantly improved the harvest index which is in contradiction with the findings of [1,10,11].

As far as GY of legumes were concerned, sole crops of green and black grams remained unmatched while their grain yields were significantly reduced by three double rows (90 cm apart) of maize with three sets of quadratic rows (20 cm apart) of green or black gram (Table 3). Comparatively higher productivity of legumes monocultures might be due to better aeration, more ground area available for nutrition uptake and less shading effect of maize strips [45–48] which ultimately slashed the GY of inter-seeded legumes. These results are in agreement with [1,10,11], who reported that in cereal-legumes intercropping, legumes remained recessive compared to cereals in terms of acquiring growth resources which led to reduction in their yields compared to solo crops. It was also suggested that added advantage of intercropping could only be achieved by ensuring compatibility of intercrops in temporal and spatial dimensions, whereby intercrops peak their requirements at different times.

Land use efficiency for intercropping systems is measured as LER which indicates added advantage of intercropping if their values are above 1 [1,11]. Our results exhibited total LER of over 1 for all intercropping systems, while three double rows (90 cm apart) of maize with three sets of quadratic rows (20 cm apart) of green gram expressed the highest LER as well as intercropping advantage which remained statistically at par to T_7 (Table 4). High LER and IA of maize and gram intercropping systems might be attributed to enhanced and efficient exploitation of available resources such as land, light, moisture and fertilizer etc. [49,50]. The LERs of all intercropping system greater than one indicated higher efficiency and more productive use of all environmental resources by gram intercrops [7,40]. Additionally, sole legumes probably intercepted more radiation compared to monoculture of maize, while the interception by intercrops remained in between monocultures of legumes and maize which led to higher IA. It was recorded by [1,51,52] that intercrop converted the intercepted radiation into grain yield more efficiently which led to higher land use efficiency by maize-legumes intercropping systems.

The correlation analyses indicated inverse association among grain yield of pulses with weeds infestation. The increase in weeds density and their biomass (fresh and dry weights) resulted in sequential decline pulses grain yield. It might be attributed that weeds flora (*Echinochloa colona, Trianthema portuclacastrum, Convulvulus arvensis* and *Convulvulus esculentus*) sliced the growth resources share of green and black gram crops as weeds hold advantage in acquiring mineral nutrients from soil solution and moisture by virtue of their superior botanical traits [53–55]. Previously, crop losses caused by weeds ranged up to 71% depending on infestation level, diversity, availability of nutrients and moisture as well as competitive potential of crop species [33–35].

5. Conclusions

The research findings were in line with the postulated hypothesis as maize intercropping with green and black gram significantly suppressed weeds infestation as indicated by low weeds interference especially by three double rows (90 cm apart) of maize with three sets of quadratic rows (20 cm apart) of black and green gram intercrops. Likewise, row spacing was also proved a vital factor which significantly affected the productivity of monocultures and intercrops along with weed flora. Solo crops of maize and gram (green and black) exhibited higher grain yield in comparison to intercropping systems. Maximal reduction in weed infestation, the highest 1000 grains weight, biological and grain yields were attained by intercropping system encompassing three double rows (90 cm apart) of maize with three sets of quadratic rows (20 cm apart) of green gram. This intercropping system is recommended for general adoption in semi-arid regions of South Asia as it seems to have high resource use efficiency. Moreover, our findings re-emphasized that maize-green gram intercropping might be developed as eco-friendly and biologically viable strategy for suppressing weeds infestation and imparting sustainability to maize production under semi-arid conditions.

Author Contributions: Conceptualization, R.N.A. and A.I.; methodology, M.A.A. and M.A.I.; formal analysis, M.A.I.; data curation, M.A.I.; writing—original draft, M.A.I., M.I. and R.N.A.; writing—review and editing, A.R., J.-T.C., M.N.A. and D.I.H.; funding acquisition, A.R., M.N.A. and D.I.H. validation, J.-T.C. All authors have read and agreed to the published version of the manuscript.

Funding: This research received no external funding.

Institutional Review Board Statement: Not applicable.

Informed Consent Statement: Not applicable.

Acknowledgments: The authors would like to extend their sincere appreciation to the Researchers Supporting Project Number (RSP-2021/180), King Saud University, Riyadh, Saudi Arabia.

Conflicts of Interest: The authors declare no conflict of interest.

References

1. Iqbal, M.A.; Hamid, A.; Imtiaz, H.; Muzammil, H.S.; Tanveer, A.; Abdul, K.; Zahoor, A. Competitive indices in cereal and legume mixtures in a South Asian environment. *Agron. J.* **2019**, *111*, 242–249. [CrossRef]
2. Maqsood, Q.; Rana, N.A.; Iqbal, M.A.; Serap, K.A.; Asif, I.; Ayman, S. Overviewing of weed management practices to reduce weed seed bank and to increase maize yield. *Planta Daninha.* **2020**, *38*, e020199716. [CrossRef]
3. Yap, V.Y.; Xaphokhame, P.; de Neergaard, A.; Bech Bruun, T. Barriers to agro-ecological intensification of smallholder upland farming systems in Lao PDR. *Agronomy* **2019**, *9*, 375. [CrossRef]

4. Madembo, O.; Blessing, M.; Christian, T. Productivity or stability? Exploring maize-legume intercropping strategies for smallholder conservation agriculture farmers in Zimbabwe. *Agric. Syst.* **2020**, *185*, 102921. [CrossRef]
5. Tadele, Z. Raising Crop Productivity in Africa through intensification. *Agronomy* **2017**, *7*, 22. [CrossRef]
6. Xu, Z.; Chunjie, L.; Chaochun, Z.; Yang, Y.; Wopke, W.; Fusuo, Z. Intercropping maize and soybean increases efficiency of land and fertilizer nitrogen use; A meta-analysis. *Field Crops Res.* **2020**, *246*, 107661. [CrossRef]
7. Harun, I.G.; Shadrack, O.N.; Solomon, K.; Nancy, N.K.; Charles, K.K.G.; Muhammad, A.R.; Sagar, M.; Elmar, S. Revisiting intercropping indices with respect to potato-legume intercropping systems. *Field Crops Res.* **2020**, *258*, 107957. [CrossRef]
8. Costa, N.D.; Carlos, A.C.C.; Paulo, C.O.T.; Cristiano, M.P.; Ciniro, C.; André, M.C.; Daniel, M.S.; João, W.B.; Marcelo, A.; Paulo, R.L.M.; et al. Recovery of 15N fertilizer in intercropped maize, grass and legume and residual effect in black oat under tropical conditions. *Agric. Eco. Environ.* **2021**, *310*, 107226. [CrossRef]
9. Ngwira, A.R.; Vernon, K.; Pacsu, S.; Kondwani, M.; Kefasi, K. Productivity and profitability of maize-legume cropping systems under conservation agriculture among smallholder farmers in Malawi. *Acta Agric. Scand. Sec. B.* **2020**, *70*, 241–251. [CrossRef]
10. Iqbal, M.A.; Hamid, A.; Tanveer, A.; Imtiaz, H.; Sajid, A.; Ansar, A.; Zahoor, A. Forage sorghum-legumes intercropping: Effect on growth, yields, nutritional quality and economic returns. *Bragantia* **2019**, *78*, 82–95. [CrossRef]
11. Iqbal, M.A.; Muzammil, H.S.; Sher, A.; Zahoor, A.; Qaiser, M.; Rana, D.K. Forage productivity of cowpea [*Vigna unguiculata* (L.) Walp] cultivars improves by optimization of spatial arrangements. *Rev. Mex. Cienc. Pec.* **2018**, *9*, 203–219. [CrossRef]
12. Amos, R.N.; Aune, J.B.; Mkwinda, S. On farm evaluation of yield and economic benefits of short term maize legume intercropping systems under conservation Agriculture in Malawi. *Field Crops Res.* **2012**, *132*, 149–157.
13. Blanchart, E.; Villenave, C.; Viallatoux, A.; Barthes, B.; Girardin, C.; Azontonde, A.; Feller, C. Long-term effect of a legume cover crop (*Mucuna pruriens* var. utilis) on the communities of soil macrofauna and nematofauna, under maize cultivation, in southern Benin. *Eur. J. Soil Biol.* **2006**, *42*, 136–144. [CrossRef]
14. Hakim, M.A.; Juraimi, A.S.; Rezaul, K.S.M.; Khan, M.S.I.; Islam, M.S.; Choudhury, M.K.; Soufan, W.; Alharby, H.; Bamagoos, A.; Iqbal, M.A.; et al. Effectiveness of herbicide to control rice weeds in diverse saline environments. *Sustainability* **2021**, *13*, 2053. [CrossRef]
15. Malik, M.S.; Norsworthy, J.K.; Culpepper, A.S.; Riley, M.B.; Bridges, B. Use of wild radish (Raphanusraphanistrum) and rye cover crops for weed suppression in sweet corn. *Weed Sci.* **2008**, *56*, 588–595. [CrossRef]
16. Weerarathne, L.V.Y.; Marambe, B.; Chauhan, B.S. Intercropping as an effective component of integrated weed management in tropical root and tuber crops: A review. *Crop. Protec.* **2017**, *95*, 89–100. [CrossRef]
17. Xu, G.; Shen, S.; Zhang, Y.; Clements, D.R.; Yang, S.; Li, J.; Dong, L.; Zhang, F.; Jin, G.; Gao, Y. Designing cropping systems to improve the management of the invasive weed Phalaris minor Retz. *Agronomy* **2019**, *9*, 809. [CrossRef]
18. Walker, J.A.; King, J.R. Above and below-ground competition between Kura clover (*Trifolium ambiguum*) and meadow bromegrass (*Bromus biebersteinii*): Greenhouse study. *Can. J. Plant. Sci.* **2009**, *89*, 21–27. [CrossRef]
19. Loureiro, I.; Santin-Montanyá, I.; Escorial, M.-C.; García-Ruiz, E.; Cobos, G.; Sánchez-Ramos, I.; Pascual, S.; González-Núñez, M.; Chueca, M.C. Glyphosate as a tool for the incorporation of new herbicide options in integrated weed management in maize: A weed dynamics evaluation. *Agronomy* **2019**, *9*, 876. [CrossRef]
20. Iqbal, M.A. Comparative performance of forage cluster bean accessions as companion crops with sorghum under varied harvesting times. *Bragantia* **2018**, *77*, 476–484. [CrossRef]
21. Iqbal, M.A.; Asif, I.; Rana, N.A. Spatio-temporal reconciliation to lessen losses in yield and quality of forage soybean (*Glycine max* L.) in soybean-sorghum intercropping systems. *Bragantia* **2018**, *77*, 283–291. [CrossRef]
22. Addis, H.; Andreas, K.; Theib, O.; Stefan, S. Linking selected soil properties to land use and hill slope- A watershed case study in the Ethiopian highlands. *Soil Water Res.* **2016**, *11*, 163–171. [CrossRef]
23. Sparks, D.L.; Page, A.L.; Helmke, P.A.; Leoppert, R.H.; Soltanpour, P.N.; Tabatabai, M.A.; Johnston, G.T.; Sumner, M.E. *Methods of Soil Analysis*; Soil Science Society of America: Madison, WI, USA, 1996.
24. Naresh, R.K.; Tomar, S.S.; Kumar, D.; Samsher, P.; Singh, S.; Dwivedi, P.; Kumar, A. Experiences with rice grown on permanent raised beds: Effect of crop establishment techniques on water use, productivity, profitability and soil physical properties. *Rice Sci.* **2014**, *21*, 170–180. [CrossRef]
25. Iqbal, M.A.; Iqbal, A.; Maqbool, Z.; Ahmad, Z.; Ali, E.; Siddiqui, M.H.; Ali, S. Revamping soil quality and correlation studies for yield and yield attributes in sorghum-legumes intercropping systems. *Biosci. J.* **2018**, *34*, 1165–1176. [CrossRef]
26. AOAC. *AOAC-Official Methods of Analysis*, 17th ed.; Association of Official Analytical Chemists, Inc.: Arlington, VA, USA, 2003.
27. Iqbal, M.A.; Brandon, J.B.; Asif, I.; Rana, N.A.; Zubair, A.; Haroon, Z.K.; Bilal, A. Agro-botanical response of forage sorghum-soybean intercropping systems under atypical spatio-temporal patterns. *Pak. J. Bot.* **2017**, *49*, 987–994.
28. Jackson, M.L. *Soil Chemical Analysis*; Prentice Hall, Inc.: Englewood Cliffs, NJ, USA, 1962.
29. Black, C.A. *Methods of Soil Analysis, Part II*; American Society of Agronomy: Madison, WI, USA, 1965.
30. Piper, C.S. *Soil and Plant. Analysis*; The University of Adelaide, Press Adelaide: Adelaide, Australia, 1950.
31. Steel, R.G.; Jhdickey, D.A. *Principles and Procedures of Statistics a Biometrical Approach*; McGraw-Hill: New York, NY, USA, 1986; pp. 223–241.
32. Nurk, L.; Grab, R.; Pekrun, C.; Wachendorf, M. Effect of sowing method and weed control on the performance of maize (*Zea mays* L.) intercropped with climbing beans (*Phaseolus vulgaris* L.). *Agriculture* **2017**, *7*, 51. [CrossRef]

33. Berdjour, A.; Dugje, I.Y.; Israel, K.D.; Nurudeen, A.R. Maize–soybean intercropping effect on yield productivity, weed control and diversity in northern Ghana. *Weeds Biol. Manag.* **2020**, *20*, 69–81. [CrossRef]
34. Rad, S.V.; Valadabadi, S.A.R.; Pouryousef, M.; Saifzadeh, S.; Zakrin, H.R.; Mastinu, A. Quantitative and qualitative evaluation of *Sorghum bicolor* L. under intercropping with legumes and different weed control methods. *Horticulturae* **2020**, *6*, 78. [CrossRef]
35. Gerhards, R. Weed suppression ability and yield impact of living mulch in cereal crops. *Agriculture* **2018**, *8*, 39. [CrossRef]
36. Verret, V.; Antoine, G.; Elise, P.; Safia, M.; David, M.; Muriel, V. Can legume companion plants control weeds without decreasing crop yield? A meta-analysis. *Field Crops Res.* **2017**, *204*, 158–168. [CrossRef]
37. Kermah, M.; Angelinus, C.F.; Samuel, A.; Benjamin, D.K.A.; Robert, C.A.; Ken, E.G. Maize-grain legume intercropping for enhanced resource use efficiency and crop productivity in the Guinea savanna of northern Ghana. *Field Crops Res.* **2017**, *213*, 38–50. [CrossRef] [PubMed]
38. Duchene, O.; Jean-François, V.; Florian, C. Intercropping with legume for agroecological cropping systems: Complementarity and facilitation processes and the importance of soil microorganisms. A review. *Agric. Ecosyst. Environ.* **2017**, *240*, 148–161. [CrossRef]
39. Zhang, X.; Huang, G.; Bian, X.; Zhao, Q. Effects of root interaction and nitrogen fertilization on the chlorophyll content, root activity, photosynthetic characteristics of intercropped soybean and microbial quantity in the rhizosphere. *Plant. Soil Environ.* **2013**, *59*, 80–88. [CrossRef]
40. Zuo, Y.; Zhang, F. Iron and zinc biofortification strategies in dicot plants by intercropping with gramineous species. A review. *Agron. Sustain. Dev.* **2009**, *29*, 63–71. [CrossRef]
41. Mariotti, M.; Masoni, A.; Ercoli, L.; Arduini, I. Above-and below-ground competition between barley, wheat, lupin and vetch in a cereal and legume intercropping system. *Grass Forage Sci.* **2009**, *64*, 401–412. [CrossRef]
42. Mushagalusa, G.N.; Ledent, J.F.; Draye, X. Shoot and root competition in potato/maize intercropping: Effects on growth and yield. *Environ. Exp. Bot.* **2008**, *64*, 180–188. [CrossRef]
43. Zuo, Y.; Zhang, F.; Li, X.A.; Cao, Y. Studies on the improvement in iron nutrition of peanut by intercropping with maize on a calcareous soil. *Plant. Soil.* **2000**, *220*, 13–25. [CrossRef]
44. Zhang, F.; Li, L. Using competitive and facilitative interactions in intercropping systems enhances crop productivity and nutrient-use efficiency. *Plant. Soil.* **2003**, *248*, 305–312. [CrossRef]
45. Schulz, V.S.; Schumann, C.; Weisenburger, S.; Müller-Lindenlauf, M.; Stolzenburg, K.; Möller, K. Row-intercropping maize (*Zea mays* L.) with biodiversity-enhancing flowering-partners-effect on plant growth, silage yield, and composition of harvest material. *Agriculture* **2020**, *10*, 524. [CrossRef]
46. Nassary, E.K.; Baijukya, F.; Ndakidemi, P.A. Assessing the productivity of common bean in intercrop with maize across agro-ecological zones of smallholder farms in the Northern Highlands of Tanzania. *Agriculture* **2020**, *10*, 117. [CrossRef]
47. Gecaitė, V.; Arlauskienė, A.; Cesevičienė, J. Competition effects and productivity in oat–forage legume relay intercropping systems under organic farming conditions. *Agriculture* **2021**, *11*, 99. [CrossRef]
48. Njira, K.O.W.; Semu, E.; Mrema, J.P. Productivity of pigeon pea, cowpea and maize under sole cropping, legume–legume and legume–cereal intercrops on Alfisols in Central Malawi. *Agroforest Syst.* **2021**, *95*, 279–291. [CrossRef]
49. Iqbal, M.A.; Iqbal, A.; Siddiqui, M.H.; Maqbool, Z. Bio- agronomic evaluation of forage sorghum-legumes binary crops on Haplic Yermosol soil of Pakistan. *Pak. J. Bot.* **2018**, *50*, 1991–1997.
50. Namatsheve, T.; Regis, C.; Marc, C.; Claire, M.; Icard-Vernière, C.; Rémi, C. Maize-cowpea intercropping as an ecological intensification option for low input systems in sub-humid Zimbabwe: Productivity, biological N2-fixation and grain mineral content. *Field Crops Res.* **2021**, *263*, 108052. [CrossRef]
51. Iqbal, M.A.; Asif, I.; Ayub, M.; Akhtar, J. Comparative study on temporal and spatial complementarity and profitability of forage sorghum-soybean intercropping systems. *Cust. Agronegocio* **2016**, *12*, 2–18.
52. Islam, A.K.M.M.; Nasir, M.; Akter Mou, M.; Yeasmin, S.; Islam, M.S.; Ahmed, S.; Anwar, M.P.; Hadifa, A.; Baazeem, A.; Iqbal, M.A.; et al. Preliminary reports on comparative weed competitiveness of Bangladeshi monsoon and winter rice varieties under puddled transplanted conditions. *Sustainability* **2021**, *13*, 5091. [CrossRef]
53. Karim, S.M.R.; Iqbal, T.M.T.; Islam, N. Relátive Yields of crops and crop losses due to weed competition in Bangladesh. *Pak. J. Sci. Ind. Res.* **1998**, *41*, 318–324.
54. Buhler, D.D.; Liebman, M.; Obrycki, J.J. Theoretical and practical challenges to an IPM approach to weed management. *Weed Sci.* **2000**, *48*, 274–280. [CrossRef]
55. Iqbal, M.A.; Sajid, A.; Sabagh, A.; Ahmad, Z.; Siddiqui, M.H. Changing climate and advances on weeds utilization as forage: Provisions, nutritional quality and implications. In *Invasive Species—Introduction Pathways, Economic Impact, and Possible Management Options*; Intechopen Ltd.: London, UK, 2020.

agronomy

MDPI

Article

Assessment of Production and Qualitative Characteristics of Different Populations of *Salvia sclarea* L. Found in Sicily (Italy)

Teresa Tuttolomondo [1,†], Giuseppe Virga [2,†], Mario Licata [1,*], Nicolò Iacuzzi [1,*], Davide Farruggia [1] and Salvatore La Bella [1]

[1] Department of Agricultural, Food and Forest Sciences, Università degli Studi di Palermo, Viale delle Scienze 13, Building 4, 90128 Palermo, Italy; teresa.tuttolomondo@unipa.it (T.T.); davide.farruggia@community.unipa.it (D.F.); salvatore.labella@unipa.it (S.L.B.)

[2] Research Consortium for the Development of Innovative Agro-Environmental Systems (Corissia), Via della Libertà 203, 90143 Palermo, Italy; giuseppe.virga@corissia.it

* Correspondence: mario.licata@unipa.it (M.L.); nicolo.iacuzzi@unipa.it (N.I.)

† These authors are equally contributed.

Abstract: *Salvia sclarea* L. is an important industrial crop, valued for its herbal-aromatic properties and high quality essential oils, that is used in food, pharmaceuticals and cosmetics. In this study, carried out from 2009 to 2010, the morphological and production characteristics and essential oil content and composition of three Sicilian populations were studied. In particular, the composition of essential oils extracted from primary and secondary inflorescences using steam distillation was assessed. Morphological, production and qualitative data from the three populations were subjected to analysis of variance and cluster analysis. Regarding the quality of the oils, only the most prevalent compounds were taken into consideration in this study. The three populations were linalyl acetate/linalool chemotypes. Highly significant variations were found for the effective local population and inflorescence type in the composition of the essential oil principal components. In particular, the primary inflorescences were found to be accumulation sites favoured by monoterpenes, and secondary inflorescences were favoured by sesquiterpenes and sclareol. Populations "S. Stefano Quisquina" and "Alcara Li Fusi" performed best on a morphological and production level, whereas populations "Prizzi"and "Alcara Li Fusi" performed best in terms of quality. Population "S. Stefano Quisquina" produced high levels of sclareol. Biotype selection from within the populations should be based on both morphological, production and quality analyses.

Keywords: *Salvia sclarea* L.; spike yield; primary and secondary inflorescences; local populations; essential oil principal components

Citation: Tuttolomondo, T.; Virga, G.; Licata, M.; Iacuzzi, N.; Farruggia, D.; Bella, S.L. Assessment of Production and Qualitative Characteristics of Different Populations of *Salvia sclarea* L. Found in Sicily (Italy). *Agronomy* **2021**, *11*, 1508. https://doi.org/10.3390/agronomy11081508

Academic Editor: Antonios Chrysargyris

Received: 19 June 2021
Accepted: 28 July 2021
Published: 29 July 2021

Publisher's Note: MDPI stays neutral with regard to jurisdictional claims in published maps and institutional affiliations.

1. Introduction

Salvia sclarea L. is a medicinal and aromatic species from the Mediterranean belonging to the *Lamiaceae* family. Clary sage is a hardy plant that grows wild in temperate areas and is xerophyte in nature. The whole plant is highly aromatic [1]. However, during flowering, which occurs during the second year, the inflorescences are covered in a dense, exceedingly aromatic resin similar in fragrance to muscat wine; hence the name "moscatella" or muscat sage [2]. It is one of the most highly valued Mediterranean species as a result of these aromatic properties.

The dried inflorescences and leaves are used in the production of spirits, herbal medicines, extracts and teas; the floral heads are used chiefly for essential oil extraction. The essential oils (EOs), characterized by an intense floral aroma and the scent of fresh grass, are used in the food, pharmaceutical and cosmetics industries. Clary sage essential oil is used to aromatize beer, tonic water, spirits and even muscat wine and vermouth [3]. In the perfume industry, the essential oil is highly prized due both to the quality of its fragrance and the fact it is an excellent fixative. Furthermore, sclareol, one of the principal components of the essential oil, is used as a base for the chemical synthesis of Ambrox. Ambrox is commonly used as an alternative to ambergris, a waxy substance produced in the digestive tract of the sperm whale and the source of one of the most prized animal-extract essences in the industry [3]. Within the perfume industry, *S. sclarea* provides the dry amber/tobacco note in oriental tobacco scents [4]. Recent studies on *S. sclarea* have shown it to have allelopathic and insecticidal properties, even acting as a biofilter in water treatment [5–8].

Scientific literature reports the traditional use of clary sage oil as an agent against gingivitis, stomatitis and mouth ulcers. Other scientific evidence demonstrates the analgesic, anti-inflammatory, antioxidant, antimicrobial, antiviral, antifungal and cytotoxic activities of its essential oils [9–20]. Clary sage is also used in aromatherapy as a highly effective relaxant for the treatment of stress, asthma, digestion and menstrual problems as well as an aid to induce childbirth [21,22]. Some authors have found that clary sage seeds are rich in polyunsaturated fatty acids, which make them ideal for use in nutraceuticals. They are also a good source of edible oils, having omega 3-linoleic acid [23,24].

Its hardy nature and essential oil profile mean that clary sage is widely grown for extraction purposes in France, Bulgaria, the post-Soviet states, the United States and Western China [25]. In Italy, however, although the species grows wild in a number of areas, it is one of many medicinal and aromatic species that are either not cultivated or are generally underused.

These aromatic species, especially Mediterranean species, have high phenotypical plasticity; they adapt well to a range of environments, such as the xerophytic conditions typical of the Mediterranean and, as a consequence, are able to change their chemical composition [26–28]. It is worth noting that percentage content and essential oil composition are important parameters in the evaluation of aromatic species, as they delineate numerous and varied properties (antioxidant, antimicrobial, etc.) that can be used to create innovative products [29].

It is widely known that the essential oils of a number of species belonging to the *Lamiaceae* family show a degree of chemical variability due to certain exogenous factors (climate, soil, altitude, latitude, agronomic techniques, post-harvest management, etc.) and endogenous factors (plant age, development stage, genetic properties, plant parts, etc.) [30–38]. There is also known to be a strict correlation between the formation of primary and secondary compounds. The latter can be affected by the amount of biomass and by the relationship between the organs of the plant and substance accumulation levels in its tissues [39,40].

For the species in this study, therefore, it is important to evaluate a number of factors that could lead to greater efficiency in terms of biomass yield, particularly with regard to inflorescence production, but also in terms of essential oil composition.

These aspects are fundamental for agronomic selection in the development of industrial crops.

Based on the above, this study compared three local Sicilian populations (LP) of *Salvia sclarea* to evaluate both quality and production aspects. Furthermore, the effects of two types of inflorescence—primary and secondary—on the principal components of the essential oils of the three populations were evaluated.

2. Materials and Methods

2.1. Site of Experiments and Treatments

The tests were carried out in the two years 2009 and 2010 at the Orleans Experimental Station, University of Palermo (Italy) (Table 1).

Table 1. Test site information.

Test Site	Province	Geographical Coordinates	Altitude (m a.s.l.)	Average Annual Rainfall (mm)	Average Annual Temperature (°C)
"Orleans" experimental station	Palermo	38°06′26.2″ N 13°20′56.0″ E	34	605	18.40

Soils in the test area were sandy clay loam (Aric Regosol, 54% sand, 23% clay, and 21% silt), with a pH of 7.6, 14 g kg^{-1} organic matter, 3.70% active carbonates, 1.32% total nitrogen, 18.1 ppm available phosphorus and 320 ppm exchangeable potassium. The hot, temperate climate is characterized by humid winters and dry, hot summers, typical of the Mediterranean. August is the hottest month of the year, with an average temperature of 26.2 °C, and January is the coldest at 12.1 °C.

Three populations of clary sage sourced from the wild from different sites in Sicily (Italy) were compared (Table 2).

Table 2. Provenance and local *Salvia sclarea* L. population code.

Provenance	Province	LP Code
S. Stefano Quisquina	Agrigento	SS
Prizzi	Palermo	PR
Alcara Li Fusi	Messina	AF

The main climatic and environmental characteristics of the test site are shown in Table 1.

The plants used to create the experimental plot were obtained from seeds taken from each population located in the plant collection field at the test site. As shown in Table 2, the plants were identified using initials linked to their provenance: SS, PR and AF. In order to assess their quantitative and qualitative characteristics, the three populations were planted in the field using a plant density of 2 plants per m^2. The plants were grown using the same organic cropping techniques in both years. Weed control was carried out mechanically without the use of herbicides or chemical fertilizers, and irrigation was not used.

2.2. Plant Measurements

Observations were carried out at the full flowering stage, which occurred the year following planting in the open field, as the species is a biennial.

The following parameters were recorded during harvesting: plant height, plant fresh weight, plant dry weight, number of branches, number of stems and inflorescence length. In addition, inflorescence as well as leaf and stem ratios (as % of total dry weight of the plant) were also measured.

Dry matter weight was calculated when constant sample weight was reached (dried in a shaded and well-aerated environment at a temperature of approximately 30 °C). Spike yield per hectare was also estimated.

2.3. Essential Oil Extraction and Oil Yield Calculation

For a sample of 500 g of dried inflorescences, the total essential oil (EO) content was determined (expressed as a % *v/w*: oil volume/sample weight in g) following steam distillation extraction. Oil yields were calculated by multiplying inflorescence yields by oil content and 0.90 (approximate specific gravity of oil) [41]. Furthermore, both the content and composition of the essential oils were determined for two types of inflorescence: primary inflorescence stem "ISP" and secondary inflorescences stem "ISS". The length of the ISP inflorescences and the ISS inflorescences was also measured.

2.4. GC and GC/MS Analyses of Essential Oils

In accordance with international guidelines [42], gas chromatographic (GC) analyses were run on a Shimadzu gas chromatograph, Model 17-A, equipped with a flame ionization detector (FID) and an operating software Class VP Chromatography Date System, version 4.3 (Shimadzu Corporation, Duisburg, Germany). Analytical conditions were as follows SPB-5 capillary column (15 m × 0.10 mm × 0.15 μm), helium as carrier gas (1 mL min^{-1}), injection in split mode (1:200), injected volume 1 μL (4% essential oil/CH2Cl2 *v/v*), injector and detector temperature 250–280 °C, linear velocity in column 19 cm s^{-1}. The oven temperature was held at 60 °C for 1 min, then programmed from 60 to 280 °C at 10 °C min^{-1} then 280 °C for 1 min. Percentages of compounds were determined from their peak areas in the GC/FID profiles. Gas chromatography mass spectrometry (GC/MS) was carried out in the fast mode on a Shimadzu GC/MS mod. GCMS-QP5050A, with the same column and the same operative conditions as used for analytical GC/FID, using operating software GC/MS solution, version 1.02 (Shimadzu). The ionization voltage was 70 eV, the electron multiplier was 900 V, and the ion source temperature was 180 °C. Mass spectra data were acquired in the scan mode in an m/z range of 40–400. The same oil solutions (1 μL) were injected with the split mode (1:96).

2.5. Identification of Components of Essential Oils

The identity of components was based on their GC retention index (relative to C9–C22 n-alkanes on the SPB-5 column), computer matching of spectral MS data with those from NIST MS libraries [43], the comparison of the fragmentation patterns with those reported in the literature [40] and, whenever possible, co-injections with authentic samples.

For each sage population three samples of essential oils were subjected to GC. The values shown in the tables are the result of the average of the 3 replicates.

2.6. Statistical Analysis

Data (two-year averages) relating to the morphology and production of the three populations were subjected to analysis of variance (one-way ANOVA) followed by cluster analysis (UPGMA).

Analysis of variance (two-way ANOVA) and cluster analysis (UPGMA) were also carried out on the essential oil compounds to assess the effects of the local populations and the type of inflorescence (TIPS)—primary (IPS) and secondary (ISS). Arcsine transformation was performed on all data percentages prior to elaboration. Variations between treatments were compared using Tukey's test, with a 5% probability level. A randomized plot design with three replications was used, and statistical analysis was conducted using the software PAST 3.

3. Results

Following analysis of variance (Table 3), the three local populations of clary sage in the study showed significant differences for most of the characteristics under examination. Statistical differences were not found for plant height, percentage incidence of stems and the number of stems per plant.

As Table 3 clearly demonstrates, populations SS and AF obtained greater spike yields at 2.76 and 2.10 Mg ha^{-1}, respectively, both statistically differing from PR, which produced 1.8 Mg ha^{-1}. Similar trends were also found regarding plant fresh weight and dry weight percentage incidence of leaves and number of branches. Fresh weight was found to be approximately 1200 g in both population AF (1232 g) and SS (1114 g), while PR was found to be considerably lower (796 g).

Table 3. Morphological and production characteristics of three LPs (Local Populations) of *Salvia sclarea* L.

Characteristic	AF	SS	PR	Significance
spike yield (Mg ha^{-1})	2.10 a	2.77 a	1.8 b	*
plant fresh weight (g)	1232 a	1114 a	768.9 b	*
plant dry weight (g)	337.2 a	356.1 a	214.5 b	*
plant height (cm)	137.9	142	139.8	n.s
inflorescence (%)	34.00 b	40.02 a	42.52 a	**
leaves (%)	25.63 a	19.49 a,b	19.18 b	**
stems (%)	40.37	40.48	37.62	n.s
no. stems	4.63	4.28	3.88	n.s
no. branches	13.18 a	13.93 a	9.8 b	**
inflorescence length (cm)	42.01 b	46.10 a,b	49.81 a	*
EO content (%)	1.29 a	0.68 c	0.91 b	**
EO yield (kg ha^{-1})	24.20 a	15.98 b	14.05 b	*
ISP length (cm)	54.42 b	58.25 a,b	61.63 a	*
ISS length (cm)	45.96 b	52.33 a	47.47 a	**
EO yield % ISP	0.98 a	0.58 b	0.98 a	**
EO yield % ISS	1.61 a	0.78 b	0.83 b	**

Local population code: PR = Prizzi; AF = Alcara Li Fusi; SS = S. Stefano Quisquina. EC = essential oil; ISP = primary inflorescences; ISS = secondary inflorescences. ** = significant at $p < 0.01$; * = significant = $p < 0.05$; n.s. = not significant. Within the same row, means followed by the same letter are not significantly different for $p \leq 0.05$ according to Tukey's test.

In a similar fashion, plant dry weights for AF (337.20 g) and SS (356.10 g) were found to be considerably higher than for PR (214.50 g). This was also true for number of branches, which was recorded at 13 for both AF and SS and 10 for PR. The AF population (25.60%) was shown to have a leaf percentage incidence significantly higher than PR (19.20%; the lowest value), and SS maintained an intermediate position (19.50%), with no statistical differences between the other two populations.

The PR population produced greater average inflorescence length (49.81 cm), ISP length (61.63 cm) and ISS length (47.47 cm) as well as inflorescence percentage incidence (42.52%) compared to the other two populations, although statistical differences were not found with SS.

Differences were only found compared to AF, which recorded the lowest values for the abovementioned parameters. The population with the greatest essential oil content compared to the others was the AF population (1.29%), followed by PR (0.91%) and finally SS (0.68%).

AF performed the best for both essential oil yields (24.20 kg ha^{-1} vs. PR and SS averages of 15 kg ha^{-1}) and oil percentage content of the primary inflorescences (ISP) (0.98%) and secondary inflorescences (ISS) (1.61%). It is also worth noting that PR obtained similar results to AF regarding essential oil % content of the ISP. The dendrogram (Figure 1), based on cluster analysis using morphological and production characteristics, shows the two main clusters. The first cluster grouped the two populations AF and SS, and the second cluster was constituted only by the population PR. These results are in accordance with data from ANOVA analysis.

Table 4 shows average values for the population characteristics in each cluster. The values are purely descriptive and are shown only to highlight the distinctive features of each cluster, as grouped by the analysis.

The two populations SS and AF located in the first cluster (Table 4) recorded greater plant fresh (1173 g vs. 760 g for PR) and dry (347 g vs. 215 g for PR) biomass production, greater inflorescence yield (2.43 vs. 1.80 Mg ha^{-1} for PR), number of branches (14 vs. 10 for PR), and percentage incidence of leaves (23 vs. 19% for PR) and stems (40 vs. 38% for PR).

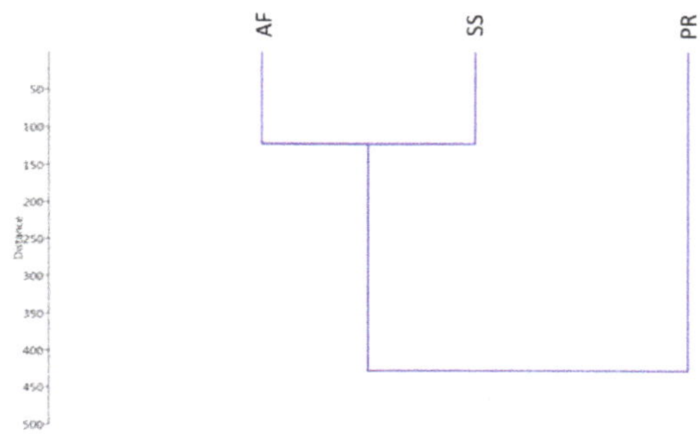

Figure 1. Cluster analysis of local populations in the study base on morphological and production data. Local population code: PR = Prizzi; AF = Alcara Li Fusi; SS = S. Stefano Quisquina.

Table 4. Average values regarding morphological and production characteristics of the populations in each cluster.

	First Cluster (SS, AF)	Second Cluster (PR)
Characteristic	Average	Average
inflor. yield (Mg ha^{-1})	2.43	1.80
EO content (%)	0.99	0.91
EO yield (kg ha^{-1})	20.10	14.05
plant height (cm)	139.96	139.83
plant fresh weight (g)	1173.05	768.86
plant dry weight (g)	346.62	214.51
inflorescence (%)	37.01	43.19
leaves (%)	22.57	19.18
stems (%)	40.42	37.62
no. branches	13.55	9.80
no. stems	4.45	3.88
inflor. length (cm)	44.05	49.81
ISP length (cm)	56.33	61.63
ISS length (cm)	49.14	47.47
EO yield % ISP	0.78	0.98
EO yield % ISS	1.20	0.83

Local population code: PR = Prizzi; AF = Alcara Li Fusi; SS = S. Stefano Quisquina. EO = essential oil; ISP = primary inflorescences; ISS = secondary inflorescences.

These same populations (SS and AF) on average also produced the greatest EO yields (20 vs. 14 kg ha^{-1} for PR) and % content (0.99 vs. 0.90% for PR), with greater incidence in ISS (1.2 vs. 0.83% for PR).

In contrast, the greatest % incidence of inflorescences per plant (43% vs. 37% for SS and AF) was found in the population PR, in addition to the greatest average inflorescence length (50 cm vs. 44 cm for SS and AF). This greater length was linked to ISP length (62 cm vs. 56 cm for SS and AF) in particular, which was further found to have a higher EO % incidence (0.98% vs. 0.78% for SS and AF) than the other two populations.

Seventy-six components emerged from GC analysis, constituting approximately 98% of the chemical profile. Regarding the compound classes, the monoterpenes were the most abundant class. Oxygenated monoterpenes, in particular (73 ÷ 79%), were far more abundant than hydrocarbons (5 ÷ 6.0%). Sesquiterpenes oscillated between 11 and 14% and diterpenes were also worthy of note, ranging between 4 and 7.0%. Finally, the content of the class named "others" was negligible, being far below 1.0% (Figure 2).

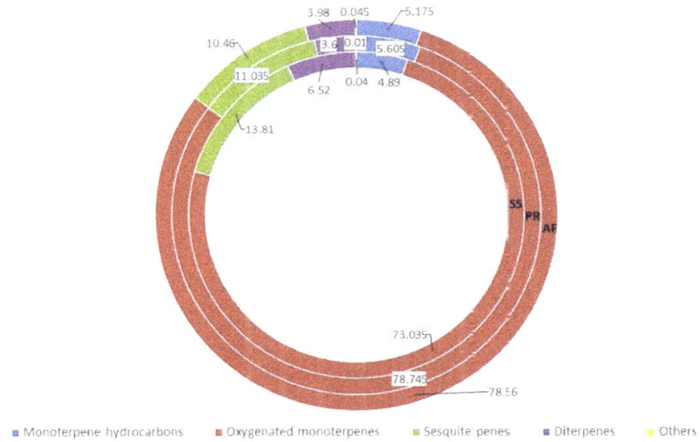

Figure 2. Class of compounds of the essential oils of the local populations in the study. Local population code: PR = Prizzi; AF = Alcara Li Fusi; SS = S. Stefano Quisquina.

Only the most abundant components were taken into consideration in this study, in decreasing order from linalyl acetate (38.73–41.63%), linalool (22.42–25.55%), α-terpineol (5.76–7.14%), germacrene D (3.61–4.50%), sclareol (2.93–5.50%), geranyl acetate (2.64–3.04%), β-caryophyllene (2.00–2.66%), valencene (1.77–2.56%), β- myrcene (1.56–1.91%), neryl acetate (1.54–1.71%), trans-ocimene (1.19–1.46%), and nerol (1.19–1.40%), constituting a little over 90% of the oil composition.

In all three populations, the only chemotype found was "linalyl acetate/linalool", as the two compounds together accounted for the highest percentage of the total, with values of 61.15% (SS), 63.50% (PR) and 65.60% (AF) (Table 5).

Table 5. Effect of the local population (LP) on the composition of the principal components of *Salvia sclarea* L. essential oils.

Component		LP		Significance
	SS	PR	AF	LP
β-myrcene	1.56 c	1.91 a	1.70 b	**
trans-Ocimene	1.19 c	1.46 a	1.29 b	**
linalool	22.42 c	25.55 a	23.97 b	**
α-Terpineol	5.76 c	7.14 a	6.48 b	**
nerol	1.19 c	1.40 a	1.25 b	**
linalyl acetate	38.73 c	38.95 b	41.63 a	**
neryl acetate	1.54 b	1.71 a	1.54 b	**
geranyl acetate	2.64 b	3.04 a	2.74 b	**
β-caryophyllene	2.66 a	2.00 c	2.23 b	**
germacrene D	4.42 b	4.49 a	3.61 c	**
valencene	2.56 a	1.77 c	1.80 b	**
sclareol	5.50 a	2.93 c	3.21 b	**

Local population code: PR = Prizzi; AF = Alcara Li Fusi; SS = S. Stefano Quisquina. ** = significant at $p < 0.01$; n.s. = not significant. Within the same row, means followed by the same letter are not significantly different for $p \leq 0.05$ according to Tukey's test.

As illustrated in Table 5, highly significant variations were found for the effect of local populations on the composition of the essential oil principal components. Population AF had the highest percentage of linalyl acetate (41.63%), followed by PR (38.95%) and SS (38.73%), while linalool varied from 25.55% in PR to 22.42% in SS, with the intermediate value of 23.97% obtained by AF.

Furthermore, following comparison of the three populations, results showed that the chemical profile of PR had a greater content of β-myrcene, trans-Ocimene, linalool, α-Terpineol, nerol, neryl acetate, geranyl acetate and germacrene D compared to AF and SS, which followed in decreasing order. It is worth noting that AF produced intermediate quantities of nearly all the components examined, although it excelled in the production of linalyl acetate and produced the lowest levels of germacrene D (3.61%). SS, though lagging behind the other two populations regarding most of the chemical components examined, pulled ahead in β-caryophyllene, valencene and, in particular, sclareol (5.50%) production; for the latter component, it obtained three percentage points more than AF and PR. As the dendrogram (Figure 3) shows, cluster analysis of the percentage composition of the essential oil principal components highlighted 2 groups. The first group included populations PR and AF, and the other group comprised only SS.

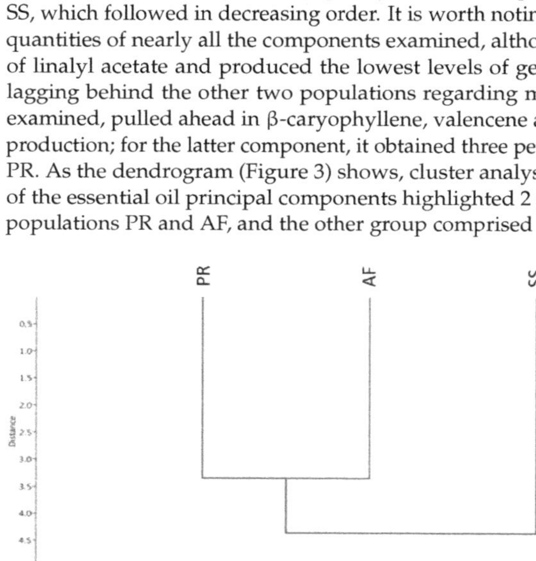

Figure 3. Cluster analysis of local populations in the study based on qualitative data. Local population code: PR = Prizzi; AF = Alcara Li Fusi; SS = S. Stefano Quisquina.

ANOVA analysis showed highly significant variations for the effect of inflorescence type on all of the principal components of the essential oils tested (Table 6, Figure 4).

Table 6. Effect of inflorescence type (TIPS) on the composition of the essential oil principal components.

Component	Inflorescence		
	ISP	**ISS**	**Significance**
β-myrcene	1.59 b	1.86 a	**
trans-Ocimene	1.23 b	1.41 a	**
linalool	23.20 b	24.78 a	**
α-Terpineol	5.87 b	7.06 a	**
nerol	1.19 b	1.37 a	**
linalyl acetate	39.78 b	39.86 a	**
neryl acetate	1.53 b	1.68 a	**
geranyl acetate	2.55 b	3.06 a	**
β-caryophyllene	2.57 a	2.02 b	**
germacrene D	4.53 a	3.82 b	**
valencene	2.33 a	1.75 b	**
sclareol	4.59 a	3.18 b	**

ISP = primary inflorescences; ISS = secondary inflorescences. ** = significant at $p < 0.01$; n.s. = not significant. Within the same row, means followed by the same letter are not significantly different for $p \leq 0.05$ according to Tukey's test.

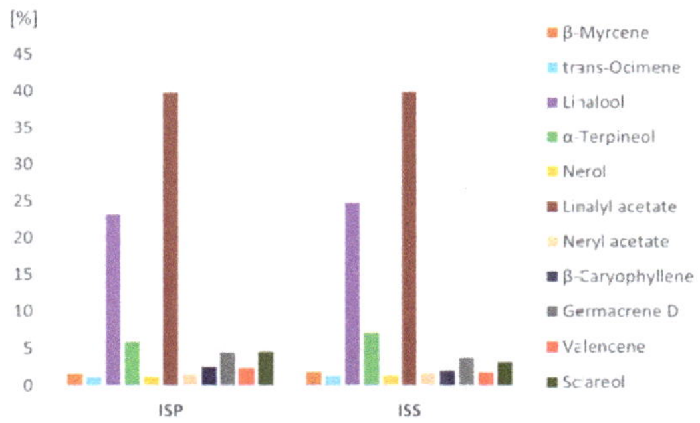

Figure 4. Principal components of the essential oils of primary inflorescences (ISP) and secondary inflorescences (ISS).

Relative to the components that characterize the chemotypes of the three populations (linalyl acetate and linalool), the most abundant contents were found in the secondary inflorescences (ISS) together with β-myrcene, trans-Ocimene, α-Terpineol, nerol, neryl acetate and geranyl acetate, therefore showing a prevalence of the monoterpene fraction. In contrast, primary inflorescences (ISP) had greater abundance of β-caryophyllene, germacrene D, valencene and sclareol—the sesquiterpene and diterpene classes.

Due to the combined effect of the two factors (Table 7, Figure 5), the primary inflorescences of population SS were statistically more abundant in germacrene D (5.08%), β-caryophyllene (3.28%), valencene (3.17%) and sclareol (7.76%), while the remaining components were found to be lower compared to the other treatments.

Table 7. Composition of the essential oils of the ISP and ISS of local *Salvia sclarea* populations—interaction of LP∗TIPS factors.

Component							**Significance**
	SSP	**PRP**	**AFP**	**SSS**	**PRS**	**AFS**	**LP∗TIPS**
β-myrcene	1.39 e	1.72 c	1.67 d	1.74 b	2.10 a	1.73 b	**
trans-Ocimene	1.07 e	1.35 b	1.28 d	1.33 c	1.59 a	1.32 c,d	**
linalool	21.42 f	23.79 c	24.40 b	23.43 e	27.31 a	23.59 d	**
α-Terpineol	4.83 e	6.40 d	6.38 d	6.69 b	7.90 a	6.58 c	**
nerol	1.07 f	1.27 d	1.23 e	1.32 b	1.52 a	1.28 c	**
linalyl acetate	35.80 e	42.46 a	40.85 d	41.71 c	35.46 f	42.42 b	**
neryl acetate	1.43 f	1.62 c	1.53 e	1.66 b	1.81 a	1.56 d	**
geranyl acetate	2.26 f	2.76 d	2.62 e	3.02 b	3.32 a	2.85 c	**
β-caryophyllene	3.28 a	2.03 d	2.39 b	2.03 d	1.97 e	2.07 c	**
germacrene D	5.08 a	4.55 b	3.95 d	3.76 e	4.42 c	3.27 f	**
valencene	3.17 a	1.93 c	1.88 d	1.94 b	1.60 f	1.72 e	**
sclareol	7.76 a	2.71 d	3.29 b	3.25 c	3.15 d	3.14 d	**

SSP = S. Stefano Quisquina Primary inflorescences; PRP = Prizzi Primary inflorescences; AFP = Alcara Li Fusi Primary inflorescences; SSS = S. Stefano Quisquina Secondary inflorescences; PRS = Prizzi Secondary inflorescences; AFS = Alcara Li Fusi Secondary inflorescences. ** = significant at $p < 0.01$; n.s. = not significant. Within the same row, means followed by the same letter are not significantly different for $p \leq 0.05$ according to Tukey's test.

In population SS, the components most frequently found in the primary inflorescences were equally as prominent in the secondary inflorescences, except for β-caryophyllene and germacrene D, which were lower. All the other components were greater in value. In the PR primary inflorescences, the highest levels of linalyl acetate (42.46%) and germacrene D (4.55%) were found, although the latter was still lower than the levels found in the SS primary inflorescences. In the PR secondary inflorescences, the statistically highest levels of linalool (27.31%), α-Terpineol (7.90%), β-myrcene (2.10%), trans-Ocimene (1.59%), nerol (1.52%), neryl acetate (1.81%) and geranyl acetate (3.32%) were found. With the exception of germacrene D (4.42%) (levels of which were among the highest but below those found in the primary inflorescences (PR)), the remaining components were among the lowest. In population AF, (except for linalool (24.40%), β-caryophyllene (2.39%) and sclareol (3.29%) from the primary inflorescences and β-myrcene (1.73%) and linalyl acetate (42.42%) from the secondary inflorescences (AF)—all of which were in a sub-apical position) all the components in both types of inflorescence were found to have medium-low values.

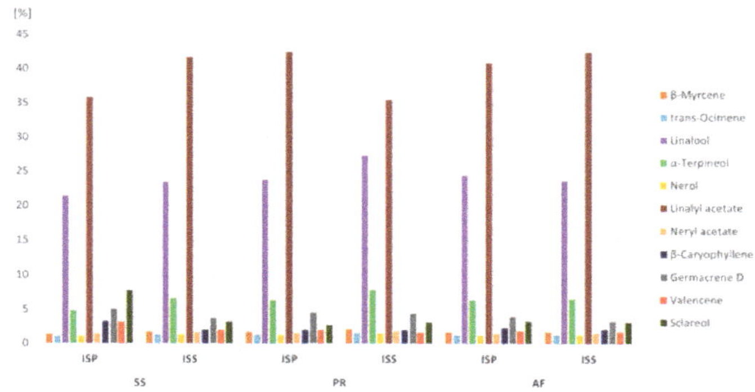

Figure 5. Principal components of the essential oils of the two inflorescence types (TIPS) of the three LPs. Local population code: PR = Prizzi; AF= Alcara Li Fusi; SS = S. Stefano Quisquina. ISP = Primary inflorescences steam; ISS = Secondary inflorescences steam.

4. Discussion

The morphological and production study provided an initial insight into the different populations in the study. Populations SS and AF demonstrated very similar characteristics, especially in terms of plant biomass production (both fresh and dry), inflorescence yields and essential oil yields. These populations performed the best in production terms, and plants were vigorous with dense leaves and inflorescence. In contrast, PR lagged in performance from this point of view. Despite the high incidence of inflorescences in the total plant weight produced by PR, both inflorescence and essential oil yields were not as high as the other two populations, as PR plants were smaller.

Population PR was different from the other two populations, with thinner plants and longer inflorescences. An analysis of the morphological characteristics is the first step towards crop improvement [44–46]. Morphological differences can be used to classify the plant material into various groups. As reported by Yaseen et al. [41], the populations in our study were classified as medium size (100–150 cm), as all the plants were a little under 150 cm in height. However, in studies carried out by Tibaldi et al. [1] in Piemonte (north Italy), tall biotypes were identified (>150 cm) and in Sicily small sizes (<100 cm) were identified [47]. Other studies on the species have shown that plant height can become an important distinguishing feature in the selection of accessions, as there is a positive correlation with a number of important production parameters, such as no. of inflorescences/plant, inflorescence length and oil yields [41]. In agreement with our study, Balmus et al. [48], while researching promising varieties in Moldovia, also noted that encouraging results in terms of quality were shown by tall plants with a height of approximately 140 cm and with an inflorescence length approximately 20 cm longer than ours. In addition to confirming a number of Yaseen's results [41], Tuttolomondo et al. [49] added that the length of the inflorescence (both primary (ISP) and secondary (ISS)), appears to be equally as interesting in terms of classifying accessions, deemed a reliable characteristic for selecting high EO-content biotypes. Furthermore, the same authors found longer ISS inflorescences were produced in the year with lower rainfall levels (compared to an exceptionally rainy year) together with a slightly higher EO content than the ISP. It emerged, therefore, that the accessions that were able to employ this production feature as an adaptation strategy in reaction to difficult environmental conditions were also those that performed best in terms of biomass production. This suggests that the length of the inflorescence, in particular the ISS, should be given particular consideration when selecting accessions for production purposes, especially in Mediterranean areas. Regarding differences in EO % content, our study recorded statistical differences between populations: the EO % content for population AF was higher than the other two populations in the study, consistent with results from the best variety identified by Balmus et al. [43]. Referring once again to Yaseen's classification [41] regarding oil yields, population AF can be classified as high yielding (>20 kg ha^{-1}), while populations SS and PR were medium yielding (10–15 kg ha^{-1}). Although a high EO-yielding variety was identified, better yields were found in the varieties studied in Moldavia by Balmus et al. [48]. However, the study of wild populations is the first step towards identifying good biotypes to be used in the development of *S. sclarea* high EO-content varieties for the Mediterranean.

Cluster analysis results regarding the morphological and production data showed, as illustrated above, that two of the three populations (SS and AF), although originating in different areas, were grouped together in the same cluster due to similarity of characteristics. The populations seem to have been little affected by differences in geographic origin, when affected at all. Furthermore, agronomic characteristics are known to be easily influenced by environmental conditions and cropping practices [50].

Population PR, however, formed a separate cluster from the other two populations, even though the test environment and cropping practices were identical. In addition to highlighting a different phenotype, this could also indicate genetic differences based on different geographical locations of origin.

Compared to the many studies in the scientific literature regarding the composition and biological activity of *S. sclarea* essential oils, there are relatively few that assess production aspects and essential oil quality together. A few studies have shown (for the most part) only differences in the chemical composition of *S. sclarea* essential oil in relation to geographical location of origin [13,16,17,19] and to different cropping and harvesting conditions. Field tests carried out in the Ukraine showed that agronomic-technical factors modify essential oil yields of *Salvia sclarea*; sowing in December rather than April, and harvesting in the cooler hours of the day produced higher yields [51]. Other studies underlined different yields from different plant parts (flowers and leaves in particular) [16]. Other variations in the production of the chemical components are linked to different ecotypes or chemotypes. It is worth noting that the cultivars produced greater yield stability in terms of chemical composition, while the effect of the environment was greater for ecotypes.

Regarding the chemical composition of *S. sclarea* essential oils, various authors have demonstrated that, in most cases, the principal volatile components belong to the terpenoids group [43,52], among which are linalyl acetate and linalool (which characterize good quality oils suited to aromatizing) [2,12,14,53]. No studies on the composition of the essential oils, in relation to the two types of inflorescence examined in this study, were found in the scientific literature. A small number of studies were found that reported that linalool, linalyl acetate and sclareol are essential oil components typical of the flowers, whereas germacrene D was found in higher proportions in the leaves [39,41,54].

These results reinforce data found in literature on possible quantitative and/or qualitative differences in essential oil components from local populations/ecotypes and different plant parts.

Regarding the qualitative aspects, population PR performed the best of the three populations as it obtained higher values in approximately 60.00% the chemical components examined. AF was found to have the greatest linalyl acetate content and the lowest germacrene D content while maintaining intermediate values for the other components. SS lagged behind the other two populations for most of the components, although it excelled in sclareol, β-caryophyllene and valencene content. These differences were highlighted by the cluster analysis, with the grouping of populations PR and AF and the separation of SS from the other two populations. Similar results regarding essential oil composition between populations in different areas within the same region were reported by Pitarokili et al. [13] in Greece. Two populations were studied, one more abundant in linalyl acetate, linalool and a-terpinol, and the other with higher sclareol content, comparable to our population SS.

We would like to underline the difference in groupings of the populations in the clusters following analysis of the morphological and production characteristics and analysis of the essential oil components. The first dataset showed similarity between AF and SS (the most productive in biomass terms); the second dataset grouped PR and AF (based on affinity of the essential oil composition). The selection of ecotype from the populations should, therefore, include a morphological, production and quality analysis.

The inflorescences, divided into primary and secondary inflorescences, are characterized by the same predominant components but in differing ratios. Primary inflorescences were found to be accumulation sites favoured by sesquiterpenes and sclareol, and secondary inflorescences by monoterpenes. This differentiation was not shown by the interaction effect of LP∗TIPS, as the levels of the different components were found to be relatively heterogeneous between the inflorescences of the three populations.

However, a closer examination of the results showed that most of the components followed the same trend. This is likely due to the fact that the populations, particularly abundant in monoterpene components or fraction (or other fractions), maintained high levels. These levels, however, were lower in one of the two inflorescence types and frequently higher than the corresponding inflorescence fractions in the other populations thereby obscuring the general accumulation trend. This principle does not hold for linalool in AF and linalyl acetate in PR.

Finally, due to considerable interest regarding this species, further studies could be conducted on more efficient propagation methods, using in vitro technologies already adopted for other typical Mediterranean species, such as capers and hops [55–57].

5. Conclusions

The three populations in this study demonstrated significant production differences in terms of both biomass and essential oil yields. Quality analysis of the essential oils of the three populations produced only one chemotype, the "linalyl acetate/linalool" type. Primary inflorescences were found to be a preferred accumulation site of sesquiterpenes and sclareol, whereas secondary inflorescences hosted monoterpenes. Cluster analysis of the morphological, production and quality characteristics, each illustrated on a dendrogram revealed two clusters for each.

Regarding the first dataset, populations SS and AF performed the best, while PR and AF excelled in the results of the quality analysis. SS was found to have the highest levels of sclareol, particularly in the ISP. These differences in characteristics within the populations can be of interest in terms of end use. AF, for example, successfully combined morphological and production parameters with quality characteristics, proving to be of interest for a number of final uses. Population SS showed good production levels but trailed in quality, particularly concerning sclareol content, one of the most valued components of essential oils for the perfume industry. PR produced high quality oils, chiefly with regard to monoterpene fractions; however, it was the least productive of the populations.

This considerable diversity among the three populations regarding most of the characteristics examined assumes different levels of importance depending on use. Knowledge of this kind is precious when selecting biotypes based on their morphological, production and/or qualitative performance as well as the intended final use.

Based on our results, this study of Sicilian *S. sclarea* populations is the first step toward identifying biotypes within populations and can contribute to the development of this crop. It is a crop is of considerable importance to the area, both in economic and agronomic terms, and may provide producers with the opportunity to grow quality crops with local plant materials.

Author Contributions: Conceptualization, T.T., G.V. and S.L.B.; methodology, T.T., G.V. and S.L.B.; software, N.I.; validation, N.I., D.F. and M.L.; formal analysis, T.T., M.L., G.V. and D.F.; investigation, M.L., N.I., D.F. and S.L.B.; resources, T.T. and S.L.B.; data curation, G.V., M.L., N.I. and D.F.; writing—original draft preparation, T.T., G.V. and S.L.B.; writing—review and editing, M.L. and N.I.; visualization, M.L., N.I. and D.F.; supervision, T.T. and S.L.B.; project administration, T.T. and S.L.B.; funding acquisition, S.L.B. All authors have read and agreed to the published version of the manuscript.

Funding: This research was funded by the Sicilian Regional Ministry of Agriculture and Food Resources (Italy), project "Environmental and plant resources in the Mediterranean: study, valorisation and defence", grant number 2309/2005.

Institutional Review Board Statement: Not applicable.

Informed Consent Statement: Not applicable.

Data Availability Statement: Data are available by contacting the authors.

Acknowledgments: The authors would like to thank Lucie Branwen Hornsby for her linguistic assistance.

Conflicts of Interest: The authors declare no conflict of interest. The funders had no role in the design of the study; in the collection, analyses, or interpretation of data; in the writing of the manuscript, or in the decision to publish the results.

References

1. Tibaldi, G.; Fontana, E.; Nicola, S. Cultivation practices do not change the *Salvia sclarea* L. essential oil but drying process does. *J. Food Agric. Environ.* **2010**, *8*, 790–794.
2. Cai, J.; Lin, P.; Zhu, X.; Su, Q. Comparative analysis of clary sage (*S. sclarea* L.) oil volatiles by GC-FTIR and GC-MS. *Food Chem.* **2006**, *99*, 401–407. [CrossRef]
3. Gonceariuc, M.; Balmuş, Z.; Cotelea, L. Genetic diversification of *Salvia sclarea* L. quality by increasing the storage capacity of the essential oil. *Oltenia-Studii si Comunicari Stiintele Naturii* **2016**, *32*, 29–36.
4. Leffingwell, J.C.; Stallings, J.W.; Sellers, F.O.; Lloyd, R.A.; Kane, F.C., Jr. Clary sage production in the Southeastern United States. In Proceedings of the 6th International Congress of Essential Oils, San Francisco, CA, USA, 8–12 September 1974.
5. Angelova, V.R.; Ivanova, R.V.; Todorov, G.M.; Ivanov, K.I. Potential of *Salvia sclarea* L. for phytoremediation of soils contaminated with heavy metals. *Int. J. Agric. Biosyst. Eng.* **2016**, *10*, 780–790.
6. Šućur, J.; Popović, A.; Petrović, M.; Anačkov, G.; Malenčić, D.; Prvulović, D. Allelopathic Effects and Insecticidal Activity of *Salvia sclarea* L. *Studia UBB Chem.* **2015**, *1*, 253–264.
7. Šućur, J.; Prvulović, D.; Anačkov, G.; Malenčić, Đ. Oxidative stress in bromus (*Bromus mollis* L.) seedlings treated with clary sage (*Salvia sclarea* L.) aqueous extract. *Biol. Nyssana* **2016**, *7*, 141–144.
8. Šućur, J.; Gvozdenac, S.; Anačkov, G.; Malenčić, Đ.; Prvulović, D. Allelopathic effect of *Clinopodium menthifolium* and *Salvia sclarea* aqueous extracts. *Matica Srpska J. Nat. Sci. Novi Sad* **2016**, *131*, 177–188. [CrossRef]
9. Kostić, M.; Kitić, D.; Petrović, M.B.; Jevtović-Stoimenov, T.; Jović, M.; Petrović, A.; Živanović, S. Anti-inflammatory effect of the *Salvia sclarea* L. ethanolic extract on lipopolysaccharide-induced periodontitis in rats. *J. Ethnopharmacol.* **2017**, *199*, 52–59. [CrossRef]
10. Gülçin, I.; Uğuz, M.T.; Oktay, M.; Beydemir, Ş.; Küfrevioğlu, Ö.İ. Evaluation of the antioxidant and antimicrobial activities of clary sage (*Salvia sclarea* L.). *Turk. J. Agric. For.* **2004**, *28*, 25–33.
11. Tuttolomondo, T.; Dugo, G.; Ruberto, G.; Leto, C.; Napoli, E.M.; Potortì, A.G.; Fede, M.R.; Virga, G.; Leone, R.; D'Anna, E.; et al. Agronomical evaluation of Sicilian biotypes of *Lavandula stoechas* L. spp. *Stoechas* and analysis of the essential oils. *J. Essent. Oil Res.* **2015**, *27*, 115–124. [CrossRef]
12. Pesic, P.Z.; Bankovi, V.M. Investigation on the essential oil of cultivated *Salvia sclarea* L. *Flavour Fragr. J.* **2003**, *18*, 228–230. [CrossRef]
13. Pitarokili, D.; Couladis, M.; Petsikos-Panayotarou, N.; Tzakou, O. Composition and antifungal activity on soil-borne pathogens of the essential oil of *Salvia sclarea* from Greece. *J. Agric. Food Chem.* **2002**, *50*, 6688–6691. [CrossRef] [PubMed]
14. Fakir, H.; Korkmaz, M.; Güller, B. Medicinal plant diversity of western Mediterranean region in Turkey. *J. Appl. Biol. Sci.* **2009**, *3*, 30–40.
15. Özdemir, E.; Alpınar, K. An ethnobotanical survey of medicinal plants in western part of central Taurus Mountains, Aladaglar (Nigde–Turkey). *J. Ethnopharmacol.* **2015**, *166*, 53–65. [CrossRef]
16. Georgiev, E.; Stoyanova, A. *A Guide for the Specialist in Aromatic Industry*; UFT Academic Publishing House: Plovdiv, Bulgaria, 2006.
17. Jirovetz, L.; Buchbauer, G.; Denkova, Z.; Slavchev, A.; Stoyanova, A.; Schmidt, E. Chemical composition, antimicrobial activities and odor descriptions of various *Salvia* sp. and *Thuja* sp. essential oils. *Nutrition* **2006**, *90*, 152–159.
18. Peana, A.T.; Moretti, M.D.; Juliano, C. Chemical composition and antimicrobial action of the essential oils of *Salvia desoleana* and *S. sclarea*. *Planta Med.* **1999**, *65*, 752–754. [CrossRef] [PubMed]

19. Jirovetz, L.; Wlcek, K.; Buchbauer, G.; Gochev, V.; Girova, T.; Stoyanova, A.; Schmidt, E.; Geissler, M. Antifungal activities of essential oils of *Salvia lavandulifolia*, *Salvia officinalis* and *Salvia sclarea* against various pathogenic *Candida* species. *J. Essent. Oil Bear. Plants* **2007**, *10*, 430–439. [CrossRef]
20. Geun, H.S.; Hyun, S.S.; Pill-Joo, K.; Hean, K.M.; Ki, H.L.; Insop, S.; Suk, S.M. Antidepressant-like effect of *Salvia sclarea* is explained by modulation of dopamine activities in rats. *J. Ethnopharmacol.* **2010**, *130*, 187–190.
21. Verma, R.S. Chemical investigation of decanted and hydrophilic fractions of *Salvia sclarea* essential oil. *Asian J. Tradit. Med.* **2010**, *5*, 102–108.
22. Mahboubi, M. Clary sage essential oil and its biological activities. *Adv. Tradit. Med.* **2020**, *20*, 517–528. [CrossRef]
23. Tulukcu, E.; Yalcin, H.; Ozturk, I.; Sagdic, O. Changes in the fatty acid compositions and bioactivities of clary sage seeds depending on harvest year. *Ind. Crop. Prod.* **2012**, *39*, 69–73. [CrossRef]
24. Taârit, M.B.; Msaada, K.; Hosni, K.; Marzouk, B. Fatty acids, phenolic changes and antioxidant activity of clary sage (*Salvia sclarea* L.) rosette leaves grown under saline conditions. *Ind. Crop. Prod.* **2012**, *38*, 58–63. [CrossRef]
25. Lawrence, B.M. *Advances in Labiate Science*; Harley, R.M., Reynolds, T., Eds.; Royal Botanical Gardens: Kew, UK, 1992; pp. 399–400.
26. Valussi, M. *Il Grande Manuale Dell'Aromaterapia. Fondamenti di Scienza degli Oli Essenziali*; Tecniche Nuove: Milan, Italy, 2005.
27. Tuttolomondo, T.; La Bella, S.; Leto, C.; Gennaro, M.C.; Calvo, R.; D'Asaro, F. Biotechnical characteristics of root systems in erect and prostrate habit Rosmarinus officinalis L. accessions grown in a Mediterranean climate. *Chem. Eng. Trans.* **2017**, *58*, 769–774.
28. La Bella, S.; Rossini, F.; Licata, M.; Virga, G.; Ruggeri, R.; Iacuzzi, N.; Leto, C.; Tuttolomondo, T. Four-Year Study on the Bio-Agronomic Response of Biotypes of *Capparis spinosa* L. on the Island of Linosa (Italy). *Agriculture* **2021**, *11*, 327. [CrossRef]
29. Barreca, S.; La Bella, S.; Maggio, A.; Licata, M.; Buscemi, S.; Leto, C.; Pace, A.; Tuttolomondo, T. Flavouring extra-virgin olive oil with aromatic and medicinal plants essential oils stabilizes oleic acid composition during photo-oxidative stress. *Agriculture* **2021**, *11*, 266. [CrossRef]
30. La Bella, S.; Virga, G.; Iacuzzi, N.; Licata, M.; Sabatino, L.; Consentino, B.B.; Leto, C.; Tuttolomondo, T. Effects of irrigation, peat-alternative substrate and plant habitus on the morphological and production characteristics of Sicilian rosemary (*Rosmarinus officinalis* L.) biotypes grown in pot. *Agriculture* **2021**, *11*, 13. [CrossRef]
31. Virga, G.; Sabatino, L.; Licata, M.; Tuttolomondo, T.; Leto, C.; La Bella, S. Effects of irrigation with different sources of water on growth, yield and essential oil compounds in oregano. *Plants* **2020**, *9*, 1618. [CrossRef]
32. Saija, A.; Speciale, A.; Trombetta, D.; Leto, C.; Tuttolomondo, T.; La Bella, S.; Licata, M.; Virga, G.; Bonsangue, G.; Gennaro, M.C.; et al. Phytochemical, ecological and antioxidant evaluation of wild sicilian thyme: *Thymbra capitata* (L.) Cav. *Chem. Biodivers.* **2016**, *13*, 1641–1655. [CrossRef] [PubMed]
33. Tuttolomondo, T.; La Bella, S.; Leto, C.; Bonsangue, G.; Leone, R.; Gennaro, M.C.; Virga, G.; Inguanta, R.; Licata, M. Effect of plant density on the number of glandular trichomes and on yield and quality of essential oils from oregano. *Nat. Prod. Res.* **2016**, *11*, 849–852. [CrossRef]
34. Tuttolomondo, T.; Martinelli, F.; Mariotti, L.; Leto, C.; Maggio, A.; La Bella, S. Agronomic, metabolomic and lipidomic characterisation of Sicilian *Origanum vulgare* (L.) ecotypes. *Nat. Prod. Res.* **2016**, *30*, 1103–1107. [CrossRef]
35. Tuttolomondo, T.; Dugo, G.; Ruberto, G.; Leto, C.; Napoli, E.M.; Cicero, N.; Gervasi, T.; Virga, G.; Leone, R.; Licata, M.; et al. Study of quantitative and qualitative variations in essential oils of Sicilian *Rosmarinus officinalis* L. *Nat. Prod. Res.* **2015**, *29*, 1928–1934. [CrossRef] [PubMed]
36. Tuttolomondo, T.; La Bella, S.; Licata, M.; Virga, G.; Leto, C.; Saija, A.; Trombetta, D.; Tomaino, A.; Speciale, A.; Napoli, E.M.; et al. Biomolecular characterization of wild Sicilian oregano: Phytochemical screening of essential oils and extracts, and evaluation of their antioxidant activities. *Chem. Biodivers.* **2013**, *10*, 411–433. [CrossRef] [PubMed]
37. Licata, M.; Tuttolomondo, T.; Dugo, G.; Ruberto, G.; Leto, C.; Napoli, E.M.; Rando, R.; Fede, M.R.; Virga, G.; Leone, R.; et al. Study of quantitative and qualitative variations in essential oils of Sicilian oregano biotypes. *J. Essent. Oil Res.* **2015**, *27*, 293–306. [CrossRef]
38. La Bella, S.; Tuttolomondo, T.; Dugo, G.; Ruberto, G.; Leto, C.; Napoli, E.M.; Potorti, A.G.; Fede, M.R.; Virga, G.; Leone, R.; et al. Composition and variability of the essential oil of the flowers of *Lavandula stoechas* from various geographical sources. *Nat. Prod. Commun.* **2015**, *10*, 2001–2004. [CrossRef] [PubMed]
39. Farkas, P.; Holla, M.; Jozef, T.; Mellen, S. Composition of the essential oils from the flowers and leaves of *Salvia sclarea* L. (Lamiaceae) cultivated in Slovak Republic. *J. Essent. Oil Res.* **2005**, *17*, 141–144.
40. Schmiderer, C.; Grassi, P.; Novak, J.; Weber, M.; Franz, C. Diversity of essential oil glands of clary sage (*Salvia sclarea* L., Lamiaceae). *Plant Biol.* **2008**, *10*, 433–440. [CrossRef] [PubMed]
41. Yaseen, M.; Singh, M.; Ram, D.; Singh, K. Production potential, nitrogen use efficiency and economics of clary sage (*Salvia sclarea* L.) varieties as influenced by nitrogen levels under different locations. *Ind. Crop. Prod.* **2014**, *54*, 86–91. [CrossRef]
42. Adams, R.P. *Identification of Essential Oil Components by Gas Chromatographic/Quadrupole Mass Spectrometry*, 4th ed.; Allured Publishing Corporation: Carol Stream, IL, USA, 2007; pp. 1–698.
43. National Institute of Standard and Technology (NIST). Mass Spectral Library, Version 1998. Available online: www.nist.gov (accessed on 25 May 2021).
44. Alquezar, J.B. *Investigación y Experimentación de Plantas Aromàticas e Medicinalesen Aragón*; Cultivo, Trasformación y Analìtica, Espanha; Governo de Aragón: Zaragoza, Spain, 2003; 262p.

45. Julia, C.C.; Waters, D.L.E.; Wood, R.H.; Rose, T.J. Morphological characterisation of Australian ex situ wild rice accessions and potential for identifying novel sources of tolerance to phosphorus deficiency. *Genet. Resour. Crop Evol.* **2016**, *63*, 327–337. [CrossRef]
46. Peratoner, G.; Seling, S.; Klotz, C.; Florian, C.; Figl, U.; Schmitt, A.O. Variation of agronomic and qualitative traits and local adaptation of mountain landraces of winter rye (*Secale cereale* L.) from Val Venosta/Vinschgau (South Tyrol). *Genet. Resour. Crop Evol.* **2016**, *63*, 261–273. [CrossRef]
47. Carrubba, A.; Torre, R.; Piccaglia, R.; Marotti, M. Characterization pf an Italian biotype of clary sage (*Salvia sclarea* L.) grown in semi-arid Mediterranean environment. *Flavour Fragr. J.* **2002**, *17*, 191–194. [CrossRef]
48. Balmush, Z.; Gonceariuc, M.; Cotelea, L.; Butnarash, V. Parfum perfect the new early variety of *Salvia sclarea* L. (clary sage). In Proceedings of the International Congress of Geneticists and Breeders from the Republic of Moldova, Chisinau, Moldova, 15–16 June 2021; p. 70.
49. Tuttolomondo, T.; Iapichino, G.; Licata, M.; Virga, G.; Leto, C.; La Bella, S. Agronomic evaluation and chemical characterization of Sicilian *Salvia sclarea* L. accessions. *Agronomy* **2020**, *10*, 1114. [CrossRef]
50. Shen, G.; Girdthai, T.; Liu, Z.Y.; Fu, Y.H.; Meng, Q.Y.; Liu, F.Z. Principal component and morphological diversity analysis of Job's-tears (*Coix lacryma-jobi* L.). *Chil. J. Agric. Res.* **2019**, *79*, 131–143. [CrossRef]
51. Ushkarenko, V.O.; Vozhehova, R.A.; Kokovikhin, S.V.; Shepel, A.V.; Chaban, V.O. Clary Sage Yield and Essential Oil Output Depending on the Studied Agro-Technical Factors. 2020. Available online: http://dspace.ksau.kherson.ua/handle/123456789/5451 (accessed on 5 June 2021).
52. Ohloff, G. *Riechstoffe und Geruchssinn: Die molekulare Welt der Düfte*; Springer: Berlin/Heidelberg, Germany; New York, NY, USA, 1990; pp. 208–214.
53. Loumerem, M.; Alercia, A. Descriptors for jute (*Corchorus olitorius* L.). *Genet. Resour. Crop Evol.* **2016**, *63*, 1103–1111. [CrossRef]
54. Yuce, E.; Yildirim, N.; Yildirim, N.C.; Paksoy, M.Y.; Bagci, E. Essential oil composition, antioxidant and antifungal activities of *Salvia sclarea* L. from Munzur Valley in Tunceli, Turkey. *Cell. Mol. Biol.* **2014**, *60*, 1–5.
55. Moulines, J.; Bats, J.P.; Lamidey, A.M.; Da Silva, N. About a practical synthesis of Ambrox from sclareol: A new preparation of a ketone key intermediate and a close look at its Baeyer-Villiger oxidation. *Helv. Chim. Acta* **2004**, *87*, 2695–2705. [CrossRef]
56. Germanà, M.A.; Lesto, F.; Lo Bosco, F.; Iacuzzi, N.; Ozudogru, E.A. Encapsulation of in vitro-derived propagules of two genotypes of *Capparis spinosa* (L.) from Pantelleria Island. *Acta Hortic.* **2020**, *1285*, 39–44. [CrossRef]
57. Rossini, F.; Virga, G.; Loreti, P.; Iacuzzi, N.; Ruggeri, R.; Provenzano, M.E. Hops (*Humulus lupulus* L.) as a novel multipurpose crop for the mediterranean region of Europe: Challenges and opportunities of their cultivation. *Agriculture* **2021**, *11*, 484. [CrossRef]

agronomy

MDPI

Article

High Nitrogen Fertilization Modulates Morpho-Physiological Responses, Yield, and Water Productivity of Lowland Rice under Deficit Irrigation

Nasr M. Abdou [1], Mohamed A. Abdel-Razek [1], Shimaa A. Abd El-Mageed [2], Wael M. Semida [3,*], Ahmed A. A. Leilah [4], Taia A. Abd El-Mageed [1], Esmat F. Ali [5], Ali Majrashi [5] and Mohamed O. A. Rady [2]

1 Soil and Water Department, Faculty of Agriculture, Fayoum University, Fayoum 63514, Egypt; nma02@fayoum.edu.eg (N.M.A.); maa27@fayoum.edu.eg (M.A.A.-R.); taa00@fayoum.edu.eg (T.A.A.E.-M.)
2 Agronomy Department, Faculty of Agriculture, Fayoum University, Fayoum 63514, Egypt; sa1944@fayoum.edu.eg (S.A.A.E.-M.); mer00@fayoum.edu.eg (M.O.A.R.)
3 Horticulture Department, Faculty of Agriculture, Fayoum University, Fayoum 63514, Egypt
4 Agronomy Department, Faculty of Agriculture, Mansoura University, Mansoura 35516, Egypt; ahmed_leila@mans.edu.eg
5 Department of Biology, College of Science, Taif University, P.O. Box 11099, Taif 21944, Saudi Arabia; a.esmat@tu.edu.sa (E.F.A.); aa.majrashi@tu.edu.sa (A.M.)
* Correspondence: wms00@fayoum.edu.eg; Tel.: +2-010-107-877-88

Citation: Abdou, N.M.; Abdel-Razek, M.A.; Abd El-Mageed, S.A.; Semida, W.M.; Leilah, A.A.A.; Abd El-Mageed, T.A.; Ali, E.F.; Majrashi, A.; Rady, M.O.A. High Nitrogen Fertilization Modulates Morpho-Physiological Responses, Yield, and Water Productivity of Lowland Rice under Deficit Irrigation. *Agronomy* **2021**, *11*, 1291. https://doi.org/10.3390/agronomy11071291

Academic Editors: Umberto Anastasi and Aurelio Scavo

Received: 10 May 2021
Accepted: 23 June 2021
Published: 25 June 2021

Abstract: Sustainability of rice production under flooding conditions has been challenged by wate shortage and food demand. Applying higher nitrogen fertilization could be a practical solution t alleviate the deleterious effects of water stress on lowland rice (*Oryza sativa* L.) in semi-arid condition For this purpose, field experiments were conducted during the summer of 2017 and 2018 season These trials were conducted as split-split based on randomized complete blocks design with soil moisture regimes at three levels (120, 100 and 80% of crop evapotranspiration (ETc), nitrogen fertilizers a two levels (N_1—165 and N_2—200 kg N ha^{-1}) and three lowland Egyptian rice varieties [V_1 (Giza$_{178}$) V_2 (Giza$_{177}$) and V_3 (Sakha$_{104}$)] using three replications. For all varieties, growth (plant height, tiller No, effective tillers no), water status ((relative water content RWC, and membrane stability inde MSI), physiological responses (chlorophyll fluorescence, Relative chlorophyll content (SPAD), and yield were significantly increased with higher addition of nitrogen fertilizer under all water regime Variety V_1 produced the highest grain yield compared to other varieties and the increases were 38% and 15% compared with V_2 and V_3, respectively. Increasing nitrogen up to 200 kg N ha^{-1} (N_2 resulted in an increase in grain and straw yields by 12.7 and 18.2%, respectively, compared with N_1. The highest irrigation water productivity (IWP) was recorded under I_2 (0.89 kg m^{-3}) compared to (0.83 kg m^{-3}) and (0.82 kg m^{-3}) for I_1 and I_3, respectively. Therefore, the new applied agro management practice (deficit irrigation and higher nitrogen fertilizer) effectively saved irrigation water input by 50–60% when compared with the traditional cultivation method (flooding system). Hence, the new proposed innovative method for rice cultivation could be a promising strategy fo enhancing the sustainability of rice production under water shortage conditions.

Keywords: *Oryza sativa*; drought stress; chlorophyll fluorescence; varieties; grain yield and wate productivity

1. Introduction

Rice (*Oryza sativa* L.) is a staple food crop for about half of the world's populatio and ranks 2nd in production after wheat [1,2]. Globally, in 2018, the productivity o rice approximately amounted to 700 million tons, which was produced from 167 millior hectares, by an average grain yield of 4.2 tons per hectare [2]. Rice (*Oryza sativa* L.) is being cultivated in various agro-ecosystems: irrigated rice, rainfed lowland rice, uplanc rice, and flood-prone rice. More than 75% of rice production is supplied by irrigatec lowland rice [3]. Generally, rice has been grown under flooded conditions, maintaining a continuous water depth by 5–10 cm [4]. Under flooded conditions, a large amount of irrigation water supply is required, which is not only used for the growth and development of rice plants but also as a management technique during rice cultivation [5,6]. In a puddled rice field, the consumption of water depends on the rates of evaporation, transpiration, and water losses by percolation, seepage, and surface runoff. The irrigation water demand for rice crops under the traditional flooded system is more than 20,000 m^3 ha^{-1} which is more than 3–4 times that of its biological needs from water, which ranges between 6000–8000 m^3 ha^{-1} [7].

With increasing water scarcity, rapid population growth, increased urbanization, and the expected potential climate change, the sustainability, food production, and ecosystem services of rice fields are threatened [7–10]. It is estimated that, by 2025, 15–20 million ha of irrigated rice will suffer from some degree of water scarcity. Rice production in the Nile Delta of Egypt consumes about 11 billion m^3 of irrigation water which represents about 20% of the whole quantity of irrigation water used in agriculture (55.5 billion^3/year). In addition, many studies conducted in Egypt concluded that the total seasonal water input to rice under continuous flooding ranges between 16,190–21,428 m^3 ha^{-1} [11]. Therefore, Egypt's policymakers annually reduce the allotted area for rice cultivation, which has decreased by 59% from 745,000 ha to 304,080 ha during the past ten years (2008–2018) [2].

Therefore, good water governance should be adopted to develop socially acceptable, economically viable and environmentally sustainable novel rice-based systems that tend to reduce water losses and enhance crop productivity challenged by high evaporative demand and severe shortages of water supply. Hence, cultivating rice aerobically in non-puddled and non-saturated soils under water-saving irrigation technique as deficit irrigation could be a promising water-saving strategy to cope with water scarcity.

The application of irrigation water below the ET demand is termed deficit irrigation, aiming at optimization in economic output when water is very limited [12]. Plants under deficit irrigation receive a lower amount of irrigation water than their full water requirements either at specific crop growth stages or during the total cultivation period [12]. Consequently, under deficit irrigation technique, plants are subjected to water stress to some extent [13,14].

However, rice is very sensitive to water stress. Water stress negatively affects the growth and productivity of crops [15,16]. Physiological functioning in rice plants [17] viz root length density, root moisture extraction, the rate of apical development, canopy size, leaf elongation rate, leaf rolling, transpiration rate, relative water content, biomass production, spikelet number, spikelet sterility, panicle development, grain size and grain yield [17,18] may be drastically reduced due to water stress, if it occurs during vegetative or reproductive stages of rice, depending upon the stress severity and cultivar tolerance.

However, optimal application of nitrogen also plays a valuable role in combating drought [19]. Nitrogen (N) is considered a key component of many organic compounds. Nitrogen is one of the most essential nutrient elements for rice growth and metabolic processes [20,21]. Nitrogen represents a vital role in improving yield production and enhancing the photosynthetic activity especially during the grain filling stage of rice crops. Hence, the efficient use and nitrogen management respecting crop production is an urgent case for maximizing crop productivity, environmental safety with increasing economic returns [22,23] also concluded that increasing nitrogen supplying dose up to 144 kg N ha^{-1} improved and significantly affected plant growth, grain yield and yield components. Ref. Ref. [24] reported that increasing nitrogen application rates from 120 to 190 kg N ha^{-1} significantly improved plant height, panicle length, filled grains by panicle and grain yields. Ref. Ref. [25] reported that the inoculation with *G. diazotrophicus* Pal5 strain was alleviated deleterious effects of drought stress on rice plants, and improved biomass and grain yield. In addition, [26] noted that there were significant increases in plant growth traits, yield parameters and grain yield due to increasing nitrogen supplying rate of 100, 200 and 300 kg N ha^{-1}. Ref. Ref. [27] observed an interaction between soil moisture deficit and N supply rates on the activity of photosynthesis and transpiration processes in rice. Application of N fertilizer resulted in a significant increase in grain yield of rainfed-lowland rice under water deficit, where observed the optimal timing of N application for continuously irrigated rice was when the rice was exposed to moderate water deficit before flowering [28]. The absorption and utilization of water and nitrogen nutrition are two coupled physiological processes [29,30]. Supplying plants with N can increase drought resistance by increasing root hydraulic conductivity (Lpr) through increased abscisic acid (ABA) and aquaporin expression [31–34].

Therefore, the current study hypothesized that exogenous application of N-fertilizer may positively affect the rice performance, irrigation water productivity, chlorophyll fluorescence, water status as well as yield of some drought-stressed lowland rice varieties. Accordingly, the recent investigation was conducted to evaluate the effect of new applied water and nutrition environment on water-saving capacity and productivity of some lowland rice varieties under water-scarce conditions in arid and semi-arid regions.

2. Materials and Methods

2.1. Experimental Design, Treatments, and Cultural Practices

The current investigation was conducted at the experimental station farm of Faculty of Agriculture, Fayoum University, Egypt, (30°56_N latitude, 75°52_ longitude). The site is described by arid climate type, hot in summer. The meteorological data in Table 1 show that the highest mean values of maximum temperature, minimum temperature and relative humidity % by 40.4 °C, 26 °C and 45%, respectively, were recorded during August. The maximum pan evaporation rate (mm day^{-1}) 7.5 occurred in June month and decreased to its lowest value 5.3 in September during the two successive seasons of rice cultivation (2017–2018).

Table 1. Meteorological data recorded at Meteorological observatory of Fayoum governorate, during crop growing seasons of 2017 and 2018.

Month	Year	Temperature °C			Relative Humidity (%)	Wind Speed (m s^{-1})	Pan Evaporation (mm day^{-1})
		Max.	Min.	Mean			
May	2017	35.2	20.9	28.1	39.6	4.2	6.5
	2018	37.3	22.6	29.9	41.0	3.9	6.6
June	2017	36.0	21.7	28.9	42.1	5.2	7.3
	2018	40.3	24.4	32.3	38.6	5.0	7.5
July	2017	37.0	21.8	29.4	35.5	4.0	6.9
	2018	39.3	23.9	31.6	37.8	3.7	6.9
August	2017	40.4	26.0	33.2	36.9	1.9	6.2
	2018	36.4	23.0	29.6	45.2	3.7	6.3
September	2017	38.3	13.8	31.0	36.6	2.0	5.5
	2018	35.3	21.0	28.0	44.3	3.5	5.3

Max, and Min are maximum, and minimum temperatures, respectively.

2.2. Soil Characteristics

Table 2 shows that the soil is clay-textured. Soil moisture content % (at 0.33 bar and at 15 bar), available water %, bulk density (g cm^{-3}) and saturated hydraulic conductivity (cm h^{-1}) were determined at the surface soil layer (0–0.25 m) and amounted to 34.33, 14.60, 19.73, 1.40 and 1.2, respectively, and by 32.19, 13.06, 19.13, 1.36 and 0.9, respectively, a subsurface (0.25–0.5 m).

According to the data represented in Table 2, organic matter content was 1.2%, available nitrogen—0.04%, available phosphorus (mg kg^{-1} soil) 5.84 and available potassium 61.9 (mg kg^{-1} soil), ECe, (dS·m^{-1}), pH, CEC (cmol$_e$ kg^{-1}) and $CaCO_3$ (%) were measured and amounted 2.62, 7.76, 14.10 and 4.81, respectively. Soil physical and chemical properties were determined according to [35,36].

Table 2. Some initial physical and chemical properties of the experimental soil.

Physical Properties										
Layer (cm)	Particle Size Distribution				Bulk Density (g cm^{-3})	K_{sat} Cm h^{-1}	FC (%)	WP (%)	AW (%)
	Sand %	Silt %	Clay %	Texture Class					
0–25	10.00	20.00	70.00	Clay	1.40	1.20	34.33	19.73	14.60
25–50	7.00	21.00	72.00	Clay	1.36	0.90	32.19	19.13	13.06
Physical Properties									
	pH	ECe (dS·m^{-1})	CEC (cmole kg^{-1})	$CaCO_3$ (%)	Organic Matter (%)	N (%)	P (mg kg^{-1} soil)		K (mg kg^{-1} soil)
0–25	7.76	2.62	14.10	4.81	1.20	0.04	5.8		61.90
25–50	7.78	2.52	14.00	4.76	1.10	0.04	5.6		60.00

K_{sat} = Hydraulic conductivity, FC = Field capacity, WP = wilting point, and AW = Available water.

2.3. Experimental Layout and Treatments

Seeds of the three studied lowland rice varieties [V_1 (Giza-178), V_2 (Giza-177) and V_3 (Sakha-104)] were sown manually on 16 May 2017 and 10 May 2018 in rows with seed rate 130 kg h^{-1} with rows spaced distance 0.25 m. The characteristics of the tested rice varieties were reported by [37] as follows:

V_1 (Giza-178): Pedigree (Giza175/Milyang 49), Properties (Indica/Japonica type—tolerant to drought—medium maturing—high yield), grain yield (10 t ha^{-1}) growth period (135 days)
V_2 (Giza-177): Pedigree (Giza 171/Yomjo No. 1//PiNo.4,) Japonica type—sensitive to drought—short stature—early duration—,grain yield (9 t ha^{-1}) and the growth period (125 days)
V_3 (Sakha-104): Pedigree (GZ4096-8-1/GZ4100-9-1), Properties (Japonica type—sensitive to drought—), grain yield (10 t ha^{-1}) growth period (135 days)

Each sub-subplot (experimental unit) involved 5 rows. Agronomic practices for crop management viz fertilizers, pesticides and herbicides applications were implemented according to the recommended practices described by [38]. Fertilizers were manually broadcast then incorporated within the basal application (35 kg P and 50 kg K per hectare). The experiments were conducted in a randomized complete block design (Split-split Plot) in three replicates including three factors. Using a surface irrigation system, three different irrigation regimes as follows: I_1 (120% of ETc), I_2 (100% of ETc) and I_3 (80% of ETc) were applied and allocated to the main plots and two N fertilization levels N_1 (100% of recommended dose (RD) by 165 kg N ha^{-1}) and N_2 (125% of RD by 200 kg N ha^{-1}), were broadcasted in three equal splits and added at basal, mid-tillering and panicle initiation of rice developing stages and allocated to the sub main plots, where the three lowland rice varieties seeded and allocated to sub-sub main plots. The total experimental area specified for each year was 1134 m^2 divided into 54 experimental plots of 21 m^2 for each. To protect against irrigation and nitrogen fertilizer treatment effects, three meters were utilized to isolate the experimental units.

2.4. Irrigation Water Requirements (IWR)

The IWR was determined according to [39] equation as follows:

$$IWR = \frac{A \times ETc \times Ii}{Ea \times 1000}$$

where, IWR: irrigation water requirements (m^3), A plot area (m^2), ETc: water consumptive use (mm day^{-1}), Ii: intervals between irrigation (day), and Ea: application efficiency (%).

To convey water for each plot plastic pipe (spiles) of 2 inch diameter was used, and the amount of water delivered through a plastic pipe was calculated according to [40].

$$Q = CA\sqrt{2gh} \times 10^{-3}$$

where: Q is the discharge of irrigation water (m^3), C is the coefficient of discharge, A is a cross-sectional area of irrigation pipe (cm^2), g is gravity acceleration ($cm\ s^{-2}$) and h is the average of the effective head of water (cm) above the pipe.

ETc: Water consumptive use (mm day^{-1})

$$ET_c = ET_0 \times K_c$$

where: ETc: crop water consumption (mm d^{-1}), ET_0 is the reference evapotranspiration (mm d^{-1}), and Kc: crop coefficient.

$$ET_0 = E_{pan} \times K_p$$

where E_{pan}: is the evaporation from a class A (mm day^{-1}) and K_p: is the pan coefficient.

2.5. Plant Physiological Measurements

2.5.1. Chlorophyll a Fluorescence (Fv/Fm) and Performance Index (PI) Values

Chlorophyll *a* fluorescence was determined by plant efficiency analyzer, Handy PEA (Hansatech Instruments Ltd., Norfolk, UK). From each plot, ten fully expanded flag leaves were randomly selected, the leaf samples were dark-adapted for 15 min before being illuminated with irradiance intensity of 3000 μmol/(m^2·s) [41,42]. While PI was measured as reported by [43].

2.5.2. Relative Water Content (RWC)

Leaf samples for RWC measurement were randomly collected in the morning (8:00 to 9:00 a.m.). RWC was estimated according to the method described by [44].

$$RWC\ (\%) = (FW - DW) \times 100/(TW - DW)$$

where: FW: Fresh weight was measured within two hours after excision of leaves. Turgid weight (TW) was computed by soaking leaves in distilled water and left at room temperature for (16–18 h) then rapidly and carefully blotted dry by tissue paper to determine turgid weight. The small leaf pieces were later oven-dried for 48 h at 70 °C to estimate the dry weight (DW).

2.5.3. Relative Chlorophyll Content (SPAD)

The SPAD meter (SPAD502, KONICAMINOLTA. Inc., Tokyo, Japan) was used to estimate the relative chlorophyll content of the rice.

2.5.4. Membrane Stability Index (MSI %)

The (MSI %) was measured by the method described by [45]. The small leaf strips (0.2 g) of equal size were prepared and taken in two sets of test tubes containing 10 mL of distilled water. The test tubes arranged in one set were maintained at 40 °C in a water bath for 30 min then the ECe of the water covering the leaf samples was estimated (C_1). While the test tubes of the other set were incubated in a bath of boiling water at 100 °C for 15 min then measured ECe (C_2). The MSI was computed: MSI = $[1 - (C_1/C_2)] \times 100$.

2.6. Growth, Yield and Yield Components Measurements

All the studied three rice varieties were harvested after 135 days from the sowing date. From each plot, ten plants were selected randomly to determine the plant growth and yield components parameters, i.e., tillers and productive tillers number, plant height (from ground level to the tip of panicle, was measured by meter-scale), length and weight of panicle, number and weight of grains per panicle and 100-grain weight. The grain and straw yield were measured by using digital balance from all plants collected from 1m^2 sampling area.

2.7. Grain N Content

Digestion process was performed for the dried grain samples with a mixture consisting of perchloric and nitric acids (at 1:3, *v*/*v*, respectively). Using the previous digestion solution, an assessment of N content was performed. Determination of N was performed using the micro-Kjeldahl apparatus (Ningbo Medical Instruments Co., Ningbo, China) following [46] methods.

2.8. Irrigation Water Productivity (IWP)

IWP was determined according to [47].

$$\text{IWP}\left(\text{Kg m}^{-3}\right) = \frac{\text{grain yield}\left(\text{Kg ha}^{-1}\right)}{\text{irrigation water applied}\left(\text{m}^3\ \text{ha}^{-1}\right)}$$

Crop water productivity (CWP): was computed according to [48].

$$\text{CWP}\left(\text{Kg m}^{-3}\right) = \frac{\text{grain yield}\left(\text{Kg ha}^{-1}\right)}{\text{water consumptive}\left(\text{m}^3\ \text{ha}^{-1}\right)}$$

2.9. Statistical Analysis

The obtained data for each variable were subjected to two-way analysis of variance (ANOVA) using GenStat statistical package (12th Ed., VSN International Ltd., Oxford, UK). In case of significant effects, the treatment means were separated using LSD test at $p \leq 0.05$ probability level.

3. Results and Discussion

3.1. Rice Water Status

Results of plant water status (RWC and MSI) in response to irrigation, nitrogen and variety are displayed in Table 3. In both seasons, the highest MSI, RWC values were recorded when rice was subjected to irrigation at 120% of ETc and received 200 kg N ha^{-1}; on the other hand, the integrative application of irrigation at 80% of ETc and plants received 165 kg N ha^{-1} recorded the lowest values over all varieties.

Regarding the effect of varieties, data enumerated in Table 3 revealed that Giza-178 (V_1) recorded the highest MSI, RWC values while Giza-177 (V_2) gave the lowest values in both seasons. Application of 200 kg N ha^{-1} to drought-stressed plants up to 20% compensated for this shortage of irrigation and recorded similar values to well-irrigated plants ($I_{120\%}$) and received 165 kg N ha^{-1}. RWC and MSI declined remarkably in both $I_{100\%}$ and $I_{80\%}$ treatment compared with well-watered treatment ($I_{120\%}$) [49,50]. We found that RWC and MSI had positive relationships with IWA irrespective of nitrogen applications. The RWC and MSI values were decreased as drought increased. Interestingly, leaf RWC and MSI in higher N treatment were 3.2 and 5.2% higher than that low N treatment, irrespective of the variety effect. In this investigation, N-supply decreased the detrimental effects of water stress on rice plants and kept their RWC and MSI values at close levels as in well-watered plants (Table 3). In the present study, the adverse changes that occurred in the health of cell membranes under drought stress were assessed. Our results revealed that higher N-application plays an important role in stabilizing membrane integrity and maintaining cell turgor of rice leaves under drought stress. In this respect, increases of tissue RWC and MSI as metabolically available water seems to maintain tissue health and may reflect on the metabolic processes in rice under water stress, which agrees with the results of [34,51], who stated that, N supply under drought stress improved RWC and MSI as well as enhanced the photosynthetic efficiency with increased grain yield of wheat plants. Our results agree with those of [48,52]. They reported that drought stress severely affected and reduced rice growth, photosynthesis, stomatal conductance. Ref. [53] reported drought adversely affects physiological responses of plants through a reduction in gas exchange especially stomatal conductance, photosynthetic pigments and overall crop water status.

3.2. Leaf Physiological Traits

For all varieties, physiological traits like relative chlorophyll content (SPAD value), chlorophyll fluorescence (Fv/Fm) and performance index (PI) were significantly improved in drought stress-treated rice plants under high N application (Table 3). Drought stress (80% of ETc) reduced the values of SPAD (18.7%), Fv/Fm (5.5%), and PI (46.1%) as compared to plants irrigated at 120% of ETc (Table 3). All physiological traits were significantly decreased under drought [17,54]. However, rice plants that received high N concentrations (200 kg ha^{-1}) significantly increased SPAD (7.3%), Fv/Fm (3.8%), and PI (45.3%), as compared to low N concentration, over all varieties as average for both seasons. Moreover, results in Table 3 showed that the highest SPAD, Fv/Fm, and PI values were recorded in $Giza_{178}$ (V_1), while the lowest values were recorded under $Giza_{177}$ (V_2) for both seasons. The best results of these traits were recorded in $Giza_{178}$ with application of 200 kg N ha^{-1} + I_{120} of ETc. Our results suggest that the application of high N concentration can mitigate the negative effects of water stress on SPAD, Fv/Fm, and PI and as a result of increasing the photosynthetic efficiency of rice plants. Similar trends were reported by [16,55–58].

3.3. Plant Growth Characteristics

Results presented in Table 4 showed that the applied irrigation regimes, nitrogen fertilization levels and their interaction significantly affected the plant height and the number of effective tillers. Applying water stress level (I 80%) significantly reduced plant height, tillers No, effective tillers no and panicle length by 9.45, 21.54, 21.12 and 10.58% compared by control (I 120%). Increasing the amount of the applied irrigation water and nitrogen dose at ($I_{120} \times N_2$) treatment gave the greatest values of plant height and effective tillers. Tillers number in particular fertile tillers is considered one of the most important components of yield. The observed increase in number of fertile tillers in the current investigation might be related to the higher availability of the nitrogen element that played a vital role in cell division. On the contrary, the lowest estimation for these parameters was observed at ($I_{80} \times N_1$) treatment. Drought stress may cause various structural and functional disruptions in reproductive organs [20]. Among the grown rice varieties ($Giza_{178}$) recorded the maximum plant height (76.14 cm) and effective tillers number (2.75) as compared with other varieties. The increases in both plant height and produced effective tillers under ($I_1 \times N_2$) treatment could be due to the availability of water and nitrogen resulting in better translocation of photosynthesis, higher cell deviation and there by favored highest yield attributes under these treatments. The obtained results were in line with [59] who observed that increasing nitrogen fertilizer resulted in drier matter accumulation. Ref. [60,61] found that the irrigation regime and N application significantly affected rice yield and yield traits. In addition, increasing the N uptake may have beneficial effects on plants grown under drought conditions, where the plant drought resistance increased with increasing N supply resulted in increasing root hydraulic conductivity (Lpr) through increased abscisic acid and aquaporin expression [30,31].

Table 3. Effect of deficit irrigation, nitrogen fertilizer varieties, and their interaction on plant water status (RWC% and MSI %), chlorophyll florescence (Fv/Fm and PI), and leaf chlorophyll content (SPAD) of rice plants in 2017 (S_I) and 2018 (S_{II}) seasons.

Source of Variation	RWC%		MSI%		SPAD		Fv/Fm		PI	
	S_I	S_{II}	S_I	S_{II}	S_I	S_{II}	S_I	S_{II}	S_I	S_{II}
Irrigation (I)	**	**	**	**	**	**	**	**	**	**
$I_{120\%}$	70.20 ± 1.1a	70.41 ± 1.2a	65.70 ± 1.6a	65.65 ± 1.5a	44.64 ± 0.63a	44.60 ± 0.46a	0.82 ± 0.00a	0.83 ± 0.00a	5.50 ± 0.30a	5.60 ± 0.28a
$I_{00\%}$	68.91 ± 1.7b	69.22 ± 1.6b	68.94 ± 2.1a	67.64 ± 1.8a	42.14 ± 0.60b	42.80 ± 0.56b	0.80 ± 0.00b	0.81 ± 0.01b	4.04 ± 0.26b	3.93 ± 0.23b
$I_{80\%}$	61.65 ± 1.4c	62.14 ± 1.3c	61.00 ± 1.6b	60.50 ± 1.7b	36.35 ± 0.82c	36.20 ± 0.72c	0.78 ± 0.01c	0.78 ± 0.01c	2.94 ± 0.35c	3.04 ± 0.32c
Nitrogen (N)	**	**	**	**	**	**	**	**	**	**
N_{165}	65.44 ± 1.2b	66.20 ± 1.1b	62.52 ± 1.4b	63.00 ± 1.3b	39.82 ± 0.69b	39.64 ± 0.72b	0.78 ± 0.00b	0.79 ± 0.00b	3.44 ± 0.27b	3.40 ± 0.26b
N_{200}	68.40 ± 1.1a	68.34 ± 1.0a	66.54 ± 1.4a	66.24 ± 1.6a	42.34 ± 0.59a	42.92 ± 0.65a	0.81 ± 0.00a	0.82 ± 0.01a	4.94 ± 0.25a	5.00 ± 0.30a
Variety (V)	**	**	**	NS	**	NS	NS	NS	**	**
V_1	69.11 ± 1.0a	69.42 ± 1.1a	66.00 ± 1.9a	65.14 ± 1.6a	42.13 ± 0.68a	41.74 ± 0.74a	0.81 ± 0.01a	0.81 ± 0.01a	4.34 ± 0.38a	4.34 ± 0.36a
V_2	66.14 ± 1.2b	66.73 ± 1.3b	65.50 ± 1.8a	65.24 ± 1.4a	40.84 ± 0.88b	41.07 ± 0.95a	0.81 ± 0.00a	0.81 ± 0.01a	4.45 ± 0.39a	4.50 ± 0.33a
V_3	65.50 ± 1.4b	65.64 ± 1.3c	62.14 ± 1.5b	63.40 ± 1.3a	40.20 ± 0.72b	41.15 ± 0.63a	0.80 ± 0.00a	0.80 ± 0.01a	3.80 ± 0.27b	3.81 ± 0.31b
I × N	**	**	NS	NS	**	**	**	**	**	**
I × V	**	**	NS	*	NS	**	**	**	**	NS
V × N	**	**	*	NS	**	**	NS	NS	NS	**
I × N × V	NS	**	*	**	NS	NS	NS	**	**	**

** and * refer to the significant difference at $p \leq 0.01$ and $p \leq 0.05$, respectively; and "ns" refers to no significant difference. Different letters next to mean values in each column indicate significant differences according to the LSD test ($p < 0.05$).

Table 4. Effect of deficit irrigation, nitrogen fertilizers varieties, and their interaction on growth characteristics of rice plants in 2017 (S_I) and 2018 (S_{II}) seasons.

Source of Variation	Plant Height (cm)		Tillers No		Effective Tillers No		Panicle Length (cm)	
	S_I	S_{II}	S_I	S_{II}	S_I	S_{II}	S_I	S_{II}
Irrigation (I)	**	**	**	*	*	*	**	**
I_{120}	74.73 ± 0.87a	75.30 ± 0.67a	3.11 ± 0.08a	3.14 ± 0.09a	2.95 ± 0.09a	2.78 ± 0.08a	20.46 ± 0.21a	20.58 ± 0.18a
I_{00}	71.87 ± 0.65b	72.26 ± 0.62b	2.56 ± 0.09b	2.79 ± 0.08b	2.45 ± 0.09b	2.54 ± 0.08b	19.41 ± 0.24b	19.38 ± 0.21b
I_{80}	67.84 ± 0.67c	68.00 ± 0.63c	2.44 ± 0.08c	2.46 ± 0.08c	2.35 ± 0.09b	2.17 ± 0.09c	18.36 ± 0.23c	18.34 ± 0.20c
Nitrogen (N)	**	**	**	NS	**	NS	**	**
N_{165}	69.92 ± 0.67b	70.01 ± 0.55a	2.54 ± 0.08b	2.81 ± 0.08a	2.44 ± 0.08b	2.49 ± 0.09a	18.87 ± 0.23b	18.89 ± 0.22b
N_{200}	73.04 ± 0.70a	73.69 ± 0.59a	2.85 ± 0.08a	2.80 ± 0.09a	2.72 ± 0.08a	2.51 ± 0.07a	19.95 ± 0.19a	19.97 ± 0.17a
Variety (V)	**	**	NS	NS	NS	*	**	**
V_1	75.79 ± 0.77a	76.49 ± 0.73a	2.86 ± 0.09a	3.01 ± 0.09a	2.70 ± 0.08a	2.80 ± 0.06a	20.57 ± 0.21a	20.56 ± 0.19a
V_2	67.31 ± 0.57c	67.47 ± 0.65b	2.65 ± 0.09b	2.76 ± 0.09b	2.59 ± 0.10b	2.44 ± 0.06b	18.54 ± 0.23c	18.57 ± 0.24c
V_3	71.34 ± 0.57b	71.59 ± 0.59b	2.59 ± 0.10b	2.64 ± 0.09b	2.44 ± 0.09c	2.26 ± 0.07b	19.11 ± 0.20b	19.14 ± 0.22b
I×N	**	**	**	**	*	*	**	**
I×V	**	**	**	*	**	**	**	**
V×N	**	**	NS	*	*	*	**	**
I×N×V	NS	NS	*	*	NS	NS	*	*

** and * refer to the significant difference at $p \leq 0.01$ and $p \leq 0.05$, respectively; and "ns" refers to no significant difference. Different letters next to mean values in each column indicate significant differences according to the LSD test ($p < 0.05$).

3.4. Yield Components

Yield components of, i.e., panicle length (cm), grains number per panicle, panicle weight, weight of grain per panicle (g) and weight of 100 grain (g), were significantly affected by irrigation and nitrogen nutrition treatments. Results in Table 5 showed that rice variety $Giza_{178}$ yielded the highest values of these traits under irrigation and nutrition level ($I_{120} \times N_2$), while the lowest values were yielded by $Giza_{177}$ under ($I_{80} \times N_1$). Among irrigation regimes (I_{120}) produced the highest averages of these traits, that gradually reduced by increasing drought level. The reduction percent amounted to 5.48, 7.78, 13.63 18.41 and 11.07% respectively, as compared with those obtained at moderate stress level (I_{100}) and by 10.57, 17.83, 29.66 32.61and 18.38% respectively compared to (I_{80}). The obtained results were in line with those reported by [52] who found that yield components of rice crop (panicle number, panicle length, 100-grain weight and grain yield) decreased with increasing water stress especially if it occurred at the stage of panicle initiation Ref. [17,62] noted that depending upon the stress severity the plant physiological responses the apical development rate, biomass production, panicle development, spikelet number and grain yield were decreased.

However, applying nitrogen fertilization at N_2 resulted in an increase in panicle length (cm), grains number per panicle, panicle weight, weight of grain per panicle (g) and weight of 100 grain (g) by 5.72, 10.81, 12.85, 11.91 and 5.21% as compared to N_1. Inter varietal comparison showed that, yield component traits were significantly differed in both seasons Results are in agreement with those revealed by [63,64] who noted that the application of N fertilizers significantly increased the yield and yield components of rice.

3.5. Grain, Straw and Biological Yield

Results presented in Table 6 show that, grain, straw, biological yield and grains N content were varied significantly among the grown rice varieties as affected by both irrigation and nitrogen nitration management. As average (2017–2018) rice variety $Giza_{17}$ gave the highest grain, straw and biological yield 7.97, 12.23 and 20.19 t ha^{-1} respectively while $Giza_{177}$ recorded the lowest yields (5.78, 8.84 and 14.61 t ha^{-1} respectively). As presented above in (Tables 3–5) the higher application of irrigation water (I_{120}) improved the plant water statues, enhanced growth and development of rice plants that contributed to achieve the maximum values for grain yield (8.11 t ha^{-1}), straw yield (11.89 t ha^{-1}) and biological yield (20.00 t ha^{-1}). Furthermore, comparing with irrigation treatment (I_{120})

moderate irrigation regime (I_{100}) reduced the grain, straw and biological yield of grown lowland rice varieties. The reductions were by 10.98, 5.55 and 7.75%, respectively, and by 34.18, 26.53 and 29.63%, respectively, under the application of deficit irrigation treatment (I_{80}). A similar trend was observed by [62,65–67].

Table 5. Effect of deficit irrigation, nitrogen fertilizer varieties, and their interaction on yield component of rice in 2017 (S_I) and 2018 (S_{II}) seasons.

Source of Variation	Grains No/Panicle		Panicle Weight		Weight of Grains/Panicle		W of 100 g	
	S_I	S_{II}	S_I	S_{II}	S_I	S_{II}	S_I	S_{II}
Irrigation (I)	**	**	**	**	**	**	**	**
I_{120}	93.56 ± 1.7a	94.45 ± 1.6a	2.50 ± 0.03a	2.49 ± 0.09a	2.24 ± 0.03a	2.27 ± 0.13a	2.39 ± 0.29a	2.40 ± 0.28a
I_{00}	86.34 ± 1.9b	87.05 ± 1.8b	2.15 ± 0.06b	2.16 ± 0.10b	1.80 ± 0.06b	1.88 ± 0.16b	2.10 ± 0.47b	2.16 ± 0.36b
I_{80}	77.16 ± 1.5c	78.18 ± 1.6c	1.75 ± 0.07c	1.76 ± 0.06c	1.46 ± 0.06c	1.58 ± 0.09c	1.89 ± 0.49c	2.02 ± 0.45c
Nitrogen (N)	**	**	**	**	**	**	**	**
N_{165}	82.99 ± 1.6b	83.63 ± 1.5b	1.98 ± 0.06b	2.00 ± 0.06b	1.72 ± 0.06b	1.81 ± 0.10b	2.07 ± 0.45b	2.16 ± 0.39b
N_{200}	88.39 ± 1.6a	89.47 ± 1.7a	2.27 ± 0.06a	2.27 ± 0.12a	1.93 ± 0.06a	2.02 ± 0.16a	2.19 ± 0.49a	2.26 ± 0.38a
Variety (V)	**	**	**	**	**	**	**	**
V_1	97.20 ± 1.6a	95.85 ± 1.8a	2.46 ± 0.04a	2.48 ± 0.11a	2.20 ± 0.04a	2.25 ± 0.08a	2.27 ± 0.26a	2.34 ± 0.36a
V_2	77.00 ± 1.2c	79.01 ± 1.3c	1.81 ± 0.07c	1.80 ± 0.09c	1.41 ± 0.07c	1.54 ± 0.11c	1.84 ± 0.65c	1.95 ± 0.54c
V_3	82.87 ± 1.3b	84.80 ± 1.3b	2.11 ± 0.07b	2.13 ± 0.08b	1.88 ± 0.12b	1.94 ± 0.09b	2.27 ± 0.67b	2.29 ± 0.57b
I × N	**	**	**	*	*	**	NS	**
I × V	**	**	**	**	**	**	**	**
V × N	**	**	NS	**	**	*	NS	**
I × N × V	NS	NS	**	**	*	**	*	**

** and * refer to the significant difference at $p \leq 0.01$ and $p \leq 0.05$, respectively; and "ns" refers to no significant difference. Different letters next to mean values in each column indicate significant differences according to the LSD test ($p < 0.05$).

However, nitrogen application gave an appositive effect and improved the productivity of all grown rice varieties. Data in Table 6 showed that increasing the nitrogen fertilization dose from N_1 to N_2 resulted in an increase in grain, straw and biological yield by 12.66, 18.20 and 15.99%, respectively. Nitrogen application could produce promoted root growth, enhanced water and nitrogen extraction from soil, resulting in better crop growth and higher yield productivity. Similar trend was noted by [68] who observed an increase in grain yield with the increasing of nitrogen application levels. The combined effect of (I × N) on grain and straw was found to be significant during both seasons except their effect on straw yield in the second year was non-significant. The effect of (V × N) during two seasons was non-significant. Meanwhile, the effect of (I × V × N) was significant.

3.6. Irrigation Water Applied and Water Productivity

Results in Table 7 show that total water applied was varied between the applied irrigation treatments. The lowest amount of irrigation water (646 mm ha^{-1}) required was at (I_{80}), while the highest (970 mm ha^{-1}) was needed for (I_{120}). According to the grain yield obtained under each watering treatment, the applied irrigation water and crop water consumption, irrigation water productivity (IWP) and crop water productivity (CWP) were significantly differed ($p \leq 0.05$) in both growing seasons. Among irrigation treatments watering at (I_{100}) gave the highest values of IWP and CWP by 0.89 and 1.28 (kg m^3), respectively, when comparing with other irrigation treatments. In addition, increasing nitrogen application dose up to N_2 resulted in higher IWP and CWP by 12.50 and 12.17%, respectively, than N_1 (as average for both seasons). Between rice varieties, $Giza_{179}$ corresponding to its high grain productivity which resulted in the highest value as averages of IWP (0.99 Kg m^{-3}) and CWP (1.41 kg m^{-3}) meanwhile, the lowest IWP and CWP amounted 0.71 and 1.02 kg m^{-3}, respectively, noted for $Giza_{177}$.

Table 6. Effect of deficit irrigation, nitrogen fertilizers varieties, and their interaction on grain yield, straw yield, biological yield, and grains N content of rice plants in 2017 (S_I) and 2018 (S_{II}) seasons.

Source of Variation	Grain Yield (t ha^{-1})		Straw Yield (t ha^{-1})		Biological Yield (t ha^{-1})		N (%)	
	S_I	S_{II}	S_I	S_{II}	S_I	S_{II}	S_I	S_{II}

Table 7. Effect of deficit irrigation, nitrogen fertilizer varieties, and their interaction on grain yield, straw yield, biological yield, irrigation water productivity and crop water productivity of rice plants in 2017 (S_I) and 2018 (S_{II}) seasons.

Source of Variation	Irrigation Water Applied ($m^3\ ha^{-1}$)		Irrigation Water Productivity ($kg\ m^{-3}$)		Crop Water Productivity ($kg\ m^{-3}$)	
	S_I	S_{II}	S_I	S_{II}	S_I	S_{II}
Irrigation (I)			**	**	**	**
$I_{120\%}$	9661	9733	0.83 ± 0.01b	0.84 ± 0.01b	1.19 ± 0.09b	1.20 ± 0.12b
$I_{00\%}$	8051	8111	0.89 ± 0.02a	0.90 ± 0.02a	1.27 ± 0.08a	1.29 ± 0.09a
$I_{80\%}$	6441	6489	0.82 ± 0.01b	0.83 ± 0.01b	1.17 ± 0.11c	1.19 ± 0.08b
Nitrogen (N)			**	**	**	**
N_{165}	8051	8111	0.79 ± 0.01b	0.81 ± 0.01b	1.13 ± 0.09c	1.16 ± 0.08c
N_{200}	8051	8111	0.90 ± 0.02a	0.90 ± 0.02a	1.29 ± 0.06a	1.29 ± 0.07a
Variety (V)			**	**	**	**
V1	8051	8111	0.98 ± 0.02a	0.99 ± 0.02a	1.40 ± 0.11a	1.42 ± 0.09a
V2	8051	8111	0.72 ± 0.00c	0.71 ± 0.01c	1.03 ± 0.09c	1.02 ± 0.0c
V3	8051	8111	0.84 ± 0.01b	0.87 ± 0.02b	1.20 ± 0.08b	1.24 ± 0.10b
I × N			**	**	**	**
I × V			**	**	**	**
V × N			NS	NS	*	**
I × N × V			NS	NS	NS	NS

** and * refer to the significant difference at $p \leq 0.01$ and $p \leq 0.05$, respectively; and "ns" refers to no significant difference. Different letters next to mean values in each column indicate significant difference according to the LSD test ($p < 0.05$).

Many studies have been summarized that rice cultivation under flooded condition consumes approximately 20,000 $m^3\ ha^{-1}$ [7]. In North Delta, Egypt [69] estimated water requirements for flooded-rice by 19,000 $m^3\ ha^{-1}$ and water utilization efficiency by 48%. Similar results reported by [70] who found that rice normally needs under traditional methods in Egypt a water application of about 20,000 $m^3\ ha^{-1}$. Studies conducted by [11] concluded that the water requirement of paddy fields in Egypt is about 1800–2200 mm ha^{-1}. Therefore, compared with previous studies, the current investigation aimed to create new environmental cultivation conditions to grow lowland rice varieties (aerobically) in non-puddled fields under non-flooded conditions and maintain a profitable grain yield could save irrigation water by 60%. Moreover, results concluded that lowland rice variety $Giza_{178}$ was observed to be more tolerant than other varieties, $Giza_{177}$ and $Sakha_{104}$, while v. $Giza_{177}$ was much affected by deficiency of irrigation water. A similar trend was reported by [3].

3.7. Correlation Analysis

Results in Tables 8 and 9 illustrated the correlation coefficients between rice grain yield and the other yield components. This type of examination could be used as an appropriate instrument to indicate which one of them is positive and greatly associated with the obtained yield of grains.

Table 8. A matrix of simple correlation coefficients between grain yield and other important traits estimated for 2017 and 2018 seasons.

	Parameter	Season	1	2	3	4	5	6	7	8	9	10
1	Grain yield	2017	1									
		2018	1									
2	Plant height	2017	0.920 **	1								
		2018	0.921 **	1								
3	Branch No	2017	0.486 **	0.469 **	1							
		2018	0.361 **	0.308 **	1							
4	No spike	2017	0.405 **	0.396 **	0.712 **	1						
		2018	0.377 **	0.313 **	0.568 **	1						
5	Spike length	2017	0.827 **	0.815 **	0.535 **	0.455 **	1					
		2018	0.819 **	0.775 **	0.306 **	0.349 **	1					
6	Grain No	2017	0.903 **	0.879 **	0.430 **	0.380 **	0.844 **	1				
		2018	0.894 **	0.849 **	0.268*	0.294 **	0.736 **	1				
7	Spike weight	2017	0.915 **	0.866 **	0.541 **	0.454 **	0.841 **	0.872 **	1			
		2018	0.899 **	0854 **	0.393 **	0.376 **	0.781 **	0.812 **	1			

Table 9. Correlation coefficient (r), coefficient of determination (R^2) and standard error of the estimates (SEE) for predicting grain yield for S_I (2017) and S_{II} (2018) seasons.

Season	r	R^2	SEE	Significance	Fitted Equation
2017	0.961	0.924	0.33	***	Grain yield = −4.1 + 0.09 plant height + 1.69 spike weight + 0.179 skillets no − 0.074 weight of 1000 grain
2018	0.961	0.924	0.33	***	Grain yield = −6.18+ 0.08 plant height + 0.72 spike weight + 0.031 grain no + 0.104 spike length

*** indicate correlation is significant at the 0.01 level.

It was observed that grain yield was positive and strongly correlated with plant height, spike length, grain no/spike then followed by spike weight, which validates their economic importance.

Data in Table 9 show that plant height and spike weight in both seasons were significantly ($p \leq 0.01$) contributed to variations in rice grain yield.

4. Conclusions

Cultivating rice aerobically under deficit irrigation has been observed to be used as an effective and efficient approach in saving irrigation water and improving IWP compared with the flooding system. Additionally, improving nitrogen supply enhanced rice crop productivity under deficit irrigation conditions. The combined effect of ($I_{120} \times N_2$) treatment achieved the greatest values of all agronomic and physiological traits, grain and straw yield of rice. Applying of irrigation water at (I_2) produced the maximum (IWP) and (CWP) by 0.89 and 1.28 kg m^{-3}, respectively, compared with other irrigation schedules. Rice variety ($Giza_{178}$, V1), gave the highest CWP and grain yield which amounted to 1.41 and 7.97, respectively. Consequently, it could be a suitable genotype for improving rice productivity under drought conditions as compared with other tested rice varieties. Thus, based on the results of the current study the incorporating deficit irrigation technique at level (100 of ETc) with nitrogen application at 120% of recommended dose could be a valuable agro-management strategy for maintaining relatively high yields of some sensitive-drought Egyptian rice varieties and saving irrigation water supply by 50–60% compared to the conventional rice cultivation method.

Author Contributions: Conceptualization, N.M.A., M.O.A.R. and T.A.A.E.-M.; methodology, N.M.A., M.O.A.R., W.M.S. and T.A.A.E.-M.; formal analysis, T.A.A.E.-M., M.O.A.R., N.M.A. and W.M.S.; investigation, T.A.A.E.-M., W.M.S., N.M.A. and M.O.A.R.; resources N.M.A., T.A.A.E.-M., W.M.S., S.A.A.E.-M., M.A.A.-R., E.F.A., A.M. and M.O.A.R.; data curation, N.M.A., M.O.A.R., T.A.A.E.-M.; E.F.A., A.M. and M.A.A.-R.; writing—original draft preparation, N.M.A., M.O.A.R. and T.A.A.E.-M.; and S.A.A.E.-M.; writing—review and editing, N.M.A., T.A.A.E.-M., W.M.S., A.A.A.L.; visualization, N.M.A., M.O.A.R., T.A.A.E.-M.; W.M.S., E.F.A., A.M., S.A.A.E.-M. and M.A.A.-R.; supervision, N.M.A., T.A.A.E.-M. and M.O.A.R. project administration, N.M.A., M.O.A.R. and T.A.A.E.-M.; All authors have read and agreed to the published version of the manuscript.

Funding: This research received no external funding.

Institutional Review Board Statement: Not applicable.

Informed Consent Statement: Not applicable.

Data Availability Statement: The data presented in this study are available upon request from the corresponding author.

Acknowledgments: The authors are thankful to Taif University Researchers Supporting Project number (TURSP-2020/110), Taif University, Saudi Arabia, for the financial support and research facilities.

Conflicts of Interest: The authors declare no conflict of interest.

References

1. Manjappa, G.U.; Shailaja, H. Identification of drought tolerant and high yielding genotypes of rice under aerobic condition. *Oryz* **2014**, *51*, 273–278.
2. FAOSTAT. *Food and Agriculture Data*; Food and Agriculture Organization: Rome, Italy, 2019.
3. Tantawi, A.; Ghanem, S.A. Water use efficiency in rice culture. In *Future of Water Management for Rice in Mediterranean Climat Areas: Proceedings of the Workshops*; CIHEAM: Montpellier, France, 1999; pp. 39–45.
4. Bouman, B.A.M. *Water Management in Irrigated Rice: Coping with Water Scarcity*; Internationa Rice Research Institute: Los Baños Philippines, 2007.
5. Brown, K.W.; Turner, F.T.; Thomas, J.C.; Deuel, L.E.; Keener, M.E. Water balance of flooded rice paddies. *Agric. Water Manag* **1977**, *1*, 277–291. [CrossRef]
6. McCauley, G.N. Sprinkler vs. flood irrigation in traditional rice production regions of southeast Texas. *Agron. J.* **1990**, *82* 677–683. [CrossRef]
7. Tuong, T.P.; Bam, B.; Mortimer, M. More Rice, Less Water—Integrated Approaches for Increasing Water Productivity in Irrigatec Rice-Based Systems in Asia. *Plant Prod. Sci.* **2005**, *3*, 231–241. [CrossRef]
8. WWAP, U. *World Water Assessment Programme: The United Nations World Water Development Report 4: Managing Water unde Uncertainty and Risk*; UNESCO: Paris, France, 2012.
9. Saxena, R.; Vanga, S.K.; Wang, J.; Orsat, V.; Raghavan, V. Millets for food security in the context of climate change: A review *Sustainability* **2018**, *10*, 2228. [CrossRef]
10. Abd El-Mageed, T.A.; Rady, M.O.A.; Semida, W.M.; Shaaban, A.; Mekdad, A.A.A. Exogenous Micronutrients Modulate Morpho physiological Attributes, Yield, and Sugar Quality in Two Salt-Stressed Sugar Beet Cultivars. *J. Soil Sci. Plant Nutr.* **2021** 1–16. [CrossRef]
11. Watanabe, T. Irrigation water requirement. In *The Japanese Society of Irrigation, Drainage and Reclamation Engineering (JSIDRE Editor*; Advanced Paddy Field Engineering; Shinzansha Sci. & Tech.: Tokyo, Japan, 1995; pp. 31–50.
12. English, M.J.; Nuss, G.S. Designing for deficit irrigation. *J. Irrig. Drain. Div.* **1982**, *108*, 91–106. [CrossRef]
13. Abd El–Mageed, T.A.; El-Sherif, A.M.A.; Ali, M.M.; Abd El-Wahed, M.H. Combined effect of deficit irrigation and potassium fertilizer on physiological response, plant water status and yield of soybean in calcareous soil. *Arch. Agron. Soil Sci.* **2017**, *63* 827–840. [CrossRef]
14. Rady, M.O.A.; Semida, W.M.; Howladar, S.M.; Abd El-Mageed, T.A. Raised beds modulate physiological responses, yield and water use efficiency of wheat (*Triticum aestivum* L.) under deficit irrigation. *Agric. Water Manag.* **2021**, *245*, 106629. [CrossRef]
15. Moustafa, E.S.A.; El-Sobky, E.-S.E.A.; Farag, H.I.A.; Yasin, M.A.T.; Attia, A.; Rady, M.O.A.; Awad, M.F.; Mansour, E. Sowing Date and Genotype Influence on Yield and Quality of Dual-Purpose Barley in a Salt-Affected Arid Region. *Agronomy* **2021** *11*, 717. [CrossRef]
16. Abd El-Mageed, T.A.; Semida, W.M.; Rady, M.M. Moringa leaf extract as biostimulant improves water use efficiency, physic biochemical attributes of squash plants under deficit irrigation. *Agric. Water Manag.* **2017**, *193*, 46–54. [CrossRef]
17. Yang, X.; Wang, B.; Chen, L.; Li, P.; Cao, C. The different influences of drought stress at the flowering stage on rice physiologica traits, grain yield, and quality. *Sci. Rep.* **2019**, *9*, 1–12. [CrossRef]
18. Palanog, A.D.; Swamy, B.M.; Shamsudin, N.A.A.; Dixit, S.; Hernandez, J.E.; Boromeo, T.H.; Kumar, A. Grain yield QTLs with consistent-effect under reproductive-stage drought stress in rice. *F. Crop. Res.* **2014**, *161*, 46–54. [CrossRef]
19. Saud, S.; Fahad, S.; Yajun, C.; Ihsan, M.Z.; Hammad, H.M.; Nasim, W.; Alharby, H. Effects of nitrogen supply on water stress and recovery mechanisms in Kentucky bluegrass plants. *Front. Plant Sci.* **2017**, *8*, 983. [CrossRef] [PubMed]
20. Noor, M.A. Nitrogen management and regulation for optimum NUE in maize–A mini review. *Cogent Food Agric.* **2017** *3*, 1348214. [CrossRef]
21. Ghoneim, A.M.; Gewaily, E.E.; Osman, M.M. Effects of nitrogen levels on growth, yield and nitrogen use efficiency of some newly released Egyptian rice genotypes. *Open Agric.* **2018**, *3*, 310–318. [CrossRef]
22. Yoshida, S. Physiological aspects of grain yield. *Annu. Rev. Plant Physiol.* **1972**, *23*, 437–464. [CrossRef]
23. Ebaid, R.A.; Ghanem, S.A. Productivity of Giza 177 rice variety grown after different winter crops and fertilized with differen nitrogen levels. *Egypt. J. Agric. Res.* **2000**, *78*, 717–731.
24. El-Batal, M.A.; Abdel-Gawad, M.H.; Abdou, F.A.; Abdel-Aziz, E.A. Uniconazole application as antilodging for rice plants fertilized with high nitrogen rate. *Zagazig J. Agric. Res.* **2004**, *31*, 473–490.
25. Filgueiras, L.; Silva, R.; Almeida, I.; Vidal, M.; Baldani, J.I.; Meneses, C.H.S.G. Gluconacetobacter diazotrophicus mitigates drought stress in *Oryza sativa* L. *Plant Soil* **2020**, *451*, 57–73. [CrossRef]
26. Yoseftabar, S. Effect nitrogen management on panicle structure and yield in rice (Oryza sativa L.). *Int. J. Agric. Crop Sci.* **2013** *5*, 1224–1227.
27. Otoo, E.; Ishii, R.; Kumura, A. Interaction of nitrogen supply and soil water stress on photosynthesis and transpiration in rice *Jpn. J. Crop Sci.* **1989**, *58*, 424–429. [CrossRef]
28. Castillo, E.G.; Buresh, R.J.; Ingram, K.T. Lowland rice yield as affected by timing of water deficit and nitrogen fertilization. *Agron J.* **1992**, *84*, 152–159. [CrossRef]
29. Shangguan, Z.P.; Shao, M.A.; Dyckmans, J. Nitrogen nutrition and water stress effects on leaf photosynthetic gas exchange and water use efficiency in winter wheat. *Environ. Exp. Bot.* **2000**, *44*, 141–149. [CrossRef]

30. Ding, L.; Lu, Z.; Gao, L.; Guo, S.; Shen, Q. Is nitrogen a key determinant of water transport and photosynthesis in higher plants upon drought stress? *Front. Plant Sci.* **2018**, *9*, 1143. [CrossRef] [PubMed]
31. Guo, S.; Chen, G.; Zhou, Y.; Shen, Q. Ammonium nutrition increases photosynthesis rate under water stress at early development stage of rice (*Oryza sativa* L.). *Plant Soil* **2007**, *296*, 115–124. [CrossRef]
32. Shatil-Cohen, A.; Attia, Z.; Moshelion, M. Bundle-sheath cell regulation of xylem-mesophyll water transport via aquaporins under drought stress: A target of xylem-borne ABA? *Plant J.* **2011**, *6*, 72–80. [CrossRef] [PubMed]
33. Ren, B.; Wang, M.; Chen, Y.; Sun, G.; Li, Y.; Shen, Q.; Guo, S. Water absorption is affected by the nitrogen supply to rice plants. *Plant Soil* **2015**, *396*, 397–410. [CrossRef]
34. Agami, R.A.; Alamri, S.A.M.; Abd El-Mageed, T.A.; Abousekken, M.S.M.; Hashem, M. Role of exogenous nitrogen supply in alleviating the deficit irrigation stress in wheat plants. *Agric. Water Manag.* **2018**, *210*, 261–270. [CrossRef]
35. Klute, A.; Dirksen, C. Hydraulic conductivity and diffusivity. Laboratory methods. *Methods Soil Anal. Part 1. Phys. Mineral. Methods* **1986**, *9*, 687–734.
36. Page, A.I.; Miller, R.H.; Keeny, D.R. Methods of Soil Analysis. In *Part II. Chemical and Microbiological Methods*, 2nd ed.; American Society of Agronomy: Madison, WI, USA, 1982; pp. 225–246.
37. Gaballah, M.M.; Metwally, A.M.; Skalicky, M.; Hassan, M.M.; Brestic, M.; El Sabagh, A.; Fayed, A.M. Genetic Diversity of Selected Rice Genotypes under Water Stress Conditions. *Plants* **2021**, *10*, 27. [CrossRef]
38. Sariam, O.; Azmi, M.; Chan, C.S.; Badrulhadza, A.; Mohd Khusairy, K.; Mohd Fitri, M.; Alicia, J.A.; Rosnani, H.; Shamsul, S.A. *Aerobic Rice Manual*; MARDI Press: Kuala Lumpur, Malaysia, 2015.
39. Allen, R.G.; Pereira, L.S.; Raes, D.; Smith, M. Crop evapotranspiration: Guidelines for computing crop requirements. *Irrig. Drain. Pap. No. 56 FAO* **1998**, *300*, D05109.
40. Israelsen, O.W.; Hansen, V.C. *Irrigation Principles and Practices*; John Wiley & Sons Inc.: New York, NY, USA, 1962.
41. Maxwell, K.; Johnson, G.N. Chlorophyll fluorescence—A practical guide. *J. Exp. Bot.* **2000**, *51*, 659–668. [CrossRef]
42. Spoustová, P.; Synková, H.; Valcke, R.; Čevrovská, N. Chlorophyll a fluorescence as a tool for a study of the Potato virus Y effects on photosynthesis of nontransgenic and transgenic Pssu-ipt tobacco. *Photosynthetica* **2013**, *51*, 191–201. [CrossRef]
43. Clark, A.J.J.; Landolt, W.; Bucher, J.B.B.; Strasser, R.J.J. Beech (Fagus sylvatica) response to ozone exposure assessed with a chlorophyll a fluorescence performance index. *Environ. Pollut.* **2000**, *109*, 501–507. [CrossRef]
44. Semida, W.M.; Abd El-mageed, T.A.; Abdelkhalik, A.; Hemida, K.A.; Abdurrahman, H.A.; Howladar, S.M.; Leilah, A.A.A.; Rady, M.O.A. Selenium Modulates Antioxidant Activity, Osmoprotectants, and Photosynthetic Efficiency of Onion under Saline. *Agronomy* **2021**, *11*, 855. [CrossRef]
45. Rady, M.M.; Taha, R.S.; Semida, W.M.; Alharby, H.F. Modulation of salt stress effects on vicia faba l. plants grown on a reclaimed-saline soil by salicylic acid application. *Rom. Agric. Res.* **2017**, *34*, 175–185.
46. AOAC. *Official Methods of Analysis*, 17th ed.; The Association of Official Analytical Chemists: Gaithersburg, MD, USA, 2000; ISBN 0935584676.
47. Michael, A.M. *Irrigation Theory and Practice-2Nd Edn: Theory and Practice*; Vikas Publishing House: Delhi, India, 2009.
48. Quampah, A.; Ren-Ming, W.; Shamsi, I.H.; Jilani, G.; Zhang, Q.; Hua, S.; Xu, H. Improving water productivity by potassium application in various rice genotypes. *Int. J. Agric. Biol.* **2011**, *13*, 9–17.
49. Abd El-Mageed, T.A.; El-Sherif, A.M.A.; Abd El-Mageed, S.A.; Abdou, N.M. A novel compost alleviate drought stress for sugar beet production grown in Cd-contaminated saline soil. *Agric. Water Manag.* **2019**, *226*, 105831. [CrossRef]
50. Desoky, E.M.; Elrys, A.S.; Mansour, E.; Eid, R.S.M.; Selem, E.; Rady, M.M.; Ali, E.F.; Mersal, G.A.M.; Semida, W.M. Application of biostimulants promotes growth and productivity by fortifying the antioxidant machinery and suppressing oxidative stress in faba bean under various abiotic stresses. *Sci. Hortic.* **2021**, *288*, 110340. [CrossRef]
51. Abid, M.; Tian, Z.; Ata-Ul-Karim, S.T.; Cui, Y.; Liu, Y.; Zahoor, R.; Dai, T. Nitrogen nutrition improves the potential of wheat (*Triticum aestivum* L.) to alleviate the effects of drought stress during vegetative growth periods. *Front. Plant Sci.* **2016**, *7*, 981. [CrossRef]
52. Zaman, N.K.; Abdullah, M.Y.; Othman, S.; Zaman, N.K. Growth and physiological performance of aerobic and lowland rice as affected by water stress at selected growth stages. *Rice Sci.* **2018**, *25*, 82–93. [CrossRef]
53. Anjum, S.A.; Xie, X.; Wang, L.; Saleem, M.F.; Man, C.; Lei, W. Morphological, physiological and biochemical responses of plants to drought stress. *Afr. J. Agric. Res.* **2011**, *6*, 2026–2032. [CrossRef]
54. Badal, E.; El-Mageed, T.A.A.A.; Buesa, I.; Guerra, D.; Bonet, L.; Intrigliolo, D.S. Moderate plant water stress reduces fruit drop of "Rojo Brillante" persimmon (Diospyros kaki) in a Mediterranean climate. *Agric. Water Manag.* **2013**, *119*, 154–160. [CrossRef]
55. Belal, H.E.E.; Rady, M.M.; Gadallah, F.M.; Semida, W.M. Application in two methods elevates drought tolerance by increasing antioxidant defense system and reducing oxidative stress biomarkers in solanum lycopersicum plant. *Fayoum J. Agric. Res. Dev.* **2020**, *34*, 130–147. [CrossRef]
56. Abd El-mageed, T.A.; Abdelkhalik, A.; Abd El-mageed, S.A.; Semida, W.M. Co-composted Poultry Litter Biochar Enhanced Soil Quality and Eggplant Productivity Under Different Irrigation Regimes. *J. Soil Sci. Plant Nutr.* **2021**, 1–17. [CrossRef]
57. Abd El-mageed, T.A.; Rady, M.M.; Taha, R.S.; Abd El Azeam, S.; Simpson, C.R.; Semida, W.M. Effects of integrated use of residual sulfur-enhanced biochar with effective microorganisms on soil properties, plant growth and short-term productivity of Capsicum annuum under salt stress. *Sci. Hortic.* **2020**, *261*, 108930. [CrossRef]

58. Semida, W.M.; El-mageed, T.A.A.; Abdalla, R.M.; Hemida, K.A.; Howladar, S.M.; Leilah, A.A.A.; Rady, M.O.A. Sequential Antioxidants Foliar Application Can Alleviate Negative Consequences of Salinity Stress in *Vicia faba* L. *Plants* **2021**, *10*, 914. [CrossRef]
59. Parasuraman, P. Integrated nitrogen management in rice (Oryza sativa) through split application of fertilizers at different levels with green manure. *Mysore J. Agric. Sci.* **2005**, *39*, 90–94.
60. Sandhu, S.S.; Mahal, S.S.; Vashist, K.K.; Buttar, G.S.; Brar, A.S.; Singh, M. Crop and water productivity of bed transplanted rice as influenced by various levels of nitrogen and irrigation in northwest India. *Agric. Water Manag.* **2012**, *104*, 32–39. [CrossRef]
61. Jiao, J.; Shi, K.; Li, P.; Sun, Z.; Chang, D.; Shen, X.; Xu, L. Assessing of an irrigation and fertilization practice for improving rice production in the Taihu Lake region (China). *Agric. Water Manag.* **2018**, *201*, 91–98. [CrossRef]
62. Kruzhilin, I.P.; Doubenok, N.N.; Ganiev, M.A.; Abdou, N.M.; Melikhov, V.V.; Bolotin, A.G.; Rodin, K.A. Water-saving technology of drip irrigated aerobic rice cultivation. *ISVESTIYA* **2015**, *3*, 47–56.
63. Chaturvedi, I. Effect of nitrogen fertilizers on growth, yield and quality of hybrid rice (*Oryza sativa*). *J. Cent. Eur. Agric.* **2005**, *6*, 611–618.
64. Singh, H.; Verma, A.; Ansari, M.W.; Shukla, A. Physiological response of rice (*Oryza sativa* L.) genotypes to elevated nitrogen applied under field conditions. *Plant. Signal. Behav.* **2014**, *9*, 1–8. [CrossRef]
65. Semida, W.M.; El-Mageed, T.A.A.; Mohamed, S.E.; El-Sawah, N.A. Combined effect of deficit irrigation and foliar-applied salicylic acid on physiological responses, yield, and water-use efficiency of onion plants in saline calcareous soil. *Arch. Agron. Soil Sci.* **2017**, *63*, 1227–1239. [CrossRef]
66. Abd El-Mageed, T.A.; Semida, W.M. Effect of deficit irrigation and growing seasons on plant water status, fruit yield and water use efficiency of squash under saline soil. *Sci. Hortic.* **2015**, *186*, 89–100. [CrossRef]
67. Abd El-Mageed, T.A.; Semida, W.M.; Taha, R.S.; Rady, M.M. Effect of summer-fall deficit irrigation on morpho-physiological, anatomical responses, fruit yield and water use efficiency of cucumber under salt affected soil. *Sci. Hortic.* **2018**, *237*, 148–155. [CrossRef]
68. Mannan, M.A.; Bhuiya, M.S.U.; Akhand, M.I.M.; Saman, M.M. Growth and yield of basmati and traditional aromatic rice as influenced by water stress and nitrogen level. *J. Sci. Found.* **2012**, *10*, 52–62. [CrossRef]
69. Abo Soliman, M.S.M.; Ghanem, S.A.; Abd El-Hafez, S.A.; El-Mowelhi, N. Effect of irrigation regimes and nitrogen levels on rice production and nitrogen losses under title drainage. *Minist. Agric. L Reclam. Agric. Res.* **1990**, *1*, 14–15.
70. Tuong, T.P.; Bouman, B.A.M. Rice Production in Water Scarce Environments. In *Water Productivity in Agriculture: Limits and Opportunities for Improvement*; Kijne, J.W., Barker, R., Molden, D., Eds.; CABI Publishing: Wallingford, UK, 2003.

agronomy

MDPI

Article

Productivity and Profitability of Kharif Rice Are Influenced by Crop Establishment Methods and Nitrogen Management in the Lateritic Belt of the Subtropical Region

Samata Mohanta [1], Mahua Banerjee [1], Ganesh Chandra Malik [1], Tanmoy Shankar [2], Sagar Maitra [2], Ismail Ahmed Ismail [3], Eldessoky S. Dessoky [3,*], Attia O. Attia [3] and Akbar Hossain [4,*]

1 Department of Agronomy, Palli Siksha Bhavana, Visva-Bharati, Sriniketan 731204, India; samata94@gmail.com (S.M.); mahua.banerjee@visva-bharati.ac.in (M.B.); ganeshmalik_2004@rediffmail.com (G.C.M.)
2 Department of Agronomy, Centurion University of Technology and Management, Odisha 761211, India; tanmoy.shankar125@gmail.com (T.S.); sagar.maitra@cutm.ac.in (S.M.)
3 Department of Biology, College of Science, Taif University, P.C. Box 11099, Taif 21944, Saudi Arabia; i.ismail@tu.edu.sa (I.A.I.); ao.attia@tu.edu.sa (A.O.A.)
4 Bangladesh Wheat and Maize Research Institute, Dinajpur 5200, Bangladesh
* Correspondence: es.dessouky@tu.edu.sa (E.S.D.); akbarhossainwrc@gmail.com (A.H.)

Citation: Mohanta, S.; Banerjee, M.; Malik, G.C.; Shankar, T.; Maitra, S.; Ismail, I.A.; Dessoky, E.S.; Attia, A.O.; Hossain, A. Productivity and Profitability of Kharif Rice Are Influenced by Crop Establishment Methods and Nitrogen Management in the Lateritic Belt of the Subtropical Region. *Agronomy* **2021**, *11*, 1280. https://doi.org/10.3390/agronomy11071280

Academic Editors: Umberto Anastasi and Aurelio Scavo

Received: 18 May 2021
Accepted: 21 June 2021
Published: 24 June 2021

Abstract: Nitrogen management is vital for economic and environmental sustainability. Asynchrony of fertilizer application with crop demand along various nitrogen losses in Eastern India leads to low fertilizer efficiency in *Kharif* rice. At the same time, direct-sowing is gaining popularity due to water and labor scarcity. In an experiment between 2017–2018 in West Bengal, India, the main plots represented establishment methods: conventional transplanting, TPR; direct-seeded rice, DSR; and drum seeded rice, DRR; while subplots represented nitrogen management options: farmer's practice (FP), the state-recommended (SR), nutrient expert-based (NE), Green seeker-based (GS) and LCC-based (LCC) in a split-plot design with three repetitions. Plant growth, productivity, and profitability were evaluated. All indicators of growth or production were affected by establishment methods and by N-management options. The yield enhancement of TPR and DSR over DRR was 21.1 and 16.8%, respectively, while it was enhanced by 19.21, 14.71, 6.49, and 2.52% by GS, NE, LCC, and SR, respectively, over FP. The highest net return and return per rupee invested were recorded with DSR, while both GS and NE had better economics. The results suggest that the combination of DSR establishment with GS or NE requires further studies to find climate-smart management techniques in *Kharif* rice.

Keywords: direct seeding; drum seeder; establishment method; green seeker; growth and yield of rice; *Kharif* rice; LCC; nitrogen supplementation; nutrient expert; production cost

1. Introduction

Nitrogen (N) is a yield-limiting nutrient for rice in India [1] and its efficient use is crucial for economic and environmental sustainability [2]. During the last five decades, the use of nitrogenous fertilizer in cropping systems has been increased [3], but a gradual reduction has also been noted in the crop yield response and fertilizer N use efficiency [4]. Inefficient utilization of nitrogen is considered to be the most critical one among various reasons for this low productivity. India is occupied by a large rice area, accounting for 43.7 Mha [5]. Around 65% population of India depends on this cereal for their food security [6,7]. Hence, in India, food security is also synonymous with rice production security. In India, the crops which are usually sown at the beginning of the monsoon season around June and harvested in October–November are known as *Kharif* crops. The primary rice-growing season in India is the "*Kharif*". The *Kharif* rice (grown between June to November) in India accounts for 89% of the total rice area and 85% of total rice production at all Indian levels [8]. The largest *Kharif* rice area is in Eastern India, such as Odisha, West Bengal, Eastern Uttar Pradesh, Bihar, and Chhattisgarh. Rice productivity in Eastern India is lower (2.0 to 3.0 tonnes ha^{-1}). The current production of *Kharif* rice is 97.10 Mt. By 2050, to feed the projected population of 1.65 billion, rice demand will increase to 197.40 Mt. If 40% increase in *Kharif* rice production alone in suitable areas, production can be increased to 104.55 Mt [9].

In the early 1970s, Indian farmers were applying only 5% of the global fertilizer N, but since 2015 it has increased to 16% [10]. Blanket application of N fertilizer is the prime cause of low nitrogen use efficiency, increased cost of cultivation, and environmental degradation [11]. Due to substantial temporal and plot-to-plot unevenness in indigenous nitrogen supply of soil, broad-based N recommendations like state-recommended N application for rice cannot be helpful [12]. Real-time nitrogen management and site-specific nitrogen management (SSNM) are recent scientific approaches that ensure both increases in rice productivity and sustainability of the rice ecosystem. Nutrient expert is a nutrient estimation support tool that uses the concept of site-specific nutrient management and assists the extension functionaries in developing fertilizer recommendations customized to a specific field [13]. An optical sensor is an important tool of SSNM which uses crop biomass and nitrogen status of standing crop for fertilizer assessment [3].

While analyzing the cause of the low productivity of rice in *Kharif* conditions, it was observed that faulty rice establishment methods have a significant impact along with low nitrogen use efficiency. Successful cropping begins with good crop establishment. Over time, many rice establishment methods have been adopted depending on farmer's willingness, input and technology availability [14]. The benefits of the conventional transplanted rice in puddled conditions are added to nutrient availability, weed control, and reduced percolation loss of water because of puddling. On the other hand, continuous conventional puddling caused soil structure damage, hard-pan formation, soil permeability reduction and decline in groundwater table [15]. "Direct seeded rice (DSR)" refers to the process of establishing a rice crop from seeds sown in the field rather than by transplanting seedlings from the nursery [16]. DSR is a possible establishment technique for future generations to combat issues, such as water scarcity, labor shortage, and greenhouse gas emission [17]. The constraint of labor requirement at the peak stage of cultural practices and the higher cost of cultivation can be redressed through direct seeding by sidestepping nursery raising, seedling uprooting, and transplanting. The availability of better weed management technologies, herbicides, and escalating labor costs are encouraging many farmers to move to direct seeding [18,19]. Many rice varieties were recognized in different countries of the world to cultivate under different establishment methods successfully [20,21]. Under DSR prescriptive N fertilizer application in two or three equal splits followed by a corrective GS guided N fertilizer application at panicle initiation stage can improve N fertilizer use efficiency without any yield loss compared with the general recommendation in North-western India [22]. A higher benefit-cost ratio was obtained in DSR than the transplanted rice using a site-specific nitrogen management tool [23].

Previous researchers have conducted many experiments to determine the yield differences between DSR and TPR under various agroclimatic regions. Some studies have indicated that due to multiple issues with TPR and comparable yield associated with higher net return in DSR [24], it became popular among farmers. However, in contrast to this, many other studies reported yield loss and instability in DSR [25,26]. Experiments conducted before on climate-smart tools like nutrient expert-based or green seeker-based N management were mainly in the high N application area, focusing on saving N with equivalent yield [27,28]. However, in West Bengal, where the *Kharif* rice is predominant farmers apply less N than optimal [29,30]. Therefore, along with climate-smart agriculture strategies, our research aimed to find out the most productive and profitable option of different management aspects suited for different rice establishment methods.

Based on the above considerations, our experiments were carried out to find the best N management options suitable under different establishment methods. Therefore, the objectives of this study were to (1) quantify the effect of establishment methods and N demand-supply on rice growth parameters, yield attributes, and yield and to (2) identify the most cost-effective establishment methods and N management tools; so that the most productive and profitable climate-smart strategies can be identified.

2. Materials and Methods

2.1. Experimental Site

The experimental site was at Chellakamarpada (Birbhum district) village in farmer's field, (23°62′ N latitude and 87°62′ E) in sandy loam soil under the red and lateritic belt (Ultisols) of West Bengal, India. The climate is subtropical. The area falls in the region of the southwest monsoon, and rain generally starts in the third week of June and receives an annual rainfall of about 1190 mm, of which about 80% is received a short duration of three months from mid-June to mid-September and the rest between October to May.

The crop received 1364.6 mm and 836.9 mm rainfall during the cropping period of 2017 and 2018. In 2017, crops received comparatively more rain during the crop establishment period that provided stress due to excess soil moisture. The meteorological data of the experimental site related to the weather conditions prevailing during crop seasons (from June 2017 to November 2017 in the first year) and (from June 2018 to November 2018 in the second year) with respect to rainfall, relative humidity, and temperature obtained from the agro-meteorological advisory services is presented in Figure 1.

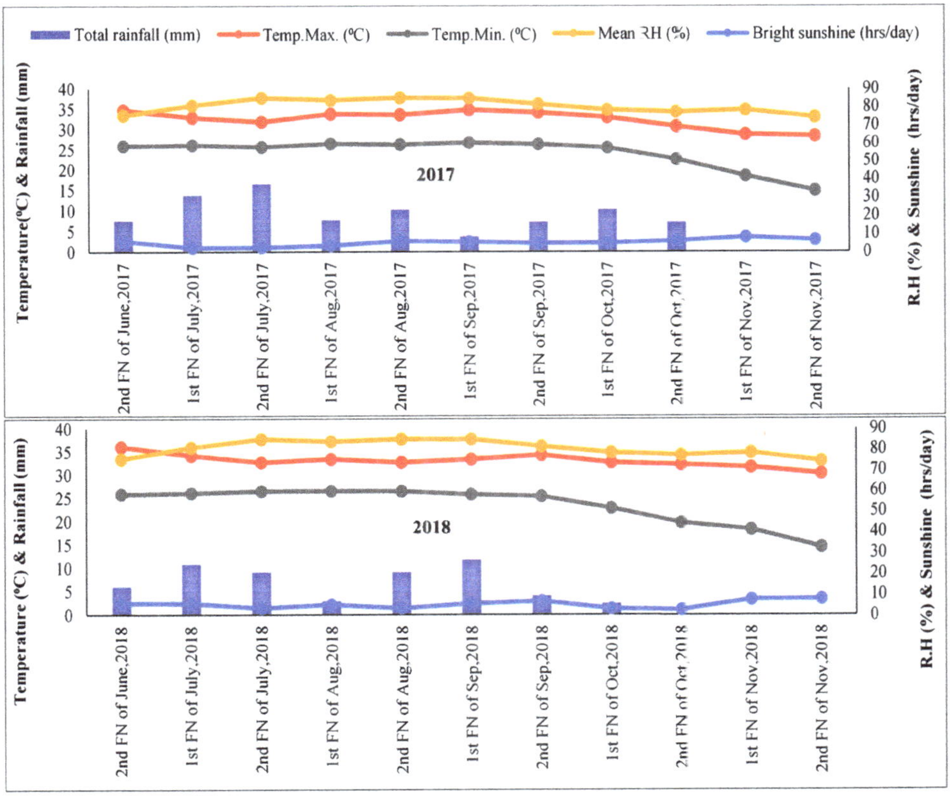

Figure 1. Fortnightly (FN) weather parameters in both crop seasons during 2017 and 2018.

The initial soil textural class and fertility status like pH (6.06), organic carbon content (0.36%), available N (185 kg ha^{-1}), available P_2O_5 (26.6 kg ha^{-1}), available K_2O (270.6 kg ha^{-1}) were determined in the laboratory at the beginning of the experiment and presented in Table 1.

Table 1. Initial soil fertility status.

Particulars	Value	Methods Followed
Soil textural classes	Sandy loam	-
Sand (%)	72.6	Hydrometer method [31]
Silt (%)	17.8	Hydrometer method [31]
Clay (%)	9.6	Hydrometer method [31]
Soil pH	6.06	Determined with the help of pH meter in 1:2.5 ratio of soil water suspension [32]
Electrical conductivity (EC) (dS m^{-1})	0.22	Using conductivity meter [32]
Organic carbon (%)	0.36	Volumetric weight combustion method [33]
Available nitrogen (kg ha^{-1})	185	Alkaline permanganate method [34]
Available phosphorus (kg ha^{-1})	26.6	Brays method No.1 [35]
Available potassium (kg ha^{-1})	270.6	Flame photometer method [36]

2.2. Experimental Treatments and Design

The experiment was carried out for two consecutive years (2017 and 2018) *in Kharif* season. The research was conducted in a split-plot design comprising fifteen treatment combinations in 5 m × 4 m net plot size and replicated thrice. Three crop establishment methods, i.e., "conventional transplanted rice (TPR)", "direct-seeded rice (DSR)", and "drum seeded rice (DRR)" were arranged in main plots and five nitrogen management options, i.e., "farmers' practice-based N management (FP)", "State recommended based N management (SR)", "Nutrient expert based N management (NE)", "Green seeker based N management (GS)" and "LCC based N management (LCC)" were taken as subplot treatments.

In this experiment, TPR was considered the standard practice among the establishment methods, which is being followed by the majority of the farmers in this area. The performance of the practices like DSR and DRR were compared with the standard practice. Similarly, in N-management, the improved N management options were compared with the farmer's practice.

2.2.1. Farmer's Practice (FP)

In the farmer's practice, fertilizer application was made as per the past 3 years' fertilizer application information of the experimental paddy plot. A questionnaire for past years' nutrient management was prepared, and information was collected from the farmer based on the questionnaire. A fertilizer dose of 68:46:37 kg ha^{-1} of N: P_2O_5:K_2O was applied (Table 2). A total amount of phosphorus and potash were used as basal, and nitrogen was used in 3 splits.

2.2.2. State Recommended Nitrogen Application (SR)

In this treatment, fertilizer was applied as per the state recommended nutrient recommendations of West Bengal, i.e., 80:40:40 kg ha^{-1} of N: P_2O_5:K_2O was applied (Table 2) with three splittings of nitrogen and two splittings of potash at different growth stages of rice [37].

2.2.3. Nutrient Expert Based Nitrogen Management (NE)

The nutrient Expert fertilizer decision tool is developed by The International Plant Nutrition Institute (IPNI), in collaboration with IIRR and other national partners, which provides field-specific fertilizer recommendations. In this present study, through a prepared questionnaire, information such as farmer's current yield, characteristics of growing environment, soil fertility indicator (soil texture, color), crop sequence in the farmer's cropping pattern, crop residue management, and fertilizer and organic manure inputs were collected from the farmer [38]. An experimental field-specific fertilizer recommendation was developed using collected information as input of Nutrient Expert® for Rice-South Asia (India) software tool. Based on the developed Rice-Recommendation sheet of the experimental plot of Kamarpara village in May 2017, fertilizers were applied in both years of the experiment. The total dose of phosphoric fertilizer and potassic fertilizers were applied at the basal with three splittings of N.

Table 2. Treatment wise N, P_2O_5 and K_2O application dose.

Treatment	Total N Applied (kg ha^{-1})	Total P_2O_5 Applied (kg ha^{-1})	Total K_2O Applied (kg ha^{-1})
	Farmer's practice (FP)		
Conventional transplanted rice (TPR)	68	46	37
Direct seeded rice (DSR)	68	46	37
Drum seeded rice (DRR)	68	46	37
	State recommended based N management (SR)		
Conventional transplanted rice (TPR)	80	40	40
Direct seeded rice (DSR)	80	40	40
Drum seeded rice (DRR)	80	40	40
	Nutrient expert based N management (NE)		
Conventional transplanted rice (TPR)	118	37	51
Direct seeded rice (DSR)	118	37	51
Drum seeded rice (DRR)	118	37	51
	Green seeker based N management (GS)		
Conventional transplanted rice (TPR)	92.9	37	51
Direct seeded rice (DSR)	95.3	37	51
Drum seeded rice (DRR)	97.6	37	51
	Leaf color chart based N management (LCC)		
Conventional transplanted rice (TPR)	79.6	40	40
Direct seeded rice (DSR)	79.6	40	40
Drum seeded rice (DRR)	79.6	40	40

2.2.4. Greenseeker Handheld Crop Sensor (GS)

The Greenseeker handheld crop sensor (GS) was developed by Trimble agriculture as an active light source optical sensor used to measure plant biomass and displayed as NDVI (normalized difference vegetation index), which is used for N prescription recommendation [39]. In Green seeker-based nitrogen management, the basal dose and first top dressing of N fertilizer were applied as per nutrient expert-based nitrogen management, because of the interference of the exposed water background in reflectance measurements when the crop's canopy was not fully developed [40,41]. The second top dressing at the panicle initiation stage was done as per the estimated fertilizer rate. Total phosphoric fertilizer and potassic fertilizers were applied as basal.

2.2.5. Leaf Color Chart (LCC)

The leaf color chart used in the experiment was developed by the Central Rice Research Institute (CRRI), Cuttack, India, which is now renamed as ICAR-National Rice Research Institute (ICAR-NRRI). Two times nitrogen was top-dressed when LCC < 3* with a basal dose of 26.5 kg N ha^{-1} as per the recommendation of customized leaf color chart for nitrogen management in rice for different ecology [42]. Phosphorus and potassic fertilizers were applied as per the state recommended.

2.2.6. Direct Seeded Rice (DSR)

In DSR, dry rice seeds were sown in line immediately after receiving favorable rain in moist but unsaturated soil before the onset of monsoon in both years of study.

2.2.7. Drum Seeded Rice (DRR)

In Drum seeded rice, direct seeding of pre-germinated paddy seeds was done through the fiber drums, which dispense seeds evenly in puddled and leveled fields.

2.2.8. Conventional Transplanting (TPR)

In the conventional transplanting (TPR) system, 21 days seedlings were transplanted manually in the puddled main field. Rice variety (HYV) "Pratikhya" was taken as an experimental crop variety. The dose of nutrients in all the treatment are presented in Table 2. In NE, attainable yield (Ya) was estimated through the Nutrient Expert fertilizer decision tool from maximum attainable yield (Ymax) for a geographic region or growing environment and farmers' actual nutrient-limited yield (Y) [38]. (Ya) was considered as target yield in both NE and GS. The amount of N fertilizer was calculated based on the grain yield targets 5 t ha^{-1} in NE and GS.

2.3. Experimental Procedure

In TPR, seedlings were grown in a wet nursery. Seeds were treated with Thiram 75% WP @ 2 gm kg^{-1} of seed to prevent fungal diseases. For nursery raising, raised bed was laid out near the main field. In TPR and DRR methods of establishment, after the final puddling, planking and leveling were done. The beds were leveled vigilantly so that water would not stagnate at any place on the bed. Well-decomposed farmyard manure (FYM) was applied on the bed as per the local farmer's practice. In DRR, after puddling and land leveling, excess water was drained out the day before drum seeding. To achieve the successful establishment of DSR, the plot was ploughed two times with a disc harrow after tillage with a cultivator and followed by one planking to make ready a fine seedbed for seeding. Sowing in the nursery bed for seedling raising in TPR and sowing in the main field in DSR and DRR was done on the same day. Under TPR of transplanting, 21–22 days old seedlings were planted in a leveled field @ 2–3 seedlings/hill at a 20 cm × 10 cm spacing. In DSR, treated dry seeds were sown in line @ 2–3 seeds/hill with the same spacing (20 cm × 10 cm). In DRR, soaked and pre-germinated seeds were sown with an eight-row seed drum, in line with spacing (20 cm × 8 cm). The seed drum was filled up to two-thirds of capacity. One of the funnel-shaped holes of the drum seeder was blocked by the cap to change the plant to plant spacing to 8 cm. The crop was sown on 25 June 2017 and 16 June 2018.

The fertilizers were applied in the plots after layout as per treatments (Table 2). An N-rich strip was maintained within the field in a small area where enough fertilizer had been applied. This N-rich strip was taken as a reference area, and the normalized difference vegetation index ($NDVI_{ref}$) value reading was recorded. By taking the $NDVI_{ref}$ NDVI value of the experimental plot, and fertilizer estimation chart, the fertilizer rate was obtained for the GS.

2.4. Measurements and Analytical Procedures

2.4.1. Growth and Yield Attributes

The height of ten plants was recorded, and the mean value was calculated and expressed in centimeters (cm). The number of tillers was counted by using 1 m × 1 m quadrate from the second row. Leaving the first row from the border of each side of a plot, destructive samples were taken from the second row to record biometric observations such as dry matter accumulation (gm^{-2}) and leaf area index (LAI). To determine the dry matter accumulation, rice plants were cut at ground level from each plot randomly as destructive samples. For leaf area, the representative green leaves were taken randomly from destructive samples, and their areas were recorded by leaf area meter. The destructive samples were dried in a hot air oven at 80 °C for 10 h until constant weights were obtained as per the standard procedure [43]. The recorded dry weights of plants and leaves were used to calculate dry matter accumulation and leaf area index. The ratio of the recorded leaf area and dry weight of these leaves was used to measure the leaf area indices, since LAI is the area of leaf surface per unit of the land surface [44,45] (Equation (1)).

$$Leaf\ area\ index = \frac{Leaf\ area}{ground\ area} \quad (1)$$

Five plants were harvested, dried and their yield attributes were recorded.

2.4.2. Yield

The grain yield obtained from each treatment in the net plot area was sundried, threshed, winnowed, and cleaned. After that, the weight of the grains per net plot was recorded at 14% moisture with the help of electronic balance. Each treatment's grain yield per hectare was calculated from the net plot yield and expressed in kg ha^{-1}.

2.4.3. Economics

The total cost of production ha^{-1} for each treatment was calculated based on the current market rate of inputs like seed, fertilizer, herbicide, pesticide. Hired machinery costs for land preparation, sowing, and threshing, labor cost, irrigation costs were were also added to production costs. Gross return was calculated based on the products' prevailing market price, and accordingly, net return was calculated. Dividing this net return by the cost of cultivation, we obtained the return per rupee invested. Based on the return per rupee (Indian currency) invested, the most beneficial treatment for the crop sequence was determined (Equation (2)).

$$Return\ per\ rupee\ invested = \frac{Net\ return}{cost\ of\ cultivation} \quad (2)$$

2.5. Calculations and Statistical Analysis

The experimental data were analyzed statistically by using analysis of variance (ANOVA). The standard error of means (SEm±) and the critical difference at a 5% probability level of significance (CD, $p \leq 0.05$) [46]. Excel software (Microsoft Office Home and Student version 2019-en-us, Microsoft Inc., Redmond, Washington, DC, (USA) was used for statistical analysis and drawing graphs and figures.

3. Results

3.1. Growth Parameter

The pooled data for two years of study are presented in Table 3 to show the impact of establishment methods and nitrogen management on plant height (cm) at harvest and tillers m^{-2} at the maximum tillering stage of the "Pratikhya" variety of rice grown in *Kharif* conditions. The observations showed that establishment methods influenced the plant height of rice in 2018, and DSR recorded significantly taller plant height than TPR and DRR at harvest. (Table 3). Pooled data for two years did not show any significant impact in increasing plant height among different establishment methods. NE and GS exhibited the highest plant height among different N management treatments, while LCC, SR, and FP presented significantly lower plant height than NE. LCC showed significantly higher plant height than FP (Table 3).

The data on the number of tillers per unit area (m^{-2}) revealed that the crop establishment methods and nitrogen management influenced the tillers production in "Pratikhya" *Kharif* rice in both the years of experimentation (Table 3). The treatment TPR being statistically on par with DSR produced a significantly greater number of tillers than the DRR method of rice establishment during both years. Pooled data for two years also showed a similar trend. Among the N management options, FP resulted in significantly lowest tillers production of "Pratikhya" *Kharif* rice over other treatments in the study. The data of individual years and pooled data showed that GS and NE treatments remained statistically on par in production of the number of tillers per unit area at 75 DAS. These two treatments were significantly superior to other treatments.

Table 3. Influence of crop establishment method and nitrogen management on plant height (cm) and the number of tillers m^{-2} of "Pratikhya" *Kharif* rice variety.

Treatments	Plant Height (cm)			Number of Tillers m^{-2}		
	At harvest			At 75 DAS		
	2017	2018	Pooled	2017	2018	Pooled
	Establishment method					
TPR	117.7 [a]	102.1 [b]	109.9	350.9 [a]	343.5 [a]	347.2 [a]
DSR	111.8 [a]	118.6 [a]	115.2	321.2 [b]	329.4 [a]	325.9 [ab]
DRR	114.4 [a]	107.0 [b]	110.6	299.5 [b]	306.5 [b]	300.0 [b]
SEm±	2.0	2.0	1.4	7.1	6.9	7.0
CD at 5%	NS	7.9	NS	27.8	27.2	27.7
	Nitrogen management					
FP	107.0 [c]	101.8 [c]	104.4 [d]	285.4 [c]	281.4 [c]	283.7 [c]
SR	111.9 [bc]	104.4 [c]	108.2 [c]	306.8 [bc]	312.6 [b]	308.6 [b]
NE	122.2 [a]	119.2 [a]	120.7 [a]	347.2 [a]	360.9 [a]	353.2 [a]
GS	118.4 [ab]	113.0 [ab]	115.7 [b]	364.7 [a]	357.0 [a]	360.0 [a]
LCC	113.7 [abc]	107.7 [bc]	110.7 [c]	315.2 [b]	320.4 [b]	316.3 [b]
SEm±	3.2	2.6	2.1	8.2	8.8	8.2
CD at 5%	9.4	7.7	5.9	23.9	25.8	23.9
Interaction effect	NS	NS	NS	NS	NS	NS

Different lowercase letters within the continuous columns are significantly different at 5% level of probability in Duncan's multiple ranges test (DMRT). Note: Conventional transplanting (TPR), direct-seeded rice (DSR) and drum seeded rice (DRR); farmer's practice (FP), the state recommended (SR), nutrient expert based (NE), Green seeker based (GS), leaf color chart (LCC) and non-significant (NS).

Year-wise and pooled data on dry matter accumulation were presented in Table 4 indicated that nitrogen management and establishment methods significantly influenced dry matter accumulation (gm^{-2}) of "Pratikhya" *Kharif* rice. Among the different rice establishment methods, TPR showed its significant superiority over other two methods, namely DSR and DRR in dry matter production in the harvest stage of "Pratikhya" *kharif* rice in 2017; but in 2018, DSR resulted in significantly more dry matter accumulation than TPR and DRR. Based on pooled data, DSR and TPR recorded significantly more dry matter production (gm^{-2}) than DRR. In N management, a similar trend was noted in individual years and the pooled data (Table 4). The N management treatments, such as NE and GS, resulted in significantly more dry matter accumulation over other treatments.

Data on leaf area index (LAI) revealed that TPR was statistically on par with DSR, registered significantly higher values of LAI at 95 DAS than DRR rice in 2017 and pooled data (Table 4). Among different N management treatments, NE registered higher values in LAI during both the years, and GS closely followed it; however, these two treatments were statistically on par as reflected in individual years' data. But pooled data registered the significant superiority of the treatment NE over other treatments. N management in FP resulted in the least values of LAI of "Pratikhya" *Kharif* rice at 95 DAS.

Table 4. Influence of crop establishment method and nitrogen management on dry matter accumulation (gm^{-2}) and leaf area index of "Pratikhya" *Kharif* rice variety.

Treatments	Dry Matter Accumulation (gm^{-2}) at Harvest			Leaf Area Index (LAI) at 95 DAS		
	2017	2018	Pooled	2017	2018	Pooled
Establishment method						
TPR	1274.1 [a]	1129.2 [b]	1201.7 [a]	4.97 [a]	4.52	4.74 [a]
DSR	1020.8 [b]	1462.4 [a]	1241.5 [a]	4.70 [a]	4.39	4.54 [a]
DRR	980.9 [b]	1093.9 [b]	1037.4 [b]	4.12 [b]	4.24	4.18 [b]
SEm±	22.1	45.9	29.6	0.08	0.13	0.08
CD at 5%	87.0	180.3	116.4	0.33	NS	0.26
Nitrogen management						
FP	948.2 [c]	1051.1 [c]	999.6 [b]	4.22 [c]	3.97 [c]	4.09 [d]
SR	1022.6 [c]	1163.0 [bc]	1092.8 [b]	4.30 [c]	4.23 [bc]	4.27 [d]
NE	1278.2 [a]	1428.5 [a]	1353.4 [a]	5.15 [a]	4.74 [a]	4.95 [a]
GS	1169.9 [b]	1315.2 [ab]	1242.5 [a]	4.82 [ab]	4.53 [ab]	4.67 [b]
LCC	1040.9 [c]	1184.8 [bc]	1112.9 [b]	4.49 [bc]	4.45 [ab]	4.47 [c]
SEm±	35.6	77.3	43.2	0.12	0.12	0.08
CD at 5%	103.9	225.5	126.2	0.34	0.35	0.24
Interaction effect	NS	NS	NS	NS	NS	NS

Different lowercase letters within the continuous columns are significantly different at 5% level of probability in Duncan's multiple ranges test (DMRT). Note: Conventional transplanting (TPR), direct-seeded rice (DSR) and drum seeded rice (DRR); farmer's practice (FP), the state recommended (SR), nutrient expert based (NE), Green seeker based (GS), leaf color chart (LCC) and non-significant (NS)

3.2. Grain Yield and Yield Attributes

The individual years and pooled data on yield attributes of "Pratikhya" rice under *Kharif* season indicated that crop establishment methods and nitrogen management influenced the yield parameters and yield (Table 5; Figures 2 and 3). The effect of treatments on yield attributes and yield are narrated in the following segment. The present study revealed that the treatment TPR being statistically on par with DSR registered significantly higher effective tillers m^{-2} than the DRR method of crop establishment as noted in individual years and pooled data. N management treatments also significantly affected effective tillers. GS-based N management being statistically on par with NE-based nutrient management resulted in significantly superior to other treatments. FP-based N management showed poor performance in comparison to others. The pooled data revealed that SR, LCC, NE, and GS increased effective tillers over farmer's practice by 4.36, 7.8, 16.40, and 22.88%, respectively.

The number of filled grains panicle^{-1} is an important yield attributing character influencing the productivity of "Pratikhya" *Kharif* rice. The results revealed that the conventional TPR resulted in significantly higher filled grains panicle^{-1} than the other two methods, viz, DSR and DRR in "Pratikhya" under *Kharif* 2017 (Table 5). DSR and DRR remained statistically on par in an expression of filled grains panicle^{-1}. Pooled data also showed a similar trend as noted in 2017. N management treatments influenced filled grains panicle^{-1} of "Pratikhya" *Kharif* rice. The treatment GS recorded significantly higher filled grains panicle^{-1} than other treatments, namely, FP, SR, LCC, and NE-based nitrogen management.

Table 5. Influence of crop establishment methods and nitrogen management on yield attributes of "Pratikhya" *Kharif* rice variety.

Treatments	Effective Tillers m^{-2}			Filled Grains/Panicle			Test Weight (g)		
	2017	2018	Pooled	2017	2018	Pooled	2017	2018	Pooled
Establishment method									
TPR	287.6 ^{a}	277.2 ^{a}	282.4 ^{a}	104.2 ^{a}	96.8 ^{a}	100.5 ^{a}	26.4	25.8	26.1
DSR	279.7 ^{a}	263.4 ^{ab}	271.5 ^{a}	86.2 ^{b}	93.3 ^{b}	89.7 ^{b}	25.7	26.0	25.9
DRR	245.2 ^{b}	240.6 ^{b}	242.9 ^{b}	83.1 ^{b}	86.6 ^{b}	84.8 ^{b}	25.0	24.9	25.0
SEm±	7.9	7.0	5.2	3.3	2.8	2.2	0.4	0.5	0.3
CD at 5%	30.8	27.3	17.1	13.0	NS	7.0	NS	NS	NS
Nitrogen Management									
FP	231.9 ^{c}	249.7 ^{b}	240.8 ^{c}	79.2 ^{b}	82.5 ^{c}	80.9 ^{d}	23.6	24.7	24.1
SR	249.4 ^{bc}	253.1 ^{b}	251.3 ^{c}	86.2 ^{b}	84.3 ^{c}	85.3 ^{c}	25.2	25.5	25.3
NE	295.4 ^{a}	265.2 ^{ab}	280.3 ^{ab}	97.4 ^{a}	97.5 ^{b}	97.5 ^{b}	26.4	26.6	26.5
GS	313.9 ^{a}	277.9 ^{a}	295.9 ^{a}	106.1 ^{a}	107.1 ^{a}	106.6 ^{a}	26.4	25.6	26.0
LCC	263.4 ^{b}	256.1 ^{b}	259.8 ^{bc}	86.9 ^{b}	89.6 ^{bc}	88.3 ^{c}	27.0	25.6	26.3
SEm±	9.3	6.6	5.7	3.2	3.1	2.2	0.9	0.7	0.6
CD at 5%	27.0	19.2	16.1	9.2	9.2	6.3	NS	NS	NS
Interaction effect	NS		NS		NS	NS		NS	NS

Different lowercase letters within the continuous columns are significantly different at 5% level of probability in Duncan's multiple ranges test (DMRT). Note: Conventional transplanting (TPR), direct-seeded rice (DSR) and drum seeded rice (DRR); farmer's practice (FP), the state recommended (SR), nutrient expert based (NE), Green seeker based (GS), leaf color chart (LCC) and non-significant (NS).

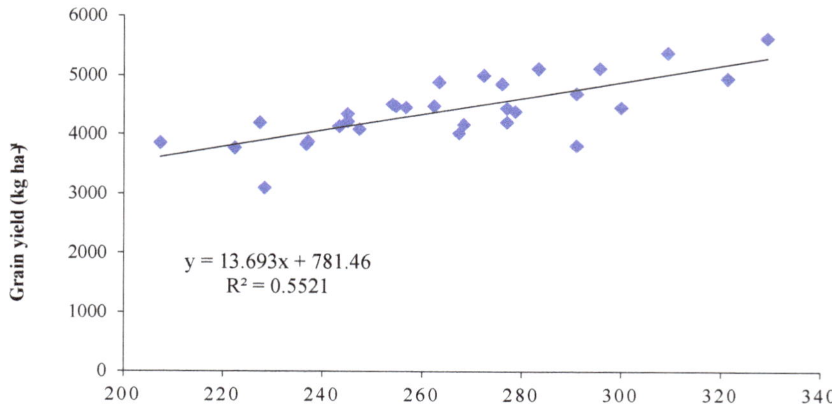

Figure 2. Linear regression between yield (kg ha^{-1}) and panicles (m^{-2}) of "Pratikhya" *Kharif* rice variety.

Grain and straw yields of "Pratikhya" *Kharif* rice were influenced by the crop establishment method and nitrogen management. The result of the treatments on grain and straw yield of rice in the present experiment was narrated in the following segment. During both years, grain yield was influenced by rice establishment methods (Table 6). In 2017 the TPR method of crop establishment produced more grain yield, and during 2018, DSR yielded more rice grains, but there was no significant difference between TPR and DSR during both years. TPR and DSR produced significantly higher grain yields than DRR (Table 6).

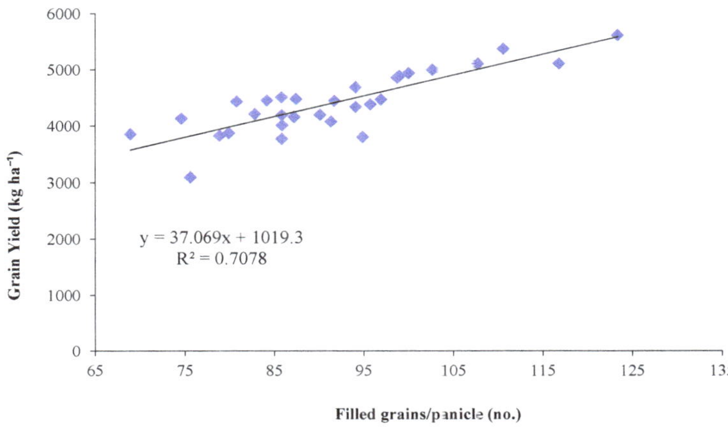

Figure 3. Linear regression between yield (kg ha^{-1}) and filled grain/panicle (no.) of "Pratikhya" *Kharif* rice variety.

The pooled data showed a similar trend as noted in 2017. Among the N management treatments, GS being statistically on par with NE, registered a significantly higher grain yield than other treatments, such as FP, SR, and LCC, as noted in individual years and pooled data (Table 6). The pooled data showed that yield enhancement in GS and NE was 19.2% and 14.7% over FP. LCC recorded a significantly lower yield than GS and NE. Crop establishment methods impacted on straw yield of "Pratikhya" under *Kharif* rice (Table 6). The maximum straw yield was recorded with TPR during both years, and pooled data also showed a similar trend. However, data of 2017 and pooled data showed that TPR produced significantly higher straw yield than the other two establishment methods, namely, DSR and DRR. But in 2018, TPR being statistically on par with DSR, produced significantly more straw yield than DRR. Among N management treatments, NE resulted in the maximum straw yield. In 2017, NE being statistically on par with GS and LCC, produced significantly higher straw yield than FP and SR. But the data of 2018 and pooled data revealed that NE being statistically on par with GS, had significantly more straw yield than other treatments, namely, FP, SR, and LCC. The percentage increase in straw yield of NE, GS, LCC, and SR over FP was 19.19, 14.94, 10.40, and 6.08%, respectively.

3.3. Economics

Based on pooled data (Table 7), the result showed that the maximum gross return per hectare (Rs. 110,121 ha^{-1}) was recorded with TPR, but no significant variation was observed between TPR and DSR. The conventional transplanting fetched Rs. 5122 ha^{-1} higher gross return than direct-seeded rice and Rs. 19,249 ha^{-1} higher gross return than drum seeded rice (Table 7).

But DSR recorded the highest net return per hectare (Rs. 63,726 ha^{-1}) due to less cost of cultivation involved in DSR in puddling and transplanting operation. DSR fetched more net return of (+ Rs. 1457 ha^{-1}) than TPR, but there was no significant variation in net return between TPR and DSR. The maximum return per rupee invested (Rs. 2.54) was recorded with DSR, and it was significantly higher than TPR (Rs. 2.30) (Table 7). GS fetched a more gross return of Rs. 17,118 ha^{-1}, Rs. 14,027 ha^{-1}, Rs. 10,241 ha^{-1}, and Rs. 2426 ha^{-1} compared to FP, SR, LCC, and NE, respectively (Table 7). The same trend was observed with net return ha^{-1}. GS fetched more net return of (+ Rs. 16,798 ha^{-1}), (+ Rs. 13,725 ha^{-1}), (+Rs. 9932 ha^{-1}) and (+Rs. 2836 ha^{-1}) than FP, SR, LCC, and NE, respectively. There was no significant difference in gross return and net return between GS and NE. The highest return per rupee invested (Rs. 2.52) was recorded in GS, which was statistically on par with NE (Rs. 2.44) (Table 7).

Table 6. Influence of crop establishment methods and nitrogen management on yield of "Pratikhya" *Kharif* rice variety.

Treatments	Grain Yield (kg ha)			Straw Yield (kg ha)		
	2017	2018	Pooled	2017	2018	Pooled
			Establishment method			
TPR	4990 a	4513 a	4751 a	6359 a	5940 a	6150 a
DSR	4491 a	4671 a	4581 a	5676 b	5594 a	5635 a
DRR	3750 b	4095 b	3923 b	5130 c	5002 b	5066 b
SEm±	136	99	84	138	126	93
CD at 5%	534.3	390.1	274.8	542.7	492.7	304.4
			Nitrogen Management			
FP	4114 b	4024 c	4069 c	5151 c	5050 d	5100 d
SR	4155 b	4188 bc	4172 bc	5543 bc	5278 cd	5411 c
NE	4682 a	4653 ab	4668 a	6191 a	5967 a	6079 a
GS	4795 a	4906 a	4851 a	5981 ab	5744 ab	5862 ab
LCC	4306 ab	4360 abc	4333b	5742 ab	5520 bc	5631 bc
SEm±	161	186	123	183	115	108
CD at 5%	470.8	541.9	349.7	533.1	336.3	307.0
Interaction effect	NS	NS	NS	NS	NS	NS

Different lowercase letters within the continuous columns are significantly different at 5% level of probability in Duncan's multiple ranges test (DMRT). Note: Conventional transplanting (TPR), direct-seeded rice (DSR) and drum seeded rice (DRR); farmer's practice (FP), the state recommended (SR), nutrient expert based (NE), Green seeker based (GS), leaf color chart (LCC) and non-significant (NS).

Table 7. Influence of crop establishment methods and nitrogen management on the economics of "Pratikhya" *Kharif* rice variety.

Treatments	Gross Return (Rs/ha)			Net Return (Rs/ha)			Return per Rupee Invested (Rs/ha)		
	2017	2018	Pooled	2017	2018	Pooled	2017	2018	Pooled
				Establishment method					
TPR	115,257 a	104,985 a	110,121 a	67,709 a	56,830 b	62,269 a	2.42 a	2.18 b	2.30 b
DSR	103,548 b	106,450 a	104,999 a	62,549 a	64,902 a	63,726 a	2.53 a	2.56 a	2.54 a
DRR	88,020 c	93,724 b	90,872 b	46,007 b	51,111 b	48,559 b	2.09 b	2.20 b	2.15 c
SEm±	2947	1919	1758	2947	1919	1758	0.07	0.05	0.04
CD at 5%	11,567	7534	5733	11,567	7534	5733	0.27	0.18	0.14
				Nitrogen management					
FP	94,656 c	92,627 b	93,642 c	51,359 b	48,730 c	50,045 c	2.19 b	2.12 c	2.16 c
SR	96,965 c	96,501 b	96,733 bc	53,651 b	52,586 bc	53,118 bc	2.23 b	2.21 bc	2.22 bc
NE	109,036 ab	107,632 a	108,334 a	65,009 a	63,004 ab	64,007 a	2.47 a	2.42 ab	2.44 a
GS	110,236 a	111,283 a	110,760 a	66,583 a	67,102 a	66,843 a	2.52 a	2.52 a	2.52 a
LCC	100,481 bc	100,556 ab	100,519 b	57,173 ab	56,649 ab	56,911 b	2.32 ab	2.30 abc	2.31 b
SEm±	3170	3569	2387	3170	3569	2387	0.07	0.08	0.05
CD at 5%	9251	10415	6785	9251	10415	6785	0.21	0.23	0.15
Interaction effect NS	NS	NS		NS	NS		NS	NS	

Different lowercase letters within the continuous columns are significantly different at 5% level of probability in Duncan's multiple ranges test (DMRT). Note: Conventional transplanting (TPR), direct-seeded rice (DSR) and drum seeded rice (DRR); farmer's practice (FP), the state recommended (SR), nutrient expert based (NE), Green seeker based (GS), leaf color chart (LCC) and non-significant (NS).

4. Discussion

4.1. Growth Parameter

Weather and the rainfall pattern during the seeding and early establishment stage greatly influenced growth variation between DSR and TPR. Similar results were reported from the previous research [23,47,48]. The highest plant height was in NE treatment, followed by the state recommended dose and farmer field practice due to site-specific balanced fertilizer application as per demand in NE [49]. All treatments received more N than FP, and NE received maximum N. Probably, such expression of plant height is due to the N factor as N is known to increase plant height in rice. NE with more N might have resulted in elongation of internodes reflected in enhanced plant height [50].

The lower tiller number in DSR or DRR might be due to the fact that net photosynthate production was lower with these establishment methods than TPR [51,52]. Nitrogen is known to promote growth and tillering, and the treatments NE and GS received more N than other treatments. Because receiving more N, NE, and GS treatments produced more tillers in "Pratikhya" *Kharif* rice. The present study results conform to earlier research which mentioned more tillering due to higher leaf N content due to more N application [53,54]. In the first year (2017), rainfall affected the establishment of DSR adversely, but in the second year (2018), the rainfall pattern favored the initial establishment of rice crops and subsequent dry matter production. A similar type of observation was also reported in the previous experiment [55]. Therefore, DSR under favorable agroclimatic conditions might result in greater biomass production [56].

Site-specific N management, namely NE and GS, showed a positive impact in improving dry matter production in *"Pratikhya" under Kharif condition* rice as these treatments received more N than the remaining treatments. NE optimized fertilization management based on the 4R concept to harmonize crop nutrient requirement and application during the growing season [57]. Similarly, GS has also known as a reliable precision tool for N management and biomass production and superior to traditional practices [58]. Nutrient expert and Green Seeker-based nutrient management treatments received a higher dose of N and N has the most noticeable effect on LAI by increasing the number of tillers and leaf size [54]. The findings corroborate with earlier research [59], where differences in LAI due to the rice establishment method were observed [60–62].

4.2. Yield and Yield Parameter

Earlier findings [63] also noted higher panicles in rice with direct seeding and transplanting. TPR and DSR also produced a more significant number of effective tillers m^{-2}. Enhancing effective tillers m^{-2} with these two treatments was probably due to proper partitioning of dry matter from source to sink. Earlier research also indicated the superiority of GS in the enhancement of effective tillers [40]. The impact of the establishment method on filled grains panicle^{-1} of rice was also proved by an earlier study [64]. The superior tiller growth with the photosynthates in functional leaves after heading in TPR may enhance the source for grain filling resulting in higher filled grain per panicle [65]. On the other hand, comparatively small sink capacity or insufficient source content and export might be the reason for low filled grain per panicle in DSR and DRR [66].

Earlier, several findings confirmed that rice grain and straw yield were influenced by the crop establishment method [67] and nitrogen management [68]. In our present study, the difference in grain yield of rice with DSR method between two years was probably due to climatic variation, i.e., higher rainfall in 2017 immediately after seeding and during the crop establishment period. In direct seeding, due to climatic factors and irregular stand establishment, rice yield was affected, as mentioned by earlier studies [69–73]. The yield variation between DSR and TPR is determined by climate and soil properties. Significant yield loss was reported in DSR due to climatic stress. DSR relative yield was −25% when unbalanced climate stress occurred, whereas it was only −7% without climate stress [74]. DSR could produce comparable yields to TPR but is more prone to yield losses due to inappropriate management practices, unsuitable soil properties, and climatic stresses. Wet seeding in puddled soil might be less suitable on soils prone to cracking and when there is a chance of rainfall after seeding unless fields are well-leveled and with good surface drainage systems [75]. Poor establishment in DSR can result from seed rotting and seedling damage if rain is immediate after sowing and continues for few days, which eventually negatively impacts growth and productivity [76,77]. Heavy rain, especially in heavier clay soils after sowing, tends to have poor distribution, germination, and emergence [78]. Although TPR and DSR were recorded statistically on par yield, numerically, TPR obtained a higher yield than DSR, which can be better explained due to the higher number of filled grain in TPR [79]. TPR resulted in significantly more filled grains panicle^{-1} than DSR and DRR due to 15% higher floret fertility in the TPR than DSR [75]. Grain yield was positively correlated with the yield attributing characters (panicle m^{-2} and number of grains panicle^{-1}) of "Pratikhya" *Kharif* rice. In our present study, a higher correlation was found between the number of filled grains per panicle with grain yield (r = 0.84**, data was not displayed in this paper), and the regression equation indicated that 70% of the total variation in yield could be explained by the linear relation between the number of filled grains per panicle and grain yield.

In the current study, both GS and NE based N management expressed higher growth parameters, namely, dry matter production, LAI, number of tillers m^{-2} and yield attributes (particularly, panicles m^{-2} and number of grains panicle^{-1}), and the impact of these characters was reflected in the productivity of rice. Regression study indicates that the number of panicle m^{-2} is correlated with yield, and 55% of the total variation in yield can be explained by the linear relation between the number of panicle m^{-2} and yield. Earlier research also evidenced GS and NE-based precision nutrient management in rice [2,80,81]. The number of filled grains per panicle was higher in GS, contributing significantly to grain yield [40,49,80,82]. By recommending a moderate amount of basal nitrogenous fertilizer at transplanting, enough N fertilizer in between active tillering and panicle initiation stage, and optical sensor-directed fertilizer N dose at panicle initiation stage, higher yield and nitrogen use efficiency can be achieved in transplanted rice [40]. Growth attributes and grain yield positively correlate with the canopy NDVI of rice [83]. GS considers leaf greenness and plant biomass compared to LCC, which determines nitrogenous fertilizer only based on leaf color, so GS may be considered a better N management tool even in not very well managed rice fields [40,58]. NE recorded a higher yield than SR and FP due to yield parameters, i.e., panicle m^{-2} [84] and filled grain per panicle [49]. The poor performance with DRR in the production of straw yield was probably due to inferior dry matter accumulation.TPR provided a uniform crop stand, and TPR superiority was observed from the earlier research [85,86]. Higher straw yields of NE and GS were probably due to more dry matter accumulation by the treatments. The results conform with other research [87,88].

4.3. Economics

Higher return per rupee invested in DSR, compared to TPR and DRR, suggests that DSR is more cost-effective than other treatments. A similar observation was recorded by previous researchers [89]. A higher gross return was obtained from the transplanted method, but a higher net return and benefit–cost ratio were noted from the DSR method [90]. Most of the cost-saving in DSR came from avoiding nursery establishment and reduced labor in crop establishment as DSR required only 9% of the total cost, while it required 23% in TPR [91]. Despite the higher costs associated with an increased dose of N in NE, a higher income was obtained from yield gain than other treatments [2]. Previous research also confirmed as compared to FP, SSNM recorded an increase in yield by 7% and profitability by 12% [92]. NE is the better option over other nutrient management regarding yield and profit [60,93]. SSNM had a positive and significant effect on the economics of the "Pratikhya" *Kharif* rice variety. In our study, the highest return per rupee invested was in GS and NE due to higher yield and net return although higher production costs in these two treatments. There are vast possibilities to practice precision farming technologies in

India using inexpensive and handy gadgets like LCCs and expensive devices like optical sensors [89].

5. Conclusions

In conclusion, the two years field study on crop establishment methods of "Pratikhya" *Kharif* rice indicated that DSR could obtain comparative growth, yield attributes, and yield with TPR in the *Kharif* rice system of the red and lateritic belt of West Bengal. Seed sowing in DSR and DRR should be done before the onset of monsoon for better establishment and a comparable yield with TPR. Poor establishment and growth attributes in drum seeding field management practices resulted in lower yield. The growth parameters, yield attributes (panicle m^{-2} and grains per panicle), and grain yield of "Pratikhya" *Kharif* rice variety were enhanced with site-specific nitrogen management like GS, NE, and LCC based N management compared to blanket application. Due to the lesser cost of cultivation and higher return per rupees invested, DSR proved to be more economically viable than TPR or DRR. GS and NE can be considered a more economical remunerative N management option in the experimental field of Birbhum. Based on present results, it could be concluded that DSR an alternative of transplanted rice and optimization of N application with site-specific nutrient management like GS and NE as a better N management option in terms of growth, productivity, and profitability of rice production in the *Kharif* season under the red and lateritic belt of sub-tropical India. Further research is needed to work out the prescriptive N management to be followed at transplanting and active tillering before applying Green seeker-based fertilizer dose at the panicle initiation stage and to evaluate the N management and crop establishment methods under various agro-climatic zones and varieties.

Author Contributions: Conceptualization, S.M. (Samata Mohanta), M.B., G.C.M., T.S. and S.M. (Sagar Maitra); methodology, S.M. (Samata Mohanta), M.B., G.C.M., T.S. and S.M. (Sagar Maitra); software, S.M. (Samata Mohanta) and M.B.; validation, S.M. (Sagar Maitra), (Samata Mohanta), M.B. and G.C.M.; formal analysis, S.M. (Samata Mohanta) and M.B.; investigation, S.M. (Samata Mohanta), M.B. and G.C.M.; resources, M.B. and G.C.M.; data curation, S.M. and M.B.; writing—original draft preparation, S.M. (Samata Mohanta), M.B., G.C.M., T.S., S.M. (Sagar Maitra) and A.H.; writing—review and editing, I.A.I., E.S.D., A.O.A. and A.H.; visualization, S.M. (Samata Mohanta) and M.B.; supervision, M.B.; project administration, M.B. and G.C.M.; Funding, I.A.I., E.S.D., A.O.A. and A.H. All authors have read and agreed to the published version of the manuscript.

Funding: The current work was funded by Taif University Researchers Supporting Project number (TURSP-2020/120), Taif University, Taif, Saudi Arabia.

Institutional Review Board Statement: Not applicable.

Informed Consent Statement: Not applicable.

Data Availability Statement: Most of the recorded data are available in all Tables in the manuscript.

Acknowledgments: The authors extend their appreciation to Taif University for funding current work by Taif University Researchers Supporting Project number (TURSP-2020/120), Taif University, Taif, Saudi Arabia.

Conflicts of Interest: Authors would hereby like to declare that there is no conflict of interests.

References

1. Panda, D.; Nayak, A.K.; Mohanty, S. Nitrogen management in rice. *Oryza* **2019**, *56*, 125–135. [CrossRef]
2. Sapkota, T.B.; Jat, M.L.; Rana, D.S.; Khatri-Chhetri, A.; Jat, H.S.; Bijarniya, D.; Sutaliya, J.M.; Kumar, M.; Singh, L.K.; Jat, R.K.; et al. Crop nutrient management using Nutrient Expert improves yield, increases farmers' income and reduces greenhouse gas emissions. *Sci. Rep.* **2021**, *11*, 1564. [CrossRef] [PubMed]
3. Bijay, S.; Singh, Y. Management and Use Efficiency of Fertilizer Nitrogen in Production of Cereals in India; Issues and Strategies. *J. Indian Nitrogen Manag.* **2017**, *10*, 149–162.
4. Lassaletta, L.; Billen, G.; Grizzetti, B.; Anglade, J.; Garnier, J. 50 year trends in nitrogen use efficiency of world cropping systems: The relationship between yield and nitrogen input to cropland. *Environ. Res. Lett.* **2014**, *9*, 105011. [CrossRef]

5. Ram, M.S.; Shankar, T.; Maitra, S.; Adhikary, R.; Swamy, G.V.V.S.N. Productivity, nutrient uptake and nutrient use efficiency of summer rice (*Oryza sativa*) as influenced by integrated nutrient management practices. *Crop Res.* **2020**, *55*, 65–72. [CrossRef]
6. GOI. *Agricultural Statistics at a Glance 2018*; Government of India Ministry of Agriculture and Farmers Welfare Department of Agriculture, Cooperation and Farmers Welfare, Directorate of Economics and Statistics, Government of India; GOI: New Delhi, India, 2018; pp. 87–89.
7. Shankar, T.; Banerjee, M.; Malik, G.C.; Dutta, S.; Maiti, D.; Maitra, S.; Alharby, H.; Bamagoos, A.; Hossain, A.; Ismail, I.A.; et al. The productivity and nutrient use efficiency of rice–rice–black gram cropping sequence are influenced by location specific nutrient management. *Sustainability* **2021**, *13*, 3222. [CrossRef]
8. Mohapatra, T.; Nayak, A.K.; Raja, R.; Shahid, M. Vision 2050; *Central Rice Research Institute*; Indian Council of Agricultural Research: Cuttack, India, 2013; p. 26.
9. Pathak, H.; Tripathi, R.; Jambhulkar, N.N.; Bisen, J.P.; Panda, B.B. *Eco-Regional Rice Farming for Enhancing Productivity, Profitability and Sustainability*; NRRI Research Bulletin No. 22; ICAR-National Rice Research Institute: Cuttack, India, 2020; p. 28.
10. FAOSTAT 2020. Available online: http://www.fao.org/faostat/en/#data/RFN; http://www.fao.org/faostat/en/#data/RL (accessed on 17 March 2020).
11. Ladha, J.K.; Pathak, H.; Krupnik, T.J.; Six, J.; Van, K.C. Effciency of fertilizer nitrogen in cereal production: Retrospect and prospects. *Adv. Agron.* **2005**, *87*, 85–156.
12. Adhikari, C.; Bronson, K.F.; Panaullah, G.M.; Regmi, A.P.; Saha, P.K.; Dobermann, A.; Olk, D.C.; Hobbs, P.R.; Pasuquin, E. On-farm N supply and N nutrition in the rice wheat system of Nepal and Bangladesh. *Field Crop. Res.* **1999**, *64*, 273–286. [CrossRef]
13. Surekha, K.; Kumar, M.R.; Nagendra, V.; Sailaja, N.; Satyanarayana, T. 4R Nitrogen management for sustainable rice production. *Better Crop. South Asia* **2016**, *10*, 16.
14. Alam, M.K.; Bell, R.W.; Hasanuzzaman, M.; Salahin, N.; Rashid, M.H.; Akter, N.; Akhter, S.; Islam, M.S.; Islam, S.; Naznin, S.; et al. Rice (*Oryza sativa* L.) Establishment techniques and their implications for soil properties, global warming potential mitigation and crop yields. *Agronomy* **2020**, *10*, 888. [CrossRef]
15. Bhatt, R.; Singh, P.; Hossain, A.; Timsina, J. Rice–wheat system in the northwest Indo-Gangetic plains of South Asia: Issues and technological interventions for increasing productivity and sustainability. *Paddy Water Environ.* **2021**, 1–21. [CrossRef]
16. Farooq, M.; Siddique, K.H.M.; Rehman, H.; Aziz, T.; Lee, D.J.; Wahid, A. Rice direct seeding: Experiences, challenges and opportunities. *Soil Tillage Res.* **2011**, *111*, 87–98. [CrossRef]
17. Pathak, H.; Tewari, A.N.; Sankhyan, S.; Dubey, D.S.; Singh, M.U.; Jain, V.K.N.; Bhatia, A. Direct-seeded rice: Potential, performance and problems—A review. *Curr. Adv. Agric. Sci.* **2011**, *3*, 77–88.
18. De Datta, S.K. Technology development and the spread of direct seeded flooded rice in Southeast Asia. *Exp. Agric.* **1986**, *22*, 417–426. [CrossRef]
19. Pandey, S.; Velasco, L.E.; Suphanchaimat, N. Economics of direct seeding in northeast Thailand. In *Direct Seeding: Research Strategies and Opportunities*; Pandey, S., Mortimer, M., Wade, L., Tuong, T.P., Lopez, K., Hardy, B., Eds.; IRRI: Manila, Philippines, 2002; pp. 139–160.
20. Thet, K.; Ko, K.M.M.; Ngwe, K.; Min, T.D.; Win, K.K. Performance of Different Varieties in Direct Seeded Rice (*Oryza sativa* L.) as Affected by Different Sowing Methods. *J. Agric. Res.* **2019**, *6*, 65–71.
21. Bhatt, R.; Kukal, S.S. Direct Seeded Rice in South Asia. In *Sustainable Agriculture Reviews*; Lichtfouse, E., Ed.; Springer International Publishing: Cham, Switzerland, 2015; pp. 217–252. [CrossRef]
22. Ali, A.M.; Thind, H.S.; Singh, V.; Singh, B. A framework for refining nitrogen management in dry direct-seeded rice using GreenSeeker™ optical sensor. *Comput. Electron. Agric.* **2015**, *110*, 114–120. [CrossRef]
23. Ali, M.A.; Ladha, J.K.; Rickman, J.; Lales, J.S. Comparison of different methods of rice establishment and nitrogen management strategies for lowland rice. *J. Crop Improv.* **2006**, *16*, 173–189. [CrossRef]
24. Ahmed, S.; Latiful, B. Performance of Aus Rice in Different Tillage Systems and Crop Establishment Method in Southwest Bangladesh. *J. Exp. Sci.* **2018**, *9*, 05–08. [CrossRef]
25. Chen, S.; Ge, Q.; Chu, G.; Xu, C.; Yan, J.; Zhang, X.; Wang, D. Seasonal differences in the rice grain yield and nitrogen use efficiency response to seedling establishment methods in the Middle and Lower reaches of the Yangtze River in China. *Field Crop. Res.* **2017**, *205*, 157–169. [CrossRef]
26. Chauhan, B.S.; Awan, T.H.; Abugho, S.B.; Evengelista, G.; Sudhir, Y. Effect of crop establishment methods and weed control treatments on weed management, and rice yield. *Field Crop. Res.* **2015**, *172*, 72–84. [CrossRef]
27. Singh, B.; Sharma, R.K.; Kaur, J.; Jat, M.L.; Martin, K.L.; Singh, Y.; Singh, V.; Chandna, P.; Chaudhary, O.P.; Gupta, R.K.; et al. Assessment of the nitrogen management strategy using an optical sensor for irrigated wheat. *Agron. Sustain. Dev.* **2011**, *31*, 589–603. [CrossRef]
28. Ali, M.A. Development of an algorithm for optimizing nitrogen fertilization in wheat using GreenSeeker proximal optical sensor. *Exp. Agric.* **2020**, *56*, 688–698. [CrossRef]
29. Ramesh, C.; Pavithra, S. Fertiliser Use and Imbalance in India Analysis of States. *Econ. Political Wkly.* **2015**, *50*, 98–104.
30. Pathak, H. Nitrogen and climate change: Interactions, impacts, mitigation and adaptation. *J. Indian Soc. Soil Sci.* **2013**, *60*, 109–119.
31. Bouyoucos, G.J. Hydrometer method improved for making particle size analysis of soils. *J. Agron.* **1951**, *54*, 464–465. [CrossRef]
32. Jackson, M.L. *Soil Chemical Analysis*; Prentice Hall of India Pvt. Ltd.: New Delhi, India, 1973; pp. 183–193.

43. Walkley, A.; Black, I.A. An examination of the Degtjareff method for determining soil organic matter and a proposed modification of the chromic acid titration method. *Soil Sci.* **1934**, *37*, 29–38. [CrossRef]
44. Subbiah, B.V.; Asija, G.L. A rapid procedure for the determination of available nitrogen in soils. *Curr. Sci.* **1956**, *25*, 259–260.
45. Bray, R.H.; Kurtz, L.T. Determinations of total, organic and available forms of phosphorus in soils. *Soil Sci.* **1945**, *59*, 39–45. [CrossRef]
46. Hanway, J.J.; Heidel, H. Soil analyses methods as used in Iowa State College Soil Testing Laboratory. *Iowa Agric.* **1952**, *57*, 1–31.
47. Adhikari, B.; Bag, M.K.; Bhowmick, M.K.; Kundu, C. Status Paper on Rice in West Bengal; Hyderabad (India): Rice Knowledge Management Portal (RKMP). Directorate of Rice Research, Rajendranagar, Hyderabad 500030, India. 2011; pp. 1–47. Available online: http://www.rkmp.co.in/sites/default/files/ris/rice-state-wise/Status%20Paper%20on%20Rice%20in%20West%20Bengal.Pdf (accessed on 28 April 2021).
48. Pampolino, M.F.; Witt, C.; Pasuquin, J.M.; Johnston, A.; Fisher, M.J. Development approach and evaluation of the Nutrient Expert software for nutrient management in cereal crops. *Comput. Electron. Agric.* **2012**, *88*, 103–110. [CrossRef]
49. Ali, A.M.; Abou-Amer, I.; Ibrahim, S.M. Using GreenSeeker active optical sensor for optimizing maize nitrogen fertilization in calcareous soils of Egypt. *Arch. Agron. Soil Sci.* **2018**, *64*, 1083–1093. [CrossRef]
50. Singh, B.; Singh, V.; Purba, J.; Sharma, R.K.; Jat, M.L.; Singh, Y.; Thind, H.S.; Gupta, R.K.; Chaudhary, O.P.; Chandna, P.; et al. Site-specific fertilizer nitrogen management in irrigated transplanted rice (*Oryza sativa*) using an optical sensor. *Precis. Agric.* **2015**, *16*, 455–475. [CrossRef]
51. Xue, L.; Li, G.; Qin, X.; Yang, L.; Zhang, H. Topdressing nitrogen recommendation for early rice with an active sensor in South China. *Precis. Agric.* **2014**, *15*, 95–110. [CrossRef]
52. Nayak, A.K.; Mohanty, S.; Raja, R.; Shahid, M.; Lal, B.; Tripathi, R.; Bhattacharyya, P.; Panda, B.B.; Gautam, P.; Thilagam, V.K.; et al. Customized Leaf Colour Chart for Nitrogen Management in Rice for Different Ecologies. *CRRI News Letter* January–March. **2013**, *34*, 14. Available online: http://www.crri.nic.in/CRRI_newsletter/crnljanmar13 pdf (accessed on 20 May 2021).
53. Roshanak, S.; Rahimmalek, M.; Goli, S.A.H. Evaluation of seven different drying treatments in respect to total flavonoid, phenolic, vitamin C content, chlorophyll, antioxidant activity and color of green tea (*Camellia sinensis* or *C. assamica*) leaves. *J. Food Sci. Technol.* **2016**, *53*, 721–729. [CrossRef] [PubMed]
54. Sestak, Z.; Catsky, J.; Jarvis, P.G. *Plant Photosynthetic Production. Manual of Methods*. Junk, W., Ed.; The Hague: The Netherlands, 1971; pp. 818–824.
55. Watson, D.J. The physiological basis of variation in yield. *Adv. Agron.* **1952**, *4*, 101–145.
56. Cochran, W.G.; Cox, G.M. *Experimental Design*; Asia Publishing House: Calcutta, India, 1977; pp. 142–181.
57. Yogeswari, D.; Porpavai, S. Effect of crop establishment methods and irrigation scheduling on growth and yield of rice. *Int. J. Chem. Stud.* **2018**, *6*, 32–35.
58. Gill, J.S.; Walia, S.S.; Gill, R.S. Direct seeded rice: An alternative rice establishment technique in north-west India—A review. *Int. J. Adv. Res.* **2014**, *2*, 375–386.
59. Gupta, G.; Shrestha, A.; Shrestha, A.; Amagain, L.P. Evaluation of different nutrient management practice in yield and growth in rice in Morang district. *Adv. Plants Agric. Res.* **2016**, *3*, 187–191. [CrossRef]
60. Chou, J.T.; Wu, Z.; Jiang, T.H.; Liu, F.; Zhang, Z.J. Effects of nitrogen reduction at different growth stages on rice population production characteristics and preliminary analysis of nitrogen reduction strategies. *J. Yangzhou Univ.* **2020**, *65*, 52–58.
61. Moldenhauer, K.A.K.; Gibbons, J.H. Rice Morphology and Development. In *Rice: Origin, History, Technology, and Production*; Smith, C.W., Dilday, R.H., Eds.; John Wiley & Sons, Inc.: Hoboken, NJ, USA, 2003; pp. 103–128.
62. Huang, M.; Zou, Y.; Jiang, P.; Xia, B.; Feng, Y.; Cheng, Z.; Mo, Y. Yield Component Differences between Direct-Seeded and Transplanted Super Hybrid Rice. *Plant Prod. Sci.* **2011**, *14*, 331–338. [CrossRef]
63. Zhong, X.; Peng, S.; Sanico, A.L.; Liu, H. Quantifying the interactive effect of leaf nitrogen and leaf area on tillering of rice. *J. Plant Nutr.* **2003**, *26*, 1203–1222. [CrossRef]
64. Fageria, N.K. Yield physiology of rice. *J. Plant Nutr.* **2007**, *30*, 843–879. [CrossRef]
65. Kumhar, B.L.; Chavan, V.G.; Rajemahadik, V.A.; Kanade, V.M.; Dhopavkar, R.V.; Tilekar, R.N. Effect of different rice establishment methods on growth, yield and different varieties during kharif season. *Int. J. Plant Ani Environ. Sci.* **2016**, *6*, 127–132.
66. Liu, H.; Hussain, S.; Zheng, M.; Peng, S.; Huang, J.; Cui, K.; Nie, L. Dry direct-seeded rice as an alternative to transplanted-flooded rice in Central China. *Agron. Sustain. Dev.* **2015**, *35*, 285–294. [CrossRef]
67. Wang, Y.; Li, C.; Li, Y.; Zhu, L.; Liu, S.; Yan, L.; Feng, G.; Gao, Q. Agronomic and environmental benefits of nutrient expert on maize and rice in Northeast China. *Environ. Sci. Pollut. Res.* **2020**, *27*, 28053–28065. [CrossRef] [PubMed]
68. Ali, A.M.; Thind, H.S.; Sharma, S.; Singh, V. Prediction of dry direct-seeded rice yields using chlorophyll meter, leaf color chart and GreenSeeker optical sensor in northwestern India. *Field Crop. Res.* **2014**, *161*, 11–15. [CrossRef]
69. Sharma, V.; Bali, A.S.; Kachroo, D. Effect of different establishment methods and sowing schedules on growth and yield of hybrid rice (*Oryza sativa*) and their after effects on succeeding wheat (*Triticum aestivum*) in rice—Wheat cropping system. *Econ. Aff.* **2016**, *61*, 487–493. [CrossRef]
70. Budhathoki, S.; Amgain, L.P.; Subedi, S.; Iqbal, M.; Nikee Shrestha, N.; Aryal, S. Assessing growth, productivity and profitability of drought tolerant rice using nutrient expert—Rice and other precision fertilizer management practices in Lamjung, Nepal. *Acta Sci. Agric.* **2018**, *2*, 153–158. [CrossRef]

61. Kumar, T.; Singh, G.; Singh, R.A.; Shahi, A.K.; Kumar, M.; Rajput, S.K.S. Effect of site-specific nutrient management on productivity and profitability of rice in low land situation. *Int. J. Chem. Stud.* **2019**, *7*, 1963–1966.
62. Pateel, V.L.; Veeresh, H.; Narayana, K.R.; Gaddi, A.K.; Basavanneppa, M.A. Use of chlorophyll meter and optical sensors for nitrogen management in direct seeded rice. *J. Farm Sci.* **2017**, *30*, 365–369.
63. Alam, M.; Sarker, S.; Momin, M.A. Profitability of rice production using a drum seeder. *J. Bangladesh Agril. Univ.* **2007**, *5*, 135–144
64. Ramesh, S.; Chandrasekaran, B. Evaluation of crop establishment methods and nitrogen management strategies on realizing yield potential of rice hybrid ADTRH 1. *Asian J. Plant Sci.* **2007**, *6*, 239–251. [CrossRef]
65. Zhu, D.; Zhang, H.; Guo, B.; Xu, K.; Dai, Q.; Wei, H.; Gao, H.; Hu, Y.; Cui, P.; Huo, Z. Effects of nitrogen level on yield and quality of japonica soft super rice. *J. Integr. Agric.* **2017**, *16*, 984–991. [CrossRef]
66. Yang, J.; Zhang, J. Crop management techniques to enhance harvest index in rice. *J. Exp. Bot.* **2010**, *61*, 3177–3189. [CrossRef]
67. Bhagavathi, M.S.; Baradhan, G.; Suresh Kumar, S.M.; Arivudainambi, S. Effect of different crop establishment methods on rice (*Oryza sativa* L.): A review. *Plant Arch.* **2020**, *20*, 3416–3422.
68. Bana, R.C.; Yadav, S.S.; Shivran, A.C.; Singh, P.; Kudi, V.K. Site-specific nutrient management for enhancing crop productivity. *Int. Res. J. Pure Appl. Chem.* **2020**, *21*, 17–25. [CrossRef]
69. Rickman, J.F.; Pyseth, M.; Bunna, S.; Sinath, P. Direct seeding of rice in Cambodia. In *Increased Lowland Rice Production in the Mekong Region*; Fukai, S., Basnayake, J., Eds.; ACIAR: Canberra, ACT, Australia, 2001; p. 101.
70. Lav, B.; Ladha, J.K.; Gupta, R.K.; Singh, S.; Tirol-Padre, A.; Saharawat, Y.S.; Gathala, M.; Pathak, H. Saving of water and labour in rice-wheat systems with no-tillage and direct seeding technologies. *Agron. J.* **2007**, *99*, 1288–1296.
71. Kaur, J.; Singh, A. Direct seeded rice: Prospects, problems/constraints and researchable issues in India. *Curr. Agric. Res. J.* **2017**, *5*, 13–32. [CrossRef]
72. Kumar, P.; Kumar, S.; Mouriya, A.K.; Kumar, V. Productivity and economics of direct seeded rice. *Int. J. Sci. Environ. Technol.* **2018**, *7*, 2033–2039.
73. Rashid, M.H.; Alam, M.M.; Khan, M.A.H.; Ladha, J.K. Productivity and resource use of direct-(drum)-seeded and transplanted rice in puddled soils in rice–rice and rice–wheat ecosystems. *Field Crop. Res.* **2009**, *113*, 274–281. [CrossRef]
74. Xu, L.; Li, X.; Wang, X.; Xiong, D.; Wang, F. Comparing the Grain Yields of Direct-Seeded and Transplanted Rice: A Meta-Analysis. *Agronomy* **2019**, *9*, 767. [CrossRef]
75. Yadav, S.; Evangelista, G.; Faronilo, J.; Humphreys, E.; Henry, A.; Fernandez, L. Establishment method effect on crop performance and water productivity of irrigated rice in the tropics. *Field Crop. Res.* **2014**, *166*, 112–127. [CrossRef]
76. Gopal, R.; Jat, R.K.; Malik, R.K.; Kumar, V.; Alam, M.M.; Jat, M.L.; Mazid, M.A.; Saharawat, Y.S.; McDonald, A.; Gupta, R. *Direct Dry Seeded Rice Production Technology and Weed Management in Rice Based Systems*; Technical Bulletin; International Maize and Wheat Improvement Center: New Delhi, India, 2010; p. 28.
77. Gupta, R.K.; Ladha, J.K.; Singh, S.; Singh, R.; Jat, M.L.; Saharawat, Y.; Singh, V.P.; Singh, S.S.; Singh, G.; Sah, G.; et al. *Production Technology for Direct Seeded Rice*; Rice–Wheat Consortium Technical Bulletin 8; Rice–Wheat Consortium for the Indo-Gangetic Plains; NASC: New Delhi, India, 2006; p. 16.
78. Bell, M.A.; Rickman, J.; Castro, J.E.C.; Actan, L.B.; McNamara, J. Precision land leveling for rice production in Asia. In *Proceedings of the International Agricultural Engineering Conference*; Salokhe, V.M., Jianxia, Z., Eds.; Asian Institute of Technology (AIT): Bangkok, Thailand, 1998; pp. 257–264.
79. Rahman, A.; Salam, M.A.; Kader, M.A. Effect of crop establishment methods on the yield of boro rice. *J. Bangladesh Agric. Univ.* **2019**, *17*, 521–525. [CrossRef]
80. Goudra, S.; Mudalagiriyappa, D.C.; Kalyana, K.N.; Murthy, P.K.; Kumar, M.P. Influence of precision nitrogen management through crop sensors on growth and yield of aerobic rice (*Oryza sativa* L.). *J. Pharm. Phytochem.* **2019**, *8*, 2409–2413.
81. Ladha, J.K.; Jat, M.L.; Stirling, C.M.; Chakraborty, D.; Pradhan, P.; Krupnik, T.J.; Sapkota, T.B.; Pathak, H.; Rana, D.S.; Tesfaye, K.; et al. Achieving the sustainable development goals in agriculture: The crucial role of nitrogen in cereal-based systems. *Adv. Agron.* **2020**, *163*, 39–116. [CrossRef]
82. Kandel, S.; Shrestha, A.; Neupane, M.P. Assessing the growth and productivity of rice under different fertilizer management practices grown under SRI Production in Kavre, Midhill of Nepal. *Ann. Rev. Res.* **2018**, *3*, 555614.
83. Liu, X.; Ferguson, R.B.; Zheng, H.; Cao, Q.; Tian, Y.; Cao, W.; Zhu, Y. Using an Active-Optical Sensor to Develop an Optimal NDVI Dynamic Model for High-Yield Rice Production (Yangtze, China). *Sensors* **2017**, *17*, 672. [CrossRef] [PubMed]
84. Shrestha, S.; Amgain, L.; Subedi, R.; Shrestha, P.; Shahi, S. Productivity and Profitability Analysis of Old Aged Hybrid Rice Seedling Using Nutrient-Rice Expert and Other Precision Nutrient Management Practices at Lamjung, Nepal. *Int. J. Appl. Sci. Biotechnol.* **2018**, *6*, 232–237. [CrossRef]
85. Hossain, M.F.; Salam, M.A.; Uddin, M.R.; Pervez, Z.; Sarkar, M.A.R. A comparative study of direct seeding versus transplanting method on the yield of aus rice. *J. Agron.* **2002**, *1*, 86–88. [CrossRef]
86. Raj, S.K.; Jose, N.; Mathew, R.; Leenakumary, S. Influence of stand establishment techniques on yield and economics of rice cultivation in Kuttanad. *Int. J. Sci. Res. Publ.* **2013**, *3*, 1–6.
87. Seth, M.; Thakurand, D.R.; Manuja, S. Effect of tillage and site-specific nutrient management on productivity of rice-wheat cropping system. *J. Crop Weed.* **2019**, *15*, 115–119.
88. Krishnakumar, S.; Haefelem, S. Integrated nutrient management and LCC based nitrogen management on soil fertility and yield of rice (*Oryza sativa* L.). *Sci. Res. Essays* **2013**, *8*, 2059–2067. [CrossRef]

89. Kumar, S.; Panwar, A.S.; Naresh, R.K.; Singh, P.; Mahajan, N.C.; Chowdhary, U.; Kumar, S.; Malik, M.; Meena, A.L.; Ghashal, P.C.; et al. Improving Rice-wheat cropping system through precision nitrogen management: A review. *J. Pharma. Phytochem.* **2018**, *7*, 1119–1128.
90. Kumar, A.; Jnanesha, A.C. Effect of Crop Establishment Methods on Growth Yield and Water Productivity of Rice. *Int. J. Agric. Sci.* **2017**, *8*, 40–45.
91. Devkota, K.P.; Khanda, C.M.; Beebout, S.J.; Mohapatra, B.K.; Singleton, G.R.; Puskur, R. Assessing alternative crop establishment methods with a sustainability lens in rice production systems of Eastern India. *J. Clean. Prod.* **2020**, *244*, 118835. [CrossRef]
92. Dobermann, A.; Witt, C.; Dawe, D.; Gines, G.C.; Nagarajan, R.; Satawathananont, S.; Son, T.T.; Tan, P.S.; Wang, G.H.; Chien, N.V.; et al. Site-specific nutrient management for intensive rice cropping systems in Asia. *Field Crop. Res.* **2002**, *74*, 37–66. [CrossRef]
93. Qureshi, A.; Singh, D.K.; Pandey, P.C.; Singh, V.P.; Raverkar, K.P. Site specific nutrient management approaches for enhancing productivity and profitability in rice and wheat under rice-wheat cropping system. *Int. J. Agric. Sci.* **2016**, *8*, 2838–2842.

agronomy

MDPI

Article

Understanding the Dynamic of Rice Farming Systems in Southern Mozambique to Improve Production and Benefits to Smallholders

Fátima Ismael [1,2,3,*], Aires A. Mbanze [2,4,5,*], Alexis Ndayiragije [6] and David Fangueiro [3]

1 Department of Rural Development, Faculty of Agricultural Sciences, Campus Universitário de Unango, Universidade Lúrio, Sanga 3302, Mozambique
2 Nova School of Business and Economics, Campus de Carcavelos, Universidade Nova de Lisboa, Rua da Holanda 1, 2775-405 Lisbon, Portugal
3 Linking Landscape, Environment, Agriculture and Food (LEAF), Instituto Superior de Agronomia (ISA), Universidade de Lisboa, Tapada da Ajuda, 1349-017 Lisbon, Portugal; dfangueiro@isa.ulisboa.pt
4 Department of Environment and Nature Conservation, Faculty of Agrarian Sciences, Campus Universitários de Unango, Universidade Lúrio, Sanga 3302, Mozambique
5 ECOSLIVE—Ecosystem Services and Livelihoods Enhancement (Serviços do Ecossistema e Melhoria dos Modos de Vida), Lichinga 3300, Mozambique
6 Internacional Rice Research Instituto (IRRI), Instituto de Investigação Agrária de Moçambique-IIAM, Avenida FPLM, Maputo 3658, Mozambique; a.ndayiragije@irri.org
* Correspondence: fatima786.ismael@gmail.com (F.I.); airesmbanze@unilurio.ac.mz (A.A.M.)

Citation: Ismael, F.; Mbanze, A.A.; Ndayiragije, A.; Fangueiro, D. Understanding the Dynamic of Rice Farming Systems in Southern Mozambique to Improve Production and Benefits to Smallholders. *Agronomy* **2021**, *11*, 1018. https://doi.org/10.3390/agronomy11051018

Academic Editors: Umberto Anastasi and Aurelio Scavo

Received: 21 March 2021
Accepted: 9 May 2021
Published: 20 May 2021

Publisher's Note: MDPI stays neutral with regard to jurisdictional claims in published maps and institutional affiliations.

Abstract: Rice farming systems (RFSs) in southern Mozambique are very heterogeneous and diversified, which has implications for smallholders' adoption of each RFS, as well as on rice production and productivity in the region. In this regard, it is important to understand: (i) which RFS typologies can be leveraged to improve rice production and productivity; (ii) the drivers for smallholder farmers' decisions to adopt an RFS; and (iii) which policies/incentives could enhance existing RFSs. The present study was based on surveys of 341 smallholder rice farmers in the Chókwè Irrigation Scheme (CIS), southern Mozambique. Data on the productivity of rice, size of the herd, and total other crop types were used to frame the RFS typologies. A multinomial logit model (MLM) and multiple linear regression (MLR) were applied to determine the driver for each RFS, and predict the constraints for production and yield. Based on cluster analysis, four typologies of RFSs were identified: the subsistence farming system (FS), specialised rice FS, mixed crops FS, and rice–livestock FS. Farms with longer experience reported applying more fertiliser and seedlings per unit hectare. The availability of labour increased the likelihood of adopting the mixed crops FS and rice–livestock FS. Older households were more likely to adopt the subsistence FS, and live closer to the farming fields. Yield of rice was positively associated with inputs such as fertilisers, pesticides, and seedlings, as well as years of experience of the household. Our results suggest that smallholder farmers need more assistance and technical support to identify and adopt more productive and less costly RFSs in this region.

Keywords: crop–livestock; farming systems; production and productivity of rice; fertilisation; smallholder farmers

1. Introduction

Global rice production reached 0.5 billion tonnes (on a milled basis) in 2018, which represents an increase of 1.4% [1], and it will continue to grow, especially in Africa, where production is far behind the global average [2]. This significant growth was driven by market demand, prices, and state subsidies [3]. The Asian region accounted for almost 80% of the increased production [4], while Sub-Saharan Africa is the only region in the world where food production per capita, including of staple cereals, has been growing slowly, leaving many people more vulnerable to food insecurity [5,6]. By 2050, the African population is predicted to reach 2.5 billion, more than double the current population [7]. Feeding this growing population will remain a great challenge for most of the African governments', requiring rapid changes to policies and agricultural technology [8].

Global cereal production is increasing dramatically, providing a platform for rural and urban economic growth [9]. In Africa, as in many other parts of the developing world, cereals such as maize, rice, and wheat are essential for the daily diet of most rural and urban households, preventing them from falling into acute food insecurity [1,10,11]. Indeed, these three staple cereals together account for 94% of all cereal consumption in Africa [10], which also helps to frame the prevailing narrative that African agriculture has lagged behind the rest of the world [9].

Even in the context of lower staple cereal production, there is a widespread agreement that the agriculture sector will remain pivotal for the development of the sub-Saharan region [9], employing many rural people; up to 80% are smallholder farmers who produce most of their regional food [11,12]. Rural smallholder farms grow a wide variety of food grains, root crops, cash crops, and livestock that support diverse food and livelihood systems in different agricultural zones, and traditionally produce modest surpluses for local or distant trade [13].

Despite the availability of lowland and wetland suitable for sustainable rice-based cropping [14], rice production and productivity in sub-Saharan Africa is hindered by low soil fertility [5,15,16], a lack of technology [11,17], poor agricultural polices [18], and a lack of adequate infrastructure [6] and skilled workers [19]. As in most sub-Saharan countries, agriculture is a key sector in Mozambique, employing 80% of the labour force and contributing approximately 20% of the GDP [20]. Rice is the main cereal, second to maize, and its production area encompasses 204,000 ha, with an average paddy yield of 1.27 t/ha (Japan International Cooperation Agency [21]). This figure is remarkably low compared to the average paddy yields of 4.2 t ha^{-1} in Asia [22]. Most of the farming plots are located in lowlands, which are seasonally rain-fed and account for 90% of the total rice area (Ministry of Agriculture, [23]), and contribute about 10–15% of the cereal caloric supply at the national level [24].

The growing human population and increase in middle-class consumers have exacerbated the demand for rice in Mozambique [24]. This increasing demand has created 300,000 t $year^{-1}$ of rice deficit, which has been covered by importation from Asian countries (National Institute of Statistics [25]). The production deficit is likely related to (i) a lack of technology (agricultural mechanization, use of chemical and organic fertilisers, herbicides, and improved rice varieties); (ii) insufficient support for smallholder farmers, who are the main rice producer in the country [26]; (iii) a lack of extension services to smallholders [27]; and (iv) high heterogeneity and diversity of farming systems (FS), which hampers the implementation of agriculture policies. The construction of specific typologies of FSs, and understanding of drivers that motivate smallholder farms to adopt each specific FS, will be a useful step forward to frame the aforementioned problems [13,28].

Although the country has a potential to reach 900,000 hectares of rice production, it is estimated that only 35% of this area is under cultivation, mostly in Gaza province (south Mozambique), Zambézia and Sofala provinces, in the centre of the country, and Nampula and Cabo Delgado provinces, in the north [23,29]. The majority of rice farming fields in the south and centre of the country are located in the Chókwè and Baixo Limpopo irrigation schemes, respectively [24]. The Chókwè Irrigation Scheme (CIS) is the largest irrigated area in Mozambique, and it has a vigorous agricultural community (including rice cultivation, horticulture, and sheep farming) in intensive and mixed FSs. However, the production volume has been remarkably low, with a lengthy stagnation since 1988, due to internal warfare, floods of the Limpopo River [21], and a lack of policies, especially around the difficulty of accessing credit [30]. Most rice producers are smallholders (>5 ha) and medium holders (5–20 ha), who also need to diversify their production to improve their livelihood and income generation. The spatial predominance of mixed FSs is dependent on the drivers and constraints. Thus, it is important to propose incentives to effectively improve the production of rice in the CIS. This will require an in-depth understanding of the existing FSs and other alternative livelihood options available in the region. The present study aimed to answer the following questions: (i) Which typologies of rice Farming Systems (RFSs) are predominant in the CIS? (ii) What are the drivers for each RFS in the area? (iii) Do different demographic patterns affect household decisions to embrace different FSs? (iv) What factors affect production and productivity for smallholder farmers in the CIS? (v) What policies/incentives can be proposed to enhance production and productivity of rice in the CIS?

To answer the above questions, a survey was carried out with smallholder rice farmers who were based adjacent to the CIS. Answering these questions was important in order to underpin development strategies, assess production constraints, prioritise research, and identify scaling potentials, which in turn will improve local food production and nutrition security, and the rice value scheme.

2. Materials and Methods

2.1. Study Area: Chókwè Irrigation Scheme (CIS)

The Chókwè Irrigation Scheme (CIS) is located in southern Mozambique, in the Chókwè District, adjacent to the Limpopo River [31,32] (Figure 1). It is the largest irrigation scheme in the country [23,24], covering 33,000 ha of land [21]. Its production potential has not yet been achieved, and is hindered by an insufficient supply of irrigation water, excessively expensive chemical fertiliser, and moderately costly labour [30]. The total population living in Chókwè district is about 212,071 people, with 56% living in rural areas (INE, 2017). The climate of the region is semi-arid [33], with an average annual temperature of 22–26 °C, and rainfall averaging between 500 and 700 mm/year [30,32]. The CIS is composed of three main hydraulic sectors: montante (upstream), sul (midstream), and rio (downstream) (Figure 1). The hydraulic structures in the irrigation scheme include the Massingir dam, Macarretane weir, and the main, secondary, and tertiary canals, as well as the drainage network [32].

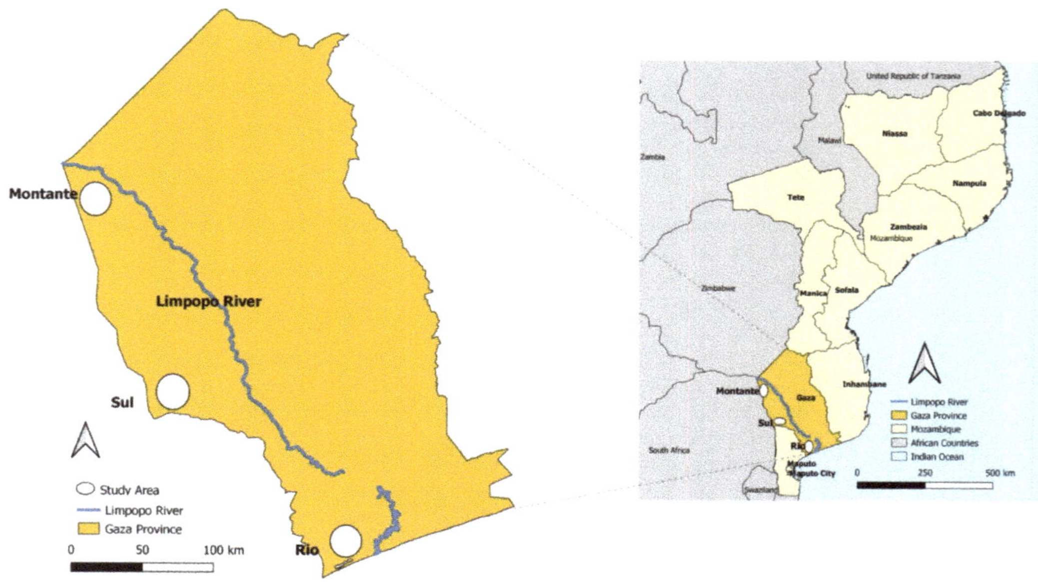

Figure 1. The location of the Chokwe Irrigation Scheme (CIS).

The predominant FS in the area is agropastoral [13], with approximately 3500 farm households, extension workers, and technical staff. Small-scale farms hold on average less than 5 ha per plot of land for rice production [21]. Most of the agricultural activities in the area include mixed crops and livestock production, in which most of the vegetables are produced on wetlands in the dry season [33]. Rice yield is on average low in small scale FSs, but has slightly increased in the last few years [21]. Few commercial rice farmers use farm machinery, fertilisers, pesticides, and improved seeds varieties [30].

2.2. Data Collection

The survey was conducted in 25 villages, selected to cover the three main regions of the CIS: upstream/montante, midstream/sul, and downstream/rio (Figure 1). The villages were selected after an exploratory field visit and consultation with key informants (government entities and the village chairperson). Most of the households in the villages were essentially rice farmers, who also practiced other activities to sustain their daily livelihoods and generate income, such as cattle herding, fishing, and small-scale informal business, and who also cultivated fresh vegetables. Upstream, we surveyed eight villages representing a total of 93 farms (27.19%), while in the midstream and downstream a total of 5 and 12 villages were sampled, covering a total of 110 farms (32.16%) and 138 farms (40.35%), respectively (Table 1). The number of households sampled was more than 10% of the total in all of the sampled villages. According to Bartlett at al. (2001) [34] and Landry et al. [35], a 5% sampling intensity is sufficient for social studies.

Table 1. Total number of smallholder rice farmers in the three main zones of the irrigation scheme, sample size and number of village samples per zone

Chókwè Scheme	Number of Villages Sampled	Total Small Rice Farms	N° and (%) of Household Sampled
Upstream/Montante	8	782	93 (11.9)
Midstream/Sul	12	1137	138 (12.1)
Downstream/Rio	5	994	110 (11.1)
Total	25	2913	341 (11.7)

The survey was conducted from February to June 2019. During this period, a total of 346 smallholder farmers were surveyed; of these, 5 surveys were not validated. The criteria used to select the 341 surveyed households was: (i) living in a village adjacent to the CIS for more than two years; and (ii) having a rice farming plot of less than 5 ha in size in the CIS.

Household Survey

The first section of the survey obtained the socio-economic characteristics of the surveyed households, and a description of their agriculture production system: (i) socio-demographic and socio-economic information of the household (gender, age, the composition of the family, and education); (ii) different sources of income; (iii) if the household held a permanent labour force or hired labour workers; and (iv) agricultural and livestock production. The second section obtained the characterisation of the RFS, including (i) type of land access; (ii) soil fertilisation; (iii) weed control; (iv) types of seeds and yield; and (v) the destination of the production, including market access. All questions included in the survey were based on a literature review, field visits, and meetings with key informants, and part of the questionnaire was from a similar study conducted in Angola, with some adjustments [36]. The survey was conducted in close collaboration with two field workers, who were native to the area and able to speak Changana, the native language, because most of the respondents either did not understand Portuguese or preferred to be interviewed in Changana. To control for potential mistranslation, the survey was translated to Changana and reverse translated to Portuguese, until a similar meaning was consistently achieved.

The questionnaire was pre-tested with five respondents from upstream and was later adjusted based on their observations.

2.3. Data Analysis

Data were coded and analysed using R software. Exploratory analysis and descriptive statistics were performed. A table of frequencies and percentages was used to represent the socio-demographic and socio-economic characteristics of the households in the CIS. The information included gender of the household head, education, the main and secondary activities, accessibility of land, the number of households who hire or have permanent workers, average distance to the farming field, the use of fertiliser and types, livestock husbandry, and destination of production output.

2.3.1. Typology of Rice Farming Systems

The prevailing RFSs were developed based on three main variables: yield of rice, expressed in ton ha^{-1}; livestock breeding; and total other types of crops (vegetables (cabbage, lettuce, tomato, etc.), maize (*Zea mays*), cowpea (*Vigna unguiculata*), common bean (*Phaseolus vulgaris*), and sweet potato (*Ipomoea batatas*), that households reported growing. Three livestock categories (cattle, swine, and goats) were considered. We used only these categories because: (i) these are the livestock categories that most households declared; (ii) households devote a considerable amount of time and effort to these activities; (iii) depending on the size of the herd, the household may need an additional labour force and it may impact resource allocation; (iv) a trade-off between livestock production and agricultural activities is also common, as suggested by [36,37]. Because the survey was focused only on rice, the yield of other crops was not assessed. Nevertheless, the annual yield of each crop was needed for construction of the FS typology [16,38,39]; an artefact was thus used to derive a proxy variable, which consisted of summing the number of all other crops the household reported growing, besides rice. Although we recognise the limitations of this approach, it was the most appropriate, in the sense that it best captured the main purpose of the work, which was to understand the basis of household decision making with regards to enterprise diversification, with an emphasis on rice. In future research, a more flexible approach to the typology of FS construction might provide further context and insight into the causes, consequences, and negotiations of farm diversity [40].

The classification of the RFS was assessed through cluster analysis of the household data on rice yield, total number of other types of crop the household reported growing, and household herd size [40,41]. We used the Minkoskwi distance as a measure of dissimilarity and ran Ward's method because our variables were mixed between continuous and discrete. A Z transformation was also used to standardise the different scales of variables, minimising the object function error [42]. To understand to what extent each variable described the FS, an analysis of variance (ANOVA) was also conducted. If a significant difference between each variable that described the RFS was detected, the post hoc test of Tukey was used to test for statistical differences of means between pairs of clusters of the RFS.

2.3.2. Crop Patterns across Farming Systems

To characterise the distribution patterns of other crops at the FS level, a cross-tabulation between the FS and each crop type was assembled and tested to verify whether the null hypothesis of similar patterns of crop distribution across the FS could be rejected. Post hoc cell-wise tests were performed to find out which crop types were above/below what would be expected by chance in each FS [39,43]. The same procedure was also used to characterise the proportion of literate households across each FS. The ANOVA and a non-parametric Kruskal–Wallis (KW) test were required to test for statistical differences of average distances households travel to reach the farming field.

2.3.3. Drivers and Predictors of Rice Yield Production and Farming Systems

In Table 2, we present all variables used to compute the models for prediction of the driver of different RFSs, and the predictors of production and yield of rice in the CIS.

Table 2. Drivers for rice farming systems (RFSs), and predictors of production and productivity of rice, in the Chókwè Irrigation Scheme.

Variable Name/Cod	Type	Unity of Measuring/Class	Min-Max	Mean (SE)
Age of the household	Numerical	NA	25–79	53.30 (±0.59)
Education	Ordinal	4 Classes	Illiterate (0)–high School (3)	1.01 (±0.35)
Distance to the farming field	Ordinal	4 Classes	<30 min (1)–>60 min (4)	2.47 (±0.58)
Labour force in the family	Numerical	NA	1–28	10.23 (±0.25)
Permanent workers	Categorical	Dummy	0–1	NA
Seeding	Numerical	Kg/ha	20–300	50.98 (±1.47)
Application of fertiliser	Categorical	dummy	0–1	NA
Quantity of fertiliser	Numerical	Kg/ha	0–200	22.58 (±2.35)
Use of pesticides	Categorical	Dummy	0–1	NA
Total rice activities	Numerical	NA	0–3	1.3 (±0.04)
Region	Categorical	Dummy	0–1	NA

Socio-economic drivers and predictors were: (i) age and education of the household. Age was a numerical variable, while education was ordinal, they were coded in the following categories: (0 = illiterate, 1 = primary school, 2 = secondary school and 3 = high school); (ii) the distance to the farming field. This was the average time that the household spent travelling to the farming area; thus, this time represents how far the field was from the place where the family lived and was coded in four categories (1 for <30 min; 2 = 30–45 min; 3 = 46–60 min; and 4 for $\geq$60 min); (iii) the availability of labour in the family. This was the number of family members of active working age, including men, women, and youth; (iv) whether the household had permanent paid workers, and whether the household used fertilisers or not, were both coded as dummy variables (yes = 1 or no = 0); (v) the amount of seedlings and fertiliser that households use per hectare of rice; (vi) The total number of activities involved in rice production. The biophysical variable only considered the region in which the farming field was located. This variable was coded as a dummy (0 = if the household farm was in the upstream and 1 = if it was in the midstream or downstream).

2.3.4. Rice Farming Systems

A multinomial logistic model (MLM) was applied to investigate the importance of each driver of the RFSs. The importance of the variables in the fitted model was detected based on the log-likelihood, likelihood ratio, and Nagelkerke and Cox and Snell pseudo R-square. Predictors were selected based on their significance in the model, and possible meaningful interpretations. The importance of each predictor included in the model was assessed at the $p \leq 0.05$ level of significance.

2.3.5. Predicting Factors Affecting Production and Yield

A multiple linear regression (MLR) was also applied to investigate the factors affecting the production and yield of rice in the CIS. The rice yield (tonnes ha^{-1}) was used as a response variable to predict the factor constraint productivity, while the extension of rice farming (ha) was used as a response variable to predict the factors that positively or negatively affect the expansion of the rice area. The importance of each predictor in the model was assessed through the significance of each coefficient. A stepwise procedure was used to select the most parsimonious sub-model, based on the Akaike information criterion (AIC). All models were selected based on the significance of the p-value, residual standard error (RSE), multiple R-squared, and Adjusted R-square.

3. Results

3.1. Socio-Economic and Demographic Characteristics

The socio-economic and demographic characteristics of the respondents are presented in Table 3 and Table S1 (Supplementary Materials). More than half of the respondent farmers were men (210, 62%), while the remaining 131 (38%) were women. The predominant age group ranged from 35 to 64 years old, comprising 80% of all sample households. Only 19 (5.6%) respondents were in a younger (25–34) age group. The smallest group was the older population (65–79 years old, at 47, or 14%). Most of the households were either illiterate (55, 17.3%) or had a primary school level of education (230, 67.4%). Most respondents (256, 75.1%) reported they do not use fertiliser for their rice crop. Urea and Urea + NPK (84, 24.2%) were the most widely used fertilisers (Supplementary Materials).

Table 3. Socio-economic and demographic characteristics of surveyed households in the CIS.

Variables	Frequency	Percentage %
	Gender	
Male	210	62
Female	131	38
Total	341	100
	Age	
25–34	19	5.6
35–49	103	30
50–64	172	50
65–79	47	14
Total	341	100
	Level of education	
Illiterate	59	17.3
Primary	230	67.4
Secondary	42	12.3
High School	10	2.9
Total	341	100
	Use of fertilisers	
No	256	75.1
Yes	85	24.9
Total	341	100

Most of the households (294, 86%) reported paying for extra labour; only 13 (3.8%) farmers had permanent workers. Livestock rearing was also common among the farmers (246, 72%). Most of the farmers (224, 66%) reported that the largest proportion of production was for both consumption and sale (Supplementary Materials).

3.2. Typologies of Different Rice Farming Systems in the CIS

Four FSs of rice were identified in the CIS (Table 4): (i) the subsistence farming system where the outputs of rice and livestock are relatively lower, but only two other crops (maize and common bean) beside rice are produced; (ii) the specialised rice FS, where rice production is on average higher and dominant. The average size of the livestock herd in this system is four animals, and the number of other crops besides rice that the household grows was slightly different from subsistence farming. The common bean is the most evident secondary crop; (iii) mixed crops, in which the rice yield is lower and statistically similar to the subsistence FS. In this system, the size of the herd of each household is relatively lower, but similar to the previous two systems, and on average each household grows more crops besides rice (vegetables, maize, cowpeas, and sweet potatoes); and (iv) crop-livestock, where livestock is clearly predominant compared to the rest of the systems, followed by rice. On average, each household reported growing four other crops besides rice; specifically, the same crops as in the previous FS.

Table 4. Farming systems in the Chókwè Irrigation Scheme, and some of their socio-economic determinants.

	Farming System						
	Subsistence	**Specialised Rice**	**Mixed Crops**	**Rice–Livestock**			
Variables	**N = 105 ǀ 30.8%**	**N = 85 ǀ 24.9%**	**N = 61 ǀ 17.9%**	**N = 90 ǀ 26.4%**	**Mean**	***p*-Value**	**Eta^2**
Rice (ton/ha)	1.64 c	3.24 a	1.88 c	2.49 b	2.31	0.000 ***	0.374
Livestock	2.95 b	3.59 b	3.61 b	18.46 a	7.32	0.000 ***	0.478
Other Crops	1.61 c	1.94 b	3.59 a	3.56 a	2.56	0.000 ***	0.729
Proportion of Households Growing Each Type of Other Crop (%)							
Vegetables	6.7	15.3	60.7	61.1	31.8(112)	0.000 ***	0.54
Adjusted residual	− − −	− −	+++	+++			
Maize	92.4	89.4	100	97.8	94.4(322)	0.015 *	0.175
Adjusted residual		−	+				
Cowpeas	26.7	37.6	100	100	61.9(211)	0.000 ***	0.705
Adjusted residual	− − −	− −	+++	+++			
Common beans	35.2	35.3	0.0	1.1	19.9(68)	0.000 ***	0.430
Adjusted residual	+++	++	− − −	− −			
Sweet potatoes	0.0	16.5	98.4	95.6	46.9(160)	0.000 ***	0.898
Adjusted residual	− − −	− − −	+++	+++			
Other Characteristics of the Farming System							
Distance to the farming field # (min)	1.06 b	1.05 b	1.98 a	1.99 a	1.47	0.000 ***	0.190
Proportion of literate households	81.0	74.1	91.8	86.7	82.7	0.028 *	0.164
Adjusted residual		−	+				

Note: α = *** and * denote significance at 0.1%, 1%, and 5%, respectively. Lowercase letters in the line indicate the difference between farming systems for each variable. Similar letters in the line are not statistically different. Proportions of households who reported growing other types of crops in each RFS were determined based on the adjusted residual: the symbols plus (+) and minus (−) indicate the existence of a relationship or no relationship between farming systems and the proportion of households who are literate, or grow each type of other crop. + ǀ −; ++ ǀ − − and +++ ǀ − − − denote significance at 5%, 1%, and 0.1%, respectively. Values within brackets represent the total number of households in all RFSs who reported growing each crop. The value of the ETA is the proportion of variance between FS for the variables, (yield of rice, herd size, and other crops) that characterise each FS. The variance is higher when the Eta is close to one.

Most smallholders (30.8%) in the CIS are subsistence farmers, followed by crop–livestock 26.4%, and 24.9% are specialised rice farmers. Mixed crops are the least adopted FS (17.9%). The proportion of literate farmers is on average higher in the mixed crops (91.8%), and lower in the specialised rice. Households who practice subsistence and specialised rice travel less distance from home to the field than those who have adopted mixed crop and crop–livestock FS.

All of the main variables that characterise the FSs (yield of rice, herd size, and other crops) are significantly ($p < 0.001$) different across FSs. The proportion of variance across the FSs is higher than in other crops; it represents more than half of the total variance ($ETA^2 = 0.73$), and is moderate for livestock and rice ($ETA^2 = 0.478$ and 0.374, respectively). All other crops, except maize ($p = 0.015$ and $ETA^2 = 0.175$) and common beans, vary considerably across FSs. Maize is the only crop which has been adopted by almost all households in all RFSs.

3.3. Farmer Choices Regarding Different Rice Farming Systems in the CIS

The estimated multinomial logistic model of choice for different FSs is represented in Table 5. The model shows that the location of a smallholder farmer in the upstream increases the likelihood of choice of subsistence and specialised rice FSs, in opposition to mixed and crop–livestock FSs, which are more likely to be located in the midstream and downstream. Meanwhile, increases in the household age reduce the likelihood of adopting the mixed crop FS and crop–livestock FS, as opposed to the subsistence and specialised FSs.

Table 5. Multinomial logistic model of farming system choice and its drivers/determinants.

Farming System	Drivers	Coefficient	Std. Error	Z-Value	Alf (α)	Exp(B)
Specialised Rice	Intercept	−1.492	1.126	1.756	0.185	
	Region = mid + downstream	−0.262	0.245	1.148	0.284	0.769
	Age	−0.022	0.016	1.803	0.179	0.978
	Total rice activities	0.114	0.326	0.123	0.726	1.121
	Distance to the farming field	−0.221	0.205	1.160	0.281	0.802
	Household labour force	0.044	0.043	1.073	0.300	1.045
	Permanent labour force = Yes	0.951	1.271	0.559	0.455	2.587
	Seeding (kg/ha)	0.027	0.008	10.515	0.001 ***	1.028
	Fertiliser application = No	−0.966	0.404	5.735	0.017 **	0.380
Mixed Crops	Intercept	−5.206	1.629	10.213	0.001 **	
	Region = mid + downstream	0.884	0.297	8.882	0.003 **	2.421
	Age	−0.073	0.022	11.028	0.001 **	0.929
	Total rice activities	1.480	0.436	11.533	0.001 **	4.392
	Distance to the farming field	0.406	0.222	3.344	0.067	1.501
	Household labour force	0.091	0.057	2.521	0.112	1.096
	Permanent labour force = Yes	3.584	1.456	6.056	0.014 **	36.010
	Seeding (kg/ha)	−0.007	0.013	0.269	0.604	0.993
	Fertiliser application = No	0.712	0.610	1.364	0.243	2.038
Rice–livestock	Intercept	−7.030	1.565	20.170	0.000 ***	
	Region = mid + downstream	0.879	0.287	9.404	0.002 **	2.408
	Age	−0.070	0.022	10.344	0.001**	0.933
	Total rice activities	0.952	0.408	5.440	0.020*	2.590
	Distance to the farming field	0.455	0.217	4.370	0.037*	1.576
	Household labour force	0.201	0.054	13.878	0.000***	1.223
	Permanent labour force = Yes	2.461	1.436	2.936	0.087	11.718
	Seeding (kg/ha)	0.022	0.010	4.687	0.030*	1.022
	Fertiliser application = No	−0.123	0.525	0.055	0.815	0.884

Note: Subsistence farming system is a reference category; α = *** is significant at 0.1%, ** = 1%, * = 5%, NS = not significant. Model fit (log-likelihood = 650.61); likelihood ratio test (Chi-square = 282.67, α = 0.000). Number of observations = 341; Pseudo R-squared (Nagelkerke = 0.60, Cox and Snell = 0.56).

Having more activities related to rice production increases the likelihood of choice of a mixed and crop–livestock FS, as opposed to a subsistence FS. Increasing the distance to the farming field increases the likelihood of a choice for crop–livestock, as opposed to a subsistence FS. The availability of permanent labour, either from the family or by hire, significantly increases the likelihood of adopting a mixed crop (α = 0.014) and crop–livestock (α = 0.000) FS, rather than subsistence farming. Increasing the seedling rate per hectares increases the likelihood of the choice to adopt a specialised rice FS (α = 0.001) and crop–livestock FS (α = 0.030), as opposed to the subsistence FS. Likewise, the use of fertiliser significantly increases the likelihood of choosing specialised rice over subsistence farming.

3.4. Production and Yield of Rice in the CIS

Table 6 presents the results of the multiple linear regression that predicts the factors that constrain or stimulate the yield of rice in the CIS. The four variables (fertiliser, seeds, pesticides, and age of the household) selected through the AIC algorithm explained 51% of the variance ($p = 0.000$). All variables selected positively affected the yield of rice in the CIS, although they were not equally significant.

Table 6. Multiple linear regression to predict rice yield in the Chókwè Irrigation Scheme.

Variables	Coefficient	Std. Error	T-Value	Alf(α)
Intercept	0.227	0.473	0.479	0.633
Fertiliser (Kg/ha)	0.007	0.003	1.938	0.056 *
Seeds (Kg/ha)	0.016	0.003	5.813	0.000 ****
Pesticide (yes/no)	0.531	0.335	1.583	0.117
Age of the household	0.021	0.009	2.338	0.022 **
R^2		0.5101		
p-value		0.000 ****		
Start AIC		−4.92		
End AIC		−12.83		

Note: Model was plotted based on the Akaike information criterion (AIC). α = ****, ** and * is significant at 0.001; 0.05 and 0.1, respectively.

Increasing the amount of fertiliser and seeds per hectare slightly increases the yield per hectare, with $p = 0.056$ and $p = 0.000$ significance, respectively. The application of pesticides has a positive effect on the rice yield, although the effect is not statistically significant. Increasing age of the household also has a positive effect on rice yield.

Table 7 presents the Multiple Linear Regression that describes the increase in production in the CIS. Only four variables (years of experience, total rice activities, and the amount of money invested in paying casual and permanent labour) were selected by the stepwise algorithm. Total rice activities were the only variable with a negative effect on production. The expansion of land for growing rice is positively affected by the years of household experience, the amount of money that the household spends to hire labour, and the availability of permanent labour for farming activities.

Table 7. Multiple linear regression to predict rice production in the Chókwè Irrigation Scheme.

Variables	Coefficient	Std. Error	T-value	Alf(α)
Intercept	0.317	0.150	2.119	0.035 **
Years of experience	0.022	0.005	4.576	0.000 ****
Total rice activities	−0.330	0.087	−3.784	0.000 ****
Amount of money paid for labour	0.009	0.002	4.890	0.000 ****
Permanent labour force	2.669	0.259	10.302	0.000 ****
R^2		0.3501		
R^2 Adjusted		0.3423		
p-value		0.000 ****		
Start AIC		−73.34		
End AIC		−82.15		

Note: Model was plotted based on the AIC criterion. α = ****, and ** is significant at 0.001; and 0.05, respectively.

More years of experience also increases the likelihood of increased production ($p = 0.035$), while hiring permanent labour and increasing the amount of money that the household spends to hire labour also significantly increase the likelihood of expanding the farming area ($p = 0.000$). Both models, the MLM of yield and production, showed no significant effect of any interaction between factors (e.g., gender, education, hiring permanent labour, and application of pesticides) or covariates (e.g., amount of fertiliser and seeds per hectare, age of the household, distance to the farming field, and the amount of money paid for labour, etc.)

4. Discussion

4.1. Drivers of Rice Farming Systems

Livestock and other crop types, including vegetables, maize, cowpeas, common beans, and sweet potatoes, appear to be indispensable components of all RFSs in the CIS. Even in specialised rice FSs, most households grow other crops and raise livestock (e.g., cattle, swine, and goats). There are several reasons for households to diversify their FSs; most importantly, it can secure their livelihood, and protect against food scarcity [35,43,44] For instance, Mota et al. (2019) [38] reported less income generation and more food insecurity for farmers with no livestock, compared to those who owned livestock. Based on the average plot size allocated to rice in the surveyed households (min = 0.25 ha, max = 5.3 ha, mean = 0.94 ha and, mode = 1 ha), one can easily infer that only a few farmers can survive by specialising in rice production. This allocation of a small rice plot is also related to the fact that most households also grow maize, which is a rice substitute, as well as the existing land restrictions. The optimal allocation of substitute crops, such as maize and cassava, has also been reported in other FSs in north Mozambique [45]. The use of a small amount of land for subsistence agriculture is a widespread practice in the rural areas of developing countries, such as Mozambique [35] and the sub-Saharan region [41,43,46,47] due to the lack of technology and capital to hire a labour force [48]. Even in the specialised rice system, the average rice output of 3.24 ton/ha is still very low when compared to other regions [49]. The adoption of livestock rearing is a traditional practice in south Mozambique [13], as it confers a comparative advantage due to the availability of lower lands and favourable climate conditions [50]. To underline the above hypothesis, moving from upstream to downstream increases the likelihood of adoption of the rice–livestock FS, as opposed to other FSs. Rice is, on average, grown over a single rainy season from October to March [51], and other crops, such as sweet potatoes, vegetables, and cowpeas are grown after harvesting the rice.

Increasing age of the household is associated with increased likelihood of adopting specialised rice and subsistence farming, as opposed to mixed and crop–livestock FSs; this is likely because either experience is important for specialised rice adoption, or because older households only grow a few other crops for subsistence (e.g., maize and common beans). Other studies conducted in Africa have also highlighted the importance of experience in rice production [52]. In contrast, Kajisa (2014) [49] argued that, for rice production technology is more important than years of experience. However, technology is still very expensive for smallholder farmers in this region, hence it was not yet being used by the surveyed households.

A study conducted in the same region found that younger people are more willing to move closer to bigger cities, such as Maputo and Xai-Xai, seeking employment opportunities and better living conditions [50]. Thus, they are less likely to embrace farming. It is also important to highlight that the Chókwé district headquarters is in the upstream region so most of the households in this region are formally employed in the government and in NGO institutions; they practice subsistence agriculture as a second occupation.

Labour availability significantly increases the likelihood of adoption of the mixed crops and rice–livestock FSs as opposed to the subsistence FS, which is probably related to the fact that, in rural areas of developing countries, farmers need to hire more workers or have a large family to overcome labour scarcity, especially when there are other off-farming activities such as animal rearing to consider [40]. A study conducted by Sraïri and Ghabiyel (2017) [51], in the Gharb Irrigation Scheme in rural Morocco, found that crop–livestock farms devote, on average, 56.41% of their annual working time to livestock raising, and the remaining time to crop-related necessities. Increasing the labour force also increases the likelihood of choosing the specialised rice FS, although not as significantly as the crop–livestock FS, which is likely because an increase in the productivity of rice requires more rice-related activities (e.g., ploughing, harrowing, fertiliser and pesticide application and weed control), hence requiring additional labour.

4.2. Promoting Factors and Constraints for Rice Productivity in the CIS

Based on the MLR, we have demonstrated that only four variables—seeds, fertilisers pesticides, and age of the household—predict 51% of the variability in the rice yields in the CIS. Thus, optimising these inputs could greatly improve the productivity of rice in the CIS. Fertiliser and seeds appear to be the most important production inputs to improve the yield of rice (see, for instance, Afolami et al., 2012) [52]. However, in this study, only 24.5% of farmers reported applying fertiliser. This relatively low number of farmers is probably related to the following factors: (i) the lack of financial incentives, such as bank loans and low import tariffs, for rice inputs such as fertiliser and improved seeds, so that smallholder farmers can afford to purchase them, and (ii) lack of information and capacity-building required to take advantage of the available animal manure [53].

Based on the MLR (see Table 7), a lack of experience and lower labour availability (either permanent or hired) also hinder farmers' decisions to expand their production area. Again, we might therefore infer that a lack of financial support constrains famers in terms of intensification and expansion of rice production in the CIS, which is also in line with conclusions from Chukwu et al. (2016) [45]. The number of activities that farmers carry out prior to harvesting the rice is negatively correlated with land expansion; this is likely because intensification requires more inputs (activities), such as weed and pest control, which in turn requires more labour and inputs, and these are out of reach for smallholder farms in the CIS. Thus, within the existing constraints that farmers are exposed to, they must choose to optimise scarce resources either by intensifying or diversifying their production. According to Ayoola and Dangbegnon (2011) [48], increasing the use of land and inputs such as fertilisers, herbicides, and labour could also increase the production and productivity of rice.

4.3. Policy Recommendations to Improve Production and Productivity for Smallholder Farmers in the CIS

The main constraints for productivity and production in the CIS are (i) land restriction; (ii) lack of inputs (e.g., fertilisers, herbicides, and labour); and (iii) poor training and lack of rural extension services. The expansion of farming areas appears to be hampered by a lack of capital for the acquisition of more land, since most households (70.4%) in our survey reported inheriting land from relatives (Supplementary Materials). To overcome this restriction, smallholder farmers need to strengthen corporativisms and intensify productivity in small plots by using more inputs [47]. However, to facilitate intensification, the government and NGOs operating in the CIS need to create incentives, such as providing loans and cheaper agrarian credits, so that farmers can better access market inputs (e.g., fertilisers, herbicides, and labour). Providing more extension services and training to smallholder farmers in the CIS should be the top priority to improve production and productivity in this area [27]. For instance, we noted that limited chemical fertiliser could be replaced by adopting animal manure, which has a great potential to contribute to the development of green agriculture in the area [53]. However, we also acknowledge the crop yield gap between organic and conventional agriculture [54]. Nevertheless, training households to adopt animal ploughing could minimise scarcity, without being constrained by the unaffordable price of human labour.

Despite the efforts towards fitting the destination of the rice output (e.g., for sale locally or for household consumption) as an FS driver, no meaningful explanation was captured. All FSs in the CIS appear to be more consumption-oriented than market-oriented (Supplementary Materials). Most of the FSs appear to have evolved towards mixed crop and rice–livestock production, which is typical of subsistence agriculture, even though the CIS was primarily designed for rice production. This remarkable dynamic is likely because there is a market opportunity for other crops and products, such as vegetables and meat, that can be sold at nearby towns such as Maputo and Xai-Xai. According to Glover and Jones (2019) [55], farms are highly selective in their locations, preferring areas close to existing infrastructure and markets. This hypothesis was highlighted by the fact that educated households are more likely to focus production on more cattle and vegetables, since they are also in a better position to see market opportunities and assess the viability of other commodities which do not demand higher production inputs. Second, rice appears not to be a profitable business for the smallholder farmer, since Asian rice (from China, Thailand, and Vietnam) is sold at a very competitive price in Mozambique. To make matters worse, maize is a competing product, and a substitute for rice that requires less water and production inputs, especially in the context of climate change.

To better advise decision-makers in this area, the hypotheses explored above need to be tested with comprehensive data on production, yield, value chain, and gross margin analysis of each crop and livestock type, in comparison to large and middle-sized farmers, since some authors have reported possible economic spillovers from commercial activities to local smallholders [55]. However, due to time and resource restrictions, this will be a focus for future research.

5. Conclusions

This study aimed to assess rice farming system typologies in the CIS to better understand the drivers and constraints for smallholder rice farmers, and proposes alternative policies for decision-makers to improve production and productivity. The results demonstrated that the use of different RFS typologies, rather than one FS for all farmers, can better capture the specific drivers for each FS in the region.

Four RFSs were identified: the subsistence FS, specialised rice FS, mixed crops FS, and rice–livestock FS. Subsistence and specialised rice are predominant in the upstream region of the CIS, while mixed crops and crop–livestock are predominant in the midstream and downstream. The households who adopted subsistence FS were on average older and had fewer resources. Specialized rice farmers had access to more resources and were driven by the household power for purchasing production inputs. Mixed crops and rice–livestock were driven by the availability of labour and possession of lower lands. In general, increased production inputs (e.g., fertiliser, pesticides, weed control, and the number of seeds per hectare) might greatly improve productivity, whilst household experience and labour availability could greatly improve production.

This research suggests that rice farmers in the region require more training opportunities to optimise the available resources, such as animal manure and animal traction, as well as to explore other more valuable trade-offs, such as potential market opportunities and the production cost of other crops and livestock in the region.

Supplementary Materials: The following are available online at https://www.mdpi.com/article/10.3390/agronomy11051018/s1, Table S1: Socio-economic and demographic characteristics of the surveyed households in the CIS. Survey to households head.

Author Contributions: Conceptualization, F.I., A.N. and D.F.; methodology, A.A.M., F.I., and D.F.; validation, A.A.M., F.I. and D.F.; formal analysis, A.A.M., F.I. and D.F.; investigation, all authors.; resources, F.I. and D.F.; data curation, A.A.M., and D.F.; writing—original draft preparation, F.I. and D.F.; writing—review and editing, all authors.; visualisation, all authors.; supervision, D.F.; project administration, F.I. and D.F.; funding acquisition, F.I. All authors have read and agreed to the published version of the manuscript.

Funding: This research received no external funding.

Institutional Review Board Statement: Not applicable.

Informed Consent Statement: Not applicable.

Data Availability Statement: Data available under request.

Acknowledgments: The authors would like to acknowledge all institutions and individuals who contributed to this work. We especially acknowledge the following institutions: Hidráulica de Chókwè, Empresa Pública (HICEP) and Serviço Distrital de Actividades Económicas (SDAE), for their support to conduct the research with the smallholders. We also acknowledge the field assistants Alexandre Quetane, Pedrito Quetane, and Argêncio Mucavel, and all of the surveyed smallholder farmers in the CIS, as well as Marina Temudo, who provided important inputs and ideas. Paula Alves carried out thorough English grammar correction. We also acknowledge the Fundação para Ciência e Tecnologia (FCT) of Portugal, which provided a scholarship to the first author (Ref no SFRH/BD/128098/2016) and support the LEAF center with the project UID/AGR/04129/2020 (LEAF).

Conflicts of Interest: The authors declare no conflict of interest.

References

1. FAO. *FAO Rice Market Monitor (RMM)*; FAO: Rome, Italy, 2018; Volume 21.
2. Karimov, A.A.; Terefe, B.; Baye, K.; Hazard, B.; Abate, G.T.; Covic, N. Wheat contribution to food and nutrition security and leveraging opportunities for sustainability, nutrition and health outcomes. In *Encyclopedia of Food Security and Sustainability*; Elsevier: Amsterdam, The Netherlands, 2019; Volume 3, pp. 270–277. ISBN 9780128126882.
3. FAO. *FAO Rice Market Monitor*; FAO: Rome, Italy, 2013; Volume 21.
4. Fahad, S.; Adnan, M.; Noor, M.; Arif, M.; Alam, M.; Khan, I.A.; Ullah, H.; Wahid, F.; Mian, I.A.; Jamal, Y.; et al. Major Constraints for Global Rice Production. In *Advances in Rice Research for Abiotic Stress Tolerance*: Elsevier Inc.: Amsterdam, The Netherlands, 2019; pp. 1–22. ISBN 9780128143322.
5. Sanchez, P.A. Soil fertility and hunger in Africa. *Science* **2002**, *295*, 2019–2020. [CrossRef]
6. Meijer, S.S.; Catacutan, D.; Ajayi, O.C.; Sileshi, G.W.; Nieuwenhuis, M. The role of knowledge, attitudes and perceptions in the uptake of agricultural and agroforestry innovations among smallholder farmers in sub-Saharan Africa. *Int. J. Agric. Sustain.* **2015**, *13*, 40–54. [CrossRef]
7. Adenle, A.A.; Wedig, K., Azadi, H. Sustainable agriculture and food security in Africa: The role of innovative technologies and international organizations. *Technol. Soc.* **2019**, *58*, 101143. [CrossRef]
8. Goldstone, J. *Africa 2050: Demographic Truth and Consequences*; Hoover: Stanford, CA, USA, 2019.
9. Karingi, S.; Mwakubo, S. *Promotion of Fertilizer Production, Cross-Border Trade and Consumption in Africa*; African Development Bank Group: Addis Ababa, Ethiopia, 2018.
10. Pretty, J.; Toulmin, C.; Williams, S. Sustainable intensification in African agriculture. *Int. J. Agric. Sustain.* **2011**, *9*, 5–24. [CrossRef]
11. Ranum, P.; Peña-Rosas, J.P.; Garcia-Casal, M.N. Global maize production, utilization, and consumption. *Ann. N. Y. Acad. Sci.* **2014**, *1312*, 105–112. [CrossRef]
12. Mutsvangwa-Sammie, E.P.; Manzungu, E.; Siziba, S. Key attributes of agricultural innovations in semi-arid smallholder farming systems in south-west Zimbabwe. *Phys. Chem. Earth* **2018**, *105*, 125–135. [CrossRef]
13. Dixon, J. *Concept and Classifications of Farming Systems*; Elsevier: Ansterdam, The Netherlands, 2019; Volume 3, ISBN 9780081005965.
14. Fagnombo, J.; Sintondji, L.O.; Malé, A.; Baggie, I.; Agbahungba, G.; Hamadoun, A.; Zwart, S.J. Predictors determining the potential of inland valleys for rice production development in West Africa. *Appl. Geogr.* **2018**, *96*, 86–97. [CrossRef]
15. Sileshi, G.; Akinnifesi, F.K.; Ajayi, O.C.; Chakeredza, S.; Kaonga, M.; Matakala, P. Contributions of agroforestry to Ecosystem Services in the Miombo eco-region of Eastern and Southern Africa. *Afr. J. Environ. Sci. Technol.* **2007**, *1*, 68–80. [CrossRef]
16. Makumba, W.; Janssen, B.; Oenema, O.; Akinnifesi, F.K.; Mweta, D.; Kwesiga, F. The long-term effects of a gliricidia-maize intercropping system in Southern Malawi, on gliricidia and maize yields, and soil properties. *Agric. Ecosyst. Environ.* **2006**, *116*, 85–92. [CrossRef]
17. Kwesiga, F.; Akinnifesi, F.K.; Mafongoya, P.L.; McDermott, M.H.; Agumya, A. Agroforestry research and development in southern Africa during the 1990s: Review and challenges ahead. *Agrofor. Syst.* **2003**, *59*, 173–186. [CrossRef]
18. Mills, G.; Obasanjo, O.; Herbst, J.; Davis, D. *Making Africa Work*, 1st ed.; Ronan, M., Moodie, G., Eds.; Tafelberg: Cape Town, South Africa, 2017; ISBN 978-0-624-08027-5.
19. Balasubramanian, V.; Sie, M.; Hijmans, R.J.; Otsuka, K. Increasing Rice Production in Sub-Saharan Africa: Challenges and Opportunities. In *Advances in Agronomy*; Elsevier: Amsterdam, The Netherlands, 2007; Volume 94, pp. 55–133. ISBN 0123741076.
20. Maclean, J.; Hardy, B.; Hettle, G. *Rice Almanac: Source Book for One of the Most Important Economic Activities on Earth*, 4th ed.; IRRI: Los Baños, Philippines, 2013.
21. Cunguara, B.; Thompson, T.; Davis, K. *Mozambique: Desk Study of Extension and Advisory Services Developing Local Extension Capacity (DLEC) Project*; Feed the Future: Washington, DC, USA, 2018.
22. JICA. *The Project for Rice Produtivity Improvement in Chokwe Irrigation Scheme*; JICA: Maputo, Mozambique, 2014.
23. Zeigler, R.S. Rice. Encyclopedia of Applied Plant Sciences. *Encycl. Appl. Plant Sci.* **2016**, *3*, 45–49. [CrossRef]
24. Ministry of Agriculture. *National Rice Development Strategy Mozambique*; Ministry of Agriculture: Maputo, Mozambique, 2009.
25. Alasia, A.; Low, J.; Soregaroli, C. *Food Demand Changes in Mozambique: Assessing the Role of the Rice Sector*; FAO: Rome, Italy, 2003.
26. Instituto Nacional de Estatística. *Anuário Estatístico*; Instituto Nacional de Estatística: Maputo, Mozambique, 2018.
27. Silici, L.; Bias, C.; Cavane, E. *Sustainable Agriculture for Small-Scale Farmers in Mozambique*; Internationa Institute for Environment and Development: London, UK, 2015.
28. Benfica, R.; Cunguara, B.; Thurlow, J. Linking agricultural investments to growth and poverty: An economywide approach applied to Mozambique. *Agric. Syst.* **2018**, *172*, 91–100. [CrossRef]
29. Dixon, J.; Gulliver, A.; Gibbon, D. *Farming Systems and Poverty: Improving Farmers' Livelihoods in a Changing Word*; FAO: Rome, Italy; Washington, DC, USA, 2001.
30. *MASA Plano Nacional de Arroz em Moçambique*; Ministério de Agricultura e Segurança Alimentar: Maputo, Mozambique, 2019.
31. Kajisa, K.; Payongayong, E. Potential of and constraints to the rice Green Revolution in Mozambique: A case study of the Chokwe irrigation scheme. *Food Policy* **2011**, *36*, 615–626. [CrossRef]
32. Abbas, M. *Chokwé: Efeitos Locais de Políticas Instáveis, Erráticas e Contraditórias*; OMR: Maputo, Mozambique, 2018.

33. De Sousa, L.S.; Wambua, R.M.; Raude, J.M.; Mutua, B.M. Assessment of Water Flow and Sedimentation Processes in Irrigation Schemes for Decision-Support Tool Development: A Case Review for the Chókwè Irrigation Scheme, Mozambique. *AgriEngineering* **2019**, *1*, 100–118. [CrossRef]
34. Bartlett, J.E.; Kotrlik, J.W. Organizational Research: Determining Appropriate Sample Size in Survey Research. *Sci. Open* **2001**, *19*, 43–50.
35. Landry, J.; Chirwa, P.W. Analysis of the potential socio-economic impact of establishing plantation forestry on rural communities in Sanga district, Niassa province, Mozambique. *Land Use Policy* **2011**, *28*, 542–551. [CrossRef]
36. *MAE Perfil do Distrito do CHÓKWE Provincia de Gaza*; Governo do Mozambique: Maputo, Mozambique, 2012.
37. Temudo, M.P.; Talhinhas, P. Dynamics of change in a 'female farming system', Mbanza Kongo/Northern Angola. *J. Peasant Stud.* **2019**, *46*, 258–275. [CrossRef]
38. Mota, A.A.; Lachore, S.T.; Handiso, Y.H. Assessment of food insecurity and its determinants in the rural households in Damot Gale Woreda, Wolaita zone, southern Ethiopia. *Agric. Food Secur.* **2019**, *8*, 1–11. [CrossRef]
39. Michalscheck, M.; Mellon-Bedi, S.; Kuivanen, K.S.; Alvarez, S.; Adjei-Nsiah, S.; Descheemaeker, K. Characterising the diversity of smallholder farming systems and their constraints and opportunities for innovation: A case study from the Northern Region, Ghana. *Wageningen J. Life Sci.* **2016**, *78*, 153–166. [CrossRef]
40. Ribeiro, P.F.; Nunes, L.C.; Beja, P.; Reino, L.; Santana, J.; Moreira, F.; Santos, J.L. A Spatially Explicit Choice Model to Assess the Impact of Conservation Policy on High Nature Value Farming Systems. *Ecol. Econ.* **2018**, *145*, 331–338. [CrossRef]
41. Legendre, P.; Legendre, L. Numerical Ecology. *Dev. Environ. Model.* **2003**, *24*, 870. [CrossRef]
42. García-Pérez, M.A.; Núñez-Antón, V.V. Cellwise Residual Analysis in Two-Way Contingency Tables. *Educ. Psychol. Meas.* **2003**, *63*, 825–839. [CrossRef]
43. Sharpe, D. Your Chi-Square Test Is Statistically Significant: Now What? Practical Assessment, Research & Evaluation. *Pract. Assessment Res. Eval.* **2015**, *20*, 1.
44. Mbanze, A.M.; Viera da Silva, C.; Ribeiro, N.S.; Silva, J.; Santos, J.L. A Livelihood and Farming System Approach for Effective Conservation Policies in Protected Areas of Developing Countries: The Case Study of the Niassa National Reserve in Mozambique. *Land Use Policy* **2020**, *99*, 1–16. [CrossRef]
45. Chukwu, V.A.; Eze, A.; Osuafor, O.O. Socio-Economic Determinants of Adoption of Improved Rice Production Technologies among Rice Farmers in Ebonyi State, Nigeria: A Logit Regression. *Agriculture* **2016**, *94*, 39900–39908.
46. Ducrot, R.; Leite, M.; Gentil, C.; Bouarfa, S.; Rollin, D.; Famba, S. Strengthening the capacity of irrigation schemes to cope with flood through improved maintenance: A collaborative approach to analySe the case of Chókwè, Mozambique. *Irrig. Drain.* **2018**. [CrossRef]
47. Menete, M.Z.L.; van Es, H.M.; Brito, R.M.L.; DeGloria, S.D.; Famba, S. Evaluation of system of rice intensification (SRI) component practices and their synergies on salt-affected soils. *Field Crop. Res.* **2008**, *109*, 34–44. [CrossRef]
48. Ayoola, J.; Dangbegnon, C. Socio-economic factors influencing rice production among male and female farmers in Northern Guinea Savanna Nigeria: Lessons for promoting gender equity in action research. *Agric. Biol. J. N. Am.* **2011**, *2*, 1010–1014. [CrossRef]
49. Kajisa, K. *Constraints on Rice Sector Development in Mozambique*; JICA: Tokyo, Japan, 2014; ISBN 8134098045.
50. Ibraimo, M. *Observador Rural*; OMR: Mozambique, Maputo, 2017.
51. Sraïri, M.T.; Ghabiyel, Y. Coping with the work constraints in crop-livestock farming systems. *Ann. Agric. Sci.* **2017**, *62*, 23–32. [CrossRef]
52. Afolami, C.A.; Obayelu, A.E.; Agbonlahor, M.U.; Lawal-Adebowale, O. Socioeconomic Analysis of Rice Farmers and Effects of Group Formation on Rice Production in Ekiti and Ogun States of South-West Nigeria. *J. Agric. Sci.* **2012**, *4*, 233–244. [CrossRef]
53. Ismael, F.; Ndayiragije, A.; Fangueiro, D. New Fertilizer Strategies Combining Manure and Urea for Improved Rice Growth in Mozambique. *Agronomy* **2021**, *11*, 783. [CrossRef]
54. De Ponti, T.; Rijk, B.; Van Ittersum, M.K. The crop yield gap between organic and conventional agriculture. *Agric. Syst.* **2012**, *108*, 1–9. [CrossRef]
55. Glover, S.; Jones, S. Can commercial farming promote rural dynamism in sub-Saharan Africa? Evidence from Mozambique. *World Dev.* **2019**, *114*, 110–121. [CrossRef]

agronomy

MDPI

Article

Maximizing Land Use Efficiency and Productivity of Soybean and Fodder Maize Intercrops through Manipulating Sowing Schedule and Maize Harvest Regime

Heba S. A. Salama * and Mahmoud H. Abdel-Moneim

Crop Science Department, Faculty of Agriculture (EL-Shatby), Alexandria University, Alexandria 21545, Egypt; mahmoudh3884@gmail.com
* Correspondence: heba.salama@alexu.edu.eg

Abstract: The incorporation of both food and forage crops in an intercropping system is receiving increasing attention, especially in developing countries with increasing populations and limited resources. In a two-year (2019–2020) field trial, conducted in Northern Egypt, productivity of soybean and fodder maize, as well as the quality of maize herbage, were investigated under three sowing schedules; soybean and maize sown together, and maize sown 15 and 30 days after soybean, in addition to soybean and fodder maize sown in pure stands, with maize harvested at green fodder maturity (GFM), and silage maturity (SM). Harvesting fodder maize at SM resulted in higher herbage yield than harvesting it at GFM, yet it negatively affected the soybean productivity. However, this negative impact was offset when fodder maize sowing was delayed 30 days after soybean sowing. Maize harvested at GFM was characterized by a higher leaf component, which was reflected in its higher crude protein content, yet the decline in quality with advanced maturity was to a great extent, counterbalanced by the presence of high-quality ears in maize harvested at SM. This was clear in its lower fiber and higher non-fiber carbohydrate contents. Land equivalent ratio (LER) demonstrated yield advantage with the delayed sowing of fodder maize (LER > 1), while the dry matter equivalent ratio (DMER) associated the yield advantage with the late harvesting of fodder maize at SM (DMER > 1), across all sowing schedules, which was more realistic for an additive intercropping model where the dry matter is the economic component. In a soybean-fodder maize intercropping system, whether fodder maize will be cultivated for green feeding or for silage production, it is recommended that sowing is delayed until 30 days after the soybean, in order to maximize yield advantage and land use efficiency.

Keywords: intercropping; soybean; fodder maize; maturity stage; sowing schedule; land equivalent ratio; dry matter equivalent ratio

Citation: Salama, H.S.A.; Abdel-Moneim, M.H. Maximizing Land Use Efficiency and Productivity of Soybean and Fodder Maize Intercrops through Manipulating Sowing Schedule and Maize Harvest Regime. *Agronomy* **2021**, *11*, 863. https://doi.org/10.3390/agronomy11050863

Academic Editors: Umberto Anastasi and Aurelio Scavo

Received: 4 April 2021
Accepted: 25 April 2021
Published: 28 April 2021

1. Introduction

The agricultural systems in developing countries are, nowadays, striving to reach sustainability in food production and food security under the existing high population pressure. Intercropping is a farming system, where two or more crops are cultivated together in the same field for a significant period of time during the growing season, even though the component crops are not necessarily sown or harvested simultaneously. It is one of the vital practices widely proposed to improve productivity and land use efficiency, especially in developing countries suffering from limited arable land and restricted agricultural inputs [1]. This is usually achieved by increasing the resource use efficiency [2].

Due to the pressing needs of the increasing populations, there has been escalating interest in the incorporation of both food and forage crops in the same farming system. Therefore, intercropping soybean (*Glycine max* [L.] Merr.), as a prominent oil crop, with fodder maize (*Zea mays* L.), as principle forage crop, increases the overall benefit from the farming practice, especially in the low input agricultural systems of the developing countries [3]. In a soybean-maize intercropping system, soybean can secure sufficient amounts of nitrogen through biological N fixation, thereby, enhancing soil quality [4]. There is evidence that nodulation and nodule longevity of soybean is generally improved in a soybean-maize intercropping system due to the improved microclimate that favors the growth of the nodular bacteria, in addition to the stimulation effect caused by the exudates produced from maize roots [4,5]. This will be positively reflected on the growth and productivity of both crops allowing them to mutually benefit from the intercropping system [3]. Soybean-maize intercropping is also encouraged in areas with limited water resources, like Egypt and other developing countries, due to its water-saving abilities [6]. This is mostly attributed to the root complementarity of both crops that increases their ability to capture soil water at different depths [7].

Previous studies have documented that maize is likely to dominate the soybean-maize intercropping system due to its higher competitive ability and relatively rapid initial growth [8], which suppresses the growth of soybean, especially when both crops are sown at the same time [9]. The early growth of the intercropped species is very important in determining their competitive abilities, which is reflected in their growth dynamics and final productivity [10]. Therefore, the interspecific competition between the intercrop components can be manipulated by adjusting the sowing schedule, i.e., varying the sowing dates of the different species (sometimes known as relay intercropping). This mechanism is expected to provide an advantage to the first sown crop by increasing its competitiveness, and thus, vigor [11]. Hence, achieving the maximum benefit from the soybean-maize intercropping system would be feasible only with the proper management of the intercropping component crops, especially in terms of sowing and harvesting adjustment, which would minimize competition and ensure complementarity in resources' utilization [12,13]. Many attempts were made to maximize the land use efficiency and productivity of intercropped soybean and maize by manipulating the row spacing [14,15], or sowing pattern and planting structure [4,5,9,16,17]. However, the variations in soybean and fodder maize productivity when different sowing/harvesting schedules are adopted is not yet exploited.

This study aimed at developing guidelines for intercropping soybean with fodder maize in Northern regions of Egypt, characterized by their arid Mediterranean climate. The main goal of the study was to develop practical recommendations about the appropriate sowing schedule for both crops in combination with the best harvest regime at which fodder maize should be removed in order to achieve optimum balance between soybean seed yield on the one hand and maize herbage productivity and quality on the other hand. It was hypothesized that consecutive sowing of both crops would interact with harvesting fodder maize at different stages of maturity in a way that alters the competition between the two crops. This would positively enhance their productivity and improve the land use efficiency. In this study, productivity of soybean and fodder maize, in addition to quality of maize herbage were investigated under variable sowing schedules, and maize harvest regimes. Land use efficiency and yield gain were also evaluated using the dry matter equivalent ratio (DMER), compared to the traditional land equivalent ratio (LER).

2. Materials and Methods

2.1. Field Trial

A two-year field trial was conducted at the experimental station of the Faculty of Agriculture, Alexandria University, Alexandria, Egypt (31°20′ N, 30° E), during 2019 and 2020 summer seasons. Texture of the experimental soil was sandy loam (54% sand, 30% silt, and 16% clay), with pH 8.15, 1.30 dS m^{-1} electrical conductivity, and 7.50% $CaCO_3$. The top 25 cm of soil contained 1.50% organic matter and 100, 4.80, and 290 mg kg^{-1} available N, P, and K, respectively. The experimental location is characterized by its hot, arid Mediterranean climate with zero precipitation during the summer season. Average monthly temperature and humidity during both experimental seasons are illustrated in Figures 1 and 2, respectively.

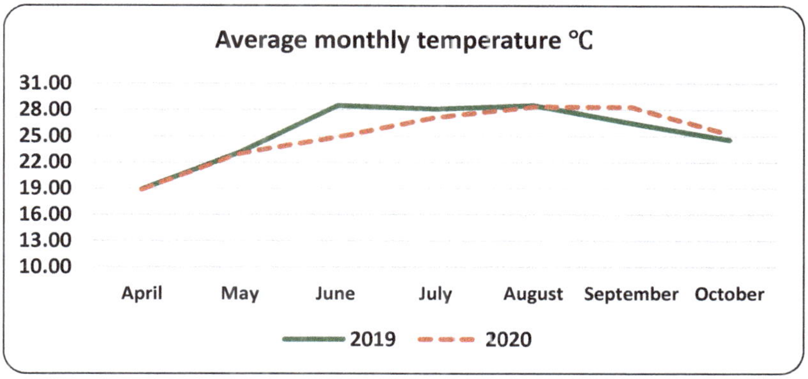

Figure 1. Average monthly temperature (°C) of the experimental site during summers 2019 and 2020.

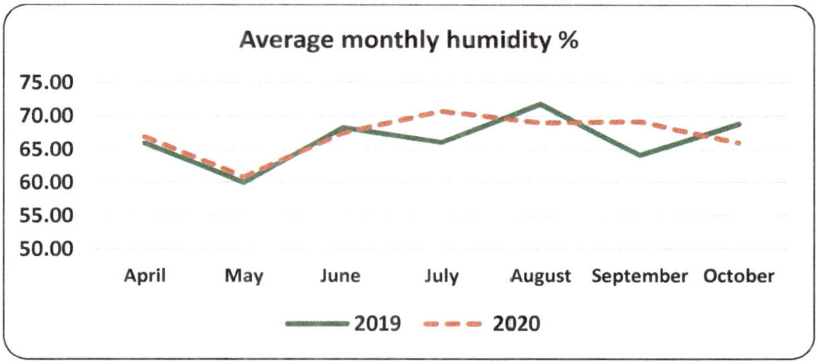

Figure 2. Average monthly humidity (%) of the experimental site during summers 2019 and 2020.

Soil preparation included plowing, disking, levelling and, finally dividing into experimental plots. Each experimental plot consisted of two adjacent wide beds, 60 cm apart. Each wide bed was 200 cm long and 120 cm wide, resulting in a total plot area of 6 m^2. On each wide bed, two border rows of maize, and three rows of soybean were sown at 30 cm intra-row spacing (Figure 3). A distance of 60 cm was left between each two experimental plots. Two seeds of the soybean and maize intercrops were sown in hills 15 and 30 cm apart, respectively. This sowing pattern was followed to maintain 75% plant density for soybean, in addition to 50% plant density for maize, in an additive intercropping model. Pure soybean and fodder maize stands were established during both seasons and were sown to 100% plant density for both crops.

A split-plot experimental design with four replications was employed, with the main plots assigned to the sowing schedule; 1. SS1: Soybean and maize sown together, 2. SS2: Maize sown 15 days after soybean, 3. SS3: Maize sown 30 days after soybean. 4. Pure stands of soybean and fodder maize. Sub-plots were dedicated to maize harvest regime; 1. HR1: Green fodder maturity (55 DAS), 2. HR2: Silage maturity (100 DAS). In both seasons, the maize three-way hybrid 368 and soybean cultivar Giza 111 were used. Soybean sowing was performed on 1 May and 20 April during 2019, and 2020, respectively, while maize sowing was done according to the investigated sowing schedules. Plant thinning was performed 21 days after sowing (DAS), by leaving 1 and 2 plants per hill for maize and soybean, respectively. To maintain adequate soil moisture and avoid induced drought stress, surface irrigation was scheduled on weekly basis. Based on the official recommendations of soybean and fodder maize production in the region, an amount of 200 kg ha^{-1} calcium monophosphate (15.5% P_2O_5) was applied once with seedbed preparation. In addition a total amount of 144 kg N ha^{-1}, in the form of ammonium nitrate was split into three equal doses and applied with sowing of soybean (side-banded), then after 30 and 60 days (top dressing). The experimental plots were sprayed with 720 g Lannate ($C_5H_{10}N_2O_2S$) dissolved in 480 L water ha^{-1}, 30 days after maize sowing to protect against maize stem borers, while, weeds were hand-hoed when necessary.

Figure 3. Experimental plot design illustrating the sowing pattern of the soybean and fodder maize on two adjacent wide beds.

2.2. Sampling and Measurements:

Maize was harvested at green fodder and silage maturity stages. Harvesting of green fodder maize was done after 55 DAS, which was supposed to provide the optimum balance between yield and fodder quality for green feeding as concluded in a previous study [18]. On the other hand, silage maturity was identified by the 1/2 milk line stage, which is considered to attain high silage nutritional quality [19]; this stage was reached at 100 DAS for the investigated cultivar. At harvesting, stalks were manually cut directly above ground level, and fresh matter yield (FMY) per plot was weighed immediately in the field. Plant height (cm), stem diameter (mm), plant weight (g), and leaf, stem and ear percentages, were determined as an average of five randomly chosen plants from each plot. To determine dry matter content (DMC) of the plant material, a subsample of approximately 1 kg was taken from each plot and oven-dried at 60 °C to constant weight. The dry matter yield (DMY) per plot was estimated based on the FMY and the DMC of the subsample. Prior to soybean harvesting, plant height was measured from the soil surface to the uppermost node with at least one pod, for 5 random plants from each plot. Plots were manually harvested and fresh biological yield (FBY) was weighed in the field; after that plants per plots were left to air-dry until constant weight was reached to determine the dry biological yield (DBY). Soybean plants were manually threshed and seeds were weighed to determine seed yield (t ha^{-1}) and then sieved to remove seed splits. Harvest index was calculated as seed yield divided by FBY and expressed as percentage. The 100-seed weight (g) was determined as an average of three random seed samples taken from each plot.

2.3. Laboratory Analyses

For maize quality analyses, the dried subsamples were milled to a 1 mm particle size. The contents of neutral detergent fiber (NDF), acid detergent fiber (ADF), and acid detergent lignin (ADL) were sequentially determined using ANKOM^{200} Fiber analyzer (ANKOM Technology, Macedon, NY, USA) as described by [20]. The nitrogen (N) content was analyzed using the Kjeldahl procedure [21], then crude protein (CP) was calculated as N × 6.25. The crude ash (CA) content was determined by incinerating the samples in a muffle oven at 550 °C for 3 h [21]. The crude fat (CF) content in maize samples and oil content in soybean seed samples were determined using the Soxhlet procedure [21]. Non-fiber carbohydrates (NFC) content (g kg^{-1}) was calculated as follows:

$$NFC = 1000 - (CP + CF + NDF + CA). \quad (1)$$

2.4. Land Use Efficiency and Yield Advantage

Land equivalent ratio (LER): Determined as the sum of the fractions of the fresh biological yield (t ha^{-1}) of soybean and maize intercrops relative to their sole crop yields [22,23]:

$$LER = \frac{Yab}{Yaa} + \frac{Yba}{Ybb}. \quad (2)$$

where, *Yab* is yield of soybean "*a*" intercropped with maize "*b*", *Yaa* is pure stand yield of soybean "*a*", *Yba* is yield of maize "*b*" intercropped with soybean "*a*", *Ybb* is pure stand yield of maize "*b*".

Dry matter equivalent ratio (DMER): Determined as the sum of the dry yield of the main soybean crop and the maize companion crop relative to the *DM* yield of the sole main soybean crop [24,25]:

$$DMER = \frac{DMYab + DMYba}{DMYaa}. \quad (3)$$

where *DMYab* is *DMY* of soybean "*a*" intercropped with maize "*b*", *DMYba* is *DMY* of maize "*b*" intercropped with soybean "*a*", *DMYaa* is pure stand *DMY* of soybean "*a*".

2.5. Statistical Analyses

Analysis of variance (ANOVA) was conducted using Proc Mixed of SAS 9.4 [26], with only replicates considered random. The investigated variables (*V*) were analysed according to the following model:

$$V_{ijk} = \mu + R_i + SS_j + (R \times SS)_{ij} + HR_k + (SS \times HR)_{jk} + e_{ijk} \quad (4)$$

where μ is the overall mean, R_i is the replication (i = 1,2,3,4), SS_j is the sowing schedule effect (j = 1,2,3,4), $(R \times SS)_{ij}$ is the experimental error "a", HR_k is the maize harvest regime effect (k = 1,2), $(SS \times HR)_{ij}$ is the effect of the interaction between the sowing schedule and maize harvest regime, and e_{ijk} is the experimental error "b". The crop was not considered as an experimental factor, and the statistical analysis was conducted separately for each crop.

Data were presented in a combined analysis for the two growing seasons (2019 and 2020) upon homogeneity of variance's error [27]. Prior to the statistical analysis of the data, the harvest index was arcsine transformed and expressed as percentage. Mean comparisons were made using the least significant difference (L.S.D) procedure, with significances declared at $p < 0.05$.

3. Results

The main effects of sowing schedule and maize harvest regime will be presented and discussed only when their interaction is not significant.

3.1. Performance of Fodder Maize

The SS exerted a significant influence on fodder maize FMY, DMY, DMC, stem diameter and plant weight, which were all, in addition to plant height, significantly affected by the maize HR and by the interaction between the SS and HR (Table 1). The means presented in Table 2 revealed that harvesting maize at SM caused a significant increase in the above-mentioned parameters, except for the stem diameter, compared to harvesting at GFM. Obviously, maize harvested at SM produced highest significant FMY and DMY with the highest significant accumulated DMC, as well as the tallest and heaviest significant plants, with the least significant stem diameter except for SS2.

Table 1. p values for fresh matter yield (FMY), dry matter yield (DMY) as t ha^{-1}, dry matter content (DMC) as g kg^{-1}, plant height (cm), stem diameter (mm), and plant weight (g) for fodder maize, combined over 2019 and 2020 growing seasons.

S.O.V.	D.F.	FMY	DMY	DMC	Plant Height	Stem Diameter	Plant Weight
SS	3	<0.0001	<0.0001	0.0380	0.1398	0.0073	0.0278
HR	1	<0.0001	<0.0001	<0.0001	<0.0001	<0.0001	<0.0001
SS * HR	3	<0.0001	<0.0001	0.0172	0.0001	0.0393	0.0150

S.O.V.: Source of variation, D.F.: Degrees of freedom, SS: Sowing schedule, HR: Harvest regime.

Table 2. Variations of the fresh matter yield (FMY), dry matter yield (DMY) as t ha^{-1}, dry matter content (DMC) as g kg^{-1}, plant height (cm), stem diameter (mm), and plant weight (g) for fodder maize, as affected by the interaction between sowing schedule (SS) and maize harvest regime (GFM and SM).

Treatment	FMY		DMY		DMC	
	GFM	**SM**	**GFM**	**SM**	**GFM**	**SM**
SS1	9.15 bB	49.95 aC	1.01 bA	17.12 aC	110.89 bA	343.67 aB
SS2	11.67 bB	39.32 aD	1.20 bA	17.63 aC	103.29 bA	447.41 aA
SS3	9.69 bB	66.03 aB	1.01 bA	28.07 aB	107.49 bA	423.74 aA
Pure	20.32 bA	88.40 aA	2.28 bA	32.17 aA	112.37 bA	363.59 aB
Treatment	**Plant height**		**Stem diameter**		**Plant weight**	
	GFM	**SM**	**GFM**	**SM**	**GFM**	**SM**
SS1	128.00 bB	259.83 aAB	24.42 aAB	16.03 bB	224.17 bB	1021.50 aA
SS2	102.08 bC	273.84 aA	22.43 aB	21.19 aA	154.17 bC	1233.83 aA
SS3	158.34 bA	246.25 aB	28.33 aA	20.78 bA	388.75 bA	979.17 aAB
Pure	131.33 bB	265.42 aAB	23.50 aB	16.07 bB	211.70 bB	608.14 aB

Means followed by different small letter(s) within the same row, and different capital letter(s) within the same column, for each studied parameter, are significantly different according to the L.S.D. test at 0.05 level of probability.

At SM, the pure maize stands were significantly superior to the tested sowing schedules concerning the FMY and DMY, with 88.40, and 32.17 t ha^{-1}, respectively, followed by SS3, with the least amount of decrease reaching 25.31 and 12.74% for FMY, and DMY, respectively. The SS2 and SS3 produced significantly higher DMC than SS1 and the pure maize stands. Moreover, SS2 was significantly superior to SS3 concerning plant height, while SS2 and SS3 produced the highest significant values for stem diameter, and SS1 and SS2 resulted in the heaviest plants. On the other hand, when maize was harvested at GFM, all the tested sowing schedules accumulated significantly similar amounts of DMC to the maize pure stands, which ranged from 103.29 to 112.37 g kg^{-1}. This was reflected on the significantly similar amounts of DMY produced, despite that the pure stands were characterized by the highest significant amount of FMY amounting to 20.32 t ha^{-1}. Nonetheless, when harvesting was done at GFM, SS3 was significantly superior to the other sowing schedules and to the maize pure stands, concerning plant height and weight and stem diameter.

Analysis of the variations in leaf, stem and ear% of fodder maize, in addition to its quality in terms of CP, fiber fractions (NDF, ADF, and ADL), and NFC, revealed that all the tested parameters, except ear percentage and ADL, were significantly affected by the SS, while, leaf and stem%, and the tested quality parameters were significantly affected by the HR (Table 3). The interaction between the two studied factors non significantly affected all the parameters. The pure maize stands where characterized by the highest significant leaf% that was significantly similar to SS1 and SS2, while SS3 was characterized by the highest significant stem% (Table 4). No significant variation was detected among the three tested sowing schedules with regard to the ear%. Maize harvested at GFM consisted of only leaves and stems, while maize harvested at SM consisted of leaves, stems and ears. Comparing both harvesting regimes revealed that early harvesting at GFM produced more leaves and stems than late harvesting at SM.

Table 3. *p* values for leaf, stem and ear percentages, crude protein (CP), fiber fractions (NDF, ADF, ADL) and non-fiber carbohydrates (NFC), expressed as g kg^{-1} for fodder maize, combined over 2019 and 2020 growing seasons.

S.O.V.	D.F.	Leaf%	Stem%	Ear%	CP	NDF	ADF	ADL	NFC
SS	3	0.0011	0.0466	0.9086	0.0331	0.0184	0.0006	0.6427	0.0016
HR	1	<0.0001	<0.0001	–	<0.0001	0.0047	<0.0001	<0.0001	<0.0001
SS * HR	3	0.0610	0.3258	–	0.2660	0.3494	0.0890	0.3564	<0.0001

S.O.V.: Source of variation, D.F.: Degrees of freedom, SS: Sowing schedule, HR: Harvest regime l.

Table 4. Variations of the leaf, stem and ear percentages, crude protein (CP), fiber fractions (NDF, ADF, ADL) and non-fiber carbohydrates (NFC) expressed as g kg^{-1} for fodder maize, as affected by the sowing schedule (SS) and maize harvest regime (GFM and SM).

Treatment	Leaf%	Stem%	Ear%	CP
Sowing schedule:				
SS1	28 30 ab	52.59 b	38.22 a	63.94 b
SS2	27 92 ab	53.41 ab	37.33 a	64.93 b
SS3	24 78 b	57.36 a	35.72 a	69.96 a
Pure	31.56 a	50.39 b	36.11 a	65.35 b
Maize harvest regime:				
GFM	40.45 a	59.55 a	-	74.65 a
SM	15.84 b	47.32 b	36.84	57.44 b
Treatment	**NDF**	**ADF**	**ADL**	**NFC**
Sowing schedule:				
SS1	618.85 a	284.05 a	40.95 a	180.97 b
SS2	617.34 a	277.49 a	40.15 a	182.74 b
SS3	613.21 a	290.88 a	41.70 a	188.08 ab
Pure	576.60 b	233.15 b	40.56 a	207.93 a
Maize harvest regime:				
GFM	622.14 a	295.51 a	47.54 a	135.65 b
SM	590.86 b	247.27 b	34.14 b	244.20 a

Means followed by different small letter(s) within the same studied factor for each parameter, are significantly different according to the L.S.D. test at 0.05 level of probability.

Regarding the fodder maize quality parameters, little, yet significant, variation was observed among the tested sowing schedules for the CP content, where SS3 was characterized by the highest significant CP content, amounting to around 7%, which was more than that produced from the pure stands, SS2 and SS1 by 0.5, 0.5, and 0.6%, respectively. On the other hand, a more pronounced variation was detected between the two maize harvest regimes, with harvesting at GFM producing around 7.5% CP, with 1.7% CP more than harvesting at SM. The three tested sowing schedules produced highest significant NDF and ADF contents compared to the maize pure stands, while no significant variation was detected for the ADL content. Similar to the CP content, harvesting at GFM produced highest significant amounts of the three fiber fractions than harvesting at SM. The difference between both harvest regimes amounted to 3.13, 4.82, and 1.34% for NDF, ADF, and ADL, respectively. The pure maize stands produced the highest significant NFC content followed by SS3, amounting to 207.93, and 188.08 g kg^{-1}, respectively. On the contrary, SS1 and SS2 were significantly inferior with 180.97, and 182.74 g kg^{-1}, respectively. As opposed to CP and fiber fractions, harvesting maize at GFM produced around half the amount of NFC that was produced when maize was harvested at SM, with 135.65 against 244.20 g kg^{-1} for GFM, and SM, respectively.

3.2. *Performance of Soybean*

Soybean fresh biological yield (FBY), dry biological yield (DBY), seed yield, plant height and 100-seed weight were significantly affected by the SS and HR, while HI was only variable among the tested sowing schedules. Meanwhile, non-significant variations were detected for the seed oil content. The interaction between the two studied factors was significant only in case of soybean FBY, DBY and 100-seed weight (Table 5). Early harvesting of the companion maize crop at GFM was accompanied with the highest significant soybean FBY and DBY for SS1 and SS2, while for SS3 difference between the two maize harvest regimes was non-significant (Table 6). The highest significant soybean FBY and DBY were produced from the pure stands, amounting to 41.94, and 21.34 t ha^{-1}, respectively. When maize was removed at GFM, no significant variation was detected between SS2 and SS3, while both were superior to SS1 for soybean FBY and DBY. However, in case of maize harvesting at SM, SS3 produced much higher soybean FBY and DBY, than SS1 and SS2. The SS3 was higher than SS1 and SS2 by 114.71, and 61.00% for FBY, respectively, and 113.56, and 54.77% for DBY, respectively. Pure soybean stands were superior to all the tested treatments in the production of the highest 100-seed weight, amounting to 17.81 g. Meanwhile, the three tested sowing schedules resulted in significantly similar 100-seed weight, yet slightly, but significantly, lower than the pure stands. At SS2 and SS3, harvesting maize at GFM and SM produced soybean with significantly similar 100-seed weight, while at SS1, soybean with highest significant 100-seed weight was produced when maize was harvested at GFM than when it was harvested at SM. As shown in Table 7, the highest significant soybean seed yield was produced from the pure stands (4.45 t ha^{-1}), followed by SS2 and SS3, while the least significant seed yield resulted from SS1 (1.78 t ha^{-1}). The HI followed the same trend of the seed yield, with the highest HI recorded for the pure stands that was at par with SS2 and SS3, while SS1 produced the least HI, with 6.67% less than the pure stands. Similarly, SS1 was accompanied with the shortest significant soybean plants, compared to the other sowing schedules and pure stands. Seed oil content was non-significantly affected by the SS, and reached 203.24 g kg^{-1}, in average for the three tested sowing schedules, against 206.17 g kg^{-1} for the soybean pure stands. When maize was harvested at GFM, soybean produced the highest significant seed yield, with the tallest significant plants compared to maize harvesting at SM. While, HI and seed oil content were not significantly variable among the two harvest regimes.

Table 5. p values for fresh biological yield (FBY) and dry biological yield (DBY) as t ha^{-1}, seed yield (t ha^{-1}), harvest index (HI%), plant height (cm), 100-seed weight (g), and seed oil content (g kg^{-1}) for soybean, combined over 2019 and 2020 growing seasons.

S.O.V.	D.F.	FBY	DBY	Seed Yield	HI	Plant Height	100-Seed Weight	Oil Content
SS	3	<0.0001	<0.0001	<0.0001	0.0240	0.0073	0.0088	0.6050
HR	1	0.0002	0.0002	0.0180	0.6493	0.0065	0.0005	0.8418
SS * HR	3	0.0079	0.0040	0.3207	0.9133	0.3288	0.0280	0.6190

S.O.V.: Source of variation, D.F.: Degrees of freedom, SS: Sowing schedule, HR: Harvest regime.

Table 6. Variations of the fresh biological yield (FBY) and dry biological yield (DBY) as t ha^{-1}, 100-seed weight (g) for soybean, as affected by the interaction between sowing schedule (SS) and maize harvest regime (GFM and SM).

Treatment	FBY		DBY		100-Seed Weight	
	GFM	SM	GFM	SM	GFM	SM
SS1	21.49 aC	14.48 bD	9.87 aC	7.45 bD	16.89 aB	16.16 bB
SS2	32.24 aB	19.31 bC	16.51 aB	10.28 bC	16.88 aB	16.81 aB
SS3	33.10 aB	31.09 aB	16.83 aB	15.91 aB	16.39 aB	16.63 aB
Pure	41.94 A	41.94 A	21.34 A	21.34 A	17.81 A	17.81 A

Means followed by different small letter(s) within the same row, and different capital letter(s) within the same column, for each studied parameter, are significantly different according to the L.S.D. test at 0.05 level of probability.

Table 7. Variations of the seed yield (t ha^{-1}), harvest index (HI%), plant height (cm), and seed oil content (g kg^{-1}) of soybean, as affected by the sowing schedule (SS) and maize harvest regime (GFM and SM).

Treatment	Seed Yield	HI	Plant Height	Oil Content
Sowing schedule:				
SS1	1.78 c	14.11 b	82.42 b	202.39 a
SS2	2.45 bc	18.87 ab	90.61 a	203.46 a
SS3	3.14 b	19.05 ab	92.50 a	203.87 a
Pure	4.45 a	20.78 a	97.29 a	206.17 a
Maize harvest regime:				
GFM	3.27 a	18.65 a	94.69 a	204.18 a
SM	2.64 b	17.75 a	86.72 b	203.77 a

Means followed by different small letter(s) within the same studied factor for each parameter, are significantly different according to the L.S.D. test at 0.05 level of probability.

3.3. Land Use Efficiency and Yield Advantage

Data of LER, presented in Table 8 and Figure 4, indicated that late sowing of fodder maize, in general, had a positive impact on land use resulting in a clear yield advantage. LER values for sowing fodder maize 15 days after soybean was 1.32 and 1.00, when harvesting was done at GFM, and SM, respectively. While, sowing maize 30 days after soybean resulted in LER values of 1.27 and 1.49 for the two respective fodder maize harvesting regimes. Determining the yield gain in terms of DMER (Table 8, Figure 4), showed an advantage only when fodder maize was harvested at SM associated with the three sowing schedules, while harvesting at GFM was accompanied with DMER values less than 1. Even though harvesting at SM caused a clear dry matter yield gain, the values of DMER progressively increased with later sowing of fodder maize, with the highest value (2.06) reached when fodder maize was sown 30 days after soybean and harvested at SM, indicating around 200% gain in the dry matter yield of the intercropping system compared to sole cropping of both crops. On the other hand, harvesting fodder maize at GFM resulted in a clear loss in the dry matter yield (DMER < 1), across all sowing schedules, with the most severe loss occurring when both crops were sown together at the same time (DMER = 0.51).

Table 8. Relative yields of the main soybean crop (La) and the companion fodder maize crop (Lb), land equivalent ratio (LER), and dry matter equivalent ratio (DMER) for the tested sowing schedules (SS) and maize harvest regimes (GFM and SM).

Sowing Schedule	Maize Harvest Regime	La	Lb	LER	DMER
SS1	GFM	0.51	0.45	0.96	0.51
	SM	0.35	0.57	0.92	1.15
SS2	GFM	0.74	0.57	1.31	0.83
	SM	0.56	0.44	1.00	1.31
SS3	GFM	0.79	0.48	1.27	0.84
	SM	0.74	0.75	1.49	2.06

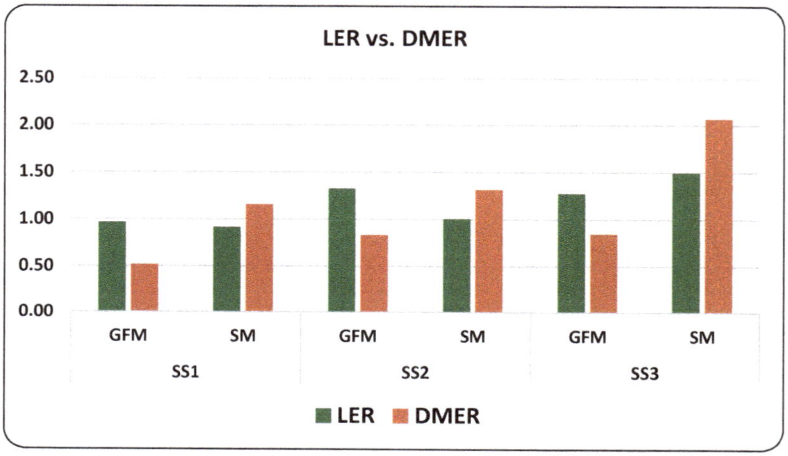

Figure 4. Variations in land equivalent ratio (LER) and dry matter equivalent ratio (DMER) in response to the tested sowing schedules (SS) and maize harvest regimes (GFM and SM).

4. Discussion

In an intercropping system, the best productivity from the component crops could be achieved if they vary in their growth duration so that their peak demand for growth resources can be reached at different periods [7,12]. The critical periods of yield definition for soybean and fodder maize occur usually at different timings along the growing season. Therefore, it is necessary to minimize the competition between both crops during these critical periods, which could be achieved by shifting the sowing/harvesting schedules of one or both crops.

The evaluated sowing schedules, in the current study, exerted a pronounced influence on the soybean and fodder maize performances, that was significantly dependent on the two tested maize harvest regimes. A stronger effect for the sowing schedule on both crops was observed when the fodder maize companion crop was harvested at SM, than when it was harvested at GFM. This was probably because harvesting fodder maize at SM acquired longer existence of maize crop neighboring soybean than harvesting it at GFM, which entailed longer period of interspecific competition between both crops. Meanwhile, each of the two crops showed different response to the sowing/harvesting treatments. While, fodder maize, harvested at SM, produced significantly higher fresh yield with higher dry matter content, resulting in higher dry matter yield than that harvested at GFM, an opposite impact was detected on soybean fresh, dry and seed yields.

A deep insight into the growth dynamics of both crops would help to explain their responses to the treatments. According to [28], the critical period of pod development and seed setting in soybean occurs between R4 and R7, which usually lasts for around 42 days in average approaching the end of the crop's growth cycle. Therefore, early harvesting of fodder maize at GFM terminates the competition between both crops before the beginning of the critical reproductive period of soybean. In addition, harvesting the fodder maize long before the canopy of the soybean matures permits light and air through the understory, which will be reflected on a healthier soybean canopy [7]. In similar studies, soybean plants were able to exhibit fast recovery growth after maize crop was harvested with good compensation to the previous severe competition that occurred during the intercropping period [29].

On the other hand, late harvesting of fodder maize at SM provides a longer period of competition between both crops during the soybean's critical period of development, which will not be in favor of the legume crop. In agreement with the current results [4], the differences in yield of soybean intercropped with maize were attributed to the stage of maturity of the maize companion crop. Cereals are generally characterized by vigorous plants with higher growth rates than legumes, thus, they often suppress the growth of accompanying legumes when intercropped together [13]. This was true for many legume-cereal intercropping systems, like soybean-maize and soybean-sorghum [30,31]. In their study of intercropping soybean and maize using variable patterns, the authors in [17] concluded that intercropping stimulated the growth of maize, which was negatively reflected on the growth of the accompanying soybean. In a similar soybean-maize intercropping system, the authors in [32] reported that fodder maize will be ready for harvesting and ensiling, while soybean is in the R7 developmental stage.

In addition to the vigorous growth nature of maize compared to soybean, the sowing pattern followed to establish the intercropping stands in the current study was in favor of fodder maize crop. Sowing fodder maize on the adjacent borders of the plots allowed it to benefit from the border-row effect [3,13,16,29] that was believed to increase sunlight capture by plants and improve photosynthesis [3], in addition to the use of the optimal intercropping arrangement of four maize rows: six soybean rows as recommended by [5]. This explains the vigorous growth and enhanced productivity of fodder maize achieved in the current study.

The negative impact of late harvesting of the companion fodder maize at SM on soybean crop was clearly offset by manipulating the sowing schedule of the companion crop. The worst impact on the productivity of both crops was achieved when they were sown together. It is well-known that the early growth of the intercrop component crops is very crucial in determining the competition dynamics between them [10]. Therefore, sowing both crops at the same time allowed the competition to begin very early in the season [7], negatively impacting both crops, with heavier impact on the legume component. This was clearly indicated by the significantly lowest soybean seed yield and HI. It was observed that late sowing of fodder maize, resulted in a soybean HI similar to that obtained from the soybean pure stands. While, sowing both crops together significantly decreased soybean HI, probably because the high competition associated with sowing both crops together at the same time significantly suppressed the ability of the soybean plant to convert the photosynthetic assimilates into the economic component, i.e., seed yield. In addition, the shortest soybean plants were produced when both crops were sown together, probably due to the high shading of the fast-growing fodder maize crop, reducing the light intensity reaching the lower soybean canopy, which resulted in stunted plants. In partial agreement with the current results, [33] reported that most soybean cultivars that grow under shade, induced by a taller neighbor plant like maize, exhibit yield reductions. They added, however, that, unlike the current study, shade might enhance stem elongation of soybean and, consequently increase the risk of lodging. On the other hand, delayed sowing of fodder maize allowed the establishment of soybean crop, increasing its competitiveness for when fodder maize was introduced. The best results arising from soybean and fodder maize yields were achieved when fodder maize was sown 30 days after soybean. Soybean plants at 30 DAS were in the third/fourth node developmental stage (V3/V4), thus, plants have already fully developed leaves beginning with the uni-foliolate nodes [28], in addition to the well-developed tap-root system, and are therefore, able to withstand the high competition associated with the introduction of fodder maize crop. Noticeably, the delayed sowing of fodder maize to 30 days after soybean sowing had also a better impact on the fodder maize fresh and dry matter yields, especially when it was harvested at SM. This result suggests that this consecutive sowing schedule ensured complementarity in resource-use in time driven by the different growth periods of both crops [29]. Yet, this delayed sowing of fodder maize resulted in taller maize plants, especially at early growth stages, probably because sowing maize 30 days after soybean (S3) encouraged the plant to strive for solar radiation by increasing stem elongation. This was clearly reflected on taller maize plants cut at GFM. On the other hand, later in the season, the speed of stem elongation slows down, ending up with maize sown early in the season (S1 and S2) and cut at SM having taller stems than late sown maize (S3).

In relation to the quality of the produced maize forage, it was observed that early harvested maize at GFM was characterized by higher significant CP content than late harvested maize at SM [34], which was directly proportional to the leaf component of the crop. Nonetheless, despite the lower leaf component of maize harvested at SM, it was characterized by the lowest significant fiber content (NDF, ADF, ADL) and highest significant NFC content. This might be attributed to the contribution of the ear to the resulting forage material, where the reduction in quality of the plant with advanced maturity is to a great extent compensated by the high quality ears [35,36]. During growth of the maize plant, carbohydrates are stored in the vegetative parts (leaves and stems) and whilst the plant is approaching maturity, the stored carbohydrates are translocated to the ear and deposited into the grains [37]. The importance of the maize grain content in determining its feeding value was well-documented in the early work of several researchers [38–41]. Little variation was detected for fodder maize quality among the evaluated sowing schedules, yet compared to the pure stands, intercropped maize was characterized with low leaf component and high stem component, especially with delayed fodder maize sowing. Maize in late sowing, was already surrounded with a 30 days old soybean canopy that obstructed light penetration into the newly emerging maize population and retarded its photosynthetic activity, resulting in the development of taller plants with smaller leaves and more stems. This was directly reflected on the higher NDF and ADF contents of the intercropped fodder maize compared to the pure stands. Meanwhile, intercropped fodder maize was characterized by higher CP content than pure maize. This result confirmed the ability of maize to benefit from the atmospheric fixed nitrogen by the soybean crop and convert it into higher protein content in the herbage [9].

In the present additive intercropping model, soybean and fodder maize were intercropped at 75% and 50% of the optimal plant densities, respectively, resulting in a total of 125% for both crops. It was, thus, obvious that the pure soybean and fodder maize stands sown at the optimal (100%) plant density for each crop, were significantly more productive compared to all the evaluated intercropping treatments, in terms of herbage and seed yields of fodder maize and soybean, respectively. These results agree with the findings of [15] who has reported higher yields for sole over intercropped soybean and maize. However, the analysis of land use efficiency and yield gain revealed that the LER values for the delayed sowing of fodder maize (15 or 30 days after sowing of soybean) were more than one which indicated the advantage of intercropping soybean and fodder maize over the sole cropping of both crops. The maximum LER value (1.49) was obtained when fodder maize was sown 30 days after soybean and harvested at SM, indicating 49% yield gain over sole cropping. On the other hand, the lowest LER values were 0.91 and 0.96, achieved in case of sowing both crops together at the same time and harvesting fodder maize at GFM, and SM, respectively. In line with the current results, in experiments involving intercropping soybean and maize, high LER (more than one) were achieved [3,15,29,42]. The achieved yield gain in terms of high LER values could be attributed to the complementarity in utilization of above- and below-ground resources and farming inputs between the intercrop component crops [5], which was enhanced by the intercropping pattern used in the current study, by late sowing and harvesting of fodder maize. This intercropping model increased the overall resource use efficiency during the part of the growing season that was occupied by both crops together. Observing the relative yields of the two intercrops revealed that the high LER values were mainly caused by the high relative intercrop soybean yields, which confirms the assumption that the sowing schedules adopted in the current study were mostly in favor of the early sown (soybean) crop. Similarly, the authors in [3], in China, reported that soybean-maize intercropping significantly improved the productivity of soybean.

In addition to the LER, the DMER was used as a key index to gauge dry matter yield gain. A pronounced intercropping advantage in terms of high dry matter yield gain was observed when fodder maize was harvested at SM, noted by DMER values higher than one. This was attributed to the high dry matter contents (34% to 45%), reflected on high dry matter yields (17 to 28 t ha^{-1}) of fodder maize harvested at SM, compared to harvesting at GFM. Coupled with the previously reported advantage of late sowing of fodder maize, the highest DMER (around 200% dry matter yield gain) was reached when fodder maize was sown 30 days after soybean and harvested at SM. Therefore, as opposed to the LER, the productivity of the companion fodder maize crop was more important in determining the DMER than the productivity of the main soybean crop. This is due to the higher dry matter content of the fodder maize, especially when harvested at SM, in addition to the higher growth rate and competitive ability of maize as a cereal crop [15]. Nonetheless, several studies reported that land use efficiency and yield advantage were mainly caused by the subordinate rather than the dominant main crop [11,15]. Notably, the values of the DMER were more realistic in describing the yield gain of the intercropping system compared to sole cropping of both crops, than the LER, which confirms the assumptions raised by the authors in [2], that DMER is more adequate in determining the expected gain, in case of an additive intercropping model, especially in case of crops where the dry matter is the main economic component [43].

5. Conclusions

It has been demonstrated that the soybean-fodder maize additive intercropping practice might be beneficial for the low input agricultural systems of the developing countries. In the current study, the reduction in productivity of the main soybean crop, accompanied with late harvesting of fodder maize companion crop at silage maturity was counterbalanced with the delayed sowing of maize to 30 days after soybean. Late harvesting of fodder maize at silage maturity was not necessarily accompanied by reductions in herbage quality due to the presence of the high-quality ears. Although intercropping reduced the productivity of soybean and fodder maize compared to their pure stands, considering the LER revealed an intercropping advantage with the delayed sowing of fodder maize (LER > 1). On the other hand, the dry matter equivalent ratio (DMER) associated the yield advantage with the late harvesting of fodder maize at SM (DMER > 1), across all sowing schedules, which was more realistic for an additive intercropping model where the dry matter is the economic component. In a soybean-fodder maize intercropping system, whether fodder maize will be cultivated for green feeding or for silage production, it is recommended to delay its sowing to 30 days after soybean in order to maximize yield advantage and land use efficiency.

Author Contributions: Conceptualization, H.S.A.S.; methodology and investigation, H.S.A.S. and M.H.A.-M.; resources, H.S.A.S. and M.H.A.-M.; laboratory analyses, H.S.A.S.; data collection and statistical analysis, H.S.A.S. and M.H.A.-M.; writing—original draft preparation, H.S.A.S. and M.H.A.-M.; writing—review and editing, H.S.A.S. and M.H.A.-M.; supervision, H.S.A.S.; project administration, H.S.A.S. and M.H.A.-M.; funding acquisition, H.S.A.S. and M.H.A.-M. Both authors have read and agreed to the published version of the manuscript.

Funding: This research received no external funding.

Institutional Review Board Statement: Not applicable.

Informed Consent Statement: Not applicable.

Conflicts of Interest: The authors declare no conflict of interest.

Abbreviations

SS	Sowing schedule
HR	Harvest regime
GFM	Green fodder maturity
SM	Silage maturity
FMY	Fresh matter yield
DMY	Dry matter yield
DMC	Dry matter content
CP	Crude protein
NDF	Neutral detergent fiber
ADF	Acid detergent fiber
ADL	Acid detergent lignin
NFC	Non-fiber carbohydrates
FBY	Fresh biological yield
DBY	Dry biological yield
HI	Harvest index

References

1. Salama, H.S.A.; Khalil, H.E.; Nawar, A.I. Utilization of thinned sunflower and soybean intercrops as forage: A useful strategy for small scale farms in intensive agricultural systems. *Int. J. Plant Prod.* **2020**, *14*, 487–499. [CrossRef]
2. Nawar, A.I.; Salama, H.S.A.; Khalil, H.E. Additive intercropping of sunflower and soybean to improve yield and land use efficiency: Effect of thinning interval and nitrogen fertilization. *Chil. J. Agric. Res.* **2020**, *80*, 142–152. [CrossRef]
3. Du, J.B.; Han, T.F.; Gai, J.Y.; Yong, T.W.; Xin, S.U.N.; Wang, X.C.; Feng, Y.; Jiang, L.; Kai, S.; Liu, W.G.; et al. Maize-soybean strip intercropping: Achieved a balance between high productivity and sustainability. *J. Integr. Agric.* **2018**, *17*, 747–754. [CrossRef]
4. Dolijanović, Ž.; Oljača, S.; Kovačević, D.; Simić, M.; Momirović, N.; Jovanović, Ž. Dependence of the productivity of maize and soybean intercropping systems on hybrid type and plant arrangement pattern. *Genetika* **2013**, *45*, 135–144. [CrossRef]
5. Zhang, Y.; Liu, J.; Zhang, J.; Liu, H.; Liu, S.; Zhai, L.; Wang, H.; Lei, Q.; Ren, T.; Yin, C. Row ratios of intercropping maize and soybean can affect agronomic efficiency of the system and subsequent wheat. *PLoS ONE* **2015**, *10*, e0129245. [CrossRef]
6. Ouda, S.A.; El Mesiry, T.; Abdallah, E.F.; Gaballah, M.S. Effect of water stress on the yield of soybean and maize grown under different intercropping patterns. *Aust. J. Basic Appl. Sci.* **2007**, *1*, 578–585.
7. Dowling, A.; Sadras, V.O.; Roberts, P.; Doolette, A.; Zhou, Y.; Denton, M.D. Legume-oilseed intercropping in mechanised broadacre agriculture—A review. *Field Crops Res.* **2021**, *260*, 107980. [CrossRef]
8. Kitonyo, O.M.; Chemining'wa, G.N.; Muthomi, J.W. Productivity of farmer-preferred maize varieties intercropped with beans in semi-arid Kenya. *Int. J. Agron. Agric. Res.* **2013**, *3*, 6–16.
9. Htet, M.N.S.; Soomro, R.N. Effect of different planting structure of maize and soybean intercropping on fodder production and silage quality. *Curr. Agric. Res.* **2016**, *4*, 125. [CrossRef]
10. Tofinga, M.P.; Paolini, R.; Snaydon, R.W. A study of root and shoot interactions between cereals and peas in mixtures. *J. Agric. Sci.* **1993**, *120*, 13–24. [CrossRef]
11. Andrade, J.F.; Cerrudo, A.; Rizzalli, R.H.; Monzon, J.P. Sunflower–soybean intercrop productivity under different water conditions and sowing managements. *Agron. J.* **2012**, *104*, 1049–1055. [CrossRef]
12. Ijoyah, M.O. Review of intercropping research: Studies on cereal-vegetable based cropping system. *Sci. J. Crop Sci.* **2012**, *1*, 55–62.
13. Belel, M.D.; Halim, R.A.; Rafii, M.Y.; Saud, H.M. Intercropping of corn with some selected legumes for improved forage production: A review. *J. Agric. Sci.* **2014**, *6*, 48. [CrossRef]
14. Kim, J.; Song, Y.; Kim, D.W.; Fiaz, M.; Kwon, C.H. Evaluating different interrow distance between corn and soybean for optimum growth, production and nutritive value of intercropped forages. *J. Anim. Sci. Technol.* **2018**, *60*, 1–6. [CrossRef]
15. Kamara, A.Y.; Tofa, A.I.; Ademulegun, T.; Solomon, R.; Shehu, H.; Kamai, N.; Omoigui, L. Maize–soybean intercropping for sustainable intensification of cereal–legume cropping systems in northern Nigeria. *Exp. Agric.* **2019**, *55*, 73–87. [CrossRef]
16. Sánchez, D.G.R.; Silva, J.E.; Gil, A.P.; Corona, J.S.S.; Wong, J.A.C.; Mascorro, A.G. Forage yield and quality of intercropped corn and soybean in narrow strips. *Span. J. Agric. Res.* **2010**, *8*, 713–721. [CrossRef]
17. Ariel, C.E.; Eduardo, O.A.; Benito, G.E.; Lidia, G. Effects of two plant arrangements in corn (*Zea mays* L.) and soybean (*Glycine max* L. Merrill) intercropping on soil nitrogen and phosphorus status and growth of component crops at an Argentinean Argiudoll. *Am. J. Agric. For.* **2013**, *1*, 22–31. [CrossRef]

8. Salama, H.S.A. Yield and nutritive value of maize (*Zea mays* L.) forage as affected by plant density, sowing date and age at harvest. *Ital. J. Agron.* **2019**, *14*, 114–122. [CrossRef]
9. Bal, M.A.; Coors, J.G.; Shaver, R.D. Impact of the maturity of corn for use as silage in the diets of dairy cows on intake, digestion, and milk production. *J. Dairy Sci.* **1997**, *80*, 2497–2503. [CrossRef]
0. Van Soest, P.J.; Robertson, J.B.; Lewis, B.A. Methods for dietary fiber, neutral detergent fiber, and nonstarch polysaccharides in relation to animal nutrition. *J. Dairy Sci.* **1991**, *74*, 3583–3597. [CrossRef]
1. Association of Official Analytical Chemists (AOAC). *Official Methods of Analysis*, 19th ed.; Association of Official Analytical Chemists (AOAC): Gaithersburg, MD, USA, 2012.
2. De Wit, C.T. On competition. *Versl. Landbouwkd. Onderz.* **1960**, *66*, 1–82.
3. De Wit, C.T.; Van den Bergh, J.P. Competition between herbage plants. *Neth. J. Agric. Sci.* **1965**, *13*, 212–221.
4. Shaalan, A.M.; Khalil, H.E.; Nawar, A.I.; El-Salamouni, M.M. Intercropping of grain and fodder maize crops under different nitrogen levels and cutting dates. *Alex. Sci. Exch. J.* **2015**, *36*, 373–380.
5. Salama, H.S.A.; El-Karamity, D.E.; Nawar, A.I. Additive intercropping of wheat, barley, and faba bean with sugar beet: Impact on yield, quality and land use efficiency. *Egypt. J. Agron.* **2016**, *38*, 413–430. [CrossRef]
6. SAS Institute, Inc. *SAS/STAT User's Guide*; Version 9.1; SAS Institute: Cary, NC, USA, 2012.
7. Winer, B.J. *Statistical Principles in Experimental Design*, 2nd ed.; McGraw-Hill Kogakusha, Ltd.: Tokyo, Japan, 1971.
8. Fehr, W.R.; Caviness, C.E. *Stages of Soybean Development. Iowa Cooperative Extension Service, Iowa Agriculture and Home Economics Experiment Station Special Report Nr 80*; Iowa State University: Ames, IA, USA, 1977; Available online: https://lib.dr.iastate.edu/specialreports/87 (accessed on 12 January 2021).
9. Raza, M.A.; Khalid, M.H.B.; Zhang, X.; Feng, L.Y.; Khan, I.; Hassan, M.J.; Ahmed, M.; Ansar, M.; Chen, Y.K.; Fan, Y.F.; et al. Effect of planting patterns on yield, nutrient accumulation and distribution in maize and soybean under relay intercropping systems. *Sci. Rep.* **2019**, *9*, 1–14. [CrossRef] [PubMed]
0. Ghosh, P.K.; Tripathi, A.K.; Bandyopadhyay, K.K.; Manna, M.C. Assessment of nutrient competition and nutrient requirement in soybean/sorghum intercropping system. *Eur. J. Agron.* **2009**, *31*, 43–50. [CrossRef]
1. Echarte, L.; Della Maggiora, A.; Cerrudo, D.; Gonzalez, V.H.; Abbate, P.; Cerrudo, A.; Sadras, V.O.; Calvino, P. Yield response to plant density of maize and sunflower intercropped with soybean. *Field Crops Res.* **2011**, 423–429. [CrossRef]
2. Batista, V.V.; Adami, P.F.; Sartor, L.R.; Silveira, M.F.; Soares, A.B.; Oligini, K.F.; Kwiecinski, D.; Ferreira, M.L.; Camana, D.; Giacomel, C.L.; et al. Forage yield and silage quality of intercropped maize+ soybean with different relative maturity cycle. *J. Agric. Sci.* **2018**, *10*, 249–261. [CrossRef]
3. Liu, W.; Deng, Y.; Hussain, S.; Zou, J.; Yuan, J.; Luo, L.; Yang, C.; Yuan, X.; Yang, W. Relationship between cellulose accumulation and lodging resistance in the stem of relay intercropped soybean [*Glycine max* (L.) Merr.]. *Field Crops Res.* **2016**, *196*, 261–267. [CrossRef]
4. Millner, J.P.; Aver, R.V.; Hardacre, A.K. The yield and nutritive value of maize hybrids grown for silage. *N. Z. J. Agric. Res.* **2005**, *48*, 101–108. [CrossRef]
5. Thom, E.R.; Dorofaeff, F.D.; Dyson, C.B. Effect of plant population and time of harvest on yield and quality of maize (*Zea mays* L.) grown for silage: I. Yield and chemical composition, and sampling procedures for large areas. *N. Z. J. Agric. Res.* **1981**, *24*, 285–292. [CrossRef]
6. Phipps, R.H.; Wilkinson, M. *Maize Silage*; Chalcombe Publications: Great Britain, UK, 1985.
7. Daynard, T.B.; Tanner, J.W.; Hume, D.J. Contribution of stalk soluble carbohydrates to grain yield in corn (*Zea mays* L.). *Crop Sci.* **1969**, *9*, 831–834. [CrossRef]
8. Bunting, E.S. The question of grain content and forage quality in maize: Comparisons between isogenic fertile and sterile plants. *J. Agric. Sci. Camb.* **1975**, *85*, 455–463. [CrossRef]
9. Bunting, E.S. Effects of grain formation on dry matter distribution and forage quality in maize. *Exp. Agric.* **1976**, *12*, 417–428. [CrossRef]
0. Phipps, R.H. A note on the effect of genotype, density and row width on the yield and quality of forage maize. *J. Agric. Sci. Camb.* **1975**, *84*, 567–569. [CrossRef]
1. Phipps, R.H.; Weller, R.F.; Fulford, R.J. The development of plant components and their effects on the composition of fresh and ensiled forage maize 3. The effect of grain content on milk production. *J. Agric. Sci. Camb.* **1979**, *92*, 493–498. [CrossRef]
2. Matusso, J.M.M.; Mugwe, J.N.; Mucheru-Muna, M. Effects of different maize (*Zea mays* L.) soybean (*Glycine max* (L.) Merrill) intercropping patterns on yields and land equivalent ratio. *J. Cereals Oilseeds* **2013**, *4*, 48–57. [CrossRef]
3. Nawar, A.I.; Salama, H.S.A.; Shaalan, A.M.; Khalil, H.E. Land Equivalent Ratio Versus Dry Matter Equivalent Ratio: Adequacy for additive intercropping. In Proceedings of the 61st Annual conference of the German Society of Agronomy, Kiel, Germany, 8–9 October 2018; Volume 30, pp. 217–218.

agronomy

MDPI

Article

The Effect of Subsurface Placement of Mineral Fertilizer on Some Soil Properties under Reduced Tillage Soybean Cultivation

Piotr Kraska, Sylwia Andruszczak *, Paweł Gierasimiuk and Hubert Rusecki

Department of Herbology and Plant Cultivation Techniques, University of Life Sciences, Akademicka 13, 20-950 Lublin, Poland; piotr.kraska@up.lublin.pl (P.K.); pawel.gierasimiuk@interia.pl (P.G.); hubert.rusecki@gmail.com (H.R.)

* Correspondence: sylwia.andruszczak@up.lublin.pl; Tel.: +48-81-445-66-89

Citation: Kraska, P.; Andruszczak, S.; Gierasimiuk, P.; Rusecki, H. The Effect of Subsurface Placement of Mineral Fertilizer on Some Soil Properties under Reduced Tillage Soybean Cultivation. *Agronomy* **2021**, *11*, 859. https://doi.org/10.3390/agronomy11050859

Academic Editors: Aurelio Scavo and Umberto Anastasi

Received: 20 March 2021
Accepted: 25 April 2021
Published: 28 April 2021

Abstract: One of the adverse effects of no-tillage is the accumulation of nutrients (in particular P and K) in the top soil layer. The subsurface application of mineral fertilizers at a depth of 10–30 cm can reduce this phenomenon and at the same time provide a relatively uniform access to soil nutrients for plant roots. Such a method of mineral fertilizer application can additionally decrease the environmental risk associated with water eutrophication because the water runoff from fields where the soil P content is high, is reduced. The aim of this research was to evaluate the effect of the subsurface application of different rates of a compound mineral fertilizer on the content of some macronutrients, soil organic carbon content (SOC), and soil pH in a field after the harvest of soybean grown under reduced tillage conditions. The field experiment was conducted during the growing seasons of 2014/2015–2016/2017 in the village of Rogów, Zamość County, Poland. It was set up as a split-plot design in four replicates. The first experimental factor included two methods of mineral fertilization application: fertilizer broadcast over the soil surface (S); fertilizer applied deep (subsurface placed) using a specially designed cultivator (Sub-S). The other factor was the rates of the mineral fertilizer (NPKS): 85 kg·ha^{-1} (F85) and 170 kg·ha^{-1} (F170). Over the successive years of the study, the SOC content was found to increase. However, neither the fertilization rate nor the method of fertilizer application caused any significant difference in organic carbon. Under subsurface fertilizer application conditions, a higher soil pH was found in treatment F85, however when the fertilizer was surface-applied, the soil in treatment F170 had a higher pH value. During the three-year study period, the P and K content in the 0–30 cm soil layer was higher than in the 30–60 cm and 60–90 cm layers. In turn, the highest Mg content was determined in the 30–60 cm layer. In the case of both mineral fertilizer application methods, a higher P content was determined in the soil fertilized at a rate of 170 kg NPKS, compared with a rate of 85 kg·ha^{-1}. The surface application of the higher rate of mineral fertilization resulted in an increase in the soil K content. On the other hand, when the mineral fertilizer was subsurface-applied, a higher soil K was determined in the treatments with lower mineral fertilization.

Keywords: soil; chemical composition; reduced tillage system; subsurface fertilization; soybean

1. Introduction

In studies addressing the effects of different agronomic practices on the productivity of agroecosystems, it is of key importance to evaluate the direction of changes that occur in soil biological, chemical, and physical properties. Currently, conventional tillage with a plow dominates in Central Europe. Such tillage helps to aerate the soil, introduce crop residues, and control weeds to prepare the final seedbed. On the other hand, a traditional plow-based system can lead to many negative changes in the soil environment, such as leaching nutrients from the soil and reducing the amount of soil organic matter (SOM). In addition, the loss of SOM has a negative effect on soil structure, water capacity, and biological activity. It also increases susceptibility to water and wind erosion [1–6]. The degradation of the soil environment associated with conventional soil tillage results in a need to use new tillage technologies that will allow the biodiversity of biocenosis to be preserved, including soil conservation [7,8]. Lahmar [9] and Wauters et al. [10] report that, during the last decade, minimum/reduced tillage systems and no-tillage systems have attracted an ever-greater interest, as parts of a sustainable agriculture. Reduced cultivation is a tillage practice that does not invert the soil, combined with 30% of crop residues left on the soil surface, while no-tillage is defined as a system in which the soil remains undisturbed from harvest to planting and the seeds are drilled into the stubble of the previous crop. Compared with conventional tillage, these cultivation methods are less labor- and energy-consuming; they also beneficially affect the biological activity of soil as well as its chemical and physical properties [1–3,11,12]. It was found that no-tillage is conducive to increasing the content of organic matter in soil. Additionally, water and wind erosion are reduced, and the risk of elements being leached outside an agricultural ecosystem diminishes substantially [6,8,13–18].

The chemical properties of soil depend on the content of the elements present in the soil, the forms in which they occur, and the changes they undergo. Furthermore, the chemical properties of soil also depend on soil fauna and vegetation, human activity, as well as cropping and soil use intensity [5,7,17,19,20]. According to Wróbel and Pabin [21], changes in soil chemistry under reduced tillage conditions adversely affect nutrient supply to plants. Mineral fertilization, as one of the elements of agronomic practices, directly impacts the availability of essential nutrients in soil.

Modern agronomic technologies allow mineral fertilizers to be placed at different depths relative to the soil surface [22–25]. Lakew [26] thinks that nutrients must be supplied at an appropriate amount, form, and time in order to provide to the greatest possible extent, proper growth, and development conditions for crops. The yield-increasing effect of various nutrient application methods largely depends on soil nutrient availability and the tillage system used. The beneficial effects of subsurface fertilization are manifested more strongly under low soil disturbance conditions; hence, this fertilization method is primarily recommended in the no-tillage system [23,24,27–29].

One of the negative effects of no-tillage is the accumulation of nutrients in the top soil layer. This applies in particular to phosphorus and potassium [4,7,16,30–33]. The deep (subsurface) application of mineral fertilizers prevents P and K from accumulating in the limited soil volume and can contribute to an increased nutrient efficiency. The deep application of fertilizers, especially P-containing ones, can reduce the concentration of this element on the field surface. In this way, the environmental risks related to water erosion and the surface runoff of water, from fields in which the level of the soil P content is high, is reduced. At the same time, the deep placement of fertilizers is thought to improve the availability of nutrients contained in them, thus enhancing the effectiveness of their application [22,25,28,34–37]. Randall and Hoeft [33] give several methods of localized subsurface fertilization, notably, deep band placement, surface band placement under the seed, and band placement of fertilizer directly with the seed. Stanisławska-Glubiak and Korzeniowska [39] are of the opinion that such application of mineral fertilizers should increase the use of nutrients by plants.

This study's hypothesis was that the deep application of mineral fertilizer, compared with its surface placement, under reduced tillage conditions, would allow soybean plants to have better availability of nutrients supplied with the mineral fertilizer. Moreover, the subsurface placement of the mineral fertilizer could contribute to more even distribution of nutrients in the soil profile.

The aim of this experiment was to evaluate the effect of subsurface application of various doses of mineral fertilizer on soil pH; soil organic carbon (SOC) and the content of P, K; and Mg in the soil after the harvest of soybean grown in crop rotation (soybean—winter wheat—maize), under the conditions of reduced tillage system.

2. Materials and Methods

2.1. Study Area and Field Experiment

This study was conducted over the period 2015–2017, based on a field experiment established in the autumn of 2014 in the village of Rogów, Municipality of Grabowiec, Zamość County [50°48′22.4″ N; 23°30′00.5″ E]. The experiment was set up on brown soil (CAMBISOLS according to the World Reference Base for Soil Resources 2014) [40,41].

Before the establishment of the experiment in the autumn of 2014, soil samples were taken to determine the availability of essential elements (P, K, Mg) in the soil and its pH in the layer from 0 to 90 cm, as well as the soil organic carbon content in the 0–30 cm layer. The properties of the initial soil are given in Table 1.

Table 1. Properties of the initial soil of the experiment site at Rogów, in 2014.

Initial Soil Properties		Value
Soil pH	0–30 cm soil depth	5.01
	30–60 cm soil depth	5.94
	60–90 cm soil depth	6.61
Available P content ($mg \cdot kg^{-1}$)	0–30 cm soil depth	18.84
	30–60 cm soil depth	10.68
	60–90 cm soil depth	16.69
Available K content ($mg \cdot kg^{-1}$)	0–30 cm soil depth	78.92
	30–60 cm soil depth	43.77
	60–90 cm soil depth	44.51
Available Mg content ($mg \cdot kg^{-1}$)	0–30 cm soil depth	64.07
	30–60 cm soil depth	69.33
	60–90 cm soil depth	65.46
SOC ($g \cdot kg^{-1}$)	0–30 cm soil depth	7.9
Particle size distribution	Sand (%)	23.6
	Silt (%)	70.6
	Clay (%)	5.8

SOC—soil organic carbon.

The study was set up as a split-plot design. The first experimental factor included two methods of mineral fertilization application under reduced tillage conditions. In one treatment, the compound mineral fertilizer was broadcast over the soil surface (S). In the other treatment, the fertilizer was placed deep, using a specially designed cultivator, evenly at a depth of 10–30 cm of the operation of the soil loosening and fertilizer spreading attachment (S-Sub). Another factor included was the different rates of the mineral fertilizer: 85 kg NPKS·ha^{-1} (F85) and 170 kg NPKS·ha^{-1} (F170). In total, the experiment consisted of four treatments, each in four replicates (16 plots per year). The area of a single plot was 175 m^2. Between the plots with the different mineral fertilization treatments, there was a 20 m wide buffer zone necessary to properly perform specific agronomic operations.

In the experiment, the soybean cultivar, 'Annushka', was grown in crop rotation with winter wheat and maize. 'Annushka', which originated from the soybean breeding company Hodowla Soi Agroyoumis Polska, was listed in the Common Catalogue of Varieties of Agricultural Plant Species (CCA) in 2009 [42]. It is recommended for cultivation across the entire country, it is a very early variety (earliness group 0000), and its growing season lasts about 100–130 days.

Before the establishment of the experiment, winter oilseed rape was grown in the field under the condition of conventional tillage and after its harvest liming was applied by spreading chalk (CaO content 39.2%; $CaCO_3$—70%) at a rate of 5 t·ha^{-1} (New Holland Tm 165 + Joskin Siroko spreader).

Soil cultivation involved disking (Terradisc 6001 T disk harrow), which was performed twice: after harvesting the previous crop and before winter. Before seed sowing, a cultivator was used (Pöttinger SYNKRO 5003 K cultivator). On the plots with surface fertilizer application, such cultivation treatment was carried out immediately after fertilizer placement (Amazone ZA TS 4200), whereas on plots with subsurface fertilizer application, the treatment was carried out during the same pass (Figure 1).

Figure 1. The sweep for deep mineral fertilizer application used in soybean cultivation technology.

Before sowing the soybean seeds, mineral fertilizer was applied in the form of Polifoska®6 NPK(S) 6-20-30(7), at a rate of 200 (F85) or 400 (F170) kg·ha^{-1}. The percentage content of all nutrients in the applied fertilizer was as follows: N—6%; P_2O_5—20%; K_2O—30%; SO_3—7%. In total, the mineral fertilization was the following (per hectare):

F85 = 12 kg N, 17.5 kg P, 50 kg K, 5.5 kg S (85 kg NPKS·ha^{-1}).

F170 = 24 kg N, 35 kg P, 100 kg K, 11 kg S (170 kg NPKS·ha^{-1}).

As soybean is a plant that fixes atmospheric nitrogen, no nitrogen top dressing was applied in the soybean crop. Moreover, the soybean plants were not irrigated during the growing season.

The surface placement of the fertilizer was carried out using an Amazone ZA TS 4200 spreader, whereas the subsurface application was performed using a rigid tine cultivator with its sweeps adapted to subsurface fertilizer placement. The sweeps were connected with a fertilizer hopper via a compressed air turbine, used to feed the fertilizer to the sweeps through the distribution mechanism. Moreover, this device places the fertilizer evenly at a depth of 10–30 cm of the operation of the soil loosening and fertilizer spreading attachment during one travel (Figure 1).

Cv. 'Annushka' soybeans were sown at a rate of 120 kg·ha^{-1}. A TERRASEM C6 seed drill was used to seed soybeans. The chemical plant protection of the soybean was as shown in Table 2.

Table 2. Chemical plant protection of the soybean during the growing seasons.

	Plant Protection Product	Dose	Application Date
Seed dressing	T75 DS/WS [thiuram (a compound from the dithiocarbamate group)—750 g·kg^{-1}]	2g·kg^{-1} seeds	Before sowing
	Nitragina	300 g·ha^{-1}	Before sowing
Herbicide	Roundup 360 SL [glyphosate (a compound from the amino phosphonic acid group) as potassium salt—360 g·L^{-1}].	1.5 L·ha^{-1}	Before emergence
	Corum 502.4 SL [bentazon (a compound from the diazine group)—480 g·L^{-1}; imazamox (a compound from the imidazolinone group)—22.4 g·L^{-1}]	1.25 L·ha^{-1}	BBCH 12–25
Adjuvant	Dash HC [methyl oleate—348.75 g·L^{-1}; fatty alcohol (alkoxylated phosphoric acid ester)—209.25 g·L^{-1}]	1.0 L·ha^{-1}.	BBCH 12–25

BBCH—scale used to identify the phenological development stages of plants [43].

The soybean crop was harvested at full maturity stage using a New Holland CR 8090 combine harvester.

2.2. Analyses

In each year of the study, soil samples were collected for analysis after the soybean harvesting, using a modified soil auger. Soil samples were taken at 10 randomly selected sites from each experimental plot, at a soil depth of 0–30 cm, 30–60 cm, and 60–90 cm. Then, the collected soil samples were combined into one aggregate sample from each plot, separately for each soil layer. The total number of samples was 48 per year. The content of phosphorus, potassium, and magnesium, as well as the pH were determined for soil layers 0–30 cm, 30–60 cm, and 60–90 cm. The organic carbon content, in turn, was determined for a layer of 0–30 cm. The chemical analyses were carried out at the accredited laboratory, Chemical and Agricultural Station in Lublin (accreditation certificate No. AB 1186 issued by the Polish Centre for Accreditation), which meets the requirements of the PN/EN ISO/IEC 17025:2018-02 standard. The organic carbon content in the soil was determined by the Tiurin method (oxidation of soil organic carbon with excess potassium dichromate in concentrated sulphuric acid) [44], total nitrogen by the Kjeldahl method [45], available phosphorus (P) and potassium (K) by the Egner-Riehm method [46,47], available magnesium (Mg) by ASA, after the extraction of 0.0125 mole $CaCl_2 \cdot dm^{-3}$ [48], and pH_{KCl} was determined potentiometrically [49].

2.3. Data Analysis

Analysis of variance (ANOVA) was used to statistically analyze the results by employing Statistica PL 13.3 (TIBCO Software Inc., Tulsa, OK, USA). Tukey's multiple comparison test was applied to determine the differences between the means for the main factors (methods of fertilizer application: MFA; fertilizer dose: FD; soil layer: SL; years: Y), whereas confidence intervals for the means of LSD (lowest significant difference; $p = 0.05$) were used to compare the means from the subclasses (interaction Y × SL; Y × MFA; Y × FD; MFA × SL; MFA × FD; FD × SL). The three-way interactions were not considered.

2.4. Characteristics of Three Growing Seasons Based on Selyaninov's Hydrothermal Coefficient

To evaluate the thermal and pluvio-thermal conditions in the three growing seasons analyzed, Selyaninov's hydrothermal coefficient was applied, following Stachowski [50] in the following form:

$$K = (P \cdot 10)/\sum t \quad (1)$$

P—sum of monthly total rainfall in mm

Σt—sum of mean daily temperatures >0 °C.

The humidity characteristics of the months and the interpretation of the hydrothermal coefficient followed Skowera and Puła [51] as well as Skowera [52], depending on the value of the coefficient k: extremely dry—k ≤ 0.4; very dry—0.4 < k ≤ 0.7; dry—0.7 < k ≤ 1.0; rather dry—1.0 < k ≤ 1.3; optimal—1.3 < k ≤ 1.6; rather humid—1.6 < k ≤ 2.0; humid—2.0 < k ≤ 2.5; very humid—2.5 < k ≤ 3.0; extremely humid—k > 3.0.

In 2015, the hydrothermal coefficient values show that water deficits occurred only in the months of June, July, and August (Table 3). The humidity index in this year demonstrates that March, April, and May were humid months.

Table 3. Selyaninov hydrothermal coefficients (K) during the growing seasons in the years of the experiment.

Months	Years		
	2015	2016	2017
March	k = 2.73 very humid	k = 4.49 extremely humid	k = 1.79 rather humid
April	k = 1.47 optimal	k = 2.40 humid	k = 2.66 very humid
Maj	k = 4.75 extremely humid	k = 1.23 rather dry	k = 1.67 rather humid
June	k = 0.30 extremely dry	k = 1.23 rather dry	k = 0.50 very dry
July	k = 0.70 very dry	k = 2.20 humid	k = 1.66 rather humid
August	k = 0.10 extremely dry	k = 0.94 dry	k = 0.65 very dry
September	k = 1.90 rather humid	k = 0.24 extremely dry	k = 2.50 very humid
October	k = 2.14 humid	k = 5.89 extremely humid	k = 3.97 extremely humid
November	k = 2.35 humid	k = 7.30 extremely humid	k = 3.11 extremely humid

In the second year of the experiment (2016), April was a humid month that had been preceded by an extremely humid March, whereas May was a rather dry month, similarly to June.

In 2017, the humidity characteristics of the analyzed months of the growing season tended toward humid periods. Only June and August were very dry months (Selyaninov's hydrothermal coefficient was k = 0.50 and k = 0 65, respectively). During the spring and summer period, the highest rainfall was recorded in May and July, which is confirmed by Selyaninov's coefficient, according to which these months were rather humid.

3. Results

Given the variance analysis, the effect of years, the method of fertilizer application, the fertilizer dose, and the soil layer, as well as the interaction of these factors on pH and the content of P, K, and Mg in the soil, were significant. In contrast, no significant interactions were found between experimental factors with regard to the content of organic carbon in the soil (Table 4).

Table 4. Effect of years, method of fertilizer application, fertilizer dose, soil layer, and interaction of experimental factors on examined features.

Feature	Y	MFA	FD	SL	Y × MFA	Y × FD	Y × SL	MFA × FD	MFA × SL	FD × SL
SOC	**	ns	ns	–	ns	ns	–	ns	–	–
pH	**	**	**	**	**	**	**	**	**	**
P	**	**	**	**	**	**	**	**	**	**
K	**	**	**	**	**	**	**	**	**	**
Mg	**	**	**	**	**	**	**	ns	**	**

Y—year; MFA—method of fertilizer application; FD—fertilizer dose; SL—soil layer; SOC—soil organic carbon [$g \cdot kg^{-1}$]; P—content of available P [$mg \cdot kg^{-1}$]; K—content of available K [$mg \cdot kg^{-1}$]; Mg—content of available Mg [$mg \cdot kg^{-1}$]; **—significant at $p = 0.05$; ns—not significant at $p = 0.05$.

The content of SOC differed significantly between the years. The highest value of SOC was found in the last year of the experiment, whereas 2015 and 2016 have similar values (Table 5). Over the three-year study period, the fertilizer application method and fertilizer rate did not significantly affect the soil organic carbon content.

Table 5. Soil organic carbon content in the 0–30 cm soil layer after soybean harvest ($g \cdot kg^{-1}$).

Method of Fertilizer Application (MFA)	Fertilizer Dose (FD)	Years (Y)		
		2015	2016	2017
S	F85	11.3	12.9	18.7
	F170	10.8	13.2	18.5
	Mean	11.1	13.1	18.6
Sub-S	F85	12.7	14.0	19.5
	F170	11.7	13.2	22.0
	Mean	12.2	13.6	20.7
Mean	F85	12.0	13.5	19.1
	F170	11.3	13.2	20.3
	Mean	11.6	13.3	19.7
LSD $_{0.05}$			Years 4.25	

S—Surface fertilizer application, Sub-S—Subsurface fertilizer application; F85—fertilizer dose 85 kg $NPKS \cdot ha^{-1}$; F170—fertilizer dose 170 kg $NPKS \cdot ha^{-1}$; LSD $_{0.05}$—the lowest significant difference at $p = 0.05$.

The soil pH value differed significantly over the years of the study. The lowest pH was found in 2015, while the highest in the second year of the experiment (Table 6). The pH value was shown to change significantly, depending on the depth. The soil in the top 0–30 cm layer exhibited the lowest pH, whereas, with the increase in depth, the pH measured in the successive soil layers increased significantly. Treatments S were found to have a higher soil pH, compared with Sub-S. Moreover, the higher rate of mineral fertilization contributed to a significant increase in pH (Table 6).

In the soil after soybean harvest, a significant increase in the P content was found in each successive year of the study (Table 6). Furthermore, in the last year of the experiment, the soil K content was shown to significantly increase, relative to the first two years of observation. In turn, the highest Mg content was determined in 2016 (Table 6).

The content of the evaluated macronutrients in the individual soil layers differed significantly. The highest amount of P was determined in the top soil layer; it was significantly lower at the level of 60–90 cm, while it was at its lowest in the 30–60 cm layer. In the case of potassium, with the increasing depth, the content of this element significantly decreased. In turn, the highest Mg content was found in the 30–60 cm soil layer; it was significantly lower in the top 0–30 cm layer, while it was at its lowest in the 60–90 cm layer (Table 6).

Table 6. Evaluation of the pH and soil content of available forms of selected macronutrients after soybean harvest.

Specification		pH (KCl)	P (mg·kg^{-1})	K (mg·kg^{-1})	Mg (mg·kg^{-1})
Years (Y)	2015	5.74	13.53	53.73	68.43
	2016	6.23	15.01	54.17	71.09
	2017	5.89	16.80	55.02	69.05
	LSD $_{0.05}$	0.006	0.293	0.673	0.842
Soil layer (SL)	0–30 cm	5.32	19.26	72.53	69.80
	30–60 cm	6.00	11.29	46.42	71.66
	60–90 cm	6.55	14.79	43.98	67.11
	LSD $_{0.05}$	0.006	0.293	0.673	0.842
Method of fertilizer application (MFA)	S	6.17	14.59	57.45	68.43
	Sub-S	5.74	15.64	51.16	70.62
	LSD $_{0.05}$	0.004	0.199	0.457	0.572
Fertilizer dose (FD)	F85	5.94	13.04	53.95	69.98
	F170	5.97	17.18	54.67	69.07
	LSD $_{0.05}$	0.004	0.199	0.457	0.572

S—Surface fertilizer application, Sub-S—Subsurface fertilizer application; F85—fertilizer dose 85 kg NPKS·ha^{-1}; F170—fertilizer dose 170 kg NPKS·ha^{-1}; LSD $_{0.05}$—the lowest significant difference at $p = 0.05$.

In the soil sampled from the plots where the fertilizer was surface-applied (S), a significantly higher K content and, at the same time, a lower P and Mg content were found, compare to those found under deep fertilizer application conditions (Sub-S). Furthermore, it was demonstrated that the increased level of mineral fertilization promoted an increase in the soil P and K content. On the other hand, in the soil taken from the plots where the lower rate of the fertilizer Polifoska®6 had been used, a higher Mg content was determined (Table 5).

The subsurface application of mineral fertilizer, compared with S treatment, significantly decreased the soil pH in each of the evaluated soil layers (0–30 cm, 30–60 cm, and 60–90 cm) (Figure 2A). The experiment confirmed that the effect of mineral fertilizer rate on soil pH is dependent on soil depth. In the 0–30 cm and 30–60 cm layers, a higher pH was found in treatment F170, whereas, in the 60–90 cm layer, the double rate of fertilizer (F170) significantly decreased the soil pH, in comparison with treatment F85 (Figure 2B).

The highest pH was found in the plots where the higher rate of surface-applied mineral fertilization was applied (SF170) (Figure 2C).

The effect of the fertilizer application method on the soil P content was dependent on soil depth. As regards the 0–30 cm soil layer, subsurface fertilization resulted in a significant increase in the content of this element; whereas, in the 30–60 cm layer, such application of fertilizer decreased the P content, compared with the plots where the fertilizer was surface-applied. In the deepest soil layer (60–90 cm), a different method of fertilizer application did not cause any significant differences in the P content in soil (Figure 3A).

In all the soil layers evaluated, fertilization with a doubled rate of NKPS (F170) resulted in a significant increase in the P content, compared with the lower dose of 85 kg NPKS·ha^{-1} (Figure 3B). The statistically proven interaction between experimental factors showed that the highest content of P in the soil was in the plot with the higher rate of mineral fertilization, applied over the soil surface (SF170) (Figure 3C).

In the 0–30 cm soil layer, the soil K content in treatment S was significantly higher, compared with SubS. A similar relationship was found for the 30–60 cm soil layer. As far as the 60–90 cm layer is concerned, the soil K concentration in treatments S and SubS was similar (Figure 4A).

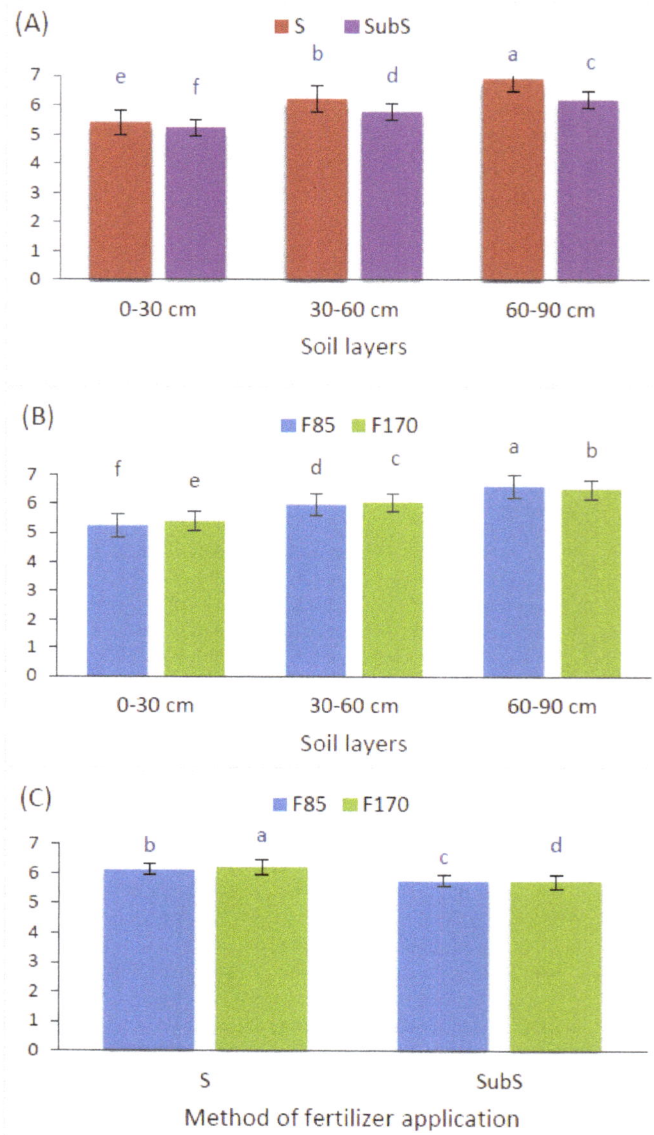

Figure 2. Effect of the interaction of the experimental factors on soil pH: (**A**) Method of fertilizer application × soil layer; (**B**) Fertilizer dose × soil layer; (**C**) Method of fertilizer application × fertilizer dose (S—Surface fertilizer application, SubS—Subsurface fertilizer application; F85—fertilizer dose 85 kg NPKS·ha^{-1}; F170—fertilizer dose 170 kg NPKS·ha^{-1}); different letters indicate significant differences (p = 0.05).

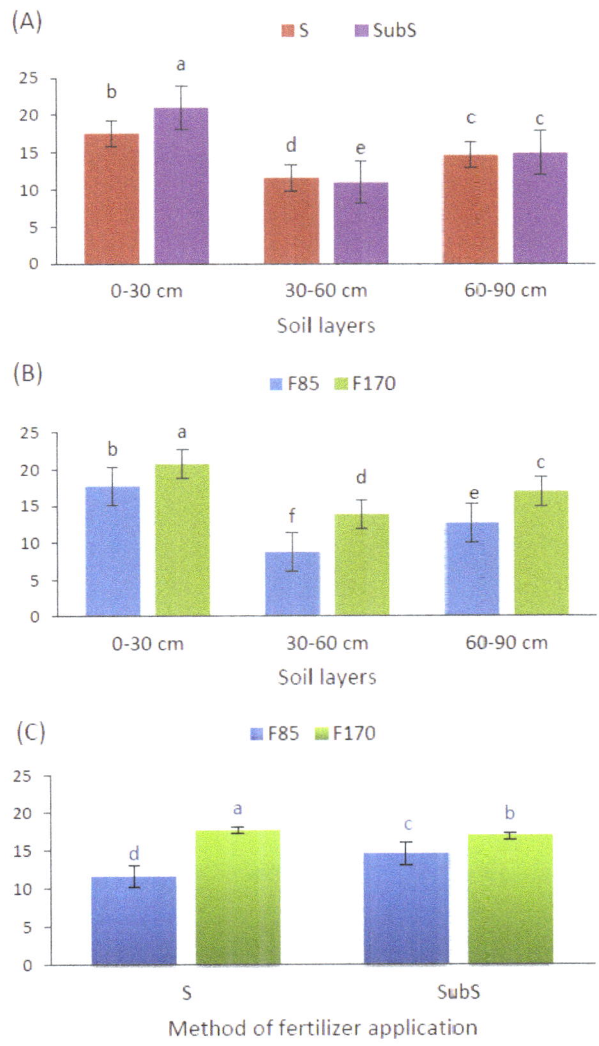

Figure 3. Effect of the interaction of the experimental factors on the soil content of available P [$mg \cdot kg^{-1}$]: (**A**) Method of fertilizer application × soil layer; (**B**) Fertilizer dose × soil layer; (**C**) Method of fertilizer application × fertilizer dose (S—Surface fertilizer application, SubS—Subsurface fertilizer application; F85—fertilizer dose 85 kg NPKS·ha^{-1}; F170—fertilizer dose 170 kg NPKS·ha^{-1}); different letters indicate significant difference ($p = 0.05$).

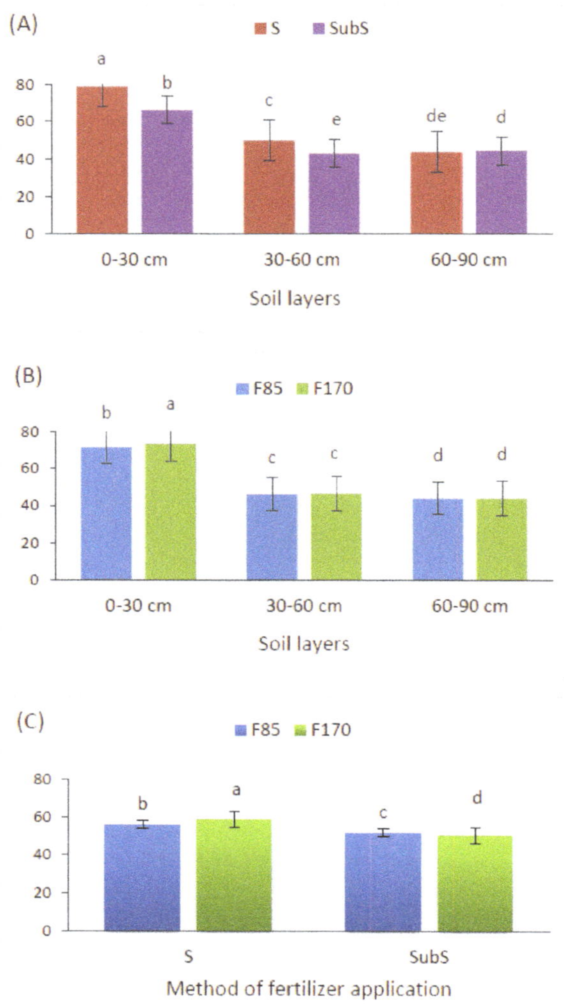

Figure 4. Effect of the interaction of the experimental factors on the soil content of available K [$mg \cdot kg^{-1}$]: (**A**) Method of fertilizer application × soil layer; (**B**) Fertilizer dose × soil layer; (**C**) Method of fertilizer application × fertilizer dose (S—Surface fertilizer application, SubS—Subsurface fertilizer application; F85—fertilizer dose 85 kg NPKS·ha^{-1}; F170—fertilizer dose 170 kg NPKS·ha^{-1}); different letters indicate significant difference ($p = 0.05$).

The proven interaction demonstrated that an increased mineral fertilization of the soybean crop significantly increased the K content in the soil only to a depth of 30 cm; whereas, in the deeper soil layers (30–60 cm and 60–90 cm), the different fertilizer rate had no impact on the K content (Figure 4B).

The surface application of the higher rate of fertilizer (SF170) resulted in an increase in the soil K content, compared with treatment F85. On the other hand, when the mineral fertilizer was subsurface-applied (SubS), a reverse relationship was found—a higher soil K content in the soil was determined in treatments F85 (Figure 4C).

The effect of the fertilizer application method on the soil Mg content was dependent on soil depth. In the 30–60 cm and 60–90 cm soil layers, a significantly higher soil Mg content was found in the treatment with subsurface fertilizer application (SubS), whereas, in the soil up to a depth of 30 cm, the Mg content did not significantly differ in both fertilization treatments (S and SubS) (Figure 5A).

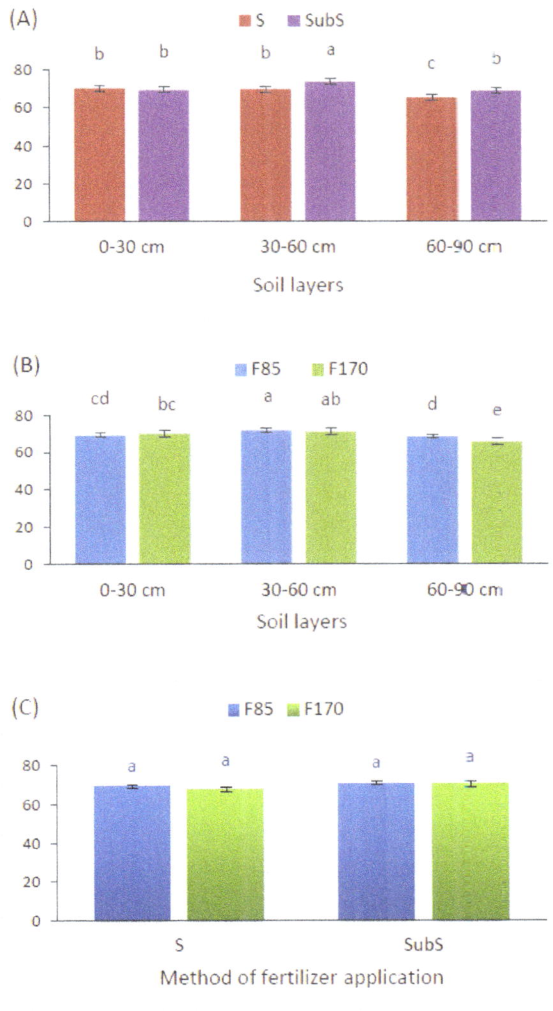

Figure 5. Effect of the interaction of the experimental factors on the soil content of available Mg [mg·kg^{-1}]: (**A**) Method of fertilizer application × soil layer; (**B**) Fertilizer dose × soil layer; (**C**) Method of fertilizer application × fertilizer dose (S—Surface fertilizer application, SubS—Subsurface fertilizer application; F85—fertilizer dose 85 kg NPKS·ha^{-1}; F170—fertilizer dose 170 kg NPKS·ha^{-1}); different letters indicate significant difference (p = 0.05).

The effect of fertilizer rate on the soil Mg content depended on soil layer. Evaluating the content of this element in the 0–30 cm and 30–60 cm layers, it was found that mineral fertilization at a higher rate (170 kg NPKS) essentially did not have any statistically proven effect on its occurrence. In turn, in the 60–90 cm soil layer, a higher content of Mg was found in plots with the lower rate of mineral fertilizer (F85) (Figure 5B). The experiment did not prove a significant interaction between the fertilizer application method and fertilizer rate in relation to the soil Mg content (Figure 5C).

4. Discussion

In the study discussed in this paper, the soil pH in the top soil layer (0–30 cm) was lower than the one at the deeper soil levels (30–60 cm and 60–90 cm). Limousin and Tessier [53], López-Fando and Pardo [31], as well as Neugschwandtner et al. [7] also report that the non-tilled upper soil layer generally has a lower value of pH. One of the reasons for the acidification of surface soil layers under reduced tillage conditions is the accumulation of decomposition products and fertilizer substances with an acidifying effect [54]. In the study carried out by Wróbel and Pabin [21], the changes in the concentration of the main nutrients in the soil in which reduced tillage was used were accompanied by a decrease in the value of pH_{KCl} in the 0–5 cm soil layer, relative to the 10–15 cm layer. Dorneles et al. [33] found a similar relationship with regard to the 5–10 cm layer. Haruna and Nkongolo [55] also obtained a lower soil pH for the 0–10 cm layer, in comparison with the soil levels of 10–20 cm, 20–40 cm, and 40–60 cm.

In the third year of the experiment, the SOC content in the top soil layer (0–30 cm) was shown to significantly increase in relation to the first and the second year of the study. It is worth noting that the value of this parameter in 2015–2017 was distinctly higher than the one determined before the establishment of the experiment (2014), by 47% in 2015 and 68% in 2016; this is while, in 2017, the value of this parameter was twice as high as in 2014. The likely cause of the increase in the soil organic carbon was the change in tillage methods: from the conventional to the reduced tillage system. Plowing, which was used in the years prior to the experiment, could accelerate the warming and drying of the soil and thus contributed to accelerating the mineralization of organic matter and reducing its content in the soil. In turn, in our experiment, a minimum/reduced tillage without plowing was used, which was conducive to increasing the SOC content. Likewise, Alam et al. [25] found that the elimination of plowing leads to a slowed-down rate of mineralization of soil organic matter and lower soil aeration, which, in turn, promotes the greater accumulation of organic carbon in the top soil layer. Ogle et al. [56], Hermle et al. [57], Chatterjee and Lal [58], as well as Erns and Emmerling [59] found that, in non-tilled soil, the amount of the accumulated organic matter in the soil layer below 10-15 cm is lower than that in the surface layers.

Under the reduced tillage system and direct drilling, nutrients are unevenly distributed due to their greater accumulation in the top soil layer [5,7,17,55,60–62]. Under such tillage conditions, the accumulation of crop residues in the surface soil layer promotes a higher concentration of P, K, and Mg, compared with conventional tillage [7,16,31,63]. This relationship was confirmed under the conditions of the present experiment.

In the 0–30 cm and 30–60 cm soil layers, a higher potassium content was found in the treatments with the surface placement of the mineral fertilizer, compared with the deep fertilization treatment. Wróbel and Pabin [21] report that the slow movement of K deeper into the soil profile is the reason for the increased K concentration in the top soil layer under no-tillage conditions. Borges and Mallarino [37], as well as Mallarino and Borges [22] found an increased uptake of P and K by soybean under the subsurface fertilizer placement conditions. In the studies carried out by Kraska [16], as well as by Woźniak and Soroka [8], no-tillage increased the potassium content in the top soil layer. Alvarez [64], Kraska et al. [65], and Van den Putte et al. [66] also found reduced tillage to promote an increased potassium content in the soil.

Kraska et al. [65], as well as Haruna and Nkongolo [55], found that the use of reduced tillage leads to an increase in the magnesium content in the top soil layer. Włodek et al. [67], on the other hand, revealed an opposite relationship—they obtained a higher soil magnesium content in conventional tillage treatments, compared with those obtained in the reduced tillage plots. Biskupski et al. [68], in turn, did not find tillage to affect the magnesium content in the 0–40 cm soil layer. Likewise, the present study found no clear trend with regard to Mg concentration in the soil profile layers based on the method of mineral fertilizer placement under reduced tillage conditions.

According to Biskupski et al. [68], the variability in the study-results regarding the soil content of available forms of elements in different tillage systems can be due to the fact that soil can exhibit a lower temperature under reduced tillage conditions, compared with the conventional tillage conditions. This, in turn, may contribute to the slowing of chemical reactions occurring in the soil. Shen et al. [69] confirm a decrease in soil temperature under reduced tillage conditions.

The diversified level of mineral fertilizer significantly influenced the content of nutrients in the soil. The higher dose of NPKS (170 kg·ha^{-1}), compared with the 85 kg NPKS·ha^{-1}, increased the P content in all tested soil layers and increased the K content in the 0–30 cm soil layer. However, in the case of Mg, a higher dose of mineral fertilization resulted in a decrease in the content of this element in the 60–90 cm soil layer. According to Bhatt et al. [70], high doses of NPK fertilizers are required to maintain soil fertility and raise crop yields. Skowrońska [71] is of the opinion that the content of the elements in soil is primarily determined, apart from mineral fertilization, by the quantity of the yields and the uptake of nutrients from an agroecosystem.

5. Conclusions

The method of application and rate of mineral fertilizer did not have a significant effect on the SOC content in the top soil layer (0–30 cm). Under the deep fertilizer application conditions, the pH was lower in all the soil layers considered, in comparison with the surface fertilization treatment. The mineral fertilizer applied at the double rate (170 kg NPKS) contributed to an increase in the pH in the surface soil layer of 0–30 cm. The P and K content in the 0–30 cm soil layer was higher than the one at deeper levels of the soil profile (30–60 cm and 60–90 cm). The subsurface application of mineral fertilizer favored an increase in the content of P and Mg in the soil and a decrease in the K content, in comparison with the surface application of mineral fertilizer. The higher level of mineral fertilization promoted an increase in the soil P and K content.

Author Contributions: P.K., P.G., and S.A. conceptualized the research; P.K. and P.G. designed the research; P.K., P.G., and H.R. performed the experiments; P.K., S.A., P.G., and H.R. analyzed the data; P.K. and S.A. wrote the paper. All authors have read and agreed to the published version of the manuscript.

Funding: This research was funded by the Ministry of Agriculture and Rural Development, Ministry of Science and Higher Education, Poland.

Institutional Review Board Statement: Not applicable.

Informed Consent Statement: Not applicable.

Data Availability Statement: Not applicable.

Acknowledgments: We would like to give a special thanks to W. Gryn, who made it possible to conduct the field experiment.

Conflicts of Interest: The authors declare no conflict of interest. The funders had no role in the design of the study; in the collection, analyses, or interpretation of data; in the writing of the manuscript; or in the decision to publish the results.

References

1. Nouri, A.; Lee, J.; Yin, X.; Tyler, D.D.; Jagadamma, S.; Arelli, P. Soil physical properties and soybean yield as influenced by long-term tillage systems and cover cropping in the Midsouth USA. *Sustainability* **2018**, *10*, 4696. [CrossRef]
2. Nouri, A.; Youssef, F.; Basaran, M.; Lee, J.; Saxton, A.M.; Erpul, G. The Effect of Fallow Tillage Management on Aeolian Soil Losses in Semiarid Central Anatolia, Turkey. *Agrosyst. Geosci. Environ.* **2018**, *1*, 1–13. [CrossRef]

3. Fonteyne, S.; Gamiño, M.-A.M.; Tejeda, A.S.; Verhulst, N. Conservation Agriculture Improves Long-term Yield and Soil Quality in Irrigated Maize-oats Rotation. *Agronomy* **2019**, *9*, 845. [CrossRef]
4. Harasim, E.; Antonkiewicz, J.; Kwiatkowski, C.A. The effects of catch crops and tillage systems on selected physical properties and enzymatic activity of loess soil in a spring wheat monoculture. *Agronomy* **2020**, *10*, 334. [CrossRef]
5. Singh, D.; Lenka, S.; Lenka, N.K.; Trivedi, S.K.; Bhattacharjya, S.; Sahoo, S.; Saha, J.K.; Patra, A.K. Effect of Reversal of Conservation Tillage on Soil Nutrient Availability and Crop Nutrient Uptake in Soybean in the Vertisols of Central India. *Sustainability* **2020**, *12*, 6608. [CrossRef]
6. Bienes, R.; Marques, M.J.; Sastre, B.; García-Díaz, A.; Esparza, I.; Antón, O.; Navarrete, L.; Hernánz, J.L.; Sánchez-Girón, V.; Sánchez del Arco, M.J.; et al. Tracking Changes on Soil Structure and Organic Carbon Sequestration after 30 Years of Different Tillage and Management Practices. *Agronomy* **2021**, *11*, 291. [CrossRef]
7. Neugschwandtner, R.W.; Liebhard, P.; Kaul, H.-P.; Wagentristl, H. Soil chemical properties as affected by tillage and crop rotation in a long-term field experiment. *Plant. Soil Environ.* **2014**, *60*, 57–62. [CrossRef]
8. Woźniak, A.; Soroka, M. Effect of crop rotation and tillage system on the weed infestation and yield of spring wheat and on soil properties. *Appl. Ecol. Environ. Res.* **2018**, *16*, 3087–3096. [CrossRef]
9. Lahmar, R. Adoption of conservation agriculture in Europe. Lessons of the KASSA project. *Land Use Policy* **2010**, *27*, 4–10. [CrossRef]
10. Wauters, E.; Bielders, C.; Poesen, J.; Govers, G.; Mathijs, E. Adoption of soil conservation practices in Belgium: An examination of the theory of planned behaviour in the agri-environmental domain. *Land Use Policy* **2010**, *27*, 86–94. [CrossRef]
11. Derpsch, R.; Friedrich, T. Global overview of conservation agriculture adoption. In Proceedings of the 4th World Congress on Conservation Agriculture, New Delhi, India, 4–7 February 2009; pp. 429–438.
12. Lal, R.; Reicosky, D.C.; Hanson, J.D. Evolution of the plow over 10,000 years and the rationale for no-till farming. *Soil Till. Res.* **2007**, *93*, 1–12. [CrossRef]
13. Holland, J.M. The environmental consequences of adopting conservation tillage in Europe: Reviewing the evidence. *Agric. Ecosys. Environ.* **2004**, *103*, 1–25. [CrossRef]
14. Berner, A.; Hildermann, I.; Fliessbach, A.; Pfiffer, L.; Niggli, U.; Mäder, P. Crop yield and soil fertility response to reduced tillage under organic management. *Soil Tillage Res.* **2008**, *101*, 89–96. [CrossRef]
15. Vogeler, I.; Rogasik, J.; Funder, U.; Schnug, E. Effect of tillage systems and P-fertilization on soil physical and chemical properties, crop yield and nutrient uptake. *Soil Tillage Res.* **2009**, *103*, 137–143. [CrossRef]
16. Kraska, P. Effect of conservation tillage and catch crops on some chemical properties of rendzina soil. *Acta Sci. Pol. Agric.* **2011**, *10*, 77–92.
17. Kraska, P. The content of some micronutrients in rendzina soil cultivated using different tillage systems and catch crops. *Polish J. Agron.* **2011**, *4*, 7–11. [CrossRef]
18. Van den Putte, A.; Govers, G.; Diels, J.; Gillijns, K.; Demuzere, M. Assessing the effect of soil tillage on crop growth: A meta-regression analysis on European crop yields under conservation agriculture. *Eur. J. Agron.* **2010**, *33*, 231–241. [CrossRef]
19. Lipiński, W. The content of available phosphorus in soils of Poland. *Fertil. Fertil.* **2005**, *2*, 49–54. (In Polish)
20. Lipiński, W. Agrochemical properties of agriculturally used soils. *Inż. Ekol.* **2019**, *20*, 1–12. (In Polish) [CrossRef]
21. Wróbel, S.; Pabin, J. Effect of tillage system on macronutrient content in soil and maize plants cultivated in monoculture. *Rocz. Glebozn.* **2008**, *59*, 226–232. (In Polish)
22. Mallarino, A.P.; Borges, R. Phosphorus and Potassium Distribution in Soil Following Long-Term Deep-Band Fertilization in Different Tillage Systems. *Soil Sci. Soc. Am. J.* **2006**, *70*, 702–707. [CrossRef]
23. Barbieri, P.A.; Sainz Rozas, H.R.; Covacevich, F.; Echeverría, H.E. Phosphorus placement effects on phosphorous recovery efficiency and grain yield of wheat under no-tillage in the humid pampas of Argentina. *Int. J. Agron.* **2014**, *2014*, 507105. [CrossRef]
24. Hansel, F.D.; Amado, T.J.C.; Ruiz Diaz, D.A.; Rosso, L.H.M.; Nicoloso, F.T.; Schorr, M. Phosphorus Fertilizer Placement and Tillage Affect Soybean Root Growth and Drought Tolerance. *Agron. J.* **2017**, *109*, 2936–2944. [CrossRef]
25. Alam, M.; Bell, R.W.; Salahin, N.; Pathan, S.; Mondol, A.T.; Alam, M.J.; Rashid, M.H.; Paul, P.L.; Hossain, M.I.; Shil, N.C. Banding of fertilizer improves phosphorus acquisition and yield of zero tillage maize by concentrating phosphorus in surface soil. *Sustainability* **2018**, *10*, 3234. [CrossRef]
26. Lakew, A. Influence of N and P fertilizer rates on yield and yield components of bread wheat (*Triticum aestivum* L.) in Sekota District of Wag-Himira Zone, North Eastern Ethiopia. *Arch. Agric. Environ. Sci.* **2019**, *4*, 8–18. [CrossRef]
27. Randall, G.W.; Vetsch, J.A.; Murrell, T.S. Corn response to phosphorus placement under various tillage practices. *Better Crops* **2001**, *85*, 12–15.
28. Randall, G.W.; Vetsch, J.A. Optimum placement of phosphorus for corn/soybean rotations in a strip-tillage system. *J. Soil Water Conserv.* **2008**, *63*, 152A–153A. [CrossRef]
29. de Andrade Costa, S.E.V.G.; Souza, E.D.; Anghinoni, I.; Flores, J.P.C.; Cao, E.G.; Holzschuh, M.J. Phosphorus and root distribution and corn growth related to longterm tillage systems and fertilizer placement. *Rev. Bras. Cienc. Solo* **2009**, *33*, 1237–1247. [CrossRef]
30. Galvani, R.; Hotta, L.F.K.; Rosolem, C.A. Phosphorus sources and fractions in an oxisol under no-tilled soybean. *Sci. Agric.* **2008**, *65*, 415–421. [CrossRef]
31. López-Fando, C.; Pardo, M.T. Changes in soil chemical characteristics with different tillage practices in a semi-arid environment. *Soil Tillage Res.* **2009**, *104*, 278–284. [CrossRef]

32. Abdi, D.; Cade-Menun, B.J.; Ziadi, N.; Parent, L.-É. Long-Term Impact of Tillage Practices and Phosphorus Fertilization on Soil Phosphorus Forms as Determined by ^{31}P Nuclear Magnetic Resonance Spectroscopy. *J. Environ. Qual.* **2014**, *43*, 1431–1441. [CrossRef] [PubMed]
33. Dorneles, E.P.; Lisboa, B.B.; Abichequer, A.D.; Bissani, C.A.; Meurer, E.J.; Vargas, L.K. Tillage, fertilization systems and chemical attributes of a Paleudult. *Sci. Agric.* **2015**, *72*, 175–186. [CrossRef]
34. Randall, G.W.; Hoeft, R.G. Placement Methods for Improved Efficiency of P and K Fertilizers: A Review. *J. Prod. Agric.* **1988**, *1*, 70–79. [CrossRef]
35. Duiker, S.W.; Beegle, D.B. Soil fertility distributions in long-term no-till, chisel/disk and moldboard plow/disk systems. *Soil Tillage Res.* **2006**, *88*, 30–41. [CrossRef]
36. Nkebiwe, P.M.; Weinmann, M.; Bar-Tal, A.; Muller, T. Fertilizer placement to improve crop nutrient acquisition and yield: A review and meta-analysis. *Field Crops Res.* **2016**, *196*, 389–401. [CrossRef]
37. Borges, R.; Mallarino, A.P. Broadcast and deep-band placement of phosphorus and potassium for soybean managed with ridge tillage. *Soil Sci. Soc. Am. J.* **2003**, *67*, 1920–1927. [CrossRef]
38. Randall, G.W.; Hoeft, R.G. Fertilizer placement methods: New wrinkles on a new face. *Crops Soils Magaz.* **1986**, *38*, 17–22.
39. Stanisławska-Glubiak, E.; Korzeniowska, J. Efficiency of deep-placed fertilization for maize and pea in conventional and no-tillage system. *Fragm. Agron.* **2010**, *27*, 160–169. (In Polish)
40. IUSS Working Group WRB. *World Reference Base for Soil Resources 2014, International Soil Classification System for Naming Soils and Creating Legends for Soil Maps*; Update 2015; World Soil Resources Reports No. 106; Food and Agriculture Organization of the United Nations: Rome, Italy, 2015.
41. Polish Society of Soil Science. Particle size distribution and textural classes of soils and mineral materials—Classification of Polish Society of Soil Science 2008. *Soil Sci. Ann.* **2009**, *60*, 5–16.
42. European Commission. *Common Catalogue of Varieties of Agricultural Plant Species*, 28th ed.; 2009/C 302 A/01; European Commission: Luxembourg, 2009.
43. Bleiholder, H.; Buhr, L.; Feller, C.; Hack, H.; Hess, M.; Klose, R.; Lancashire, P.D.; Meier, U.; Stauss, R.; Van den Boom, T.; et al. Growth stages of mono- and dicotyledonous plants. In *BBCH Monograph*; Uwe Meier Julius Kühn-Institut (JKI), Ed.; Open Agrar Repositorium: Berlin, Germany, 2018; ISBN 978-3-95547-071-5. [CrossRef]
44. Józefowska, A.; Miechówka, A. Comparison of the results of organic carbon determination in soils of the Carpathian Foothills obtained by the Tiurin and dry combustion methods (thermal method). *Rocz. Gleb.* **2011**, *1*, 65–69. (In Polish)
45. ISO 11261. *Soil Quality—Determination of Total Nitrogen—Modified Kjeldahl Method*; ISO: Geneva, Switzerland, 2002.
46. PN-R-04023. Analiza Chemiczno-Rolnicza Gleby. Oznaczanie Zawartości Przyswajalnego Fosforu w Glebach Mineralnych./Chemical and Agricultural Analysis of Soil. Determination of Available Phosphorus Content in Mineral Soils. 1996. Available online: https://www.iso.org/standard/19239.html (accessed on 27 April 2021).
47. PN-R-04022:1996/Az1. Analiza Chemiczno-Rolnicza Gleby. Oznaczanie Zawartości Przyswajalnego Potasu w Glebach Mineralnych./Chemical and Agricultural Analysis of Soil. Determination of Available Potassium in Mineral Soils. 2002.
48. PN-R-04020:1994/Az1. Analiza Chemiczno-Rolnicza Gleby. Oznaczanie Zawartości Przyswajalnego Magnezu./Chemical and Agricultural Analysis of Soil. Determination of Available Magnesium Content. 2004.
49. ISO 10390. International Standard Organization. Soil Quality—Determination of pH. 2002. Available online: https://www.iso.org/standard/40879.html (accessed on 27 April 2021).
50. Stachowski, P. Assessment of meteorological droughts on the postmining areas in the Konin Region. Środkowo-Pomorskie Towarzystwo Naukowe Ochrony Środowiska. *Rocz. Ochr. Środowiska* **2010**, *12*, 587–606.
51. Skowera, B.; Puła, J. Pluviometric extreme conditions in spring season in Poland in the years 1971–2000. *Acta Agrophys.* **2004**, *3*, 171–177.
52. Skowera, B. Changes of hydrothermal conditions in the polish area (1971–2010). *Fragm. Agron.* **2014**, *31*, 74–87.
53. Limousin, G.; Tessier, D. Effect of no-tillage on chemical gradients and topsoil acidification. *Soil Till. Res.* **2007**, *92*, 167–174. [CrossRef]
54. Lilienfein, J.; Wilcke, W.; Vilela, L.; Lima, S.D.; Thomas, R.; Zech, W. Effect of no-tillage and conventional tillage systems on the chemical composition of soil. *J. Plant. Nutr. Soil Sci.* **2000**, *163*, 411–419. [CrossRef]
55. Haruna, S.I.; Nkongolo, N.V. Tillage, cover crop and crop rotation effects on selected soil chemical properties. *Sustainability* **2019**, *11*, 2770. [CrossRef]
56. Ogle, S.M.; Breidt, F.J.; Paustian, K. Agricultural management impacts on soil organic carbon sto-rage under moist and dry climatic conditions of temperate and tropical regions. *Biogeochemistry* **2005**, *72*, 87–121. [CrossRef]
57. Hermle, S.; Anken, T.; Leifeld, J.; Weisskopf, P. The effect of the tillage system on soil organic carbon content under moist, cold-temperate conditions. *Soil Till. Res.* **2008**, *98*, 94–105. [CrossRef]
58. Chatterjee, A.; Lal, R. On farm assessment of tillage impact on soil carbon and associated soil quality parameters. *Soil Till. Res.* **2009**, *104*, 270–277. [CrossRef]
59. Ernst, G.; Emmerling, C. Impact of five different tillage systems on soil organic carbon content and the density, biomass, and community composition of earthworms after a ten year period. *Eur. J. Soil Biol.* **2009**, *45*. 247–251. [CrossRef]
60. Thomas, G.A.; Dalal, R.C.; Standley, J. No-till effect on organic matter, pH, cation exchange capacity and nutrient distribution in a Luvisol in the semi-arid subtropics. *Soil Till. Res.* **2007**, *94*, 295–304. [CrossRef]

61. Shi, X.H.; Yang, X.M.; Drury, C.F.; Reynolds, W.D.; McLaughlin, N.B.; Zhang, X.P. Impact of ridge tillage on soil organic carbon and selected physical properties of a clay loam in southwestern Ontario. *Soil. Till. Res.* **2012**, *120*, 1–7. [CrossRef]
62. Hausherr Lüder, R.-M.; Qin, R.; Richner, W.; Stamp, P.; Streit, B.; Noulas, C. Effect of tillage systems on spatial variation in soil chemical properties and winter wheat (*Triticum aestivum* L.) performance in small fields. *Agronomy* **2019**, *9*, 182. [CrossRef]
63. Martin-Rueda, I.; Munoz-Guerra, L.M.; Yunta, F.; Esteban, E.; Tenorio, J.L.; Lucena, J.J. Tillage and crop rotation effects on barley yield and soil nutrients on a *Calciortidic Haploxeralf*. *Soil Till. Res.* **2007**, *92*, 1–9. [CrossRef]
64. Alvarez, R. A review of nitrogen fertilizer and conservation tillage effects on soil organic carbon storage. *Soil Use Manag.* **2005**, *21* 38–52. [CrossRef]
65. Kraska, P.; Palys, E.; Jedruszczak, M. Conservation tillage system and chemical properties of sandy soil under crops in crop rotation. In Proceedings of the 17th Triennial Conference Sustainability–its Impact on Soil Management and Environment, Kiel Germany, 28 August–3 September 2006; pp. 566–570.
66. Van den Putte, A.; Govers, G.; Diels, J.; Langhans, C.; Clymans, W.; Vanuytrecht, E.; Merckx, R.; Raes, D. Soil functioning and conservation tillage in the Belgian Loam Belt. *Soil Till. Res.* **2012**, *122*, 1–11. [CrossRef]
67. Włodek, S.; Hryńczuk, B.; Pabin, J.; Biskupski, A. The contents of carbon and nutrients in the soil after many years' use of different cultivation systems. *Zesz. Prob. Post. Nauk Rol.* **2003**, *493*, 727–732. (In Polish)
68. Biskupski, A.; Sienkiewicz-Cholewa, U.; Włodek, S.; Pabin, J. Differentiation in the contents of carbon and nutrients in experiment with many years simplifications in cultivation. *Rocz. Gleb.* **2009**, *60*, 5–11. (In Polish)
69. Shen, Y.; McLaughlin, N.; Zhang, X.; Xu, M.; Liang, A. Effect of tillage and crop residue on soil temperature following planting for a Black soil in Northeast China. *Sci. Rep.* **2018**, *8*, 4500. [CrossRef]
70. Bhatt, M.K.; Labanya, R.; Joshi, H.C. Influence of long-term chemical fertilizers and organic manures on soil fertility—A review *Univ. J. Agric. Res.* **2019**, *5*, 177–188. [CrossRef]
71. Skowrońska, M. Nutrient balance under differentiated maize fertilization and sowing. *Przem. Chem.* **2018**, *95*, 1595–1598 (In Polish) [CrossRef]

agronomy

MDPI

Article

Interseeding Camelina and Rye in Soybean with Varying Maturity Provides Soil Cover without Affecting Soybean Yield

Kory L. Johnson, Hans J. Kandel, Dulan P. Samarappuli and Marisol T. Berti *

Department of Plant Sciences, North Dakota State University, Fargo, ND 58104, USA; johnsonkory14@gmail.com (K.L.J.); Hans.Kandel@ndsu.edu (H.J.K.); Dulan.Samarappuli@ndsu.edu (D.P.S.)
* Correspondence: Marisol.Berti@ndsu.edu; Tel.: +1-701-231-6110

Abstract: Low adoption to utilize cover crops interseeded into soybean (*Glycine max* (L.) Merr.), in the northern Plains in the USA, is due to a short growing season and a few adapted winter-hardy species. The objective was to evaluate the impact of interseeded winter camelina (*Camelina sativa* (L.) Crantz) and winter rye (*Secale cereale* L.) using different soybean relative maturities on soybean yield, canopy coverage, spring cover crop biomass, and subsequent wheat (*Triticum aestivum* L.) yield. Cover crops interseeded into early-maturing (0.4–0.8) soybean cultivars had more fall coverage compared with the 0.9 maturity cultivar, but the spring biomass was similar for all maturities. The soybean yield of the 0.9 cultivar was significantly higher, 2365 kg ha^{-1} compared with 2037 kg ha^{-1} for the 0.4 cultivar. Rye outperformed winter camelina and had higher fall canopy cover (15 vs. 7%), spring canopy cover (16% vs. 4%), and higher spring biomass (313 vs. 100 kg ha^{-1} dry matter). Spring wheat, after rye, yielded 90% of the check. It is not recommended to plant spring wheat following winter rye, but there was no negative yield effect from winter camelina. Interseeding cover crops into soybean in the northern Plains is possible but needs further research to optimize interseeding systems.

Keywords: cover crop; canopy cover; wheat; winter survival

Citation: Johnson, K.L.; Kandel, H.J.; Samarappuli, D.P.; Berti, M.T. Interseeding Camelina and Rye in Soybean with Varying Maturity Provides Soil Cover without Affecting Soybean Yield. *Agronomy* **2021**, *11*, 353. https://doi.org/10.3390/agronomy11020353

Academic Editor: Umberto Anastasi and Aurelio Scavo
Received: 24 January 2021
Accepted: 13 February 2021
Published: 16 February 2021

1. Introduction

Soil erosion is a major problem in soybean production regions as crop residue is limited [1], leaving the soil with limited cover during the winter. The Red River of the North Valley of North Dakota and Minnesota is flat and has few natural wind barriers. The importance in finding a solution to loss of topsoil due to erosion, especially following soybean production in conventionally tillage systems, is vital in sustaining soil health, as continued topsoil loss from wind and water erosion will have detrimental effects on crop productivity of regional soils and eventually will increase fertilizer inputs if current management practices continue [2].

Cover crops can provide protection to the soil by reducing soil particle removal due to wind and water erosion. Cover crops can be winter grasses, forbs, or legumes that are typically planted in the fall and overwinter until the spring. Cover crops are used for erosion control, improving soil structure, moisture, and nutrient content, increasing beneficial soil biota, suppressing weeds, providing habitat for beneficial predatory insects, facilitating crop pollinators, providing wildlife habitat, and as forage for farm animals [3]. Cover crops may provide benefits to the soil, the ecosystem, and potentially increase grain yield of the following crop by increasing diversity of microorganisms, providing soil coverage, enriching soil organic matter, and enhancing the nutrient cycling [4,5]. Diverse plant species promote the soil microbial community differently, and this may result in greater soil microbial diversity [6]. Soil organic matter is a major contributing factor in soil productivity and can be enhanced by incorporating cover crops [7].

Although cover crop utilization has increased in the Corn Belt and the northern Great Plains Region (North of latitude 44°, including eastern Montana, north-eastern Wyoming, most of North and South Dakota, and the Canadian Prairies), adoption has been slower than average for overwintering cover crops. This is primarily due to timing of when soil moisture is available for successful cover crop stand establishment at time of seeding late summer or early fall and shorter growing season to establish sufficient growth for overwintering for winter annuals. Overwintering cover crops are planted after or between the primary crop with the goals of surviving the winter and to resume growth in the spring. Cover crops must produce enough biomass in the fall and or spring for benefits to be expressed [8]. Following soybean harvest, there is an extremely short remaining growing period before a fall frost poses challenges for establishing a cover crop with conventional seeding methods. Alternative methods of planting are needed for successful biomass growth and soil coverage.

Most producers evaluate cropping systems based on the economics of grain yield and short-term profitability, not on the value of soil health and long-term sustainability. Several studies have suggested current conventional cropping systems are less sustainable because of limited benefits to the ecosystem [9–11].

Camelina (*Camelina sativa* (L.) Crantz) is a short-season annual oilseed crop in the Brassicaceae family with agronomic low-input features that has been produced for the oil in Europe for over 3000 years [12]. Camelina has two biotypes, summer and winter [13,14]. The winter annual biotype is winter-hardy and has a high level of tolerance to drought and low-temperature stress, and has the ability to adapt across a wide range of environments [14,15]. Because of winter camelina's desirable agronomic traits, further research is being conducted to improve its adoption of cultivation and cover crops use.

North Dakota farmers need winter annual biotypes of camelina that are proven to be winter-hardy and suitable for the northern Plains Region [13,16]. Fall-seeded camelina will remain in the rosette stage throughout the winter, with growth resuming in the spring [17].

Rye is the most common and reliable winter annual cover crop utilized in the upper Midwest (Iowa, Michigan, Minnesota, North Dakota, South Dakota, and Wisconsin). It is one of the few cover crops that can establish successfully when planted late in the growing season. It is winter-hardy throughout the region and accumulates biomass before spring planting of the subsequent crop [18–20]. Rye has an extensive root system that can lead to reduction of nitrate leaching [21]. Rye can germinate at temperatures as low as 1 °C and vegetative growth can begin at 3 °C [22]. With vegetative growth still active at near freezing temperatures, winter rye has a longer time in the field compared with non-winter-hardy cover crops to produce biomass and canopy coverage, which is an important factor in North Dakota. With a prolonged growing season, winter rye can be a good weed suppressor in the fall and in the spring when soil canopy coverage increases rapidly [23].

Several researchers have reported on interseeding cover crops into soybean at different stages of growth [17,24,25]. Interseeding involves planting of the cover crop by drilling the seed into the soil or broadcasting it before soybean matures. The advantages of interseeding include not needing to seed after soybean harvest (during the busy harvest season), providing more time for cover crop establishment, improved cover crop growth, and increased winter survival [26]. Interseeding usually requires special or modified equipment that is able to leave established soybean plants undamaged. Research has shown that weeds can be suppressed effectively without yield reduction of the main crop by interseeding cover crops in organic farming systems [27,28].

Berti et al. [17] reported that the camelina plants have difficulty competing with the dense canopy of soybean, and interseeding should occur during soybean reproductive stages [17,25]. Establishment of winter camelina and winter rye by aerial broadcasting is mainly dependent on timely rainfall after sowing [29] and seeding rates need to be increased by a minimum of 50% [22].

Winter rye as a cover crop can be integrated into existing corn (*Zea mays* L.)–soybean production systems and has been recommended as a cost-effective strategy for improving environmental stewardship [21]. Rye is superior among cool-season cereal cover crops for absorbing unused soil NO_3-N. It has a fast-growing fibrous root system, which helps scavenge for residual NO_3-N throughout the soil profile. Where rye has been interseeded into soybean in August, leaching losses from September to May were less than 5.6 kg of N ha^{-1} [30]. Rye has the ability to access K from lower in the soil profile [31].

Although several studies have been done on interseeded cover crops in soybean [17,24,25,32,33], this research is unique as it is evaluating the effect of soybean maturity on the establishment of winter camelina and rye and the following hard red spring wheat (HRSW) (*Triticum aestivum* L.) crop. The objectives of this research were to evaluate cover crop development and biomass production when interseeded into soybean cultivars with different relative maturity and to evaluate the effect of cover crop growth on soybean and HRSW grain yield.

2. Materials and Methods

2.1. Experimental Sites

The experiments were established at North Dakota State University's (NDSU) experiment field (46.932124° N, 96.858941° W) located near Fargo, ND, between 2016 and 2018. The soil at the experimental site is a mixture of Fargo (fine, smectitic, frigid Typic Epiaquerts) and Ryan (fine, smectitic, frigid Typic Natraquerts) silty clay, naturally poorly or very poorly drained and slowly permeable. The parent material of the soil is clayey glaciolacustrine deposits [34]. The crop grown before soybean seeding was corn in 2015 and HRSW in 2016. Conventional tillage management practices were used before the establishment of the experiment. No-till management was used for the first time at the research site in the spring of 2016 and has been continued during the subsequent seasons. Weather data for the 2016, 2017, and 2018 growing seasons were obtained from the North Dakota Agricultural Weather Network [35] at the Fargo weather station located in Fargo, ND.

2.2. Experimental Design and Management

Two cover crop experiments were conducted during each of the 2016 and 2017 growing seasons with data collection on spring wheat in 2017 and 2018. Each of the experiments were considered a separate environment. The method of establishment was not an objective in this trial. Therefore, in one experiment in each year, the cover crop was direct-planted. In the second experiment, simulated air seeding was used to represent possible cover crop establishment methods.

The experimental design was a randomized complete block with a factorial arrangement. There were four replicates per experiment and each replicate consisted of 20 experimental units. The experimental unit size was 1.52 × 7.62 m. Treatments included soybean relative maturity (cultivar), cover crop species, and cover crop seeding rate. Soybean relative maturities included 0.4, 0.5, 0.8, and 0.9. The 0.4 is the earliest maturing cultivar. Soybean cultivars are listed in Table 1. All soybean cultivars were glyphosate-tolerant (Roundup Ready 2 Yield), carried resistance to soybean cyst nematode (*Heterodera glycines* Ichinohe) (except AG0434), had *Phytophthora* resistance, and were pre-treated by the seed company (Asgrow; Bayer, Monheim, Germany) with Acceleron (a.i. pyraclostrobin and metalaxyl) seed treatment. Acceleron seed treatment is a fungicide combination providing protection from seed and soil borne diseases such as but not limited to; *Pythium irregulare*, *Phytophthora sojae*, *Fusarium solani*, and *Rhizoctonia solani*. The cultivars were inoculated with Vault SP (*Bradyrhizobium japonicum*) inoculum (BASF, Ludwigshafen, Germany) at a rate of 1.8 g kg^{-1} soybean seed on the day of planting to encourage nodulation. The same cultivars were used in both growing seasons.

Soybean was planted as soon as field conditions were favorable in early to mid-May, with four soybean rows spaced 30.5-cm apart and using a seeding rate of 469,300 live seeds ha^{-1}. The plots were planted with a Hege 1000 no-till planter (Hege Company, Waldenberg, Germany). Seeds were planted to a depth of approximately 3 cm.

Table 1. Details of soybean cultivars used in experiments in 2016 and 2017 at Fargo, ND, USA.

Cultivar	Company	Maturity	IDC †	SCN ‡	Canopy	Plant Height
AG0434	Asgrow	0.4	2.0	None	Medium bushy	Medium
AG0536	Asgrow	0.5	1.6	R	Medium bushy	Medium tall
AG0835	Asgrow	0.8	1.8	R	Bushy	Medium tall
AG0934	Asgrow	0.9	2.1	R	Medium bushy	Medium short

† IDC = iron deficiency chlorosis. IDC scored on 1–5 scale (1 = Green, 3 = Yellow, 5 = Dead) [36]; ‡ SCN = soybean cyst nematode. R = resistant SCN using PI347654 source.

Cover crop treatments were none (control), winter camelina, and rye. Cover crop seeding rate treatments were 100% of seeding rate and 75% of seeding rate. Winter camelina cultivar "Joelle" was planted at 6.72 kg ha^{-1} live seeds for the 100% seeding rate treatments and 5.04 kg ha^{-1} for 75% rate treatments to a depth of 1.3 cm. The quantity of winter camelina seeds per kg can be upwards of 770,000 seeds kg^{-1} compared with 39,000 seeds kg^{-1} for rye [12]. The rye cultivar "Rymin" was planted at 67.2 kg ha^{-1} for the 100% seeding rate and 50.4 kg ha^{-1} for 75% rate, to the depth of 2.5 cm. Germination testing was conducted before planting. For both 2016 and 2017 growing seasons, a 95% germination rate was determined for rye and 90% for camelina. Seeding rate was adjusted based on germination.

All cover crops were interseeded into established soybean at the R7 growth stage of the 0.4 maturity cultivar. Staging of soybean was based on NDSU Soybean Production Field Guide, which defines R7 as beginning maturity—one normal pod on the main stem that has reached its mature pod color [37]. In one experiment, the cover crops were planted in a single furrow in the center of all soybean rows, 15.3-cm from each corresponding row resulting in three cover crop rows per experimental unit. Furrows were made to the depth of 1.3 cm for camelina and 2.5 cm for rye using a standard garden hoe. No furrows were made in the control plot (without cover crops). In the other experiment, the cover crop seed was broadcasted, to simulate seeding by airplane.

Weeds in soybean plots were controlled twice in 2016 and once in 2017, prior to the planting of the cover crops using (a.i. 48.7% glyphosate, N-(phosphonomethyl) glycine in potassium salt form) Roundup PowerMAX (Monsanto Co., St. Louis, MO, USA and (12.6% (E)-2-[1-[[(3-chloro-2-propenyl)oxy]imino]propyl]-5-[2-(ethylthio)propyl]-3-hydroxy-2-cyclohexen-1-one) and SelectMax (Valent U.S.A. Corporation, Walnut Creek CA, USA). The herbicides were applied using TeeJet 8001 XR nozzle at a rate of 1.6 in 94 L ha^{-1} water and a spray pressure of 200 kPa. Cover crops were terminated in the spring using Roundup WeatherMAX.

In 2017 and 2018, (a.i. 9.15% S-cyano(3-phenoxyphenyl)methyl (+/−)-cis/trans-3-(2,2-dichloethenyl)-2,2-dimethylcyclopropanecarboxylate) Mustang Maxx (FMC Corporation Philadelphia, PA, USA) was applied at a rate of 1.75 L ha^{-1} to both soybean and HRSW as soybean aphid (*Aphis glycines* Matsumura) levels in soybean and grasshopper (Orthoptera Acrididae) thresholds in HRSW surpassed thresholds as described by NDSU [37,38]. Important field operation and measuring dates are provided in Table 2.

Wheat was planted after cover crops were terminated. Fertilizer was broadcast-applied during the spring before the HRSW at a rate 112 kg per ha^{-1} of N using urea (46-0-0). In both years, the HRSW cultivar "Glenn" was used. All HRSW plots were planted as soon as field conditions were favorable in early May, with a Great Plains 3P605NT no-till planter (Great Plains Ag, Salina, KS, USA). Experimental units had seven rows spaced 18.3-cm apart. The seeding rate was 2,739,000 live seeds ha^{-1} and seeding depth approximately 2 cm. Weeds were controlled using Wolverine Advanced (4.56% fenoxaprop-p-ethyl 1.5% pyrasulfotole, 6.13% bromoxynil octanoate, 5.93% bromoxynil heptanoate) (Bayer CropScience LP, Research Triangle Park, NC, USA) to control selective postemergent grassy and broadleaf weeds.

Table 2. Dates of important measurements and field operations between 2016 and 2018 at Fargo, ND, USA.

Measurement/Operation	Date	
	2016	2017
Soybean planting	6 May	6 May
First herbicide application	9 June	9 June
Second herbicide application	30 June	-
Cover crop planting	22 Aug.	22 Aug.
Cover crop Canopeo † reading	15 Nov.	31 Oct.
Soybean harvest	27 Sept.	6 Oct.
	2017	2018
Spring cover crop Canopeo reading	1 May	13 May
Spring cover crop biomass	1 May	13 May
Cover crop termination	6 May	16 May
Wheat planting	6 May	16 May
Wheat Canopeo reading	9 June	9 June
Wheat harvest	22 Aug.	16 Aug.

† Canopeo, a mobile application developed to measure crop canopy coverage.

2.3. Evaluations

Soybean plant density was determined shortly after emergence (VE) by randomly selecting one linear m near the center of the plot. Then, counting all plants within the linear m in both inner two rows.

Cover crop canopy coverage, defined as a percentage of green plant matter, which covers the soil, was measured using the mobile phone application "Canopeo" developed by the Oklahoma State University Department of Plant and Soil Sciences, following cover crop emergence, before the first killing frost, and before termination in the spring. Canopeo measures the fractional green canopy cover through an image processed through the Canopeo application providing a green canopy coverage percentage [39]. Canopy coverage data was collected from pictures used taken in the center of each plot at a height of 1 m, allowing 15 cm from the outside of last soybean row. Picture data were then processed using Canopeo application, which resulted in a percentage of green tissue within the area of the picture.

Cover crop biomass was collected in the spring preceding termination and subsequent HRSW planting. Biomass was sampled from an area within a 30.5 × 50 cm plastic square (0.1525 m^2). The square was randomly tossed into each half of the lengthwise portion of each experimental unit, creating two samples per plot. An average of the two samples was used for the biomass calculation. Biomass samples were created by cutting all cover crop plants within the square at the soil level. Samples were then place in a dryer at a temperature of 40 °C until biomass sample showed no difference in weight during 24 h. Samples were then individually placed on a tray where foreign material was removed before weighing the sample using a Mettler Toledo XS6001S scale (Mettler-Toledo, LLC, Columbus, OH, USA).

The soybean and HRSW plots were harvested after physiological maturity [40,41], at harvestable moisture content using a Wintersteiger Classic plot combine (Wintersteiger Ag. Ried, Austria). Seed samples were cleaned using a Clipper seed cleaner (Ferrell-Ross, Bluffton, IN, USA), and seed samples were then weighed for yield. Moisture and test weight were determined using a GAC 2100 moisture tester (DICKEY-John Corp., Minneapolis, MN, USA), and observations were corrected to 13% and 13.5% moisture content for soybean and HRSW, respectively. Soybean oil and protein contents were not significantly different between treatments and are not reported in this paper.

2.4. Statistical Analysis

Statistical analysis was conducted using a randomized complete block design with a two-factor factorial arrangement. All dependent variables were analyzed with a mixed model (PROC MIXED) on SAS 9.3 [42]. Cultivars and cover crops were considered fixed variables, and environment was considered a random variable. Cover crop treatments, winter camelina and cereal rye, and 100% and 75% seeding rate were combined across soybean cultivars during statistical analysis to make five treatments (Camelina100, Camelina75, Rye100, Rye75, and Check, without cover crops).

Homogeneity of variance tests was done to determine if environments (defined as the combination of location-year) could be combined. If homogeneous, a combined analysis across four environments was conducted. Treatment means were separated using Fisher's protected least significant difference (LSD) at the 95% level of confidence ($p \leq 0.05$).

3. Results and Discussion

3.1. Weather Data

The production years 2016 and 2017 differed for total precipitation and air temperature as observed by NDAWN weather stations (Tables 3 and 4). During the 2016 interseeding of the cover crops, below average precipitation during seeding was followed by above average precipitation in September and October. Differences were also observed between growing seasons for spring data collection. The spring of 2017 had below average precipitation, yet with soil moisture levels greater than that of 2018 (due to lower precipitation during the 2017) (Table 3).

Table 3. Monthly total rainfall in 2016, 2017, and 2018 compared with the 30-year average at Fargo, ND, USA.

	Total Rainfall			
Month	**2016**	**2017**	**2018**	**Historical Avg.** †
			mm	
April	59	25	6	35
May	33	26	44	71
June	69	57	123	99
July	132	23	81	71
August	48	58	101	65
September	80	70	64	65
October	64	20	58	55
Total	485	279	477	461

† Historical data represents a 30-year average from 1981–2010 [35].

The 2017 growing season only had 279 mm of precipitation compared with 485 and 477 mm for 2016 and 2018, respectively (Table 3). This difference was the leading factor in lower cover crop germination rates, irregular germination, and difficulties during the cover crop establishment phase. In addition, below average temperatures as compared with the historical average during the months of March and April of 2018 (Table 4) negatively affected the already inhibited cover crops, resulting in low biomass growth. Solar radiation was higher during the 2017–2018 cover crop growing season, compared with the 2016–2017 season.

3.2. Cultivar and Cover Crops

3.2.1. Cultivar

The analysis of variance and significance levels are provided in Table 5. On average soybean established plant density was 440,000 plants ha^{-1} and not significantly different among cultivars. Interseeding cover crops into different cultivars at the R7 stage of the early maturity cultivar produced no soybean yield reductions comparing soybean yield with cover crops (camelina or rye) to the check plot of each cultivar, which is consistent with previous research [17,24,25].

Table 4. Monthly average air temperature and solar radiation for 2016, 2017, and 2018, and historical data for average air temperature at Fargo, ND, USA.

Month	Average Air Temperature				Average Solar Radiation †		
	2016	2017	2018	Historical ‡	2016	2017	2018
			°C			Langley §	
March	3.3	−1.5	−3.0	−2.3	280	308	325
April	6.3	7.6	1.7	6.8	337	417	479
May	15.5	14.0	18.0	14.0	509	465	500
June	20.0	19.8	21.4	19.0	561	565	551
July	22.0	22.3	21.7	21.5	525	571	555
August	21.1	19.3	20.6	20.7	477	440	440
September	16.8	16.5	15.0	15.1	325	323	329
October	9.7	8.7	4.4	7.5	188	230	186

† No historical data available. ‡ Historical data represent a 30-year average from 1981 to 2010 [35]. § Total incident solar radiation flux density is measured in Watts m^{-2} at approximately 2 m above the soil surface with a pyranometer. The solar radiation energy units reported are Langleys (Ly) per day or MJ m^{-2} day^{-1}. One Ly = 1 calorie cm^{-2}.

Table 5. Analysis of variance and mean squares for cover crops (CC) and cultivars (Cul) across four environments (Env) in Fargo, ND, USA, 2016–2018

SOV	df	Wheat Canopy Coverage	Soybean Yield	Wheat Yield	df †	Fall Canopy Coverage	Spring Canopy Coverage	Spring CC Biomass
Env	3	2.232	505990	21717067	3	0.5369	0.3736	1556563
Rep (Env)	12	0.007	542972	853373	12	0.0597	0.0177	81544
Cul	3	0.001	1614987 *	95644	3	0.0239 *	0.0035	35460
Env × Cul	9	0.003	117854	67174	9	0.0121	0.0019	23671
CC	4	0.273 *	70538	1102703 *	3	0.1413 *	0.3046 *	973545
Env × CC	12	0.096	41174	1045313	9	0.0223	0.0717	267847
Cul × CC	12	0.001	23022	52859	9	0.0039	0.0018	20706
Env × Cul × CC	36	0.002	32093	95382	27	0.0026	0.0018	17095
Error	228	0.003	26699	105005	180	0.0060	0.0026	14737

* Significant at the 0.05 probability level. † Cover crop check plots removed from data.

Cover crop fall canopy coverage percentages followed expected outcomes with the greatest value associated with the 0.4 soybean maturity and lowest with 0.9 (Table 6). These differences were expected due to 0.4 maturity soybean cultivar entering plant senescence much quicker than the 0.9 cultivar, allowing for greater light penetration and decreased competition of the soybean with the interseeded cover crops. Despite increased canopy coverage percentages from the cover crops in the earlier maturity soybean group, cover crop biomass differences were not observed due to cultivar maturity differences (Table 6).

Table 6. Mean fall and spring canopy coverage and cover crop biomass for four soybean cultivars across four environments in Fargo, ND, USA, from 2016, 2017, and 2018.

Cultivar	Fall Canopy Coverage		Spring Canopy Coverage	Cover Crop Biomass
		%		kg ha^{-1}
AG0434	12.6	a †	10.7	234
AG0536	11.5	a	10.3	193
AG0835	10.0	a	9.4	216
AG0934	8.2	b	9.1	182
LSD 0.05			ns	ns

† Within a column, mean followed by a different letter is significantly different at $p \leq 0.05$. ns = not significant.

If seeding the cover crop earlier increases soil coverage, early-maturing soybean cultivars may have the advantage over late-maturing cultivars. However, cultivars with later relative maturities had higher soybean yield (Table 7), as was also found by [43].

Table 7. Mean soybean yield, wheat yield, and wheat canopy coverage readings for four soybean cultivars averaged across four environments in Fargo, ND, USA, from 2016, 2017, and 2018.

Cultivar	Soybean Yield		Wheat Yield	Wheat Canopy Coverage
	kg ha^{-1}			%
AG0434	2037	d [†]	2715	47.8
AG0536	2154	c	2631	48.0
AG0835	2270	b	2665	47.8
AG0934	2365	a	2677	48.4
LSD 0.05			ns	ns

[†] Within a column, mean followed by a different letter are significantly different at $p \leq 0.05$. ns = not significant.

The greatest yield difference averaged across all environments was shown between the AG0434 (2340 kg ha^{-1}) and AG0934 (2675 kg ha^{-1}), which equals to a monetary difference of \$17.37 ha^{-1}, using a \$0.294 kg^{-1} soybean price. This negative return for the benefit of reduced competition of the soybean plant (for AG0434) needs to be considered by agricultural producers to determine the best economic return for an interseeded-cover crop system. This research did not analyze the benefit of the increased biomass for potential reduced fertilizer application in future crops, herbicide cost reduction, and potential long term soil health benefits. Several studies have been conducted about economic returns on cover crops [44–46], yet further research is suggested to improve grower decision making of maximum cover crop economic benefit to improve sustainability of interseeding cover crops into soybean.

Further research needs to be conducted to show the economic return resulting from the additional cover crop growth achieved by interseeding into an early-maturing soybean cultivar compared with the lower yield and monetary loss associated with not planting a later maturing cultivar. The wheat grain yield was not influenced by the soybean maturity of the cultivar (Table 7).

3.2.2. Cover Crops by Seeding Rate

Nearly all cover crop treatments metrics were significantly different when comparing winter rye and winter camelina (Table 8). Rye had higher cover percent and biomass. No significant differences were found between seeding rate treatments within rye or camelina although all 100% seeding rate treatments produced larger values. The lower seeding rates would allow for reduced cover crop seed expense. Despite no differences between seeding rates in this study, several studies have suggested positive results for 100% seeding rate treatments [47].

Table 8. Mean fall and spring canopy coverage and cover crop biomass for cover crops across four environments in Fargo, ND, USA, from 2016, 2017, and 2018.

Cultivar	Fall Canopy Coverage		Spring Canopy Coverage		Spring Cover Crop Biomass	
	%				kg ha^{-1}	
Camelina100	7.0	b [†]	4.3	b	103	b
Camelina75	6.4	b	3.7	b	97	b
Rye100	16.2	a	16.1	a	321	a
Rye75	12.8	a	15.5	a	304	a

[†] Within a column, mean followed by a different letter is significantly different at $p \leq 0.05$.

Rye treatments coverage percent and biomass values observed upwards of three times those of the winter camelina values (Table 8). These values were consistent across all environments and similar to trends found in other research [24,25]. No economical or soil nutrient analysis was done in this study to show the economic impact of these differences yet based on this study's data, rye was superior compared with camelina.

Soybean yield was not different for soybean interseeded with camelina or rye compared with soybean without a cover crop (Table 9). For the HRSW growing seasons o

2017 and 2018, the termination of the cover crops was conducted using an application of Roundup WeatherMAX applied the same day as HRSW planting. Ten days after termination, winter camelina was visually eliminated with limited competition with HRSW. Winter rye took 30 to 45 d to become eliminated, and by this time, the HRSW had nearly 35% canopy coverage and was beginning to tiller.

Table 9. Mean soybean (2016–2017) and wheat yield (2017–2018) and wheat canopy coverage for five cover crops across four environments in Fargo, ND, USA.

Cultivar	Soybean Yield	Wheat Yield		Wheat Canopy Coverage	
	kg ha^{-1}			%	
Camelina100	2208	2718	a †	52.0	a
Camelina75	2193	2767	a	51.7	a
Rye100	2197	2507	b	40.1	b
Rye75	2175	2562	b	41.9	b
Check	2262	2808	a	51.9	a
LSD 0.05	ns				

† Within a column, mean followed by a different letter are significantly different at $p \leq 0.05$. ns = not significant.

An advantage of chemical elimination of the cover crop is protection to the soil from wind erosion and excess sunlight resulting in preventing moisture loss or crusting, compared with tillage. The wheat after cereal rye plots were significantly inhibited in growth (Table 9) and vigor as expected, which is constant with previous research [48].

This wheat yield difference after rye (Table 9) was caused by the substantial biomass growth produced by cereal rye, 313 kg ha^{-1} average of both seeding rates, as compared with biomass of winter camelina at 100 kg ha^{-1}, average of both seeding rates, respectively (Table 8). These biomass differences compounded by late termination of cover crops, canopy coverage differences, and slower herbicide (glyphosate) action in rye resulted in the significant differences of the wheat cover percentage and yield (Table 9).

The substantial wheat biomass growth inhibition due to the late termination of the cover crops was exacerbated by the no-till tillage system as crop residue was high. Since the rye showed canopy cover percentages averaging above 38% at termination, germinating HRSW plants were covered by the dying cereal rye plants. This difference between rye and camelina or check plots was easily observed, with the cereal rye plots expressing stunting, chlorosis, and poor vigor.

The economic loss using wheat yield data between check plots (2808 kg ha^{-1}) and rye plots (2535 kg ha^{-1}) (Table 9) was about \$ 60.33 ha^{-1} using a price of \$ 0.22 kg^{-1} for wheat. With this amount of economic loss, planting rye before growing HRSW is not recommended when rye is chemically terminated at the same time as wheat planting. Further research is needed to investigate if other termination timings will have different results and evaluate the economic cost or benefits.

Producers are interested in including cover crops in their farming systems to increase soil protection and soil health benefits. Additional research will be needed to evaluate the long-term benefits of cover crops after soybean.

4. Conclusions

Earlier maturing soybean cultivars produced increased cover crop growth resulting in increased canopy coverage. However, the opportunity cost of planting an earlier maturing cultivar may be larger due to reduction of soybean yield compared with the later maturing cultivar. The early maturing cultivar with 0.4 maturity had higher cover crop soil cover percent later in the fall and early in the spring, with 53.7% more canopy coverage in the fall compared with the 0.9 maturity cultivar when cover crops were planted at the R7 growth stage of the early maturing cultivar.

Cover crop seeding rates did not increase cover crop biomass production. Interseeded cover crops into different cultivars at the R7 stage of the early maturity cultivars did not reduce soybean yield compared with the check plot, which is consistent with previous research. Growing HRSW after interseeded cereal rye into soybean resulted in reduced yields compared with winter camelina and the check plots. This was expected, as the HRSW cover percentage after rye was significantly lower compared with camelina and check plots, and the visual stress observed during the summer months was obvious. Further research needs to be conducted to show the economic return resulting from the cover crop grow after soybean.

Author Contributions: K.L.J.: Conceptualization, formal analysis, investigation, methodology, writing—original draft, writing—review and editing. H.J.K.: Conceptualization, investigation, supervision, writing—review and editing. D.P.S.: Writing—review and editing. M.T.B.: Formal analysis, funding acquisition, investigation, project administration, resources, writing—review and editing. All authors have read and agreed to the published version of the manuscript.

Funding: This project was funded by USDA-NIFA, Award no. 2016-69004-24784

Acknowledgments: The authors would like to thank Chad Deplazes and Darin Eisinger for their technical assistance.

Conflicts of Interest: The authors declare no conflict of interest.

References

1. Dickey, E.C.; Shelton, D.P.; Jasa, P.J.; Peterson, T.R. Soil erosion form tillage systems used in soybean and corn residues. *Trans. ASAE* **1985**, *28*, 1124–1130. [CrossRef]
2. Franzen, D.; Goos, R.J.; Kandel, H.; Augustin, C.; Buetow, R.; Teboh, J.; Forster, S.; Endres, G. *Soybean Soil Fertility*; NDSU Ext. Serv. SF1164 (Revised Nov. 2019); North Dakota Agric. Exp. Station: Fargo, ND, USA, 2019.
3. NRCS (Natural Resources Conservation Service). Cover Crop Termination Guidelines Version 4. USDA Risk Management Agency. 2021. Available online: https://www.rma.usda.gov/Topics/Cover-Crops (accessed on 16 January 2021).
4. Robertson, G.P.; Gross, K.L.; Hamilton, S.K.; Landis, D.A.; Schmidt, T.M.; Snapp, S.S.; Swinton, S.M. Farming for ecosystem services: An ecological approach to production agriculture. In *The Ecology of Agricultural Landscapes: Long-Term Research on the Path to Sustainability*; Hamilton, S.K., Doll, J.E., Robertson, G.P., Eds.; Oxford Univ. Press: New York, NY, USA, 2015; pp. 33–53.
5. Delgado, J.A.; Gantzer, C.J. The 4Rs for cover crops and other advances in cover crop management for environmental quality. *J. Soil Water Conserv.* **2015**, *70*, 142A–145A. [CrossRef]
6. Garbeva, P.; van Veen, J.A.; van Elsas, J.D. Microbial diversity in soil: Selection of microbial populations by plant and soil type and implications for disease suppressiveness. *Ann. Rev. Phytopathol.* **2004**, *42*, 243–270. [CrossRef] [PubMed]
7. Gasch, C.; DeJong-Hughes, J. *Soil Organic Matter Does Matter*; SF1942; North Dakota State Univ. Ext. Pub.: Fargo, ND, USA, 2019.
8. Lu, Y.C.; Watkins, K.B.; Teasdale, J.R.; Abdul-Baki, A.A. Cover crops in sustainable food production. *Food Rev. Intern.* **2000**, *16*, 121–157. [CrossRef]
9. O'Brien, P.L.; Hatfield, J.L.; Dold, C.; Kistner-Thomas, E.J.; Wacha, K.M. Cropping pattern changes diminish agroecosystem services in North and South Dakota, USA. *Agron. J.* **2020**, *112*, 1–24. [CrossRef]
10. Syswerda, S.P.; Robertson, G.P. Ecosystem services along a management gradient in Michigan (USA) cropping systems. *Agric. Ecosyst. Environ.* **2014**, *189*, 28–35. [CrossRef]
11. Schipanski, M.E.; Barbercheck, M.; Douglas, M.R.; Finney, D.M.; Haider, K.; Kayne, J.P.; Kemaian, A.R.; Mortensen, D.A.; Ryan, M.R.; Tooker, J.; et al. A framework for evaluating ecosystem services provided by cover crops in agroecosystems. *Agric. Syst.* **2014**, *125*, 12–22. [CrossRef]
12. Berti, M.; Gesch, R.; Eynck, C.; Anderson, J.; Cermak, S. Camelina uses, genetics, genomics, production, and management. *Ind. Crops Prod.* **2016**, *94*, 690–710. [CrossRef]
13. Anderson, J.V.; Horvath, D.P.; Doğramacı, M.; Dorn, K.M.; Chao, W.S.; Watkin, E.; Marks, M.D.; Gesch, R. Expression of FLOWERING LOCUS C and a frame shift mutation of this gene on chromosome 20 differentiate a summer- and winter-annual biotype of *Camelina sativa*. *Plant Direct* **2018**, *2*, 1–14. [CrossRef]
14. Wittenberg, A.; Anderson, J.V.; Berti, M.T. Winter and summer annual biotypes of camelina have different morphology and seed characteristics. *Ind. Crops Prod.* **2019**, *135*, 230–237. [CrossRef]
15. Gesch, R.W.; Archer, D.; Berti, M.T. Dual cropping winter camelina with soybean in the northern Corn Belt. *Agron. J.* **2014**, *106*, 1735–1745. [CrossRef]
16. Gesch, R.; Cermak, S. Sowing date and tillage effects on fall-planted camelina in the northern Corn Belt. *Agron. J.* **2011**, *103*, 980–987. [CrossRef]
17. Berti, M.; Samarappuli, D.; Johnson, B.L.; Gesch, R.W. Integrating winter camelina into maize and soybean cropping systems. *Ind. Crops Prod.* **2017**, *107*, 595–601. [CrossRef]
18. Snapp, S.; Swinton, S.; Labarta, R.; Mutch, D.; Black, J.; Leep, R.; Nyiraneza, J.; O'Neil, K. Evaluating cover crops for benefits, costs and performance within cropping system niches. *Agron. J.* **2005**, *97*, 322–332.

19. Wilson, M.L.; Baker, J.M.; Allan, D.L. Factors affecting successful establishment of aerially seeded winter rye. *J. Environ. Qual.* **2013**, *33*, 1010. [CrossRef]
20. Crowley, K.A.; Van Es, H.M.; Gomex, M.I.; Ryan, M.R. Trade-offs in cereal rye management strategies prior to organically managed soybean. *Agron. J.* **2018**, *110*, 1492–1504. [CrossRef]
21. Kladivko, E.; Kaspar, T.; Jaynes, D.; Malone, R.; Singer, J.; Morin, X.; Searchinger, T. Cover crops in the upper Midwest USA: Potential adoption and reduction of nitrate leaching in the Mississippi River Basin. *J. Soil Water Conserv.* **2014**, *69*, 279–291. [CrossRef]
22. SARE (Sustainable Agriculture Research and Education). *Managing Cover Crops Profitably*, 3rd ed.; Handbook Series 9; SARE: College Park, MD, USA, 2012; pp. 98–105.
23. Weil, R.; Kremen, A. Thinking across and beyond disciplines to make cover crops pay. *J. Sci. Food Agric.* **2007**, *87*, 551–557. [CrossRef]
24. Mohammed, Y.A.; Matthees, H.L.; Gesch, R.W.; Patel, S.; Forcella, F.; Aasand, K.; Steffl, N.; Johnson, B.L.; Wells, M.S.; Lenssen, A.W. Establishing winter annual cover crops by interseeding into maize and soybean. *Agron. J.* **2020**, 1–14. [CrossRef]
25. Peterson, A.T.; Berti, M.T.; Samarappuli, D. Intersowing cover crops into standing soybean in the US Upper Midwest. *Agronomy* **2019**, *9*, 264. [CrossRef]
26. Midwest Cover Crops Council. Integrating Cover Crops in Soybean Rotations: Challenges and Recommendations for the North Central Region. 2015. Available online: http://mccc.msu.edu/wp-content/uploads/2016/10/ALL_2015_Integrating-cover-crops-in-soybean-rotations.pdf (accessed on 16 January 2021).
27. Uchino, H.; Iwama, K.; Jitsuyama, Y.; Ichiyama, K.; Sugiura, E.; Yudate, T.; Nakamura, S.; Gopal, J. Effect of interseeding cover crops and fertilization on weed suppression under an organic and rotational cropping system: 1. Stability of weed suppression over years and main crops of potato, maize and soybean. *Field Crops Res.* **2012**, *127*, 9–16. [CrossRef]
28. Masilionyte, L.; Maiksteniene, S.; Kriauciuniene, Z.; Jablonskyte-Rasce, D.; Zou, L.; Sarauskis, E. Effect of cover crops in smothering weeds and volunteer plants in alternative farming systems. *Crop Prot.* **2017**, *91*, 74–81. [CrossRef]
29. Fisher, K.A.; Momen, B.; Kratochvil, R.J. Is broadcasting seed an effective winter cover crop planting method? *Agron. J.* **2011**, *103*, 472–478. [CrossRef]
30. Parkin, T.B.; Kaspar, T.C.; Cambardella, C.A. Small grain cover crops to manage nitrogen in the Midwest. In Proceedings of the Cover Crops, Soil Quality, and Ecosystems Conference, Sacramento, CA, USA, 12–14 March 1997.
31. Eckert, D.J. Chemical attributes of soils subjected to no-till cropping with rye cover crops. *Soil Sci. Soc. Am. J.* **1991**, *55*, 405–409. [CrossRef]
32. Hively, W.D.; Cox, W.J. Interseeding cover crops into soybean and subsequent corn yields. *Agron. J.* **2001**, *93*, 308–313. [CrossRef]
33. Koehler-Cole, K.; Elmore, R.W. Seeding rates and productivity of broadcast interseeded cover crops. *Agronomy* **2020**, *10*, 1723. [CrossRef]
34. USDA (United States Department of Agriculture). Web Soil Survey. 2021. Available online: https://websoilsurvey.sc.egov.usda.gov/App/HomePage.htm (accessed on 16 January 2021).
35. NDAWN. *North Dakota Agricultural Weather Network*; North Dakota State Univ.: Fargo, ND, USA, 2020; Available online: http://ndawn.ndsu.nodak.edu (accessed on 16 January 2021).
36. Helms, T.C.; Scot, R.A.; Schapaugh, W.T.; Goos, R.J.; Franzen, D.W.; Shlegel, A.J. Soybean iron-deficiency chlorosis tolerance and yield decrease on calcareous soils. *Agron. J.* **2010**, *102*, 492–498. [CrossRef]
37. Kandel, H.; Endres, G. *Soybean Production Field Guide*; A1172; North Dakota State Univ. Ext. Pub.: Fargo, ND, USA, 2019.
38. Knodel, J.; Beauzay, P.; Boetel, M.; Prochaska, T.J.; Lubenow, L. *North Dakota Field Crop Insect Management Guide*; NDSU Ext. E1143-20; North Dakota Agric. Exp. Station: Fargo, ND, USA, 2020.
39. Patrignani, A.; Ochsner, T.E. Canopeo: A powerful new tool for measuring fractional green canopy cover. *Agron. J.* **2015**, *107*, 2312–2320. [CrossRef]
40. Fehr, W.R.; Caviness, C.E.; Burmood, D.T.; Pennington, J.S. Stage of development descriptions for soybeans, *Glycine max* (L.) Merrill. *Crop Sci.* **1971**, *11*, 929–931. [CrossRef]
41. Zadoks, J.C.; Chang, T.T.; Konzak, C.F. A decimal code for the growth stages of cereals. *Weed Res.* **1974**, *14*, 415–421. [CrossRef]
42. SAS Institute. *SAS User's Guide*; Statistics; SAS Institute: Cary, NC, USA, 2020.
43. Egli, D.B. Cultivar maturity and potential yield of soybean. *Field Crops Res.* **1993**, *32*, 147–158. [CrossRef]
44. Plastina, A.; Liu, F.; Sawadgo, W.; Miguez, F.; Carlson, F. Partial budgets for cover crops in Midwest row crop farming. *J. ASFMRA* **2018**, 90–106. [CrossRef]
45. Bergtold, J.; Ramsey, S.; Maddy, L.; Williams, J. A review of economic considerations for cover crops as a conservation practice. *Renew. Agric. Food Syst.* **2019**, *34*, 62–76. [CrossRef]
46. Myers, R.; Weber, A.; Tellatin, S. Cover Crop Economics: Opportunities to Improve Your Bottom Line in Row Crops. SARE Technical Bulletin. 2019. Available online: https://extension.umd.edu/sites/extension.umd.edu/files/_docs/SARE%20Cover%20Crop%20Economics%202019.pdf (accessed on 16 January 2021).
47. Gesch, R.W.; Matthees, H.L.; Alvarez, A.L.; Gardner, R.D. Winter camelina: Crop growth, seed yield, and quality response to cultivar and seeding rate. *Crop Sci.* **2018**. [CrossRef]
48. Nielsen, D.C.; Lyon, D.J.; Higgins, R.K.; Hergert, G.W.; Holman, J.D.; Vigil, M.F. Cover crop effect on subsequent wheat yield in the central Great Plains. *Agron. J.* **2016**, *108*, 243–256. [CrossRef]

agronomy

MDPI

Article

Potato Growth and Yield Characteristics under Different Cropping System Management Strategies in Northeastern U.S. †

Robert P. Larkin [1,*], C. Wayne Honeycutt [2], Timothy S. Griffin [3], O. Modesto Olanya [4] and Zhongqi He [5]

1 USDA-ARS, New England Plant, Soil, and Water Laboratory, University of Maine, Orono, ME 04469, USA
2 Soil Health Institute, Morrisville, NC 27560, USA; whoneycutt@soilhealthinstitute.org
3 Friedman School of Nutrition Science and Policy, Tufts University, Boston, MA 02111, USA; Timothy.Griffin@tufts.edu
4 USDA-ARS, Eastern Regional Research Center, Wyndmoor, PA 19038, USA; modesto.olanya@usda.gov
5 USDA-ARS, Southern Regional Research Center, New Orleans, LA 70124, USA; zhongqi.he@usda.gov
* Correspondence: bob.larkin@usda.gov; Tel.: +1-207-581-3367

† Mention of trade names, proprietary products, or specific equipment does not constitute a guarantee or warranty by the U.S. Department of Agriculture and does not imply approval to the exclusion of other products that may be suitable.

Abstract: Cropping systems and management practices that improve soil health may greatly enhance crop productivity. Four different potato cropping systems designed to address specific management goals of soil conservation (SC), soil improvement (SI), disease suppression (DS), and a status quo (SQ) standard rotation, along with a non-rotation (PP) control, were evaluated for their effects on potato crop growth, nutrient, and yield characteristics under both irrigated and non-irrigated (rainfed) conditions in field trials in Maine, USA, from 2004 to 2010. Both cropping system and irrigation significantly ($p < 0.05$) affected most potato crop parameters associated with growth and yield. All rotations increased tuber yield relative to the non-rotation PP control, and the SI system, which included yearly compost amendments, resulted in overall higher yields and a higher percentage of large-size tubers than all other systems with no irrigation (increases of 14 to 90%). DS, which contained disease-suppressive green manures and cover crops, produced the highest yields overall under irrigation (increases of 11 to 35%). Irrigation increased tuber yields in all cropping systems except SI (average increase of 27–37%). SI also resulted in significant increases in leaf area duration and chlorophyll content (as indicators of photosynthetic potential) and root and shoot biomass relative to other cropping systems, particularly under non-irrigated conditions. SI also resulted in higher shoot and tuber tissue concentrations of N, P, and K, but not most micronutrients. Overall, cropping systems that incorporate management practices such as increased rotation length and the use of cover crops, green manures, reduced tillage, and particularly, organic amendments, can substantially improve potato crop growth and yield. Irrigation also substantially increased growth and yield under normal field conditions in Maine, but SI, with its large organic amendments, was essentially a substitute for irrigation, producing comparable results without irrigation.

Keywords: compost amendment; cover crops; crop production; green manure; leaf area duration; soil health; tuber yield

Citation: Larkin, R.P.; Honeycutt, C.W.; Griffin, T.S.; Olanya, O.M.; He, Z. Potato Growth and Yield Characteristics under Different Cropping System Management Strategies in Northeastern U.S. *Agronomy* **2021**, *11*, 165. https://doi.org/10.3390/agronomy11010165

Received: 17 December 2020
Accepted: 13 January 2021
Published: 16 January 2021

1. Introduction

Sustainability of crop production systems is dependent on many factors, from the cost/benefit of the many operations involved to the inputs and outputs obtained to the continued health of the soil and overall agroecosystem. Probably the most important single attribute to growers is crop productivity, usually measured by yield. Crop yield is the final result, but numerous aspects of crop development and growth may be involved in or responsible for the resultant yield observed, and may give indications of where production problems may be occurring. Soil health, defined as the continued capacity of soil to function as a vital living system to sustain biological productivity; maintain environmental quality; and promote plant, animal, and human health [1–3] is a critical component of agricultural productivity, sustainability, and ecosystem function. Incorporation of soil and crop management practices that promote soil health, such as crop rotations, cover crops and green manures, organic amendments, and conservation tillage, into improved cropping systems may help maintain and/or improve soil health and enhance productivity, sustainability, economic vitality, and environmental quality [3,4].

Potato (*Solanum tuberosum* L.), an important crop in the US and throughout the world, can be particularly hard on soils due to the intensive tillage operations and cropping patterns used. Potato production in the northeast U.S., as well as in other potato growing regions, has been characterized by short (2 y) rotations, extensive tillage, minimal crop residue return, and minimal crop diversity, often taking a toll on soil health and crop productivity over time [5]. Increasing rotation length from 2 years to 3 or more years between potato crops has been shown to improve productivity, as well as reduce soilborne diseases in multiple studies [6–10]. Other practices, such as the addition of cover crops and green manures [11,12], amendments of compost or animal manure [13–15], and reduced tillage [6,16], have all shown promise for having positive effects on tuber yield and quality, as well as other benefits to various soil properties and soil health in potato systems. However, most previous research has focused on the assessment of individual practices or rotations, and not necessarily on the combined effects of multiple different practices in integrated cropping systems for the total system effects on productivity and plant and soil properties.

In this research, which builds upon our previous work with improving potato cropping systems [17–21], we assessed the effects of cropping systems incorporating multiple soil health management practices focused on specific soil and crop management goals, on various plant characteristics of the potato crop itself, including growth, productivity, and nutrient concentrations. In 2004, we established field trials for long-term evaluation of different potato cropping systems to better determine what factors were most limiting to potato production in Northeastern U.S., and how these limitations could be addressed through cropping systems [20]. Three specific cropping systems were established to address the crop and soil management goals of soil conservation, soil improvement, and disease suppression, and these were compared to a standard rotation and a non-rotation control.

Previously, we characterized the effects of these cropping systems on various management concerns, such as soilborne potato diseases and soil microbiology [20–23], soil health (represented by various soil physical, chemical, and biological properties) [24,25], and soil nutrient-related enzyme activities and P status [26–28]. In the present research, we examined cropping system effects on crop productivity using such crop characteristics as photosynthetic potential (measured as leaf area index and chlorophyll content), tuber yield and quality (total and various size class distributions, misshapenness, and specific gravity), biomass production (both above- and below-ground), and plant tissue nutrient concentrations.

2. Materials and Methods

2.1. Cropping Systems

Cropping systems consisted of five different systems designed to address specific management goals of soil conservation, soil improvement and disease suppression, as well as a system representing a typical standard rotation currently used in the Northeast U.S., and a non-rotation control of continuous potato. An overview of the cropping systems and their features is provided in Table 1, which have been previously described [20]. In brief, the standard or "status quo" (SQ) rotation consisted of a 2 y rotation of barley (*Hordeum vulgare* L.) underseeded with red clover (*Trifolium pretense* L.) as a cover crop, followed by potato the following year, and includes regular spring and fall tillage each year. The soil conserving (SC) system consisted of a 3 y rotation of barley underseeded with the forage grass timothy (*Phleum pratense* L), which would overwinter and be allowed to continue undisturbed for a full year (2nd y), and then followed by potato in the third year. In this system, tillage was also greatly reduced, with no tillage except as needed for maintenance and harvest in the potato crop year, thus substantially improving soil conservation. In addition, straw mulch (2 Mg/ha) was applied after potato harvest to further conserve soil resources. The soil improving (SI) system consisted of the same basic rotation as SC (3 y, barley/timothy-timothy-potato, limited tillage, straw mulch), but with yearly additions of compost (composted dairy manure added at 45 Mg/ha fresh wt [~18 Mg/ha dry wt]), to provide abundant organic matter to improve soil quality. The disease-suppressive (DS) system was designed to make use of multiple strategies for suppressing soilborne diseases, and included the use of disease-suppressive rotation crops, a longer rotation period, crop diversity, green manures, and fall cover crops. The DS system consisted of a 3 y rotation with the disease-suppressive *Brassica* "Caliente 119" Mustard Blend (blend of oriental and white mustard seeds, *Brassica juncea* L. and *Sinapis alba* L.) grown as a green manure, followed by a fall cover crop of rapeseed (*Brassica napus* L."Dwarf Essex") in the first year. In the second year, a disease-suppressive Sorghum–Sudangrass hybrid (*Sorghum bicolor* x *S. bicolor* var. *sudanense* L.) was grown as a green manure, followed by a fall cover crop of winter rye (*Secale cereale* L.), with potato in the third year. For this study, green manure refers to a crop whose full biomass was incorporated into the soil while fresh and green, whereas cover crop refers to a crop that is left in the field to overwinter unplowed and uncut. Continuous potato (PP) was the non-rotation control consisting of a potato crop planted in the same plots each year (spring and fall tillage). All cropping systems were evaluated under both irrigated and non-irrigated management. Irrigation was applied with a lateral, overhead sprinkler system when soil tensiometer readings at the 10–15 cm depth exceeded 50 KPa. Each irrigation event consisted of application of 1.3 cm of water.

Table 1. Names, descriptions, and features of the cropping systems used to address specific management goals in these studies.

Cropping System Parameters				
Name	**Abbreviation**	**Length**	**Rotation Description**	**Features**
Status Quo	SQ	2 y	Barley/Clover, Potato	Typical rotation (Industry standard)
Soil Conserving	SC	3 y	Barley/Timothy, Timothy, Potato	Additional year of forage, limited tillage, straw mulch after potato
Soil Improving	SI	3 y	Barley/Timothy, Timothy, Potato	SC plus yearly compost amendments
Disease-Suppressive	DS	3 y	Mustard GM/Rapeseed cover, Sudangrass GM/Rye cover, Potato	Biofumigation crops, green manures, cover crops, and increased crop diversity
Continuous Potato	PP	1 y	Potato, Potato	Non-rotation control

2.2. Field Set-Up and Management

Long-term research plots were established in 2004 at the USDA-ARS New England Plant, Soil and Water Laboratory Field Experimental Site in Presque Isle, Maine, USA, as a split-block design with 5 replicate blocks, with irrigation (Irr) and cropping system (CS) as the main and split factors, respectively. Soil type was a Caribou sandy loam (Fine-loamy, isotic, frigid Typic Haplorthods). Each rotation entry point (representing each possible rotation crop for all years) was included in each block, so that each full rotation was represented each year (SQ, 2 entry points; SC, SI, and DS, 3 entry points; and PP, 1 entry point), resulting in 12 treatment plots (6 × 15 m each) per block for each of the irrigated and non-irrigated components. Average soil properties measured at the time of initial planting (with no significant differences among treatment plots) were as follows: pH 5.88, total soil C 22.5 g kg^{-1} soil, soil N 1.7 g kg^{-1}, P 17.7 mg kg^{-1}, K 139 mg kg^{-1}, Ca 607 mg kg^{-1}, Mg 158 mg kg^{-1}, and CEC 5.58. For potato planting, seed tubers of the potato variety "Russet Burbank" were cut to seedpieces of ~50–60 g each 7 to 10 days prior to planting and stored at 8 °C until 48 h prior to planting, when they were stored at room temperature until planted. Seedpieces were planted by hand in furrows in each plot (four rows, 0.9 m centers, with a 35 cm spacing between plants). Potato plots were fertilized with the equivalent of 224 kg ha^{-1} N and 249 kg ha^{-1} P_2O_5 and K_2O. Fertilizer rate was based on years of previous research for this region in similar soils establishing 150–200 kg N ha^{-1} as optimal for potato production [29–33]. Fertilizer applications were purposely applied to be equal across all systems and above optimal rates for crop nutritional needs, so that nutrition would not be limiting in any cropping system and observed system effects would likely be related to factors other than fertility. Further details of the planting and management of the crops and rotations have been described previously [20].

Site environmental conditions, including air and soil temperature, relative humidity, and rainfall were monitored throughout each growing season using a CR10X datalogger (Campbell Scientific Inc., Logan, UT, USA) outfitted with temperature probes and a tipping bucket rain gauge. Data were recorded every hour and converted to daily minimum, maximum, and average values as well as total daily rainfall. For ease of presentation, temperature and rainfall data were summarized as average monthly values, and are presented along with the number of individual irrigation events for each year of the study (Table 2).

Table 2. Average daily temperature, total rainfall, and number of irrigation events for the months of May through September at the Presque Isle research site for 2006 to 2010 compared with long-term (30 year) average conditions.

	Environmental Parameters					
	2006	2007	2008	2009	2010	Long-Term Avg
Average Daily Temperature (°C)						
May	12.7	10.7	10.3	11.3	13.1	11.4
June	18.1	16.7	15.9	14.7	16.2	16.4
July	20.3	19.0	20.4	13.4	20.8	19.0
August	16.1	17.3	17.7	15.9	18.8	18.2
September	13.1	14.1	13.6	11.7	14.6	13.2
Season avg	16.1	15.6	15.6	13.4	16.7	15.6
Rainfall (cm)						
May	11.3	6.1	5.3	12.5	6.5	8.7
June	10.9	5.1	11.6	8.6	13.0	8.6
July	11.7	9.8	8.2	12.2	7.2	9.4
August	6.3	12.0	11.2	5.9	3.3	10.0
September	7.1	4.3	7.9	3.8	7.2	8.7
Season total	47.3	37.3	44.2	43.0	37.2	45.4
Irrigation events (no.)	3	6	4	0	6	

2.3. Tuber Yield and Quality Assays

In October of each year, potatoes were harvested from the full-length of the center two rows from each potato plot. Total weight of the harvested tubers was used to determine total yield on a Mg/ha basis. A subset of the harvested tubers, amounting to a total of 20–25 kg/plot and taken from multiple randomly selected plot sections, were washed, graded, and sized into 4 categories from small to extra large (small, <114 g; medium, 114–227 g; large, 228–342 g; and extra large >342 g). Marketable yield was calculated as the total weight of tubers of a size greater than 114g each. Tuber specific gravity was assessed on the graded subset of marketable tubers for each plot using standard weight in air vs. weight in water calculations. The weight of severely misshaped tubers (due to knobs, irregular shapes) from all size classes was also assessed for each plot.

2.4. Crop Growth Assays

2.4.1. Leaf Area Index, Duration, and Chlorophyll Content

Potato canopy density and light interception were estimated by the leaf area index (LAI), which is a measure of leaf area per unit of ground area, and leaf area duration (LAD), which is a measure of LAI over time. LAI was measured using the SunScan Canopy Analysis System (Dynamax Inc., Houston, TX, USA) which assesses the photosynthetically active radiation above and below the canopy and calculates LAI. Starting around 60 days after planting (DAP) and continuing once each week for 9–10 weeks thereafter, LAI was estimated in each plot, with a minimum of 12 below canopy readings made/plot at each sampling. LAD was determined from plots of LAI over time as the area under the LAI progress curve, and calculated as the integration of 2nd order polynomial functions generated through linear regression of LAI versus time (in DAP).

Leaf chlorophyll content was measured at the same time and same weekly schedule as LAI readings using the Minolta SPAD-502 chlorophyll meter (Spectrum Technologies, Plainfield, IL, USA). For each plot, 12 leaflets were sampled (third terminal leaflet from top of plant) from random plants at each sampling date. SPAD values indicated relative chlorophyll content, and were analyzed for individual sampling dates, as well as averaged over all sampling dates to give an overall season average for comparison among cropping systems.

2.4.2. Root, Tuber, and Shoot Biomass

Potato plant samples for biomass determinations were collected in mid-August (80–85 DAP) in the years 2007, 2008, and 2009. Four whole potato plants (2 each from the 2 center rows), including full root systems and developing tubers, were randomly selected in each plot, and removed intact from the soil. Plants were separated into root, tuber, and shoot components and weighed fresh, and then brought back to the lab. Roots and tubers were washed (to remove soil and adhering debris), and all plant parts were oven-dried (65 °C for 1 week in a drying oven) and weighed again. Dry weights per plot were then converted to biomass on an area basis (Mg/ha).

2.4.3. Root, Tuber, and Shoot Tissue Composition

In 2008 only, subsamples from plant samples collected for biomass determination were also analyzed for full elemental composition to compare nutritional qualities of the plant material among treatments. Subsamples were dried and ground in a Wiley mill (1-mm screen). Plant tissue C and N were determined by dry combustion using an elemental analyzer. Plant tissue concentrations of Ca, K, Mg, P, Al, B, Fe, Mn, Zn, and Cu were analyzed using inductively coupled plasma optical emission spectroscopy (ICP-OES) by the University of Maine Analytical Lab (Orono, ME, USA). Values were expressed as mg kg^{-1} dry weight of tissue.

2.5. Statistical Analyses

Data were analyzed using standard analysis of variance (ANOVA) with factorial treatment structure and interactions appropriate for a split-block design. Data from each crop year were analyzed separately, and then data from multiple seasons were also combined and analyzed (with year as additional factor, with interactions) together to evaluate cumulative and multi-year effects of the cropping systems. Correlation analyses were conducted (using Pearson's product-moment correlation coefficients) among crop growth, yield, and other soil property parameters (from soil properties determined as part of a previously published study [25]). Significance was evaluated at $p < 0.05$ for all tests. Mean separation was accomplished with Fisher's protected LSD test. All analyses were conducted using the Statistical Analysis Systems ver. 9.4 (SAS Institute, Cary, NC, USA).

3. Results

Cropping system significantly affected virtually all aspects of potato crop growth, nutrient, and yield characteristics. Effects also varied somewhat from year to year, depending on environmental conditions, but generally showing consistent trends by cropping system over time. The interaction of irrigation (whether irrigated or not irrigated) and cropping system was significant ($p < 0.05$) for most measured parameters, so for those cases, results are presented separately for irrigated and non-irrigated treatments. When irrigation by cropping system interaction was not significant, data are presented for cropping systems over both irrigation regimes.

3.1. Tuber Yield

Under non-irrigated (rainfed) conditions, the soil improving (SI) system, which included yearly compost amendments, resulted in the highest total and marketable yields of all cropping systems, showing significantly higher yields than the standard 2 y (SQ) rotation and non-rotation (PP) controls in all years, with values ranging from 34 to 44 Mg ha^{-1} and 26–35 Mg ha^{-1}, for total and marketable yield, respectively, which represented increases of 14 to 60% and 15 to 93% over those from SQ and PP systems (Table 3). The disease-suppressive (DS) system also resulted in higher total and marketable yields than PP in most years, and the SC and SQ systems in some years, whereas PP consistently resulted in the lowest overall yields of all systems. Although yields for SC remained relatively low in the early years of the study (representing the first full rotation cycle, 2006–2008), by the second rotation cycle (2009–2010), yields for SC averaged greater than both SQ and PP systems, by 12 to 23%. When averaged over all five years, all cropping systems significantly increased both total and marketable yield over PP, and DS also increased yield relative to SC and SQ, but SI produced significantly higher overall yields than all other systems, with increases in total yield averaging 41% higher than PP and 30% higher than SC and SQ (Figure 1).

Table 3. Effect of different cropping systems on total and marketable (tubers >114 g each) potato tuber yield over five field seasons (2006–2010) under irrigated (Irr) and non-irrigated (Non-irr) conditions.

	Tuber Yield (Mg/ha)									
	2006		2007		2008		2009		2010	
Treatment [x]	**Non-Irr**	**Irr**	**Non-Irr**	**Irr**	**Non-Irr**	**Irr**	**Non-Irr**	**Irr**	**Non-Irr**	**Irr**
Total yield										
SI	41.5 a [y]	43.3 ab	44.2 a	43.0 ab	33.9 a	33.9 ab	34.1 a	32.0 c	35.1 a	40.9 a
DS	36.6 ab	44.7 a	35.6 b	44.9 a	31.4 ab	36.5 a	28.1 bc	39.3 a	30.6 b	39.4 ab
SC	27.9 c	37.1 c	33.1 b	39.0 bc	30.5 ab	37.1 a	30.6 ab	36.7 b	25.5 c	37.7 abc
SQ	34.5 b	44.4 a	31.4 bc	39.3 bc	29.7 b	32.7 bc	26.7 c	32.0 c	22.0 d	35.4 c
PP	32.4 bc	38.1 bc	27.8 c	34.0 c	22.8 c	29.6 c	25.1 c	33.7 c	25.0 cd	36.5 bc
LSD (p = 0.05)	5.1	5.7	4.8	5.3	3.4	3.8	3.8	2.3	3.5	3.6
Avg.	34.6	41.5 * [z]	34.4	40.0*	29.5	34.0 *	28.8	34.7 *	27.6	38.0 *
Marketable yield										
SI	32.0 a	34.3 a	35.4 a	33.9 ab	28.1 a	27.2 a	27.9 a	26.8 b	26.3 a	30.7 ab
DS	29.3 ab	34.7 a	25.6 b	35.9 a	24.8 ab	29.3 a	22.3 bc	33.4 a	23.3 a	33.4 a
SC	21.2 c	29.5 a	22.9 b	28.6 bc	23.0 c	27.7 a	24.9 ab	28.5 b	17.6 b	33.2 a
SQ	25.4 bc	35.7 a	23.7 b	30.7 bc	24.4 bc	25.5 ab	21.7 bc	27.4 b	13.6 b	28.2 b
PP	22.8 c	29.1 a	18.8 c	25.2 c	16.2 d	21.6 c	17.9 c	25.8 b	16.5 b	29.1 b
LSD (p = 0.05)	5.9	7.3	3.8	6.3	3.5	4.5	5.4	3.7	4.8	3.2
Avg.	26.2	32.7 *	25.5	30.9 *	23.3	26.3	22.9	28.4 *	19.4	30.9 *

[x] SI = soil improving, SC = soil conserving, DS = disease suppressive, SQ = status quo, PP = continuous potato systems. [y] Values within columns for each yield type followed by the same letter are not significantly different from each other based on ANOVA and Fisher's protected LSD test ($p < 0.05$). [z] Mean values for irrigated treatments (Irr) followed by an asterisk are significantly greater than their corresponding non-irrigated (Non-irr) mean value within each year based on ANOVA and Fisher's protected LSD test.

Under irrigated conditions, yield (both total and marketable) in all cropping systems (except SI) increased compared to non-irrigated conditions. DS resulted in the numerically highest yields in most years, ranging from 36 to 45 Mg/ha and 29 to 36 Mg/ha for total and marketable yield, respectively, although values were generally comparable to SI or SC in individual years (Table 3). SI and SC increased total yield relative to PP in most years and SQ in some years. When averaged over all five years, again, all cropping systems increased yield relative to PP, but now DS resulted in significantly higher yields than all other systems, averaging increases of 19 and 26% over PP and 11 and 13% over SQ for total and marketable yield, respectively (Figure 1). The combined effect of irrigation and cropping system is realized when noting that DS with irrigation increased total and marketable yield by an average of 54 and 82%, respectively, over non-irrigated PP and by 42 and 54%, respectively, over non-irrigated SQ (Figure 1).

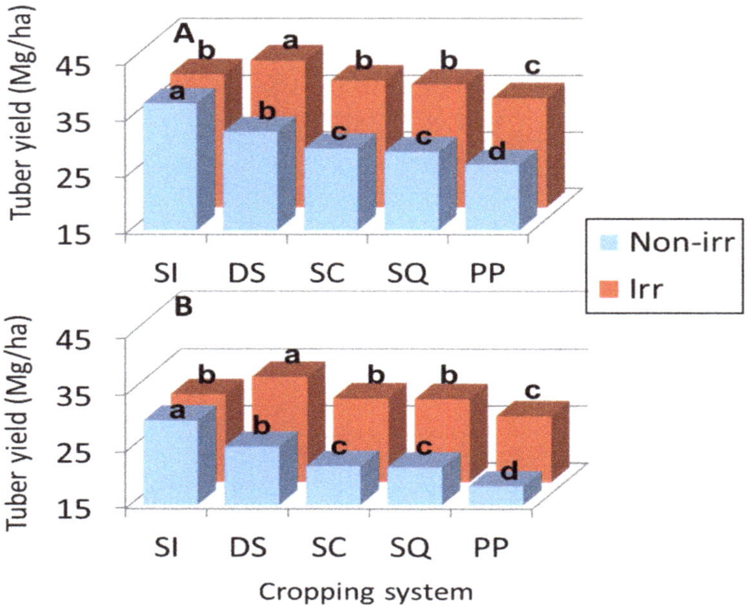

Figure 1. Effects of cropping system (SQ—standard rotation, SC—soil conserving, SI—soil improving, DS—disease suppressive, and PP—nonrotation control) and irrigation (Non-irr = Non-irrigated and Irr = Irrigated) on average (**A**) total and (**B**) marketable tuber yield over a 5 year period (2006–2010). Bars topped by the same letter within each irrigation regime are not significantly different from each other based on ANOVA and Fisher's protected LSD test ($p < 0.05$).

3.2. Tuber Size and Quality

On average, under non-irrigated conditions, SI resulted in the largest percentage of tubers in the large (228–342 g) and extra-large (>342 g) size classes, accounting for a combined 51% of all tubers compared to 29 to 35% in all other cropping systems, representing an increase of 60% relative to PP (Figure 2A). The largest individual size class for all but the SI system was the medium size class (115–227 g), accounting for 37 to 44% of the total (compared to 29% for SI). PP also resulted in the largest percentage of small-sized tubers (<114 g), comprising 31% of the total, compared to 26% for SC and SQ, and 20–22% for DS and SI systems. This resulted in PP having the lowest overall percentage of marketable tubers (comprising all size classes greater than small), at 68% for PP vs. 80 and 78% for SI and DS, and 74% for SC and SQ (Figure 2A).

Under irrigated conditions, the relative proportion of large and extra-large tubers increased for all cropping systems (except SI) relative to non-irrigated conditions, with all cropping systems still demonstrating significantly greater percentages of both size classes combined than PP (48 to 54% vs. 43% for PP) (Figure 2B). The large size class also constituted the largest individual size class for all cropping systems (36 to 39% of total). Once again, PP also resulted in the overall greatest proportion of small sized tubers, accounting for 23% of the total, which was significantly greater than the 18% in DS. Overall, percentage of marketable tubers was highest for DS (82%), significantly greater than PP, with the lowest percentage (77%), and in-between for the other cropping systems (79–81%) (Figure 2B).

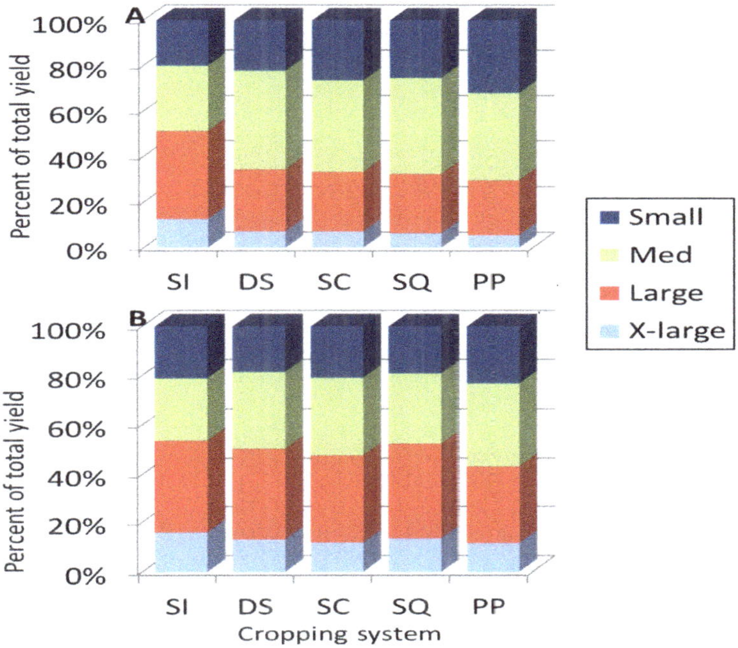

Figure 2. Effect of cropping system on average tuber size class distribution under (**A**) non-irrigated and (**B**) irrigated conditions as averaged over five cropping seasons (2006–2010).

Under non-irrigated conditions, the proportion of severely misshapen tubers was significantly lower for all cropping systems than the non-rotation control PP when averaged over all cropping years (9 to 14% vs. 20% for PP), but was not statistically different among cropping systems (Figure 3). Under irrigated conditions, DS maintained the overall lowest percentage of misshapen tubers (12.1%), significantly lower than SI and PP (20 to 22%), and SC and SQ were also lower than PP (at 15%) (Figure 3).

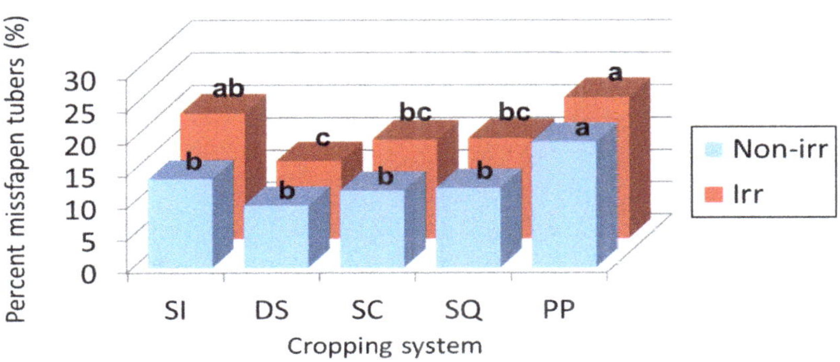

Figure 3. Effect of cropping system (SQ—standard rotation, SC—soil conserving, SI—soil improving, DS—disease suppressive, and PP—non-rotation control) and irrigation (Non—Non-irrigated and Irr—Irrigated) on average percentage of severely misshapen tubers as averaged over a 5 year period (2006–2010). Bars topped by the same letter within each irrigation regime are not significantly different from each other based on ANOVA and Fisher's protected LSD test ($p < 0.05$).

Tuber specific gravity varied somewhat from year to year, with generally lower values in 2010 and higher values in 2009 than other years. Although, generally, there were no significant effects due to irrigation, the interaction between irrigation and cropping system was significant ($p < 0.05$), so results are presented separately for irrigated and non-irrigated conditions. Under non-irrigated conditions, there was no effect of cropping system on specific gravity in 2007 and 2009, but in 2008 and 2010, SI resulted in lower specific gravity than all other cropping systems (Table 4). SC also showed lower specific gravity than PP and SQ in 2010. Over all years, SI and SC averaged slightly lower specific gravity than the other cropping systems. Under irrigated conditions, PP resulted in higher specific gravity than all other systems in 2008, 2009, and 2010, and both PP and SQ demonstrated higher specific gravity than SI in 2007. SI generally resulted in lower specific gravity than most cropping systems. Averaged over all years, PP resulted in higher specific gravity and SI lower specific gravity than all other cropping systems (Table 4).

Table 4. Effect of different cropping systems on tuber specific gravity over four field seasons (2007–2010) under irrigated (Irr) and non-irrigated (Non-irr) conditions.

	Tuber Specific Gravity									
	2007		2008		2009		2010		Mean	
Treatment [y]	Non-Irr	Irr	Non-Irr	Irr	Non-Irr	Irr	Non-Irr	Irr	Non-Irr	Irr
SI	1.083 a [z]	1.082 b	1.082 b	1.083 c	1.088 a	1.087 b	1.074 c	1.071 c	1.082 b	1.081 c
DS	1.085 a	1.086 ab	1.087 a	1.085 b	1.086 a	1.092 ab	1.079 ab	1.075 b	1.084 a	1.085 b
SC	1.082 a	1.086 ab	1.085 a	1.085 b	1.086 a	1.090 b	1.077 b	1.074 bc	1.082 b	1.084 b
SQ	1.085 a	1.087 a	1.086 a	1.084 c	1.087 a	1.089 b	1.080 a	1.075 b	1.084 a	1.084 b
PP	1.082 a	1.087 a	1.087 a	1.088 a	1.087 a	1.096 a	1.080 a	1.078 a	1.084 a	1.087 a
LSD	0.004	0.005	0.004	0.002	0.003	0.005	0.002	0.003	0.002	0.002

[y] SI = soil improving, SC = soil conserving, DS = disease suppressive, SQ = status quo, PP = continuous potato systems. [z] Values within columns followed by the same letter are not significantly different from each other based on ANOVA and Fisher's protected LSD test ($p < 0.05$).

3.3. *Crop Growth Assays*

3.3.1. Leaf Area Duration and Chlorophyll Content

Emergence was uniformly high across years and cropping systems (94 to 98%), with no significant effects due to irrigation or among cropping systems (data not shown). Leaf area index values collected over time within each cropping season were converted to leaf area duration (LAD) to provide an overall measure of leaf area and biomass production for each cropping system. Under non-irrigated conditions, SI produced significantly greater LAD in 2007–2009 than all other cropping systems, with both SI and DS producing higher LAD than others in 2006 (Table 5). In 2008, PP resulted in lower LAD than all other systems. Averaged over all years, SI resulted in higher LAD than all other systems, and DS in higher LAD than the remaining systems, with SI averaging 60 to 66% higher than SQ and PP values. Irrigation significantly increased LAD values in all years and in all cropping systems except SI. Under irrigated conditions, there were fewer differences among cropping systems, but SI resulted in greater LAD than SC and PP in 2006, and PP resulted in lower LAD than all other systems in 2008. When averaged over all years, SI and DS resulted in significantly higher LAD than SC and PP (by about 13%), and LAD for PP was also lower than for SQ (Table 5).

Table 5. Effect of different cropping systems on leaf area duration and chlorophyll content (SPAD assessment) over four field seasons (2006–2009) under irrigated (Irr) and non-irrigated (Non-irr) conditions.

	2006		2007		2008		2009		Mean (2006–2009)	
Treatment [x]	**Non-Irr**	**Irr**	**Non-Irr**	**Irr**	**Non-Irr**	**Irr**	**Non-Irr**	**Irr**	**Non-Irr**	**Irr**
Leaf area duration										
SI	194.1 a [y]	228.2 a	244.2 a	216.6 a	224.3 a	221.8 a	184.6 a	176.9 a	211.7 a	210.9 a
DS	166.5 a	204.2 ab	173.6 b	222.7 a	174.6 b	232.9 a	120.0 b	185.9 a	158.2 b	211.4 a
SC	117.2 b	181.2 b	149.1 b	190.6 a	152.1 c	216.9 a	115.8 b	170.5 a	133.5 c	189.8 bc
SQ	129.6 b	214.3 ab	141.0 b	213.1 a	152.2 c	237.3 a	108.1 b	158.7 a	132.7 c	205.8 ab
PP	123.2 b	178.5 b	169.5 b	226.4 a	118.0 d	187.4 b	98.7 b	156.0 a	127.3 c	187.1 c
LSD	42.6	46.4	30.9	42.9	18.9	28.9	41.6	31.2	14.8	18.0
Avg.	146.1	201.3 * [z]	175.5	214.8 *	163.8	219.3 *	125.4	169.6 *	152.6	201.1 *
Chlorophyll content (SPAD)										
SI	39.9 a	39.0 a	42.6 a	42.6 a	37.0 a	37.1 ab	38.8 a	40.4 ab	39.6 a	39.8 a
DS	39.5 a	38.2 a	39.9 b	41.8 ab	34.6 bc	36.0 bc	37.8 ab	40.2 ab	38.0 b	39.1 b
SC	38.4 a	38.1 a	38.5 b	41.7 ab	33.9 c	36.4 ab	37.2 b	40.3 ab	37.0 c	39.2 b
SQ	39.6 a	39.0 a	40.0 b	42.7 a	35.6 b	37.3 a	38.0 ab	40.7 a	38.3 b	39.9 a
PP	39.4 a	38.6 a	38.7 b	40.8 b	34.2 c	34.9 c	37.4 b	39.6 b	37.4 c	38.5 c
LSD	1.0	1.1	1.4	1.2	1.2	1.2	1.3	1.1	0.6	0.5
Avg.	39.4	38.6 *	39.9	41.9 *	35.1	36.4 *	37.8	40.2 *	38.0	39.3 *

[x] SI = soil improving, SC = soil conserving, DS = disease suppressive, SQ = status quo, PP = continuous potato systems. [y] Values within columns for each parameter followed by the same letter are not significantly different from each other based on ANOVA and Fisher's protected LSD test ($p < 0.05$). [z] Mean values for irrigated treatments followed by an asterisk are significantly greater than their corresponding non-irrigated mean value within each year based on ANOVA and Fisher's protected LSD test.

Leaf chlorophyll content as estimated by SPAD determinations was also affected by both cropping system and irrigation in all years but 2006 (Table 5). Under non-irrigated conditions, SI resulted in higher chlorophyll content than all other cropping systems in 2007 and 2008, and higher than SC and PP in 2009. Irrigation generally increased chlorophyll content across cropping systems. Under irrigated conditions, PP resulted in lower chlorophyll content than SI and SQ in 2007; SI, SC, and SQ in 2008; and SQ in 2009. Averaged over all years, SI and SQ resulted in higher chlorophyll content and lower PP content than all other cropping systems (Table 5).

3.3.2. Root, Shoot, and Tuber Biomass

Biomass of above- and below-ground plant parts, such as root, shoot, and tuber biomass, collected in August of each year demonstrated some differences due to both cropping system and irrigation, although there was no significant effect on root biomass or tuber biomass in 2007. Under non-irrigated conditions, SI resulted in greater root biomass than PP in 2008 and greater than all other cropping systems in 2009, as well as greater than all cropping systems when averaged over all three years (Table 6). DS also resulted in greater root biomass than PP over all years combined. Under irrigated conditions, root biomass increased in 2008 and 2009 overall, relative to no irrigation, as well as across all years when averaged together. SC resulted in greater root biomass than SQ and PP in 2008, as well as greater root biomass than PP when averaged over all years. Shoot biomass showed the greatest differences among cropping systems under non-irrigated conditions, with SI resulting in greater shoot biomass than all other cropping systems in all three years, as well as when averaged over all years (Table 6). Under irrigated conditions, SI resulted in greater shoot biomass than PP in 2008 and all other cropping systems in 2009, as well as greater shoot biomass than all other cropping systems when averaged over all years. PP also resulted in lower shoot biomass than SI, DS, and SC when averaged over all years (Table 6). Tuber biomass was only significantly affected by cropping system under irrigated conditions in 2008 and under non-irrigated conditions in 2009, with DS resulting in higher tuber biomass than SI and SQ in 2008 and PP resulting in higher biomass than SI in 2009 (Table 6). Averaged over all three years, DS resulted in higher tuber biomass than SI under both irrigated and non-irrigated conditions, and higher tuber biomass than SQ under irrigation.

Table 6. Effect of different cropping systems on potato plant root and shoot biomass (dry wt) over three field seasons (2007–2009) under irrigated and non-irrigated conditions.

	Biomass (Mg dry wt/ha)							
	2007		2008		2009		Mean (2007–2009)	
System [y]	**Non-Irr**	**Irrigated**	**Non-Irr**	**Irrigated**	**Non-Irr**	**Irrigated**	**Non-Irr**	**Irrigated**
Root Biomass								
SI	0.136 a [z]	0.106 a	0.168 a	0.198 ab	0.188 a	0.190 a	0.164 a	0.165 ab
DS	0.118 a	0.098 a	0.140 ab	0.211 ab	0.134 b	0.176 a	0.131 b	0.161 ab
SC	0.104 a	0.120 a	0.130 ab	0.236 a	0.138 b	0.180 a	0.124 bc	0.179 a
SQ	0.086 a	0.090 a	0.124 ab	0.168 b	0.136 b	0.194 a	0.115 bc	0.151 ab
PP	0.076 a	0.108 a	0.100 b	0.144 b	0.136 b	0.166 a	0.104 c	0.139 b
LSD ($p = 0.05$)	0.056	0.042	0.051	0.064	0.025	0.041	0.024	0.027
Shoot biomass								
SI	3.37 a	3.56 a	3.36 a	3.71 a	3.62 a	4.00 a	3.44 a	3.76 a
DS	2.55 b	3.45 a	2.03 c	2.73 ab	1.98 b	2.93 b	2.19 b	3.04 b
SC	2.16 b	3.25 a	1.96 c	3.01 ab	1.88 b	2.85 b	1.97 b	3.04 b
SQ	2.26 b	3.21 a	2.64 b	2.79 ab	1.76 b	2.43 b	2.26 b	2.81 bc
PP	1.86 b	3.06 a	1.89 c	2.34 b	2.12 b	2.30 b	1.96 b	2.56 c
LSD ($p = 0.05$)	0.68	0.91	0.49	0.98	0.44	0.67	0.33	0.41
Tuber biomass								
SI	3.19 a	3.62 a	2.44 a	1.88 bc	3.35 b	3.85 a	2.99 b	3.12 b
DS	3.58 a	4.00 a	3.06 a	3.11 a	4.51 ab	4.87 a	3.72 a	3.99 a
SC	3.69 a	3.75 a	3.14 a	2.80 ab	4.03 ab	4.43 a	3.62 ab	3.66 ab
SQ	3.07 a	4.24 a	2.81 a	1.79 c	3.92 ab	3.75 a	3.26 ab	3.26 b
PP	3.53 a	3.64 a	2.72 a	2.41 abc	4.63 a	5.02 a	3.63 ab	3.69 ab
LSD ($p = 0.05$)	1.08	1.71	1.04	0.91	1.10	1.32	0.65	0.59

[y] SI = soil improving, SC = soil conserving, DS = disease suppressive, SQ = status quo, PP = continuous potato systems. [z] Values within columns for each parameter followed by the same letter are not significantly different from each other based on ANOVA and Fisher's protected LSD test ($p < 0.05$).

3.3.3. Plant Tissue Elemental Analyses

Shoot tissue concentrations were significantly affected by cropping system for all nutrient elements measured (except Boron), but were generally not affected by irrigation (and no irrigation by cropping system interaction), except for slight increases in P and B, and a decrease in Zn observed in irrigated vs. non-irrigated systems (data not shown). SI resulted in higher shoot tissue concentrations of P and K, and SI and SQ for N, than all other cropping systems, whereas PP tended to have the lowest concentrations of N and P (Table 7). However, PP and SC tended to have higher and SI lower shoot tissue concentrations of Ca and Mg. For Mn, DS averaged the highest and SI the lowest shoot tissue concentration, and for the metals Fe, Al, Cu, and Zn, SI averaged lower and PP higher concentrations than most other cropping systems (Table 7).

Overall, cropping system and irrigation did not significantly affect root tissue nutrient concentrations for all elements measured, except for a slight increase in P for SI relative to the other cropping systems, and higher K, Mg, and Zn levels in non-irrigated vs. irrigated systems (data not shown). For tuber concentrations, irrigation only significantly affected N, Mg, and B concentrations, with irrigation resulting in slightly lower N and Mg concentrations, and higher B concentration than non-irrigated (data not shown). Overall, tuber tissue concentrations were substantially lower than shoot tissue concentrations for all elements, except for P, which were higher in tuber tissue.

Table 7. Effect of different cropping systems on shoot tissue elemental composition (2008 data).

Treatment [y]	Elemental Composition									
	N	P	K	Ca	Mg	Mn	Fe	B	Cu	Zn
	%					ppm				
Shoot tissue										
SI	3.95 a [z]	0.235 a	5.47 a	1.21 c	0.834 c	262.9 c	165.8 b	27.1 a	7.77 b	113.4 b
DS	3.58 b	0.185 cd	4.67 b	1.24 bc	0.893 bc	422.1 a	199.0 ab	25.4 a	8.88 b	127.0 ab
SC	3.53 bc	0.198 bc	4.68 b	1.25 bc	0.861 bc	359.9 b	181.8 b	25.2 a	9.94 ab	129.8 a
SQ	3.83 a	0.199 b	4.32 b	1.35 a	0.929 ab	389.6 ab	240.0 a	26.2 a	9.25 ab	128.8 ab
PP	3.32 c	0.181 d	4.39 b	1.32 ab	0.949 a	355.5 b	207.0 ab	24.9 a	11.27 a	130.9 a
LSD	0.23	0.014	0.38	0.088	0.075	54.0	50.0	2.6	2.05	14.9
Tuber tissue										
SI	1.74 a	0.260 a	2.49 a	0.043 a	0.107 ab	17.2 ab	19.5 ab	5.50 a	8.00 a	17.5 a
DS	1.66 a	0.210 d	2.24 b	0.032 c	0.105 ab	19.6 a	24.1 a	4.76 a	8.14 a	15.1 c
SC	1.70 a	0.246 ab	2.29 b	0.033 c	0.104 ab	18.6 ab	18.9 b	4.97 a	8.44 a	16.1 bc
SQ	1.67 a	0.222 cd	2.15 b	0.038 b	0.102 b	18.7 ab	18.2 b	5.18 a	7.98 a	15.4 c
PP	1.72 a	0.230 bc	2.24 b	0.035bc	0.111 a	15.1 b	19.6 ab	4.86 a	8.32 a	16.7ab
LSD	0.09	0.017	0.13	0.004	0.006	3.4	4.7	0.75	0.49	1.0

[y] SI = soil improving, SC = soil conserving, DS = disease suppressive, SQ = status quo, PP = continuous potato systems. [z] Values within columns for each parameter followed by the same letter are not significantly different from each other based on ANOVA and Fisher's protected LSD test ($p < 0.05$).

Cropping system effects on tuber tissue concentration included higher concentrations of P, K, and Ca for SI than all other cropping systems, although there was no effect on N among cropping systems (Table 7). DS averaged higher tuber tissue concentrations of Mn than PP and higher concentrations of Fe and Al than SC and SQ. SI registered the highest tuber concentration of Zn and PP the highest concentration of Mg among cropping systems.

3.4. Parameter Correlations

Total and marketable tuber yields were highly correlated ($p < 0.001$) with crop growth parameters such as LAD, chlorophyll content, shoot biomass, and total biomass, and weakly correlated ($p < 0.05$) with the soil physical/chemical parameter soil C/N ratio across all samples (irrigated and non-irrigated conditions) and all years. Total and marketable yields were also correlated ($p < 0.05$) with soil moisture, total soil C, soil N, and soil ammonium (NH_4) concentration sampled in spring under non-irrigated, but not under irrigated, conditions. Marketable yield was also correlated with potentially mineralizable N and negatively correlated with bulk density under non-irrigated, but not under irrigated, conditions. The percentage of large and extra-large tubers was correlated with LAD, chlorophyll content, root and shoot biomass, soil moisture, aggregate stability, active C, soil C and N, POM C and N, potentially mineralizable N, and NO_3 and NH_4 concentrations, and negatively correlated with bulk density under both irrigated and non-irrigated conditions across all years.

4. Discussion

In this research, multiple individual soil health management practices were combined into cropping systems with specific management goals of soil conservation, soil improvement, and disease suppression, and effects on crop growth, nutrition, and tuber yield and quality were assessed over up to five full cropping seasons (and after cropping systems in place for 3 to 7 years) under both non-irrigated (rainfed) and irrigated conditions. Combined data from all five seasons demonstrated that cropping system significantly affected virtually all of the crop and plant characteristics measured, ranging from crop growth (photosynthetic potential and biomass) and tissue nutrient concentration to potato tuber yield and quality, with the soil improving (SI) system, which included compost amendments, cover crops, and reduced tillage in a 3 y rotation, producing the greatest overall effects and improvements in these crop production parameters, particularly without irrigation. Irrigation effects were also significant for most parameters. This research demonstrated that improved cropping systems can substantially enhance characteristics associated with potato crop productivity.

Concurrent research on these same cropping systems over the same years documented the system effects on soil physical, chemical, and biological properties, and that effects tended to increase over time [24,25]. In these studies, all rotations increased aggregate stability, water availability, microbial biomass C, and total C and N compared to no rotation (PP), and the 3 y systems (SI, SC, DS) increased aggregate stability relative to the 2 y system (SQ). Additionally, the 3 y systems with reduced tillage (SI and SC) increased water availability and reduced bulk density relative to the other systems. However, the SI system resulted in greater increases in total and particulate organic matter (POM) C and N; active C; microbial biomass C; water availability; CEC; concentrations of P, K, Ca, Mg, and S; and lower bulk density than all other cropping systems [24,25]. SI was also shown to increase microbial activity and greatly affect soil microbial community characteristics, whereas PP showed the lowest microbial activity, with the others in between [20,21]. These changes all constitute parameters associated with improved soil health.

In the present study, under non-irrigated conditions, all crop rotations increased total and marketable tuber yields over no rotation (PP), but the SI system resulted in the highest tuber yield of all systems (both total and marketable), averaging 30 to 40% higher than SQ and PP systems over all years. Yield differences were greatest in the drier years (2007 and 2010), when SI yields were 40–90% higher than SQ and PP. In addition, SI resulted in the highest percentage of large and extra-large size-class tubers, and fewer small or under-sized tubers. It is also noteworthy that with irrigation, all cropping systems, with the exception of SI, produced substantially higher yields than their non-irrigated counterpart, with total and marketable yields averaging 27 and 37% higher, respectively, demonstrating that only SI produced comparable (and high) yields under both irrigated and non-irrigated conditions. These yield effects indicate the importance of adequate soil water in potato production and, as has also been previously demonstrated, that in most years, supplemental irrigation is needed in Maine to increase productivity [29,31]. However, the data also strongly suggest that the yield increases observed in SI are related to soil health improvements associated with increased water-holding capacity and plant-available water. Thus, the improvements in soil characteristics, and particularly the increased organic matter and ability to store and hold available water provided by the compost amendments, apparently enabled SI to produce higher yields when not irrigated than all other cropping systems. Essentially, under these conditions, the improvements resulting from the compost amendments -effectively substituted for water additions through irrigation in these studies. This aspect was noted and explored in previous research examining the economics of potentially using compost amendments as an alternative to irrigation [34]. In other research, compost amendments have been shown to provide similar increases in organic matter, water availability, various soil quality parameters, and generally higher tuber yields [5,14,35,36], although in some cases, tuber yields were not significantly increased with compost amendments even when there was substantial improvement in soil quality parameters [37,38]. Organic matter amendments have been shown to improve soil structural stability primarily through increases in aggregate stability, as well as improvements in bulk density, aeration, porosity, and water movement [39–42].

There are many aspects and changes in soil characteristics involved with the compost amendments and other factors within the SI system, and thus the specific cause of the yield increases observed cannot be conclusively determined. However, based on the results observed and the characteristics of these systems, it is apparent that improvements in properties associated with the ability to store and hold soil water were at least partially, if not primarily, responsible for the yield increases observed in the SI system, rather than such aspects as nutritional improvements. First, all systems were supplied with adequate (above optimal NPK) fertilization so as not to limit productivity based on numerous studies in this area [29–33]. Additional NPK fertility above these levels provided by the compost amendments would not be expected to increase yield further, as studies have indicated depressed yields, not increased yields, with above optimal fertilizer additions [29–33]. Most importantly, SI had higher yields than other systems when not irrigated, but no improvement with irrigation, even though all other systems showed increased yields when irrigated. If yield increases were primarily related to increased nutrition, then we would expect to observe an irrigation effect, as with all other systems, but in SI, comparable yields were produced under irrigated and non-irrigated conditions. Additionally, under irrigated conditions, SI would be expected to produce higher yields than the other systems, but again, this was not observed, thus further indicating that added nutrition was not the primary cause of increased yields with SI. Overall, differences in water availability appeared to explain a substantial part of the cropping system yield differences observed. Under irrigated conditions, DS produced the highest overall yields, while PP still resulted in lower yields than all other cropping systems. Interestingly, these irrigation effects on yield were observed even in 2009, a year in which no irrigation treatments were applied (as were not needed), yet effects were still observed, possibly a result of cumulative beneficial effects due to a history of previous irrigation.

The DS system also resulted in overall significant increases in total and marketable yield relative to SC, SQ, and PP under non-irrigated conditions. These increases were presumably due to the beneficial effects of the added green manure and cover crops in reducing potential pathogens and soilborne diseases, and maintaining various soil health parameters, as has been observed in other potato systems [12,43,44]. As previously reported, DS resulted in lower incidence and severity of multiple soilborne diseases (including stem canker, black scurf, and common scab), as well as significant effects on soil microbial community characteristics, but more modest effects on soil chemical and biological parameters [20,21]. The SC system, however, despite increased rotation length, use of cover crops, and reduced tillage, resulted in comparable tuber yield to the standard 2 y SQ rotation through most years of this study, although it did show indications of higher yields than SQ in the later years (following the second full rotation cycle). Other researchers have also noted that significant effects due to increased rotation length and cover crops alone may take several years to develop [6,45,46], and this was also indicated in the overall comparable soil properties observed for SC and SQ through the early years of the study [24,25].

Overall, average yield values for DS, SC, and SQ under non-irrigated conditions were comparable to average state-wide values for commercial production in Maine for this period (~32 Mg/ha, 2006–2010) [47], whereas SI averaged higher, and PP lower than average, as the majority of commercial production in Maine is not irrigated. However, under irrigated conditions, all cropping systems resulted in yields above the state-wide averages (by 8 to 28%).

Specific gravity is an important quality characteristic for processing potatoes, as it represents the dry matter content of tubers. Higher specific gravity means higher dry matter content, which produces lighter color, absorbs less oil, and requires fewer tubers and less time to produce the same yield of finished product (thus less costly to produce) [48]. Specific gravity varied somewhat among cropping systems, with PP resulting in the overall highest values, and SI resulting in lower values. However, acceptable specific gravity values for Russet Burbank of 1.082 or higher (representing total solids content of >21.5%) were observed for all cropping systems in all years except 2010, which was the warmest and driest summer, with higher than normal temperatures and lower than normal rainfall observed throughout July, August, and September. High temperatures and water stress are known to depress specific gravities, as well as excessive water and/or fertilization [49,50]. The slightly lower specific gravity observed in SI is probably due to the higher organic matter content and lower bulk density of those soils, as organic amendments may also reduce specific gravity [36]. Although there was no overall irrigation effect on specific gravity, there was a significant interaction between cropping system and irrigation, and it appears that for most of the systems, there was a slight increase in specific gravity associated with irrigation for most systems, but a slight decrease in SI, resulting in no overall effect.

Under non-irrigated conditions, SI also resulted in the greatest photosynthetic potential, as represented by the leaf area index, leaf area index over time (leaf area duration—LAD), and leaf chlorophyll content. Yields are closely associated with the ability of a plant to intercept solar radiation and its efficiency in accumulating dry matter. LAD has been shown to be more closely related to yield than LAI and other indicators of leaf area [51]. SPAD readings are closely related to actual chlorophyll content and have been used as an indicator of leaf N content, but recent research indicates that the relationship between SPAD readings and Leaf N content can be greatly affected by environmental conditions and crop species [52]. SI also resulted in greater overall root and shoot biomass than the other cropping systems, demonstrating the impact of the improved soil quality parameters for SI on all aspects of crop growth dynamics. Previous research has also demonstrated that large additions of organic matter can dramatically affect these growth parameters [36,53]. Surprisingly, however, SI resulted in overall lower tuber biomass than DS, but it must be taken into account that the biomass measurements were made in early August, when tubers were first developing, and do not represent any potential effects on yield. Although DS resulted in overall greater LAD than the remaining systems, and greater chlorophyll content than SQ and PP, there were fewer differences among the other cropping systems for biomass measurements under non-irrigated conditions, although PP generally resulted in lower values for most parameters. Irrigation resulted in overall increases in LAD and chlorophyll content for all systems, but irrigation effects on biomass were inconsistent. Although averages over all three years of biomass data indicated overall increases due to irrigation, individual years varied. There were some differences among cropping systems overall, including greater root biomass in SC than PP, greater shoot biomass in SI than all systems, and greater tuber biomass in DS than SI and SQ, but again, effects were variable between years. These differences reflect the generally more favorable conditions for biomass growth under irrigated conditions.

SI also resulted in generally higher levels of N, P, and K in above-ground shoot and tuber tissues, but lower concentrations of Ca, Mg, and Mn in shoot tissue, as well as generally lower concentrations of Fe, Cu, and Zn, relative to most other cropping systems. This observation indicated that SI management did not always increase the levels of these micronutrients, even though soil levels of these nutrients were generally increased in SI [24,25]. However, this observation was consistent with studies of other cropping systems amended by organic fertilizers. For example, application with poultry litter resulted in a greater concentration of extractable soil P, K, Ca, Mg, Cu, Zn, and Na. However, these increases did not always result in greater concentrations of these elements in cotton plant parts [54,55]. SQ and PP tended to have higher Ca and Mg concentrations in shoot tissue, as well as higher Fe, Cu, and Zn, than most other systems. Overall, tissue concentrations for all major nutrient elements (N, P, K, Ca, Mg) for all cropping systems were within the normal (sufficient level) ranges previously observed and reported for potato leaf and tuber tissues [56,57].

Although only one potato variety, Russet Burbank, which is the predominant processing variety grown commercially in the northeast, was used in this study, observed results should be generally applicable to other potato varieties as well. Previous studies in this region have indicated similar responses to rotations, amendments, and fertilization in multiple different potato varieties [29–32,36], and improvements in soil health have been associated with increases in yield across not only different potato varieties but many different crops as well [4,6,12–15,44].

Overall, this study demonstrated that incorporating soil health management practices into integrated cropping systems can greatly affect crop growth and productivity parameters, and can be used to improve crop growth and yield, in addition to benefits in soil health and other soil properties. The integration of practices such as extending crop rotations, use of cover crops and green manures, reduced tillage, and, particularly, organic amendments, into existing, modified, and enhanced potato cropping systems may provide the basis for greater sustainability and productivity in potato production systems. This study also demonstrated that development of improved cropping systems can substantially enhance productivity from the standard cropping system currently used throughout Northeastern US for potato production. The SI system, which incorporated large organic amendments along with a longer rotation period, use of cover crops, and reduced tillage, resulted in substantial effects and improvements in crop growth and yield, particularly under non-irrigated conditions. Organic matter affects and influences many different soil physical, chemical, and biological properties in various ways, and has often been cited as the single most important aspect of soil health [4,58]. Characterization of the water-extractable organic matter samples within the cropping systems suggested that these management practices stimulated the decomposition of the humic fraction in the soil organic matter pool, implying healthier soil conditions with these practices than in continuous potato growth [59]. The current study further revealed that large organic matter amendments had the most immediate and substantial effects of all the cropping systems. This research also emphasized the importance of soil water, and that under normal environmental conditions during cropping seasons in Maine, irrigation provides a definite yield benefit under most cropping systems, but also that one of the benefits of the large organic amendments in SI was that it could be an effective substitute for irrigation and produce high yields without irrigation, at least under the conditions occurring during this study. Although SI and the effects of organic amendments appeared to provide the most substantial effects, the DS system, which included disease-suppressive rotation crops, green manures, crop diversity, and increased tillage (for incorporation of cover crops and green manures), also resulted in high yields (highest under irrigation) throughout, demonstrating the impact of reduced disease levels and other benefits provided by green manure crops and crop diversity. However, for the relatively short duration of this study, a small increase in rotation length, cover crops, and reduced tillage, as provided in the SC system was not sufficient to produce an overall increase in yield and growth parameters relative to the standard 2 y rotation, although by the second rotation cycle, indications of higher SC yields were evident. Additional time, or more aggressive changes, appear to be necessary to achieve enhanced productivity in this system. However, all these approaches still may provide some benefits in contributing to the overall goals of maintaining and/or improving soil health. Research is continuing to integrate the principles of these systems into more productive and economically viable enhanced cropping systems for growers in the northeast and elsewhere.

Author Contributions: Project conceptualization, methodology, and investigation, R.P.L., C.W.H., T.S.G., O.M.O. and Z.H.; formal analysis and writing original draft, R.P.L.; project administration and supervision, C.W.H.; reviewing and editing by all authors. All authors have read and agreed to the published version of the manuscript.

Funding: This research received no external funding.

Institutional Review Board Statement: Not applicable.

Informed Consent Statement: Not applicable.

Data Availability Statement: The data presented in this study are available upon request from the corresponding author.

Acknowledgments: Many people have worked on or contributed to this research in various ways over the years of these studies, including summer workers and student aides, field crews, technicians, and colleagues. Most importantly, we would like to thank D. Cowperthwaite for managing and maintaining the field site, as well as J. Hunt, P. Pinette, E Champaco, L. Matthiesen, G. Trusty, E. Mallory, B. LeGasse, and D. Torrey for additional technical support.

Conflicts of Interest: The authors declare no conflict of interest.

References

1. Doran, J.W.; Sarrantonio, M.; Leibig, M. Soil health and sustainability. In *Advances in Agronomy*; Sparks, D.L., Ed.; Academic Press San Diego, CA, USA, 1996; pp. 1–54.
2. Doran, J.W.; Zeiss, M.R. Soil health and sustainability: Managing the biotic component of soil quality. *Appl. Soil Ecol.* **2000** *15*, 3–11. [CrossRef]
3. Larkin, R.P. Soil health paradigms and implications for disease management. *Annu. Rev. Phytopathol.* **2015**, *53*, 19–221. [CrossRef] [PubMed]
4. Magdoff, F.; van Es, H. *Building Soils for Better Crops*, 3rd ed.; Sustainable Agriculture Research and Education: Waldorf, MD USA, 2009.
5. Grandy, A.S.; Porter, G.A.; Erich, M.S. Organic amendment and rotation crop effects on the recovery of soil organic matter and aggregation in potato cropping systems. *Soil Sci. Soc. Am. J.* **2002**, *66*, 1311–1319. [CrossRef]
6. Carter, M.R.; Sanderson, J.B. Influence of conservation tillage and rotation length on potato productivity, tuber disease and soil quality parameters on a fine sandy loam in eastern Canada. *Soil Tillage Res.* **2001**, *63*, 1–13. [CrossRef]
7. Mohr, R.M.; Volkmar, K.; Derksen, D.A.; Irvine, R.B.; Khakbazan, M.; McLaren, D.L.; Monreal, M.A.; Moulin, A.P.; Tomasiewicz D.J. Effect of rotation on crop yield and quality in an irrigated potato system. *Am. J. Potato Res.* **2011**, *88*, 346–359. [CrossRef]
8. Peters, R.D.; Sturz, A.V.; Carter, M.R.; Sanderson, J.B. Influence of crop rotation and conservation tillage practices on the severity of soil-borne potato diseases in temperate humid agriculture. *Can. J. Soil Sci.* **2004**, *84*, 397–402. [CrossRef]
9. Scholte, K. Causes of differences in growth pattern, yield and quality of potatoes (*Solanum tuberosum* L.) in short rotations on sandy soil as affected by crop rotation, cultivar, and application of granular nematicides. *Potato Res.* **1990**, *33*, 181–190. [CrossRef]
10. Wright, P.J.; Falloon, R.E.; Hedderly, D. A long-term vegetable crop rotation study to determine effects on soil microbial communities and soilborne diseases of potato and onion. *N. Z. J. Crop Hortic. Sci.* **2017**, *45*, 29–54. [CrossRef]
11. Essah, S.Y.C.; Delgado, J.A.; Dillon, M.; Sparks, R. Cover crops can improve potato tuber yield and quality. *HortTechnology* **2012** *22*, 185–190. [CrossRef]
12. N'Dayegamiye, A.; Nyiraneza, J.; Grenier, M.; Bippfubusa, M.; Drapeau, A. The benefits of crop rotation including cereals and green manure on potato yield and nitrogen nutrition and soil properties. *Adv. Crop Sci. Tech.* **2017**, *5*, 279. [CrossRef]
13. Bernard, E.; Larkin, R.P.; Tavantzis, S.; Erich, M.S.; Alyokhin, A.; Gross, S. Rapeseed rotation, compost, and biocontrol amendments reduce soilborne diseases and increase tuber yield in organic and conventional potato production systems. *Plant Soil* **2014** *374*, 611–627. [CrossRef]
14. Blanchet, G.; Gavazov, K.; Bragazza, L.; Sinaj, S. Responses of soil properties and crop yields to different inorganic and organic amendments in a Swiss conventional farming system. *Agric. Ecosyst. Environ.* **2016**, *230*, 116–126. [CrossRef]
15. Ninh, H.T.; Grandy, A.S.; Wickings, K.; Snapp, S.S.; Kirk, W.; Hao, J. Organic amendment effects on potato productivity and quality are related to soil microbial activity. *Plant Soil* **2015**, *386*, 223–236. [CrossRef]
16. Carter, M.R.; Noronha, C.; Peters, R.D.; Kimpinski, J. Influence of conservation tillage and crop rotation on the resilience of an intensive long-term potato cropping system. *Agric. Ecosyst. Environ.* **2009**, *133*, 32–39. [CrossRef]
17. Larkin, R.P.; Griffin, T.S.; Honeycutt, C.W. Rotation and cover crop effects on soilborne potato diseases, tuber yield, and soil microbial communities. *Plant Dis.* **2010**, *94*, 1491–1502. [CrossRef]
18. Larkin, R.P.; Halloran, J.M. Management effects of disease-suppressive rotation crops on potato yield and soilborne disease and their economic implications in potato production. *Am. J. Potato Res.* **2014**, *91*, 429–439. [CrossRef]
19. Larkin, R.P.; Honeycutt, C.W. Effects of different 3-year cropping systems on soil microbial communities and Rhizoctonia diseases of potato. *Phytopathology* **2006**, *96*, 68–79. [CrossRef]
20. Larkin, R.P.; Honeycutt, C.W.; Griffin, T.S.; Olanya, O.M.; Halloran, J.M.; He, Z. Effects of different potato cropping system approaches and water management on soilborne diseases and soil microbial communities. *Phytopathology* **2011**, *101*, 58–67 [CrossRef]
21. Larkin, R.P.; Honeycutt, C.W.; Griffin, T.S.; Olanya, O.M.; He, Z.; Halloran, J.M. Cumulative and residual effects of different potato cropping system management strategies on soilborne diseases and soil microbial communities over time. *Plant Pathol.* **2017**, *66*, 437–449. [CrossRef]
22. Olanya, O.M.; Honeycutt, C.W.; He, Z.; Larkin, R.P.; Halloran, J.M.; Frantz, J.M. Early and late blight potential on Russet Burbank potato as affected by microclimate, cropping systems, and irrigation management in Northeastern United States. In *Sustainable Potato Production: Global Case Studies*; He, Z., Larkin, R.P., Honeycutt, C.W., Eds.; Springer: Amsterdam, The Netherlands, 2012 pp. 43–60.
23. Olanya, O.M.; Larkin, R.P.; Halloran, J.M.; He, Z. Relationships of crop and soil management systems to meteorological variables and potato diseases on a Russet Burbank cultivar. *J. Agric. Meteorol.* **2014**, *70*, 91–104. [CrossRef]
24. Larkin, R.P. Cumulative and residual effects of potato cropping system management strategies on soil physical, chemical, and biological properties. *Phytopathology* **2015**, *105*, S4.
25. Larkin, R.P.; Griffin, T.S.; Honeycutt, C.W.; Olanya, O.M.; He, Z. Potato cropping system management strategy impacts soil physical, chemical, and biological properties over time. *Soil Tillage Res.* **2021**. submitted.

26. He, Z.; Honeycutt, C.W.; Griffin, T.S.; Larkin, R.P.; Olanya, O.M.; Halloran, J.M. Increases in phosphatase and urease activities in potato fields by cropping rotation practices. *J. Food Agric. Environ.* **2010**, *8*, 1112–1117 [CrossRef]
27. He, Z.; Honeycutt, C.W.; Olanya, O.M.; Larkin, R.P.; Halloran, J.M. Soil test phosphorus and microbial biomass phosphorus in potato fields. *J. Food Agric. Environ.* **2011**, *9*, 540–545. [CrossRef]
28. He, Z.; Honeycutt, C.W.; Olanya, O.M.; Larkin, R.P.; Halloran, J.M.; Frantz, J.M. Comparison of phosphorus status and organic matter composition in potato fields with different crop rotation systems. In *Sustainable Potato Production: Global Case Studies*; He, Z., Larkin, R.P., Honeycutt, C.W., Eds.; Springer: Amsterdam, The Netherlands, 2012; pp. 61–79.
29. Belanger, G.; Walsh, J.R.; Richards, J.E.; Milburn, P.H.; Ziadi, N. Yield response of two potato cultivars to supplemental irrigation and N fertilization in New Brunswick. *Am. J. Potato Res.* **2000**, *77*, 11–21. [CrossRef]
30. Honeycutt, C.W.; Clapham, W.M.; Leach, S.S. Crop rotation and N fertilization effects on growth, yield, and disease incidence in potato. *Am. Potato J.* **1996**, *73*, 45–61. [CrossRef]
31. Porter, G.A.; Sisson, J.A. Response of Russet Burbank and Shepody potatoes to nitrogen fertilizer in two cropping systems. *Am. Potato J.* **1991**, *68*, 428–463. [CrossRef]
32. Zaeen, A.A.; Sharma, L.K.; Jasim, A.; Bali, S.; Buzza, A.; Alyokhin, A. Yield and quality of three potato cultivars under series of nitrogen rates. *Agrosystems Geosci. Environ.* **2020**, *3*, e20062. [CrossRef]
33. Zebarth, B.J.; Bélanger, G.; Cambouris, A.N.; Ziadi, N. Nitrogen fertilization strategies in relation to potato tuber yield, quality, and crop N recovery. In *Sustainable Potato Production: Global Case Studies*; He, Z., Larkin, R.P., Honeycutt, C.W., Eds.; Springer: Amsterdam, The Netherlands, 2012; pp. 165–186.
34. Halloran, J.M.; Larkin, R.P.; DeFauw, S.L.; Olanya, O.M.; He, Z. Economic potential of compost amendment as an alternative to irrigation in Maine potato production systems. *Am. J. Plant Sci.* **2013**, *4*, 238–245. [CrossRef]
35. Griffin, T.S.; Porter, G.A. Altering soil carbon and nitrogen stocks in intensively tilled two-year rotations. *Biol. Fertil. Soils* **2004**, *39*, 366–374.
36. Porter, G.A.; Opena, G.B.; Bradbury, W.B.; McBurnie, J.C.; Sisson, J.A. Soil management and supplemental irrigation effects on potato: I. Soil properties, tuber yield, and quality. *Agron. J.* **1999**, *91*, 416–425. [CrossRef]
37. Wilson, C.; Zebarth, B.J.; Burton, D.L.; Goyer, C. Short-term effects of diverse compost products on soil quality in potato production. *Soil Sci. Soc. Am. J.* **2018**, *82*, 889–900. [CrossRef]
38. Wilson, C.; Zebarth, B.J.; Burton, D.L.; Goyer, C.; Moreau, G.; Dixon, T. Effect of diverse compost products on potato yield and nutrient availablity. *Am. J. Potato Res.* **2019**, *96*, 272–284. [CrossRef]
39. Abiven, S.; Menasseri, S.; Chenu, C. The effects of organic inputs over time on soil aggregate stability. *Soil Biol. Biochem.* **2008**, *41*, 1–12. [CrossRef]
40. Albaich, R.; Canet, R.; Pomares, F.; Ingelmo, F. Organic matter components and aggregate stability after application of different amendments to a horticultural soil. *Bioresour. Technol.* **2001**, *76*, 125–129. [CrossRef]
41. Diacono, M.; Montemurro, F. Long-term effects of organic amendments on soil fertility. A review. *Agron. Sustain. Dev.* **2010**, *30*, 401–422. [CrossRef]
42. He, Z.; Tazisong, I.A.; Senwo, Z.N.; Zhang, D. Soil properties and macro cations impacted by long-term applied poultry litter. *Commun. Soil Sci. Plant Anal.* **2008**, *39*, 858–872. [CrossRef]
43. Sexton, P.; Plant, A.; Johnson, S.B. Effect of a mustard green manure on potato yield and disease incidence in a rainfed environment. *Crop Manag.* **2007**. [CrossRef]
44. Sharifi, M.; Lynch, D.H.; Hammermeister, A.; Burton, D.L.; Messiga, A.J. Effect of green manure and supplemental fertility amendments on selected soil quality parameters in an organic potato rotation in Eastern Canada. *Nutr. Cycl. Agroecosyst.* **2014**, *100*, 135–146. [CrossRef]
45. Carter, M.R.; Holmstrom, D.; Sanderson, J.B.; Ivany, J.A.; DeHaan, R. Comparison of conservation with conventional tillage for potato production in Atlantic Canada: Crop productivity, soil physical prcerties and weed control. *Can. J. Soil Sci.* **2005**, *85*, 453–461. [CrossRef]
46. Griffin, T.S.; Larkin, R.P.; Honeycutt, C.W. Delayed tillage and cover crop effects on potato systems. *Am. J. Potato Res.* **2009**, *86*, *79–87*. [CrossRef]
47. National Agricultural Statistics Service (NASS). New England Agricultural Statistics 2011. Available online: https://www.nass.usda.gov/Statistics_by_State/New_England_includes/Publications/Annual_Statistical_Bulletin/2011/fallpots2011.pdf (accessed on 20 August 2020).
48. Lulai, E.C.; Orr, P.H. Influence of specific gravity on yield and oil content of chips. *Am. Potato J.* **1977**, *56*, 379–389. [CrossRef]
49. Eldredge, E.P.; Holmes, Z.A.; Mosley, A.R.; Shock, C.C.; Stieber, T.D. Effects of transitory water stress on potato tuber stem-end reducing sugar and fry color. *Am. J. Potato Res.* **1996**, *73*, 517–530. [CrossRef]
50. Kincaid, D.C.; Westermann, D.T.; Trout, T.J. Irrigation and soil temperature effects on Russet Burbank quality. *Am. J. Potato Res.* **1993**, *70*, 711–723. [CrossRef]
51. Boyd, N.S.; Gordon, R.; Martin, R.C. Relationship between leaf area index and ground cover in potato under different management conditions. *Potato Res.* **2002**, *45*, 117–129. [CrossRef]

52. Xiong, D.; Chen, J.; Yu, T.; Gao, W.; Ling, X.; Li, Y.; Peng, S.; Huang, J. SPAD-based leaf nitrogen estimation is impacted by environmental factors and crop leaf characteristics. *Sci. Rep.* **2015**, *5*, 13389. [CrossRef]
53. Opena, G.B.; Porter, G.A. Soil management and supplemental nutrition effects on potato II. Root growth. *Agron. J.* **1999**, *91*, 426–431. [CrossRef]
54. He, Z.; Shankle, M.; Zhang, H.; Way, T. R.; Tewolde, H.; Uchimaya, M. Mineral composition of cottonseed is affected by fertilization management practices. *Agron. J.* **2013**, *105*, 341–350. [CrossRef]
55. Tewolde, H.; Adeli, A.; Sistani, K.R.; Rowe, D.E. Mineral nutrition of cotton fertilized with poultry litter or ammonium nitrate. *Agron. J.* **2011**, *103*, 1704–1711. [CrossRef]
56. Walworth, J.L.; Muniz, J.E. A compendium of tissue nutrient concentrations for field-grown potatoes. *Am. Potato J.* **1993**, *70*, 579–597. [CrossRef]
57. Warman, P.R.; Havard, K.A. Yield, vitamin and mineral contents of organically and conventionally grown potatoes and sweet corn. *Agric. Ecosys. Environ.* **1998**, *68*, 207–216. [CrossRef]
58. Magdoff, F.; Weil, R.R. *Soil Organic Matter in Sustainable Agriculture*; CRC Press: Boca Raton, FL, USA, 2004.
59. He, Z.; Zhang, M.; Zhao, A.; Olanya, O.M.; Larkin, R.P.; Honeycutt, C.W. Quantity and nature of water-extractable organic matter from sandy loam soils with potato cropping management. *Agric. Environ. Lett.* **2016**, *1*, 6. [CrossRef]

agronomy

MDPI

Article

Response of Milling and Appearance Quality of Rice with Good Eating Quality to Temperature and Solar Radiation in Lower Reaches of Huai River

Nianbing Zhou, Haiyan Wei and Hongcheng Zhang *

Jiangsu Key Laboratory of Crop Genetics and Physiology/Jiangsu Co-Innovation Center for Modern Production Technology of Grain Crops, Yangzhou University, Yangzhou 225009, China; iceazhou@163.com (N.Z.); wei_haiyan@163.com (H.W.)
* Correspondence: hczhang@yzu.edu.cn

Abstract: The effects of temperature and solar radiation on milling and appearance quality of rice (*Oryza sativa* L.) were evaluated to find the optimal temperature and solar radiation for optimizing milling and appearance quality of rice in the lower reaches of Huai River. Field experiments were conducted with two medium-maturing *japonica* soft rice varieties (SMR), two late-maturing *japonica* soft rice varieties (SLR) and two late-maturing *japonica* non-soft rice varieties (LR) as experimental materials. Seeds were sown on 10 May (T1), 17 May (T2), 24 May (T3), 31 May (T4), 7 June (T5), 14 June (T6), and 21 June (T7) in 2017 and 2018. Compared with solar radiation, temperature was the main environmental factor affecting the milling and appearance quality of rice in the lower reaches of Huai River. Under the condition of ensuring relatively high-yield, the milling quality of SMR and SLR can reach the second grade of China's national standard of high quality paddy. The mean daily temperature (T_{mean}) range were 20.2–22.7 °C and 20.4–22.0 °C respectively. The temperature range for LR to obtain a relatively high-yield, good milling and appearance quality was 20.4–20.7 °C. The optimal sowing dates of SMR, SLR and LR were 15 May to 1 June, 15 May to 20 May and 15 May to 20 May, respectively.

Keywords: rice quality; temperature; solar radiation; sowing date

Citation: Zhou, N.; Wei, H.; Zhang, H. Response of Milling and Appearance Quality of Rice with Good Eating Quality to Temperature and Solar Radiation in Lower Reaches of Huai River. *Agronomy* **2021**, *11*, 77. https://doi.org/10.3390/agronomy11010077

Received: 9 December 2020
Accepted: 29 December 2020
Published: 31 December 2020

1. Introduction

The lower reaches of Huai River are located in the north of Jiangsu Province. This region is an important rice production area in Jiangsu Province. The total production and planting area of rice in the lower reaches of Huai River have increased since 1980, accounting for 42.1% and 43.5% of Jiangsu Province, respectively [1]. With a rapid development in the economy and improvement of living standards, the demand for high-quality rice has been increasing in China. The grain yield and quality are influenced by numerous factors such as varietal differences, agronomic practices, and climatic conditions [2–6]. Most studies have identified that the quality of field-grown rice strongly depends on the temperature and solar radiation throughout the grain filling period [4,5], and ameliorating environmental conditions during the rice growing period by adjusting the sowing date is a practical and simple agronomic methods to improve the quality of rice [6,7].

Rice is primarily consumed as an intact kernel and the appearance quality reflects the ability to attract consumers. The milling and appearance quality are the foremost indicators for evaluating rice quality [8,9]. The milling appearance is an essential parameter of the final quality of rice, wheat, and other cereals final products, since that the milling process is able to generates the greatest and deepest changes in the final products [10–12], even higher than other essential processes such as kneading and baking [13,14]. Several studies have argued that high temperature at the heading–maturity stage of rice will shorten the filling time of rice, reduce the plumpness of grains and reduce the rate of brown rice, milled rice and head rice [15], In addition, a higher temperature will also cause the chalky grains and chalkiness degree to increase [16–18]. The low temperature during the grain filling stage will reduce the accumulation and transportation of assimilates, increase the "green rice rate", reduce the milling quality of rice [4,19], and increase the chalky grain and chalkiness degree [20,21]. The weak light environment at the filling stage of rice has also been reported to cause deterioration of rice milling and appearance quality [22,23]. Previous studies have suggested that the optimum temperature in the rice filling stage was 21.7–26.7 °C [16]. Although there has been much research on the influence of temperatures or solar radiation on rice quality, the optimal range of temperatures or solar radiation for new rice varieties with good eating quality are not clear, and the adaptability of good eating quality rice in the lower reaches of Huai River is rarely reported. Therefore, it is necessary to study the response of milling and appearance quality to temperature and solar radiation in this region, and then the optimal sowing dates can be recommended for high-yielding and good quality production. Six rice varieties with good eating quality were selected as raw materials. Seven different temperature and solar radiation environments were established by setting different sowing dates in the lower reaches of Huai River. The objectives of the study were: (1) to reveal the rice requirement of temperature and solar radiation for high milling and appearance quality production in the lower reaches of Huai River; and (2) to propose an optimal range of sowing dates for high-yield, good milling and appearance quality production in this area.

2. Materials and Methods

2.1. Plant Materials and Experimental Design

The Field experiment was conducted during the rice cropping season in 2017 and repeated in 2018 in the same experimental field at Lingqiao township, Huai'an city, Jiangsu Province, China (N 33°35′, E 118°51′). Huai'an city, which is located in the lower reaches of Huai River, it has a typical transitional monsoon climate in the north subtropical north warm temperate zone. The annual average temperature is about 14 °C, the annual precipitation is about 960 mm, the annual sunshine hours are about 2358.4 h, and the frost free period is 239 days. The soil properties determined from the upper 20 cm layer were organic matter 21.42 g kg^{-1}, total N 1.59 g kg^{-1}, available phosphorus 48.22 mg kg^{-1}, and available potassium 98.28 mg kg^{-1}.

The treatments were arranged in a split plot design with sowing dates as main plots and varieties as subplots, and the range of sowing dates in this study was designed to create contrasting environmental conditions that represent a wide range of situations for rice growth and development. Seven sowing dates were used, and six varieties were arranged in three replications within each sowing date. Planting dates were as follows: 10 May (T1), 17 May (T2), 24 May (T3), 31 May (T4), 7 June (T5), 14 June (T6), and 21 June (T7). Two medium-maturing *japonica* soft rice (SMR) varieties (Amylose content < 15%) "Nangeng 2728" and "Nangeng 505", two late-maturing *japonica* soft rice (SLR) varieties (Amylose content < 15%) "Nangeng 9108","Fenggeng 1606", and two late-maturing *japonica* non-soft rice (LR) varieties (amylose content > 15%) "Fenggeng 3227", "Wuyungeng 80" were used in 2017 and 2018. These six good eating quality varieties were chosen as they are currently the most widely cultivated in the lower reaches of Huai River. The varieties were raised in plastic plates and the seedlings were transplanted to the field 20 days after sowing at a hill spacing of 12 cm × 30 cm.

The total N application rate was 270 kg ha^{-1}. N was applied in three splits: 35% as basal fertilizer, 35% at tillering initiation, and 30% at panicle initiation. Nitrogen was applied as urea (46.4% N). For each plot, calcium superphosphate (P_2O_5 content: 12%) was applied as a basal fertilizer at the rate of 135 kg P_2O_5 ha^{-1}. Similarly, potassium chloride (K_2O content: 60%) was applied at a rate of 135 kg K_2O ha^{-1} as both basal fertilizer and at panicle initiation. The experimental field was flooded post-transplant and remained flooded until 7 days before maturity. Insects, diseases, and weeds were intensively controlled by chemicals to avoid losses in rice quality and yield.

2.2. Sample and Data Collection

All rice plants were hand harvested. The final grain yield was adjusted to 14% moisture content. The China national standard of high-quality paddy (GB/T17891-2017) was an evaluation standard for rice quality promulgated by the National Food Administration Standard Quality Center, which has the general function of judging the quality of high quality paddy in China. According to the China national standard of high quality paddy, the grading index for milling quality and appearance of *japonica* rice is the head rice rate and chalkiness degree. The head rice rate should be equal or greater than 67%, 61% and 55%, respectively, and the chalkiness degree should be equal or lesser than 2%, 4% and 6%, respectively, when the milling and appearance quality reaches the first, second or third grade of China's national standard of high quality paddy. Rice quality analysis was performed according to the GB/T17891-2017 in this study. The brown rice, milled rice and head milled rice rate were expressed as percentages of the total grain weights, chalkiness was evaluated on 100 milled grains per plot. Chalkiness size was expressed as percentage of the total area of the kernel.

$$\text{Chalkiness degree (\%)} = \text{Chalkiness rate} \times \text{Chalkiness size},$$

The dates of heading and maturity were observed and recorded for each treatment. The daily air temperature and number of sunshine hours during the rice growing season in both experimental years were collected from a local weather observation point at the Huai'an Meteorological Station (Jiangsu Province, China).

2.3. Calculation Methods and Statistical Analysis

The effective accumulated temperature (EAT) in the determined growth duration expressed as °C d was calculated as:

$$\text{EAT} = \sum (\text{T} - \text{T}_0) \times \text{Growth duration},$$

where T and T_0 (10 °C for japonica rice varieties) are the mean daily temperature and the biological zero temperature, respectively [19].

The environmental data for the period 2007–2016 in Huai'an City were collected from the National Meteorological Information Center of the China Meteorological Administration. The Angstrom–Prescott (AP) model was used to calculate daily global solar radiation from sunshine duration, because solar radiation could not be directly recorded at the meteorological station. It was calculated as follows:

$$\frac{Q}{Q_0} = a + b \times \frac{S}{S_0}$$

where Q (MJ m^{-2} d^{-1}) is global solar radiation, Q_0 (MJ m^{-2} day^{-1}) is extraterrestrial solar radiation and total solar radiation of the ideal atmosphere, S is the actual sunshine hours in a day, and S_0 is the potential sunshine hours in a day. The constitute climatology coefficients *a* and *b* (Table 1), were described by Chen et al. as the extraterrestrial solar radiation and total solar radiation [24].

Table 1. The coefficients *a* and *b* for each month in the Angstrom-Prescott model.

Coefficient	May	June	July	August	September	October	November
a	0.211	0.239	0.303	0.272	0.304	0.290	0.206
b	0.712	0.624	0.529	0.576	0.487	0.567	0.679

The cumulative solar radiation (CSR) in the determined growth duration expressed as MJ m^{-2} was calculated as:

$$\text{CSR} = \sum \text{Q} \times \text{Growth duration},$$

Q (MJ $m^{-2}d^{-1}$) is the daily global solar radiation

$$\text{Relative grain yield} = \frac{\text{Yield}_{\text{Ti}}}{\sum \text{Yield}_{\text{Tn}}},$$

Yield_{Ti} represents the yield of rice under Ti treatment, Yield_{Tn} represents the yield of the treatment that the rice can mature normally, SMR: $n = 7$, SLR: $n = 4$, LR: $n = 4$.

Data were analyzed using analysis of variance (ANOVA) with SPSS 13.0. Means were compared by the least significant difference (LSD) test at the 0.05 probability level. In addition, the graphs were prepared with Microsoft Excel.

3. Results

3.1. Effects of Sowing Date on Milling and Appearance Quality

The six tested varieties experienced different temperature and solar radiation during their ripening phase due to a wide range in sowing dates. The late-maturing *japonica* soft rice(SLR) and late-maturing *japonica* non-soft rice(LR) cannot mature in T5, T6, and T7, and the harvest time (November 8) was taken as the deadline for rice growth, and it was used to calculate the effective accumulated temperature (EAT), mean daily temperature (T_{mean}), cumulative solar radiation (CSR), and mean daily solar radiation (R_{mean}). With the delay of sowing date, the EAT, T_{mean}, CSR, and R_{mean} of six rice varieties showed a decreasing trend at the stage from heading to maturity (Figures 1 and 2). The temperature under the same sowing date had similar values in the two year experiment, and the CSR, and R_{mean} in 2018 are slightly higher than those in 2017. The seven temperature and solar radiation treatments with significant differences were established for each rice variety by setting seven sowing dates in the same area.

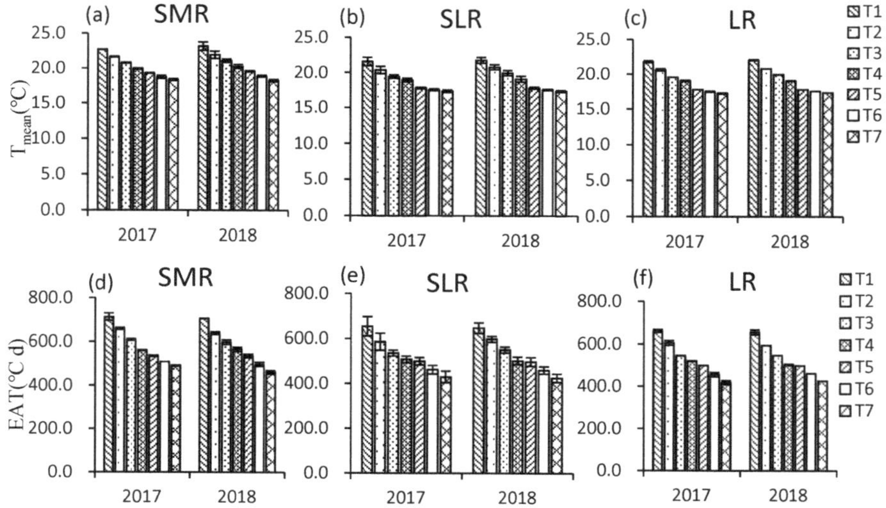

Figure 1. Differences in mean daily temperature (T_{mean}, °C) and effective accumulated temperature (EAT, °C d) of rice at the stage from heading–maturity in the seven environmental condition treatments. (**a**–**c**) respesent the T_{mean} of SMR, SLR and LR at the stage from heading-maturity in the seven environmental condition treatments. (**d**–**f**) respesent the EAT of SMR, SLR and LR at the stage from heading-maturity in the seven environmental condition treatments. T1, T2, T3, T4, T5, T6, and T7 represent the sowing dates 10 May, 17 May, 24 May, 31 May, 7 June, 14 June, and 21 June. SMR: medium-maturing *japonica* soft rice, SLR: late-maturing *japonica* soft rice, LR: late-maturing *japonica* non-soft rice.

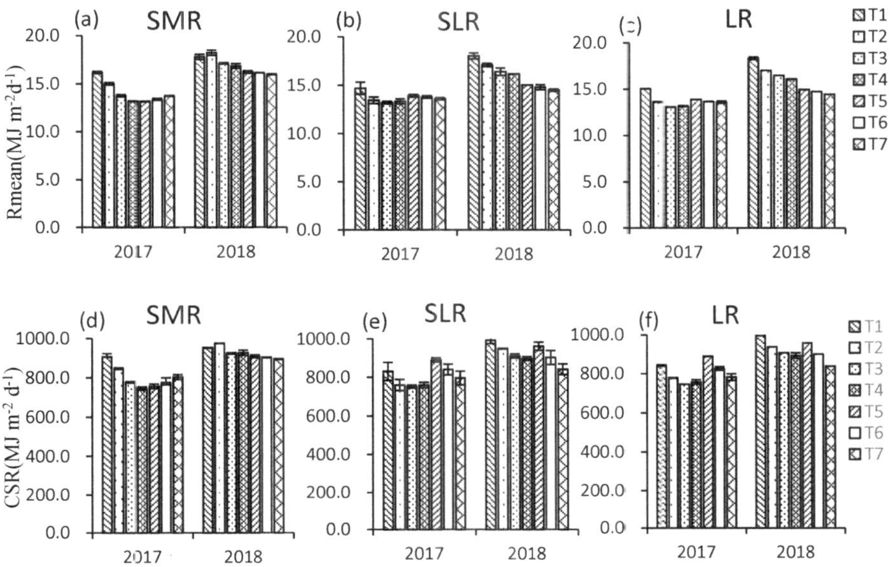

Figure 2. Differences in mean daily solar radiation (R_{mean} MJ m^{-2} d^{-1}) and cumulative solar radiation (CSR MJ m^{-2}) of rice at the stage from heading–maturity in the seven environmental condition treatments. (**a–c**) respesent the R_{mean} of SMR, SLR and LR at the stage from heading-maturity in the seven environmental condition treatments. (**d–f**) respesent the CSR of SMR, SLR and LR at the stage from heading-maturity in the seven environmental condition treatments.

In addition to the head milled rice rate, the milling quality had extremely significant differences among the years (Y), types (T), varieties (V) and sowing dates (S) (Table 2). The reason for this insignificant interaction of Y × S was due to the difference in milling quality in two years and the different changing trends of the six varieties under the conditions of the sowing date.

Table 2. Analysis of variance table for milling and appearance quality of rice among years, sowing dates and varieties.

Analysis of Variance	*df*	Brown Rice Rate	Milled Rice Rate	Head Milled Rice Rate	Chalky Grains	Chalkiness Degree
Year (Y)	1	50.956 **	11.653 **	3.004 NS	28.403 **	319.798 **
Type (T)	2	11.481 **	26.513 **	36.675 **	3488.960 **	55.969 **
Sowing date (S)	6	28.068 **	31.464 **	48.648 **	459.804 **	444.033 **
Variety (V)	1	92.966 **	19.921 **	36.472 **	437.967 **	306.443 **
Y × T	2	67.557 **	31.659 **	5.760 **	92.896 **	108.934 **
Y × S	6	0.380 NS	1.105 NS	0.324 NS	29.270 **	2.974 *
Y × V	1	2.183 NS	35.045 **	1.692 NS	15.430 **	485.505 **
T × S	12	58.571 **	76.606 **	47.115 **	141.089 **	59.200 **
T × V	2	73.169 **	13.829 **	16.992 **	320.189 **	248.186 **
S × V	6	2.276 *	2.229 *	0.613 NS	9.717 **	4.856 **
Y × T × S	12	1.372 NS	2.633 **	0.581 NS	10.609 **	0.690 NS
Y × T × V	2	3.730 *	24.066 **	2.411 NS	11.225 **	119.044 **
Y × S × V	6	1.328 NS	2.191 NS	0.206 NS	2.757 *	1.477 NS
T × S × V	12	3.660 **	2.264 *	0.389 NS	10.793 **	5.388 **
Y × T × S × V	12	0.627 NS	1.346 NS	0.291 NS	6.278 **	6.801 **

** and * indicate significant difference at $p = 0.01$ and $p = 0.05$ levels, respectively, NS means not significant at the $p = 0.05$ level.

A wide range in the milling quality of six tested varieties was observed across seven sowing dates. The milling quality of medium-maturing *japonica* soft rice (SMR) had been improved with the decrease in temperature and solar radiation. On the contrary, the SLR and LR showed a deterioration trend (Table 3). The brown rice rate, milled rice rate and head milled rice rate of SMR in T7 were 0.67–4.09%, 0.80–5.50% and 0.71–5.23% higher than those in T1, T2, T3, T4, T5 and T6, respectively. The brown rice rate, milled rice rate and head milled rice rate of SLR in T1 were 0.50–5.78%, 0.43–6.85% and 0.83–10.30% higher than those in T2, T3, T4, T5, T6 and T7, respectively. The T1 of LR were 0.29–3.36%, 0.42–6.86% and 0.72–8.93% higher than those in T2, T3, T4, T5, T6 and T7, respectively. Notably, a significant decrease was observed in the milling quality of SLR and LR in T5, T6 and T7.

Table 3. Differences in milling quality of rice under different temperature and solar radiation conditions.

Variety	Treatment	Brown Rice Rate (%)		Milled Rice Rate (%)		Head Milled Rice Rate (%)	
		2017	2018	2017	2018	2017	2018
Nangeng2728	T1	83.17 ^{d}	82.45 ^{b}	70.57 ^{d}	71.41 ^{b}	63.54 ^{b}	64.47 ^{c}
	T2	83.59 ^{c,d}	82.66 ^{b}	71.40 ^{d}	71.34 ^{b}	64.35 ^{b}	64.84 ^{b,c}
	T3	84.06 ^{c}	83.02 ^{a,b}	71.71 ^{c,d}	71.66 ^{b}	64.70 ^{a,b}	65.25 ^{b,c}
	T4	84.90 ^{b}	83.43 ^{a,b}	72.63 ^{b,c}	72.53 ^{a}	64.98 ^{a,b}	66.09 ^{a,b,c}
	T5	85.37 ^{b}	83.53 ^{a,b}	73.09 ^{a,b}	72.66 ^{a}	65.83 ^{a,b}	66.33 ^{a,b}
	T6	85.62 ^{b}	83.67 ^{a,b}	73.74 ^{a,b}	72.92 ^{a}	66.60 ^{a}	66.63 ^{a,b}
	T7	86.65 ^{a}	84.22 ^{a}	74.25 ^{a}	73.24 ^{a}	67.29 ^{a}	67.29 ^{a}
Nangeng505	T1	83.82 ^{c}	82.06 ^{c}	70.47 ^{d}	70.10 ^{d}	64.09 ^{c}	64.88 ^{b}
	T2	84.61 ^{b,c}	82.12 ^{c}	70.66 ^{d}	70.81 ^{c,d}	64.17 ^{c}	65.41 ^{a,b}
	T3	84.61 ^{b,c}	82.56 ^{c}	71.63 ^{c,d}	71.27 ^{c,d}	64.78 ^{b,c}	65.85 ^{a,b}
	T4	85.72 ^{a,b}	83.18 ^{b,c}	72.69 ^{b,c}	71.77 ^{b,c}	65.75 ^{a,b,c}	66.30 ^{a,b}
	T5	86.10 ^{a,b}	83.51 ^{a,b,c}	73.25 ^{b,c}	72.19 ^{a,b,c}	65.81 ^{a,b,c}	66.74 ^{a,b}
	T6	86.29 ^{a}	84.07 ^{a,b}	72.97 ^{a,b}	72.88 ^{a,b}	66.52 ^{a,b}	67.16 ^{a}
	T7	87.17 ^{a}	84.66 ^{a}	74.53 ^{a}	73.72 ^{a}	67.03 ^{a}	67.45 ^{a}
Nangeng9108	T1	85.31 ^{a}	85.25 ^{a}	73.77 ^{a}	73.65 ^{a}	67.63 ^{a}	67.68 ^{a}
	T2	84.89 ^{a}	85.09 ^{a}	73.20 ^{a}	73.25 ^{a}	67.23 ^{a}	67.25 ^{a}
	T3	84.76 ^{a}	84.84 ^{a}	73.12 ^{a}	73.18 ^{a}	66.58 ^{a}	66.78 ^{a}
	T4	83.75 ^{a}	83.70 ^{b}	73.05 ^{a}	73.07 ^{a}	66.17 ^{a}	66.23 ^{a}
	T5	81.08 ^{b}	81.83 ^{c}	71.36 ^{b}	71.62 ^{b}	64.37 ^{b}	63.67 ^{b}
	T6	80.35 ^{b}	81.35 ^{c}	70.43 ^{b}	70.69 ^{b}	63.25 ^{b,c}	62.77 ^{b,c}
	T7	80.00 ^{b}	79.01 ^{d}	68.53 ^{c}	68.60 ^{c}	62.36 ^{c}	61.61 ^{c}
Fenggeng1606	T1	86.72 ^{a}	86.83 ^{a}	73.87 ^{a}	73.57 ^{a}	67.81 ^{a}	66.58 ^{a}
	T2	86.29 ^{a,b}	86.14 ^{a,b}	73.59 ^{a}	73.34 ^{a}	66.88 ^{a,b}	65.92 ^{a}
	T3	85.22 ^{a,b,c}	86.58 ^{a,b}	73.25 ^{a}	72.97 ^{a}	65.90 ^{a,b}	65.53 ^{a}
	T4	84.99 ^{b,c}	85.57 ^{b}	72.25 ^{b}	72.49 ^{a,b}	64.57 ^{b,c}	64.71 ^{a}
	T5	84.61 ^{c,d}	84.27 ^{c}	71.66 ^{b}	71.45 ^{b,c}	62.50 ^{c,d}	61.69 ^{b}
	T6	83.32 ^{d}	84.03 ^{c}	70.30 ^{c}	71.07 ^{c,d}	61.03 ^{d}	61.28 ^{b}
	T7	83.16 ^{d}	83.66 ^{c}	69.63 ^{c}	69.87 ^{d}	60.43 ^{d}	60.20 ^{b}
Fenggeng3227	T1	84.85 ^{a}	84.63 ^{a}	75.86 ^{a}	75.64 ^{a}	67.49 ^{a}	67.54 ^{a}
	T2	84.68 ^{a,b}	84.27 ^{a,b}	72.06 ^{b}	75.32 ^{a}	66.94 ^{a}	67.15 ^{a}
	T3	84.14 ^{a,b,c}	83.98 ^{a,b,c}	71.66 ^{b,c}	74.98 ^{a}	66.23 ^{a,b}	66.87 ^{a}
	T4	83.97 ^{a,b,c}	83.57 ^{a,b,c}	71.06 ^{b,c}	74.47 ^{a,b}	65.20 ^{b}	66.19 ^{a,b}
	T5	83.19 ^{a,b,c}	83.77 ^{a,b,c}	70.89 ^{b,c}	73.57 ^{b,c}	63.49 ^{c}	64.50 ^{b,c}
	T6	82.88 ^{b,c}	83.20 ^{b,c}	71.64 ^{b,c}	73.17 ^{b,c}	62.31 ^{c,d}	64.06 ^{c}
	T7	82.68 ^{c}	82.80 ^{c}	72.65 ^{c}	72.96 ^{c}	61.82 ^{d}	62.99 ^{c}
Wuyungeng80	T1	85.40 ^{a}	84.82 ^{a}	74.48 ^{a}	73.61 ^{a}	66.84 ^{a}	66.66 ^{a}
	T2	84.94 ^{a}	84.69 ^{a}	73.86 ^{a,b}	73.30 ^{a}	66.43 ^{a}	65.47 ^{a,b}
	T3	84.50 ^{a,b}	83.81 ^{b}	72.98 ^{b,c}	72.85 ^{a,b}	65.04 ^{b}	64.78 ^{b}
	T4	83.46 ^{b,c}	83.53 ^{b}	72.50 ^{c}	72.13 ^{a,b,c}	64.65 ^{b}	63.92 ^{b,c}
	T5	83.13 ^{b,c}	82.91 ^{c}	71.90 ^{c}	71.58 ^{b,c}	62.67 ^{c}	62.81 ^{c,d}
	T6	82.79 ^{c}	82.43 ^{c,d}	70.52 ^{d}	71.04 ^{c}	61.72 ^{c,d}	62.61 ^{c,d}
	T7	82.03 ^{c}	82.71 ^{d}	70.05 ^{d}	70.95 ^{c}	61.50 ^{d}	61.31 ^{d}

Values followed by different lowercase letters within a column are significantly different at the $p = 0.05$ level. T1, T2, T3, T4, T5, T6, and T7 represent the sowing dates 10 May, 17 May, 24 May, 31 May, 7 June, 14 June, and 21 June.

The milling quality of the same type of rice varieties was similar on the same sowing date. The milling quality of SLR and LR was better than that of SMR in T1, T2 and T3. While in T5, T6 and T7, the milling quality of SMR was better than that of SLR and LR.

The chalky grain rate and chalkiness degree of SMR decreased with the reduction in temperature and solar radiation. The chalky grain and chalkiness degree of T1 were 0.37–146.23% and 1.97–187.67% higher than those in other treatments, respectively (Table 4).

The chalky grain and chalkiness degree of SLR and LR decreased first and then increased with the reduction of temperature and solar radiation. The significant increase in chalky grain and chalkiness degree of SLR and LR were related to the incomplete maturity in T5, T6 and T7.

Table 4. Differences in appearance quality of rice under different temperature and solar radiation conditions.

Variety	Treatment	Chalky Grains (%)		Chalkiness Degree (%)	
		2017	2018	2017	2018
Nangeng2728	T1	53.61 ^{a}	53.89 ^{a}	7.00 ^{a}	6.68 ^{a}
	T2	53.31 ^{a}	53.69 ^{a}	6.86 ^{b}	6.36 ^{a,b}
	T3	53.07 ^{a}	48.11 ^{b}	6.27 ^{c}	6.03 ^{b}
	T4	38.26 ^{b}	44.89 ^{c}	5.76 ^{d}	5.52 ^{c}
	T5	30.09 ^{c}	44.42 ^{c}	5.46 ^{e}	5.24 ^{c}
	T6	26.63 ^{d}	38.00 ^{d}	5.32 ^{f}	4.72 ^{d}
	T7	21.77 ^{e}	29.20 ^{e}	5.18 ^{g}	4.46 ^{d}
Nangeng505	T1	52.08 ^{a}	52.44 ^{a}	6.36 ^{a}	8.56 ^{a}
	T2	50.11 ^{a,b}	49.80 ^{a,b}	5.69 ^{b}	7.73 ^{a,b}
	T3	47.39 ^{b}	47.49 ^{b}	4.65 ^{c}	7.22 ^{b,c}
	T4	40.07 ^{c}	41.43 ^{c}	3.24 ^{d}	6.44 ^{c,d}
	T5	35.96 ^{c,d}	37.96 ^{c,d}	3.09 ^{d}	6.14 ^{d,e}
	T6	32.54 ^{d}	35.73 ^{d}	2.86 ^{e}	5.67 ^{d,e}
	T7	28.23 ^{e}	26.18 ^{e}	2.21 ^{f}	5.30 ^{e}
Nangeng9108	T1	46.58 ^{a}	46.70 ^{a}	6.46 ^{a}	7.84 ^{a}
	T2	44.95 ^{a}	39.04 ^{b}	5.70 ^{b}	6.34 ^{b}
	T3	30.41 ^{c}	32.15 ^{c}	4.40 ^{d}	4.87 ^{c}
	T4	23.39 ^{d}	21.02 ^{e}	3.54 ^{e}	3.89 ^{c}
	T5	28.83 ^{c}	27.81 ^{d}	4.06 ^{d}	4.44 ^{d}
	T6	40.24 ^{b}	33.71 ^{c}	5.01 ^{c}	5.61 ^{e}
	T7	42.45 ^{a,b}	37.97 ^{b}	5.41 ^{b,c}	5.88 ^{f}
Fenggeng1606	T1	36.95 ^{a}	30.05 ^{a}	7.43 ^{a}	7.82 ^{a}
	T2	30.16 ^{b}	23.90 ^{b}	6.54 ^{b}	7.14 ^{b}
	T3	20.87 ^{c,d}	19.93 ^{c}	4.56 ^{d}	6.06 ^{d,e}
	T4	18.38 ^{d}	19.58 ^{c}	3.55 ^{e}	5.51 ^{f}
	T5	20.71 ^{c,d}	19.84 ^{c}	4.25 ^{d}	5.82 ^{e,f}
	T6	21.80 ^{c,d}	20.99 ^{b,c}	5.52 ^{c}	6.55 ^{s,d}
	T7	24.97 ^{c}	21.62 ^{b,c}	6.21 ^{b}	7.02 ^{b,c}
Fenggeng3227	T1	33.89 ^{a}	23.33 ^{a}	6.99 ^{a}	5.29 ^{a}
	T2	28.68 ^{b}	20.92 ^{a,b}	5.60 ^{b}	4.89 ^{a,b}
	T3	17.32 ^{d}	18.07 ^{c,d}	3.94 ^{d}	3.60 ^{c,d}
	T4	14.51 ^{d}	15.53 ^{d}	3.18 ^{e}	2.87 ^{e}
	T5	14.91 ^{d}	17.36 ^{c,d}	3.55 ^{d,e}	3.37 ^{d,e}
	T6	22.05 ^{c}	18.71 ^{b,c}	4.53 ^{c}	3.99 ^{d,e,c}
	T7	24.61 ^{c}	18.85 ^{b,c}	5.47 ^{c}	4.68 ^{b}
Wuyungeng80	T1	33.88 ^{a}	22.96 ^{a}	7.32 ^{a}	8.37 ^{a}
	T2	27.50 ^{b}	19.63 ^{b}	6.80 ^{b}	7.20 ^{b}
	T3	16.11 ^{d}	16.44 ^{c}	5.20 ^{e}	5.40 ^{e}
	T4	14.27 ^{d}	16.21 ^{c}	4.11 ^{g}	4.24 ^{f}
	T5	14.78 ^{d}	16.38 ^{c}	4.79 ^{f}	5.21 ^{e}
	T6	21.90 ^{c}	17.38 ^{b,c}	5.65 ^{d}	6.19 ^{d}
	T7	26.44 ^{b}	18.66 ^{b,c}	6.26 ^{c}	6.83 ^{c}

Values followed by different lowercase letters within a column are significantly different at the $p = 0.05$ level.

For different types of varieties, the LR had the best appearance quality under the same sowing date.

3.2. Correlation between Rice Milling Quality, Appearance Quality and Temperature or Solar Radiation

The milling quality of the three types of rice showed a significantly correlation with T_{mean} or EAT at the stage from heading to maturity. The correlation coefficients of milling quality with R_{mean} or CSR were smaller than the correlation coefficients of milling quality with T_{mean} or EAT (Table 5). The T_{mean} and, EAT at the stage from heading to maturity showed a positive correlation with chalky grain and chalkiness degree. However, marked differences were observed in correlations between solar radiation and appearance quality in two years. These results indicated that the influence of temperature on rice milling and appearance quality was greater than that of solar radiation.

Table 5. Correlation analysis between rice quality and environmental factors at the stage from heading to maturity.

Type	Rice Quality	T_{mean}		EAT		R_{mean}		CSR	
		2017	2018	2017	2018	2017	2018	2017	2018
SMR	brown rice rate (%)	−0.939 **	−0.933 **	−0.909 **	−0.938 **	−0.716 **	−0.867 **	−0.538 *	−0.844 **
	milled rice rate (%)	−0.963 **	−0.904 **	−0.967 **	−0.912 **	−0.765 **	−0.824 **	−0.654 *	−0.803 **
	head milled rice (%)	−0.962 **	−0.981 **	−0.948 **	−0.974 **	−0.717 **	−0.921 **	−0.609 *	0.853 **
	Chalky grains (%)	0.924 **	0.943 **	0.921 **	0.945 **	0.690 **	0.881 **	0.568 *	0.846 **
	Chalkiness degree (%)	0.737 **	0.793 **	0.692 **	0.803 **	0.617 *	0.716 **	0.440	0.711 **
SLR	brown rice rate (%)	0.719 **	0.683 **	0.621*	0.628 *	0.045	0.686 **	0.045	0.686 **
	milled rice rate (%)	0.862 **	0.864 **	0.837 **	0.860 **	0.083	0.875 **	0.083	0.875 **
	head milled rice (%)	0.917 **	0.931 **	0.905 **	0.909 **	0.130	0.931 **	0.130	0.931 **
	Chalky grains (%)	0.475	0.456	0.531	0.494	0.537 *	0.440	0.537*	0.440
	Chalkiness degree (%)	0.440	0.413	0.387	0.388	0.451	0.396	0.451	0.396
LR	brown rice rate (%)	0.962 **	0.927 **	0.964 **	0.932 **	0.369	0.916 **	−0.137	0.706 **
	milled rice rate (%)	0.881 **	0.675 **	0.895 **	0.642 *	0.634 *	0.673 **	0.127	0.423
	head milled rice (%)	0.956 **	0.859 **	0.945 **	0.839 **	0.300	0.859 **	−0.221	0.601 *
	Chalky grains (%)	0.553 *	0.664 **	0.489	0.677 **	0.730 **	0.648 *	0.114	0.532
	Chalkiness degree (%)	0.496	0.352	0.442	0.369	0.667 **	0.313	0.119	0.261

** and * respectively represent extremely significant correlation and significant correlation. $r_{0.01} = 0.661$; $r_{0.05} = 0.533$.

Under conditions of complete maturity, the head milled rice rate and chalkiness degree showed a significant correlation with EAT or T_{mean} (Figures 3–6). The result showed that, to obtain the second grade of milling and appearance quality of China's national standard GB/T 17891-2017, the demand of temperature at the stage from heading to maturity for SMR were lower than those of SLR and LR (Tables 6 and 7).

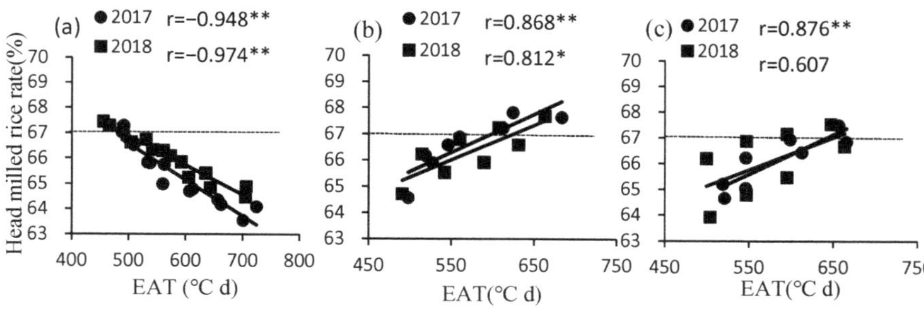

Figure 3. The correlation between head milled rice rate and the EAT of rice at stage of heading–maturity (**a**): SMR, $n = 14$, (**b**): SLR, $n = 8$, (**c**): LR, $n = 8$, the (immature treatment including T5, T6 and T7 was removed from SLR and LR). * and ** indicate $p < 0.05$ and $p < 0.01$, respectively.

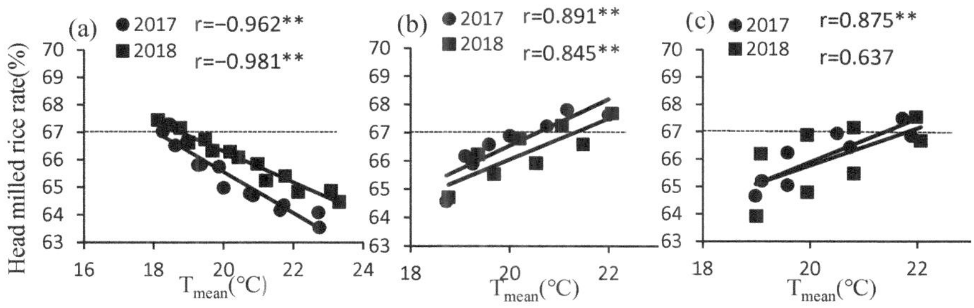

Figure 4. The correlation between head milled rice rate and the T_{mean} of rice at stage of heading–maturity (**a**): SMR, $n = 14$, (**b**): SLR, $n = 8$, (**c**): LR, $n = 8$, (the immature treatment including T5, T6 and T7 was removed from SLR and LR). ** indicate $p < 0.01$.

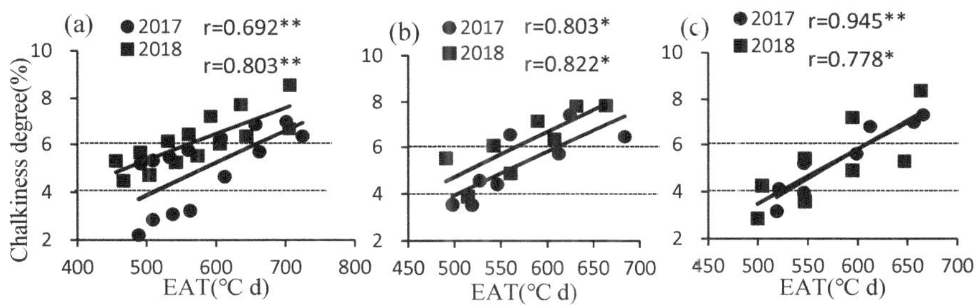

Figure 5. The correlation between chalkiness degree and the EAT of rice at stage of heading-maturity (**a**): SMR, $n = 14$, (**b**): SLR, $n = 8$, (**c**): LR, $n = 8$, (the immature treatment including T5, T6 and T7 was removed from SLR and LR). * and ** indicate $p < 0.05$ and $p < 0.01$, respectively.

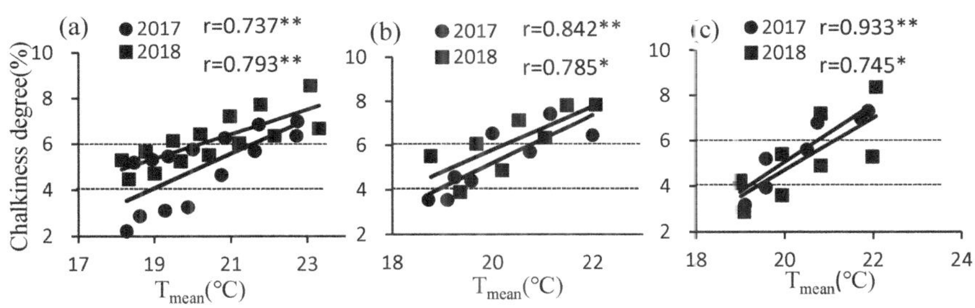

Figure 6. The correlation between chalkiness degree and the T_{mean} of rice at stage of heading-maturity (**a**): SMR, $n = 14$, (**b**): SLR $n = 8$, (**c**): LR, $n = 8$, the immature treatment including T5, T6 and T7 was removed from SLR and LR). * and ** indicate $p < 0.05$ and $p < 0.01$, respectively.

Table 6. Characteristics of EAT at stage from heading to maturity of good eating quality rice.

Type	SMR		SLR		LR	
	2017	2018	2017	2018	2017	2018
Relatively high-yield	580.6–724.9 °C	574.6–606.3 °C	576.1–683.9 °C	575.3–663.6 °C	576.1–683.9 °C	575.3–663.6 °C
GBMI	450.3–470.4 °C	455.4–490.6 °C	599.6–683.9 °C	623.8–663.6 °C	642.0–665.9 °C	647.3–663.6 °C
GBMII	450.3–713.9 °C	455.4–708.6 °C	431.8–683.9 °C	450.7–663.6 °C	488.6–665.9 °C	400.6–663.6 °C
GBAII	488.6–508.3 °C	-	498.1–502.2 °C	-	518.0–528.2 °C	499.7–520.7 °C
GBAIII	488.6–654.3 °C	455.4–560.6 °C	498.1–609.1 °C	491.0–566.2 °C	519.0–607.9 °C	499.7–606.9 °C

GBMI: The milling quality reaches the first grade of China's national standard of high quality paddy (head rice rate ≥ 67%), **GBMII**: The milling quality reaches second grade of China's national standard of high quality paddy (head rice rate ≥ 61%), **GBAII** The appearance quality reaches second grade of China's national standard of high quality paddy (chalkiness degree ≤ 4.0%), **GBAIII**: The appearance quality reaches third grade of China's national standard of high quality paddy (chalkiness degree ≤ 6.0%); "-"indicated that the appearance quality did not reach the second grade of China's national standard of high quality paddy.

Table 7. Characteristics of T_{mean} at stage from heading to maturity of good eating quality rice.

Type	SMR		SLR		LR	
	2017	2018	2017	2018	2017	2018
Relatively high-yield	20.2–22.7 °C	20.5–23.3 °C	20.2–22.0 °C	20.4–22.1 °C	20.2–22.0 °C	20.4–22.1 °C
GBMI	17.7–18.1 °C	18.1–18.8 °C	20.6–22.0 °C	21.3–22.1 °C	21.4–21.9 °C	21.8–22.1 °C
GBMII	17.7–22.7 °C	18.1–23.8 °C	17.6–22.0 °C	18.2–22.1 °C	18.4–21.9 °C	17.3–22.1 °C
GBAII	18.3–18.9 °C	-	18.7–18.9 °C	-	19.0–19.2 °C	19.0–19.4 °C
GBAIII	18.3–21.6 °C	18.1–20.2 °C	18.7–20.7 °C	18.8–20.2 °C	19.0–20.7 °C	19.0–21.1 °C

GBMI: The milling quality reaches first grade of China's national standard of high quality paddy (head rice rate ≥ 67%), **GBMII**: The milling quality reaches second grade of China's national standard of high quality paddy (head rice rate ≥ 61%), **GBAII** The appearance quality reaches second grade of China's national standard of high quality paddy (chalkiness degree ≤ 4.0%), **GBAIII**: The appearance quality reaches third grade of China's national standard of high quality paddy (chalkiness degree ≤ 6.0%); "-" indicated that the appearance quality did not reach the second grade of China's national standard of high quality paddy.

3.3. Optimum Sowing Date for Good Milling and Appearance Quality Production of Rice

Under conditions of normal maturity, if the relative grain yield of a given variety is bigger than one, the grain yield of that variety in a certain planting condition is higher than the average yield of that variety in all treatments, indicating a relatively higher grain yield of that variety [19]. The ranges of EAT and T_{mean} at the stage from heading to maturity for relative high yields production of SMR, SLR and LR are listed in Tables 5 and 6 [25]. The high temperature at the stage from heading to maturity under early sowing conditions was beneficial to increase yield and milling quality of SLR and LR, while it deteriorated the appearance quality of SMR, SLR and LR. The EAT at the stage from heading to maturity of SMR to obtain high yield and good milling quality are 580.6–713.9 °C and 574.6–606.3 °C in 2017 and 2018, and the T_{mean} was 20.2–22.7 °C and 20.5–23.3 °C. The EAT of SLR was 576.1–683.9 °C and 575.3–663.6 °C, and the T_{mean} was 20.2–22.0 °C and 20.4–22.1 °C respectively. The EAT of LR to obtain high yield, good milling and appearance quality was 576.1–665.9 °C and 575.3–663.6 °C, and the T_{mean} was 20.2–20.7 °C and 20.4–22.1 °C respectively.

EAT is often used to evaluate the accumulation of heat resources of a certain rice variety under certain cultivation conditions [26]. The EAT was taken as the index to use for statistical analysis of the best sowing times in the recent years in this study. It is important to highlight that the daily minimum temperature stably passed 10 °C was the same as the earliest sowing date for the formation of high yield, good milling, and appearance quality (Table 8). During the years 2007–2016, the date when the daily minimum temperature stably passed 10 °C in 2011 was significantly later than that in other years, which led to the earliest optimal sowing date in 2011 being more than 7 d later than that in the other years. The EATs in 2014 and 2015 were 6.04% and 6.31% lower than the average EAT in 2007–2018, which resulted in the latest optimal sowing dates in 2014 and 2015 being more than 9 d earlier than that in the other years [25]. Therefore, compared with the perennial climate the climate in 2011, 2014 and 2015 was abnormal and the optimal sowing date selected by the remaining seven years was representative in the lower reaches of Huai River. The earliest optimal sowing date for three types of rice obtaining high yield and good quality in the lower reaches of the Huai River was 15 May, and the latest optimal sowing dates for SMR, SLR and LR were 1 June, 20 May and 20 May, respectively.

Table 8. The optimal sowing dates for rice to obtain high yield and good quality.

Year	EOS	SMR		SLR		LR	
		LOS		LOS		LOS	
		2017	2018	2017	2018	2017	2018
2007	5/10	5/30	6/1	5/21	5/21	5/21	5/22
2008	5/15	5/31	6/2	5/21	5/22	5/21	5/22
2009	4/28	6/1	6/3	5/22	5/22	5/22	5/24
2010	5/14	6/4	6/7	5/23	5/24	5/23	5/25
2011	5/23	5/26	5/28	-	-	-	-
2012	4/18	5/30	6/1	5/18	5/21	5/18	5/21
2013	4/27	6/4	6/7	5/24	5/25	5/24	5/25
2014	5/6	5/19	5/21	5/6	5/8	5/6	5/8
2015	4/23	5/18	5/22	5/6	5/6	5/5	5/6
2016	4/26	6/1	6/3	5/18	5/20	5/18	5/20

EOS: earliest optimal sowing date; LOS: latest optimal sowing date; "-" indicated that there is no suitable sowing date.

4. Discussion

4.1. Response of Milling Quality to Temperature and Solar Radiation

The filling stage of rice is generally considered to be the key period affecting rice quality [15,27]. A low coefficient of correlation was observed between solar radiation and milling quality in this study. Therefore, we believe that the solar radiation resource in this region was abundant, and it was not a limiting factor that affects the formation of good milling quality. The EAT and T_{mean} of the three types of rice at the stage from heading to maturity showed a downward trend with the delay of the sowing date. However, the brown rice rate, milled rice rate and head milled rice rate of SMR increased by 0.67–4.09%, 0.80–5.50% and 0.71–5.23% respectively. The milling quality of SMR showed a significant negative correlation with EAT and T_{mean} at the stage from heading to maturity. Previous studies on the effect of temperature during the rice filling stage on rice milling quality believed that high temperature would increase the amount of broken rice leading to poor milling quality [15,28,29]. We supposed that the milling quality of SMR is sensitive to high temperature. It was found that lower temperatures (17.7–18.8 °C) were favorable for SMR forming the first grade of China's national standard of high quality paddy. In contrast, the milling quality of SLR and LR showed significant positive correlation with EAT and T_{mean}. The temperature requirement of SLR and LR cannot be lower than 20.6 °C and 21.4 °C in order to constitute the first grade of China's national standard of high quality paddy. Thus, it appears that the temperature requirements are different for different types of rice, which is consistent with the views of Li et al. [30]. Increasing temperature in an appropriate range was beneficial to improve the milling quality of the late-maturing rice [15,31]. The T_{mean} ranges were 20.2–23.3 °C (SMR), 20.2–22.1 °C (SLR) and 20.2–22.1 °C (LR), respectively, when the rice obtained relatively high-yields, and the milling quality reached the second grade of China's national standard of high quality paddy. It was considered that late sowing was beneficial for SMR to obtain good milling quality, and early sowing was conducive to improving the milling quality of SLR and LR.

4.2. Response of Appearance Quality of Good Eating Quality Rice to Temperature and Solar Radiation

Chalkiness rate and chalkiness degree are the main indexes to evaluate rice appearance quality. The appearance quality of rice is often affected by environmental factors. Many studies believe that the high temperature during the grain filling stage will cause the grain refractive index to decrease and formation of chalkiness [32,33]. Compared with solar radiation, a higher correlation coefficient was observed between temperature and chalkiness degree. With the decline in temperature, the chalkiness rate and chalkiness degree of SMR decreased. Interestingly, with the decline in temperature, the chalky grains and chalkiness degree of SLR and LR decreased first and then increased. Gong et al. summarized the effect of low temperature on the appearance and quality of rice and pointed out that the chalkiness is caused by low temperature hindering the division of endosperm cells and reducing the volume of amyloplasts [15]. We believe that low temperature during rice filling stage was beneficial to improve the appearance quality of SLR and LR, but too low temperature is not conducive to the formation of good appearance quality of late-maturing *japonica* rice.

Under the condition of fully maturity, the appearance quality of the three types of rice failed to reach the first grade of China's national standard of high quality paddy. A very small range of T_{mean} during the grain filling stage for SMR, SLR and LR to reach the second grade of appearance quality of China National Standard were found, which were 18.3–18.9 °C, 18.7–18.9 °C and 19.0–19.4 °C, respectively. The results show that compared with the milling quality, the appearance quality of rice had more stringent requirements in regard to temperature.

Previous research has suggested that a poor transparency of rice with low amylose content was caused by the cavities of starch granules, and the cavity size was negatively correlated with amylose content [34,35]. The appearance quality of the three types of rice had a greater range of changes under different sowing dates and the seven sowing dates represented actual field growing conditions. However, under the condition of obtaining a relatively high yield, except for LR, the temperature of SMR and SLR at the stage from heading to maturity cannot meet the appearance quality to reach the third grade of China's national standard of high quality paddy. In addition to the higher T_{mean} during grain filling stage, this may also be related to the low amylose content of SMR and SLR [8]. Hence, how to improve the yield and milling quality of SMR and SLR through cultivation measures or variety improvement under the low temperatures needs further research.

4.3. Recommending an Optimal Sowing Date Range for Good Eating Quality Rice in the Lower Reaches of Huai River

The rice-wheat rotation is the main planting mode in the lower reaches of Huai River. The annual harvest time for winter wheat in this area is from 1 June to 15 June [36,37]. Considering the time of harvesting, land preparation, and other agricultural consumption, the earliest sowing date and latest maturity date for rice in the lower reaches of Huai River is 16 May and 5 November, respectively. The sowing date ranges for SMR, SLR and LR under the conditions of the rice–wheat double-cropping system were 16 May–1 June, 16 May–20 May and 16 May–20 May, respectively. The planting area of rice-vegetable, rice-rape and rice-green manure rotations account for 5% of the double-cropped planting area along the lower reaches of Huai River, and the harvest times are 10–15 days earlier than that of wheat [38]. The optimal sowing dates for the three types of rice were 15 May to 1 June, 15 May to 20 May and 15 May to 20 May, respectively.

5. Conclusions

The temperature and solar radiation of six rice varieties showed a decreasing trend at the stage from heading to maturity with the delay of sowing date. Compared with solar radiation, temperature was the main environmental factor affecting the milling and appearance quality of good eating quality rice in the lower reaches of Huai River. Under the condition of obtaining a relatively high yield, the three types of rice can obtain good milling quality. The average temperature and cumulative temperature from May to November in different years have similar changes, although it is difficulties to find every year the same temperature every year during the same sowing days. According to the temperature requirements of different types of rice and the meteorological conditions in the past 10 years, we think that the optimal sowing dates for high yield, good milling and appearance quality production of LR was 15 May to 20 May, the optimal sowing dates for high yield, good milling quality production of SMR and SLR were 15 May to 1 June and 15 May to 20 May, respectively. Proposing an appropriate range for the sowing period is beneficial to improve the milling quality and appearance quality of rice in the lower Huaihe River and similar ecological areas. At the same time, finding a suitable growth temperature for rice with good eating-quality is helpful to cope with the decline in yield and quality caused by future warmer climates. However, more experiments are needed, including research on the physiological mechanism of temperature on rice milling and appearance quality.

Author Contributions: Conceptualization, H.W.; methodology, H.Z.; validation, N.Z. and H.W.; formal analysis, N.Z.; investigation, N.Z.; resources, H.W.; data curation, N.Z.; writing—original draft preparation, N.Z.; writing—review and editing, N.Z.; supervision, H.Z.; project administration, H.W.; funding acquisition, H.W. All authors have read and agreed to the published version of the manuscript.

Funding: This research was funded by the National Key Research Program (grant number 2016YFD0300503); the National Rice Industry Technology System (grant number CARS0127); the National Natural Science Foundation of China (grant number 31971841); the Key Research Program of Jiangsu Province (grant number BE2018355); the Earmarked Fund for Jiangsu Agricultural Industry Technology System, China (grant number JATS[2019]444); and the Project Funded by the Priority Academic Program Development of Jiangsu Higher Education Institutions, China.

Institutional Review Board Statement: The research material is rice and does not require ethical approval.

Informed Consent Statement: Not applicable.

Data Availability Statement: Data is contained within the article.

Acknowledgments: We fully appreciate the editors and all anonymous reviewers for their constructive comments on this manuscript.

Conflicts of Interest: The authors declare no conflict of interest. The funders had no role in the design of the study; in the collection, analyses, or interpretation of data; in the writing of the manuscript, or in the decision to publish the results.

References

1. Du, Y.L.; Zhang, W.J.; Wu, X.R.; Li, G.H.; Wang, S.H.; Liu, Z.H.; Tang, S.; Ding, Y.F. The characteristics of spatial and temporal change of rice yield in Jiangsu Province. *J. Nanjing Agric. Univ.* **2014**, *37*, 7–12.
2. Guerrini, L.; Napoli, M.; Mancini, M.; Masella, P.; Cappelli, A.; Parenti, A.; Orlandini, S. Wheat Grain Composition, Dough Rheology and Bread Quality as Affected by Nitrogen and Sulfur Fertilization and Seeding Density. *Agronomy* **2020**, *10*, 233. [CrossRef]
3. Cappelli, A.; Oliva, N.; Cini, E. A Systematic Review of Gluten-Free Dough and Bread: Dough Rheology, Bread Characteristics, and Improvement Strategies. *Appl. Sci.* **2020**, *10*, 6559. [CrossRef]
4. Zhang, C.X.; Guo, B.W.; Tang, J.; Xu, F.F.; Xu, K.; Hu, Y.J.; Xing, Z.P.; Zhang, H.C.; Dai, Q.G.; Huo, Z.Y.; et al. Combined effects of low temperature and weak light at grain-filling stage on rice grain quality. *Acta. Agron. Sin.* **2019**, *45*, 1208–1220.
5. Wu, H.B.; Liu, D.H.; Zhong, M.; Wang, Y.Y. Research progress of climate factor on quality formation and influence mechanism in rice. *Hubei Agric. Sci.* **2019**, *58*, 13–18.
6. Cheng, F.M.; Zhong, L.J. Variation of rice quality traits under different climate conditions and its main affected factors. *Chin. J. Rice Sci.* **2001**, *15*, 187–191.
7. Cheng, F.M.; Hu, D.W.; Ding, Y.S. Dynamic change of chalkiness and observation of grain endosperm structure with scanning electron microscope under controlled temperature condition. *Chin. J. Rice Sci.* **2000**, *14*, 83–87.
8. Ge, G.K.; Zhang, Z.; Shi, C.H.; Wu, J.G.; Ye, Z.H. Impacts of amylose content or protein content on genetic correlations between the weight of brown rice and appearance quality traits of rice. *Chin. J. Rice Sci.* **2007**, *21*, 44–50.
9. Cooper, N.T.W.; Siebenmorgen, T.J.; Counce, P.A. Effects of Nighttime Temperature During Kernel Development on Rice Physicochemical Properties. *Cereal Chem. J.* **2008**, *85*, 276–282. [CrossRef]
10. Cappelli, A.; Mugnaini, M.; Cini, E. Improving roller milling technology using the break, sizing, and reduction systems for flour differentiation. *LWT* **2020**, *133*, 110067. [CrossRef]
11. Cappelli, A.; Oliva, N.; Cini, E. Stone milling versus roller milling: A systematic review of the effects on wheat flour quality, dough rheology, and bread characteristics. *Trends Food Sci. Technol.* **2020**, *97*, 147–155. [CrossRef]

12. Cappelli, A.; Guerrini, L.; Parenti, A.; Palladino, G.; Cini, E. Effects of wheat tempering and stone rotational speed on particle size dough rheology and bread characteristics for a stone-milled weak flour. *J. Cereal Sci.* **2020**, *91*, 102879. [CrossRef]
13. Cappelli, A.; Canessa, J.; Cini, E. Effects of CO2 snow addition during kneading on thermoregulation, dough rheologica properties, and bread characteristics: A focus on ancient and modern wheat cultivars. *Int. J. Refrig.* **2020**, *117*, 52–60. [CrossRef]
14. Cappelli, A.; Bettaccini, L.; Cini, E. The kneading process: A systematic review of the effects on dough rheology and resulting bread characteristics, including improvement strategies. *Trends Food Sci. Technol.* **2020**, *104*, 91–101. [CrossRef]
15. Gong, J.L.; Zhang, H.C.; Hu, Y.J.; Long, H.Y.; Chang, Y.; Wang, Y.; Xing, Z.P.; Huo, Z.Y. Effects of air temperature during rice grain filling period on the formation of rice grain yield and its quality. *Chin. J. Ecol.* **2013**, *32*, 482–491.
16. Patindol, J.A.; Siebenmorgen, T.J.; Wang, Y.-J. Impact of environmental factors on rice starch structure: A review. *Starch Stärke* **2015**, *67*, 42–54. [CrossRef]
17. Ishimaru, T.; Horigane, A.K.; Ida, M.; Iwasawa, N.; San-Oh, Y.A.; Nakazono, M.; Nishizawa, N.K.; Masumura, T.; Kondo, M. Yoshida, M. Formation of grain chalkiness and changes in water distribution in developing rice caryopses grown under high temperature stress. *J. Cereal Sci.* **2009**, *50*, 166–174. [CrossRef]
18. Tang, S.; Chen, W.Z.; Liu, W.Z.; Zhou, Q.Y.; Zhang, H.X.; Wang, S.H.; Ding, Y.F. Open-field warming regulates the morphologica structure, protein synthesis of grain and affects the appearance quality of rice. *J. Cereal Sci.* **2008**, *84*, 20–29. [CrossRef]
19. Xing, Z.-P.; Wu, P.; Zhu, M.; Qian, H.-J.; Hu, Y.-J.; Guo, B.-W.; Wei, H.-Y.; Xu, K.; Huo, Z.-Y.; Dai, Q.-G.; et al. Temperature and solar radiation utilization of rice for yield formation with different mechanized planting methods in the lower reaches of the Yangtze River, China. *J. Integr. Agric.* **2017**, *16*, 1923–1935. [CrossRef]
20. Yuan, L.M.; Chang, E.H.; Xu, W.; Wang, Z.Q.; Yang, J.C. Effects of low temperature during grain filling on the structure of endosperm in hybrid rice. *Acta. Agron. Sin.* **2006**, *32*, 96–102.
21. Chu, C.Y.; Wang, J.D.; Cheng, Y.; Zhang, Y.; Sun, G.Y. Effect of low temperature treatment in booting and filling stage on yield components and quality of main rice cultivars in Sanjiang Plain. *Chin. J. Agrom.* **2018**, *39*, 751–761.
22. Li, T.; Ryu, O.; Tohru, Y.; Haruto, S. Effects of weak light on rice starch accumulation and starch synthesis enzyme activities at grain filling stage. *Chin. J. Rice Sci.* **2005**, *19*, 545–550.
23. Ren, W.J.; Yang, W.Y.; Xu, J.W.; Fan, G.Q.; Ma, Z.H. Effect of low light on grains growth and quality in rice. *Acta. Agron. Sin.* **2003** *29*, 785–790.
24. Chen, R.; Ersi, K.; Yang, J.; Lu, S.; Zhao, W. Validation of five global radiation models with measured daily data in China *Energy Convers. Manag.* **2004**, *45*, 1759–1769. [CrossRef]
25. Zhou, N.B.; Zhang, J.; Fang, S.L.; Wei, H.Y.; Zhang, H.C. The effects of temperature and solar radiation on yield in good eating-quality rice grown in the lower reaches of the Huai River Basin. *J. Integr. Agric.* under review.
26. Wang, W.T.; Zhang, T.F.; Ma, P.L.; Pu, J.Y.; Lü, F.P.; Huang, B. Response of abnormal growth of winter wheat to climate warming in the Loess Plateau of Longdong, Gansu Province. *Chin. J. Ecol.* **2008**, *27*, 1492–1497.
27. Song, Z.H.; Zhang, Y.G.; Lu, K.D. Influence of temperature on the super rice polished rice rate and the size of chalkiness in the grouting period. *Seed* **2009**, *28*, 66–68.
28. Rehmani, M.I.A.; Wei, G.; Hussain, N.; Ding, C.; Li, G.; Liu, Z.; Wang, S.; Ding, Y. Yield and quality responses of two indica rice hybrids to post-anthesis asymmetric day and night open-field warming in lower reaches of Yangtze River delta. *Field Crop. Res* **2014**, *156*, 231–241. [CrossRef]
29. Zhang, G.F.; Wang, S.H.; You, J.; Wang, Q.S.; Ding, Y.F.; Ji, Z.J. Effect of higher temperature in different filling stages on rice qualities. *Acta Agron. Sin.* **2006**, *32*, 283–287.
30. Li, L.; Sha, G.D.; Lu, J.H. Effect of temperature and light factors on rice quality during rice filling stage. *Chin. J. Agrom.* **1989**, *10* 33–38.
31. Dou, Z.; Tang, S.; Chen, W.; Zhang, H.; Li, G.; Liu, Z.; Ding, C.; Chen, L.; Wang, S.; Zhang, H.; et al. Effects of open-field warming during grain-filling stage on grain quality of two japonica rice cultivars in lower reaches of Yangtze River delta. *J. Cereal Sci.* **2018** *81*, 118–126. [CrossRef]
32. Zhang, G.L.; Zhang, S.T.; Wang, L.; Xiao, Y.H.; Tang, W.B.; Chen, G.H.; Chen, L.Y. Effects of high temperature at different times during the heading and filling periods on rice quality. *Sci. Agric. Sin.* **2013**, *46*, 2869–2879.
33. Yang, T.T.; Sun, Y.N.; Zeng, Y.H.; Huang, S.; Zhang, J.; Tan, X.M.; Zeng, Y.J.; Pan, X.H. Effect of post-anthesis warming on the grain yield and quality of double-cropped high-quality rice cultivars. *J. Nucl. Agric. Sci.* **2019**, *33*, 0583–2591.
34. Zhang, L.; Zhao, L.; Zhang, J.; Cai, X.-L.; Liu, Q.; Wei, C. Relationships between transparency, amylose content, starch cavity, and moisture of brown rice kernels. *J. Cereal Sci.* **2019**, *90*, 102854. [CrossRef]
35. Lu, Y.; Zhang, X.M.; Qi, Y.; Zhang, C.Q.; Ling, Y.P.; Liu, Q.Q. Scanning electron microscopic analysis of grain cross-section from rice with different transparency. *Chin. J. Rice Sci.* **2018**, *32*, 189–199.
36. Ji, H.J.; Shi, G.Y. Characteristics and High-yielding cultivation techniques of super late sowing wheat in northern Jiangsu Province *Bull Agric. Sci. Technol.* **2018**, *2*, 178–180.
37. Jiang, S.Q.; Xue, D.H.; Jiang, Z.Q. Obstacle factors and countermeasures of high and stable yield of wheat in Yancheng area of northern Jiangsu Province. *Mod. Agric. Sci.* **2016**, *20*, 36–38.
38. Shi, G.Y.; Cao, J.H.; Min, H.X.; Gao, D.F. High yield and high efficiency cultivation techniques of rice in greenhouse vegetable stubble. *Bull. Agric. Sci. Technol.* **2014**, *3*, 172–174.

agronomy

MDPI

Review

Perspectives and Advances in Organic Formulations for Agriculture: Encapsulation of Herbicides for Weed Control

Francisco J. Rodríguez-Mejías [1,2,*], Aurelio Scavo [3], Nuria Chinchilla [1], José M. G. Molinillo [1], Stefan Schwaiger [2], Giovanni Mauromicale [3] and Francisco A. Macías [1,*]

1 Allelopathy Group, Department of Organic Chemistry, Institute of Biomolecules (INBIO), Campus CEIA3, School of Science, University of Cádiz, C/República Saharaui, 7, Puerto Real, 11510 Cádiz, Spain; nuria.chinchilla@uca.es (N.C.); chema.gonzalez@uca.es (J.M.G.M.)
2 Institute of Pharmacy/Pharmacognosy, Center for Molecular Biosciences Innsbruck (CMBI), University of Innsbruck, CCB, Innrain 80/82, 6020 Innsbruck, Austria; stefan.schwaiger@uibk.ac.at
3 Department of Agriculture, Food and Environment (Di3A), University of Catania, 95123 Catania, Italy; aurelio.scavo@unict.it (A.S.); gmauromi@unict.it (G.M.)
* Correspondence: francisco.rodriguez-mejias@uibk.ac.at (F.J.R.-M.); famacias@uca.es (F.A.M.)

Abstract: This article offers a critical analysis of the evolution of encapsulation methods for herbicides and natural products, with a main focus on organic formulations. It extols the possibilities presented by these micro- and nanomaterials, such as their slow release, stability, bioavailability, water solubility, and stability for classical and natural herbicides from their origins to the present.

Keywords: nanoencapsulation; allelochemical; organic nanoparticle; weed management; bioherbicides; allelopathy; polymeric nanoparticles; formulation

Citation: Rodríguez-Mejías, F.J.; Scavo, A.; Chinchilla, N.; Molinillo, J.M.G.; Schwaiger, S.; Mauromicale, G.; Macías, F.A. Perspectives and Advances in Organic Formulations for Agriculture: Encapsulation of Herbicides for Weed Control. *Agronomy* **2023**, *13*, 1898. https://doi.org/10.3390/agronomy13071898

Academic Editor: Ilias Travlos

Received: 5 June 2023
Revised: 11 July 2023
Accepted: 13 July 2023
Published: 18 July 2023

1. Introduction

"There is need for the development of safe and effective controlled release formulations of pesticides. [. . .] formulations could make it possible to use smaller amounts of pesticides and perhaps even improve performance efficiency [1]." These were the words of Richard G. Sinclair, the author of the first research paper on the encapsulation of agrochemicals in 1973. The paper was entitled 'Polymers of Lactic and Glycolic Acids as Ecologically Beneficial, Cost-Effective Encapsulating Materials' and was based on the pillars established by the Pennwalt Corporation in 1972, whose development of Penncap-M shook up the agricultural field. This first agro-material was based on polyamide spheres in which methyl parathion was encapsulated, and the spheres were spread by spraying an aqueous suspension. This was a broad-spectrum pesticide that was mainly used to control insects, such as caterpillars, beetles, and grasshoppers [2]. However, it was the starting point for the encapsulation of agrochemicals, and these techniques were recently applied for herbicides. The commercial pesticide Penncap-M® (O,O-dimethyl-O-p-nitrophenyl phosphorothioate) is currently being recalled, as it is a health risk for humans and is banned from sale and import in nearly all countries around the world [3,4].

Other renowned enterprises, such as 3M Corp and National Cash Register (currently known as NCR Corporation), also began large-scale field trials in 1973. The systems tested include 'biodegradable plastic compositions' and 'proteinaceous films' [2]. These companies started a race to develop the safest, cheapest, and most profitable encapsulation system, and this race continues today.

Scientific research on the encapsulation of agrochemicals has been influenced by market demands, but this is always with some delay. The scientific community is focused on advancing knowledge and humanity, and studies have been carried out to identify natural products as alternatives to classical herbicides and to replace field-persistent encapsulation structures with ecologically sound materials. In this respect, the number of studies focused on this topic has markedly increased over the last decade and, particularly, in the last four years, i.e., 2018–2022 (Figure 1); this has maintained the high impact of this subject.

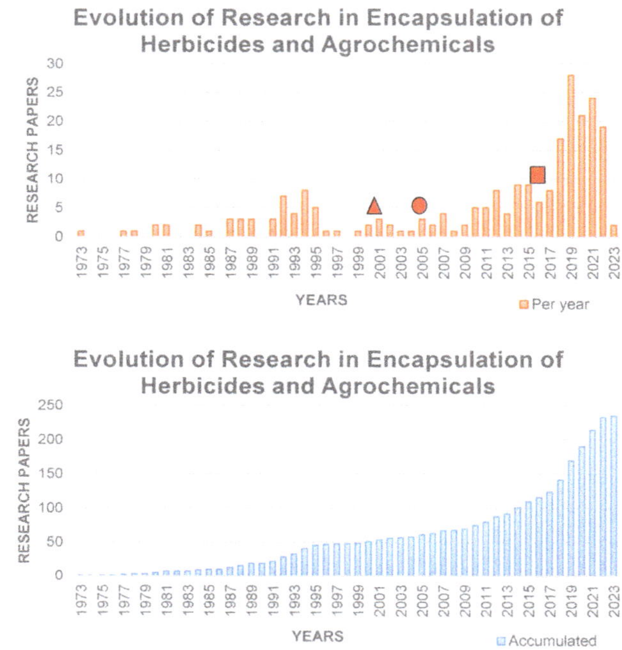

Figure 1. Evolution of research on the encapsulation of herbicides/bioherbicides according to Scifinder® (Keywords: Encapsulation, Herbicide, Agrochemicals, Nanoecapsulation, Natural Products). (Triangle) First publication about atrazine contamination of drinking water in the USA (2001) [5]. (Circle) The use of Atrazine and Alachlor in the European Union is banned (2004) [6]. (Square) Plan to ban glyphosate in most countries of the European Union within two years, and some states of the USA start to evaluate the adverse effects (2016) [7]. (**Top**) Papers published per year. (**Bottom**) Cumulative papers published per year.

The use of encapsulation has been very successful in terms of property modification, the application of smaller amounts of herbicides, and enhancements in stability. These modifications also result in higher water solubility, lower soil and environmental pollution, and more targeted products. Since the 1970s, this approach has been applied to classical herbicides. Such chemicals have very limited pollution control and little specificity in terms of their mode of action. This undisciplined approach has led to a rapid increase in herbicide-resistant weed species worldwide, which has led to higher herbicide application rates and the use of other active principles with longer environmental persistence [8]. Furthermore, Hulme stated that the number of herbicide-resistant weeds is probably underestimated and that agronomic drivers suggest that, in many countries, the number of resistant weeds will increase [9]. As a consequence, in recent decades, the use of natural alternatives for weed control, crop protection, and increased production has been promoted. In this respect, organic encapsulation has been successfully applied to these new natural and nature-inspired options.

The benefits associated with bioherbicides/allelochemicals can be summarized as follows: natural origin of the chemical compounds, low impact on the environment, new modes of action against weeds, and public acceptance [10–12]. However, there are still barriers that limit the use of these systems under natural conditions, and these include their low water solubility, rapid biodegradation in the environment, and high cost of synthesis, among others [13]. Roberts et al. stated that, in order to be successfully integrated within weed management systems, bioherbicides should have a suitable formulation, be economically sustainable, cause a high mortality rate for target plants, and lead to very limited or zero impact on the surrounding natural environment and human health [14]. Current examples include the encapsulation of phytotoxic sesquiterpene lactones in organic nanotubes that show activity against *Phalaris arundinacea* L., *Lolium perenne* L., and *Portulaca oleracea* L. weeds or monoterpenes encapsulated in organoclays for the prevention of volatilization [15,16]. The release rate is another important factor; in general, larger sizes facilitate a gradual and prolonged release of the active substances, while smaller particles allow a more homogeneous dispersion, increase the release rate, and facilitate the transport and absorption of the substances. This results in a controlled release of the active substance. For this reason, among others, different technologies that allow increasingly smaller encapsulation sizes have been developed. In general, one can speak of microencapsulation when the particles are between 1 and 1000 μm and of nanoencapsulation when the particle size is smaller, down to 10 nm [17,18].

The aim of this paper is to provide a perspective on how encapsulation systems have evolved and discuss the experimental results that have been obtained in field studies. The main focus is on the most relevant and promising organic encapsulation systems that have been studied to develop safer, non-persistent, and ecological agrochemicals.

2. Perspectives and Analysis of Organic Encapsulation Systems Employed for Weed Control

2.1. Cyclodextrins and Macrocycles

A large number of compounds encapsulated with cyclodextrins (CDs) have been used in the field of medicinal chemistry, but the use of these systems for weed control and crop enhancement is very limited. Szejtli was the first to report the safe application of CDs to plants by analyzing the physiological effect of this macrocycle on seeds from crops of interest [19]. He studied the phytotoxic effect of β-CD, and two years later, in 1985, he applied the encapsulation method to several herbicides (e.g., molinate, dichlobenil, and benthiocarb, among others), pesticides, and fungicides [20]. Since then, several studies have focused on the complexation of a range of CDs and herbicides, although these have only concerned supramolecular properties, such as solubility or soil stability, and biological applications have not been considered [21,22]. Lezcano et al. reported the complexation of fungicides with these macrocycles—specifically, with the three natural CDs (α, β, and γ) [23,24]. However, only complex production and characterization were described, without reference to biological applications. Comparable results were published by Benfeito et al., who used 2-hydroxypropyl-β-cyclodextrin to host Oxadiargyl (5-tert-butyl-3-[2,4-dichloro-5-(prop-2-ynyloxy)phenyl]-1,3,4-oxadiazol-2(3*H*)-one) [25]. Interestingly, CDs were also tested as soil remediators, but these were not combined with applications for crop protection or enhancement purposes [26,27].

The first experimental application of this formulation method was not reported until 2017. Cala et al. encapsulated three sesquiterpene lactones (Figure 2), and these showed phytotoxic effects against parasitic plants (*Orobanche cumana* Wallr., *Orobanche minor* Sm., *Orobanche ramosa* L. (syn.: *Phelipanche ramosa* (L.) Pomel), *Orobanche aegyptiaca* Pers. (syn.: *Phelipanche aegyptiaca* (Pers.) Pomeland) *Striga hermonthica* (Delile) Benth.) of the Fabaceae and Asteraceae families, but also tomato, maize, and sugar cane. This study revealed that β-CD encapsulation improved the water solubility of these allelochemicals and enhanced their bioactivity when compared to that of free sesquiterpenes, and it also highlighted this as a potential pre-emergence herbicide for food production [28]. Another sesquiterpene lactone, Inuloxin A (Figure 2), was also tested against *Orobanche ramosa* L. (syn.: *Phelipanche ramosa* (L.) Pomel) after complexation with β-CD [29].

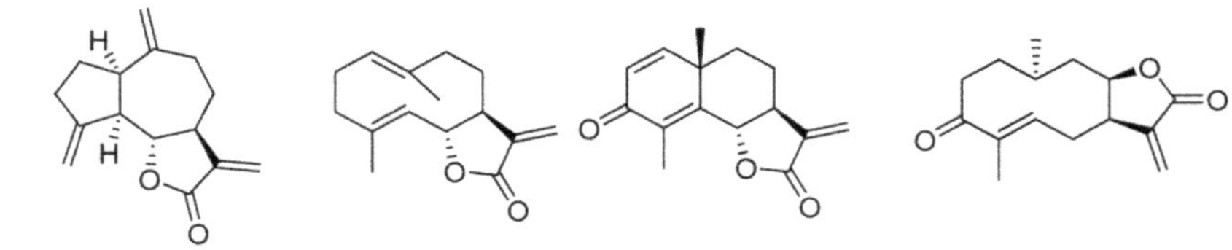

Dehydrocostuslactone Costunolide α-Santonin Inuloxin A

Figure 2. Sesquiterpene lactones encapsulated with cyclodextrins to fight parasitic plants.

In addition to those used against parasitic plants, formulations for combatting *Echinochloa crus-galli* (L.) P.Beauv. are the second main focus for the agrochemical application of cyclodextrin formulations in crop protection. Atrazine [30], butachlor [31], cyanazine [32], and diuron [33] have been complexed with β-CD or its 2-hydroxypropyl derivative. In all cases, the inhibition of different plant parameters was higher when compared to that obtained with free herbicides. For example, plant height, root length, and fresh weight were directly affected by encapsulation with CDs. HP-β-CD and γ-CD were recently employed to encapsulate 2,2′-disulfanediyldianiline (DiS-NH_2), an aminophenoxazinone mimic, to target *Portulaca oleracea* L., *Plantago lanceolata* L., and *Lolium rigidum* Gaudin, which are problematic weeds in rice, wheat, and barley cultures, respectively. The results showed an enhancement of the water solubility and bioactivity for the γ-CD complex, with inhibition values higher than 80% with respect to the control for germination, shoot length, and root formation of *P. lanceolata* [34].

Interestingly, there is a significant gap in the information concerning the use of CDs in field experiments and the more dominant in vitro tests. The results in the literature support the application of these systems in field experiments, but there is a lack of further research focused on this area. Furthermore, most research has focused on β-CD, and the other natural CDs have been largely overlooked (Figure 3). β-CD is approved by the EFSA (Food code: E459), and this fact has encouraged research on crops for human consumption [35]. However, β-CD has the lowest water solubility of the CD family. The inclusion in the structure of 2-hydroxypropyl substituents improves solubility, and this explains why there are some research papers on this macrocycle. Many authors seem to be attracted by new nanostructures, and natural formulation methods are often overlooked, though we should, in fact, seek to rediscover them. γ-CD, which allows the generation of higher-order complexes (1:2) with respect to the guest, seems to be a particularly economically interesting option due to the lower amount of cyclodextrin that is required.

Other macrocycles have recently been studied for weed control, but these are synthetic materials. One example is cucurbit[*n*]urils (CB*n*), whose main structural motifs are glycoluril units, and these can usually be obtained with 5–8 subunits. Most of the studies on herbicide encapsulation concern physicochemical characterization, as in the cases of ametryn [36], atrazine, and imazapyr [37,38], but their biological activity was not described. Nevertheless, the encapsulation of natural phytotoxic aminophenoxazinones and their sulfur mimics by complexation with CB7 has recently been reported, and these displayed improved phytoactivity in the growth of wheat (*Triticum aestivum* L.) models when compared with the free compounds [39].

The formulation process for using CDs as host materials is rather simple, and no extra adjuvants or steps are needed, apart from mixing the correct concentrations once the binding constant is known. They are also natural products, so this is a green approach for formulations. Current biotechnological production makes their obtention cheap. Furthermore, the main units of CDs are glucose units, which have been demonstrated to enhance the bioavailability of the drugs/herbicides encapsulated.

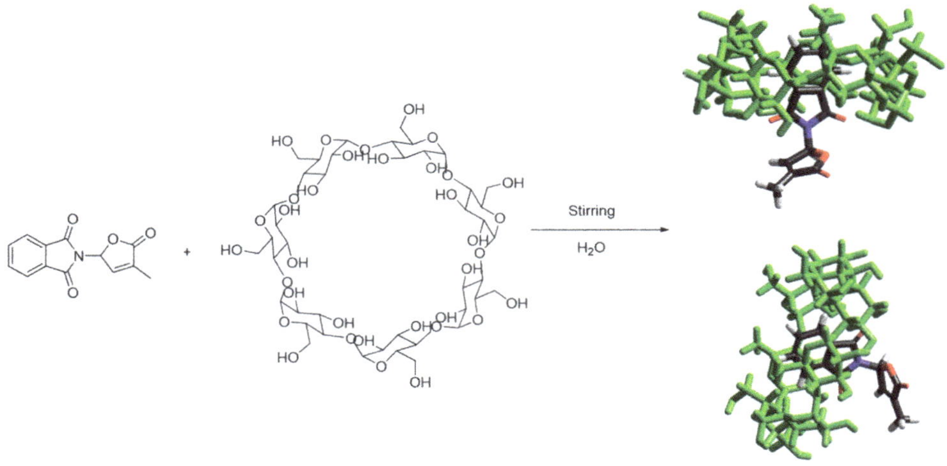

Figure 3. Scheme of the procedure of encapsulation with cyclodextrins and herbicides. This example includes the formation of PL01@βCD. Reprinted/adapted with permission from Ref. [36]. Copyright 2023, copyright owner Royal Society of Chemistry.

2.2. Clays

Clays have been extensively studied in an agronomic context, as they are porous materials that are present in soil. There are several research papers on the adsorption of herbicides onto soil clays and on how this reduces the efficacy of herbicides. In contrast, the use of clays as carriers has not been widely investigated, but this has changed over the last decade. The main advantage of this approach is the biocompatibility of the material with the medium in which the crop and weeds grow. In most cases, this is a green approach because the encapsulating or carrier material is already present as part of the soil.

The first applications of clays were reported in 1984, i.e., around ten years after the first use of encapsulation in agrochemistry. Connick et al. employed a kaolin clay to adsorb 2,6-dichlorobenzonitrile and studied its properties as a carrier to control common purslane (*Portulaca oleracea* L.), broadleaf signalgrass (*Brachiaria platyphylla* (Munro ex C.Wright) Nash), goosegrass (*Eleusine indica* Gaertn.), and large crabgrass (*Digitaria sanguinalis* (L.) Scop.) in vitro [40]. These weeds occur in corn, cotton, soybean, rice, and wheat cultures, and they cause yield losses of up to 20% [41–44]. Further research on clays for herbicide/weed control was not carried out until 1994, when Carr et al. developed an interesting method with montmorillonite to support starch with encapsulated metolachlor and atrazine [45]. However, these formulations were not applied in the field or in vitro. Montmorillonite has also been used to encapsulate chloridazon and metribuzin [46], glyphosate [47], paraquat [48,49], and picloram [50], but these studies are limited to the characterization of the encapsulated agrochemical compound in terms of release, stability, and water solubility. Generally, the encapsulation method involves the preparation of the clay in the presence of the herbicide to enhance the probabilities of capture in the clay pores, as observed in Figure 4.

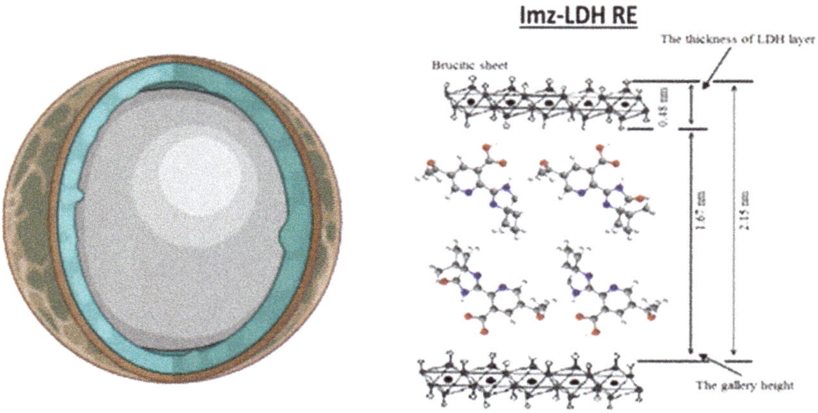

Figure 4. Encapsulation method of imazamox in a cationic nanoclay (Cloisite 10A). Partially reproduced from [51] with permission from Elsevier (license number: 5522980963504).

Mixtures of starch with different clays/organoclays, inspired by the work of Carr et al., have also been reported in recent years. These materials showed interesting properties. For example, isoproturon encapsulated in sodium montmorillonite with carboxymethyl starch-based micro-particles gave a reduction of around 90% in the herbicide released per irrigation of the soil [52]. This enabled the long-term delivery of the herbicide and, thus, reduced the pollution effect. In addition, a similar starch/montmorillonite nanocomposite with encapsulated ametryne displayed an interesting photoprotective effect on the herbicide [53]. This prolonged the action time for weed control, thus avoiding an extra application of the herbicide for days.

Chitosan has been employed as a matrix to be dispersed on the surface of clays. For example, the herbicide imazamox has been encapsulated in a chitosan matrix and adsorbed on sodium-enriched montmorillonite. This system showed good in vitro phytotoxicity for standard target species, such as cauliflower (*Brassica oleracea* var. botrytis L.) [54]. Similar results were obtained with imazamox encapsulated in cloisite clay and a modified quaternary alkylammonium montmorillonite clay in combating the invasive plant *Brassica nigra* W.D.J.Koch [51,55].

In terms of nanomaterials mixed with clays, additives other than starch have been employed, and these include phosphatidylcholine vesicles. In this case, atrazine and alachor were encapsulated in vesicles and supported on sodium montmorillonite. The resulting materials were tested in vitro against green foxtail (*Setaria viridis* (L.) P.Beauv.) germination [56]. This weed affects late-seeded wheat (*Triticum aestivum* L.), sugarbeet (*Beta vulgaris* L.), and maize (*Zea mays* L.) [57]. An experiment was designed to determine the content of the herbicide and its efficacy. The authors prepared a soil column and added the nanocomposite, which was then eluted with water. Green foxtail seeds were distributed at different heights in the column, and germination was evaluated to assess the release of the herbicide. A similar technique was employed for atrazine and imidacloprid encapsulated in chitosan and supported on bentonite clay [58], as well as for sulfosulfuron encapsulated in montmorillonite to target green foxtail [59]. This approach provided interesting data about the release profile from the clay. Other cases of mixed nanomaterials have been published, and these include encapsulation of the herbicide in micelles with subsequent adsorption on clays. Research on alkylpolyglucosides, ethoxylated amines [60], and octadecyltrimethylammonium bromide micelles (ODTMs) [61–64] has been published, with sepiolite and sodium montmorillonite acting as carriers. Pendimethalin was encapsulated with ODTMs and montmorillonite, and it was shown to be effective in reducing the root penetration of tomato (*Solanum lycopersicum* L.) into greenhouse drippers, thus enhancing the yield of this fruit.

Natural bioherbicides based on allelochemicals have been encapsulated in clays. *S*-Carvone, a monoterpene that is usually isolated from spearmint (*Mentha spicata* L.) or caraway (*Carum carvi* L.), was encapsulated in an organobentonite clay modified with dimethyl, benzyl, and hydrogenated alkyl tallow quaternary ammonium salts [15]. Its bioactivity was tested in vitro on standard target species (specifically, *Lactuca sativa* L.) in terms of general phytotoxicity, and it was found that the formulation improved the inhibition of shoots and germination when compared to the free compound. Similar results were obtained by Galán-Pérez et al. when encapsulating scopoletin in montmorillonite clays with the same modifications as those outlined for the previous organobentonite. This formulation also showed phytotoxic effects on *Lactuca sativa* L. germination and root length, and the results were better than those for free scopoletin [65].

More biological studies have been carried out on clay encapsulation than on macrocycle complexation, and there is, therefore, more knowledge on these systems for both classical herbicides and allelochemicals. However, despite the ecofriendly nature of this approach, field experiments have not been widely employed. The most remarkable results were reported by Galán-Jiménez et al. on the encapsulation of the herbicide mesotrione in sepiolite clays. These materials were applied post-emergence on a maize (*Zea mays* L.) crop to target broadleaf weeds between maize rows. The authors performed the experiment on an area of 0.216 ha and observed better results in terms of maize yield when compared to the positive control mesotrione/atrazine. The formulation was applied by directly spraying on the weeds [66]. The potential applications of this encapsulation technique remain unexploited when compared with currently available systems. Novel encapsulation methods could be interesting, but the biocompatibility of clay particles with the soil is a key factor in terms of a green approach, and these carriers have shown interesting properties for slow release and in-depth soil applications.

2.3. Matrices from Starch to Hybrids

Starch matrices were among the most relevant systems for encapsulation in the early research into this approach in agriculture. The modification of starch with xanthates or alkali chlorides generates microporous organic materials that are useful for the encapsulation of herbicides. The earliest system was developed with butylate and diazinon as bioactive compounds in the fight against foxtail (*Hordeum murinum* L.), which infests barley crops [67]. Other herbicides, such as EPTC [68,69] and trifluralin [70–72], were later encapsulated. Starch is readily available and cheap, and methods for chemical modification are well established. It is noteworthy that the application of this method leads to an enhancement in the persistence of the herbicide as the volatility is decreased. The increased interest in starch has allowed more in-depth characterization, and authors have studied how different levels of amylose/amylopectin in starch improve herbicide release in soil [73,74]. Bioassays were carried out, especially via field testing, with trials on encapsulated trifluralin against *Echinochloa crus-galli* L., which infests soybean (*Glycine max* L.) [75], and against foxtail (*Hordeum murinum* L.) [76]. In the latter case, different ions were evaluated, and it was found that calcium and borate were the best combination for achieving slow release.

There were reports about the environmental risks of trifluarin [77,78], and research over the following decade focused more on other classical herbicides, e.g., atrazine [79–83] and alachlor [82,84–86]. Strategies other than adduct formation were studied, e.g., twin-screw extrusion. However, the use of these techniques to produce starch for herbicide encapsulation generates slurries that, despite showing promise in vitro, were ruled out in subsequent research papers due to their problematic soil distribution in field experiments [87–90]. Ion adducts with starch were produced by Fleming et al. [91] and Reed et al. [92], who obtained interesting results through the encapsulation of alachlor/metribuzin with a starch–borate matrix and EPTC/butylate with a starch–iron ($FeCl_3$) matrix, respectively. In the former case, the encapsulated system led to an enhancement in soybean crop yield and protection against large crabgrass (*Digitaria sanguinalis* (L.) Scop.), foxtail millet (*Setaria italica* (L.) P.Beauv.), and longspine sandbur (*Cenchrus longispinus* (Hack.) Fernald). The application of starch–iron inhibited several weeds, such as johnsongrass (*Sorghum halepense* Pers.), giant foxtail (*Setaria faberi* R.A.W.Herrm.), and redroot pigweed (*Amaranthus retroflexus* L.), due to the enhanced release of the herbicides. However, this approach could cause high iron accumulation in the soil. Green approaches should take precedence over the efficacy of the formulation, and fortunately, this is the current trend.

In recent years, interest in starch as an encapsulation system for agrochemicals has decreased due to increased research on new materials, such as nanoparticles, new polymers or biomaterials that offer different physicochemical properties. In terms of applications starch materials are still very interesting due to their low cost, biocompatibility, and low soil pollution. However, this material does suffer from some drawbacks, such as low thermal stability and strong retention of the encapsulated bioactive compound. Researchers have therefore, studied the hybridization of starch to enhance these properties. One such example is the use of starch-coated clay (montmorillonite) to encapsulate ametryn [93] and a mixed starch–alginate matrix to encapsulate 2,4-D [94,95]. However, the biological efficacy of these hybrids was not studied.

The first application of allelochemicals encapsulated in starch is an interesting example. This system was developed by Alipour et al. in 2019 and involved the encapsulation of rosemary essential oil (*Rosmarinus officinalis* L.) to control weeds such as amaranth (*Amaranthus retroflexus* L.), common purslane (*Portulaca oleracea* L.), and knapweed (*Acroptilon repens* (L.) DC.) [96]. The oil was trapped in the starch matrix, and this allowed its application as a solid. The same strategy was employed with savory (*Satureja hortensis* L.) essential oils, albeit encapsulated in a different type of matrix, namely, an Arabic gum matrix and apple pectin. This approach also led to high growth inhibition in the pre-emergence mode for amaranth weed [97]. Further studies on matrices in agriculture for weed control were carried out last year. Carboxymethyl chitosan [58,98], carboxymethyl [99], or ethyl-cellulose [100,101] and lignin [102,103] are the most interesting materials for herbicide and bioherbicide encapsulation on the basis of properties such as their release delivery, and stability. These matrices were used in conjunction with metolachlor, 2,4-D and atrazine, amongst others, but biological results were not obtained in vitro or in the field to demonstrate their efficacy.

Other matrices are currently under investigation, as they are readily available from natural sources and they show appropriate physicochemical properties a priori. Examples include β-CD nanosponges, which are obtained by crosslinking cyclodextrins [104], and biochars, which are stable carbon-rich materials formed through the pyrolysis of biomass under oxygen-limited conditions [105]. These materials were used to encapsulate the post-emergence herbicide nicosulfuron and natural coumarins, respectively. Only in the case of biochar@coumarin was phytotoxicity evaluated in *Lactuca sativa* L. models.

Similarly to the encapsulation method for clay systems, current formulation systems with matrices apply an in situ method to keep the bioactive component inside. The polymeric grid or structure is self-assembled while the herbicide is dispersed in the media. This increases the encapsulation efficiency and conveniently reduces the number of steps in the formulation. Figure 5 shows an example of the methodology for the encapsulation of agrochemicals with new polymers based on polyethylene glycol.

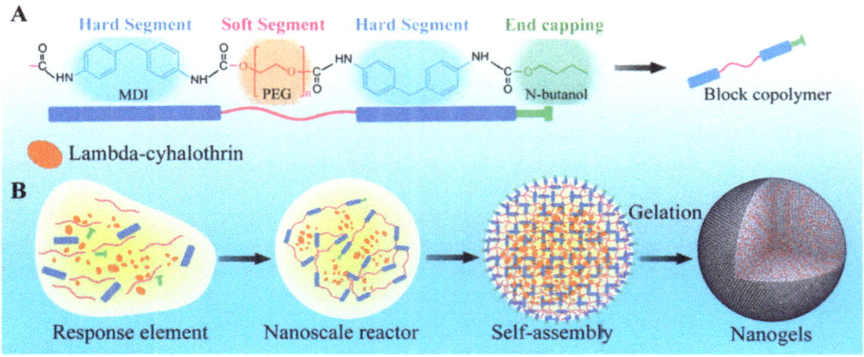

Figure 5. (**A**) Synthesis of block copolymers through the self-assembly of reaction elements. (**B**) Schematic of one-step synthesis of herbicide-loaded flexible nanogels. Reprinted (adapted) with permission from [106]. Copyright 2021 American Chemical Society.

Matrices, particularly starch, are still of great interest for field applications in weed control. The possibility of combining matrices with new biomaterials that are under development could improve properties and applications, especially in the case of allelochemicals. However, many more biological studies on the new matrices are required and are a prerequisite for future applications.

2.4. Micro- and Nanoparticles

The relevance of organic micro- and nanoparticles can be seen in Figure 6. These types of particle are the major contributors in the representation of the most widely employed methods for encapsulation. These contributions have undergone exponential growth in the last 15 years, and this is much more than any other formulation method for weed control. This increase is due to improvements in characterization techniques, such as electron microscopy, and the boost in polymer engineering.

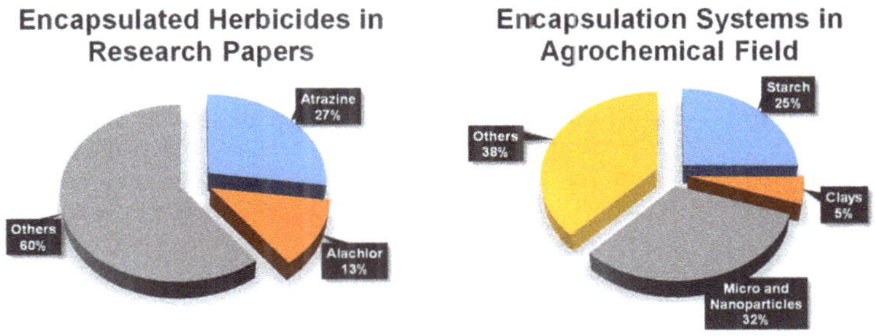

Figure 6. (**Left**) Most commonly studied herbicides for encapsulation. (**Right**) Most widely used systems for encapsulating agrochemicals.

The first use of microparticles for weed control involved the encapsulation of chlorpropham to target several grass weeds that infect tomatoes, safflowers, and onions [107]. It is interesting to note that this encapsulation was inspired by the volatilization issues associated with this herbicide. Therefore, the intention of the authors was to improve the persistence of the bioactive compound, as in the case of the early starch encapsulation approach. This idea contrasts with the current approach of nano- and microparticle encapsulation of various kinds of herbicides. However, Petersen and Shea exploited this idea for slow release and established the modern concepts of encapsulated herbicides for crop protection. Polyurea polymers were used to encapsulate alachlor, and the efficacy was demonstrated on *Triticum aestivum* L., which was protected for a longer time than with free alachlor [108].

Researchers subsequently employed different organic polymers, such as polylactic acid [109–111], polyvinyl alcohol [112], chitosan [112–115], poly(hydroxyvalerate) [116,117], and ethyl cellulose [118–122], for encapsulation in weed control. Norfluorazon, alachlor, and 2,4-D are the most widely studied herbicides, but it is worth highlighting the study by Chang et al., which is one of the first field studies on the bioactivity of organic nanoparticles/microparticles without an encapsulated bioactive compound [109]. These authors showed that the carriers alone can also stimulate the growth and yield of soybeans. This finding established the interesting pillars of new encapsulation models that address the dual effect of the phytotoxicity of the core and the synergistic properties of the shell.

The work by Quiñones et al. is worth highlighting, as it is the first report on allelochemical encapsulation with this system [114]. Brassinosteroids, which are usually isolated from *Brassica napus* L., were encapsulated in chitosan microparticles. The resulting materials were characterized, but they were not biologically tested. A similar approach was employed by Cho et al. [123] with the encapsulation of a vitamin B1 derivate in lecithin nanoparticles. However, the biological evaluation only showed good results against fungal infection prevention on white radish (*Raphanus sativus* L.), and relevant activities against weeds were not observed.

In the last ten years, nanoparticle encapsulation in agrochemistry has been improved by using new polymers that had already been tested for biological purposes. Poly(ε-caprolactone) [124–129] or alginate polymers [130–132] have attracted attention, and, for example, they have been applied against invasive plants such as *Brassica* spp. It is curious, however, that the increase in the number of publications about the encapsulation of herbicides with these structures does not necessarily correlate with a higher number of in vitro or field experiments [125,132–134]. Several papers were only concerned with the characterization or physicochemical properties, and any enhancement in weed control activity was only assumed. However, this trend changed dramatically around 2018 with the new requirements for publications, and most of the papers published later contain data from biological evaluations. As a consequence, more papers have been published on the encapsulation of new commercially available herbicides. Polymers have been explored in greater detail, and they have been tested on a variety of weeds and invasive plants. For example, poly(methylmethacrylate) has been employed to encapsulate haloxyfop and Gallant® in the fight against duckweed (*Lemna minor* L.) and greater duckweed (*Spirodela polyrhiza* (L.) Schleid.), both of which are invasive aquatic plants that particularly affect crops that have a high water demand [135–138]. The published papers describe the better efficacy of the herbicide in the encapsulated version and a reduction in water pollution. Several interesting mechanistic studies have been described in which the delivery processes in plant cells were examined to understand the mode of transport. One such example is atrazine encapsulated by poly(ε-caprolactone) nanoparticles, which were tested for the control of *Brassica juncea* (L.) Czern., which infests spring grain crops. The authors discovered that the formulation allowed penetration into the leaf tissue, with the formulation reaching the mesophyll through the stomata. This encapsulation improved the efficacy of the herbicide more than ten-fold, and side effects due to the capsule were not observed. In the same context, Falsini et al. explored the delivery mechanism of gibberellic acid encapsulated in lignin nanoparticles. This represents the first application of natural polymers for encapsulating a natural product, and the authors showed how the lignin nanoparticles entered the root of the seedling through cortical cells to enhance the growth of tomatoes (*Solanum lycopersicum* L.) and arugula (*Eruca vesicaria* (L.) Cav.) [139].

The trend in the application of allelochemicals has subsequently increased, but the isolation and synthesis of natural products still limit industrial approaches. For this reason, natural extracts are more commonly encapsulated with organic nanoparticles than with pure compounds. For example, Synowiec et al. employed maltodextrin nanoparticles to encapsulate caraway (*Carum carvi* L.) essential oil and obtained good results against

Echinochloa crus-galli (L.) P.Beauv. and *Galinsoga parviflora* Cav. Weeds, which infect rice and potato crops [140]. Taban et al. also encapsulated essential oil for agrochemical application, but this was sourced from savory (*Satureja hortensis* L.) and encapsulated with Arabic gum nanoparticles. This agro-nanomaterial showed high specificity in the control of *Amaranthus retroflexus* L. in post-emergence treatment without harming tomato crops [141]. This new strategy facilitates the desired green approach in agriculture for the replacement of classical herbicides, and in vitro and field experiments are currently supporting fully organic bioherbicides from the core (allelochemical) to the shell (formulation).

2.5. Metal–Organic Systems

In the past, metal–organic systems for encapsulation were inspired by the use of metalloids such as boron in starch–borate systems for butylate and *S*-ethyl dipropylthiocarbamate [142,143]. Currently, organometallic approaches have also been applied in formulations, especially in recent years since the discovery of metal–organic frameworks (MOFs). These materials are synthesized with zinc [144–146], iron [147–149], or gadolinium [150] as metal cores, and they display interesting properties in terms of their stability, delivery, and pH-responsiveness. Wang et al. tested 2,4-D encapsulated in Fe-MOFs in vitro against *Cichorium intybus* L. and found improved growth inhibition in comparison with that of the free herbicide. Similar phytotoxicity results were obtained with Zn-MOFs in which disulfide herbicides were encapsulated in tests against *Lolium rigidum* Gaudin, *Echinochloa crus-galli* (L.) P.Beauv., and *Amaranthus blitum* L. (syn.: *Amaranthus viridis* All.). These weeds mainly affect rice, corn, and potato crops, and the aforementioned formulation method led to a reduction in the root formation of the weeds that was twice as good as that of commercial herbicides and 5–10 times better than that of the non-encapsulated compound (Figure 7).

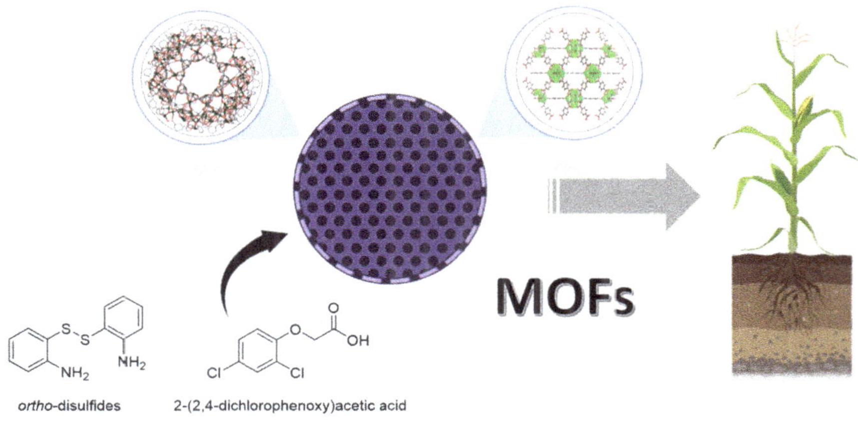

Figure 7. Scheme of encapsulation with *o*-disulfides and 2,4-D in metal–organic frameworks based on Fe and Zn, respectively [144,146]. Those agromaterials displayed phytotoxicity against weeds and protective effects on the crops.

Copper and silver are the metals that are most widely employed to generate encapsulation systems after those employed for MOFs. Copper can be found in agrochemical applications in stabilizers with biological polymers such as alginate [151] and incorporated into other nanoparticles to enhance their properties [152] or to enhance delivery to the surface of 2D graphene materials [153]. In a copper alginate carrier, this system was employed to encapsulate sodium selenate, which improved cherry radish (*Raphanus sativus* L.) yield and showed inhibitory effects on the fungus *Fusarium oxysporum* Schltdl [151]. Silver nanoparticles have been used to support paraquat encapsulated in chitosan polymer, and this nanomaterial was tested against the invasive plant *Eichhornia crassipes* (Mart.) Solms with enhanced results. The authors also tested its phytotoxicity in crops of interest, such as black gram (*Vigna mungo* (L.) Hepper), but inhibitory effects were not observed [154].

Different organometallic nanomaterials, such as metallacrowns [155], sandwich nanohybrid complexes [156], and organosilica vesicles [157], have been considered for weed control or other agrochemical purposes. In reality, the use of metal cores increases the cost of formulations and increases the environmental risk of soil and water pollution. Researchers have clarified the potential use of these systems and obtained good results even when using trace metals that are essential for plant development in the nanomaterial design. However the lack of field experiments with these formulation methods is the best explanation for the limited use of this approach.

2.6. New Trends

New encapsulation methods in medicinal chemistry have been exploited to develop new formulations in agriculture in recent years, especially in the last decade. Applications in agrochemistry require low-cost and large-scale production not only for bioactive compounds, but also for carriers. However, it is important to note that a good formulation method can decrease the concentration of the herbicide/bioherbicide required in the field. Enhancements in water solubility, stability, or targeting could decrease the amount required for weed control by 10–50 times according to current research papers [11].

One of the main encapsulation techniques reported in the scientific literature involves the use of nanotubes. The first use of nanotubes for agrochemical purposes was in 2014 with the application of carbon nanotubes containing a polycitric acid surface shell. This matrix was adsorbed onto the surface of the nanotubes, followed by encapsulation of zineb and mancozeb, two pesticides that act against the fungus *Alternaria alternate* (Fr.) Keissl (Fr.), which infects most cereal plants [158]. However, some level of toxicity has been associated with carbon nanotubes, and this approach does not seem to represent a green method [159]. It was not until 2019 that the first application of nanotubes in phytotoxicity studies was reported. In this case, nanotubes were formed with lithocholic acid, a natural product that is produced by the human body, and these nanotubes were employed to encapsulate disulfide herbicides [160] and natural sesquiterpene lactones (Figure 8) [16]. The authors demonstrated an enhancement in water solubility and in vitro efficacy against *Phalaris arundinacea* L., *Lolium perenne* L., and *Portulaca oleracea* L. The bioactivity was higher than for the free compounds and the positive control (Logran®) at higher concentrations (1000–300 μM) of the allelochemicals (aguerin B, cynaropicrin, and grosheimin). More specifically, the activity was mainly observed in the root formation of the weeds, and this system was more active against dicotyledons [16]. The data obtained—as well as the method itself—are of great interest for future field applications, particularly in the case of the natural sesquiterpene lactones due to their encapsulation with nanotubes generated by natural products. This would represent a green approach to weed control and food enhancement. In terms of natural/biological encapsulation systems, other interesting methods have been reported, and these include polymers generated by coumarin moieties for the encapsulation of 2,4-D [161]. This method was tested in vitro in *Cucurbita maxima* Duchesne models, and a boost in the activity was observed in comparison with the non-encapsulated herbicide. In addition to the idea of 'natural product carriers', apple pectin and Arabic gum have been employed [162]. There are other interesting ideas, such as the use of plant virus nanoparticles to deliver herbicides. Chariou et al. employed the icosahedral cowpea mosaic virus and the physalis mosaic virus to encapsulate nematocidal abamectin inside a virus capsule [163]. The results showed better soil mobility when compared to other encapsulation methods (e.g., silica nanoparticles) and a higher loading capacity.

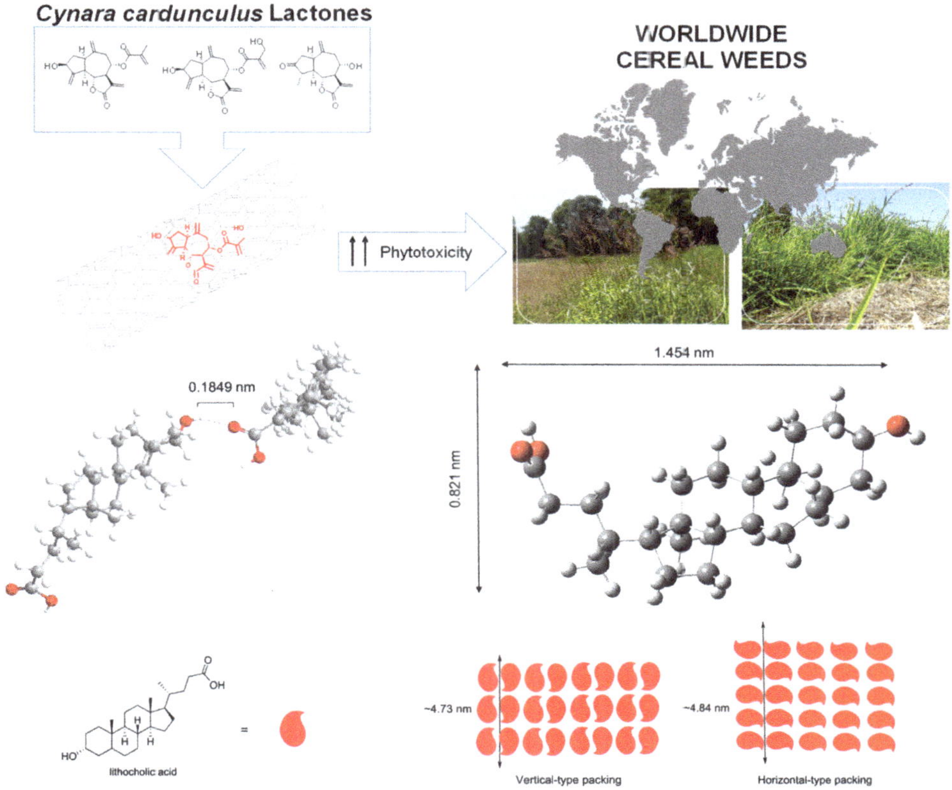

Figure 8. Nanotubes generated with lithocholic acid, a natural bile acid, to encapsulate *Cynara cardunculus* sesquiterpene lactones. Adapted with permission from [16,160]. Copyright 2019 and 2022, American Chemical Society.

In the last few years, allelopathy has gained some momentum, and natural products (allelochemicals) are seen as valid options for weed control. However, as outlined above, a good formulation without structural modification is important for retaining the role of a 'natural herbicide'. Some of the new methods presented here are promising in terms of formulation, but real applications in the field are still underexplored. It is our belief that the possibilities offered by organic encapsulation systems will meet with success, especially those employing other natural components as carriers for their formulation.

3. Conclusions

The most relevant advances in the encapsulation and formulation of herbicides and allelochemicals for weed control have been presented. Several methods have been successfully applied since this method was established in 1973. Some of these approaches have been extensively studied, e.g., that using starch, but they are now less widely studied due to new advances in nanotechnology and polymers. These advances have allowed the emergence of nanoparticle encapsulation, as well as the use of new materials, such as nanotubes and metal–organic hybrids. However, there is a lack of biological studies on these materials, and they must be analyzed in vitro and in the field before their large-scale application. Most of the knowledge on the encapsulating materials presented here has been applied to classical herbicides, with enhanced results being obtained for their physicochemical and biological properties. Nevertheless, in the future, it is expected that this technique will be applied to natural products/allelochemicals to achieve green approaches in agriculture. In the last five years, advances have been made in this respect, but challenges remain in terms of formulation and before industrial applications are developed. The authors suggest that the methods presented here indicate that applications using organic nanoparticles are very promising due to their biodegradability, ecological materials, slow-release properties, and greater potential for surface functionalization. In general, nanoparticles have three dimensions at the nanoscale, which offers more options for bioavailability compared to microstructures or 2D nanomaterials. Recognition, assimilation, and transport by and through plant cells are easier for 3D nanomaterials.

Author Contributions: F.J.R.-M., A.S. and J.M.G.M.: Conceptualization, methodology, investigation, resources, and writing—original draft preparation, N.C., S.S., G.M. and F.A.M.: conceptualization, resources, writing—review and editing, funding, and design and verification of all figures. All authors have read and agreed to the published version of the manuscript.

Funding: This research was funded by the Agencia Estatal de Investigación, Ministerio de Ciencia e Innovacion (grant number: PID2020-15747RB-I00/AEI/10.13039) and Consejo de Seguridad Nuclear (grant number: IA2-RACOHN-CSN-2022).

Data Availability Statement: Not applicable.

Acknowledgments: F.J.R.M. thanks the Universidad de Cádiz for the postdoctoral support with a Margarita-Salas fellowship funded by the European Union (NextGenerationEU), as well as the University of Innsbruck for the use of their facilities.

Conflicts of Interest: The authors declare no conflict of interest.

References

1. Sinclair, R.G. Slow-Release Pesticide System. Polymers of Lactic and Glycolic Acids as Ecologically Beneficial, Cost-Effective Encapsulating Materials. *Environ. Sci. Technol.* **1973**, *7*, 955–956. [CrossRef]
2. Pennwalt to Test Encapsulated Pesticide. *Chem. Eng. News Arch.* **1972**, *50*, 68–69. [CrossRef]
3. Jaga, K.; Dharmani, C. Methyl Parathion: An Organophosphate Insecticide Not Quite Forgotten. *Rev. Environ. Health* **2006**, *21*, 57–68. [CrossRef] [PubMed]
4. CNN News Monsanto Pleads Guilty to Illegally Spraying Banned Pesticide in Maui. Available online: https://edition.cnn.com/2019/11/22/us/monsanto-maui-pesticide-guilty-plea/index.html (accessed on 2 June 2023).
5. Gilliom, R.J.; Barbash, J.E.; Crawford, C.G.; Hamilton, P.A.; Martin, J.D.; Nakagaki, N.; Nowell, L.H.; Scott, J.C.; Stackelberg, P.E.; Thelin, G.P.; et al. Occurrence and Distribution in Streams and Ground Water. In *Pesticides in the Nation's Streams and Ground Water, 1992–2001*; Geological Survey (USGS): Reston, VA, USA, 2001; pp. 41–66.
6. Law, E.U. Commission Decision of 10 March 2004 Concerning the Non-Inclusion of Atrazine in Annex I to Council Directive 91/414/EEC and the Withdrawal of Authorisations for Plant Protection Products Containing This Active Substance. Available online: https://eur-lex.europa.eu/eli/dec/2004/248/oj (accessed on 12 June 2023).
7. Hedlund, B. Flemish Government Approves Belgium Glyphosate Ban for Individuals. Available online: https://www.baumhedlundlaw.com/belgium-glyphosate-ban-individuals/ (accessed on 12 June 2023).
8. Peterson, M.A.; Collavo, A.; Ovejero, R.; Shivrain, V.; Walsh, M.J. The Challenge of Herbicide Resistance around the World: A Current Summary. *Pest Manag. Sci.* **2018**, *74*, 2246–2259. [CrossRef] [PubMed]
9. Hulme, P.E. Weed Resistance to Different Herbicide Modes of Action Is Driven by Agricultural Intensification. *Field Crops Res.* **2023**, *292*, 108819. [CrossRef]
10. Scavo, A.; Mauromicale, G. Crop Allelopathy for Sustainable Weed Management in Agroecosystems: Knowing the Present with a View to the Future. *Agronomy* **2021**, *11*, 2104. [CrossRef]
11. Macías, F.A.; Mejías, F.J.; Molinillo, J.M. Recent Advances in Allelopathy for Weed Control: From Knowledge to Applications. *Pest Manag. Sci.* **2019**, *75*, 2413–2436. [CrossRef]
12. Macías, F.A.; Durán, A.G.; Molinillo, J.M.G. Allelopathy: The Chemical Language of Plants. In *Progress in the Chemistry of Organic Natural Products*; Springer: Berlin/Heidelberg, Germany, 2020; Volume 112, pp. 1–84.
13. Harding, D.P.; Raizada, M.N. Controlling Weeds with Fungi, Bacteria and Viruses: A Review. *Front. Plant Sci.* **2015**, *6*, 659. [CrossRef]
14. Roberts, J.; Florentine, S.; Fernando, W.G.D.; Tennakoon, K.U. Achievements, Developments and Future Challenges in the Field of Bioherbicides for Weed Control: A Global Review. *Plants* **2022**, *11*, 2242. [CrossRef]
15. Gámiz, B.; Celis, R. S-Carvone Formulation Based on Granules of Organoclay to Modulate Its Losses and Phytotoxicity in Soil. *Agronomy* **2021**, *11*, 1593. [CrossRef]

16. Mejías, F.J.R.; Fernández, I.P.; Rial, C.; Varela, R.M.; Molinillo, J.M.G.; Calvino, J.J.; Trasobares, S.; Macías, F.A. Encapsulation of *Cynara Cardunculus* Guaiane-Type Lactones in Fully Organic Nanotubes Enhances Their Phytotoxic Properties. *J. Agric. Food Chem.* **2022**, *70*, 3644–3653. [CrossRef] [PubMed]
17. Suganya, V.; Anuradha, V. Microencapsulation and Nanoencapsulation: A Review. *Int. J. Pharm. Clin. Res.* **2017**, *9*, 233–239. [CrossRef]
18. Scavo, A.; Mejías, F.J.R.; Chinchilla, N.; Molinillo, J.M.G.; Schwaiger, S.; Lombardo, S.; Macías, F.A.; Mauromicale, G. Wheat Response and Weed-Suppressive Ability in the Field Application of a Nanoencapsulated Disulfide (DiS-NH_2) Bioherbicide Mimic. *Agronomy* **2023**, *13*, 1132. [CrossRef]
19. Szejtli, J. Physiological Effects of Cyclodextrins on Plants. *Starch* **1983**, *35*, 433–438. [CrossRef]
20. Szejtli, J. Cyclodextrins in Pesticides. *Starch* **1985**, *37*, 382–386. [CrossRef]
21. Gines, J.M.; Perez-Martinez, I.; Arias, M.J.; Moyano, J.R.; Morillo, E.; Ruiz-Conde, A.; Sachez-Soto, P.J. Inclusion of the Herbicide 2,4-Dichlorophenoxyacetic Acid (2,4-D) with β-Cyclodextrin by Different Processing Methods. *Chemosphere* **1996**, *33*, 321–334. [CrossRef]
22. Hosangadi, B.; Asgaonkar, A. Inclusion of Acaricides by Complexation with β-Cyclodextrin. *J. Incl. Phenom. Mol. Recognit. Chem.* **1995**, *23*, 35–39. [CrossRef]
23. Lezcano, M.; Novo, M.; Al-Soufi, W.; Rodríguez-Núñez, E.; Tato, J.V.; Vázquez Tato, J. Complexation of Several Fungicides with β-Cyclodextrin: Determination of the Association Constants and Isolation of the Solid Complexes. *J. Agric. Food Chem.* **2003**, *51*, 5036–5040. [CrossRef]
24. Lezcano, M.; Al-Soufi, W.; Novo, M.; Rodríguez-Núñez, E.; Vázquez Tato, J. Complexation of Several Benzimidazole-Type Fungicides with α- and β-Cyclodextrins. *J. Agric. Food Chem.* **2002**, *50*, 108–112. [CrossRef]
25. Benfeito, S.; Rodrigues, T.; Garrido, J.; Borges, F.; Garrido, E.M. Host-Guest Interaction between Herbicide Oxadiargyl and Hydroxypropyl-β-Cyclodextrin. *Sci. World J.* **2013**, *2013*, 1–6. [CrossRef]
26. Perez-Martinez, J.I.; Morillo, E.; Gines, J.M. β-CD Effect on 2,4-D Soil Adsorption. *Chemosphere* **1999**, *39*, 2047–2056. [CrossRef]
27. Cai, X.; Liu, W.; Chen, S. Environmental Effects of Inclusion Complexation between Methylated β-Cyclodextrin and Diclofop-Methyl. *J. Agric. Food Chem.* **2005**, *53*, 6744–6749. [CrossRef] [PubMed]
28. Cala, A.; Molinillo, J.M.G.G.; Fernández-Aparicio, M.; Ayuso, J.; Álvarez, J.A.; Rubiales, D.; Macías, F.A.; Delavault, P. Complexation of Sesquiterpene Lactones with Cyclodextrins: Synthesis and Effects on Their Activities on Parasitic Weeds. *Org. Biomol. Chem.* **2017**, *15*, 6500–6510. [CrossRef]
29. Moeini, A.; Masi, M.; Zonno, M.C.; Boari, A.; Cimmino, A.; Tarallo, O.; Vurro, M.; Evidente, A. Encapsulation of Inuloxin A, a Plant Germacrane Sesquiterpene with Potential Herbicidal Activity, in β-Cyclodextrins. *Org. Biomol. Chem.* **2019**, *17*, 2508–2515. [CrossRef] [PubMed]
30. Zhang, J.; Li, B.; Bi, H.; Zhang, P. Synthesis of Atrazine-HPCD Inclusion and Its Bioactivity. *Trans. Tianjin Univ.* **2014**, *20*, 350–357. [CrossRef]
31. Geng, Q.; Xie, J.; Wang, X.; Cai, M.; Ma, H.; Ni, H. Preparation and Characterization of Butachlor/(2-Hydroxypropyl)-β-Cyclodextrin Inclusion Complex: Improve Soil Mobility and Herbicidal Activity and Decrease Fish Toxicity. *J. Agric. Food Chem.* **2018**, *66*, 12198–12205. [CrossRef] [PubMed]
32. Gao, S.; Jiang, J.; Li, X.; Liu, Y.; Zhao, L.; Fu, Y.; Ye, F. Enhanced Physicochemical Properties and Herbicidal Activity of an Environment-Friendly Clathrate Formed by β-Cyclodextrin and Herbicide Cyanazine. *J. Mol. Liq.* **2020**, *305*, 112858. [CrossRef]
33. Gao, S.; Jiang, J.Y.; Liu, Y.Y.; Fu, Y.; Zhao, L.X.; Li, C.Y.; Ye, F. Enhanced Solubility, Stability, and Herbicidal Activity of the Herbicide Diuron by Complex Formation with β-Cyclodextrin. *Polymers* **2019**, *11*, 1396. [CrossRef]
34. Mejías, F.J.R.; Carrasco, Á.; Durán, A.G.; Molinillo, J.M.G.; Macías, F.A.; Chinchilla, N. On the Formulation of Disulfide Herbicides Based on Aminophenoxazinones: Polymeric Nanoparticle Formulation and Cyclodextrin Complexation to Combat Crop Yield Losses. *Pest Manag. Sci.* **2022**, *79*, 1547–1556. [CrossRef]
35. Mortensen, A.; Aguilar, F.; Crebelli, R.; Di Domenico, A.; Dusemund, B.; Frutos, M.J.; Galtier, P.; Gott, D.; Gundert-Remy, U.; Leblanc, J.C.; et al. Re-Evaluation of β-Cyclodextrin (E 459) as a Food Additive. *EFSA J.* **2016**, *14*, e04628. [CrossRef]
36. Mokhtar, M.S.; Suliman, F.O.; Elbashir, A.A. Investigation of Inclusion Complexes of Ametryne and Atrazine with Cucurbit[n]Urils (n = 6–8) Using Experimental and Theoretical Techniques. *J. Incl. Phenom. Macrocycl. Chem.* **2019**, *94*, 31–43. [CrossRef]
37. Mokhtar, M.S.; Suliman, F.O.; Elbashir, A.A. The Binding Interaction of Imazapyr with Cucurbit[n]Uril (n = 6–8): Combined Experimental and Molecular Modeling Study. *Spectrochim. Acta Part A Mol. Biomol. Spectrosc.* **2018**, *194*, 67–75. [CrossRef]
38. Mokhtar, M.S.; Elbashir, A.A.; Suliman, F.O. Spectroscopic and Molecular Simulation Studies on the Interaction of Imazaquin Herbicide with Cucurbiturils (n = 6–8). *J. Mol. Struct.* **2023**, *1274*, 134444. [CrossRef]
39. Mejías, F.J.R.; He, S.; Varela, R.M.; Molinillo, J.M.G.; Barba-Bon, A.; Nau, W.M.; Macías, F.A. Stability and PKa Modulation of Aminophenoxazinones and Their Disulfide Mimics by Host–Guest Interaction with Cucurbit[7]Uril. Direct Applications in Agrochemical Wheat Models. *J. Agric. Food Chem.* **2023**, *71*, 480–487. [CrossRef] [PubMed]
40. Connick, W.J.; Bradow, J.M.; Wells, W.; Steward, K.K.; Van, T.K. Preparation and Evaluation of Controlled-Release Formulations of 2,6-Dichlorobenzonitrile. *J. Agric. Food Chem.* **1984**, *32*, 1199–1205. [CrossRef]
41. Qasem, J.R. Nutrient Accumulation by Weeds and Their Associated Vegetable Crops. *J. Hortic. Sci.* **1992**, *67*, 189–195. [CrossRef]
42. Chauhan, B.S.; Johnson, D.E. Germination Ecology of Goosegrass (*Eleusine indica*): An Important Grass Weed of Rainfed Rice. *Weed Sci.* **2008**, *56*, 699–706. [CrossRef]

43. Zhang, S.-Z.; Li, Y.-H.; Kong, C.-H.; Xu, X.-H. Interference of Allelopathic Wheat with Different Weeds. *Pest Manag. Sci.* **2016**, *72*, 172–178. [CrossRef]
44. Zhou, B.; Kong, C.H.; Li, Y.H.; Wang, P.; Xu, X.H. Crabgrass (*Digitaria sanguinalis*) Allelochemicals That Interfere with Crop Growth and the Soil Microbial Community. *J. Agric. Food Chem.* **2013**, *61*, 5310–5317. [CrossRef]
45. Carr, M.E.; Wing, R.E.; Doane, W.M. Clay as a Carrier in Starch Encapsulated Herbicides Prepared by Extrusion Processing. *Starch-Stärke* **1994**, *46*, 9–13. [CrossRef]
46. Flores Céspedes, F.; Villafranca Sánchez, M.; Pérez García, S.; Fernández Pérez, M. Modifying Sorbents in Controlled Release Formulations to Prevent Herbicides Pollution. *Chemosphere* **2007**, *69*, 785–794. [CrossRef] [PubMed]
47. Rivoira, L.; Frassati, S.; Cordola, S.; Castiglioni, M.; Onida, B.; Ronchetti, S.; Ingrando, I.; Bruzzoniti, M.C. Encapsulation of the Glyphosate Herbicide in Mesoporous and Soil-Affine Sorbents for Its Prolonged Release. *Chem. Eng. J.* **2022**, *431*, 134225. [CrossRef]
48. Han, Y.-S.; Lee, S.-Y.; Yang, J.-H.; Soo Hwang, H.; Park, I. Paraquat Release Control Using Intercalated Montmorillonite Compounds. *J. Phys. Chem. Solids* **2010**, *71*, 460–463. [CrossRef]
49. Rashidzadeh, A.; Olad, A.; Hejazi, M.J. Controlled Release Systems Based on Intercalated Paraquat onto Montmorillonite and Clinoptilolite Clays Encapsulated with Sodium Alginate. *Adv. Polym. Technol.* **2017**, *36*, 177–185. [CrossRef]
50. Marco-Brown, J.L.; Undabeytia, T.; Torres Sánchez, R.M.; dos Santos Afonso, M. Slow-Release Formulations of the Herbicide Picloram by Using Fe–Al Pillared Montmorillonite. *Environ. Sci. Pollut. Res.* **2017**, *24*, 10410–10420. [CrossRef]
51. Khatem, R.; Celis, R.; Hermosín, M.C. Cationic and Anionic Clay Nanoformulations of Imazamox for Minimizing Environmental Risk. *Appl. Clay Sci.* **2019**, *168*, 106–115. [CrossRef]
52. Wilpiszewska, K.; Spychaj, T.; Paździoch, W. Carboxymethyl Starch/Montmorillonite Composite Microparticles: Properties and Controlled Release of Isoproturon. *Carbohydr. Polym.* **2016**, *136*, 101–106. [CrossRef]
53. Giroto, A.S.; De Campos, A.; Pereira, E.I.; Ribeiro, T.S.; Marconcini, J.M.; Ribeiro, C. Photoprotective Effect of Starch/Montmorillonite Composites on Ultraviolet-Induced Degradation of Herbicides. *React. Funct. Polym.* **2015**, *93*, 156–162. [CrossRef]
54. Cabrera, A.; Celis, R.; Hermosín, M.C. Imazamox-Clay Complexes with Chitosan- and Iron(III)-Modified Smectites and Their Use in Nanoformulations. *Pest Manag. Sci.* **2016**, *72*, 1285–1294. [CrossRef]
55. Cal-IPC (California Invasive Plant Council). *Cal-IPC Plant Assessment Form for Brassica Nigra*; Cal-IPC: Berkeley, CA, USA, 2004.
56. Sánchez-Verdejo, T.; Undabeytia, T.; Nir, S.; Villaverde, J.; Maqueda, C.; Morillo, E. Environmentally Friendly Formulations of Alachlor and Atrazine: Preparation, Characterization, and Reduced Leaching. *J. Agric. Food Chem.* **2008**, *56*, 10192–10199. [CrossRef]
57. Wall, D.A.; Friesen, G.H. Effect of Duration of Green Foxtail (*Setaria viridis*) Competition on Potato (*Solanum tuberosum*) Yield. Effect of Duration of Green Foxtail (*Setaria viridis*) Competition on Potato (*Solanum tuberosum*) Yield. *Weed Technol.* **1990**, *4*, 539–542. [CrossRef]
58. Li, J.; Yao, J.; Li, Y.; Shao, Y. Controlled Release and Retarded Leaching of Pesticides by Encapsulating in Carboxymethyl Chitosan/Bentonite Composite Gel. *J. Environ. Sci. Health Part B Pestic. Food Contam. Agric. Wastes* **2012**, *47*, 795–803. [CrossRef] [PubMed]
59. Mishael, Y.G.; Undabeytia, T.; Rabinovitz, O.; Rubin, B.; Nir, S. Sulfosulfuron Incorporated in Micelles Adsorbed on Montmorillonite for Slow Release Formulations. *J. Agric. Food Chem.* **2003**, *51*, 2253–2259. [CrossRef] [PubMed]
60. Galán-Jiménez, M.D.C.; Mishael, Y.-G.; Nir, S.; Morillo, E.; Undabeytia, T. Factors Affecting the Design of Slow Release Formulations of Herbicides Based on Clay-Surfactant Systems. A Methodological Approach. *PLoS ONE* **2013**, *8*, e59060. [CrossRef] [PubMed]
61. Zait, Y.; Segev, D.; Schweitzer, A.; Goldwasser, Y.; Rubin, B.; Mishael, Y.G. Development and Employment of Slow-Release Pendimethalin Formulations for the Reduction of Root Penetration into Subsurface Drippers. *J. Agric. Food Chem.* **2015**, *63*, 1682–1688. [CrossRef]
62. Goldreich, O.; Goldwasser, Y.; Mishael, Y.G. Effect of Soil Wetting and Drying Cycles on Metolachlor Fate in Soil Applied as a Commercial or Controlled-Release Formulation. *J. Agric. Food Chem.* **2011**, *59*, 645–653. [CrossRef]
63. Hermosin, M.C.; Calderón, M.J.; Aguer, J.P.; Cornejo, J. Organoclays for Controlled Release of the Herbicide Fenuron. *Pest Manag. Sci.* **2001**, *57*, 803–809. [CrossRef]
64. Mishael, Y.G.; Undabeytia, T.; Rabinovitz, O.; Rubin, B.; Nir, S. Slow-Release Formulations of Sulfometuron Incorporated in Micelles Adsorbed on Montmorillonite. *J. Agric. Food Chem.* **2002**, *50*, 2864–2869. [CrossRef]
65. Galán-Pérez, J.A.; Gámiz, B.; Celis, R. Granulated Organoclay as a Sorbent to Protect the Allelochemical Scopoletin from Rapid Biodegradation in Soil. *Environ. Technol. Innov.* **2022**, *28*, 102707. [CrossRef]
66. del Carmen Galán-Jiménez, M.; Morillo, E.; Bonnemoy, F.; Mallet, C.; Undabeytia, T. A Sepiolite-Based Formulation for Slow Release of the Herbicide Mesotrione. *Appl. Clay Sci.* **2020**, *189*, 105503. [CrossRef]
67. Doane, W.M.; Shasha, B.S.; Russell, C.R. Encapsulation of Pesticides within a Starch Matrix. *Control. Release Pestic.* **1977**, *53*, 74–83.
68. Schreiber, M.M.; White, M.D. Effect of Formulation Technology on the Release of Starch Encapsulated EPTC. *Proc. Br. Crop Prot. Conf. Weeds* **1980**, *15*, 225–229.
69. Schreiber, M.M.; White, M.D. Granule Structure and Rate of Release with Starch-Encapsulated Thiocarbamates. *Weed Sci.* **1980**, *28*, 685–690. [CrossRef]

70. Shasha, B.S.; Trimnell, D.; Otey, F.H. Encapsulation of Pesticides in a Starch-Calcium Adduct. *J. Polym. Sci. Part A Polym. Chem.* **1981**, *19*, 1891–1899. [CrossRef]
71. Trimnell, D.; Shasha, B.S.; Doane, W.M. Release of Trifluralin from Starch Xanthide Encapsulated Formulations. *J. Agric. Food Chem.* **1981**, *29*, 637–640. [CrossRef]
72. Trimnell, D.; Shasha, B.S.; Otey, F.H. The Effect of α-Amylases upon the Release of Trifluralin Encapsulated in Starch. *J. Control. Release* **1985**, *1*, 183–190. [CrossRef]
73. Wing, R.E.; Doane, W.M. The Role of Retrogradation on Starch Encapsulation and Release of Active Agents. *Polym. Prepr.* **1987**, *28*, 108–109.
74. Wing, R.E.; Maiti, S.; Doane, W.M. Effectiveness of Jet-cooked Pearl Cornstarch as a Controlled Release Matrix. *Starch-Stäke* **1987**, *39*, 422–425. [CrossRef]
75. White, M.D.; Schreiber, M.M. Herbicidal Activity of Starch Encapsulated Trifluralin. *Weed Sci.* **1984**, *32*, 387–394. [CrossRef]
76. Trimnell, D.; Shasha, B S. Autoencapsulation: A New Method for Entrapping Pesticides within Starch. *J. Control. Release* **1988**, *7*, 25–31. [CrossRef]
77. Grover, R.; Wolt, J.; Cessna, A.; Schiefer, B. Environmental Fate of Trifluralin. In *Reviews of Environmental Contamination and Toxicology*; Springer: Berlin/Heidelberg, Germany, 1997; pp. 1–64.
78. European Union European Union-Final Regulatory Action. Available online: http://archive.pic.int/view_displayFRA.php?id=1186&back=viewB_FRAchems.php?sort=chemical (accessed on 7 June 2023).
79. Schreiber, M.M.; Hickman, M.V.; Vail, G.D. Starch-encapsulated Atrazine: Efficacy and Transport. *J. Environ. Qual.* **1993**, *22*, 443–453. [CrossRef]
80. Vail, G.D.; Hickman, M.V.; Schreiber, M.M. Atrazine Dissipation and Carryover from Commercial and Starch-Encapsulated Atrazine Formulations. *Weed Sci.* **1997**, *45*, 842–847. [CrossRef]
81. Fleming, G.F.; Wax, L.M.; Simmons, F.W. Leachability and Efficacy of Starch-Encapsulated Atrazine. *Weed Technol.* **1992**, *6*, 297–302. [CrossRef]
82. Buhler, D.D.; Schreiber, M.M.; Koskinen, W.C. Weed Control with Starch-Encapsulated Alachlor, Metolachlor, and Atrazine. *Weed Technol.* **1994**, *8*, 277–284. [CrossRef]
83. Williams, C.F.; Nelson, S.D.; Gish, T.J. Release Rate and Leaching of Starch-Encapsulated Atrazine in a Calcareous Soil. *Soil Sci. Soc. Am. J.* **1999**, *63*, 425–432. [CrossRef]
84. Gish, T.G.; Shirmohammadi, A.; Wienhold, B.J. Field-Scale Mobility and Persistence of Commercial and Starch-Encapsulated Atrazine and Alachlor. *J. Environ. Qual.* **1994**, *23*, 355–359. [CrossRef]
85. Wienhold, B.J.; Sadeghi, A.M.; Gish, T.J. Effect of Starch Encapsulation and Temperature on Volatilization of Atrazine and Alachlor. *J. Environ. Qual.* **1993**, *22*, 162–166. [CrossRef]
86. Wienhold, B.J.; Gish, T.J. Effect of Formulation and Tillage Practice on Volatilization of Atrazine and Alachlor. *J. Environ. Qual.* **1994**, *23*, 292–298. [CrossRef]
87. Wing, R.E.; Maiti, S.: Doane, M. Amylose Content of Starch Control the Release of Encapsulated Bioactive Agents. *J. Control. Release* **1988**, *7*, 33–37. [CrossRef]
88. Trimnell, D.; Wing, R.E.; Carr, M.E.; Doane, W.M. Encapsulation of EPTC in Starch by Twin-screw Extrusion. *Ind. Crops Prod.* **1991**, *43*, 146–151. [CrossRef]
89. Wing, R.E.; Carr, M.E.; Trimnell, D.; Doane, W.M. Comparison of Steam Injection Cooking versus Twin-Screw Extrusion of Pearl Cornstarch for Encapsulation of Chloroacetanilide Herbicides. *J. Control. Release* **1991**, *16*, 267–277. [CrossRef]
90. Carr, M.E.; Wing, R.E.; Doane, W.M. Encapsulation of Atrazine within a Starch Matrix by Extrusion Processing. *Cereal Chem.* **1991**, *68*, 262–266.
91. Fleming, G.F.; Wax, L.M.; Simmons, F.W.; Felsot, A.S. Movement of Alachlor and Metribuzin from Controlled Release Formulations in a Sandy Soil. *Weed Sci.* **1992**, *40*, 606–613. [CrossRef]
92. Reed, J.P.; Hall, F.R.; Trimnell, D. Effect of Encapsulating Thiocarbamate Herbicides within Starch for Overcoming Enhanced Degradation in Soils. *Starch-Stärke* **1989**, *41*, 184–186. [CrossRef]
93. Giroto, A.S.; de Campos, A.; Pereira, E.I.; Cruz, C.C.T.; Marconcini, J.M.; Ribeiro, C. Study of a Nanocomposite Starch-Clay for Slow-Release of Herbicides: Evidence of Synergistic Effects between the Biodegradable Matrix and Exfoliated Clay on Herbicide Release Control. *J. Appl. Polym. Sci.* **2014**, *131*, 41188. [CrossRef]
94. Riyajan, S.A. Physical Property Testing of a Novel Hybrid Natural Rubber-Graft-Cassava Starch/Sodium Alginate Bead for Encapsulating Herbicide. *Polym. Test.* **2017**, *58*, 300–307. [CrossRef]
95. Riyajan, S.A. A Novel Hybrid 2,4-Dichlorophenoxy Acetate Bead from Modified Cassava Starch and Sodium Alginate with Modified Natural Rubber Coating. *J. Polym. Environ.* **2018**, *26*, 1950–1961. [CrossRef]
96. Alipour, M.; Saharkhiz, M.J.; Niakousari, M.; Seidi Damyeh, M. Phytotoxicity of Encapsulated Essential Oil of Rosemary on Germination and Morphophysiological Features of Amaranth and Radish Seedlings. *Sci. Hortic.* **2019**, *243*, 131–139. [CrossRef]
97. Taban, A.; Saharkhiz, M.J.; Naderi, R. A Natural Post-Emergence Herbicide Based on Essential Oil Encapsulation by Cross-Linked Biopolymers: Characterization and Herbicidal Activity. *Environ. Sci. Pollut. Res.* **2020**, *27*, 45844–45858. [CrossRef]
98. Dong, H.; Li, F.; Li, J.; Li, Y. Characterizations of Blend Gels of Carboxymethylated Polysaccharides and Their Use for the Controlled Release of Herbicide. *J. Macromol. Sci. Part A Pure Appl. Chem.* **2012**, *49*, 235–241. [CrossRef]

99. Li, J.; Jiang, M.; Wu, H.; Li, Y. Addition of Modified Bentonites in Polymer Gel Formulation of 2,4-D for Its Controlled Release in Water and Soil. *J. Agric. Food Chem.* **2009**, *57*, 2868–2874. [CrossRef] [PubMed]
100. Flores-Céspedes, F.; Daza-Fernández, I.; Villafranca-Sánchez, M.; Fernández-Pérez, M. Use of Ethylcellulose To Control Chlorsulfuron Leaching in a Calcareous Soil. *J. Agric. Food Chem.* **2009**, *57*, 2856–2861. [CrossRef] [PubMed]
101. Fernández-Pérez, M.; Villafranca-Sánchez, M.; Flores-Céspedes, F.; Daza-Fernández, I. Ethylcellulose and Lignin as Bearer Polymers in Controlled Release Formulations of Chloridazon. *Carbohydr. Polym.* **2011**, *83*, 1672–1679. [CrossRef]
102. Fernández-Pérez, M.; Villafranca-Sánchez, M.; Flores-Céspedes, F.; Daza-Fernández, I. Lignin-Polyethylene Glycol Matrices and Ethylcellulose to Encapsulate Highly Soluble Herbicides. *J. Appl. Polym. Sci.* **2014**, *132*, 41422. [CrossRef]
103. Yiamsawas, D.; Kangwansupamonkon, W.; Kiatkamjornwong, S. Lignin-Based Nanogels for the Release of Payloads in Alkaline Conditions. *Eur. Polym. J.* **2021**, *145*, 110241. [CrossRef]
104. Liu, X.; Li, W.; Xuan, G. Preparation and Characterization of β-Cyclodextrin Nanosponges and Study on Enhancing the Solubility of Insoluble Nicosulfuron. *IOP Conf. Ser. Mater. Sci. Eng.* **2020**, *774*, 012108. [CrossRef]
105. Gámiz, B.; López-Cabeza, R.; Velarde, P.; Spokas, K.A.; Cox, L. Biochar Changes the Bioavailability and Bioefficacy of the Allelochemical Coumarin in Agricultural Soils. *Pest Manag. Sci.* **2021**, *77*, 834–843. [CrossRef]
106. Luo, J.; Gao, Y.; Liu, Y.; Huang, X.; Zhang, D.; Cao, H.; Jing, T.; Liu, F.; Li, B. Self-Assembled Degradable Nanogels Provide Foliar Affinity and Pinning for Pesticide Delivery by Flexibility and Adhesiveness Adjustment. *ACS Nano* **2021**, *15*, 14598–14609. [CrossRef] [PubMed]
107. Turner, B.C.; Glotfelty, D.E.; Taylor, A.W.; Watson, D.R. Volatilization of Microencapsulated and Conventionally Applied Chlorpropham in the Field. *Agron. J.* **1978**, *70*, 933–937. [CrossRef]
108. Petersen, B.B.; Shea, P.J. Microencapsulated Alachlor and Its Behavior on Wheat (*Triticum aestivum*) Straw. *Weed Sci.* **1989**, *37*, 719–723. [CrossRef]
109. Chang, Y.N.; Mueller, R.E.; Iannotti, E.L. Use of Low MW Polylactic Acid and Lactide to Stimulate Growth and Yield of Soybeans. *Plant Growth Regul.* **1996**, *19*, 223–232. [CrossRef]
110. Stloukal, P.; Kucharczyk, P.; Sedlarik, V.; Bazant, P.; Koutny, M. Low Molecular Weight Poly(Lactic Acid) Microparticles for Controlled Release of the Herbicide Metazachlor: Preparation, Morphology, and Release Kinetics. *J. Agric. Food Chem.* **2012**, *60*, 4111–4119. [CrossRef]
111. Boehm, A.L.; Martinon, I.; Zerrouk, R.; Rump, E.; Fessi, H. Nanoprecipitation Technique for the Encapsulation of Agrochemical Active Ingredients. *J. Microencapsul.* **2003**, *20*, 433–441. [CrossRef] [PubMed]
112. Yeom, C.K.; Oh, S.B.; Rhim, J.W.; Lee, J.M. Microencapsulation of Water-Soluble Herbicide by Interfacial Reaction. I. Characterization of Microencapsulation. *J. Appl. Polym. Sci.* **2000**, *78*, 1645–1655. [CrossRef]
113. Wang, X.; Zhao, J. Encapsulation of the Herbicide Picloram by Using Polyelectrolyte Biopolymers as Layer-by-Layer Materials. *J. Agric. Food Chem.* **2013**, *61*, 3789–3796. [CrossRef]
114. Quiñones, J.P.; García, Y.C.; Curiel, H.; Covas, C.P. Microspheres of Chitosan for Controlled Delivery of Brassinosteroids with Biological Activity as Agrochemicals. *Carbohydr. Polym.* **2010**, *80*, 915–921. [CrossRef]
115. Yeom, C.K.; Kim, Y.H.; Lee, J.M. Microencapsulation of Water-Soluble Herbicide by Interfacial Reaction. II. Release Properties of Microcapsules. *J. Appl. Polym. Sci.* **2002**, *84*, 1025–1034. [CrossRef]
116. Grillo, R.; de Melo, N.F.S.; de Lima, R.; Lourenço, R.W.; Rosa, A.H.; Fraceto, L.F. Characterization of Atrazine-Loaded Biodegradable Poly(Hydroxybutyrate-Co-Hydroxyvalerate) Microspheres. *J. Polym. Environ.* **2010**, *18*, 26–32. [CrossRef]
117. Lobo, F.A.; De Aguirre, C.L.; Silva, M.S.; Grillo, R.; De Melo, N.F.S.; De Oliveira, L.K.; De Morais, L.C.; Campos, V.; Rosa, A.H.; Fraceto, L.F. Poly(Hydroxybutyrate-Co-Hydroxyvalerate) Microspheres Loaded with Atrazine Herbicide: Screening of Conditions for Preparation, Physico-Chemical Characterization, and in Vitro Release Studies. *Polym. Bull.* **2011**, *67*, 479–495. [CrossRef]
118. Pérez-Martínez, J.I.; Morillo, E.; Maqueda, C.; Ginés, J.M. Ethyl Cellulose Polymer Microspheres for Controlled Release of Norfluazon. *Pest Manag. Sci.* **2001**, *57*, 688–694. [CrossRef]
119. Dailey, O.D.J. Volatilization of Alachlor from Polymeric Formulations. *J. Agric. Food Chem.* **2004**, *52*, 6742–6746. [CrossRef] [PubMed]
120. Dowler, C.C.; Dailey, O.D.; Mullinix, B.G. Polymeric Microcapsules of Alachlor and Metolachlor: Preparation and Evaluation of Controlled-Release Properties. *J. Agric. Food Chem.* **1999**, *47*, 2908–2913. [CrossRef] [PubMed]
121. Elbahri, Z.; Taverdet, J.L. Optimization of an Herbicide Release from Ethylcellulose Microspheres. *Polym. Bull.* **2005**, *54*, 353–363. [CrossRef]
122. Sopeña, F.; Cabrera, A.; Maqueda, C.; Morillo, E. Controlled Release of the Herbicide Norflurazon into Water from Ethylcellulose Formulations. *J. Agric. Food Chem.* **2005**, *53*, 3540–3547. [CrossRef]
123. Cho, J.; Seo, Y.; Yim, T.; Lee, H. Effect of Nanoencapsulated Vitamin B1 Derivative on Inhibition of Both Mycelial Growth and Spore Germination of *Fusarium oxysporum* f. sp. Raphani. *Int. J. Mol. Sci.* **2013**, *14*, 4283–4297. [CrossRef] [PubMed]
124. Clemente, Z.; Grillo, R.; Jonsson, M.; Santos, N.Z.P.; Feitosa, L.O.; Lima, R.; Fraceto, L.F. Ecotoxicological Evaluation of Poly(ε-Caprolactone) Nanocapsules Containing Triazine Herbicides. *J. Nanosci. Nanotechnol.* **2014**, *14*, 4911–4917. [CrossRef]
125. Grillo, R.; Rosa, A.H.; Fraceto, L.F. Poly(ε-Caprolactone) Nanocapsules Carrying the Herbicide Atrazine: Effect of Chitosan-Coating Agent on Physico-Chemical Stability and Herbicide Release Profile. *Int. J. Environ. Sci. Technol.* **2014**, *11*, 1691–1700. [CrossRef]

26. Pereira, A.E.S.; Grillo, R.; Mello, N.F.S.; Rosa, A.H.; Fraceto, L.F. Application of Poly(Epsilon-Caprolactone) Nanoparticles Containing Atrazine Herbicide as an Alternative Technique to Control Weeds and Reduce Damage to the Environment. *J. Hazard. Mater.* **2014**, *268*, 207–215. [CrossRef]
27. Diyanat, M.; Saeidian, H. The Metribuzin Herbicide in Polycaprolactone Nanocapsules Shows Less Plant Chromosome Aberration than Non-Encapsulated Metribuzin. *Environ. Chem. Lett.* **2019**, *17*, 1881–1888. [CrossRef]
28. Preisler, A.C.; Pereira, A.E.S.; Campos, E.V.R.; Dalazen, G.; Fraceto, L.F.; Oliveira, H.C. Atrazine Nanoencapsulation Improves Pre-Emergence Herbicidal Activity against Bidens Pilosa without Enhancing Long-Term Residual Effect on Glycine Max. *Pest Manag. Sci.* **2020**, *76*, 141–149. [CrossRef]
29. Sousa, B.T.; Santo Pereira, A.D.E.; Fraceto, L.F.; de Oliveira, H.C.; Dalazen, G. Effectiveness of Nanoatrazine in Post-Emergent Control of the Tolerant Weed Digitaria Insularis. *J. Plant Prot. Res.* **2020**, *60*, 185–192. [CrossRef]
30. Babaei, S.; Kahrizi, D.; Nosratti, I.; Karimi, N.; Arkan, E.; Tahir, M.B. Preparation and Characterization of Chloridazon-Loaded Alginate/Chitosan Nanocapsules. *Cell. Mol. Biol.* **2022**, *68*, 34–42. [CrossRef] [PubMed]
31. Artusio, F.; Casà, D.; Granetto, M.; Tosco, T.; Pisano, R. Alginate Nanohydrogels as a Biocompatible Platform for the Controlled Release of a Hydrophilic Herbicide. *Processes* **2021**, *9*, 1641. [CrossRef]
32. Faria, D.M.; Dourado Júnior, S.M.; Nascimento, J.P.L.D.; Nunes, E.D.S.; Marques, R.P.; Rossino, L.S.; Moreto, J.A. Development and Evaluation of a Controlled Release System of TBH Herbicide Using Alginate Microparticles. *Mater. Res.* **2016**, *20*, 225–235. [CrossRef]
33. Kumar, S.; Bhanjana, G.; Sharma, A.; Sarita; Sidhu, M.C.; Dilbaghi, N. Herbicide Loaded Carboxymethyl Cellulose Nanocapsules as Potential Carrier in Agrinanotechnology. *Sci. Adv. Mater.* **2015**, *7*, 1143–1148. [CrossRef]
34. Maruyama, C.R.; Guilger, M.; Pascoli, M.; Bileshy-José, N.; Abhilash, P.C.; Fraceto, L.F.; De Lima, R. Nanoparticles Based on Chitosan as Carriers for the Combined Herbicides Imazapic and Imazapyr. *Sci. Rep.* **2016**, *6*, 19768–19781. [CrossRef]
35. Gerardo, R.; de Lima, I.P. Monitoring Duckweeds (*Lemna minor*) in Small Rivers Using Sentinel-2 Satellite Imagery: Application of Vegetation and Water Indices to the Lis River (Portugal). *Water* **2022**, *14*, 2284. [CrossRef]
36. Torbati, S.; Mahmoudian, M.; Alimirzaei, N. Nanocapsulation of Herbicide Haloxyfop-R-Methyl in Poly(Methyl Methacrylate): Phytotoxicological Effects of Pure Herbicide and Its Nanocapsulated Form on Duckweed as a Model Macrophyte. *Turk. J. Chem.* **2018**, *42*, 132–145. [CrossRef]
37. Torbati, S.; Mahmoudian, M.; Alimirzaei, N. Toxicological Effects of a Post Emergent Herbicide on Spirodela Polyrhiza as a Model Macrophyte: A Comparison of the Effects of Pure and Nano-Capsulated Form of the Herbicide. *Iran. J. Toxicol.* **2018**, *12*, 45–54. [CrossRef]
38. Mahmoudian, M.; Torbati, S.; AliMirzayi, N.; Nozad, E.; Kochameshki, M.G.; Shokri, A. Preparation and Investigation of Poly(methylmethacrylate) Nano-Capsules Containing Haloxyfop-R-Methyl and Their Release Behavior. *J. Environ. Sci. Health Part B* **2019**, *55*, 301–309. [CrossRef]
39. Falsini, S.; Clemente, I.; Papini, A.; Tani, C.; Schiff, S.; Salvatici, M.C.; Petruccelli, R.; Benelli, C.; Giordano, C.; Gonnelli, C.; et al. When Sustainable Nanochemistry Meets Agriculture: Lignin Nanocapsules for Bioactive Compound Delivery to Plantlets. *ACS Sustain. Chem. Eng.* **2019**, *7*, 19935–19942. [CrossRef]
40. Synowiec, A.; Lenart-Boroń, A.; Bocianowski, J.; Lepiarczyk, A.; Kalemba, D. How Soil-Applied Maltodextrin with Caraway (*Carum carvi* L.) Oil Affects Weed and Soil Microbiological Activity in Maize (*Zea mays* L.) Stands. *Pol. J. Environ. Stud.* **2019**, *29*, 817–826. [CrossRef] [PubMed]
41. Taban, A.; Saharkhiz, M.J.; Khorram, M. Formulation and Assessment of Nano-Encapsulated Bioherbicides Based on Biopolymers and Essential Oil. *Ind. Crops Prod.* **2020**, *149*, 112348. [CrossRef]
42. Wing, R.E.; Maiti, S.; Doane, M. Factors Affecting Release of Butylate from Calcium Ion-Modified Starch-Borate Matrices. *J. Control. Release* **1987**, *5*, 79–89. [CrossRef]
43. Shasha, B.S.; Trimnell, D.; Otey, F.H. Starch–Borate Complexes for EPTC Encapsulation. *J. Appl. Polym. Sci.* **1984**, *29*, 67–73. [CrossRef]
44. Mejías, F.J.R.; Trasobares, S.; Varela, R.M.; Molinillo, J.M.G.; Calvino, J.J.; Macías, F.A. One-Step Encapsulation of Ortho-Disulfides in Functionalized Zinc MOF. Enabling Metal–Organic Frameworks in Agriculture. *ACS Appl. Mater. Interfaces* **2021**, *13*, 7997–8005. [CrossRef] [PubMed]
45. Raju, P.; Natarajan, S. Investigation of Pesticidal and Anti-Biofilm Potential of Calotropis Gigantea Latex Encapsulated Zeolitic Imidazole Nanoframeworks. *J. Inorg. Organomet. Polym. Mater.* **2022**, *32*, 2771–2780. [CrossRef]
46. Ma, S.; Wang, Y.; Yang, X.; Ni, B.; Lü, S. MOF Hybrid for Long-Term Pest Management and Micronutrient Supply Triggered with Protease. *ACS Appl. Mater. Interfaces* **2022**, *14*, 17783–17793. [CrossRef] [PubMed]
47. Wang, Y.; Ma, S.; Yang, X.; Li, Y.; Lü, S. Facile Synthesis of the Dual Pesticide-Loaded Metal–Organic Framework Hybrid for PH-Responsive Release. *ACS Agric. Sci. Technol.* **2022**, *2*, 1267–1275. [CrossRef]
48. Shan, Y.; Xu, C.; Zhang, H.; Chen, H.; Bilal, M.; Niu, S.; Cao, L.; Huang, Q. Polydopamine-Modified Metal–Organic Frameworks, NH_2-Fe-MIL-101, as PH-Sensitive Nanocarriers for Controlled Pesticide Release. *Nanomaterials* **2020**, *10*, 2000. [CrossRef]
49. Dong, J.; Chen, W.; Feng, J.; Liu, X.; Xu, Y.; Wang, C.; Yang, W.; Du, X. Facile, Smart, and Degradable Metal–Organic Framework Nanopesticides Gated with Fe III-Tannic Acid Networks in Response to Seven Biological and Environmental Stimuli. *ACS Appl. Mater. Interfaces* **2021**, *13*, 19507–19520. [CrossRef]

150. Lunn, R.D.J.; Tocher, D.A.; Sidebottom, P.J.; Montgomery, M.G.; Keates, A.C.; Carmalt, C.J. Encapsulation of Aromatic Compounds and a Non-Aromatic Herbicide into a Gadolinium-Based Metal–Organic Framework via the Crystalline Sponge Method. *Cryst. Growth Des.* **2020**, *20*, 7238–7245. [CrossRef]
151. Ma, X.; Zhang, S.; Yang, Y.; Tong, Z.; Shen, T.; Yu, Z.; Xie, J.; Yao, Y.; Gao, B.; Li, Y.C.; et al. Development of Multifunctional Copper Alginate and Bio-Polyurethane Bilayer Coated Fertilizer: Controlled-Release, Selenium Supply and Antifungal. *Int. J. Biol. Macromol.* **2023**, *224*, 256–265. [CrossRef] [PubMed]
152. Liang, Y.; Song, J.; Dong, H.; Huo, Z.; Gao, Y.; Zhou, Z.; Tian, Y.; Li, Y.; Cao, Y. Fabrication of PH-Responsive Nanoparticles for High Efficiency Pyraclostrobin Delivery and Reducing Environmental Impact. *Sci. Total Environ.* **2021**, *787*, 147422. [CrossRef] [PubMed]
153. Sharma, S.; Singh, B.; Bindra, P.; Panneerselvam, P.; Dwivedi, N.; Senapati, A.; Adholeya, A.; Shanmugam, V. Triple-Smart Eco-Friendly Chili Anthracnose Control Agro-Nanocarrier. *ACS Appl. Mater. Interfaces* **2021**, *13*, 9143–9155. [CrossRef] [PubMed]
154. Karthick Raja Namasivayam, S.; Aruna, A. Gokila Evaluation of Silver Nanoparticles-Chitosan Encapsulated Synthetic Herbicide Paraquate (AgNp-CS-PQ) Preparation for the Controlled Release and Improved Herbicidal Activity against *Eichhornia crassipes*. *Res. J. Biotechnol.* **2014**, *9*, 19–27.
155. Dendrinou-Samara, C.; Alevizopoulou, L.; Iordanidis, L.; Samaras, E.; Kessissoglou, D.P. 15-MC-5 Manganese Metallacrowns Hosting Herbicide Complexes. Structure and Bioactivity. *J. Inorg. Biochem.* **2002**, *89*, 89–96. [CrossRef]
156. Hussein, M.Z.; Hashim, N.; Yahaya, A.H.; Zainal, Z. Synthesis of an Herbicides-Inorganic Nanohybrid Compound by Ion Exchange-Intercalation of 3(2-Chlorophenoxy)Propionate into Layered Double Hydroxide. *J. Exp. Nanosci.* **2010**, *5*, 548–558. [CrossRef]
157. Liang, Y.; Gao, Y.; Wang, W.; Dong, H.; Tang, R.; Yang, J.; Niu, J.; Zhou, Z.; Jiang, N.; Cao, Y. Fabrication of Smart Stimuli Responsive Mesoporous Organosilica Nano-Vehicles for Targeted Pesticide Delivery. *J. Hazard. Mater.* **2020**, *389*, 122075. [CrossRef]
158. Sarlak, N.; Taherifar, A.; Salehi, F. Synthesis of Nanopesticides by Encapsulating Pesticide Nanoparticles Using Functionalized Carbon Nanotubes and Application of New Nanocomposite for Plant Disease Treatment. *J. Agric. Food Chem.* **2014**, *62*, 4833–4838. [CrossRef]
159. Francis, A.P.; Devasena, T. Toxicity of Carbon Nanotubes: A Review. *Toxicol. Ind. Health* **2018**, *34*, 200–210. [CrossRef] [PubMed]
160. Mejías, F.J.R.; Trasobares, S.; López-Haro, M.; Varela, R.M.; Molinillo, J.M.G.; Calvino, J.J.; Macías, F.A. In Situ Eco Encapsulation of Bioactive Agrochemicals within Fully Organic Nanotubes. *ACS Appl. Mater. Interfaces* **2019**, *11*, 41925–41934. [CrossRef] [PubMed]
161. Atta, S.; Paul, A.; Banerjee, R.; Bera, M.; Ikbal, M.; Dhara, D.; Singh, N.P. Photoresponsive Polymers Based on a Coumarin Moiety for the Controlled Release of Pesticide 2,4-D. *RSC Adv.* **2015**, *5*, 99968–99975. [CrossRef]
162. Taban, A.; Saharkhiz, M.J.; Kavoosi, G. Development of Pre-Emergence Herbicide Based on Arabic Gum-Gelatin, Apple Pectin and Savory Essential Oil Nano-Particles: A Potential Green Alternative to Metribuzin. *Int. J. Biol. Macromol.* **2021**, *167*, 756–765. [CrossRef]
163. Chariou, P.L.; Dogan, A.B.; Welsh, A.G.; Saidel, G.M.; Baskaran, H.; Steinmetz, N.F. Soil Mobility of Synthetic and Virus-Based Model Nanopesticides. *Nat. Nanotechnol.* **2019**, *14*, 712–718. [CrossRef]

agronomy

MDPI

Review

Crop Allelopathy for Sustainable Weed Management in Agroecosystems: Knowing the Present with a View to the Future

Aurelio Scavo * and Giovanni Mauromicale

Department of Agriculture, Food and Environment (Di3A), University of Catania, 95123 Catania, Italy; g.mauromicale@unict.it

* Correspondence: aurelio.scavo@unict.it; Tel.: +39-09-54-78-34-15

Abstract: In the face of yield losses caused by weeds, especially in low-input agricultural systems, and environmental pollution due to the excessive use of synthetic herbicides, sustainable weed management has become mandatory. To address these issues, allelopathy, i.e., the biochemical phenomenon of chemical interactions between plants through the release of secondary metabolites into the environment, is gaining popularity. Although many important crops are known for their allelopathic potential, farmers are still reluctant to use such knowledge practically. It is therefore important to assist advisors and farmers in assessing whether allelopathy can be effectively implemented into an eco-friendly weed management strategy. Here, we aim to give a comprehensive and updated review on the herbicidal potential of allelopathy. The major findings are the following: (1) Crops from different botanical families show allelopathic properties and can be cultivated alone or in combination with other non-allelopathic crops. (2) Many allelopathic tools can be adopted (crop rotation, intercropping, cover cropping as living or dead mulches, green manuring, use of allelochemical-based bioherbicides). (3) These methods are highly flexible and feature increased efficiency when combined into an integrated weed management strategy. (4) Recent advances in the chemistry of allelopathy are facilitating the use of allelochemicals for bioherbicide production. (5) Several biotechnologies, such as stress induction and genetic engineering techniques, can enhance the allelopathic potential of crops or introduce allelopathic traits de novo. This review shows how important the role of allelopathy for sustainable weed management is and, at the same time, indicates the need for field experiments, mainly under an integrated approach. Finally, we recommend the combination of transgenic allelopathy with the aforementioned allelopathic tools to increase the weed-suppressive efficacy of allelopathy.

Keywords: allelopathy; weed management; crop rotation; cover crops; intercropping; bioherbicides

Citation: Scavo, A.; Mauromicale, G. Crop Allelopathy for Sustainable Weed Management in Agroecosystems: Knowing the Present with a View to the Future. *Agronomy* **2021**, *11*, 2104. https://doi.org/10.3390/agronomy11112104

Academic Editor: Silvia Pampana

Received: 4 October 2021
Accepted: 19 October 2021
Published: 20 October 2021

1. Introduction

Weeds are considered the most serious biotic constraint on crop production, with yield losses ranging from 45–95%, depending on environmental conditions and agronomic practices [1]. The reduction of yields is not the only harmful effect associated with weeds; a decrease in the quality and market value of agricultural products is also reported. For many years, weed control in agroecosystems has been pursued almost exclusively mechanically and chemically, keeping weed pressure and management costs low and crop productivity high. Nowadays, agriculture is facing the negative effects derived from the improper use of tillage and synthetic herbicides: depletion of soil organic matter, disruption of soil structure, acceleration of soil erosion, increase in herbicide-resistant weeds, development of a substitution weed flora, herbicides persistence in vegetables, and contamination of food and the environment. Furthermore, in organic farming, an exponentially increasing sector throughout Europe (~13.7 million ha) and the world (~71.4 million ha of total organic area) [2], chemical control is avoided and both hand weeding and mechanical tools alone are not agronomically or economically sufficient. As a consequence, the search for alternative and sustainable weed control practices has become an imperative. Among the low-input and environmentally-friendly available methods for weed management, the manipulation of allelopathic mechanisms plays a central role, as demonstrated by the increasing interest of the scientific community [3]. In the past 11 years (late 2020-early 2010), 3257 articles have been published in the field of allelopathy (this number refers to the Scopus® database using the keyword "allelopathy"), 53% of which were focused on weed control.

Allelopathy is a biochemical phenomenon with ecological implications involving any harmful or beneficial effect, either direct or indirect, by one plant (donor) on another (target) through the production of chemical compounds that escape into the environment [4]. Allelochemicals, i.e., the defensive secondary metabolites involved in the allelopathic interactions, can have negative effects on conspecific (autoallelopathy or autotoxicity) and/or heterospecific species (heterotoxicity). They encompass a very wide range of chemical classes, the most representative of which are phenolic compounds (simple phenols, flavonoids, quinones, coumarins, etc.), terpenoids (mono-, di- and triterpenes, sesquiterpenes and steroids) and compounds containing a nitrogen atom (e.g., benzoxazinoids) [5]. The mechanisms of action of allelochemicals are the most varied, considering that the visible effects on target plants (e.g., inhibition of seed germination, reduction of seedling growth) are often secondary signs of primary changes (inhibition of cell division and elongation, interference with cell membrane permeability, enzymatic activities, respiration and photosynthesis, etc.) [3]. Furthermore, in field conditions, the allelopathic effects are generally caused by the joint action of mixtures of allelochemicals. The role of allelochemicals in acting as biopesticides in agricultural pest management against weeds, insects and diseases has been examined and reviewed [6,7]. In this review, only the detrimental effects of allelopathy and the plant–plant interactions will be considered, with special reference to crop–crop and weed–crop allelopathic interference. Nevertheless, even in this case, there is a large literature on how allelopathy could be exploited for weed control [8,9]. However, in the past 5 years more than 1000 papers have been published on this topic, thus an updated review of them is needed. After discussing the allelopathic behavior of the main crops and the allelochemicals involved, this article reviews the recent advances in weed management through allelopathy by reporting practical applications of crop rotation, cover cropping and bioherbicides, also under an integrated approach. The last chapter focuses on the role of modern biotechnologies in plant allelopathy. The goal of the review is to find new possible applicative solutions in the allelopathy field by using the acquired knowledge to make weed management more sustainable.

2. Crop Allelopathy

Although the ability of certain plants to negatively affect other plants is an ancient concept, and was well-documented 2000 years ago (e.g., by Demokritus, Theophrastus, Pliny the Elder, Columella, etc.) [10], allelopathy in the narrow sense has been demonstrated only in the past four decades [11,12]. Over these years, several eminent scientists in the field of allelopathy have proposed guidelines for the suspected cases of crop allelopathy [13,14]. Summarizing, an allelopathic crop should present (i) vegetation patterns around itself, (ii) affect the growth of other crops or the same crop growth when cultivated in succession, (iii) cause problems of soil sickness due to the build-up of allelochemicals in the soil, (iv) synthesize and release into the environment bioactive allelochemicals. A large number of plant species are known to possess allelopathic properties, mainly weeds. The latest estimation found ~240 allelopathic weeds [15], but many other weed species were found in the past 20 years to show allelopathic effects. In addition to weeds, many herbaceous and woody crops have allelopathic traits both on other crops and weeds [16]. Most allelopathic crops belong to the Asteraceae and Poaceae families, but Brassicaceae and Fabaceae are also well-represented (Table 1).

Table 1. List of main crops showing allelopathic properties with corresponding allelochemicals involved.

Botanical Family	Binomial Name	Allelochemicals	References
Amaryllidaceae	*Allium cepa* L. *A. sativum* *A. ursinum*	S-containing compounds (alliin, isoalliin, methiin, allicin, ajoene, sulfenic acid, methyl propenyl disulfate, methylpropyl trisulfate) and phenolic acids (ferulic, *p*-coumaric, *p*-hydroxybenzoic, syringic, vanillic)	[17]
Asteraceae	*Artemisia absinthium* L.	Tannins, terpenes and alkaloids (absinthium)	[18]
	Carthamus tinctorius L.	Sesquiterpene lactones (dehydrocostuslactone, costunolide) and strigolactones (solanacol, GR24 and abacyl acetate)	[19]
	Cichorium intybus L.	Sesquiterpene lactones (8α-angeloyloxycichoralexin, lactupicrin) and guaianolides (cichoralexin, 10α-hydroxycichopumilide)	[20]
	Cynara cardunculus L.	Sesquiterpene lactones (cynaropicrin, deacylcynaropicrin, 11,13-dihydro-deacylcynaropicrin, grosheimin, 11,13-dihydroxi-8-deoxygrosheimin, aguerin B, cynaratriol), pinoresinol and polyphenols (caffeoylquinic and dicaffeoylquinic acids, luteolin and apigenin derivatives)	[21–23]
	Helianthus annuus L.	Sesquiterpene lactones (helivypolide D, leptocarpin, helivypolide E, annuolide F, annuolide H, helivypolides F, helivypolides H, helivypolides J, helieudesmanolide A, 8β-angeloiloxicumambranolide), heliannuoles (heliannuol J), bisnorsesquiterpenes (annuionone D, (+)-dehydrovomifoliol), flavonoids (heliannone A, kukulkanine B, heliannone B, tambuline) and (+)-loliolide	[24]
	H. tuberosus	Sesquiterpene lactones (1,10-epoxidized heliangolides, 1-keto-2,3-unsaturated-furanoheliangolides, 4,15-isoatriplicolide angelate, 4,15-isoatriplicolide methylacrylate), diterpenes (*ent*-17-oxokaur-15(16)-en-19-oic acid, *ent*-17-hydroxykaur-15(16)-en-19-oic acid, *ent*-15β-hydroxykaur-16(17)-en-19-oic acid methyl ester and *ent*-15-*nor*-14-oxolabda-8(17),12E-dien-18-oic acid), phenolic compounds (*p*-hydroxybenzoic acid, *p*-hydroxybenzaldehyde, salicylic acid, coumarin, *o*-coumarinic acid, and *p*-coumaric acid) and (+)- pinoresinol	[25,26]
	Lactuca sativa L.	Phenolic acids (coumarin, *trans*-cinnamic acid, *o*-coumaric acid, *p*-coumaric acid and chlorogenic acid)	[27]
Brassicaceae	*Brassica juncea* (L.) Czern. *B. napus* *B. nigra* *B. oleracea*	Glucosinolates [isothiocyanates (allyl-ITC, 2-phenylethyl, 3-butenyl, 4-pentenyl, 4-methylthiobutyl, 5-methylthiopentyl), nitriles (5-methylthiopentanenitrile, 6-methylthiohexanenitrile), oxazolidinethione (goitrin)] and brassinosteroids (brassinolide, 24-epibrassinolide, 28-homobrassinolide)	[28]
	Capparis spinosa L.	Flavonoids (quercetin-3-*O*-β-D-glucopyranoside, quercetin, kaempferol 3-*O*-β-D-glucopyranoside, rhamnetin, isorhamnetin, rhamnozin, thomnocitirin), carotenoids (β-carotene, lutein, neoxanthin and violaxanthin), tocopherols (α- and γ-tocopherol) and glucosinates	[29,30]

Table 1. *Cont.*

Botanical Family	Binomial Name	Allelochemicals	References
Cannabaceae	*Cannabis sativa* L.	Glycosides, alkaloids, flavonoids, flavones, steroids, tannins, phenols and saponins	[31]
Convolvulaceae	*Ipomea batatas* (L.) Lam.	Polyphenols (coumarin, trans-cinnamic acid, hydroxycinnamic acid, *p*- and *o*-coumaric acid, caffeic acid, chlorogenic acid)	[32]
Cucurbitaceae	*Citrullus lanatus* (Thunb.) Matsum and Nakai	Phenolic acids (ρ-hydrobenzoic, vanillic, syringic, ρ-coumaric and frulic acids)	[33]
Cucurbitaceae	*Cucumis sativus* L.	Phenolic acids (benzoic, *p*-hydroxybenzoic, 2,5-dihydroxybenzoic, 3-phenylpropionic, cinnamic, *p*-hydroxycinnamic, gallic, vanillic, caffeic, hydrocaffeic, *p*-coumaric, ferulic, sinapic, *p*-thiocyanatophenol and 2-hydroxybenzothiazole) and fatty acids (myristic, palmitic, and stearic)	[34]
Fabaceae	*Arachis hypogaea* L.	Phenolic acids (*p*-coumaric and benzoic) and fatty acids (tetradecanoic, hexadecanoic, octadecanoic)	[35]
	Glycyrrhiza uralensis Fisch.	Flavonones (liquiritin, isoliquiritigenin) and triterpenes (glycyrrhizic acid, dodecanoic acid)	[36]
	Medicago sativa L.	Phenolic compounds (salicylic acid, *p*-hydroxybenzoic acid, *trans*-cinnamic acid, *o*-coumaric acid, *p*-coumaric acid, ferulic acid, vanillic acid, chlorogenic acid, caffeic acid, coumarin, rutin, quercetin, scopoletin, medicarpin, sativan, 4-methoxymedicarpin, 5-methoxy-sativan)	[37,38]
	Phaseolus vulgaris L.	Phenolic acids (benzoic, salicylic, and malonic)	[39]
	Pisum sativum L.	Phenolic acids (benzoic, cinnamic, *p*-hydroxybenzoic, 3,4-dihydrobenzoic, vanillic, *p*-coumaric, sinapic) and pisatin	[40]
	Vicia faba L.	Phenolic acids (lactic, benzoic, *p*-hydroxybenzoic, vanillic, adipic, succinic, malic, glycolic, *p*-hydroxyphenylacetic, salicylic)	[39]
	Tamarindus indica L.	Phenolic acids (caffeic), methyl-2,3,4-trihydroxyhexanoate and organic acids (citric, malic, oxalic, and tartaric)	[41,42]
Juglandaceae	*Juglans nigra* L. *J. regia*	Naphthoquinones (juglone, 1,4-naphthoquinone, plumbagin, 2-methyl-1,4-naphthoquinone), triterpenoids (lupenone, lupeol, squalene) fatty acids (n-hexadecanoic, 9,12-octadecadienoic, 8-octadecenoic, palmitic, stearic), phenolic acids (chlorogenic, *p*-coumaric, *o*-coumaric, ferulic, tannic, caffeic, vanillic, syringic), flavonoides (catechin, epicatechin and myricetin), flavonoids (quercetin and quercetin derivatives), hydroxybenzoic acids (gallic, ellagic, protocatechuic), and steroids (γ-sitosterol, sitostenone)	[43,44]
Labiateae	*Rosmarinus officinalis* Schleid.	Monoterpenoids (α-pinene, myrcene, α-terpinene, β-cymene, 1,8-cineole, camphene, α-limonene, sabinene) and polyphenols (caffeic, ferulic, gallic, rosmarinic, carnosic, and chlorogenic acids)	[45,46]

Table 1. *Cont.*

Botanical Family	Binomial Name	Allelochemicals	References
Moraceae	*Ficus carica* L.	Phenolic acids (3-*O*-caffeoylquinic acid, 5-*O*-caffeoylquinic acid, dihydroxybenzoic acid, caffeoylmalic acid) and flavonoids (rutin, isoquercetin, catechin and astragalin)	[47]
Myrtaceae	*Eucalyptus globulus* Labill. *E. urograndis* *E. urophylla*	Phenolic compounds (chlorogenic acid, ρ-coumaric acid, ellagic acid, hyperoside, rutin, quercitrin and kaempferol 3-*O*-glucoside) and organic acids (citric, malic, shikimic, succinic and fumaric)	[48]
Poaceae	*Avena sativa* L.	Flavonoids (2-*O*-glucoside, isovitexin 2″-*O*-arabinoside), phenolic acids (caffeic, ferulic, coumaric, salicylic, coumarin, cynnamic and derivatives) and saponins (avenacoside A, avenacoside B, 26-desglucoavenacoside A, 26-desglucoavenacoside B)	[49,50]
Poaceae	*Hordeum vulgare* L.	Phenolic acids (benzoic, caffeic, chlorogenic, coumaric, coumarin, ferulic, *p*-hydroxybenzoic, ferulic, gentisic, salicylic, sinapic, syringic, vanillic, cinnamic, hydroxycinnamic, scopoletin), benzoxazinoids (DIBOA, DIMBOA), alkaloids (gramine, hordenine), flavonoids (saponarin, apigenin, lutonarin, catechin, cyanidin, isovitexin) and cyanoglucosides (heterodendrin, epidermin, epiheterodendrin, sutherlandin, osmaronin, dihydroosmaronin)	[51]
	Oryza sativa L.	Diterpenes (momilactones and oryzalexins), phenolic acids (caffeic, ferulic, coumaric, salicylic, syringic, *p*-hydroxybenzoic, coumarin, cynnamic and derivatives), flavones (5,7,4′-trihydroxy-3′,5′-dimethoxyflavone) and cyclohexenones (3-isopropyl-5-acetoxycyclohexene-2-one-1)	[13,49,52]
	Secale cereale L.	Benzoxazinoids (DIBOA, HMBOA, BOA, DIMBOA, MBOA) and phenolic acids (caffeic, ferulic, coumaric, salicylic, coumarin, cynnamic and derivatives)	[49]
	Sorghum bicolor (L.) Moench *S. bicolor* × *S. sudanense* *S. halepense*	Benzoxazinoids (DIBOA, HMBOA, BOA, DIMBOA, MBOA), benzoquinones (sorgoleone), cyanogenic glycosides (dhurrin), phenolic acids (*p*-hydroxybenzoic, *p*-hydroxybenzaldehyde, coumaric, ferulic)	[53]
	Triticum aestivum L. *T. durum*	Benzoxazinoids (DIBOA, DIMBOA-Glc, HMBOA, BOA, DIMBOA, MBOA), phenolic acids (*trans*-ρ-coumaric, *cis*-ρ-coumaric ferulic, vanillic, syringic, ρ-hydroxybenzoic), fatty acids (acetic, propionic and butyric), triterpenoids (cycloart-5-ene-3β,25-diol and cycloart-3β,25-diol) and steroids (cholesterol, ergosterol, campesterol, stigmasterol, sitosterol, spinasterol and stigmastanol)	[54–56]
Rubiaceae	*Coffea arabica* L.	Alkaloids (caffeine, theobromine, theophylline, paraxanthine), coumarins (scopoletin) and phenolic acids (chlorogenic, ferulic, *p*-coumaric, *p*-hydroxybenzoic, caffeic and vanillic)	[57]

Table 1. *Cont.*

Botanical Family	Binomial Name	Allelochemicals	References
Solanaceae	*Capsicum annuum* L.	Phthalate esters (N-phenyl-2-naphthylamine, dibutyl phthalate, butyl cyclohexyl ester), dicarboxylic acids (phthalic acid, 1,2-benzenedicarboxylic acid) and phenols anilines (diphenylamine, 4,4′-(1-methylethylidene) bis-phenol, 1-naphthalenamine, n-phenyl-1-naphthylamine)	[58,59]
	Nicotiana tabacum L.	Alkaloids (nicotine), sucrose esters (fluoro derivatives of sucrose) and diterpenes (duvatrienediol and duvatrienediol derivatives)	[60]
	Solanum lycopersicum L.	Alkaloids (α-tomatine), steroids (stigmasterol), furocoumarins (bergapten) and strigolactones (7-oxoorobanchylacetate, solanacol, orobanchol, strigol, fabacyl acetate, orobanchyl acetate and 5-deoxysrtrigol)	[61]

As regards the Asteraceae, the most studied allelopathic crop is sunflower (*Helianthus annuus* L.). In Asteraceae members including sunflower, the main allelochemicals are sesquiterpenes, especially heliannuoles, sesquiterpene lactones and bisnorsesquiterpenes, in addition to triterpenes and flavonoids [24]. Its allelopathic effects have been tested on both other crops and weeds, in field conditions and in vitro bioassays [62]. In recent years, the allelopathic potential of *Cynara cardunculus* L., an herbaceous perennial species belonging to the Mediterranean basin [63], was assessed on seed germination and seedling growth of some weeds and target crops [21,64,65]. Allelochemicals responsible for *C. cardunculus* allelopathy are the sesquiterpene lactones cynaropicrin, deacylcynaropicrin, 11,13-dihydro-deacylcynaropicrin, aguerin B, grosheimin, 11,13-dihydroxy-8-deoxygrosheimin and cynaratriol, as well as polyphenols such as caffeoylquinic and dicaffeoylquinic acids, luteolin and apigenin derivatives [21–23]. Recently, Rial et al. [19] demonstrated the phytotoxicity of safflower (*Carthamus tinctorius* L.), a thistle-like herbaceous plant cultivated in regions with arid or semiarid climate for industrial applications (oil production, pigments and human consumption), indicating the sesquiterpene lactones dehydrocostuslactone and costunolide and several strigolactones as the main allelochemicals released by root exudation.

Allelopathy in Poaceae plants has been widely described. Rice (*Oryza sativa* L.), rye (*Secale cereale* L.), common (*Triticum aestivum* L.) and durum wheat (*T. durum*), sorghum (*Sorghum* spp.), barley (*Hordeum vulgare* L.) and oat (*Avena sativa* L.) are probably the most studied allelopathic crops. The spectrum of their allelochemicals has been investigated in depth, with benzoxazinoids (DIBOA, HMBOA, BOA, DIMBOA, MBOA), phenolic acids, flavonoids and terpenoids recognized as major allelochemicals [49,53,55]. Moreover, the biosynthetic pathways of some of these allelochemicals have been sequenced, such as in the case of sorgoleone [66]. Given the considerable knowledge of this family and chemicals involved, the recent research has focused on the utilization of their allelopathic mechanisms for weed control.

The Brassicaceae family comprises more than 3200 species, of which the *Brassica* genus includes several highly allelopathic crops such as canola (*Brassica napus* L.), Indian mustard (*B. juncea*), black mustard (*B. nigra*) and cabbage (*B. oleracea*). Rehman et al. [28] reviewed the use of *Brassica* allelopathy for weed management and documented that glucosinolates (mainly isothiocyanates and nitriles) and the endogenous steroidal compounds brassinosteroids (e.g., brassinolide, 24-epibrassinolide, 28-homobrassinolide) are responsible for their phytotoxicity. Significant evidence of allelopathic effects has been reported in some leguminous crops. The best known example is alfalfa (*Medicago sativa* L.), commonly used as living or dead mulch for weed management, and widely studied also as a plant model in autoallelopathy [37,38]. Other examples of allelopathic Fabaceae plants are the common bean (*Phaseolus vulgaris* L.), faba bean (*Vicia faba* L.), peanut (*Arachis hypogaea* L.) and, recently, liquorice (*Glycyrrhiza uralensis* Fisch.) [35,36,39]. Fabaceae allelopathy is mainly due to phenolic acids such as benzoic, cinnamic, *p*-hydroxybenzoic, vanillic, coumaric, ferulic, caffeic, salicylic, etc.

The Solanaceae family is gaining in interest for the allelopathic potential shown by some important members. Rial et al. [61], for example, investigated the allelopathic traits of tomato (*Solanum lycopersicum* L.) and identified its major root allelochemicals as the alkaloid α-tomatine, the steroid stigmasterol, the furocoumarin bergapten and the strigolactones solanacol, orobanchol, strigol, etc. Important phytochemical advances were also made in the discovery and identification of tobacco (*Nicotiana tabacum* L.) [60] and red pepper (*Capsicum annuum* L.) allelochemicals [59].

3. Allelopathic Practices Involved in Weed Management

Allelopathic crops can be employed to manage weeds in agroecosystems by (i) including them in rotational sequences or (ii) intercropping in close proximity with a cash crop, (iii) cover cropping as living or dead mulches, (iv) crop residue incorporation into the soil and (v) by using their allelochemicals as bioherbicides [6–9]. The adoption of allelopathy for weed management is highly flexible, varying site by site depending on the specific characteristics of the context: pedo-climatic conditions, weed species, agricultural practic used, economic constraints and farmer's expectations. Allelopathy can be exploited i different cropping systems, but it certainly plays more beneficial roles in organic farmin and in conservative, minimum and no-tillage agricultural systems, where weed control often problematic. Moreover, the abovementioned allelopathic weed control tools can b individually applied or combined into an integrated weed management strategy (IWM to increase their efficiency [67].

3.1. Crop Rotation

Planting different crops in sequence in the same field is a traditional agricultur tactic allowing many benefits in cropping systems: weed and pest control, reduction autoallelopathy or soil sickness related to monoculture, reduction of nutrient leachin improvement of soil organic matter and soil microorganisms, enhancement of soil fertili and crop yields [68]. Crop rotation offers the best results when considered within an IWM as prevention against weed establishment and reduction of the soil seedbank [67]. Indee such techniques avoid the development of specialized and invasive weed flora, while lowing a multifaceted weed community characterized by low densities for each species [6 It does not eradicate troublesome weeds, but limits their reproduction and reduces th impact of subsequent direct control methods. Including an allelopathic crop within ro tional sequences can help to control weeds both in the current and next crops by releasi allelochemicals into the soil through root exudation, decomposition of plant residues a leaching from plant foliage. Once released into the rhizosphere, allelochemicals can direct or indirectly—by microbial transformation into more active, less active or entirely inacti compounds—inhibit seed germination and reduce weed density and biomass [70]. Th positive effects of allelopathic crops within crop rotations are often exacerbated in conserv tive agricultural systems. For example, studying the impact of five tillage systems and fi crop rotations, Shahzad et al. [71] found that sorghum-wheat rotation had the stronge weed-suppressive effect in terms of density and dry biomass reduction in all tillage system especially during the second year, thanks to the accumulation of sorghum allelochemic (sorgoleone) in the soil. In another study, Scherner et al. [72] examined the combin 11-year-long effects of tillage and crop rotations on weed flora and reported that, amo the four crop sequences under study, the most diversified sequence (winter wheat–spri barley–peas rotation) had the lowest density of grass weeds, while tillage effects did n differ within rotations. Similarly, Hunt et al. [73] compared the environmental and ag nomic impact of three crop rotation systems and integrated mechanical–chemical we control tactics. A 4-year rotation based on corn–soybean–oat/alfalfa–alfalfa combin with mechanical and chemical weed control returned similar results to a conventiona managed less diverse system in terms of grain yields and weed suppression, while al achieving a significant reduction in herbicide use and water contamination. Overall, t scientific literature agrees in promoting the diversification of crop rotations to suppr weeds while limiting the adoption of herbicides. In this regard, Weisberger et al. [conducted a meta-analysis across studies involving simple and more diverse crop rotatio They found that diversifying crop rotations, often involving allelopathic crops such wheat, oat, corn, alfalfa, sunflower, etc., reduced weed density by 49% compared to sim sequences, while no significant effects were observed on weed biomass, likely due to lower number of studies. Furthermore, they indicated that the effect of crop rotation on size and structure of weed communities was markedly influenced by tillage systems. particular, crop rotation showed the best results in zero-tillage systems where, in addit to weed control, it reduced or eliminated the yield losses and improved soil conservati Therefore, the authors demonstrated the high synergism between zero-tillage and cr rotation for weed management.

Recently, Scavo et al. [75] suggested the inclusion of cultivated cardoon or gl artichoke for 2–3 years in Mediterranean crop rotations to reduce the size of the we seedbank and stimulate the soil eubacterial communities. Sometimes, the combinat of crop allelopathy and crop rotation can produce negative effects on the cash crop. For instance, Karkanis et al. [76] indicated that the inclusion of spearmint (*Mentha* × *piperita*) or peppermint (*Mentha spicata* L.) in rotational sequences decreased the dry biomass, photosynthetic rate and grain yield of corn as the succeeding crop. Demonstrating the direct allelopathic effect of crop rotation in field experiments is very complex, since it is often confused with the role of competition and other indirect disturbances. For this reason, a multidisciplinary approach in such experiments is needed.

3.2. Cover Cropping

Cover cropping is the mono- or intercropping of herbaceous plants either for part of or an entire year with the aim of enhancing yields [77]. Cover crops are grown for their numerous ecosystem services: protection from soil erosion, reduction of nutrient leaching (especially nitrates), enhancement of soil organic matter levels and microbial activities, improvement of soil structure and hydraulic properties, conservation of soil moisture, pest management and weed control [78,79]. Altogether, these benefits result in higher yields for the subsequent crops. Cover cropping, mulching, intercropping and green manuring are often indicated as distinct and separate techniques, but they can be considered as synonymous [67,79]. Indeed, cover crops can be used as living mulches when intercropped with the cash crop, as dead mulches by leaving plant residues on the soil surface or green manures by incorporating the residues into the soil (Table 2).

Table 2. Exploitation of cover cropping for the allelopathic management of weeds.

Cover Cropping Type	Allelopathic Cover Crop	Main Crop	Target Weeds	References
Dead mulching	*Helianthus annuus* L., *Zea mays* L., *Oryza sativa* L., *Sorghum bicolor* L.	Wheat	*Phalaris minor* Retz.	[80]
	Trifolium subterraneum L.	Apricot	Several monocots and dicots	[81]
	Fagopyrum esculentum Moench., *Sinapis alba* L., *T. subterraneum*, *H. annuus*, *Linum usitatissimum* L., *Raphanus sativus* L., *Vicia sativa* L., *Avena strigosa* Schreb., *Cannabis sativa* L. and mixtures	Sugar beet	*Stellaria media* (L.) Vill., *Chenopodium album* L., *Matricaria chamomilla* L.	[82]
Green manuring	*Eucalyptus globulus* Labill.	Corn	*Digitaria sanguinalis* (L.) Scop., *C. album*	[83]
	S. alba and *S. alba* + *F. esculentum*	Red clover, wheat, pea, barley rotation with red clover as a undersown crop	*Cirsium arvense* (L.) Scop., *Sonchus arvensis* L., *Galium aparine* L., *Lamium purpureum* L., *Fallopia convolvulus* (L.) Á. Löve, *C. album*, *S. media*	[84]
	V. faba	Corn	*A. retroflexus*, *C. album*, *Solanum nigrum*, *D. sanguinalis*, *Cyperus rotundus* L.	[85]
	Hordeum vulgare L., *V. sativa*	Corn, sunflower	*Xanthium spinosum* L. and other broadleaf species	[86]
Intercropping	*Crotalaria juncea* L.	Cotton	*C. rotundus*, *Alternanthera paronychioides* A. St.-Hil.	[87]
	Trigonella foenum-graecum L.	Coriander	Several monocots and dicots	[88]
	F. esculentum, *Lens culinaris* Medik., *S. bicolor*, *H. annuus*	Soybean	*C. album*, *Polygonum persicaria* L.	[89]
	T. repens	Wheat	*A. fatua*, *S. media*, *M. recutica*	[90]

As observed for crop rotation, cover cropping is commonly adopted in low-input agriculture and organic farming, mainly in IWMS to control the soil weed seedbank and prevent weed emergence [67]. The weed-suppressive ability of cover crops is a result of physical and allelopathic effects, which in the field often act in synergism [91].

Root development, rapid growth, length of biological cycle and biomass production are factors determining the competitive capacity of cover crops for light, water, nutrients and space. Generally, the greater the biomass production and biological length, the higher the cover crop phytotoxicity [92]. Both cover crops and mulches can indirectly affect weed density by favouring predators that eat seeds [93]. Moreover, surface-applied or soil-incorporated mulches obstruct weed seed germination and reduce weed emergence by acting as physical barriers [78]. In addition to such kinds of physical interference, living and dead mulches of allelopathic species directly exude into the rhizosphere or release allelochemicals by residues' decomposition that, once present in the soil solution, must reach the target plants to exert phytotoxic effects [70]. The efficacy of this allelopathic weed suppression is mediated by climatic conditions, soil properties, cover crop genotype, quantity, duration and placement of cover crop residues, and biological characteristics of the weed communities (Figure 1).

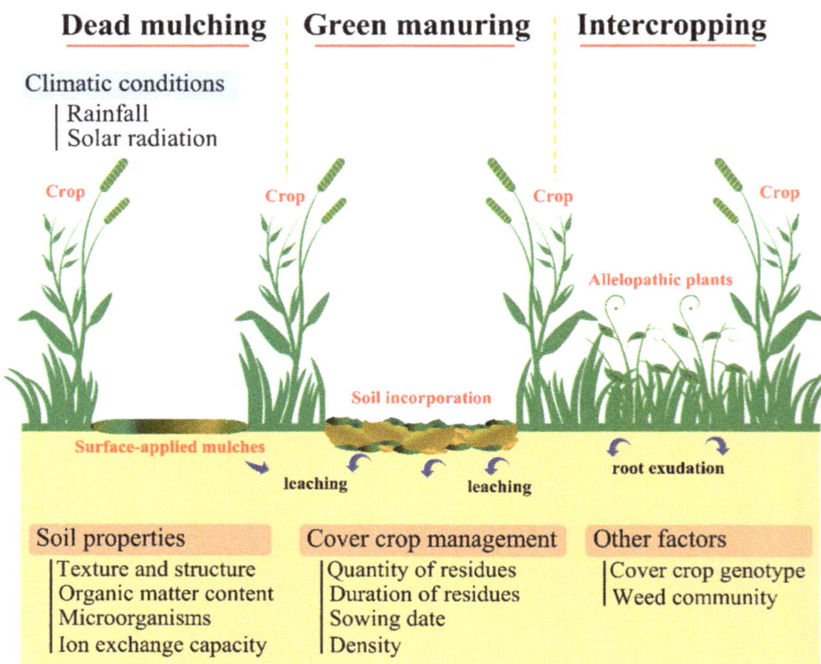

Figure 1. Practical applications of allelopathic cover cropping by intercropping, surface-applicatio (dead mulching) and soil incorporation (green manuring) of crop residues. Allelopathic effects a due to the release of active allelochemicals into the soil through root exudation from living mulche and leaching from decomposing residues. The level of phytotoxicity is closely influenced by climat conditions (rainfall, solar radiation), soil properties (texture, structure, organic matter content, pH cation exchange capacity and microorganisms), characteristics of the weed community, cover cro genotype and management (quantity and duration of residues, sowing date and density).

For example, Kruidhof et al. [94] reported an increased allelopathic activity of lucerr (*M. sativa*), winter oilseed rape (*B. napus*) or winter rye (*S. cereale*) with increasing rainfa level, probably due to the higher leaching of allelochemicals. The same authors also ind cated that the larger the weed seed size, the higher its resistance and detoxification capaci especially with respect to surface-applied dead mulch, highlighting the importance of wee community composition. The role of soil properties (e.g., texture, structure, pH, organ matter level, ion exchange capacity, etc.) in affecting allelochemicals retention, transpo and availability in the soil has been examined in depth by Scavo et al. [70]. Concerning cover crop management, the extent of an effective weed control increases with increasing seeding rate, amount and duration of plant residues [8,79]. These aspects of cover cropping allelopathy are exacerbated in no-tillage systems. Some important cover crops with allelopathic potential include the cereals rye, sorghum, wheat, barley, oat, buckwheat (*Fagopyrum esculentum* Moench.), the Brassicaceae black mustard, field mustard (*B. rapa*), rapeseed (*B. napus*) and white mustard (*Sinapis alba* L.), and the legumes alfalfa, hairy vetch (*Vicia villosa* Roth), common vetch (*V. sativa*), velvet bean (*Mucuna pruriens* ((L.)) DC.), cowpea (*Vigna unguiculata* ((L.)) Walp.), crimson clover (*Trifolium incarnatum* L.), subterranean clover (*T. subterraneum*), red clover (*T. pratense*) and Egyptian clover (*T. alexandrinum*).

3.2.1. Dead Mulching

Surface-applied mulches are recognized to act as preventive weed control by affecting the soil weed seedbank, weed emergence and establishment [67]. In a two-year field trial, Abbas et al. [80] noted a significant decay of the soil weed seedbank and an increased wheat yield after the application of sunflower, rice, corn and sorghum mulches. Moreover, the integrated use of allelopathic crop mulches and post-emergence herbicide mixtures at low doses provided an effective control of *Phalaris minor* Retz., thus reducing the herbicide-resistance development in *P. minor* and the herbicide selection pressure. Sturm et al. [82] evaluated ten different cover crop mulches and mixtures in a sugar beet cultivation, obtaining different results in relation to cover crop type and target weed. In particular, a significant reduction of weed density by all cover crops compared to untreated control was observed, with the highest reduction caused by cover crop mixtures, likely due to a synergistic effect of allelochemicals. The release of allelochemicals into the soil largely depends on the decomposability of the residues (C/N ratio) and residue management, particularly pre-treatment [95]. Dead mulching is often less efficient in suppressing weeds than incorporating plant residues into the soil, likely due to the higher release rate mediated by soil microorganisms and to the longer persistence of released allelochemicals [91]. Scavo et al. [81] studied the effect of *T. subterraneum* cover cropping, with or without burying dead mulches into the soil, on the quali-quantitative composition of the soil weed seedbank in an apricot (*Prunus armeniaca* L.) orchard. *Trifolium subterraneum* green manuring was more effective than dead mulching in decreasing the size of the soil weed seedbank, although both have significantly reduced it compared to spontaneous flora cover cropping and conventional apricot management. In contrast, Kruidhof et al. [95] reported a genotype-dependent effect, where winter rye dead mulching inhibited weed emergence more so than green manuring, while an opposite trend was observed for winter oilseed rape.

3.2.2. Green Manuring

In recent years, several reports have been made on incorporating allelopathic plant residues into the soil (i.e., green manuring) for weed control under field conditions. Puig et al. [83] evaluated the allelopathic potential of *Eucalyptus globulus* Labill. leaf green manure in corn fields for two seasons and in two different locations. *Eucalyptus* green manure significantly reduced weed biomass, especially that of *Digitaria sanguinalis* (L.) Scop. and *Chenopodium album* L., and mainly at early stages of corn establishment, while corn was not negatively affected. Soil pH, cation exchange capacity and microbial biomass carbon were also increased by *E. globulus* green manure. In previous research, Puig et al. [48] reported that *Eucalyptus* leaf green manure continuously releases different phenolic acids (chlorogenic acid, ellagic acid, hyperoside and rutin) and volatile compounds (the monoterpenes α-pinene, β-pinene, α-phellandrene, eucalyptol, β-cis-ocimene, γ-terpinene, terpinen-4-ol, terpineol and the sesquiterpene longifolene and β-caryophyllene) during a 30-day period of decomposition. In another study, Álvarez-Iglesias et al. [85] reported that faba bean (*V. faba*) green manure suppressed both density (from −14.8% to −69.8%) and biomass (from −46.9% to −78.5%) of some dicotyledon and monocotyledon weeds in correspondence with the early critical period of corn, thus reducing the need for post-emergence herbicides.

Investigating the weed-suppressive capacity of different cover crops cultivated for gree manure in a 6-year field experiment, Masilionyte et al. [84] found that white mustard (*alba*), especially when combined with buckwheat (*F. esculentum*), showed a higher reductic of the number of weeds and weed biomass than narrow-leafed lupine (*Lupinus angustifoli* L.) in a mixture with oil radish (*Raphanus sativus* L.), thanks to its biomass production an release of allelochemicals into the soil. Alonso-Ayuso et al. [86] studied barley (*H. vulga* and vetch (*V. sativa*) green manure in a long-term field experiment involving two cas crops: corn and sunflower. Overall, replacing a winter fallow by cover crops had positi effects on weed density, diversity and soil seedbank, with both cover crops showing weed density reduction of 51–63% in spring, while barley green manure suppressed wint weeds better than vetch. However, the weed seedbank size was not affected after 10 yea of cover cropping. Additionally, only during 1 year did the weed control efficacy increa as the cover crop termination date was delayed.

3.2.3. Intercropping

Intercropping, i.e., growing multiple crop species together in the same field during growing season, has been widely used throughout history to maximize ecosystem servic per unit area per unit of time. It still remains a common agricultural practice in sm farms, conservative agriculture and resource-limited agricultural systems, although it increasingly used also in modern intensive agriculture. Many studies describe intercrc ping as a means to address weed control in an economical and environmentally friend way [96]. The genotypes of both cash and cover crop, plant density, plant arrangement, e closely affect the level of weed suppression. On this point, it is possible to distinguish thr main types of intercropping: fully mixed (without a specific arrangement), relay (with temporal separation between crops) and strip (two or more crops are separated in the spa by cultivating them in strips) [96]. Strip intercropping is the most adopted in allelopat field experiments, since it allows more interactions between crops and facilitates th cultivation. Allelopathic crops, when included in intercropping systems, release alle chemicals into the environment through root exudation, volatilization from abovegrou plant parts and leaching from rainfall or decay of plant debris [3]. Intercropping is report to enhance the allelopathic weed–cover crop interactions and, consequently, the phytoto effects by improving soil microbial diversity and facilitating allelochemicals' transp into the soil [96]. Barto et al. [97] described the ability of common mycorrhizal netwo in acting as 'superhighways' directly connecting the plants belowground and deliveri allelochemicals to target plants.

Cereal–legume intercropping is the most common example of allelopathic intercr ping due to the high numbers of allelopathic crops suitable for cover cropping both in Poaceae and Fabaceae families. Rad et al. [98] investigated the effect of *S. bicolor* intercr ping with different ratios of hairy vetch (*V. villosa*) and lathyrus (*Lathyrus sativus* L.), a three weed management strategies (no weed control, full weed control, hand-weedir The selection of appropriate intercropping ratios played a key role in enhancing the we control and improving the quali-quantitative traits of sorghum forage. Sorghum with 10 lathyrus showed the highest weeding efficacy, but good results in terms of forage yie were obtained by adding a minimum of 33–66% of hairy vetch to intercropping un no weeding conditions. Analyzing the trade-off between wheat yield, protein cont and weed control, Vrignon-Brenas et al. [90] indicated that combining simultaneous wh clover (*T. repens*) intercropping with high N availability significantly increased cover c biomass, decreased weed shoots' dry matter and improved N accumulation while ma taining high wheat yields and protein content. Intercropping has also been applied for control of parasitic weeds by using allelopathic species as trap crops. For instance, a 3 years of field experimentation, Fernández-Aparicio et al. [99] suggested intercropp grain legumes such as faba bean and pea with Egyptian clover (*T. alexandrinum*) to red the infection of the holoparasitic plant *Orobanche crenata* Forssk.

Research involving other cover crop species has also been published in recent years. Blaise et al. [87], evaluating twelve different intercrops over 5 years in cotton and found an average reduction of 43–71% of weed emergence and a 91–96% reduction of weed biomass compared to the control. Among intercrops, sun hemp (*Crotalaria juncea* L.) on one hand, showed the highest phytotoxicity with a significant suppression of purple nutsedge (*Cyperus rotundus* L.) and smooth joyweed (*Alternanthera paronychioides* A. St.-Hil.), while on the other, it allowed the highest cotton yield levels. Cheriere et al. [89] compared the combinations of different allelopathic crop species and spatial arrangements on grain production and weed control in soybean. A trade-off between soybean production and weed control was found, with sunflower allowing the lowest yield but, at the same time, the highest weed control level. The authors concluded that this trade-off could be managed by farmers by combining associated species choice and spatial arrangement; for example, planting alternate rows in sorghum and buckwheat intercrops. Intercropping density can also affect the weed–crop allelopathic interactions and the levels of weed control. For instance, Pouryousef et al. [88] studied five fenugreek (*Trigonella foenum-graecum* L.) densities of intercropping with coriander (*Coriandrum sativum* L.) and reported an increased reduction in weed biomass at increasing density, even if the weed control level was not entirely adequate, likely due to the low allelopathic potential of this cover crop. The authors suggested that this issue could be solved by using mixtures of different cover crops.

Much evidence has been reported about the increase in weed control when using mixtures of cover crops [100]. Kunz et al. [101], for instance, evaluated the weed suppressive effects of four cover crops in single and two mixed cultivations in three different locations. On average, both cover crop mixtures performed better than single cover crops and reduced weed density by 66% during the fallow period. The authors hypothesized the synergisms between allelochemicals (glucosinolates) and the additive allelopathic + competitive effects as an explanation. Similar results were also obtained by Florence et al. [100], according to which cover crop mixtures are able to compensate for the temporal and spatial variations in growing conditions, thus outperforming single species in the long term. Therefore, the combination of both allelopathic and competitive traits in cover crop species may help in increasing their weed-suppressive capacity. Several attempts were carried out with the goal of shifting allelopathy from competition in cover cropping experiments [78]. Using active carbon for allelochemicals' immobilization in a glasshouse experiment, Sturm et al. [102] found that allelopathic effects were species-specific, with the weed *Stellaria media* (L.) Vill. showing a greater sensitivity to allelopathy than *Alopecurus myosuroides* Huds. and volunteer wheat (*T. aestivum*). Allelopathy played an important role on overall weed suppression, although a greater contribution was played by competition. The authors concluded that an allelopathic cover crop should have competitive prerequisites (rapid germination, fast development, dense canopy and high soil coverage) to enhance the efficiency of its weed control.

3.3. Bioherbicides

Bioherbicides are broadly defined as natural products of biological origin derived either from living organisms or their secondary metabolites to suppress target weed populations without harming the environment [103]. For the purpose of this review, only plant-derived allelochemicals will be considered. With the aim of reducing the use of synthetic herbicides, overcoming weed-resistance phenomena and minimizing their environmental impact, plant-based allelochemical bioherbicides are gaining in popularity by virtue of the numerous advantages provided: water solubility, environmentally-friendly chemical structure (low amounts of 'heavy atoms', absence of 'unnatural' rings, high number of oxygen-, nitrogen- and sp^3-hybridized carbon molecules), high degradability in soil and water, possibility of new molecular targets in weeds, public acceptance [104]. Some of these benefits, however, in certain situations could represent a drawback. Their structural complexity, indeed, generates more stereocenters than synthetic molecules, making them unstable and rapidly degradable, thus reducing their environmental half-life.

Nevertheless, the chemical characteristics of allelochemicals lead to an increase in the costs for their synthesis, also considering that their discovery and set-up as bioherbicides is more complicated than that of synthetic herbicides [105] (Figure 2). In this regard, the recent advances in metabolomic techniques and chemical analytic instrumentations (e.g., liquid or gas chromatography combined with mass spectrometry) are simplifying the rediscovery of allelochemicals and their identification and quantification directly in crude plant extracts. As a result, tens of articles have been published in the past 10 years on the chemistry of crop allelopathy and, nowadays, we have a wide knowledge of the secondary metabolites involved in the phytotoxic effects of the most important crop species [5].

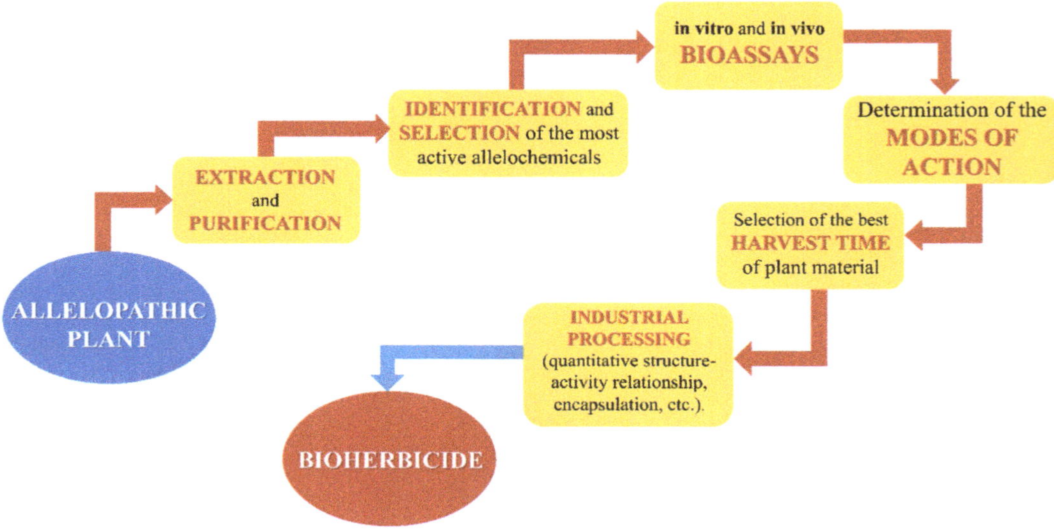

Figure 2. Steps for producing a commercial bioherbicide. After allelopathic potential is established and the best allelopathic genotype selected, allelochemicals need to be extracted, purified and identified. Then, allelopathic effects need to be checked both in vitro and in field conditions on a wide range of different weeds. Once the choice of the harvest time (i.e., maximum concentration of allelochemicals in plant tissues) has been made, a number of industrial processes such as quantitative structure-activity relationship (QSAR), chemical stabilization, encapsulation, etc., complete the whole process.

The application of plant extracts from various plant parts for weed control is well documented, especially under laboratory conditions by using organic solvents for allelochemicals' extraction and detection (commonly methanol, ethanol, ethyl acetate and dichloromethane) [22] (Table 3).

Table 3. Application of plant water extracts for weed management in field conditions.

Donor Plant (Dose)	Extract Concentration	Main Crop (Yield)	Target Weeds	Weed Control	References
Sorghum ($0.0006\ L\ m^{-2}$)	10%	Wheat (+39%)	*Avena fatua* L., *Phalaris minor* Retz.	−31% and −32% of DW	[106]
Sorghum + sunflower ($0.0012\ L\ m^{-2}$)		Wheat (+49.5%)		−52% and −45.5% of DW	
Sorghum + sunflower ($0.0006\ L\ m^{-2}$)		Wheat (+62%)		−31.5% and −32.5% of DW	
Sunflower ($0.1\ L\ m^{-2}$)	10%	Wheat (no yield losses)	*Chenopodium album* L.	−70% of biomass	[107]
Chinese cabbage ($0.002\ L\ m^{-2}$)	10%	Mung bean	*Trianthema portulacastrum* L., *Cyperus rotundus* L.	−14.6% of density and DW	[108]

Table 3. *Cont.*

Donor Plant (Dose)	Extract Concentration	Main Crop (Yield)	Target Weeds	Weed Control	References
Tree wormwood (4 L m^{-2})	13.82%	Wheat (−52.9%)	Several monocots and dicots, mainly *A. fatua* and *P. paradoxa*	~30% of weed suppression	[109]
Sicilian sumac (4 L m^{-2})	8.75%	Wheat (+9%)		50.8% of weed suppression	
Common thyme (4 L m^{-2})	22.33%	Wheat (−7.2%)		~35% of weed suppression	
Common lantana (4 L m^{-2})	6.14%	Wheat (+16.5%)		16% of weed suppression	
Mediterranean spurge (4 L m^{-2})	2.27%	Wheat (−2.3%)		~40% of weed suppression	
Tree of heaven (0.001–0.002 g L^{-1})	20%	Sage, rosemary, carnation	*Lepidium sativum* L., *Raphanus sativus* L.	0% weed presence in sage and rosemary, ~24% in carnation	[110]

Several studies have examined the allelopathic effects of plant water extracts under field conditions in important crops such as wheat, corn, cotton, etc. [7], with phytotoxic results expressed in terms of weed density and biomass reduction. Sorghum and sunflower are the best known examples of plant water extracts applied in field trials for weed control. Anjum and Bajwa [107] studied the bioherbicidal activity of sunflower leaf extracts (at 100 mL m^{-2}) applied three times in post-emergence at 7-day intervals, reporting that extracts at the highest concentration decreased lambsquarters (*C. album*) by 70% and increased wheat biomass and harvest index, compared to a weedy control, thanks to the overcoming of weed–crop competition. Jamil et al. [106] tested the water extracts of sorghum alone (at 12 L ha^{-1}) and in combination with sunflower, Chinese cabbage, eucalyptus, tobacco and sesame (at 6 L ha^{-1}) for weed control in wheat fields, following the hypothesis of a greater effectiveness of different allelochemicals when acting in synergism. Among treatments under study, sorghum + sunflower at 12 L ha^{-1} was the most effective in reducing wild oat (*Avena fatua* L.) and canary grass (*Phalaris minor* Retz.) dry weight, while sorghum + sunflower at 6 L ha^{-1} was the most economically viable treatment. In another 2-year field experiment carried out by Carrubba et al. [109] on wheat, aqueous extracts from *Rhus coriaria* L., *Lantana camara* L., *Thymus vulgaris* L. and *Euphorbia characias* L. were used in post-emergence as bioherbicides. The authors reported season-dependent results, with none of the tested extracts showing an eradication capacity of weeds. Overall, *R. coriaria* showed the most positive effect on wheat yield and weed suppression, although total weed biomass was not correlated to grain production. Application of *B. campestris* water extracts at 20 L ha^{-1} was found to significantly reduce horse purslane (*Trianthema portulacastrum* L.) and purple nutsedge (*C. rotundus*) density and dry weight in mung bean (*V. radiata*), with a 10.5% increase of yield [108]. In a nursery production system involving three horticultural crops (i.e., *Salvia officinalis* L., *S. rosmarinus* and *Dianthus caryophyllus* L.), *Ailanthus altissima* (Mill.) Swingle extracts at 100 and 200 mg L^{-1} were evaluated on the weeds *Lepidium sativum* L. and *R. sativus*. Use of *A. altissima* leaf extracts as post-emergence bioherbicides eradicated the two indicator weeds in *S. officinalis* and *S. rosmarinus*, while increased the percentage of weed presence in *D. caryophyllus*. The authors suggested applying the extract directly to the soil or growth media in order to alleviate phytotoxicity on the cash crops [110]. The chemical effects caused by the application of plant extracts were investigated in detail and can be summarized in: increase of reactive oxygen species (ROS), inhibition of gibberellin pathway and accumulation of abscisic, salicylic and jasmonic acid, alteration of cell membrane permeability and deregulation of nutrient uptake (Ca, K, Mg, Fe), alteration of photosynthesis and respiration [3,111]. Radhakrishnan et al. [112] reviewed the effects of plant-based bioherbicides on weed physiology, highlighting significant metabolic processes, resulting in the inhibition of seed germination and seedling growth.

Another strategy consists in the combined application of plant water extracts an synthetic herbicides with the aim of reducing herbicide doses. The reviews by Alsaadaw et al. [113], Farooq et al. [7] and Soltys et al. [105] involved a compendium of man articles prepared under this integrated approach, while in the past 5 years no significan steps forward have been made. Bioherbicides and synthetic herbicides can be combined b applying them at the same time or with different times of application: one in pre-emergen and the other one in post-emergence. Encouraging results were obtained in wheat, ric corn and cotton without affecting crop yields, but further studies are necessary to evalua the synergism between bioherbicides and synthetic herbicides.

Despite these findings on allelopathic water extracts, there is still a lack of know edge on the specific role of environmental factors on allelochemicals' bioavailability ar effectiveness. Air and mainly soil are the media for the transport of allelochemicals, b pedoclimatic conditions are well-known to greatly affect their movement and retention [7 Moreover, dosage, rate, time of application (in post- or pre-emergence), frequency of a plication, spectrum of target weeds (mono- or dicotyledons), persistence into the soil, et are issues needing to be addressed before producing a commercial formulation. For the reasons, the scientific community is called on to improve its efforts for the set-up of fie experiments focused on the application of plant-based bioherbicides, also under an IWM

4. Biotechnologies in Crop Allelopathy

In addition to these agronomic techniques, allelopathic traits of crops can be manag to obtain weed-suppressive cultivars and improve their allelopathic potential, since t level of crop allelopathy is often insufficient to provide effective weed management in t field. The basic principle is to develop crops able to control weeds on their own through t synthesis and release of active allelochemicals. Table 4 summarizes the main agricultu biotechnologies, apart from the use of bioherbicides which were already discussed, th could be used for the enhancement of crop allelopathy.

4.1. Screening and Selection of Allelopathic Crop Cultivars

It is well known that the allelopathic potential of crops closely depends on the ge type [16]. Many studies have pointed out how cultivars differ from each other in th allelochemical concentrations and allelopathic activities. The most studied allelopat crops showing significant allelochemicals variations (momilactones, benzoxazinoids a phenolic acids) in relation to the cultivar are rice, wheat, rye, barley and sorghum [8 Considerable effort has been made by the research community in recent years on this to Recently, metabolomic and phytotoxic differences have been observed among six canola *napus*) and two rye (*S. cereale*) cultivars [114,115]. Screening twelve barley accessions the content of the alkaloids gramine, hordenine and its direct precursor *N*-methyltyrami Maver et al. [116] found remarkable differences not only based on plant parts, but a between wild relatives and modern genotypes, thus providing important progress the breeding of this crop. Similarly, Ladhari et al. [47] screened thirteen fig (*Ficus car* L.) cultivars for their allelopathic activity and allelochemical concentration. The auth reported that the degree of inhibition was cultivar-dependent, as well as the phytochemi profiles, according to which fig cultivars were clustered into three groups. Scavo et al. [1 investigated the phytotoxicity and the quali-quantitative sesquiterpene lactone profile six *C. cardunculus* genotypes belonging to its three botanical varieties, i.e., globe artich (var. *scolymus*), and cultivated (var. *altilis*) and wild cardoon (var. *sylvestris*). Ultra-hi performance liquid chromatography in tandem with mass spectrometry highlighted t wild cardoon showed the highest levels of sesquiterpene lactones, in accordance w similar research reporting higher amounts of allelochemicals in ancestor ecotypes, wh the globe artichoke—i.e., the domesticated form—contained both the lowest concentrati and phytotoxic activity.

Table 4. Main biotechnologies involved in plant allelopathy.

Biotechnology	Main Effect	Description	Reference
Genotype selection	Screening allelopathic cultivars	Crop genotypes differ from each other in their allelochemicals' concentration and allelopathic activity. Screening and selecting genotypes allow obtaining a more allelopathic crop.	[117]
Stress induction	Increase in allelochemicals production	Induction of biotic and abiotic stress factors, or a combination of them, stimulates the synthesis of allelochemicals in donor plants.	[118]
Tissue culture	Increase in allelochemicals production Isolation from external factors during the study of allelopathic effects	Plant organ cultures such as hairy root cultures, both via normal callogenesis or using *Agrobacterium* spp. strains, may be applied to increase some competitive traits (e.g., rooting ability) and the production of allelochemicals, as well as to facilitate allelopathic studies.	[119]
Traditional breeding	Increase of crops' allelopathic potential or introduction of allelopathy de novo	Breeding programs can improve the allelopathic potential of crops just as they improved crop yields. However, poligeneticity and the low economic added value make this approach very difficult.	[120]
QTL analysis	Identification of genetic markers encoding allelopathic-related traits	The genetic analysis of quantitative trait loci (QTL) is very useful to identify the genes encoding the synthesis of allelochemicals.	[121]
Green chemistry	Increase in allelochemicals production	Improving allelochemicals' biotransformation by overexpressing the nitroreductase enzyme NfsB in *Escherichia coli* strains as a whole-cell biocatalyst.	[122]

4.2. Stress Induction

One of the main problems associated with allelopathic weed management is the low amount of allelochemicals' concentration in the donor plant and relative synthesis of these compounds for commercial use. A solution to this issue may derive from the exploitation of stress factors. In their 'stress hypothesis of allelopathy', Reigosa et al. [118] stated that plants' allelopathic potential is closely influenced by environmental changes, increasing when plants are under stress. Consequently, a stress condition generally enhances the synthesis of allelochemicals in the donor plant and the sensitivity of the target plant [3]. This is likely because when a plant recognizes a stress at cellular level, it usually starts a signal transduction leading to gene expression and to metabolic responses in terms of increased synthesis of secondary metabolites [118]. Different kinds of abiotic (light, drought, temperature, salinity, mineral availability) and biotic (pathogens, diseases, plant density) stress factors are known to increase the production of allelochemicals. For example, Oueslati et al. [123] demonstrated an increase of barley autotoxicity in drought conditions. In another study, 60% of plant shading was found to raise the concentration of sesquiterpene lactones in cultivated cardoon leaf extracts and their allelopathic activity [124]. Xuan et al. [125] reported that rice reacts to drought and salinity by enhancing the production of momilactones A and B, well-known rice allelochemicals, as a defensive mechanism. Under field conditions, generally different stress factors act in synergism, resulting in a further increase of allelochemicals' synthesis [3]. Although it is not certain if this behavior corresponds to a heightened release of allelochemicals into the environment by donor plants, stress induction can still be manipulated to obtain adequate amounts of allelochemicals and increase their concentration for the production of bioherbicides.

4.3. Genetic Engineering

In the past 20 years, biotechnology applications in weed management have focused on the development of transgenic allelopathy in crops through genetic engineering (GE techniques. Before this approach, research attempted to use traditional breeding programs to enhance the natural allelopathic potential of crops or introduce allelopathy de novo, in the same way as breeding was used to improve crop production [126]. However, breeding methods have not succeeded for two main reasons: on one side the low economic added valued provided by allelopathy, compared to yield, and on the other side its polygenetic ity [120]. Allelopathy, indeed, is a polygenetic characteristic weakly correlated to yield, thus needing the manipulation of more than one gene to encode the synthesis of allelochemicals This aspect has been observed, for instance, in the case of benzoxazinoids such as DIMBOA and DIBOA in Poaceae members [127].

To overcome these difficulties, several GE tools (e.g., recombinant DNA, polymerase chain reaction, metabolic engineering, overexpression of genes, etc.) are currently under evaluation to better understand the metabolic pathways, enzymes and genes involved in the synthesis of allelochemicals [16,105]. Being a quantitative and polygenetic trait, one of the most promising GE approaches is provided by the analysis of quantitative trait loc (QTLs) based on restriction fragment length polymorphism (RFLP), amplified fragmen length polymorphism (AFLP) and microsatellite (SSRs) markers to identify the geneti markers conferring crop allelopathic activity. QTL maps associated with allelopathi properties have already been developed in rice [128], wheat [129] and sorghum [121], bu the knowledge of crop genomes can allow extending this approach to other crops. Fo example, despite its high heterozygosity, the whole genome sequence of *C. cardunculu* was published [130] and this was the first case for the Compositae family, one of th most important allelopathic families. Furthermore, Eljounaidi et al. [131] identified th P450 genes (CYP71AV9 and CYP71BL5) and the enzymes—the germacrene A synthas (GAS), the germacrene A oxidase (GAO) and the costunolide synthase (COS)—involve in the biosynthetic pathway of sesquiterpene lactones in *C. cardunculus*. Combining th genome sequence on one hand, and the knowledge of genes and enzymes encoding th synthesis of allelochemicals on the other hand, is it possible to develop a transgenic *C cardunculus* genotype with improved allelopathy by genes' overexpression. Such a G approach can also be applied to other allelopathic Compositae members by isolating th pools of mRNAs expressing an allelopathic trait, creation of an expression sequence ta (EST) library and transfer to the desirable crop. However, even though we know th biosynthetic pathways of some allelopathic phenolic compounds and terpenoids, as well a the enzymes involved in the biosynthesis of secondary metabolites, our actual knowledg on this topic is limited and what is more, both the enzymes and genes are species-specifi so that it is necessary to directly isolate them from the allelopathic plant [120]. Anoth recent and important advance in bioherbicide production by GE was achieved by de Calle et al. [122], who improved the biosynthesis of D-DIBOA (2-deoxy-DIBOA) with 100 molar yield by *Escherichia coli* strains overexpressing the nitroreductase NfsB as biocataly

Transgenic allelopathy on its own would unlikely provide a satisfactory weed ma agement level in the field. For this reason, the latest GE future perspective is the creatic of commercial cultivars with incorporated or introduced allelopathic traits together wi competitive components (fast seedling emergence, high growth rate, early vigour, ro development, wide leaf area).

5. Conclusions

In this review, we have pointed out the increasing importance of allelopathy as a ne tool to make weed management more sustainable, both in conventional and organic ag culture. This field of research, which embraces different sciences, has gained in importan in recent years and, step by step, is becoming ever more common among farmers. Th important advances in analytic chemistry, metabolomics, biotechnology and genetics ha enabled the identification, isolation and purification of new allelochemicals, as well as th creation of transgenic allelopathic cultivars with marked allelopathic traits. Agronomic research has evaluated this broad and recent knowledge for its application as a weed control practice. To this end, promising results derive from the inclusion of allelopathic crops in rotational sequences, as living or dead mulches and by applying plant extracts as pre- or post-emergence bioherbicides. Despite these efforts, many allelopathic studies are still limited to laboratory conditions and most allelochemicals' modes of actions are unknown.

We recommend (1) more rigorously testing the agronomic performances of allelopathic crop rotation, cover cropping and bioherbicidal application; (2) a focus on the setup of field trials with involvement of biotic and abiotic factors, with the dual aim of considering both direct, indirect and synergistic allelopathic effects and acquiring a complete overview of allelopathy; (3) investigation of the different types of combination between allelopathic methods and traditional agricultural practices; (4) a focus on industrial processing for the development of commercial bioherbicides, given the high amounts of allelochemical-based candidates; (5) expanding the GE approach to the many well-known allelopathic crops and explore new GE tools for the biofortification of allelopathic crops. These recommendations will help to improve the efficiency of allelopathy as an environmentally friendly tool for weed management in agroecosystems, increase its diffusion among farmers and stakeholders, and reduce the use of synthetic herbicides and thus the development of weed-resistant ecotypes.

Author Contributions: Conceptualization, A.S. and G.M.; resources, data curation, writing—original draft preparation, illustrations and tables preparation, A.S.; review and editing, A.S. and G.M.; supervision, G.M. All authors have read and agreed to the published version of the manuscript.

Funding: This research received no external funding.

Institutional Review Board Statement: Not applicable.

Conflicts of Interest: The authors declare no conflict of interest.

ferences

Oerke, E.-C. Crop losses to pests. *J. Agric. Sci.* **2006**, *144*, 31–43. [CrossRef]
Willer, H.; Schlatter, B.; Trávníček, J.; Kemper, L.; Lernoud, J. *The World of Organic Agriculture Statistics and Emerging Trends 2020*; Research Institute of Organic Agriculture (FiBL): Frick, Switzerland, 2020; p. 337.
Scavo, A.; Restuccia, A.; Mauromicale, G. Allelopathy: General principles and basic aspects for agroecosystem control. In *Sustainable Agriculture Reviews*; Gaba, S., Smith, B., Lichtfouse, E., Eds.; Springer: Cham, Switzerland, 2018; Volume 28, pp. 47–101. [CrossRef]
Rice, E.L. *Allelopathy*, 2nd ed.; Academic Press: Orlando, FL, USA, 1984.
Macías, F.A.; Mejías, F.J.R.; Molinillo, J.M.G. Recent advances in allelopathy for weed control: From knowledge to applications. *Pest Manag. Sci.* **2019**, *75*, 2413–2436. [CrossRef] [PubMed]
Khanh, T.D.; Chung, M.I.; Xuan, T.D.; Tawata, S. The exploitation of crop allelopathy in sustainable agricultural production. *J. Agron. Crop. Sci.* **2005**, *191*, 172–184. [CrossRef]
Farooq, M.; Jabran, K.; Cheema, Z.A.; Wahid, A.; Siddique, K.H.M. The role of allelopathy in agricultural pest management. *Pest Manag. Sci.* **2011**, *67*, 493–506. [CrossRef] [PubMed]
Bhowmik, P.C.; Inderjit. Challenges and opportunities in implementing allelopathy for natural weed management. *Crop. Prot.* **2003**, *22*, 661–671. [CrossRef]
Jabran, K.; Mahajan, G.; Sardana, V.; Chauhan, B.S. Allelopathy for weed control in agricultural systems. *Crop. Prot.* **2015**, *72*, 57–65. [CrossRef]
Willis, R.J. *The History of Allelopathy*; Springer: Dordrecht, The Netherlands, 2007. [CrossRef]
Weidenhamer, J.D.; Romeo, J.T. Allelopathic properties of *Polygonella myriophylla*. *J. Chem. Ecol.* **1989**, *15*, 1957–1970. [CrossRef]
Macías, F.A.; Oliveros-Bastidas, A.; Marín, D.; Chinchilla, N.; Castellano, D.; Molinillo, J.M.G. Evidence for an allelopathic interaction between rye and wild oats. *J. Agric. Food Chem.* **2014**, *62*, 9450–9457. [CrossRef]
Macías, F.A.; Molinillo, J.M.G.; Varela, R.M.; Galindo, J.C. Allelopathy—A natural alternative for weed control. *Pest Manag. Sci.* **2007**, *63*, 327–348. [CrossRef]
Duke, S.O. Proving allelopathy in crop–weed interactions. *Weed Sci.* **2015**, *63*, 121–132. [CrossRef]
Qasem, J.R.; Foy, C.L. Weed allelopathy, its ecological impacts and future prospects: A review. *J. Crop. Prot.* **2001**, *4*, 43–119. [CrossRef]
Tesio, F.; Ferrero, A. Allelopathy, a chance for sustainable weed management. *Int. J. Sustain. Develop. World Ecol.* **2010**, *17*, 377–389. [CrossRef]

17. Sharangi, A.B. In search of allelopathy from common alliaceae crops for managing weeds in coriander: An overview. *Int. J. Agr. Res.* **2011**, *6*, 209–217. [CrossRef]
18. Funke, G.L. The influence of *Artemisia absinthium* on neighbouring plants. *Blumea* **1943**, *5*, 281–293.
19. Rial, C.; Tomé, S.; Varela, R.M.; Molinillo, J.M.G.; Macías, F.A. Phytochemical study of safflower roots (*Carthamus tinctorius*) on th induction of parasitic plant germination and weed control. *J. Chem. Ecol.* **2020**, *46*, 871–880. [CrossRef] [PubMed]
20. Nishimura, H.; Satoh, A. Antimicrobial and nematicidal substances from the root of chicory (*Cichorium intybus*). In *Allelochemica Biological Control of Plant Pathogens and Diseases. Disease Management of Fruits and Vegetables*; Inderjit, Mukerji, K., Eds.; Springe Dordrecht, The Netherlands, 2006; Volume 2, pp. 177–180. [CrossRef]
21. Rial, C.; Novaes, P.; Varela, R.M.; Molinillo, J.M.G.; Macías, F.A. Phytotoxicity of cardoon (*Cynara cardunculus*) allelochemicals standard target species and weeds. *J. Agric. Food Chem.* **2014**, *62*, 6699–6706. [CrossRef]
22. Scavo, A.; Rial, C.; Molinillo, J.M.G.; Varela, R.M.; Mauromicale, G.; Macías, F.A. The extraction procedure improves th allelopathic activity of cardoon (*Cynara cardunculus* var. *altilis*) leaf allelochemicals. *Ind. Crop. Prod.* **2019**, *128*, 479–487. [CrossRe
23. Scavo, A.; Pandino, G.; Restuccia, A.; Mauromicale, G. Leaf extracts of cultivated cardoon as potential bioherbicide. *Sci. Hort* **2020**, *261*, 109024. [CrossRef]
24. Macías, F.A.; Fernández, A.; Varela, R.M.; Molinillo, J.M.G.; Torres, A.; Alves, P.L.C.A. Sesquiterpene lactones as allelochemica *J. Nat. Prod.* **2006**, *69*, 795–800. [CrossRef]
25. Pan, L.; Sinden, M.R.; Kennedy, A.H.; Chai, H.; Watson, L.E.; Graham, T.L.; Kinghorn, A.D. Bioactive constituents of *Helianth tuberosus* (Jerusalem artichoke). *Phytochem. Lett.* **2009**, *2*, 15–18. [CrossRef]
26. Tesio, F.; Weston, L.A.; Ferrero, A. Allelochemicals identified from Jerusalem artichoke (*Helianthus tuberosus* L.) residues and th potential inhibitory activity in the field and laboratory. *Sci. Hortic.* **2011**, *129*, 361–368. [CrossRef]
27. Chon, S.U.; Jang, H.G.; Kim, D.K.; Kim, Y.M.; Boo, H.O.; Kim, Y.J. Allelopathic potential in lettuce (*Lactuca sativa* L.) plants. S *Hortic.* **2005**, *106*, 309–317. [CrossRef]
28. Rehman, S.; Shahzad, B.; Bajwa, A.A.; Hussain, S.; Rehman, A.; Cheema, S.A.; Abbas, T.; Ali, A.; Shah, L.; Adkins, S.; et Utilizing the allelopathic potential of *Brassica* species for sustainable crop production: A review. *J. Plant Growth Regul.* **2018**, 343–356. [CrossRef]
29. Tlili, N.; Nasri, N.; Saadaoui, E.; Khaldi, A.; Triki, S. Carotenoid and tocopherol composition of leaves, buds and flowers *Capparis spinosa* grown wild in Tunisia. *J. Agric. Food Chem.* **2009**, *57*, 5381–5385. [CrossRef] [PubMed]
30. Ladhari, A.; Omezzine, F.; Dellagreca, M.; Zarrelli, A.; Haouala, R. Phytotoxic activity of *Capparis spinosa* L. and its discove active compounds. *Allelopath. J.* **2013**, *32*, 175–190.
31. Ahmed, M.; Ji, M.; Peiwen, Q.; Liu, Y.; Gu, Z. Phytochemical screening, total phenolic and flavonoids contents and antioxida activities of *Citrullus colocynthis* L. and *Cannabis sativa* L. *Appl. Ecol. Environ.* **2019**, *17*, 6961–6979. [CrossRef]
32. Soni, B.; Tseng, T.P.M.; Yue, Z. Identification and quantification of allelochemicals from selected sweet potato (*Ipomoea batatas* Lam.) cultivars. *Am. J. Plant Sci.* **2019**, *10*, 12. [CrossRef]
33. Hao, Z.P.; Wang, Q.; Christie, P.; Li, X.L. Allelopathic potential of watermelon tissues and root exudates. *Sci. Hortic.* **2007**, *1* 315–320. [CrossRef]
34. Yu, J.Q.; Ye, S.F.; Zhang, M.F.; Hu, W.H. Effects of root exudates and aqueous root extracts of cucumber (*Cucumis sativus*) a allelochemicals, on photosynthesis and antioxidant enzymes in cucumber. *Biochem. Syst. Ecol.* **2003**, *31*, 129–139. [CrossRef]
35. Li, X.; Ding, C.; Hua, K.; Zhang, T.; Zhang, Y.; Zhao, L. Soil sickness of peanuts is attributable to modifications in soil micro induced by peanut root exudates rather than to direct allelopathy. *Soil Biol. Biochem.* **2014**, *78*, 149–159. [CrossRef]
36. Ren, X.; Yan, Z.; He, X.; Li, X.; Qin, B. Allelochemicals from rhizosphere soils of *Glycyrrhiza uralensis* Fisch: Discovery of autotoxic compounds of a traditional herbal medicine. *Ind. Crop. Prod.* **2017**, *97*, 302–307. [CrossRef]
37. Chon, S.U.; Kim, J.D. Biological activity and quantification of suspected allelochemicals from alfalfa plant parts. *J. Agron. C Sci.* **2002**, *188*, 281–285. [CrossRef]
38. Chon, S.U.; Jennings, J.A.; Nelson, C.J. Alfalfa (*Medicago sativa* L.) autotoxicity: Current status. *Allelopath. J.* **2006**, *18*, 58–80.
39. Asaduzzaman, M.; Asao, T. Autotoxicity in beans and their allelochemicals. *Sci. Hortic.* **2012**, *134*, 26–31. [CrossRef]
40. Kato-Noguchi, H. Isolation and identification of an allelopathic substance in *Pisum sativum*. *Phytochemistry* **2003**, *62*, 1141–1 [CrossRef]
41. Biswas, S.M.; Chakraborty, N. Strong sour tamarind flavor of methyl-2, 3, 4-trihydroxyhexanoate, a new compound isolated fr leaves of *Tamarindus indica* L. plays a role in plant. *Ann. Trop. Res.* **2012**, *34*, 26–44. [CrossRef]
42. Syed, S.; Ahmed, Z.I.; Al-Haq, M.I.; Mohammad, A.; Fujii, Y. The possible role of organic acids as allelochemicals in *Tamarin indica* L. leaves. *Acta Agric. Scand.* **2014**, *64*, 511–517. [CrossRef]
43. Solar, A.; Colarič, M.; Usenik, V.; Stampar, F. Seasonal variations of selected flavonoids, phenolic acids and quinones in ann shoots of common walnut (*Juglans regia* L.). *Plant Sci.* **2006**, *170*, 453–461. [CrossRef]
44. Wang, Q.; Xu, Z.; Hu, T.; Rehman, H.; Chen, H.; Li, Z.; Ding, B.; Hu, H. Allelopathic activity and chemical constituents of wal (*Juglans regia*) leaf litter in walnut–winter vegetable agroforestry system. *Nat. Prod. Res.* **2014**, *28*, 2017–2020. [CrossRef]
45. Chen, F.; Peng, S.; Chen, B.; Ni, G.; Liao, H. Allelopathic potential and volatile compounds of *Rosmarinus officinalis* L. aga weeds. *Allelopath. J.* **2013**, *32*, 57–66.
46. Appiah, K.S.; Mardani, H.K.; Omari, R.A. Involvement of carnosic acid in the phytotoxicity of *Rosmarinus officinalis* leaves. *To* **2018**, *10*, 498. [CrossRef] [PubMed]

47. Ladhari, A.; Gaaliche, B.; Zarrelli, A.; Ghannem, M.; Mimoun, M.B. Allelopathic potential and phenolic allelochemicals discrepancies in *Ficus carica* L. cultivars. *S. Afr. J. Bot.* **2020**, *130*, 30–44. [CrossRef]
48. Puig, C.G.; Reigosa, M.J.; Valentão, P.; Andrade, P.B.; Pedrol, N. Unravelling the bioherbicide potential of *Eucalyptus globulus* Labill: Biochemistry and effects of its aqueous extract. *PLoS ONE* **2018**, *13*, e0192872. [CrossRef] [PubMed]
49. Chon, S.U.; Kim, Y.M. Herbicidal potential and quantification of suspected allelochemicals from four grass crop extracts. *J. Agron. Crop. Sci.* **2004**, *190*, 145–150. [CrossRef]
50. De Bertoldi, C.; De Leo, M.; Braca, A.; Ercoli, L. Bioassay-guided isolation of allelochemicals from *Avena sativa* L.: Allelopathic potential of flavone C-glycosides. *Chemoecology* **2009**, *19*, 169–176. [CrossRef]
51. Kremer, R.J.; Ben-Hammouda, M. Allelopathic Plants. 19. Barley (*Hordeum vulgare* L.). *Allelopath. J.* **2009**, *24*, 225–242.
52. Kong, C.H.; Li, H.B.; Hu, F.; Xu, X.H.; Wang, P. Allelochemicals released by rice roots and residues in soil. *Plant Soil* **2006**, *288*, 47–56. [CrossRef]
53. Weston, L.A.; Alsaadawi, I.S.; Baerson, S.R. Sorghum allelopathy-from ecosystem to molecule. *J. Chem. Ecol.* **2013**, *39*, 142–153. [CrossRef]
54. Wu, H.; Pratley, J.; Lemerle, D.; Haig, T. Allelopathy in wheat (*Triticum aestivum*). *Ann. Appl. Biol.* **2001**, *139*, 1–9. [CrossRef]
55. Stochmal, A.; Kus, J.; Martyniuk, S.; Oleszek, W. Concentration of benzoxazinoids in roots of field-grown wheat (*Triticum aestivum* L.) varieties. *J. Agric. Food Chem.* **2006**, *54*, 1016–1022. [CrossRef]
56. Fragasso, M.; Iannucci, A.; Papa, R. Durum wheat and allelopathy: Toward wheat breeding for natural weed management. *Front. Plant Sci.* **2013**, *4*, 375. [CrossRef] [PubMed]
57. Chou, C.H.; Waller, G.R. Possible allelopathic constituents of *Coffea arabica*. *J. Chem. Ecol.* **1980**, *6*, 643–654. [CrossRef]
58. Geng, G.D.; Zhang, S.Q.; Cheng, Z.H. Allelopathy and allelochemicals of root exudates in hot pepper. *Acta Hortic. Sin.* **2009**, *36*, 873–878.
59. Sun, H.; Wang, Y. Potential allelopathic effects of allelochemicals in aqueous extracts of leaves and root exudates of *Capsicum annuum* on vegetable crops. *Allelopath. J.* **2015**, *35*, 11–22.
60. Baek, J.M.; Kawecki, O.J.; Lubin, K.D.; Song, J.; Wiens, O.A.; Wu, F. Plant warfare: Allelopathic effects of *Nicotiana tabacum* on the germination of *Vigna radiata* and *Triticum aestivum*. *WURJ Health Nat. Sci.* **2017**, *8*. [CrossRef]
61. Rial, C.; Gómez, E.; Varela, R.M.; Molinillo, J.M.G.; Macías, F.A. Ecological relevance of the major allelochemicals in *Lycopersicon esculentum* roots and exudates. *J. Agric. Food Chem.* **2018**, *66*, 4638–4644. [CrossRef] [PubMed]
62. Rawat, L.S.; Maikhuri, R.K.; Bahuguna, Y.M.; Jha, N.K.; Phondani, P.C. Sunflower allelopathy for weed control in agriculture systems. *J. Crop. Sci. Biotech.* **2017**, *20*, 45–60. [CrossRef]
63. Pandino, G.; Mauromicale, G. Globe artichoke and cardoon forms between traditional and modern uses. *Acta Hortic.* **2020**, *1284*, 1–18. [CrossRef]
64. Scavo, A.; Restuccia, A.; Pandino, G.; Onofri, A.; Mauromicale, G. Allelopathic effects of *Cynara cardunculus* L. leaf aqueous extracts on seed germination of some Mediterranean weed species. *Ital. J. Agron.* **2018**, *13*, 119–125. [CrossRef]
65. Scavo, A.; Pandino, G.; Restuccia, A.; Lombardo, S.; Pesce, G.R.; Mauromicale, G. Allelopathic potential of leaf aqueous extracts from *Cynara cardunculus* L. on the seedling growth of two cosmopolitan weed species. *Ital. J. Agron.* **2019**, *14*, 78–83. [CrossRef]

Dayan, F.E.; Rimando, A.M.; Pan, Z.; Baerson, S.R.; Gimsing, A.L.; Duke, S.O. Sorgoleone. *Phytochemistry* **2010**, *71*, 1032–1039. [CrossRef]

Scavo, A.; Mauromicale, G. Integrated weed management in herbaceous field crops. *Agronomy* **2020**, *10*, 466. [CrossRef]

Snapp, S.S.; Swinton, S.M.; Labarta, R.; Mutch, D.; Black, J.R.; Leep, R.; Nyiraneza, J.; O'Neil, K. Evaluating cover crops for benefits, costs and performance within cropping system niches. *Agron. J.* **2005**, *97*, 322–332. [CrossRef]

Buhler, D.D.; Liebman, M.; Obrycki, J.J. Theoretical and practical challenges to an IPM approach to weed management. *Weed Sci.* **2000**, *48*, 274–280. [CrossRef]

Scavo, A.; Abbate, C.; Mauromicale, G. Plant allelochemicals: Agronomic, nutritional and ecological relevance in the soil system. *Plant Soil* **2019**, *442*, 23–48. [CrossRef]

Shahzad, M.; Farooq, M.; Jabran, K.; Hussain, M. Impact of different crop rotations and tillage systems on weed infestation and productivity of bread wheat. *Crop. Prot.* **2016**, *89*, 161–169. [CrossRef]

Scherner, A.; Melander, B.; Kudsk, P. Vertical distribution and composition of weed seeds within the plough layer after eleven years of contrasting crop rotation and tillage schemes. *Soil Tillage Res.* **2016**, *161*, 135–142. [CrossRef]

Hunt, N.D.; Hill, J.D.; Liebman, M. Reducing freshwater toxicity while maintaining weed control, profits, and productivity: Effects of increased crop rotation diversity and reduced herbicide usage. *Environ. Sci. Technol.* **2017**, *51*, 1707–1717. [CrossRef]

Weisberger, D.; Nichols, V.; Liebman, M. Does diversifying crop rotations suppress weeds? A meta-analysis. *PLoS ONE* **2019**, *14*, e0219847. [CrossRef]

Scavo, A.; Restuccia, A.; Abbate, C.; Mauromicale, G. Seeming field allelopathic activity of *Cynara cardunculus* L. reduces the soil weed seed bank. *Agron. Sustain. Dev.* **2019**, *39*, 41. [CrossRef]

Karkanis, A.; Alexiou, A.; Katsaros, C.; Petropoulos, S. Allelopathic activity of spearmint (*Mentha spicata* L.) and peppermint (*Mentha* × *piperita* L.) reduces yield, growth, and photosynthetic rate in a succeeding crop of maize (*Zea mays* L.). *Agronomy* **2019**, *9*, 461. [CrossRef]

Mauromicale, G.; Occhipinti, A.; Mauro, R. Selection of shade-adapted subterranean clover species for cover cropping in orchards. *Agron. Sustain. Dev.* **2010**, *30*, 473–480. [CrossRef]

78. Gerhards, R.; Schappert, A. Advancing cover cropping in temperate integrated weed management. *Pest Manag. Sci.* **2020**, *76*, 42–49. [CrossRef] [PubMed]
79. Mennan, H.; Jabran, K.; Zandstra, B.H.; Pala, F. Non-chemical weed management in vegetables by using cover crops: A review. *Agronomy* **2020**, *10*, 257. [CrossRef]
80. Abbas, T.; Nadeem, M.A.; Tanveer, A.; Ali, H.H.; Farooq, N. Role of allelopathic crop mulches and reduced doses of tank-mixed herbicides in managing herbicide-resistant *Phalaris minor* in wheat. *Crop. Prot.* **2018**, *110*, 245–250. [CrossRef]
81. Scavo, A.; Restuccia, A.; Lombardo, S.; Fontanazza, S.; Abbate, C.; Pandino, G.; Anastasi, U.; Onofri, A.; Mauromicale, G. Improving soil health, weed management and nitrogen dynamics by *Trifolium subterraneum* cover cropping. *Agron. Sustain. Dev.* **2020**, *40*, 18. [CrossRef]
82. Sturm, D.J.; Kunz, C.; Gerhards, R. Inhibitory effects of cover crop mulch on germination and growth of *Stellaria media* (L.) Vill., *Chenopodium album* L. and *Matricaria chamomilla* L. *Crop. Prot.* **2016**, *90*, 125–131. [CrossRef]
83. Puig, C.G.; Revilla, P.; Barreal, M.E.; Reigosa, M.J.; Pedrol, N. On the suitability of Eucalyptus globulus green manure for field weed control. *Crop. Prot.* **2019**, *121*, 57–65. [CrossRef]
84. Masilionyte, L.; Maiksteniene, S.; Kriauciuniene, Z.; Jablonskyte-Rasce, D.; Zou, L.; Sarauskis, E. Effect of cover crops in smothering weeds and volunteer plants in alternative farming systems. *Crop. Prot.* **2017**, *91*, 74–81. [CrossRef]
85. Álvarez-Iglesias, L.; Puig, C.G.; Revilla, P.; Reigosa, M.J.; Pedrol, N. Faba bean as green manure for field weed control in maize. *Weed Res.* **2018**, *58*, 437–449. [CrossRef]
86. Alonso-Ayuso, M.; Gabriel, J.L.; García-González, I.; Del Monte, J.P.; Quemada, M. Weed density and diversity in a long-term cover crop experiment background. *Crop. Prot.* **2018**, *112*, 103–111. [CrossRef]
87. Blaise, D.; Manikandan, A.; Verma, P.; Nalayini, P.; Chakraborty, M.; Kranthi, K.R. Allelopathic intercrops and its mulch as an integrated weed management strategy for rainfed Bt-transgenic cotton hybrids. *Crop. Prot.* **2020**, *135*, 105214. [CrossRef]
88. Pouryousef, M.; Yousefi, A.R.; Oveisi, M.; Asadi, F. Intercropping of fenugreek as living mulch at different densities for weed suppression in coriander. *Crop. Prot.* **2015**, *69*, 60–64. [CrossRef]
89. Cheriere, T.; Lorin, M.; Corre-Hellou, G. Species choice and spatial arrangement in soybean-based intercropping: Levers that drive yield and weed control. *Field Crop. Res.* **2020**, *256*, 107923. [CrossRef]
90. Vrignon-Brenas, S.; Celette, F.; Piquet-Pissaloux, A. Intercropping strategies of white clover with organic wheat to improve the trade-off between wheat yield, protein content and the provision of ecological services by white clover. *Field Crop. Res.* **2018**, *224*, 160–169. [CrossRef]
91. Lemessa, F.; Wakjira, M. Mechanisms of ecological weed management by cover cropping: A review. *J. Biol. Sci.* **2014**, *14*, 452–45 [CrossRef]
92. Restuccia, A.; Scavo, A.; Lombardo, S.; Pandino, G.; Fontanazza, S.; Anastasi, U.; Abbate, C.; Mauromicale, G. Long-term effect of cover crops on species abundance and diversity of weed flora. *Plants* **2020**, *9*, 1506. [CrossRef] [PubMed]
93. Westerman, P.R.; Luijendijk, C.D.; Wevers, J.D.A.; van der Werf, W. Weed seed predation in a phenologically late crop. *Weed Res.* **2011**, *51*, 157–164. [CrossRef]
94. Kruidhof, H.M.; Gallandt, E.R.; Haramoto, E.R.; Bastiaans, L. Selective weed suppression by cover crop residues: Effect of seed mass and timing of species sensitivity. *Weed Res.* **2011**, *51*, 177–186. [CrossRef]
95. Kruidhof, H.M.; Bastiaans, L.; Kropff, M.J. Cover crop residue management for optimizing weed control. *Plant Soil* **2009**, *31*, 169–184. [CrossRef]
96. Brooker, R.W.; Bennett, A.E.; Cong, W.F.; Daniell, T.J.; George, T.S.; Hallett, P.D.; Hawes, C.; Iannetta, P.P.; Jones, H.G.; Karley, A.J.; et al. Improving intercropping: A synthesis of research in agronomy, plant physiology and ecology. *New Phytol.* **2015**, *20*, 107–117. [CrossRef]
97. Barto, E.K.; Weidenhamer, J.D.; Cipollini, D.; Rillig, M.C. Fungal superhighways: Do common mycorrhizal networks enhance below ground communication? *Trends Plan Sci.* **2012**, *17*, 633–637. [CrossRef] [PubMed]
98. Rad, S.V.; Valadabadi, S.A.R.; Pouryousef, M.; Saifzadeh, S.; Zakrin, H.R.; Mastinu, A. Quantitative and qualitative evaluation of *Sorghum bicolor* L. under intercropping with legumes and different weed control methods. *Horticulturae* **2020**, *6*, 78. [CrossRef]
99. Fernández-Aparicio, M.; Emeran, A.A.; Rubiales, D. Inter-cropping with berseem clover (*Trifolium alexandrinum*) reduces infection by *Orobanche crenata* in legumes. *Crop. Prot.* **2010**, *29*, 867–871. [CrossRef]
100. Florence, A.M.; Higley, L.G.; Drijber, R.A.; Francis, C.A.; Lindquist, J.L. Cover crop mixture diversity, biomass productivity, weed suppression, and stability. *PLoS ONE* **2019**, *14*, e0206195. [CrossRef] [PubMed]
101. Kunz, C.; Sturm, D.J.; Varnholt, D.; Walker, F.; Gerhards, R. Allelopathic effects and weed suppressive ability of cover crops. *Plant Soil Environ.* **2016**, *62*, 60–66. [CrossRef]
102. Sturm, D.J.; Peteinatos, G.; Gerhards, R. Contribution of allelopathic effects to the overall weed suppression by different cover crops. *Weed Res.* **2018**, *58*, 331–337. [CrossRef]
103. Hoagland, R.E.; Boyette, C.D.; Weaver, M.A.; Abbas, H.K. Bioherbicides: Research and risks. *Toxin Rev.* **2007**, *26*, 313–3 [CrossRef]
104. Dayan, F.E.; Owens, D.K.; Duke, S.O. Rationale for a natural products approach to herbicide discovery. *Pest Manag. Sci.* **2012**, 519–528. [CrossRef]
105. Soltys, D.; Krasuska, U.; Bogatek, R.; Gniazdowska, A. Allelochemicals as bioherbicides—Present and perspectives. In *Herbicides: Current Research and Case Studies in Use*; Price, A.J., Kelton, J.A., Eds.; IntechOpen: London, UK, 2013. [CrossRef]

6. Jamil, M.; Cheema, Z.A.; Mushtaq, M.N.; Farooq, M.; Cheema, M.A. Alternative control of wild oat and canary grass in wheat fields by allelopathic plant water extracts. *Agron. Sustain. Dev.* **2009**, *29*, 475–482. [CrossRef]

7. Anjum, T.; Bajwa, R. The effect of sunflower leaf extracts on *Chenopodium album* in wheat fields in Pakistan. *Crop. Prot.* **2007**, *26*, 1390–1394. [CrossRef]

8. Ullah, R.; Aslam, Z.; Maitah, M.; Bashir, S.; Hassan, W. Sustainable weed control and enhancing nutrient use efficiency in crops through brassica (*Brassica campestris* L.) allelopathy. *Sustainability* **2020**, *12*, 5763. [CrossRef]

9. Carrubba, A.; Labruzzo, A.; Comparato, A.; Muccilli, S.; Spina, A. Use of plant water extracts for weed control in durum wheat (*Triticum turgidum* L. subsp. *durum* Desf.). *Agronomy* **2020**, *10*, 364. [CrossRef]

0. Caser, M.; Demasi, S.; Caldera, F.; Dhakar, N.K.; Trotta, F.; Scariot, V. Activity of *Ailanthus altissima* (Mill.) Swingle extract as a potential bioherbicide for sustainable weed management in horticulture. *Agronomy* **2020**, *10*, 965. [CrossRef]

1. Barros, V.M.D.S.; Pedrosa, J.L.F.; Gonçalves, D.R.; De Medeiros, F.C.L.; Carvalho, G.R.; Gonçalves, A.H.; Teixeira, P.V.V.Q. Herbicides of biological origin: A review. *J. Hortic. Sci. Biotech.* **2021**, *96*, 288–296. [CrossRef]

2. Radhakrishnan, R.; Alqarawi, A.A.; Abd Allah, E.F. Bioherbicides: Current knowledge on weed control mechanism. *Ecotoxicol. Environ. Saf.* **2018**, *158*, 131–138. [CrossRef]

3. Alsaadawi, I.S.; Khaliq, A.; Farooq, M. Integration of allelopathy and less herbicides effect on weed management in field crops and soil biota: A Review. *Plant Arch.* **2020**, *20*, 225–237.

4. Asaduzzaman, M.; Pratley, J.E.; An, M.; Luckett, D.J.; Lemerle, D. Metabolomics differentiation of canola genotypes: Toward an understanding of canola allelochemicals. *Front. Plant Sci.* **2015**, *5*, 765. [CrossRef]

5. Ghimire, B.K.; Yu, C.Y.; Ghimire, B.; Seong, E.S.; Chung, I.M. Allelopathic potential of phenolic compounds in *Secale cereale* cultivars and its relationship with seeding density. *Appl. Sci.* **2019**, *9*, 3072. [CrossRef]

5. Maver, M.; Miras-Moreno, B.; Lucini, L.; Trevisan, M.; Pii, Y.; Cesco, S.; Mimmo, T. New insights in the allelopathic traits of different barley genotypes: Middle Eastern and Tibetan wild-relative accessions vs. cultivated modern barley. *PLoS ONE* **2020**, *15*, e0231976. [CrossRef]

7. Scavo, A.; Rial, C.; Varela, R.M.; Molinillo, J.M.G.; Mauromicale, G.; Macías, F.A. Influence of genotype and harvest time on the *Cynara cardunculus* L. sesquiterpene lactone profile. *J. Agric. Food Chem.* **2019**, *67*, 6487–6496. [CrossRef]

Reigosa, M.J.; Souto, X.C.; González, L. Effect of phenolic compounds on the germination of six weeds species. *Plant Growth Regul.* **1999**, *28*, 83–88. [CrossRef]

Stanišić, M.; Ćosić, T.; Savić, J.; Krstić-Milošević, D.; Mišić, D.; Smigocki, A.; Ninković, S.; Banjac, N. Hairy root culture as a valuable tool for allelopathic studies in apple. *Tree Physiol.* **2019**, *39*, 888–905. [CrossRef]

Duke, S.O.; Scheffler, B.E.; Boyette, C.D.; Dayan, F.E. *Biotechnology in Weed Control. Kirk-Othmer Encyclopedia of Chemical Technology*; John Wiley & Sons, Inc.: New York, NY, USA, 2015. [CrossRef]

Shehzad, T.; Okuno, K. Genetic analysis of QTLs controlling allelopathic characteristics in sorghum. *PLoS ONE* **2020**, *15*, e0235896. [CrossRef] [PubMed]

De la Calle, M.E.; Cabrera, G.; Cantero, D.; Valle, A.; Bolivar, J. A genetically engineered *Escherichia coli* strain overexpressing the nitroreductase NfsB is capable of producing the herbicide D-DIBOA with 100% molar yield. *Microb. Cell Factories* **2019**, *18*, 86. [CrossRef] [PubMed]

Oueslati, O.; Ben-Hammouda, M.; Ghorbal, M.H.; Guezzah, M.; Kremer, R.J. Barley autotoxicity as influenced by varietal and seasonal variation. *J. Agron. Crop. Sci.* **2005**, *191*, 249–254. [CrossRef]

Scavo, A.; Rial, C.; Molinillo, J.M.G.; Varela, R.M.; Mauromicale, G.; Macías, F.A. Effect of shading on the sesquiterpene lactone content and phytotoxicity of cultivated cardoon leaf extracts. *J. Agric. Food Chem.* **2020**, *68*, 11946–11953. [CrossRef]

Xuan, T.D.; Minh, T.N.; Khanh, T.D. Allelopathic momilactones A and B are implied in rice drought and salinity tolerance, not weed resistance. *Agron. Sustain. Dev.* **2016**, *36*, 52. [CrossRef]

Tranel, P.J.; Horvath, D.P. Molecular biology and genomics: New tools for weed science. *Bioscience* **2009**, *59*, 207–215. [CrossRef]

Frey, M.; Chomet, P.; Glawischnig, E.; Stettner, C.; Grün, S.; Winklmair, A.; Eisenreich, W.; Bacher, A.; Meeley, R.B.; Briggs, S.P.; et al. Analysis of a chemical plant defense mechanism in grasses. *Science* **1997**, *277*, 696–699. [CrossRef]

Jensen, L.B.; Courtois, B.; Shen, L.S.; Li, Z.K.; Olofsdotter, M.; Mauleon, R.P. Locating genes controlling allelopathic effects against barnyardgrass in upland rice. *Agron. J.* **2001**, *93*, 21–26. [CrossRef]

Wu, H.; Pratley, J.; Ma, W.; Haig, T. Quantitative trait loci and molecular markers associated with wheat allelopathy. *Theor. Appl. Genet.* **2003**, *107*, 1477–1481. [CrossRef]

Scaglione, D.; Reyes-Chin-Wo, S.; Acquadro, A. The genome sequence of the outbreeding globe artichoke constructed de novo incorporating a phase-aware low-pass sequencing strategy of F_1 progeny. *Sci. Rep.* **2016**, *6*, 19427. [CrossRef] [PubMed]

Eljounaidi, K.; Cankar, K.; Comino, C.; Moglia, A.; Hehn, A.; Bourgaud, F.; Bouwmeester, H.; Menin, B.; Lanteri, S.; Beekwilder, J. Cytochrome P450s from *Cynara cardunculus* L. CYP71AV9 and CYP71BL5, catalyze distinct hydroxylations in the sesquiterpene lactone biosynthetic pathway. *Plant Sci.* **2014**, *223*, 59–68. [CrossRef] [PubMed]

MDPI
St. Alban-Anlage 66
4052 Basel
Switzerland
www.mdpi.com

Agronomy Editorial Office
E-mail: agronomy@mdpi.com
www.mdpi.com/journal/agronomy

www.ingramcontent.com/pod-product-compliance
Lightning Source LLC
Chambersburg PA
CBHW061449220926
51850CB00004B/50
9783725800636